saur

International Federation of Film Archives (FIAF)

International Directory of Cinematographers Set- and Costume Designers in Film

Vol. 11: Part A
Film Titles

General Index Vol. 1–10

Albania, Bulgaria, Czechoslovakia, Denmark, Finland, France, German Democratic Republic, Germany (to 1945), Greece, Hungary, Italy, Norway, Poland, Portugal, Romania, Spain, Yugoslavia

Edited by Alfred and Hille Krautz

K · G · Saur
München · London · New York · Paris 1992

Die Deutsche Bibliothek - CIP-Einheitsaufnahme

International directory of cinematographers, set- and costume designers in film / International Federation of Film Archives (FIAF). Ed. by Alfred Krautz. - München ; New York ; London ; Paris : Saur.
NE: Krautz, Alfred [Hrsg.]; International Federation of Film Archives

Vol. 11. General index vol. 1-10.
Pt. A. Film title. - 1992
ISBN 3-598-21444-8

Gedruckt auf säurefreiem Papier / Printed on acid-free paper

Part of Reed International P.L.C.
Printed in the Federal Republic of Germany

Druck/Bindung: Druck Partner Rübelmann, Hemsbach

ISBN 3-598-21442-1 (Gesamtwerk in 2 Bänden)

Preface

As former leader of the working group on documentation at FIAF (the International Federation of Film Archives) I was commissioned in 1980 to put together an "International Directory of Cinematographers, Set- and Costume Designers in Film". I turned then to K. G. Saur Publishers with this ambitious project, namely to index and publish by country the holdings of various national film archives.

Since then, with the help of friends and colleagues in the FIAF archives, the holdings of a total of 17 countries' national film archives have been published in 10 volumes: German Democratic Republic, Poland, France, Albania, Bulgaria, Greece, Romania, Yugoslavia, Germany (to 1945), Denmark, Finland, Norway, Italy, Portugal, Spain, Hungary und Czechoslovakia. Each of these volumes lists film cinematographers, set- and costume designers with their dates of birth and death (as appropriate) and a title directory of film directors and year of production. Each volume also provides Film Director and Title Indexes.

In accordance with the many requests received from users of this International Directory, K. G. Saur Publishers and the editors decided to produce a General Index for the first ten columes of the work:

Vol. 11: Part A: Index of Film Titles (35,110 entries)
Vol. 11: Part B: Index of Film Directors (16,270 entries)

The first ten volumes are therefore now available with a complete and user-friendly alphabetical index in two parts.

Berlin, August 1992 Alfred Krautz

Index of Film Titles

A

A.-B. GIFTA BORT BARON OLSON **Vol. 5** 891, 925
A BALADA FA PRAIA DOS CAES **Vol. 8** 612
A BELLES DENTS **Vol. 2** 773
A BJAHME MLADI **Vol. 3** 56 **Vol. 6** *page 26, page 29*
A BOUT DE SEXE **Vol. 2** 442
A BOUT DE SOUFFLE **Vol. 2** 208, 465
A BUEN JUEZ, MEJOR TESTIGO **Vol. 8** 264
A CAUSE A CAUSE D'UNE FEMME **Vol. 2** 398
A CAUSE DE L'HOMME A LA VOITURE BLANCHE **Vol. 2** 283
A CHACUN SON ENFER **Vol. 2** 185, 293
A CHACUN SON PARADIS **Vol. 2** 208
A CHE SERVONO QUESTI QUATTRINI ? **Vol. 7** 792, 1036
A CHI TOCCA ... TOCCA ! **Vol. 7** 780
A COEUR JOIE **Vol. 2** 598, 781
A CONTRATIEMPO **Vol. 8** 373
A COSTUREIRINHA DA SE **Vol. 8** 51, 159
A COUTEAUX TIRES **Vol. 2** 570, 738
A CUARENTA Y CINCO REVOLUCIOES POR MINUTO **Vol. 8** 524, 667
A CULPA **Vol. 8** 12, 157
A DOPPIA FACCIA **Vol. 7** 910, 1084
A DOS GRADOS DEL ECUADOR **Vol. 8** 268
A DOUBLE TOUR **Vol. 2** 224, 285, 700, 770
A ETE PERDUE UNE MARIEE, IL **Vol. 2** 153, 573
A FIL DI SPADA **Vol. 7** 16
A FLEUR DE PEAU **Vol. 2** 196
A FLOR DO MAR **Vol. 8** 7
A FORCE ON S'HABITUE **Vol. 6** *page 113, page 115*
A FOST PRIETENUL MEU **Vol. 3** 268, 303, 317
A FUGA **Vol. 8** 2
A HIERRO MUERE **Vol. 8** 276, 558, 725
A ILHA DE MORAES **Vol. 8** 56
A LA BELLE FREGATTE **Vol. 2** 32, 722
A LA DYNAMITE **Vol. 8** 732
A LA GUERRE COMME A LA GUERRE **Vol. 2** 335
A LA JAMAIQUE **Vol. 2** 368, 627
A LA MANIERE DE SHERLOCK HOLMES **Vol. 2** 627, 872
A LA PALIDA LUZ DE LA LUNA **Vol. 8** 297, 438
A LEGGE **Vol. 7** 829
A L'HORIZON DU SUD **Vol. 2** 440
A LOS PIES DE USTED **Vol. 8** 611
A LUZ VEM DO ALTO **Vol. 8** 96, 130, 155
A MALA NOVA **Vol. 7** 829
A MANTILHA DE BEATRIZ **Vol. 8** 44, 88, 106, 178, 207, 475
A MARECHIARE 'NCE STA NA FENESTA **Vol. 7** 565, 829
A ME MI PIACE **Vol. 7** 383
A ME NO ME MIRE USTED **Vol. 7** 587 **Vol. 8** 475
A MENINA DA RADIO **Vol. 8** 122, 133
A MI-CHEMIN DU CIEL **Vol. 2** 603
A MI, LA LEGION ! **Vol. 8** 357, 400
A MI, LAS MUJERES NI FU NI FA **Vol. 8** 331, 553
A MINUIT LE 7 **Vol. 2** 191, 273, 454, 595
A MORGADINHA DOS CANAVIAIS **Vol. 8** 10, 116
A MORT L'ARBITRE ! **Vol. 6** *page 94, page 130, page 134, page 140*
A MOSCA CIECA **Vol. 7** 493
A MOURA ENCANTADA **Vol. 8** 32
A NOI PIACE FREDDO ..! **Vol. 7** 372
A NOIVA DO BRASIL **Vol. 8** 107, 122
A NOS AMOURS **Vol. 6** *page 86, page 107, page 126*
A NOUS DEUX **Vol. 2** 879
A NOUS DEUX, MADAME LA VIE **Vol. 2** 8, 193, 362, 863
A NOUS LA LIBERTE **Vol. 2** 573, 656, 708, 837
A NOUS, LES PETITES ANGLAISES ! **Vol. 2** 335
A NOUS LES PETITES SALOPES ISABELLE ET OLGA **Vol. 6** *page 108, page 142*
A NOUS QUATRE, CARDINAL **Vol. 2** 738
A PARIS TOUS LES DEUX **Vol. 2** 593
...A PATY JEZDEC JE STRACH **Vol. 10** 229, 369
A PEGGIO OFFESA **Vol. 7** 424
A PIED, A CHEVAL ET EN SPOUTNIK **Vol. 2** 47
A PIED, A CHEVAL ET EN VOITURE **Vol. 2** 344, 386
A PIEDIGROTTA **Vol. 7** 829
A PLEINE BOUCHE **Vol. 2** 735
A PLEINES MAINS **Vol. 2** 205
A PORTE CHIUSE **Vol. 7** 798
... A POZDRAVUJI VLASTOVKY **Vol. 10** 87, 395, 517
A PRINCESINHA DAS ROSAS **Vol. 8** 78, 165
A PROMESSA **Vol. 8** 35, 169
A PROPOS DE LA FEMME **Vol. 2** 814
A PROPOS DE NICE **Vol. 2** 425
A PUNTA DE LATIGO **Vol. 8** 279
A QUALCUNO PIACE CALVO **Vol. 7** 514
A QUELQUES JOURS PRES **Vol. 2** 769

A REBROUSSE-POIL Vol. 2 270, 318, 807
A REVOLUCAO DE MAIO Vol. 8 18, 79, 190, 199
A RIBEIRA DA SAUDADE Vol. 8 130
A ROSA DO ADRO Vol. 8 20, 79
A SAGRADA FAMILIA Vol. 8 7
A SAN FRANCISCO Vol. 7 264, 701
A SANGRE FRIA Vol. 8 614
A SANTA ALIANCA Vol. 8 164, 183, 185
A SANTA LUCIA Vol. 7 965
A SANTANOTTE Vol. 7 829
A SEVERA Vol. 8 50
A SOUL ADRIFT Vol. 2 384
A SUD NIENTE DI NUOVO Vol. 7 1140
A TIRO LIMPIO Vol. 8 523, 732
A TOI DE FAIRE MIGNONNE Vol. 2 615, 659
A TOI DE JOUER CALLAGHAN Vol. 2 740 Vol. 7 982
A TOPE Vol. 8 465, 666
A TOUT CASSER Vol. 2 377, 512
A TU PER TU Vol. 7 383
... A TUTTE LE AUTO DELLA POLIZIA Vol. 7 1022
A ULTIMA PEGA Vol. 8 54, 160, 176
A VARANDA, DOS ROUXINOIS Vol. 8 18, 50
A VASEN VITEZI Vol. 10 201, 290, 584
A VENDRE Vol. 6 *page 108, page 142*
A VENISE UNE NUIT Vol. 2 344, 533, 775
Ä VI GIFTA ? Vol. 5 849, 1121, 1205
A VIDA E BELA ... ? Vol. 8 164, 169
A VIZINHA DO LADO Vol. 8 18, 200
A VOLTA DE JOSE DO TELHADO Vol. 8 96, 121
A VOS ORDRES, MADAME Vol. 2 9, 395
A VOZ DO SANGUE Vol. 8 54, 176
... A ZASE TA LUCIE ! Vol. 10 70, 577
A 008: OPERAZIONE STERMINIO Vol. 7 1121
A ZIVOT JDE DAL Vol. 10 34, 167, 256, 379, 507, 524
AALTOSKA ORKANISEERAA Vol. 5 374, 443
AATAMIN PUVUSSA JA VÄHAN EEVANKIN Vol. 5 349, 356, 428, 494, 549, 639
AB MITTERNACHT Vol. 4 410
ABAJO ESPERA LA MUERTE Vol. 8 350, 450
ABAJO LOS HOMBRES Vol. 8 285, 485
ABANDERADO, EL Vol. 8 716, 752
ABANDERADOS DE LA PROVIDENCIA, LOS Vol. 8 732, 753
ABASSO LA MISERIA ! Vol. 7 701
ABAT - JOUR Vol. 7 1174
ABBA - THE MOVIE Vol. 5 900, 1072
ABBANDONO Vol. 7 453, 954, 1096
ABBANDONO DI PAPA Vol. 7 1189
ABBASSO IL CAMBIO ! Vol. 7 1211
ABBASSO LA FORTUNA Vol. 7 487
ABBASSO LA RICCHEZZA Vol. 7 1136
ABBASSO TUTTI VIVA NOI Vol. 7 901
ABBE CONSTANTIN, L' Vol. 2 503, 598, 840
ABBIAMO VINTO ! Vol. 7 1170
ABBRACCIO DELLA VERGINE DI FERRO, L' Vol. 7 91, 980
A.B.C. OF LOVE Vol. 2 384
A.B.C... ... ...ZH Vol. 3 9
A.B.C.MILOSCI Vol. 1 278, 308, 404
ABECEDA STRAHA Vol. 3 421
ABEL MIT DER MUNDHARMONIKA Vol. 4 30, 325, 445, 567
ABEL, TWOJ BRAT Vol. 1 243, 305 Vol. 6 *page 245*
ABELE FRATRICIDA Vol. 7 1189
ABELHA NA CHUVA, UMA Vol. 8 185
ABEMENNESKET Vol. 5 83
ABEND - NACHT - MORGEN Vol. 4 240, 479
ABENTEUER DER MARQUISE VON KÖNIGSMARK, DIE Vol. 4 213, 356
ABENTEUER DER SIBYLLE BRANT, DAS Vol. 4 213, 626
ABENTEUER DER THEA ROLAND, DAS Vol. 4 28, 584
ABENTEUER DES DR. KIRCHEISEN, DAS Vol. 4 660
ABENTEUER DES WERNER HOLT, DIE Vol. 1 47, 111, 125
ABENTEUER EINER NACHT Vol. 4 471, 560
ABENTEUER EINES ERMORDETEN, DIE Vol. 4 309, 780
ABENTEUER EINES JUNGEN HERRN IN POLEN Vol. 4 54, 340, 373, 466
ABENTEUER EINES ZEHNMARKSCHEINES, DIE Vol. 4 28, 43, 397, 546
ABENTEUER GEHT WEITER, DAS Vol. 4 85, 620
ABENTEUER IM ENGADIN Vol. 4 15, 192, 422, 620, 738
ABENTEUER IM SÜDEXPRESS Vol. 4 132, 230
ABENTEUER IN BAMSDORF Vol. 1 26, 65, 73
ABENTEUER IN DER KARNEVALSNACHT Vol. 4 617
ABENTEUER IN MAROKKO Vol. 4 292
ABENTEUER IN WARSCHAU Vol. 4 340, 361, 466
ABENTEUER VON SAGOSSA, DAS Vol. 4 255, 327, 344, 545
ABENTEUERIN VON MONTE CARLO, DIE Vol. 4 240

ABENTEUERIN VON TUNIS, DIE Vol. 4 321, 479, 634
ABENTEUERLICHE HOCHZEIT, DIE Vol. 4 25, 212, 544, 726
ABENTEUR DER SCHÖNEN DORETTE, DIE Vol. 4 239, 294
ABENTEUR IM GRANDHOTEL Vol. 4 18, 44
ABENTEUR IM NACHTEXPRESS Vol. 4 356, 471, 560, 779
ABENTEURER, DER Vol. 4 492, 584, 634
ABENTEURER G. M. B. H., DIE Vol. 4 73, 626, 747
ABENTEURER VON PARIS, DER Vol. 4 536, 558
ABER DOKTOR ! Vol. 6 *page 161, page 162, page 168*
ABER MEIN LIEBER HERR NEUMANN Vol. 4 25, 367
ABER VATI! Vol. 1 11, 98, 122
ABERGLAUBE Vol. 4 743
ABETE FULMINATO, L' Vol. 7 420
ABGEFUNDEN Vol. 6 *page 159, page 166, page 175*
ABGRUND DER SEELEN, DER Vol. 4 217, 329
ABGRÜNDE DER LIEBE Vol. 4 98, 112
ABIGEL Vol. 9 193, 203, 329
ABISMOS DA MEIA-NOITE, OS Vol. 8 169
ABITO NERO DA SPOSA, L' Vol. 7 589, 910, 1136
ABITO NUOVO, L' Vol. 7 1189
ABOGADO, EL ALCALDE Y EL NOTARIO, EL Vol. 8 426, 732
ABOMINABLE HOMBRE DE LA COSTA DEL SOL, EL Vol. 8 524
ABOMINABLE HOMME DES DOUANES, L' Vol. 2 503, 598, 840
ABORTO CRIMINAL Vol. 8 261
ABRACADABRA Vol. 7 792, 1063
ABRE TU FOSA AMIGO ... LIEGA SABATA PRESAGIO Vol. 8 629, 768
ABRIGO A CUADROS, UN Vol. 8 692, 772
ABSAGE AN VIKTORIA Vol. 1 19, 34, 104
ABSCHIED Vol. 1 7, 50, 73 Vol. 4 339, 629
ABSCHIED VOM FRIEDEN Vol. 1 30, 73, 114
ABSCHIEDSSYMPHONIE Vol. 4 354, 614
ABSCHIEDSWALZER Vol. 4 89, 246
ABSEITS VOM GLÜCK Vol. 4 178
ABSENCES REPETEES Vol. 2 754
ABSTURZ, DER Vol. 4 51, 213, 361, 651, 800
ABTRÜNNIGE, DER Vol. 4 119
ABUELO, EL Vol. 8 655
ABUELO MADE IN SPAIN Vol. 8 524
ABUELO TIENE UN PLAN, EL Vol. 8 681, 749
ABUNA MESSIAS Vol. 7 466, 476, 926, 1136
ABUS DE CONFIANCE Vol. 2 153
ABWEGE Vol. 4 149, 660, 663
ABYSSES, LES Vol. 2 126
ACAPULCO, PRIMA SPIAGGIA A SINISTRA Vol. 7 239, 241, 901
ACASA Vol. 6 *page 269, page 271, page 282*
ACCADDE AL COMMISSARIATO Vol. 7 476
ACCADDE AL PENITENZIARO Vol. 7 333
ACCADDE TRA LE SBARRE Vol. 7 302
ACCALMIE, L' Vol. 2 804
ACCATONE Vol. 7 333, 781
ACCIAIO Vol. 7 756, 1036, 1116
ACCIDENT Vol. 3 211, 233, 235
ACCIDENT, L' Vol. 2 49, 97
ACCIDENTE 703 Vol. 8 524, 725
ACCIDENTI ALLE TASSE Vol. 7 333
ACCIDIA, L' Vol. 7 441, 701
ACCORD FINAL Vol. 2 427, 439
ACCORDO IN MINORE, L' Vol. 7 824, 825
ACCOUPLEMENTS POUR VOYEURS Vol. 2 735
ACCROCHE-COEUR, L' Vol. 2 84, 249
ACCUSATEUR, L' Vol. 2 257
ACCUSED Vol. 2 32
ACCUSEE, LEVEZ-VOUS Vol. 2 32, 84, 197
ACECHO, EL Vol. 8 732
ACES HIGH Vol. 2 299
ACH, DU FRÖHLICHE Vol. 1 13, 48, 107
ACH TE SPODNIE Vol. 1 174, 188
ACHATMURMEL, DIE Vol. 1 36, 92, 103
ACHILLESFERSE Vol. 1 1, 17, 33
ACHILLESHAELEN ER MIT VÄBEN Vol. 5 46, 99
ACHT MÄDELS IM BOOT Vol. 4 49, 316
ACHT UHR DREIZEHN Vol. 4 479, 634
ACHTE GEBOT, DAS. DU SOLLST NICHT FALSCH ZEUGNIS REDEN... Vol. 4 396, 423
ACHTE MIR AUFS GAKEKI Vol. 4 115, 361
ACHTEN SIE AUF MEYER Vol. 4 558, 747
ACHTUNG ! – AUTODIEBE ! Vol. 4 119, 479
ACHTUNG, BANDITI Vol. 7 401, 1167
ACHTUNG ! FEIND HÖRT MIT ! Vol. 4 79, 115, 745, 774
ACHTUNG HARRY ! AUGEN AUF ! Vol. 4 270, 471, 779
ACHTUNG KURVE Vol. 4 525, 546
ACHTUNG ! LIEBE ! LEBENSGEFAHR ! Vol. 4 84, 452

ACHTUNG ! WER KENNT DIESE FRAU ? **Vol. 4** 344, 647
ACHTZEHNJÄHRIGEN, DIE **Vol. 4** 424, 527
ACOMPANAME **Vol. 8** 749, 762
ACORDARTI CIRCUMSTANTE ATENUANTE ? **Vol. 6** *page 271, page 280, page 281*
ACOSADA **Vol. 8** 639
ACQUA, ACQUA, FUOCO, FUOCO **Vol. 7** 598
ACQUA CHETA **Vol. 7** 493
ACQUA E SAPONE **Vol. 7** 383, 1008, 1166
ACQUE AMARE **Vol. 7** 17
ACQUE DI PRIMAVERA **Vol. 7** 792, 1063
ACROBATE, L' **Vol. 2** 32, 128, 329, 332, 513
ACTE D'AMOUR, UN **Vol. 2** 626, 824, 837
ACTEON **Vol. 8** 492, 495, 581
ACTION **Vol. 7** 315, 585, 590
ACTION IMMEDIATE **Vol. 2** 496, 515
ACTO DE PRIMAVERA **Vol. 8** 197
ACTO DOS FEITOS DA GUINE **Vol. 8** 34, 152
ACTORUL SI SALBATICII **Vol. 3** 291, 319, 332
AD OCCHI BENDATI **Vol. 7** 1189
AD OGNI COSTO **Vol. 8** 517, 732
AD OVEST DI PAPERINO **Vol. 7** 13, 303
ADA **Vol. 6** *page 307*
ADA SE UCI JEZDIT **Vol. 10** 45, 58
ADA, TO NIE WYPADA **Vol. 1** 278, 308, 404
ADALEN 31 **Vol. 5** 1096, 1176
ADALENS POESI **Vol. 5** 924, 978, 1042
ADAM A EVA **Vol. 10** 2, 154, 243, 536
ADAM A OTKA **Vol. 10** 327, 515, 568
ADAM EST EVE **Vol. 2** 480
ADAM ET EVE **Vol. 6** *page 114*
ADAM, EVA UND DER APFEL **Vol. 4** 361, 558
ADAM OCH EVA **Vol. 5** 869, 881
ADAM OG EVA **Vol. 5** 31, 179, 241, 248
ADAM UND EVA **Vol. 4** 85, 208, 373, 638, 646, 755
ADAMO ED EVA **Vol. 7** 1136
ADAMSSON I SVERIGE **Vol. 5** 881, 940, 1081
ADAN Y EVA, LA PRIMERA HISTORIA DE AMOR **Vol. 8** 375, 376
ADAPTACIJA **Vol. 3** 117 **Vol. 6** *page 35*
ADDIO ALLE ARMI **Vol. 7** 756
ADDIO, AMORE **Vol. 7** 390, 756, 796
ADDIO GIOVINEZZA ! **Vol. 7** 456, 729, 756, 796, 1060, 1136
ADDIO, KIRA **Vol. 7** 3, 117, 250
ADDIO MIA BELLA, ADDIO **Vol. 7** 829
ADDIO MIA BELLA NAPOLI **Vol. 7** 843, 910, 1027
ADDIO, MIA BELLA SIGNORA ! **Vol. 7** 1063
ADDIO MIMI **Vol. 7** 756
ADDIO PER SEMPRE **Vol. 7** 196
ADDIO, PICCOLA MIA **Vol. 1** 18, 48, 80
ADDIO ULTIMO UOMO **Vol. 7** 273
ADDITION, L' **Vol. 6** *page 78, page 80*
ADDYS AEGTESKAB **Vol. 5** 9
ADELA **Vol. 6** *page 269, page 271, page 276* **Vol. 8** 205, 234, 455
ADELA JESTE NEVECERELA **Vol. 10** 292, 299, 369, 425
ADELAIDE **Vol. 2** 689
ADELHEID **Vol. 10** 139, 425, 567
ADEMAI AU MOYEN-AGE **Vol. 2** 122, 538
ADEMAI AU POITEAU-FRONTIERE **Vol. 2** 229, 598
ADEMAI AVIATEUR **Vol. 2** 122, 778
ADEMAI BANDIT D'HONNEUR **Vol. 2** 57, 197
ADERFIA MOU, ALETES POULIA **Vol. 3** 165
ADERPOS ANNA **Vol. 3** 141
ADEUS PORTUGUES, UM **Vol. 8** 7, 91, 97, 118
ADHEMAR LAMPIOT **Vol. 2** 660, 761
ADHEMAR OU LE JOUET DE LA FATALITE **Vol. 2** 671, 706
ADIEU CHERIE **Vol. 2** 388, 488
ADIEU L'AMI **Vol. 2** 263, 814
ADIEU LEONARD **Vol. 2** 251, 826
ADIEU LES COPAINS **Vol. 2** 603
ADIEU MASCOTTE **Vol. 4** 164, 584, 707
ADIEU NU, L' **Vol. 2** 265, 323, 337
ADIEU PARIS **Vol. 2** 471, 661
ADIEU PHILIPPINE **Vol. 2** 564
ADIEU POULET **Vol. 2** 196, 783
ADIEU VIENNE **Vol. 2** 530
ADIKE KATARA **Vol. 3** 152, 164
ADIO, DRAGA NELA ! **Vol. 3** 276, 285
ADIOS, CIGUENA, ADIOS **Vol. 8** 348
ADIOS CORDERA **Vol. 8** 338, 611
ADIOS GRINGO **Vol. 8** 306, 719
ADIOS MIMI POMPOM **Vol. 8** 207, 214
ADIOS, MUCACOS **Vol. 3** 41 **Vol. 6** *page 35, page 53*
ADIOS PEQUENA **Vol. 8** 373, 709
ADIOS TEXAS **Vol. 8** 626
ADJ, KIRALY, KATONAT **Vol. 9** 107, 167, 238, 240, 279
ADJO SOLIDARITET **Vol. 5** 667, 737, 774
ADJUNKT VRBA **Vol. 10** 167, 453
ADJUTANT, DER **Vol. 1** 2, 26, 85
ADJUTANT DES ZAREN, DER **Vol. 4** 149, 660, 707
ADJUTANT SEINER HOHEIT, DER **Vol. 4** 257
ADOLESCENCE PERVERSE **Vol. 2** 215

ADOLESCENTE, L' **Vol. 2** 265, 323, 337 **Vol. 7** 474
ADOLESCENTES A DEPUCELER **Vol. 2** 294
ADOLESCENTES AU PENSIONNAT **Vol. 6** *page 127*
ADOLESCENTES, LAS **Vol. 8** 465, 616
ADOLESCENTES TROP CURIEUSES **Vol. 6** *page 118, page 119*
ADOLESCENTS, LES **Vol. 7** 300, 634, 778, 910
ADOLESCENTS POUR VICIEUX OU SCRABBLE PARTOUZE **Vol. 2** 294
ADOLF ARMSTARKE **Vol. 5** 878, 927
ADOLF I ELD OCH LAGOR **Vol. 5** 923, 1205
ADOLF KARAR SKIVAN **Vol. 5** 923, 1205
ADOLPHE OU L'AGE TENDRE **Vol. 2** 175
ADOLPHE OU LE L'EDUCATION AMOUREUSE **Vol. 2** 285
ADOM OU LE SANG D'ABEL **Vol. 2** 78, 754
ADORABILI E BUGIARDE **Vol. 7** 333
ADORABLE MENTEUSE **Vol. 2** 398, 481
ADORABLES CREATURES **Vol. 2** 226, 388, 566
ADORABLES DEMONS **Vol. 2** 128, 581
ADRESSATIN VERSTORBEN **Vol. 4** 183
ADRESSE UNBEKANNT **Vol. 4** 273, 623
ADRIANA LECOUVREUR **Vol. 7** 86, 965, 1057
ADRIANOPELS HEMMELIGHED **Vol. 5** 70
ADRIEN **Vol. 2** 332, 824
ADUA E LE COMPAGNE **Vol. 7** 817, 1035
ADULATERA, L' **Vol. 7** 1189
ADULTERIO DECENTE, UN **Vol. 8** 207
ADULTERIO LEGALE **Vol. 7** 1189
ADULTERO, EL **Vol. 8** 207, 581, 749, 773
ADULTERO LUI, ADULTERA LEI **Vol. 7** 514, 634, 1142
ADVENT **Vol. 10** 184, 486, 564
ADVERSIDAD **Vol. 8** 661
ADVOKAT CHUDYCH **Vol. 10** 142, 345, 532
ADVOKATKA **Vol. 10** 155, 263, 483
ADVOKATKA VERA **Vol. 10** 256, 412
AEDLE SKRAEDDER, DEN **Vol. 5** 83
AEGTEMAND FOR EN TIME **Vol. 5** 83
AEGTESKAB OG PIGESJOV **Vol. 5** 9
AEGTESKABETS TORNEFULDE VEJ **Vol. 5** 81
AEGYPTEREN RADAMAS **Vol. 5** 203
AE'KKE TE A TRU, DET **Vol. 5** 726
AENNCHEN VON THARAU **Vol. 4** 232
AERA **Vol. 3** 163, 192
AERELOSE, DEN **Vol. 5** 58
AEREN TABT – ALT TABT **Vol. 5** 97
AEREOPORTO **Vol. 7** 910
AERESOPREJSNING, EN **Vol. 5** 58
AEROGUAPAS, LAS **Vol. 8** 353 393, 400, 668, 677, 749
AEROPUERTO **Vol. 8** 614, 749
AETHER-BRETTL **Vol. 4** 367
AEVENTYRERSKEN **Vol. 5** 9
AF ELSKOVS NAADE **Vol. 5** 9
AFACEREA PROTEA **Vol. 3** 231, 266, 317, 332
AFAN – EVU **Vol. 8** 625, 692
AFBRUDTE BRYLLUPSNAT, DEN **Vol. 5** 203
AFERA AHIL **Vol. 6** *page 314*
AFERA KRYSTYNY **Vol. 1** 404
AFERA PLUKOVNIKA REDLA **Vol. 10** 29, 528, 586
AFERA V GRANDHOTELU **Vol. 10** 34
AFERY ME ZENY **Vol. 10** 184, 193, 296, 612
AFFAEREN BIRTE **Vol. 5** 36, 89, 144, 224
AFFAEREN I MOLLEBY **Vol. 5** 61, 125, 195
AFFAIRE BLAIREAU, L' **Vol. 2** 380, 381, 387, 469, 731
AFFAIRE BLUM **Vol. 1** 6, 39, 90
AFFAIRE COQUELET, L' **Vol. 2** 336, 566
AFFAIRE CRAZY CAPO, L' **Vol. 2** 241
AFFAIRE DE LA RUE DE LOURCINE, L' **Vol. 2** 388
AFFAIRE DES POISONS, L' **Vol. 2** 279, 601, 680
AFFAIRE D'HOMMES, UNE **Vol. 6** *page 83, page 96*
AFFAIRE DOMINICI, L' **Vol. 2** 35, 704
AFFAIRE DU COLLIER DE LA REINE, L' **Vol. 2** 27, 251, 402
AFFAIRE DU COURRIER DE LYON, L' **Vol. 2** 427, 439
AFFAIRE DU GRAND-HOTEL, L' **Vol. 2** 6, 328, 471
AFFAIRE D'UNE NUIT, L' **Vol. 2** 185, 232, 488
AFFAIRE EST DANS LE SAC, L' **Vol. 2** 837
AFFAIRE LAFARGE, L' **Vol. 2** 81, 526, 644, 646
AFFAIRE LESURGUES, L' **Vol. 2** 427
AFFAIRE MAURIZIUS, L' **Vol. 2** 232, 251, 488
AFFAIRE NINA B., L' **Vol. 2** 279, 427
AFFAIRE ROEDERN, DIE **Vol. 4** 525
AFFAIRES ETRANGERES **Vol. 2** 74
AFFAIRES SONT LES AFFAIRES, LES **Vol. 2** 122, 722
AFFAMEES, LES **Vol. 6** *page 119*
AFFÄRE HEYDE-SAWADE, DIE **Vol. 1** 21, 73, 125
AFFARE SI COMPLICA, L' **Vol. 7** 460, 941

AFFICHE, L' **Vol. 2** 100, 237
AFFICHE ROUGE, L' **Vol. 2** 754
AFFITTACAMERE, L' **Vol. 7** 1202
AFFRESCO DI POMPEI, L' **Vol. 7** 986
AFFREUX, LES **Vol. 2** 293, 294, 598
AFGRUNDEN **Vol. 5** 188
AFRICA DOLCE E SELVAGGIA **Vol. 7** 273
AFRICA SOTTO I MARI **Vol. 7** 591
AFRICAIN, L' **Vol. 6** *page 85, page 91*
AFRIKA EXPRESS **Vol. 7** 514
AFRIKAI VÖLEGENY **Vol. 9** 58, 270
AFRIKANEREN **Vol. 5** 754, 772, 799
AFRODITE, DEA DELL' AMORE **Vol. 7** 1027
AFSKEDENS TIME **Vol. 5** 52, 125, 247
AFSPORET **Vol. 5** 196, 264
AFTENLANDET **Vol. 5** 56, 57, 221
AFTENSCENE, EN **Vol. 5** 9, 197
AFTER THE VERDICT **Vol. 6** *page 184, page 198*
AFYON – OPPIO **Vol. 7** 869
AGACHATE QUE DISPARAN **Vol. 8** 317, 405
AGAPE GIA PANTA **Vol. 3** 153, 170, 205
AGAPE KAI AIMA **Vol. 3** 158, 205
AGAPES KAI KAEMOL **Vol. 3** 130, 132, 155
AGAPESA ENAN ALETE **Vol. 3** 161
AGAPESA ENAN PRODOTE **Vol. 3** 155
AGAPI MOU OUA OUA **Vol. 3** 153, 205
AGATHA OU LES LECTURES ILLIMITEES **Vol. 6** *page 89, page 128*
AGATHE ET MARTHA **Vol. 6** *page 79*
AGE D'OR, L' **Vol. 2** 277, 395, 558, 617, 775
AGE EN FLEUR, L' **Vol. 2** 244
AGE INGRAT **Vol. 2** 197
AGENCE MATRIMONIALE **Vol. 2** 251, 309
AGENT 69 JENSEN **Vol. 5** 127, 158, 257
AGENT NR. 1 **Vol. 1** 281, 311 **Vol. 6** *page 248, page 258*
AGENT O,5 OCH KVARTEN – FATTARUVÄL ! **Vol. 5** 887
AGENTE 3 S 3 – MASSACRO AL SOLE **Vol. 8** 438
AGENTE 3 S 3 – PASAPORTE PARA EL INFIERNO **Vol. 8** 732
AGENTE 3 S 3 – PASSAPORTE PER L'INFERNO **Vol. 8** 732
AGENTE END **Vol. 8** 732
AGENTE SIGMA 3 – MISION GOLDWATER **Vol. 8** 407
AGENTE SIGMA 3 – MISSIONE GOLDWATER **Vol. 8** 407
AGENTE SO 3, OPERACION ATALANTIDA **Vol. 8** 708
AGENTE SO 3, OPERAZIONE ATLANDITE **Vol. 8** 708
AGENTE X – 1 – 7, OPERACION OCEANO **Vol. 8** 752
AGENTE Z – 55, MISION HONG – KONG **Vol. 8** 280, 732
AGENTE Z – 55, MISSIONE DISPERATA **Vol. 8** 280
AGENTENS TVILLINGER **Vol. 5** 83
AGENTES DEL QUINTO GRUPO, LOS **Vol. 8** 768
AGENTTI OOO JA KUOLEMAN KURVIT **Vol. 5** 314
AGENTUL STRANIU **Vol. 3** 277, 327, 335
AGENZIA RICCARDO FINZI ... PRATICAMENTE DETECTIVE **Vol. 7** 310
ÄGGET AR LÖST ! **Vol. 5** 852, 947, 1170
AGGUATO A TANGERI **Vol. 7** 817, 871, 910
AGGY, ISTEN BIRI ! **Vol. 9** 28, 71
AGI MURAD, IL DIAVOLO BIANCO **Vol. 7** 106, 1019
AGIOS NEKTARIOS **Vol. 3** 199
AGITESE ANTES DE USARLA **Vol. 8** 681, 770
AGNE TOU LIMANIOU **Vol. 3** 168
AGNES DE RIEN **Vol. 2** 254, 833
AGNESE VA A MORIRE, L' **Vol. 7** 19, 909, 1153
AGNIESZKA 46 **Vol. 1** 179, 243
AGONIA **Vol. 3** 180
AGONIA GIA TON EROTA **Vol. 3** 132
AGONIE DE JERUSALEM, L' **Vol. 2** 387, 672, 824
AGONIE DES AIGLES, L' **Vol. 2** 131, 209, 217, 308, 496, 727
AGONIZANDO EN EL CRIMEN **Vol. 8** 249, 772
AGORA E QUE SAO ELAS **Vol. 8** 172, 192
AGOROKORITSO **Vol. 3** 161
AGOSTINO O LA PERDITA DELL' INNOCENZA **Vol. 7** 1136
AGOSTO **Vol. 8** 7
AGOSTO, DONNE MIE NON VI CONOSCO **Vol. 7** 877
AGRESSION, L' **Vol. 2** 642
AGRIA OIKOGENEIA **Vol. 3** 189
AGROLSZAKADT URILANY **Vol. 9** 58, 76, 196, 270
AGUA EN EL SUELO, EL **Vol. 8** 244, 268, 270
AGUAS BAJAN NEGRAS **Vol. 8** 207
AGUAS SANGRIENTAS **Vol. 8** 606
AGUILAS DE ACERO **Vol. 8** 606
AGUSTINA DE ARAGON **Vol. 8** 246, 298, 606, 607
AGYU ES HARANG **Vol. 9** 337
AH ! CHE RAGAZZI ! **Vol. 7** 1189
AH, EN SA'N ADVOKAT **Vol. 5** 843, 923

AH, EN SA'N GRABB ! Vol. 5 840, 841
AH, EN SA'N NATT Vol. 5 923
AH ! LES BELLES BACCHANTES Vol. 2 137, 193
AH, QUELLA DORY ! Vol. 7 148
AH ! QUELLE EQUIPE Vol. 2 270, 501, 669
AH ! QUELLE GARCE! Vol. 2 646
AH SI ? ... E IO LO DICO ZZZORRO ! Vol. 7 1179
AHASVER Vol. 4 124, 232 Vol. 6 *page 104* Vol. 10 45
AHI VA OTRO RECLUTA Vol. 8 214
AHNUNGSLOSE ENGEL, DER Vol. 4 302, 344, 647, 689
AHOI, AHOI Vol. 4 623
AI MARGINI DELLA METROPOLI Vol. 7 60, 1167
AI VOSTRI ORDINI, SIGNORA ! Vol. 7 493
AIDA Vol. 7 923
AIDANKAATAJAT Vol. 5 324, 606, 630
ÄIDITTÖMAT Vol. 5 443, 643
AIGLE A DEUX TETES, L' Vol. 2 85, 282, 566, 863
AIGLE ET LA COLOMBE, L' Vol. 2 76
AIGLON, L' Vol. 2 93, 153, 579, 676, 711, 833
AIGLONNE, L' Vol. 2 452
AIGRETTE, L' Vol. 7 150, 965
AIGUILLE ROUGE, L' Vol. 2 47, 625, 863
AIKA HYVÄ IHMISEKSI Vol. 5 381, 459, 545
AIKALAINEN Vol. 5 434, 589
AILA, POHJOLAN TYTÄR Vol. 5 334, 656
AILE OU LA CUISSE, L' Vol. 2 142, 409, 719
AILES BLANCHES, LES Vol. 2 330, 407, 540, 603
AILES DE LA COLOMBE, LES Vol. 6 *page 123*
AIMEE Vol. 2 806 Vol. 6 *page 91, page 104*
AIMEZ-VOUS BRAHMS? Vol. 2 824
AIMEZ-VOUS LES FEMMES? Vol. 2 285, 322, 854
AIMEZ VOUS LES UNS LES AUTRES...MAIS PAS TROP Vol. 2 74
AINAMA Vol. 2 406 Vol. 6 *page 86*
AINE DES FERCHAUX, L' Vol. 2 224, 376
AINSI FINIT LA NUIT Vol. 2 9, 247
AIR DE PARIS, L' Vol. 2 96, 402
AIUTAMI A SOGNARE Vol. 7 333, 593
AJANDEK EZ A NAP Vol. 9 50, 51, 167
AJAX Vol. 7 209
AJOLÄHTÖ Vol. 5 532, 648, 653
A.K.A. KNUD RASMUSSENS GRONLANDSFILM Vol. 5 263, 288
AKADEMIA PANA KLEKSA - PRZYDGODA KSIECIA MATEUSZA Vol. 6 *page 244, page 251, page 252*
AKADEMIA PANA KLEKSA - TAJEMNICA GOLARZA FILIPA Vol. 6 *page 251*
AKATON MIES Vol. 5 477, 639
AKCE B Vol. 10 413, 580
AKCE BORORO Vol. 10 110, 153, 173, 517
AKCE KALIMANTAN Vol. 10 299, 577
AKCE V ISTANBULU Vol. 10 277, 286, 408
AKCIA EDELSTEIN Vol. 10 72, 148, 544
AKCIE P.Z.H. Vol. 10 536, 591
AKCIJA Vol. 3 345, 395
AKCIJA STADION Vol. 3 371
AKCJA „BRUTUS" Vol. 1 211, 281 Vol. 6 *page 240, page 248*
AKCJA POD ARSENALEM Vol. 1 156, 178, 374, 393
AKE OCH HANS VÄRLD Vol. 5 850, 1035, 1096
AKELARRE Vol. 8 219, 580
AKIK ELETET CSERELTEK Vol. 9 77
AKIKET ELKAP AZ AR Vol. 9 126, 315, 317
AKLI MIKLOS Vol. 9 158, 265, 285
AKO NE IDE VLAK Vol. 3 60, 78
AKORD SMRTI Vol. 10 122, 198
AKROBAT SCHÖ-Ö-ÖN Vol. 4 97, 215, 228
AKROGIALI TOU EROTA Vol. 3 132
AKSELI JA ELINA Vol. 5 324, 613, 635, 639
ÄKTENSKAPSBROTTAREN Vol. 5 878, 1052, 1053, 1130
ÄKTENSKAPSLEKEN Vol. 5 840, 909
AKTIE BOLAGET HÄLSANS GAVA Vol. 5 944, 966
AKTIVISTIT Vol. 5 330, 471
AKTÖREN Vol. 5 923, 928
AKTORZY PROWINCJONALNI Vol. 6 *page 220, page 221, page 252*
AKWARELE Vol. 1 276, 293 Vol. 6 *page 227*
AL AMOR EMPIEZA A MEDIANOCHE Vol. 8 681
AL BAR DELLO SPORT Vol. 7 599, 957
AL BUIO INSIEME Vol. 7 48, 466, 756
AL CHIAROR DEI LAMPI Vol. 7 26
AL CONFINE DELLA MORTE Vol. 7 984
AL DI LA DEL BENE E DEL MALE Vol. 7 62, 817, 1142
AL DI LA DELLA VITA Vol. 7 914, 1209
AL ESTE DEL OESTE Vol. 8 728, 770
AL HOLLYWOOD MADRILENO Vol. 8 516
AL LIMITE CIOE NON GLIELO DICO Vol. 7 987, 1022
AL MARGEN DE LA LEY Vol. 8 396
AL OESTE DE RIO GRANDE Vol. 8 673

AL OTRO LADO DE LA CIUDAD **Vol. 8** 276, 732
AL OTRO LAO DEL ESPEJO **Vol. 8** 552, 770
AL PATRULEA STOL **Vol. 3** 320 **Vol. 6** *page 267*
AL PIACERE DI RIVEDERLA **Vol. 7** 567
AL ROMBO DEL CANNON ! **Vol. 7** 341
AL SERVICIO DE LA MUJER ESPANOLA **Vol. 8** 728, 749
AL SERVIZIO DELL' IMPERATORE **Vol. 7** 693
AL TREILEA SALT MORTAL **Vol. 3** 262, 308 **Vol. 6** *page 265*
ÄL' YLI PÄÄSTÄ PERHANAA **Vol. 5** 375, 594
ALA - ARRIBA ! **Vol. 8** 18, 50, 67, 68, 171
ÄLÄ NUOLASE... **Vol. 5** 443, 549
ALABAMA **Vol. 6** *page 224, page 255*
ALALA **Vol. 8** 653
ALAMBRADAS DE VIOLENCIO **Vol. 8** 495
ALARCOSBAL **Vol. 9** 288
ALARIDO, EL **Vol. 8** 606
ALARM **Vol. 4** 198, 292, 422 **Vol. 5** 7
ALARM AUF STATION III **Vol. 4** 8, 564
ALARM IM ZIRKUS **Vol. 1** 9, 100, 101
ALARM IN PEKING **Vol. 4** 49, 108, 270
ALARMA IN DELTA **Vol. 3** 208, 321
ALARMA IN MUNTI **Vol. 3** 224, 297, 310
ALARMSTUFE V **Vol. 4** 288, 619, 660, 728
ALAS DE JUVENTUD **Vol. 8** 298, 524, 716
ALASKA STORY **Vol. 3** 437
ALASKAFÜCHSE **Vol. 1** 68, 74, 102
ALASTON MALLI KARKUTEILLÄ **Vol. 5** 451, 592
ALAZATOSAN JELENTEM **Vol. 9** 67, 149, 285
ALBA DE AMERICA **Vol. 8** 298, 400
ALBA DI LIBERTA **Vol. 7** 390
ALBA DI SANGUE **Vol. 7** 81
ALBA NUZIALE **Vol. 7** 961
ALBA REGIA **Vol. 9** 117, 165, 259
ALBA ROSSA **Vol. 7** 1005
ALBA TORBIDA **Vol. 7** 932
ALBA, UN' **Vol. 7** 15, 429
ALBATROS, L' **Vol. 2** 300, 866
ALBERGO A ORE **Vol. 7** 609
ALBERGO DEGLI ASSENTI, L' **Vol. 7** 476
ALBERGO DELLA FELICITA, L' **Vol. 7** 48
ALBERGO LUNA, CAMERA TRENTA QUATTRO **Vol. 7** 796
ALBERGO MODERNO **Vol. 7** 1189
ALBERO DEGLI ZOCCOLI, L' **Vol. 7** 1144, 1212
ALBERO DI ADAMO, L' **Vol. 7** 453, 493
ALBUM POLSKI **Vol. 1** 144, 156, 325, 360 **Vol. 6** *page 230*
ALCALDE DEZALAMENA, EL **Vol. 8** 214, 306, 614
ALCAMIR **Vol. 8** 376, 672
ALCHIMISTEN **Vol. 1** 111, 128
ALCOVA, L' **Vol. 7** 349
ALCOVA MUTA, L' **Vol. 7** 932, 1104
ALDEA MALDITA, LA **Vol. 8** 207, 246, 449
ALDEBARAN **Vol. 3** 183 **Vol. 7** 48, 453, 796, 1036, 1116
ALDEIA DA ROUPA BRANCA **Vol. 8** 18, 122
ALDO ET JUNIOR **Vol. 6** *page 70, page 101*
ALDOZAT, AZ **Vol. 9** 20, 41, 311
ALDRI ANNET ENN BRAK **Vol. 5** 762, 831
ALDRIG I LIVET **Vol. 5** 533
ALDRIG MED MIN KOFOT **Vol. 5** 1033
ALEGRA CARAVANA, LA **Vol. 2** 646
ALEGRE CARAVANA, LA **Vol. 8** 298, 607, 705
ALEGRE DIVORCIADO, EL **Vol. 8** 313, 333
ALEGRE JUVENTUD **Vol. 8** 450, 546
ALEGRE VOY **Vol. 8** 718
ALEGRES BRIBONES, LES **Vol. 8** 257, 474, 757
ALEGRES CHICAS DE COLSADA, LAS **Vol. 8** 207, 214
ALEGRIA DE LA HUERTA, LA **Vol. 8** 263, 430
ALEGRIA DEL BATALLON, LA **Vol. 8** 422
ALEGRIA QUE PASA, LA **Vol. 8** 531
ALEKSA DUNDIC **Vol. 3** 383
ALENA **Vol. 10** 142, 256, 412
ALENE I VERDEN **Vol. 5** 97
ALERTA EN EL CIELO **Vol. 8** 400, 450
ALERTE AU DEUXIEME BUREAU **Vol. 2** 309, 376, 743
ALERTE AU SUD **Vol. 2** 401, 419, 777
ALERTE AUX CANARIES **Vol. 2** 131, 703
ALERTE EN MEDITERRANEE **Vol. 2** 533
ALERTEZ LES BEBES **Vol. 2** 157
ALESSANDRO, SEI GRANDE ! **Vol. 7** 16, 487, 756, 1099
ALESSIA ... UN VULCANO SOTTO LA PELLE **Vol. 7** 949
ALEXANDER DEN STORE **Vol. 5** 937, 997
ALEXANDER DER KLEINE **Vol. 6** *page 167, page 173*
ALEXANDRA **Vol. 4** 202, 223
ALEXANDRA SI INFERNUL **Vol. 3** 250, 268, 272, 334 **Vol. 6** *page 270*
ALEXANDRE LE BIENHEUREUX **Vol. 2** 564, 642
ALEXANDRINEKVADRILLE **Vol. 5** 97
ALEXIS, GENTLEMAN-CHAUFFEUR **Vol. 2** 273, 727
ALFA ROMEO ES JULIA **Vol. 9** 67, 138, 180

ALFA TAU ! Vol. 7 250
ALFAMA, VELHA LISBOA Vol. 8 105
ALFONS ZITTERBACKE Vol. 1 11, 54, 62, 87
ALFRED VON INGELHEIMS LEBENSDRAMA Vol. 4 113, 780
ALGO AMARGO EN LA BOCA Vol. 8 706, 773
ALGOL Vol. 4 213, 328, 546
ALGUNAS LECCIONES DE AMOR Vol. 8 396
ALHAMBRA Vol. 8 359, 628, 704, 715
ALHUCEMAS Vol. 8 692
ALI BABA ET LES QUARANTE VOLEURS Vol. 2 232, 488, 863
ALI DELLA VENDETTA, LE Vol. 7 320
ALI, LE Vol. 7 52
ALI ROZSAKERTJE Vol. 9 2
ALI SPEZZATE Vol. 7 787
ALIBI Vol. 2 526, 646
ALIBI, L' Vol. 2 581, 676
ALIBI NA VODE Vol. 10 193, 296, 433
ALIBI POUR UN MEURTRE Vol. 2 270, 479
ALIBISVINDLERNE 1 Vol. 5 81
ALICE OU LA DERNIERE FUGUE Vol. 2 700, 783
ALICJA Vol. 6 *page 217, page 227*
ALIEN 2 SULLA TERRA Vol. 7 474, 784
ALIKE STO NAUTIKO Vol. 3 144
ALIKE TES SIGHRONES GENIAS Vol. 3 201
ALIMA Vol. 7 1185
ALIMENTE Vol. 4 73, 247
ALKALOM Vol. 9 26, 117, 133
ALKOHOL Vol. 4 124
ALKOHOL UND STEUERRAD Vol. 4 18, 558
ALKYMISTENS DATTER Vol. 5 97
ALL A BAL Vol. 9 71, 322
ALL JORDENS FRÖJD Vol. 5 871, 933, 1073
ALL' OMBRA DELLA GLORIA Vol. 7 402, 551, 809, 1155
ALL' OMBRA DELLA GRANDE QUERCIA Vol. 7 74, 383, 719
ALLA CONQUISTA DI UN TRONO Vol. 7 731
ALLA TIDERS 91 : AN KARLSSON Vol. 5 930, 1066
ALLA TIDERS KARLSSON Vol. 5 923, 1181
ALLA VI BARN I BULLERBYN Vol. 5 977
ALLAMI ARUHAZ Vol. 9 83, 237, 326
ALLE GAAR RUNDT OG FORELSKER SIG Vol. 5 31, 32, 55, 70, 101, 208
ALLE KNEB GAELDER Vol. 5 92
ALLE MAN PA POST Vol. 5 870, 923
ALLE MAND PAA DAEK ! Vol. 5 55, 70, 196
ALLE MEINE MÄDCHEN Vol. 6 *page 158, page 164, page 166*
ALLE TAGE IST KEIN SONNTAG Vol. 4 361, 558
ALLE TIDERS KUPP Vol. 5 682, 717, 772, 814
ALLEGRI MASNADIERI Vol. 7 725, 1000, 1172
ALLEGRO BARBARO Vol. 9 20, 154, 318
ALLEGRO CANTATE, L' Vol. 7 792, 796, 902
ALLEGRO FANTASMA, L' Vol. 7 756, 1063
ALLEGRO NON TROPPO Vol. 7 741
ALLEGRO SQUADRONE O ALBERTO IL MARMITTONE Vol. 7 915
ALLEIN IM URWALD Vol. 4 219, 340, 683
ALLENATORE NEL PALLONE, L' Vol. 7 350, 1202
ALLER LIEBE ANFANG Vol. 1 12, 17, 70
ALLER RETOUR Vol. 3 367, 410, 421
ALLER SIMPLE, UN Vol. 2 351, 770
ALLERSEELEN Vol. 4 113
ALLES AUS LIEBE Vol. 4 30
ALLES FÜR ANITA Vol. 4 115
ALLES FÜR GELD Vol. 4 242, 407, 760
ALLES FÜR GLORIA Vol. 4 292
ALLES FÜR VERONIKA Vol. 4 139, 723
ALLES HÖRT AUF MEIN KOMMANDO Vol. 4 119, 340, 466
ALLES SCHWINDEL Vol. 4 135, 615, 623, 660
ALLES UM EINE FRAU Vol. 4 344, 558
ALLES WEG'N DEM HUND Vol. 4 230, 361, 458
ALLEZ FRANCE Vol. 2 548, 835
ALLIANCE, L' Vol. 2 234, 599
ALLJON MEG A MENET Vol. 9 20, 167, 318
ALLMÄCHTIGE DOLLAR, DER Vol. 4 192, 780
ALLO JE T'AIME Vol. 2 595, 627
ALLO, MADEMOISELLE Vol. 2 833
ALLONS Z'ENFANTS Vol. 6 *page 82, page 96*
ALLOTRIA Vol. 4 82, 504, 614
ALLTÄGLICHE GESCHICHTE, EINE Vol. 4 97, 298, 727
ALLVARSAMMA LEKEN, DEN Vol. 5 684, 843, 852, 892, 923, 1068, 1109, 1186
ALLZUVIEL IST UNGESUND Vol. 4 18, 461
ALMA ARAGONESA Vol. 8 551, 749
ALMA BATURRA Vol. 8 614
ALMA CANARIA Vol. 8 357
ALMA DE DIOS Vol. 8 281, 657

ALMA DE LA COLPA, EL Vol. 8 298, 541
ALMA DE SERENA, EL Vol. 8 493, 749
ALMA MATER Vol. 7 560
ALMA RIFENA Vol. 8 655
ALMA SCHAFFT ALLE Vol. 6 *page 163, page 165, page 176*
ALMAS EN PELIGRO Vol. 8 396, 503
ALMATLAN EVEK Vol. 9 134, 183, 314
ALMENRAUSCH UND EDELWEISS Vol. 4 135, 774
ALMODO IFJUSAG Vol. 9 253, 287, 303
ALMODOZASOK KORA Vol. 9 306
ALO, AMERICA ! Vol. 3 249
ALO, ATERIZEAZA STRABUNICA ! Vol. 6 *page 272, page 277*
ALO ? ! ... ATI GRESIT NUMARUL Vol. 3 248, 266, 281
ALOHA, LE CHANT DES ILES Vol. 2 9, 338, 339, 826
ALOIS VYHRAL LOS Vol. 10 167, 198
ALOISE Vol. 2 655
ALOMKERINGÖ Vol. 9 132, 236
ALOMSARKANY Vol. 9 94, 164
ALORS, HEUREUX ! Vol. 2 640, 690 Vol. 6 *page 90, page 92*
ALOYSE Vol. 2 265, 289
ALPAGUEUR, L' Vol. 2 285, 655
ALPENGLÜHEN Vol. 4 47
ALPENTRAGÖDIE Vol. 4 10, 14, 208, 788
ALPEROSEN Vol. 5 216
ALPHAVILLE Vol. 2 208
ALPONERSE EL SOL Vol. 8 258, 338
ALRAUNE Vol. 4 149, 255, 359, 546, 626, 627, 660
ALRAUNE, DIE HENKERSTOCHTER, GENANNT DIE ROTE HANNE Vol. 4 232, 300
ALRAUNE UND DER GOLEM Vol. 4 646
ALS DIE ROSEN ERBLÜHTEN Vol. 4 514
ALS ICH WIEDERKAM Vol. 4 558, 724
ALS MARTIN 14 WAR Vol. 1 7, 11, 47
ALS UNKU EDES FREUNDIN WAR Vol. 6 *page 159, page 160, page 169*
ALSACE Vol. 2 153, 167
ÄLSKANDE PAR Vol. 5 880, 973, 1080
ÄLSKARINNAN Vol. 5 869, 1052, 1053
ÄLSKLING, JAG GER MIG Vol. 5 840, 909
ÄLSKLING PA VAGEN Vol. 5 1080, 1156, 1164
ÄLSKLING PA VIFT Vol. 5 852, 919
ALT DETTE – OG ISLAND MED Vol. 5 115, 378, 395, 616
ALT DETTE – OG ISLAND OGSA Vol. 5 115, 378, 395, 616, 721
ALT FOR KARRIEREN Vol. 5 55, 70, 196
ALT FOR KVINDEN Vol. 5 128, 199, 201, 246
ALT-HEIDELBERG Vol. 4 452, 646
ALT-HEIDELBERG, DU FEINE Vol. 4 99, 124
ALT PA ET BRAET ! Vol. 5 35, 165, 257
ALT PAA ET KORT Vol. 5 97
ALTA COSTURA Vol. 8 449, 616, 630
ALTA INFEDELTA Vol. 7 567, 1167
ALTA TENSION Vol. 8 388, 569
ALTAIR Vol. 7 373
ALTALENE DELLA VITA, L' Vol. 7 131
ALTAR MAYOR Vol. 8 359, 437
ALTAS VARIEDADES Vol. 8 602, 732
ALTE BALLHAUS, DAS Vol. 4 634, 648
ALTE FRITZ, DER Vol. 4 247, 461
ALTE GESETZ, DAS Vol. 4 294, 316, 318, 663
ALTE GOSPODAR, DER Vol. 4 71, 341, 366
ALTE KAMERADEN Vol. 4 97, 546, 644
ALTE KLEIDER Vol. 4 82
ALTE LIEBE, EINE Vol. 1 73, 88
ALTE LIED, DAS Vol. 4 49, 164, 385, 584, 623
ALTE RECHT, DAS Vol. 4 361, 626
ALTE UND DER JUNGE KÖNIG, DER Vol. 4 437, 536, 738
ALTE VAGABUND UND SEIN HÜNDCHEN, DER Vol. 4 642
ALTEMER LE CYNIQUE Vol. 2 378, 688
ALTER KAHN UND JUNGE LIEBE Vol. 1 54, 58, 92
ALTER SCHÜTZT VOR TORHEIT NICHT Vol. 4 514
ALTERNATIVA, LA Vol. 8 450, 493
ALTES HERZ WIRD WIEDER JUNG Vol. 4 298, 727, 729
ALTID BALLADE Vol. 5 31, 241, 248
ALTITUDE 3200 Vol. 2 439, 626, 824
ALTO EN EL CAMINO, UN Vol. 8 246
ALTRA DONNA, L' Vol. 7 819, 908
ALTRA FACCIA DEL PADRINO, L' Vol. 7 910
ALTRA GELOSIA, L' Vol. 7 139
ALTRA, L' Vol. 7 923
ALTRA META DEL CIELO, L' Vol. 7 599
ALTRA ONESTA, L' Vol. 7 343
ALTRA RAZZA, L' Vol. 7 277
ALTRI DESIDERI PARTICOLARI Vol. 7 1179
ALTRI TEMPI Vol. 7 283, 332, 634, 796, 910, 1116
ALTRIMENTO CI ARRABBIAMO Vol. 8 595, 659
ALTRO DIO, L' Vol. 7 1145
ALTRO INFERNO, L' Vol. 7 123
ALTRO IO, L' Vol. 7 369
ALTRO PERICOLO, L' Vol. 7 37, 266, 545
ALTURA Vol. 7 923
ALUCINEMA Vol. 8 215, 372

ALVAJARO, AZ **Vol. 9** 323
ALVARO PIUTOSTO CORSARO **Vol. 7** 338, 1027
ALVILAG PROFESSZORA, AZ **Vol. 9** 83, 259
ALWIN AUF DER LANDSTRASSE **Vol. 1** 49, 92, 116
ALWIN DER LETZTE **Vol. 1** 8, 88
ALYSE ET CHLOE **Vol. 2** 294, 532
AM ABEND AUF DER HEIDE **Vol. 4** 30, 340
AM ABEND NACH DER OPER **Vol. 4** 8, 54
AM ABGRUND **Vol. 6** *page 194*
AM AMBOSS DES GLÜCKS **Vol. 4** 159, 488
AM ANFANG WAR ES SÜNDE **Vol. 3** 387
AM ENDE DER WELT **Vol. 1** 15, 17, 127 **Vol. 4** 13, 614
AM FOST 16 **Vol. 3** 250, 268, 321
AM GRAUEN STRAND, AM GRAUEN MEER **Vol. 6** *page 171, page 174, page 175*
AM KREUZWEG DER LEIDENSCHAFT **Vol. 4** 232
AM O IDEE **Vol. 6** *page 277, page 278, page 279, page 281*
AM RANDE DER GROSSTADT **Vol. 4** 124, 290
AM RANDE DER SAHARA **Vol. 4** 751
AM RANDE DER WELT **Vol. 4** 479, 729
AM ROTEN KLIFF **Vol. 4** 35, 359
AM RÜDESHEIMER SCHLOSS STEHT EINE LINDE **Vol. 4** 43, 796
AM SEIDENEN FADEN **Vol. 4** 298, 742
AM TELEFON WIRD GEWÜNSCHT **Vol. 4** 361, 461
AM WEBSTUHL DER ZEIT **Vol. 4** 89, 548
AM WEIBE ZERSCHELLT **Vol. 4** 756
AMA ROSA **Vol. 8** 752
AMADOR **Vol. 8** 243, 719
AMALIA CATENA **Vol. 7** 424
AMAMI, ALFREDO ! **Vol. 7** 196, 453, 1000
AMANDUS **Vol. 3** 387, 390
AMANECER ENPUERTA OSCURA **Vol. 8** 611, 725
AMANHA COMO HOJE **Vol. 8** 21, 79, 153
AMANT DE BORNEO, L' **Vol. 2** 32, 197
AMANT DE CINQ JOURS, L' **Vol. 2** 285, 655
AMANT DE LADY CHATTERLEY, L' **Vol. 2** 156, 454, 656, 837 **Vol. 6** *page 80*
AMANT DE MADAME VIDAL, L' **Vol. 2** 279, 407, 644
AMANT DE PAILLE, L' **Vol. 2** 332, 427
AMANT DE POCHE, L' **Vol. 2** 513
AMANTE BISEX, L' **Vol. 7** 311
AMANTE DEL VAMPIRO, L' **Vol. 7** 77
AMANTE DELLA LUNA, L' **Vol. 7** 104
AMANTE DI PARIDE, L' **Vol. 7** 978
AMANTE INCATENATA, L' **Vol. 7** 764
AMANTE SEGRETA, L' **Vol. 7** 427, 453, 1177
AMANTES DE VERONA, LOS **Vol. 8** 772
AMANTES DEL DESIERTO, LOS **Vol. 8** 253, 261, 298
AMANTES DEL DIABLO, LAS **Vol. 8** 332, 346
AMANTES DO TEJO, OS **Vol. 8** 87
AMANTI DE RAVELLO, GLI **Vol. 7** 113
AMANTI DEL PASSATO **Vol. 7** 603
AMANTI DI DOMANI, GLI **Vol. 7** 619
AMANTI MIEI **Vol. 7** 649
AMANTI SENZA AMORE **Vol. 7** 522, 796
AMANTS DE BRAS MORT, LES **Vol. 2** 402 **Vol. 7** 583
AMANTS DE DEMAIN, LES **Vol. 2** 309, 475
AMANTS DE MINUIT, LES **Vol. 2** 427, 832
AMANTS DE TERUEL, LES **Vol. 2** 271, 719
AMANTS DE TOLEDE, LES **Vol. 2** 59, 427
AMANTS DE VERONE, LES **Vol. 2** 13, 226, 615
AMANTS DU PONT SAINT-JEAN, LES **Vol. 2** 15, 501
AMANTS DU TAGE, LES **Vol. 2** 279, 402 **Vol. 8** 87
AMANTS ET VOLEURS **Vol. 2** 623, 657, 726
AMANTS, LES **Vol. 2** 224, 285, 770, 856
AMANTS MAUDITS, LES **Vol. 2** 460
AMANTS TERRIBLES, LES **Vol. 2** 332, 626, 824
AMARAMENTE **Vol. 7** 1121
AMARCORD **Vol. 7** 388, 537, 664, 988
AMARSI MALE **Vol. 7** 1179
AMARSI UN PO' ... **Vol. 7** 318, 340
AMARTI E' IL MIO DESTINO **Vol. 7** 33
AMARTI E' IL MIO VECCATO **Vol. 7** 838
AMATOR **Vol. 6** *page 222, page 230, page 254*
AMATORDETEKTIVEN **Vol. 5** 83
AMATÖRFILMEN **Vol. 5** 849, 1183
AMAYA **Vol. 8** 207, 630
AMAZONE, DIE **Vol. 4** 479, 535
AMAZONKE **Vol. 3** 403
AMAZZONE MACABRA, L' **Vol. 7** 277
AMAZZONE MASCHERATA, L' **Vol. 7** 965
AMAZZONE NERA, L' **Vol. 7** 529
AMAZZONI BIANCHE **Vol. 7** 730, 822, 1121
AMBASADOR **Vol. 6** *page 322, page 323*
AMBASCIATORE, L' **Vol. 7** 725, 1029, 1060

AMBASSADEURS, LES **Vol. 2** 742
AMBASSADORS DAUGHTER, THE **Vol. 2** 59
AMBICIOSA **Vol. 8** 261, 385
AMBITIEUSE **Vol. 2** 863
AMBIZIONE SBAGLIATE, LE **Vol. 7** 593, 1093
AMBIZIOSE, LE **Vol. 7** 46
AMBROSIUS **Vol. 5** 121
AME D'ARTISTE **Vol. 2** 440, 521, 833
AME DE CLOWN **Vol. 2** 42, 279, 404, 708
AME DE PIERRE, L' **Vol. 2** 277, 410
AMEDEE **Vol. 2** 197, 652
AMELIE OU LE TEMPS D'AIMER **Vol. 2** 571, 835
AMENAZA BLACK BOX **Vol. 8** 221, 708
AMERE VICTOIRE **Vol. 2** 279, 320, 427, 663
AMERICA RUGIENTE **Vol. 8** 311, 402
AMERICAIN A PARIS, UN **Vol. 2** 593
AMERICAIN, L' **Vol. 2** 870
AMERICAN FEVER **Vol. 7** 633
AMERICANO A ROMA, UN **Vol. 7** 796
AMERICANO EN TOLEDO, UN **Vol. 8** 396, 772
AMERICANO IN VACANZA, UN **Vol. 7** 1177
AMERICKY SOUBOJ **Vol. 10** 363
AMERIKA CIGARETTA **Vol. 9** 70, 167, 280
AMERIKA-EUROPA IM LUFTSCHIFF **Vol. 4** 402
AMERIKAI ANZIX **Vol. 9** 192
AMERIQUE INSOLITE **Vol. 2** 368
AMERYKANSKA AWANTURA **Vol. 1** 176, 205, 402
AMES CORSES **Vol. 2** 54
AMES D'ORIENT **Vol. 2** 798
AMETRALLADORA, LA **Vol. 8** 299, 523
ÄMGLAR, FINNS DOM... **Vol. 5** 880, 933
AMI DE LA FAMILLE, L' **Vol. 2** 386, 559, 661
AMI DE VINCENT, L' **Vol. 6** *page 72, page 83*
AMI D'ENFANCE **Vol. 2** 37
AMI FRITZ, L' **Vol. 2** 167, 279, 614, 744
AMI VIENDRA CE SOIR, UN **Vol. 2** 388, 488
AMIAMOCI COSI **Vol. 7** 792, 1036
AMICA DI SONIA, L' **Vol. 7** 311
AMICA, L' **Vol. 7** 5, 6, 719, 1144
AMICHE, LE **Vol. 7** 763, 915, 1167
AMICI DI NICK HEZARD, GLI **Vol. 7** 514
AMICI MIEI **Vol. 7** 80, 599
AMICI MIEI ATTO II **Vol. 7** 80, 542, 834
AMICI MIEI ATTO III **Vol. 7** 318, 908
AMICI PER LA PELLE **Vol. 7** 634, 910
AMICIZIA **Vol. 7** 460, 1177
AMICO DELL' UOMO, L' **Vol. 7** 1189
AMICO DELLA VEDOVA, L' **Vol. 7** 1189
AMICO DELLE DONNE, L' **Vol. 7** 493, 750, 1080
AMICO DI CASA, L' **Vol. 7** 1189
AMICO, L' **Vol. 7** 369
AMIG HOLNAP LESZ **Vol. 9** 117, 183, 332
AMIGAS, LAS **Vol. 8** 524
AMINATA **Vol. 2** 427
AMINTIRI DIN COPILARIE **Vol. 3** 241, 321
AMIRES DE DON JUAN, LOS **Vol. 8** 598
AMIS, LES **Vol. 2** 300, 738
AMITIES PARTICULIERES, LES **Vol. 2** 722
AMLETO **Vol. 7** 254, 454, 1002
AMLETO DI MENO, UN **Vol. 7** 741
AMLETO E IL SUO CLOWN **Vol. 7** 45, 436, 569
AMMAZZALI TUTTI E TORNA SOLO **Vol. 8** 626, 762
AMMENKÖNIG, DER **Vol. 4** 92, 196, 536, 626
AMMIRATORE DI BUONAPARTE, UN **Vol. 7** 1189
AMMITRATRICE, L' **Vol. 7** 474
AMMUTINAMENTO, L' **Vol. 7** 915
AMNESTIA **Vol. 6** *page 217, page 239, page 244*
AMO NON AMO **Vol. 7** 853
AMO TE SOLA **Vol. 7** 756, 796, 1060
AMO UN ASSASSINO **Vol. 7** 476
AMOK **Vol. 2** 206, 573 **Vol. 3** 159
AMOKFUTAS **Vol. 9** 20, 167, 318
AMOR A LA ESPANOLA **Vol. 8** 492
AMOR A TODO GAS **Vol. 8** 261, 495
AMOR AM STEUER **Vol. 4** 560, 735
AMOR AUF SKI **Vol. 4** 344, 544
AMOR BAJO CERA **Vol. 8** 214, 762
AMOR BRUJO, EL **Vol. 8** 214, 358, 449, 523, 644, 771
AMOR CASI ... LIBRE **Vol. 8** 299, 773
AMOR CHE A NULLA AMATO ... **Vol. 7** 531
AMOR CHE FA MORIRE **Vol. 7** 806
AMOR CHE TACE **Vol. 7** 731
AMOR CHE TUTTO VINCE **Vol. 7** 493
AMOR DE DON JUAN, EL **Vol. 8** 298
AMOR DE LOS AMORES, EL **Vol. 8** 298, 753
AMOR DE PERDICAO **Vol. 8** 10, 18, 35, 71, 90, 171, 185
AMOR DESCEU EM PARA - QUEDAS, O **Vol. 8** 176
AMOR DI LADRO **Vol. 7** 264
AMOR DI MADRE **Vol. 7** 1189
AMOR EMPIEZA A MEDIANOCHE, EL **Vol. 8** 681, 749
AMOR EMPIEZA EN SABADO, EL **Vol. 8** 493, 732
AMOR EN EL AIRE **Vol. 8** 261, 749

AMOR EN MANIOBRAS **Vol. 8** 437
AMOR ES VENENO **Vol. 8** 549, 728
AMOR FATAL **Vol. 3** 236
AMOR HOI ! **Vol. 5** 379
AMOR HOLT SICH NASSE FÜSSE **Vol. 1** 69, 70
AMOR I TELEFONEN **Vol. 5** 2, 23
AMOR MIEI **Vol. 7** 524
AMOR MIO ! **Vol. 7** 454
AMOR MIO NON MUORE, L' **Vol. 7** 822, 890
AMOR PAA KROGVEJE **Vol. 5** 295
AMOR PAA SPIL **Vol. 5** 97
AMOR QUE YO TE DI, EL **Vol. 8** 298, 400
AMOR SOBRE RUEDAS **Vol. 8** 261, 616, 630
AMOR TOSSICA **Vol. 7** 801
AMOR TUO LI REDIME, L' **Vol. 7** 1038
AMORE **Vol. 2** 85, 234 **Vol. 7** 196, 725, 1136, 1172
AMORE A PRIMA VISTA **Vol. 7** 468
AMORE A ROMA, UN **Vol. 7** 798
AMORE A TUTTE LE ETA **Vol. 7** 1189
AMORE AL FOTOGRAFO **Vol. 7** 1189
AMORE CANTA, L' **Vol. 7** 639, 1097, 1189
AMORE CHE UCCIDE OVVERO CHI MI DARA L'OBZIO SENZA MORIRE ? **Vol. 7** 1038
AMORE COMINCIA DOMANI, L' **Vol. 7** 158, 196
AMORE COSI FRAGILE, COSI VIOLENTO, UN **Vol. 7** 514
AMORE DI BARBARO **Vol. 7** 843
AMORE DI LOREDANA, L' **Vol. 7** 155, 508, 529
AMORE DI NORMA, L' **Vol. 7** 476, 701
AMORE DI USSARO **Vol. 7** 40, 50, 796
AMORE DIFFICILE – IL SERPENTE **Vol. 7** 257, 763
AMORE DIFFICILE – L'AVARO **Vol. 7** 257, 763
AMORE DIFFICILE – L'AVVENTURA DI UN SOLDATO **Vol. 7** 257, 763
AMORE DIFFICILE – LE DONNE **Vol. 7** 257, 763
AMORE E CHIACCHIERE **Vol. 7** 332, 910
AMORE E DESTINO **Vol. 7** 454
AMORE E GINNASTICA **Vol. 7** 95, 502
AMORE E IL CODICILLO, L' **Vol. 7** 369
AMORE E LA MASCHERA, L' **Vol. 7** 968
AMORE E MORTE **Vol. 7** 314
AMORE E PATRIA **Vol. 7** 1189
AMORE E RABBIA – LA SEQUENZA DE FIORE DI CARTA **Vol. 7** 1007
AMORE E SMARRIMENTO **Vol. 7** 501
AMORE FACILE **Vol. 7** 347, 798
AMORE – FRAMMENTI DI STORIE D'AMORE **Vol. 7** 601
AMORE GRANDE, AMORE LIBERO **Vol. 7** 1015
AMORE IMPERIALE **Vol. 2** 100 **Vol. 7** 16, 149
AMORE IN CITTA – GLI ITALIANI SI VOLTANO **Vol. 7** 915, 1167
AMORE IN CITTA – L'AMORE CHE SI PAGA **Vol. 7** 915, 1167
AMORE IN CITTA – PARADISO PER QUATTRO ORE **Vol. 7** 915, 1167
AMORE IN CITTA – STORIA DI CATERINA **Vol. 7** 915, 1167
AMORE IN CITTA – TENTATO SUICIDIO **Vol. 7** 915, 1167
AMORE IN CITTA – UNA AGENZIA MATRIMONIALE **Vol. 7** 915, 1167
AMORE IN FUGA **Vol. 7** 110
AMORE IN MAREMMA **Vol. 7** 932
AMORE IN PRIMA CLASSE, UN **Vol. 7** 109, 134, 605
AMORE INSIDIOSE **Vol. 7** 1189
AMORE MIO **Vol. 7** 627, 738
AMORE MIO, AIUTAMI **Vol. 7** 853
AMORE MIO, SPOGLIATI ... CHE POI TI SPIEGO **Vol. 7** 1022
AMORE NASCE ROMA, L' **Vol. 7** 17
AMORE NEL LACCIO **Vol. 7** 731, 936
AMORE PIOMBO E FURORE **Vol. 7** 988
AMORE ROSSO **Vol. 7** 16, 301
AMORE SENZA FINE, UN **Vol. 7** 17
AMORE SENZA STIMA **Vol. 7** 369
AMORE SI FA COSI, L' **Vol. 7** 16, 756
AMORE STANCO **Vol. 7** 932
AMORE TOSSICO **Vol. 7** 854
AMORE VOUL DIR GELOSIA **Vol. 7** 901
AMORES DE DON JUAN, LOS **Vol. 8** 407, 600
AMORES DE JUVENTUD **Vol. 8** 546
AMORES DIFICILES, LOS **Vol. 8** 249, 749
AMORI DI ERCOLE, GLI **Vol. 7** 1062
AMORI DI MANON LESCAUT, GLI **Vol. 7** 196
AMORI E VELENI **Vol. 7** 798
AMORI PERICOLOSI **Vol. 7** 333
AMOROSA AVVENTURA, L' **Vol. 7** 168
AMOROSE NOTTI DI ALI BABA, LE **Vol. 7** 1152
AMORS SPILOPPER **Vol. 5** 67
AMORURILE UNEI PRINTESE **Vol. 3** 234
AMOUR **Vol. 5** 35, 80, 257, 268, 307
AMOUR A LA BOUCHE, L' **Vol. 2** 721
AMOUR A LA CHAINE, L' **Vol. 2** 294
AMOUR A LA MER, L' **Vol. 2** 732
AMOUR A L'AMERICAINE, L' **Vol. 2** 402, 780
AMOUR AUTOUR DE LA MAISON, L' **Vol. 2** 34, 652

AMOUR AUX SPORTS D'HIVER, L' **Vol. 6** *page 69*
AMOUR AUX TROUSSES, L' **Vol. 2** 738
AMOUR AVEC DES SI..., L' **Vol. 2** 196, 317
AMOUR AVEC DEUX M, L' **Vol. 2** 600
AMOUR C'EST GAI? L'AMOUR C'EST TRISTE, L' **Vol. 2** 742
AMOUR CHEZ LES POIDS LOURDS, L' **Vol. 2** 553
AMOUR DE LOUVE **Vol. 2** 37
AMOUR DE PLUIE, UN **Vol. 2** 225, 873
AMOUR DE POCHE **Vol. 2** 97, 190, 385, 808
AMOUR DE SWANN, UN **Vol. 6** *page 90, page 96*
AMOUR D'EMMERDEUSE, UN **Vol. 2** 368 **Vol. 6** *page 112*
AMOUR DESCEND DU CIEL, L' **Vol. 2** 376, 497
AMOUR D'UNE FEMME **Vol. 2** 185, 644
AMOUR EN ALLEMAGNE, UN **Vol. 6** *page 254, page 259*
AMOUR EN CAGE, L' **Vol. 2** 397
AMOUR EN FUITE, L' **Vol. 2** 17, 356, 434
AMOUR EN HERBE, L' **Vol. 2** 806
AMOUR EN QUESTION, L' **Vol. 2** 49, 185, 746
AMOUR EST EN JEU, L' **Vol. 2** 156, 454, 837, 875
AMOUR ET CARBURATEUR **Vol. 2** 42, 197
AMOUR ET CIE **Vol. 2** 197
AMOUR ET DISCIPLINE **Vol. 2** 297, 205, 488
AMOUR ET LA VEINE, L' **Vol. 2** 205, 279, 731
AMOUR FOU, L' **Vol. 2** 74, 513
AMOUR, L' **Vol. 2** 808
AMOUR L'APRES-MIDI, L' **Vol. 2** 17, 701
AMOUR, MADAME..., L' **Vol. 2** 185, 407
AMOUR N'EST PAS UN PECHE, L' **Vol. 2** 322, 738
AMOUR NU, L' **Vol. 6** *page 76*
AMOUR TOUJOURS L'AMOUR, L' **Vol. 2** 48, 131
AMOUR TROP FORT, L' **Vol. 2** 164 **Vol. 6** *page 109*
AMOUR VEILLE, L' **Vol. 2** 273, 460, 807
AMOUR VIOLE, L' **Vol. 2** 60, 351
AMOUR...AMOUR **Vol. 2** 330
AMOUREUSE AVENTURE, L' **Vol. 2** 290, 388
AMOUREUX DE LA CAISSIERE, LES **Vol. 2** 159, 286, 330
AMOUREUX DE MARIANNE, LES **Vol. 2** 322, 395
AMOUREUX DU FRANCE, LES **Vol. 2** 732
AMOUREUX SONT SEULS AU MONDE, LES **Vol. 2** 15, 824
AMOURS CELEBRES **Vol. 2** 630, 863
AMOURS DE CASANOVA, LES **Vol. 2** 538
AMOURS DE LADY HAMILTON, LES **Vol. 2** 661
AMOURS DE MINUIT, LES **Vol. 2** 402
AMOURS DE PARIS, LES **Vol. 2** 383, 540, 738
AMOURS DE PERGOLESE, LES **Vol. 7** 1088
AMOURS, DELICES ET ORGUES **Vol. 2** 48, 57, 388
AMOURS D'HERCULE, LES **Vol. 2** 107
AMOURS FINISSENT A L'AUBE, LES **Vol. 2** 13, 96
AMPELOPEDE, L' **Vol. 2** 76
AMPHITRYON **Vol. 4** 82, 266, 574, 729 **Vol. 6** *page 187*
AMPRENTA **Vol. 3** 258, 317
AMURGUL FINTINILOR **Vol. 6** *page 275, page 277*
AN 01, L' **Vol. 2** 685
AN 40, L' **Vol. 2** 871
AN DER SCHÖNEN BLAUEN DONAU **Vol. 4** 14, 186, 584
AN DER WESER **Vol. 4** 221, 270
ÄN EN GANG GÖSTA EKMAN **Vol. 5** 958
AN FRANZÖSISCHEN KAMINEN **Vol. 1** 17, 42, 48
AN HEILIGEN WASSERN **Vol. 4** 306, 344
ÄN LEVA DE GAMLA GUDAR **Vol. 5** 928, 1032
AN-MAGRITT **Vol. 5** 773, 797
ANA **Vol. 8** 7, 169
ANA DICE SI **Vol. 8** 546
ANA MARIA **Vol. 8** 449
ANA SI „HOTUL“ **Vol. 6** *page 273, page 276*
ANA Y LOS LOBOS **Vol. 8** 584
ANADIOMENE **Vol. 7** 968
ANAMESA SE DYO GYNAIKES **Vol. 3** 137
ANAMNESE **Vol. 6** *page 162, page 166, page 173*
ANAPARASTASE **Vol. 3** 135
ANARKISTENS SVIGERMODER **Vol. 5** 97, 178
ANARKISTER OMBORD **Vol. 5** 97
ANASTASIA, DIE FALSCHE ZARENTOCHTER **Vol. 4** 247, 461
ANASTASIA MIO FRATELLO **Vol. 7** 443, 834
ANATOLE CHERI **Vol. 2** 6
ANATOMIA MILOSCI **Vol. 1** 148, 202, 227, 289
ANATOMIE D'UN RAPPORT **Vol. 2** 312
ANATRA ALL' ARANCIA, L' **Vol. 7** 524

ANCHE GLI ANGELI MANIAGNI FAGIOLI **Vol. 8** 523, 732
ANCHE I LADRI HANNO UN SANTO **Vol. 7** 281
ANCHE LEI FUMAVA IL SIGARO **Vol. 7** 181, 832
ANCHE L'INFERNO TREMA **Vol. 7** 1147
ANCHETA **Vol. 3** 208, 239
ANCIENS DE SAINT-LOUP, LES **Vol. 2** 185, 644
ANCORA UNA VOLTA ... A VENEZIA **Vol. 7** 611
ANDALOUSIE **Vol. 2** 332, 826
ANDALUSISCHE NÄCHTE **Vol. 4** 373, 626
ANDATA ALLA FUCINA, L' **Vol. 7** 1189
ANDEL BLAZENE SMRTI **Vol. 10** 30, 353, 428
ANDEL NA HORACH **Vol. 10** 313, 433, 460
ANDEL S DABLEM V TELE **Vol. 10** 254, 299, 327
ANDEL SVADI DABLA **Vol. 10** 251, 299, 327
ANDELICKARKA **Vol. 10** 45, 345
ANDERE, DER **Vol. 4** 87, 158
ANDERE, DIE **Vol. 4** 288, 306
ANDERE FRONT, DIE **Vol. 1** 7, 65, 90
ANDERE ICH, DAS **Vol. 4** 49, 149, 188
ANDERE LÄNDER, ANDERE SITTEN **Vol. 4** 119, 517, 558
ANDERE MANN, DER **Vol. 4** 247, 461
ANDERE SEITE, DIE **Vol. 4** 124, 204
ANDERE STRASSE, DIE **Vol. 6** *page 313*
ANDERE – STUDENT VON PRAG, DER **Vol. 4** 247
ANDERE WELT **Vol. 4** 368
ANDERS ALS DIE ANDERN **Vol. 4** 159
ANDERSSON, PETTERSSON OCH LUNDSTRÖM **Vol. 5** 857, 923
ANDERSSONS KALLE PA NYA UPPTAG **Vol. 5** 996
ANDERSSONSKANS KALLE **Vol. 5** 840, 870, 909, 1003, 1033, 1042
ANDERSSONSKANS KALLE I BUSFORM **Vol. 5** 1033
ANDOR **Vol. 9** 77
ANDORRA OU LES HOMMES D'AIRAIN **Vol. 2** 595, 717
ANDRA DANSEN **Vol. 5** 947, 1071
ANDRANIK **Vol. 2** 378
ANDRAS **Vol. 9** 24, 196
ANDRE CORNELIS **Vol. 2** 167, 205, 339, 584
ANDRE FOLKS BORN **Vol. 5** 179, 248
ANDRE SKIFTET, DET **Vol. 5** 684, 746
ANDREA CHENIER **Vol. 7** 923
ANDREA DORIA – 74 **Vol. 7** 749
ANDREAS HOFER **Vol. 4** 25, 419 **Vol. 6** *page 181*
ANDREAS SCHLÜTER **Vol. 4** 24, 119, 265, 266
ANDREINA **Vol. 7** 264, 701
ANDRINE OG KJELL **Vol. 5** 723, 800
ANDROCLES ET LE LION **Vol. 2** 792
ANDROKLUS UND DER LÖWE **Vol. 1** 74, 92
ANDULA VYHRALA **Vol. 10** 92, 256, 412
ANDULA ZARLI **Vol. 10** 363
ANE DE BURIDAN, L' **Vol. 2** 6, 55, 332
ANELLO DI CONGIUNZIONE, L' **Vol. 7** 1113
ANEMA E CORE **Vol. 7** 16
ANEMA NOVA **Vol. 7** 553, 993
ANFANG AM ENDE DER WELT **Vol. 1** 12, 53, 55
ANFLUG ALPHA 1 **Vol. 1** 11, 34, 114
ANG.: LONE **Vol. 5** 259
ANGE AU PARADIS, UN **Vol. 2** 481
ANGE DE LA MAISON, L' **Vol. 2** 798
ANGE DE LA NUIT, L' **Vol. 2** 9, 48, 627
ANGE DU FOYER, L' **Vol. 2** 197, 339
ANGE GARDIEN, L' **Vol. 2** 137, 560, 573, 595
ANGE PASSE, UN **Vol. 2** 331
ANGE QUE J'AI VENDU, L' **Vol. 2** 59, 407, 581, 657
ANGE QU'ON M'A DONNE, L' **Vol. 2** 368, 469
ANGE ROUGE, L' **Vol. 2** 254, 309
ANGEBOT AUS SCHENECTADY **Vol. 1** 63, 87, 100
ANGEL, EL **Vol. 8** 258, 299
ANGEL ESTA EN LA CUMBRE, EL **Vol. 8** 753
ANGEL GRIS, EL **Vol. 8** 677
ANGEL PASO POR BROOKLYN, UN **Vol. 8** 449, 725
ANGEL TUVO LA CULPA, UN **Vol. 8** 611
ANGELA **Vol. 7** 86
ANGELA DAVIS, ENCHAINEMENT **Vol. 2** 119
ANGELA ES ASI **Vol. 8** 396
ANGELA MERGE MAI DEPARTE **Vol. 6** *page 266, page 273, page 283*
ANGELAN SOTA **Vol. 5** 384, 522, 606
ANGELAS KRIG **Vol. 5** 384, 522
ANGELE **Vol. 2** 143, 485, 871
ANGELES DEL VOLANTE, LOS **Vol. 8** 509, 721
ANGELES GORDOS **Vol. 8** 438, 681
ANGELES SIN CIELO **Vol. 8** 602
ANGELFREUNDE, DIE **Vol. 4** 366
ANGELI BIANCHI ... ANGELI NERI **Vol. 7** 946
ANGELI CUSTODI, GLI **Vol. 7** 436, 559
ANGELI DALLE MANI BENDATE, GLI **Vol. 7** 509

ANGELI DEL QUARTIERE, GLI **Vol. 7** 1063
ANGELI E DEMONI **Vol. 7** 668
ANGELICA **Vol. 2** 29, 775 **Vol. 7** 491, 701, 838
ANGELIKA **Vol. 4** 292, 317, 422
ANGELINA SUPERPORNO **Vol. 7** 600
ANGELIQUE ET LE ROI **Vol. 2** 347
ANGELIQUE ET LE SULTAN **Vol. 2** 347, 659
ANGELIQUE, MARQUISE DES ANGER **Vol. 2** 615, 659
ANGELO BIANCO, L' **Vol. 7** 125, 296, 333, 734, 1054, 1199
ANGELO DEL CREPUSCOLO, L' **Vol. 7** 890, 994
ANGELO DELLA FAMIGLIA, L' **Vol. 7** 1189
ANGELO DELLE ALPI, L' **Vol. 7** 493
ANGELO DER RIESE **Vol. 4** 482, 598
ANGELO E IL DIAVOLO, L' **Vol. 7** 1036, 1063, 1116
ANGELO E SCESO A BROOKLYN, UN **Vol. 7** 571
ANGELO FERITO, L' **Vol. 7** 578
ANGELO REDENTORE **Vol. 7** 1189
ANGELO TRA LA FOLLA **Vol. 7** 113
ANGELO, TYRAN FRA PADUA **Vol. 5** 97
ANGELUS, L' **Vol. 7** 1113
ANGENEHME RUHE **Vol. 4** 119, 517, 558
ANGES DU PECHE, LES **Vol. 2** 8, 722
ANGES, LES **Vol. 2** 276
ANGES NOIRS, LES **Vol. 2** 152, 812
ANGI VERA **Vol. 9** 107, 167, 311
ANGOSCIA DI SATANA, L' **Vol. 7** 843
ANGST **Vol. 4** 536, 626 **Vol. 5** 767, 780, 824
ANGSTHASE **Vol. 4** 361, 558
ANGUSTIA **Vol. 8** 207, 488
ANGYALI ÜDVÖZLET **Vol. 9** 147, 171
ANGYALOK FÖLDJE **Vol. 9** 67, 265, 285
ANICKA JDE DO SKOLY **Vol. 10** 70, 380, 385
ANICKO, VRAT SE ! **Vol. 10** 167, 453
ANIELLO A' FFEDE **Vol. 7** 39
ANIKI - BOBO **Vol. 8** 121, 158
ANIKINA VREMENA **Vol. 3** 425, 436
ANILLO DE NIEBLA, EL **Vol. 8** 379
ANIMA ALLEGRA **Vol. 7** 264, 701
ANIMA DEL MARE, L' **Vol. 7** 711
ANIMA DI BANDITA **Vol. 7** 729
ANIMA DI BARBARO **Vol. 7** 1169
ANIMA NERA **Vol. 7** 1147
ANIMA PERSA **Vol. 7** 333, 964
ANIMAL, L' **Vol. 2** 588, 719
ANIMALI PAZZI **Vol. 7** 587, 1004
ANIMATRICE **Vol. 6** *page 100, page 102*
ANIMAUX, LES **Vol. 2** 60
ANIME BUIE **Vol. 7** 369
ANIME ERRANTI **Vol. 7** 212
ANIME FIERE **Vol. 7** 398
ANIME GEMELLE **Vol. 7** 943, 1103, 1169
ANIME IN TUMULTO **Vol. 7** 40, 796
ANIME INQUIETE **Vol. 7** 39
ANIME RIBELLI **Vol. 7** 91
ANIME TRAVIATE **Vol. 7** 1189
ANITA **Vol. 5** 919, 1213
ANITA JO **Vol. 4** 724
ANITA O IL ROMANZO D' AMORE DELL' EROE DEI DUE MONDI **Vol. 7** 218, 665, 725, 730
ANITA V RAJI **Vol. 10** 167, 256
ANJA, TULE KOTIIN **Vol. 5** 386, 471
ÄNKEMAN JARL **Vol. 5** 841, 1032
ANKETA **Vol. 3** 38, 69 **Vol. 6** *page 33*
ANLAUF **Vol. 1** 18, 19, 128
ANN AND EVA **Vol. 3** 437
ANN OCH EVE - DE EROTISKA **Vol. 5** 1138, 1217
ANNA **Vol. 5** 434, 1011 **Vol. 7** 443, 531, 667, 725, 1079, 1147 **Vol. 9** 6, 152, 202
ANNA BOLEYN **Vol. 4** 294, 560, 663
ANNA CHRISTIE **Vol. 4** 117
ANNA-CLARA OCH HENNES BRÖDER **Vol. 5** 1210
ANNA DA SAN CELSO **Vol. 7** 1113
ANNA DI BROOKLYN **Vol. 7** 756, 988
ANNA FAVETTI **Vol. 4** 82, 340, 466
ANNA I WAMPIR **Vol. 6** *page 213, page 246, page 250*
ANNA KARENINA **Vol. 4** 208, 390 **Vol. 7** 965
ANNA KARENINE **Vol. 2** 13, 23
ANNA LANS **Vol. 5** 923, 1205
ANNA-LIISA **Vol. 5** 351, 355, 356, 386, 406, 628
ANNA MAGNANI **Vol. 7** 910
ANNA MARIA **Vol. 4** 135, 628
ANNA PERDONAMI **Vol. 7** 113
ANNA PETROVNA **Vol. 7** 621
ANNA PROLETARKA **Vol. 10** 175, 296, 522
ANNA RODITE **Vol. 3** 154, 182
ANNA, SESTRA JANY **Vol. 10** 173, 517, 577
ANNA SUSANNA **Vol. 1** 31, 107, 109, 117
ANNA UND ELISABETH **Vol. 4** 51, 437, 742
ANNAMARIA **Vol. 9** 76, 132, 164
ÄNNCHEN VON THARAU **Vol. 4** 115, 354
ANNE DU BAC, L' **Vol. 2** 322, 873
ANNE-LIESE VON DESSAU, DIE **Vol. 4** 124, 733
ANNE, MA SOEUR ANNE, NE VOIS-TUS RIEN VENIR **Vol. 2** 45
ANNE-MARIE **Vol. 2** 309, 440, 657

ANNEE DERNIERE A MARIENBAD, L' Vol. 2 770, 854
ANNEE PROCHAINE SI TOUT VA BIEN, L' Vol. 6 *page 97, page 101, page 145*
ANNEE SAINTE A ROME, L' Vol. 2 703
ANNEE SAINTE, L' Vol. 2 97, 808
ANNEES LUMIERE, LES Vol. 2 739
ANNELIE Vol. 4 246, 365
ANNEMARIE Vol. 4 108, 132, 270
ANNEMARIE, DIE BRAUT DER KOMPAGNIE Vol. 4 239, 270
ANNEMARIE UND IHR ULAN Vol. 4 223, 424
ANNETTE ET LA DAME BLONDE Vol. 2 332, 401, 488
ANNETTE IM PARADIES Vol. 4 257 Vol. 10 167, 256
ANNI DIFFICILI Vol. 7 101 , 796
ANNI FACILI Vol. 7 522, 1136
ANNI RUGGENTI Vol. 7 257, 913
ANNIBALE Vol. 7 739
ANNO 79. D. C. Vol. 3 395
ANNO ZERO - GUERRA NELLO SPAZIO Vol. 7 474
ANNONSERA Vol. 5 840, 878
ANNY - ELLER EN GADEPIGES ROMAN Vol. 5 677
ANO DE LAS LUCES, EL Vol. 8 232, 684
ANOIHTE EPISTOLE Vol. 3 131, 172
ANONIMA COCOTTES Vol. 7 693
ANONIMA DE ASESINOS Vol. 8 311
ANONIMA ROYLOTT, L' Vol. 7 1107, 1116
ANONYME BRIEF, DER Vol. 4 517, 525, 558
ANOTIMPURI Vol. 3 208, 297, 298, 325
ANSCHLAG AUF BAKU Vol. 4 28, 298, 354, 541, 633, 727
ANSCHLAG AUF SCHWEDA Vol. 4 135, 382, 517
ANSCHLUSS UM MITTERNACHT Vol. 4 97, 208, 558
ANSIGTET I FLODEN Vol. 5 242
ANSIGTET LYVER II Vol. 5 256
ANSIKTET Vol. 5 940, 1000, 1036, 1052
ANSTÄNDIGE FRAUEN Vol. 4 724
ANSTÄNDIGT LIV, ETT Vol. 5 1008, 1039, 1051
ANTE Vol. 5 706, 767, 785, 789
ANTEK CWANIAK Vol. 1 174, 317
ANTEK POLICMAJSTER Vol. 1 278, 308, 404
ANTENATO, L' Vol. 7 725, 1162
ANTENNENDRAHT, DER Vol. 4 25
ANTES A SORTE QUE TAL MORTE Vol. 8 34, 57, 60
ANTES DE ANOCHECER Vol. 8 317
ANTES DE ENTRAR DEJEN SALIR Vol. 8 359, 437
ANTES DO ADEUS Vol. 8 56
ANTES LLEGA LA MUERTE Vol. 8 603, 626
ANTHRACITE Vol. 2 534 Vol. 6 *page 124*
ANTICA FIAMMA, L' Vol. 7 1162
ANTICASANOVA Vol. 6 *page 310, page 322*
ANTIGONE Vol. 3 133, 169
ANTIKE MÜNZE, DIE Vol. 1 2, 104
ANTIQUAR VON STRASSBURG, DER Vol. 4 202
ANTIQUARIO BURUATO, L' Vol. 7 1189
ÄNTLIGEN ! Vol. 5 926, 1019
ANTOINE ET ANTOINETTE Vol. 2 330
ANTOINE ET SEBASTIEN Vol. 2 450, 851
ANTOINETTE SABRIER Vol. 2 308, 648, 762, 785
ANTON Vol. 5 684, 799, 808
ANTON DER LETZTE Vol. 4 124, 232, 375
ANTON DER ZAUBERER Vol. 1 18, 42, 75
ANTON SPELEC, OSTROSTRELEC Vol. 10 167, 453
ANTONIA, ROMANCE HONGROISE Vol. 2 32, 197, 626
ANTONIETA Vol. 8 358, 736
ANTONIO DI PADOVA, IL SANTO DEI MIRACOLI Vol. 7 460, 569, 792
ANTONIO GRAMSCI Vol. 7 411, 910, 916
ANTONIO MEUCCI Vol. 7 204, 786, 978
ANTONY AND CLEOPATRA Vol. 8 215, 221, 603
ANTONYHO SANCE Vol. 10 258, 495, 497
ANTOS KOMBINATOR Vol. 1 174, 188
ANTOS PIERWSZY RAZ W WARSZAWIE Vol. 1 258
ANTREAS JA SYNTINEN JOLANDA Vol. 5 379, 430
ANTROPOPHAGUS Vol. 7 152
ANTTI PUUHAARA Vol. 5 465, 527, 554, 589
ANTYKI Vol. 1 186, 269, 300
ANU JA MIKKO Vol. 5 552, 592, 613
ANUSCHKA Vol. 4 110, 524, 544
ANWALT DES HERZENS, DER Vol. 4 166, 536
ANYAMASSZONY KATONAJA Vol. 9 58, 71
ANYASZIV, AZ Vol. 9 337
AOOM Vol. 8 523, 768
APA Vol. 9 215, 263
APACHE KID Vol. 8 252, 296
APACHEN Vol. 1 8, 96, 102
APACHEN, DIE Vol. 4 124, 514
APACHEN VON PARIS, DIE Vol. 4 368, 442, 707
APACHENLORD, DER Vol. 4 390, 611
APACHEPIGENS HAEVN Vol. 5 198, 203

APAGA Y VAMONOS Vol. 8 358
APAGOGE TES NYFES Vol. 3 203
APAM NEHANY BOLDOG EVE Vol. 9 6, 152, 259
APARECIDOS, LOS Vol. 8 246
APARTADO DE CORREOS 1001 Vol. 8 382, 394, 493
APARTAMENTO DE LA TENTACION, EL Vol. 8 603, 626
APARTMENTZAUBER Vol. 3 395
APASZKA Vol. 1 215, 326
APE REGINA, UNA STORIA MODERNA Vol. 7 567
APEN FRAMTID Vol. 5 667, 737, 774
APENBARINGEN Vol. 5 728, 784
APHRODITE Vol. 4 124 Vol. 6 *page 76, page 99* Vol. 9 315
APINAN VUOSI Vol. 5 387, 585, 633
APO LAHTARA SE LAHTARA Vol. 3 198
APO POU PANE GIA TE HABOUZA Vol. 3 180
APOBEZIM AZ NA KRAJ SVETA Vol. 10 82, 228, 548
APOCALIPSIS CANIBAL Vol. 8 303
APOCALISSE, L' Vol. 7 1136
APOCALISSE SUL FIUME GIALLO Vol. 7 1062
APOCALPSIS JOE Vol. 8 598
APOCALYPSE, L' Vol. 2 742
APOI S-A NASCUT LEGENDA Vol. 3 239, 286, 319
APOKALYPSE Vol. 4 186, 202, 560
APOLLON DES ROCHES NOIRES, L' Vol. 2 798
APOPLANESI Vol. 3 132
APOSTOLITE Vol. 3 63, 114 Vol. 6 *page 23, page 47*
APPARENCE FEMININE Vol. 6 *page 116*
APPARIZIONE Vol. 7 443, 528
APPARTEMENT DES FILLES, L' Vol. 2 481
APPASSIONATA Vol. 5 843, 923
APPASSIONATAMENTE Vol. 7 344, 604
APPASSIONNATA, L' Vol. 2 339, 413
APPEL DE LA CHAIR, L' Vol. 2 37
APPEL DU BLED, L' Vol. 2 48, 657
APPEL DU DESTIN, L' Vol. 2 488, 611
APPEL DU SILENCE, L' Vol. 2 363, 595
APPEL, L' Vol. 2 740
APPELEZ-MOI MATHILDE Vol. 2 368, 457
ÄPPELKRIGET Vol. 5 852, 1093, 1170
APPETIT OG KAERLIGHED Vol. 5 262
APPRENTI SALAUD, L' Vol. 2 481
APPRENTIS SORCIERS, LES Vol. 2 734
APPREZZATO PROFESSIONISTA DI SICURO AVVENIRE Vol. 7 257
APPRODO, L' Vol. 7 1073
APPUNTAMENTA O RENDEZ-VOUS, L' Vol. 7 534
APPUNTAMENTO AD ISCHIA Vol. 7 514, 1044
APPUNTAMENTO IN RIVIERA Vol. 7 1044
APPUNTAMENTO, L' Vol. 7 185
APRENDIENDO A MORIR Vol. 8 400, 616, 630
APRENDIZ DE CLOWN, EL Vol. 8 232, 405
APRENDIZ DE MALO, EL Vol. 8 298
APRES L'AMOUR Vol. 2 32, 128, 332, 613, 824
APRES L'ORAGE Vol. 2 249, 460
APRES „MEIN KAMPF", MES CRIMES Vol. 2 187
APRES VOUS, DUCHESSE Vol. 2 344, 627
APRIL Vol. 5 23, 248
APRIL, APRIL Vol. 4 329, 774
APRIL HAT 30 TAGE, EIN Vol. 1 42, 96, 116
APRILISI RIADO Vol. 9 265, 297, 314
APROAPE DE SOARE Vol. 3 208, 250, 277
AQUEL HOMBRE DE TANGER Vol. 8 214, 692
AQUEL MADITO DIA Vol. 8 629
AQUEL VIEJO MOLINO Vol. 8 614
AQUELLA CASA EN LAS AFUERAS Vol. 8 215, 681
AQUELLA JOVEN DE BLANCO Vol. 8 708, 772
AQUELLAS PALABRAS... Vol. 8 396, 524, 732
AQUELLOS TIEMPOS DEL CUPLE Vol. 8 276
AQUI ESTAN LOS VICETIPLES Vol. 8 214, 600
AQUI HA FANTASMAS Vol. 8 54
AQUI HAY PETROLEO Vol. 8 214, 541
AQUI MANDO YO Vol. 8 338, 598
AQUI, PORTUGAL Vol. 8 80, 92
AQUI ROBAMOS TODOS Vol. 8 306
AQUI VA OTRO RECLUTA Vol. 8 600
AQUILA, L' Vol. 7 621
AQUILA NERA Vol. 7 404, 638, 639
AQUILE ROMANE Vol. 7 843
AR ALDRIG FÖR SENT, DET Vol. 5 871, 1073
ÄR DU INTE RIKRIGT KLOK ? Vol. 5 953, 1033, 1046, 1173
ÄR HOS MIG HAN HAR VARIT, DET Vol. 5 959, 1130, 1142, 1217
ÄR MIN MODELL, DET Vol. 5 840, 909
ÄR MIN MUSIK, DET Vol. 5 840, 940, 1028
ARABELLA Vol. 1 215, 326 Vol. 4 211, 247, 452, 720

ARABIAN BOY IN PARIS **Vol. 2** 759 **Vol. 6** *page 96*
ARAGOSTA A COLAZIONE **Vol. 7** 29, 257, 263
ARAIGNEE D'EAU, L' **Vol. 2** 359, 532
ARAIGNEE, L' **Vol. 2** 678
ARAMÜTES **Vol. 9** 6, 259
ARANA NEGRA **Vol. 8** 438, 600
ARANYEMBER **Vol. 9** 71, 164, 172, 200, 322
ARANYEMBER, AZ **Vol. 9** 67, 83, 84
ARANYFEJ, AZ **Vol. 9** 122, 302, 332
ARANYHAJU SZFINX, AZ **Vol. 9** 301
ARANYKESZTYÜ LOVAGJAI, AZ **Vol. 9** 67, 84, 122
ARANYORA **Vol. 9** 30, 132, 312
ARANYPAVA **Vol. 9** 26, 117
ARANYSARKANY **Vol. 9** 134, 259, 265
ARAUCANA, LA **Vol. 8** 602, 749
ARBEJDET ADLER **Vol. 5** 295
ARBRE DE GUERNICA, L' **Vol. 2** 806
ARBRE DE NOEL, L' **Vol. 2** 13, 21
ARC **Vol. 9** 152, 253, 316
ARCHA BLAZNU ANEB VYPRAVENI Z KONCE ZIVOTA **Vol. 10** 139, 289, 471
ARCHE DE NOE, L' **Vol. 2** 34, 128, 427, 832
ARCHE, DIE **Vol. 4** 178, 237
ARCHI SEXE, L' **Vol. 2** 793
ARCHIMEDE LE CLOCHARD **Vol. 2** 197, 644
ARCHIMEDOV ZAKON **Vol. 10** 287, 451, 576
ARCHIV DES TODES **Vol. 1** 45 **Vol. 6** *page 158, page 159, page 164*
ARCIDIAVOLO, L' **Vol. 7** 507, 792, 978
ARCIERE DE FUOCO, L' **Vol. 8** 214
ARCIERE NERO, L' **Vol. 7** 556
ARCNELKÜLI VAROS, AZ **Vol. 9** 83, 149, 332
ARCOBALENO **Vol. 7** 2, 910
ARCOLAIO DI BARBERINA, L' **Vol. 7** 45, 1162
ARCTIC FURY **Vol. 5** 334
ARDITI CIVILI **Vol. 7** 167, 595, 792
ARDOISE, L' **Vol. 2** 835
ARENA HUMSTI BUMSTI **Vol. 4** 110
ARENDAS ZSIDO, AZ **Vol. 9** 28
ARENES JOYEUSES **Vol. 2** 347, 460, 488, 581, 803, 872
ARGENT DE POCHE, L' **Vol. 2** 351, 370, 434
ARGENT DES AUTRES, L' **Vol. 2** 668, 787
ARGENT DU MINISTRE, L' **Vol. 2** 740
ARGENT, L' **Vol. 2** 58, 70, 440, 511, 566, 573, 688 **Vol. 6** *page 83, page 87, page 141*
ARGINE, L' **Vol. 7** 40, 955, 1177
ARGYRO E PRODÖMENE TSELIDGO-POULA **Vol. 3** 161
ARIA DEL CONTINENTE, L' **Vol. 7** 665, 730
ARIA DI PAESE **Vol. 7** 460, 730, 792
ARIA DI PARIGI **Vol. 7** 583
ARIA DLA ATLETY **Vol. 6** *page 219, page 234, page 243*
ARIADNE IN HOPPEGARTEN **Vol. 4** 73, 97, 132
ARIANE **Vol. 4** 609, 788
ARIANE, JEUNE FILLE RUSSE **Vol. 2** 440, 726
ARISTO, L' **Vol. 2** 84, 279, 388, 726
ARISTOCRATES, LES **Vol. 2** 128, 559, 661
ARISTOCRATICA PERVERSA, L' **Vol. 7** 311
ARIZONA BILL – SERIE **Vol. 2** 159, 286, 330
ARIZONA SI SCATENO ... E LI FECE FUORI TUTTI **Vol. 8** 551
ARIZONA VUELVE **Vol. 8** 551
ARKEN **Vol. 5** 924
ARLESIENNE, L' **Vol. 2** 48, 96, 153, 388, 644
ARLETTE ET L'AMOUR **Vol. 2** 32, 347
ARLETTE ET SES PAPAS **Vol. 2** 6, 332
ARM WIE EINE KIRCHENMAUS **Vol. 4** 379, 626
ARMA BIANCA **Vol. 7** 725, 1029
ARMA, L' **Vol. 7** 19
ARMA, L'ORA, IL MOVENTE, L' **Vol. 7** 310
ARMADNI DVOJCATA **Vol. 10** 532, 621
ARMAGUEDON **Vol. 2** 738
ARMANDO **Vol. 3** 59, 119
ARMAS CONTRA LA LEY **Vol. 8** 602, 749
ARMATA AZZURA, L' **Vol. 7** 652, 756, 796, 1036, 1116
ARMATA BRANCALEONE, L' **Vol. 7** 522
ARMATI E IL MIO DESTINO **Vol. 7** 1152
ARMATURA DI CARLOMAGNO, L' **Vol. 7** 131
ARMBAANDET **Vol. 5** 97
ARME A GAUCHE, L' **Vol. 2** 722, 875
ARME JENNY, DIE **Vol. 4** 124
ARME KLEINE COLOMBINE **Vol. 4** 25, 255
ARME KLEINE EVA **Vol. 4** 233, 239
ARME KLEINE INGE **Vol. 4** 582
ARME KLEINE SIF **Vol. 4** 287, 291, 565
ARME MARIE **Vol. 4** 87
ARME MILLIONÄR, DER **Vol. 4** 616
ARME PRINZESSIN **Vol. 4** 514
ARME REICHE, DER **Vol. 4** 15, 230
ARME REICHE HUBERT B., DER **Vol. 1** 9, 17

ARME, SYNDIGE MENNESKE Vol. 5 716, 755, 824
ARME THEA Vol. 4 660, 704
ARME VIOLETTA Vol. 4 729
ARMEE DES OMBRES, L' Vol. 2 517, 588
ARMES DE LA FEMME, LES Vol. 2 798
ARMES GLÜCK Vol. 4 514
ARMES KLEINES MÄDCHEN Vol. 4 239, 568
ARMIAMOCI E ... PARTITE ! Vol. 7 131
ARMOIRE VOLANTE, L' Vol. 2 15, 395
ARNAUD, LES Vol. 2 872
ARNLJOT Vol. 5 870, 1084
ARNO STARKS KRAFT IM KAMPF ZWISCHEN HANDWERK UND MASCHINEN Vol. 4 212, 594
ARNYEK A SATORON Vol. 9 257
ARPA NERA, L' Vol. 7 961
ARQUERO DE SHERWOOD, EL Vol. 8 214
ARRANGIATEVI Vol. 7 257
ARRAPAHO Vol. 7 123, 1161
ARRASTAO, LES AMANTS DE LA MER Vol. 2 172
ARREBATO Vol. 8 373
ARREGLO, EL Vol. 8 274, 299
ARRESTATION D'UN IVROGNE, L' Vol. 2 678
ARRESTATION MED „EKSTRAFOR PLEJNING" Vol. 5 97
ARRET AU MILIEU, L' Vol. 2 296
ARRETE TON CHAR, BIDASSE Vol. 2 600
ARRETEZ LE MASSACRE Vol. 2 419, 515
ARRETEZ LES TAMBOURS Vol. 2 293, 475
ARRIBA HAZANA Vol. 8 610, 754
ARRIBA LAS MUJERES ! Vol. 8 317, 504
ARRIBADA FORZOSA Vol. 8 716, 752
ARRIGOGOLO, L' Vol. 7 48, 254, 640
ARRIVA LA BANDA Vol. 7 852
ARRIVA LA ZIA D'AMERICA Vol. 7 891
ARRIVANO I BERSAGLIERI Vol. 7 383, 778
ARRIVANO I DOLLARI Vol. 7 1027
ARRIVANO I GATTI Vol. 7 633, 908
ARRIVANO I MIEI Vol. 7 544, 698
ARRIVANO I NOSTRI Vol. 7 16
ARRIVANO I TITANI Vol. 7 347, 1054
ARRIVEDERCI PAPA Vol. 7 1054, 1177
ARRIVEDERCI ROMA Vol. 7 333, 1063
ARRIVIAMO NOI ! Vol. 7 2, 48
ARRIVISTE, L' Vol. 2 378
ARRIVISTES, LES Vol. 2 59, 146, 549
ARS AMANDI – L'ARTE DI AMARE Vol. 7 208, 719
ARSENE LUPIN CONTRE ARSENE LUPIN Vol. 2 185, 661
ARSENE LUPIN DETECTIVE Vol. 2 217, 254
ART. 519 CODICE PENALE Vol. 7 196, 709
ART D'AIMER, L' Vol. 6 *page 146* Vol. 7 208, 719
ARTATLAN GYILKOSOK Vol. 9 67, 273, 327
ARTATLAN VAGYOK Vol. 9 323
ARTE DE CASARSE, EL Vol. 8 406, 681
ARTE DE NO CARSARSE, EL Vol. 8 406, 681
ARTE DE VIVIR, EL Vol. 8 406, 748
ARTE DI ARRANGIARSI, L' Vol. 7 161, 299, 498, 1044
ARTE NEL SUO MISTERO, L' Vol. 7 277
ARTEFICE DELL' AMORE Vol. 7 940
ARTHUR Vol. 2 138, 339, 413
„ARTHUR" – GROTESKFILM-SERIE Vol. 4 64, 459
ARTHUR RUBINSTEIN, L'AMOUR DE LA VIE Vol. 2 715
ARTHURS FORBRYTELSE Vol. 5 674
ARTIGLI D'ACCIAIO, GLI Vol. 7 81
ARTISTA, AURUL SI ARDELENII Vol. 3 255
ARTISTA, DOLARII SI ARDELENII Vol. 3 239, 314
ARTISTEN Vol. 4 119, 270, 400, 733
ARTISTENTREUE Vol. 4 113, 780
ARTUR A LEONTYNA Vol. 10 460, 621
ARTUR BECKER Vol. 1 27, 38, 85
ARVACSKA Vol. 9 263, 312, 318
ARVEN Vol. 5 684, 730, 731, 808
ARVETANTEN Vol. 5 92
ARVINGEN Vol. 5 89, 281
ARVINGEN TIL KRAGSHOLM Vol. 5 97
ARVOTTOMAT Vol. 5 352, 436, 591, 642
ARZT AUS LEIDENSCHAFT Vol. 4 361, 424, 544
ARZT VON BOTHENOW, DER Vol. 1 22, 50
ÄRZTE Vol. 1 22, 48, 100
ÄRZTINNEN Vol. 6 *page 163, page 166, page 176*
AS DE PIC Vol. 8 732
AS DU TURF, LES Vol. 2 41, 603
AS ME KRINOUN OI GYNAIKES Vol. 3 164
AS PUPILAS DO SENHOR REITOR Vol. 8 24, 44, 75, 130, 170, 182
AS YOU LIKE IT Vol. 2 573
ASA E VIATA Vol. 3 315
ASA – HANNA Vol. 5 863, 1032
ASA – NISSE Vol. 5 908, 923
ASA – NISSE BLAND GEVAR OCH BARONER Vol. 5 1041
ASA – NISSE FLYGER I LUFTEN Vol. 5 834
ASA – NISSE I AGENTFORM Vol. 5 914
ASA – NISSE I FULL FART Vol. 5 883

ASA – NISSE I KRONANS KLÄDER Vol. 5 1029
ASA – NISSE I POPFORM Vol. 5 852, 905
ASA – NISSE I RAKETFORM Vol. 5 843
ASA – NISSE I REKORDFORM Vol. 5 1143
ASA – NISSE JUBILERAR Vol. 5 1029
ASA – NISSE OCH DEN STORA KALABALIKEN Vol. 5 914, 1142
ASA – NISSE OCH TJOCKA SLÄKTEN Vol. 5 905
ASA – NISSE ORDNAR ALLT Vol. 5 834, 923
ASA – NISSE PA HAL IS Vol. 5 834
ASA – NISSE PA JAKSTIGEN Vol. 5 913, 923
ASA – NISSE PA MALLORCA Vol. 5 903
ASA – NISSE PA NYA ÄVENTYR Vol. 5 923, 1029
ASA – NISSE PA SEMESTER Vol. 5 923, 1029
ASA – NISSE SLAR TILL Vol. 5 843, 1142
ASA – NISSE SOM POLIS Vol. 5 883, 1073
ASALTO ALBANCO CENTRAL Vol. 8 670
ASCENSEUR POUR L'ECHAFAUD Vol. 2 224, 548, 598, 856
ASCHENBRÖDEL Vol. 4 213, 342 Vol. 6 *page 182*
ASCHERMITTWOCH Vol. 4 208, 236, 239, 270, 356, 423
ASCOLTAMI Vol. 7 603
ASEDIUL Vol. 3 218, 230, 291
ASES BUSCAN LA PAZ, LOS Vol. 8 515, 753
ASESINATO EN COMITE CENTRAL Vol. 8 219, 229
ASESINO DE DÜSSELDORF, EL Vol. 8 214
ASESINO DE MUNECAS, EL Vol. 8 419, 677
ASESINO NO ESTA SOLO, EL Vol. 8 245
ASESINO NO ESTA SUELA, EL Vol. 8 252
ASESSORIN NAISHUOLET Vol. 5 356, 428 Vol. 5 356, 528, 451, 462, 634
ASFALTTILAMPAAT Vol. 5 516, 520
ASI ES MADRID Vol. 8 449, 616, 630
ASINO DI BURIDANO, L' Vol. 7 229
ASINO POLIZIOTTO, L' Vol. 7 1189
ASPHALT Vol. 4 295, 325, 567 Vol. 6 *page 189*
ASPHALTE Vol. 2 315, 348, 402
ASPHALTROSE, DIE Vol. 4 762 Vol. 6 *page 201*
ASPRO MAURO Vol. 3 176
ASSALTO AI PESCICANI Vol. 7 548
ASSASSIN A PEUR LA NUIT, L' Vol. 2 156, 205
ASSASSIN CONNAIT LA MUSIQUE, L' Vol. 2 309, 650
ASSASSIN EST A L'ECOUTE, L' Vol. 2 6, 423, 475
ASSASSIN EST DANS L'ANNUAIRE, L' Vol. 2 128, 661
ASSASSIN HABITE AU 21, L' Vol. 2 23, 824
ASSASSIN MUSICIEN, L' Vol. 2 633
ASSASSIN N'EST PAS COUPABLE, L' Vol. 2 48, 70
ASSASSIN QUI PASSE, UN Vol. 2 633 Vol. 6 *page 139*
ASSASSIN VIENDRA CE SOIR, L' Vol. 2 3, 737
ASSASSINAT DU DUC DE GUISE, L' Vol. 2 93
ASSASSINAT DU PERE NOEL, L' Vol. 2 332, 824
ASSASSINIO AL CIMITERO ETRUSCO Vol. 7 426, 504
ASSASSINIO DEL CORRIERE DI LIONE, L' Vol. 7 886
ASSASSINIO DEL JOKEY, L' Vol. 7 968
ASSASSINIO DI TROTSKY, L' Vol. 7 1025
ASSASSINIO SUL TEVERE Vol. 7 310, 315
ASSASSINO E AL TELEFONO, L' Vol. 7 742
ASSASSINO E ANCORA TRA NOI, L' Vol. 7 127, 450
ASSASSINO E COSTRETTO AD UCCIDERE ANCORA Vol. 7 852
ASSASSINO, L' Vol. 7 401, 853
ASSASSINO MADE IN ITALY Vol. 8 214, 602
ASSASSINS DE L'ORDRE, LES Vol. 2 49, 598
ASSASSINS DU DIMANCHE, LES Vol. 2 124, 650, 743
ASSASSINS ET VOLEURS Vol. 2 205, 249, 259
ASSAUT, L' Vol. 2 41, 460, 579
ASSEDIO DELL' ALCAZAR, L' Vol. 7 587, 756, 1096
ASSEDIO DI SIRACUSA, L' Vol. 7 257, 1054
ASSEDIO DI TROJA, L' Vol. 7 186
ASSEDIO E LA CADUTA DI TROJA, L' Vol. 7 186
ASSENZA INGIUSTIFICATA Vol. 7 756, 1000, 1177
ASSI ALLA RIBALTA Vol. 7 606
ASSIGNATURA PENDIENTE Vol. 8 214
ASSISTENTE, L' Vol. 7 122
ASSISTENTE SOCIALE TUTTO PEPE, L' Vol. 7 122
ASSO Vol. 7 1144
ASSOCIE, L' Vol. 2 97, 810
ASSOLUTO NATURALE, L' Vol. 7 567

ASSOMMOIR, L' **Vol. 2** 47, 64, 131, 357, 404, 831
ASSUNTA SPINA **Vol. 7** 264, 454, 701, 910
ASSZONY A TELEPEN **Vol. 9** 36, 263, 318
ASSZONYBOSSZU **Vol. 9** 220
ASSZONYFALO **Vol. 9** 220
ASSZONYI ESKÜ **Vol. 9** 77
ASTA, MEIN ENGELCHEN **Vol. 6** *page 159, page 167, page 168*
ASTA SEARA DANSAM IN FAMILIE **Vol. 3** 230, 291
ASTEPTAREA **Vol. 3** 225, 237, 332
ASTEPTIND UN TREN **Vol. 6** *page 268, page 269, page 276*
ASTERÖ **Vol. 3** 154, 169
ASTRAGALE, L' **Vol. 2** 728, 787
ASTRAPOGIANNOS **Vol. 3** 173, 197, 199
ASTRID **Vol. 7** 390
ASTRONAUTA, EL **Vol. 8** 524
ASTUZIA DI DONNA **Vol. 7** 131
ASZANTKA **Vol. 1** 215
AT DERE TOR ! **Vol. 5** 750, 780, 808
AT ZIJE NEBOZTIK **Vol. 10** 256, 524
AT ZIJE REPUBLIKA **Vol. 10** 235, 289, 516
AT ZIJI DUCHOVE ! **Vol. 10** 299, 371, 425
ATA ISHIN KATER **Vol. 3** 15, 26
ATAKUALPA **Vol. 1** 346
ATALANTE, L' **Vol. 2** 86, 420, 425
ATANTAS KNALDSUCCES **Vol. 5** 83
ATAQUE DE LOS KURDOS **Vol. 8** 523, 732
ATAVISMO DELL' ANIMA, L' **Vol. 7** 961
ATENTAT **Vol. 10** 182, 313, 352
ATERIZARE FORTATA **Vol. 3** 296, 307
ATHENA E KLOPE TES ODOU STADIOU **Vol. 3** 200
ATHLETE AUX MAINS NUES, L' **Vol. 2** 848
ATHLETEN **Vol. 4** 242, 316
ATIUNEA ZUZUC **Vol. 6** *page 274, page 281*
ATLANTAS BORTFORELSE **Vol. 5** 291
ATLANTÄVENTYRET **Vol. 5** 840, 841
ATLANTIDE, L' **Vol. 2** 62, 612, 614, 639, 778, 798
ATLANTIK **Vol. 4** 578
ATLANTIS **Vol. 2** 751 **Vol. 5** 9
ATLAS **Vol. 7** 774, 1116
ATOK VARA **Vol. 9** 308
ATOLL K **Vol. 2** 698, 824
ATOMIQUE MONSIEUR PLACIDO, L' **Vol. 2** 191
ATOMOVA KATEDRALA **Vol. 10** 173, 291, 466
ATOUT-COEUR **Vol. 2** 9, 42, 122, 152
ATOUT COEUR A TOKYO POUR OSS 117 **Vol. 2** 251, 368
ATOUTS DE M. WENS, LES **Vol. 2** 228
ATRACADORES, LOS **Vol. 8** 492, 732
ATRACO A LAS TRES **Vol. 8** 725, 762
ATRACO A SEXO ARMADO **Vol. 8** 438
ATRACO DE IDA Y VUELTA, UN **Vol. 8** 517
ATRACO EN LA COSTA AZUL **Vol. 8** 312, 595
ATRAKCE SVEDSKEHO ZAJEZDU **Vol. 10** 139, 414, 466, 487
ATRAPADO **Vol. 8** 302, 369, 639
ATRAPADOS EN EL MIEDO **Vol. 8** 414
ATSALUT PADER **Vol. 7** 667, 947
ATT ÄLSKA **Vol. 5** 880, 1053, 1080
ATT ANGÖRA EN BRYGGA **Vol. 5** 878, 881, 1093
ATTANASIO, CAVALLO VANESIO **Vol. 7** 1027
ATTENTAT **Vol. 5** 11, 25, 165, 306
ATTENTAT, L' **Vol. 2** 35, 235
ATTENTAT U SARAJEVU **Vol. 3** 343
ATTENTATO, L' **Vol. 7** 490
ATTENTI A QUEI DUE NAPOLETANI **Vol. 7** 556
ATTENTI A QUEI P 2 **Vol. 7** 426, 1127
ATTENTI A QUELLE DUE ... NINFOMANI **Vol. 7** 311
ATTENTI AL BUFFONE ! **Vol. 7** 347, 907
ATTENTI RAGAZZI ... CHI ROMPE PAGA **Vol. 7** 692
ATTENTION FILLETTES **Vol. 6** *page 129*
ATTENTION LES ENFANTS REGARDENT **Vol. 2** 719, 746
ATTENTION LES YEUX ! **Vol. 2** 782
ATTENTION, UNE FEMME PEUT EN CACHER UNE AUTRE **Vol. 6** *page 69, page 77, page 81*
ATTENZIONE, L' **Vol. 7** 381, 585, 1083
ATTI IMPURI ALL' ITALIANA **Vol. 7** 509
ATTICO, L' **Vol. 7** 284, 502
ATTILA **Vol. 7** 1102, 1136, 1147
ATTILA FLAGELLO DI DIO **Vol. 7** 257, 1165
ATTILAS '74 **Vol. 3** 176
ATTIMO CHE UCCIDE, L' **Vol. 7** 369, 377
ATTO DI ACCUSA **Vol. 7** 693
ATTONDE DAGEN, DEN **Vol. 5** 1052, 1096
ATTORE SCOMPARSO, L' **Vol. 7** 902, 1036
ATUNCI I-AM CONDAMNAT PE TOTI LA MOARTE **Vol. 3** 218, 235, 254
ATVALTOZAS **Vol. 9** 238, 249
AU BONHEUR LES DAMES **Vol. 2** 23, 180, 226, 387, 824
AU BOUT DU BOUT DU BANC **Vol. 2** 810
AU COEUR DE LA CASBAH **Vol. 2** 224, 571, 856
AU COEUR DE LA VIE **Vol. 2** 110, 651, 815

AU COEUR DE L'ORAGE **Vol. 2** 57, 777, 832
AU-DELA DE CETTE LIMITE VOTRE TICKET N'EST PLUS VALABLE **Vol. 6** *page 116*
AU-DELA DE LA PEUR **Vol. 2** 661
AU-DELA DES GRILLES **Vol. 2** 644
AU DIABLE LA VERTU **Vol. 2** 247, 615
AU FIL DE L'EAU **Vol. 2** 566
AU FIL DES ONDES **Vol. 2** 496
AU FIL DES ROUTES ET DES ANS **Vol. 2** 811
AU FOND DU GOUFFRE **Vol. 2** 798
AU FRAIS DE LA PRINCESSE **Vol. 2** 139
AU GRAND BALCON **Vol. 2** 395, 722
AU GRE DES FLOTS **Vol. 2** 375
AU HASARD, BALTHAZAR **Vol. 2** 170, 190
AU LONG DE RIVIERE FANGO **Vol. 2** 177
AU NOM DE LA LOI **Vol. 2** 84, 152, 197
AU NOM DE TOUS MIENS **Vol. 6** *page 75, page 93*
AU PAN COUPE **Vol. 2** 732
AU PAYS DE GEORGE SAND **Vol. 2** 656
AU PAYS DES CIGALES **Vol. 2** 131, 479
AU PAYS DES TENEBRES **Vol. 2** 415
AU PAYS DES TENEBRES OU LA MINE **Vol. 2** 24
AU PAYS DU SOLEIL **Vol. 2** 344, 394, 471, 871
AU PETITE BONHEUR **Vol. 2** 197, 440
AU P'TIT ZOUAVE **Vol. 2** 254, 368
AU RENDEZ-VOUS DE LA MORT JOYEUSE **Vol. 2** 185, 190
AU REVOIR MONSIEUR GROCK **Vol. 2** 676, 833
AU REVOIR...A LUNDI **Vol. 2** 695, 696
AU ROYAUME DES CIEUX **Vol. 2** 32, 615
AU SECOURS! **Vol. 2** 798
AU SERVICE DU TZAR **Vol. 2** 70, 833
AU SEUIL DE HAREM **Vol. 2** 440, 470
AU SOLEIL DE MARSEILLE **Vol. 2** 413, 460
AU SON DES GUITARES **Vol. 2** 460, 579, 722
AU VOLEUR **Vol. 2** 559, 661, 663
AUBE DE SANG, L' **Vol. 2** 104
AUBERGE DE L'ABIME, L' **Vol. 2** 6, 70
AUBERGE DU PECHE, L' **Vol. 2** 8, 67, 270
AUBERGE DU PETIT DRAGON, L' **Vol. 2** 153
AUBERGE FLEURIE, L' **Vol. 2** 243
AUBERGE, L' **Vol. 2** 134, 257
AUBERGE ROUGE, L' **Vol. 2** 42, 47, 251, 402
AUCA DEL SENOR ESTEVE, L' **Vol. 8** 422
AUDACE COLPO DEI SOLITI IGNOTI **Vol. 7** 401, 514
AUDIENCIA PUBLICA **Vol. 8** 207
AUF BEFEHL DER POMPADOUR **Vol. 4** 270, 384, 471
AUF DEM FELDE DER EHRE **Vol. 4** 247
AUF DEM FLUG NACH HAVANNA **Vol. 1** 44, 71, 111
AUF DEM SPRUNG **Vol. 6** *page 156, page 160, page 176*
AUF DEN TRÜMMERN DES PARADIESES **Vol. 4** 575
AUF DER FLUCHT ERSCHOSSEN **Vol. 1** 72, 73, 130
AUF DER REEPERBAHN NACHTS UM HALB EINS **Vol. 4** 157, 461, 742
AUF DER SONNENSEITE **Vol. 1** 46, 72, 115
AUF EIGENE FAUST **Vol. 4** 323, 367
AUF GEFÄHRLICHEN SPUREN **Vol. 4** 471, 560, 779
AUF LEBEN UND TOD **Vol. 4** 340, 608, 691
AUF WIEDERSEHEN, FRANZISKA **Vol. 4** 270, 582
AUFENTHALT, DER **Vol. 6** *page 153, page 161, page 165*
AUFFORDERUNG ZUM TANZ **Vol. 4** 82, 119, 340, 409, 466, 713
AUFMACHEN, KRIMINALPOLIZEI **Vol. 4** 18, 558
AUFRUHR DER HERZEN **Vol. 4** 54, 232, 774
AUFRUHR DES BLUTES **Vol. 4** 757 **Vol. 6** *page 199*
AUFRUHR IM DAMENSTIFT **Vol. 4** 18, 317, 558
AUFRUHR IM JUNGGESELLENHEIM **Vol. 4** 239, 255
AUFRUHR IN DAMASKUS **Vol. 4** 79, 115, 566, 623
AUFSTIEG **Vol. 6** *page 187, page 188, page 197*
AUFSTIEG DER KLEINEN LILIAN, DER **Vol. 4** 35, 290
AUFTRAG FÜR M & S **Vol. 1** 32, 87, 91
AUFTRAG HÖGLERS, DER **Vol. 1** 78, 97, 101
AUFTRAG: ÜBERLEBEN **Vol. 1** 50, 73
AUGE DES BUDDHA, DAS **Vol. 4** 239
AUGEN **Vol. 4** 291, 471
AUGEN ALS ANKLÄGER, DIE **Vol. 4** 570, 780
AUGEN DER MUMIE MA, DIE **Vol. 4** 242, 560
AUGEN DER WELT, DIE **Vol. 6** *page 189*
AUGEN DES OLE BRANDIS, DIE **Vol. 4** 646
AUGEN IM WALDE, DIE **Vol. 4** 71, 479

AUGENBLICK IM PARADIES, EIN Vol. 4 385
AUGENZEUGEN Vol. 4 18
AUGH ! AUGH ! Vol. 7 909
AUGURI E FIGLI MASCHI Vol. 7 693
AUGUST DER STARKE Vol. 4 536, 544
AUGUST JÄRJESTÄÄ KAIKEN Vol. 5 349, 566
AUGUSTAS LILLA FELSTEG Vol. 5 840, 878
AUGUSTE Vol. 2 368, 515
AUKTIONSNUMMER 117 Vol. 5 97
AUNQUE LA HORMONA SE VISTA DE SEDA ... Vol. 8 331, 627
AURAIS DU FAIRE GAFFE, LE CHOC EST TERRIBLE Vol. 2 301
AUREL VLAICU Vol. 3 214, 232, 286
AURELIA STEINER Vol. 2 267, 517
AURI SACRA FAMES Vol. 4 232, 673
AURINKOTUULI Vol. 5 377, 410, 434, 437, 500, 522, 546, 585
AURORA Vol. 7 514, 893
AURORA DE ESPERANZA Vol. 8 648
AURORA ROSSA Vol. 7 621
AURORA SUL MARE Vol. 7 16, 941
AUS DEM LEBEN EINES TAUGENICHTS Vol. 1 5, 26, 52
AUS DEM TAGEBUCH EINER MINDERJÄHRIGEN Vol. 1 26, 89, 92
AUS DEM TAGEBUCH EINES JUNGGESELLEN Vol. 4 340, 407, 608, 782
AUS DEN AKTEN EINER ANSTÄNDIGEN FRAU Vol. 4 35, 385
AUS DEN ERINNERUNGEN EINES FRAUENARZTES Vol. 4 233, 270, 461, 593
AUS DEN MEMOIREN EINER FILMSCHAUSPIELERIN Vol. 4 208, 390 Vol. 6 *page 193*
AUS DES RHEINLANDS SCHICKSALSTAGEN Vol. 4 424, 724
AUS DEUTSCHLANDS RUHMESTAGEN 1870/71 Vol. 4 89
AUS EIGENER KRAFT Vol. 4 146
AUS EINER KLEINEN RESIDENZ Vol. 4 232, 239, 270
AUS ERSTER EHE Vol. 4 246, 729
AUS MEINER KINDHEIT Vol. 1 37, 124
AUS SCHWERER ZEIT Vol. 4 187
AUS TAUSEND METER HÖHE Vol. 4 165
AUS UNSERER ZEIT - DAS DUELL Vol. 1 8, 47
AUS UNSERER ZEIT - DER COMPUTER SAGT NEIN Vol. 1 5, 10, 111
AUS UNSERER ZEIT - DIE ZWEI SÖHNE Vol. 1 30, 80, 85
AUS UNSERER ZEIT - GEWÖHNLICHE LEUTE Vol. 1 26, 80, 85
AUSGEKOCHTER JUNGE, EIN Vol. 4 270, 409, 774
AUSGELÖSTE SCHWIEGERPAPA, DER Vol. 6 *page 186*
AUSGESTOSSENEN, DIE Vol. 4 230, 350, 527
AUSGESTOSSENER, EIN Vol. 4 239, 396, 634
AUSSENSEITER, DER Vol. 4 119, 123, 247, 788
AUSSI LOIN QUE L'AMOUR Vol. 2 60
AUSSI LONGUE ABSENCE, UNE Vol. 2 194, 866
AUSTERIA Vol. 6 *page 228, page 251, page 252*
AUSTERLITZ Vol. 2 13, 249, 423, 669, 786
AUSTERNLILI, DIE Vol. 4 373, 437, 740
AUSTERNPRINZESSIN, DIE Vol. 4 560, 663
AUSTREIBUNG, DIE Vol. 4 115, 178, 202
AUTANT EN EMPORTE LE GANG Vol. 2 829
AUTE E GE EINAI DIKE MAS Vol. 3 161
AUTHENTIQUE PROCES DE CARL-EMMANUEL JUNG, L' Vol. 2 804
AUTO Vol. 9 20, 152, 167
AUTO DA FLORIPES Vol. 8 63
AUTO-SCAT DI ROBINET Vol. 7 837
AUTO-STOPPEUSES EN CHALEUR Vol. 2 291
AUTO UND KEIN GELD, EIN Vol. 4 292
AUTOBUS NR. 2 Vol. 4 462, 748 Vol. 6 *page 195*
AUTOBUS ODJEZDZA 6.20 Vol. 1 179, 334, 397 Vol. 6 *page 211*
AUTOBUS S Vol. 4 79, 123, 796
AUTOFAHRT UNTER DER ERDE, DIE Vol. 4 385
AUTOGRAMM LILI Vol. 9 7
AUTOI POU MILESAN ME TO THANATO Vol. 3 135
AUTOMARTIRIO Vol. 7 1005
AUTOMAT NA PRANI Vol. 10 250, 462
AUTOMNE, L' Vol. 2 390
AUTOMOBILE DELLA MORTE, L' Vol. 7 1189
AUTOMOBILE ERRANTE, L' Vol. 7 144
AUTONOSUYAS, LAS Vol. 8 207, 770
AUTOPSIA Vol. 8 558, 603
AUTOPSIA DE UN CRIMINAL Vol. 8 517, 749
AUTOPSIE D'UN MONSTER Vol. 2 293
AUTOSTOP ROSSO SANGUE Vol. 7 524, 1007
AUTOTYTÖT Vol. 5 452
AUTOUR D'UN BERCEAU Vol. 2 688, 727
AUTRE AILE, L' Vol. 2 192, 217

AUTRE DEVOIR, L' **Vol. 2** 798
AUTRE FEMME, L' **Vol. 2** 819 **Vol. 8** 276, 611
AUTRE FRANCE, L' **Vol. 2** 175
AUTRE HOMME, UNE AUTRE CHANCE, UN **Vol. 2** 491, 616
AUTRE MONDE, UN **Vol. 2** 359
AUTRES, LES **Vol. 2** 35
AUTUAS EVERSTI **Vol. 5** 443, 643
AUTUNNO DELL' AMORE **Vol. 7** 37
AUTUOMO, L' **Vol. 7** 949
AUVARSAMMA LEKEN, DEN **Vol. 5** 808
AUVERGNAT ET L'AUTOBUS, L' **Vol. 2** 570, 815
AUX DEUX COLOMBES **Vol. 2** 475, 706
AUX JARDINS DE MURCIE **Vol. 2** 6, 395, 744, 874
AUX PORTES DE PARIS **Vol. 2** 131, 166, 205
AUX URNES, CITOYENS ! **Vol. 2** 330, 407, 824
AUX YEUX DU SOUVENIR **Vol. 2** 488, 722
AV HJÄRTANS LUST **Vol. 5** 840, 1028, 1033
AVALANCHE **Vol. 2** 36, 351, 557
AVALANSA **Vol. 3** 263, 277, 278, 297
AVALEUSES, LES **Vol. 2** 859
AVANT LA BATAILLE **Vol. 6** *page 233*
AVANT LE DELUGE **Vol. 2** 124, 197, 801
AVANTAZ **Vol. 3** 110 **Vol. 6** *page 32, page 55*
AVANTI ! **Vol. 7** 1042
AVANTI C'E POSTO... **Vol. 7** 1030, 1036, 1063
AVANTURE BORIVOJA SURDILOVICA **Vol. 3** 408
AVARE, L' **Vol. 2** 97, 728 **Vol. 6** *page 77*
AVARIA **Vol. 3** 213, 309, 327
AVARIZIA, L' **Vol. 7** 442, 701
AVARO, L' **Vol. 7** 1189
AVARUUSRAKETILLA RAKKAUTEEN **Vol. 5** 333, 443, 455, 634
AVE CAESAR **Vol. 9** 200
AVE DE ARRIBACAO **Vol. 8** 50
AVE MARIA **Vol. 4** 246, 462 **Vol. 7** 508, 513, 837
AVE MARIA DE SCHUBERT **Vol. 2** 679
AVE MARIA DI GOUNOD, L' **Vol. 7** 1189
AVE MARIA, L' **Vol. 7** 311
AVEC ANDRE GIDE **Vol. 2** 661
AVEC LA PEAU DES AUTRES **Vol. 2** 110, 279
AVEC LE SANG DES AUTRES **Vol. 2** 621
AVEC LE SOURIRE **Vol. 2** 158, 273, 824
AVENTURA **Vol. 8** 268, 541
AVENTURA DE CINE, UNA **Vol. 8** 246, 751
AVENTURA DE GIL BLAS, UNA **Vol. 8** 261, 298
AVENTURA DE HULARIO MARTINEZ, UNA **Vol. 8** 572
AVENTURA EN EL PALACIO VIEJO **Vol. 8** 719
AVENTURA ORIENTAL, UNA **Vol. 8** 718
AVENTURAS DE EDUARDINI **Vol. 8** 561
AVENTURAS DE ENRIQUE Y ANA, LAS **Vol. 8** 297, 616
AVENTURAS DE GIL BLAS DE SANTILLANE, LAS **Vol. 8** 261, 298
AVENTURAS DE JUAN DE MAIRENA **Vol. 8** 268
AVENTURAS DE JUAN LUCAS **Vol. 8** 214, 336, 611
AVENTURAS DE ZIPI Y ZAPE, LAS **Vol. 8** 456, 473
AVENTURAS DEL BARBERO DE SEVILLA **Vol. 8** 261, 298
AVENTURAS DEL OESTE **Vol. 8** 603, 626
AVENTURE A PARIS **Vol. 2** 8, 427, 526
AVENTURE A PIGALLE **Vol. 2** 595, 698
AVENTURE C'EST L'AVENTURE, L' **Vol. 2** 196
AVENTURE COMMENCE DEMAIN, L' **Vol. 2** 344, 756
AVENTURE DE BILLY LE KID, UNE **Vol. 2** 301
AVENTURE DE CABASSOU, L' **Vol. 2** 347, 807
AVENTURE DE LUNA-PARK, L' **Vol. 2** 277
AVENTURE DE MR. SMITH, L' **Vol. 2** 159, 286, 330
AVENTURE DI BIJOU, LE **Vol. 7** 441
AVENTURE EST AU COIN DE LA RUE, L' **Vol. 2** 401, 719
AVENTURE EST SUR LA ROUTE, L' **Vol. 2** 259, 309
AVENTURE STRAORDINARIE DI SATURNINO FARANDOLA, LE **Vol. 7** 656
AVENTURERO DE GUAYNAS, EL **Vol. 8** 493
AVENTURES D'ARSENE LUPIN, LES **Vol. 2** 190, 517, 549, 598, 781
AVENTURES DE CASANOVA, LES **Vol. 2** 439, 807
AVENTURES DE GIL BLAS DE SANTILLANE, LES **Vol. 2** 294, 401, 402, 857
AVENTURES DE GUIDON FUTE, LES **Vol. 2** 294
AVENTURES DE HOLLY AND WOOD, LES **Vol. 2** 854

AVENTURES DE RABBI JACOB, LES **Vol. 2** 224, 588
AVENTURES DE ROBERT MACAIRE, LES **Vol. 2** 378, 582, 623
AVENTURES DE SALAVIN, LES **Vol. 2** 149
AVENTURES DE TILL L'ESPIEGLE, LES **Vol. 2** 59, 226, 248, 566
AVENTURES DES PIEDS NICKELES, LES **Vol. 2** 596
AVENTURES DES QUEUES NICKELEES, LES **Vol. 2** 217, 627 **Vol. 2** 596
AVENTURES DU ROI PAUSOLE, LES **Vol. 2** 563, 775
AVENTURES EXTRA – CONJUGALES **Vol. 6** *page 94*
AVENTUREUSE, L' **Vol. 2** 64
AVENTURI LA MAREA NEAGRA **Vol. 3** 208, 286, 330
AVENTURIER, L' **Vol. 2** 9, 584, 824, 871
AVENTURIERE DU TSCHAD, L' **Vol. 2** 373, 740
AVENTURIERS DE L'AIR, LES **Vol. 2** 747, 858
AVENTURIERS DES CHAMPS-ELYSEES, LES **Vol. 2** 243, 249
AVENTURIERS DU MEKONG, LES **Vol. 2** 243, 475
AVENTURIERS, LES **Vol. 2** 110, 642
AVENTURILE LUI BABUSCA **Vol. 3** 272, 321
AVENTURILE LUI TOM SAWYER **Vol. 3** 260, 317, 323
ÄVENTYR I PYJAMAS **Vol. 5** 996
ÄVENTYR PA HOTELL **Vol. 5** 863, 1042
ÄVENTYRARE, EN **Vol. 5** 888, 1015, 1032, 1181
ÄVENTYRET **Vol. 5** 840, 997
AVERE VENT' ANNI **Vol. 7** 514
AVES SIN RUMBO **Vol. 8** 652
AVEU, L' **Vol. 2** 208, 285
AVEUX LES PLUS DOUX, LES **Vol. 2** 208
AVIOLIITTOYHTIÖ **Vol. 5** 357, 471
AVION DE MINUIT, L' **Vol. 2** 70, 727
AVISDRENGEN **Vol. 5** 242
AVISMANDEN **Vol. 5** 2, 23, 228
AVOCAT, L' **Vol. 2** 79, 498
AVOIR VINGT ANS DANS LES AURES **Vol. 2** 186, 839
AVORTEMENT CLANDESTIN **Vol. 2** 859
AVOVETEEN **Vol. 5** 552, 592
AVSKEDET **Vol. 5** 651, 850, 1053
AVTOSTOP **Vol. 3** 31, 70
AVVENIRE IN AGGUATO, L' **Vol. 7** 1189
AVVENTURA A CAPRI **Vol. 7** 1177
AVVENTURA AMOROSA **Vol. 7** 1189
AVVENTURA DI ANNABELLA, L' **Vol. 7** 709
AVVENTURA DI COLLEGIO **Vol. 7** 26
AVVENTURA DI FRACASSA, L' **Vol. 7** 1116
AVVENTURA DI SALVATOR ROSA, UN' **Vol. 7** 709, 1060, 1177
AVVENTURA DI UN VIVEUR, L' **Vol. 7** 91, 806
AVVENTURA IN CITTA **Vol. 7** 587
AVVENTURA, L' **Vol. 7** 132, 599, 913, 1045
AVVENTURE DEI TRE MOSCHETTIERI, LE **Vol. 7** 877
AVVENTURE DEL MAGGIORE **Vol. 7** 1189
AVVENTURE DELL' INCREDIBILE ERCOLE, L' **Vol. 7** 381, 1083, 1086
AVVENTURE DI CARTOUCHE, LE **Vol. 7** 372
AVVENTURE DI FANTASIO NUVOLA, LE **Vol. 7** 1116
AVVENTURE DI GIACOMO CASANOVA, LE **Vol. 7** 106
AVVENTURE DI MANDRIN, LE **Vol. 7** 453, 798, 830
AVVENTURE DI PINOCCHIO, LE **Vol. 7** 57, 65
AVVENTURE DI ROBI E BUCK, LE **Vol. 7** 556
AVVENTURE DI ROBINET **Vol. 7** 1189
AVVENTURE DI ROBINSON CRUSOE **Vol. 7** 968
AVVENTURE DI UN GIGANTE **Vol. 7** 1189
AVVENTURE DI UN PROVINCIALE, LE **Vol. 7** 1189
AVVENTURE E GLI AMORE DI SCARAMOUCHE, L' **Vol. 7** 124
AVVENTURE, LE **Vol. 7** 1121
AVVENTURE NELL' ARCIPELAGO **Vol. 7** 46
AVVENTURI DI UN UBRIACO **Vol. 7** 837
AVVENTURIERA DEL PIANO DI SORPA, L **Vol. 7** 42, 332, 480, 747, 1027
AVVENTURIERI DEI TROPICI, GLI **Vol. 7** 487
AVVENTURIERI DI SAN JUANO, GLI **Vol. 7** 968
AVVENTURIERO DI CALIFORNIA, L' **Vol. 7** 148
AVVERTIMENTO, L' **Vol. 7** 347, 363, 667
AVVISO IN QUARTA PAGINA, UN **Vol. 7** 341
AVVOCATO DIFENSORE, L' **Vol. 7** 730
AVVOLTOIO, L' **Vol. 7** 212
AWANS **Vol. 1** 182, 324 **Vol. 6** *page 253*
AWANTURA O BASIE **Vol. 1** 250, 359, 400
AWANTURY MILOSNE PANNY D. **Vol. 1** 404
AWOL **Vol. 5** 1142
AXEL BREIDAHL MORER SIG **Vol. 5** 97

AXEL BREIDAHL SOM HYPNOTISOR Vol. 5 97
AXEL BREIDAHLS LOTTERIGEVINST Vol. 5 97
AYOUPA Vol. 3 198
AZ DO KONCE Vol. 10 28, 495, 570
AZ PRIJDE KOCOUR Vol. 10 41, 292
AZ SE VRATIS Vol. 10 503, 532, 569
AZAIS Vol. 2 279, 488, 824
AZAR SE DIVIERTE, EL Vol. 8 217, 576
AZERT IS MARADOK Vol. 9 26, 117, 133
AZONOSITAS Vol. 9 190, 259, 318
AZUL, AZUL Vol. 8 32, 118
AZUREXPRESS Vol. 9 71, 164
AZYL Vol. 1 202, 230, 289, 324, 379 Vol. 6 *page 248*

B

B. A. BA Vol. 2 342
B. COMME BEATRICE Vol. 2 721
B. D. INTRA IN ACTIUNE Vol. 3 244, 283, 297, 317
B. D. LA MUNTE SI LA MARE Vol. 3 283, 297, 317
BABA Vol. 9 220
BABA YAGA Vol. 7 869
BABETTE S'EN VA-T-EN GUERRE Vol. 2 21, 423, 824
BABICKA Vol. 10 45, 92, 109, 122, 569
BABICKA I, II Vol. 10 299, 473, 569
„BABICKY DOBIJEJTE PRESNE !" Vol. 10 41, 97, 180
BABINSKY Vol. 10 45, 453
BABLJE LJETO Vol. 3 445
BABY Vol. 2 388, 397 Vol. 4 257
BABY LOVE Vol. 2 82
BABY PAA EVENTYR Vol. 5 2, 228
BABY PAA LAANEKONTORET Vol. 5 97
BABY SITTER Vol. 7 1086
BACANALES ROMANAS, II PARTE Vol. 8 455
BACCANALI DI TIBERIO, I Vol. 7 1027
BACCANTI, LE Vol. 7 877
BACCARA Vol. 2 8, 427, 526, 626
BACCARAT Vol. 4 162, 522 Vol. 9 77
BACCHANALE DES TODES ODER DAS OPFER EINER GROSSEN LIEBE, DIE Vol. 4 46, 124, 695
BACCHANALES SEXUELLES Vol. 2 76
BACCHANTIN, DIE Vol. 4 192, 648
BACH DETECTIVE Vol. 2 404
BACH EN CORRECTIONNELLE Vol. 2 48, 538, 708, 833
BACH MILLIONNAIRE Vol. 2 357, 387, 404
BACHELIERES EN CHALEUR, LES Vol. 6 *page 93*
BACIAMO LE MANI Vol. 7 502, 1009
BACIO A FIOR D'ACQUA, UN Vol. 7 725, 1029
BACIO DEL SOLE, IL Vol. 7 528
BACIO DI CIRANO, IL Vol. 7 45, 569
BACIO DI EMMA, IL Vol. 7 1189
BACIO DI MORTE Vol. 7 493
BACIO DI SALOME, IL Vol. 7 91
BACIO DI UN RE, IL Vol. 7 1067
BACIO NEL DESERTO, IL Vol. 7 1116
BACIO NEL SOGNO, UN Vol. 7 806
BAD AUF DER TENNE, DAS Vol. 4 41, 517, 730
BADARANII Vol. 3 265, 297
BADARNA Vol. 5 1033
BADEHOTELLET Vol. 5 70
BADEMAUS, DIE Vol. 4 247
BADIA DI MONTENERO, LA Vol. 7 146
BAEGERSVINGEREN Vol. 5 83
BAG DE RODE PORTE Vol. 5 150, 179
BAG FILMENS KULISSER Vol. 5 24
BAG KOBENHAVNS KULISSER Vol. 5 2, 24, 250
BAG STORSTADSTEATRETS KULISSER Vol. 5 203
BAGARRES Vol. 2 427, 676
BAGATELLER FOR EN SJOMAN Vol. 5 760
BAGLIORI DI FIAMMA Vol. 7 1189
BAGNARD, LE Vol. 2 173, 460
BAHIA BLANCA Vol. 8 731
BAHIA DE PALMA Vol. 8 377, 470
BAHNO PRAHY Vol. 10 58, 345
BAHNWÄRTER THIEL Vol. 6 *page 163, page 165*
BAHUT VA CRAQUER, LE Vol. 2 49
BAIE DES ANGES, LA Vol. 2 285, 700
BAIE DU DESIR, LA Vol. 2 259
BAIETII NOSTRI Vol. 3 256, 266, 276
BAILA GUAPA Vol. 7 99
BAILARIN Y EL TRABAJADOR, EL Vol. 8 268, 751, 752
BAILE, EL Vol. 8 207, 214
BAIN DE LA PARISIENNE Vol. 2 678
BAISERS Vol. 2 2, 208
BAISERS VOLES Vol. 2 188, 674
BAISEUSES, LES Vol. 2 859
BAISEZ-MOI Vol. 2 2
BAIXANT DE LA FONT DEL GAT Vol. 8 228
BAJ GANJO Vol. 3 98
BAJADERE, DIE Vol. 4 187
BAJADSER Vol. 5 9, 197
BAJECNI MUZI S KLIKOU Vol. 10 172, 425, 516
BAJKA O RYBAKU I RYBCE Vol. 1 298
BAJO EL CIELO ANDALUZ Vol. 8 306, 396

BAJO EL CIELO DE ASTURIAS **Vol. 8** 365, 677
BAJO EL CIELO DE ESPANA **Vol. 8** 725, 752
BAJO EN NICOTINA **Vol. 8** 684, 700
BAJO LAS NIEBLAS DE ASTURIAS **Vol. 8** 619
BAJTARSAK **Vol. 9** 132, 236
BAKARUHABAN **Vol. 9** 11, 312, 326
BAKO, L'AUTRE RIVE **Vol. 2** 484
BAKOM JALUSIN **Vol. 5** 1093
BAKONJA FRA BRNE **Vol. 3** 399, 444
BAKSMÄLLA **Vol. 5** 606
BAKSMÄLLA – BITAR AV ETT ÄKTENSKAP **Vol. 5** 391
BAKTERION **Vol. 7** 124
BAL A SAVOYBAN **Vol. 9** 71, 309, 322
BAL CUPIDON **Vol. 2** 231
BAL DE NUIT **Vol. 2** 347, 581, 809
BAL DES ESPIONS, LE **Vol. 2** 376, 601
BAL DES PASSANTS, LE **Vol. 2** 407, 538
BAL DES POMPIERS, LE **Vol. 2** 48, 627
BAL DES VOYOUS, LE **Vol. 2** 738
BAL DU COMTE D'ORGEL, LE **Vol. 2** 377
BAL DU VIOL, LE **Vol. 6** *page 108*
BAL, LE **Vol. 2** 573, 824 **Vol. 6** *page 71* **Vol. 7** 29, 964, 1143
BAL NA VODI **Vol. 6** *page 311, page 320*
BAL PARE **Vol. 4** 13, 574
BALADA DA PRAIA DOS CAES **Vol. 8** 7
BALADA DE JOHNNY RINGO, LA **Vol. 8** 629, 768
BALADA O SVIREPOM **Vol. 3** 391
BALADA O TROBENTI IN OBLAKU **Vol. 3** 406, 449
BALADA PENTRU MARIUCA **Vol. 3** 244, 297, 329, 331
BALADA PRO BANDITU **Vol. 10** 80, 318, 466, 622
BALADE POUR UN VOYOU **Vol. 2** 49, 642
BALANCE, LA **Vol. 6** *page 90, page 100*
BALANCOIRES **Vol. 2** 180
BALAO OU DES PAS AU PLAFOND **Vol. 2** 24
BALARRASA **Vol. 8** 276, 717
BALCON DE LA LUNA, EL **Vol. 8** 298, 762
BALCON EN FORET, UN **Vol. 2** 505
BALCON SOBRE EL INFIERNO, UN **Vol. 8** 276, 732
BALDEVINS BRÖLLOP **Vol. 5** 1032
BALDEVINS BRYLLUP **Vol. 5** 689, 803
BALDIRI DE LA COSTA, EL **Vol. 8** 348, 732
BALETT PRIMADONNAN **Vol. 5** 937, 997
BALEYDIER **Vol. 2** 402, 780, 797
BALGARO – UNGARSKA RAPSODIJA **Vol. 9** 35, 164
BALILLA DEL' 48, UN **Vol. 7** 277
BALKAN EKSPRES **Vol. 6** *page 308, page 315, page 322*
BALKANSKI SPIJUN **Vol. 6** *page 308, page 314, page 318*
BALKEZES ANGYAL **Vol. 9** 315, 322
BALL, DER **Vol. 2** 290, 824 **Vol. 4** 158
BALL IM METROPOL **Vol. 4** 110, 437, 738
BALLAADI **Vol. 5** 349, 481, 566
BALLAD **Vol. 5** 313, 338
BALLADE **Vol. 4** 703
BALLADE A BLANC **Vol. 6** *page 78, page 114*
BALLADE DE LA FECONDUCTRICE **Vol. 6** *page 115*
BALLADE I NYHAVN **Vol. 5** 228
BALLADE PA BULLERBORG **Vol. 5** 14
BALLADE PA CHRISTIANSHAVN **Vol. 5** 16, 61, 272
BALLADE POUR UN CHIEN **Vol. 2** 689
BALLADE VOM ROTEN MOHN **Vol. 1** 17, 47, 111
BALLADE VON DER GEIGE, DIE **Vol. 1** 26, 47, 106
BALLADEN OM CARL-HENNING **Vol. 5** 125, 130
BALLADEN OM MESTERTYVEN OLE HOYLAND **Vol. 5** 693, 718, 817
BALLAGAS **Vol. 9** 151, 197, 316
BALLAGO IDÖ **Vol. 9** 122, 152, 259
BALLANDO BALLANDO **Vol. 7** 29, 964, 1143
BALLATA DEI FIORI, LA **Vol. 7** 1116
BALLATA DEL DESTINO, LA **Vol. 7** 342
BALLATA DELLA STREGA, LA **Vol. 7** 1189
BALLATA DI EVA **Vol. 7** 208, 387, 762
BALLATA TRAGICA **Vol. 7** 16
BALLE DANS LE CANON, UNE **Vol. 2** 481, 560
BALLE PER BALLE **Vol. 3** 20, 24
BALLE SUFFIT, UNE **Vol. 2** 866
BALLERINA **Vol. 2** 388, 488, 593, 869
BALLERINA E IL BUON DIO, LA **Vol. 7** 1035, 1062
BALLERINE **Vol. 7** 390, 665, 709, 1172, 1177
BALLES PERDUES **Vol. 6** *page 87, page 109*
BALLET CHINOIS **Vol. 2** 678
BALLET DES PETITS INCROYABLES **Vol. 2** 678
BALLETDANSERINDEN **Vol. 5** 97
BALLETS DE PARIS **Vol. 2** 13
BALLETTEN DANSER **Vol. 5** 2, 24
BALLHAUS-ANNA, DIE **Vol. 4** 247
BALLHAUS GOLDENER ENGEL **Vol. 4** 239
BALLMUTTER **Vol. 4** 525, 546

BALLO AL CASTELLO **Vol. 7** 253, 1054, 1177
BALLON ROUGE, LE **Vol. 2** 781
BALLONAT **Vol. 3** 11, 20
BALLONGEN **Vol. 5** 840, 878, 1173
BALLSKANDAL **Vol. 4** 598
BALOANE DE CURCUBEU **Vol. 6** *page 267, page 272, page 273*
BALOCCHI E PROFUMI **Vol. 7** 106
BALTAGUL **Vol. 3** 218
BALTHAZAR **Vol. 2** 32, 197
BALTIK 66 **Vol. 1** 49
BALTUTLÄMMINGEN **Vol. 5** 917, 1020, 1149
BALUL DE SIMBATA SEARA **Vol. 3** 230, 291
BALVANY **Vol. 9** 117, 259
BAMBINI CI GUARDANO, I **Vol. 7** 250, 1160
BAMBOLA E IL GIGANTE, LA **Vol. 7** 268, 1123
BAMBOLA E L'AMORE, LA **Vol. 7** 390
BAMBOLA INFRANTA, LA **Vol. 7** 277
BAMBOLA SALUATRICE, LA **Vol. 7** 1189
BAMBOLA VIVENTE, LA **Vol. 7** 529
BAMBOLE, LE **Vol. 7** 86, 514, 567, 798
BAMBU **Vol. 2** 427 **Vol. 8** 298
BAMSE **Vol. 5** 869, 954, 1025, 1130
BANANA JOE **Vol. 7** 199, 599
BANANES MECANIQUES **Vol. 2** 294
BANANHEJKERINGÖ **Vol. 9** 6, 215, 316
BANAT ASSZONYA, A **Vol. 9** 86
BANCA DI MONATE, LA **Vol. 7** 910
BANCALS **Vol. 6** *page 137, page 141*
BANCO A BANGKOK POUR OSS 117 **Vol. 2** 503, 615
BANCO DE PRINCE **Vol. 2** 118, 153
BANDA **Vol. 1** 232, 355 **Vol. 6** *page 240*
BANDA CASAROLI, LA **Vol. 7** 401, 408
BANDA DE LOS OCHO, LA **Vol. 8** 207, 221
BANDA DE LOS TRES CRISANTEMOS, LA **Vol. 8** 261, 768
BANDA DEGLI ONESTI, LA **Vol. 7** 451
BANDA DEI ROSSI, LA **Vol. 7** 940
BANDA DEL GOBBO, LA **Vol. 7** 1202
BANDA DEL PECAS, LA **Vol. 8** 217, 405
BANDA DEL TRUCIDO, LA **Vol. 7** 335
BANDA DELLE CIFRE, LA **Vol. 7** 369
BANDE A BONNOT, LA **Vol. 2** 513
BANDE A BOUBOULE, LA **Vol. 2** 339, 648
BANDE A PAPA, LA **Vol. 2** 515, 661
BANDE A PART **Vol. 2** 208
BANDE DU REX, LA **Vol. 2** 301 **Vol. 6** *page 139*
BANDE VON HOHENECK, DIE **Vol. 4** 88
BANDERA, LA **Vol. 2** 309, 439, 440
BANDERA NEGRA **Vol. 8** 438, 738
BANDIDO DE LA SIERRA, EL **Vol. 8** 270, 676
BANDIDO DE MADRID, EL **Vol. 8** 287
BANDIDOS **Vol. 8** 396, 626
BANDITEN VON ASNIERES, DIE **Vol. 4** 385
BANDITI A ORGOSOLO **Vol. 7** 1068
BANDITO DAGLI OCCHI AZZURRI, IL **Vol. 7** 208, 1086
BANDITO, IL **Vol. 7** 187, 1136
BANDOLERO STANCO, IL **Vol. 7** 1027
BANG ! **Vol. 5** 955, 1002, 1192
BANG-BANG **Vol. 2** 322
BANG BANG **Vol. 8** 299, 517
BANG BANG KID **Vol. 8** 299, 517
BANGKOK, CITA CON LA MUERTE **Vol. 8** 731
BANK BAN **Vol. 9** 77
BANKETT FÜR ACHILLES **Vol. 1** 57, 69, 128
BANKETTEN **Vol. 5** 1042, 1114, 1185
BANKKRACH UNTER DEN LINDEN, DER **Vol. 4** 186, 340
BANOVIC STRAHINJA **Vol. 6** *page 313, page 314*
BANQUE NEMO, LA **Vol. 2** 55, 573, 826
BANQUET DES FRAUDEURS, LE **Vol. 2** 778
BANQUET, LE **Vol. 2** 427
BANQUIERE, LA **Vol. 2** 879 **Vol. 6** *page 75, page 118, page 132*
BANTER **Vol. 8** 322, 742
BANYA TITKA, A **Vol. 9** 237
BANZAI **Vol. 6** *page 98*
BAR DE LA FOURCHE, LE **Vol. 2** 142
BAR DU SUD **Vol. 2** 6, 249
BAR DU TELEPHONE **Vol. 2** 56, 534
BAR DU TELEPHONE, LE **Vol. 6** *page 124*
BAR - EL - MANACH **Vol. 4** 98, 112
BARA DI VETRO, LA **Vol. 7** 441, 442
BARA EN DANSERSKA **Vol. 5** 925, 1103, 1136
BARA EN KVINNA **Vol. 5** 841, 923
BARA EN KYPARE **Vol. 5** 878, 1052, 1177
BARA EN MOR **Vol. 5** 878, 1173
BARA EN TRUMPETARE **Vol. 5** 840, 878, 927
BARA PER LO SCERIFFO, UNA **Vol. 8** 598, 626
BARA ROLIGT I BULLERBYN **Vol. 5** 469
BARABBA **Vol. 7** 1136
BARABBAS **Vol. 5** 956, 1042, 1080, 1164
BARACCA E BURATINI **Vol. 7** 17
BARAKA A BEYROUTH **Vol. 2** 208
BARAKA, LA **Vol. 6** *page 103, page 111*
BARAKA SUR X - 13 **Vol. 8** 732
BÄRANDE HAV **Vol. 5** 1052, 1164
BARAONDA, LA **Vol. 7** 347, 860

BARAQUE N.1 Vol. 2 228
BARATIN Vol. 2 32, 376
BARATINEURS, LES Vol. 2 131, 303
BARATSAGOS ARCOT KEREK ! Vol. 9 71, 322
BARBA ZVANE Vol. 3 365
BARBABLU Vol. 7 16
BARBACKA Vol. 5 863, 1142
BARBAGIA Vol. 7 365
BARBARA Vol. 8 48, 120, 132
BARBARA RADZIWILLOWNA Vol. 1 278, 308, 346
BARBARA, WO BIST DU ? Vol. 4 13, 24
BARBARELLA Vol. 2 305, 719 Vol. 7 498
BARBE-BLEUE Vol. 2 566, 863
BARBEROUSSE Vol. 2 153
BARBIER DE SEVILLE, LE Vol. 2 193, 388, 475, 533
BARBIERE DI SIVIGLIA, IL Vol. 7 902, 1116
BARBORA HLAVSOVA Vol. 10 34, 92, 122
BARBORA RADI Vol. 10 167, 256
BARBOSUL Vol. 3 249
BARBOUZES, LES Vol. 2 293 Vol. 8 62
BARCA SIN PESCADOR, LA Vol. 8 217, 470
BARCAIOLO DI AMALFI, IL Vol. 7 1012
BARCAIUOLO D'AMALFI, IL Vol. 7 493, 829
BARCAMENANDOCI Vol. 7 649
BARCAROLE Vol. 4 49, 266, 574
BARCELONA CONNECTION Vol. 8 297, 432
BARCELONA SUR Vol. 8 310, 621
BARCO SIN RUMBO Vol. 8 551
BARCOS DE PAPEL Vol. 8 320
BARDAME Vol. 4 115, 547, 729, 735
BARE DET NORSKE AVSNITT Vol. 5 721
BARE ET LIV Vol. 5 724, 797
BARE ET LIV – HISTORIEN OM FRIDJOF NANSEN Vol. 5 763
BÄRENBURGER SCHNURRE Vol. 1 13, 115
BARIERA Vol. 1 233, 401 Vol. 3 230, 281, 306 Vol. 6 *page 219*
BARIERATA Vol. 3 114 Vol. 6 *page 51*
BARIOLE Vol. 2 424
BARKEN MARGRETHE AF DANMARK Vol. 5 24
BARLANGLAKOK, A Vol. 9 337
BARMAIDS A JOUIR Vol. 2 294
BARMHERZIGE LÜGE, DIE Vol. 4 115, 536
BARN AV SOLEN Vol. 5 672
BARN I KIRKE Vol. 5 97
BARNA FRAN BLASJÖFJÄLLET Vol. 5 933, 1023, 1052
BARNABE Vol. 2 58, 652, 658
BARNEHJERTE, ET Vol. 5 183
BARNEHJERTS HELTEMOD Vol. 5 228
BARNEN FRAN FROSTMOFJÄLLET Vol. 5 1042, 1078, 1080
BARNENS Ö Vol. 5 850, 1160
BARNET Vol. 5 31, 32, 55, 97, 101, 208
BARNET FRA OPFOSTRINGSHUSET Vol. 5 81
BARNET SOM VELGORER Vol. 5 97
BARNETS MAGT Vol. 5 58
BARNFÖRBJUDET Vol. 5 955, 1160, 1170
BARNVAGNEN Vol. 5 1192
BAROCCO Vol. 2 311, 633
BARON DE L'ECLUSE, LE Vol. 2 644, 722
BARON FANTOME, LE Vol. 2 402, 439, 786
BARON OLSON Vol. 5 1154
BARON PRASIL Vol. 10 154, 385, 462, 550, 621
BARON TZIGANE, LE Vol. 2 733 Vol. 4 567
BARONE DI CORBO, IL Vol. 7 44, 480, 792, 974, 1036
BARONESSA DARIA, LA Vol. 7 399
BARONESSEN FRA BENZINTANKEN Vol. 5 29, 104, 199, 210
BARONI, I Vol. 7 738
BARONIN KAMMERJUNGFER Vol. 4 603
BAROUD Vol. 2 153, 863
BARRABAS Vol. 2 169, 330, 613
BARRACA, LA Vol. 8 546
BARRAGE CONTRE LE PACIFIQUE Vol. 2 178
BARRANCO, LTD Vol. 2 407
BARRANDOVSKE NOCTURNO ANEB JAK FILM TANCIL A ZPIVAL Vol. 10 229, 327, 359, 568
BARRAQUES, LES Vol. 8 723
BARRERA, LA Vol. 8 207, 749
BARRICADE DU POINT DU JOUR, LA Vol. 2 682, 806
BARRIERA A SETTENTRIONE Vol. 7 1036
BARRIERE UMANE Vol. 7 279
BARRILOT SI PORTA A DEPUTATO Vol. 7 1189
BARRIO Vol. 8 298, 436, 449, 611
BARRIOS BAJOS Vol. 8 270, 294
BARRY Vol. 2 571, 698, 807
BARTEK ZWYCIEZCA Vol. 1 191
BARTHOLOMAIOS Vol. 3 131
BARTLEBY Vol. 2 737, 828
BARUFFA Vol. 7 731
BARUTEN BUKVAR Vol. 3 60 Vol. 6 *page 42, page 51*
BARWY OCHRONNE Vol. 1 138, 144, 204, 304, 403
BARWY WALKI Vol. 1 180, 211 Vol. 6 *page 246*

BARYTON Vol. 6 *page 235, page 254*
BAS-FONDS, LES Vol. 2 48, 122, 526
BAS MAJSTOR"T NA EKSKURZIJA Vol. 6 *page 25, page 26, page 62*
BASEJN"T Vol. 3 119 Vol. 6 *page 35*
BASES KAI BASOULA Vol. 3 153
BASILISCHI, I Vol. 7 1167
BASSIFONDI DI SHANGHAI, I Vol. 7 968
BASTA CHE NON SI SAPPINA IN GIRO Vol. 7 599, 950, 1007
BASTA MI BOJADZIJATA Vol. 3 117 Vol. 6 *page 33, page 54*
BASTARD Vol. 3 234 Vol. 5 678, 714, 726, 738, 760
BASTARD, DER Vol. 4 84, 340, 584, 724
BASTARDO, IL Vol. 7 296
BASTIEN, BASTIENNE Vol. 2 685 Vol. 6 *page 87, page 92*
BASTYASETANY 74. Vol. 9 253, 259, 303
BASURA ESTA EN EL ATICO, LA Vol. 8 471
BATAILLE DE L'EAU LOURDE, LA Vol. 2 866
BATAILLE DE SAN SEBASTIAN, LA Vol. 2 185, 824
BATAILLE DU CHILI, LA Vol. 2 622
BATAILLE DU FEU, LA Vol. 2 67, 197
BATAILLE DU RAIL, LA Vol. 2 13
BATAILLE, LA Vol. 2 37, 257, 402, 676, 699
BATAILLE SILENCIEUSE, LA Vol. 2 70, 623, 644
BATAILLON DU CIEL, LE Vol. 2 395, 407, 627
BATALION Vol. 10 34, 256, 345, 460
BATALLA DEL DOMINGO, LA Vol. 8 298, 546
BATALLA DEL PORRO, LA Vol. 8 641, 660
BATALLA DEL ULTIMO PANZER, LA Vol. 8 749
BATALLON DE LAS SOMBRAS, EL Vol. 8 298, 524
BATARD, LE Vol. 6 *page 75, page 127*
BATEAU A SOUPE, LE Vol. 2 440, 469
BATEAU DE VERRE, LE Vol. 2 387, 774
BATEAU D'EMILE, LE Vol. 2 194, 423
BATEAU SUR L'HERBE, LE Vol. 2 74, 285
BATELIERS DE LA VOLGA, LES Vol. 2 122, 676, 775, 824
BATHMAN DAL PIANETA PORNO Vol. 7 728
BATISSEURS, LES Vol. 2 72, 90, 739
BATOUK Vol. 2 112
BATTAGLIA DEL DESERTO, LA Vol. 7 1202
BATTAGLIA DI ALGERI, LA Vol. 7 502
BATTAGLIA DI MARATONA, LA Vol. 7 106
BATTAGLIA, LA Vol. 7 101
BATTAGLIE DELLA VITA Vol. 7 917
BATTANT, LE Vol. 6 *page 89, page 99*
BATTELIERI DEL VOLGA, I Vol. 7 798
BATTEMENT DE COEUR Vol. 2 59, 488, 657
BATTICUORE Vol. 7 196, 487, 756
BAUDELAIRE EST MORT EN ETE Vol. 2 188
BAULÖWE, DER Vol. 6 *page 160, page 170*
BAXTER, VERA BAXTER Vol. 2 854
BAZA LUDZI UMARLYCH Vol. 1 227, 382 Vol. 6 *page 216*
BAZAR DELLE IDEE, IL Vol. 7 733, 1116
B'DI BLAGOSLOVENA Vol. 3 73, 101
B'DI STASTLIVA, ANI ! Vol. 3 66, 82 Vol. 6 *page 26*
BEARN Vol. 8 297, 616
BEATA Vol. 1 198, 213, 220
BEATA ES AZ ÖRDÖG Vol. 9 315, 322
BEATES FLITTERWOCHEN Vol. 4 25, 347, 370
BEATRICE Vol. 3 218, 313 Vol. 7 441, 701
BEATRICE CENCI Vol. 7 48, 404, 453, 491, 640, 763, 910, 1019, 1060, 1096
BEATRICE DEVANT LE DESIR Vol. 2 339, 460, 863
BEAU MARIAGE, LE Vol. 6 *page 87, page 107, page 148*
BEAU MASQUE Vol. 2 194, 527
BEAU MEC, LE Vol. 2 1
BEAU MONSTRE, UN Vol. 2 241, 377
BEAU-PERE Vol. 2 655 Vol. 6 *page 89, page 99*
BEAU SERGE, LE Vol. 2 224
BEAU VOYAGE, LE Vol. 2 158, 566
BEAUJOLAIS NOUVEAU EST ARRIVE, LE Vol. 2 12, 779
BEAUTE DU DIABLE, LA Vol. 2 59, 427, 571
BEAUX JOURS DU ROI MURAT, LES Vol. 2 187, 574
BEAUX JOURS, LES Vol. 2 67, 279, 427, 573
BEBAWI - IL DELITTO DI VIA LAZIO Vol. 7 280, 947
BEBE DE L'ESCADRON, LE Vol. 2 407, 526, 626, 824
BEBENDE BERG, DER Vol. 4 25
BEBERT ET LE OMNIBUS Vol. 2 47, 185
BEBES A GOGO Vol. 2 249, 777
BECASSINE Vol. 2 318, 858, 871
BECERRADA, LA Vol. 8 762, 768
BECSÜLET ES DICSÖSEG Vol. 9 71, 237, 241

BECZKA AMONTILLADO Vol. 1 303
BEDINGUNG - KEIN ANHANG ! Vol. 4 646
BEDNATA ULICA Vol. 3 32, 38, 48, 112 Vol. 6 *page 32*
BEDNIJAT LUKA Vol. 3 105 Vol. 6 *page 29, page 57*
BEDRE ENN SITT RYKTE Vol. 5 762, 831
BEDSTEMOR GAAR AMOK Vol. 5 89, 224
BEETHOVEN - TAGE AUS EINEM LEBEN Vol. 1 37, 57, 92
BEFEHL IST BEFEHL Vol. 4 230, 232, 292
BEFFA DELLA VITA, LA Vol. 7 531, 1079
BEFFE, LICENZE ET AMORI DEL DECAMERON SEGRETO Vol. 7 290
BEFREITE HÄNDE Vol. 4 123, 288, 544, 619
BEGÄR Vol. 5 1156, 1164
BEGEGNUNG IM ZWIELICHT Vol. 6 *page 177*
BEGEGNUNGEN Vol. 1 2, 12, 17, 70
BEGEGNUNGEN IM ZWIELICHT Vol. 1 72
BEGIERDE Vol. 4 35, 310, 720
BEGUIN DE LA GARNISON, LE Vol. 2 37, 469, 660
BEGUNEC Vol. 3 395, 417
BEGYNDTE OMBORD, DET Vol. 5 24
BEHAGELIG SPADSERETUR, EN Vol. 5 97
BEHAJTANI TILOS Vol. 9 26, 117, 133
BEHEXTE NEPTUN, DER Vol. 4 95, 124, 428
BEI DER BLONDEN KATHREIN Vol. 4 344, 647
BEICHTE EINER GEFALLENEN, DIE Vol. 4 219, 781
BEIDEN GATTEN DER FRAU RUTH, DIE Vol. 4 190
BEIDEN RIVALEN, DIE Vol. 4 734
BEIDEN SCHWESTERN, DIE Vol. 4 51, 108, 623
BEIDEN SEEHUNDE, DIE Vol. 4 361, 461
BEIL VON WANDSBEK, DAS Vol. 1 107, 129
BEIM RECHTSANWALT Vol. 4 167
BEJARANA, LA Vol. 8 655
BEJLEREN Vol. 5 11, 46, 273
BEKEIDÖ Vol. 9 238
BEKÖTÖTT SZEMMER Vol. 9 259, 285
BEKSTVA Vol. 3 367, 374
BEL-AGE, LE Vol. 2 190, 517, 854
BEL AMI Vol. 2 59, 395, 646 Vol. 4 265, 504, 614 Vol. 5 1133
BEL AMOUR Vol. 2 401
BEL INDIFFERENT, LE Vol. 2 285
BELARMINO Vol. 8 22
BELE TRAVE Vol. 3 395, 417
BELIJAT KON Vol. 3 59, 96
BELL' ANTONIO, IL Vol. 7 498, 817
BELLA ADDORMENTATA, LA Vol. 7 453, 796, 1060
BELLA ANTONIA, PRIMA MONICA E POI DIMONIA, LA Vol. 7 1027
BELLA CORSARA, LA Vol. 7 390, 490, 1071
BELLA DI GIORNO, MOGLIE DI NOTTE Vol. 7 333
BELLA DI LODI, LA Vol. 7 333, 388
BELLA DI ROMA, LA Vol. 7 493
BELLA E LA BESTIA, LA Vol. 7 143, 965
BELLA GIARDINIERA, LA Vol. 7 91
BELLA LOLA, LA Vol. 8 569
BELLA MIMI, LA Vol. 8 583, 772
BELLA MUGNAIA, LA Vol. 7 1062
BELLA NON PIANGERE Vol. 7 1147
BELLA, RICCA, LIEVE DIFETTO FISICO, CERCA ANIMA GEMELLA Vol. 7 528, 761
BELLA SALAMANDRA, LA Vol. 7 72, 428
BELLADONNA Vol. 5 61, 99
BELLE Vol. 2 190
BELLE AFFAIRE, LA Vol. 2 154, 368
BELLE AMERICAINE, LA Vol. 2 9, 190
BELLE AUX CHEVEUX D'OR, LA Vol. 2 798
BELLE AVENTURE, LA Vol. 2 96, 153, 156
BELLE CAPTIVE, LA Vol. 6 *page 70, page 116*
BELLE CAROSSERIE, UNE Vol. 6 *page 79*
BELLE DAME SANS MERCI, LA Vol. 2 636
BELLE DE CADIX, LA Vol. 2 8, 59, 540
BELLE DE JOUR Vol. 2 185, 854
BELLE DE MONTPARNASSE, LA Vol. 2 270
BELLE DE NUIT Vol. 2 152, 437, 451, 530
BELLE DELL' ARIA, LE Vol. 7 468 Vol. 8 353, 393, 400, 668, 677, 749
BELLE EQUIPE, LA Vol. 2 309, 439, 440
BELLE ET LA BETE, LA Vol. 2 13, 85, 158, 282, 615
BELLE ET LE TZIGANE, LA Vol. 2 226, 395, 846 Vol. 9 63, 183, 241, 312, 320
BELLE ET L'EMPEREUR, LA Vol. 2 566
BELLE ETOILE Vol. 2 128, 174, 402, 863
BELLE FAMIGLIE, LE Vol. 7 869
BELLE FILLE COMME MOI, UNE Vol. 2 351, 434
BELLE GARCE, UNE Vol. 2 37, 84, 131, 332
BELLE HUMEUR Vol. 2 196, 330, 613
BELLE IMAGE, LA Vol. 2 401, 419
BELLE LIMONADIERE, LA Vol. 2 731
BELLE MA POVERE Vol. 7 333
BELLE MADAME HEBERT, LE Vol. 7 730
BELLE MARINIERE, LA Vol. 2 563, 579
BELLE MENTALITE Vol. 2 595, 627

BELLE MEUNIERE, LA Vol. 2 347, 871
BELLE NIVERNAISE, LA Vol. 2 246, 378, 387
BELLE O BRUTTE SI SPOSAN TUTTE... Vol. 7 16, 453
BELLE OTERO, LA Vol. 2 248, 388, 427
BELLE QUE VOILA, LA Vol. 2 251, 824
BELLE REVANCHE, LA Vol. 2 259, 698
BELLE VIE, LA Vol. 2 110, 457
BELLES, BLONDES ET BRONZEES Vol. 6 *page 79*
BELLES DE NUIT, LES Vol. 2 59, 226, 423, 824
BELLES MANIERES, LES Vol. 2 804
BELLEZZA DEL DIAVOLO, LA Vol. 7 1131
BELLEZZA DEL MONDO, LA Vol. 7 369
BELLEZZA IN BICICLETTA Vol. 7 792
BELLEZZE A CAPRI Vol. 7 201
BELLEZZE IN MOTOSCOOTER Vol. 7 501
BELLISSIMA Vol. 7 915, 923, 992, 1142
BELLISSIME GAMBE DI SABRINA, LE Vol. 7 693
BELLO DI MAMMA Vol. 7 854
BELLO MIO BELLEZZA MIA Vol. 7 381, 775, 988
BELLO RECUERDO Vol. 8 298, 400
BELONNINGEN Vol. 5 728, 732, 824
BELOV'RHA VITOSA Vol. 3 49
... BELPAESE, IL Vol. 7 567
BELPHEGOR Vol. 2 33, 489, 724, 731
BELVA COL MITRA, LA Vol. 7 130
BEMATA TES FOTIAS Vol. 3 167
BEN ET BENEDICT Vol. 2 72
BEN HUR Vol. 7 621, 792, 796, 923, 978
BENE CONTRO MALE Vol. 7 69
BENEDEK HAZ Vol. 9 196, 274
BENEFIZ-VORSTELUNG DER VIER TEUFEL, DIE Vol. 4 535, 651
BENGASI Vol. 7 40, 423, 1136
BENI ECEN VETE Vol. 3 5, 11 Vol. 6 *page 13*
BENILDE OU A VIRGEM-MAE Vol. 8 103, 169
BENINGO HERMANO MIO Vol. 8 276, 495
BENITO CERENO Vol. 2 35
BENITOU Vol. 2 672
BENJAMIN Vol. 2 190, 674
BENNO MACHT GESCHICHTEN Vol. 6 *page 163, page 168, page 174*
BENNO STEHKRAGEN Vol. 4 223, 287
BENSAA SUONISSA Vol. 5 541
BENVENUTA Vol. 6 *page 91*
BERCEAU DE CRISTAL, LE Vol. 2 331
BERCEAU DE DIEU, LE Vol. 2 579, 767
BERCSENYI HUSZAROK Vol. 9 26, 132
BEREDSKAPSPOJKAR Vol. 5 843, 907, 923
BERENICE Vol. 2 72
BERG DES SCHICKSALS, DER Vol. 4 10, 73, 156, 240, 493, 620
BERG-EJVIND OCH HANS HUSTRU Vol. 5 937, 997
BERG RUFT, DER Vol. 4 10, 54, 215, 432, 531, 541, 564
BERGADLER, DER Vol. 4 544, 545, 721
BERGE IN FLAMMEN Vol. 4 10, 54, 73, 725
BERGENSTOGET PLYNDRET I NATT Vol. 5 673
BERGET PA MANENS BAKSIDA Vol. 5 882, 990
BERGFÜHRER VON ZAKOPANE, DER Vol. 4 565, 725
BERGKATZE, DIE Vol. 4 663, 681
BERGSLAGSFOLK Vol. 5 928
BERLIN – ALEXANDERPLATZ Vol. 4 85, 158, 200
BERLIN BLUES Vol. 8 358, 771
BERLIN. DIE SINFONIE DER GROSSTADT Vol. 4 28, 325, 373, 599
BERLIN ECKE SCHÖNHAUSER Vol. 1 26, 31, 88
BERLIN KAPUTT Vol. 6 *page 306, page 315, page 319*
BERLIN W Vol. 4 424
BERLINER RANGE, DIE Vol. 4 385, 495
BERLINER ROMANZE, EINE Vol. 1 31, 105, 125
BERLINER UNTER SICH Vol. 4 646
BERLINGOT ET CIE Vol. 2 84, 722, 726
BERLINGUER TI VOGLIO BENE Vol. 7 1105
BERNARDA LA DOTTORESSA DELLE NINFOMANI Vol. 7 99
BERÖRINGEN Vol. 5 1052, 1053, 1080
BERSAGLIO UMANO Vol. 7 548
BERSAGLIO VIVENTE Vol. 7 1189
BERTILLONSKE SYSTEM, DET Vol. 5 92
BERTOLDO, BERTOLDINO E CACASENNO Vol. 7 80, 109, 378, 460, 542, 941, 995
BERTRAND, COEUR DE LION Vol. 2 224, 627
BERU ET CES DAMES Vol. 2 475, 815
BERÜHMTE FRAU, DIE Vol. 4 246, 321, 752
BESA E KUQE Vol. 6 *page 11, page 12*
BESAETTELSE Vol. 5 55, 196
BESAME MONSTRUO Vol. 8 465, 775
BESAME TONTA Vol. 8 219, 580
BESATZUNG DORA Vol. 4 484, 469
BESCHREIBUNG EINES SOMMERS Vol. 1 46, 92, 100
BESEJREDE PEBERSVENDE, DE Vol. 5 81
BESESSENE ODER DAS FRÄULEIN VON SCUDERI, DER Vol. 4 551
BESHMRTNA MLADOST Vol. 3 349, 436

BESIEGTE SIEGER Vol. 4 247
BESO DE JUDAS, EL Vol. 8 214, 400, 748
BESO EN EL PUERTO, UN Vol. 8 311, 402
BESONDERE KENNZEICHEN Vol. 4 221, 340, 608
BESONDERE KENNZEICHEN: KEINE Vol. 1 2, 115, 120
BESSERER HERR, EIN Vol. 4 344, 544
BESSERER HERR SUCHT ANSCHLUSS Vol. 4 13, 24
BESSERWISSER, DER Vol. 4 329, 525
BESTEN JAHRE, DIE Vol. 1 47, 60
BESTIA Vol. 1 174, 215, 390
BESTIA NELLO SPAZIO, LA Vol. 7 474
BESTIA Y LA ESPADA MAGICA, LA Vol. 8 296, 580
BESTIAIRE D'AMOUR, LE Vol. 2 266
BESTIALITA Vol. 7 127
BESTIE, DIE Vol. 4 344
BESTIE IM MENSCHEN, DIE Vol. 4 36, 83, 690
BESTIJE Vol. 3 407
BESUCH AM ABEND Vol. 4 28, 340, 466
BESUCH BEI ONKEL EMIL Vol. 4 212, 617
BESUCH BEIM BETTELSTUDENT, DER Vol. 4 367
BESUCH IN DER ABENDSTUND Vol. 4 110
BESUCH UM MITTERNACHT Vol. 4 270, 409, 462
BESZALLASOLAS Vol. 9 71, 322
BESZELÖ KÖNTÖS, A Vol. 9 26, 117, 121, 133, 149, 259
BESZTERCE OSTROMA Vol. 9 71, 222, 312
BETE A L'AFFUT, LA Vol. 2 566, 662
BETE AUX SEPT MANTEAUX, LA Vol. 2 249, 603, 708, 871
BETE ERRANTE, LA Vol. 2 332
BETE HUMAINE, LA Vol. 2 206, 526, 625, 719
BETE, LA Vol. 2 215, 642
BETE, MAIS DISCIPLINE Vol. 2 151, 779
BETE NOIRE, LA Vol. 6 *page 111, page 120*
BETE TRAQUEE, LA Vol. 2 227, 614
BETHSABEE Vol. 2 395, 579, 722
BETJENTENS MIDDAGSLUR Vol. 5 97
BETRAGEN UNGENÜGEND Vol. 4 257
BETROGEN BIS ZUM JÜNGSTEN TAG Vol. 1 24, 35, 125
BETROGENE KALIF, DER Vol. 4 517, 558, 796
BETTELGRÄFIN VOM KURFÜRSTEN-DAMM, DIE Vol. 4 5, 584
BETTELPRINZESSIN, DIE Vol. 4 5, 565
BETTELSTUDENT - AUFTRITT DES OBERST OLLENDORF, DER Vol. 4 135, 514
BETTELSTUDENT, DER Vol. 4 119, 287, 384, 437, 458, 531, 644, 646, 738
BETTLER VOM KÖLNER DOM, DER Vol. 4 239, 340
BETTLER VON SAUERN, DER Vol. 4 366
BEUNRUHIGUNG, DIE Vol. 6 *page 156, page 161, page 167*
BEUTE DER ERINNYEN, DIE Vol. 4 232, 239
BEVOR DER BLITZ EINSCHLÄGT Vol. 1 58, 100, 129
BEYOND THE BORDER Vol. 2 84
BEZ Vol. 3 402
BEZ, AT TI NEUTECE Vol. 10 256, 460
BEZ KONCA Vol. 6 *page 254*
BEZ MILOSCI Vol. 6 *page 225, page 239, page 260*
BEZ SVATOZARE Vol. 10 184, 550
BEZ ZNIECZULENIA Vol. 1 204, 339 Vol. 6 *page 254*
BEZDETNA Vol. 10 167, 453
BEZDOMNI Vol. 1 194, 384
BEZIMIENNI BOHATEROWIE Vol. 1 278, 346
BEZIRKSVERTRETER GESUCHT Vol. 4 13, 24
BEZKRESNE LAKI Vol. 1 142, 197, 244
BEZKR"STNI GROBOVE Vol. 3 71
BIALA TRUCIZNA Vol. 1 278, 404
BIALY MAZUR Vol. 1 198, 254, 395, 396
BIALY MURZYN Vol. 1 194, 384
BIALY NIEDZWIEDZ Vol. 1 254, 321, 358
BIALY SLAD Vol. 1 228
BIANCA Vol. 7 658, 801, 1145
BIANCANEVE E CO Vol. 7 492
BIANCO E NERO Vol. 7 1055
BIANCO, IL GIALLO, IL NERO, IL Vol. 8 346
BIANCO, ROSSO E ... Vol. 7 32, 347, 931
BIANCO, ROSSO E VERDONE Vol. 7 1077, 1145
BIANCO, ROSSO, GIALLO, ROSA Vol. 7 772
BIBBIA, LA Vol. 7 988
BIBERPELZ, DER Vol. 1 23, 77, 107 Vol. 4 124, 232, 361, 422, 648, 743
BIBERSPUR Vol. 6 *page 160, page 163, page 167*
BIBI FRICOTIN Vol. 2 833
BIBI LA PUREE Vol. 2 488, 613
BIBLIA EN PASTA, LA Vol. 8 658, 681
BICE SKORO PROPAST SVETA Vol. 3 350, 408
BICHES, LES Vol. 2 91, 700
BICHON Vol. 2 48, 388, 475, 726

BICICLETAS SON PARA EL VERANO, LAS **Vol. 8** 616, 759
BICIKLISTI **Vol. 3** 373, 433
„BICOTS NEGRES" VOS VOISINS, LES **Vol. 2** 110, 161
BICYCLISTE, UNE **Vol. 2** 678
BICZ BOZY **Vol. 1** 169, 180, 303
BIDASSES AU PENSIONNAT, LES **Vol. 2** 294
BIDASSES AUX GRANDES MANOEUVRES, LES **Vol. 6** *page 79*
BIDASSES EN FOLIE, LES **Vol. 2** 40, 114
BIDASSES EN VADROUILLE, LES **Vol. 2** 692, 746
BIDASSES S'EN VONT EN GUERRE, LES **Vol. 2** 114
BIDON D'OR, LE **Vol. 2** 394, 672
BIDONE, IL **Vol. 7** 283, 539, 725
BIE E IL BA, IL **Vol. 7** 315, 542, 909
BIELA STUZKA V TVOJICH VLASOCH **Vol. 10** 263, 269, 544
BIEN **Vol. 5** 97
BIEN AU FOND DU PETIT TROU **Vol. 6** *page 120*
BIEN PAGADA, LA **Vol. 8** 270, 751
BIENE MAJA UND IHRE ABENTEUER, DIE **Vol. 4** 747
BIENFAITEUR, LE **Vol. 2** 440, 676
BIENVENIDO MISTER MARSHALL **Vol. 8** 276, 311, 748
BIENVENIDO PADRE MURRAY **Vol. 8** 752
BIETUL IOANIDE **Vol. 3** 288, 291
BIFFEN OCH BANANEN **Vol. 5** 909, 940, 1173
BIFONE E LA MASCHERA DAI DENTI NERI **Vol. 7** 621
BIFUR 3 **Vol. 2** 270, 407
BIGAMIE **Vol. 4** 306, 724
BIGAMISTKA **Vol. 1** 317
BIGAMO, IL **Vol. 7** 798
BIGLIETTO DA MILLE, IL **Vol. 7** 1189
BIGLIETTO DI FAVORE **Vol. 7** 1189
BIGORNE, CAPORAL DE FRANCE, LA **Vol. 2** 249, 517, 801, 866
BIJELI DJAVO **Vol. 3** 378, 403, 451
BIJOUTIERS DU CLAIR DE LUNE, LES **Vol. 2** 21, 824
BIJOUX DE FAMILLE, LES **Vol. 2** 640
BIJUTERII DE FAMILIE **Vol. 3** 287, 310, 317
BILA JACHTA VE SPLITU **Vol. 10** 168, 564
BILA NEMOC **Vol. 10** 167, 256
BILA OBLAKA **Vol. 10** 313, 395, 577
BILA PANI **Vol. 10** 359, 572
BILA SAM JACA **Vol. 3** 423, 455
BILA SPONA **Vol. 10** 369, 460
BILA TMA **Vol. 10** 193, 503
BILA VRANA **Vol. 10** 256, 460
BILANS KWARTALNY **Vol. 1** 183, 403 **Vol. 6** *page 236*
BILBAO **Vol. 8** 256
BILD DER AHNFRAU, DAS **Vol. 4** 5
BILDER DES ZEUGEN SCHATTMANN, DIE **Vol. 1** 22, 27, 104
BILDERBOGEN DER EHE **Vol. 4** 256, 605
BILDNIS DES DORIAN GRAY, DAS **Vol. 4** 159, 488
BILET POWROTNY **Vol. 6** *page 222, page 243, page 252*
BILINCSBEVERT FOLYAM **Vol. 9** 2
BILITIS **Vol. 2** 215, 787
BILJET TILL ÄVENTYR **Vol. 5** 357
BILJETT TILL PARADISET **Vol. 5** 840, 946, 959
BILLE DE CLOWN **Vol. 2** 153
BILLET DE LOGEMENT, LE **Vol. 2** 330
BILLET DE MILLE, L' **Vol. 2** 388, 566, 708
BILLET MRK. **Vol. 5** 7, 168, 222
BILLI SCHLÄGT SICH DURCHS LEBEN **Vol. 4** 624
BILLY V PRAZE **Vol. 10** 335
BILY RAJ **Vol. 10** 167, 536
BIM **Vol. 2** 811
BIM BUM BAM **Vol. 7** 1114
BIMBA DI SATANA, LA **Vol. 7** 609
BIMBI D'ITALIA SON TUTTI BALILLA, I **Vol. 7** 1195
BIMBI LANTONI **Vol. 7** 150
BING BANG **Vol. 3** 216
BINGO BONGO **Vol. 7** 208, 347
BINGUEROS, LOS **Vol. 8** 297
BIO-GRAFIA **Vol. 3** 175
BIONDA FRAGOLA **Vol. 7** 13, 134
BIONDA SOTTOCHIAVE **Vol. 7** 796, 1181
BIONDINA, LA **Vol. 7** 560
BIOTAXIA **Vol. 8** 317, 470
BIQUEFARRE **Vol. 6** *page 100, page 127*
BIRBANTE DI UN CANE **Vol. 7** 1189
BIRIBI **Vol. 2** 142, 219
BIRIBI, IL PICCOLO POLIZIOTTO TORINESE **Vol. 7** 398
BIRICCHINATE DI UN RAGAZZO, LE **Vol. 7** 1189
BIRICCHINO DI PAPA, IL **Vol. 7** 462, 756, 1027
BIRICCHINO DI PARIGI **Vol. 7** 965
BIRICCHINO DI TRISTESE, IL **Vol. 7** 131
BIRKE DA OBEN, DIE **Vol. 6** *page 161, page 174, page 176*
BIS DASS DER TOD EUCH SCHEIDET **Vol. 1** 14, 71
BIS SPÄTER – ICH MUSS MICH ERSCHIESSEN **Vol. 5** 325, 331, 343
BISACCIA, LA **Vol. 7** 891
BISARCA, LA **Vol. 7** 891

BISBETICA DOMATA, LA **Vol. 7** 756
BISBETICO DOMATO, IL **Vol. 7** 30, 347, 476
BISCOT SE TROMPE D'ETAGE **Vol. 2** 432
BISKOPPEN **Vol. 5** 55, 98, 101, 208
BISMARCK **Vol. 4** 5, 270, 283, 420, 423, 424, 462, 479, 788
BISSCHEN LIEBE FÜR DICH, EIN **Vol. 4** 208, 452
BIST DU DER WEIHNACHTSMANN ? **Vol. 4** 247
BISTURI, LA MAFIA BIANCA **Vol. 7** 781, 1007
BITKA NA NERETVI **Vol. 3** 371, 372, 421, 444
BITTE EIN AUTOGRAMM **Vol. 4** 361, 558
BITTERE KRÄUTER **Vol. 3** 349, 359, 391
BITTERE PILLEN **Vol. 1** 42, 68, 121
BITVA O HEDVIKU **Vol. 10** 352, 369, 385
BITWA O KOZI DWOR **Vol. 1** 156, 180, 386, 389 **Vol. 6** *page 237*
BIVIO DELLA MORTE, IL **Vol. 7** 1189
BIZALOM **Vol. 9** 167, 206, 259
BJAGAJ...OBICAM TE ! **Vol. 3** 31, 42 **Vol. 6** *page 28*
BJAGASTA PO V"LNITE **Vol. 3** 45 **Vol. 6** *page 59*
BJAGSTVO V ROPOTAMO **Vol. 3** 114 **Vol. 6** *page 39*
BJALA MAGIJA **Vol. 6** *page 35, page 43, page 60*
BJALATA, ODISEJA **Vol. 3** 81, 103
BJALATA STAJA **Vol. 3** 38, 69 **Vol. 6** *page 50*
BJÖRN MÖRK **Vol. 5** 986
BJORNEJAGT **Vol. 5** 97
BJORNEJAGT I RUSLAND **Vol. 5** 193
BJORNEJAGTEN **Vol. 5** 97
BJORNEPATRULJEN **Vol. 5** 756
BJORNETAEMMEREN **Vol. 5** 188
BJURRA **Vol. 5** 724, 782, 817
BLA HIMMEL **Vol. 5** 840, 843, 884
BLAA BLOD, DET **Vol. 5** 86
BLAA DRENGE, DE **Vol. 5** 55, 101, 208
BLAA VIDUNDER, DET **Vol. 5** 81
BLAAVAND MELDER FIRE **Vol. 5** 7
BLAAVAND MELDER STORM **Vol. 5** 136
BLACK ARROW **Vol. 8** 215, 303
BLACK CAT **Vol. 7** 222, 1015
BLACK JACK **Vol. 2** 21, 826 **Vol. 8** 408, 728
BLACK MAGIC **Vol. 2** 27
BLACK MOON **Vol. 2** 634, 841
BLACK STORY **Vol. 8** 306, 681
BLACK VENUS **Vol. 8** 214, 296
BLADE AF SATANS BOG **Vol. 5** 265
BLADKONGEN **Vol. 5** 81
BLAENDVAERK **Vol. 5** 14
BLAGO ACTEKA **Vol. 6** *page 312*
BLAGO U SREBRNOM JEZERU **Vol. 3** 366, 371
BLAGUE DANS LE COIN **Vol. 2** 194, 487
BLAHOVE DEVCE **Vol. 10** 345, 585
BLAHOVY SEN **Vol. 10** 122, 154, 460, 566
BLAJACKOR **Vol. 5** 862, 940, 1025, 1173
BLANANDE HAV **Vol. 5** 1080
BLANC COMME NEIGE **Vol. 2** 308, 460
BLANC ET LE NOIR, LE **Vol. 2** 402
BLANC ET NOIR **Vol. 1** 174, 268
BLANCA PALOMA, LA **Vol. 8** 276, 359
BLANCA POR FUERA, ROSA POR DENTRO **Vol. 8** 299, 681
BLANCHE **Vol. 2** 275, 642
BLANCHETTE **Vol. 2** 248
BLANCO, EL AMARILLO Y EL NEGRO, EL **Vol. 8** 346, 348
BLANDT BYENS BORN **Vol. 5** 83
BLANDT KOBENHAVNS APACHER **Vol. 5** 121
BLANDT SAMFUNDETS FJENDER **Vol. 5** 295
BLASONE MALEDETTO, IL **Vol. 7** 843
BLASTFIGHTER **Vol. 7** 99
BLATAGUL **Vol. 3** 281
BLATT AUS DEM BUCHE DES LEBENS, EIN **Vol. 4** 135, 514
BLAUE DIAMANT, DER **Vol. 4** 384, 558, 796
BLAUE ENGEL, DER **Vol. 4** 246, 298, 567, 620
BLAUE HELM, DER **Vol. 6** *page 156, page 159, page 169*
BLAUE LATERNE, DIE **Vol. 4** 773
BLAUE LICHT, DAS **Vol. 1** 5, 69, 116 **Vol. 4** 73, 620
BLAUE MAUS, DIE **Vol. 4** 49, 87, 734, 584
BLAUE VOM HIMMEL, DAS **Vol. 4** 192, 584
BLAUE ZIMMER, DAS **Vol. 1** 85
BLAUEN SCHWERTER, DIE **Vol. 1** 25, 56, 129
BLAUFUCHS, DER **Vol. 4** 614, 742
BLAUFUCHS, SCHRECKEN DES KILIMANDSCHARO, DER **Vol. 4** 25
BLAUVOGEL **Vol. 6** *page 163, page 171*
BLAZEN PURYMOWY **Vol. 1** 346, 384
BLAZNI A DEVCATKA **Vol. 10** 183, 278, 383
BLAZNI, VODNICI A PODVODNICI **Vol. 10** 395, 425, 496
BLAZNOVA KRONIKA **Vol. 10** 193, 207, 462
BLBEC Z XEENEMÜNDE **Vol. 10** 153, 296, 425
BLE EN HERBE, LE **Vol. 2** 232, 251, 488, 625

BLECHTROMMEL, DIE **Vol. 3** 437
BLED, LE **Vol. 2** 533, 623
BLEKA GREVEN **Vol. 5** 923, 1181
BLEKITNY KRZYZ **Vol. 1** 331
BLEPE LOUKIANOS **Vol. 3** 191
BLESSURE, LA **Vol. 7** 264, 701, 725
BLEUS DE LA MARINE, LES **Vol. 2** 731 **Vol. 2** 771
BLEUS DE L'AMOUR, LES **Vol. 2** 863
BLEUS DU CIEL, LES **Vol. 2** 48, 99, 626, 658
B"LGARAN E GALANT **Vol. 3** 50
B"LGARI OT STARO VREME **Vol. 3** 79
B"LGARO-UNGARSKA RAPSODIJA **Vol. 3** 90
B"LGARSKI ORLI **Vol. 3** 64
BLICK IN DEN ABGRUND, DER **Vol. 4** 114
BLICK IN DEN ABGRUND, EIN **Vol. 4** 514
BLICK ZURÜCK, EIN **Vol. 4** 614, 666
BLIND **Vol. 5** 97
BLIND ALARM **Vol. 5** 97
BLIND MAKKER **Vol. 5** 46, 125, 283
BLIND PASSAGER **Vol. 5** 58
BLINDE LIEBE **Vol. 4** 798
BLINDE PASSAGIER, DER **Vol. 4** 511, 648
BLINDE PASSAGIERE **Vol. 4** 119, 558
BLINDEKUH **Vol. 4** 560
BLINDER EIFER **Vol. 4** 18
BLINDER EIFER SCHADET NUR **Vol. 4** 110, 246
BLINKENDE FENSTER, DAS **Vol. 4** 718
BLITZZUG DER LIEBE **Vol. 4** 162, 288
BLIXT OCH DUNDER **Vol. 5** 840, 841
BLIZKE DIALAVY **Vol. 10** 72, 470, 592
BLIZNA **Vol. 1** 183, 210, 294
BLOCKERAT SPAR **Vol. 5** 933, 1073
BLOD OCH ELD **Vol. 5** 863, 1032
BLODETS RÖST **Vol. 5** 997
BLODVEIEN **Vol. 5** 800
BLOMSTERPIGEN **Vol. 5** 97
BLOMSTERPIGENS HAEVN **Vol. 5** 183
BLOMSTERTID..., DEN **Vol. 5** 863, 1032
BLOMSTRANDE TIDER **Vol. 5** 882, 935, 952
BLONDE CARMEN, DIE **Vol. 4** 123, 373, 788
BLONDE CHRISTEL, DIE **Vol. 4** 344, 647
BLONDE COMME CA, UNE **Vol. 2** 9, 368
BLONDE DE PEKIN, LA **Vol. 2** 481, 663
BLONDE DES TROPIQUES, UNE **Vol. 2** 131, 872
BLONDE FRAU ÜBERN KURZEN WEG **Vol. 4** 314, 700
BLONDE GEISHA, DIE **Vol. 4** 239, 479
BLONDE HANNELE, DAS **Vol. 4** 25, 219
BLONDE HUMIDES **Vol. 2** 596
BLONDE NACHTIGALL, DIE **Vol. 4** 89, 270
BLONDE VERHÄNGNIS, DAS **Vol. 4** 320, 423, 454
BLONDEN MÄDELS VOM LINDENHOF, DIE **Vol. 4** 385
BLONDER MANN ÜBERN WEG **Vol. 4** 97, 738, 788
BLONDER TRAUM, EIN **Vol. 4** 30, 325, 567, 713
BLONDES GIFT **Vol. 4** 240, 424, 628
BLONDIE, BIFFEN OCH BANANEN **Vol. 5** 878, 1173
BLONDIN I FARA **Vol. 5** 533
BLONDINE **Vol. 2** 193, 539
BLONDY **Vol. 2** 49
BLOUDENI **Vol. 10** 87, 191, 395, 505
BLOUDENI ORIENTACNIHO BEZCE **Vol. 10** 283, 471, 495
BLOW – UP **Vol. 7** 853
BLU GANG **Vol. 7** 1101
BLUDICKA **Vol. 10** 45, 72, 155, 270
BLUDNE DUSE **Vol. 10** 257
BLUE ECSTASY **Vol. 2** 294
BLUE EROTIC CLIMAX **Vol. 7** 742
BLUE JEANS **Vol. 2** 39
BLUE MOVIE **Vol. 7** 289
BLUE NUDE **Vol. 7** 478
BLUES ENTRE LES DENTS, LE **Vol. 2** 325, 793
BLUES METROPOLITANO **Vol. 7** 607, 933
BLUES PRO EFB **Vol. 10** 389, 466, 622
BLUFF **Vol. 4** 329, 365 **Vol. 6** *page 123*
BLUFF STOP **Vol. 5** 1035, 1097, 1170
BLUFF – STORIE DI TRUDDE E DI IMBROGLIONI **Vol. 7** 502
BLUFFEUR, LE **Vol. 2** 259
BLUME VON HAWAI, DIE **Vol. 4** 373, 626
BLUMEN FÜR DEN MANN IM MOND **Vol. 1** 33, 55, 79
BLUMENFRAU VOM POTSDAMER PLATZ, DIE **Vol. 4** 321, 626
BLUMENMÄDCHEN VOM GRAND-HOTEL, DAS **Vol. 4** 292, 340, 466
BLUMENWUNDER, DAS **Vol. 4** 203, 732
BLUSENKÖNIG, DER **Vol. 4** 560
BLUSZCZ **Vol. 6** *page 208, page 231, page 244*
BLUT, DAS **Vol. 4** 113, 399
BLUT DER AHNEN, DAS **Vol. 4** 239, 734
BLUT DER SCHWESTER, DAS **Vol. 4** 135, 187
BLUTSBRÜDER **Vol. 1** 46, 91, 96, 126 **Vol. 4** 340, 466, 471, 566
BLUTSBRÜDERSCHAFT **Vol. 4** 270, 378
BLUTSCHANDE PARAGRAPH 173 ST. G. B. **Vol. 4** 247, 255
BLYGE ANTON **Vol. 5** 930, 1032

BOB AND MARY **Vol. 4** 479, 747
BOB GRAY **Vol. 4** 71, 780
BOB HERCEG **Vol. 9** 67, 71, 84, 268, 274
BOB LE FLAMBEUR **Vol. 2** 131, 224, 577
BOBBY **Vol. 4** 329, 355
BOBBY DEERFIELD **Vol. 2** 224
BOBBY, DER BENZINJUNGE **Vol. 4** 242, 461
BOBBY GEHT LOS **Vol. 4** 119, 340
BOBBYS KRIG **Vol. 5** 712, 817, 824
BOBO JACCO **Vol. 6** *page 121*
BOBO, O **Vol. 8** 32, 118
BOBOSSE **Vol. 2** 170, 294
BOBY NESMI KOURIT **Vol. 10** 198
BOCAGE **Vol. 8** 16, 18, 43, 50, 82, 122, 162, 170, 180
BOCCA CHIUSA, LA **Vol. 7** 277
BOCCA DELL' INFERNO, LA **Vol. 7** 729
BOCCA GOLOSA **Vol. 7** 742
... BOCCA MI BACIO TUTTA TREMANTE **Vol. 7** 1174
BOCCA SUGGELLATA, LA **Vol. 7** 342
BOCCA SULLA STRADA, LA **Vol. 7** 493, 1094
BOCCACCESCA **Vol. 7** 390, 508, 1071
BOCCACCIO **Vol. 4** 298, 608, 713 **Vol. 7** 224, 402, 665, 1116
BOCCACCIO '70 – IL LAVORO **Vol. 7** 498, 988, 1142
BOCCACCIO '70 – LA RIFFA **Vol. 7** 355, 725
BOCCACCIO '70 – LE TENTAZIONI DEL DOTTOR ANTONIO **Vol. 7** 725, 1214
BOCCACCIO '70 – RENZO E LUCIANA **Vol. 7** 522, 817
BOCCONCINO, IL **Vol. 7** 130
BOCET VESEL **Vol. 6** *page 271, page 273, page 281*
BOCHORNO **Vol. 8** 221, 546
BOCK I ÖRTAGARD **Vol. 5** 909, 959, 1052
BOCKBIERFEST **Vol. 4** 246, 379
BOCKSHORN **Vol. 6** *page 161, page 165, page 170*
BOCNI VETROVI **Vol. 6** *page 312*
BODA ACCIDENTADA **Vol. 8** 640
BODA DE QUINITA FLORES, LA **Vol. 8** 400
BODA DEL SENOR CURA, LA **Vol. 8** 207
BODA EN EL INFIERNO **Vol. 8** 359, 449
BODA ERA A LAS DOCE, LA **Vol. 8** 470, 492
BODA, LA **Vol. 8** 450, 517
BODAKUNGEN **Vol. 5** 870, 1211
BODAS DE BLANCA, LAS **Vol. 8** 245, 338
BODAS DE SANGRE **Vol. 8** 358, 610
BODEGA, LA **Vol. 2** 9, 205, 277
BODEGA VON LOS CUERROS, DIE **Vol. 4** 113, 780
BODES DER FOR – , DET **Vol. 5** 55, 101, 208
BODY BODY A BANGKOK **Vol. 6** *page 134*
BOETHEIA ERHETAI O BEGGOS **Vol. 3** 149
BOF **Vol. 2** 495, 854
BOF, ANATOMIE D'UNE LIVREUR **Vol. 2** 377
BOFFEN OG BANANEN **Vol. 5** 70
BOG JE UMRO UZALUD **Vol. 3** 374, 383
BOGANCS **Vol. 9** 259, 285, 326
BOGDAN STIMOFF **Vol. 4** 213
BOGRA **Vol. 10** 115, 260
BOGURODZICA **Vol. 1** 278, 308, 372
BOHAC A FALON **Vol. 9** 67, 218, 339
BOHATEROWIE SYBIRU **Vol. 1** 278, 308, 404
BOHATERSTWO POLSKIEGO SKAUTA **Vol. 1** 174, 176
BOHEME **Vol. 4** 121, 232, 339, 385, 673
BOHEMIOS **Vol. 8** 422
BOHUS BATALJON **Vol. 5** 1052, 1080
BOIA DI LILLA, IL **Vol. 7** 372, 1063
BOIS DES AMANTS, LE **Vol. 2** 251, 625
BOIS SACRE, LE **Vol. 2** 338, 339, 395
BOITE A BAC, LA **Vol. 2** 114
BOITE AUX REVES, LA **Vol. 2** 124, 156, 173, 863
BOITE DE NUIT **Vol. 2** 15, 65, 727 **Vol. 7** 968
BOJ O MOSKVU **Vol. 10** 71, 181, 356, 406
BOJ POSLEDEN **Vol. 3** 41 **Vol. 6** *page 37, page 51*
BOJ SA SKONCI ZAJTRA **Vol. 10** 287, 351
BOJANSKIJAT MAJSTOR **Vol. 6** *page 45, page 48*
BOJKA **Vol. 3** 79
BOKHANDLAREN SOM SULTADE BADA **Vol. 5** 933, 1042, 1093
BOKSER **Vol. 1** 198, 330 **Vol. 6** *page 232*
BOKSERI IDU U RAJ **Vol. 3** 403, 436
BOKSERNES KONGE **Vol. 5** 83
BOLA AO CENTRO **Vol. 8** 20, 106, 167
BOLDOG IDÖK **Vol. 9** 132, 270
BOLDOG SZÜLETESNAPOT, MARILYN **Vol. 9** 60, 51, 141, 259
BOLDOGGA TESZLEK **Vol. 9** 133, 157, 274
BOLDOGTALAN KALAP **Vol. 9** 107, 174, 279
BOLDYN **Vol. 6** *page 214, page 241, page 255*
BOLEK I LOLEK **Vol. 1** 278, 308, 364
BOLERO **Vol. 2** 9, 32
BOLESLAW SMIALY **Vol. 1** 198, 354 **Vol. 6** *page 215*

BOLETTES BRUDEFAERD **Vol. 5** 55, 101, 208
BOLICHE **Vol. 8** 422
BOLJE JE UMETI **Vol. 3** 409
BOLLA DI SAPONE **Vol. 7** 940
BOLLE DI SAPONE **Vol. 7** 1189
BOLLENTI SPIRITI **Vol. 7** 29, 585
BOLO RAZ PRIATELSTVO **Vol. 10** 77, 273
BOLOGNESE, LA **Vol. 7** 201
BOLOND APRILIS **Vol. 9** 75, 285, 326
BOLONDOS VAKACIO **Vol. 3** 244, 267, 329 **Vol. 9** 131, 152
BOLSCEVISMO ??!! **Vol. 7** 732
BOMAN FAR SNURREN **Vol. 5** 840, 863
BOMAN PA UTSTÄLLNINGEN **Vol. 5** 891, 966
BOMANS POJKE **Vol. 5** 840, 841
BOMB AT 10, 10, THE **Vol. 3** 374, 427
BOMBA **Vol. 10** 296, 532
BOMBA U 10, 10 **Vol. 3** 374, 427
BOMBAS PARA LA PAZ **Vol. 8** 546
BOMBASI **Vol. 3** 370, 402, 415
BOMBEN **Vol. 5** 97, 976, 1110
BOMBEN AUF MONTE CARLO **Vol. 4** 325, 567, 713
BOMBENANGST, DIE **Vol. 4** 514
BOMBENGESCHÄFT, EIN **Vol. 4** 352, 700
BOMBENIDEE, DIE **Vol. 4** 339, 525
BOMBER **Vol. 7** 102, 875
BOMBI BITT OCH JAG **Vol. 5** 923, 1181
BOMSALVA **Vol. 5** 915, 1024, 1160, 1201
BON CHIC, BON GENRE MAIS SALOPES **Vol. 6** *page 129*
BON DIEU SANS CONFESSION, LE **Vol. 2** 47, 251, 625
BON ET LES MECHANTS, LE **Vol. 2** 499
BON PETIT DIABLE, UN **Vol. 6** *page 95, page 140, page 142*
BON PLAISIR, LE **Vol. 6** *page 85, page 91*
BON ROI DAGOBERT, LE **Vol. 2** 722 **Vol. 7** 434, 817, 893
BON SANS, INSPEKTORE **Vol. 6** *page 30, page 35, page 64*
BON WEEKEND **Vol. 2** 322
BONANCA & C. **Vol. 8** 4, 98, 130, 146
BONGOLO ET LA PRINCESSE NOIRE **Vol. 2** 294
BONHEUR DES AUTRES, LE **Vol. 2** 211
BONHEUR DU JOUR, LE **Vol. 2** 65, 357
BONHEUR EN LOCATION **Vol. 2** 118, 423
BONHEUR EST POUR DEMAIN, LE **Vol. 2** 655
BONHEUR, LE **Vol. 2** 72, 332, 599, 700, 803
BONHEUR TOI-MEME **Vol. 2** 754
BONIFACE SOMNAMBULE **Vol. 2** 128
BONJOUR JEUNESSE **Vol. 2** 131, 309, 740 **Vol. 7** 982
BONJOUR LA CHANCE **Vol. 2** 57, 717
BONJOUR SOURIRE **Vol. 2** 515
BONJOUR TOUBIB **Vol. 2** 97, 205, 517, 808
BONJOUR TRISTESSE **Vol. 2** 656
BONNE A TUER **Vol. 2** 232, 279, 488
BONNE ANNEE, LA **Vol. 2** 196
BONNE AVENTURE, LA **Vol. 2** 237, 357, 466, 469
BONNE CHANCE **Vol. 2** 48, 388, 726
BONNE CHANCE CHARLIE **Vol. 2** 427
BONNE ETOILE, LA **Vol. 2** 205, 270
BONNE HOTESSE, LA **Vol. 2** 798
BONNE, LA **Vol. 7** 109, 297, 358
BONNE OCCASE, LA **Vol. 2** 294
BONNE SOUPE, LA **Vol. 2** 402, 770
BONNE TISANE, LA **Vol. 2** 97, 581, 808
BONNES CAUSES, LES **Vol. 2** 548, 824
BONNES FEMMES, LES **Vol. 2** 224, 578
BONNIE E CLYDE ALL' ITALIANA **Vol. 7** 29, 524
BONNIE PRINCE CHARLIE **Vol. 2** 81
BONS BAISERS DE DINARD **Vol. 2** 224
BONS BAISERS DE HONG KONG **Vol. 2** 151, 244, 851, 875
BONS BAISERS...A LUNDI **Vol. 2** 241
BONS DE BARRAS, LES **Vol. 2** 135
BONS VIVANTS, LES **Vol. 2** 128, 293
BONSOIR MESDAMES, BONSOIR MESSIEURS **Vol. 2** 719, 722
BONSOIR PARIS, BONJOUR L'AMOUR **Vol. 2** 226, 279, 409, 427, 557
BONY A KLID **Vol. 10** 258, 495, 497
BONZESSE, LA **Vol. 2** 359
BOOM, IL **Vol. 7** 483
BOOT IM WALDE, DAS **Vol. 1** 57, 76
BOR, A **Vol. 9** 71, 288
BOR BORSON **Vol. 5** 670, 708, 751
BOR BORSON II **Vol. 5** 787, 817, 824
BOR BORSON JR. **Vol. 5** 674, 817, 824
BOR FORELSKE DEM, DE **Vol. 5** 228
BORBA ZA STASTIE **Vol. 3** 106
BORCSA AMERIKABAN **Vol. 9** 178, 237
BORDEL, LE **Vol. 2** 259, 818
BORDELLA **Vol. 7** 763
BORDELLET **Vol. 5** 12, 33
BORDON Y LA ESTRELLA, EL **Vol. 8** 681
BORGHESE PICCOLO PICCOLO, UN **Vol. 7** 1153, 1194
BORGIA, I **Vol. 7** 626, 785, 796, 1209
BORGSLAEGTENS HISTORIE **Vol. 5** 183
BORIS GODUNOV **Vol. 10** 10, 219, 378, 394
BORIS I **Vol. 6** *page 61*
BORISEK – MALY SERZANT **Vol. 10** 12, 60, 105, 456, 595
BORNENS SYND **Vol. 5** 58
BORNEVENNERNE **Vol. 5** 58
BORNHOLMERURET **Vol. 5** 97

BOROTALCO Vol. 7 363, 567
BORRASCA DE CELOS Vol. 8 614
BORS ISTVAN Vol. 9 132, 237, 270
BORSA E LA VITA, LA Vol. 7 1103
BORSALINI, LES Vol. 2 202
BORSALINO Vol. 2 457, 814
BORSALINO ET CIE Vol. 2 457, 814
BORSENS OFFER Vol. 5 81
BORTFORELSE Vol. 5 97
BORTFORTE BRUD, DEN Vol. 5 81
BORTREIST PA UBESTEMT TID Vol. 5 704, 748
BÖRZEKIRALY, A Vol. 9 28
BOSCO D'AMORE Vol. 7 207, 1173
BÖSE GEIST LUMPACI VAGABUNDUS, DER Vol. 4 190, 329
BÖSEWICHTER MÜSSEN DRAN, DIE Vol. 1 9, 15, 114
BOSKO BUHA Vol. 3 341, 370
BOSQUE DEL LOBO, EL Vol. 8 492
BOSS, IL Vol. 7 1179
BOSSIERE Vol. 2 48, 726
BOSSU, LE Vol. 2 27, 368, 503, 515, 526, 566, 581, 646, 676, 722, 871
BOSTONVILLEI KALAND, A Vol. 9 2
BOSZORKANYSZOMBAT Vol. 9 154, 243
BOTA JMENEM MELICHAR Vol. 10 41, 153, 515
BOTON DE ANCLA Vol. 8 214, 217, 276, 625
BOTOSTROJ Vol. 10 547, 580, 621
BOTSCHAFTER MORDEN NICHT Vol. 1 19, 57, 75
BOTTA E RISPOSTA Vol. 7 443, 1136
BOTTE I FARTEN Vol. 5 939, 948
BOTTEGA DELL' ANTIQUARIO, LA Vol. 7 529
BOTY PLNE VODY Vol. 10 233, 317, 496
BOUBOUROCHE Vol. 2 388
BOUBULE I., ROI NEGRE Vol. 2 167, 574, 744
BOUCHE COUSUE Vol. 2 318, 347, 807
BOUCHE-TROU, LE Vol. 2 868
BOUCHER, LA STAR ET L'ORPHELINE, LE Vol. 2 176, 190, 477, 527
BOUCHER, LE Vol. 2 700
BOUCHES LASCIVES ET PORNO Vol. 2 291
BOUDU SAUVE DES EAUX Vol. 2 37, 153, 160, 422, 469, 533
BOUGNOUL, LE Vol. 2 74
BOUIF ERRANT, LE Vol. 2 387, 824
BOULANGER DE VALORGUE, LE Vol. 2 347, 807
BOULE DE SUIF Vol. 2 59, 226, 566
BOULEVARD Vol. 2 118, 247, 503
BOULEVARD DES ASSASSINS Vol. 6 *page 87, page 129*
BOULEVARD DU CRIME Vol. 2 70, 243, 259
BOULEVARD DU RHUM Vol. 2 251
BOULOT AVIATEUR Vol. 2 595
BOUM 2, LA Vol. 6 *page 96, page 99*
BOUM, LA Vol. 2 781 Vol. 6 *page 74*
BOUM SUR PARIS Vol. 2 48, 131, 479, 866
BOUM TARATATZOUM Vol. 3 142, 197
BOUNTY KILLER A TRINITA, UN Vol. 7 742
BOUQUET DE JOIE Vol. 2 671
BOURGEOIS GENTIL MEC, LE Vol. 2 475, 503
BOURGEOIS GENTILHOMME Vol. 2 13, 185, 453, 517
BOURGEOIS GENTILHOMME, LE Vol. 6 *page 85, page 119, page 145*
BOURGEOISE ET LE LOUBARD, LA Vol. 2 352
BOURGEOISE ET ... PUTE Vol. 6 *page 129*
BOURLIVE VINO Vol. 10 41, 197, 425
BOURRASQUE Vol. 2 407, 698, 824
BOURREAU ATTENDRA, LE Vol. 2 207, 777
BOURREAU DES COEURS, LE Vol. 6 *page 85, page 140*
BOURREAUX D'ENFANTS Vol. 2 661
BOURRIN, LE Vol. 2 7
BOURSE ET LA VIE, LA Vol. 2 141, 244, 598, 835
BOUT-DE-CHOU Vol. 2 388, 404
BOUT DE LA ROUTE, LE Vol. 2 32, 717
BOUTALIN AERONAUTA Vol. 7 1189
BOUTALIN BOTANICO Vol. 7 1189
BOUTALIN SPAZZACAMINO Vol. 7 1189
BOUTIQUE AUX ILLUSIONS, LA Vol. 2 530
BOUTIQUE, LA Vol. 8 467, 578
BOXER A SMRT Vol. 10 32, 273
BOXERBRAUT, DIE Vol. 4 115, 663
BOY Vol. 8 449
BOYKOTT Vol. 4 344, 544
BOZI MLYNY Vol. 10 58, 256, 345
BOZSKA EMA Vol. 10 139, 283, 398
BOZYSZCZE Vol. 1 317, 371
BRA MENNESKER Vol. 5 678, 726
BRACCIA APERTE, LE Vol. 7 34
BRACCIALETTO AL PIEDE, IL Vol. 7 1055
BRACHA ZA VSECHNY PENIZE Vol. 10 165, 380, 394
BRACONNIER DE DIEU, LE Vol. 6 *page 78, page 80, page 91*
BRAENDENDE SPORGSMAAL, DET Vol. 5 89, 136
BRAENDT A' Vol. 5 83
BRAENDTE VINGER Vol. 5 242
BRAGHE DEL PADRONE, LE Vol. 7 257

BRAND - BORGE RYKKER UD Vol. 5 114, 306
BRAND IM OSTEN Vol. 4 378
BRAND IM OZEAN Vol. 4 329, 728
BRAND IN DER OPER Vol. 4 626, 729
BRANDMAL DER VERGANGENHEIT, DAS Vol. 4 89, 307
BRANDMANDENS DATTER Vol. 5 256
BRANDOS COSTUMES Vol. 8 7, 198
BRANDSTELLEN Vol. 1 27, 104, 111
BRANDY Vol. 8 546
BRANKAR BYDLI V NASI ULICI Vol. 10 179, 305, 407, 502
BRANNEN Vol. 5 790, 799, 817
BRÄNNINGAR Vol. 5 840, 997
BRANQUIGNOL Vol. 2 124, 137
BRÄNT BARN Vol. 5 974, 1017, 1080
BRAS DE LA NUIT, LES Vol. 2 475
BRASIER ARDENT, LE Vol. 2 100, 521, 623, 775, 833
BRAT DOKTORA HOMERA Vol. 3 356, 366, 441
BRATELE AFRODITEI Vol. 3 226, 235, 269
BRATIA Vol. 10 287, 576
BRAUT DES ENTMÜNDIGTEN, DIE Vol. 4 113, 780
BRAUT IM MOND, DIE Vol. 4 79, 115, 742
BRAUT IN EILE GESUCHT, EINE Vol. 4 361, 558
BRÄUTIGAM DER BABETTE BOMBERLING, DER Vol. 4 321, 584
BRÄUTIGAMSWITWE, DIE Vol. 4 192, 462
BRAVE SÜNDER, DER Vol. 4 85, 359
BRAVISSIMO ! Vol. 7 634, 1044
BRAVO DI VENEZIA, IL Vol. 7 2, 170, 491, 552, 725
BRAVO, KLEINER THOMAS Vol. 4 243, 740
BRAVO MAESTRO Vol. 3 454 Vol. 6 *page 304*
BRAZDAT Vol. 3 5, 11 Vol. 6 *page 16*
BRAZZA OU L'EPOPEE DU CONGO Vol. 2 254, 595
BRECHE D'ENFER, LA Vol. 2 838
BREITE WEG, DER Vol. 4 309
BRELAN D'AS Vol. 2 344, 615
BREME Vol. 3 382
BRENNENDE ACKER, DER Vol. 4 178, 202, 729
BRENNENDE GRENZE Vol. 4 49, 316
BRENNENDE HERZ, DAS Vol. 4 38, 113, 444
BRENNENDE KUGEL, DIE Vol. 4 270, 400, 762
BRENNENDE RUHR Vol. 1 7, 73, 85
BRENNENDE SCHIFF, DAS Vol. 4 211, 634
BRENNENDES GEHEIMNIS Vol. 4 15, 28, 69, 124
BRENNENDES LAND Vol. 4 487, 583
BRENNER I NATT, DET Vol. 5 672
BRENNO IL NEMICO DI ROMA Vol. 7 1152
BRENT JORD Vol. 5 718, 762, 799
BRESCIA, LEONESSA D'ITALIA Vol. 7 64
BREVE AMORE Vol. 7 255
BREVE RENCONTRE A PARIS Vol. 2 224
BREVE STAGIONE, UNA Vol. 7 1056
BREVE VACANZA, UNA Vol. 7 550, 567
BREVET 95-75 Vol. 2 152
BREVET FRA AFDODE Vol. 5 2, 7
BREVI AMORI A PALMA DI MAJORCA Vol. 7 693
BREZA Vol. 3 421, 437
BREZEL, DIE Vol. 4 228, 471
BRICOLEURS, LES Vol. 2 512, 598
BRIDE SUR LE COU, LA Vol. 2 185, 232, 488, 593
BRIEFE, DIE IHN NICHT ERREICHTEN Vol. 4 14, 186, 340
BRIEFTRÄGER, DER Vol. 4 617, 647
BRIERE, LA Vol. 2 79
BRIGADA CRIMINAL Vol. 8 677
BRIGADA DE LOS CONDENADOS, LA Vol. 8 762
BRIGADA DIVERSE IN ALERTA Vol. 3 244, 283, 297
BRIGADA LUI IONUT Vol. 3 231, 259, 266, 330
BRIGADE ANTI-GANGS Vol. 2 347, 659
BRIGADE CRIMINELLE Vol. 2 48, 61
BRIGADE DES MOEURS Vol. 2 727
BRIGADE EN FOLIE, LA Vol. 2 661
BRIGADE EN JUPONS, LA Vol. 2 32, 67, 322
BRIGADE, LA Vol. 2 51
BRIGADE MONDAINE Vol. 2 2
BRIGADE MONDAINE: LA SECTE DE MARRAKECH Vol. 2 814 Vol. 6 *page 70, page 101*
BRIGADE MONDAINE: VAUDOU AUX CARAIBES Vol. 2 2
BRIGADE SAUVAGE, LA Vol. 2 427
BRIGAND GENTILHOMME, LE Vol. 2 717
BRIGANTE DI TACCA DEL LUPO, IL Vol. 7 86, 401
BRIGANTE, IL Vol. 7 817
BRIGANTE MUSOLINO, IL Vol. 7 781, 1136
BRIGANTENLIEBE Vol. 4 186, 293, 773
BRIGANTENRACHE Vol. 4 321
BRIGANTI, I Vol. 7 540, 965
BRIGANTI IN SARDEGNA Vol. 7 1189
BRIGANTI ITALIANI, I Vol. 7 1214
BRIGANTIN VON NEW YORK, DIE Vol. 4 119, 340
BRIGATA FIRENZE Vol. 7 493

BRIGGEN TRE LILJOR **Vol. 5** 869, 1052
BRILLANTE PORVENIR **Vol. 8** 470, 492
BRILLANTEN **Vol. 4** 13, 574
BRILLANTEN DER MORANOWS, DIE **Vol. 4** 329, 355
BRILLANTENMIEZE, DIE **Vol. 4** 495, 593
BRILLANTRING, DER **Vol. 4** 514
BRILLE UND BOMBE - BEI UNS LIEGEN SIE RICHTIG **Vol. 3** 395
BRINDIS A MANOLETE **Vol. 8** 359, 449
BRINDIS AL CIELO **Vol. 8** 438, 583
BRINNER EN ELD, DET **Vol. 5** 840, 909
BRINNESO **Vol. 7** 39
BRISEUR LE CHAINES, LE **Vol. 2** 9, 566
BRISTET LYKKE **Vol. 5** 9
BRITA I GROSSHANDLARHUSET **Vol. 5** 863, 1032
BRIVIDI ... **Vol. 7** 968
BRIVIDO **Vol. 7** 603, 862
BRÖDER EMELLAN **Vol. 5** 876, 923
BRODER GABRIELSEN **Vol. 5** 772
BRODER MOD BRODER **Vol. 5** 295
BRÖDERNA LEJONHJÄRTA **Vol. 5** 933, 1052, 1061
BRÖDERNA ÖSTERMANS BRAVADER **Vol. 5** 834
BRÖDERNA ÖSTERMANS HUSKORS **Vol. 5** 334, 870, 875, 923, 1183
BRÖDERNAS KVINNA **Vol. 5** 843, 923
BRODERNE PA UGLEGAARDEN **Vol. 5** 76, 223, 261
BROGADA DIVERSE SI LA MARE **Vol. 3** 317
BROGLIACCIO D'AMORE **Vol. 7** 221
BROKIGA BLAD **Vol. 5** 891, 909
BRÖLLOPET **Vol. 5** 869, 975, 1024
BRÖLLOPET I BRÄNNA **Vol. 5** 891, 966
BRÖLLOPET PA SOLÖ **Vol. 5** 334, 875, 1156
BRÖLLOPSBESVÄR **Vol. 5** 881, 933, 1130
BRÖLLOPSDAGEN **Vol. 5** 1042, 1053, 1130, 1217
BRÖLLOPSNATT PA STJÄRNEHOV, EN **Vol. 5** 857, 923
BRÖLLOPSNATTEN **Vol. 5** 1032, 1095, 1151, 1218
BRÖLLOPSRESAN **Vol. 5** 840, 997
BRONCE Y LUNA **Vol. 8** 396, 405
BRONTE **Vol. 3** 374
BRONTOSAURUS **Vol. 10** 344, 425, 505
BRONZES FONT DU SKI, LES **Vol. 2** 739 **Vol. 6** *page 91*
BRONZES, LES **Vol. 2** 642
BRONZOVIJAT KLJUC **Vol. 6** *page 50, page 65*
BROR MIN OCH JAG **Vol. 5** 913, 923
BROT UND ROSEN **Vol. 1** 13, 79, 115
BROTHER CARL **Vol. 5** 918, 933, 1093
BROTT, ETT **Vol. 5** 871, 1156
BROTT I PARADISET **Vol. 5** 869, 1052, 1177
BROTT I SOL **Vol. 5** 907, 1073
BROTT OCH STRAFF **Vol. 5** 865, 954, 1164
BROTTMALSDOMAREN ELLER HJÄRTATS SYND **Vol. 5** 966, 976
BROWN EYE - EVIL EYE **Vol. 3** 343, 346, 374
BRUBAKER **Vol. 2** 633
BRUCE CONTRE ATTAQUE **Vol. 6** *page 79*
BRUCIA, RAGAZZO, BRUCIA **Vol. 7** 1179
BRUCIATI DA COCENTE PASSIONE **Vol. 7** 514
BRÜCKE, DIE **Vol. 1** 35, 118, 129
BRÜCKE ZWISCHEN GESTERN UND MORGEN **Vol. 1** 9, 26, 86
BRUDAR OCH BOLLAR **Vol. 5** 840
BRUDEBUKETTEN **Vol. 5** 674
BRUDEFERDEN I HARDANGER **Vol. 5** 692, 770
BRUDEGAVEN **Vol. 5** 97
BRUDEKJOLEN **Vol. 5** 203
BRUDEKRONEN **Vol. 5** 760
BRUDEN FRA DRAGSTRUP **Vol. 5** 146, 180, 199, 201
BRUDEN KOM GENOM TAKET **Vol. 5** 843, 923
BRÜDER **Vol. 4** 59, 258, 318, 537 **Vol. 6** *page 182*
BRUDER BERNHARD **Vol. 4** 344, 545
BRÜDER KARAMASOFF, DIE **Vol. 4** 660, 704
BRÜDER LAUTENSACK, DIE **Vol. 1** 30, 60, 75
BRÜDER NOLTENIUS, DIE **Vol. 4** 325, 373
BRÜDER SCHELLENBERG, DIE **Vol. 4** 211, 247, 318 **Vol. 6** *page 194*
BRÜDERCHEN UND SCHWESTERCHEN **Vol. 6** *page 181, page 190*
BRÜDERLEIN FEIN **Vol. 4** 85, 234, 620
BRUIT QUI COURT, UN **Vol. 6** *page 80, page 126, page 143*
BRUJA, LA **Vol. 8** 246, 281
BRUJA SIN ESCOBA, UNA **Vol. 8** 583, 772
BRUJAS MAGICAS **Vol. 8** 346, 407, 627
BRUMES **Vol. 2** 388
BRUMES DE PARIS **Vol. 2** 603, 753
BRUNA INDIAVOLATA, UNA **Vol. 7** 1027
BRUNE QUE VOILA, LA **Vol. 2** 9, 488
BRUNET WIECZOROWA PORA **Vol. 1** 340, 408 **Vol. 6** *page 236*
BRUNNEN DES WAHNSINNS, DER **Vol. 4** 329, 527, 745
BRUSCOLO **Vol. 7** 1005
BRUSTEN HIMMEL **Vol. 5** 869, 1097, 1195
BRUTO PARA PATRICIA, UN **Vol. 8** 517

BRUTTI, SPORCHI E CATTIVI **Vol. 7** 854, 964, 1166
BRUTTO SOGNO DI UNA SARTINA, IL **Vol. 7** 1189
BRYGGERENS DATTER **Vol. 5** 154
BRYLANTY PANI ZUZY **Vol. 1** 389, 401
BRYLLUPSAFTEN, EN **Vol. 5** 70
BRYLLUPSAFTEN PAA HOTEL, EN **Vol. 5** 97
BRYLLUPSREJSE, EN **Vol. 5** 97
BRZEZINA **Vol. 1** 304, 314 **Vol. 6** *page 258*
BÜBAJOSOK **Vol. 9** 190, 215, 316
BUBASINTER **Vol. 3** 362, 418
BUBE U GLAVI **Vol. 3** 402
BUBNY **Vol. 10** 353, 394, 425
BUCH DES LASTERS, DAS **Vol. 4** 288
BUCH ESTHER, DAS **Vol. 4** 488
BUCHHOLZENS ABENTEUER IM HOCHGEBIRGE **Vol. 4** 308
BÜCHSE DER PANDORA, DIE **Vol. 2** 23 **Vol. 4** 14, 273, 359
BUCKLIGE UND DIE TÄNZERIN, DER **Vol. 4** 178, 479
BUCURESTI **Vol. 3** 286, 295, 317
BUD PRIPRAVEN **Vol. 10** 198
BUDAI CUKRASZDA **Vol. 9** 71, 322
BUDAI SZENZACIOS RABLOGYILKOSSAG **Vol. 9** 28
BUDAPESTI MESEK **Vol. 9** 152, 259, 263
BUDAPESTI TALALKOZO **Vol. 9** 226
BUDAPESTI TAVASZ **Vol. 9** 134, 183, 312
BUDDENBROOKS, DIE **Vol. 4** 451, 679, 735
BUDJENJE PACOVA **Vol. 3** 367, 368
BÜDÖSVIZ **Vol. 9** 149, 241, 314
BUDSKAB TIL NAPOLEON PAA ELBA, ET **Vol. 5** 97
BUEK **Vol. 9** 50, 51, 339
BUEN AMOR, EL **Vol. 8** 258
BUEN VIAJE PABLO **Vol. 8** 217, 503
BUENAS NOCHES, SENOR MONSTRUO **Vol. 8** 368, 681
BUENAS NOTICIAS **Vol. 8** 259, 752
BUENO PARA NADA **Vol. 8** 306, 681
BUENOS DIAS CONDESITA **Vol. 8** 749, 762
BUENOS DIAS PERDIDOS, LOS **Vol. 8** 207, 214
BUFERA **Vol. 7** 731, 815
BUFERE **Vol. 7** 798
BUFFALO E BILL **Vol. 7** 91
BUFFALO, L'EROE DEL FAR WEST **Vol. 7** 372
BUFFET FROID **Vol. 2** 588, 655
BUFFETJOMFRUEN FRA LUNDEHUSSOEN **Vol. 5** 203
BUGERGIARIA **Vol. 7** 1005
BUGIARDA, LA **Vol. 8** 311
BUIO NELLA VALLE **Vol. 7** 1144 **Vol. 7** 358, 834
BUIO OMEGA **Vol. 7** 742
BUITRE SOBRE LA CIUDAD **Vol. 8** 299, 728
BUITRES DE LA ALDEA, LOS **Vol. 8** 647
BUKSESKORTET **Vol. 5** 203
BULDOCI A TRESNE **Vol. 10** 299, 327, 622
BULECZKA **Vol. 1** 213, 294, 313
BULETIN DE BUCURESTI **Vol. 6** *page 274, page 282, page 283*
BULEVARD **Vol. 6** *page 29, page 61, page 65*
BUMERANG **Vol. 1** 206, 281, 389 **Vol. 3** 41 **Vol. 6** *page 23, page 56, page 248*
BUMMELGENIE, DAS **Vol. 4** 339, 525
BUMMELSTUDENTEN **Vol. 4** 178, 514
BUMS, DER SCHEIDUNGSGRUND **Vol. 4** 329, 373
BUNA **Vol. 3** 69 **Vol. 6** *page 43*
BUNA SEARA, IRINA **Vol. 3** 281 **Vol. 6** *page 268*
BUND DER DREI, DER **Vol. 4** 89, 270
BUNDFALD **Vol. 5** 89, 281
BUNICUL SI DOI DELINCVENTI MINORI **Vol. 3** 240, 254, 311
BUNKEBRYLLUP **Vol. 5** 295
BÜNOS VAGYOK **Vol. 9** 164, 196
BUNT KRWI I ZELAZA **Vol. 1** 278, 407
BUNTE PALETTE, DIE **Vol. 4** 367
BUNTKARIERTEN, DIE **Vol. 1** 6, 39, 107
BUNT"T NA ROBITE **Vol. 3** 35
BUON ANNO **Vol. 7** 1189
BUON LADRONE, IL **Vol. 7** 108
BUON NATALE ! **Vol. 7** 420
BUON POSTO, UN **Vol. 7** 1189
BUON RIMEDIO PER LA GASTRICA, UN **Vol. 7** 1189
BUON SAMARITANO, IL **Vol. 7** 1187
BUON SANGUE NON MENTE **Vol. 7** 442
BUON SOLDATO, IL **Vol. 7** 13, 142
BUONA FIGLIOLA, LA **Vol. 7** 168
BUONA FORTUNA, LA **Vol. 7** 734, 1036
BUONA GIORNATA PER ROBINET **Vol. 7** 1189
BUONA IDEA SE NE VA, UNA **Vol. 7** 1189
BUONA SERA ALLA FINESTRA **Vol. 7** 1189
BUONA SERA BAMBINA E BURATTINO **Vol. 7** 1189
BUONA SERA, FIRULI **Vol. 7** 1189
BUONA SERA FOTOGRAFO ! **Vol. 7** 1189
BUONA SERA PITTORE ! **Vol. 7** 1189
BUONA SERA SIGNORINA BONELLI **Vol. 7** 1189

BUONE NOTIZIE Vol. 7 411, 746, 819, 1166
BUONGIORNO ELEFANTE Vol. 7 196, 522
BUONGIORNO, MADRID ! Vol. 7 332, 587, 968
BUONGIORNO PRIMO AMORE ! Vol. 7 693
BURDUS Vol. 3 362, 391
BUREBISTA Vol. 3 254 Vol. 6 *page 270*
BÜRGERMEISTER ANNA Vol. 1 3, 56, 97, 117
BÜRGSCHAFT, DIE Vol. 6 *page 189*
BÜRGSCHAFT FÜR EIN JAHR Vol. 6 *page 155, page 158, page 166*
BURIDAN Vol. 2 688
BURIDAN, HEROS DE LA TOUR DE NESLE Vol. 2 243, 717
BURJA NA MLADOSTTA Vol. 3 71
BURLACI SA VOLGE Vol. 6 *page 312, page 315*
BURSCHENLIED AUS HEIDELBERG, EIN Vol. 4 266, 288, 574
BUSCA, LA Vol. 8 331, 681
BUSCANDO A PERICO Vol. 8 466, 519
BUSCO AMANTE PARA UN DIVORCIO Vol. 8 299
BUSCO TONTA PARA FIN DE SEMANA Vol. 8 261, 687
BUSINESS Vol. 2 181, 727
BUSQUEME A ESA CHICA Vol. 8 214, 261
BUSSEN Vol. 5 23, 248, 674, 724
BUSTENSKJOLD Vol. 5 674, 826, 831
BUSTER BROWN Vol. 5 203
BUSTERS VERDEN Vol. 5 177, 219, 266
BUSTINO ROSA, IL Vol. 7 1189
BUTASAGOM TÖRTENETE Vol. 9 67, 84, 117
BUTNSKALA Vol. 6 *page 303*
BUZAVIRAG Vol. 9 71
BUZDUGANUL CU TREI PECETI Vol. 3 237, 239, 297
BY OG LAND, HAND I HAND Vol. 5 760
BYDEN SAR AMOK Vol. 5 275
BYE BYE BARBARA Vol. 2 481, 674
BYENS HERKULES Vol. 5 83
BYL JAZZ Vol. 6 *page 235*
BYL JEDOU JEDEN KRAL Vol. 10 460, 514, 621
BYL PRVNI MAJ Vol. 10 45
BYLO CTVRT A BUDE PUL Vol. 10 286, 433, 505, 621
BYLO NAS DESET Vol. 10 369, 427, 438, 503
BYLO TO V MAJI Vol. 10 352, 503
BYTTE ROLLER Vol. 5 81
BYZANTINE RAPSODIA Vol. 3 151, 159

C

C. A K. POLNI MARSALEK Vol. 10 167, 450, 586
C... DE MARLINE, LE Vol. 6 *page 148*
C NAVCERIETO Vol. 3 38, 97
CA, C'EST DU SPORT ! Vol. 2 47, 122, 273
CA FAIT DU BIEN Vol. 2 814
CA FAIT TILT Vol. 2 600, 663
CA GLISSE AU PAYS DES MERVEILLES Vol. 2 575, 592, 721
CA-N FILME Vol. 6 *page 277, page 278, page 284*
CA N'ARRIVE QU'AUX AUTRES Vol. 2 555, 694
CA N'ARRIVE QU'AUX VIVANTS Vol. 2 401, 559, 661
CA PEUT TOUJOURS SERVIR Vol. 2 427
CA VA BARDER ! Vol. 2 194, 501
CA VA, CA VIENT Vol. 2 406
CA VA ETRE TA FETE Vol. 2 747
CA VA FAIRE MAL Vol. 6 *page 98, page 101*
CA VA PAS ETRE TRISTE Vol. 6 *page 88*
CA VA PAS LA TETE Vol. 2 198
CABALGANDO HACIA LA MUERTE Vol. 8 603
CABALLERO ANDALUZ, UN Vol. 8 261, 611, 616, 630
CABALLERO DEL DRAGON, EL Vol. 8 580
CABALLERO FAMOSO, UN Vol. 8 396
CABALLEROS DE LA ANTORCHA, LOS Vol. 8 697, 732
CABALLEROS DEL AIRE Vol. 8 396, 640
CABALLO BLANCO, EL Vol. 8 729
CABANE AUX SOUVENIRS, LA Vol. 2 322, 368
CABARET Vol. 8 276, 620, 704
CABARET DU GRAND LARGE, LA Vol. 2 70, 529
CABEZA DE HIERRO Vol. 8 437
CABEZAS CORTADAS Vol. 8 317, 768
CABEZAS QUEMADAS Vol. 8 221
CABEZOTA, EL Vol. 8 626, 627
CABINET DES DR. CALIGARI, DAS Vol. 4 239, 546, 574, 734
CABIRIA Vol. 7 100, 304, 305, 380, 471, 584, 1132
CABO BLANCO Vol. 8 346, 756
CABRIOLA Vol. 8 214, 261
CABRITA QUE TIRA AL MONTE Vol. 8 516, 751
CACADA DO MALHADEIRO, A Vol. 8 167, 187
CACCIA AI MILIONI Vol. 7 22
CACCIA AL LUPO Vol. 7 1162

CACCIA AL MARITO **Vol. 7** 1121
CACCIA ALLE ANITRE **Vol. 7** 100
CACCIA ALL'OMBRA **Vol. 7** 732
CACCIA ALL'UOMO **Vol. 7** 408, 443
CACCIA TRAGICA **Vol. 7** 401, 547, 725
CACCIATORE DI SQUALI, IL **Vol. 7** 646 **Vol. 8** 626, 627
CACCIATORE MILLANTATORE **Vol. 7** 1189
CACCIATORI DEL COBRA D'ORO, I **Vol. 7** 692
CACORKA **Vol. 10** 154, 453, 460
CADA VEZ QUE ... **Vol. 8** 232
CADAVERE DI MARMO, IL **Vol. 7** 277
CADAVERE IMBELLETTATO, IL **Vol. 7** 91
CADAVERE PER SIGNORA **Vol. 7** 408
CADAVERE SCOMPARSO, IL **Vol. 7** 131
CADAVERE VIVENTE, IL **Vol. 7** 1132
CADAVERI ECCELLENTI **Vol. 7** 363, 1025
CADEAU, LE **Vol. 6** *page 81* **Vol. 7** 411
CADET ROUSSELLE **Vol. 2** 158, 368, 516, 857
CADETS DE L'OCEAN, LES **Vol. 2** 100, 826
CADETTI DI GUASCOGNA, I **Vol. 7** 1136
CADUTA DEGLI ANGELI RIBELLI, LA **Vol. 7** 76, 901
CADUTA DEGLI DEI **Vol. 7** 736, 817, 931, 990, 1025, 1142
CADUTA DI TROIA, LA **Vol. 7** 186, 1132
CAFE CHANTANT **Vol. 7** 526
CAFE, COCA Y PURO **Vol. 8** 519
CAFE DE CHINITAS **Vol. 8** 614, 749
CAFE DE LA MARINA, EL **Vol. 8** 652
CAFE DE PARIS **Vol. 2** 423, 439, 566 **Vol. 8** 437
CAFE DE PUERTO **Vol. 8** 298, 762
CAFE DU CADRAN, LE **Vol. 2** 15, 501
CAFE DU PORT, LE **Vol. 2** 191, 330, 454, 479, 533
CAFE EXPRESS **Vol. 7** 318, 533, 1153
CAFE LUNCHRASTEN **Vol. 5** 1003, 1073
CAFE MOSZKVA **Vol. 9** 90, 132, 237
CAFE PARADIS **Vol. 5** 89, 118
CAFE POD MINOGA **Vol. 1** 225, 250 **Vol. 6** *page 256*
CAFE X **Vol. 5** 752, 794
CAFRE, EL **Vol. 8** 241, 441
CAGE AUX FILLES, LA **Vol. 2** 368, 516, 722
CAGE AUX ROSSIGNOLS, LA **Vol. 2** 205, 254
CAGE AUX SOURIS, LA **Vol. 2** 404
CAGE, LA **Vol. 2** 760, 770, 875
CAGLIOSTRO **Vol. 2** 27, 237, 440, 573
CAGLIOSTRO. LIEBE UND LEBEN EINES GROSSEN ABENTEURERS **Vol. 6** *page 185, page 195*
CAGLIOSTROS TOTENHAND **Vol. 4** 247, 340
CAID DE CHAMPIGNOL, LE **Vol. 2** 347, 872
CAID, LE **Vol. 2** 423, 615
CAIDS, LES **Vol. 2** 110, 251, 252
CAILLES SUR CANAPE **Vol. 2** 814
CAIN, AVENTURES DES MERS DU SUD **Vol. 2** 595
CAIN DE NULLE PART **Vol. 2** 300
CAINA O L'ISOLA E IL CONTINENTE **Vol. 7** 301
CAINO **Vol. 7** 971
CAIO GIULIO CESARE **Vol. 7** 369
CAIS DO SODRE **Vol. 8** 16, 18, 46, 92, 106
CAJUS JULIUS CAESAR **Vol. 7** 369, 640
CAL **Vol. 6** *page 234*
CALA NAPRZOD **Vol. 1** 179, 338, 343, 389
CALABUCH **Vol. 8** 306, 719
CALABUIG **Vol. 7** 1058
CALAFURIA **Vol. 7** 360, 910
CALAIS-DOUVRES **Vol. 2** 679
CALAMITA, LA **Vol. 7** 621, 1189
CALAMO **Vol. 7** 585, 852
CALANDRIA, LA **Vol. 7** 585
CALCULATORUL MARTURISESTE **Vol. 6** *page 275, page 281, page 282*
CALCUTTA **Vol. 2** 74
CALDA VITA, LA **Vol. 7** 514, 781
CALDE NOTTI DEL DECAMERONE, LE **Vol. 7** 649
CALDE NOTTI DI CALIGOLA, LE **Vol. 7** 1026
CALDE NOTTI DI POPPEA, LE **Vol. 7** 1121
CALDERON **Vol. 7** 842, 1105
CALDO PROFUMO DI VERGINE **Vol. 7** 742
CALDURA **Vol. 3** 285, 296, 332
CALE **Vol. 8** 469, 672
CALEA VICTORIEI **Vol. 3** 232, 307
CALIDO VERANO DEL SENOR RODRIGUEZ, EL **Vol. 8** 541, 749
CALIENTE NINA JULIETTA, LA **Vol. 8** 471
CALIGOLA **Vol. 7** 388, 585
CALIGOLA E MESSALINA **Vol. 7** 311, 761
CALIGULA ET MESSALINE **Vol. 6** *page 117*
CALINO **Vol. 2** 159, 286, 330
CALL OF THE WILD **Vol. 5** 686
CALLAGHAN REMET CA **Vol. 2** 740
CALLE MAYOR **Vol. 2** 427 **Vol. 8** 214, 484
CALLE OG PALLE **Vol. 5** 2, 229
CALLE P. **Vol. 5** 880, 951, 1035
CALLE SIN SOL, LA **Vol. 8** 400
CALMA DI CIELO E TEMPESTA D'ANIME **Vol. 7** 316

CALMOS Vol. 2 21, 719
CALUMNIA, LA Vol. 8 228
CALUMNIADA, LA Vol. 8 716, 772
CALVAIRE D'AMOUR Vol. 2 122
CALVAIRE DE CIMIEZ, LE Vol. 2 388
CALVAIRE DE DONA PIA, LE Vol. 2 387, 467
CALVARIO DI UNA MADRE Vol. 7 1189
CALVARIODI UN MAESTRO, IL Vol. 7 1189
CALZOLAIO HA GUADAGNATO AL LOTTO, IL Vol. 7 1189
CALZONAZOS, EL Vol. 8 681, 749
CAMARADES Vol. 2 351
CAMAROTE DE LUJO Vol. 8 214, 400, 598
CAMBIALE, LA Vol. 7 693
CAMERA ALBA Vol. 3 208, 281, 285
CAMERA D'ALBERGO Vol. 7 80, 333
CAMERA OBSCURA Vol. 4 159, 759
CAMERIERA, BELLA PRESENZA, OFFRESI Vol. 7 792, 1036
CAMERIERA NERA, LA Vol. 7 311
CAMERIERA SEDUCE I VILLEGGIANTI, LA Vol. 7 188, 649
CAMERIERA SENZA MALIZIA Vol. 7 414
CAMERIERE BURLATO Vol. 7 1189
CAMERIERE, LE Vol. 7 333
CAMERIERE MUSICOMANE, IL Vol. 7 1189
CAMERINO SIN BIOMBO Vol. 8 396
CAMICIA DELLA FELICITA, LA Vol. 7 28
CAMICIA NERA Vol. 7 16, 106, 359, 554, 621, 1004, 1159
CAMICIE ROSSE Vol. 7 86, 522, 792, 1044
CAMIERE, LE Vol. 7 915
CAMILA Vol. 8 224, 245
CAMILLA Vol. 7 910, 915
CAMILLE Vol. 2 579
CAMILLO AVVELENATO SUO MALGRADO Vol. 7 131
CAMILLO EMULO DI SHERLOK HOLMES Vol. 7 26, 701
CAMILLO, LA FIGLIA E L'ALTRO Vol. 7 26
CAMILLO UCCISORO DE LEONI Vol. 7 131
CAMINO CORTADO Vol. 8 437, 470
CAMINO DE BABEL, EL Vol. 8 611
CAMINO DE LA VERDAD Vol. 8 402, 708, 749
CAMINO DEL AMOR, EL Vol. 8 276
CAMINO DEL ROCIO Vol. 8 207, 214
CAMINO DORADO, EL Vol. 8 701
CAMINO, EL Vol. 8 495, 603
CAMINO SOLITARIO Vol. 8 222
CAMION BLANC, LE Vol. 2 249, 833
CAMION, LE Vol. 2 633
CAMISARDS, LES Vol. 2 188, 468
CAMMEO DI LILY, IL Vol. 7 435
CAMMINA CAMMINA Vol. 7 835
CAMMINO DELLA SPERANZA, IL Vol. 7 86, 966
CAMMINO DELLE STELLE, IL Vol. 7 277
CAMÖES Vol. 2 775 Vol. 7 587 Vol. 8 41, 43, 46, 88, 101, 146, 153, 162, 178, 199
CAMORRA Vol. 7 19
CAMORRA, LA Vol. 7 1189
CAMORRA NAPOLETANA Vol. 7 1189
CAMPA CAROGNA ... LA TAGLIA CRESCE Vol. 8 252, 600
CAMPAGNA DELLA TEPPA, LA Vol. 7 2, 106, 552, 1116
CAMPAGNOLA BELLA Vol. 7 1024
CAMPANA DEI CADUTI A ROVERETO, LA Vol. 7 341
CAMPANA DEL INFIERNO Vol. 8 681, 749
CAMPANA DELLA CATTEDRALE, IL Vol. 7 1078
CAMPANA DELLO SCANDALO, LA Vol. 7 277
CAMPANADAS A MEDIANOCHE Vol. 8 450
CAMPANE A MARTELLO Vol. 7 522, 796
CAMPANE DI SAN LUCIO, LE Vol. 7 196
CAMPANILE D'ORO, IL Vol. 7 693
CAMPEMENT 13 Vol. 2 200, 833
CAMPEONES ! Vol. 8 268, 752
CAMPING Vol. 7 910
CAMPING-CAMPING Vol. 1 11, 70, 122
CAMPIONE, IL Vol. 7 938, 966, 1136
CAMPISTA EM APUROS, UM Vol. 8 130, 176
CAMPO BRAVO Vol. 8 753
CAMPO DE FIORI Vol. 7 603, 1030
CAMPO DI MAGGIO Vol. 7 16, 919, 1121, 1159
CAMPO MAMULA Vol. 3 436, 455
CAN CANNES Vol. 7 103
CANAGLIA DORATA Vol. 7 144
CANAILLES, LES Vol. 2 548, 738, 861
CANAL GRANDE Vol. 7 326, 1060
CANALE DEGLI ANGELI, IL Vol. 7 652, 665, 1182
CANARD EN FER BLANC, LE Vol. 2 248 Vol. 8 551, 732
CANARUL SI VISCOLUL Vol. 3 268, 291
CANAS Y BARRO Vol. 8 207, 298, 336
CANCAO DA SAUDADE Vol. 8 23, 135
CANCAO DA TERRA, A Vol. 8 79, 122
CANCAO DE LISBOA, A Vol. 8 14, 172
CANCELLI DELLA MORTE, I Vol. 7 52, 493
CANCHA VASCA Vol. 8 541, 616
CANCION ANDALUZA, LA Vol. 8 600
CANCION DE AIXA, LA Vol. 8 662, 716

CANCION DE CUNA **Vol. 8** 677, 772
CANCION DE JUVENTUD **Vol. 8** 261, 749
CANCION DE LA MALIBRAN, LA **Vol. 2** 440 **Vol. 8** 486, 776
CANCION DE MEDIANOCHE **Vol. 8** 436, 711
CANCION MORTAL **Vol. 8** 677, 732
CANCIONERA **Vol. 8** 246
CANCRE, LE **Vol. 2** 655
CANDICE **Vol. 6** *page 79*
CANDICE CANDY **Vol. 2** 804
CANDIDE **Vol. 2** 249
CANDIDE OU L'OPTIMISME AU XX. SIECLE **Vol. 2** 488
CANDIDE SALOPE **Vol. 2** 52
CANDIDO EROTICO **Vol. 7** 633
CANE ACCUSATORE, IL **Vol. 7** 1189
CANE ARRABBIATO **Vol. 7** 99
CANE DEL PICCOLO MOZZO, IL **Vol. 7** 1189
CANE DELINQUENTE, IL **Vol. 7** 1189
CANE POLIZIOTTO, IL **Vol. 7** 1189
CANE SENTINELLA **Vol. 7** 1189
CANELITA EN RAMA **Vol. 8** 276
CANGACEIRO, O **Vol. 8** 331, 762
CANI DI GERUSALEMME, I **Vol. 7** 593, 1173, 1175
CANI E GATTI **Vol. 7** 1044
CANICHE **Vol. 8** 256
CANICULE **Vol. 6** *page 73, page 78*
CANNABIS **Vol. 2** 441
CANNE MOZZE **Vol. 7** 13
CANNIBAL FEROX **Vol. 7** 124
CANNIBAL HOLOCAUST **Vol. 7** 834
CANNIBALES, LES **Vol. 2** 393
CANNIBALI DOMANI **Vol. 7** 269, 1023
CANNIBALI, I **Vol. 7** 19
CANTAMI BUONGIORNO TRISTEZZA **Vol. 7** 1152
CANTANDO A LA VIDA **Vol. 8** 306, 348
CANTANDO SOTTO LA STELLE **Vol. 7** 603
CANTANTE DELL'OPERA, LA **Vol. 7** 361, 1036, 1116
CANTANTE E IL CAMPIONE, IL **Vol. 7** 311
CANTANTE MISTERIOSO, IL **Vol. 7** 17, 1159
CANTASTORIE DI VENEZIA, IL **Vol. 7** 390, 490, 792, 795
CANTATE CON ME ! **Vol. 7** 1027, 1054
CANTATE CON NOI **Vol. 7** 500
CANTEMIR **Vol. 3** 254, 256, 281, 306
CANTERBURY TALES **Vol. 7** 333, 388, 434
CANTICO **Vol. 8** 232, 245, 299, 583
CANTICO DELLA TERRA **Vol. 7** 659
CANTICO FINAL **Vol. 8** 54
CANTIGA DA RUA **Vol. 8** 96, 122, 184
CANTINIER DE LA COLONIALE, LE **Vol. 2** 158, 193
CANTO DE LA CIGARRA, EL **Vol. 8** 299, 728
CANTO DEL GALLO, EL **Vol. 8** 299, 728
CANTO DELLA FEDE, IL **Vol. 7** 69
CANTO DELLA VITA, IL **Vol. 7** 756
CANTO DELL'AGONIE, IL **Vol. 7** 108
CANTO DELL'EMIGRANTE, IL **Vol. 7** 603
CANTO DI CIRCE, IL **Vol. 7** 630, 1079
CANTO MA SOTTOVOCE **Vol. 7** 16
CANTO PARA TI **Vol. 8** 677
CANTO PER TE **Vol. 7** 17
CANTONIERA N. 13, LA **Vol. 7** 1189
CANTOR DE MEXICO, EL **Vol. 8** 298
CANTOR E A BAILARINA, O **Vol. 8** 96, 167
CANZONE APPASSIONATA **Vol. 7** 796
CANZONE D'AMORE **Vol. 7** 891
CANZONE DEL CUORE, LA **Vol. 7** 17
CANZONE DEL DESTINO **Vol. 7** 603
CANZONE DEL FANTASMA, LA **Vol. 7** 390
CANZONE DEL SOLE, LA **Vol. 7** 1121
CANZONE DELL'AMORE, LA **Vol. 7** 756, 792, 1116
CANZONE DELLE ROSE, LA **Vol. 7** 296
CANZONE PIU BELLA, LA **Vol. 7** 958
CANZONE PROIBITA **Vol. 7** 17
CANZONE RUBATA, LA **Vol. 7** 472, 603, 994
CANZONI A DUE VOCI **Vol. 7** 493
CANZONI A TEMPO DI TWIST **Vol. 7** 1152
CANZONI, CANZONI, CANZONI **Vol. 7** 1044
CANZONI DI MEZZO SECOLO **Vol. 7** 1044
CANZONI DI TUTTA ITALIA **Vol. 7** 451
CANZONI NEL MONDO **Vol. 7** 1035
CAO E DOIS DESTINOS, UM **Vol. 8** 54, 160, 186
CAO INSPEKTORE **Vol. 6** *page 306, page 309, page 318*
CAO – XA **Vol. 8** 245
CAP AU LARGE **Vol. 2** 533
CAP CANAILLE **Vol. 6** *page 73, page 87*
CAP DE L'ESPERANCE, LE **Vol. 2** 388
CAPANNA DELLO ZIO TOM, LA **Vol. 7** 493
CAPAS NEGRAS **Vol. 8** 18, 20
CAPCANA **Vol. 3** 291, 292, 319
CAPCANA MERCENARILOR **Vol. 6** *page 274, page 276, page 282*
CAPELLO DA PRETE, IL **Vol. 7** 493
CAPESTRO DEGLI ASBURGO, IL **Vol. 7** 264, 701
CAPINERA **Vol. 7** 336
CAPITAINE ARDANT **Vol. 2** 226, 322, 419

CAPITAINE BENOIT, LE **Vol. 2** 9, 191
CAPITAINE BLOMET **Vol. 2** 47, 226, 439, 652
CAPITAINE CRADDOCK, LE **Vol. 2** 733
CAPITAINE DE QUINZE ANS, UN **Vol. 2** 796
CAPITAINE FRACASSE, LE **Vol. 2** 1, 83, 84, 368, 383, 395, 539, **Vol. 7** 580, 676
CAPITAINE JAUNE, LE **Vol. 2** 122, 775
CAPITAINE PANTOUFLE **Vol. 2** 57, 611
CAPITAINE RASCASSE, LE **Vol. 2** 42, 805
CAPITAINE SANS PEUR **Vol. 2** 173
CAPITAINE SINGRID **Vol. 2** 768 **Vol. 8** 177
CAPITALI CULTURALI - VATICAN **Vol. 6** *page 238*
CAPITAN DE LOYOLA, EL **Vol. 2** 646 **Vol. 8** 214, 336, 607, 748
CAPITAN FRACASSA **Vol. 7** 168, 711, 1096, 1107, 1172
CAPITAN FUOCO **Vol. 7** 17
CAPITAN INTREPIDO, EL **Vol. 8** 214, 517
CAPITAN, LE **Vol. 2** 32, 368, 515, 722
CAPITAN TEMPESTA **Vol. 7** 2, 170, 491, 552, 1116
CAPITAN VENENO, EL **Vol. 8** 339, 524, 630
CAPITANO BIANCO, IL **Vol. 7** 374
CAPITANO DEGLI USSARI, IL **Vol. 7** 639, 792
CAPITANO DI VENEZIA, IL **Vol. 7** 1035, 1167
CAPITANO NERO, IL **Vol. 7** 16, 796
CAPITAO SINGRID **Vol. 8** 177
CAPKOVY POVIDKY **Vol. 10** 256, 460, 561
CAPODANNO DI ROBINET, IL **Vol. 7** 1189
CAPOLAVORO DI SAETTA, IL **Vol. 7** 548
CAPOLAVORO, IL **Vol. 7** 493
CAPORAL EPINGLE, LE **Vol. 2** 480
CAPORAL SAETTA **Vol. 7** 548
CAPORAL SIMON, IL **Vol. 7** 442
CAPORALE DI GIORNATA **Vol. 7** 739, 792
CAPOTE DE PASEO, EL **Vol. 8** 246
CAPPELLO A TRE PUNTE, IL **Vol. 7** 443, 1060, 1116
CAPPELLO DA PRETE, IL **Vol. 7** 1060, 1080
CAPPELLO NELLA MINESTRA, IL **Vol. 7** 1189
CAPPOTTO DI ASTRAKAN, IL **Vol. 7** 270, 363, 567
CAPPOTTO DI LEGNO, IL **Vol. 7** 895
CAPPOTTO, IL **Vol. 7** 73, 283, 798, 915
CAPRICCI **Vol. 7** 289
CAPRICCI D'AMORE **Vol. 7** 148
CAPRICCIO **Vol. 4** 13, 574
CAPRICCIO DI DAMA **Vol. 7** 1189
CAPRICCIO FATALE **Vol. 7** 829
CAPRICE DE CAROLINE CHERIE, UN **Vol. 2** 439, 826
CAPRICE DE PRINCESSE **Vol. 2** 679
CAPRICES **Vol. 2** 23, 440
CAPRICES DE MARIE, LES **Vol. 2** 655
CAPRICES D'UNE SOURIS, LES **Vol. 6** *page 119*
CAPRICIOSA **Vol. 5** 97
CAPRIOLEN **Vol. 4** 265, 481, 527, 614
CAPTAIN HORATIO HORNBLOWER **Vol. 2** 173
CAPTIF, LE **Vol. 2** 207
CAPTURE GENANTE **Vol. 2** 159, 286, 330
CAPULITO DE ALHELI **Vol. 8** 536
CAR CRASH **Vol. 7** 315
CAR DYMITR SAMOZWANIEC **Vol. 1** 404
CAR I GENERAL **Vol. 3** 38, 96, 101 **Vol. 6** *page 41*
CAR IVAN SISMAN **Vol. 6** *page 48, page 54*
CARA A CARA **Vol. 8** 396, 603
CARA DE ACELGA **Vol. 8** 580, 738
CARA DEL TERROR, LA **Vol. 8** 207
CARA DOLCE NIPOTE **Vol. 7** 1179
CARA SPOSA **Vol. 7** 1007
CARABBIMATTI, I **Vol. 7** 124, 226
CARABINIERI, I **Vol. 7** 538, 818, 1016
CARABINIERI SI NASCE **Vol. 7** 694
CARABINIERS, LES **Vol. 2** 208, 288
CARADUROS, LOS **Vol. 8** 770
CARAMBOLAGES **Vol. 2** 47, 548
CARAPATE, LA **Vol. 2** 21, 353, 781
CARAQUE BLONDE, LA **Vol. 2** 836, 866
CARAT I JEGO SLUGI **Vol. 1** 174, 215, 326
CARAVAGGIO **Vol. 7** 1096 **Vol. 7** 40, 332, 1136
CARAVAGGIO, IL PITTORE MALEDETTO **Vol. 7** 480, 521
CARAVAN PETROL **Vol. 7** 603
CARAVANA DE ESCLAVOS **Vol. 8** 400, 772
CARAY CON EL DIVORCIO **Vol. 8** 237, 773
CARAY, QUE PALIZAS ! **Vol. 8** 346, 551
CARBANICE **Vol. 10** 148, 454, 493
CARCASSE ET LE TORD-COU, LA **Vol. 2** 247, 401
CARCEL DE CRISTAL, LA **Vol. 8** 636, 753
CARCELERAS **Vol. 8** 516
CARCERATO **Vol. 7** 222, 474
CARDINALE LAMBERTINI, IL **Vol. 7** 638, 725, 993
CARE AMICHE MIE **Vol. 7** 728
CARESSES A DOMICILE **Vol. 2** 291
CARESSES INAVOUABLES **Vol. 2** 291
CARESSEUSES EXPERTES **Vol. 6** *page 119*
CAREVO NOVO RUHO **Vol. 3** 399

CARGA DE LA POLICIA MONTADA, LA Vol. 8 306, 752
CARGAISON BLANCHE Vol. 2 402, 627, 637
CARGAISON CLANDESTINE Vol. 2 15, 727 Vol. 7 968
CARGO Vol. 6 *page 69, page 101, page 112*
CARI GENITORI Vol. 7 854
CARI MOSTRI DEL MARE Vol. 7 749
CARICA DEGLI EROI, LA Vol. 7 725, 1054
CARICA DEL 7. CAVALLEGGERI, LA Vol. 7 759 Vol. 8 541
CARICA DELLE PATATE, LA Vol. 7 502, 1202
CARICA EROICA Vol. 7 113, 443
CARILLON DE MINUIT, LE Vol. 2 55
CARILLONS SANS JOIE Vol. 2 190, 650
CARINO MIO Vol. 8 214, 611
CARIOCA TIGRE Vol. 7 287
CARL ALSTRUP OG HANS TVILLINGEBRODER Vol. 5 81
CARL ALSTRUPS KAERLIGHED PAA AKTIER Vol. 5 92
CARL GUSTAV, GJENGEN OG PARKERINGSBANDITTENE Vol. 5 750, 765, 784
CARL PETERS Vol. 4 2, 206, 344, 416, 437
CARL UND CARLA Vol. 4 743
CARL VON OSSIETZKY Vol. 1 22, 26, 115
CARL XII : S KURIR Vol. 5 870, 891
CARLI CAPLIN NA VITOSA Vol. 3 71
CARLOS SCHÖNSTES ABENTEUER Vol. 4 361
CARLOS UND ELISABETH Vol. 4 247, 436, 663, 720, 752
CARLOTA Vol. 8 477
CARLOTTA Vol. 8 549, 728
CARMELA Vol. 7 95, 360, 910
CARMELA E'UNA BAMBOLA Vol. 7 798, 915
CARMELA, LA SARTINA DI MONTESANTO Vol. 7 829, 1082
CARMEN Vol. 2 29, 159, 237, 282, 286, 330, 388, 573, 648, 715, 863 Vol. 4 242, 294, 424, 560 Vol. 5 1160 Vol. 7 48, 592, 1025 Vol. 8 358, 580
CARMEN BOOM Vol. 8 346, 583
CARMEN FRA I ROSSI Vol. 7 453, 587, 1096
CARMEN, LA Vol. 8 438, 681
CARMEN LA DE RONDA Vol. 8 214, 261
CARMEN LA DE TRIANA Vol. 8 487
CARMEN PROHIBITA Vol. 7 476
CARMEN VON ST. PAULI, DIE Vol. 4 49, 316
CARMILLA Vol. 5 909, 1142
CARMINA BURANA Vol. 10 6, 482, 573
CARMINA, FLOR DE GALICIA Vol. 8 774
CARNALITA MORBOSA Vol. 7 311
CARNAVAL Vol. 2 344, 347
CARNAVAL DE LAS BESTIAS, EL Vol. 8 762, 770
CARNAVAL DES VERITES, LE Vol. 2 44, 261, 528
CARNE DE HORCA Vol. 8 725
CARNE E L'ANIMA, LA Vol. 7 16, 149, 1116
CARNE INQUIETA Vol. 7 46, 827
CARNE PER FRANKENSTEIN Vol. 7 599
CARNEGIE HALL Vol. 2 778
CARNET DE BAL, UN Vol. 2 8, 195, 249, 514, 676
CARNET DU MAJOR THOMPSON, LE Vol. 2 57, 454, 566, 676
CARNEVALE DI VENEZIA, IL Vol. 7 48, 443, 453, 476, 640, 1116
CARNEVALE TRAGICO Vol. 7 424
CARNEVALESCA Vol. 7 560
CARO MICHELE Vol. 7 80, 333
CARO PAPA Vol. 7 841, 964
CARODEJ Vol. 10 45
CAROLA DE DIA, CAROLA DE NOCHE Vol. 8 214, 261
CAROLA LAMBERTI - EINE VOM ZIRKUS Vol. 1 35, 67, 78
CAROLINA, LA NINA DE PLATA Vol. 8 246
CAROLINA REDIVIVA Vol. 5 937, 1201
CAROLINE CHERIE Vol. 2 21, 57, 439, 770, 854
CAROS DOD Vol. 5 97
CAROSELLO DI CANZONI Vol. 7 1121
CAROSELLO NAPOLETANO Vol. 7 299, 923
CAROSSE D'OR, LE Vol. 2 178, 719
CAROTTES SONT CUITES, LES Vol. 2 259, 344, 475
CAROVANA DEL PECCATO, LA Vol. 7 603
CAROVANA DI CANZONI Vol. 7 1054, 1063
CAROVNE DEDICTVI Vol. 10 173, 233, 466, 485
CAROVNE OCI Vol. 10 243, 536
CARRE DE DAMES POUR UN AS Vol. 8 331, 546
CARRE DE VALETS Vol. 2 460, 627
CARREFOUR Vol. 2 153, 279, 322, 714, 832
CARREFOUR DES ENFANTS PERDUS, LES Vol. 2 270, 833
CARREFOUR DU CRIME Vol. 2 153, 158
CARRIERE Vol. 4 240, 644
CARRO ARMATO DELL' 8 SETTEMBRE, IL Vol. 7 388, 585
CARRO SULLA MONTAGNA, IL Vol. 7 399

CARSKA FAWORYTA **Vol. 1** 174, 176, 326
CARSKA MILOST **Vol. 3** 59, 107 **Vol. 6** *page 35*
CARSKA PIESA **Vol. 6** *page 33, page 56, page 60*
CARTA A SARA **Vol. 8** 400, 749
CARTA A UNA MUJER **Vol. 8** 390, 753
CARTA AL CIELO **Vol. 8** 583, 725
CARTAGALHA, REINE DES GITANS **Vol. 2** 254, 339
CARTAGINE IN FIAMME **Vol. 7** 923
CARTAS BOCA ARRIBA **Vol. 8** 517, 775
CARTAS NA MESA **Vol. 8** 169
CARTEA PRIVIRILOR **Vol. 6** *page 266, page 269*
CARTES SUR TABLE **Vol. 2** 279
CARTIERUL VESELIEI **Vol. 3** 244, 268, 274, 329
CARTOUCHE **Vol. 2** 248, 249, 395, 457, 566
CARTOUCHE, ROI DE PARIS **Vol. 2** 205, 226
CARUTA CU MERE **Vol. 6** *page 269, page 281*
CAS DE CONSCIENCE **Vol. 2** 55, 131, 698
CAS DU DOCTEUR GALLOY, LE **Vol. 2** 661
CAS DU DOCTEUR LAURENT, LE **Vol. 2** 13, 676, 709
CAS JERABIN **Vol. 10** 313, 572
CAS LASKY A NADEJE **Vol. 10** 254, 380, 394
CAS PRACUJE PRO VRAHA **Vol. 10** 286, 385, 408
CAS SLUHU **Vol. 10** 43, 149, 304
CASA CHE BRUCIA, LA **Vol. 7** 965
CASA CON LA SCALA AL BUIO, LA **Vol. 7** 99
CASA DALLE FINESTRE CHE RIDONO, LA **Vol. 7** 66, 810
CASA DE LA LLUVIA, LA **Vol. 8** 449
CASA DE LA MIEZUL NOPTII **Vol. 3** 254, 256, 306
CASA DE LA TROYA, LA **Vol. 8** 214, 484, 516, 628, 707
CASA DE LAS CHIVAS, LA **Vol. 8** 221, 402
CASA DE LAS MIL MUNECAS, LA **Vol. 8** 546, 595
CASA DE LAS PALOMAS, LA **Vol. 8** 245, 438
CASA DE LAS SONRISAS, LA **Vol. 8** 640, 732
CASA DE LOS MARTINEZ, LA **Vol. 8** 298, 493
CASA DEGLI SCAPOLI, LA **Vol. 7** 53
CASA DEI PULCINI, LA **Vol. 7** 751
CASA DEL PARAISO, LA **Vol. 8** 297, 355
CASA DEL PECCATO, LA **Vol. 7** 756, 812
CASA DEL SANTO, LA **Vol. 7** 266, 701
CASA DEL TAPPETO GIALLO, LA **Vol. 7** 12, 546, 908, 1112
CASA DELLE ORCHIDEE, LA **Vol. 7** 1022
CASA DELLO SCANDALO, LA **Vol. 7** 940
CASA DI BAMBOLA **Vol. 7** 973, 1116
CASA DI VETRO, LA **Vol. 7** 301
CASA DINTRE CIMPURI **Vol. 3** 275, 318, 319
CASA GRANDE, LA **Vol. 8** 616, 719
CASA IN ROVINA, LA **Vol. 7** 1057
CASA, LA **Vol. 8** 450
CASA LONTANA **Vol. 7** 902
CASA MIA, DONNA MIA **Vol. 7** 940
CASA NETERMINATA **Vol. 3** 235, 251, 303
CASA RICORDI **Vol. 7** 988, 1044
CASA SENZA TEMPO, LA **Vol. 7** 136, 171
CASA SOTTO LA NEVE, LA **Vol. 7** 393, 1116
CASA SPERDUTA NEL PARCO, LA **Vol. 7** 504, 834
CASA STREGATA, LA **Vol. 7** 30, 567
CASABLANCA **Vol. 2** 562, 595, 821
CASABLANCA, CASABLANCA **Vol. 7** 405, 524, 908
CASABLANCA, NID D'ESPIONS **Vol. 2** 11
CASANOVA **Vol. 2** 42, 100, 122, 153, 521, 538, 833 **Vol. 7** 208, 388, 537, 988, 1130, 1168 **Vol. 9** 315
CASANOVA '70 **Vol. 7** 1136
CASANOVA & COMPANY **Vol. 7** 881
CASANOVA FAREBBE COSI ! **Vol. 7** 638, 792
CASANOVA HEIRATET **Vol. 4** 49, 234, 738
CASANOVA WIDER WILLEN **Vol. 4** 659
CASANOVAS ERBE **Vol. 4** 166, 208
CASCO BLANCO, EL **Vol. 8** 493
CASE OF 44'S, THE **Vol. 5** 179, 281
CASERNE EN FOLIE, LA **Vol. 2** 731, 771
CASI JUGANDO **Vol. 8** 249
CASI UN CABALLERO **Vol. 8** 524, 725
CASIMIR **Vol. 2** 128, 344
CASINISTA, IL **Vol. 7** 426, 882, 1127
CASINO DE PARIS **Vol. 2** 13, 385, 516, 517, 549, 615, 808
CASO ALMERIA, EL **Vol. 8** 219, 671
CASO CERRADO **Vol. 8** 338
CASO DE LAS DOS BELLEZAS, EL **Vol. 8** 465, 775
CASO DI COSCIENZA, UN **Vol. 7** 556
CASO ELETTRICO **Vol. 7** 1189
CASO HALLER, IL **Vol. 7** 196, 460
CASO PISCIOTTA, IL **Vol. 7** 763
CASO RAOUL, IL **Vol. 7** 62, 514
CASOOTO **Vol. 7** 333, 434
CASQUE D'OR **Vol. 2** 279, 488, 540
CASSANDRA CROSSING **Vol. 7** 567

CASSE-COU MADEMOISELLE **Vol. 2** 368, 401
CASSE, LE **Vol. 2** 719, 770
CASSE-PIEDS, LES **Vol. 2** 57, 153, 158
CASSE-TETE CHINOIS POUR LE JUDOKA **Vol. 2** 548, 815
CASSIERA, LA **Vol. 7** 1189
CASSURE, LA **Vol. 6** *page 103, page 121, page 131*
CASTA DIVA **Vol. 7** 487, 988, 1044, 1116, 1172
CASTA E PURA **Vol. 7** 29, 1086
CASTA SUSANA, LA **Vol. 8** 214, 261
CASTAGNARA, LA **Vol. 7** 1189
CASTANUELA **Vol. 8** 207, 298
CASTELANII **Vol. 3** 285, 307, 331
CASTELLI IN ARIA **Vol. 7** 35, 453, 586, 756
CASTELLO DALLE CINQUANTASETTE LAMPADE, IL **Vol. 7** 796
CASTELLO DEI MORTI VIVI, IL **Vol. 7** 1136
CASTELLO DEL RAGNO, IL **Vol. 7** 37, 454
CASTELLO DELLA MALINCONIA, IL **Vol. 7** 545, 1057
CASTELLO DELL'URAGANO, IL **Vol. 7** 91
CASTELLO DI BRONZO, IL **Vol. 7** 108
CASTELUL CONDAMNATILOR **Vol. 3** 218, 260, 281, 313
CASTELUL DIN CARPATI **Vol. 6** *page 269, page 275, page 277*
CASTIGADOR, EL **Vol. 8** 217, 317, 470
CASTIGADORES, LOS **Vol. 8** 280, 732
CASTIGLIONE, LA **Vol. 2** 204, 249, 479, 661
CASTIGO **Vol. 7** 52
CASTILLO DE FU MANCHU, EL **Vol. 8** 546, 595
CASTILLO DE NAIPES **Vol. 2** 427 **Vol. 8** 484
CASTILLO DE ROCHAL, EL **Vol. 8** 572, 677
CATALINA DE INGLATERRA **Vol. 8** 753
CATEMBE **Vol. 8** 22, 185
CATENA, LA **Vol. 7** 1190
CATENE **Vol. 7** 796, 798, 1054
CATENE INVISIBILI **Vol. 7** 196, 1054
CATERINA **Vol. 7** 266
CATERINA SFORZA, LEONESSA DI ROMAGNA **Vol. 7** 71
CATETO A BABOR **Vol. 8** 438, 627
CATHERINE **Vol. 2** 132, 659
CATHERINE ET CIE **Vol. 2** 457, 808
CATHERINE, IL SUFFIT D'UN AMOUR **Vol. 2** 251, 659
CATHERINE OU UNE VIE SANS JOIE **Vol. 2** 48, 346
CATHERINE THE GREAT **Vol. 2** 656
CATHY FILLE SOUMISTE **Vol. 2** 793
CATOLAY **Vol. 8** 214
CATTIVA STELLA, LA **Vol. 7** 796
CATTIVI PENSIERI **Vol. 7** 347
CATTIVI VANNO IN PARADISO, I **Vol. 7** 348
CATTIVO SOGGETTO, UN **Vol. 7** 756, 796
CAUCHEMAR **Vol. 2** 806
CAUCHEMARS **Vol. 2** 814
CAUSA KRALIK **Vol. 10** 395, 516, 517
CAUSE E EFFETTI **Vol. 7** 229
CAUSE TOUJOURS, MON LAPIN **Vol. 2** 185, 668
CAUSE TOUJOURS, TU M'INTERESSES **Vol. 2** 151, 389
CAV. COSTANTE NICOSIA DEMONIACO OVVERO DRACULA IN BRIANZA, IL **Vol. 7** 1015
CAVALCADE D'AMOUR **Vol. 2** 27, 488
CAVALCADE DES HEURES, LA **Vol. 2** 6, 88
CAVALCATA ARDENTE, LA **Vol. 7** 390, 456, 569
CAVALCATA DEL CAPRICCIO, LA **Vol. 7** 1125
CAVALCATA SELVAGGIA **Vol. 7** 1121
CAVALE, LA **Vol. 2** 251, 835
CAVALEUR, LE **Vol. 2** 616, 779
CAVALIER DE CROIX-MORT, LE **Vol. 2** 652, 698
CAVALIER LAFLEUR, LE **Vol. 2** 67, 460, 775
CAVALIER NOIR, LE **Vol. 2** 309, 698
CAVALIER PETAGNA, IL **Vol. 7** 968
CAVALIERE DAI CENTO VOLTI, IL **Vol. 7** 113
CAVALIERE DALLE SPADA NERA, IL **Vol. 7** 1121
CAVALIERE DEL CASTELLO MALEDETTO, IL **Vol. 7** 1121
CAVALIERE DELLA LIETA FIGURA, IL **Vol. 7** 108
CAVALIERE DI KRUJA, IL **Vol. 7** 224, 838, 966, 1136
CAVALIERE DI MAISON ROUGE, IL **Vol. 7** 493
CAVALIERE DI SAN MARCO, IL **Vol. 7** 792, 1036, 1060
CAVALIERE LA MORTE E IL DIAVOLO, IL **Vol. 7** 8
CAVALIERE MISTERIOSO, IL **Vol. 7** 443, 638, 830
CAVALIERE SENZA NOME **Vol. 7** 332, 476, 1107, 1136, 1172
CAVALIERE SENZA PAURA **Vol. 7** 1211
CAVALIERE SENZA TERRA, IL **Vol. 7** 196
CAVALIERI DEL DESERTO, I **Vol. 7** 591, 709

CAVALIERI DEL DIAVOLO, I Vol. 7 1147
CAVALIERI DEL FERRO D'ORO, I Vol. 7 732
CAVALIERI DEL POKER, I Vol. 7 531, 1079
CAVALIERI DELLA MASCHERA NERA, I Vol. 7 923
CAVALIERI DELLA REGINA, I Vol. 7 877, 1044
CAVALIERI DELLE TENEBRE, I Vol. 7 1116
CAVALIERS DE L'ORAGE, LES Vol. 6 *page 75, page 77, page 110*
CAVALLERIA Vol. 7 587, 756, 1060, 1177
CAVALLERIA RUSTICANA Vol. 2 6 Vol. 7 53, 621, 665, 756, 852, 965, 1060, 1116
CAVALLINA STORNA Vol. 7 251
CAVE SE REBIFFE, LE Vol. 2 197, 644
CAVE, UN Vol. 2 185, 815
CAVES DU MAJESTIC, LES Vol. 2 332, 601
CAVIAR-MÄUSCHEN, DAS Vol. 4 628
CAZA, LA Vol. 8 348, 589
CAZADOR DE RECOMPENSAS Vol. 8 406
CAZADOR DE TIBURONES, EL Vol. 8 626, 627
CAZADORES, LOS Vol. 8 245, 616
CDN Vol. 6 *page 239*
CE CHER VICTOR Vol. 2 450
CE COCHON DE MORIN Vol. 2 833, 863
CE COQUIN D'ANATOLE Vol. 2 243, 717
CE CORPS TANT DESIRE Vol. 2 559, 661
CE JOLI MONDE Vol. 2 137, 248, 395
CE MAL ETRANGE Vol. 2 517
CE N'EST PAS MOI Vol. 2 6, 676
CE PAUVRE CHERI Vol. 2 42
CE QUE LES FLOTS RACONTENT Vol. 2 153
CE REPONDEUR NE PREND PAS DE MESSAGE Vol. 2 739
CE SACRE AMEDEE Vol. 2 65
CE SACRE GRAND-PERE Vol. 2 732
C'E SEMPRE UN MA ! Vol. 7 443, 487
CE SIECLE A CINQUANTE ANS Vol. 2 268
CE SOIR LES JUPONS VOLENT Vol. 2 293, 294, 376, 401
CE SOIR OU JAMAIS Vol. 2 310, 398, 481
C'E UN FANTASMA NEL CASTELLO Vol. 7 460, 606
C'E UN SENTIERO NEL CIELO Vol. 7 603
CEATA Vol. 3 227, 272
CEBO, EL Vol. 8 421
CEBO PARA UNA ADOLESCENTE Vol. 8 627, 749
CECH PANEN KUTNOHORSKYCH Vol. 10 142, 256, 460
CECILE EST MORTE Vol. 2 332, 601
CECILIA Vol. 5 804 Vol. 6 *page 17*
CECILIA, SANTA Vol. 7 320
CEKANI NA DEST Vol. 10 87, 222, 568
CEKANI NA PATRIKA Vol. 10 28, 589, 594
CEKANY Vol. 10 460, 621
CELA S'APPELLE L'AURORE Vol. 2 232, 251, 488
CELEBRITA Vol. 7 1202
CELEBRUL 702 Vol. 3 248, 306, 307
CELESTE Vol. 2 12, 309, 348
CELESTINA, LA Vol. 8 627
CELESTINA P. R., LA Vol. 7 1152
CELESTINE, BONNE A TOUT FAIRE Vol. 2 750
CELLE QUI DOMINE Vol. 2 32, 413, 871
CELLE QUI N'A PAS DIT SON NOM Vol. 2 357
CELLES QU'ON N'A PAS EUES Vol. 6 *page 69, page 92*
CELOS, AMOR Y MARCADO COMUN Vol. 8 553
CELOS Y EL DUENDE, LOS Vol. 8 402, 749
CELUI QUI DOIT MOURIR Vol. 2 204, 251, 625 Vol. 3 202
CELULOZA Vol. 1 174, 225, 227
CELULOZA – POD GWIAZDA FRYGIJSKA Vol. 6 *page 236*
CENA BURSZTYNU Vol. 6 *page 245*
CENA DE MATRIMONIOS Vol. 8 493, 732
CENA DELLE BEFFE, LA Vol. 7 359, 709, 1060
CENA MEDU Vol. 10 28, 223, 495
CENA ODVAHY Vol. 10 454, 491, 493
CENATA NA GRADOT Vol. 3 384, 396
CENCIAIUOLO DELLA SANITA, IL Vol. 7 314
CENCIAIUOLO DI PARIGI, IL Vol. 7 886
CENDRILLON DE PARIS Vol. 2 42, 330, 412, 648
CENERE Vol. 7 525, 711
CENERE E VAMPE Vol. 7 965
CENERENTOLA Vol. 7 16, 1032, 1071
CENERENTOLA '80 Vol. 7 142, 541, 1093
CENERENTOLA E IL SIGNORE BUONAVENTURA Vol. 7 360
CENICIENTA Y ERNESTO, LA Vol. 8 493, 597
CENICIENTO, EL Vol. 8 217, 768
CENT BRIQUES ET DES TUILES Vol. 2 322, 427
CENT DIX A ZERO Vol. 2 198
CENT FRANCS PAR SECONDE Vol. 2 347, 807
CENT MILLE DOLLARS AU SOLEIL Vol. 2 185, 368

CENT MILLE FRANCS DANS LE PAS D'UN CHEVAL **Vol. 2** 104
CENT MILLE FRANCS POUR UN BAISER **Vol. 2** 533
CENTAURO, EL **Vol. 8** 311, 606, 674
CENTAURO, IL **Vol. 7** 529
CENTESIMO AL SECONDO, UN **Vol. 7** 909
CENTINELA, ALERTA ! **Vol. 8** 270
CENTO ANNI D'AMORE **Vol. 7** 1136
CENTO CAVALIERI, I **Vol. 7** 717 **Vol. 8** 438, 523
CENTO CHILOMETRI, LA **Vol. 7** 739
CENTO DI QUESTI GIORNI **Vol. 7** 48, 756
CENTO GIORNI A PALERMO **Vol. 7** 474, 1184
CENTO GIORNI ALL'OMBRA **Vol. 7** 277
CENTO GIORNI, I **Vol. 7** 807, 984
CENTO LETTERE D'AMORE **Vol. 7** 603, 966
CENTO PICCOLE MAMME **Vol. 7** 838
CENTOCELLE **Vol. 7** 1079
CENTOMILA DOLLARI **Vol. 7** 487, 589
CEPO, EL **Vol. 8** 658
C'ERA UNA VOLTA... **Vol. 7** 150
C'ERA UNA VOLTA ANGELO MUSSO **Vol. 7** 737, 1152
C'ERA UNA VOLTA IN AMERICA **Vol. 7** 203, 333, 893, 1077
CERA VIRGEN, LA **Vol. 8** 276, 354
C'ERANO UNA VOLTA DUE LADRI **Vol. 7** 1189
CERASELLA **Vol. 7** 758, 1177
CERCA A LAS ESTRELLAS **Vol. 8** 732
CERCA DE LA CIUDAD **Vol. 8** 259, 400, 749
CERCA DEL CIELO **Vol. 8** 261, 311
CERCASI BIONDA BELLA PRESENZA **Vol. 7** 460, 1062
CERCASI GESU **Vol. 7** 329, 882, 1105
CERCATORI D'ORO **Vol. 7** 91
CERCEL DE CRISTAL, LA **Vol. 8** 515
CERCLE DES PASSIONS, LE **Vol. 6** *page 75* **Vol. 8** 536, 682
CERCLE ROUGE, LE **Vol. 2** 124, 224, 588
CERCLE VICIEUX, LE **Vol. 2** 344, 662
CERCO DE TERROR **Vol. 8** 331
CERCO DEL DIABLO, EL **Vol. 8** 207, 417, 449
CERCO, EL **Vol. 8** 515, 753
CERCO MOGLIE **Vol. 7** 1189
CERCO, O **Vol. 8** 7
CERCUL MAGIC **Vol. 3** 276, 291
CEREMONIA SANGRIENTA **Vol. 8** 245
CERENTOLA E IL SIGNOR BONAVENTURA **Vol. 7** 136
CEREREA IN CASATORIE **Vol. 3** 315
CERESOVA GRADINA **Vol. 3** 122 **Vol. 6** *page 23, page 43*
CERF-VOLANT DU BOUT DU MONDE, LE **Vol. 2** 13, 517, 597
CERIMONIA DEI SENSI, LA **Vol. 7** 221
CERNA DYNASTIE **Vol. 10** 313, 569, 572
CERNA PUNCOCHA **Vol. 10** 208, 389, 550
CERNA SOBOTA **Vol. 10** 70, 179, 182
CERNATA REKA **Vol. 3** 59, 121
CERNE OCI, PROC PLACETE ... ? **Vol. 10** 58, 168
CERNI MYSLIVCI **Vol. 10** 53, 310
CERNITE ANGELI **Vol. 3** 114 **Vol. 6** *page 41, page 51*
CERNITE LEBEDI **Vol. 6** *page 43*
CERNY PETR **Vol. 10** 70, 371
CERNY PLAMEN **Vol. 10** 316, 345, 524
CERNY PRAPOR **Vol. 10** 352, 621
CERNY VLK **Vol. 10** 395, 550, 569
CERO DELLA VITA, IL **Vol. 7** 1189
CERO EN CONDUCTA **Vol. 8** 417, 583, 721
CERRADO POR ASESINATO **Vol. 8** 551, 749
CERRO DE LOS LOCOS, EL **Vol. 8** 214
CERRO DOS ENFORCADOS, O **Vol. 8** 172
CERROMAIOR **Vol. 8** 2, 35
CERT NESPI **Vol. 10** 210, 576
CERTAIN MONSIEUR JO, UN **Vol. 2** 479
CERTAIN MONSIEUR, UN **Vol. 2** 36, 158, 395
CERTAINES CHATTES N'AIMENT PAS LE MOU **Vol. 2** 39
CERTAINES NOUVELLES **Vol. 6** *page 106, page 145*
CERTAINS L'AIMENT FROIDE **Vol. 2** 318, 627, 875
CERTICKATA **Vol. 3** 31, 42
CERTIFICADO, EL **Vol. 8** 317
CERTISKO **Vol. 10** 45, 363
CERTO, CERTISSIMO, ANZI... PROBABILE **Vol. 7** 347
CERTOSA DI PARMA, LA **Vol. 7** 109, 213, 299, 884, 1054, 1060
CERTOVA STENA **Vol. 10** 351, 460, 478
CERUL INCEPE LA ETAJUL III **Vol. 3** 269, 281
CERUL N-ARE GRATII **Vol. 3** 270, 281, 330
CERVANTES **Vol. 8** 214
CERVEAU, LE **Vol. 2** 21, 409, 824
CERVENA JESTERKA **Vol. 10** 154, 283, 552, 621
CERVENA KARKULA **Vol. 10** 198, 536
CERVENA KULNA **Vol. 10** 173, 227, 532
CERVENE VINO **Vol. 10** 32, 263, 483
CERVNOVE DNY **Vol. 10** 428, 572
CES DAMES AUX CHAPEAUX VERTS **Vol. 2** 48, 99, 193, 656, 722
CES DAMES PREFERENT LE MAMBO **Vol. 2** 501, 615

CES DAMES S'EN MELENT **Vol. 2** 481, 570
CES FLICS ETRANGES VENUS D'AILLEURS **Vol. 2** 76, 130
CES GARCONS QUI VENAIENT DU BRESIL **Vol. 2** 224
CES MESSIEURS DE LA FAMILLE **Vol. 2** 475, 815
CES MESSIEURS DE LA GACHETTE **Vol. 2** 475, 503
CES MESSIEURS DE LA SANTE **Vol. 2** 197, 206
CES SACREES ANGLAISES **Vol. 2** 291
CES SACREES VACANCES **Vol. 2** 131, 243, 259
CESAR **Vol. 2** 143, 871, 872
CESAR ET ROSALIE **Vol. 2** 110
CESARE BIROTTEAU **Vol. 7** 764
CESKE HRADY A ZAMKY **Vol. 10** 45
CESKE NEBE **Vol. 10** 220
CESKOSLOVENSKY JEZISEK **Vol. 10** 167, 198
CESSEZ LE FEU **Vol. 2** 48, 388
CEST A SLAVA **Vol. 10** 96, 264, 394, 473
C'EST ARRIVE A 36 CHANDELLES **Vol. 2** 16, 423, 738
C'EST ARRIVE A ADEN **Vol. 2** 226, 368, 615
C'EST ARRIVE A PARIS **Vol. 2** 124, 517, 776
C'EST DEMENT...MAIS ON Y VA **Vol. 2** 600
C'EST DUR POUR TOUT LE MONDE **Vol. 2** 495
C'EST ENCORE LOIN L'AMERIQUE ? **Vol. 2** 74, 674
C'EST FACILE ET CA PEUT RAPPORTER ... 20 ANS **Vol. 6** *page 146*
C'EST JEUNE ET CA SAIT TOUT **Vol. 2** 39
C'EST LA FAUTE A PAPA **Vol. 2** 51
C'EST LA FAUTE D'ADAM **Vol. 2** 247, 401, 857
C'EST LA VIE **Vol. 2** 804 **Vol. 6** *page 104*
C'EST LA VIE PARISIENNE **Vol. 2** 118, 826
C'EST PAS MOI, C'EST L'AUTRE **Vol. 2** 347, 738
C'EST PAS MOI, C'EST LUI ! **Vol. 2** 7, 682 **Vol. 6** *page 92*
C'EST PAS PARCE QU'ON A RIEN A DIRE QU'IL FAUT FERMER SA GUEULE ! **Vol. 2** 68, 791
C'EST PLUS FACILE DE GARDER LA BOUCHE OUVERTE **Vol. 2** 309
C'EST UNE FILLE DE PANAME **Vol. 2** 475, 604, 872
CESTA DO HLUBIN STUDAKOVY DUSE **Vol. 10** 2, 412
CESTA DUGA GODINU DANA **Vol. 3** 452 **Vol. 7** 1044
CESTA HLUBOKYM LESEM **Vol. 10** 154, 433
CESTA KE STESTI **Vol. 10** 122, 411
CESTA KOLEM ME HLAVY **Vol. 10** 106, 473, 541
CESTA, LA **Vol. 8** 546
CESTA NA JIHOZAPAD **Vol. 10** 206, 250, 508
CESTA ZENY **Vol. 10** 72, 320, 483
CESTA ZPATKY **Vol. 10** 296, 564, 569
CESTOU KRIZOVOU **Vol. 10** 532, 621
CESTY K VYSINAM **Vol. 10** 167, 536
CESTY MUZU **Vol. 10** 283, 286, 550
CET AGE EST SANS PITIE **Vol. 2** 191
CET HOMME EST DANGEREUX **Vol. 2** 279, 557, 866
CET IMBECILE DE RIMOLDI **Vol. 2** 661
CET OBSCUR OBJET DU DESIR **Vol. 2** 377, 728 **Vol. 8** 214
C'ETAIT UN MUSICIEN **Vol. 2** 676
CETATEA FERMECATA **Vol. 3** 333
CETATEA NEAMTULUI **Vol. 3** 234
CETIRI DANA DO SMRTI **Vol. 3** 381, 400
CETIRI KILOMETRA NA SAT **Vol. 3** 349, 374
CETIRIMATA OT VAGONA **Vol. 3** 65 **Vol. 6** *page 25, page 48*
CETTE NUIT-LA **Vol. 2** 153, 168, 702, 708, 775
CETTE SACREE GAMINE **Vol. 2** 168
CETTE VIEILLE CANAILLE **Vol. 2** 23, 206
CETVRTI SUPUTNIK **Vol. 3** 421, 437
CEUX DE DEMAIN **Vol. 2** 152, 722
CEUX DE LA TERRE **Vol. 2** 159, 286, 330
CEUX DU BALLON ROND **Vol. 2** 314
CEUX DU CIEL **Vol. 2** 249, 388, 833
CEUX DU RIVAGE **Vol. 2** 88, 124
CEZAR A DETEKTIVI **Vol. 10** 107, 268, 544
C.F.R. O SIMFONIE A MUNCII **Vol. 1** 278
CHA – CHA – CHA **Vol. 9** 103, 199, 316
CHACAL **Vol. 2** 835
CHACUN SA CHANCE **Vol. 2** 32, 197
CHACUN SON TOUR **Vol. 2** 460, 627
CHAIMITE **Vol. 8** 44, 167, 172
CHAIR ARDENTE **Vol. 2** 63, 394, 761, 871
CHAIR DE L'ORCHIDEE, LA **Vol. 2** 517, 653
CHAIR DE POULE **Vol. 2** 153, 457
CHAIR ET LE DIABLE, LA **Vol. 2** 427, 832, 863
CHAISE VIDE, LA **Vol. 2** 60
CHALEUR DU SEIN, LA **Vol. 2** 197
CHALEURS **Vol. 2** 300

CHALEURS D'ETE **Vol. 2** 703
CHALLENGE, THE **Vol. 2** 656
CHAM **Vol. 1** 278, 404
CHAMADE, LA **Vol. 2** 263, 517
CHAMÄLEON, DAS **Vol. 4** 385
CHAMBRE 13 **Vol. 2** 871
CHAMBRE ARDENTE, LA **Vol. 2** 294, 773
CHAMBRE EN VILLE, UNE **Vol. 6** *page 78, page 91*
CHAMBRE ROUGE, LA **Vol. 2** 124
CHAMBRE VERTE, LA **Vol. 2** 17, 434
CHAMBRE VOISINE, LA **Vol. 2** 517
CHAMBRES D'AMIS TRES „PARTICULIERES" **Vol. 6** *page 79*
CHAMPAGNE CAPRICE **Vol. 7** 104, 144
CHAMPAGNE E FAGIOLI **Vol. 7** 281
CHAMPAGNE IN PARADISO **Vol. 7** 909
CHAMPAGNE ROSE IS DEAD **Vol. 5** 946
CHAMPAGNEFLASKEN **Vol. 5** 97
CHAMPAGNEGALOPP **Vol. 5** 946, 1024, 1093
CHAMPAGNEGALOPPEN **Vol. 5** 55, 101, 208
CHAMPIGNON, LE **Vol. 2** 154, 198
CHAMPION DU REGIMENT, LE **Vol. 2** 330, 381, 387
CHAMPIONS DE FRANCE **Vol. 2** 152
CHANCE ET L'AMOUR, LA **Vol. 2** 513, 514
CHANCE, LA **Vol. 2** 41, 469, 579
CHANCHA VASCA **Vol. 8** 630
CHANEL SOLITAIRE **Vol. 6** *page 71, page 77, page 82, page 96*
CHANGE PAS DEMAIN **Vol. 2** 804
CHANS **Vol. 5** 878, 1052
CHANSON D'ARMOR, LA **Vol. 2** 530, 531
CHANSON DE ROLAND, LA **Vol. 2** 301, 765
CHANSON DES NATIONS, LA **Vol. 2** 387, 688
CHANSON DU LIN, LA **Vol. 2** 603
CHANSONS DE PARIS **Vol. 2** 48, 59, 131
CHANT DE L'AMOUR, LE **Vol. 2** 217, 279, 412, 831
CHANT DE L'AMOUR TRIOMPHANT, LE **Vol. 2** 122, 179, 521, 623
CHANT DE L'EXILE, LE **Vol. 2** 6, 88
CHANT DU DEPART, LE **Vol. 2** 301
CHANT DU MARIN, LE **Vol. 2** 676, 679
CHANT DU MONDE, LE **Vol. 2** 503
CHANTAGE **Vol. 2** 9, 37, 308, 368, 540
CHANTAJE **Vol. 8** 209, 268
CHANTAJE A UN ASESINO **Vol. 8** 207
CHANTAJE A UN TORERO **Vol. 8** 207, 214
CHANTECOQ **Vol. 2** 167
CHANTEUR DE MEXICO, LE **Vol. 2** 419, 676, 808
CHANTEUR DE MINUIT, LE **Vol. 2** 566, 719, 863
CHANTEUR INCONNU, LE **Vol. 2** 59, 206, 330, 676, 826
CHANTONS QUAND MEME **Vol. 2** 601, 871
CHANTONS SOUS L'OCCUPATION **Vol. 2** 752
CHAP-LA **Vol. 2** 210
CHAPEAU DE PAILLE D'ITALIE, UN **Vol. 2** 237, 347, 573, 753, 871
CHAQUE JOUR A SON SECRET **Vol. 2** 294, 383, 559
CHAQUE MINUTE COMPTE **Vol. 2** 433
CHARCUTIER DE MACHONVILLE, LE **Vol. 2** 224, 314
CHARGE DE CUIRASSIERS **Vol. 2** 678
CHARGE DE DRAGONS A SATORY **Vol. 2** 678
CHARGE OBLIQUE **Vol. 2** 678
CHARGE OF THE GAUCHOS **Vol. 2** 84
CHARGEURS DE DECOMBRES **Vol. 2** 678
CHARLEMAGNE **Vol. 2** 6, 193, 197, 332
CHARLES ET LUCIE **Vol. 2** 763
CHARLES' TANTE **Vol. 5** 29, 180, 199, 201
CHARLESTON **Vol. 7** 102, 798 **Vol. 8** 261, 732
CHARLEYS TANT **Vol. 5** 857, 891, 1031
CHARLEYS TANTE **Vol. 4** 132, 626
CHARLIE BRAVO **Vol. 2** 291
CHARLIE ET SES DEUX MENETTES **Vol. 2** 198, 682
CHARLOT ESPANOL, EL **Vol. 8** 357, 655
CHARLOTS CONNECTION **Vol. 6** *page 98*
CHARLOTS CONTRE DRACULA, LES **Vol. 2** 806 **Vol. 6** *page 97*
CHARLOTS EN DELIRE **Vol. 2** 491 **Vol. 6** *page 139*
CHARLOTS FONT L'ESPAGNE, LES **Vol. 2** 655
CHARLOTT ETWAS VERRÜCKT **Vol. 4** 239, 276, 626
CHARLOTTE ET SON JULES **Vol. 2** 465
CHARLOTTE LÖWENSKOLD **Vol. 5** 891, 997 **Vol. 5** 933, 1035
CHARLOTTE MOUILLE TA CULOTTE **Vol. 6** *page 100*
CHARLY OG STEFFEN **Vol. 5** 61, 169, 299
CHARMANTS GARCONS **Vol. 2** 185, 601
CHARME DISCRET DE LA BOURGEOISIE, LE **Vol. 2** 377, 728
CHARME PERVERS **Vol. 2** 693
CHARMES DE L'EXISTENCE, LES **Vol. 2** 652
CHARNELLES, LES **Vol. 2** 39
CHARRETTE FANTOME, LA **Vol. 2** 439, 440

CHARTREUSE DE PARME, LA **Vol. 2** 27, 140, 279, 395
CHASSE A L'HOMME, LA **Vol. 2** 457, 873
CHASSE AU LION A L'ARC, LA **Vol. 2** 752
CHASSE - CROISE **Vol. 6** *page 95*
CHASSE ROYALE, LA **Vol. 2** 49
CHASSEUR DE CHEZ MAXIM'S, LE **Vol. 2** 6, 32, 137, 205, 226, 237, 293, 387, 400, 413, 736
CHASTE SUZANNE, LA **Vol. 2** 279
CHAT ET LA SOURIS, LE **Vol. 2** 196
CHAT, LE **Vol. 2** 770, 875
CHAT NOIRS HOSTREVY 1974 **Vol. 5** 674, 681, 712, 772
CHATEAU DE LA DERNIERE CHANCE, LE **Vol. 2** 533, 558
CHATEAU DE LA MORT LENTE, LE **Vol. 2** 346
CHATEAU DE VERRE, LE **Vol. 2** 59, 488
CHATEAU DES QUATRE OBESES, LE **Vol. 2** 395
CHATEAU EN ENFER, UN **Vol. 2** 224
CHATEAU EN SUEDE **Vol. 2** 21, 824
CHATEAUX EN ESPAGNE **Vol. 2** 8, 540 **Vol. 8** 214
CHATELAINE DU LIBAN, LA **Vol. 2** 9, 62, 131, 332, 333, 419, 538, 566, 598, 824
CHATOLLETS HEMMELIGHED **Vol. 5** 183
CHATOUILLEUSES, LES **Vol. 2** 750
CHATS DE GOUTTIERES, LES **Vol. 2** 219
CHATTE, LA **Vol. 2** 9
CHATTE SORT SES GRIFFES, LA **Vol. 2** 9, 601
CHATTE SUR UN DOIGT BRULANT, LA **Vol. 2** 752
CHATTES EN CHALEUR **Vol. 2** 291
CHAUD LAPIN, LE **Vol. 2** 618
CHAUDES ADOLESCENTES **Vol. 6** *page 129*
CHAUFFEUR ANTOINETTE **Vol. 4** 115, 132
CHAUFFEUR DE MADEMOISELLE, LE **Vol. 2** 346, 357
CHAUSSEE DES GEANTS, LA **Vol. 2** 6, 33, 145, 193, 333, 381
CHAUVE-SOURIS, LA **Vol. 2** 9, 397
CHAVALA, LA **Vol. 8** 228
CHCEME ZIT **Vol. 10** 122, 193, 443
CHCEMY MEZA **Vol. 1** 174, 176, 326
CHCIALBYM LECZ NIE WOLNO **Vol. 1** 317
CHCIALBYM SIE ZGUBIC ... **Vol. 6** *page 236, page 244, page 254*
CHE ? **Vol. 7** 502, 1007
CHE CASINO... CON PIERINO ! **Vol. 7** 311
CHE C'ENTRIAMO NOI CON LA RIVOLUZIONE ? **Vol. 8** 595, 762
CHE COSA SONO LE NUVOLE ? **Vol. 7** 333, 581
CHE DISINTA FAMIGLIA **Vol. 7** 196, 756
CHE FARESTE VOI ? **Vol. 7** 621
CHE FEMMINA ... E CHE DOLLARI **Vol. 7** 1147
CHE FINE HA FATTO TOTO BABY ? **Vol. 7** 834
CHE GIOIA VIVERE ! **Vol. 7** 1214
CHE NOTTE QUELLA NOTTE **Vol. 7** 1114
CHE NOTTE RAGAZZI ! **Vol. 8** 306, 600
CHE, QUE LOCO ! **Vol. 8** 298, 400
CHE TEMPI ! **Vol. 7** 1054
CHECK UP **Vol. 2** 537
CHEMAREA DRAGOSTEI **Vol. 3** 216
CHEMIE UND LIEBE **Vol. 1** 39, 56, 77
CHEMIN DE DAMAS, LE **Vol. 2** 332, 778
CHEMIN DE LA DROGUE, LE **Vol. 2** 230
CHEMIN DE LA MAUVAISE ROUTE, LE **Vol. 2** 188
CHEMIN DE L'ABIME, LE **Vol. 2** 490
CHEMIN DE L'HONNEUR, LE **Vol. 2** 9, 193, 538
CHEMIN DE RIO, LE **Vol. 2** 9, 339
CHEMIN D'ERNOA, LE **Vol. 2** 346
CHEMIN DES ECOLIERS, LE **Vol. 2** 59, 566
CHEMIN DU BONHEUR, LE **Vol. 2** 425, 864
CHEMIN DU PARADIS, LE **Vol. 2** 679
CHEMIN PERDU, LE **Vol. 2** 854 **Vol. 6** *page 136*
CHEMINEAU, LE **Vol. 2** 623, 688, 726, 727
CHEMINEAU RINCKEUR, LE **Vol. 2** 159, 286, 330
CHEMINS DE KATMANDOU, LES **Vol. 2** 185, 873
CHEMINS SANS LOI **Vol. 2** 538, 595
CHEMISES ROUGES, LES **Vol. 2** 53
CHENE D'ALLOUVILLE, LE **Vol. 6** *page 137*
CHEQUE AU PORTEUR **Vol. 2** 32, 698
CHERCHANT UNE POSITION **Vol. 2** 678
CHERCHEZ LA FEMME **Vol. 2** 294, 401
CHERCHEZ L'ERREUR **Vol. 2** 60 **Vol. 6** *page 125, page 131*
CHERCHEZ L'IDOLE **Vol. 2** 322, 503
CHERE INCONNUE **Vol. 2** 190, 285 **Vol. 6** *page 121*
CHERE LOUISE **Vol. 2** 542
CHERI **Vol. 2** 254, 384, 833
CHERI-BIBI **Vol. 2** 200, 826 **Vol. 7** 798
CHERI DE SA CONCIERGE, LE **Vol. 2** 67, 70
CHERI, FAIS-MOI PEUR **Vol. 2** 47, 137
CHEVAL DE FER, LE **Vol. 2** 51, 358
CHEVAL D'ORGUEIL, LE **Vol. 2** 700 **Vol. 6** *page 87*

CHEVAL POUR DEUX, UN **Vol. 2** 149, 675
CHEVALIER DE GABY, LE **Vol. 2** 708
CHEVALIER DE LA NUIT, LE **Vol. 2** 388, 402
CHEVALIER DE MAISON ROUGE, LE **Vol. 2** 501
CHEVALIER DE PARDAILLAN, LE **Vol. 2** 615
CHEVALIER SANS ARMURE, LE **Vol. 2** 81, 573
CHEVALIERS DE LA MONTAGNE, LES **Vol. 2** 679
CHEVEUX DE L'AMIE, LES **Vol. 2** 159, 286, 330
CHEVRE D'OR, LA **Vol. 2** 193, 615
CHEVRE, LA **Vol. 6** *page 75, page 138*
CHEWINGUM **Vol. 7** 664
CHEZ NUOS **Vol. 5** 858, 900, 975
CHI DICE DONNA DICE ... DONNA **Vol. 7** 502
CHI E PIU FELICE DI ME ! **Vol. 7** 40, 725
CHI E SENZA PECCATO **Vol. 7** 638, 1054
CHI LAVORE A PERDUTO **Vol. 7** 87, 1050
CHI L'HA UCCISA ? **Vol. 7** 1189
CHI L'HA UCCISO **Vol. 7** 39
CHI L'HA VISTA MORIRE ? **Vol. 7** 524
CHI L'HA VISTO ? **Vol. 7** 1036, 1054
CHI MI AIUTA ? **Vol. 7** 347, 697
CHI NON CREDE ALL'AMORE **Vol. 7** 39, 250
CHI SEI TU ? **Vol. 7** 16, 647
CHI SI FERMA E PERDUTO **Vol. 7** 1044
CHI TROPPO VUOLE ... **Vol. 7** 26
CHI TROVA UN AMICO TROVA UN TESORO **Vol. 7** 381, 599
CHIAMATE 22 – 22 TENENTE SHERIDAN **Vol. 7** 347
CHIAMATE: 6969 TAXI PER SIGNORA **Vol. 7** 609
CHIAMAVAN CAPINERA ..., LA **Vol. 7** 17
CHIAMAVANO BULLDOZER, LO **Vol. 7** 524
CHIAMAVANO 'COSETTA', LA **Vol. 7** 569
CHIAMAVANO TANGO, LA **Vol. 7** 298
CHIAMAVANO „TRESETTE" ... GIOCAVA SEMPRE COL MORTO **Vol. 7** 743
CHIAMAVANO „VERITA", LO **Vol. 7** 247
CHIAMEREMO ANDREA, LO **Vol. 7** 567
CHIARINA LA MODISTA **Vol. 7** 829
CHIAVE D'ORO E CHIAVE DI FERRO **Vol. 7** 837
CHIAVE, LA **Vol. 7** 142, 585
CHICA CASI DECENTE, UNA **Vol. 8** 249, 406
CHICA CASI FORMAL, UNA **Vol. 8** 236, 725
CHICA DE CHICAGO, UNA **Vol. 8** 207, 600, 749
CHICA DE LOS ANUNCIOS, LA **Vol. 8** 523
CHICA DE OPERETA, UNA **Vol. 8** 394, 396
CHICA DE VIA CONDOTTI, LA **Vol. 8** 595
CHICA DEL AUTO-STOP, LA **Vol. 8** 562, 768
CHICA DEL BARRIO, LA **Vol. 8** 261, 298
CHICA DEL MOLINO ROJO, LA **Vol. 8** 245, 616
CHICA DEL TREBOL, LA **Vol. 8** 438, 602
CHICA DELGATO, LA **Vol. 8** 396, 541, 655
CHICA LLAMADA MARILYNE, UNA **Vol. 8** 299, 728
CHICA PARA DOS, UNA **Vol. 8** 402, 772
CHICA PARA TODO **Vol. 8** 396, 450
CHICA QUE CAYO DEL CIELO, LA **Vol. 8** 683
CHICA Y UN SENOR, UNA **Vol. 8** 524
CHICAS DEL BINGO, LAS **Vol. 8** 667, 728
CHICAS DEL TANGA, LAS **Vol. 8** 731
CHICO, CHICA, BOOM ! **Vol. 8** 261, 768
CHICO O CHICA ? **Vol. 8** 298, 524
CHICOS CON LAS CHICAS, LOS **Vol. 8** 438
CHICOS DE LA ESCUELA, LOS **Vol. 8** 228, 559
CHICOS DEL PREU, LOS **Vol. 8** 524, 725
CHICOS, LOS **Vol. 8** 719
CHIEDO ASILO **Vol. 7** 405, 695, 947
CHIEN ANDALOU, UN **Vol. 2** 255, 277, 775
CHIEN DANS UN JEU DE QUILLES, UN **Vol. 2** 512 **Vol. 6** *page 69, page 114*
CHIEN DE PIQUE **Vol. 2** 16
CHIEN FOU, LE **Vol. 2** 814
CHIEN JAUNE, LE **Vol. 2** 780, 833
CHIEN QUI RAPPORTE, UN **Vol. 2** 65, 330, 538
CHIENNE, LA **Vol. 2** 402, 780, 797
CHIENNES, LES **Vol. 2** 818
CHIENS DE JERUSALEM, LES **Vol. 7** 593, 1173, 1175
CHIENS, LES **Vol. 2** 74, 690
CHIENS PERDUS SANS COLLIER **Vol. 2** 601, 722, 832
CHIFFONNIERS D' EMMAUS, LES **Vol. 2** 124, 475, 801
CHIFFRIERT AN CHEF – AUSFALL NR. 5 **Vol. 6** *page 157, page 159, page 172*
CHIKWEMBO ! **Vol. 8** 80, 184
CHIMERA **Vol. 7** 442
CHIMERE **Vol. 7** 37
CHINGACHGOOK – DIE GROSSE SCHLANGE **Vol. 1** 37, 68, 102
CHINOIS A PARIS, LES **Vol. 2** 110
CHINOISE, LA **Vol. 2** 208
CHINOS Y MINIFALDAS **Vol. 8** 406, 541
CHIODO, IL **Vol. 7** 1189
CHIPEE **Vol. 2** 270, 395

CHIQUE Vol. 2 6, 193, 197
CHIROMANTE, IL Vol. 7 735, 838
CHIRURGUS JOHANN PAUL SCHROTH Vol. 6 *page 159, page 167, page 175*
CHISPITA Y SUS GORILLAS Vol. 8 627, 770
CHISSA PERCHE ... CAPITANO TUTTE A ME Vol. 7 524, 875, 1010
CHLAP JAK HORA Vol. 10 41, 352, 457
CHLAPI PRECE NEPLACOU Vol. 10 160, 161, 255
CHLOPCI Vol. 1 206, 268, 276, 282, 313, 327
CHOBIZENESSE Vol. 2 450, 588
CHOBOTNICE Z II. PATRA Vol. 10 139, 165, 496
CHOC EN RETOUR Vol. 2 55, 249, 727
CHOC, LE Vol. 6 *page 78, page 82*
CHOCOLATE Vol. 8 299, 303
CHOD A NELUC SA Vol. 10 228, 280, 592
CHODNIK CEZ DUNAJ Vol. 10 145, 178, 188
CHOIX D'ASSASSINS, UN Vol. 2 513, 520
CHOIX DES ARMES, LE Vol. 2 351 Vol. 6 *page 84*
CHOLERA V PRAZE Vol. 10 201
CHOMEUR DE CLOCHEMERLE, LE Vol. 2 347, 807,
CHORAL VON LEUTHEN, DER Vol. 4 74, 235, 323, 527, 626
CHOSES DE LA VIE, LES Vol. 2 110, 675
CHOTARD ET CIE Vol. 2 160, 623, 726
CHOUANS, LES Vol. 2 270, 719
CHOURINETTE Vol. 2 152, 330, 437
CHRISTA Vol. 2 872 Vol. 5 5, 179
CHRISTA HARTUNGEN Vol. 4 178
CHRISTIAN ASMUS VON DER HELLING Vol. 1 127
CHRISTIAN SCHRODER SOM DON JUAN Vol. 5 97
CHRISTIAN SCHRODER SOM LEJETJENER Vol. 5 295
CHRISTIAN WAHNSCHAFFE Vol. 4 124, 239
CHRISTINA Y LA RECONVERSION SEXUAL Vol. 8 243, 762
CHRISTINE Vol. 2 226, 279, 566
CHRISTINE SE MARIE Vol. 2 197
CHRISTINE UND DIE STÖRCHE Vol. 1 22
CHRISTOPH COLUMBUS Vol. 4 25, 112, 219, 240, 427
CHRISTUS O LA SFINGE DELLO JONIO Vol. 7 534
CHRISTUS VON OBERAMMERGAU, DER Vol. 4 25, 652
CHRONIK EINES MORDES Vol. 1 40, 115
CHRONIQUE D'UN ETE Vol. 2 135, 208, 609, 814
CHUDA HOLKA Vol. 10 345, 524
CHUDY I INNI Vol. 1 206, 355 Vol. 6 *page 259*
CHULO, EL Vol. 8 438, 719
CHULOS, LOS Vol. 8 368, 773
CHUNG KUO – CINA Vol. 7 1145
CHUT Vol. 2 866
CHUTE DE LA MAISON USHER, LA Vol. 2 346, 426, 430, 528, 530
CHUTE D'UN CORPS, LA Vol. 2 7, 8
CHYTTE HO ! Vol. 10 167, 453
CI SPOSEREMO A CAPRI Vol. 7 827
CI TROVIAMO IN GALLERIA Vol. 7 1044
CIANURA SI PICATURA DE PLOAIE Vol. 3 269, 286
CIAO CIALTRONI ! Vol. 7 222, 275
CIAO CIAO BAMBINA Vol. 7 798, 915
CIAO LES MECS ! Vol. 2 49
CIAO MARZIANO Vol. 7 257
CIAO MASCHIO Vol. 7 434, 1145
CIAO NEMICO Vol. 7 13, 315
CIAO NI ! Vol. 7 1015
CIAO PAIS ... Vol. 7 85
CIBLE, LA Vol. 2 521, 623, 833
CIBOULETTE Vol. 2 67, 206, 573, 837
CICALA, LA Vol. 7 383, 931, 971, 997
CICATRICES, LAS Vol. 8 719
CICCA, LA Vol. 7 1189
CICCIO, IL PIZZAIUOLO DEL CARMINE Vol. 7 829
CICCIOBOMBA Vol. 7 257, 315
CICERUACCHIO Vol. 7 369
CICHA NOC, SWIETA NOC Vol. 1 148
CICILIA, LA Vol. 2 502
CID CABREADOR, EL Vol. 8 414
CIECA DEL MOLO, LA Vol. 7 1174
CIECA DI SORRENTO, LA Vol. 7 493, 501, 912
CIECO, IL Vol. 7 701
CIEL BLEU Vol. 2 742
CIEL EST A VOUS, LE Vol. 2 251, 644
CIEL ET LA BOUE, LE Vol. 2 768
CIEL SUR LA TETE, LE Vol. 2 781, 811
CIELO BRUCIA, IL Vol. 7 196
CIELO E ROSSO, IL Vol. 7 404, 1177
CIELO NEGRO Vol. 8 276, 704
CIELO SULLA PALUDE Vol. 7 21, 423, 749
CIEMNA RZEKA Vol. 1 142, 203 Vol. 6 *page 222*
CIEN Vol. 1 242, 250 Vol. 6 *page 256*
CIEN CABALLEROS, LOS Vol. 8 438, 523
CIEN JUZ NIEDALEKO Vol. 6 *page 255*
CIERNA MINUTA Vol. 10 155, 263
CIERPKIE GLOGI Vol. 1 276, 381 Vol. 6 *page 240*
CIFRADO ESPECIAL Vol. 8 306
CIFTI I LUMTUR Vol. 3 4, 15
CIGANIN BARON Vol. 3 356

CIGANKA **Vol. 3** 440, 448
CIGANY, A **Vol. 9** 33, 71, 157
CIGARETPIGEN **Vol. 5** 58
CIGARETTE, LA **Vol. 2** 167
CIGARETTE POUR UN INGENU, UNE **Vol. 2** 347
CIGARETTES, WHISKY ET P'TITES PEPEES **Vol. 2** 419, 598
CIGARRA, LA **Vol. 8** 298, 611
CIGULI MIGULI **Vol. 3** 445
CIKAN JURA **Vol. 10** 257, 536
CIKANI **Vol. 10** 257, 536
CIKANSKA LASKA **Vol. 10** 403, 468
CIKLAMEN **Vol. 9** 77, 323
CIKLOP"T **Vol. 3** 46 **Vol. 6** *page 36, page 59*
CIMELII VERDIANI, I **Vol. 7** 308
CIMITERO DEI GIUSTIZIATI, IL **Vol. 7** 91, 980
CIMZETT ISMERETLEN **Vol. 9** 15, 322
CINCI OAMENI LA DRUM **Vol. 3** 286, 317, 325
CINCO ADVERTENCIAS DE SATANAS, LAS **Vol. 8** 640, 749
CINCO DE LA VENGANZA, LOS **Vol. 8** 562, 732
CINCO LOBITOS **Vol. 8** 53, 75, 119, 122, 188, 449
CINCO PISTOLAS DE TEXAS **Vol. 8** 562, 768
CINCO TENEDORES **Vol. 8** 243, 738
CIND PRIMAVARA E FIERBINTE **Vol. 3** 266, 273, 332
CINDERELLA **Vol. 2** 249, 425
CINE-GIRL **Vol. 2** 359, 514
CINE IUBESTE SI LASA **Vol. 6** *page 278, page 283*
CINE MA STRIGA ? **Vol. 3** 311 **Vol. 6** *page 272*
CINE VA DESCHIDE USA **Vol. 3** 285, 307, 331
CINEMA, CHE PASSIONE ! **Vol. 7** 1136
CINEMA D'ALTRI TEMPI **Vol. 7** 1044
CINEMA DE PAPA, LE **Vol. 2** 655, 770
CINEMATOGRAFO 1900 **Vol. 8** 237, 513, 524
CINGHA DELLA MORTE, LA **Vol. 7** 961
CINICO, L'INFAME, IL VIOLENTO, IL **Vol. 7** 1202
CINQ ET LA PEAU **Vol. 6** *page 77, page 100*
CINQ FILLES EN FURIE **Vol. 2** 259
CINQ GARS POUR SINGAPOUR **Vol. 2** 175, 552
CINQ GENTLEMEN MAUDITS, LES **Vol. 2** 573, 824 **Vol. 6** *page 89, page 98*
CINQ MILLIONS COMPTANT **Vol. 2** 595, 627, 726
CINQ POUR CENT DE RISQUE **Vol. 6** *page 104*
CINQ SOUS DE LAVAREDE, LES **Vol. 2** 187, 254, 360
CINQ TULIPES ROUGES **Vol. 2** 197, 368
CINQUE A ZERO **Vol. 7** 665, 730
CINQUE DELLA VENDETTA, I **Vol. 8** 562, 732
CINQUE DELL'ADAMELLO, I **Vol. 7** 501
CINQUE GIORNATE, LE **Vol. 7** 98
CINQUE PER L'INFERNO **Vol. 7** 692
CINQUE POVERI IN AUTOMOBILE **Vol. 7** 16, 161
CINQUIEME EMPREINTE, LA **Vol. 2** 193, 579, 646, 722
CINTAMANI A PODVODNIK **Vol. 10** 55, 302, 621
CINTEC PENTRU FIUL MEU **Vol. 3** 214, 262, 308
CINTECELE MARII **Vol. 3** 286, 330
CINTURA DELLE AMAZZONI, LA **Vol. 7** 774, 813, 972
CIOCIARA, LA **Vol. 7** 756, 910
CIOCOII **Vol. 3** 216
CIOCOLATA CU ALUNE **Vol. 3** 321, 335
CIOSY **Vol. 6** *page 206, page 220, page 226*
CIPELICE NA ASFALTU **Vol. 3** 344, 356, 383, 388, 426, 428
CIPRIA E SANGUE **Vol. 7** 525
CIPRIAN PORUMBESCU **Vol. 3** 225, 254, 276
CIRANO DI BERGERAC **Vol. 7** 254, 751
CIRCLE MODERNA – O – VENERE ORGIASTRA **Vol. 7** 1038
CIRCO EQUESTRE ZA-BUM **Vol. 7** 796
CIRCONSTANCES ATTENUANTES **Vol. 2** 197, 726
CIRCULEZ ! **Vol. 2** 833
CIRCULEZ Y'A RIEN A VOIR **Vol. 6** *page 80, page 90*
CIRCULO DE PASIONES **Vol. 8** 536, 682
CIRCUS MAXIMUS **Vol. 9** 51, 263, 316
CIRENO, IL **Vol. 7** 971, 1013
CIRESARII **Vol. 3** 311 **Vol. 6** *page 266, page 270*
CIRKU NE FSHAT **Vol. 3** 11, 22 **Vol. 6** *page 13*
CIRKUS **Vol. 5** 55, 899
CIRKUS BIMBINI **Vol. 5** 841
CIRKUS BUDE **Vol. 10** 412, 556, 566
CIRKUS BUSTER **Vol. 5** 31, 248, 272
CIRKUS CASABLANCA **Vol. 5** 48, 62
CIRKUS FANDANGO **Vol. 5** 672, 831
CIRKUS JEDE **Vol. 10** 131
CIRKUS V CIRKUSE **Vol. 10** 249, 292, 568
CIRKUSREVYEN 67 **Vol. 5** 179, 195, 254, 292
CIS CABREADOR, EL **Vol. 8** 407

CISARUV PEKAR PEKARUV CISAR Vol. 10 524, 547, 561, 621
CISTA REKA Vol. 10 18, 313, 517
CISTE RUKY Vol. 10 268, 320
CITA CON MI VIEJO CORAZON Vol. 8 753
CITA EN MI CORAZON Vol. 8 572, 753
CITA EN NAVARRA Vol. 8 249, 772
CITA IMPOSIBLE Vol. 8 753
CITADEL, THE Vol. 2 573
CITADELA SFARIMATA Vol. 3 219, 259, 272, 297
CITADELATA OTGOVORI Vol. 3 105 Vol. 6 *page 59*
CITADELLE DU SILENCE, LA Vol. 2 23, 332, 626, 824
CITE DE L'ESPERANCE Vol. 2 197, 368
CITE DE L'INDICIBLE PEUR, LA Vol. 2 778
CITE DES LUMIERES, LA Vol. 2 273, 360, 871
CITE FOUDROYEE, LA Vol. 2 216, 388
CITLIVA MISTA Vol. 10 70, 408, 465
CITOVA VYCHOVA JEDNEJ DASE Vol. 10 108, 263, 592
CITTA CANORA, LA Vol. 7 587
CITTA DEL SOLE, LA Vol. 7 819
CITTA DELL'AMORE, LA Vol. 7 1162
CITTA DELLE DONNE, LA Vol. 7 434, 449, 537, 893, 988, 1130
CITTA DI MIRIAM, LA Vol. 7 801, 842, 1146
CITTA DI NOTTE Vol. 7 106
CITTA DI VETRO, LA Vol. 7 730
CITTA DOLENTE, LA Vol. 7 333
CITTA SCONVOLTA: CACCIA SPIETATA AI RAPITORI Vol. 7 763
CITTA SI DIFENDE, LA Vol. 7 796
CIUADAD DE LOS SUENOS, LA Vol. 8 524
CIUADAD NO ES PARA MI, LA Vol. 8 524, 725
CIUDAD DE FUEGO, LA Vol. 8 359, 602, 630, 753
CIUDAD DE LOS MUNECOS, LA Vol. 8 512, 542, 561
CIUDAD LLAMADA BASTARDA, UNA Vol. 8 276, 558
CIUDAD PERDIDA, LA Vol. 8 214, 404
CIULEANDRA Vol. 6 *page 270, page 277, page 281*
CIULINII BARAGANULUI Vol. 3 220, 306
CIVIL A PALYAN Vol. 9 134, 237
CIZIM VSTUP POVOLEN Vol. 10 116, 508, 553
CLAIM NA HLUCHEM POTOKU Vol. 10 327
CLAIR DE FEMME Vol. 2 35 Vol. 6 *page 74, page 76, page 97*
CLAIR DE LUNE Vol. 2 237, 347, 379
CLAIR DE LUNE A MAUBEUGE, UN Vol. 2 686
CLAIR DE TERRE Vol. 2 754
CLAIRE Vol. 4 36, 270
CLAN DE LOS NAZARENOS, EL Vol. 8 348, 626
CLAN DES SICILIENS, LE Vol. 2 224, 770
CLANDESTINES, LES Vol. 2 294, 801
CLANDESTINO A TRIESTE Vol. 7 171, 1063
CLANDESTINO, EL Vol. 8 545, 598
CLANDESTINS, LES Vol. 2 401, 475, 595
CLARA DE MONTARGIS Vol. 2 719, 722
CLARA ES EL PRECIO Vol. 8 402, 732
CLARA ET LES CHICS TYPES Vol. 2 450 Vol. 6 *page 70*
CLARA ET LES MECHANTS Vol. 2 16, 661, 863
CLARETTA Vol. 7 29, 122, 684
CLARINES DEL MIEDO, LOS Vol. 8 261
CLARINES Y CAMPANAS Vol. 8 391, 630
CLARISSA Vol. 4 247, 329
CLARK Vol. 5 46, 129, 221, 283
CLASH Vol. 6 *page 100, page 144*
CLASS DE ASEN, LA Vol. 7 341
CLASSE DI FERRO Vol. 7 739
CLASSE DU SEXE, LA Vol. 2 215
CLASSE MISTA Vol. 7 1202
CLASSE TOUS RISQUES Vol. 2 190, 598
CLAUDIA Vol. 1 24, 35
CLAUDINE A L'ECOLE Vol. 2 200, 439, 871
CLAUDINE ET LE POUSSIN Vol. 2 308
CLAVELES DE LA VIRGEN, LOS Vol. 8 246, 357, 542
CLAVELES, LOS Vol. 8 217, 270, 584, 777
CLAVO, EL Vol. 8 214, 357, 400
CLE DE VOUTE, LA Vol. 2 33
CLEF SUR LA PORTE, LA Vol. 2 157
CLEMENTE VII E IL SACCO DI ROMA Vol. 7 785
CLEMENTINE CHERIE Vol. 2 165, 344
CLEMENTINE TANGO Vol. 6 *page 103, page 143*
CLEO DE 5 A 7 Vol. 2 285, 700
CLEO, ROBES ET MANTEAUX Vol. 7 730
CLEOPATRA REGINA D'EGITTO Vol. 7 124
CLERAMBARD Vol. 2 185, 564
CLIMATS Vol. 2 168, 854
CLINIQUE POUR SOINS TRES SPECIAUX Vol. 2 735
CLINT EL SOLITARIO Vol. 8 562, 732
CLIPA Vol. 3 309, 319, 321, 325

CLIPELE DINAINTEA SINGURATATII **Vol. 6** *page 267, page 270*
CLNY PROTI PRUDU **Vol. 10** 263, 269, 399
CLOCHARD **Vol. 2** 330, 394, 871
CLOCHARD MILLIARDAIRE, LE **Vol. 2** 67, 131, 856
CLOCHEMERLE **Vol. 2** 488, 558
CLODO **Vol. 2** 139
CLODOCHE **Vol. 2** 193, 479, 688, 698
CLOSERIE DES GENETS, LA **Vol. 2** 227, 467
CLOVEK NA MOSTE **Vol. 10** 36, 72, 210
CLOVEK NENI SAM **Vol. 10** 97, 313, 532
CLOVEK PROTI ZKAZE **Vol. 10** 233, 313, 516
CLOWN BUX, LE **Vol. 2** 67, 279, 646
CLOWN, DER **Vol. 4** 365, 608
CLOWN FERDINAND UND DER KOFFER **Vol. 1** 93
CLOWN FERDINAND UND DIE CHEMIE **Vol. 1** 93
CLOWNS, I **Vol. 7** 388, 544, 854
CLUB DE FEMMES **Vol. 2** 9, 440, 559, 661
CLUB DE SOLTEROS **Vol. 8** 207, 749
CLUB DEGLI IMPICCATI, IL **Vol. 7** 148
CLUB DEGLI OSSESSI, IL **Vol. 7** 1005
CLUB DEGLI STRAVAGANTI, IL **Vol. 7** 940
CLUB DEI SUICIDI, IL **Vol. 7** 886
CLUB DELLE DONNE, IL **Vol. 7** 548
CLUB DER HAZARDEURE, DER **Vol. 4** 67
CLUB DES ARISTOCRATES, LES **Vol. 2** 32, 197, 407
CLUB DES FADAS, LE **Vol. 2** 55
CLUB DES QUATRE CENTS COUPS, LE **Vol. 2** 496
CLUB DES SOUPIRANTS, LE **Vol. 2** 153, 863
CMA **Vol. 6** *page 230, page 253, page 260*
CNOSTNY METOD **Vol. 10** 72, 210, 483
CO DZIEN BLIZEJ NIEBA **Vol. 6** *page 224, page 247, page 250*
CO JE DOMA, TO SE POCITA, PANOVE ... **Vol. 10** 139, 325, 430
CO JE VAM, DOKTORE ? **Vol. 10** 258, 389, 570
CO MI ZROBISZ JAK MNIE ZLAPIESZ **Vol. 1** 233, 275, 339, 340
CO MOJ MAZ ROBI W NOCY **Vol. 1** 278, 308, 404
CO REKNE ZENA **Vol. 10** 301, 445
CO SE SEPTA **Vol. 10** 256, 412
COARTADA EN DISCO ROJO **Vol. 8** 311
COBRA, EL **Vol. 8** 331, 747
COBRA, IL **Vol. 8** 331, 747
COBRARDES, LOS **Vol. 8** 551
COCCO DI MAMMA, IL **Vol. 7** 693
COCHECITO, EL **Vol. 8** 214, 258
COCHONNES, LES **Vol. 2** 365
COCKTAIL **Vol. 5** 2, 55, 208
COCKTAIL MOLOTOV **Vol. 2** 754 **Vol. 6** *page 87*
COCO LA FLEUR **Vol. 2** 210
COCORICO ! MONSIEUR POULET **Vol. 2** 752
COCU MAGNIFIQUE, LE **Vol. 2** 228
CODE CRIMINEL **Vol. 2** 388
CODIN **Vol. 3** 218, 297
CODINE **Vol. 2** 866
CODO CON CODO **Vol. 8** 243, 562
COEUR A L'ENVERS, LE **Vol. 2** 713 **Vol. 6** *page 114*
COEUR ARDENT **Vol. 2** 159, 286, 330
COEUR BATTANT, LE **Vol. 2** 566, 638, 673
COEUR DE COQ **Vol. 2** 347, 871
COEUR DE GUEUX **Vol. 7** 16, 1159
COEUR DE LILAS **Vol. 2** 206, 676
COEUR DE PARIS, LE **Vol. 2** 387, 469
COEUR DE POUPEE **Vol. 2** 798
COEUR DES GUEUX, LE **Vol. 2** 50
COEUR FIDELE **Vol. 2** 246, 380, 805
COEUR FROID, LE **Vol. 2** 739
COEUR SUR LA MAIN, LE **Vol. 2** 460, 627
COEUR SUR MER **Vol. 2** 344, 475
COEURS JOYEUX **Vol. 2** 778
COEURS VERTS, LES **Vol. 2** 49
COFANETTO DEI MILIONI, IL **Vol. 7** 131
COFFRET DE JADE, LE **Vol. 2** 511
COFFRET DE LAQUE, LE **Vol. 2** 205
COGAGNE **Vol. 2** 402, 650
COIFFEUR POUR DAMES **Vol. 2** 226, 347, 807 **Vol. 7** 53
COIN TRANQUILLE, LE **Vol. 2** 131, 344
COINCIDENCES **Vol. 2** 194, 407
COLCHICO E LA ROSA, IL **Vol. 7** 441
COLECCIONISTA DE CADAVERES, EL **Vol. 8** 616, 719
COLEGAS **Vol. 8** 297, 349, 660
COLEI CHE SI DEVE SPOSARE **Vol. 7** 26, 701
COLEI CHE TUTTO SOFFRE **Vol. 7** 254, 420, 1038
COLERA DEL VIENTO, LA **Vol. 8** 292, 428
COLERE DES DIEUX, LA **Vol. 2** 70, 153
COLERE FROIDE **Vol. 2** 419, 698
COLIBRI **Vol. 4** 190, 290, 584
COLINA DE LOS PEQUENOS DIABLOS, LA **Vol. 8** 396, 772
COLLANA DEL MILIONE, LA **Vol. 7** 493
COLLANTS NOIRS, LES **Vol. 2** 13, 184
COLLECTION MENARD **Vol. 2** 270, 833
COLLECTIONNEUSE, LA **Vol. 2** 17
COLLECTIONS PRIVEES **Vol. 2** 315
COLLEGE **Vol. 7** 30, 287, 383, 1127
COLLEGE EN FOLIE, LE **Vol. 2** 131, 871
COLLEGIALE, LA **Vol. 7** 996, 1189

COLLEGIENNES, LES **Vol. 2** 97, 158, 205, 516, 808
COLLEURS D'AFFICHES **Vol. 2** 678
COLLIER DE CHANVRE, LE **Vol. 2** 273, 533
COLLIER DE LA REINE, LE **Vol. 2** 48, 81, 438
COLLIER DE MIMI PINSON, LE **Vol. 2** 798
COLLINA DEGLI STIVALI, LA **Vol. 7** 738
COLLINES DU PLAISIR, LES **Vol. 6** *page 110, page 135*
COLMENA, LA **Vol. 8** 297, 438
COLOCADAS, LAS **Vol. 8** 524
COLOMBA **Vol. 2** 124, 407, 451, 540, 717 **Vol. 4** 159, 681
COLONA INFAME, LA **Vol. 7** 19
COLONEL CHABERT, LE **Vol. 2** 197, 488
COLONEL DURAND, LE **Vol. 2** 247, 558
COLONEL EST DE LA REVUE, LE **Vol. 2** 254, 496
COLONIALSKANDAL **Vol. 4** 626, 634, 720, 734
COLONNELLO BRIDEAU, IL **Vol. 7** 1004, 1103
COLONNELLO CHABERT, IL **Vol. 7** 436, 569
COLORIN **Vol. 8** 628
COLOSO DE RODAS, EL **Vol. 8** 396, 438, 692
COLOSSO DI RODI, IL **Vol. 8** 396, 438, 692
COLOSSO DI ROMA, IL **Vol. 7** 1121
COLOSSO VENDICATORE, IL **Vol. 7** 108
COLPA DEL MORTO, LA **Vol. 7** 668
COLPA DI UNA MADRE, LA **Vol. 7** 528
COLPA ED ESPIAZIONE **Vol. 7** 731
COLPA VENDICA LA COLPA, LA **Vol. 7** 250
COLPE DEI PADRI, LE **Vol. 7** 1189
COLPEVOLI, I **Vol. 7** 915
COLPI DI LUCE **Vol. 7** 834
COLPI DI TIMONE **Vol. 7** 16, 403, 756
COLPIRE AL CUORE **Vol. 7** 381, 819, 1112
COLPITA DA IMPROVVISO BENESSERE **Vol. 7** 1086
COLPO DI FULMINE **Vol. 7** 70, 97, 664
COLPO DI PISTOLA, UN **Vol. 7** 120, 750, 756, 1116
COLPO DI SCENA, UN **Vol. 7** 938
COLPO DI VENTO, UN **Vol. 7** 63, 147, 1121, 1159
COLPO GROSSO ... GROSSISSIMO ... ANZI PROBABILE **Vol. 8** 354, 611
COLPO, IL **Vol. 8** 438
COLPO MAESTRO AL SERVIZIO DI SUA MAESTA BRITANNICA **Vol. 8** 306, 708
COLPO ROVENTE **Vol. 7** 1025
COLPO SEGRETO DI D'ARTAGNAN, IL **Vol. 7** 1054
COLT E LA MIA LEGGE, LA **Vol. 8** 495, 541
COLT POR CUATRO CIRIOS, UN **Vol. 8** 261, 687
COLUI CHE HA TUTTO PERDUTO **Vol. 7** 531
COLUI CHE SEPPE AMARE **Vol. 7** 574, 1004
COLUMBINE **Vol. 4** 320, 454
COLUMNA **Vol. 3** 297, 306, 319
COMANCHE BLANCHO **Vol. 8** 402, 706
COMANDO AL INFIERNO **Vol. 8** 772
COMANDO DE ASESINOS **Vol. 8** 602, 772
COMANDO SUICIDA **Vol. 8** 749
COMBAT DANS L'ILE **Vol. 2** 285, 517
COME CANI ARRABBIATA **Vol. 7** 13
COME CI SPOSAMMO **Vol. 7** 413
COME CONCLUDE AMORE **Vol. 7** 1055
COME DONNA OMBROGLIA, COSI SBROGLIA **Vol. 7** 26
COME FILO DI SPADA **Vol. 7** 320
COME IN QUEL GIORNO **Vol. 7** 1038
COME IO VI AMO **Vol. 7** 787
COME LE FOGLIE **Vol. 7** 37, 1116
COME PERDERE UNA MOGLIE ... E TROVARE UN'AMANTE **Vol. 7** 1007
COME PERSI LA GUERRA **Vol. 7** 1136
COME PRESI MOGLIE **Vol. 7** 1189
COME SCOPERSI L'AMERICA **Vol. 7** 16
COME SI AMA **Vol. 7** 1189
COME SI FA UN CANDIDATO **Vol. 7** 1189
COME TE MOVI TE FULMINO ! **Vol. 7** 1136
COME TI CHIAMI, AMORE MIO ? **Vol. 7** 139
COME TRANZUILLO ENTRO IN SOCIETA **Vol. 7** 277
COME UNA ROSA AL NASO **Vol. 7** 585
COMEDIA D'AMOUR **Vol. 2** 49
COMEDIA FANTASTICA **Vol. 3** 218, 265, 270, 286
COMEDIE DU BONHEUR, LA **Vol. 2** 237, 538, 615
COMEDIEN, LE **Vol. 2** 722, 833
COMEDIENS, LES **Vol. 2** 224, 457
COMICOS **Vol. 8** 259, 295, 752
COMINCERA TUTTO UN MATTINO: IO DONNA, TU DONNA **Vol. 7** 289
COMISAR ACUZA, UN **Vol. 3** 218, 272
COMISARIO, EL **Vol. 8** 299
COMISSARIATO DI NOTTURNA **Vol. 7** 949
COMISSARIO DE POLICIA, O **Vol. 8** 18, 45, 96
COMIZI D'AMORE **Vol. 7** 128, 130, 333, 335, 457

COMMANDO DE DESTRUCTION **Vol. 2** 423
COMMANDO DES CHAUDS LAPINS, LE **Vol. 2** 484
COMMARE SECCA, LA **Vol. 7** 820, 1083
COMME DEUX GOUTTES D'EAU **Vol. 2** 208
COMME LA LUNE **Vol. 2** 198
COMME UN BOOMERANG **Vol. 2** 351, 400
COMME UN CHEVEU SUR LA SOUPE **Vol. 2** 137, 205, 808
COMME UN ECLAIR **Vol. 2** 221, 860
COMME UN POISSON DANS L'EAU **Vol. 2** 571, 668
COMME UN POT DE FRAISES **Vol. 2** 208
COMME UNE FEMME **Vol. 2** 309
COMMEDIA DAL MIO PALCO, LA **Vol. 7** 390
COMMENT DRAGUER TOUTES LES FILLES **Vol. 6** *page 72*
COMMENT EPOUSER UN PREMIER MINISTRE **Vol. 2** 285
COMMENT LES SEDUIRE **Vol. 2** 447
COMMENT ON DEVIENT UN ENNEMI DE L'INTERIEUR **Vol. 2** 186, 502, 844
COMMENT PASSER SON PERMIS DE CONDUIRE **Vol. 2** 594
COMMENT QU'ELLE EST **Vol. 2** 615
COMMENT REUSSIR DANS LA VIE...QUAND ON EST C...ET PLEURNICHARD **Vol. 2** 293
COMMENT REUSSIR EN AMOUR **Vol. 2** 457
COMMENT SE FAIRE REFORMER **Vol. 2** 76, 130
COMMENT SE FAIRE VIRER DE L'HOSTO **Vol. 2** 721
COMMENT TROUVEZ-VOUS MA SOEUR ? **Vol. 2** 548, 668
COMMESSA, LA **Vol. 7** 611
COMMISSAIRE SAN ANTONIO **Vol. 2** 642, 815
COMMISSARIO, IL **Vol. 7** 1189
COMMISSARIO PEPE, IL **Vol. 7** 318
COMMISSARIO VERRAZZANO, IL **Vol. 7** 909
COMMUNALE, LA **Vol. 2** 268, 836
COMMUNION SOLENNELLE, LA **Vol. 2** 738, 739
COMO CASARSE EN SIETE DIAS **Vol. 8** 553, 749
COMO DOS GOTAS DE AGUA **Vol. 8** 214, 261
COMO ESTA EL SERVICIO ! **Vol. 8** 553, 749
COMO LA TIERRA **Vol. 8** 449, 704
COMO ROBAR UN QUINTAL DE DIAMANTES EN RUSIA **Vol. 8** 598, 626
COMO SOIS LAS MUJERES **Vol. 8** 524, 725
COMO TE AMO ! **Vol. 8** 611, 768
COMOARA **Vol. 6** *page 267, page 281*
COMOARA DIN VADUL VECHI **Vol. 3** 244, 250, 329
COMOARA FURATA **Vol. 3** 236
COMPAGNA DI BIANCO, LA **Vol. 7** 947
COMPAGNA DI VIAGGIO, LA **Vol. 7** 728
COMPAGNES DE LA NUIT, LES **Vol. 2** 9, 402
COMPAGNI, I **Vol. 7** 498, 988
COMPAGNIA DEI MATTI, LA **Vol. 7** 162, 1116
COMPAGNONS DE LA MARGUERITE, LES **Vol. 2** 475
COMPANEROS, LOS **Vol. 8** 331, 762
COMPANYS PROCES A CATALUNYA **Vol. 8** 611
COMPARTIMENT DE DAMES SEULES **Vol. 2** 579, 644, 722, 803
COMPARTIMENT TUEURS **Vol. 2** 598
COMPERES, LES **Vol. 6** *page 69, page 114*
COMPITO DI PERINO, IL **Vol. 7** 1189
COMPLICATO INTRIGO DI DONNE, VICOLI E DELITTI **Vol. 7** 592, 607, 887
COMPLICE MUTA, LA **Vol. 7** 37
COMPLOTTO CONTRO ROBINET, UN **Vol. 7** 1189
COMPRESSO EROTICO, IL **Vol. 7** 800
COMPTES A REBOURS **Vol. 2** 20, 21, 835
COMTE DE MONTE-CRISTO, LE **Vol. 2** 32, 185, 226, 251, 407, 423, 625, 722, 738
COMTE KOSTIA, LE **Vol. 2** 79, 528
COMTE OBLIGADO, LE **Vol. 2** 339, 688, 775
COMTESSE DODDY **Vol. 4** 560, 663
COMUNE SENSO DEL PUDORE, IL **Vol. 7** 599, 913, 1007
CON AMORE **Vol. 1** 202, 221, 232
CON EL CORAZON EN LA GARGANTA **Vol. 8** 598
CON EL CULO AL AIRE **Vol. 8** 297, 730
CON EL VIENTO SOLANO **Vol. 8** 258
CON ELLA LLEGO EL AMOR **Vol. 8** 306, 461
CON LA MUSICA A OTRA PARTE **Vol. 8** 600
CON LA RABBIA AGLI OCCHI **Vol. 7** 834
CON LA VIDA HICIERON FUEGO **Vol. 8** 477, 772
CON LA ZIA NON E PECCATO **Vol. 7** 1152
CON L'AMORE E CON L'ALA **Vol. 7** 339, 993
CON LAS BRAGAS EN LA MANO **Vol. 8** 366, 632

CON LE DONNE NON SI SCHERZA **Vol. 7** 792, 1036
CON LOS OJOS DEL ALMA **Vol. 8** 628, 772
CON QUALE AMORE, CON QUANTO AMORE **Vol. 7** 524
CON UNAS Y DIENTES **Vol. 8** 249, 269
CONCENTRATION, LA **Vol. 2** 312, 331
CONCERTO **Vol. 7** 505
CONCERTO ME POLYBOLO **Vol. 3** 205
CONCIERGE, LE **Vol. 2** 368
CONCIERTO MAGICO **Vol. 8** 493, 732
CONCORDE AFFAIRE '79 **Vol. 7** 1202
CONCORRENTE, IL **Vol. 7** 1114
CONCURS **Vol. 6** *page 271, page 278*
CONDAMNE A MORT S'EST ECHAPPE, UN **Vol. 2** 153, 170, 702
CONDAMNES, LES **Vol. 2** 15, 501
CONDANNATELO ! **Vol. 7** 603
CONDE DE MARAVILLAS, EL **Vol. 8** 516
CONDE DRACULA, EL **Vol. 8** 546
CONDE SANDORF, EL **Vol. 8** 611
CONDE, UN **Vol. 2** 732
CONDENADOS **Vol. 8** 276, 418
CONDENADOS A VIVIR **Vol. 8** 348
CONDESA MARIA, LA **Vol. 8** 357, 436
CONDOTTIERI **Vol. 4** 54, 215, 253, 531, 541 **Vol. 7** 709, 906
CONDUISEZ-MOI, MADAME **Vol. 2** 573, 708
CONDUITE A GAUCHE **Vol. 2** 515, 668
CONFEDERACAO, A **Vol. 8** 134, 169
CONFESSION DE MINUIT, LA **Vol. 2** 770
CONFESSIONE **Vol. 7** 360, 910
CONFESSIONE DI UN FIGLIO DEL SECOLO, LE **Vol. 7** 598, 679
CONFESSIONI DI UNA DONNA, LE **Vol. 7** 665, 978
CONFESSIONS TRES INTIMES D'UNE PETITE FILLE **Vol. 2** 294
CONFIDENCES D'UNE PETITE CULOTTE, LES **Vol. 2** 294
CONFIDENCES EROTIQUES D'UN LIT TROP ACCUEILLANT **Vol. 2** 154, 503, 818
CONFIDENCES POUR CONFIDENCES **Vol. 2** 455
CONFIDENCIA **Vol. 8** 298, 484
CONFIDENCIAS DE UN MARIDO **Vol. 8** 258, 749
CONFIDENT DE CES DAMES, LE **Vol. 2** 807
CONFIDENTIAL REPORT ON MR. ARKADIN **Vol. 2** 124
CONFLICTO INESPERADO **Vol. 8** 721
CONFLIT **Vol. 2** 47, 194, 646, 863
CONFORMISTA, IL **Vol. 7** 1101
CONFUSIONE **Vol. 7** 607
CONGIUNTURA, LA **Vol. 7** 408
CONGIURA DEI BORGIA, LA **Vol. 7** 557
CONGIURA DEI FIESCHI, LA **Vol. 7** 1032, 1071
CONGIURA DI SAN MARCO, LA **Vol. 7** 491, 879, 1190
CONGO VIVO **Vol. 7** 46, 990
CONGREGA DEI VENTIQUATTRO, LA **Vol. 7** 510
CONGRES DES BELLES-MERES, LE **Vol. 2** 404
CONGRES S'AMUSE, LE **Vol. 2** 399
CONGRESO EN SEVILLA **Vol. 8** 214, 611
CONNOTATI DEL BANDITO, I **Vol. 7** 1189
CONQUERANTS SOLITAIRE, LES **Vol. 2** 124, 347
CONQUEST **Vol. 7** 623
CONQUETE DE DON JUAN, LA **Vol. 2** 159, 286, 330
CONQUISTA DE ALBANIA, LA **Vol. 8** 299, 539
CONQUISTA DE LA TIERRA PERDIDA, LA **Vol. 8** 762
CONQUISTA DEFICIL, UNA **Vol. 8** 628
CONQUISTA DEI DIAMANTI, LA **Vol. 7** 796
CONQUISTA DEL PACIFICO, LA **Vol. 8** 583, 772
CONQUISTA DELL'ARIA, LA **Vol. 7** 359, 709
CONQUISTADORES, LES **Vol. 2** 51
CONQUISTATORE D'ORIENTE **Vol. 7** 1147
CONQUISTATORI D'ANIME **Vol. 7** 133, 1189
CONQUISTATORI DEL MONDO, I **Vol. 7** 196
CONQUISTATORI, I **Vol. 7** 787
CONSEJERO, EL **Vol. 8** 603, 695
CONSIGLIORI, I **Vol. 8** 603, 695
CONSIGLIORI, IL **Vol. 7** 742
CONSIGNA : MATAR AL COMANDANTE EN JEFE **Vol. 8** 749
CONSIGNA : TANGER 67 **Vol. 8** 306, 772
CONSOMMATEUR MALADROIT, LE **Vol. 2** 678
CONSORTE SCACCIATO, IL **Vol. 7** 1078, 1090
CONSPIRATIA **Vol. 3** 269, 281, 319
CONSTANCE AUX ENFERS **Vol. 2** 819 **Vol. 8** 276, 732
CONSTANS **Vol. 6** *page 206, page 238, page 259*
CONSUELITA **Vol. 7** 701, 725
CONSULTARE A MR. BROWN **Vol. 8** 311, 579
CONTACT **Vol. 7** 763
CONTACT – CONTRATTO CARNALE **Vol. 7** 486

CONTACTOS **Vol. 8** 27, 49, 169
CONTAMINATION – ALIEN ARRIVA SULLA TERRA **Vol. 7** 901
CONTE AQUILA, IL **Vol. 7** 196
CONTE DI BRECHARD, IL **Vol. 7** 650, 709, 1177
CONTE DI MATERA, IL **Vol. 7** 1121
CONTE DI MONTECRISTO, IL **Vol. 7** 1189
CONTE MAX, IL **Vol. 7** 798
CONTE TACCHIA, IL **Vol. 7** 381, 549, 599
CONTE UGOLINO, IL **Vol. 7** 161, 750, 891
CONTES DE LA FONTAINE, LES **Vol. 2** 823
CONTES DES MILLE ET UNE NUITS, LES **Vol. 2** 361, 521
CONTES IMMORAUX **Vol. 2** 215, 275, 850, 880
CONTES PERVERS **Vol. 2** 234
CONTESSA AZZURRA, LA **Vol. 7** 910, 1131
CONTESSA CASTIGLIONE, LA **Vol. 7** 453, 796, 910, 1060
CONTESSA DI PARMA, LA **Vol. 7** 725, 748, 864, 876
CONTESSA FEDRA, LA **Vol. 7** 420
CONTESSA SARA, LA **Vol. 7** 264, 701, 725
CONTESSA SCALZA, LA **Vol. 7** 506
CONTESSINA **Vol. 7** 493
CONTESSINA CHIMERA, LA **Vol. 7** 306
CONTESTAZIONE GENERALE **Vol. 7** 1007
CONTINENTAL CIRCUS **Vol. 2** 350, 411, 589
CONTO CORRENTE **Vol. 7** 1189
CONTO E CHIUSO, IL **Vol. 7** 335
CONTRA LA PARED **Vol. 8** 331, 379, 481
CONTRABBANDIERI DEL MARE, I **Vol. 7** 1063
CONTRASTO D'ANIME **Vol. 7** 69
CONTRE-ENQUETE **Vol. 2** 67, 137, 440, 615
CONTREBANDIERES, LES **Vol. 2** 818
CONTRO CORRENTE **Vol. 7** 932
CONTRO L'ALTRO ... PRATICAMENTE AMICI, UNO **Vol. 7** 310
CONTRO QUATTRO BANDIERE **Vol. 7** 20
CONTROLLORE DEI VAGONI-LETTO, IL **Vol. 7** 48, 520
CONTROSESSO **Vol. 7** 86, 567
CONTROSESSO – COCAINA DI DOMENICA **Vol. 7** 401
CONTROSESSO – IL PROFESSORE **Vol. 7** 246, 514
CONTROSESSO – UNA DONNA D'AFFARI **Vol. 7** 241
CONVERSA ACABADA **Vol. 8** 7, 89
CONVOIUL **Vol. 6** *page 271, page 283*
COPAINS, LES **Vol. 2** 47, 185
COPAK JE TO ZA VOJAKA ... **Vol. 10** 408, 495, 497
COPIA CERO **Vol. 8** 376, 672
COPIE CONFORME **Vol. 2** 388, 826
COPKILLER **Vol. 7** 208, 901
COPLA ANDALUZA, LA **Vol. 8** 744
COPLAN AGENT SECRET FX 18 **Vol. 2** 347, 372 **Vol. 8** 258
COPLAN AU SA PEAU **Vol. 2** 234, 517
COPLAN FX 18 CASSE TOUT **Vol. 2** 570, 659
COPLAN OUVRE LE FEU A MEXICO **Vol. 2** 340, 795
COPLAN PREND DES RISQUES **Vol. 2** 548, 661
COPPELIUS **Vol. 2** 528
COQ DU REGIMENT, LE **Vol. 2** 63, 731
COQ EN PATE **Vol. 2** 32, 332
COQUECIGROLE **Vol. 2** 279, 407, 824
COQUELUCHE **Vol. 8** 249, 406
COQUELUCHE, LA **Vol. 2** 517
COQUILLE ET LE CLERGYMAN, LA **Vol. 2** 648
COQUIN, UN **Vol. 2** 104
COR 'E FRATE **Vol. 7** 829
CORAGGIO, IL **Vol. 7** 451
CORAGGIO SPRECATO **Vol. 7** 1189
CORALIE ET CIE **Vol. 2** 153, 279
CORAZON **Vol. 8** 655
CORAZON DE CRISTAL **Vol. 8** 215, 762
CORAZON DE PAPEL **Vol. 8** 616, 681
CORAZON DE REINA **Vol. 8** 278
CORAZON DEL BOSQUE, EL **Vol. 8** 358, 580
CORAZON SOLITARIO **Vol. 8** 219, 688
CORAZZIERE, IL **Vol. 7** 693
CORBEAU, LE **Vol. 2** 23, 395
CORBILLARD DE JULES, LE **Vol. 6** *page 94*
CORDA AL COLLO, LA **Vol. 7** 91, 980
CORDA D'ACCIAIO, LA **Vol. 7** 1063
CORDA DELL'ARCO, LA **Vol. 7** 1189
CORDE AU COU, LA **Vol. 2** 251, 377
CORDE RAIDE, LA **Vol. 2** 348, 372
CORDE, UN COLT, UNE **Vol. 2** 548, 659
CORDON-BLEU **Vol. 2** 41, 579, 871
CORE E MAMMA ! **Vol. 7** 298
CORE FURASTIERO **Vol. 7** 298
CORE MIO **Vol. 7** 762
CORE 'NGRATO **Vol. 7** 16
CORIGENTA DOMNULUI PROFESOR **Vol. 3** 241, 244, 260, 281, 313, 329
CORINNA SCHMIDT **Vol. 1** 58, 78, 129
CORISTA, LA **Vol. 8** 583
CORKA GENERALA PANKRATOWA **Vol. 1** 278, 308, 346
CORKA KANTORA **Vol. 1** 317
CORKA PANI X **Vol. 1** 174, 176, 326

CORLEONE Vol. 7 122
CORNETTI ALLA CREMA Vol. 7 116, 426
CORNIAUD, LE Vol. 2 224, 347
CORNUTO Vol. 7 1189
COROLLA DI SANGUE, LA Vol. 7 490
CORONA DI FERRO, LA Vol. 7 359, 709, 1060, 1177
CORONA NEGRA, LA Vol. 8 214, 261, 477
CORPI NUDI Vol. 7 713
CORPI PRESENTANO TRACCE DI VIOLENZA CARNALE, I Vol. 7 208, 426
CORPO DELLA RAGAZZA, IL Vol. 7 29, 1007
CORPS A COEUR Vol. 2 804
CORPS A CORPS Vol. 2 735
CORPS A SES RAISONS, LES Vol. 2 692
CORPS BRULANTS Vol. 2 210
CORPS DE MON ENNEMI, LE Vol. 2 457, 655
CORRE GITANO Vol. 8 297, 373, 376
CORREO DE INDIAS Vol. 2 646 Vol. 8 298, 607, 692
CORREO DEL REY Vol. 8 515, 661
CORRERIAS DEL VIZCONDE ARNAU, LAS Vol. 8 317, 732
CORRI COME IL VENTO KIKO Vol. 7 601
CORRIDA POUR UN ESPION Vol. 2 11, 294, 773 Vol. 8 214
CORRIDOR OF MIRRORS Vol. 2 826
CORRIERE DEL RE, IL Vol. 7 1054
CORRIERE DELL'IMPERATORE, IL Vol. 7 1189
CORRIERE DI FERRO, IL Vol. 7 1063
CORRUZIONE, LA Vol. 7 86, 300
CORSA AL SEPOLCRO, LA Vol. 7 1116
CORSA AL TRONO, LA Vol. 7 264, 701
CORSA ALL'ABISSO Vol. 7 796, 807
CORSAIRE, LE Vol. 2 676
CORSAIRES DU BOIS DE BOULOGNE, LES Vol. 2 661
CORSARA, LA Vol. 7 399
CORSARIO, EL Vol. 8 396
CORSARIO NEGRO, EL Vol. 8 317, 732
CORSARIOS DEL CARIBE, LOS Vol. 8 311, 475
CORSARIOS, LOS Vol. 8 603, 749
CORSARO DELLA MEZZA LUNA, IL Vol. 7 17
CORSARO, IL Vol. 7 751
CORSARO NERO, IL Vol. 7 88, 91, 258, 283, 650, 1086, 1172, 1177, 1195
CORSE A PERDICUORE Vol. 7 139, 842
CORSICA Vol. 2 473, 490
CORTE D'ASSISE Vol. 7 48, 756, 1116
CORTE DE FARAON, LA Vol. 8 219, 591
CORTE NOTE DI FARFALEA, LA Vol. 3 437
CORTEGE IMPERIAL DU TZAR Vol. 2 678
CORTILE Vol. 7 48, 361, 792, 1116, 1121
CORTOCIRCUITO Vol. 7 187, 443, 838
COSA BUFFA, LA Vol. 7 524
COSACCHI, I Vol. 7 372
COSCIENZA Vol. 7 131
COSCIENZA DEL DIAVOLO, LA Vol. 7 279
COSE DA PAZZI Vol. 7 106
COSE DELL'ALTRO MONDO Vol. 7 16, 756
COSETTA ... COSETTA Vol. 7 298
COSI COME SEI Vol. 7 20, 1035
COSI DOLCE, COSI PERVERSA Vol. 7 694
COSI E LA VITA Vol. 7 621
COSI PARLO BELLAVISTA Vol. 7 750, 1093, 1165
COSI SIA Vol. 7 426
COSMIN FIUL ZIMBRULUI Vol. 3 311
COSMOPOLIS Vol. 7 1057
COSMOS MORTAL Vol. 8 237, 756
COSTA AZZURRA Vol. 7 17, 1054
COSTA DE AFRICA, O Vol. 8 20, 80, 131
COSTA DO CASTELO, O Vol. 8 25, 61, 67, 122
COSTANTINO IL GRANDE Vol. 7 372
COSTANZA DELLA RAGIONE, LA Vol. 7 567
COSTAUD DES BATIGNOLLES, LE Vol. 2 205, 776
COSTAUD DES EPINETTES, LE Vol. 2 42, 547, 805, 825
COSTAUD DES P.T.T., Le Vol. 2 833
COTALAY Vol. 8 602
COTE D'AMOUR, LA Vol. 6 *page 95, page 132*
COTO DE CAZA Vol. 8 349, 580
COUCHE DE LA MARIEE, LE Vol. 2 70, 727
COUCHE-MOI DANS LE SABLE ET FAIS JAILLIR TON PETROLE Vol. 2 692
COUCHER DE LA MARIEE, LE Vol. 2 416, 678
COUER EBLOUI, LE Vol. 2 6, 249
COULEES D'AMOUR Vol. 2 396
COULIN E MARIETTA Vol. 7 1189
COUP DE BAMBOU Vol. 2 347, 566
COUP DE FEU DANS LA NUIT Vol. 2 560, 595
COUP DE FOUDRE Vol. 2 74 Vol. 6 *page 75, page 87*
COUP DE MISTRAL, UN Vol. 2 63, 404
COUP DE ROUGE, UN Vol. 2 604, 708
COUP DE SIROCCO, LE Vol. 2 739
COUP DE TELEPHONE, UN Vol. 2 290, 526, 573, 775
COUP DE TETE Vol. 2 7, 197, 568

COUP DE TORCHON **Vol. 2** 351 **Vol. 6** *page 99*
COUP DE VENT **Vol. 7** 63, 147, 1121
COUP DU PARAPLUIE, LE **Vol. 2** 224 **Vol. 6** *page 70, page 95*
COUP DU SINGE, LE **Vol. 2** 301
COUP DUR CHEZ LES MOUS **Vol. 2** 118, 848
COUPABLE ? **Vol. 2** 15, 36
COUPABLE, LE **Vol. 2** 59, 159, 488, 657
COUPLE CHERCHE ESCLAVE SEXUEL **Vol. 2** 735
COUPLE DEBUTANT CHERCHE COUPLE INITIE **Vol. 2** 52
COUPLE IDEAL, LE **Vol. 2** 270
COUPLE MODERNE, UN **Vol. 2** 2
COUPLE TEMOIN, LE **Vol. 2** 754
COUPLES COMPLAISANTS **Vol. 2** 500
COUPLES COMPLICES **Vol. 2** 349
COUPS DE FEU **Vol. 2** 41, 407, 540
COUPS DE ROULIS **Vol. 2** 205, 488
COUPS POUR RIEN, LES **Vol. 2** 760
COURAGE FUYONS **Vol. 2** 434, 450, 478
COURRIER, BAIN ET COUCHER DE LA PARISIENNE **Vol. 2** 678
COURRIER D'ASIE **Vol. 2** 37, 706
COURRIER-SUD **Vol. 2** 58, 59, 488, 644
COURS APRES MOI QUE JE T'ATTRAPE **Vol. 2** 275
COURS DU SOIR POUR MONSIEUR SEUL **Vol. 2** 210
COURSE A LA VERTU, LA **Vol. 2** 330, 871
COURSE A L'AMOUR, LA **Vol. 2** 54, 159, 286, 330
COURSE A L'ECHALOTTE, LA **Vol. 2** 142, 224, 791
COURSE DE TAUREAUX, LA **Vol. 2** 224
COURSE DU BAT, LA **Vol. 2** 739
COURSE DU FLAMBEAU, LA **Vol. 2** 216
COURSE DU LIEVRE A TRAVERS LES CHAMPS **Vol. 2** 377, 728
COURSE EN TETE, LA **Vol. 2** 51, 522
COURTE-TETE **Vol. 2** 197, 247, 726
COURTS-CIRCUITS **Vol. 2** 534 **Vol. 6** *page 86*
COUSIN COUSINE **Vol. 2** 504
COUSIN PONS, LE **Vol. 2** 528
COUSINE BETTE, LA **Vol. 2** 316, 381
COUSINS, LES **Vol. 2** 224, 285, 700, 770
COUTEAU DANS LA PLAIE, LE **Vol. 2** 13, 837
COUTEAU SOUS LA GORGE, LE **Vol. 2** 131, 407, 709
COUTURIER DE CES DAMES, LE **Vol. 2** 318, 347, 807
COUTURIERE DE LUNEVILLE, LA **Vol. 2** 563, 579
COVEK I ZVER **Vol. 3** 371, 374
COVEK IZ HRASTOVE SUME **Vol. 3** 350, 391
COVEK NE S"M UBIVAL **Vol. 6** *page 31, page 60, page 63*
COVEK NIJE TICA **Vol. 3** 367, 418
COVEK SA CETIRI NOGE **Vol. 6** *page 316*
COVEK"T, KOJTO ZABRAVI BOGA **Vol. 3** 71
COVEK"T V SJANKA **Vol. 3** 31, 51, 75
COVER-GIRL **Vol. 2** 513
COVER GIRL **Vol. 8** 249, 699
COVIK OD SVITA **Vol. 3** 421, 444
COVJEK KOGA TREBA UBITI **Vol. 3** 350, 366
COVJEK SA FOTOGRAFIJE **Vol. 3** 391, 425
COWBOY-MILLIARDÄR, DER **Vol. 4** 628
COWBOYMILLIONAEREN **Vol. 5** 228
COX ES BOX **Vol. 9** 28
COYOTE, EL **Vol. 8** 749, 752
COZ TAKHLE DAT SI SPENAT **Vol. 10** 41, 425, 567
... COZES TY ZA PANI ... **Vol. 6** *page 223, page 230, page 258*
CRABE-TAMBOUR, LE **Vol. 2** 208
CRAC Y COMPANIA **Vol. 8** 516
CRACK DOS, EL **Vol. 8** 368, 681
CRACK, EL **Vol. 8** 580, 681
CRACKS, LES **Vol. 2** 19
CRAINQUEBILLE **Vol. 2** 48, 131, 153, 297, 308, 344, 615
CRAN D'ARRET **Vol. 2** 732
CRANATIERE DI POMERANIA, IL **Vol. 7** 679
CRANEUR, LE **Vol. 2** 294, 401
CRASH **Vol. 5** 716
CRAVACHE, LA **Vol. 2** 640
CRAZY HORSE DE PARIS **Vol. 2** 686
CREATORI DELL'IMPOSSIBILE **Vol. 7** 575
CREATURE DELLA NOTTE **Vol. 7** 493, 665
CREATURES, LES **Vol. 2** 441, 674
CREDITORES DISGRAZIATO **Vol. 7** 1189
CREMA CIOCCOLATO E PA ... PRIKA **Vol. 7** 207, 953, 1194
CREPUSCOLO TRISTE **Vol. 7** 1189
CRESCETE E MOLTIPLICATEVI **Vol. 7** 408
CRESUS **Vol. 2** 402, 819
CRETINETTI E GLI STIVALI DEL BRASILIANO **Vol. 7** 305
CREVE-COEUR **Vol. 2** 224
CRI DE LA CHAIR, LE **Vol. 2** 131, 198
CRI DU COEUR, LE **Vol. 2** 142, 513
CRI DU CORMORAN, LE SOIR, AU-DESSUS DES JONQUES, LES **Vol. 2** 279, 661
CRI DU DESIR, LE **Vol. 2** 442
CRIME AU CONCERT MAYOL **Vol. 2** 243, 259

CRIME D'AMOUR, LE **Vol. 6** *page 94, page 127*
CRIME DE ALDEIA VELHA, EL **Vol. 8** 127, 146, 173
CRIME DE DAVID LEVINSTEIN, LE **Vol. 2** 60
CRIME DE LORD ARTHUR SAVILLE, LE **Vol. 2** 614
CRIME DE MONSIEUR LANGE, LE **Vol. 2** 48, 160, 388
CRIME DE SIMAO BOLANDAS, O **Vol. 8** 96, 130
CRIME DE SYLVESTRE BONNARD, LE **Vol. 2** 407, 824
CRIME DES HOMMES, LE **Vol. 2** 136
CRIME DES JUSTES, LE **Vol. 2** 595
CRIME DU BOUIF, LE **Vol. 2** 407, 469, 603, 627, 661
CRIME DU CHEMIN ROUGE, LE **Vol. 2** 277, 407, 540
CRIME ET CHATIMENT **Vol. 2** 70, 96, 193, 526, 623, 719
CRIME, LA **Vol. 6** *page 78, page 82*
CRIME NE PAIE PAS, LE **Vol. 2** 863
CRIME OVER LONDON **Vol. 2** 32
CRIMEN **Vol. 7** 1167 **Vol. 8** 562, 768
CRIMEN DE CUENCA, EL **Vol. 8** 297, 508, 698
CRIMEN DE DOBLE FILO **Vol. 8** 725, 748
CRIMEN DE LA CALLE DE BORDADORES, EL **Vol. 8** 268, 298
CRIMEN DE PEPE CONDE, EL **Vol. 8** 400
CRIMEN EN EL ENTREACTO **Vol. 8** 438, 752
CRIMEN EN FAMILIA **Vol. 8** 414, 466
CRIMEN IMPERFECTO **Vol. 8** 524
CRIMEN IMPOSIBLE ? **Vol. 8** 214, 276
CRIMEN PARA RECIEN CASADOS **Vol. 8** 438, 762
CRIMEN TAMBIEN JUEGA, EL **Vol. 8** 611
CRIMENES DE PETIOT, LOS **Vol. 8** 555
CRIMES DE L'AMOUR - NINA DE VANGHEL, LES **Vol. 2** 571, 778
CRIMES DE L'AMOUR - RIDEAU CRAMOISI, LES **Vol. 2** 571, 778
CRIMINAL, LE **Vol. 2** 104
CRIMINE A DUE **Vol. 7** 1152
CRIMINEL **Vol. 2** 217, 466, 727
CRIN BLANC **Vol. 2** 781
CRIPTA E L'INCUBO, LA **Vol. 8** 772
CRIPTA, LA **Vol. 8** 362, 588, 621
CRISANTEMO MACCHIATO DI SANGUE, IL **Vol. 7** 133
CRISE EST FINIE, LA **Vol. 2** 200, 778
CRISI DEGLI ALLOGGI, LA **Vol. 7** 369
CRISI, LA **Vol. 7** 1057
CRISIS MORTAL **Vol. 8** 465, 775
CRISIS MUNDIAL **Vol. 8** 522, 548
CRISPINO E LA COMARE **Vol. 7** 360, 650, 941
CRISTIANA, MONACA INDEMONIATA **Vol. 7** 666
CRISTINA **Vol. 8** 732, 753
CRISTINA GUZMAN **Vol. 8** 261, 357, 436, 749, 772
CRISTINA Y LA RECONVERSION SEXUEL **Vol. 8** 243, 762
CRISTO **Vol. 8** 524
CRISTO DEL OCEANO, EL **Vol. 8** 277
CRISTO E'PASSATO SULL'AIA **Vol. 7** 657
CRISTO NEGRO **Vol. 8** 626, 752
CRISTO PROIBITO **Vol. 7** 722, 910, 1108
CRISTO SI E FERMATO A EBOLI **Vol. 7** 336, 1009, 1025
CRISTOBAL COLON DE OFICIO ... DESCUBRIDOR **Vol. 8** 338, 762
CRISTOFORO COLOMBO **Vol. 7** 299, 482, 524, 661, 750
CRITICO, IL **Vol. 7** 1189
CRITS SORDS **Vol. 8** 337
CRNE PTICE **Vol. 3** 348, 364
CRNI BISERI **Vol. 3** 338, 342
CRNI KOPLJANICI **Vol. 3** 391
CRNO SEME **Vol. 3** 384, 419
CROAZIERA **Vol. 3** 282
CROCE SENZA NOME, UNA **Vol. 7** 106
CROCETTA D'ORO **Vol. 7** 559
CROCIATA DEGLI INNOCENTI, LA **Vol. 7** 442, 493, 806
CROISADE, LA **Vol. 2** 37
CROISEE DES CHEMINS, LA **Vol. 2** 347, 826
CROISIERE JAUNE, LA **Vol. 2** 595
CROISIERE, LA **Vol. 2** 706
CROISIERE POUR COUPLES EN CHALEUR **Vol. 6** *page 79*
CROISIERE POUR L'INCONNU **Vol. 2** 8, 330
CROISIERES SIDERALES **Vol. 2** 407, 539
CROIX DE BOIS, LES **Vol. 2** 440, 657
CROIX DES VIVANTS, LA **Vol. 2** 47, 59
CROIX ET LA BANNIERE, LA **Vol. 2** 223, 259, 309
CROLLO, IL **Vol. 7** 968
CRONACA DI UN AMORE **Vol. 7** 443, 1031, 1045, 1062
CRONACA DI UN DELITTO **Vol. 7** 1121
CRONACA FAMILIARE **Vol. 7** 781, 988
CRONACA NERA **Vol. 7** 1116
CRONACHE DI POVERI AMANTI **Vol. 7** 60, 1167
CRONICA DE NUEVE MESES **Vol. 8** 553
CRONICA DE UN ATRACO **Vol. 8** 426, 732
CRONICA DE UN INSTANTE **Vol. 8** 556, 738

CRONICA DOS BONS MALANDROS **Vol. 8** 185
CRONICA SENTIMENTAL EN ROJO **Vol. 8** 754
CRONICON, EL **Vol. 8** 219, 299
CROQUE LA VIE **Vol. 6** *page 73, page 78*
CROQUEMITOUFLE **Vol. 2** 204, 501
CROQUEUSES, LES **Vol. 2** 139
CROSS COUNTRY **Vol. 3** 373, 433
CROSS OF IRON, THE **Vol. 6** *page 312*
CROSSWINDS, THE **Vol. 6** *page 312*
CROULANTS SE PORTENT BIEN, LES **Vol. 2** 249, 559, 661
CRUCERO BALEARES, EL **Vol. 8** 475, 625, 716
CRUCERO DE VERANO **Vol. 8** 207, 298
CRUDELE ATTESA **Vol. 7** 1189
CRUDELI, I **Vol. 8** 626
CRUELES, LAS **Vol. 8** 232, 768
CRUZ DE FERRO, A **Vol. 8** 130
CRUZ DE MAYO **Vol. 2** 775
CRUZ DEL DIABLO, LA **Vol. 8** 245, 742
CRUZADA EN LA MAR **Vol. 8** 249
CRVENA ZEMLJA **Vol. 3** 356, 418, 441
CRVENI CVET **Vol. 3** 391, 448
CRVENI I CRNI **Vol. 6** *page 308, page 320*
CRVENI SAL POVRATAK **Vol. 3** 347, 373, 376, 423
CRVENI UDAR **Vol. 3** 370, 402
CRVENIOT KONJ **Vol. 6** *page 312, page 317*
CRYSTAL CAGE **Vol. 8** 215, 762
CSABITO, A **Vol. 9** 315
CSAK EGY KISLANY VAN A VILAGON ! **Vol. 9** 308, 322
CSAK EGY MOZI **Vol. 9** 171, 253, 281
CSAK EGY TELEFON **Vol. 9** 67, 83, 183
CSAK NOVEL NE ! **Vol. 9** 175, 308
CSAK SEMMI PANIK ... **Vol. 9** 41, 54, 62
CSAKO ES KALAP **Vol. 9** 26, 117
CSALADI POTLEK **Vol. 9** 71, 322
CSALADI TÜZFESZEK **Vol. 9** 216, 238
CSALADUNK SZEGYENE **Vol. 9** 274, 290, 315
CSALODAS, A **Vol. 9** 26, 117, 133
CSAPLARNE A BETYART SZERETTE **Vol. 9** 77
CSARDASKIRALYNŐ **Vol. 9** 37, 71, 152, 316
CSASZAR KATONAI, A **Vol. 9** 220
CSASZAR PARANCSARA, A **Vol. 9** 11, 274, 326
CSEMPESZEK **Vol. 9** 117, 326, 332
CSEND ES KIALTAS **Vol. 9** 20, 154, 318
CSENDES OTTHON **Vol. 9** 26, 71, 215
CSEPLO GYURI **Vol. 9** 6
CSEREBERE **Vol. 9** 26, 117
CSEREPEK **Vol. 9** 179, 190
CSIGALEPCSÖ **Vol. 9** 83, 165, 274
CSIKOS, A **Vol. 9** 172
CSIKOSBARONESS, DIE **Vol. 4** 164, 208, 471, 584
CSILLAG A MAGLYAN **Vol. 9** 134, 215, 259
CSILLAGOSOK **Vol. 9** 134, 241, 274
CSILLAGOSOK, KATONAK **Vol. 9** 47, 273, 310
CSILLAGSZEMÜ, A **Vol. 9** 41, 67, 327
CSINOM PALKO **Vol. 9** 67, 199, 285
CSODA VEGE, A **Vol. 9** 170, 191, 204
CSODACSATAR, A **Vol. 9** 87, 199
CSODADOKTOR, A **Vol. 9** 26, 241, 308
CSODAGYEREK, A **Vol. 9** 172
CSOK, ANYU **Vol. 9** 152, 253, 259
CSOKOLJ MEG EDES ! **Vol. 9** 71, 322
CSONTVARY **Vol. 9** 140, 215, 316
CSUDAPEST **Vol. 9** 173, 211, 332
CSUNYA LANY, A **Vol. 9** 15, 322
CSUTAK ES A SZÜRKE LO **Vol. 9** 36, 122
CTRNACTY U STOLU **Vol. 10** 412, 621
CTYRI VRAZDY STACI, DRAHOUSKU **Vol. 10** 173, 425, 550
CU MIINILE CURATE **Vol. 3** 222, 223, 235
CUADRILATERO **Vol. 8** 706, 773
CUANDO ALMANZOR PERDIO EL TAMBOR **Vol. 8** 368, 474
CUANDO ESTALLO LA PAZ **Vol. 8** 298, 681
CUANDO LLEGUE LA NOCHE **Vol. 8** 276, 772
CUANDO LOS ANGELES DUERMEN **Vol. 8** 661, 721
CUANDO PASA EL AMOR **Vol. 8** 276, 400
CUANDO TU NO ESTAS **Vol. 8** 258, 338
CUANTO COBRA UN ESPIA ? **Vol. 8** 731
CUARENTA ANOS DE NOVIOS **Vol. 8** 544, 578
CUARENTA GRADOS A LA SOMBRA **Vol. 8** 553, 779
CUARENTA Y OCHO ORAS **Vol. 8** 400
CUARTA CARABELA, LA **Vol. 8** 551, 749
CUARTA VENTANA, LA **Vol. 8** 732, 753
CUATREROS **Vol. 8** 495, 752
CUATRO BALAZOS **Vol. 8** 752
CUATRO BODAS DE MARISOL, LAS **Vol. 8** 214, 261
CUATRO BODAS Y PICO **Vol. 8** 311, 461
CUATRO BUDAS DE KRIMINAL, LOS **Vol. 8** 396
CUATRO CABALGARON **Vol. 8** 603, 616
CUATRO DE FORT APACHE, LOS **Vol. 8** 252, 600
CUATRO DOLARES DE VENGANZA **Vol. 8** 562, 732
CUATRO EN LA FRONTERA **Vol. 8** 444, 470, 677, 697

CUATRO IMPLACABLES, LOS **Vol. 8** 551, 626
CUATRO MUJERES **Vol. 8** 276, 417
CUATRO MUJERES Y UN LIO **Vol. 8** 296, 770
CUATRO NOCHES DE BODA **Vol. 8** 331, 553
CUATRO ROBINSONES, LOS **Vol. 8** 278, 716, 724
CUATRO SALVAJES, LOS **Vol. 8** 598
CUATRO VERDADES, LAS **Vol. 8** 719, 766
CUBANA EN ESPANA, UNA **Vol. 8** 325, 717
CUBANA IN ESPANA, UNA **Vol. 2** 775
CUCA E MALEVE **Vol. 3** 1, 9, 10
CUCCAGNA, LA **Vol. 7** 150
CUD NAD WISLA **Vol. 1** 174, 176
CUDNA DEVOJKA **Vol. 3** 405, 455
CUDO NEVIDJENO **Vol. 6** *page 314*
CUDOTVORNI MAC **Vol. 3** 440
CUDOVITI PRAH **Vol. 3** 395, 417
CUENTO DE HADAS **Vol. 8** 298, 449
CUENTOS DE LA ALHAMBRA **Vol. 8** 311, 449
CUENTOS EROTICOS **Vol. 8** 245, 373, 459, 728
CUENTOS PARA UNA ESCAPADA **Vol. 8** 358, 738
CUERDA DE PRESOS **Vol. 8** 298
CUERPO A CUERPO **Vol. 8** 455, 672
CUERVOS, LOS **Vol. 8** 732
CUGINA D'ALCANTARA, LA **Vol. 7** 726
CUGINA, LA **Vol. 7** 1132
CUIBUL SALAMANDRELOR **Vol. 3** 214, 232, 246
CUIDADO CON LAS PERSONAS FORMALES **Vol. 8** 221, 753
CUIDADO CON LAS SENORAS **Vol. 8** 551
CUISINE AU BEURRE **Vol. 2** 402, 598
CUISSES EN CHALEUR **Vol. 2** 868
CUISSES INFERNALES **Vol. 2** 442
CUISSES OUVERTES **Vol. 2** 735
CUJ PETELA **Vol. 3** 121 **Vol. 6** *page 30, page 55*
CUKROVA BOUDA **Vol. 10** 87, 172, 517
CULASTRISCE NOBILE VENEZIANO **Vol. 7** 257
CULO E CAMICIA - IL TELEVEGGENTE **Vol. 7** 259, 450, 1007
CULO E CAMICIA - UN UOMO, UN UOMO E ... EVIVA UNA DONNA **Vol. 7** 29, 426, 1142
CULOTTES ROUGES, LES **Vol. 2** 515, 655
CULPA DEL OTRO, LA **Vol. 8** 396
CULPA FUE DE EVA, LA **Vol. 8** 276, 630
CULPABLE **Vol. 8** 614
CULPABLE PARA UN DELITO **Vol. 8** 562, 772
CULPABLES **Vol. 8** 602
CULPABLES, LAS **Vol. 8** 217, 470
CUMPANA **Vol. 6** *page 268*
CUMPARISTA, LA **Vol. 8** 495, 602
CUOR DI BAMBINO E CUOR DI SOLDATO - O - COSA TROSTE E LA GUERRA ! **Vol. 7** 274
CUOR DI FERRO E CUOR D'ORO **Vol. 7** 110
CUOR DI POETA **Vol. 7** 1189
CUOR DI SOLDATO **Vol. 7** 1189
CUOR DI VAGABONDO **Vol. 7** 16, 1159
CUORE **Vol. 7** 340, 599, 796, 1034
CUORE DELL'ALTRA, IL **Vol. 7** 148
CUORE DI CANE **Vol. 7** 221, 916, 931, 1069, 1141
CUORE DI MAMMA **Vol. 7** 16
CUORE DI MUSETTE, IL **Vol. 7** 131
CUORE DI ROMA, IL **Vol. 7** 250
CUORE DI SPIA **Vol. 7** 302
CUORE DI VAGABONDO **Vol. 7** 1189
CUORE DI ZINGARA **Vol. 7** 1206
CUORE E DOVERE **Vol. 7** 1189
CUORE NEL MONDO, UN **Vol. 7** 53
CUORE SEMPLICE, UN **Vol. 7** 1076, 1205
CUORE SOTTO IL MAGLIO, IL **Vol. 7** 26
CUORE, UN **Vol. 7** 940
CUORE, UN PUGNALE, UN CERVELLO, UN **Vol. 7** 940
CUORI E CASTE **Vol. 7** 825
CUORI E SENSI **Vol. 7** 731
CUORI E TUFFI **Vol. 7** 731
CUORI NELLA TORMENTA **Vol. 7** 408, 466, 697, 1136
CUORI SENZA FRONTIERE **Vol. 7** 796
CUORI SOLITARI **Vol. 7** 754
CUORICINO D'ORO **Vol. 7** 1187
CUPIDO CONTRABANDISTA **Vol. 8** 306, 483
CURA DE ALDEA **Vol. 8** 606
CURA DE ALDEA, EL **Vol. 8** 522
CURA YA TIENE HIJO, EL **Vol. 8** 728, 770
CURE DE SAINT-AMOUR, LE **Vol. 2** 243, 717
CUREE, LA **Vol. 2** 21, 719
CUREVOS, LOS **Vol. 8** 523
CURIOSA, LA **Vol. 8** 331, 627
CURIOSO IMPERTINENTE, EL **Vol. 8** 504, 772
CURITA CANON, UN **Vol. 8** 207, 284
CURPO A CUERPO **Vol. 8** 519
CURRA VELETA **Vol. 8** 207, 306
CURRANTE, EL **Vol. 8** 770, 773
CURRITO DE LA CRUZ **Vol. 8** 207, 214, 281, 357, 394, 449,
CURRO VARGAS **Vol. 8** 572, 657
CURSA **Vol. 3** 288 **Vol. 6** *page 267*
CUSTER OF THE WEST **Vol. 2** 526

CUSTODE SELVAGGIO, IL **Vol. 7** 1189
CUVAR PLAZE U ZIMSKOM PERIODOU **Vol. 3** 367, 415, 418
CUVARI ZAMKA **Vol. 3** 391
CVETJE V JESENI **Vol. 3** 395, 447
CVETNIKOV KONREC **Vol. 6** *page 305, page 316*
CYANKALI **Vol. 4** 359, 626
CYGANKA AZA **Vol. 1** 167, 404 **Vol. 4** 159
CYPRIEN NEURASTHENIQUE **Vol. 2** 159, 286, 330
CYRANO DE BERGERAC **Vol. 2** 48, 722
CYRANO ET D'ARTAGNAN **Vol. 2** 117, 249 **Vol. 8** 595
CYRANO Y D'ARTAGNAN **Vol. 8** 595
CYTRI V KRUHU **Vol. 10** 24, 313, 517
CZARCI ZLEB **Vol. 1** 169, 250
CZARDASFÜRSTIN, DIE **Vol. 2** 399 **Vol. 4** 113, 266, 288, 357
CZARNA KSIAZKA **Vol. 1** 215, 326
CZARNA KSIEZNA **Vol. 1** 176
CZARNA PERLA **Vol. 1** 278, 308, 404
CZARNE DIAMENTY **Vol. 1** 241, 290, 308, 365
CZARNE SKRZYDLA **Vol. 1** 281, 382 **Vol. 6** *page 242, page 247*
CZARODZIEJSKIE ZIARNO **Vol. 1** 298
CZAS DOJRZEWANIA **Vol. 6** *page 245, page 246, page 255*
CZAS PRZESZLY **Vol. 1** 179, 249, 398 **Vol. 6** *page 215*
CZATY **Vol. 1** 174, 404
CZEKAM W MONTE CARLO **Vol. 1** 220, 303, 305, 330
CZERWONE BERETY **Vol. 1** 356, 391 **Vol. 6** *page 255*
CZERWONE CIERNIE **Vol. 1** 197, 198, 203, 221 **Vol. 6** *page 242*
CZERWONE I BIALE **Vol. 1** 153, 293, 401, 409
CZERWONE I ZLOTE **Vol. 1** 179, 237
CZERWONE WEZE **Vol. 6** *page 227, page 237, page 247*
CZERWONY BLAZEN **Vol. 1** 346, 371
CZLOWIEK, KTORY ZDEMORALIZOWAL HADLEYBURG **Vol. 1** 305, 311, 388
CZLOWIEK NA TORZE **Vol. 1** 223, 250 **Vol. 6** *page 237*
CZLOWIEK O BLEKITNEJ DUSZY **Vol. 1** 252, 309
CZLOWIEK Z M-3 **Vol. 1** 303, 321, 389
CZLOWIEK Z MARMURU **Vol. 1** 204, 224, 313, 340 **Vol. 6** *page 254*
CZLOWIEK Z ZELAZA **Vol. 6** *page 241, page 254*
CZTEREJ PANCERNI I PIES **Vol. 1** 223, 227, 323
CZUJE SIE SWIETNIE **Vol. 6** *page 214*
CZULE MIEJSCA **Vol. 6** *page 234, page 237, page 244*
CZWARTKI UBOGICH **Vol. 6** *page 214, page 230*
CZY LUCYNA TO DZIEWCZYNA? **Vol. 1** 176, 278, 308

D

D-ALE CARNAVALULUI **Vol. 3** 265, 297, 306
D III 88 **Vol. 4** 311, 361, 422, 461
D-ZUG 13 HAT VERSPÄTUNG **Vol. 4** 89, 270, 409
DA HÄLT DIE WELT DEN ATEM AN **Vol. 4** 316, 340, 535
DA ICARO A DE PINEDO **Vol. 7** 1004
DA IZJADES JAB'LKATA **Vol. 3** 122
DA LI JE UMRO DOBAR COVJEK **Vol. 3** 366
DA LORD A DETECTIVE **Vol. 7** 304
DA QUI ALL'EREDITA **Vol. 7** 16
DA ROMA AL NIAGARA **Vol. 7** 390
DA SARTINA A BARONESSA **Vol. 7** 1189
DA STIMMT WAS NICHT **Vol. 4** 79, 115, 462
DA SVANTE FORSVANDT **Vol. 5** 179
DA TOFFELHELTEN GENERALSTREJKEDE **Vol. 5** 83
DA UN PAESE LONTANO : PAPA GIOVANNI PAOLO **Vol. 6** *page 236, page 238, page P253*
DAARSKAB DYD OG DRIVERTER **Vol. 5** 24, 83
DABELSKE LIBANKY **Vol. 10** 144, 359, 400
DABLOVA PAST **Vol. 10** 352, 425, 503
DACH ÜBERM KOPF **Vol. 6** *page 169, page 174, page 175*
DACII **Vol. 3** 227, 254, 306
DACTYLO **Vol. 2** 397
DACTYLO SE MARIE **Vol. 2** 697
DADA, A **Vol. 9** 73
D'AEKKE TE A TRU ! **Vol. 5** 676
DAFNIS KAI HLOE **Vol. 3** 142, 178
DAG I IVAN DENISSOVITSJ' LIV, EN **Vol. 5** 773, 797
DAG SKA GRY, EN **Vol. 5** 543, 570
DAGE I MIN FAR'S HUS **Vol. 5** 19, 76, 123
DAGER FRA 1000 AR – HVIS DU BAERER PA EN HJERTESORG **Vol. 5** 660, 680, 758, 817
DAGER FRA 1000 AR – MIRA **Vol. 5** 767, 817

DAGER FRA 1000 AR - VOKSE OPP Vol. 5 680, 797
DAGFIN Vol. 4 132, 397, 536, 629, 636, 788, 796
DAGLI APPENNINI ALLE ANDE Vol. 7 253, 420, 874, 1027, 1054, 1189
DAGNY Vol. 1 136, 303, 314 Vol. 5 664, 790
DAGOBERT Vol. 7 434, 817, 893
DAGS SIMPELT FAENGSEL, EN Vol. 5 97
DAGUERREOTYPES Vol. 2 45
DAHINTEN IN DER HEIDE Vol. 4 97, 230, 773
DAI FRANTUMI DELL'IDOLO Vol. 7 629, 630, 941
DAINAH LA METISSE Vol. 2 656
DAISY Vol. 4 439, 704
DAJ STO DAS Vol. 3 352
DAKOTA 308 Vol. 2 158, 344
DAKRYA GIA ENAN ALETE Vol. 3 142
DALEKA JEST DROGA Vol. 1 171, 237, 321, 323
DALEKO JE DO NEBA Vol. 10 32, 72, 86
DALEKO JE SUNCE Vol. 3 356, 374
DALEKO NEBO Vol. 6 *page 315, page 317*
DALIBOR Vol. 10 296, 412, 569
DALILA Vol. 7 298
DALL 16 AL 23 NOVEMBRE Vol. 7 454
DALLA COLPA ALL'AMORE Vol. 7 1189
DALL'ALBA AL TRAMONTO Vol. 7 731
DALL'ALBA AL TRAMONTO EFFETTI DI LUCE. L'INFERNO Vol. 7 993
DALLE CATENE ALLE MORTE Vol. 7 1078
DALLE CINQUE GIORNATE ALLA BRECCIA DI PORTA PIA Vol. 7 1004
DALL'ITALIA ALL'EQUATORE Vol. 7 1113, 1116
DALMATINSKA SVADBA Vol. 3 390
DALOLVA SZEP AZ ELET Vol. 9 117, 237
DAMA BIANCA, LA Vol. 7 443, 493
DAMA DAL PROFUMO D'AMBRA ROSA, LA Vol. 7 980, 1195
DAMA DE BEYRUT, LA Vol. 8 538, 732
DAMA DE CHEZ MAXIM'S, LA Vol. 7 559
DAMA DEL ALBA, LA Vol. 8 515, 564
DAMA DEL ARMINO, LA Vol. 8 214, 276
DAMA D'ONORE Vol. 7 1189
DAMA E IL MISTERO, LA Vol. 7 441, 560
DAMA ERRANTE, LA Vol. 7 787
DAMA NA KOLEJICH Vol. 10 41, 153, 498
DAMA ROSSA UCCIDE SETTE VOLTE, LA Vol. 7 1086
DAMA S MALOU NOZKOU Vol. 10 198
DAMA S RUZI Vol. 10 310, 331, 393
DAMA SPATHI Vol. 3 130
DAMALS Vol. 4 235, 742
DAMALS IN PARIS Vol. 1 56, 58, 115
DAME AUS BERLIN, DIE Vol. 4 613, 712
DAME AUX CAMELIAS, LA Vol. 2 8, 27, 59, 226, 402, 440, 540, 579, 722, 803
DAME DANS L'AUTO AVEC DES LUNETTES ET UN FUSIL, LA Vol. 2 400, 719
DAME DE CHEZ MAXIM'S, LA Vol. 2 231, 279, 514, 656
DAME DE HAUT-LE-BOIS, LA Vol. 2 81, 131, 533
DAME DE MALACCA, LA Vol. 2 309, 439, 440, 837
DAME DE MONSOREAU, LA Vol. 2 614, 672
DAME DE PIQUE, LA Vol. 2 513, 574, 626, 824, 833
DAME DE VITTEL, LA Vol. 2 9, 407
DAME, DER TEUFEL UND DIE PROBIERMAMSELL, DIE Vol. 4 560
DAME D'ONZE HEURES, LA Vol. 2 401
DAME EN GRIS, LA Vol. 7 886, 1079
DAME IN GRAU, DIE Vol. 4 747
DAME IN SCHWARZ, DIE Vol. 4 25, 135, 190, 422, 503, 560, 648
DAME MASQUEE, LA Vol. 2 361, 521, 623, 833
DAME MIT DEM TIGERFELL, DIE Vol. 4 213, 681
DAME MIT DEN SMARAGDEN, DIE Vol. 4 232
DAME MIT DER MASKE, DIE Vol. 4 115, 132
DAME NERE, LE Vol. 7 1189
DAME UN POCO DE AMOOOR ! Vol. 8 438, 719
DAME UND DER BLINDE, DIE Vol. 1 2, 86
DAME UND DER LANDSTREICHER, DIE Vol. 4 37, 71
DAME UND IHR CHAUFFEUR, DIE Vol. 4 149, 242, 660
DAMEN I SVART Vol. 5 1042, 1080
DAMEN MED DE LYSE HANDSKER Vol. 5 31, 32, 55, 101, 208
DAMEN MED KAMELIORNA Vol. 5 891, 966
DAMERNES BLAD Vol. 5 97
DAMERNES RIDDER Vol. 5 83
DAMERNES VEN Vol. 5 199, 223, 306
DAMES DU BOIS DE BOULOGNE, LES Vol. 2 8, 16, 251
DAMI KANJAT Vol. 3 122
DAMIGELLA DI BARD, LA Vol. 7 196, 756, 1060
DAMINA DI PORCELLANA, LA Vol. 7 399
DAMNED, THE Vol. 7 736, 817, 931, 990, 1025, 1142
DÄMON BLUT Vol. 4 192

DÄMON DER WELT **Vol. 4** 113, 193, 270, 309, 565
DÄMON DES HIMALAYA, DER **Vol. 4** 15, 154, 182, 437, 738
DÄMON DES MEERES **Vol. 4** 281
DÄMON VON KOLNO, DER **Vol. 4** 35, 514
DÄMON ZIRKUS **Vol. 4** 293, 306, 735
DÄMONISCHE TREUE **Vol. 4** 162, 223
D'AMORE SI MUORE **Vol. 7** 910, 1076
D'AMORE SI VIVE **Vol. 7** 9
D'AMOUR ET D'EAU FRAICHE **Vol. 2** 597, 781
DAN CETRNAESTI **Vol. 3** 347, 383, 388
DAN DUZI OD GODINE **Vol. 3** 356, 401
DAN, TANT OCH LILLA FRÖKEN SÖDERLUND **Vol. 5** 996
DANARO DI GIUDA, IL **Vol. 7** 1189
DANCA DOS DIAMANTES **Vol. 8** 156
DANCE FEVER **Vol. 7** 414
DANCE MUSIC **Vol. 7** 315
DANCE OF THE HERON, THE **Vol. 2** 854
DANCING W KWATERZE HITLERA **Vol. 1** 148, 156, 237, 325
DANDIN GYÖRGY VAGY A MECSUFOLT FERJ **Vol. 9** 75, 215, 241
DANGER DE MORT **Vol. 2** 191, 788
DANGER GIRL, THE **Vol. 2** 84
DANI **Vol. 3** 418, 429 **Vol. 9** 26, 117, 183
DANI OD SNOVA **Vol. 3** 357, 366 **Vol. 6** *page 307*
DANIEL UND DER WELTMEISTER **Vol. 1** 21, 44
DANIELE CORTIS **Vol. 7** 522, 1036, 1060, 1177
DANKA **Vol. 3** 56, 66, 92
DANKO PISTA **Vol. 9** 71, 299, 322
DANMARK ER LUKKET **Vol. 5** 195, 247
DANN SCHON LIEBER LEBERTRAN **Vol. 4** 149, 536, 629, 660 **Vol. 6** *page 198*
DANNATI, I **Vol. 7** 554
DANNAZIONE DI CAINO, LA **Vol. 7** 1189
DANS LA GUEULE DU LOUP **Vol. 2** 348, 372
DANS LA POUSSIERE DU SOLEIL **Vol. 2** 808
DANS LA VIE TOUT S'ARRANGE **Vol. 2** 15
DANS L'EAU QUI FAIT DES BULLES **Vol. 2** 484, 627
DANS LES RUES **Vol. 2** 23, 563, 626
DANS L'OMBRE DU HAREM **Vol. 2** 413, 727, 871
DANS PA ROSOR **Vol. 5** 1113, 1156, 1164
DANS UNE ILE PERDUE **Vol. 2** 384
DANSA MIN DOCKA **Vol. 5** 1173
DANSE DE MORT, LA **Vol. 2** 16, 488, 863
DANSE ESPAGNOLE **Vol. 2** 678
DANSE MUSIC **Vol. 7** 102
DANSEGALSKAB **Vol. 5** 93
DANSEN PAA KOLDINGHUS **Vol. 5** 97
DANSERINDENS HAEVN **Vol. 5** 58
DANSES SACREES **Vol. 2** 13
DANSEUR INCONNU, LE **Vol. 2** 42
DANSEUSE DE MARRAKECH, LA **Vol. 2** 67, 131
DANSEUSE NUE, LA **Vol. 2** 67, 131
DANSEUSE ORCHIDEE, LA **Vol. 2** 41, 153, 579
DANSEUSE ROUGE, LA **Vol. 2** 13, 833
DANSSALONGEN **Vol. 5** 862, 1042
DANTE AKTA'RE FOR HAJEN ! **Vol. 5** 862
DANTE E BEATRICE **Vol. 7** 1189
DANTE NELLA VITA E NEI TEMPI SUOI **Vol. 7** 169, 272, 796, 888
DANTE NO ES UNICAMENTE SEVERO **Vol. 8** 492, 768
DANTES MYSTERIER **Vol. 5** 332, 840, 909
DANTON **Vol. 2** 153, 837 **Vol. 4** 85, 131, 158, 724 **Vol. 6** *page 130, page 141, page 254*
DANUBE BLEU, LE **Vol. 2** 273, 727
DANZA COME SAI DANZARE TU ... **Vol. 7** 39
DANZA DE FUEGO **Vol. 8** 625, 716
DANZA DE LOS DESEOS, LA **Vol. 2** 775 **Vol. 8** 717
DANZA DEI GIOIELLI, LA **Vol. 7** 806
DANZA DEI MILIONI, LA **Vol. 7** 639, 1110, 1188, 1211
DANZA DEL CORAZON, LA **Vol. 8** 365, 400, 677
DANZA DEL FUOCO, LA **Vol. 7** 101, 638
DANZA DELLA VITA E DELLA MORTE, LA **Vol. 7** 1103
DANZA DELLE LANCETTE, LA **Vol. 7** 796, 902
DANZA DELLE ORE, LA **Vol. 7** 277
DANZA MACABRA, LA **Vol. 7** 852
DANZA SULL'ABISSO, LA **Vol. 7** 1156
DANZATRICE CIECA, LA **Vol. 7** 91, 1057
DANZATRICE DI TANGO, LA **Vol. 7** 81
DANZATRICE IGNOTA, LA **Vol. 7** 805
DAPHNE UND DER DIPLOMAT **Vol. 4** 124, 232, 742
DÄR FYREN BLINKAR **Vol. 5** 857, 891
DÄR MOLLORNA GA... **Vol. 5** 840, 1033
DAR SVATEBNI NOCI **Vol. 10** 34, 345
DARBUJAN A PANDRHOLA **Vol. 10** 380, 396, 547
DARCLEE **Vol. 3** 244, 248, 329
DARF ICH PETRUSCHKA ZU DIR SAGEN **Vol. 6** *page 159, page 176*
DARK JOURNEY **Vol. 2** 656

DARK MIRROR Vol. 2 778
DARO UN MILIONE Vol. 7 156, 725, 796, 1116
D'ARTAGNAN CONTRO I TRE MOSCHETTIERI Vol. 7 1152
DARWIN Vol. 4 283
DAS IST DER KRIEG ! Vol. 4 239
DAS WAR IN HEIDELBERG IN BLAUER SOMMERNACHT Vol. 4 125, 270
DAS WAR MEIN LEBEN Vol. 4 340, 378, 638
DATTERSEN SOLGT Vol. 5 97
DAVANTI A LUI TREMAVA TUTTA ROMA Vol. 7 756
DAVID COPPERFIELD Vol. 5 136, 157
DAVID E GOLIA Vol. 7 17
DAVID GOLDER Vol. 2 573, 656, 824
DAVID OG JONATHAN Vol. 5 81
DAVITELJ PROTIV DAVITELJA Vol. 6 *page 304, page 311, page 312*
DAY OF THE JACKAL, THE Vol. 2 400
DAY THEY ROBBED THE BANK OF ENGLAND, THE Vol. 2 656
DAYS OF HEAVEN Vol. 2 17
DAZDNIK SVATEHO PETRA Vol. 9 134, 176, 183 Vol. 10 196, 273, 609
DCERUSKA K POHLEDANI Vol. 10 168, 564
DCERY EVINY Vol. 10 138, 167
DE-AS FI HARAP ALB ! Vol. 3 270, 298
DE BABORD A TRIBORD Vol. 2 566
DE BARRO Y ORO Vol. 8 258, 495
DE BUNA VOIE SI NESILIT DE NIMENI Vol. 3 238, 245, 252, 286
DE CAIN, LAS Vol. 8 298
DE CAMISA VIEJA A CHAQUETA NUEVA Vol. 8 207
DE CE TRAG CLOPOTELE, MITICA ? Vol. 6 *page 274, page 278, page 279, page 282*
DE COLOR MORENO Vol. 8 395, 528
DE CRIADA A SENORA Vol. 8 626, 627
DE CUERPO PRESENTE Vol. 8 298, 348
DE DRAGUL TAU, ANCA Vol. 6 *page 280, page 283*
DE ESPALDAS A LA PUERTA Vol. 8 449
DE HOMBRE A HOMBRE Vol. 8 414
DE LA CANEBIERE, UN Vol. 2 279, 833
DE LA CAVALERIE, UNE Vol. 2 270
DE LA LEGION, UN Vol. 2 775
DE LA PART DES COPAINS Vol. 2 746
DE LA PIEL DEL DIABLO Vol. 8 583
DE L'AMOUR Vol. 2 728, 787
DE MADRID AL CIELO Vol. 8 214, 484
DE MAYERLING A SARAJEVO Vol. 2 206, 279
DE MECS EN MECS Vol. 6 *page 96*
DE MENDEZ, LAS Vol. 8 281, 364
DE MUJER A MUJER Vol. 8 400, 717
DE PICOS PARDOS A LA CIUDAD Vol. 8 261, 768
DE PROFESION POLIGAMO Vol. 8 299, 773
DE PROFESION SOSPECHOSOS Vol. 8 544, 578
DE PROFESION, SUS LABORES Vol. 8 318
DE PROFUNDIS Vol. 4 773
DE QUOI TU TE MELES DANIELA ! Vol. 2 259, 344, 532
DE TREI ORI BUCURESTI Vol. 3 254, 260, 286, 295, 296, 307
DE TRIPAS CORAZON Vol. 8 232
DE TROYA EN PALMAR, LA Vol. 8 366
DEADLINE Vol. 5 130
DEBANDADE , LA Vol. 2 534
DEBATIK Vol. 3 1, 4, 6, 8
DEBAUCHE, LA Vol. 2 818
DEBELLI I MRSAVI Vol. 6 *page 304, page 308, page 310*
DEBER DE ESPOSA Vol. 8 268, 359, 725
DEBER, EL Vol. 8 649
DEBITO DELL'IMPERATORE, IL Vol. 7 1189
DEBITO D'ODIO Vol. 7 1211
DEBIUTANTKA Vol. 6 *page 220, page 226, page 260*
DEBLA LA VIRGEN GITANA Vol. 8 214
DEBORDEMENTS DE PLAISIR Vol. 2 365
DEBOUT DES CRABES, LA MER MONTE Vol. 6 *page 106, page 142*
DEBOUT, LA-DEDANS ! Vol. 2 404, 771
DEBOUT LES MORTS Vol. 2 167, 798
DEBUREAU Vol. 2 706,722
DEBUTS DE CHIFFONNETTE, LES Vol. 2 678
DECA IGRAJAT V"N Vol. 3 51, 76 Vol. 6 *page 38*
DECA VOJVODE SMITA Vol. 3 418
DECADE PRODIGIEUSE, LA Vol. 2 705
DECAK I VIOLINA Vol. 3 367, 415
DECAK JE VIKAO UBISTVO Vol. 3 391
DECAK MITA Vol. 3 349, 394
DECAMERON, IL Vol. 7 333, 388, 434
DECAMERON N. 2: LE ALTRE NOVELLE DEL BOCCACCIO Vol. 7 666
DECAMERON N. 3: LE PIU BELLE DONNE DEL BOCCACCIO Vol. 7 901
DECAMERON PROIBITISSIMO Vol. 7 364
DECAMERONE NERO, IL Vol. 7 514
DECAMERONE PROIBITO, IL Vol. 7 852
DECAMEROTICUS Vol. 7 184
DECATA NA BALKANA Vol. 3 74
DECEMBRE Vol. 2 175, 642
DECENTE, LA Vol. 8 438, 493
DECHEANCE Vol. 7 264
DECIMA VITTIMA, LA Vol. 7 913, 1167

DECKO KOJI OBECAVA Vol. 6 *page 312, page 318*
DECLICET DES CLAQUES Vol. 6 *page 73, page 88*
DECOLAREA Vol. 3 225, 265
DECULOTTEZ – VOUS LES STARLETTES Vol. 6 *page 69*
DECYZJA Vol. 1 148, 281, 389 Vol. 6 *page 24*
DEDE Vol. 2 200
DEDE LA MUSIQUE Vol. 2 318, 460, 807
DEDECEK AUTOMOBIL Vol. 10 179, 286, 311, 502
DEDECEK, KYLIJAN A JA Vol. 10 283, 369, 577
DEDECKEM PROTI SVE VULI Vol. 10 345, 524
DEDEE D'ANVERS Vol. 2 124, 194, 863
DEDELGETETT KEDVENCEINK Vol. 9 238
DEDICATO A UNA STELLA Vol. 7 406
DEDICATORIA Vol. 8 269, 358
DEDISCINA Vol. 6 *page 320*
DEDO DEL DESTINO, EL Vol. 8 243, 517
DEDO EN EL GATILLO, EL Vol. 8 517
DEFA 70 Vol. 1 9, 17, 130
DEFA-DISCO 77 Vol. 1 9, 76
DEFENSE D'AIMER Vol. 2 332, 875
DEFENSE DE SAVOIR Vol. 2 527
DEFENSEUR, LE Vol. 2 279, 871
DEFILE DE CUIRASSIERS Vol. 2 678
DEFIZIT, DAS Vol. 4 288
DEFONCEUSES, LES Vol. 2 721
DEFROQUE, LE Vol. 2 128, 833
DEGENHARDTS, DIE Vol. 4 97, 416, 437
DEGOURDIS DE LA 11EME, LES Vol. 2 533
DEHORS-DEDANS Vol. 2 283
DEI SVARTE HESTANE Vol. 5 672, 723
DEIN BEGEHREN IST SÜNDE Vol. 4 255, 344
DEIN HERZ IST MEIN Vol. 4 232, 673
DEIN IST MEIN HERZ Vol. 4 361, 558
DEIN LEBEN GEHÖRT MIR Vol. 4 51, 108, 623
DEIN UNBEKANNTER BRUDER Vol. 6 *page 162, page 168, page 170*
DEINES BRUDERS WEIB Vol. 4 673
DEIXEM – ME AO MENOS SUBIR AS PALMEIRAS... Vol. 8 102
DEJEUNER DE PIERROTS Vol. 2 678
DEJEUNER DE SOLEIL, UN Vol. 2 279, 626, 824
DEJEUNER SUR L'HERBE, LE Vol. 2 239, 480
DEJLIG AEGTEMAND, EN Vol. 5 81
DEJLIG DAG, EN Vol. 5 92
DEJTE SE OMLADIT Vol. 10 5, 68
DEKAMERON 40 Vol. 1 144
DEKAMERON-NÄCHTE Vol. 4 23, 663
DEKATOS TRITOS Vol. 3 185, 207
DEL AMOR Y OTRAS SOLEDADES Vol. 8 348 Vol. 8 406
DEL BRAZO POR LA CALLE Vol. 8 544, 578
DEL ROSA AL AMERILLO Vol. 8 402
DEL SOTO DEL PARRAL, LA Vol. 8 423
DELE COLOR AL DIFUNTO Vol. 8 681, 749
DELEKA CESTA Vol. 10 283, 407, 532, 563
DELENDA CARTHAGO ! Vol. 7 1190
DELFINI, I Vol. 7 207, 1167
DELIBAB MINDEN MENNYSEGBEN Vol. 9 49, 122
DELIBABOK ORSZAGA Vol. 9 139, 152, 202
DELICES DU TOSSING, LES Vol. 6 *page 129*
DELICES D'UN SEXE CHAUD ET PROFOND Vol. 6 *page 120*
DELICIOSAMENTE TONTOS Vol. 8 214, 400
DELIJE Vol. 3 418
DELIKATESSEN Vol. 4 208, 479, 600 Vol. 6 *page 185*
DELINCUENTES Vol. 8 492, 537
DELIRES PORNO Vol. 2 294
DELIRES SEXUELS Vol. 2 294
DELIRIO Vol. 7 910
DELIRIOS DE AMOR Vol. 8 560, 645, 672, 754
DELIRIOS DE GRANDEZA Vol. 8 749
DELIRIUM Vol. 8 672
DELIRO Vol. 2 626
DELIT DE FUITE Vol. 2 615, 719, 873
DELITTI, AMORE E GELOSIA Vol. 7 502
DELITTO A PORTA ROMANA Vol. 7 310, 315
DELITTO AL BLUE GAY Vol. 7 315, 738
DELITTO AL CIRCOLO DEL TENNIS Vol. 7 1101
DELITTO AL RISTORANTE CINES Vol. 7 310, 381
DELITTO ALLO SPECCHIO Vol. 7 739
DELITTO CARNALE Vol. 7 130
DELITTO DEL COMMENDATORE, IL Vol. 7 131
DELITTO DELLA BRUGHIERA Vol. 7 1189
DELITTO DELLA PICCINA, IL Vol. 7 613
DELITTO DELLA VILLA SOLITARIA, IL Vol. 7 1116
DELITTO DI BEINASCO, IL Vol. 7 1189
DELITTO DI CASTEL GIUBILEO, IL Vol. 7 531
DELITTO DI GIOVANNI EPISCOPO, IL Vol. 7 453, 1060, 1136

DELITTO DI UN PESCATORE, IL Vol. 7 1189
DELITTO IN FORMULA, UNO Vol. 7 315, 585
DELITTO MATTEOTTI, IL Vol. 7 854, 1153
DELITTO NON MIO, IL Vol. 7 1189
DELITTO SULL'AUTOSTRADA Vol. 7 257, 315
DELITTO, UN Vol. 7 544, 915
DELIZIE EROTICHE Vol. 7 742
DELKA POLIBKU DEVADESAT Vol. 10 254, 387, 428
D'ELMORO Vol. 1 172, 259
DELO BEILISA Vol. 2 833
DELO NO. 205/1913 Vol. 6 *page 35, page 64*
DELPHINE Vol. 2 564, 664
DELTÖL HAJNALIG Vol. 9 67, 121, 318
DEM TEUFEL VERSCHRIEBEN Vol. 4 447
DEMAIN EST UN AUTRE JOUR Vol. 2 224
DEMAIN LES MOMES Vol. 2 289, 742
DEMAIN, NOUS DIVORCONS Vol. 2 423, 722
DEMANTY NOCI Vol. 10 41, 289, 292
DEMASIADAS MUJERES PARA LAYTON Vol. 8 331, 546
DEMI-VIERGES, LES Vol. 2 84, 248, 249
DEMIMONDE-HEIRAT, EINE Vol. 4 208, 232
DEMOISELLE ET SON REVENANT, LA Vol. 2 153, 249
DEMOISELLES DE PENSIONNAT, LES Vol. 2 735
DEMOISELLES DE ROCHEFORT, LES Vol. 2 190, 285
DEMOKRATI Vol. 10 72, 82, 544
DEMON DANS L'ILE, LE Vol. 6 *page 71, page 107*
DEMON RODU HALKENU Vol. 10 201, 290, 492
DEMONE DEL FUOCO, IL Vol. 7 490
DEMONE GIALLO, IL Vol. 7 984, 1079
DEMONE GLI DISSE... , UN Vol. 7 304
DEMONE, IL Vol. 7 148
DEMONI Vol. 7 97, 99, 683
DEMONIAQUES, LES Vol. 2 721
DEMONIO CON ANGEL, UN Vol. 8 217, 768
DEMONIOS DE ALCACER-KIBIR, OS Vol. 8 56, 118
DEMONIOS EN EL JARDIN Vol. 8 219, 591
DEMONS DE L'AUBE, LES Vol. 2 124, 863
DEMONS DE MIDI Vol. 2 60, 495
DEMONS DE MINUIT, LES Vol. 2 144, 768
DEMÜTIGE UND DIE SÄNGERIN, DER Vol. 4 89, 752
DEN, KTORY NEUMRIE Vol. 10 82, 273, 280
DEN PRO MOU LASKU Vol. 10 144, 299, 327
DEN SEDMY - OSMA NOC Vol. 10 154, 313
DEN SLNOVRATU Vol. 10 72, 459, 474
DEN ZAHN LASS DIR MAN ZIEHEN Vol. 4 361, 367
DENJAT NE SI LICI PO ZARANTA Vol. 6 *page 63, page 65*
...DENN ALLE SCHULD RÄCHT SICH AUF ERDEN Vol. 4 5, 592
DENONCIATION, LA Vol. 2 377, 702
DENOVI NA ISKUSENIE Vol. 3 339, 384
DENTE PER DENTE Vol. 7 443, 734
DENTELLIERE, LA Vol. 2 110, 284, 798
DENTS LONGUES, LES Vol. 2 185, 423
DENUNZIA, LA Vol. 7 1066
DENVER Vol. 8 455
DEPART A ZERO Vol. 2 826
DEPART DE L'AMBASSADE POUR L'EGLISE RUSSE Vol. 2 678
DEPART DES SOUVERAINS RUSSES POUR VERSAILLES Vol. 2 678
DEPARTE DE TIPPERARY Vol. 3 269, 281, 319
DEPORTATE DELLA SEZIONE SPECIALE SS, LE Vol. 7 834
DEPRAVES DU PLAISIR, LES Vol. 2 109
DEPRISA, DEPRISA Vol. 8 269, 358
DEPS Vol. 3 414, 437
DEPUCELAGES Vol. 2 735
DER ADJUTANT SEINER HOHEIT Vol. 10 118, 167, 169
DER BRAENDER EN ILD Vol. 5 89, 281
DER ER ET YNDIGT LAND Vol. 5 11, 46, 249
DER FINDES INGEN TAMME DYR Vol. 5 214
DER KNUTE ENTFLOHEN Vol. 4 124, 232 Vol. 6 *page 194*
DER KOM EN DAG Vol. 5 180, 199, 304
DER KOM EN SOLDAT Vol. 5 76, 306
DER LIEBE LEID Vol. 4 129
DER LIEBE LUST UND LEID Vol. 4 424, 724
DER LIEBE PILGERFAHRT Vol. 4 190, 773
DER MA VAERE EN SENGEKANT Vol. 5 156, 283, 296
DER MENSCHHEIT ANWALT Vol. 4 546, 574
DER VAR ENGANG Vol. 5 97, 128, 178, 179, 189, 232, 240, 265
DER VAR ENGANG EN GADE Vol. 5 2, 23
DER VAR ENGANG EN KRIG Vol. 5 16, 61, 195

DER VAR ENGANG EN VICEVAERT **Vol. 5** 7
DERACINES, LES **Vol. 2** 738
DERBY **Vol. 4** 132, 603, 636, 788, 796
DERECHOS DE LA MUJER, LOS **Vol. 8** 214, 681
DERELITTA, LA **Vol. 6** *page 76, page 100*
DERELITTI DI VALCOURT, I **Vol. 7** 711
DERNIER AMANT ROMANTIQUE, LE **Vol. 2** 315, 682
DERNIER AMOUR **Vol. 2** 401, 798
DERNIER ATOUT **Vol. 2** 251, 395
DERNIER BAISER, LE **Vol. 2** 234
DERNIER CHOC, LE **Vol. 2** 167, 388, 412
DERNIER COMBAT, LE **Vol. 6** *page 80, page 122, page 128, page 146*
DERNIER DES SIX, LE **Vol. 2** 23, 488
DERNIER DOMICILE CONNU **Vol. 2** 74, 163
DERNIER ETE **Vol. 6** *page 72, page 138*
DERNIER HEURE EDITION SPECIALE **Vol. 2** 197, 344
DERNIER HOMME, LE **Vol. 2** 517
DERNIER METRO, LE **Vol. 2** 17, 131, 339 **Vol. 6** *page 84, page 140*
DERNIER MILLARDAIRE, LE **Vol. 2** 9, 158, 563, 626
DERNIER QUART D'HEURE, LE **Vol. 2** 9, 481
DERNIER REFUGE **Vol. 2** 153, 197
DERNIER ROBIN DES BOIS, LE **Vol. 2** 128, 595
DERNIER SAUT, LE **Vol. 2** 49, 165
DERNIER SOU, LE **Vol. 2** 23, 67
DERNIER TIERCE, LE **Vol. 2** 663
DERNIER TOURNANT, LE **Vol. 2** 194, 566, 719, 863
DERNIERE AVENTURE **Vol. 2** 122, 270
DERNIERE BOURREE A PARIS, LA **Vol. 2** 20, 219
DERNIERE CHEVAUCHEE, LA **Vol. 2** 423, 627
DERNIERE DES PUTES, LA **Vol. 6** *page 112*
DERNIERE FEMME, LA **Vol. 2** 142
DERNIERE HEURE **Vol. 2** 604
DERNIERE JEUNESSE **Vol. 2** 29, 775 **Vol. 7** 48, 838
DERNIERE NUIT, LA **Vol. 2** 864
DERNIERE SORTIE AVANT ROISSY **Vol. 2** 527
DERNIERE VALSE, LA **Vol. 2** 153, 388
DERNIERES VACANCES, LES **Vol. 2** 8, 59
DERNIERS JOURS DE POMPEI, LES **Vol. 2** 402
DEROBADE, LA **Vol. 2** 164 **Vol. 6** *page 109, page 125*
DERRAPAGEM **Vol. 8** 130
DERRIERE LA FACADE **Vol. 2** 9, 423
DERRIERE LE MIRIOR SANS TAIN **Vol. 6** *page 94*
DERRRIERE LA GRANDE MURAILLE **Vol. 2** 655
DERVIS I SMRT **Vol. 3** 356, 374
DERYNE **Vol. 9** 71, 75, 87, 199
DERYNE HOL VAN ? **Vol. 9** 20, 167, 265
DES CHRISTS PAR MILLIERS **Vol. 2** 112, 512
DES DRACHENS GRAUER ATEM **Vol. 1** 38, 70, 91
DES ENFANTS GATES **Vol. 2** 513
DES FEMMES DISPARAISSENT **Vol. 2** 423, 515
DES FRISSONS PARTOUT **Vol. 2** 475, 659
DES GARCONS ET DES FILLES **Vol. 2** 163
DES GENS SANS IMPORTANCE **Vol. 2** 185, 644
DES HENKERS BRUDER **Vol. 1** 26, 45, 62
DES JEUNES FILLES DANS LA NUIT **Vol. 2** 48, 197
DES JOURNEES ENTIERES DANS LES ARBRES **Vol. 2** 17
DES JUNGEN DESSAUERS GROSSE LIEBE **Vol. 4** 49, 325, 445
DES KAISERS ALTE KLEIDER **Vol. 4** 135, 334
DES KLEINEN LOCKFÜHRERS GROSSE FAHRT **Vol. 1** 33, 121, 128
DES KÖNIGS BEFEHL **Vol. 4** 223, 270, 444, 798
DES LEBENS RUTSCHBAHN **Vol. 4** 565
DES LEBENS UND DER LIEBE WELLEN **Vol. 4** 565
DES LEBENS WÜRFELSPIEL **Vol. 4** 183, 287, 644
DES MEERES UND DER LIEBE WELLEN **Vol. 4** 187
DES PISSENLITS PAR LA RACINE **Vol. 2** 168, 293
DES QUINTUPLES AU PENSIONNAT **Vol. 2** 70, 193
DES SÄNGERS FLUCH **Vol. 6** *page 189*
DES TEUFELS ADVOKAT **Vol. 4** 232, 673
DES TEUFELS PUPPE **Vol. 4** 471
DES VACANCES EN OR **Vol. 2** 340 **Vol. 8** 306, 426
DESAFIO EN RIO BRAVO **Vol. 8** 243
DESAFIOS, LOS **Vol. 8** 348, 406
DESAMPARADOS, LOS **Vol. 8** 551, 732
DESANT NA DRVAR **Vol. 3** 366 **Vol. 3** 403
DESAPARECIDO, EL **Vol. 8** 652
DESARROI **Vol. 2** 205, 698
DESAT PERCENT NADEJE **Vol. 10** 228, 399, 474

DESCENDEZ ON VOUS DEMANDE **Vol. 2** 344, 598
DESCRIPTION D'UN COMBAT **Vol. 2** 190
DESEJADO OU AS MONTANHAS DA LUA, O **Vol. 8** 139
DESERT DE PIGALLE, LE **Vol. 2** 270
DESERTER USA **Vol. 5** 1051
DESERTEUR, LE **Vol. 2** 344, 488
DESERTO DEI TARTARI, IL **Vol. 7** 95, 1145
DESERTO ROSSO **Vol. 7** 41, 675, 853, 854, 913, 1086
DESERTOR, EN **Vol. 5** 7
DESERTOREN **Vol. 5** 130
DESESPERADOS, LOS **Vol. 8** 719
DESET DNI NEPLATENI **Vol. 3** 63, 110
DESETI BRAT **Vol. 6** *page 311, page 316*
DESETINA SA NAVARONA **Vol. 3** 350
DESFASURAREA **Vol. 3** 244, 299, 322, 329
DESHMORET E MONUMENTEVE **Vol. 3** 2, 7, 28 **Vol. 6** *page 13*
DESIDERI D'ESTATE **Vol. 7** 1179
DESIDERIA – LA VITA INTERIORE **Vol. 7** 318, 538
DESIDERIO **Vol. 7** 434, 636, 638, 639, 988
DESIDERIO E SOLE **Vol. 7** 603
DESIR ET L'AMOUR, LE **Vol. 2** 427, 722
DESIR, LE **Vol. 2** 87, 294, 672
DESIR MENE LES HOMMES, LE **Vol. 2** 224, 475, 856
DESIRADE, LA **Vol. 2** 641, 860
DESIRE **Vol. 2** 48, 657
DESIRELLA **Vol. 2** 494
DESIRS SOUS LES TROPIQUES **Vol. 6** *page 69*
DESNUDA CHICA DEL RELAX, LA **Vol. 8** 471
DESORDRE A VINGT ANS, LE **Vol. 2** 702
DESORDRE ET LA NUIT, LE **Vol. 2** 16, 118, 644, 832
DESORDRE, LE **Vol. 2** 53
DESPEDIDA DE SOLTERO **Vol. 8** 551
DESPERADO VON PANAMA, DER **Vol. 4** 247
DESPERTAR, EL **Vol. 8** 492, 732
DESPERTO SU CORAZON **Vol. 8** 486, 732
DESPIADADOS, LOS **Vol. 8** 626
DESPRE O ANUME FERICIRE **Vol. 3** 265, 270, 311, 321
DESPUE DEL DILUVIO **Vol. 8** 232
DESPUES DE LOS NUEVE MESES **Vol. 8** 553, 749
DESPUES DE OTRO, UNO **Vol. 8** 214, 602
DESPUES DEL GRAN ROBO **Vol. 8** 426, 732
DESSERT DE PIGALLE, LE **Vol. 2** 47
DESSOUS DES CARTES, LE **Vol. 2** 615, 824
DESTIN **Vol. 2** 698
DESTIN DE JULIETTE, LE **Vol. 6** *page 127, page 142*
DESTIN EST MAITRE, LE **Vol. 2** 584
DESTIN EXECRABLE DE GUILLEMENTE BABIN, LE **Vol. 2** 205, 226, 538
DESTIN FABULEUX DE DESIREE CLARY, LE **Vol. 2** 48, 197
DESTIN S'AMUSE, LE **Vol. 2** 67, 332
DESTINATIA MAHMUDIA **Vol. 6** *page 274, page 277, page 282*
DESTINAZIONE PIOVAROLA **Vol. 7** 451
DESTINAZIONE SAN REMO **Vol. 7** 372
DESTINE ROMANTICE **Vol. 6** *page 266, page 277, page 279, page 281*
DESTINEE **Vol. 2** 845, 871
DESTINEES **Vol. 2** 232, 279, 488, 566, 676 **Vol. 7** 1054
DESTINI DI DONNE **Vol. 7** 359, 691, 739, 1054
DESTINO : BARAJAS **Vol. 8** 598, 749
DESTINO DI UN TROVATELLO, IL **Vol. 7** 488
DESTINO : ESTAMBUL **Vol. 8** 732
DESTINO, IL **Vol. 7** 264, 701
DESTINO IN TASCA, IL **Vol. 7** 792, 1036
DESTINO MARCA A HORA, O **Vol. 8** 31, 150, 167
DESTINO SE DISPUTA, EL **Vol. 8** 711, 716
DESTINS **Vol. 2** 344, 401
DESTIVY DEN **Vol. 10** 41, 292
DESZCZOWY LIPIEC **Vol. 1** 225, 250 **Vol. 6** *page 233*
DETASAMENTUL „CONCORDIA“ **Vol. 3** 211, 320
DETECTIVE CON FALDAS **Vol. 8** 493, 515
DETECTIVE, UN **Vol. 7** 514
DETECTIVES DU DIMANCHE, LES **Vol. 2** 309, 475
DETEKTIV DES KAISERS **Vol. 4** 49, 626
DETEKTIVDUELL, DAS **Vol. 4** 503
DETEKTIVENS BARNEPIGE **Vol. 5** 183
DETI NA ZAZAZKU **Vol. 10** 532, 621
DETI OSUDU **Vol. 10** 53, 122
DETI VELKI LASKY **Vol. 10** 34, 345
DETI ZITRKU **Vol. 10** 161, 255, 462
DETOURNEMENT DE MINHEURES **Vol. 2** 131, 433
DETRAS DEL SILENCIO **Vol. 8** 207, 299
DETRUIRE, DIT-ELLE **Vol. 2** 655
DETTE DE HAINE **Vol. 2** 167
DETYRE E POSACME **Vol. 3** 4, 8, 16 **Vol. 6** *page 17*
DEUTSCHE FRAUEN – DEUTSCHE TREUE **Vol. 4** 232, 292
DEUTSCHE HELDEN **Vol. 4** 356, 778
DEUTSCHE HERZEN AM DEUTSCHEN RHEIN **Vol. 4** 208, 270

DEUTSCHE LIED, DAS **Vol. 4** 232, 421, 717
DEUTSCHE MUTTERHERZ, DAS **Vol. 4** 212, 571, 726
DEUX AMOURS **Vol. 2** 128,344
DEUX ANGLAISES ET LE CONTINENT, LES **Vol. 2** 17, 142
DEUX ANS DE VACANCES **Vol. 3** 244, 307
DEUX BALLES AU COEUR **Vol. 2** 533
DEUX BILLETS POUR MEXICO **Vol. 2** 429
DEUX CANARDS, LES **Vol. 2** 308, 439, 731
DEUX COMBINARDS, LES **Vol. 2** 330, 460
DEUX DE LA RESERVE **Vol. 2** 70, 601, 833
DEUX DE L'ESCADRILLE **Vol. 2** 128, 496
DEUX ESPIONNES AVEC UN PETIT SLIP A FLEURS **Vol. 6** *page 74*
DEUX FOIS VINGT ANS **Vol. 2** 113
DEUX FONT LA PAIRE, LES **Vol. 2** 595, 627, 726
DEUX GAMINES, LES **Vol. 2** 37, 67, 131, 169, 330, 613
DEUX GOSSES, LES **Vol. 2** 48, 724, 874
DEUX GRANDES FILLES DANS UN PYJAMA **Vol. 2** 97, 808
DEUX HEURES A TUER **Vol. 2** 106, 514
DEUX HEURES MOINS LE QUART AVANT JESUS-CHRIST **Vol. 6** *page 89, page 118, page 148*
DEUX HOMMES DANS LA VILLE **Vol. 2** 163, 814
DEUX HOMMES DANS MANHATTAN **Vol. 2** 376, 395, 577
DEUX IMBECILES HEUREUX **Vol. 2** 39
DEUX LIONS AU SOLEIL **Vol. 2** 534
DEUX MAINS, LES **Vol. 6** *page 93, page 114*
DEUX MAMANS, LES **Vol. 2** 104
DEUX MONDES, LES **Vol. 2** 751
DEUX „MONSIEUR" DE MADAME, LES **Vol. 2** 153, 412, 775
DEUX MONSIEURS DE MADAME, LES **Vol. 2** 67, 776
DEUX ORPHELINES, LES **Vol. 2** 9, 27, 84, 570 **Vol. 6** *page 99*
DEUX OU TROIS CHOSES QUE JE SAIS D'ELLE **Vol. 2** 208
DEUX POULAINS DE LUCETTE, LES **Vol. 2** 712
DEUX SOLDATS, LES **Vol. 2** 216
DEUX SOUS DE VIOLETTES **Vol. 2** 57, 59
DEUX TIMIDES, LES **Vol. 2** 8, 65, 96, 573, 753
DEUXIEME BUREAU CONTRE INCONNU **Vol. 2** 309, 376, 743
DEUXIEME BUREAU CONTRE KOMMANDANTUR **Vol. 2** 248, 249, 395
DEUXIEME BUREAU **Vol. 2** 37, 70, 191, 533
DEUXIEME BUREAU CONTRE TERRORISTES **Vol. 2** 198
DEUXIEME SOUFFLE, LE **Vol. 2** 198, 288
DEVADESAT MINUT PREKVAPENI **Vol. 10** 166
DEVATE JMENO **Vol. 10** 197, 313, 582
DEVATE SRDCE **Vol. 10** 144, 299, 327
DEVCATA, NEDEJTE SE ! **Vol. 10** 167, 256
DEVCATA, VDAVEJTE SE ! **Vol. 10** 257, 536
DEVCATKO, NERIKEJ NE ! **Vol. 10** 138, 524
DEVCATKO Z VENKOVA **Vol. 10** 256, 460
DEVCE Z HOR **Vol. 10** 243, 345, 586
DEVCE Z PODSKALI **Vol. 10** 243, 536
DEVCE Z PREDMESTI **Vol. 10** 256, 460, 612
DEVCE Z TABAKOVE TOVARNY **Vol. 10** 125, 453
DEVCE ZA VYKLADEM **Vol. 10** 256, 460
DEVCE ZE STRIBRNE HRANICE **Vol. 10** 122, 198, 272
DEVCICA Z BESKYD **Vol. 10** 412, 503
DEVET KRUHU PEKLA **Vol. 10** 74, 176, 231, 283
DEVETI KRUG **Vol. 3** 390, 452
DEVETO CUDO NA ISTOKU **Vol. 3** 351
DEVIAZIONE DI GOOLF STREAM, LA **Vol. 7** 390
DEVOJACKI MOST **Vol. 3** 367, 402 **Vol. 6** *page 308*
DEVOJKA **Vol. 3** 362, 416
DEVOJKA SA KOSMAJA **Vol. 3** 381
DEVOUEMENT **Vol. 2** 731
DEVOYES, LES **Vol. 2** 104, 723
DEZERTER **Vol. 1** 227, 331, 398
D'HOMME A HOMMES **Vol. 2** 388, 566
DI CHE SEGNO SEI ? **Vol. 7** 318
DI NOTTE ALL'ARIA APERTA **Vol. 7** 26
DI PADRE IN FIGLIO **Vol. 7** 318, 544, 909
DI PIETRO L'ARETINO SI RACCONTA... **Vol. 7** 1213
DI QUA DI LA PIAVE **Vol. 7** 891
DI QUELLE, UNA **Vol. 7** 910
DI TROPPO, UNA **Vol. 7** 426
DIA CON SERGIO, UN **Vol. 8** 237, 577, 603
DIA DE LOS ENAMORADES, EL **Vol. 8** 214, 762
DIA DE MANANA, EL **Vol. 8** 402, 589
DIA DE VIDA, UM **Vol. 8** 130, 160
DIA DESPUES DE AGOSTO, UN **Vol. 8** 249, 384
DIA EN EL TRIANGULO, UN **Vol. 8** 514, 672
DIA ES UN DIA, UN **Vol. 8** 214, 484

DIA PERDIDO, UN **Vol. 8** 611, 704
DIA TRAS DIA **Vol. 8** 260, 524
DIABEL **Vol. 6** *page 238, page 249*
DIABLE AU COEUR, LE **Vol. 2** 9, 44, 79, 511, 879
DIABLE AU CORPS, LE **Vol. 2** 251, 427
DIABLE BOITEUX, LE **Vol. 2** 722, 758, 833
DIABLE DANS LA BOITE, LE **Vol. 2** 854
DIABLE ET LES DIX COMMANDEMENTS, LE **Vol. 2** 294, 457
DIABLE LA VILLE, LE **Vol. 2** 805
DIABLE PAR LA QUEUE, LE **Vol. 2** 655
DIABLE SOUFFLE, LE **Vol. 2** 13, 249
DIABLO BAJO LA ALMOHADA, UN **Vol. 8** 611
DIABLO COJUELO, EL **Vol. 8** 214, 681
DIABLO EN VACACIONES, EL **Vol. 8** 583, 772
DIABLO TAMBIEN LLORA, EL **Vol. 8** 214, 517
DIABLO TOCA LA FLAUTA, EL **Vol. 8** 611
DIABLOS DE LA GUERRA, LOS **Vol. 8** 317, 688
DIABLOS DEL MAR, LOS **Vol. 8** 237, 524, 689
DIABLOS ROJOS, LOS **Vol. 8** 249
DIABO DESCEU A VILA, O **Vol. 8** 130, 173
DIABO ERA OUTRO, O **Vol. 8** 4, 98, 130
DIABO SAO ELAS ..., O **Vol. 8** 53, 75, 119, 122, 188
DIABOLICAMENTE ... LETIZIA **Vol. 7** 364
DIABOLICI, I **Vol. 7** 751, 1211
DIABOLICIA MALICIA **Vol. 8** 348, 714
DIABOLIQUEMENT VOTRE **Vol. 2** 59, 224
DIABOLIQUES, LES **Vol. 2** 59, 423, 738, 824
DIABOLO MENTHE **Vol. 2** 754
DIADALMAS ELET **Vol. 9** 7, 175
DIADEM DER ZARIN, DAS **Vol. 6** *page 200*
DIADIKASIA **Vol. 3** 135
DIAGNOSA X **Vol. 10** 256, 524
DIAGNOZA SMRTI **Vol. 10** 172, 430, 593
DIAGOGE MEDEN **Vol. 3** 154
DIAGONALE DU FOU, LA **Vol. 6** *page 132*
DIAKOPES STO VIETNAM **Vol. 3** 144, 205
DIALOG 20 – 40 – 60 **Vol. 10** 262, 283, 369, 459, 575, 576,
DIALOGOS DE LA PAZ **Vol. 8** 600, 749
DIALOGUE DES CARMELITES, LE **Vol. 2** 47, 194, 540, 549
DIAMANT DE CENT SOUS, LE **Vol. 2** 32, 270
DIAMANT DES ZAREN, DER **Vol. 4** 298, 321, 703, 738
DIAMANT NOIR **Vol. 2** 122, 538, 558, 584
DIAMANT VERT, LE **Vol. 2** 871
DIAMANTEN DES ZAREN, DIE **Vol. 4** 190
DIAMANTENKONKURRENZ, DIE **Vol. 4** 162, 208
DIAMANTES A GO – GO **Vol. 8** 517, 732
DIAMANTI **Vol. 7** 40, 1136
DIAMANTI E LAGRIME **Vol. 7** 277
DIAMANTI SPORCHI DI SANGUE **Vol. 7** 514
DIAMANTIA STO GYMNO SOU SOMA **Vol. 3** 196
DIANA **Vol. 9** 7
DIANA L'AFFASCINATRICE **Vol. 7** 264, 701
DIANA SOREL **Vol. 7** 212
DIANE **Vol. 4** 14, 49, 203 **Vol. 6** *page 199*
DIARIO DE INVIERNO **Vol. 8** 232, 237
DIARIO DI UNA STELLA, IL **Vol. 7** 337, 1071, 1177
DIARIO SEGRETO DI UN CARCERE FEMMINILE **Vol. 7** 996
DIARY OF A CHAMBERMAID, THE **Vol. 2** 24, 526
DIAS DE ANGUSTIA **Vol. 8** 762
DIAS DE CABIRIO, LOS **Vol. 8** 297, 749
DIAS DE FERIA **Vol. 8** 517
DIAS DE VIEJO COLOR **Vol. 8** 299, 681
DI'ASEMANTON AFORMEN **Vol. 3** 156
DIAVOLI DELLA GUERRA, I **Vol. 8** 317
DIAVOLO A PARIGI, IL **Vol. 7** 1113
DIAVOLO BIANCO, IL **Vol. 3** 378, 403, 451 **Vol. 7** 638
DIAVOLO E L'ACQUASANTA, IL **Vol. 7** 102, 315
DIAVOLO, IL **Vol. 7** 1136
DIAVOLO IN CONVENTO, IL **Vol. 7** 86
DIAVOLO IN CORPO **Vol. 7** 363, 607, 1112
DIAVOLO NEL CERVELLO, IL **Vol. 7** 1045
DIAVOLO NERO, IL **Vol. 7** 693, 755
DIAVOLO SI FA EREMITA, IL **Vol. 7** 1189
DIAVOLO VA IN COLLEGIO **Vol. 7** 756
DIAVOLO ZOPPO, IL **Vol. 7** 1189
DIBNCOLO DE NISIPURI **Vol. 3** 280
DICE WOMAN, THE **Vol. 2** 84
DICH HAB ICH GELIEBT **Vol. 4** 186, 287, 291, 458
DICHIARAZIONE DI ROBINET, UNA **Vol. 7** 1189
DICHTERLIEBE **Vol. 1** 95
DICIOTTENNI, LE **Vol. 7** 1044
DICKE TILLA, DIE **Vol. 6** *page 159, page 163, page 169*
DICKE UND ICH, DER **Vol. 6** *page 167, page 170*
DICKSCHÄDEL, DER **Vol. 4** 79, 115, 354
DIDONE ABBANDONATA **Vol. 7** 1189
DIE AUS DER 12B **Vol. 1** 34, 58, 129

DIE DA UNTEN **Vol. 4** 132, 584
DIE DAS LICHT SCHEUEN **Vol. 4** 718
DIE HEUTE ÜBER 40 SIND **Vol. 1** 34, 43
DIE IHR GLÜCK VERKENNEN **Vol. 4** 300
DIE IM SCHATTEN GEHEN **Vol. 4** 223, 424
DIE NICHT HEIRATEN DÜRFEN **Vol. 4** 424, 742
DIE – ODER KEINE **Vol. 4** 113, 626
... DIE SICH VERKAUFEN **Vol. 4** 255, 344, 545
DIE VOM ANDEREN UFER **Vol. 4** 339, 377
DIE VOM NIEDERRHEIN **Vol. 4** 339, 492, 544, 584, 646
DIE VOM RUMMELPLATZ **Vol. 4** 164, 200, 257, 584
DIE VOM SCHICKSAL VERFOLGTEN **Vol. 4** 124, 648, 747
DIE VOM ZIRKUS **Vol. 4** 522, 565
DIE VON DER LIEBE LEBEN **Vol. 4** 232, 300
DIE VON DER SCHOLLE SIND **Vol. 4** 273, 359
DIE VON DER WATERKANT **Vol. 4** 132, 584
DIEB UND WEIB **Vol. 4** 425, 433
DIEB VON SAN MARENGO, DER **Vol. 1** 74, 75, 100
DIEBE **Vol. 4** 471, 565
DIECI CANZONI D'AMORE DA SALVARE **Vol. 7** 171
DIECI COMANDAMENTI, I **Vol. 7** 891
DIECI GIORNI CHE SCONVOLSERO IL MONDO **Vol. 7** 719
DIECI MERAVIGLIE DELL AMORE, LE **Vol. 7** 1213
DIECI MINUTI DI VITA **Vol. 7** 644, 1136
DIEGO CORRIENTES **Vol. 8** 516, 572, 657, 732, 753
DIENER LASSEN BITTEN **Vol. 4** 361, 424, 689
DIENST AM KUNDEN **Vol. 4** 18, 517, 558
DIENST IST DIENST **Vol. 4** 212, 239, 270
DIESEL **Vol. 4** 325, 361
DIETER, DER MENSCH UNTER STEINEN **Vol. 4** 424, 742
DIETRO LE QUINTE **Vol. 7** 1189
DIEU A BESOIN DES HOMMES **Vol. 2** 488, 722
DIEU A CHOISI PARIS **Vol. 2** 512
DIEUX DU DIMANCHE, LES **Vol. 2** 401
DIEVCA Z JAZERA **Vol. 10** 72, 210, 483,
DIEZ DIAS MILIONARIA **Vol. 8** 262
DIEZ FUSILES ESPERAN **Vol. 8** 438, 719
DIEZ HORCAS PARA UN PISTOLERO **Vol. 8** 677
DIFENDIMI DALLA NOTTE **Vol. 7** 383
DIFENDO IL MIO AMORE **Vol. 7** 1167
DIFERENTE **Vol. 8** 221, 517
DIFFICILE MORIRE **Vol. 7** 607
DIFFICULTE D'ETRE INFIDELE, LA **Vol. 2** 208, 787
DIFUNTO ES UN VIVO, EL **Vol. 8** 396, 400, 437, 470
DIGA SUL PACIFICO, LA **Vol. 7** 725
DIGAN LO QUE DIGAN **Vol. 8** 258, 338
DIGTEREN OG BASUNBLAESEREN **Vol. 5** 83
DIH **Vol. 6** *page 305, page 307*
DIK VALECNEHO SIROTKA **Vol. 10** 525
DIKE ENOS ATHOU **Vol. 3** 141
DIKTATOR KALEI THANASE **Vol. 3** 205
DIKTATUR DER LIEBE, DIE **Vol. 4** 626
DILEMA **Vol. 8** 437, 725
DILEMMA **Vol. 3** 416 **Vol. 5** 179, 182
DILIGENCEKUSKEN FRA ST. HILO **Vol. 5** 154
DILIGENCIA DE LOS CONDENADOS, LA **Vol. 8** 768
DILIZANSA SNOVA **Vol. 3** 349, 374
DIM **Vol. 3** 408, 429
DIMANCHE A LA CAMPAGNE, UN **Vol. 6** *page 125, page 133*
DIMANCHE DE FLIC, UN **Vol. 6** *page 103, page 135*
DIMANCHE DE LA VIE, LE **Vol. 2** 263
DIMANCHES DE VILLE-D'AVRAY, LES **Vol. 2** 224, 285
DIMENSIONE VIOLENZA **Vol. 7** 99
DIMENTICA IL MIO PASSATO **Vol. 8** 546, 749
DIMENTICARE VENEZIA **Vol. 7** 13, 1008, 1035
DIMINETILE UNUI BAIAT CUMINTE **Vol. 3** 269, 307, 319
DIMITROVGRADCI **Vol. 3** 55, 75, 97 **Vol. 6** *page 41*
DIMMI CHE FAI TUTTO PER ME **Vol. 7** 524
DIMRI I FUNDIT **Vol. 3** 22, 24
DIN FORTID ER GLEMT **Vol. 5** 2, 228
DIN GROZAVIILE LUMII **Vol. 3** 249
DIN NOU IMPREUNA **Vol. 3** 307, 311
DIN VIATA LUI PACALA **Vol. 3** 212
DINA E DJANGO **Vol. 8** 7
DINAMITA ESTA SERVIDA, LA **Vol. 8** 331, 681
DINAMITA JOE **Vol. 8** 546, 772
DINAMITEROS, LOS **Vol. 8** 524, 645
DINCOLO DE BARIERA **Vol. 3** 281, 295, 330
DINCOLO DE BRAZI **Vol. 3** 248, 285
DINCOLO DE NISIPURI **Vol. 3** 324, 326
DINCOLO DE ORIZONT **Vol. 3** 229, 292
DINCOLO DE POD **Vol. 3** 239, 254, 255
DINDON, LE **Vol. 2** 204, 395, 581, 776, 856

DINERO NEGRO Vol. 8 439, 447, 670
DINERO TIENE MIEDO, EL Vol. 8 524
DING IM SCHLOSS, DAS Vol. 1 53, 87, 113
DINHEIRO DOS POBRES, O Vol. 8 44, 172
DINU SI RODICA Vol. 3 304
DIO LI FA POL LI ACCOPPIA Vol. 7 408, 697
DIO PERDONA ... IO NO Vol. 8 770
DIO PERDONI LA MIA PISTOLA Vol. 7 743
DIO SEI PROPRIO UN PADRETTERNO ! Vol. 7 1136
DIONISIA Vol. 7 250
DIOS ELIGIO SUS VIAJEROS Vol. 8 450, 600
DIOSA SALVAJE, LA Vol. 8 317, 515
DIPLE THYSIA Vol. 3 146
DIPLOMES DU DERNIER RANG, LES Vol. 6 *page 85, page 140*
DIPUTADO, EL Vol. 8 349
DIR GEHÖRT MEIN HERZ Vol. 7 453, 493
DIR ZULIEBE Vol. 4 85, 582, 678
DIRECT AU COEUR Vol. 2 113, 209, 318, 727
DIRECTORUL NOSTRU Vol. 3 231, 248, 261, 286, 330
DIREKTION VERLOBT SICH, DIE Vol. 4 743
DIREKTOR, DER Vol. 6 *page 166, page 168, page 175*
DIRETTISSIMI DI MEZZANOTTE, IL Vol. 7 548
DIRETTISSIMO DELLE 19 Vol. 7 1189
DIRITTO DI UCCIDERE Vol. 7 1189
DIRITTO DI UCCIDERE, IL Vol. 7 1038
DIRNENTRAGÖDIE Vol. 4 329, 646
DIRTY STORY Vol. 5 522, 946, 1074
DIS BONJOUR A LA DAME Vol. 2 600
DIS-MOI QUE TU M'AIMES Vol. 2 335
DIS-MOI QUI TUER Vol. 2 163, 702
DISAJ, COVECE Vol. 6 *page 24, page 33, page 43*
DISCO ROJO Vol. 8 125, 149, 551, 749
DISCO VOLANTE, IL Vol. 7 87
DISCOPRIBEH Vol. 10 159, 344, 511
DISCOTECA, LA Vol. 7 123, 222
DISERTORE, IL Vol. 7 25, 766
DISFATTA DELL'ERINNI, LA Vol. 7 369
DISFATTA DI SHERLOCK HOLMES, LA Vol. 7 560
DISGRAZIA DI UN UBRIACO, LE Vol. 7 1189
DISKENSPRINGEREN Vol. 5 83
DISKRET OPHOLD Vol. 5 150, 248
DISKRETION - EHRENSACHE Vol. 4 108, 270, 623
DISONESTI, I Vol. 7 316
DISONORATA Vol. 7 71
DISORDINE, IL Vol. 7 86, 498
DISPARU DE L'ASCENSEUR, LE Vol. 2 237
DISPARUS DE SAINT-AGIL, LES Vol. 2 344, 533, 775
DISPERATO ABBANDONO Vol. 7 304
DISPERATO ADDIO Vol. 7 373
DISPOSTA A TUTTO Vol. 7 692
DISPUTADO VOTO DEL SENOR CAYO, EL Vol. 8 610, 762
DISPUTE DE COCHERS Vol. 2 678
DISQUE 413, LE Vol. 2 217
DISSONANZEN DES LEBENS Vol. 4 603
DISTINTA FAMIGLIA, UNA Vol. 7 756
DISTRACTIONS, LES Vol. 2 742
DISTRAIT, LE Vol. 2 142, 860
DISTRATTO, IL Vol. 7 1189
DISTRAZIONI, LE Vol. 7 386
DISUBBEDIENZA, LA Vol. 7 498, 1083, 1093
DISZELÖADAS Vol. 9 87, 117, 199
DISZMAGYAR Vol. 9 117, 237
DIT VINDARNA BÄR Vol. 5 907, 971, 972, 1059
DITA DI FATA Vol. 7 1211
DITA E PARE E EMERIMIT Vol. 6 *page 14, page 15*
DITA SAXOVA Vol. 10 254, 292, 299
DITE PERIFERIE Vol. 10 58, 125, 345
DITES-LE AVEC DES FLEURS Vol. 2 11, 190
DITES-LUI QUE JE L'AIME Vol. 2 517, 536
DITIRAMBO Vol. 8 232, 768
DITO NELLA PIAGA, IL Vol. 7 692
DITTE MENNESKEBARN Vol. 5 150, 248
DIVA Vol. 6 *page 87, page 95*
DIVA BARA Vol. 10 142, 154, 563, 621
DIVERGENZE PARALLELE Vol. 7 1149
DIVERZANTI Vol. 3 343, 346, 401, 415
DIVIETO DI SOSTA Vol. 7 701, 733, 910
DIVINA CREATURA Vol. 7 988
DIVINAS PALABRAS Vol. 8 771
DIVINE Vol. 2 348, 362, 388, 402, 450
DIVINE CROISIERE, LA Vol. 2 217
DIVINE SPARK, THE Vol. 7 487, 1172
DIVISIONE FOLGORE Vol. 7 634, 1147
DIVKA NA KOSTETI Vol. 10 41, 388, 425
DIVKA S MUSLI Vol. 10 18, 165, 369
DIVKA S TREMI VELBLOUDY Vol. 10 22, 522, 621
DIVKA V MODREM Vol. 10 460, 621
DIVLJE ANDJELI Vol. 3 430
DIVLJE SENKE Vol. 3 367, 418
DIVOCH Vol. 10 256, 460
DIVOKA MARYNA Vol. 10 198
DIVOKA SRDCE Vol. 10 165, 344, 511

DIVOKY KONIK RYN **Vol. 10** 233, 313, 371
DIVORCE HEUREUX, UN **Vol. 2** 674 **Vol. 5** 93, 179, 244
DIVORCEMENT, LE **Vol. 2** 450, 828 **Vol. 6** *page 144*
DIVORCIO A LA ANDALUZA **Vol. 8** 495, 667
DIVORCIO QUE VIENE, EL **Vol. 8** 438, 762
DIVORZIO ALL'ITALIANA **Vol. 7** 86, 401
DIVOTVORNY KLOBOUK **Vol. 10** 142, 286, 532, 543
DIX-HUIT HEURES D'ESCALE **Vol. 2** 293, 294, 401
DIX-SEPTIEME CIEL, LE **Vol. 2** 60, 376, 548
DIXIEME SYMPHONIE, LA **Vol. 2** 153
DIZZENGOFF 99 **Vol. 2** 110
DJAEVELENS DATTER **Vol. 5** 9
DJAEVLESPILLET I ZOOLOGISK HAVE **Vol. 5** 97
DJANGO **Vol. 8** 311
DJANGO IL BASTARDO **Vol. 7** 1024
DJAVOLSKOTO OR"ZIE **Vol. 6** *page 37, page 53, page 64*
DJAVOL"T SOFIA **Vol. 3** 98
DJÄVULLENS ÖGA **Vol. 5** 940, 1052, 1053
DJÄVULSFORSEN **Vol. 5** 310, 634
DJEVOJKA I HRAST **Vol. 3** 451
DJUNGELSAGA, EN **Vol. 5** 1167
DJURGARDSNÄTTER **Vol. 5** 857
DJURGARDSVÄLLAR **Vol. 5** 997, 1173
DLA CIEBIE POLSKO **Vol. 1** 317
DLA SZCZESCIA **Vol. 1** 176, 326
DLUGA NOC **Vol. 6** *page 251*
DNES NAPOSLED **Vol. 10** 296, 564
DNES NEORDINUJI **Vol. 10** 256, 532
DNES O PUL JEDENACTE **Vol. 10** 92, 407
DNES VECER VSECHNO SKONCI **Vol. 10** 292, 566
DNESKA PRISEL NOVY KLUK **Vol. 10** 116, 161, 462
DNEVNA SVETLINA **Vol. 3** 76, 103 **Vol. 6** *page 48*
DNY ZRADY **Vol. 10** 283, 313, 516
DO GORY NOGAMI **Vol. 6** *page 225, page 228, page 239*
DO GRADA E BLIZO **Vol. 3** 54 **Vol. 6** *page 59*
DO KRWI OSTATNIEJ **Vol. 1** 360, 395 **Vol. 6** *page 238*
DO OUTRO LADA DO ESPELHO - ATLANTIDA **Vol. 8** 47
DO PANSKEHO STAVU **Vol. 10** 345, 586
DO POBEDATA I PO NEA **Vol. 3** 384
DO POSLEDNEHO DYCHU **Vol. 10** 263, 459, 483
DO WIDZENIA, DO JUTRA **Vol. 1** 148, 233, 250, 298
DO ZBRANE, KURUCI ! **Vol. 10** 32, 263, 483
DO ZUBU A DO SRDICKA **Vol. 10** 116, 246, 508
DOBAR VETAR PLAVA PTICO **Vol. 3** 391, 441
DOBBELTE MAND, DEN **Vol. 5** 46, 283
DOBBELTGAENGEREN **Vol. 5** 97
DOBRA KOB **Vol. 3** 401
DOBRE SITUOVANY PAN **Vol. 10** 564, 621
DOBRE SVETLO **Vol. 10** 378, 511
DOBRI HOLUBI SE VRACEJI **Vol. 10** 129, 313, 577
DOBRI STARI PIANINO **Vol. 3** 345, 387
DOBRO MORJE **Vol. 3** 345, 395
DOBRODRUZSTVI JOE FOCKA **Vol. 10** 201
DOBRODRUZSTVI NA LABI **Vol. 10** 250
DOBRODRUZSTVI NA ZLATE ZATOCE **Vol. 10** 229, 286
DOBRODRUZSTVI S BLASIEM **Vol. 10** 38, 105, 199, 428
DOBRODRUZSTVI S NAHYM KLUKEM **Vol. 10** 131, 395
DOBROVOLJCI **Vol. 6** *page 310, page 314, page 317*
DOBRY DEN, MESTO **Vol. 10** 173, 233, 577
DOBRY TRAMP BERNASEK **Vol. 10** 167, 256
DOBRY VOJAK SVEJK **Vol. 10** 132, 138, 167, 359, 453, 522, 524
DOC, MANOS DE PLATE **Vol. 8** 732
DOCE HOMBRES Y UNA MUJER **Vol. 8** 394, 648
DOCE HORAS DE VIDA **Vol. 8** 396, 732
DOCE LUNAS DE MIEL **Vol. 8** 692
DOCI I OSTATI **Vol. 3** 341, 350
DOCKHEM, ETT **Vol. 5** 840, 843
DOCTEUR FRANCOISE GAILLAND **Vol. 2** 256, 719
DOCTEUR JUSTICE **Vol. 2** 427, 445
DOCTEUR LAENNEC **Vol. 2** 330, 719
DOCTEUR POPAUL **Vol. 2** 520, 567, 700
DOCTOR CANAMON, EL **Vol. 8** 359, 475
DOCTOR, ESTOY BUENA ? **Vol. 8** 624, 770
DOCTOR JEKYLL Y EL HOMBRE LOBO **Vol. 8** 221, 402
DOCTOR, ME GUSTAN LAS MUJERES, ES GRAVE ? **Vol. 8** 407, 603
DOCTOR ROJO, EL **Vol. 8** 424
DOCTORUL POENARU **Vol. 3** 253, 289 **Vol. 6** *page 272*
DOCUMENTEUR **Vol. 6** *page 71*
DOCUMENTO, IL **Vol. 7** 493, 756, 1000

DOCUMENTO Z - 3 Vol. 7 910
DOCUMENTS SECRETS Vol. 2 273, 833
DOD I SKONHED, EN Vol. 5 295
DODE SKIB, DET Vol. 5 242
DODEK NA FRONCIE Vol. 1 278, 308, 404
DODEN ER ET KJAERTEGN Vol. 5 676, 717, 800
DODEN I GATENE Vol. 5 672, 695, 720, 786
DODEN KOMMER TIL MIDDAG Vol. 5 31, 248, 272
DODES HALSBAAND, DEN Vol. 5 97
DODES O, DE Vol. 5 86
DÖDES SKUGGA, DEN Vol. 5 1099
DODES TJERN, DE Vol. 5 724, 800
DÖDKYSSEN Vol. 5 997
DODSANGSTENS MASKESPIL Vol. 5 97
DODSBOKSEREN Vol. 5 55, 83
DODSDROMMEN Vol. 5 97
DÖDSPOLARE Vol. 5 361
DODSSEJLEREN Vol. 5 188
DODSVARSLET Vol. 5 203
DOG, THE Vol. 2 633
DOGADJAJ Vol. 3 437, 451
DÖGKESELYÜ Vol. 9 51, 171, 253
DOGN MED ILSE, ET Vol. 5 261
DOGN UDEN LOGN, ET Vol. 5 207, 294
DOI BAIETI CA PIINEA CALDA Vol. 3 248, 277, 332
DOI BARBATI PENTRU O MOARTE Vol. 3 261, 306, 319
DOI VECINI Vol. 3 281, 299, 330
DOIGTS DANS LA TETE, LES Vol. 2 276, 450
DOIS DIAS NO PARAISO Vol. 8 126, 130
DOKTOR Vol. 6 *page 313, page 316*
DOKTOR GLAS Vol. 5 843, 923
DOKTOR JUDYM Vol. 1 203, 276 Vol. 6 *page 244*
DOKTOR KOVACS ISTVAN Vol. 9 26, 117, 133
DOKTOR MINORKA VIDOR NAGY NAPJA Vol. 9 107, 210, 215
DOKTOR MLADEN Vol. 3 343, 351, 370, 415
DOKTOR MUREK Vol. 1 308, 346
DOKTOR MURSKI Vol. 1 317
DOKTOR SATANSOHN Vol. 4 366
DOKTOR UR Vol. 9 77
DOKUD MAS MAMINKU Vol. 10 256, 586
DOLAR DE FUEGO, UN Vol. 8 562, 768
DOLAR DE RECOMPENSA, UN Vol. 8 312
DOLAR Y UNA TUMBA, UN Vol. 8 708, 781
DOLCE CHE SCAPPA Vol. 7 1189
DOLCE E SELVAGGIO Vol. 7 325
DOLCE GOLA Vol. 7 828
DOLCE VELENO, IL Vol. 7 441
DOLCE VITA, LA Vol. 7 522, 725, 1205
DOLCE ZIE, LE Vol. 7 1213
DOLCH DES MALAYEN, DER Vol. 4 223, 560
DOLCI INGANNI, I Vol. 7 300, 634, 778, 910
DOLINA Vol. 10 82, 273, 548
DOLINA ISSY Vol. 6 *page 214, page 236, page 245*
DOLINA MIRA Vol. 3 439
DOLINA MIRU Vol. 3 449
DOLINA SZCZESCIA Vol. 6 *page 213, page 256*
DOLLAR Vol. 5 840, 909
DOLLARI E FRAKS Vol. 7 825
DOLLARIMILJOONA Vol. 5 576
DOLLARKIRALY LANYA Vol. 9 217
DOLLARMILLIONEN Vol. 5 891, 1031
DOLLARO DI FIFA, UN Vol. 7 1027
DOLLARO PER SETTE VIGLIACCHI, UN Vol. 8 681
DOLLARPAPA Vol. 9 71, 312, 326
DOLLARPRINSESSEN Vol. 5 121
DOLLARPRINZESSIN UND IHRE SECHS FREIER, DIE Vol. 4 239, 479
DOLLIA E DASMES SIME Vol. 3 9, 18 Vol. 6 *page 13*
DOLLY Vol. 2 32
DOLLY MACHT KARRIERE Vol. 4 28, 164, 584, 729
DOLLY TAR CHANSEN Vol. 5 840, 909, 1122, 1123, 1189
DOLMEN TRAGIQUE, LE Vol. 2 67, 131
DOLORES, LA Vol. 8 449
DOLOROSA, LA Vol. 8 565
DOLOVAI NABOB LEANYA, A Vol. 9 77
DOM Vol. 1 148
DOM BEZ OKIEN Vol. 1 156, 220, 354 Vol. 6 *page 205*
DOM KALLAR OSS MODS Vol. 5 1039
DOM NA DVE ULICI Vol. 3 101, 126
DOM NA PUSTKOWIU Vol. 1 250, 397
DOM NA RAZCESTI Vol. 10 273, 320, 478
DOM ROBERTO Vol. 8 54
DOM SWIETEGO KAZIMIERZA Vol. 6 *page 240, page 255, page 259*
DOM WARIATOW Vol. 6 *page 231*
DOM ZA NEZNI DUSI Vol. 6 *page 44, page 49*
DOMANI E TROPPO TARDI Vol. 7 21, 359, 443, 476
DOMANI MI SPOSO Vol. 7 102
DOMANI PASSO A SALUTARE LA TUA VEDOVA...PAROLA DI EPIDEMIA Vol. 7 426
DOMANI SI BALLA Vol. 7 38, 103
DOMAREN Vol. 5 1028, 1042, 1080
DOMATIO 19 Vol. 3 170

DOMEK Z KART **Vol. 1** 133, 305 **Vol. 6** *page 223*
DÓMEN ICKE **Vol. 5** 996
DOMENICA **Vol. 2** 395
DOMENICA D'AGOSTO **Vol. 7** 86, 1036
DOMENICA DELLA BUONA GENTE, LA **Vol. 7** 17
DOMENICA E SEMPRE DOMENICA **Vol. 7** 693
DOMICILE CONJUGAL **Vol. 2** 17, 548
DOMINATIONS **Vol. 2** 365
DOMINATORE DEL DESERTO, IL **Vol. 7** 528
DOMINGO A TARDE **Vol. 8** 169
DOMINGO DE CARNAVAL **Vol. 8** 268, 311, 725
DOMINIQUE **Vol. 2** 460, 671
DOMINO **Vol. 2** 158, 407
DOMINO AZZURRO, IL **Vol. 7** 1189
DOMINO VERT, LE **Vol. 2** 733
DOMMENS DAG **Vol. 5** 9
DOMMER IKKE **Vol. 5** 9, 197
DOMMERENS HUSTRU **Vol. 5** 81
DOMPTEUR, LE **Vol. 2** 197, 626, 824
DON BOSCO **Vol. 7** 1, 493, 1036
DON BUONAPARTE **Vol. 7** 359, 360, 1198
DON CAMILLO **Vol. 7** 251, 498, 524, 580, 709, 736
DON CAMILLO E GIOVANE D'OGGI **Vol. 7** 318
DON CAMILLO E L'ONOREVOLE PEPPONE **Vol. 7** 196, 709
DON CESARE DI BAZAN **Vol. 7** 359, 750, 756, 1060
DON CHISCOTTE **Vol. 7** 660, 1173
DON CIPOTE DE LA MANGA **Vol. 8** 551
DON D'ADELE, LE **Vol. 2** 243, 717
DON ERRE QUE ERRE **Vol. 8** 214, 493
DON FLORIPONDO **Vol. 8** 726
DON GABRIEL **Vol. 1** 211, 281 **Vol. 6** *page 211, page 248*
DON GIOVANNI **Vol. 2** 299 **Vol. 7** 170, 442, 552, 701, 725, 960, 1011
DON GIOVANNI IN SICILIA **Vol. 7** 514, 646, 1087
DON JOSE, PEPE Y PEPITO **Vol. 8** 546, 616, 630
DON JUAN **Vol. 2** 395, 863 **Vol. 4** 170, 320, 452, 454 **Vol. 6** *page 161, page 165, page 170, page 196, page 199* **Vol. 8** 298, 400, 438, 567
DON JUAN 73 **Vol. 2** 21, 224, 873
DON JUAN DE SERRALLONGA **Vol. 8** 515, 721
DON JUAN ET FAUST **Vol. 2** 44, 330
DON JUAN HEIRATET **Vol. 4** 135, 514
DON JUAN IN DER MÄDCHENSCHULE **Vol. 4** 340, 407, 608, 782
DON JUAN LEGUTOLSO KALANDJA **Vol. 9** 67, 117, 199
DON JUAN TENORIO **Vol. 8** 583
DON JUANS OVERMAND **Vol. 5** 83
DON LORENZO **Vol. 7** 196
DON LUCIO Y EL HERMANO PIO **Vol. 8** 214, 400
DON LUIGI STURZO **Vol. 7** 312, 411, 910
DON MANUEL, DER BANDIT **Vol. 4** 565, 725
DON MILANI **Vol. 7** 406
DON OLSEN KOMMER TIL BYEN **Vol. 5** 23, 275, 281
DON PABLO IL BANDITO **Vol. 7** 453, 796
DON PASQUALE **Vol. 7** 796, 902, 1060
DON QUICHOTTE **Vol. 2** 23, 290, 688
DON QUIJOTE **Vol. 8** 450, 583, 602
DON QUIJOTE CABALGA DE NUEVO **Vol. 8** 616, 719
DON QUIJOTE DE LA MANCHA **Vol. 8** 214, 400
DON QUINTIN EL AMARGAO **Vol. 8** 228, 270
DON QUIXOTE **Vol. 5** 24, 83, 138
DONA FRANCISQUITA **Vol. 8** 261, 298, 449
DONA JUANITA **Vol. 4** 28, 178, 203, 325, 509, 609
DONA MARIA LA BRAVA **Vol. 8** 298
DONANTE, EL **Vol. 8** 770
DONATELLA **Vol. 7** 333
DONAUMELODIEN **Vol. 4** 247, 400, 544
DONAUSCHIFFER **Vol. 4** 247, 391
DONAUWALZER **Vol. 4** 287, 458, 646
DONCEL DE LA REINA, EL **Vol. 8** 298
DONCELLA DE LA DUQUESA, LA **Vol. 8** 692
DONDE ESTARA MI NINO ? **Vol. 8** 303, 413
DONDE HAY PATRON ... **Vol. 8** 331, 396
DONDE PONGO ESTE MUERTO ? **Vol. 8** 217, 470
DONDE TU ESTES **Vol. 8** 271, 350
DONDE VAS ALFONSO XII ? **Vol. 8** 207
DONDE VAS TRISTE DE TI ? **Vol. 8** 207, 214
DONKOSAKENLIED, DAS **Vol. 4** 382, 492, 584 **Vol. 6** *page 199*
DONNA A 1 DIMENSIONE, LA **Vol. 7** 798
DONNA ABBANDONATA, LA **Vol. 7** 150
DONNA AL BAGNO **Vol. 7** 1189
DONNA ALLA FINESTRA, UNA **Vol. 7** 1136
DONNA ALLO SPECCHIO, UNA **Vol. 7** 318, 358, 770
DONNA BIANCA, LA **Vol. 7** 978
DONNA CARNEFICE NEL PAES DELL'ORO **Vol. 7** 454

DONNA CHE FU MOLTO AMATA, LA Vol. 7 159
DONNA CHE INVENTO L'AMORE Vol. 7 196
DONNA CHE NON EBBE CUORE, LA Vol. 7 965
DONNA CHE VENNE DAL MARE, LA Vol. 7 1062
DONNA CHIAMATA APACHE, UNA Vol. 7 1003
DONNA COSA SI FA PER TE Vol. 7 1179
DONNA DAI CAPELLI D'ORO, LA Vol. 7 655, 1113
DONNA D'ALTRI TEMPI, UNA Vol. 7 508, 529, 767
DONNA DEI FARAONI, LA Vol. 7 877
DONNA DEL FIUME, LA Vol. 7 725, 781
DONNA DEL GIORNO, LA Vol. 7 817
DONNA DEL MARE, LA Vol. 7 279, 296, 364, 863
DONNA DEL PECCATO, LA Vol. 7 453, 476
DONNA DELLA DOMENICA, LA Vol. 7 32, 1145
DONNA DELLA MONTAGNA, LA Vol. 7 756, 1116
DONNA DELLE MERAVIGLIE, LA Vol. 7 498, 1007
DONNA DI CUORI, LA Vol. 7 150
DONNA DI GHIACCIO, LA Vol. 7 1152
DONNA DI NOTTE, UNA Vol. 7 692
DONNA DI SECUNDA MANO, UNA Vol. 7 819
DONNA DI TRENT'ANNI, LA Vol. 7 783
DONNA DI UNA NOTTE, LA Vol. 7 978
DONNA E I BRUTI, LA Vol. 7 53
DONNA E IL CADAVERE, LA Vol. 7 1211
DONNA E L'ARMATURA, LA Vol. 7 196
DONNA E L'UOMO, LA Vol. 7 560
DONNA E MOBILE, LA Vol. 7 487
DONNA E UNA COSA MERAVIGLIOSA, LA Vol. 7 725, 1167
DONNA HA UCCISO, UNA Vol. 7 17
DONNA INVISIBILE, LA Vol. 7 585
DONNA LIBERA, UNA Vol. 7 500
DONNA LISA Vol. 7 900
DONNA NUDA, LA Vol. 7 264, 701, 725
DONNA PERDUTA, LA Vol. 7 48, 665, 1103
DONNA PIU BELLA, LA Vol. 7 106
DONNA PREGA, UNA Vol. 7 17
DONNA SCIMMIA, LA Vol. 7 498, 1136
DONNA TRA DUE MONDI, UNA Vol. 7 48, 620
DONNA, UNA Vol. 7 621
DONNA, UNA MUMMIA, UN DIPLOMATICO, UNA Vol. 7 26
DONNE CI TENGONO ASSAI, LE Vol. 7 603
DONNE E BRIGANTI Vol. 7 798, 1054
DONNE E SOLDATI Vol. 7 1167
DONNE-MOI TES YEUX Vol. 2 122, 579, 698
DONNE PROIBITE Vol. 7 196
DONNE SOLE Vol. 7 528
DONNE SONO DEBOLI, LE Vol. 7 619
DONNER, BLITZ UND SONNENSCHEIN Vol. 4 430, 558, 796
DONNEZ-MOI DIX HOMMES DESESPERES Vol. 2 137, 268
DONNEZ-MOI MA CHANCE Vol. 2 185
DONO DEL MATTINO, LE Vol. 7 570, 1189
DONOGOO TONGA Vol. 4 49, 298, 608
DÖNTÖ PILLANAT, A Vol. 9 126, 184, 315
DOPIO VOLTO, IL Vol. 7 107
DOP"LNENIE K"M ZAKONA ZA ZASTITA NA D"RZAVATA Vol. 3 38, 60, 101 Vol. 6 *page 46*
DOPO DIVORZIEREMO Vol. 7 252, 443, 487, 618
DOPO IL PECCATO Vol. 7 369
DOPO IL PERDONE Vol. 7 886, 1079
DOPO IL SUICIDO Vol. 7 277
DOPO IL VEGLIONE Vol. 7 796
DOPPELBRÄUTIGAM, DER Vol. 4 758
DOPPELGÄNGER, DER Vol. 4 123, 135, 323, 503, 508, 788
DOPPELSELBSTMORD Vol. 4 110, 558
DOPPELT ODER NICHTS Vol. 1 29, 92, 102
DOPPELTE SPIEL, DAS Vol. 1 53, 60
DOPPIA FERITA, LA Vol. 7 796
DOPPIO DELITTO Vol. 7 599
DORA ET LA LANTERNE MAGIQUE Vol. 2 46
DORA JELENTI ... Vol. 9 121, 152, 259
DORA, LA FRENESIE DU PLAISIR Vol. 2 291
DORA NELSON Vol. 2 579, 722, 727 Vol. 7 196, 1181
DORA O LE SPIE Vol. 7 264, 701
DORFBARBIER, DER Vol. 4 119, 517, 558
DORIAN GRAYS PORTRAET Vol. 5 203
DORNENWEG EINER FÜRSTIN Vol. 4 329, 634
DORNRÖSCHEN Vol. 1 2, 5, 29 Vol. 4 242, 261, 396, 652, 697
DOROTEJ Vol. 6 *page 312, page 315*
DOROTHEA Vol. 2 301
DOROTHEAS GAVE Vol. 5 81
DOROTHEE CHERCHE L'AMOUR Vol. 2 249, 395
DOROTHY Vol. 7 1113
DOROZKARZ NR. 13 Vol. 1 384, 404

DORRIT BEKOMMT 'NE LEBENSSTELLUNG **Vol. 4** 342
DORRITCHENS VERGNÜGUNGSREISE **Vol. 4** 342
DORRITS CHAUFFEUR **Vol. 4** 342
DORRITS EHEGLÜCK **Vol. 4** 342
DORS BONHOMME **Vol. 2** 39
DORTE **Vol. 5** 2, 60, 228
DORTOIR DES GRANDES **Vol. 2** 232, 488, 722
DORUCAK SA DJAVOLOM **Vol. 3** 382, 427
DOS **Vol. 8** 373
DOS ALAS **Vol. 8** 311, 402
DOS ANOS DE VACACIONES **Vol. 8** 221, 753
DOS AU MUR, LE **Vol. 2** 232, 488, 515 **Vol. 7** 619
DOS CAMINOS **Vol. 8** 207, 431, 630
DOS CHICAS LOCAS LOCAS **Vol. 8** 338, 762
DOS CONTRA AL CAPONE **Vol. 8** 306, 697
DOS COSMONAUTAS A LA FUERZA **Vol. 8** 331
DOS CRUCES EN DANGER PASS **Vol. 8** 396
DOS CUENTOS PARA DOS **Vol. 8** 396, 772
DOS DE LA MAFIA **Vol. 8** 598, 772
DOS DE MAYO, EL **Vol. 8** 281
DOS GEMELAS ESTUPENDAS **Vol. 8** 395, 532
DOS GOLFILLOS, LOS **Vol. 8** 524
DOS MEJOR QUE UNO **Vol. 8** 373, 489
DOS MIL DOLARES POR COYOTE **Vol. 8** 677, 772
DOS MUJERES EN LA NIEBLA **Vol. 8** 752
DOS MUJERES Y UN DON JUAN **Vol. 8** 285, 516
DOS MUJERES Y UN ROSTRO **Vol. 8** 272
DOS PILOTOS Y PICO **Vol. 8** 471
DOS PISTOLAS GENELAS **Vol. 8** 603
DOS PISTOLEROS **Vol. 8** 258, 438
DOS RIVALES, LOS **Vol. 8** 396
DOS VECES JUDAS **Vol. 8** 523, 732
DOS VIDAS **Vol. 8** 357, 658
DOS VIVALES EN FUERTE ALAMO **Vol. 8** 437, 732
DOS Y DOS, CINCO **Vol. 8** 438, 681
DOS Y MEDIA Y VENENO, LAS **Vol. 8** 546, 558, 725
DOSLO DOBA DA SE LJUBAV PROBA **Vol. 3** 410, 424
DOSPELACI MUZOU VSECHNO **Vol. 10** 250, 462
DOSSIER 1413 **Vol. 2** 2131, 318, 433
DOSSIER 51, LE **Vol. 2** 323, 481
DOSSIER NOIR, LE **Vol. 2** 124, 197, 801
DOSSIER PROSTITUTION **Vol. 2** 769
DOST DOBRI CHLAPI **Vol. 10** 82, 263, 320
DOSTIH **Vol. 10** 389, 425, 571
DOTKNIECIE NOCY **Vol. 1** 189, 220, 401
DOTTER FÖDD, EN **Vol. 5** 864, 1205
DOTTOR ANTONIO, IL **Vol. 7** 570, 1116
DOTTOR JEKYLL E GENTILE SIGNORA **Vol. 7** 567, 1015
DOTTORESSA ... CI STA COL COLONELLO **Vol. 7** 116, 1194
DOTTORESSA DI CAMPAGNA, LA **Vol. 7** 289
DOTTORESSA PREFERISCE I MARINAI **Vol. 7** 811, 1202
DOTYKY **Vol. 10** 204, 389, 496
D'OU VIENS-TU JOHNNY ? **Vol. 2** 194, 875
DOUA LOZURI **Vol. 3** 265, 297, 330
DOUAUMONT **Vol. 4** 97, 204
DOUBLE AMOUR, LE **Vol. 2** 237, 426
DOUBLE CRIME SUR LA LIGNE MAGINOT **Vol. 2** 395, 858
DOUBLE DESTIN **Vol. 2** 47
DOUBLE ENIGME DU DR. MORART, LA **Vol. 2** 428
DOUBLE EXISTENCE DE LORD SAMSEY, LA **Vol. 2** 688
DOUCE **Vol. 2** 8, 439, 786
DOUCE ENQUETE SUR LA VIOLENCE **Vol. 6** *page 75, page 121, page 145*
DOUCE VIOLENCE **Vol. 2** 309, 532
DOUCEMENT LES BASSES **Vol. 2** 457, 814
DOUCEUR D'AIMER, LA **Vol. 2** 871
DOUGLAS **Vol. 5** 660, 712
DOULEUR, LA **Vol. 2** 217, 238, 724
DOULOREUSE, LA **Vol. 7** 1211
DOULOS, LE **Vol. 2** 376, 395
DOUZE HEURES DE BONHEUR **Vol. 2** 185, 309
DOUZE HEURES D'HORLOGE **Vol. 2** 13, 248, 671
DOV E LA LIBERTA ? **Vol. 7** 43, 333, 438, 1136
DOV' E LA MIA VITA ! **Vol. 7** 298
DOVE ANDIAMO, SIGNORA ? **Vol. 7** 890
DOVE VAI IN VACANZA ? **Vol. 7** 383
DOVE VAI IN VACANZA ? – SARO TUTTA PER TE **Vol. 7** 1145
DOVE VAI IN VACANZA ? – VACANZA INTELLIGENTI **Vol. 7** 834
DOVE VAI TUTTA NUDA ? **Vol. 7** 514
DOVE VOLANO I CORVI D'ARGENTO **Vol. 7** 141
DOVERIJE **Vol. 5** 399, 477, 504, 506
DOVIDENIA V PEKLE, PRIATELIA ! **Vol. 10** 36, 200, 228
DOVOLENA S ANDELEM **Vol. 10** 407, 433, 524
DPBAR VETAR PLAVA PTICO **Vol. 3** 356

DR **Vol. 3** 349, 374
DR. BESSELS VERWANDLUNG **Vol. 4** 213, 422, 648
DR. BLUFF **Vol. 4** 361
DR. CRIPPEN AN BORD **Vol. 4** 147, 167, 232
DR. GAR EL HAMA I **Vol. 5** 97
DR. GAR EL HAMAS FLUGT **Vol. 5** 97
DR. GLAS **Vol. 5** 74, 191, 273, 832, 933, 1042, 1053
DR. JEKYLL ET LES FEMMES **Vol. 6** *page 99, page 106*
DR. MABUSE, DER SPIELER **Vol. 4** 288, 298, 325, 547, 665, 727
DR. MED. SOMMER II **Vol. 1** 32, 127
DR. MONNIER UND DIE FRAUEN **Vol. 4** 315, 479
DR. NICOLA **Vol. 5** 97
DR. SACROBOSCO, DER GROSSE UNHEIMLICHE **Vol. 4** 643, 765
DR. SCHLÜTER **Vol. 1** 7, 22, 75, 111
DR. STEFFENS' SELTSAMER FALL **Vol. 4** 25, 329, 745
DR. WISLIZENUS **Vol. 4** 124, 747
DR. X **Vol. 5** 295
DRACULA CERCA SANGUE DI VERGINE E ... MORI DI SETE !!! **Vol. 7** 599
DRACULA, PERE ET FILS **Vol. 2** 151, 513
DRAGA IRENA **Vol. 6** *page 307*
DRAGA MOJA IZA **Vol. 3** 360
DRAGEE HAUTE, LA **Vol. 2** 199, 559
DRAGEES AU POIVRE **Vol. 2** 224, 322, 631
DRAGHFY EVA **Vol. 9** 260
DRAGON RAPIDE **Vol. 8** 232, 580
DRAGONERLIEBCHEN **Vol. 4** 287, 384, 458, 646
DRAGOSTE LA ZERO GRADE **Vol. 3** 230, 281, 291, 306
DRAGOSTE LUNGA DE-O SEARA **Vol. 3** 306, 319
DRAGOSTEA INCEPE VINERI **Vol. 3** 239, 256
DRAGOSTEA MARINARULUI **Vol. 3** 234
DRAGOSTEA MEA CALATOARE **Vol. 3** 301
DRAGOSTEA SI REVOLUTIA **Vol. 6** *page 279, page 281, page 282*
DRAGUEURS, LES **Vol. 2** 250, 251, 374, 781, 856
DRAHE TETY A JA **Vol. 10** 144, 359, 398
DRAK SA VRACIA **Vol. 10** 273, 459
DRAKULA **Vol. 9** 93
DRAMA FRA RIDDERTIDEN, ET **Vol. 5** 97
DRAMA NUEVO, UN **Vol. 8** 298, 311, 436
DRAMA PAA SLOTTET **Vol. 5** 55, 196
DRAMA V KABARE FUTURISTOV N.13 **Vol. 2** 833
DRAMA VON GLOSSOW, DAS **Vol. 4** 232, 673
DRAME AU CARLTON CLUB, UN **Vol. 2** 104
DRAME AU VEL' D'HIV' **Vol. 2** 637, 855
DRAME DANS UN MIROIR **Vol. 2** 279, 604
DRAME DE SHANGHAI, LE **Vol. 2** 23, 27, 206, 332, 644, 778
DRAMES DIVERS **Vol. 2** 330, 375, 432
DRAMMA AL MULINO **Vol. 7** 1189
DRAMMA BORGHESE, UN **Vol. 7** 347, 1059
DRAMMA DEL MACCHINISTA, IL **Vol. 7** 1189
DRAMMA DEL MULINO, IL **Vol. 7** 296
DRAMMA DELL'AMORE, IL **Vol. 7** 369
DRAMMA DI VITTORIANO, UN **Vol. 7** 454
DRAMMA IGNORATO, UN **Vol. 7** 277
DRAMMA IN CASERMA **Vol. 7** 1189
DRAMMA IN MONTAGNA, UN **Vol. 7** 454
DRAMMA IN WAGON-LITS, UN **Vol. 7** 304
DRAMMA NELL'OLIMPO, UN **Vol. 7** 399
DRAMMANELLA CASBAH **Vol. 7** 16
DRAMMI DELLA UITA **Vol. 7** 1189
DRAMMI DELL'ALCOLISMO **Vol. 7** 1189
DRANKERSKEN **Vol. 5** 183
DRAPEAU NOIR FLOTTE SUR LA MARMITE, LE **Vol. 2** 279, 408
DRATENICEK **Vol. 10** 45, 167, 290, 453
DRAUFGÄNGER, DER **Vol. 4** 192, 458, 462, 584
DRAVCI **Vol. 10** 92, 503
DREI BLAUEN TIGER, DIE **Vol. 3** 395
DREI BLONDE JUNGS – EIN BLONDES MÄDEL **Vol. 4** 344, 424
DREI FRAUEN VON URBAN HELL, DIE **Vol. 4** 239, 306
DREI GRÜNEN HUNDE, DIE **Vol. 3** 395
DREI HASELNÜSSE FÜR ASCHENBRÖDEL **Vol. 1** 102, 114
DREI KAISERJÄGER **Vol. 4** 287, 462, 644
DREI KAPITEL GLÜCK **Vol. 1** 46, 104, 107
DREI KRIEGE – HINTER DEN FRONTEN **Vol. 1** 26, 32, 86
DREI KRIEGE – IN BERLIN **Vol. 1** 30, 32, 115
DREI KRIEGE – TAUROGGEN **Vol. 1** 17, 32, 86
DREI KUCKUCKSUHREN, DIE **Vol. 4** 306, 729
DREI MÄDCHEN IM ENDSPIEL **Vol. 1** 22, 103, 115
DREI MÄDELS UND IHRE FREIER **Vol. 4** 540, 797
DREI MÄDERL UM SCHUBERT **Vol. 4** 292, 437, 738

DREI MANNEQUINS, DIE **Vol. 4** 306, 535
DREI MARIEN UND DER HERR VON MARANA, DIE **Vol. 4** 89, 342, 752, 793
DREI NÄCHTE **Vol. 4** 218
DREI NIEMANDSKINDER, DIE **Vol. 4** 85, 700
DREI PORTIERMÄDEL, DIE **Vol. 4** 124, 242
DREI SMARAGDE, DIE **Vol. 4** 18, 558
DREI TAGE AUF LEBEN UND TOD **Vol. 4** 76, 204, 424
DREI TAGE LIEBE **Vol. 4** 306, 321
DREI TAGE MITTELARREST **Vol. 4** 98, 246, 379
DREI TANTEN, DIE **Vol. 4** 660, 704
DREI TÄNZE DER MARY WILFORD, DIE **Vol. 4** 190
DREI TOLLE MÄDELS **Vol. 4** 292 **Vol. 7** 756
DREI TOLLE TAGE **Vol. 4** 110, 124, 517
DREI UM CHRISTINE, DIE **Vol. 4** 344, 647, 689
DREI UM EDITH, DIE **Vol. 4** 49, 316
DREI UNTEROFFIZIERE **Vol. 4** 234, 365, 608
DREI VÄTER UM ANNA **Vol. 4** 82, 184
DREI VON DER KAVALLERIE **Vol. 4** 255, 373
DREI VON DER STEMPELSTELLE **Vol. 4** 323, 558, 646
DREI VON DER STRASSE, DIE **Vol. 4** 93, 329
DREI VON DER TANKSTELLE, DIE **Vol. 4** 298, 527
DREI WUNDERSCHÖNE TAGE **Vol. 4** 110, 123, 544
DREIGROSCHENOPER, DIE **Vol. 2** 23 **Vol. 4** 14, 729 **Vol. 6** *page 79*
DREIKLANG **Vol. 4** 123, 365, 544
DREIKLANG DER NACHT **Vol. 4** 132, 157, 270
DREIMÄDERLHAUS, DAS **Vol. 4** 159
DREIMAL EHE **Vol. 4** 329, 774
DREIMAL HOCHZEIT **Vol. 4** 265, 614, 774
DREIMAL KOMÖDIE **Vol. 4** 619
DREIZEHN AUS STAHL, DIE **Vol. 4** 648, 747
DREIZEHN MANN UND EINE KANONE **Vol. 4** 247, 647
DREIZEHN STÜHLE **Vol. 4** 85, 292
DREIZEHNTE WELTWUNDER, DAS **Vol. 4** 115, 323
DRENGE **Vol. 5** 46
DRENGEN DER FORSVANDT **Vol. 5** 11, 219, 296
DRENGEN MED DEN SJETTE SANS **Vol. 5** 97
DREPTATE IN LANTURI **Vol. 6** *page 265, page 271*
DRESZCZE **Vol. 6** *page 234*
DREVENA DEDINA **Vol. 10** 268, 273, 349
DREWNIANY ROZANIEC **Vol. 1** 254, 281, 377 **Vol. 6** *page 247*
DREYFUS **Vol. 4** 32, 49, 626, 734
DREYFUS, L'INTOLERABLE VERITE **Vol. 2** 60, 713
DRITAT E QYTEZES **Vol. 6** *page 14, page 15, page 16, page 18*
DRITTE, DER **Vol. 1** 18, 36, 50 **Vol. 4** 247
DRITTE ESKADRON, DIE **Vol. 4** 149, 660, 715
DRITTE, LE **Vol. 7** 891
DRIVER DAGG, FALLER REGN **Vol. 5** 840, 878
... DROGA DALEKA PRZED NAMI ... **Vol. 6** *page 206, page 225, page 252*
DROGA DI SATANA, LA **Vol. 7** 548
DROGA LLAMADA HELEN, UNA **Vol. 8** 299
DROGA NA ZACHOD **Vol. 1** 227, 254 **Vol. 6** *page 218*
DROGA W SWIETLE KSIEZYCA **Vol. 1** 303
DROGUE DU VICE, LA **Vol. 2** 728
DROHENDE FAUST, DIE **Vol. 4** 290
DROIT A LA VIE, LE **Vol. 2** 153
DROIT D'AIMER, LE **Vol. 2** 224, 353
DROIT DE CUISSAGE, LE **Vol. 2** 291
DROIT DE L'ENFANT, LE **Vol. 2** 167, 328, 496
DROLE DE BOURRIQUE, UNE **Vol. 2** 661
DROLE DE COLONEL, UN **Vol. 2** 97, 872
DROLE DE DIMANCHE, UN **Vol. 2** 194, 625
DROLE DE DRAME **Vol. 2** 13, 644, 778, 837
DROLE DE JEU **Vol. 2** 480, 624
DROLE DE NOCE **Vol. 2** 15, 32
DROLE DE PAROSSIEN, UN **Vol. 2** 840
DROLE DE SHOW **Vol. 6** *page 69*
DROLES DE DIAM'S **Vol. 2** 534, 682
DROLES DE PHENOMENES **Vol. 2** 97, 581, 808
DROLES DE ZEBRES **Vol. 2** 60
DROLESSE, LA **Vol. 2** 754
DRÖM IM FRIHET, EN **Vol. 5** 496
DRÖM POJKEN **Vol. 5** 923
DRÖMMARES VANDRING, EN **Vol. 5** 565, 572
DROMME STOJER IKKE NAR DE DOR **Vol. 5** 46, 61
DRÖMMEN OM AMERIKA **Vol. 5** 882, 900, 1153

DROMMEN OM DET HVIDE SLOT Vol. 5 23, 248
DRÖMMEN OM ETT ANNAT LIV Vol. 5 856
DROMMEN OM STENALDEREN Vol. 5 81
DROMOS ME TIS AKAKIES Vol. 3 144, 205
DROMOS TON EROON Vol. 3 142
DRÖMPOJKEN Vol. 5 946, 1046
DRÖMSEMESTER Vol. 5 928, 1066
DRONNER GJENNOM DALEN, DET Vol. 5 760
DRONNINGENS VAGTMESTER Vol. 5 2, 139, 275
DROPS WIRD FLIEGER Vol. 4 186
DROSKE 519 Vol. 5 97
DROSO Vol. 3 174
DROTOSTOT, A Vol. 9 337
DROTTINGEN AV PELLAGONIEN Vol. 5 891, 1031
DRRAPA – KUL GREJ PA VÄG TILL GÖTET Vol. 5 887
DRSNA PLANINA Vol. 10 41, 385, 466
DRUG PREDSEDNIK CENTARFOR Vol. 3 337, 349
DRUGA MLODOSC Vol. 1 278, 308, 404
DRUGA STRANA MEDALJE Vol. 3 374, 437
DRUGARCINE Vol. 3 367, 407 Vol. 6 *page 306*
DRUGI BRZEG Vol. 1 149, 303, 355
DRUGI CZLOWIEK Vol. 1 202, 223, 401
DRUGOTO STASTIE Vol. 3 93, 100
DRUHA SMENA Vol. 10 34, 621
DRUHE MLADI Vol. 10 345, 585
DRUHY TAH PESCEM Vol. 10 173, 233, 258
DRUM IN PENUMBRA Vol. 3 239, 256
DRUM NOU Vol. 3 285, 295, 303
DRUM, THE Vol. 2 656
DRUMUL OASELOR Vol. 3 240, 297, 301
DRUMURI IN CUMPANA Vol. 3 214, 240, 272
DRUNTER UND DRÜBER Vol. 4 292, 517, 558
DRUSTVENA IGRA Vol. 3 367, 415
DRUZBA PERE KVRZICE Vol. 3 44, 451
DRUZINA CERNEHO PERA Vol. 10 380, 515, 568
DRUZINSKI DNEVNIK Vol. 3 447
D"RVO BEZ KOREN Vol. 3 114 Vol. 6 *page 36, page 41*
DRVOSTEP Vol. 10 167, 272, 536
DRZANJE ZA VAZDUH Vol. 6 *page 309, page 310, page 318*
DRZWI W MURZE Vol. 1 141, 203, 206, 403
DS -70 NEVYJIZDI Vol. 10 460, 503
DSCHUNGEL RUFT, DER Vol. 4 97, 212, 215, 720, 727
DU ÄR INTE KLOK MADICKEN Vol. 5 1001, 1052, 1096
DU ÄR MITT ÄVENTYR Vol. 5 940, 1052
DU BIST DAS LEBEN Vol. 4 232, 673
DU BIST ENZÜCKEND, ROSMARIE ! Vol. 4 354, 558
DU BIST MEIN GLÜCK Vol. 4 344, 647, 689
DU BIST SO SCHÖN, BERLINERIN Vol. 4 323, 329
DU BLE EN LISSES OU L'ARRIVISTE Vol. 2 409, 796
DU CIEL PLEIN LE COEUR Vol. 2 72
DU COTE DES TENNIS Vol. 2 278
DU COTE D'OROUET Vol. 2 618
DU ER IKKE ALENE Vol. 5 120, 283
DU GAMLA DU FRIA Vol. 5 1032, 1181
DU GEHÖRST ZU MIR Vol. 4 325, 378
DU GRABUGE CHEZ LES VEUVES Vol. 2 729, 824
DU GUERRE AUX ALLEMANDS...? Vol. 2 39
DU GUESCLIN Vol. 2 439, 833
DU HAR LOVET MEG EN KONE Vol. 5 803
DU HAUT EN BAS Vol. 2 778
... DU KANNST ARBEITEN, LEBEN ...! Vol. 6 *page 191*
DU KANNST NICHT TREU SEIN Vol. 4 344, 647, 689
DU LOVTE MIG EN KONE Vol. 5 770
DU MOU DANS LA GACHETTE Vol. 2 131, 172
DU MOURON POUR LES PETITS OISEAUX Vol. 2 625, 770
DU PIU DUE UGUALE A SETTE Vol. 7 621
DU RIFFIFI A PANAME Vol. 2 185, 875
DU RIFFIFI CHEZ LES FEMMES Vol. 2 248, 598, 601
DU RIFFIFI CHEZ LES HOMMES Vol. 2 8, 156, 540
DU SKAL AERE Vol. 5 9
DU SKAL AERE DIN HUSTRU Vol. 5 69, 265
DU SKAL ELSKE DIN NAESTE Vol. 5 9
DU SOLEIL PLEIN LES YEUX Vol. 2 732
DU SOLLST DER KAISER MEINER SEELE SEIN Vol. 4 644, 747
DU SOLLST NICHT BEGEHREN Vol. 4 82, 325
DU SOLLST NICHT STEHLEN Vol. 4 245, 292, 584
DU UND ICH Vol. 4 265, 462, 614
DU UND ICH UND KLEIN-PARIS Vol. 1 63, 85, 121

DU UND ICKE UND BERLIN **Vol. 1** 59, 79, 89
DUAJE EMRIN TEND **Vol. 6** *page 16, page 17*
DUAS CAUSAS **Vol. 8** 44, 130, 131
DUBARRY VON HEUTE, EINE **Vol. 4** 729, 752
DUBARY, DIE **Vol. 4** 514
DUBBIO, IL **Vol. 7** 212
DUBLJOR"T **Vol. 3** 48, 72 **Vol. 6** *page 34*
DÜBÖRGÖ CSEND **Vol. 9** 107, 190, 318
DUCA NERO, IL **Vol. 8** 517, 569, 772
DUCH Z CANTERVILLE **Vol. 1** 211, 302, 321
DUCHACEK TO ZARIDI **Vol. 10** 167, 621
DUCHESSA DEL BAL-TABARIN, LA **Vol. 7** 806
DUCHESSA DI NALA, LA **Vol. 7** 548
DUCHESSA DI SANTA LUCIA, LA **Vol. 7** 693
DUCHESSA MISTERO, LA **Vol. 7** 730
DUCHESSE DE LANGEAIS, LA **Vol. 2** 27, 566, 676
DUCHESSINA **Vol. 7** 398
DUCHINO, IL **Vol. 7** 1189
DUDA **Vol. 8** 493, 732
DUDA, LA **Vol. 8** 207, 214
DUDOSA VIRILIDAD DE CRISTOBAL, LA **Vol. 8** 261
DUDU, EIN MENSCHENSCHICKSAL **Vol. 4** 321, 648
DUE BARBIERI, I **Vol. 7** 460, 665, 789
DUE ... BOCCHE DI MARTINA, LE **Vol. 7** 99
DUE BUONI AMICI **Vol. 7** 1189
DUE CARABINIERI, I **Vol. 7** 383, 1166
DUE CATENE, LE **Vol. 7** 825
DUE COLLEGIALI, I **Vol. 7** 1189
DUE COLONELLI, I **Vol. 7** 1027
DUE COMPARI, I **Vol. 7** 528
DUE CROCIFISSI, I **Vol. 7** 1132
DUE CUORI **Vol. 7** 187, 454, 606
DUE CUORI FRA LE BELVE **Vol. 7** 161, 639, 830
DUE CUORI SOTTO SEQUESTRO **Vol. 7** 638, 756
DUE CUORI, UNA CAPPELLA **Vol. 7** 1136
DUE CURORI FELICI **Vol. 7** 193, 508, 756
DUE DEPUTATI, I **Vol. 7** 852
DUE DERELITTI, I **Vol. 7** 1136
DUE ESISTENZE, LE **Vol. 7** 1032, 1071
DUE EVASI DI SING-SING, I **Vol. 7** 17
DUE FIGLI DI TRINITA, I **Vol. 7** 324
DUE FOSCARI, I **Vol. 7** 170
DUE GATTONI A NOVE CODE ... E MEZZA AD AMSTERDAM, I **Vol. 7** 324
DUE GIORNI DI FELICITA **Vol. 7** 1189
DUE GLADIATORI, I **Vol. 7** 877
DUE LACRIME **Vol. 7** 891
DUE LADRI ECCEZIONALI **Vol. 7** 1189
DUE LETTERE ANONIME **Vol. 7** 756, 1116
DUE MADRI, LE **Vol. 7** 196, 212, 756, 1162, 1189
DUE MAFIOSI **Vol. 8** 598, 772
DUE MAGNUM 38 PER UNA CITTA DI CAROGNE **Vol. 7** 949
DUE MARIE, LE **Vol. 7** 965
DUE MILLIONI PER UN SORRISO **Vol. 7** 16, 200, 472, 603
DUE MISANTROPI, I **Vol. 7** 196, 902, 1060
DUE MOGLI SONO TROPPE **Vol. 7** 915
DUE MOZZI, I **Vol. 7** 711
DUE NEMICI, I **Vol. 7** 988
DUE NOTTI CON CLEOPATRA **Vol. 7** 852, 1102
DUE OCCHI PER NON VEDERE **Vol. 7** 792, 1036
DUE ORFANELLE DI TORINO, LE **Vol. 7** 454
DUE ORFANELLE, LE **Vol. 2** 27 **Vol. 7** 50, 196, 453, 503, 709
DUE ORFANELLI, I **Vol. 7** 756, 1026, 1027, 1096
DUE PERROT, I **Vol. 7** 1113
DUE PEZZI DI PANE **Vol. 7** 333, 964
DUE ROSE, LE **Vol. 7** 796
DUE SAMURAI PER CENTO GEISHE **Vol. 7** 1152
DUE SERGENTI DEL GENERALE CUSTER **Vol. 8** 437, 732
DUE SERGENTI, I **Vol. 7** 196, 372, 493, 709, 1121
DUE SOGNI AD OCCHI APERTI **Vol. 7** 436, 1162
DUE SOLDI DI FELICITA **Vol. 7** 587
DUE SOLDI DI SPERANZA **Vol. 7** 493, 817
DUE SORELLE AMANO **Vol. 7** 792
DUE SORELLE, LE **Vol. 7** 1189
DUE SOSIA IN ALLEGRIA **Vol. 7** 1036
DUE STRADE, LE **Vol. 7** 984
DUE STRANI PAPA **Vol. 7** 1202
DUE SUL PIANEROTTOLO **Vol. 7** 528
DUE SUPERPIEDI QUASI PIATTI, I **Vol. 7** 318
DUE TIGRI, LE **Vol. 7** 792, 796, 1036
DUE TORERI, I **Vol. 7** 1147
DUE TRISTEZZE: „UN AMORE" **Vol. 7** 296
DUE VERITA, LE **Vol. 7** 1035, 1062
DUE VITE DI MATTIA PASCAL, LE **Vol. 7** 80, 109, 542
DUE VOLTE GIUDA **Vol. 8** 523, 732
DUE VOLTI DI NUNU, I **Vol. 7** 37
DUE ZOCCOLETTI, I **Vol. 7** 965
DUECENTO ALL'ORA **Vol. 7** 277
DUEL **Vol. 2** 167

DUEL 77 **Vol. 3** 75 **Vol. 6** *page 45*
DUEL ADAKAR **Vol. 2** 270, 661
DUEL I HESHTUR **Vol. 3** 14, 26
DUEL, LE **Vol. 2** 99, 423, 566
DUELLE **Vol. 2** 527, 787
DUELLEN **Vol. 5** 31, 97, 179, 248
DUELLO NELL'OMBRA, UN **Vol. 7** 531
DUELLO SENZA ONORE **Vol. 7** 1054
DUELO A MUERTE **Vol. 8** 465, 549
DUELO DE PASIONES **Vol. 8** 614, 749
DUELO EN EL AMAZONAS **Vol. 8** 298, 461
DUELO EN LA CANADA **Vol. 8** 298, 524
DUELUL **Vol. 6** *page 265, page 269, page 270, page 275*
DUENDE DE JEREZ, EL **Vol. 8** 261, 732
DUENDE Y EL REY, EL **Vol. 8** 298, 583
DUENDE Y MISTERIO DEL FLAMENCO **Vol. 8** 298
DUENDES DE ANDALUCIA, LOS **Vol. 8** 477
DUERME, DUERME MI AMOR **Vol. 8** 243, 627
DUETT FÖR KANNIBALER **Vol. 5** 1170
DUETTO VAGABONDO **Vol. 7** 16
DUGA ULICA-POSLIJE PREDSTAVE **Vol. 3** 445
DUGI BRODOVI **Vol. 3** 455 **Vol. 6** *page 312, page 323*
DUHUL AURULUI **Vol. 3** 223, 237, 316, 324
DUIOS ANASTASIA TRECEA **Vol. 3** 288 **Vol. 6** *page 265*
DUKE KERKUAR 5 ORESHIN **Vol. 3** 3, 22 **Vol. 6** *page 19*
DULCE NOMBRE **Vol. 8** 358, 677
DULCES HORAS **Vol. 8** 269, 358
DULCINEA **Vol. 8** 214, 276, 600, 725
DULCINEA DEL TOBOSO **Vol. 8** 583, 772
DULSCY **Vol. 1** 180, 314, 367
DUM NA ORECHOVCE **Vol. 10** 197, 369, 433
DUM NA PORICI **Vol. 10** 165, 394, 415
DUM NA PREDMESTI **Vol. 10** 256, 412, 524, 586
DUM PRO DVA **Vol. 10** 28, 172, 246
DUM ZTRACENEHO STESTI **Vol. 10** 92, 453
DUM ZTRACENYCH DUSI **Vol. 10** 283, 369, 577
DUMA VEZ POR TODAS **Vol. 8** 27, 47
DUMBOM **Vol. 5** 1073, 1164
DUMBRAVA MINUNATA **Vol. 3** 321, 332, 335
DUMINICA LA ORA 6 **Vol. 3** 267, 269, 332
DUMME AUGUST DES ZIRKUS ROMANELLI, DER **Vol. 4** 149, 321, 535, 660
DUMME MANDFOLK, DE **Vol. 5** 9
DUMMKOPF, DER **Vol. 4** 270, 743, 772
DUNAI HAJOS, A **Vol. 9** 161, 180, 327
DUNAPARTI RANDEVU **Vol. 9** 178, 237
DUNKELKLUMPEN **Vol. 5** 896
DUNKLE GASSEN **Vol. 4** 213, 412, 436, 479
DUNKLE PUNKT, DER **Vol. 4** 167, 517, 558
DUNKLE TAG, DER **Vol. 4** 110, 524, 544
DUNUNGEN **Vol. 5** 849, 927, 937, 997
DUOS SUR CANAPE **Vol. 2** 818
DUPLIZITÄT DER EREIGNISSE, DIE **Vol. 4** 131
DUPONT-BARBES **Vol. 2** 67, 131
DUPONT LAJOIE **Vol. 2** 522, 642
DUQUE NEGRO, EL **Vol. 8** 517, 569, 772
DUQUESA DE BENAMEJI, LA **Vol. 8** 607, 717, 748
DUQUESA DE BENAMYI, LA **Vol. 2** 775
DUR, DUR **Vol. 6** *page 93, page 114*
DURAND BIJOUTIER **Vol. 2** 9, 48, 322
DURAND CONTRO DURAND **Vol. 7** 1189
DURANTON, LES **Vol. 2** 627, 726
DURCH ALLE HÖLLEN **Vol. 4** 25
DURCH DIE WÜSTE **Vol. 4** 340, 466, 471, 566
DURCH HÖHERE GEWALT **Vol. 4** 514
DURCH LIEBE ERLÖST **Vol. 4** 232, 673
DURCH NACHT ZUM LICHT **Vol. 6** *page 187, page 198*
DURCHGÄNGERIN, DIE **Vol. 4** 89, 788
DURCHLAUCHT AMÜSIERT SICH **Vol. 4** 132, 391, 424
DURCHLAUCHT RADIESCHEN **Vol. 4** 192, 462, 560
DURCHS BRANDENBURGER TOR **Vol. 4** 437, 691 **Vol. 6** *page 191*
DURCHS WILDE KURDISTAN **Vol. 8** 523, 732
DÜRFEN WIR SCHWEIGEN ? **Vol. 4** 84, 558, 715
DURS A CUIRE, LES **Vol. 2** 249, 481
DÜSTERE SCHATTEN, STRAHLENDES GLÜCK **Vol. 4** 219, 424
DUSZE W NIEWOLI **Vol. 1** 174, 407
DUVAD **Vol. 9** 67, 183, 285
DUVOD K ROZVODU **Vol. 10** 167, 256
DVA DIOPTRA DALEKOGLEDSTVO **Vol. 3** 116 **Vol. 6** *page 38, page 42*
DVA KLUCI V PALBE **Vol. 10** 233, 371, 395
DVA MUZI HLASI PRICHOD **Vol. 10** 41, 197, 425
DVA NA KONI, JEDEN NA OSLU **Vol. 10** 173, 283, 466
DVA OHNE **Vol. 10** 407, 532
DVA PEKELNE DNY **Vol. 10** 275, 345

DVA SAGAPA PARA **Vol. 7** 984
DVA TYDNY STESTI **Vol. 10** 345, 532
DVA TYGRI **Vol. 10** 96, 179, 299
DVA Z ONOHO SVETA **Vol. 10** 235, 382, 460
DVA ZRNA GROZDJA **Vol. 3** 405
DVAASEDMDESATKA **Vol. 10** 256, 412, 569
DVACATY DEVATY **Vol. 10** 97, 369, 505
DVADESET I CETIRI CASA D"ZD **Vol. 6** *page 62*
DVAERGEN **Vol. 5** 11, 38, 302
DVAMA POD NEBETO **Vol. 3** 38, 58, 112 **Vol. 6** *page 32*
DVANACT **Vol. 10** 235, 433, 572
DVANACT KRESEL **Vol. 10** 376, 604
DVANAEST STOLICA **Vol. 3** 409
DVAVOLSKOTO OR"ZIE **Vol. 6** *page 53*
DVE MATKY **Vol. 10** 121, 122
DVE NOCI U JEDNOM DANU **Vol. 3** 389, 437
DVE POBEDI **Vol. 3** 51, 53, 85, 95 **Vol. 6** *page 58, page 62*
DVE VECI PRO ZIVOT **Vol. 10** 173, 233, 577
DVERE DOKORAN **Vol. 10** 32, 228, 592
DVIJE POLOVINE SRCA **Vol. 6** *page 306*
DVOBOJ ZA JUZNU PRUGU **Vol. 3** 404, 428
DVOJE **Vol. 3** 390, 455
DVOJI SVET HOTELU PACIFIK **Vol. 10** 369, 398, 425, 617
DVOJI ZIVOT **Vol. 10** 243, 345, 412, 586
DVOJNIK"T **Vol. 6** *page 30, page 38, page 55*
DVORAKOVA HUMORESKA **Vol. 10** 243
DVORAKUV SLOVANSKY TANEC C. 1 **Vol. 10** 243
DVOSTRUKI OBRUC **Vol. 3** 421, 444
DVOSTRUKI UDAR **Vol. 6** *page 308*
DWA DNI W RAJU **Vol. 1** 278, 308, 404
DWA ZEBRA ADAMA **Vol. 1** 148, 202, 232
DWAJ PANOWIE „N“ **Vol. 1** 148, 343, 355
DWANASCIE KRZESEL **Vol. 1** 278, 380
DWIE BRYGADY **Vol. 1** 169, 229
DWIE GODZINY **Vol. 1** 305, 397
DWIE JOASIE **Vol. 1** 176, 278, 308
DWIE URNY **Vol. 1** 404
DWOJE Z WIELKIEJ RZEKI **Vol. 1** 227, 254 **Vol. 6** *page 239*
DYANE **Vol. 7** 600
DYBUK **Vol. 1** 278, 308, 404
DYCKERPOTTS ERBEN **Vol. 4** 246, 292, 752
DYDEN GAR AMOK **Vol. 5** 108, 199, 304
DYDSDRAGONEN **Vol. 5** 183, 295
DYM BRAMBOROVE NATE **Vol. 10** 395, 567
DYMKY **Vol. 10** 25, 577, 621
DYNAMIITTITYTTÖ **Vol. 5** 552, 600
DYNAMIT **Vol. 5** 97, 1073, 1181 **Vol. 10** 176, 359, 448
DYNAMITE JACK **Vol. 2** 347, 402
DYO FEGGARIA TON AUGOUSTO **Vol. 3** 156, 187
DYO KOTHONIA STO NAUTIKO **Vol. 3** 203
DYO PODIA SE ENA PAPOUTSI **Vol. 3** 142, 205
DYPLOMATYCZNA ZONA **Vol. 1** 205, 290, 372
DYRLAEGENS PLEJEBORN **Vol. 5** 76, 207
DYRYGENT **Vol. 6** *page 238, page 254*
DZIADEK DO ORZECHOW **Vol. 1** 171, 368
DZIECI KAPITANA GRANTA **Vol. 1** 298
DZIECINNE PYTANIA **Vol. 6** *page 241* **Vol. 6** *page 245*
DZIECIO **Vol. 6** *page 246*
DZIECIOL **Vol. 1** 314, 403
DZIEJE GRZECHU **Vol. 1** 136, 174, 188, 308, 314, 346
DZIEN KOLIBRA **Vol. 6** *page 211, page 221, page 248*
DZIEN OCZYSZCZENIA **Vol. 1** 242, 281 **Vol. 6** *page 248*
DZIEN WIELKIEJ PRZYGODY **Vol. 1** 278, 308, 346
DZIEN WISLY **Vol. 6** *page 223, page 225, page 226*
DZIESIEC PROCENT DLA MNIE **Vol. 1** 278, 346
DZIESIECIU Z PAWIAKA **Vol. 1** 174, 176, 410
DZIEWCYNA I CHLOPAK **Vol. 6** *page 223, page 244, page 255*
DZIEWCZETA Z NOWOLIPEK **Vol. 1** 278, 308, 346
DZIEWCZYNA I GOLEBIE **Vol. 1** 313
DZIEWCZYNA SZUKA MILOSCI **Vol. 1** 174, 194, 278, 308, 372
DZIEWCZYNA Z DOBREGO DOMU **Vol. 1** 243, 403 **Vol. 6** *page 237*
DZIKIE POLA **Vol. 1** 199, 278, 285, 383
DZIKUSKA **Vol. 1** 252, 278, 371
DZINGIS KAN **Vol. 6** *page 312*
DZIS W NOCY UMRZE MIASTO **Vol. 1** 232, 237, 305
DZIURA W ZIEMI **Vol. 1** 233, 355 **Vol. 6** *page 223*
DZUNGLE VELKOMESTA **Vol. 10** 138, 586
DZUSOVY ROMAN **Vol. 10** 44, 56, 495
DZWONY WIECZORNE **Vol. 1** 137, 181

E

E AGAPE MAS **Vol. 3** 137, 161, 193
E ALIKE DIKTATOR **Vol. 3** 153
E ALIKE STO NAUTIKO **Vol. 3** 172
E ARHONTISSA KAI O ALETES **Vol. 3** 135
E ARPAGI TES PERSEFONES **Vol. 3** 141
E ARRIVATA LA PARISINA **Vol. 7** 693, 1054
E ARRIVATO L'ACCORDATORE **Vol. 7** 476, 1054
E ARRIVATO MIO FRATELLO **Vol. 7** 30, 358, 834
E ARRIVATO SLIM CALLIGHAN **Vol. 7** 982
E ATHENA META TA MESANYHTE **Vol. 3** 168
E ATIT DE APROAPE FERICIREA **Vol. 3** 146, 249, 279, 300 **Vol. 6** *page 272*
E CADUTA UNA DONNA **Vol. 7** 2, 149, 170, 552, 891, 960
E DASKALA ME TA XANTHA MALLIA **Vol. 3** 135
E, DER SCHARLACHROTE BUCHSTABE **Vol. 4** 213
E DI SHAUL E DEI SICARI SULLE VIE DI DAMASCO **Vol. 7** 128
E DIKE TON DIKASTON **Vol. 3** 151
E DIO DISS A CAINO **Vol. 7** 852, 1147
... E DOPO ? **Vol. 7** 503
E EPISTROFE TES MEDEIAS **Vol. 3** 186, 197
E FONE TES KARDIAS **Vol. 3** 179
E FORTE UN CASINO ! **Vol. 7** 127
... E I RETTILI FURONO VINTI ! **Vol. 7** 963, 1055
E KOINONIA MAS ADIKISE **Vol. 3** 132
E KOMISSA TES FAMPRIKAS **Vol. 3** 153
E KOMISSA TES KERKYRAS **Vol. 3** 205
E KORE TIS PENTAYOITISSAS **Vol. 3** 152
E KORE TOU ELIOU **Vol. 3** 136, 205
... E LA CIVETTA CANTO **Vol. 7** 243, 651
E LA NAVE VA ... **Vol. 7** 434, 775, 988
... E LA VITA CONTINUA **Vol. 7** 408, 627, 697, 1010
... E L'ALTARE CROLLO ! **Vol. 7** 148
E L'AMOR CHE MI ROVINA **Vol. 7** 453, 798
E LIMNE TON PATHON **Vol. 3** 162
E LITANEIA TON EROON **Vol. 3** 141
E LIZA KAI E ALLE **Vol. 3** 172
E MAGIKE POLIS **Vol. 3** 202
E MAHE TES KRETES **Vol. 3** 185
E MARIA TES SIOPES **Vol. 3** 199
E MESOGEIOS FLEGETAI **Vol. 3** 159
E MOIRA MIAS GYNAIKAS **Vol. 3** 170
E MPORA **Vol. 3** 154
.... E NAPOLI CANTA **Vol. 7** 891
E NERAIDA KAI TO PALLEKARI **Vol. 3** 135
... E NOI NON FAREMO KARIKIRI **Vol. 7** 762
E ORA TES DIKAIOSINES **Vol. 3** 174
E ORAIA TOU KOUREA **Vol. 3** 135
E PARANOME **Vol. 3** 204
E PASSATA UNA NUVOLA **Vol. 7** 968
E PERIGOSO DEBRUCAR - SE ... **Vol. 8** 76
E PERMESSO MARESCIALLO ? **Vol. 7** 739
E PIU CHE UN CAMMELLO ... **Vol. 7** 796
... E POI IO CHIAMARONO IL MAGNIFICO **Vol. 7** 528
E PRIMAVERA **Vol. 7** 1027
E PROEDRINA **Vol. 3** 163
E PROIKA TES ANNOULAS **Vol. 3** 157
E PROKLESE **Vol. 3** 151, 199
E RODA **Vol. 3** 128
E SBARCATO UN MARINAIO **Vol. 7** 734, 838, 966
E SMYRNIA **Vol. 3** 170
E STEFANIA STO ANAMORFOTERIO **Vol. 3** 159, 205
... E TANTA PAURA **Vol. 7** 524
E TELEUTAIA APOSTOLIE **Vol. 3** 157
E THYSIA MIAS GYNAIKAS **Vol. 3** 180
E THYSIA TOU ABRAAM **Vol. 3** 159, 207
E TORNATO CARNEVALE **Vol. 7** 756, 792, 1116
... E TU VIVRAI NEL TERRORE, ALDILA **Vol. 7** 623, 1015
E VENNE UN UOMO **Vol. 7** 771, 923
E VENNERO IN 4 ... PER UCCIDERE SARTANA **Vol. 7** 1145
EAU A LA BOUCHE, L' **Vol. 2** 294
EAU DU NIL, L' **Vol. 2** 89, 227, 387, 794, 824
EAU VIVE **Vol. 2** 48, 530
EAU VIVE, L' **Vol. 2** 795, 819
EAUX PROFONDES **Vol. 6** *page 85, page 132*
EAUX TROUBLES, LES **Vol. 2** 247
EBBEROD BANK **Vol. 5** 136, 142
EBBERÖDS BANK **Vol. 5** 840, 841, 891, 1031, 1045, 1202
EBBREZZA DEL CIELO, L' **Vol. 7** 1063, 1199
EBRANLEES, LES **Vol. 2** 139
EBREO ERRANTE, L' **Vol. 7** 1177
EBREO FASCISTA, L' **Vol. 7** 358, 1213
ECATERINA TEODOROIU **Vol. 3** 212, 233, 289, 320
ECCE BOMBO **Vol. 7** 901
ECCE HOMO HOMOLKA **Vol. 10** 70, 402
ECCEZZIUNALE ... VERAMENTE **Vol. 7** 823, 1086

ECCO LA FELICITA ! **Vol. 7** 106, 1116
ECCO LA RADIO ! **Vol. 7** 978
ECCO NOI PER ESEMPIO **Vol. 7** 988
ECHAFAUD PEUT ATTENDRE, L' **Vol. 2** 254
ECHAME LA CULPA **Vol. 8** 602
ECHANGES DE PARTENAIRES **Vol. 2** 294
ECHAPPEMENT LIBRE **Vol. 2** 781, 863
ECHEC AU PORTEUR **Vol. 2** 388, 501, 557
ECHEC AU ROY **Vol. 2** 13, 488, 698
ECHEC ET MAT **Vol. 2** 9, 122
ECHI DI SQUILLI E TROFEI DI VITTORIA **Vol. 7** 993
ECHIPAJ PENTRU SINGAPORE, UN **Vol. 6** *page 268, page 277, page 282*
ECHO **Vol. 1** 206, 398, 403
ECLIPSE DE LUNA **Vol. 8** 546
ECLIPSE SUR UN ANCIEN CHEMIN VERS COMPOSTELLE **Vol. 2** 854 **Vol. 6** *page 99*
ECLISSE, L' **Vol. 7** 913, 1025, 1167
ECOLE BUISSONNIERE, L' **Vol. 2** 131, 268, 309, 652
ECOLE DES COCOTTES, L' **Vol. 2** 9, 67, 197, 368, 395, 427, 503, 516
ECOLE DES CONTRIBUABLES, L' **Vol. 2** 41, 727
ECOLE DES JOURNALISTES, L' **Vol. 2** 362, 871
ECOLE DES PETITES BAISEUSES, L' **Vol. 2** 291
ECOLE EST FINIE, L' **Vol. 2** 536, 879 **Vol. 6** *page 87, page 88*
ECOLE SAUVAGE, L' **Vol. 2** 110
ECOLIERES TRES POLISSONNES **Vol. 6** *page 144*
ECONOMICAMENTE DEBILES, LOS **Vol. 8** 258
ECTRANO AMANECER **Vol. 2** 775
ECUME DES JOURS, L' **Vol. 2** 263, 643, 742
ECUYERE, L' **Vol. 2** 102, 380, 603
EDDIE OCH SUZANNE **Vol. 5** 744, 920
EDDY POLO IM WESPENNEST **Vol. 4** 84, 211
EDDY POLO MIT PFERD UND LASSO **Vol. 4** 221, 565
EDELWEISSKÖNIG, DER **Vol. 4** 30, 135, 370
EDELWILD **Vol. 4** 212
EDEN A POTOM **Vol. 10** 273, 323, 455
EDERA, L' **Vol. 7** 283, 499, 792, 1044, 1162
EDERA SENZA QUERICA, L' **Vol. 7** 316
EDES A BOSSZU ! **Vol. 9** 71, 322
EDES ANNA **Vol. 9** 75, 183, 285
EDES ELLENFEL **Vol. 9** 26, 132
EDES ES KESERÜ **Vol. 9** 215, 285, 332
EDES MOSTOHA **Vol. 9** 34, 288
EDIN MIG SVOBODA **Vol. 3** 31, 51, 69, 76, 109
EDIN SNIMACEN DEN **Vol. 6** *page 61, page 64*
EDINSTENNAJA DOROGA **Vol. 6** *page 309*
EDIPO RE **Vol. 7** 388, 415, 1007, 1035
EDITH **Vol. 2** 777 **Vol. 5** 203
EDITH ET MARCEL **Vol. 6** *page 73, page 75, page 128*
EDITIE SPECIALA **Vol. 3** 244, 253, 288, 301
EDLE BLUT, DAS **Vol. 4** 36, 379, 424
EDNA B"LGARKA **Vol. 3** 61, 85
EDNA ZENA NA 33 **Vol. 6** *page 31, page 49, page 64*
EDONE KAI PATHOS **Vol. 3** 170
EDOUARD ET CAROLINE **Vol. 2** 197, 488
EDUCABDO A PAPA **Vol. 8** 541, 704
EDUCANDE DI SAINT - CYR, L' **Vol. 7** 792, 1036
EDUCANDO A UNA IDIOTA **Vol. 8** 583
EDUCATION ANGLAISE **Vol. 6** *page 89*
EDUCATION DE PRINCE **Vol. 2** 99, 153, 384, 402, 832
EDUCATION SENTIMENTALE, L' **Vol. 2** 49, 770
EDUCATRICE, L' **Vol. 6** *page 144*
EDVARD MUNCH **Vol. 5** 701, 724, 787, 797
EEN BLANDT MANGE **Vol. 5** 82
EEN PIGE OG 39 SOMAEND **Vol. 5** 108, 199, 201, 275, 304
EFFIE BRIEST **Vol. 1** 73, 85, 110
EFFRACTION **Vol. 6** *page 75, page 95*
EFTERLYST **Vol. 5** 928, 1042
EGAREMENTS, LES **Vol. 2** 208
EGEN INGANG **Vol. 5** 940, 954
EGER A PALOTANBAN **Vol. 9** 26, 76, 117, 133
EGER, AZ **Vol. 9** 146, 148
EGESZEGES EROTIKA **Vol. 9** 92, 147, 275
EGI BARANY **Vol. 9** 20, 318
EGI MADAR **Vol. 9** 83, 165, 312
EGIGERÖ FÜ **Vol. 9** 20, 83, 333
EGIPT - PALESTYNA **Vol. 1** 317
EGKLEMA STA PARASKENIA **Vol. 3** 168
EGKLIMA STO KAVOURI **Vol. 3** 163
EGLANTINE **Vol. 2** 234, 787
EGOISMUS **Vol. 3** 144
EGOISTEN, DIE **Vol. 4** 297, 549, 780
EGON UND DAS ACHTE WELTWUNDER **Vol. 1** 32, 44, 86
EGOUTS DU PARADIS, LES **Vol. 2** 51, 663
EGRENYILO ABLAK **Vol. 9** 122, 261
EGRI CSILLAGOK **Vol. 9** 26, 181, 215, 239, 283, 285, 3
EGY ASSZONY ELINDUL **Vol. 9** 117, 237, 277

EGY ASSZONY VISSZANEZ Vol. 9 126, 196
EGY AZ EGGYEL Vol. 9 7
EGY BARANY Vol. 9 154
EGY BOLOND SZAZAT CSINAL Vol. 9 33, 196, 274
EGY CSOK ES MAS SEMMI Vol. 9 182, 315, 322
EGY EJ VELENCEBEN Vol. 9 40, 71, 322
EGY EJSZAKA ERDELYBEN Vol. 9 26, 71, 222
EGY EMBER, AKI NINCS Vol. 9 117, 314
EGY EMBER TRAGEDIAJA Vol. 9 132, 142, 222
EGY ERKÖLCSÖS EJSZAKA Vol. 9 51, 297, 316
EGY FEJEDELMI NAP Vol. 9 315
EGY FIUNAK A FELE Vol. 9 7, 196, 274
EGY GEP NEM TER VISSZA Vol. 9 228
EGY KALANDOR NAPLOJA Vol. 9 200
EGY KICSIT EN ... EGY KICSIT TE ... Vol. 9 20, 238, 280
EGY KIS HELY A NAP ALATT Vol. 9 253
EGY KRAJCAR TÖRTENETE Vol. 9 28
EGY LANY ELINDUL Vol. 9 319, 322
EGY MAGYAR NABOB Vol. 9 122, 181, 332
EGY NAP A VILAG Vol. 9 117
EGY ÖRÜLT EJSZAKA Vol. 9 154, 316
EGY PIKOLO VILAGOS Vol. 9 71, 274
EGY POFON, EGY CSOK Vol. 9 132
EGY SRAC FEHER LOVON Vol. 9 83, 129, 161
EGY SZERELEM HEROM EJSZAKAJA Vol. 9 183, 259, 273
EGY SZIV MEGALL Vol. 9 58, 196
EGY SZOKNYA, EGY NADRAG Vol. 9 132, 164, 290
EGY TAL LENCSE Vol. 9 71, 274
EGY TELJES NAP Vol. 9 100
EGYENLÖSEG Vol. 9 220
EGYHUSZASOS LANY, AZ Vol. 9 71, 113
EGYIPTOMI TÖRTENET Vol. 9 183, 241, 259
EGYMASRA NEZVE Vol. 9 6, 51, 316
EGYMILLIO FONTOS BANKO, AZ Vol. 9 323
EGYSZEREGY Vol. 9 154, 224, 318
EHE AUS HASS, DIE Vol. 4 558
EHE DER FÜRSTIN DEMIDOFF, DIE Vol. 4 291 Vol. 6 *page 193*
EHE DER LUISE ROHRBACH, DIE Vol. 4 178
EHE, DIE Vol. 4 318, 710
EHE EINER ACHTZEHNJÄHRIGEN, DIE Vol. 4 25
EHE IM SCHATTEN Vol. 1 6, 23, 58
EHE IN DOSEN Vol. 4 51, 108, 623
EHE IN NOT Vol. 4 49, 626
EHE MAN EHEMANN WIRD Vol. 4 18, 558
EHE M.B.H. Vol. 4 132, 149, 660
EHEFERIEN Vol. 4 245, 584, 720
EHEKONFLIKTE Vol. 4 329, 646
EHEPARADIES Vol. 6 *page 186*
EHESACHE LORENZ Vol. 1 47, 73
EHESANATORIUM, DAS Vol. 4 108, 270, 703
EHESCHULE, DIE Vol. 6 *page 186*
EHESKANDAL IM HAUSE FROMONT JUN. UND RIESLER SEN. Vol. 4 119, 270, 305, 313, 354, 644
EHESTIFTER, DER Vol. 4 212
EHESTREIK Vol. 4 132, 157, 325, 445, 565
... EHI, AMICO, C'E SABATA ... HAI CHIUSO ! Vol. 7 692
EHRE DEINE MUTTER Vol. 4 73, 267, 743
EHRE DES HAUSES, DIE Vol. 4 603
EHRENGARD Vol. 7 411, 607, 801
EHRENPOSTEN, DER Vol. 4 18, 230
EHRENSCHULD Vol. 4 479, 535
EHRENWORT, EIN Vol. 4 124, 183, 247
EHTHROS TOU LAOU Vol. 3 144
EI AUTA ITKU MARKKINOILLA Vol. 5 355, 420, 472, 476
EI ENÄÄ EILISPÄIVÄÄ Vol. 5 363, 452
EI RUUMIITA MAKUUHUONEESEEN Vol. 5 443, 639
EI SE MITÄÄN, SANOI EEMEL Vol. 5 443, 549
EID DES STEPHAN HULLER, DER Vol. 4 423
EIDECHSE, DIE Vol. 4 583
EIFERSUCHT Vol. 4 211, 247
EIGENE HAUT, DIE Vol. 1 87, 95, 102
1860 Vol. 7 196, 219, 238, 652, 830
1812 Vol. 4 67, 424
80 HUSZAR Vol. 9 152, 259, 263
EIMAI ATHOOS Vol. 3 142
EINAR SCHANKES GLEDESHUS Vol. 5 674, 681, 712, 772
EINÄUGIGE, DER Vol. 4 135, 223
EINBRECHER Vol. 4 325, 567, 713
EINBRUCH Vol. 4 255, 527
EINBRUCH IM BANKHAUS REICHENBACH Vol. 4 270, 409, 471
EINBRUCH IN DER VILLA HOWARTH Vol. 4 125, 354, 367
EINE VON ALLEN Vol. 10 168
EINE VON UNS Vol. 4 132, 306
EINE VON VIELEN Vol. 4 124
EINER FRAU MUSS MAN ALLES VERZEIH'N Vol. 4 97, 204, 558
EINER GEGEN ALLE Vol. 4 255, 373, 595
EINER MUSS DIE LEICHE SEIN Vol. 1 5, 52, 116

EINER MUTTER OPFER **Vol. 4** 183
EINER UND EIN QUARTETT **Vol. 1** 37, 60, 85, 103
EINER VOM RUMMEL **Vol. 6** *page 155, page 169, page 175*
EINER VON UNS **Vol. 1** 37, 54, 115
EINER ZUVIEL AN BORD **Vol. 4** 28, 149, 660
EINES GROSSEN MANNES LIEBE **Vol. 4** 660, 704
EINES TAGES **Vol. 4** 24, 558, 631
EINFACH BLUMEN AUFS DACH **Vol. 1** 59, 104
EINGEBILDETE KRANKE, DER **Vol. 4** 329, 517, 558, 774, 796
EINGEWEICHTE DON JUAN, DER **Vol. 6** *page 186*
EINLADUNG ZUM NACHTESSEN **Vol. 4** 354, 422
EINMAL DER LIEBE HERRGOTT SEIN **Vol. 4** 524, 544, 678
EINMAL EINE GROSSE DAME SEIN **Vol. 4** 89, 149, 660
EINMAL IST KEINMAL **Vol. 1** 9, 100, 115
EINMAL KEHR' ICH WIEDER **Vol. 3** 390
EINMAL MÖCHT ICH KEINE SORGEN HABEN **Vol. 4** 230, 558
EINMAL UM MITTERNACHT **Vol. 4** 290, 367
EINMAL UND NIE WIEDER **Vol. 6** *page 186*
EINMAL UNTEN, EINMAL OBEN **Vol. 4** 261
EINMAL WERD' ICH DIR GEFALLEN **Vol. 4** 323, 340, 462, 466
EINMALEINS DER LIEBE, DAS **Vol. 4** 13, 149, 660
EINQUARTIERUNG BEI KLAWUNDE **Vol. 4** 525
EINS PLUS EINS GLEICH DREI **Vol. 4** 558, 796
EINS ZU EINS **Vol. 4** 314, 375
EINSAME INSEL, DIE **Vol. 4** 240, 260
EINSAME VILLA, DIE **Vol. 4** 361, 461
EINSAME WRACK, DAS **Vol. 4** 254
EINSTEIN KONTRA BABINSKY **Vol. 10** 182, 359, 564
EISBRAUT, DIE **Vol. 4** 124, 247
EISENBAHNKÖNIG, DER **Vol. 4** 567
EISENBAHNMARDER, DER **Vol. 4** 71, 488
EISERNE ACHT, DIE **Vol. 4** 270, 720
EISERNE BRAUT, DIE **Vol. 4** 247, 424
EISERNE JUNGFRAU, DIE **Vol. 4** 379, 748
EISERNE UND DAS ROTE KREUZ, DAS **Vol. 4** 366
EISMEER RUFT, DAS **Vol. 6** *page 155, page 160, page 176*
EISZEIT **Vol. 5** 812, 824
EJ ES VIRRADAT **Vol. 9** 237
EJFELI GYORS **Vol. 9** 269, 315
EJFELI KERINGÖ **Vol. 9** 132, 236
EJFELI TALALKOZAS **Vol. 9** 77
EJFELKOR **Vol. 9** 117, 314, 326
EJFELRE KIDERÜL **Vol. 9** 26, 117, 133
EJJELI ZENE **Vol. 9** 68, 71
EJSZAKA LANYA, AZ **Vol. 9** 117, 126, 133
EKEINES POU DEN XEROUN NA AGAPOUN **Vol. 3** 141
EKEINO TO KALOKAIRI **Vol. 3** 199
EKEINOS KAI EKEINI **Vol. 3** 158, 205
EKEL **Vol. 2** 778
EKEL, DAS **Vol. 4** 110, 149, 370, 629, 660, 751 **Vol. 6** *page 149*
EKEZET **Vol. 9** 154, 318
EKIPAZ"T NA NADEZDA **Vol. 3** 38, 44, 61, 75, 126 **Vol. 6** *page 58*
EKKO AF ET SKUD **Vol. 5** 38, 109
EKSPEDITRICEN FRA OSTERGADE **Vol. 5** 242
EKSPRESSENS MYTERIUM **Vol. 5** 183
EKTEMANN ALENE **Vol. 5** 674, 799
EL AGENTES DEL QUINTO GRUPO, LOS **Vol. 8** 677
EL ALAMEIN **Vol. 7** 1147
EL CANTOR **Vol. 1** 46, 57, 96
EL GRECO **Vol. 7** 86
EL HOMBRE DE LA LEGION **Vol. 7** 951, 1111
EL LEON DE DAMASCO **Vol. 7** 1116
EL TRIUNFO DEL AMOR **Vol. 7** 968
EL VERDUGO **Vol. 4** 71
ELA STO THEIO **Vol. 3** 157
ELADO BIRTOK **Vol. 9** 132, 222, 312
ELÄKÖÖN ITSEMURHAAJA **Vol. 5** 325, 331, 343
ELÄMÄN MAANTIELLÄ **Vol. 5** 406, 502
ELCSERELT EMBER **Vol. 9** 71, 208, 322
ELD OMBORD **Vol. 5** 954, 1152, 1164
ELDORADO **Vol. 2** 330, 476, 528 **Vol. 9** 147
ELEFANTER PA LOFTET **Vol. 5** 199, 304
ELEGANTES PACK **Vol. 4** 321, 626
ELEGIA **Vol. 6** *page 249, page 258, page 259*
ELEGIDO, EL **Vol. 8** 416, 642
ELEGIDOS, LOS **Vol. 8** 207, 214
ELEGIJA **Vol. 6** *page 43, page 48, page 55*
ELEKTRA **Vol. 3** 136, 172, 202
ELEKTRISCHE AUSSTELLUNG UND DAS MAGNETISCHE KABEL, DIE **Vol. 4** 308
ELEKTRISCHE FUNKE, DER **Vol. 4** 247
ELEMENT OF CRIME, THE **Vol. 5** 73, 125, 199, 249, 304
ELENA ET LES HOMMES **Vol. 2** 21, 680, 719

ELENDEN DER STRASSE, DIE Vol. 4 603, 648
ELENDEN, DIE Vol. 1 105 Vol. 2 676
ELEPHANT CA TROMPE ENORMEMENT, UN Vol. 2 434, 564
ELET HALAT, SZERELEM Vol. 9 26, 71
ELET HIDJA, AZ Vol. 9 134, 183, 312
ELET KIRALYA, AZ Vol. 9 288
ELET MUZSIKAJA, AZ Vol. 9 45, 51, 85, 121, 312
ELETBE TANCOLTATOTT LANY, AZ Vol. 9 20, 285, 329
ELETJEL Vol. 9 117, 237
ELETRE ITELTEK Vol. 9 24, 132
ELEUTHERIOS VENIZELOS Vol. 3 135, 151
ELEVAZIONE Vol. 7 503
11.35 AB HAMBURG Vol. 4 718
ELF SCHILLSCHEN OFFIZIERE, DIE Vol. 4 292, 323, 558 Vol. 4 340, 407
ELF TEUFEL, DIE Vol. 4 291, 339, 377
ELFTE GEBOT, DAS Vol. 4 598
ELHAGYOTTAK, AZ Vol. 9 28
ELI SJURSDOTTER Vol. 5 726, 735, 793, 987, 1134
ELIA KAZAN OUTSIDER Vol. 6 *page 74*
ELIAS REKEFISKER Vol. 5 698, 807
ELINAN SURMA Vol. 5 420
ELISA Vol. 2 402, 439, 786
ELISABET Vol. 5 1088 Vol. 8 629, 768
ELISABETH UND DER NARR Vol. 4 135, 546, 742
ELISABETH VON ÖSTERREICH Vol. 4 186, 626
ELISE OU LA VRAIE VIE Vol. 2 878
ELISIR D'AMORE, L' Vol. 7 756, 1063
ELITELT, AZ Vol. 9 337
ELIXIERE DES TEUFELS, DIE Vol. 1 1, 57, 80 Vol. 4 160, 205, 283
ELIXIR DO DIABO, O Vol. 8 74, 129, 130
ELIXIRUL TINERETII Vol. 3 208, 218, 323
ELIXIRY DABLA Vol. 10 1, 372
ELKESETT LEVEL Vol. 9 48, 196, 322
ELLA Vol. 8 681, 749
ELLA, EL Y SUS MILIONES Vol. 8 214, 357
ELLA, ELLOS Y LA LEY Vol. 8 354, 611
ELLA NON TRADI Vol. 7 145
ELLA Y EL Vol. 8 437
ELLA Y EL MIEDO Vol. 8 221, 461
ELLA Y LOS VETERANOS Vol. 8 749, 752
ELLA Y SUS MILLONES Vol. 8 436
ELLE BOIT PAS, ELLE FUME PAS, ELLE DRAGUE PAS MAIS...ELLE CAUSE Vol. 2 279, 661
ELLE CAUS PLUS...ELLE FLINGUE Vol. 2 661
ELLE DISAIT PRAGUE... Vol. 2 45
ELLE EN VEUT Vol. 2 210
ELLE ET MOI Vol. 2 185, 644
ELLE VOIT DES NAINS PARTOUT Vol. 6 *page 110, page 146*
ELLENIKE RAPSODIA Vol. 3 178
ELLES ETAIENT DOUZE FEMMES Vol. 2 23, 32
ELLOPOTT SZERDA, AZ Vol. 9 272
ELLY PETERSEN Vol. 5 89, 224
ELNEMULT HARANGOK Vol. 9 71, 274, 337
ELNÖK KISASSZONY Vol. 9 15, 322
ELOGIO DELLA PAZZIA Vol. 7 607
ELOISA ESTA DEBAJO DE UN ALMENDRO Vol. 8 214, 400
ELOKUU Vol. 5 347, 516
ELRABOLT SZERENCSE Vol. 9 3
ELSIETETT HAZASSAG Vol. 9 67, 84, 273
ELSK... DIN NAESTE Vol. 5 23, 76, 307
ELSKERE Vol. 5 718, 754, 762, 799
ELSKOVS GAEKKERI Vol. 5 295
ELSKOVS OPFINDSOMMED Vol. 5 97
ELSKOVSLEG Vol. 5 9, 58
ELSÖ, AZ Vol. 9 26, 117, 133
ELSÖ ESZTENDÖ, AZ Vol. 9 83, 215, 332
ELSÖ FECSKEK Vol. 9 71, 274
ELSÖ KETSZAZ EVEM Vol. 9 20, 198, 265
ELTERNLOS Vol. 4 366, 367
ELTÜSSZENTETT BIRODALOM Vol. 9 134, 168, 312
ELUS DE LA MER, LES Vol. 2 217, 724
ELVARAZSOLT DOLLAR, AZ Vol. 9 22, 125, 134
ELVAVOZOTT NAP Vol. 9 20, 149, 273
ELVE VAGY HALVA Vol. 9 259, 265, 279
ELVERHOJ Vol. 5 2, 24, 203
ELVESZETT ILLUZIOK Vol. 9 139, 259, 281
ELVESZETT PARADICSOM Vol. 9 67, 149, 285
ELVIRA MADIGAN Vol. 5 213, 1032, 1096, 1125, 1181
ELVIS ! ELVIS ! Vol. 5 1030, 1127
ELYSIUM Vol. 9 152, 316, 330
EM LEGITIMA DEFESA Vol. 8 52, 130, 167
EMANUELLE BIANCA E NERA Vol. 7 289
EMANUELLE NERA Vol. 7 257
EMANUELLE NERA N. 2 Vol. 7 694
EMBAJADORES EN EL INFIERNO Vol. 8 261, 438
EMBARAZADOS, LOS Vol. 8 426
EMBASSY Vol. 2 208
EMBER A HID ALATT Vol. 9 132, 178, 237
EMBER NEHA TEVED, AZ Vol. 9 94, 322
EMBEREK A HAVASON Vol. 9 76, 274
EMBERRABLAS MAGYAR MODRA Vol. 9 67, 273, 327

EMBRASSE-MOI **Vol. 2** 139
EMBRASSEZ-MOI **Vol. 2** 339
EMBRAYE BIDASSE, CA FUME **Vol. 2** 294
EMBRIOK **Vol. 9** 108, 253, 264
EMBRUJO **Vol. 8** 753, 761
EMBUSCADE, L' **Vol. 2** 722, 726
EMEROLOGIO TES SARKAS MOU **Vol. 3** 132
EMFYLIOS LOGOS **Vol. 3** 194
EMIGRADO, EL **Vol. 8** 625
EMIGRANTE **Vol. 7** 304, 1190
EMIGRANTE, EL **Vol. 8** 217
EMIGRANTE HAENDELSER, EN **Vol. 5** 97
EMIGRANTE, L' **Vol. 2** 13, 330, 688, 778 **Vol. 7** 208, 536 **Vol. 8** 426
EMIGRANTES **Vol. 7** 923
EMIL OCH GRISEKNOEN **Vol. 5** 862, 1027, 1093
EMIL SCHIELE **Vol. 3** 437
EMIL UND DIE DETEKTIVE **Vol. 4** 89, 614
EMILE L'AFRICAIN **Vol. 2** 57, 322
EMILIA GALOTTI **Vol. 1** 22, 35, 107
EMILIA PARADA Y FONDA **Vol. 8** 581
EMILIE HÖGQVIST **Vol. 5** 840, 927, 997, 1189
EMILIENNE **Vol. 2** 525, 600
EMINENZA GRIGIA, L' **Vol. 7** 316, 986
EMIR CAVALLO DA CIRCO **Vol. 7** 390
EMIR PREFERE LES BLONDES, L' **Vol. 6** *page 79*
EMISIA CONTINUA **Vol. 6** *page 266, page 279, page 280*
EMISSARIOS DE ICHALOM, OS **Vol. 8** 34
EMMA III **Vol. 4** 115, 354
EMMAHU, DER SCHRECKEN AFRIKAS **Vol. 4** 251
EMMANUELLE **Vol. 2** 315, 682
EMMANUELLE 4 **Vol. 6** *page 82, page 92*
EMMANUELLE II, L'ANTI-VIERGE **Vol. 2** 315, 457
EMMANUELLE, IO **Vol. 7** 276
EMMENEZ-MOI AU RITZ **Vol. 2** 480
EMMERDEUR, L' **Vol. 2** 208, 457
EMMY **Vol. 9** 71, 90, 322
EMOZIONE IN PIU, UN' **Vol. 7** 762
EMPECINADO, EL **Vol. 8** 228
EMPEREUR DES PAUVRES, L' **Vol. 2** 380, 731
EMPEZO EN BODA **Vol. 8** 692
EMPIRE DE LA NUIT, L' **Vol. 2** 185, 427
EMPIRE OF DIAMONDS **Vol. 2** 384
EMPREINTE DES GEANTS, L' **Vol. 2** 326, 815
EMPREINTE DU DIEU, L' **Vol. 2** 397
EMPREINTE ROUGE, L' **Vol. 2** 191, 273, 595
EMPRISE, L' **Vol. 2** 388
EMSER DEPESCHE **Vol. 1** 1, 69, 122
EN AMOUR, CA VA, CA VIENT **Vol. 2** 396
EN ANDALUCIA NACIO EL AMOR **Vol. 8** 450, 482
EN BORDEE **Vol. 2** 65, 387, 475
EN BUSCA DE UNA CANCION **Vol. 8** 246
EN BUSCA DEL HUEVO PERDIDO **Vol. 8** 773
EN CAS DE MALHEUR **Vol. 2** 248, 251, 625
EN DETRESSE **Vol. 2** 167
EN EFFEUILLANT LA MARGUERITE **Vol. 2** 156, 644, 669, 837
EN EL OESTE SE PUEDE HACER ... AMIGO **Vol. 8** 749
EN ES A NAGYAPAM **Vol. 9** 4, 71
EN ETES-VOUS BIEN SUR ? **Vol. 2** 249, 871
EN HAUT DES MARCHES **Vol. 6** *page 98, page 104*
EN LA RED DE MI CANCION **Vol. 8** 331, 553
EN LANYOM NEM OLYAN, AZ **Vol. 9** 319, 322
EN LAS RUINAS DE BABIBOLNIA **Vol. 8** 306, 752
EN LEGITIME DEFENSE **Vol. 2** 318, 627, 875
EN MARGE **Vol. 2** 381
EN MIL PEDAZOS **Vol. 8** 728, 770
EN OCH EN **Vol. 5** 1080, 1093, 1196
EN ONOMATI TOU NOMOU **Vol. 3** 144, 205
EN PENUMBRA **Vol. 8** 216, 756
EN PLEIN CIRAGE **Vol. 2** 293, 475
EN PLONGEE **Vol. 2** 192, 418
EN QUE LIO ME HAN METIDO ! **Vol. 8** 426
EN RADE **Vol. 2** 1, 745
EN UN LUGAR DE LA MANGA **Vol. 8** 331, 396
EN UN MUNDO NUEVO **Vol. 8** 261, 331
EN UN RINCON DE ESPANA **Vol. 8** 382, 394, 437, 677
EN VAGYOK, JEROMOS **Vol. 9** 36, 134
EN VOLTAM ! **Vol. 9** 71, 322
EN VOTRE AME ET CONSCIENCE **Vol. 2** 207, 481, 570
ENA ASTEIO KORITSI **Vol. 3** 193, 205
ENA KORITSI ALLIOTIKO **Vol. 3** 142, 191
ENA KORITSI POU TA THELEI OLA **Vol. 3** 158
ENA KORITSI SE PERIMENEI **Vol. 3** 128
ENA MPOUZOUKI ALLIOTIKO APO T'ALLA **Vol. 3** 201
ENA TANKS STRO KREBATI MOU **Vol. 3** 153, 205

ENAS ANDRAS ME SYNEIDESE **Vol. 3** 161
ENAS GERMANOS STA KALABRYTA **Vol. 3** 142
ENAS IPPOTES GIA BASOULA **Vol. 3** 149, 205
ENAS NOMOTAGES POLITES **Vol. 3** 153
ENAS NTELEIKANES **Vol. 3** 193
ENCLOS, L' **Vol. 2** 423, 518
ENCONTRO COM A MORTE **Vol. 8** 52, 130, 167
ENCONTRO COM A VIDA **Vol. 8** 25, 130, 168
ENCRUCIJADA, LA **Vol. 8** 400, 503
ENCRUCIJADA PARA UNA MONJA **Vol. 8** 626, 644
ENDA NATT, EN **Vol. 5** 840, 841
ENDE DES ABENTEURERS PAOLO DE CASPADO, DAS **Vol. 4** 25, 135
ENDE DES KUNSTREITERS, DAS **Vol. 4** 514
ENDE EINES MÄZENS, DAS **Vol. 6** *page 164, page 166, page 176*
ENDE VOM LIED, DAS **Vol. 1** 127 **Vol. 4** 535, 780 **Vol. 6** *page 164*
ENDELIG ALENE **Vol. 5** 58
ENDEMONIADA, LA **Vol. 8** 442, 553
ENDEMONIADAS, LAS **Vol. 8** 602, 749
ENDER MED BRYLLUP, DET **Vol. 5** 89
ENDERR PER NJE KARRIGE **Vol. 6** *page 11, page 17*
ENDSTATION **Vol. 4** 119, 135, 270
ENE I VERDEN **Vol. 5** 228
ENEIDA **Vol. 3** 403
ENEK A BUZAMEZÖKRÖL **Vol. 9** 117, 196, 312
ENEMIGOS **Vol. 8** 622
ENERGETICOS, LOS **Vol. 8** 547
ENES DÖD..., DEN **Vol. 5** 1023, 1030, 1126
ENFALDIGE MÖRDAREN, DEN **Vol. 5** 882, 1096, 1153
ENFANCE NUE, L' **Vol. 2** 72
ENFANT DANS LA FOULE, UN **Vol. 2** 256, 537
ENFANT DANS LA TOURMENTE, UNE **Vol. 2** 404
ENFANT DE L'AMOUR, L' **Vol. 2** 9, 32, 131, 368 **Vol. 7** 369
ENFANT DE MA SOEUR, L' **Vol. 2** 330, 381, 387
ENFANT DE PARIS, L' **Vol. 2** 798
ENFANT DES HALLES, L' **Vol. 2** 339, 731
ENFANT DES NEIGES, L' **Vol. 2** 469, 661
ENFANT DU CARNAVAL, L' **Vol. 2** 122
ENFANT DU MIRACLE, L' **Vol. 2** 237, 357, 469
ENFANT ET LA BOUTEILLE, L' **Vol. 2** 286, 330
ENFANT ET LE CHIEN, L' **Vol. 2** 159, 286, 330
ENFANT-ROI, L' **Vol. 2** 42, 452
ENFANT SAUVAGE, L' **Vol. 2** 17, 548, 674
ENFANT SECRET, L' **Vol. 6** *page 126*
ENFANTASME **Vol. 7** 567
ENFANTS DE CAIN, LES **Vol. 2** 740
ENFANTS DE L'AMOUR, LES **Vol. 2** 423, 598, 738
ENFANTS DE NEANT, LES **Vol. 2** 135
ENFANTS DE PARIS **Vol. 2** 217, 387, 603, 727
ENFANTS D'EDOUARD, LES **Vol. 2** 237
ENFANTS DU PARADIS, LES **Vol. 2** 59, 322, 402, 571, 837
ENFANTS DU PLACARD, LES **Vol. 2** 12, 517
ENFANTS TERRIBLES, LES **Vol. 2** 224, 577
ENFER DES ANGES, L' **Vol. 2** 57, 128, 279, 719, 826
ENFILADES **Vol. 2** 766
ENFILEES, LES **Vol. 6** *page 94*
ENGEL IM FEGEFEUER **Vol. 1** 84, 100, 130
ENGEL IM SEPAREE – MÄDCHEN IN GEFAHR **Vol. 4** 296, 495
ENGEL IM TAXI, EIN **Vol. 6** *page 169, page 170, page 171*
ENGEL IM VISIER, DER **Vol. 1** 26, 46, 85
ENGEL MIT DEM SAITENSPIEL, DER **Vol. 4** 119, 270
ENGEL MIT KLEINEN FEHLERN **Vol. 4** 230, 232, 796
ENGELEIN **Vol. 4** 178, 213, 651
ENGELEINS HOCHZEIT **Vol. 4** 178, 213, 651
ENGELSKA ME JE VOLELA **Vol. 6** *page 315*
ENGLANDSFARERE **Vol. 5** 723, 738
ENGLEN I SORT **Vol. 5** 14, 29
ENGLER I SNEEN **Vol. 5** 666, 701, 772, 824
ENGLISCHE HEIRAT, DIE **Vol. 4** 49, 298
ENGRENAGE, L' **Vol. 2** 169, 213, 330, 349, 613
ENHÖRNINGEN **Vol. 5** 878, 1052
ENIGMA **Vol. 6** *page 91, page 110, page 127* **Vol. 7** 731, 937
ENIGMA DEL BAULE ROSSO, L' **Vol. 7** 69
ENIGMA DEL YETE, EL **Vol. 8** 213
ENIGMA DELLA CASA BIANCA, L' **Vol. 7** 1196
ENIGMA DELLE PIRAMIDI, L' **Vol. 7** 548
ENIGMA ROSSO **Vol. 7** 257
ENIGME AUX FOLIES BERGERES **Vol. 2** 287, 856

ENIGME, L' **Vol. 2** 169, 330, 613
ENKEL MELODI, EN **Vol. 5** 869, 1009
ENLEVEMENT DES SABINES, L' **Vol. 2** 722
ENLEVEZ-MOI **Vol. 2** 48, 55, 332
ENNEMI DANS L'OMBRE, L' **Vol. 2** 570, 578, 703
ENNEMI PUBLIC N.1, L' **Vol. 2** 347, 626, 824
ENNEMI SANS VISAGE, L' **Vol. 2** 205, 538
ENNEMIS, LES **Vol. 2** 546
ENNUI ET SA DIVERSION : L'EROTISME, L' **Vol. 2** 343
ENQUANTO HA GUERRA, HA ESPERANCA **Vol. 8** 138, 148
ENQUETES **Vol. 6** *page 93*
ENQUIQUINEURS, LES **Vol. 2** 139, 423
ENREDO DE FAMILIA, UN **Vol. 8** 396
ENRICO CARUSO **Vol. 7** 171, 1027
ENRICO IV **Vol. 7** 96, 607, 750, 796, 1043, 1080, 1112
ENSAYO GENERAL PARA LA MUERTE **Vol. 8** 207, 616, 630
ENSLINGEN I BLASVÄDER **Vol. 5** 840, 1033
ENSLINGEN JOHANNES **Vol. 5** 840, 1033
ENSOM KVINDE, EN **Vol. 5** 183
ENSOMME VANDRER, DEN **Vol. 5** 106
ENTDECKUNG DES JULIAN BÖLL, DIE **Vol. 1** 36, 44, 86
ENTEHRT **Vol. 6** *page 194*
ENTENTE CORDIALE **Vol. 2** 309, 332, 646
ENTFESSELTE MENSCHHEIT, DIE **Vol. 4** 270, 380, 535, 634
ENTFÜHRTE PRINZ, DER **Vol. 6** *page 159, page 162, page 176*
ENTFÜHRUNG, DIE **Vol. 1** 89, 121, 126 **Vol. 4** 89, 246
ENTGLEIST **Vol. 4** 192, 482
ENTIERRO DE UN FUNCIONARIO EN PRIMAVERA **Vol. 8** 677
ENTLASSEN AUF BEWÄHRUNG **Vol. 1** 102, 111, 115 **Vol. 6** *page 177*
ENTLASSUNG, DIE **Vol. 4** 298, 531, 729
ENTOURLOUPE, L' **Vol. 2** 163, 351 **Vol. 6** *page 82, page 139*
ENTR'ACTE **Vol. 2** 87
ENTRAINEUSE, L' **Vol. 2** 733 **Vol. 4** 485
ENTRE BARRACAS **Vol. 8** 396, 772
ENTRE DIOS Y EL DIABLO **Vol. 8** 677
ENTRE DOS AMORES **Vol. 8** 331, 396
ENTRE HERMANOS **Vol. 8** 220
ENTRE LAS REDES **Vol. 8** 426, 732
ENTRE ONZE HEURES ET MINUIT **Vol. 2** 15, 395
ENTRE TINIEBLAS **Vol. 8** 373, 570
ENTRECHATTES **Vol. 2** 721
ENTRECUISSES **Vol. 2** 721
ENTREE DES ARTISTES **Vol. 2** 423, 439, 566, 837
ENTREZ DANS LA DANSE **Vol. 2** 706
ENTREZ VITE...VITE, JE MOUILLE **Vol. 2** 421
ENTSCHEIDUNG DES DR. AHRENDT, DIE **Vol. 1** 24, 34, 35
ENTSCHLEIERTE MAJA, DIE **Vol. 4** 499
ENVERS DU PARADIS, L' **Vol. 2** 153, 249, 501
ENVERS DU SOMMEIL, L' **Vol. 2** 527
ENVOI DE FLEURS **Vol. 2** 254, 309
ENYEM VAGY **Vol. 9** 274, 315
EOLOMEA **Vol. 1** 52, 62, 79, 87
EPAFE NO.2 **Vol. 3** 190
EPAIS MANTEAU DE SANG, UN **Vol. 2** 513
EPAVE, L' **Vol. 2** 460
EPERON D'OR **Vol. 2** 566
EPERVIER, L' **Vol. 2** 33, 145, 388, 440, 615, 726
E'PICCIRELLA **Vol. 7** 829
EPIHEIRESE APOLLON **Vol. 3** 151
EPILOG NORYMBERSKI **Vol. 1** 232, 253
EPILOGO **Vol. 8** 299, 738
EPINGLE ROUGE, L' **Vol. 2** 43, 245, 257
EPISIMI AGAPEMENE **Vol. 3** 167
EPISODE **Vol. 5** 693
EPISODEN VOM GLÜCK **Vol. 1** 12, 44, 70
EPISTROFE **Vol. 3** 194
EPITAFIUM DLA BARBARY RADZIWILLOWNY **Vol. 6** *page 211, page 250, page 251*
EPOPEA NAPOLEONICA, L' **Vol. 7** 314
EPOPEE DU RAMAYANA, L' **Vol. 2** 120
EPOUSE A TOUT FAIRE, UNE **Vol. 6** *page 94*
EPOUX SCANDALEUX, L' **Vol. 2** 247, 698, 771
EQUATEUR **Vol. 6** *page 84, page 86*
EQUATION A DEUX INCONNUS **Vol. 2** 2
EQUATORE **Vol. 7** 572, 639, 1111
EQUILIBRIUM **Vol. 5** 772
EQUIPAGE, L' **Vol. 2** 9, 153, 159, 824
ER DET HJERTE, DET **Vol. 5** 81
ER GEHT RECHTS – SIE GEHT LINKS **Vol. 4** 560, 724
ER GING ALLEIN **Vol. 1** 38, 59
ER IKKE APPELSINER, DET – DET ER HESTE **Vol. 5** 76, 195
ER NAT MED FRU KNUDSEN, DET **Vol. 5** 3, 125, 169
ER ODER ICH **Vol. 4** 119, 479
ER SA SYND FOR FARMAND, DET **Vol. 5** 261, 294
ER SA YNDIGT AT FOLGES AD, DET **Vol. 5** 31, 241, 248, 272

ER, SIE, ES **Vol. 1** 52, 87, 100
ER SOLL DEIN HERR SEIN ODER IN DER EIGNEN SCHLINGE GEFANGEN **Vol. 4** 385
ER SPLITTERGALE, DE **Vol. 5** 83
ER UND SEIN DIENER **Vol. 4** 292
ER UND SEINE SCHWESTER **Vol. 4** 257, 339, 727 **Vol. 10** 117, 167
ER WEISS, WAS ER WILL **Vol. 4** 647, 689, 787
ERA DI VENERDI 17 **Vol. 7** 530, 580
ERA LEI CHE LO VOLEVA **Vol. 7** 852
ERA LUI, SI, SI ... **Vol. 7** 333, 1063
ERA NOTTE A ROMA **Vol. 7** 257, 781
ERA UNA NOTTE BUIA E TEMPETUSA ... **Vol. 7** 631, 909
ERAM DUZENTOS IRMAOS **Vol. 8** 46, 50, 122, 130, 167
ERÄMAAN TURVISSA **Vol. 5** 340
ERAMOS SIETE ALA MESA **Vol. 8** 606, 674
ERAN TRECENTO **Vol. 7** 1063
ERAS ELÄMÄN MURHENÄY-TELMÄ **Vol. 5** 344
ERASTES TOU ONEIROU **Vol. 3** 153
ERAVAMO SETTE SORELLE **Vol. 7** 443, 493
ERAVAMO SETTE VEDOVE **Vol. 7** 443, 796
ERBE, DAS **Vol. 4** 424
ERBE DER VAN DIEMEN, DER **Vol. 4** 288, 558
ERBE VON CARLINGTON, DAS **Vol. 4** 113, 780
ERBE VON PRETORIA, DAS **Vol. 4** 247, 647
ERBFÖRSTER, DER **Vol. 4** 78, 580
ERBIN DES GRAFEN VON MONTE CHRISTO, DIE **Vol. 4** 232
ERBIN VOM ROSENHOF, DIE **Vol. 4** 588, 633
ERBIN VON TORDIS, DIE **Vol. 4** 400, 733
ERBSCHAFT, DIE **Vol. 4** 302
ERBSCHAFTSTAUMEL **Vol. 4** 167, 339
ERCOLE AL CENTRO DELLA TERRA **Vol. 7** 106, 634
ERCOLE ALLA CONQUISTA DI ATLANTIDE **Vol. 7** 257, 634
ERCOLE CONTRO I FIGLI DEL SOLE **Vol. 7** 323
ERCOLE CONTRO I TIRANNI DI BABILONIA **Vol. 7** 1121
ERCOLE CONTRO MOLOCH **Vol. 7** 1121
ERCOLE CONTRO ROMA **Vol. 7** 1121
ERCOLE E LA REGINA DI LIDIA **Vol. 7** 106
ERCOLE, SANSONE, MACISTE, URSUS : GLI INVINCIBILI **Vol. 7** 113
ERCOLI CONTRO I FIGLI DEL SOLE **Vol. 8** 598, 772
ERDEDITA DEL LEBBROSO, L' **Vol. 7** 91
ERDEKHAZASSAG **Vol. 9** 3
ERDELYI KASTELY **Vol. 9** 26, 132, 299
ERDGEIST **Vol. 4** 479
ERDGIFT **Vol. 4** 320, 358
ERDÖ SZEPE **Vol. 9** 337
EREDITA DELLA LAGUNA, L' **Vol. 7** 229
EREDITA DELLO ZIO BIONANIMA, L' **Vol. 7** 196, 756, 796
EREDITA DI CAINO, L' **Vol. 7** 1057
EREDITA FERRAMONTI, L' **Vol. 7** 567, 1035
EREDITA IN CORSA, L' **Vol. 7** 902, 1136
EREDITIERE SUPERPORNO **Vol. 7** 742
EREHTYNEET SYDÄMET **Vol. 5** 428, 472, 634
EREMIT, DER **Vol. 4** 36
ERENDIRA **Vol. 6** *page 75, page 76*
ERES UN CASO **Vol. 8** 422
ERESZD EL A SZAKALLAMAT **Vol. 9** 138, 229, 339
ERETICO, L' **Vol. 7** 693
ERGASTOLO **Vol. 7** 603, 792
ERICH KUBAK **Vol. 1** 26, 82, 90
ERIK A BUZAKALASZ **Vol. 9** 178, 237
ERIK A FENY **Vol. 9** 259, 273
ERIK EJEGODS PILGRIMSFAERD **Vol. 5** 2, 7
ERIK EL VIKINGO **Vol. 8** 626
ERIK IL VIKINGO **Vol. 8** 626
ERIK XIV **Vol. 5** 1112
ERINNERUNGEN EINER NONNE **Vol. 4** 344, 544
ERINNERUNGEN EINES ILLEGALEN **Vol. 1** 73
ERKEL **Vol. 9** 71, 75, 168
ERLEBNIS EINER NACHT **Vol. 4** 292
ERLEBNIS EINER NACHT, DAS **Vol. 4** 132, 734
ERLEBNISSE DER BERÜHMTEN TÄNZERIN FANNY ELSSLER, DIE **Vol. 4** 208, 390
ERLEBNISSE EINER KAMMERZOFE, DIE **Vol. 4** 461, 522
ERLKÖNIG, DER **Vol. 4** 45
ERLKÖNIGS TOCHTER **Vol. 4** 646
ERLOSCHENE AUGEN **Vol. 4** 124, 232 **Vol. 6** *page 194*
ERLOSCHENES LICHT **Vol. 4** 183
ERMA BIFRONTE, L' **Vol. 7** 212
ERMORDETE GREIFT EIN, DER **Vol. 1** 35, 74, 125
ERNEST LE REBELLE **Vol. 2** 6, 488, 775
ERNESTO **Vol. 7** 29, 109
ERNST IST DAS LEBEN **Vol. 4** 774
ERNST SCHNELLER **Vol. 1** 7, 38, 70

ERNST THÄLMANN – FÜHRER SEINER KLASSE **Vol. 1** 13, 23, 54, 90, 101
ERNST THÄLMANN – SOHN SEINER KLASSE **Vol. 1** 23, 54, 90, 101
ERNSTE SPIELE **Vol. 6** *page 162, page 176*
ERNSTES LEBEN, EIN **Vol. 1** 63
ERO E LEANDRO **Vol. 7** 1189
ERÖD, AZ **Vol. 9** 37, 152, 316
ERODE IL GRANDE **Vol. 7** 372, 1037
EROE DEI NOSTRI TEMPI, UN **Vol. 7** 401, 1027
EROE DEL NOSTRO TEMPO, UN **Vol. 7** 1036
EROE SONO IO, L' **Vol. 7** 1044, 1111
EROGENA ZONA **Vol. 6** *page 307, page 311, page 320*
EROI DEL DOPPIO GIOCO, GLI **Vol. 7** 1121
EROI DEL WEST, GLI **Vol. 7** 1027
EROI DELLA DOMENICA, GLI **Vol. 7** 106, 443
EROI E BRIGANTI **Vol. 7** 798
EROI, GLI **Vol. 7** 257, 363, 875 **Vol. 8** 306
EROI NELLA SVENTURA **Vol. 7** 899
EROICA **Vol. 1** 179, 377, 398 **Vol. 3** 172, 202
EROINA **Vol. 7** 728
EROISMO DI DONNA **Vol. 7** 1113
EROS IN KETTEN **Vol. 6** *page 192, page 193, page 200*
EROS PERVERSION **Vol. 7** 780
EROTAS KAI KYMATA **Vol. 3** 154
EROTAS STEN KAUTE AMMO **Vol. 3** 165
EROTAS STOUS AMMOLOFOUS **Vol. 3** 153
EROTIC FAMILY **Vol. 8** 426
EROTIC FLASH **Vol. 7** 609
EROTICO 2000 **Vol. 7** 624
EROTICO ENMASCARADO, EL **Vol. 8** 237, 249
EROTIKES STIGMES **Vol. 3** 132
EROTIKON **Vol. 5** 937, 957, 996 **Vol. 10** 40, 147, 586
EROTIKOS PAROXYSMOS **Vol. 3** 167
EROTISMO Y LA INFORMATICA, EL **Vol. 8** 702
EROTISSIMO **Vol. 2** 732
ERPRESSER **Vol. 4** 648, 779
ERPRESSERTRICK, EIN **Vol. 4** 190, 773
ERRANTE, L' **Vol. 7** 1104
ERREUR JUDICIAIRE **Vol. 2** 322, 595
ERROE TELEFONICO **Vol. 7** 1189
ERSCHEINEN PFLICHT **Vol. 6** *page 154, page 159, page 172*
ERSCHIESSUNG DES SPANISCHEN REBELLEN FRANCISCO FERRER GUARDIA, DIE **Vol. 4** 102, 247
ERSTE FRÜHLING, DER **Vol. 1** 63
ERSTE KUSS, DER **Vol. 4** 257, 558
ERSTE RECHT DES KINDES, DAS **Vol. 4** 527, 600
ERSTENS KOMMT ES ANDERS **Vol. 4** 115, 774
ERUPTIA **Vol. 3** 217, 221, 270, 332
ERWACHEN DES WEIBES, DAS **Vol. 4** 208, 560
ERZEKENY BUCSU A FEJEDELEMTÖL **Vol. 9** 20, 140, 215
ERZGAUNER **Vol. 4** 290, 651
ERZHERZOG OTTO UND DAS WÄSCHERMÄDEL **Vol. 6** *page 200*
ERZIEHER MEINER TOCHTER, DER **Vol. 4** 208, 479 **Vol. 6** *page 197*
ERZIEHERIN GESUCHT **Vol. 4** 736, 742
ERZIEHUNG VOR VERDUN **Vol. 1** 7, 36, 50
ERZSEBET KIRALYNE **Vol. 9** 13, 132, 164, 322
ES A VAKOK LATNAK **Vol. 9** 76
ES AKKOR A PASAS ... **Vol. 9** 67, 122, 183
ES ASESINO DE PEDRALBES **Vol. 8** 621
ES BLASEN DIE TROMPETEN **Vol. 4** 211, 242
ES FIEL EIN REIF IN DER FRÜHLINGSNACHT **Vol. 4** 467
ES FING SO HARMLOS AN **Vol. 4** 110, 740
ES FLÜSTERT DIE NACHT... **Vol. 4** 287, 384, 458, 646
ES GEHT NICHT OHNE LIEBE **Vol. 1** 33, 86, 103
ES GEHT UM ALLES **Vol. 4** 379, 479, 600
ES GEHT UM MEIN LEBEN **Vol. 4** 108, 270, 373
ES GIBT EINE FRAU, DIE DICH NIEMALS VERGISST **Vol. 4** 149, 219, 660
ES GIBT NOCH KAVALIERE **Vol. 4** 535
ES GIBT NUR EINE LIEBE **Vol. 4** 115, 774
ES HAT NICHT SOLLEN SEIN **Vol. 4** 385
ES IST EINE ALTE GESCHICHTE **Vol. 1** 80, 122, 126
ES IST NICHTS SO FEIN GESPONNEN **Vol. 4** 247, 461
ES KANN DER BESTE NICHT IN FRIEDEN LEBEN **Vol. 4** 323, 461
ES KNALLT **Vol. 4** 745
ES KOMMT ALLE TAGE VOR... **Vol. 4** 186, 734
ES KOMMT DER TAG **Vol. 4** 634
ES LEBE DER KÖNIG ! **Vol. 4** 239
ES LEBE DIE LIEBE **Vol. 4** 85, 110
ES LEUCHTEN DIE STERNE **Vol. 4** 361, 424, 462

ES LEUCHTET MEINE LIEBE **Vol. 4** 190, 546
ES MI HOMBRE **Vol. 8** 207, 214, 472, 548, 604
ES PELIGROSO ASOMARSE AL EXTERIOR ... **Vol. 8** 76, 422
ES PELIGROSO CASARSE A LOS 60 **Vol. 8** 368, 627
ES SOLLTE EWIG SONNTAG SEIN **Vol. 6** *page 160, page 168, page 174*
ES STEHT EIN WIRTSHAUS AN DER LAHN **Vol. 4** 97, 232
ES TUT SICH WAS UM MITTERNACHT **Vol. 4** 115, 323
ES USTED MI PADRE ? **Vol. 8** 214, 261, 549
ES WAR EINE RAUSCHENDE BALLNACHT **Vol. 4** 531, 626
ES WAR EINMAL EIN MUSIKUS **Vol. 4** 373, 626
ES WAR EINMAL EIN TREUER HUSAR **Vol. 4** 221, 340
ES WAR EINMAL EIN WALZER **Vol. 4** 192, 584
ES WAREN ZWEI JUNGGESELLEN **Vol. 4** 432
ES WAREN ZWEI KÖNIGSKINDER... **Vol. 4** 48, 270, 321, 720, 799
ES WERDE LICHT ! **Vol. 4** 137, 159, 565
ES WIRD GEHEIRATET **Vol. 4** 76
ES WIRD NICHTS SO HEISS GEGESSEN **Vol. 4** 239
ES WIRD SCHON WIEDER BESSER **Vol. 4** 28, 85, 729
ES ZOGEN DREI BURSCHEN **Vol. 4** 84, 560
ESA MUJER **Vol. 8** 338, 538
ESA PAREJA FELIZ **Vol. 8** 259, 436
ESA PICARA PELIRROJA **Vol. 8** 583, 772
ESALON DOKTORA M **Vol. 3** 378, 423
ESCADRILA ALBA **Vol. 3** 210
ESCADRILEI ALBE **Vol. 7** 402
ESCADRILLE DE LA CHANCE, L' **Vol. 2** 41, 205, 727
ESCADRON BLANC **Vol. 2** 833
ESCALA EN HI-FI **Vol. 8** 450, 719
ESCALA EN TENERIFE **Vol. 8** 396, 772
ESCALE **Vol. 2** 439, 451, 595
ESCALE A ORLY **Vol. 2** 347
ESCALIER DE SERVICE **Vol. 2** 423, 454, 676
ESCALIER SANS FIN, L' **Vol. 2** 249, 566
ESCALOFRIO **Vol. 8** 274
ESCALOFRIO DIABOLICO **Vol. 8** 583
ESCANDALO, EL **Vol. 8** 221, 484
ESCANDALO EN LA FAMILIA **Vol. 8** 467, 578
ESCAPADA **Vol. 6** *page 268*
ESCAPADA FINAL **Vol. 8** 381, 455, 554
ESCAPADE **Vol. 2** 661, 863
ESCAPADE, L' **Vol. 2** 284
ESCAPADES **Vol. 2** 559
ESCAPE ME NEVER **Vol. 2** 656
ESCARGOT DANS LA TETE, UN **Vol. 6** *page 69, page 116, page 133*
ESCLAVA BLANCA, LA **Vol. 8** 222
ESCLAVA DEL PARAISO, LA **Vol. 8** 299, 557
ESCLAVA TE DOY **Vol. 8** 249
ESCLAVAS DEL CRIMEN **Vol. 8** 731, 734
ESCLAVE BLANCHE, L' **Vol. 2** 13, 23, 332, 427, 866
ESCLAVE DE PHIDIAS, L' **Vol. 2** 798
ESCLAVE, L' **Vol. 2** 368, 615, 857
ESCONDITES, LOS **Vol. 8** 518
ESCOPETA NACIONAL, LA **Vol. 8** 610, 738
ESCRITO EN LOS CIELOS **Vol. 8** 386
ESCUADRILLA **Vol. 7** 587 **Vol. 8** 357, 400, 475, 516
ESCUADRILLA DE VUELO **Vol. 8** 280, 732
ESCUADRON DE LA MUERTE, EL **Vol. 8** 396
ESCUCHA MI CANCION **Vol. 8** 298, 524
ESCUELA DE ENFERMERAS **Vol. 8** 495, 551
ESCUELA DE PERIODISMO **Vol. 8** 405
ESCUELA DE SEDUCTORAS **Vol. 8** 461, 772
ESE OBSCURO OBJETO DEL DESEO **Vol. 8** 214
ESECUTORI, GLI **Vol. 7** 869
ESENA MONO AGAPO **Vol. 3** 161
ESENNO SL"NCE **Vol. 6** *page 32, page 57*
ESERCITO DI CINQUE UOMINI, UN **Vol. 7** 85
ESERCITO PIU PAZZO DEL MONDO, L' **Vol. 7** 1202
ESFINGE DE CRISTAL, EL **Vol. 8** 243, 555
ESFINGE MARAGATA, LA **Vol. 8** 583
ESKAPADE **Vol. 4** 49, 148, 149, 660
ESKIMO **Vol. 5** 55, 689
ESÖS VASARNAP **Vol. 9** 168, 304, 332
ESPACE D'UN MATIN, L' **Vol. 2** 433
ESPADA NORMANDA, LA **Vol. 8** 214
ESPALIER DE LA MARQUISE, L' **Vol. 2** 798
ESPANA OTRA VEZ **Vol. 8** 348, 497
ESPANJANKÄVIJÄT **Vol. 5** 437, 531, 548, 577
ESPANOLAS EN PARIS **Vol. 8** 438
ESPANOLEAR **Vol. 8** 492, 732
ESPANOLITO QUE VIENES AL MUNDO **Vol. 8** 446, 585
ESPANTO SURGE DE LA TUMBA, EL **Vol. 8** 237, 546

ESPARTACO Y LOS DIEZ GLADIADORES Vol. 8 732
ESPERANDO A PAPA Vol. 8 303, 331
ESPERIAME EN EL CIELO Vol. 8 681
ESPERIENZE DI MARIA ROSA, CAMERIERA CURIOSA, LE Vol. 7 649
ESPIA ENAMORADA, UNA Vol. 8 546
ESPIA...NDO Vol. 8 558, 631
ESPIAS MATAN EN SILENCIO, LOS Vol. 8 598
ESPIAZIONE Vol. 7 782
ESPION, L' Vol. 2 208, 377
ESPION LEVE-TOI Vol. 6 *page 73, page 78*
ESPIONNAGE OU LA GUERRE SANS ARMES Vol. 2 222, 864
ESPIONNE AUX YEUX NOIRS, L' Vol. 2 42
ESPIONNE, L' Vol. 2 528
ESPIONS, LES Vol. 2 248, 566, 722
ESPOIR Vol. 2 644, 826
ESPOIRS Vol. 2 744
ESPONTANEO, EL Vol. 8 493, 581, 732
ESPRIT DE FAMILLE, L' Vol. 2 781
ESPRONCEDA Vol. 8 339, 611
ESSAYEUSE, L' Vol. 2 442
EST CHARMANT, IL Vol. 2 722, 803
EST-CE BIEN RAISONNABLE ? Vol. 2 224 Vol. 6 *page 81*
ESTA QUE LO ES Vol. 8 627, 749
ESTA TIERRA... Vol. 8 249
ESTABA ESCRITO Vol. 8 396
ESTAMBUL 65 Vol. 8 426, 732
ESTAMPAS DE AYER Vol. 8 207
ESTATE VIOLENTA Vol. 7 246, 283, 1027
ESTE CURA Vol. 8 544, 578
ESTELYIRUHA KÖTELEZŐ Vol. 9 26, 117, 133
ESTERINA Vol. 7 514
ESTHER E IL RE Vol. 7 106
ESTIVANTES POUR HOMME SEUL Vol. 2 793
ESTOS LOCOS CUATREROS Vol. 8 219, 535
ESTOUFFADE A LA CARAIBE Vol. 2 368, 791
ESTOY EN CRISIS Vol. 8 373, 666
ESTRADA DA VIDA Vol. 8 128, 130
ESTRANERA, L' Vol. 7 825
ESTRANGEIRA, A Vol. 8 7
ESTRASBURGO DE CHILE Vol. 6 *page 88*
ESTRELLA DE SIERRA MORENA Vol. 8 214, 276, 705
ESTRELLAS, LAS Vol. 8 217, 228
ESTRELLITA Vol. 7 1189
ESTUDIANTES Y MODISTILLAS Vol. 8 270, 655, 751
ESTUDIO AMUEBLADO 2 P Vol. 8 232, 299
ESZKIMO ASSZONY FAZIK Vol. 9 10, 169, 205
ESZMELES Vol. 9 204, 316, 318
ESZTERLANC Vol. 9 177, 280
ET, AVEC LES OREILLES ? QU'EST-CE QUE VOUS FAITES ? Vol. 2 814
ET DIX DE DER Vol. 2 70, 871
...ET EN PLUS IL Y A DES CONS QUI COUPENT LES ARBRES Vol. 2 39
ET LA TENDRESSE ?...BORDEL !... Vol. 2 39
ET MOI J'TE DIS QU ELLE T'A FAIT DE L'OEIL Vol. 2 205, 439, 579, 871
ET MOURIR DE DESIR Vol. 2 872
...ET MOURIR DE PLAISIR Vol. 2 21, 719, 873
ET PAR ICI LA SORTIE Vol. 2 740 Vol. 7 982
ET POURTANT ELLE TOURNE Vol. 6 *page 82, page 108*
ET QU'CA SAUTE Vol. 2 642, 815
ET SATAN CONDUIT LE BAL Vol. 2 280
ET TA SOEUR Vol. 2 16, 131, 488
ET VIVE LA LIBERTE Vol. 2 814
ETA CRITICA, L' Vol. 7 559
ETA DA SBALLO, UN Vol. 7 610
ETA DELL'AMORE, L' Vol. 7 798
ETALON, L' Vol. 2 300, 866
ETAPPENHASE, DER Vol. 4 232, 323, 727
ETAT-CIVIL FRANCOIS Vol. 2 72
ETAT DE SIEGE Vol. 2 351, 642
ETAT SAUVAGE, L' Vol. 2 163, 517
ET...DIEU CREA LA FEMME Vol. 2 21, 626, 824
ETE DE NOS QUINZE ANS, L' Vol. 6 *page 70, page 77*
ETE D'ENFER, UN Vol. 8 358, 549
ETE MEURTRIER, L' Vol. 6 *page 72, page 81*
E.T.E. Y EL OTO, EL Vol. 8 670
ETEEPÄIN – ELÄMÄÄN Vol. 5 349, 485
ETERNEL CONFLIT Vol. 2 16, 59, 566
ETERNEL ESPOIR Vol. 2 16, 496
ETERNEL RETOUR, L' Vol. 2 27, 402, 863
ETES-VOUS FIANCEE A UN MARIN GREC OU A UN PILOTE DE LIGNE ? Vol. 2 208, 787
ETES-VOUS JALOUSE ? Vol. 2 425, 722
ETIENNE Vol. 2 32, 197
ETINCELLE, L' Vol. 6 *page 81, page 100*
ETOILE AU SOLEIL, UNE Vol. 2 330, 407
ETOILE DU NORD, L' Vol. 6 *page 70, page 82, page 128*
ETOILE DU SUD, L' Vol. 2 208, 819
ETOILE ROUGE, L' Vol. 2 183
ETOILE SANS LUMIERE Vol. 2 205, 279
ETOILES DE MIDI, LES Vol. 2 804, 849

ETRANGE AFFAIRE, UNE **Vol. 6** *page 70, page 72*
ETRANGE AMAZONE, L' **Vol. 2** 205
ETRANGE AVENTURE, L' **Vol. 2** 388, 614
ETRANGE DESIR DE MONSIEUR BARD, L' **Vol. 2** 173, 671
ETRANGE DESTIN **Vol. 2** 70, 153, 226
ETRANGE MADAME X, L' **Vol. 2** 254, 644
ETRANGE MONSIEUR STEVE, L' **Vol. 2** 376, 501, 517
ETRANGE NUIT DE NOEL, L' **Vol. 2** 395, 583, 807
ETRANGE SUZY, L' **Vol. 2** 460, 501
ETRANGE VOYAGE, UN **Vol. 2** 739
ETRANGERE, L' **Vol. 2** 153, 573
ETRANGERS, LES **Vol. 2** 368
ETRANGLEUR, L' **Vol. 2** 804
ETRE OU NE PAS ETRE **Vol. 2** 731
ETTARO DI CIELO, UN **Vol. 7** 915, 1167
ETTORE FIERAMOSCA **Vol. 7** 16, 50, 72, 491, 734, 830, 922, 1054, 1177, 1193
ETTORE FIERAMOSCA O LA DISFIDA DI BARLETTA **Vol. 7** 711
EU, TU SI OVIDIU **Vol. 3** 269, 280, 284
EUCH WERD ICH'S ZEIGEN **Vol. 1** 100, 111, 113, 127
EUDOKIA **Vol. 3** 175
EUGEN ONEGIN **Vol. 4** 496, 760
EUGENIA DE MONTIJO **Vol. 8** 606, 674, 711
EUGENIE GRANDET **Vol. 2** 194 **Vol. 7** 522, 756, 1060, 1177
EUGENIE REDRESSE-TOI **Vol. 2** 159, 286, 330
EUGENIUSZ ONEGIN **Vol. 1** 317
EULALIA TÄTI **Vol. 5** 334, 346
EULE, DIE **Vol. 4** 221, 367
EULENHAUS, DAS **Vol. 4** 146
EURIDIKE **Vol. 3** 183
EUROPA '51 **Vol. 7** 709, 1136
EUROPA CANTE **Vol. 8** 772
EUROPA DI NOTTE **Vol. 7** 781, 910
EUROPA NEM VALASZOL **Vol. 9** 126, 132, 133
EUROPA POSTLAGERND **Vol. 4** 159
EUSEBE, DEPUTE **Vol. 2** 460
EUSEBIO, A PANTERA NEGRA **Vol. 8** 7
EUSEBIO, LA PANTERA NEGRA **Vol. 8** 7
EUTANASIA DI UN AMORE **Vol. 7** 502
EVA **Vol. 3** 202 **Vol. 4** 183 **Vol. 5** 1173 **Vol. 6** *page 304* **Vol. 7** 1005, 1035 **Vol. 9** 28, 175
EVA 63 **Vol. 8** 311, 541
EVA – DEN UTSTÖTTA **Vol. 5** 1217
EVA EN LA SELVA **Vol. 8** 546
EVA GAR OMBORD **Vol. 5** 928, 982
EVA IN SEIDE **Vol. 4** 247, 424
EVA NERA **Vol. 7** 7, 46
EVA, QUE HAVE ESE HOMBRE EN TU CAMA ? **Vol. 8** 462
EVA TROPI HLOUPOSTI **Vol. 10** 154, 621
EVA UND ADAM **Vol. 1** 68, 104
EVA UND ADAM ODER DRUM PRÜFE! **Vol. 1** 42
EVA UND ADAM ODER GEFECHTE MIT NAPOLEON **Vol. 1** 42
EVA UND ADAM ODER PRIVAT NACH VEREINBARUNG **Vol. 1** 42
EVA UND ADAM ODER WIEVIEL STERNE HAT DER HIMMEL **Vol. 1** 42
EVADAREA **Vol. 3** 213, 225, 279
EVADEE, L' **Vol. 2** 579
EVADES, LES **Vol. 2** 185, 309, 743
EVAKKO **Vol. 5** 451, 592
EVANGELIEMANDENS LIV **Vol. 5** 58
EVANGELIMANN, DER **Vol. 4** 5, 186, 287
EVAS HEMD **Vol. 4** 135
EVAS TÖCHTER **Vol. 4** 257, 670, 712
EVASE, LE **Vol. 7** 383
EVASION DU BARON DE CROZE, L' **Vol. 2** 798
EVASIONE DI DIAVOLO NERO, L' **Vol. 7** 837
EVASIONE DI ROBINET, L' **Vol. 7** 1189
EVE ET L'AMOUR...SOUS TOUTES SES FORMES **Vol. 2** 442
EVE ET LE SERPENT **Vol. 2** 34, 191
EVE ET LES BONNES POMMES **Vol. 2** 795
EVEIL DE L'AMOUR, L' **Vol. 2** 620
EVEIL, L' **Vol. 2** 136
EVENEMENT LE PLUS IMPORTANT DEPUIS QUE L'HOMME A MARCHE SUR LA LUNE **Vol. 2** 285
EVENTAIL, L' **Vol. 2** 67, 332
EVENTYR OM TRE, ET **Vol. 5** 2, 23, 228
EVENTYR PA MALLORCA **Vol. 5** 23, 76
EVENTYR PAA FODREJSEN **Vol. 5** 83
EVENTYREJSEN **Vol. 5** 235, 248
EVFORDULO **Vol. 9** 71, 322
EVIGA LÄMKAR **Vol. 5** 843, 923, 1044
EVIGE EVA, DEN **Vol. 5** 760, 830
EVIGE FLAMMER, DE **Vol. 5** 81
EVIN HRICH **Vol. 10** 115
EVINRUDE, DIE GESCHICHTE EINES ABENTEURERS **Vol. 4** 124, 646
EVO, VDEJ SE ! **Vol. 10** 66, 149, 258
EVROPA TANCILA VALCIK **Vol. 10** 18, 139, 283
EVTL. SPÄTERE HEIRAT NICHT AUSGESCHLOSSEN **Vol. 4** 325, 742
EWA CHCE SPAC **Vol. 1** 250, 254 **Vol. 6** *page 239*
EWIG DEIN **Vol. 4** 796
EWIGE FLUCH, DER **Vol. 4** 291, 734
EWIGE KAMPF, DER **Vol. 4** 479

EWIGE KLANG, DER **Vol. 4** 232, 369
EWIGE MASKE, DIE **Vol. 4** 623
EWIGE MÖNCH IM BANNE DER MUSIK, DER **Vol. 4** 118, 507
EWIGE NACHT, DIE **Vol. 4** 178, 213, 651
EWIGE QUELL, DER **Vol. 4** 302, 344, 524, 544
EWIGE RÄTSEL, DAS **Vol. 4** 239, 329
EWIGE TRAUM, DER **Vol. 4** 15, 481, 614
EWIGER STROM **Vol. 4** 71, 479
EXCELLENSEN **Vol. 5** 840, 878, 929, 1052
EXCES PORNOGRAPHIQUES **Vol. 2** 291
EXECUTION, L' **Vol. 2** 190, 241
EXEKUTOR V KABARETU **Vol. 10** 34, 58, 345
EXERCICE DU POUVOIR, L' **Vol. 2** 633
EXHIBITION **Vol. 2** 294
EXHIBITION 79 **Vol. 2** 294
EXHIBITION II **Vol. 2** 294
EXHIBITIONS DANOISES **Vol. 2** 365
EXIT **Vol. 5** 16, 195, 759
EXO FTOHIA **Vol. 3** 154
EXODOS MESSOLOGGIOU **Vol. 3** 186
EXORCISMO **Vol. 8** 515, 708
EXORISTOS STEN KENTRIKE LEOFORO **Vol. 3** 176
EXPEDITRICEN **Vol. 5** 97
EXPERICENCIA PREMATRIMONIAL **Vol. 8** 524
EXPERIMENT **Vol. 10** 122, 460
EXPERIMENT DES PROF. MITHRANY, DAS **Vol. 4** 495, 724
EXPERIMENT EVA **Vol. 10** 44, 173, 291
EXPLICADOR DE MATEMATICA, O **Vol. 8** 130
EXPLOSION, L' **Vol. 2** 208
EXPLOZIA **Vol. 3** 283, 307, 317
EXPONERAD **Vol. 5** 946
EXPRES Z NORIMBERKA **Vol. 10** 70, 428, 532, 621
EXPRESO DE ANDALUCIA **Vol. 8** 438, 713
EXPRESS MATRIMONIAL, L' **Vol. 2** 798
EXPRESUL DE BUFTEA **Vol. 3** 235, 332
EXQUISES SALOPES **Vol. 2** 291
EXTASE **Vol. 10** 168, 524
EXTASES IMPUDIQUES **Vol. 2** 814
EXTERIEUR NUIT **Vol. 2** 351
EXTERMINADOR DE LA CARRETERA, EL **Vol. 8** 389, 762
EXTRA-CONIUGALE **Vol. 7** 333, 347, 763
EXTRA-DRY **Vol. 7** 131
EXTRAMUROS **Vol. 8** 358, 610
EXTRANA AVENTURA DE LUIS CANDELAS, UNA **Vol. 8** 516
EXTRANGER - OH ! DE LA CALLE DELSUR, EL **Vol. 8** 580, 728, 756
EXTRANJERA, LA **Vol. 8** 217
EXTRANO AMANECER, UN **Vol. 8** 207, 311
EXTRANO AMOR DE LOS VAMPIROS, EL **Vol. 8** 221, 551
EXTRANO CASO DE LUCY CORNELL, EL **Vol. 8** 461
EXTRANO CASO DEL DOCTOR FAUSTO, EL **Vol. 8** 738
EXTRANO VIAJE, EL **Vol. 8** 207, 298
EXTRASENSORIAL **Vol. 7** 13, 134
EXTRAVAGANTE MISSION, L' **Vol. 2** 96, 156, 719
EXTRAVAGANTE THEODORA, L' **Vol. 2** 131, 514
EXTREMENOS SE TOCAN, LOS **Vol. 8** 681
EXZELLENZ UNTERROCK **Vol. 4** 319, 598
EYE FOR EYE **Vol. 2** 579
EZ A VILLA ELADO **Vol. 9** 71, 322
EZ TÖRTENT BUDAPESTEN **Vol. 9** 71, 274
EZEK A FIATALOK **Vol. 9** 20, 304, 318
EZOP **Vol. 10** 18, 289, 359, 446
EZREDES, AZ **Vol. 9** 71

F

F. COMME FAIRBANKS **Vol. 2** 240
F. P. I. ANTWORTET NICHT **Vol. 4** 30, 262, 325, 567, 713
FABBRICA DELL' IMPREVISTO, LA **Vol. 7** 796, 1132
FABBRO DEL CONVENTO, IL **Vol. 7** 279
FABBRO DEL CONVENTO OVVERO, IL **Vol. 7** 16, 1035
FABEL **Vol. 5** 778
FABIAN BALINT TALALKOZASA ISTENNEL **Vol. 9** 134, 152, 316
FABIAN HAR TANDPINE **Vol. 5** 97
FABIAN HENTER JORDEMODER **Vol. 5** 97
FABIAN KORER I SKOVEN **Vol. 5** 97
FABIAN ORDNER GARDINSTANG **Vol. 5** 97
FABIAN PAA KAERLIGHEDSSTIEN **Vol. 5** 97
FABIAN PAA ROTTEJAGT **Vol. 5** 97
FABIAN RENSER KAKKELOVN **Vol. 5** 97
FABIAN SOM AFHOLDSMAND **Vol. 5** 97
FABIANS SKOVTUR **Vol. 5** 97
FABIOLA **Vol. 7** 332, 359, 465, 584, 621, 634, 712, 1131
FÄBODJÄNTAN **Vol. 5** 1040
FABRICANTE DE SUICIDIOS **Vol. 8** 422
FABRICANTES DE PANICO **Vol. 8** 773
FABRIK-MARIANNE **Vol. 4** 159
FABULEUSE AVENTURE DE MARCO POLO, LA **Vol. 2** 770, 824
FABULOSOS DE TRINIDAD, LOS **Vol. 8** 261, 687

FACCIA A FACCIA **Vol. 8** 396, 603
FACCIA DA MASCALZONE **Vol. 7** 16, 877
FACCIA DA SCHIAFFI **Vol. 7** 514
FACCIA VIOLENTA DI NEW YORK, LA **Vol. 7** 763
FACE A L'OCEAN **Vol. 2** 731
FACE AU DESTIN **Vol. 2** 55, 131, 858
FACE CACHEE D'HITLER, LA **Vol. 2** 692
FACEREA LUMII **Vol. 3** 228, 260, 332
FACKELTRÄGER, DER **Vol. 1** 22, 34, 82
FACTEUR S'EN VA-T-EN GUERRE, LE **Vol. 2** 347, 368
FACULTAD DE LETRAS **Vol. 8** 438, 524
FADDJA **Vol. 7** 827
FADER OG SON **Vol. 5** 295
FADEREN **Vol. 5** 97
FADERN **Vol. 5** 1042, 1146
FADNI ODPOLEDNE **Vol. 10** 41, 292
FADO **Vol. 8** 37, 44, 88, 146, 184
FADO CORRIDO **Vol. 8** 130
FAEDRENES SYND **Vol. 5** 9
FAENGSLENDE FERIEDAGE **Vol. 5** 162, 165, 306
FAERGEKROEN **Vol. 5** 115, 180, 199, 201
FAETRENE PA TORNDAL **Vol. 5** 11, 175, 306
FAGER ER LIEN **Vol. 5** 671
FAGYÖNGYÖK **Vol. 9** 135
FAGYOSSZENTEK **Vol. 9** 67, 149, 285
FAHNDUNGSAKTE D. V. C. 452 **Vol. 4** 167, 339
FAHNE VON KRIWOJ ROG, DIE **Vol. 1** 1, 19, 36, 53
FAHNENTRÄGER VON SEDAN, DER **Vol. 4** 124, 291
FAHRENDES VOLK **Vol. 4** 131, 302, 345, 437, 724, 740
FAHRERFLUCHT **Vol. 4** 115, 239
FÄHRMANN MARIA **Vol. 4** 422, 742
FAHRRAD, DAS **Vol. 6** *page 157, page 161, page 176*
FAHRT INS ABENTEUER **Vol. 4** 340, 623, 638
FAHRT INS ABENTEUER, DIE **Vol. 4** 113, 162
FAHRT INS BLAUE **Vol. 1** 46, 50
FAHRT INS BLAUE, DIE **Vol. 4** 190
FAHRT INS GLÜCK **Vol. 4** 325, 742
FAHRT INS GLÜCK, DIE **Vol. 4** 584, 779
FAHRT INS GRÜNE, DIE **Vol. 4** 339, 544, 646
FAHRT INS LEBEN **Vol. 4** 2, 110, 615
FAHRT INS VERDERBEN, DIE **Vol. 4** 124, 247
FAHRT NACH BAMSDORF, DIE **Vol. 1** 26, 36, 82
FAIBLES FEMMES **Vol. 2** 21, 232, 488
FAIR PLAY **Vol. 3** 238, 317, 326
FAIS GAFFE A LA GAFFE **Vol. 2** 337 **Vol. 6** *page 74*
FAIS M'EN PLUS **Vol. 2** 294
FAISEUR, LE **Vol. 2** 152, 330
FAISON UN REVE **Vol. 2** 84, 388, 726
FAITES DONC PLAISIR AUX AMIS **Vol. 2** 661
FAITES-MOI CONFIANCE **Vol. 2** 197, 259, 309
FAITES SAUTER LA BANQUE **Vol. 2** 97, 344
FAITS DIVERS A PARIS **Vol. 2** 314
FAKIR, DER **Vol. 4** 743
FAKIR DU GRAND HOTEL, LE **Vol. 2** 153, 330
FAKIR IM FRACK, DER **Vol. 4** 219, 558
FALAISER, LA **Vol. 2** 54
FALAK **Vol. 9** 134, 183, 259
FALAKROS MATHETES **Vol. 3** 180
FALANSTERUL **Vol. 3** 208, 277
FALBALAS **Vol. 2** 251, 395, 501
FALCE, LA **Vol. 7** 671
FALCO D'ORO, IL **Vol. 7** 693
FALCO E LA COLOMBA, IL **Vol. 7** 450, 834
FALCO E L'ALLODOLA, IL **Vol. 7** 196, 357, 563, 568
FALCO ROSSO, IL **Vol. 7** 796
FALDGRUBEN **Vol. 5** 198, 203
FALENA, LA **Vol. 7** 545, 560
FALESNA KOCICKA **Vol. 10** 167, 256, 453, 460
FALEZE DE NISIP **Vol. 6** *page 271, page 278*
FALFURO **Vol. 9** 100, 171, 275
FALKEDRENGEN **Vol. 5** 97
FALL BACHMEIER, DER **Vol. 6** *page 239*
FALL BRENKEN, DER **Vol. 4** 247, 340, 466
FALL DERUGA, DER **Vol. 4** 82, 123, 544
FALL DES GENERALSTABS-OBERST REDL, DER **Vol. 4** 292, 758 **Vol. 10** 29, 528, 586
FALL DES STAATSANWALTES M..., DER **Vol. 4** 84, 124, 158
FALL DR. WAGNER, DER **Vol. 1** 58, 115, 120
FALL GLEIWITZ, DER **Vol. 1** 47, 54
FALL HAARMANN, DER **Vol. 1** 42
FALL MOLANDER, DER **Vol. 4** 266
FALL O'NEIL, DER **Vol. 4** 780
FALL RAINER, DER **Vol. 4** 65, 287, 729
FALL ROSENTOPF, DER **Vol. 4** 560
FALLE, DIE **Vol. 1** 111, 127 **Vol. 4** 798
FALLEN IDOL, THE **Vol. 2** 656
FALLET INGEGERD BREMSSEN **Vol. 5** 863, 1032
FALLIMENTO DI SATANA, IL **Vol. 7** 133
FALLITEN **Vol. 5** 97

FALLO DELL'ISTITUTRICE **Vol. 7** 212
FALOSNY PRINC **Vol. 10** 205, 228, 548, 578
FALSA AMANTE, LA **Vol. 7** 560
FALSARI, I **Vol. 7** 1036
FALSCHE ADMIRAL, DER **Vol. 4** 119, 517, 558
FALSCHE BRAUT, DIE **Vol. 4** 135, 619, 647
FALSCHE DIMITRY, DER **Vol. 4** 397, 546
FALSCHE EHEMANN, DER **Vol. 4** 266, 288, 574
FALSCHE FELDMARSCHALL, DER **Vol. 4** 257
FALSCHE PRINZ, DER **Vol. 4** 424, 535
FALSCHE SCHAM **Vol. 4** 95, 190, 261
FALSCHE SCHEIN, DER **Vol. 4** 329
FALSCHE TENOR, DER **Vol. 4** 270, 409, 536
FALSCHE WALDEMAR, DER **Vol. 4** 366
FALSCHER FUFFZIGER, EIN **Vol. 4** 123, 132, 788
FALSCHER FÜNFZIGER, EIN **Vol. 4** 361, 558
FALSCHER START **Vol. 4** 385
FALSCHMÜNZER **Vol. 4** 445, 525
FALSIFIKATOR"T OT „CERNIJA KOS" **Vol. 6** *page 25, page 29, page 62*
FALSKA GRETA **Vol. 5** 982
FALSKA MILLIONÄREN **Vol. 5** 840, 841, 855, 878, 891
FALSO ALLARME, UN **Vol. 7** 1189
FALSO HEREDERO, EL **Vol. 8** 395, 546
FALSTAFF IN WIEN **Vol. 4** 458, 461, 462, 690
FALU ALATT FOLYIK EGY PATAK **Vol. 9** 35
FALU ROSSZA **Vol. 9** 94, 319, 337
FALUSI KISLANY PESTEN, A **Vol. 9** 7
FALUSI MADONNA **Vol. 9** 77
FAMIGLIA BRAMBILLA IN VACANZA, LA **Vol. 7** 792, 978
FAMIGLIA IMPOSSIBILE, UNA **Vol. 7** 196, 453
FAMIGLIA PASSAGUAI FA FORTUNA, LA **Vol. 7** 106
FAMILIA BIEN GRACIAS, LA **Vol. 8** 438, 762
FAMILIA PROVISIONAL **Vol. 8** 438, 493
FAMILIA VILA, LA **Vol. 8** 294, 640, 677
FAMILIA ... Y UNO MAS, LA **Vol. 8** 524, 725
FAMILIARI DELLE VITTIME NON SARANNO AVVERTITI, I **Vol. 7** 742
FAMILIE AUF BESTELLUNG **Vol. 4** 119, 517, 558
FAMILIE BENTHIN **Vol. 1** 3, 129
FAMILIE BUCHHOLZ **Vol. 4** 28, 235
FAMILIE SCHIMECK **Vol. 4** 79, 132, 558, 584, 774
FAMILIEN GELINDE **Vol. 5** 2, 7
FAMILIEN GYLDENKAL **Vol. 5** 35, 59, 179
FAMILIEN GYLDENKAL SPRAENGER BANKEN **Vol. 5** 35, 59, 179
FAMILIEN GYLDENKAL VINDER VALGET **Vol. 5** 165, 274, 296
FAMILIEN MED DE 100 BORN **Vol. 5** 166
FAMILIEN OLSEN **Vol. 5** 7
FAMILIEN PA BORGAN **Vol. 5** 726, 760
FAMILIEN PILLE SOM SPEJDERE **Vol. 5** 81
FAMILIEN SCHMIDT **Vol. 5** 150, 241, 248
FAMILIEN SWEDENHIELM **Vol. 5** 28, 39, 89, 119, 224
FAMILIENANSCHLUSS **Vol. 4** 381, 445, 582
FAMILIENBANDE **Vol. 6** *page 154, page 158, page 168*
FAMILIENPARADE **Vol. 4** 354, 437, 738
FAMILIENTAG IM HAUSE PRELLSTEIN **Vol. 4** 113, 777
FAMILJEN ANDERSSON **Vol. 5** 840, 878
FAMILJEN BJÖRCK **Vol. 5** 1156, 1205
FAMILJEN SOM VAR EN KARUSELL **Vol. 5** 928, 1032
FAMILJENS HEMLIGHET **Vol. 5** 840, 909
FAMILJENS TRADITIONER **Vol. 5** 891, 1110
FAMILLE CUCUROUX, LA **Vol. 2** 404, 717
FAMILLE DURATON, LA **Vol. 2** 47, 623
FAMILLE FENOUILLARD, LA **Vol. 2** 47, 128
FAMILLE HERNANDEZ, LA **Vol. 2** 427
FAMILLE NOMBREUSE **Vol. 2** 152, 330, 437, 454, 479
FAMILLE PONT-BIQUET, LA **Vol. 2** 871
FAMILY HONOR, THE **Vol. 2** 74
FAMILY ROCK **Vol. 6** *page 76*
FAMOSA LUZ MARIA, LA **Vol. 8** 752
FAMOSO CARBALLEIRA, EL **Vol. 8** 246
FAMOZNIJAT KILIM **Vol. 3** 106
FANATIQUES, LES **Vol. 2** 153, 650, 702
FANATISME **Vol. 2** 131, 708
FANCIULLA DELLA NEVE, LA **Vol. 7** 1189
FANCIULLA DELL'ALTRA RIVA, LA **Vol. 7** 86, 966
FANCIULLA DELL'ALTRO MONDO, LA **Vol. 7** 756, 1036, 1116
FANCIULLA DELLE ACQUE, LA **Vol. 7** 1189
FANCIULLA DI PORTICI, LA **Vol. 7** 16, 443, 1060
FANCIULLA D'UNA VOLTA, LA **Vol. 7** 654, 918, 1018

FANCIULLA, IL POETA E LA LAGUNA, LA **Vol. 7** 569
FANCIULLE DI LUSSO **Vol. 7** 923
FANCIULLO DEL WEST, IL **Vol. 7** 307, 552, 891, 1054
FANDANGO **Vol. 2** 173, 147, 249
FANDY O FANDY **Vol. 10** 87, 428, 570
FANEFLUKT **Vol. 5** 718, 780, 808
FANFAN LA TULIPE **Vol. 2** 282, 339, 388, 566, 731
FANFARE D'AMOUR **Vol. 2** 48
FANFULLA DA LODI **Vol. 7** 224, 1054, 1177
FANGAD AV EN RÖST **Vol. 5** 840, 923, 1022, 1205
FANGE HAR RYMT, EN **Vol. 5** 907, 923
FANGE NO. 113 **Vol. 5** 58
FANGE NO. 19 **Vol. 5** 58
FANGE NR. 1 **Vol. 5** 79, 183
FANGELSE **Vol. 5** 1052, 1164
FANGEN 53 **Vol. 5** 894, 1031
FANGEN FRA ERIE COUNTRY **Vol. 5** 9
FANGEN PA KARLSTENS FÄSTING **Vol. 5** 1157
FANGENS SON **Vol. 5** 81
FANGO BOLLENTE **Vol. 7** 19
FANGO CHE TRAVOLGE **Vol. 7** 1189
FANGO E LE STELLE, IL **Vol. 7** 48
FANNY **Vol. 2** 83, 84, 209, 217, 308, 385, 402, 549, 780, 833, 872 **Vol. 7** 756, 1036, 1116
FANNY ELSSLER **Vol. 4** 325
FANNY ET ALEXANDRE **Vol. 6** *page 90*
FANNY HILL **Vol. 5** 881, 1033, 1104
FANNY OCH ALEXANDER **Vol. 5** 850, 1080, 1196
FANNY PELOPAJA **Vol. 8** 232, 656
FÄNRIK STALS SÄGNER **Vol. 5** 891, 925
FANT **Vol. 5** 669
FANTAISIE D'UN JOUR **Vol. 2** 247,322
FANTASIA **Vol. 7** 250, 493
FANTASIA ... 3 **Vol. 8** 342, 749
FANTASIA BIANCA **Vol. 7** 250, 927
FANTASIA CHEZ LES PLOUCS **Vol. 2** 728
FANTASIA 'E SURDATE **Vol. 7** 829
FANTASIA EROTICA IN CONCERTO **Vol. 7** 624
FANTASIA ESPANOLA **Vol. 8** 396, 503
FANTASIAS DE CUNY, LAS **Vol. 8** 728
FANTASIAS SEXUELLES **Vol. 2** 735
FANTASMA D'AMORE **Vol. 7** 697
FANTASMA DEI LAGHI, IL **Vol. 7** 806
FANTASMA DELLA MEZZANOTTE, IL **Vol. 7** 1189
FANTASMA, IL **Vol. 7** 829
FANTASMA LLAMADO AMOR, UN **Vol. 8** 495, 752
FANTASMA Y DONA JUANITA, EL **Vol. 8** 214, 448
FANTASMAS EN LA CASA **Vol. 8** 396
FANTASMI A ROMA **Vol. 7** 988
FANTASMI DEL MARE **Vol. 7** 113
FANTASMI E LADRI **Vol. 7** 891
FANTASMI IN CINECITTA **Vol. 7** 1121
FANTASMO D'AMORE **Vol. 7** 333
FANTASTERNE **Vol. 5** 261, 278
FANTASTICA **Vol. 2** 655, 695
FANTASTICA STORIA DELLA MIA VITA **Vol. 7** 1189
FANTASTICO MUNDO DEL DR. COPELIUS, EL **Vol. 8** 611, 616
FANTASTIQUE HISTORIE VRAIE D'EDDIE CHAPMAN, LA **Vol. 2** 13
FANTE – ANNE **Vol. 5** 770
FANTEGUTTEN **Vol. 5** 700, 788
FANTOM MORRISVILLU **Vol. 10** 428, 498, 550
FANTOMAS **Vol. 2** 128, 197, 205, 330, 368, 375, 402, 780
FANTOMAS CONTRA FANTOMAS **Vol. 2** 57, 322, 375
FANTOMAS CONTRE SCOTLAND YARD **Vol. 2** 251, 368
FANTOMAS DU CHAPELIER, LES **Vol. 6** *page 93*
FANTOMAS SE DECHAINE **Vol. 2** 251, 503
FANTOME A VENDRE **Vol. 2** 755
FANTOME DE LA LIBERTE, LE **Vol. 2** 377, 728
FANTOME DU MOULIN ROUGE, LE **Vol. 2** 87, 167, 388
FANTOMELE SE GRABESC **Vol. 3** 256, 291
FANTOZZI CONTRO TUTTI **Vol. 7** 318, 1153
FANTOZZI SUBISCE ANCORA **Vol. 7** 88, 1086
FAPADOS SZERELEM **Vol. 9** 39, 117, 305, 326
FAPT DIVERS **Vol. 6** *page 268, page 273, page 276*
FAR BETALER **Vol. 5** 2, 228
FAR JAG LANA DIN FRU ? **Vol. 5** 1042, 1053, 1080 **Vol. 53** 1042, 1053, 1080
FAR JAG LOV, MAGISTERN ! **Vol. 5** 634, 863, 1091
FAR LAVER SOVSEN **Vol. 5** 96, 201, 248, 275, 292
FAR OCH FLYG **Vol. 5** 1033
FAR SKAL GIFTES **Vol. 5** 89
FAR TAPETSERER **Vol. 5** 97
FAR TIL FIRE **Vol. 5** 89, 281
FAR TIL FIRE I BYEN **Vol. 5** 89, 281
FAR TIL FIRE I HOJT HUMOR **Vol. 5** 260, 306

FAR TIL FIRE I SNEEN Vol. 5 89, 281
FAR TIL FIRE MED FULD MUSIK Vol. 5 89, 281
FAR TIL FIRE OG ONKEL SOFUS Vol. 5 89, 281
FAR TIL FIRE OG ULVEUNGERNE Vol. 5 89, 281
FAR TIL FIRE PA BORNHOLM Vol. 5 89, 281
FAR TIL FIRE PA LANDET Vol. 5 89, 281
FAR TIL SOL OCH VAR Vol. 5 878, 1052
FARANDOLE Vol. 2 401, 824
FARANDULA, LA Vol. 8 357
FARAON Vol. 1 246, 303, 313, 321, 398
FARARUV KONEC Vol. 10 139, 516, 517
FARBROR BLAS NYA BAT Vol. 5 843
FARBROR FRANS Vol. 5 891, 1031
FARCE DANS UNE CHAMBREE A LA CASERNE Vol. 2 678
FARCEUR, LE Vol. 2 517, 655, 770
FARE FOR SAMFUNDET, EN Vol. 5 295
FARENDE FOLK Vol. 5 806
FAREWELL TO ARMS, A Vol. 7 756
FARFALLA DELLA MORTE, LA Vol. 7 105
FARFALLA NERA E ARGENTO, LA Vol. 7 1195
FARFALLINO Vol. 7 26, 701, 725
FARFUI Vol. 7 729
FARI NELLA NEBBIA Vol. 7 756, 1136
FARIAHO Vol. 6 *page 156, page 160, page 176*
FARINA DEL DIAVOLO, LA Vol. 7 878
FARLIG FORBRYDER, EN Vol. 5 9, 97
FARLIG FRIHET Vol. 5 1181, 1206
FARLIG KURVA Vol. 5 840, 994
FARLIG SOMMER Vol. 5 238
FARLIG UNGDOM Vol. 5 89, 281
FARLIG VAR Vol. 5 1114, 1173
FARLIGA LEKEN, DEN Vol. 5 1042, 1205
FARLIGA VÄGAR Vol. 5 841, 923
FARLIGE ALDER, DEN Vol. 5 97
FARLIGE KYS Vol. 5 52, 156
FARLIGE LEKEN, DEN Vol. 5 676, 726
FARLIGT FRIERI, ETT Vol. 5 897, 944, 976
FARLIGT LÖFTE Vol. 5 933, 1073
FARMACIA DE GUARDIA Vol. 8 401
FARMER AUS TEXAS, DER Vol. 4 132, 176, 396, 437
FARMORS REVOLUTION Vol. 5 899, 979, 1010
FARO DA PADRE, LE Vol. 7 221, 916, 931
FARO N. 13, IL Vol. 7 1078
FARO ROSSO, IL Vol. 7 86
FARREBIQUE Vol. 2 217
FARSANTES DEL AMOR, LOS Vol. 8 768
FARSANTES, LOS Vol. 8 753, 768
FART PAA Vol. 5 97
FARTFEBER Vol. 5 907, 923
FARULLI SI ARRUOLA Vol. 7 1073
FASCHING Vol. 4 149, 344, 390, 647
FASCHINGSFEE, DIE Vol. 4 49, 164, 584
FASCHINGSKÖNIG, DER Vol. 4 140, 270, 387, 634
FASCHINGSPRINZ, DER Vol. 4 287, 384, 458, 646
FASCHINGSZAUBER Vol. 4 239, 449, 458, 584
FASCINANTE AMAZONIE Vol. 2 456
FASCINATION Vol. 6 *page 120*
FASCINO Vol. 7 941
FASCINO D'ORO Vol. 7 843
FASCISTA, DONA PURA Y EL FOLLON DE LA ESCULTURA, EL Vol. 8 426
FASHION MOVIE Vol. 7 728
FASSADENGESPENST Vol. 4 71, 644
FASSE DICH KURZ Vol. 4 76, 600
FASTERS MILLIONER Vol. 5 840, 841
FÄSTMAN I TAGET, EN Vol. 5 871, 1156
FÄSTMÖ UTHYRES Vol. 5 909, 1173
FATA MORGANA Vol. 3 214, 262 Vol. 4 35, 212 Vol. 7 1190 Vol. 8 406, 492
FATAL DOMINIO Vol. 8 278
FATALE BELEZZA Vol. 7 724
FATALISTA Vol. 1 237
FATALITA Vol. 7 1060
FATALNA GODZINA Vol. 1 174, 188
FATALNA KLATWA Vol. 1 317
FATALNATA ZAPETAJA Vol. 6 *page 27, page 29, page 45*
FATICHE DI ERCOLE, LE Vol. 7 106
FATIMA, TERRA DE FE ! Vol. 8 44, 172
FATTIG MILJONÄR, EN Vol. 5 871, 1042, 1045
FATTIGA RIDDARE Vol. 5 1045
FATTIGDRENGEN Vol. 5 83
FATTO DI CRONACA Vol. 7 827, 1189
FATTO DI SANGUE FRA DUE UOMINI PER CAUSA DI UNA VEDOVA Vol. 7 333, 592, 887
FATTO SU MISURA Vol. 7 909
FAUBOURG-MONTMARTRE Vol. 2 197, 440
FAUCON, LE Vol. 6 *page 100, page 126, page 127*
FAUN Vol. 9 172, 200
FAUNO DI MARMO, IL Vol. 7 168
FAUNO, IL Vol. 7 1190
FAUNOVO VELMI POZDNI ODPOLEDNE Vol. 10 172, 293, 334
FAUSSE ALERTE Vol. 2 48, 273, 526
FAUSSE IDENTITE Vol. 2 249, 496, 661
FAUSSE MAITRESSE, LA Vol. 2 23, 488
FAUSSES NOUVELLES Vol. 2 573
FAUST Vol. 2 27, 64 Vol. 4 266, 574 Vol. 6 *page 190* Vol. 10 201

FAUST DES SCHICKSALS, DIE **Vol. 6** *page 189*
FAUST E MARGHERITA O LA LEGGENDA DI FAUST **Vol. 7** 1177
FAUST ET MARGUERITE **Vol. 2** 159, 286, 330
FAUST IM DUNKEL, DIE **Vol. 4** 240, 494
FAUST XX **Vol. 3** 265, 270, 286, 313
FAUSTINA **Vol. 8** 400, 438
FAUSTINE ET LE BEL ETE **Vol. 2** 190, 674
FAUT CE QU'IL FAUT **Vol. 2** 48, 863
FAUT-IL LES MARIER ? **Vol. 2** 397
FAUT PAS PRENDRE LES ENFANTS DU BON DIEU POUR DES CANARDS SAUVAGES **Vol. 2** 60, 279
FAUT REPARER SOPHIE **Vol. 2** 37, 648, 837
FAUT S'LES FAIRE ! ... CES LEGIONNAIRES **Vol. 6** *page 72*
FAUTE DE L'ABBE MOURET, LA **Vol. 2** 314, 588
FAUTE DE MONIQUE, LA **Vol. 2** 6, 722
FAUTE DES AUTRES, LA **Vol. 2** 277
FAUTE D'ODETTE MARECHAL, LA **Vol. 2** 636
FAUTE D'ORTHOGRAPHIE, LA **Vol. 2** 330
FAUTEUIL 47, LE **Vol. 2** 55, 122, 722
FAUTEUIL DE SARAH, LE **Vol. 2** 210
FAUVE EST LACHE, LE **Vol. 2** 207, 661
FAUVES, LES **Vol. 6** *page 102, page 137*
FAUX-CUL, LE **Vol. 2** 513
FAUX PAS, LE **Vol. 2** 60
FAVOLA DI FONTAINE, LA **Vol. 7** 559, 679
FAVORIT DER KAISERIN, DER **Vol. 4** 246, 623
FAVORIT DER KÖNIGIN, DER **Vol. 4** 25, 330, 527, 545
FAVORITA, LA **Vol. 7** 372
FAVORITI E VINCENTI **Vol. 7** 99
FAVORITTEN **Vol. 5** 295
FAYETTE, LA **Vol. 2** 194, 402, 719
FAYETTE, WE COME, LA **Vol. 2** 384
FE, LA **Vol. 8** 214, 400
FEBBRE **Vol. 7** 587, 968
FEBBRE DA CAVALLO **Vol. 7** 633
FEBBRE DI GLORIA **Vol. 7** 965
FEBBRE DI VIVERE **Vol. 7** 407, 1062, 1152
FECONDITE **Vol. 2** 145, 242, 277, 446, 775
FEDELTA DEI VEDOVI, LA **Vol. 7** 1189
FEDERICA D'ILLIRIA **Vol. 7** 454
FEDÖNEVE : LUKACS **Vol. 9** 154, 316
FEDORA **Vol. 2** 22, 41, 299, 579 **Vol. 3** 234 **Vol. 4** 190, 232, 663, 704 **Vol. 7** 16, 603, 701, 1054, 1060
FEDRA **Vol. 8** 276, 616
FEE PAS COMME LES AUTRES, UNE **Vol. 2** 293, 294
FEE VON SAINT MENARD, DIE **Vol. 4** 113, 780
FEENHÄNDE **Vol. 4** 319
FEENS ROSE **Vol. 5** 97
FEHER EJSZAKAK **Vol. 9** 323
FEHER GALAMBOK FEKETE VAROSBAN **Vol. 9** 220
FEHER ROZSA **Vol. 9** 172, 200, 288
FEHER VONAT **Vol. 9** 58, 71
FEHERVARI HUSZAROK **Vol. 9** 178
FEHLTRITT, EIN **Vol. 4** 183
FEIND IM BLUT **Vol. 4** 63, 864
FEIND IM LAND, DER **Vol. 4** 183
FEINDE **Vol. 4** 85, 285, 729
FEINDLICHEN VÄTER, DIE **Vol. 4** 370
FEITICO DO IMPERIO, O **Vol. 8** 79, 190, 199
FEJESA E BLERTES **Vol. 6** *page 16, page 18*
FEJLÖVES **Vol. 9** 339
FEKETE GYEMANTOK **Vol. 9** 134, 178, 237, 316, 323, 327
FEKETE HAJNAL **Vol. 9** 186, 196
FEKETE KAPITANY **Vol. 9** 146, 148
FEKETE SZEM EJSZAKAJA **Vol. 8** 63, 183, 241, 312, 320 **Vol. 9** 63, 183, 241, 312, 320
FEKETE SZIVARVANY, A **Vol. 9** 28
FEKETE VAROS, A **Vol. 9** 57, 149, 203
FEL A FEJJEL **Vol. 9** 134, 199, 237
FELAGA **Vol. 6** *page 59*
FELDARZT, DER **Vol. 6** *page 189*
FELDHERRNHÜGEL, DER **Vol. 4** 544, 720
FELDMARSCHALL, DER **Vol. 4** 565, 566
FELDOBOTT KÖ **Vol. 9** 67, 154, 263
FELESEG, A **Vol. 9** 77
FELFELE A LEJTÖN **Vol. 9** 71, 165, 233, 332
FELHÖJATEK **Vol. 9** 20, 232, 265
FELIANA L'ESPIONNE **Vol. 2** 136, 816
FELICES 60, LOS **Vol. 8** 426, 732
FELICES PASCUAS **Vol. 8** 611, 616, 630
FELICIE NANTEUIL **Vol. 2** 156, 644
FELICITA COLOMBO **Vol. 7** 792, 796, 902
FELICITA, LA **Vol. 7** 731
FELICITAS GROLANDIN **Vol. 4** 36, 75, 294, 340
FELICITE **Vol. 2** 450
FELINS, LES **Vol. 2** 21, 224
FELIX **Vol. 5** 48, 61, 770, 788
FELIX SI OTILIA **Vol. 3** 222, 223, 250, 268, 306
FELIZ AL FRACASAR **Vol. 8** 606, 674
FELLA **Vol. 2** 2
FELLINI ROMA **Vol. 7** 388, 664, 988

FELLINI SATYRICON **Vol. 7** 388, 664, 988, 1035
FELMEGYEK A MINISZTERHEZ **Vol. 9** 183, 259, 285
FELSZABADULT FÖLD **Vol. 9** 134, 237
FELTEKENYSEG **Vol. 9** 26, 117
FELTHERRENS HAEVN **Vol. 5** 97
FELUTON **Vol. 9** 195, 215, 261
FEM DOGN I AUGUST **Vol. 5** 765, 792
FEM KOPIER **Vol. 5** 9
FEME **Vol. 4** 119, 340
FEMEIA DIN URSA MARE **Vol. 6** *page 272, page 280*
FEMENINO SINGULAR **Vol. 8** 247, 414
FEMIJET E SAJ **Vol. 3** 17
FEMINA RIDENS **Vol. 7** 6
FEMININ, FEMININ **Vol. 2** 215
FEMME A L'ORCHIDEE, LA **Vol. 2** 740 **Vol. 7** 982
FEMME A PASSE, UNE **Vol. 2** 493
FEMME AU VOLANT, UNE **Vol. 2** 563, 626
FEMME CHIPEE, UNE **Vol. 2** 197
FEMME COUPEE EN MORCEAUX, LA **Vol. 2** 173, 460, 670, 671
FEMME D'A COTE, LA **Vol. 6** *page 84, page 87*
FEMME DANS LA NUIT, UNE **Vol. 2** 173, 249
FEMME DE JEAN, LA **Vol. 2** 60, 351
FEMME DE L'AVIATEUR, LA **Vol. 2** 534
FEMME DE MES REVES, LA **Vol. 2** 424
FEMME DE MON PITE, LA **Vol. 6** *page 89, page 91*
FEMME DE NULLE PART, LA **Vol. 2** 330, 346, 420, 528
FEMME DISPARAIT, UNE **Vol. 2** 279, 427, 581
FEMME DOUCE, UNE **Vol. 2** 170, 190
FEMME DU BOULANGER, LA **Vol. 2** 84, 485
FEMME DU BOUT DU MONDE, LA **Vol. 2** 90, 205, 727
FEMME D'UNE NUIT, LA **Vol. 2** 153, 775, 833
FEMME EN BLEU, LA **Vol. 2** 481
FEMME EN HOMME, LA **Vol. 2** 573, 656
FEMME EN ROUGE, LA **Vol. 2** 158, 601
FEMME-ENFANT, LA **Vol. 6** *page 77, page 96*
FEMME EST UNE FEMME, UNE **Vol. 2** 208, 285
FEMME ET LE PANTIN, LA **Vol. 2** 388, 402, 863
FEMME ET LE ROISSIGNOL, LA **Vol. 2** 6, 9, 55, 193
FEMME FATALE, LA **Vol. 2** 9, 807
FEMME FIDELE, UNE **Vol. 2** 21, 719
FEMME FLIC, LA **Vol. 2** 522, 783
FEMME IDEALE, LA **Vol. 2** 279, 407, 540, 603, 604
FEMME INCONNUE, LA **Vol. 2** 153, 681
FEMME INFIDELE, LA **Vol. 2** 700
FEMME INTEGRALE, LA **Vol. 6** *page 124*
FEMME INVISIBLE, LA **Vol. 2** 167, 573, 744
FEMME, LA **Vol. 5** 97
FEMME LIBRE, UNE **Vol. 2** 814
FEMME MARIEE, UNE **Vol. 2** 208, 632
FEMME NUE, LA **Vol. 2** 6, 153, 193, 413, 427, 573, 579, 627
FEMME OBJET, LA **Vol. 6** *page 69*
FEMME PAR JOUR, UNE **Vol. 2** 332, 807
FEMME PERDUE, LA **Vol. 2** 193, 698
FEMME PUBLIQUE, LA **Vol. 6** *page 97, page 100, page 248*
FEMME QUE J'AI ASSASSINEE, LA **Vol. 2** 193, 254
FEMME QUE J'AI LE PLUS AIMEE, LA **Vol. 2** 402, 722
FEMME QUI PLEURE, LA **Vol. 2** 450
FEMME QUI SE PARTAGE, UNE **Vol. 2** 270, 308
FEMME REVEE, LA **Vol. 2** 648
FEMME SANS IMPORTANCE, UNE **Vol. 2** 423, 439, 440
FEMME SANS PASSE **Vol. 2** 193, 254
FEMME SPECIALE, UNE **Vol. 6** *page 79*
FEMME, UN JOUR..., UNE **Vol. 2** 35, 787
FEMMES **Vol. 2** 871 **Vol. 6** *page 96, page 107*
FEMMES AU SOLEIL **Vol. 2** 17, 142
FEMMES COLLANTES, LES **Vol. 2** 249, 871
FEMMES D'ABORD, LES **Vol. 2** 475, 661
FEMMES DE PARIS **Vol. 2** 318, 347, 807
FEMMES DE PERSONNE **Vol. 6** *page 70, page 99*
FEMMES DES AUTRES, LES **Vol. 2** 388
FEMMES ENTRE HOMMES **Vol. 6** *page 93, page 129*
FEMMES, FEMMES **Vol. 2** 804
FEMMES IMPUDIQUES **Vol. 2** 565
FEMMES, LES **Vol. 2** 444, 481, 732
FEMMES PAR DERRIERE **Vol. 6** *page 120*
FEMMES PRETES A TOUT **Vol. 6** *page 120*
FEMMES S'EN BALANCENT, LES **Vol. 2** 501, 615, 709
FEMMES SEULES POUR DRAGUEURS **Vol. 6** *page 94*
FEMMES SONT DES ANGES, LES **Vol. 2** 368, 663
FEMMES SONT FOLLES, LES **Vol. 2** 57, 158
FEMMES SONT MARRANTES, LES **Vol. 2** 97, 205, 516, 702

FEMMES VICIEUSES Vol. 2 309, 588, 721
FEMMINA Vol. 7 456, 1132
FEMMINA INCATENATA Vol. 7 16
FEMMINA SENZA CUORE Vol. 7 587
FEMMINE DI LUSSO Vol. 7 1027
FEMMINE INSAZIABILI Vol. 7 834
FEMMINE TRE VOLTE Vol. 7 333
F.E.N. Vol. 8 358
FENESTA CA LUCIVE ... Vol. 7 424
FENIX Vol. 10 263, 269, 544
FENSTER IM 2. STOCK, DAS Vol. 4 325, 742
FENY A REDÖNY MÖGÖTT Vol. 9 36, 122, 149
FENY ES ARNYEK Vol. 9 71, 87, 156, 194, 196, 219
FENYES SZELEK Vol. 9 20, 273, 302
FERDINAND LASALLE Vol. 4 747
FERDINAND LE NOCEUR Vol. 2 67, 460, 775
FERDINANDO I RE DI NAPOLI Vol. 7 798
FERENC SE ZENI Vol. 10 393
FERFI EGESZEN MAS, A Vol. 9 67, 285, 318
FERFI MIND ÖRÜLT, A Vol. 9 71, 322
FERFIHÜSEG Vol. 9 68, 132
FERIA EN SEVILLA Vol. 8 477, 495
FERIEDRENDGEN Vol. 5 97
FERIEN VOM ICH Vol. 4 323
FERIENKIND, DAS Vol. 4 13, 317, 699
FERJET KERESEK Vol. 9 71, 322
FERJHEZ MENNI TILOS Vol. 9 207, 268, 305
FERME AUX LOUPS, LA Vol. 2 23, 824
FERME DES SEPT PECHES, LA Vol. 2 401, 419
FERME DU PENDU, LA Vol. 2 251, 826
FERMIN GALAN Vol. 8 281, 707
FERMO CON LE MANI ! Vol. 7 665, 724, 1159
FERNAND Vol. 6 *page 72, page 87*
FERNAND CLOCHARD Vol. 2 419, 515
FERNAND COW-BOY Vol. 2 57, 515, 517
FERNER LIEFEN Vol. 4 361, 558
FERNGESPRÄCH MIT HAMBURG Vol. 4 18, 558
FEROCE SALADINO, IL Vol. 7 792, 796
FEROZ Vol. 8 358, 771
FERRAGOSTO IN BIKINI Vol. 7 1121
FERRAGUS Vol. 2 371 Vol. 7 729
FERRAVILLA NELLE SUE PIU CARATTERISTICHE INTERPRETAZIONI Vol. 7 341
FERREOL Vol. 4 35, 366 Vol. 7 442
FERRO DI CAVALLO Vol. 7 144, 206, 815
FERRO, IL Vol. 7 965
FERRORE CON GLI OCCHI STORTI, IL Vol. 7 1007
FERROVIERE, IL Vol. 7 86, 401, 803
FESAK HUBERT Vol. 10 183, 517, 571
FESCHE ERZHERZOG, DER Vol. 4 113, 329
FESCHE HUSAR, DER Vol. 4 246, 292, 752
FESSEE, LA Vol. 2 237, 442
FEST AUF HADERSLEVHUUS, EIN Vol. 4 119, 126, 545
FESTA DI LAURA Vol. 7 96, 947, 1091
FESTA PERDUTA, LA Vol. 7 623, 901
FESTFORESTILLING I SNOLDEROD Vol. 5 97
FESTIVAL Vol. 8 602, 630
FESTIVAL EN BENIDORM Vol. 8 311, 752
FESTIVAL EROTIQUE Vol. 2 462
FESTIVAL PORNO Vol. 2 766
FESTIVITETSSALONGEN Vol. 5 880, 933, 1035
FESTSTELLUNG, DIE Vol. 1 9, 101, 103
FETE A HENRIETTE, LA Vol. 2 279, 402
FETE A JULES, LA Vol. 2 66
FETE ESPAGNOLE, LA Vol. 2 503, 615, 648
FETE SAUVAGE, LA Vol. 2 879
FETES GALANTES, LES Vol. 2 863
FETITIA MINCINOASA Vol. 3 259, 261, 312
FETYSZ Vol. 6 *page 206, page 240, page 242*
FETZERS FLUCHT Vol. 1 9, 65, 102
FEU ! Vol. 2 167, 388, 488, 863
FEU A LA MINETTE, LE Vol. 2 52
FEU AU SEXE, LE Vol. 2 349
FEU AU VENTRE, LE Vol. 2 76
FEU AUX LEVRES, LE Vol. 2 482, 713
FEU AUX PUDRES, LE Vol. 2 601
FEU DANS LA PEAU, LE Vol. 2 407, 557, 559
FEU DE DIEU Vol. 2 207
FEU DE PAILLE, LE Vol. 2 191, 330, 454, 479, 533
FEU FOLLET, LE Vol. 2 190, 285
FEU MATHIAS PASCAL Vol. 2 87, 122, 153, 162, 379, 387, 511, 573
FEU NICOLAS Vol. 2 205, 251
FEU SACRE Vol. 2 153, 173, 249, 441
FEU TOUPINEL Vol. 2 63, 731
FEUER, DAS Vol. 4 178, 213, 651
FEUER UNTER DECK Vol. 1 52, 53, 59
FEUERDRACHEN Vol. 6 *page 156, page 161, page 171*
FEUERPROBE, DIE Vol. 4 24
FEUERTÄNZERIN, DIE Vol. 4 36, 270
FEUERTEUFEL, DER Vol. 4 54, 215, 286, 541

FEUERZANGENBOWLE, DIE **Vol. 4** 119, 270
FEUERZEUG, DAS **Vol. 1** 36, 92, 103
FEUX DE JOIE **Vol. 2** 579
FEUX DE LA CHANDELEUR, LES **Vol. 2** 448, 814
FF. SS. CIOE „... CHE MI HAI PORTATO..." **Vol. 7** 1083, 1105, 1165
FIA JANSSON FRAN SÖDER **Vol. 5** 993, 1045
FIACCOLA SOTTO IL MOGGIO, LA **Vol. 7** 1189
FIACCOLA UMANA, LA **Vol. 7** 598
FIACCOLE **Vol. 7** 463, 1156
FIACRE 13 **Vol. 2** 6, 799
FIACRE N. 13, IL **Vol. 7** 756, 1026, 1027, 1096, 1190
FIAKER NR. 13 **Vol. 4** 84, 396, 715
FIAKERLIED **Vol. 4** 246, 292
FIAMMA ! **Vol. 7** 1132
FIAMMA CHE NON SI SPEGNE, LA **Vol. 7** 17, 1054
FIAMMA NERA **Vol. 7** 196
FIAMMA SACRA, LA **Vol. 7** 940
FIAMMA SIMBOLICA **Vol. 7** 567
FIAMMA VELATA **Vol. 7** 900
FIAMMATA, LA **Vol. 7** 569, 621, 750, 796, 900
FIAMME ABISSINE **Vol. 7** 1185
FIAMME E LA CENERE, LA **Vol. 7** 399
FIAMME E SPENTA, LA **Vol. 7** 843
FIAMME FUNESTE **Vol. 7** 148
FIAMME INDOMABILI **Vol. 7** 86
FIAMME SUL MARE **Vol. 7** 756
FIAMME SULLA LAGUNA **Vol. 7** 1170
FIAMMELLA SPENTA **Vol. 7** 1189
FIANCAILLES ROUGES, LES **Vol. 2** 79, 380
FIANCEE DES TENEBRES, LA **Vol. 2** 402, 439
FIANCEE DU DIABLE, LA **Vol. 2** 798
FIANCEE DU PIRATE, LA **Vol. 2** 49, 458
FIANCEE QUI VENAIT DU FROID, LA **Vol. 6** *page 69, page 72*
FIANCES DE L'AIR, LES **Vol. 2** 798
FIASKO, EN **Vol. 5** 83
„FIAT VOLUNTAS DEI" **Vol. 7** 493, 665, 1136
FIBRA DEL DOLORE, LA **Vol. 7** 150, 1103
FICCANASO, IL **Vol. 7** 310, 315, 617
FICHISSIMI, I **Vol. 7** 408, 450
FICKLE FINGER OF FATE, THE **Vol. 8** 243, 517
FICO D'INDIA **Vol. 7** 257, 340
FIDALGOS DE CASA MOURISCA, OS **Vol. 8** 20, 79, 122
FIDANZATI DELLA MORTE, I **Vol. 7** 528
FIDANZATI DI MIA MOGLIE, IL **Vol. 7** 639, 1107
FIDANZATI, I **Vol. 7** 221, 635
FIDANZATO INDEGNO **Vol. 7** 1189
FIDDLER ON THE ROOF **Vol. 6** *page 312*
FIDELE BAUER, DER **Vol. 4** 270, 292, 725
FIDELE HERRENPARTIE, DIE **Vol. 4** 287, 384, 458, 646
FIDELE RAZZIA, DIE **Vol. 4** 204, 565
FIDELES BÜRO, EIN **Vol. 4** 361, 558
FIDELES GEFÄNGNIS, EIN **Vol. 4** 560
FIDLOVACKA **Vol. 10** 122, 453, 586
FIEBRE **Vol. 7** 587, 968 **Vol. 8** 674, 692, 717
FIEBRE DE DANZA **Vol. 8** 536
FIEL INFANTERIE, LA **Vol. 8** 276, 749
FIELES SERVIENTES, LAS **Vol. 8** 249, 473, 684
FIERA DEI DESIDERI, LA **Vol. 7** 971
FIERAS SIN JAULA **Vol. 8** 392, 706
FIERECILLA DOMADA, LA **Vol. 8** 261, 298
FIEREZZA ITALICA **Vol. 7** 196, 548
FIESKO **Vol. 4** 288
FIESTA BRAVA **Vol. 7** 7, 46
FIESTA SIGUE, LA **Vol. 8** 614
FIETE IM NETZ **Vol. 1** 36, 99, 100
FIEVRE **Vol. 2** 73, 330, 346, 528, 775
FIEVRES **Vol. 2** 205, 538, 558
FIEVRES D'ETE **Vol. 6** *page 119*
FIFI LA PLUME **Vol. 2** 293, 661
15 BAJO LA LONA **Vol. 8** 616, 630, 719
FIFTY FIFTY **Vol. 5** 716
FIFTY-FIFTY **Vol. 5** 780 **Vol. 6** *page 104*
56, RUE PIGALLE **Vol. 2** 460
FIGARO E LA SUA GRAN GIORNATA **Vol. 7** 756, 885, 1036, 1116
FIGARO QUA, FIGARO LA **Vol. 7** 16
FIGAROS HOCHZEIT **Vol. 1** 39, 58, 90, 107 **Vol. 4** 24, 423, 644, 681
FIGHT FOR SURVIVAL **Vol. 3** 272, 291
FIGLI CHIEDONO PERCHE, I **Vol. 7** 242
FIGLI DEL MARCHESE LUCERA, I **Vol. 7** 40, 1116
FIGLI DELLA NOTTE, I **Vol. 7** 40
FIGLI DI NESSUNO, I **Vol. 7** 53, 638, 1054
FIGLI DI SANSONIA, I **Vol. 7** 1186
FIGLI NON SI TOCCANO, I **Vol. 7** 130
FIGLI NON SI VENDONO, I **Vol. 7** 1027
FIGLI ... SO' PEZZI 'E CORE **Vol. 7** 222, 474
FIGLI SPERDUTI **Vol. 7** 806
FIGLIA DEL CAPITANO, LA **Vol. 7** 443, 1136
FIGLIA DEL CONDANNATO, LA **Vol. 7** 1090
FIGLIA DEL CONTRABBANDIERE, LA **Vol. 7** 1189

FIGLIA DEL CORSARO VERDE, LA **Vol. 7** 443, 910, 1096
FIGLIA DEL DESERTO, LA **Vol. 7** 359
FIGLIA DEL DIAVOLO, LA **Vol. 7** 852
FIGLIA DEL FORZATO, LA **Vol. 7** 827
FIGLIA DEL MENDICANTE, LA **Vol. 7** 978
FIGLIA DEL PECCATO, LA **Vol. 7** 1063
FIGLIA DEL REGGIMENTO, LA **Vol. 7** 1027
FIGLIA DELLA MADONNA, LA **Vol. 7** 1063
FIGLIA DELLA TEMPESTA, LA **Vol. 7** 34, 131, 437
FIGLIA DELLE ONDE, LA **Vol. 7** 865
FIGLIA DELL'ORO, LA **Vol. 7** 531
FIGLIA DI IORIO, LA **Vol. 7** 442
FIGLIA DI JORIO, LA **Vol. 7** 1189
FIGLIA DI MATA HARI, LA **Vol. 7** 910
FIGLIA UNICA, LA **Vol. 7** 26, 701
FIGLIE DEL MARE, LE **Vol. 7** 196, 254
FIGLIE DELL'AVARO, LE **Vol. 7** 1162
FIGLIO DEL CORSARO, IL **Vol. 7** 86, 963
FIGLIO DEL CORSARO NERO, IL **Vol. 7** 1195
FIGLIO DEL CORSARO ROSSO, IL **Vol. 7** 257, 809, 838, 897
FIGLIO DELLA STRADA, IL **Vol. 7** 729
FIGLIO DELLE SELVE, IL **Vol. 7** 1189
FIGLIO DELLE STELLE **Vol. 7** 1146
FIGLIO DELLO SCEICCO, IL **Vol. 7** 1007
FIGLIO DELL'UOMO **Vol. 7** 1152
FIGLIO DI CLEOPATRA, IL **Vol. 7** 17
FIGLIO DI CORALIE, IL **Vol. 7** 250
FIGLIO DI D'ARTAGNAN, IL **Vol. 7** 161, 891
FIGLIO DI LA GARDERE, IL **Vol. 7** 493
FIGLIO DI MADAME SANS GENE **Vol. 7** 37, 584, 730
FIGLIO DI SPARTACUS, IL **Vol. 7** 1054
FIGLIO, IL **Vol. 7** 782, 783, 879
FIGLIO MIO INFINITAMENTE CARO **Vol. 7** 599
FIGLIO MIO SONO INNOCENTE ! **Vol. 7** 1022
FIGLIOCCIO DEL PADRINO, IL **Vol. 7** 515, 1027
FIGLIOCCIO DI RIRETTE, IL **Vol. 7** 1162
FIGURANTE, LA **Vol. 7** 26
FIGURE DE PROUE **Vol. 2** 388
FIGURKY ZE SMANTU **Vol. 10** 28, 233, 389, 408
FIJE QE PRITEN **Vol. 3** 14, 20
FIL A LA PATTE, UN **Vol. 2** 41, 270, 388, 579, 646, 661
FILANDON, EL **Vol. 8** 254, 315, 749
FILAVA, FILAVA... **Vol. 7** 1174
FILESE ME MARITSA **Vol. 3** 154
FILESE ME PRIN FYGEIS GIA PANTA **Vol. 3** 161
FILIBUS **Vol. 7** 454, 755
FILIGRANA **Vol. 8** 298, 400
FILIP CEL BUN **Vol. 3** 281, 288, 291
FILIP Z KONOPI **Vol. 6** *page 236, page 245, page 246*
FILLE A CROQUER, UNE **Vol. 2** 6, 194, 344
FILLE A LA DERIVE, UNE **Vol. 2** 545
FILLE A L'ENVERS, LA **Vol. 2** 734
FILLE A PAPA, UNE **Vol. 2** 193, 579, 722, 727
FILLE AU FOUET, LA **Vol. 2** 270, 309
FILLE AUX YEUX D'OR, LA **Vol. 2** 642, 651
FILLE AUX YEUX GRIS, LA **Vol. 2** 395, 615
FILLE BIEN GARDEE, UNE **Vol. 2** 169, 330, 613
FILLE COUSUE DE FIL BLANC, UNE **Vol. 2** 335, 414
FILLE D'AMERIQUE, LA **Vol. 2** 377, 728
FILLE DANS LE SOLEIL, UNE **Vol. 2** 8, 347
FILLE DE FEU, LA **Vol. 2** 131, 407
FILLE DE HAMBOURG, LA **Vol. 2** 383, 824
FILLE DE JEPHTE, LA **Vol. 2** 237
FILLE DE LA MADELON, LA **Vol. 2** 228, 404, 722
FILLE DE L'EAU, LA **Vol. 2** 48, 346
FILLE DE MADAME ANGOT, LA **Vol. 2** 249
FILLE DE MATA-HARI, LA **Vol. 2** 550
FILLE DE PRAGUE AVEC UN SAC TRES LOURD, LA **Vol. 2** 45
FILLE D'EN FACE, LA **Vol. 2** 689
FILLE DU BOCHE, LA **Vol. 2** 167
FILLE DU DIABLE **Vol. 2** 627, 824
FILLE DU GARDE-BARRIERE, LA **Vol. 2** 176, 477
FILLE DU PUISATIER, LA **Vol. 2** 143, 871
FILLE DU REGIMENT, LA **Vol. 2** 397
FILLE ELISA, LA **Vol. 7** 986
FILLE EPATANTE, UNE **Vol. 2** 294, 401
FILLE ET LE GARCON, LA **Vol. 2** 399
FILLE NOMMEE AMOUR, UNE **Vol. 2** 241, 475
FILLE PERDUE, LA **Vol. 2** 404
FILLE PRODIGUE, LA **Vol. 6** *page 84, page 86*
FILLE PUR L'ETE, UNE **Vol. 2** 124, 173, 385, 515
FILLE SUR LA ROUTE, UNE **Vol. 2** 128, 309
FILLES DE COPACABANA, LES **Vol. 2** 393
FILLES DE FERME, LES **Vol. 6** *page 119*

FILLES DE GRENOBLES, LES **Vol. 6** *page 76, page 94, page 138*
FILLES DE LA CONCIERGE, LES **Vol. 2** 388, 427, 863
FILLES DE LA ROCHELLE, LES **Vol. 2** 433
FILLES DE MADAME CLAUDE, LES **Vol. 2** 234
FILLES DE MALEMORT, LES **Vol. 2** 804
FILLES DE NUIT **Vol. 2** 347, 581
FILLES DU CAMPING, LES **Vol. 6** *page 79*
FILLES DU FAUBORG, LES **Vol. 2** 450
FILLES DU GOLDEN SALOON, LES **Vol. 2** 396
FILLES DU REGIMENT, LES **Vol. 2** 201, 445
FILLES DU RHONE, LES **Vol. 2** 70, 153
FILLES SEMENT LE VENT, LES **Vol. 2** 205
FILM CU O FATA FERMECATOARE, UN **Vol. 3** 296, 330
FILM D'AMORE E D'ANARCHIA **Vol. 7** 592, 988
FILM, FLIRT OG FORLOVELSE **Vol. 5** 83
FILM-KATHI, DIE **Vol. 4** 217
FILM VON DER KÖNIGIN LUISE, AUS PREUSSENS SCHWERER ZEIT, DER **Vol. 4** 89
FILM VON DER KÖNIGIN LUISE, DIE KÖNIGIN DER SCHMERZEN, DER **Vol. 4** 89
FILMEMACHER **Vol. 1** 42, 68, 100
FILMENS DATTER **Vol. 5** 183
FILMENS HELTE **Vol. 5** 24, 55
FILMENS VIDUNDERLIGE VERDEN **Vol. 5** 666
FILMPRIMADONNA, DIE **Vol. 4** 178, 213, 651
FILMREGENY **Vol. 9** 167, 238
FILMS HISTORIQUES **Vol. 2** 330, 375
FILMSKUESPILLEREN **Vol. 5** 97
FILO D'ARIANNA, IL **Vol. 7** 266
FILO DEL MIEDO, EL **Vol. 8** 450, 482
FILO D'ERBA O HA DA VENI DON CALOGERO, IL **Vol. 7** 756
FILON DU BOUIF, LE **Vol. 2** 731
FILOSOFKA MAJA **Vol. 10** 45, 85
FILOSOFSKA HISTORIE **Vol. 10** 167, 256, 412
FILOUS ET COMPAGNIE **Vol. 2** 205, 401
FILS D'AMERIQUE, UN **Vol. 2** 42, 206, 387, 658, 676, 697
FILS D'AMR EST MORT, LE **Vol. 6** *page 72*
FILS DE CAROLINE CHERIE, LE **Vol. 2** 57, 439, 454
FILS DE FRANCE **Vol. 2** 322, 344
FILS DU FLIBUSTIER, LE **Vol. 2** 169, 330, 613, 877
FILS DU SOLEIL, LES **Vol. 2** 380, 452, 498
FILS IMPROVISE, LE **Vol. 2** 646
FILS, LE **Vol. 2** 146, 770
FILS PUNI, LE **Vol. 2** 71, 854
FILTRO DI CIRCE, IL **Vol. 7** 980
FILUMENA MARTURANO **Vol. 7** 86, 443
FIM-DE-SEMANA COM A MORTE **Vol. 8** 17, 115, 144
FIMPEN **Vol. 5** 952, 1061, 1086, 1160
FIN DE CURSO **Vol. 8** 640
FIN DE SEMANA **Vol. 8** 311, 493
FIN DE SEMANA CON LA MUERTE **Vol. 8** 17, 115, 144
FIN DES PYRENEES, LA **Vol. 2** 721
FIN DU JOUR, LA **Vol. 2** 125, 423, 439, 566, 824
FIN DU MONDE, LA **Vol. 2** 308, 402, 440, 573, 657, 753
FINAL DE UNA LEYENDA, EL **Vol. 8** 515, 661
FINALE DER LIEBE **Vol. 4** 329, 527
FINALMENTE LIBERO ! **Vol. 7** 838
FINALMENTE MIO !... **Vol. 7** 886
FINALMENTE SI **Vol. 7** 333, 679
FINALMENTE SOLI ! **Vol. 7** 332, 603
FINANCE NOIRE **Vol. 2** 273, 395
FINANZEN DES GROSSHERZOGS, DIE **Vol. 4** 115, 119, 178, 202, 527, 626
FINCHE C'E GUERRA C'E SPERANZA **Vol. 8** 138, 148
FINE DEL MONDO NEL NOSTRO SOLITO LETTO IN UNA NOTTE PIENA DI PIOGGIA **Vol. 7** 988
FINE DELL'AMORE, LA **Vol. 7** 559
FINE DELL'INNOCENZA, LA **Vol. 7** 335
FINESTRA SUL LUNA PARK, LA **Vol. 7** 817
FINGER ON THE TRIGGER **Vol. 8** 517
FINIS TERRAE **Vol. 2** 62
FINISCE SEMPRE COSI **Vol. 7** 493
FINNS INGA GUDAR PA JORDEN, DET **Vol. 5** 944, 976
FINNSKOGENS FOLK **Vol. 5** 933, 1073
FINO ALLA TENEBRA **Vol. 7** 250
FINSKY NUZ **Vol. 10** 87, 502
FIOCCA LA NEVE **Vol. 7** 39
FIOLE D'AMOUR **Vol. 6** *page 69*
FIOR D'AMORE **Vol. 7** 266
FIOR DI LEVANTE **Vol. 7** 264, 701, 725
FIOR DI MORTE **Vol. 7** 1189
FIOR DI SIEPE **Vol. 7** 729
FIORAIA DI COMO, LA **Vol. 7** 730
FIORDALISI D'ORO POUR LA REINE **Vol. 7** 16, 1159
FIORE D'AUTUNNO **Vol. 7** 1038
FIORE DEL CAUCASO, IL **Vol. 7** 277

FIORE DEL DESTINO, IL Vol. 7 357
FIORE DEL SILENZIO, IL Vol. 7 968
FIORE DELLE MILLE E UNA NOTTE, IL Vol. 7 388, 434, 1007
FIORE DI PRATO Vol. 7 731, 936
FIORE NEL FANGO, UN Vol. 7 83
FIORE SELVAGGIO Vol. 7 212, 1164
FIORE SOTTO GLI OCCHI, IL Vol. 7 359
FIORENZA MIA ! Vol. 7 69
FIORI D'AMORE ... FIORI DI MORTE Vol. 7 131
FIORI D'ARANCIO Vol. 7 266
FIORI DI SANT'ANTONIO, I Vol. 7 1189
FIORINA LA VACCA Vol. 7 763
FIORITURES Vol. 2 153
FIRBENEDE BARNEPIGE, DEN Vol. 5 83
FIRBENEDE PLEJEBARN, DET Vol. 5 83
FIRBENEDE SHERLOCK HOLMES, DEN Vol. 5 83
FIRE DJAELVE, DE Vol. 5 188
FIRE KAERESTER, DE Vol. 5 97
FIRMA, DIE Vol. 1 17, 71
FIRMA HEIRATET, DIE Vol. 4 158, 452, 743
FIRMAFESTEN Vol. 5 869, 975, 1149
FIRMASKOVTUREN Vol. 5 68, 117, 195
FIRMIN, LE MUET DE SAINT-PATACLET Vol. 2 152, 812
FIRNENRAUSCH Vol. 4 71, 724
FIRST OF THE FEW, THE Vol. 2 656
FIRULI'HA PERSO LABALIA Vol. 7 1189
FISCHERMÄDCHEN VON SKAGEN, DAS Vol. 4 628
FISCHZÜGE Vol. 1 26, 59, 69
FISCIO DELLA SIRENA, IL Vol. 7 1189
FISKEBYN Vol. 5 937, 996
FISKER I HANSTHOLM Vol. 5 46
FISKEREN OG HANS BRUD Vol. 5 97
FISKERLEJETS DATTER Vol. 5 228
FISKERLIV I NORDEN Vol. 5 97
FIU VAGY LANY ? Vol. 9 132, 186
FIUK A TERRÖL Vol. 9 183, 259, 304
FIUL MUNTILOR Vol. 3 240, 321
FIUME DEL GRANDE CAIMANO, IL Vol. 7 426
5 A 7 TRES PARTICULIERS Vol. 6 *page 79*
5 I FEDTEFADET, DE Vol. 5 261
5 LODDRETT Vol. 5 674, 717, 742, 826
5 MAND OG ROSA Vol. 5 128, 199, 304
5 OG SPIONERNE, DE Vol. 5 76, 261
5 – ÖS SZAMU ÖRHAZ, AZ Vol. 9 26, 117
5 RASKE PIGER Vol. 5 24
500 KR. INDEN LORDAG Vol. 5 58
FIZESSEN NAGYSAD ! Vol. 9 153, 288, 334
FJELDEVENTYRET Vol. 5 685, 771
FJELDPIGEN Vol. 5 183
FJERDE DAME, DEN Vol. 5 81
FJOLS TIL FJELLS Vol. 5 674, 831
FLAADENS BLAA MATROSER Vol. 5 2, 24
FLACHSMANN ALS ERZIEHER Vol. 4 242, 471
FLADENS FRISKE FYRE Vol. 5 2, 23
FLAGELADOS DO VENTO LESTE, OS Vol. 8 65, 166
FLAGERMUSEN Vol. 5 96, 108, 176, 199, 201, 275
FLAGRANT DELIRE Vol. 2 491
FLAGRANT DELIT Vol. 2 733
FLAGRANTE DELITTO Vol. 7 1189
FLAKLYPA GRAND PRIX Vol. 5 687, 779
FLAM Vol. 10 70, 352, 620
FLAMBEE, LA Vol. 2 533
FLAMBEUSE, LA Vol. 6 *page 97, page 127*
FLAMEN NAD JADRANOM Vol. 3 340, 351
FLAMENCOS, LOS Vol. 8 207
FLAMMAN Vol. 5 967, 1041
FLAMME, DIE Vol. 4 242, 560, 663, 681 Vol. 6 *page 190*
FLAMME, LA Vol. 2 279, 387, 407, 467
FLAMMEN LÜGEN, DIE Vol. 4 213, 626
FLAMMES Vol. 2 30
FLAMMES SUR L'ADRIATIQUE Vol. 2 742 Vol. 3 340, 351
FLASH BACK Vol. 2 351
FLASH LOVE Vol. 2 213
FLECHA NEGRA Vol. 8 215, 303
FLECK „AUF DER EHR", DER Vol. 6 #190, page 193
FLECKEN AUF DER EHRE Vol. 4 99
FLEDERMAUS, DIE Vol. 4 242, 246, 266, 257, 462, 681, 774
FLEDERMÄUSE VON RONDSHILL, DIE Vol. 4 135
FLEMMING OG KVIK Vol. 5 248, 272
FLEMMING PA KOSTSKOLE Vol. 5 107, 248
FLEUR D'AMOUR Vol. 2 228, 824
FLEUR DES RUINES, LA Vol. 2 153
FLEUR D'ORANGER, LA Vol. 2 32, 332
FLEUR D'OSEILLE Vol. 2 279, 293
FLEUR LAFONTAINE Vol. 1 63, 128
FLEURS DU MIEL, LES Vol. 2 732
FLIBUSTIER, DIE Vol. 4 747
FLIC OU VOYOU Vol. 2 224, 746
FLIC STORY Vol. 2 462, 588, 814
FLIC, UN Vol. 2 270, 588, 595, 875
FLICKA FÖR MEJ, EN Vol. 5 863, 1028, 1032
FLICKA I KASERN Vol. 5 871, 1003, 1073
FLICKA KOMMER TILL STAN, EN Vol. 5 958, 1121
FLICKA MED MELODIE Vol. 5 840, 1033

FLICKA OCH HYACINTER **Vol. 5** 1042, 1164
FLICKA PA HALSEN, EN **Vol. 5** 869, 932, 963
FLICKAN ÄR ETT FYND **Vol. 5** 864, 1073
FLICKAN FRAN BACKAFALL **Vol. 5** 840, 884, 928
FLICKAN FRAN FJÄLLBYN **Vol. 5** 1091
FLICKAN FRAN PARADISET **Vol. 5** 870, 923
FLICKAN FRAN TREDJE RADEN **Vol. 5** 1042, 1164
FLICKAN FRAN VÄRMLAND **Vol. 5** 891, 978
FLICKAN FRAN VARUHUSET **Vol. 5** 928
FLICKAN I FÖNSTRET MITTEMOT **Vol. 5** 840, 878
FLICKAN I FRACK **Vol. 5** 1053, 1073, 1080, 1210
FLICKAN I REGNET **Vol. 5** 933, 1073
FLICKAN OCH DJÄVULEN **Vol. 5** 871, 1052, 1220
FLICKAN UTAN NAMN **Vol. 5** 933
FLICKOR I HAMN **Vol. 5** 1032, 1181
FLICKOR PA FABRIK **Vol. 5** 863, 1042
FLICKORNA **Vol. 5** 918, 933, 1153
FLICKORNA FRAN GAMLA STAN **Vol. 5** 863, 1042, 1141
FLICKORNA GYRKOVICS **Vol. 5** 891, 988
FLICKORNA I SMALAND **Vol. 5** 928, 1156
FLICKORNA PA SOLVIK **Vol. 5** 1037
FLICKORNA PA UPPAKRA **Vol. 5** 930, 1032
FLICKORNAS ALFRED **Vol. 5** 840, 841
FLICS DE CHOC **Vol. 6** *page 79, page 102*
FLIEGENDE AUTO, DAS **Vol. 4** 471
FLIEGENDE HOLLÄNDER, DER **Vol. 1** 36, 50, 54 **Vol. 4** 232, 646, 796
FLIEGENDE TOD, DER **Vol. 4** 241, 598
FLIEGENDEN BRIGANTEN, DIE **Vol. 4** 634, 734
FLIEGER VON GOERZ, DER **Vol. 4** 663
FLIMMERSTERNE **Vol. 4** 99, 180, 247
FLINK BARNEPIGE, EN **Vol. 5** 97
FLINTESONNERNE **Vol. 5** 89, 281
FLIPPER **Vol. 7** 695, 1173
FLIRT **Vol. 7** 599, 697
FLIRT SE SLECNOU STRIBRNOU **Vol. 10** 87, 96, 299
FLITTER - DÖRTJE **Vol. 4** 247
FLITTERWOCHEN **Vol. 4** 49, 123, 212, 560, 788, 798
FLITTERWOCHEN OHNE EHEMANN **Vol. 1** 34, 86, 111
FLOFLOCHE **Vol. 2** 47, 404
FLOH IM OHR **Vol. 4** 292, 489, 517
FLOJTESPILLEREN **Vol. 5** 89, 281
FLOR DE ESPANA **Vol. 8** 281
FLOR DE LAGO **Vol. 8** 207, 311
FLOR DE SANTIDAD **Vol. 8** 245, 688
FLOR SALVAJE **Vol. 8** 306, 708
FLORA Y MARIANA **Vol. 8** 268, 357
FLORAEA REGINEI **Vol. 3** 210, 260, 299
FLORENC 13.30 **Vol. 10** 433, 524, 621
FLORENCE **Vol. 6** *page 119*
FLORENCE EST FOLLE **Vol. 2** 158, 824,
FLORENTINER 73 **Vol. 1** 34, 68, 75, 95
FLORENTINER HUT, DER **Vol. 4** 288, 410, 660, 728
FLORES DEL MIEDO, LAS **Vol. 8** 207
FLORES SILVESTRES **Vol. 8** 774
FLORETTE E PATAPON **Vol. 7** 369, 1038, 1162
FLORIAN **Vol. 1** 241, 278, 308 **Vol. 10** 111, 153, 235
FLORINE, LA FLEUR DU VALOIS **Vol. 2** 245, 346
FLORINTINISCHE NÄCHTE **Vol. 4** 131
FLORISTA DE LA REINA, LA **Vol. 2** 646 **Vol. 8** 607, 692
FLOSS DER TOTEN, DAS **Vol. 4** 35, 212, 219
FLOSSIE **Vol. 5** 946
FLÖTENKONZERT VON SANSSOUCI **Vol. 4** 266, 288, 574
FLOTTA DEGLI EMIGRANTI, LA **Vol. 7** 1005
FLOTTANS GLADA GOSSAR **Vol. 5** 878, 1173
FLOTTANS KAVALJERER **Vol. 5** 909, 1173
FLOTTANS LILLA FÄSTMÖ **Vol. 5** 870, 891, 909
FLOTTANS MUNTERGÖKAR **Vol. 5** 834, 923
FLOTTANS ÖVERMAN **Vol. 5** 878, 1052
FLOTTARE MED FÄRG **Vol. 5** 923, 1029
FLUCH DER BÖSEN TAT, DER **Vol. 4** 25, 255
FLUCH DER MENSCHHEIT, DER **Vol. 4** 270, 570
FLUCH DER SCHÖNHEIT **Vol. 4** 87, 124
FLUCH DER SONNE, DER **Vol. 4** 124
FLUCH DER VERERBUNG, DER **Vol. 4** 73, 428, 663
FLUCH DES NURI, DER **Vol. 4** 217, 495
FLUCH DES SCHWEIGENS, DER **Vol. 4** 36, 660
FLUCH DES SPIELS, DER **Vol. 4** 747
FLUCHT AN DIE ADRIA **Vol. 10** 168, 460
FLUCHT AUS DEM HEERE DER HEIMATLOSEN, DIE **Vol. 4** 778
FLUCHT AUS DER HÖLLE **Vol. 1** 2, 17 **Vol. 4** 166, 747
FLUCHT, DIE **Vol. 1** 57, 80, 128
FLUCHT DURCH FLAMMEN, DIE **Vol. 4** 240

FLUCHT IN DEN ZIRKUS, DIE Vol. 4 211, 219, 565, 663
FLUCHT IN DIE EHE, DIE Vol. 4 743
FLUCHT IN DIE FREMDENLEGION Vol. 4 157, 213, 716
FLUCHT IN DIE NACHT, DIE Vol. 4 113, 724, 734
FLUCHT INS DUNKEL Vol. 4 270, 606, 623
FLUCHT INS SCHWEIGEN Vol. 1 102, 111, 114
FLUCHT NACH NIZZA Vol. 4 321, 626, 703
FLUCHT UND VERFOLGUNG DES RAUBMÖRDERS RUDOLF HENNIG ÜBER DIE DÄCHER VON BERLIN, DIE Vol. 4 308
FLUCHT VOR BLOND Vol. 4 270, 471
FLUCHT VOR DER KRONE. DER SCHRECKEN VON SCHLOSS WOOD, DIE Vol. 4 213
FLUCHT VOR DER LIEBE, DIE Vol. 4 270, 527
FLÜCHTLING AUS CHIKAGO, DER Vol. 4 344, 647, 788
FLÜCHTLINGE Vol. 4 266, 574, 729
FLUG IN DEN TOD, DER Vol. 4 288, 766
FLUG IN DIE SONNE, DER Vol. 4 247
FLUG UM DEN ERDBALL, DER Vol. 4 89, 149, 660
FLUG UMS LEBEN, DER Vol. 4 583
FLUG ZUR WESTGRENZE, DER Vol. 4 239
FLUGA GÖR INGEN SOMMAR, EN Vol. 5 1032, 1091
FLUGTEN Vol. 5 46, 125
FLUGTEN FRA LIVET Vol. 5 58
FLUGTEN FRA MILLIONERNE Vol. 5 79, 183
FLUGTEN FRA SERAILLET Vol. 5 97
FLUKT FRA PARADISET Vol. 5 723, 760
FLUKTEN FRA DAKAR Vol. 5 722, 738
FLY MEJ EN GREVE Vol. 5 1042, 1130, 1217
FLYG - BOM Vol. 5 878, 1173
FLYGANDE HOLLÄNDAREN Vol. 5 1210
FLYGNIVA 450 Vol. 5 946, 1072, 1170
FLYGPLAN SAKNAS Vol. 5 914, 1217
FLYVENDE CIRKUS, DEN Vol. 5 188
FLYVENDE HOLLAENDER, DEN I-IV Vol. 5 181
FOCA, LA Vol. 7 102
FOCURI SUB ZAPADA Vol. 3 333
FODBOLDPRAESTEN Vol. 5 89, 118
FODSELSDAGSGAVEN Vol. 5 9
FODSPORET Vol. 5 203
FOGADALOM, A Vol. 9 239
FOGADO AZ ÖRÖK VILAGOSSAGHOZ Vol. 9 50, 152, 154
FOGJUK MEG ES VIGYETEK Vol. 9 19, 38
FOGLIO DI VIA Vol. 7 493
FOGO ! Vol. 8 53, 75
FÖHN Vol. 4 423
FOIRE AUX CANCRES, LA Vol. 2 502, 650, 655
FOIRE AUX CHIMERES, LA Vol. 2 279, 601
FOIRE AUX FEMMES, LA Vol. 2 185, 581, 726
FÖLDINDULAS Vol. 9 71, 312
FOLGE EINEM STERN Vol. 1 47, 89
FOLGEN EINER UNGLÜCKLICHEN EHE, DIE Vol. 4 232
FOLGORE Vol. 7 886, 1079
FOLIE DES GRANDEURS, LA Vol. 2 224, 863 Vol. 8 749
FOLIE DES VAILLANTS, LA Vol. 2 308, 648, 833
FOLIE DOUCE Vol. 2 231, 661
FOLIE DU DOCTEUR TUBE, LA Vol. 2 153
FOLIES-BERGERE Vol. 2 197, 601
FOLIES BOURGEOISES Vol. 2 700, 783
FOLIES D'ELODIE Vol. 6 *page 109, page 114*
FOLKESAGN, ET Vol. 5 97
FOLKET I FÄLT Vol. 5 933, 1073
FOLKET I SIMLANGSDALEN Vol. 5 870, 923, 1042, 1164
FOLKET PA HÖGBOGARDEN Vol. 5 846
FOLKETS VEN Vol. 5 183
FOLLE A TUER Vol. 2 110, 783
FOLLE AVENTURE, LA Vol. 2 679
FOLLE NUIT, LA Vol. 2 65, 330, 595
FOLLES NUITS DE NATHALIE, LES Vol. 6 *page 119*
FOLLIA D'AMORE Vol. 7 1189
FOLLIA DEL GIUOCO, LA Vol. 7 288
FOLLIA DI NORETTA, LA Vol. 7 1116
FOLLIE Vol. 7 399, 914, 1209
FOLLIE DEL SECOLO Vol. 7 472, 725
FÖLTAMADOTT A TENGER Vol. 9 117, 199, 237
FONISSA Vol. 3 156, 207
FONTAINE DES AMOURS, LA Vol. 2 103
FONTAMARA Vol. 7 227, 1035, 1194
FONTANA DI TREVI Vol. 8 396, 515
FONTANA INSANGUINATA, LA Vol. 7 189
FONTANA PRE ZUZANU Vol. 10 210, 228, 578
FOOLE VIRGIN, THE Vol. 2 24
FOR AABENT TAEPPE Vol. 5 97
FÖR ATT INTE TALA OM ALLA DESSA KVINNOR Vol. 5 1052, 1053, 1080
FOR BARNETS SKYLD Vol. 5 295
FOR EYES ONLY Vol. 1 21, 54, 90
FOR FRIHED OG RET Vol. 5 2, 252, 272, 290

FOR FROSTNETTENE Vol. 5 754, 772, 799
FÖR HEMMET OCH FLICKAN Vol. 5 996
FOR HENDES SKYLD Vol. 5 295
FOR HENNES SKULL Vol. 5 891, 962, 991, 997
FÖR KATTEN Vol. 5 1147
FÖR MIN HETA UNGDOMS SKULL Vol. 5 1042, 1130, 1181
FOR SIN DRENG Vol. 5 295
FOR SIN FADERS SKYLD Vol. 5 58
FÖR SIN KÄRLEKS SKULL Vol. 5 925, 997
FOR SIT LANDS AERE Vol. 5 9
FÖR TAPPERHET I TÄLT Vol. 5 843, 967
FOR TORS SKYLD Vol. 5 732, 824, 827
FÖR VÄNSKAPS SKULL Vol. 5 940, 1052, 1053
FORAERING TIL MIN KONE, EN Vol. 5 97
FORAJIDOS IMPLACABLES Vol. 8 602, 626
FORASTERA, LA Vol. 8 493, 732
FORBIDDEN FRUIT Vol. 8 74, 129, 130
FORBIDDEN STREET, THE Vol. 2 656
FORBIDDEN WATERS Vol. 2 84
FORBRYDELSENS ELEMENT Vol. 5 73, 125, 249
FORBRYDER, EN Vol. 5 2, 7
FORBRYDERISK ELSKOV Vol. 5 16, 195
FORBRYDERKONGENS DATTER Vol. 5 256
FORBRYDERS LIV OG LEVNED, DEN Vol. 5 81
FORCE 8 Vol. 2 405
FORELDRELOSE, DE Vol. 5 711
FORELLE, DIE Vol. 1 8, 26, 85
FORELLEN Vol. 5 55, 196
FORELSKET I KOBENHAVN Vol. 5 241, 248
FORET DE L'ADIEU, LA Vol. 2 661, 698
FORFAITURE Vol. 2 778
FORFALDNE HUSLEJE, DEN Vol. 5 97
FORFLOJEN AEGTEMAND, EN Vol. 5 83
FORFOLGELSEN Vol. 5 684
FÖRFÖLJELSEN Vol. 5 852, 892, 1186
FORFRISKENDE BAD, ET Vol. 5 97
FORFULGTE BRUDGOM, DEN Vol. 5 83
FORGERONS Vol. 2 678
FORHEKSEDE STOLVER, DE Vol. 5 83
FORJA DE ALMAS Vol. 8 214, 716, 752
FORJA DE AMIGOS Vol. 8 722, 773
FORMULA UNO FEBBRE DELLA VELOCITA Vol. 7 383, 567
FORMYNDERE Vol. 5 780, 784, 824
FORNARETTO DI VENEZIA, IL Vol. 7 48, 67, 257, 587, 1035, 1096,
FORNARINA, LA Vol. 7 224, 603, 709
FORO NEL PARABREZZA, UN Vol. 7 909, 1083
FORRAEDERNE Vol. 5 111, 164, 259
FORRO MEZÖK Vol. 9 117, 237
FORRO VIZET A KOPASZRA ! Vol. 9 339
FORRYKTE KOMPONIST, DEN Vol. 5 58
FORSE CHI SI, FORSE CHE NO Vol. 7 796
FORSE ERI TU L'AMORE Vol. 7 792, 1036
FÖRSEGLADE LÄPPAR Vol. 5 891, 997
FÖRSÖK INTE MEJ Vol. 5 876, 1073
FÖRSTA DIVISIONEN Vol. 5 928, 1156
FÖRSTAD PRÄSTEN Vol. 5 944, 976
FORSTAERKNINGSMANDEN Vol. 5 203
FORSTE KAERLIGHED, DEN Vol. 5 9
FÖRSTERCHRISTEL, DIE Vol. 4 14, 186, 340
FÖRSTERCHRISTL, DIE Vol. 4 157, 164, 219, 584
FORSTYRRET MIDDAG Vol. 5 97
FÖRSUMMAD AV SIN FRU Vol. 5 871, 1042, 1146
FORSVUNDNE FULDMAEGTIG, DEN Vol. 5 25, 179, 298
FORSVUNDNE POLSEMAKER, DEN Vol. 5 676, 723
FORT 13 Vol. 6 *page 238, page 249, page 256*
FORT DE LA SOLITUDE Vol. 2 57, 322, 722
FORT-DOLORES Vol. 2 425
FORT DU FOU Vol. 2 128, 661
FORT SAGANNE Vol. 6 *page 84, page 90, page 115, page 125*
FORTE TETE Vol. 2 339, 560
FORTEZZE VUOTE Vol. 7 141
FORTICHES, LE Vol. 2 207, 479
FORTIDENS SKYGGE Vol. 5 97
FÖRTJUSANDE FRÖKEN, EN Vol. 5 843, 923, 1044
FÖRTROLLAD VANDRING Vol. 5 871, 1042, 1130, 1181
FORTSETZUNG FOLGT Vol. 4 325, 713
FORTUNA Vol. 7 196, 902
FORTUNA DI ESSERE DONNA, LA Vol. 7 725
FORTUNA DI ZANZE, LA Vol. 7 730, 1189
FORTUNA VIENE DAL CIELO, LA Vol. 7 476, 896
FORTUNAT Vol. 2 661, 776
FORTUNATA Y JACINTA Vol. 8 299
FORTUNATO Vol. 4 293, 321, 380 Vol. 8 298, 606, 674
FORTUNE CARREE Vol. 2 395, 615
FORTUNE DE MARSEILLE Vol. 2 67, 131
FORTUNELLA Vol. 7 1136
FORVANDLEDE DON JUAN, DEN Vol. 5 83
FÖRVANDLINGEN Vol. 5 1041, 1187
FORVISTE, DE Vol. 5 58

FORZA BRUTA, LA **Vol. 7** 493, 638, 639, 756, 1060
FORZA DEL DESTINO, LA **Vol. 7** 756
FORZA DELL'INTESTINO, LA **Vol. 7** 1189
FORZA DI UN RICORDA, LA **Vol. 7** 1125
FORZA IRRESISTIBILE **Vol. 7** 1189
FORZATO DELL'AMORE, IL **Vol. 7** 1067
FOSCARI, I **Vol. 7** 48, 520
FOSSA DEGLI ANGELI, LA **Vol. 7** 16, 495, 941
FOTBAL **Vol. 10** 70, 380
FOTO HABER **Vol. 9** 67, 122, 149
FOTO PROIBITE DE UNA SIGNORA PER BENE, LE **Vol. 8** 762
FOTOGENICO, EL **Vol. 8** 233, 253, 440, 546, 749
FOTOGRAFANDO PATRIZIA **Vol. 7** 297, 880, 1093
FOTOGRAFIA **Vol. 9** 253
FOTOGRAFO BURLATO **Vol. 7** 1189
FOTOROMANZO **Vol. 7** 1202
FOU D'AMOUR **Vol. 2** 187, 722
FOU DANSE LA MENAGERIE OU LE JUGEMENT DU FAUVE, LE **Vol. 2** 159, 286, 330
FOU DE GUERRE, LE **Vol. 7** 102, 617, 697
FOU DE LA FALAISE, LE **Vol. 2** 153
FOU DE MAI, LE **Vol. 2** 600
FOU DU LABO 4, LE **Vol. 2** 503, 791
FOU DU ROI, LE **Vol. 6** *page 82, page 127*
FOUR FEATHERS, THE **Vol. 2** 656
FOUR HORSEMEN OF THE APOCALYPSE **Vol. 2** 656
4 RAGAZZE SOGNANO **Vol. 7** 179
...4 ... 3... 2 ...1... MORTE **Vol. 8** 546, 626
4 X 4 **Vol. 5** 1192
4 X 4 – POURQUOIS **Vol. 5** 452
FOURBERIES DE SCAPIN, LES **Vol. 2** 481 **Vol. 6** *page 95*
FOURCHAMBAULT, LES **Vol. 2** 37, 286 491 **Vol. 5** 940, 1052, 1130
40ES RUGISSANTS, LES **Vol. 6** *page 77, page 91*
40 ANOS SIN SEXO **Vol. 8** 641
40 GRADI ALL'OMBRA DEL LENZUOLO **Vol. 7** 426
48 PESETAS DE TAXI **Vol. 8** 281
47 : AN LÖKEN **Vol. 5** 887
47, MORTO CHE PARLA **Vol. 7** 16
FOUS DU STADE, LES **Vol. 2** 114
FOYER PERDU **Vol. 2** 193, 475
FRA BAGDAD **Vol. 5** 97
FRA DEN GAMLE KOBMANDSGAARD **Vol. 5** 14, 180, 199, 252
FRA DET MORKE KOBENHAVN **Vol. 5** 203
FRA DIAVOLO **Vol. 2** 100, 290 **Vol. 4** 158, 452 **Vol. 7** 101, 729, 1088, 1189 **Vol. 8** 719, 725
FRA FRYSTE TIL KNEJPEVAERT **Vol. 5** 58
FRA I DUE LITIGANTI **Vol. 7** 1189
FRA MORKE TIL LYS **Vol. 5** 242
FRA PIAZZA DEL POPOLO **Vol. 5** 136, 157, 183
FRA PUPPE TIL SOMMERFUGL **Vol. 5** 97
FRACASSA E L'ALTRO **Vol. 7** 1116
FRACCHIA CONTRO DRACULA **Vol. 7** 258, 1145
FRACCHIA, LA BELVA UMANA **Vol. 7** 32, 208, 1086
FRACHT VON BALTIMORE **Vol. 4** 79, 115, 536
FRAENDELOS **Vol. 5** 155
FRAGMENTOS DE UM FILME – ESMOLA **Vol. 8** 7
FRAIRIE, LA **Vol. 2** 618
FRAK E UN APACHE, UN **Vol. 7** 824
FRAKTURA **Vol. 6** *page 12, page 18*
FRAM FÖR LILLA MÄRTA **Vol. 5** 871, 1120
FRAM ÖVER FRAMGANG **Vol. 5** 840, 878
FRA'MANISCO CERCA GUAI **Vol. 7** 756, 1152
FRÄMLINGEN FRAN SKYN **Vol. 5** 1003, 1073, 1166
FRÄMMANDE HAMN **Vol. 5** 924, 1052
FRANCAIS, SI VOUS SAVIEZ **Vol. 2** 350
FRANCAISE ET L'AMOUR, LA **Vol. 2** 9, 232, 388, 488
FRANCE, MERE PATRIE **Vol. 2** 685
FRANCE, SOCIETE ANONYME **Vol. 2** 351, 434
FRANCESCA DA RIMINI **Vol. 7** 176, 376, 807, 1156
FRANCESCO D'ASSISI **Vol. 7** 1007
FRANCESCO GIULLARE DI DIO **Vol. 7** 709, 725
FRANCISCA **Vol. 8** 10, 35, 81, 169
FRANCISCAIN DE BOURGES, LE **Vol. 2** 251, 427
FRANCO, CICCIO E IL PIRATA BARBANERA **Vol. 7** 1027
FRANCO DE PORT **Vol. 2** 249, 604, 871
FRANCOIS I. **Vol. 2** 344, 533, 775
FRANCOIS VILLON **Vol. 2** 251, 644
FRANCTIREUR **Vol. 4** 514
FRANGINES, LES **Vol. 2** 404
FRANKENSTEIN ALL'ITALIANA **Vol. 7** 46
FRÄNKISCHE LIED, DAS **Vol. 4** 467
FRANSKILD **Vol. 5** 909, 1173
FRANSSON DEN FÖRSKRÄCKLIGE **Vol. 5** 907, 923
FRANTOIO, IL **Vol. 7** 316, 730

FRANZ Vol. 2 513
FRATE FRANCESCO Vol. 7 317, 354, 490, 508, 978, 1071
FRATE SOLE Vol. 7 164, 230, 584, 965
FRATELLI CASTEGLIONE, I Vol. 7 453, 587, 1121, 1177
FRATELLI DINAMITE Vol. 7 220
FRATELLI D'ITALIA Vol. 7 333
FRATELLI KARAMAZOFF, I Vol. 7 603
FRATELLO CARNEFICE, IL Vol. 7 768
FRATELLO DI SATANA, IL Vol. 7 209
FRATELLO, IL Vol. 7 141, 856
FRATELLO MARE Vol. 7 396, 1044
FRATER LORAND Vol. 9 132, 199, 270
FRATII Vol. 3 307, 317, 321, 326
FRATII JDERI Vol. 3 307
FRAU AM SCHEIDEWEGE, DIE Vol. 4 139, 246
FRAU AM STEUER Vol. 4 82, 325
FRAU AUF DER FOLTER, DIE Vol. 4 246, 321, 703, 752
FRAU, DIE JEDER LIEBT, BIST DU !, DIE Vol. 4 268, 340, 535, 608
FRAU – DIE NACHTIGALL, DIE Vol. 4 7, 40, 111 Vol. 6 *page 72*
FRAU, DIE NICHT „NEIN" SAGEN KANN, DIE Vol. 4 97, 149, 208, 660
FRAU, DIE WEISS, WAS SIE WILL, EINE Vol. 4 292, 666 Vol. 10 300, 524, 611
FRAU EVA Vol. 4 178
FRAU EVA WIRD MONDÄN Vol. 4 361, 558
FRAU FÜR 24 STUNDEN, DIE Vol. 4 560, 634
FRAU FÜR DREI TAGE, EINE Vol. 4 72, 365
FRAU HOLLE Vol. 1 36, 54, 62, 87
FRAU IM DELPHIN ODER 30 TAGE AUF DEM MEERESGRUND, DIE Vol. 4 178, 558
FRAU IM DOKTORHUT, DIE Vol. 4 660, 704
FRAU IM FEUER, DIE Vol. 4 149, 212, 221, 660, 753
FRAU IM HIMMEL, DIE Vol. 4 648
FRAU IM KÄFIG, DIE Vol. 4 71, 479
FRAU IM MOND, DIE Vol. 2 206 Vol. 4 113, 173, 246, 298, 321, 713, 727
FRAU IM SCHRANK, DIE Vol. 4 89, 115
FRAU IM STROM Vol. 4 247, 391
FRAU IM TALAR, DIE Vol. 4 306, 379
FRAU IN DEN WOLKEN, DIE Vol. 4 192
FRAU IN GOLD, DIE Vol. 4 356 Vol. 6 *page 92*
FRAU KOMMT IN DIE TROPEN, EINE Vol. 4 132, 149, 481, 660
FRAU LEHMANNS TÖCHTER Vol. 4 232, 270, 471
FRAU LUNA Vol. 4 51, 108, 378
FRAU MEINER TRÄUME, DIE Vol. 4 325, 713
FRAU MIT DEM ETWAS, DIE Vol. 4 192, 462, 560
FRAU MIT DEM SCHLECHTEN RUF, DIE Vol. 4 288, 306
FRAU MIT DEM WELTREKORD, DIE Vol. 4 49, 124
FRAU MIT DEN KARFUNKELSTEINEN, DIE Vol. 4 146
FRAU MIT DEN MILLIONEN, DIE Vol. 4 11, 131, 495, 724
FRAU MIT DEN ORCHIDEEN, DIE Vol. 6 *page 189*
FRAU MIT VERGANGENHEIT, EINE Vol. 4 288, 356, 522
FRAU, NACH DER MAN SICH SEHNT, DIE Vol. 4 113, 479 Vol. 6 *page 197*
FRAU NACH MASS Vol. 4 270, 525
FRAU OHNE BEDEUTUNG, EINE Vol. 4 119, 148, 149, 660
FRAU OHNE GELD, DIE Vol. 4 373
FRAU OHNE HERZ, DIE Vol. 4 300
FRAU OHNE NAMEN, DIE Vol. 4 626, 634, 720, 734
FRAU OHNE NERVEN, DIE Vol. 4 239, 486, 546
FRAU OHNE SEELE, DIE Vol. 4 124, 223, 479
FRAU OHNE VERGANGENHEIT, DIE Vol. 4 391, 774
FRAU SCHLANGE Vol. 4 495, 511
FRAU SIXTA Vol. 4 347, 370, 620
FRAU SORGE Vol. 4 10, 431, 479
FRAU SYLVELIN Vol. 4 298, 536, 608
FRAU VENUS UND IHR TEUFEL Vol. 1 46, 72, 92
FRAU, VON DER MAN SPRICHT, DIE Vol. 4 287, 644, 646
FRAU VON FORMAT, EINE Vol. 4 306, 724
FRAU VON VIERZIG JAHREN, EINE Vol. 4 267, 396, 437, 663
FRAU WIE DU..., EINE Vol. 4 85, 285
FRAU WIE DU, EINE Vol. 4 344, 373, 536, 545
FRAUDEUR, LE Vol. 2 228, 273
FRAUEN AM ABGRUND Vol. 4 255, 527
FRAUEN DER LEIDENSCHAFT Vol. 4 724
FRAUEN DER WARDINS, DIE Vol. 1 26, 76, 113
FRAUEN, DIE DEN WEG VERLOREN Vol. 4 230, 366
FRAUEN, DIE DURCH DIE HÖLLE GEHEN Vol. 8 243
FRAUEN, DIE MAN OFT NICHT GRÜSST Vol. 4 186

FRAUEN, DIE NICHT HEIRATEN SOLLTEN Vol. 4 232, 300
FRAUEN, DIE NICHT LEBEN DÜRFEN Vol. 4 212, 571, 726, 753
FRAUEN, DIE SICH OPFERN Vol. 4 277, 695
FRAUEN FÜR GOLDEN HILL Vol. 4 340, 365, 466
FRAUEN IN LEIDENSCHAFT Vol. 4 124
FRAUEN SIND DOCH BESSERE DIPLOMATEN Vol. 4 325, 382, 713
FRAUEN SIND KEINE ENGEL Vol. 4 85, 666
FRAUEN UND BANKNOTEN Vol. 4 356, 373
FRAUEN VOM GNADENSTEIN, DIE Vol. 4 298, 733
FRAUEN VOM TANNENHOF, DIE Vol. 4 344
FRAUEN VON FOLIES BERGERES, DIE Vol. 4 725, 734
FRAUEN WOLLEN BETROGEN SEIN Vol. 4 247, 461
FRAUEN ZWEIER JUNGGESELLEN, DIE Vol. 4 25, 255
FRAUENARZT DR. SCHÄFER Vol. 4 255, 471
FRAUENBEICHTE Vol. 4 159, 461
FRAUENBRIEFE Vol. 4 731
FRAUENDIPLOMAT, DER Vol. 4 132, 149, 660
FRAUENGASSE VON ALGIER, DIE Vol. 4 288, 306
FRAUENHAUS VON BRESCIA, DAS Vol. 4 131, 628
FRAUENHAUS VON RIO, DAS Vol. 4 149, 527, 660
FRAUENKÖNIG, DER Vol. 4 290
FRAUENLIEBE Vol. 4 125, 210
FRAUENLIEBE - FRAUENLEID Vol. 4 149, 354, 660
FRAUENMORAL Vol. 4 471, 780
FRAUENOPFER Vol. 4 396
FRAUENRAUB Vol. 4 254
FRAUENRAUB IN MAROKKO Vol. 4 149, 219, 660
FRAUENSCHICKSAL Vol. 4 495
FRAUENSCHICKSALE Vol. 1 3, 23, 43, 78
FRÄULEIN Vol. 3 437, 454 Vol. 4 28, 298, 727
FRÄULEIN AUS ARGENTINIEN, DAS Vol. 4 221, 340, 608
FRÄULEIN BARONIN Vol. 4 385
FRÄULEIN CHAUFFEUR Vol. 4 306, 724
FRÄULEIN DOKTOR Vol. 2 676, 688, 778 Vol. 7 207, 299, 599, 750
FRÄULEIN ELSE Vol. 4 28, 178, 325, 609
FRÄULEIN FÄHNRICH Vol. 4 379, 565, 643
FRÄULEIN - FALSCH VERBUNDEN Vol. 4 208, 232, 479 Vol. 7 848
FRÄULEIN FRAU Vol. 4 115, 774
FRÄULEIN HOFFMANNS ERZÄHLUNGEN Vol. 4 257, 432, 788
FRÄULEIN JOSETTE - MEINE FRAU Vol. 4 321, 734
FRÄULEIN JULIE Vol. 4 36, 266, 574
FRÄULEIN KITTY Vol. 7 982
FRÄULEIN LAUSBUB Vol. 4 213, 558
FRÄULEIN LEUTNANT Vol. 4 743
FRÄULEIN LIESELOTT Vol. 4 115, 774
FRÄULEIN MAMA Vol. 4 212, 571, 726
FRÄULEIN MUTTER Vol. 4 565
FRÄULEIN PFIFFIKUS Vol. 4 366
FRÄULEIN PICCOLO Vol. 4 356, 779
FRÄULEIN RAFFKE Vol. 4 192, 220, 319, 584
FRÄULEIN RECHTSANWALT Vol. 4 5
FRÄULEIN SEIFENSCHAUM Vol. 4 560
FRÄULEIN VOM AMT, DAS Vol. 4 115, 729
FRÄULEIN VON BARNHELM, DAS Vol. 4 288, 524, 544, 619
FRÄULEIN VON DER KASSE 12, DAS Vol. 4 245, 560, 720
FRÄULEIN VON SCUDERIE, DAS Vol. 1 58, 78, 129
FRAY DOLLAR Vol. 8 749
FRAY ESCOBA Vol. 8 595, 752
FRAY TORERO Vol. 8 438, 681
FRECCIA D'ORO Vol. 7 108, 756, 796, 941
FRECCIA NEL CUORE, LA Vol. 7 53
FRECCIA NEL FIANCO, LA Vol. 7 200, 756, 827, 1116
FRECH UND VERLIEBT Vol. 4 678
FRECHDACHS, DER Vol. 4 82, 270, 409, 713
FREDDY Vol. 2 12, 76
FREDDY KLARAR BIFFEN Vol. 5 1130, 1142, 1217
FREDEK USZCZESLIWIA SWIAT Vol. 1 194, 278
FREDERICIA Vol. 2 419, 558
FREDERIK BUCH SOM DEKORATOR Vol. 5 97, 295
FREDI NA DIVIJEM ZAPADU Vol. 3 350
FREDLOS Vol. 5 55, 97, 359, 867, 899
FREDLOS I SKOGEN Vol. 5 753
FREDY, EL CROUPIER Vol. 8 237, 274
FREEDOM Vol. 5 405
FREGOLI Vol. 7 95, 382, 770, 1194
FREI LUIS DE SOUSA Vol. 8 10, 77, 122, 131
FREIE LIEBE Vol. 4 98, 686
FREIES LAND Vol. 1 4 Vol. 4 30
FREIES VOLK Vol. 4 291

FREIHEIT, GLEICHHEIT, BRÜDERLICHKEIT Vol. 4 232
FREIHEIT IN FESSELN Vol. 4 471, 734
FREISPRUCH MANGELS BEWEISES Vol. 1 34, 42, 50
FREITAG, DER 13. Vol. 4 167, 232
FREIWILD Vol. 4 213, 339
FRELSENDE FILM, DEN Vol. 5 58
FRELSERPIGEN Vol. 5 203
FREMDE AUS DER ELSTERGASSE, DIE Vol. 4 35, 113
FREMDE, DER Vol. 1 24, 88
FREMDE, DIE Vol. 4 39, 288 Vol. 6 *page 75*
FREMDE FRAU, DIE Vol. 4 167, 340
FREMDE FÜRST, DER Vol. 4 186, 202
FREMDE HAND, DIE Vol. 4 110, 517
FREMDE LEBEN, DAS Vol. 4 51, 108, 525
FREMDE MIT DER TEUFELSFRATZE, DER Vol. 4 718
FREMDE VOGEL, DER Vol. 4 646
FREMDENHEIM FILODA Vol. 4 149, 292, 660
FREMDENLEGIONÄR, DER Vol. 4 344, 544
FREMMED BANKER PA, EN Vol. 5 41, 132, 137
FREMMEDE TJENER, DEN Vol. 5 97
FREMMELIGT BARN, ET Vol. 5 83
FRENCH-CANCAN Vol. 2 226, 251, 427, 832
FRENCH ERECTION Vol. 2 291
FRENCH LOVE, THE Vol. 2 818
FRENCH POST-CARDS Vol. 2 633
FRENESIA Vol. 7 756, 796
FRENETIS Vol. 3 153
FRENTE DE LOS SUSPIROS, EL Vol. 8 449
FRERE JACQUES Vol. 2 64, 614
FRERES BOUQUINQUANT, LES Vol. 2 96, 644
FRERES CORSES, LES Vol. 2 159, 368
FRERES KARAMAZOFF, LES Vol. 2 290, 657
FRERES ZEMGANNO, LES Vol. 2 727
FRESKE FRASPARK Vol. 5 724, 796, 874
FRESTELSE Vol. 5 1032, 1181
FREUDLOSE GASSE, DIE Vol. 4 149, 379, 492, 660
FREUND ODER FEIND Vol. 4 558, 747
FREUND RIPP Vol. 4 626, 724
FREUNDE Vol. 4 374, 620
FREUNDIN EINES GROSSEN MANNES, DIE Vol. 4 292, 325, 445
FREUNDIN SO GOLDIG WIE DU, EINE Vol. 4 164, 257, 584
FREUT EUCH DES LEBENS Vol. 4 17, 232, 713
FRIARANNONSEN Vol. 5 1033
FRIAREN FRAN LANDSVÄGEN Vol. 5 996
FRIARNA PA LANGESTEN Vol. 5 1118
FRIC-FRAC Vol. 2 294, 626, 722, 824
FRIC-FRAC EN DENTELLES Vol. 2 475, 777
FRIC, LE Vol. 2 294, 347
FRICOT AI BAGNI Vol. 7 1189
FRICOT AMOROSO Vol. 7 1189
FRICOT BEVE LA MEDICINA Vol. 7 1189
FRICOT CANTANTE MUNICIPALE Vol. 7 1189
FRICOT DIVENTA LIBERTINO Vol. 7 1189
FRICOT E LA STATUA Vol. 7 1189
FRICOT HA FREDDO Vol. 7 1189
FRICOT HA SMARRITO IL BOTTONE Vol. 7 1189
FRICOT IMPARA UN MESTIERE Vol. 7 1189
FRICOT SOTTO LE ARMI Vol. 7 1189
FRICOT TRASLOCA Vol. 7 1189
FRICOT VA IN COLLEGIO Vol. 7 1189
FRIDAG, EN Vol. 5 97
FRIDAS VISOR Vol. 5 891, 997
FRIDERICUS Vol. 4 148, 149, 462, 660
FRIDERICUS REX Vol. 4 131, 174, 232, 275, 420, 452, 548, 646
FRIDOLF I LEJONKULAN Vol. 5 1092, 1205
FRIDOLF STICKER OPP Vol. 5 1042, 1217
FRIDOLFS FARLIGA ALDER Vol. 5 1042, 1137
FRIE FUGLE Vol. 5 181
FRIEDEMANN BACH Vol. 4 147, 232, 525
FRIEDENSREITER Vol. 4 762
FRIEDERIKE Vol. 4 82, 89, 556, 734
FRIEDHOF DER LEBENDEN Vol. 4 461, 746
FRIEDL VOM HOCHLAND, DER Vol. 4 499
FRIEDRICH SCHILLER Vol. 4 35, 625, 626, 704, 729
FRIEDRICH WERDERS SENDUNG Vol. 4 288
FRIERSCENEN Vol. 5 81
FRIESENBLUT Vol. 4 82, 270, 725
FRIESENNOT Vol. 4 10, 124, 644
FRIHED, LIGHED – OG LOUISE Vol. 5 89, 224
FRIHEDENS PRIS Vol. 5 2, 7
FRIHETENS MURAR Vol. 5 423, 1007, 1050, 1201
FRINE CORTIGIANA D'ORIENTE Vol. 7 16
FRIOS OJOS DEL MIEDO, LOS Vol. 8 261, 306
FRIQUET Vol. 2 32 Vol. 7 53
FRISCHER WIND AUS KANADA Vol. 4 325, 445, 462 Vol. 6 *page 189*
FRISEE AUX LARDONS, LA Vol. 2 68

FRISSON DES VAMPIRES, LE Vol. 2 721
FRISTLOS ENTLASSEN Vol. 4 352, 700
FRITIOFS SAGA Vol. 5 934, 1112
FRITZE BOLLMANN WOLLTE ANGELN Vol. 4 49, 517, 580
FRIVOLITES Vol. 2 153
FRK. KIRKEMUS Vol. 5 264
FRK. MOLLERS JUBILAEUM Vol. 5 7
FRK. VILDKAT Vol. 5 89, 196, 246
FRÖHLICHE WEINBERG, DER Vol. 4 321, 752
FRÖHLICHES HAUS, EIN Vol. 4 688, 736
FRÖKEN APRIL Vol. 5 840, 843, 923, 1104
FRÖKEN BLIR PUGA Vol. 5 1205
FRÖKEN CHIC Vol. 5 878, 984, 1052
FRÖKEN FOB Vol. 5 870
FROKEN HVAD VI VIL Vol. 5 81
FRÖKEN JULIE Vol. 5 1042, 1164
FROKEN KYRKRATTA Vol. 5 928, 929, 1156
FROKEN NITOUCHE Vol. 5 29, 104, 199, 201, 246
FRÖKEN PA BJÖRNEBORG Vol. 5 870
FROKEN STATSADVOKAT Vol. 5 770
FROKEN THEODOR Vol. 5 83
FRÖKEN VILDKATT Vol. 5 840, 909
FRÖKENS FÖRSTA BARN Vol. 5 871, 1042
FROM A FAR COUNTRY Vol. 6 *page 238*
FROMME LÜGE, DIE Vol. 4 79, 115, 247
FROMONT JEUNE ET RISLER AINE Vol. 2 339, 413
FRONA Vol. 10 41, 321, 388
FRONDA D'ULIVO Vol. 7 277
FRONTERA DE DIOS, LA Vol. 8 681, 749
FRONTERA DEL MIEDO, LA Vol. 8 311, 753
FRONTERA EL SUR Vol. 8 306
FRONTIERE Vol. 7 1189
FRONTTHEATER Vol. 4 79, 270, 623
FROSINA Vol. 3 339, 448
FROU-FROU Vol. 2 13, 96, 709 Vol. 4 99, 124
FROU FROU Vol. 7 264, 701
FROUFROU Vol. 2 33
FRU INGER TIL OSTRAT Vol. 5 664, 748, 787
FRU MARIANNES FRIARE Vol. 5 1088
FRU POTIFAR Vol. 5 97
FRUCHTBARKEIT Vol. 4 318
FRUCTE DE PADURE Vol. 6 *page 269, page 279*
FRÜH ÜBT SICH Vol. 4 329, 713, 774
FRÜHER SOMMER Vol. 6 *page 157, page 174, page 176*
FRÜHERE VERHÄLTNISSE Vol. 4 287, 291, 420, 565
FRÜHJAHRSPARADE Vol. 4 139
FRÜHLING BRAUCHT ZEIT, DER Vol. 1 34, 59, 98
FRÜHLINGS ERWACHEN Vol. 4 292, 339
FRÜHLINGSERWACHEN Vol. 4 35, 219
FRÜHLINGSFLUTEN Vol. 4 3, 153, 634, 722
FRÜHLINGSLUFT Vol. 4 215, 531, 774
FRÜHLINGSMÄRCHEN Vol. 4 373, 626
FRÜHLINGSRAUSCHEN Vol. 4 230, 681, 691
FRÜHSTÜCK IM BETT Vol. 6 *page 156, page 170, page 171*
FRUIT DEFENDU, LE Vol. 2 13, 598
FRUITS AMERS Vol. 2 293
FRUITS AMERS-SOLEDAD Vol. 3 350, 359
FRUITS DE L'ETE, LE Vol. 2 59, 488
FRUITS SAUVAGES, LES Vol. 2 231, 581
FRUMOASELE VACANTE Vol. 3 244, 267, 329 Vol. 9 131
FRUN TILL HANDA Vol. 5 863, 1032
FRUSTATION Vol. 2 804
FRUTTO ACERBO Vol. 7 48, 745, 756
FU COSI CHE... Vol. 7 984
FU MANCHU Y EL BESO DE LA MUERTE Vol. 8 546
FU MATTIA PASCAL, IL Vol. 7 63, 452, 639, 1060
FU SIGNOR ME STESSO, IL Vol. 7 1125
FUCHA Vol. 6 *page 211, page 244*
FUCHS, DU HAST DIE GANS GESTOHLEN Vol. 4 18, 558
FUCHS VON GLENARVON, DER Vol. 4 123, 149, 729
FUEGO Vol. 8 276, 449, 725
FUEGO CRUZADO Vol. 8 296, 376
FUEGO EN LA SANGRE Vol. 8 503, 677
FUEGO ETERNO Vol. 8 210, 771
FUENTE ENTERRADA, LA Vol. 8 484, 732
FUENTE MAGICA, LA Vol. 8 221, 752
FUENTEOVEJUNA Vol. 8 298, 311, 449
FUERA DE LA LEY Vol. 8 461, 630
FUERTE PERDIDO Vol. 8 677, 772
FUERZA DEL ARRASTRARSE, LA Vol. 8 572, 657
FUERZA DEL DESEO, LA Vol. 8 304, 673
FUERZA MORTAL Vol. 8 261, 408, 590
FUGA A DUE VOCI Vol. 7 638, 792, 1063
FUGA DAL BRONX Vol. 7 544, 623
FUGA DALL'ARCIPELAGO MALEDETTO Vol. 7 852
FUGA DE CEYLAN Vol. 8 493
FUGA DEGLI AMANTI, LA Vol. 7 796
FUGA DEI DIAMANTI, LA Vol. 7 796
FUGA DEL GATTO, LA Vol. 7 829
FUGA DESESPERADA Vol. 8 493
FUGA DI SOCRATE, LA Vol. 7 1211
FUGA IN FRANCIA Vol. 7 522, 817, 1036
FUGA IN RE MAGGIORE Vol. 7 824

FUGA, LA Vol. 7 62, 502, 817
FUGA NELLA TEMPESTA Vol. 7 910
FUGA O FUGA SCABROSAMENTE PROIBITA Vol. 7 240, 692
FUGA SENZA FINE Vol. 7 61, 632
FÜGEFALEVEL Vol. 9 84, 117, 259
FUGGE LA GLORIA Vol. 7 968
FUGGITIVA, LA Vol. 7 645, 734, 1097, 1211
FUGITIF, LE Vol. 2 153, 439
FUGITIVO DE AMBERES, EL Vol. 8 390, 493
FUGLESKRAEMSLET Vol. 5 83, 97
FUGUE DE MONSIEUR PERLE, LA Vol. 2 270
FUGUTIFS, LES Vol. 2 266
FÜHRER EN FOLIE, LE Vol. 2 779
FUITE EN AVANT, LA Vol. 6 *page 136*
FULDENT GENTLEMAN, EN Vol. 5 142, 197
FULL UTRYKNING Vol. 5 736
FULMINE SULLA CAPANNA, IL Vol. 7 1070
FUMEE BLONDE Vol. 2 247, 259, 401
FUMEES Vol. 2 131
FUMERIA D'OPPIO, LA Vol. 7 79, 200, 487
FUMO Vol. 7 971
FUNCION DE NOCHE Vol. 8 358
FUNDI I NJE GJAKMARRJE Vol. 6 *page 14, page 15, page 17*
FUNDVOGEL Vol. 4 495, 536, 646, 734 Vol. 6 *page 200*
FUNEBRAK Vol. 10 117, 138, 167, 524
FUNERAL DO PATRAO, O Vol. 8 34
FÜNF BANGE TAGE Vol. 4 149, 157, 535, 660
FÜNF FRANKFURTER, DIE Vol. 4 71, 162, 479
FÜNF MILLIONEN SUCHEN EINEN ERBEN Vol. 4 108, 119, 270
FÜNF MINUTEN ZU SPÄT Vol. 4 423
FÜNF PATRONENHÜLSEN Vol. 1 17, 48, 73
FÜNF PERSONEN SUCHEN ANSCHLUSS Vol. 4 13, 24
FÜNF TAGE – FÜNF NÄCHTE Vol. 1 82, 107
FÜNF VERFLUCHTEN GENTLEMEN, DIE Vol. 6 *page 89, page 98*
FÜNF VON DER JAZZBAND Vol. 4 115, 373
FÜNFTAUSEND MARK BELOHNUNG Vol. 4 619, 647, 660
FÜNFTE JUNI, DER Vol. 4 525
FÜNFTE STRASSE, DIE Vol. 4 112, 283, 634, 663
FÜNFUHRTEE IN DER ACKERSTRASSE Vol. 4 149, 527, 660
FÜNFZEHN MINUTEN NACH MITTERNACHT Vol. 4 525, 638
FÜNFZIGMARKSCHEIN, DER Vol. 4 329, 525
FUNICULI – FUNICULA Vol. 4 308
FUNICULI ... FUNICULA ... Vol. 7 829
FUNKEN UNTER DER ASCHE Vol. 4 514
FUNKZAUBER Vol. 4 213, 340, 608
FUOCCO TRITTICO Vol. 7 305
FUOCHI D'ARTIFICIO Vol. 7 792, 796, 1036
FUOCO, IL Vol. 7 305
FUORILEGGE DEL MATRIMONIO, I Vol. 7 763
FUORILEGGE, I Vol. 7 897, 1044
FÜR DEN RUHM DES GELIEBTEN Vol. 4 124
FÜR DIE KATZ Vol. 4 378, 445
FÜR DIE LIEBE NOCH ZU MAGER ? Vol. 1 5, 63, 79
FÜR MORD KEIN BEWEIS Vol. 1 49, 85, 103
FURBERIE DI ROBINET, LE Vol. 7 1189
FURCHT VOR DEM WEIBE, DIE Vol. 4 35, 514
FÜRE LEPNI SZABAD Vol. 9 67, 134, 183
FURET, LE Vol. 2 595, 698
FURIA Vol. 7 923, 965
FURIA A BAHIA POUR OSS 117 Vol. 2 128, 368
FURIA DE JOHNNY KID Vol. 8 772
FURIA DEI BARBARI, LA Vol. 7 1063
FURIA DEL HOMBRE LOBO, LA Vol. 8 383, 773
FURIA ESPANOLA Vol. 8 688, 754
FURIES SEXUELLES Vol. 2 291
FÜRST DER DIEBE UND SEINE LIEBE, DER Vol. 4 36
FÜRST ODER CLOWN Vol. 4 567, 626
FÜRST SEPPL Vol. 4 344, 544, 787
FÜRST VON PAPPENHEIM, DER Vol. 4 192, 462, 584
FÜRST WORONZEFF Vol. 4 30, 325, 445, 567
FÜRSTENKIND, DAS Vol. 4 292, 584
FÜRSTENLIEBE Vol. 4 247
FÜRSTIN DER RIVIERA, DIE Vol. 4 212, 571, 726
FÜRSTIN VON BERANIEN, DIE Vol. 4 71, 488
FÜRSTIN WORONZOFF, DIE Vol. 4 131
FÜRSTLICHES BLUT Vol. 4 366
FURTIVOS Vol. 8 348, 599
FURTO DEL SENTIMENTO, IL Vol. 7 81
FURTUNA Vol. 3 219, 281, 317
FUSEE, LA Vol. 2 419, 708
FUSHE E BLERTE, FUSHE E KUQE Vol. 6 *page 13, page 14, page 15, page 18*

FUSIL CHARGE **Vol. 2** 742
FUSILLE A L'AUBE **Vol. 2** 67, 226, 538, 856
FUSS, HOGY UTOLERJENEK **Vol. 9** 67, 84, 122
FUSSBALL, DER **Vol. 4** 24, 525
FÜSSCHEN MIT DER GOLDENEN SCHLANGE, DAS **Vol. 4** 603
FUSSPUR, DIE **Vol. 4** 199, 747
FÜSZER ES CSEMEGE **Vol. 9** 164, 182, 315
FUTBOL, AMOR Y TOROS **Vol. 8** 246
FUTOHOMOK **Vol. 9** 133, 157
FUTOTÜZ **Vol. 9** 132, 236
FUTUR AUX TROUSSES, LE **Vol. 2** 234, 258
FUTURES VEDETTES **Vol. 2** 423, 454, 627
FUTURO E DONNA, IL **Vol. 7** 333, 405, 434
FUTURO IMPERFECTO **Vol. 8** 601, 627
FYNTANAKI M. **Vol. 3** 130
FYRA GANGER KÄRLEK **Vol. 5** 378, 395, 616
FYRSTIDENS SKAEBNE **Vol. 5** 265
FYRTOJET **Vol. 5** 97, 126, 147
FYRVAKTERNES DOTTER **Vol. 5** 976, 1157

G

G **Vol. 5** 1089
GA PA VATTNET OM DU KAN **Vol. 5** 852, 917, 1098
GAA MED MIG HJEM **Vol. 5** 32, 55, 101, 208
GAADEFULDE VAESEN, DET **Vol. 5** 295
GAARDMANDSDATTEREN **Vol. 5** 203
GAARDMANDSSON OG HUSMANDSDATTER **Vol. 5** 97
GABBIA DORATA, LA **Vol. 7** 1057
GABBIA, LA **Vol. 7** 29 **Vol. 8** 232, 297, 770
GABBIANI VOLANO BASSO, I **Vol. 7** 1026
GABI CSINYJEI **Vol. 9** 300
GABOR DIAK **Vol. 9** 11, 199, 312
GABRIEL, TULE TAKAISIN **Vol. 5** 379, 579
GABRIELA **Vol. 10** 154, 345
GABRIELE DAMBRONE **Vol. 4** 15, 147, 485
GABRIELE EINS, ZWEI, DREI **Vol. 4** 373, 626
GABRIELE IL LAMPIONARO DI PORTO **Vol. 7** 829
GABRIELLE **Vol. 5** 940, 1173
GADE UDEN ENDE **Vol. 5** 238
GADO BRAVO **Vol. 8** 75, 99, 136
GAELDER LIVET, DET **Vol. 5** 2, 23, 228
GAELDER OS ALLE, DET **Vol. 5** 2, 228
GAELLE, MALOU ET VIRGINE **Vol. 2** 76, 721
GAESTEARBEJDERE **Vol. 5** 12
GAESTESPILLET **Vol. 5** 97
GAEVE RIDDER, EN **Vol. 5** 256
GAFE, EL **Vol. 8** 493
GAGE, LE **Vol. 2** 54
GAGNANT, LE **Vol. 2** 495, 787
GAGNE TA VIE **Vol. 2** 440
GAIETES DE LA FINANCE, LES **Vol. 2** 460, 807
GAIETES DE L'ESCADRON, LES **Vol. 2** 6, 193, 197
GAIETES DE L'EXPOSITION, LES **Vol. 2** 366, 425
GAIETES DU PALACE, LES **Vol. 2** 595, 698
GAJANEH **Vol. 1** 2, 63, 92
GALA UNTER DEN LINDEN **Vol. 1** 12
GALAJSLAVEN **Vol. 5** 97
GALANTE KÖNIG, DER **Vol. 4** 132, 406, 512, 773
GALAOR CONTRO GALAOR **Vol. 7** 548
GALAX, OMUL PAPUSA **Vol. 6** *page 267, page 272, page 274*
GALAXIE **Vol. 2** 649
GALAZIA KERIA **Vol. 3** 203
GALE DANSKER, DEN **Vol. 5** 22, 278
GALE DIGTER, DEN **Vol. 5** 81
GALEERENSTRÄFLING, DER **Vol. 4** 186, 560 **Vol. 6** *page 199*
GALERIE DES MONSTRES, LA **Vol. 2** 123, 162, 798
GALERIES LEVY ET CIE, LES **Vol. 2** 6, 55
GALES FERIE, DEN **Vol. 5** 97
GALETS D'ETRETAT, LES **Vol. 2** 241
GALETTES DE PONT-AVEN, LES **Vol. 2** 198, 455
GALGENBRAUT, DIE **Vol. 4** 540, 742
GALGENHUMOR **Vol. 4** 479, 536
GALGENTONI, DIE **Vol. 6** *page 187, page 190*
GALGMANNEN **Vol. 5** 840, 909, 1125
GALIA **Vol. 2** 279, 293
GALILÄER, DER **Vol. 4** 1, 532, 724
GALILEO GALILEI **Vol. 3** 105 **Vol. 7** 1189
GALLI DI MARE, I **Vol. 7** 792
GALLOS DE LA MADRUGADA, LOS **Vol. 8** 331, 553
GALLOS DE PELEA **Vol. 8** 453, 583
GALOPPO DELLA MORTE, IL **Vol. 7** 932
GALOSE STASTIA **Vol. 10** 299, 465, 493
GALSKAP **Vol. 5** 808
GAMBE D'ORO **Vol. 7** 113
GAMBERGE, LA **Vol. 2** 661, 770
GAMBERROS, LOS **Vol. 8** 501, 677

GAMELIES PERIPETEIES **Vol. 3** 165
GAMIN DE PARIS, LE **Vol. 2** 131, 169, 243, 330, 404, 475, 613, 877
GAMINES A TOUT FAIRE **Vol. 2** 291
GAMINES OUVERTES **Vol. 6** *page 94*
GAMLA GATANS KARNEVAL **Vol. 5** 841
GAMLA HERRGARDEN, DEN **Vol. 5** 1037
GAMLE GULD, DET **Vol. 5** 89, 118
GAMLE MAJORS UNGDOMSKAERLIGHED, DEN **Vol. 5** 97
GAMLE MOLLE PA MOLS, DEN **Vol. 5** 180, 199, 201
GAMLE PRAEST, DEN **Vol. 5** 2, 228
GAMOS A LA ELLENIKA **Vol. 3** 149, 153, 163
GANG DEL PARIGINO, LA **Vol. 7** 585
GANG DES OTAGES, LE **Vol. 2** 208, 475
GANG DES TRACTIONS-ARRIERE, LE **Vol. 2** 193, 475
GANG IN DIE NACHT, DER **Vol. 4** 423, 558
GANG, LE **Vol. 2** 542, 588
GANG NACH DEM EISENHAMMER, DER **Vol. 6** *page 189*
GANGSTER LLEGADO DE BROOKLYN, UN **Vol. 8** 626
GANGSTER MALGRE LUI **Vol. 2** 113, 152, 330, 437, 454, 812
GANGSTERENS LAERLING **Vol. 5** 25, 129
GÄNGSTERFILMEN **Vol. 5** 852, 1096
GANGSTERS DU CHATEAU D'IF, LES **Vol. 2** 279, 833
GANGSTERZY I FILANTROPI **Vol. 1** 148, 242, 401
GANOVENEHRE **Vol. 4** 119
GANS VON SEDAN, DIE **Vol. 2** 510, 676
GÄNSEHIRTIN AM BRUNNEN, DIE **Vol. 6** *page 156, page 159, page 170*
GÄNSEKNÖCHLEIN **Vol. 4** 365, 608
GANSKE ALMINDELIG PIGE, EN **Vol. 5** 49, 89
GANTELET VERT, LE **Vol. 2** 719, 837
GANTS BLANCS DU DIABLE, LES **Vol. 2** 68, 72
GANZ GROSSEN TORHEITEN, DIE **Vol. 4** 527, 626
GANZER KERL, EIN **Vol. 4** 149, 167, 344, 647, 660, 689
GAR EL HAMA III **Vol. 5** 295
GAR EL HAMA IV **Vol. 5** 295
GAR EL HAMA V **Vol. 5** 295
GARAGET **Vol. 5** 1052, 1093
GARBANZA NEGRA QUE ENPAZ DESCANSE **Vol. 8** 249, 495
GARCA E A SERPENTE, A **Vol. 8** 25, 50, 70, 130, 146
GARCES, LES **Vol. 2** 508
GARCON ! **Vol. 6** *page 70, page 73*
GARCON COMME TOI ET MOI, UN **Vol. 2** 759
GARCON SAUVAGE, LE **Vol. 2** 488, 722
GARCONNE, LA **Vol. 2** 9, 67, 368, 402, 516, 724
GARCONNIERE, LA **Vol. 7** 514, 1054
GARDARNA RUNTS SJÖN **Vol. 5** 871, 1073
GARDE A VUE **Vol. 2** 633 **Vol. 6** *page 90*
GARDE-DIVA, DIE **Vol. 4** 73, 535 **Vol. 6** *page 188*
GARDE DU CORPS, LE **Vol. 2** 97 **Vol. 6** *page 135, page 142*
GARDENIA **Vol. 7** 1003
GARDIEN DU FEU, LE **Vol. 2** 42, 805
GARDIEN, LE **Vol. 2** 193, 270
GARDIENS DE PHARE **Vol. 2** 58, 656
GARE...SONNE, LA **Vol. 2** 792
GARGOUSSE **Vol. 2** 193, 579
GARIBALDI E I SUOI TEMPI **Vol. 7** 397, 440, 796, 1004
GARIBALDINO AL CONVENTO, UN **Vol. 7** 332, 487
GARNISON AMOUREUSE, LA **Vol. 2** 332, 658, 679
GAROFANO ROSSO **Vol. 7** 381, 1205
GARRAGAN **Vol. 4** 51, 247, 373, 646
GARRAS DE LORELEI, LAS **Vol. 8** 551
GARRINGO **Vol. 8** 243
GARY COOPER QUE ESTAS EN LOS CIELOS **Vol. 8** 738
GARZONLAKAS KIADO **Vol. 9** 164, 196
GAS-OIL **Vol. 2** 197, 601, 832
GASMANN, DER **Vol. 4** 235, 373
GASPARDS, LES **Vol. 2** 835
GASPARONE **Vol. 4** 184, 325, 713
GASRORET **Vol. 5** 97
GÄSSCHEN ZUM PARADIES, DAS **Vol. 4** 515 **Vol. 10** 8, 168, 412
GASSE DER LIEBE UND DER SÜNDE, DIE **Vol. 4** 356
GASSENHAUER **Vol. 4** 28, 479, 629
GÄST I EGET HUS **Vol. 5** 1073, 1080
GAST IM EIGENEN HEIM **Vol. 4** 18, 430
GASTHAUS ZUR EHE, DAS **Vol. 4** 149, 321, 660
GASTMAHL DES SATANS, DAS **Vol. 4** 277, 598
GASTONE **Vol. 7** 1136
GASTONE E ROBINET VOGLIONO AMMOGLIARSI **Vol. 7** 1189
GASTSPIEL IM PARADIES **Vol. 4** 344, 614
GATA, LA **Vol. 8** 214, 524
GATAN **Vol. 5** 907, 1052
GATANS BARN **Vol. 5** 996
GATAS TIENEN FRIO, LAS **Vol. 8** 219
GATEGUTTER **Vol. 5** 792, 800

GATES TO PARADISE Vol. 3 349
GATO MONTES, EL Vol. 8 516
GATOS NEGROS, LOS Vol. 8 583, 772
GATTA CI COVA Vol. 7 792, 796
GATTA DA PELARE, LA Vol. 7 257, 834, 882, 1127
GATTA IN CALORE, LA Vol. 7 742
GATTIN, DIE Vol. 4 246, 742
GATTO DAGLI OCCHI DI GIADA, IL Vol. 7 1194
GATTO DI BROOKLYN ASPIRANTE DETECTIVE, IL Vol. 7 952, 1147
GATTO, IL Vol. 7 567
GATTO MAMMONE, IL Vol. 7 347
GATTO NERO, IL Vol. 7 940
GATTOPARDO, IL Vol. 7 498, 988, 1142
GAUDEAMUS IGITUR Vol. 3 218, 256
GAUKLER Vol. 4 85, 208
GAUKLER DER STRASSE Vol. 4 673
GAUKLER VON PARIS, DER Vol. 4 321, 632, 704, 706, 799
GAULOISES BLEUES, LES Vol. 2 513, 520
GAUNER IM FRACK Vol. 4 97, 119, 340, 517, 558
GAUNER ODER GESELLSCHAFT Vol. 4 380
GAUNERLIEBCHEN Vol. 4 73, 431
GAVTYVE Vol. 5 97
GAY CLUB Vol. 8 245, 438
GAY SALOME Vol. 7 1202
GAYARRE Vol. 8 762
GAZ MORTELS, LES Vol. 2 153
GAZDATLAN ASSZONY Vol. 9 196, 236
GAZIJA Vol. 6 *page 313, page 320*
GAZOLAS Vol. 9 117, 237
GDZIE JEST GENERAL ? Vol. 1 198, 343 Vol. 6 *page 218*
GDZIE JEST TRZECI KROL ? Vol. 1 180, 213, 313
GDZIE WODA CZYSTA I TRAWA ZIELONA Vol. 1 271, 321
...GDZIEKOLWIEK JESTES PANIE PREZYDENCIE Vol. 1 219, 303, 360 Vol. 6 *page 251*
GEÄCHTET Vol. 4 183
GEÄCHTETEN, DIE Vol. 4 270, 535
GEBIETERIN VON ST. TROPEZ, DIE Vol. 4 219
GEBIETERISCHE RUF, DER Vol. 4 13
GEBOREN UNTER SCHWARZEN HIMMELN Vol. 1 22, 34, 50
GEBOT DER LIEBE, DAS Vol. 4 113, 780
GEDANKENSÜNDEN Vol. 4 297, 549, 780
GEEL Vol. 2 105
GEFÄHRDETE MÄDCHEN Vol. 4 291, 424
GEFAHREN DER LIEBE Vol. 4 204, 558
GEFÄHRLICHE ALTER, DAS Vol. 4 183, 340, 428
GEFÄHRLICHE FAHNDUNG Vol. 1 5, 29, 100
GEFÄHRLICHE FRACHT Vol. 1 30, 94, 115
GEFÄHRLICHE FREUNDSCHAFT Vol. 4 179, 673
GEFÄHRLICHE REISE Vol. 1 12, 71, 91
GEFÄHRLICHER FRÜHLING Vol. 4 378, 574
GEFÄHRLICHES SPIEL Vol. 4 234, 462, 734
GEFÄHRTIN MEINES SOMMERS Vol. 4 247, 391, 557
GEFANGENE DES KÖNIGS, DER Vol. 4 344, 647
GEFANGENE SEELE Vol. 4 178, 319
GEFANGENE VON SHANGHAI, DIE Vol. 4 212, 400
GEFÄNGNIS AUF DEM MEERESGRUND, DAS Vol. 4 270, 471, 476
GEFESSELTE POLO, DER Vol. 4 211, 407
GEFOLTERTE HERZEN Vol. 4 290, 584
GEFRORENEN BLITZE, DIE Vol. 1 17, 42, 104
GEFUNDENE BRAUT, DIE Vol. 4 318, 537, 567
GEGE BELLAVITA Vol. 7 585
GEHEIMAGENT, DER Vol. 4 119, 340
GEHEIMAKTE WB 1 Vol. 4 135, 269, 344, 416, 422, 437
GEHEIMAKTEN SOLVAY Vol. 1 35, 54, 58
GEHEIMARCHIV AN DER ELBE Vol. 1 60, 104
GEHEIMBUND DER FALKEN Vol. 4 503
GEHEIME AGENT, DER Vol. 4 186, 788
GEHEIME KURIER, DER Vol. 4 149, 660, 743
GEHEIME MACHT, DIE Vol. 4 49, 584
GEHEIME SPUREN Vol. 1 57, 70, 111
GEHEIMKOMMANDO BUMERANG Vol. 1 7, 12, 70
GEHEIMKOMMANDO CIUPAGA Vol. 1 50, 76
GEHEIMKOMMANDO SPREE Vol. 1 44, 46, 70
GEHEIMNIS DER 17 Vol. 1 17, 47
GEHEIMNIS DER ALTEN MAMSELL, DAS Vol. 4 14, 146, 186, 340
GEHEIMNIS DER ALTEN TRUHE, DAS Vol. 4 385
GEHEIMNIS DER AMERIKA-DOCKS, DAS Vol. 4 124, 514
GEHEIMNIS DER ANDEN, DAS Vol. 1 38, 53, 59
GEHEIMNIS DER CHRYSANTHEMEN, DAS Vol. 4 322, 522
GEHEIMNIS DER GRÜNEN VILLA, DAS Vol. 4 320, 454

GEHEIMNIS DER MARTHA LÜDERS, DAS **Vol. 6** *page 193*
GEHEIMNIS DER MUMIE, DAS **Vol. 4** 132, 560
GEHEIMNIS DER ROTEN KATZE, DAS **Vol. 4** 270, 409, 462
GEHEIMNIS DER SANTA MARGHERITA, DAS **Vol. 4** 340, 366
GEHEIMNIS DER SCHWARZEN WITWE, DAS **Vol. 8** 438, 600
GEHEIMNIS DER SECHS SPIELKARTEN, DAS **Vol. 4** 380, 529, 565
GEHEIMNIS DER SPIELHÖLLE VON SEBASTOPOL, DAS **Vol. 4** 270, 634
GEHEIMNIS DER TOTEN, DAS **Vol. 4** 634
GEHEIMNIS DES ABBE X, DAS **Vol. 4** 215, 432, 486, 681
GEHEIMNIS DES BLAUEN ZIMMERS **Vol. 4** 323, 340, 409
GEHEIMNIS DES ZIRKUS BARRE, DAS **Vol. 4** 535
GEHEIMNIS EINES ALTEN HAUSES **Vol. 4** 323, 340, 466
GEHEIMNIS UM BETTY BONN, DAS **Vol. 4** 30, 123, 544
GEHEIMNIS UM JOHANN ORTH, DAS **Vol. 4** 149, 634, 660
GEHEIMNIS VON BOMBAY, DAS **Vol. 4** 266, 574, 747
GEHEIMNIS VON BRINKENHOF, DAS **Vol. 4** 36, 51, 190, 554
GEHEIMNIS VON CHATEAU RICHMOND, DAS **Vol. 4** 135, 503
GEHEIMNIS VON D 14, DAS **Vol. 4** 779
GEHEIMNIS VON GENF, DAS **Vol. 4** 544, 774
GEHEIMNIS VON LONDON, DAS **Vol. 4** 321
GEHEIMNIS VON MONTE CARLO, DAS **Vol. 4** 247, 646
GEHEIMNIS VON ST. PAULI, DAS **Vol. 4** 124, 239
GEHEIMNISSE DES ORIENTS **Vol. 2** 100, 833 **Vol. 4** 66, 104, 113, 413, 442, 707
GEHEIMNISSE EINER SEELE **Vol. 4** 379, 452, 492, 646
GEHEIMNISSE VON BERLIN, DIE **Vol. 4** 223
GEHEIMNISVOLLE GEWALTEN **Vol. 4** 467
GEHEIMNISVOLLE MISTER X, DER **Vol. 4** 340, 466, 471, 566
GEHEIMNISVOLLE PERSER, DER **Vol. 4** 97, 230, 239
GEHEIMNISVOLLE SPIEGEL, DER **Vol. 4** 79, 288
GEHEIMNISVOLLE UHR, DIE **Vol. 4** 684
GEHEIMNISVOLLE VILLA, DIE **Vol. 4** 159
GEHEIMNISVOLLE WANDERER, DER **Vol. 4** 213
GEHEIMNISVOLLE WRACK, DAS **Vol. 1** 35, 81
GEHEIMNISVOLLE ZEICHEN **Vol. 4** 779
GEHEIMNISVOLLEN PIRATEN, DIE **Vol. 4** 747
GEHEIMPOLIZISTEN **Vol. 4** 340, 608, 691
GEHEIMTRESOR, DER **Vol. 4** 221, 495
GEHEIMZEICHEN LB 17 **Vol. 4** 79, 115, 361
GEHENNA **Vol. 1** 78, 308, 404
GEHETZTE FRAUEN **Vol. 4** 340, 796
GEHETZTE MÄDCHEN **Vol. 4** 340, 608 **Vol. 6** *page 199*
GEHETZTE MENSCHEN **Vol. 4** 70, 119, 329, 435, 527, 600, 734 **Vol. 10** 91, 479, 524, 586
GEIER-WALLY, DIE **Vol. 4** 247, 396, 724
GEIERWALLY, DIE **Vol. 4** 15
GEIGE LOCKT, DIE **Vol. 4** 329, 713
GEIGER VON FLORENZ, DER **Vol. 4** 115, 609, 724, 752
GEISHA PARISIENNE, LA **Vol. 2** 215
GEISHA UND DER SAMURAI, DIE **Vol. 4** 217, 594
GEISTERJAGD, DIE **Vol. 4** 71, 488
GEISTERSEHER, DER **Vol. 4** 124, 747
GEISTERZUG, DER **Vol. 4** 321, 752
GEJAGT BIS ZUM MORGEN **Vol. 1** 46, 47, 78
GELBE FLAGGE, DIE **Vol. 4** 344, 544
GELBE FRATZE, DIE **Vol. 4** 232
GELBE HAUS DES KING-FU, DAS **Vol. 4** 89
GELBE RASSE, DIE **Vol. 4** 239
GELBE SCHATTEN, DER **Vol. 4** 218
GELBE SCHEIN, DER **Vol. 4** 300, 560
GELBE TOD, DER **Vol. 4** 213
GELBE WÜRGER, DER **Vol. 4** 135, 652
GELBSTERN **Vol. 4** 239
GELD AUF DER STRASSE, DAS **Vol. 6** *page 189*
GELD FÄLLT VOM HIMMEL **Vol. 4** 248, 517, 558
GELD INS HAUS **Vol. 4** 678
GELDTEUFEL, DER **Vol. 4** 388, 423
GELIEBTE DES CHINESEN, DIE **Vol. 4** 247
GELIEBTE DES GOUVERNEURS, DIE **Vol. 4** 247, 339
GELIEBTE DES KÖNIGS, DIE **Vol. 6** *page 190, page 193*
GELIEBTE, DIE **Vol. 4** 53, 208, 211, 373, 390, 445, 724
GELIEBTE ROSWOLSKYS, DIE **Vol. 4** 132, 479, 773
GELIEBTE RUF, DER **Vol. 4** 614

GELIEBTE SEINER HOHEIT, DIE **Vol. 4** 270, 634
GELIEBTE STIMME, DIE **Vol. 4** 167, 230
GELIEBTE TOTE, DIE **Vol. 4** 213
GELIEBTE WEISSE MAUS **Vol. 1** 42, 62, 87
GELIEBTE WELT **Vol. 4** 135, 347, 370, 742
GELIEBTER SCHATZ **Vol. 4** 65, 369, 422, 582
GELOSA, LA **Vol. 7** 1189
GELOSIA, LA **Vol. 7** 86, 401, 493, 730, 1060, 1080
GELÖSTE KETTEN **Vol. 4** 178
GELOZIA BAT-O VINA **Vol. 3** 244
GELÜBDE, DAS **Vol. 4** 660, 704
GELÜBDE DER KEUSCHHEIT, DAS **Vol. 4** 247, 340
GEMELAS, LAS **Vol. 8** 207, 298
GEMELLA EROTICA, LA **Vol. 7** 289
GEMELLI DE TEXAS, I **Vol. 8** 461
GEMELOS DE TEXAS, LOS **Vol. 8** 461
GEMINIS **Vol. 8** 455, 536
GEMISCHTE FRAUENCHOR, DER **Vol. 4** 560
GENBOERNE **Vol. 5** 7, 290
GENDARME A NEW YORK, LE **Vol. 2** 97, 781
GENDARME DE CHAMPIGNOL, LE **Vol. 2** 309, 548
GENDARME DE ST. TROPEZ, LE **Vol. 2** 97, 309
GENDARME EN BALADE, LE **Vol. 2** 97, 601
GENDARME ET LES EXTRA-TERRESTRES, LE **Vol. 2** 97, 368, 815
GENDARME ET LES GENDARMETTES, LE **Vol. 6** *page 73*
GENDARME SE MARIE, LE **Vol. 2** 97, 368
GENDRE DE MONSIEUR PORIER, LE **Vol. 2** 99, 395, 485, 871, 872
GENERAL CAMILLO, IL **Vol. 7** 26, 701
GENERAL IDI AMIN DADA: AUTOPORTRAIT **Vol. 2** 17
GENERAL VON BERNING **Vol. 4** 603
GENERAL VON DÖBELN **Vol. 5** 871, 959, 1156
GENERALDIREKTOREN **Vol. 5** 97
GENERALE DELLA ROVERE, IL **Vol. 7** 257, 1214
GENERALE DORME IN PIEDI, IL **Vol. 7** 854
GENERALEN **Vol. 5** 1180
GENERAL...NOUS VOILA ! **Vol. 2** 368
GENERALPROBE **Vol. 6** *page 163, page 168, page 174*
GENESUNG **Vol. 1** 9, 72, 101
GENEVIEVE **Vol. 2** 79, 595
GENGAELD **Vol. 5** 2, 23
GENGAELDELSENS RET **Vol. 5** 9
GENGHIS KAAN **Vol. 6** *page 312*
GENIA EROON **Vol. 3** 198
GENIO ALEGRE, EL **Vol. 8** 449, 602, 606, 772
GENIO, DUE COMPARI, UN POLLO, UN **Vol. 7** 1007
GENIO EN APUROS, UN **Vol. 8** 310, 426
GENITORI IN BLUE-JEANS **Vol. 7** 693
GENIUS **Vol. 10** 107, 263, 548
GENIUSZ SCENY **Vol. 1** 278, 404
GENNARIELLO, IL FIGLIO DEL GALEOTTO **Vol. 7** 829
GENNARIELLO POLIZZIOTTO **Vol. 7** 829
GENNEM DE MANGE TIL EN **Vol. 5** 97
GENNEM KAMP TIL SEJR **Vol. 5** 97
GENNEM LIVETS SKOLE **Vol. 5** 97
GENNEM MORKE TIL LYS **Vol. 5** 9
GENOU DE CLAIRE, LE **Vol. 2** 17
GENOVA A MANO ARMATA **Vol. 7** 1202
GENOVEFA DI BRABANTE **Vol. 7** 360, 743, 796, 839
GENOVEVA DE BRABANTE **Vol. 8** 598, 772
GENS DU VOYAGE, LES **Vol. 2** 81, 279
GENS NOVA **Vol. 7** 668
GENSDARM MÖBIUS **Vol. 4** 124, 247
GENTE BRAVA **Vol. 8** 281, 657
GENTE DELL'ARIA **Vol. 7** 792, 796
GENTE DI RISPETTO **Vol. 7** 567, 1035
GENTLEMAN ATT HYRA **Vol. 5** 840, 878
GENTLEMAN AUF ZEIT **Vol. 4** 132, 270
GENTLEMAN DE COCODY, LE **Vol. 2** 661
GENTLEMAN D'EPSOM, LE **Vol. 2** 197, 644
GENTLEMAN FOR EN TIME **Vol. 5** 83
GENTLEMAN-GAUNER **Vol. 4** 704, 773
GENTLEMAN JO **Vol. 8** 523, 732
GENTLEMAN JO ... UCCIDI **Vol. 8** 523, 732
GENTLEMANNAGANGSTERN **Vol. 5** 840, 909
GENTLEMANSEKRETAEREN **Vol. 5** 9
GENTRYFESZEK **Vol. 9** 33, 117
GENUINE **Vol. 4** 239, 333
GEOMETRIE, LA **Vol. 2** 703
GEORGE BULLY **Vol. 4** 12, 159, 759
GEORGE QUI ? **Vol. 2** 551
GEORGES BRAQUE OU LE TEMPS DIFFERENT **Vol. 2** 835
GEORGIA, GEORGIA **Vol. 5** 860
GEPÄCKSCHEIN 712 **Vol. 4** 430, 558, 796
GEPEITSCHT **Vol. 4** 212, 594
GEPPO IL FOLLE **Vol. 7** 347, 1135
GERACITE **Vol. 3** 31, 93, 126 **Vol. 6** *page 26*
GERECHTEN VON KUMMEROW, DIE **Vol. 6** *page 170, page 171, page 172*

GERICHTET BEI NACHT Vol. 1 37, 86, 120
GERLA DI PAPA MARTIN, LA Vol. 7 493, 756, 1190
GERLOVSKA ISTORIJA Vol. 3 31, 96, 109
GERMANIA, ANNO ZERO Vol. 2 423 Vol. 7 443, 444
GERMANIN Vol. 4 666, 714, 736
GERMINAL Vol. 2 9, 124, 650 Vol. 9 43, 117, 183
GERMOGLIO Vol. 7 82
GERN HAB' ICH DIE FRAUEN GEKÜSST Vol. 4 329, 366
GERN HAB' ICH DIE FRAU'N GEKÜSST Vol. 4 119, 232, 270
GEROITE NA SIPKA Vol. 3 95 Vol. 6 *page 26, page 45, page 57*
GEROLSTEINI KALAND Vol. 9 36, 122, 199
GERTRUD Vol. 5 23, 225, 248, 249
GERUSALEMME LIBERATA, LA Vol. 7 621, 785 Vol. 7 638
GERVAISE Vol. 2 96, 423
GESANGSTUNDE, DIE Vol. 4 774
GESANGVEREIN SORGENFREI Vol. 4 292, 339
GESCHÄFT BLÜHT, DAS Vol. 4 361, 558
GESCHEITERT Vol. 4 247
GESCHENK, DAS Vol. 4 115, 354
GESCHENK DES INDERS, DAS Vol. 4 135
GESCHENKTE LOGE, DIE Vol. 4 164, 354
GESCHICHTE DES GRAUEN HAUSES, DIE Vol. 4 35, 359, 362
GESCHICHTE VOM ARMEN HASSAN, DIE Vol. 1 81, 92, 103
GESCHICHTE VOM KLEINEN MUCK, DIE Vol. 1 3, 107, 129
GESCHICHTEN JENER NACHT - DER GROSSE UND DER KLEINE WILLI Vol. 1 21, 60
GESCHICHTEN JENER NACHT - DIE PRÜFUNG Vol. 1 50, 100, 112
GESCHICHTEN JENER NACHT - MATERNA Vol. 1 48, 80, 102
GESCHICHTEN JENER NACHT - PHÖNIX Vol. 1 21, 98
GESCHIEDENE FRAU, DIE Vol. 4 132, 584
GESCHLECHT DERER VON RINGWALL, DAS Vol. 4 178, 773
GESCHLECHT IN FESSELN Vol. 4 339, 379, 437
GESCHLOSSENE KETTE, DIE Vol. 4 729, 773
GESCHMINKTE JUGEND Vol. 4 247, 424
GESCHÖPF, DAS Vol. 4 219, 560
GESCHWADER FLEDERMAUS Vol. 1 2, 90, 105, 107
GESCHWISTER Vol. 1 11, 62
GESETZ DER LIEBE, DAS Vol. 4 660, 678
GESETZ DER SCHWARZEN BERGE, DAS Vol. 4 643
GESPAY, FANTINO E GENTILUOMO Vol. 7 264
GESPENSTER VON GARDEN HALL, DIE Vol. 4 277, 747
GESPRENGTE FESSELN Vol. 6 *page 187, page 198*
GESTÄNDNIS DER DREI, DAS Vol. 4 124, 725
GESTÄNDNIS DER GRÜNEN MASKE, DAS Vol. 4 219, 558
GESTES INTERDITS Vol. 6 *page 130*
GESTOHLENE GESICHT, DAS Vol. 4 82, 614, 629
GESTOHLENE HOTEL, DAS Vol. 4 218
GESTOHLENE MILLIONENREZEPT, DAS Vol. 4 329
GESTOHLENE SCHLACHT, DIE Vol. 1 37, 55, 102
GESTÖRTE FLITTERWOCHEN Vol. 4 558, 747
GESTÖRTE HOCHZEITSNACHT, DIE Vol. 4 342
GESTOS E FRAGMENTOS Vol. 8 7, 34, 198
GESTRANDETE MENSCHEN Vol. 4 221, 340
GESTREIFTE DOMINO, DER Vol. 4 159
GESU DI NAZARETH Vol. 7 817
GESUCHT WIRD DIE FREUNDLICHE WELT Vol. 1 12
GESUNKENEN, DIE Vol. 4 584, 663
GETEILTE HIMMEL, DER Vol. 1 9, 34, 48
GEVATTER TOD Vol. 6 *page 162, page 170, page 172*
GEWALT GEGEN RECHT Vol. 4 366, 518
GEWEHR ÜBER, DAS Vol. 4 422, 588 Vol. 6 *page 181*
GEWISSEN IN AUFRUHR Vol. 1 13, 21, 37, 85, 102, 112
GEWISSER HERR GRAN, EIN Vol. 4 149, 292, 660
GEWITTER IM MAI Vol. 4 25, 135, 370
GEWITTERFLUG ZU CLAUDIA Vol. 4 28, 311, 329
GEZEICHNETEN, DIE Vol. 4 403, 743
GEZHOJA E VJETER Vol. 3 21, 25, 27 Vol. 6 *page 12, page 18*
GHETTO EXPERIMENTAL, LE Vol. 2 157
GIA POION HTYPA E KOUDOUNA Vol. 3 143
GIA TEN AGAPE MAS Vol. 3 178
GIA TEN KARDIA TES ORAIAS ELENES Vol. 3 152
GIA TEN TIME KAI TON ERÖTA Vol. 3 161
GIACCA VERDE, LA Vol. 7 593, 841, 854

GIACOMO L'IDEALISTA **Vol. 7** 328, 589, 827, 862, 1060
GIAKOUMES MIA ELLINIKI KARDIA **Vol. 3** 161, 207
GIALLO **Vol. 7** 756, 1116
GIALLO NAPOLETANO **Vol. 7** 599
GIAN BURRASCA **Vol. 7** 332, 902, 978, 1126, 1202
GIARABUB **Vol. 7** 170, 250
GIARDINI D'ARMIDA, I **Vol. 7** 399, 574
GIARDINO DEL MISTERO, IL **Vol. 7** 1196
GIARDINO INCANTALO, IL **Vol. 7** 369, 377, 395
GIB MICH FREI **Vol. 4** 179, 673
GIBELTA NA ALEKSAND"R VELIKI **Vol. 3** 73, 101 **Vol. 6** *page 27*
GIBIER DE POTENCE **Vol. 2** 439
GIBIGIANNA, LA **Vol. 7** 711, 1189
GIBRALTAR **Vol. 2** 581, 646, 863 **Vol. 8** 311, 611
GIDINI ZA LJUBOV **Vol. 6** *page 45*
GIFLE, LA **Vol. 2** 196, 377
GIFLEE **Vol. 7** 1005
GIFT **Vol. 5** 195
GIFT DER MEDICI, DAS **Vol. 4** 212
GIFT UND LIEBE **Vol. 4** 297, 549, 780
GIFTAS **Vol. 5** 840, 843, 891, 966, 1161
GIFTASVUXNA DÖTTRAR **Vol. 5** 840, 997
GIFTES – NEJ TAK ! **Vol. 5** 24
GIFTGAS **Vol. 4** 157, 565
GIFTPILEN **Vol. 5** 183
GIGANTE DELLE DOLOMITI, IL **Vol. 7** 491, 640, 1116
GIGANTE, I SERPENTI E LA FORMICA, IL **Vol. 7** 1067
GIGANTES Y CABEZUDOS **Vol. 8** 246
GIGANTI DELLA TESSAGLIA, I **Vol. 7** 533, 634, 739, 1177
GIGANTI DI ROMA, I **Vol. 7** 1213
GIGETTA AL REGGIMENTO **Vol. 7** 1189
GIGETTA SI VENDICA DI ROBINET **Vol. 7** 1189
GIGGI IL BULLO **Vol. 7** 1202
GIGI **Vol. 2** 254, 516, 658
GIGOLETTE **Vol. 2** 249, 395
GIGOLETTE, LA **Vol. 7** 968
GIGOLO **Vol. 4** 738, 788
GIGOLO, LE **Vol. 2** 118, 247, 503
GILIAP **Vol. 5** 850, 890, 1061, 1086
GILLEKOP **Vol. 5** 9
GILLY POPRVE V PRAZE **Vol. 10** 45, 329
GINERE FARA VOIE **Vol. 3** 304
GINEVRA DEGLI ALMIERI **Vol. 7** 48, 756, 1000
GINGER E FRED **Vol. 7** 333, 388, 434, 567
GIOCARE D'AZZARDO **Vol. 7** 1173
GIOCATTOLO, IL **Vol. 7** 567, 1035
GIOCHI CARNALI **Vol. 7** 152
GIOCHI D'ESTATE **Vol. 7** 102
GIOCHI EROTICI NELLA 3 A GALASSIA **Vol. 7** 692
GIOCO D'AZZARDO **Vol. 7** 16, 443
GIOCO PERICOLOSO **Vol. 7** 493, 792
GIOCONDA FARA SURIS **Vol. 3** 239, 256, 274
GIOCONDA, LA **Vol. 7** 1189
GIOGO, IL **Vol. 7** 1057, 1116
GIOIE DEL FOCOLARE, LE **Vol. 7** 1103
GIOIELLO RECUPERATO, IL **Vol. 7** 837
GIORGINA **Vol. 7** 1156
GIORNALE, IL **Vol. 7** 1190
GIORNATA BALORDA, UNA **Vol. 7** 401, 1045
GIORNATA DI „BUSCABOTTE", LA **Vol. 7** 1189
GIORNATA DI FRETTA, UNA **Vol. 7** 1189
GIORNATA PARTICOLARE, UNA **Vol. 7** 964, 1009, 1025
GIORNI CONTATI, I **Vol. 7** 284, 567, 854, 908
GIORNI D'AMORE **Vol. 7** 942
GIORNI DEL SOLE, I **Vol. 7** 524
GIORNI DELLA CHIMERA, IL **Vol. 7** 514
GIORNI DI GLORIA **Vol. 7** 938, 1116
GIORNI FELICI **Vol. 7** 443, 1027
GIORNI PIU BELLI, I **Vol. 7** 798
GIORNO A MADERA, UN **Vol. 7** 940
GIORNO A PALERMO, UN **Vol. 7** 1211
GIORNO ALLA FINE D'OTTOBRE, UN **Vol. 7** 703
GIORNO DA LEONI, UN **Vol. 7** 401, 502
GIORNO D'AMORE **Vol. 7** 725
GIORNO DEL COBRA, IL **Vol. 7** 124, 840
GIORNO DELL'ASSUNTA, UN **Vol. 7** 741
GIORNO DI NOZZE **Vol. 7** 16, 493, 756
GIORNO IN PRETURA, UN **Vol. 7** 1044
GIORNO NELLA VITA, UN **Vol. 7** 40, 359
GIORNO PER GIORNO DISPERATAMENTE **Vol. 7** 401, 869
GIOVANE CANAGLIA **Vol. 7** 257 **Vol. 8** 732
GIOVANE NORMALE, IL **Vol. 7** 408
GIOVANEZZA, GIOVANEZZA **Vol. 7** 1101
GIOVANEZZA, GIOVANEZZA, PRIMAVERA BELEZZA **Vol. 7** 341
GIOVANI, BELLE ... PROBABILMENTE RICCHE **Vol. 7** 32, 901
GIOVANI MARITI **Vol. 7** 817
GIOVANNA I D'ANGIO, REGINA DI NAPOLI **Vol. 7** 493, 1104
GIOVANNA LA PALLIDA **Vol. 7** 654
GIOVANNI DA UNA MADRE ALL'ALTRA **Vol. 7** 12, 116, 173, 967
GIOVANNI DALLE BANDE NERE **Vol. 7** 693
GIOVANNI EPISCOPO **Vol. 7** 621

GIOVANNI SENZAPENSIERI **Vol. 7** 111, 334
GIOVANNINO **Vol. 7** 1205
GIOVANNONA LOSCIALUNGA CON ONORE **Vol. 7** 823
GIOVE IN DOPPIO PETTO **Vol. 7** 1062
GIOVENTU ALLA SBARRA **Vol. 7** 333
GIOVENTU PERDUTA **Vol. 7** 754, 796
GIOVINEZZA **Vol. 7** 1027
GIOVINEZZA DEL DIAVOLO, LA **Vol. 7** 264, 701, 725
GIPFELSTÜRMER **Vol. 4** 25, 149, 660
GIPSY **Vol. 7** 806
GIRANDOLA DI FUOCO, LA **Vol. 7** 268
GIRASOLI, I **Vol. 7** 988
GIRL VON DER REVUE, DAS **Vol. 4** 192, 422, 462, 648
GIRLS **Vol. 2** 7, 862
GIRO GIROTONDO ... CON IL SESSO E BELLO IL MONDO **Vol. 7** 509
GIROTONDO DI UNDICI LANCIERI, IL **Vol. 7** 45, 1162
GIROVAGHI, I **Vol. 7** 693
GISELA MAY SINGT BRECHT **Vol. 1** 14
GISSEMANDS KAERLIGHEDSHISTORIE **Vol. 5** 256
GITA IN CAMPAGNA **Vol. 7** 1189
GITA SCOLASTICA, UNA **Vol. 7** 96, 286, 947, 1043, 1134
GITAN, LE **Vol. 2** 400, 814
GITANA **Vol. 8** 495, 541
GITANA Y EL REY, LA **Vol. 8** 272
GITANA Y EL CHARRO, LA **Vol. 8** 532, 546, 722
GITANES **Vol. 2** 167
GITANILLA, LA **Vol. 2** 346, 731 **Vol. 8** 298, 336, 449
GITTA ENTDECKT IHR HERZ **Vol. 4** 113, 626
GIU IL SIPARIO **Vol. 7** 750, 1099
GIUDA **Vol. 7** 209
GIUDICATEMI ! **Vol. 7** 372
GIUDITTA **Vol. 7** 782
GIUDITTA E OLOFERNE **Vol. 7** 48, 640, 1116
GIUDITTA E OLOFERNE **Vol. 7** 531, 877, 1037
GIUDIZIO UNIVERSALE, IL **Vol. 7** 910, 990
GIULIA DI TRECOEUR **Vol. 7** 26, 701
GIULIA FIGLIA DI CLAUDIO **Vol. 7** 1032, 1071
GIULIANO DE'MEDICI **Vol. 7** 487, 1060, 1107, 1172
GIULIANO L'APOSTATA **Vol. 7** 230, 301, 584
GIULIETTA DEGLI SPIRITI **Vol. 7** 208, 522, 964, 1025, 1106, 1167
GIULIETTA E ROMEO **Vol. 7** 443, 445, 587, 597, 910, 1080, 1180 **Vol. 8** 772
GIULIO CESARE CONTRO I PIRATI **Vol. 7** 1063
GIUNGLA **Vol. 7** 453, 827
GIURAMENTO **Vol. 7** 222, 474
GIURAMENTO D'AMORE **Vol. 7** 827
GIUSEPPE FAVA : SICILIANO COME ME **Vol. 7** 700
GIUSEPPE VERDI **Vol. 7** 89, 161, 453, 870, 1000, 1027, 1116
GIUSEPPE W WARSZAWIE **Vol. 1** 179, 283 **Vol. 6** *page 216*
GIUSTA VENDETTA **Vol. 7** 1189
GIUSTIZIERE DELLA STRADA, IL **Vol. 7** 599
GIUSTIZIERE SFIDA LA CITTA, IL **Vol. 7** 1202
GIV GUD EN CHANCE OM SONDAGEN **Vol. 5** 52
GIVRES, LES **Vol. 2** 72
GJENERAL GRAMAFONI **Vol. 3** 6, 9, 23 **Vol. 6** *page 13*
GJEST BAARDSEN **Vol. 5** 688, 714, 726, 738
GJURMA **Vol. 3** 10, 14 **Vol. 6** *page 18*
GJURME NE DBORE **Vol. 6** *page 12, page 17*
GJURME NE KALTERSI **Vol. 6** *page 11, page 12*
GJURME TE BARDHA **Vol. 6** *page 309*
GKOLFO **Vol. 3** 177
GLACE A TROIS FACES, LA **Vol. 2** 286, 426
GLACE AVEC DEUX BOULES, UNE **Vol. 6** *page 105*
GLAD DIG I DIN UNGDOM **Vol. 5** 840, 909
GLAD GUTT, EN **Vol. 5** 692, 770
GLADA KALASET, DET **Vol. 5** 634, 863, 1032
GLADE ENKE, DEN **Vol. 5** 97
GLADE ENKE I TRANGVIK, DEN **Vol. 5** 685, 760
GLADE SKOMAKAREN, DEN **Vol. 5** 862, 1042
GLADE SKRÄDDAREN, DEN **Vol. 5** 1032, 1181
GLADE VRINSK **Vol. 5** 762
GLADIADOR INVENCIBLE, EL **Vol. 8** 541, 595
GLAEDENS DAG **Vol. 5** 256
GLAIVE ET LA BALANCE, LE **Vol. 2** 294, 598
GLAMADOR **Vol. 2** 545
GLANZ GEGEN GLÜCK **Vol. 4** 291, 423, 479

GLANZ UND ELEND DER KURTISANEN Vol. 4 149, 527, 660
GLAS WASSER, EIN Vol. 4 38, 115, 359, 734, 735
GLAS WASSSER, DAS Vol. 1 46, 47, 72
GLASAM ZA LJUBAV Vol. 3 343, 351
GLASBERGET Vol. 5 909, 1052
GLÄSERNE KUGEL, DIE Vol. 4 167, 230, 458
GLATZKOPFBANDE, DIE Vol. 1 49, 50, 72
GLAUBE UND HEIMAT Vol. 4 60, 292
GLEICHE STRECKE, DIE Vol. 1 8, 44, 86
GLEISDREIECK Vol. 4 79, 115, 536
GLEMSOMME PROFESSOR, DEN Vol. 5 83
GLI ULTIMI GIORNI DI POMPEI Vol. 2 32
GLINENI GOLUB Vol. 3 338, 351
GLISSANDO Vol. 6 *page 267, page 276*
GLISSEMENTS PROGRESSIFS DU PLAISIR Vol. 2 450
GLOBUL DE CRISTAL Vol. 3 313
GLÖCKCHEN DES GLÜCKS, DAS Vol. 4 190
GLOCKE, DIE Vol. 4 219, 366
GLOCKENKÄUFER, DER Vol. 6 *page 279*
GLÖCKERL VON BIRKENSTEIN, DAS Vol. 4 25
GLOMDALSBRUDEN Vol. 5 775, 818
GLORIA Vol. 2 251 Vol. 4 186, 600 Vol. 8 357
GLORIA AI CADUTI Vol. 7 829
GLORIA DEL MONCAYO Vol. 8 422
GLORIA DI SANGUE Vol. 7 621
GLORIA MAIRENA Vol. 8 259, 611, 616
GLORIA NU CINTA Vol. 3 247, 276, 280, 300
GLOS MA PROKURATOR Vol. 1 143, 355, 389 Vol. 6 *page 220*
GLOS PUSTYNI Vol. 1 278, 346
GLOS Z TAMTEGO SWIATA Vol. 1 171, 403 Vol. 6 *page 242*
GLOSY Vol. 6 *page 205, page 213, page 233, page 243*
GLOWY PELNE GWIAZD Vol. 6 *page 212, page 249*
GLU, LA Vol. 2 48, 191, 369, 371, 452, 595, 697
GLÜCK BEI FRAUEN Vol. 4 85, 666
GLÜCK IM HINTERHAUS Vol. 6 *page 161, page 165, page 176*
GLÜCK IM SCHLOSS Vol. 4 132, 230
GLÜCK IM WINKEL Vol. 4 18, 430, 540, 797
GLÜCK MUSS MAN HABEN Vol. 4 18
GLÜCK ÜBER NACHT Vol. 4 192, 255
GLÜCK UND GLAS Vol. 4 110, 517
GLÜCK UNTERWEGS Vol. 4 121, 269, 740
GLÜCK WOHNT NEBENAN, DAS Vol. 4 18, 292, 517, 558
GLÜCKLICHE FINDER, DER Vol. 4 124, 232, 471
GLÜCKLICHE REISE Vol. 4 626, 747
GLÜCKLICHER MENSCH, EIN Vol. 4 298, 486, 727, 729
GLÜCKSFALLE, DIE Vol. 4 36, 660
GLÜCKSKINDER Vol. 4 325, 713
GLÜCKSPERLEN Vol. 1 22, 75, 91
GLÜCKSPILZE Vol. 4 28, 424, 544
GLÜCKSSCHMIED, DER Vol. 4 118, 367
GLÜHENDE BERGE - FLAMMENDES HERZ Vol. 4 25 Vol. 6 *page 182*
GLÜHENDE GASSE, DIE Vol. 4 599, 648
GLÜHENDE KAMMER, DIE Vol. 4 522, 558
GLUSTIZIERE DEI MARI, IL Vol. 7 1131
GLUTNICATA Vol. 3 51, 76 Vol. 6 *page 38*
G.M.B.H. TENOR, DER Vol. 4 560, 663
GNESELLA Vol. 7 829
GNEVNO P"TUVANE Vol. 3 60, 63
GNIAZDO Vol. 1 202, 227, 282
GOBBO, IL Vol. 7 46, 86, 299, 988, 1136, 1189
GOBSECK Vol. 4 169, 377
GOCCE D'AMORE Vol. 7 311
GODA VÄNNER OCH TROGNA GRANNAR Vol. 5 864, 1130, 1217
GODA VÄNNER TROGNA GRANNAR Vol. 5 1042
GODE OG DET OMDE, DET Vol. 5 52
GODELUREAUX, LES Vol. 2 353, 385, 700
GODINA OT PONEDELNICI Vol. 3 38, 96 Vol. 6 *page 30, page 44*
GODINI ZA LJUBOV Vol. 3 55, 78, 97 Vol. 6 *page 45*
GODISNJA DOBA Vol. 3 430
GODNOSC Vol. 6 *page 219, page 234, page 247*
GODSEJEREN Vol. 5 83
GODSFORVALTEREN Vol. 5 183
GODSJERENS DATTER Vol. 5 97
GODURIA Vol. 7 311
GODVAKKER - MAREN Vol. 5 692, 760, 816
GODZINA PASOWEJ ROZY Vol. 1 211, 323, 403 Vol. 6 *page 220*
GODZINA SZCZYTU Vol. 1 156, 201, 289, 353
GODZINA ZA GODZINA Vol. 1 148, 202, 289
GODZINY NADZIEI Vol. 1 171, 179, 323
GOETHE-FILME DER UFA Vol. 4 89, 667
GOGLERBANDENS ADOPTIVDATTER Vol. 5 295
GOGLERENS ELSKOV Vol. 5 203
GOHA Vol. 2 124, 436, 873
GÖINGEHÖVDINGEN Vol. 5 840, 884

GOLA, LA **Vol. 3** 395 **Vol. 7** 442, 701
GOLA S"VEST **Vol. 3** 73 **Vol. 6** *page 23, page 54*
GOLD **Vol. 4** 30, 82, 298, 567
GOLD IN NEW FRISCO **Vol. 4** 30, 123, 288, 544
GOLDBLONDES MÄDCHEN, ICH SCHENK' DIR MEIN HERZ - ICH BIN JA SO VERLIEBT... **Vol. 4** 32, 316, 774
GOLDENE ABGRUND, DER **Vol. 4** 14, 26, 166, 219, 634
GOLDENE FESSEL, DIE **Vol. 4** 374, 620
GOLDENE FLIEGE, DIE **Vol. 4** 99, 247, 646
GOLDENE FRIEDELCHEN, DAS **Vol. 4** 33, 124
GOLDENE GANS, DIE **Vol. 1** 75, 90, 125
GOLDENE GLETSCHER, DER **Vol. 4** 432, 745
GOLDENE HAAR, DAS **Vol. 4** 385, 584
GOLDENE JURTE, DIE **Vol. 1** 36, 115
GOLDENE KALB, DAS **Vol. 4** 149, 219, 535, 660
GOLDENE KRONE, DIE **Vol. 4** 190, 319
GOLDENE KUGEL, DIE **Vol. 4** 89, 400
GOLDENE LÜGE, DIE **Vol. 4** 113, 780
GOLDENE MASKE, DIE **Vol. 4** 49, 424, 788
GOLDENE NETZ, DAS **Vol. 4** 479, 535
GOLDENE PEST, DIE **Vol. 4** 208, 287
GOLDENE POL, DER **Vol. 4** 747
GOLDENE SCHMETTERLING, DER **Vol. 4** 84, 396, 715
GOLDENE SEE, DER **Vol. 4** 298, 329, 634, 734
GOLDENE SKORPION, DER **Vol. 4** 283
GOLDENE SPINNE, DIE **Vol. 4** 65, 167, 422
GOLDENE STADT, DIE **Vol. 4** 424, 462, 788
GOLDENE ZEITEN - FEINE LEUTE **Vol. 1** 53, 71, 98
GOLDFELDER VON JACKSONVILLE, DIE **Vol. 4** 366
GOLDGRUBE, DIE **Vol. 4** 361, 558
GOLDSNAKE ANONIMA KILLERS **Vol. 8** 396, 772
GOLEM **Vol. 10** 8, 256, 524, 586
GOLEM, DER **Vol. 4** 202, 646 **Vol. 6** *page 187, page 206, page 232, page 251, page 256*
GOLEM, LE **Vol. 2** 23, 799
GOLEM UND DIE TÄNZERIN, DER **Vol. 4** 202, 425
GOLEM, WIE ER IN DIE WELT KAM, DER **Vol. 4** 178, 533 **Vol. 6** *page 187*
GOLEMANOV **Vol. 3** 31, 44, 126 **Vol. 6** *page 41, page 59*
GOLFO DE VIZCAYA **Vol. 8** 210, 489
GOLFO, EL **Vol. 8** 258, 308
GOLFOS, LOS **Vol. 8** 214, 258
GOLGOTA **Vol. 3** 230, 297, 317
GOLGOTHA **Vol. 2** 309, 423, 440, 657, 726
GOLI COVIK **Vol. 3** 350, 374
GOLIA ALLA CONQUISTA DI BAGDAD **Vol. 7** 1121
GOLIA CONTRO I GIGANTI **Vol. 8** 762
GOLIA E IL CAVALIERE MASCHERATO **Vol. 7** 1121
GOLIAT CONTRA LOS GIGANTES **Vol. 8** 762
GOLJAMATA POBEDA **Vol. 3** 77 **Vol. 6** *page 30, page 31*
GOLJAMATA SKUKA **Vol. 3** 69 **Vol. 6** *page 30*
GOLJAMOTO NOSTNO K"PANE **Vol. 3** 122 **Vol. 6** *page 35, page 54*
GOLOWIN GEHT DURCH DIE STADT **Vol. 4** 85, 285, 288, 619
GOLPE DE MANO **Vol. 8** 602, 732
GOLPE DE MIL MILLONES, UN **Vol. 8** 306
GOLYAKALIFA, A **Vol. 9** 172
GOMA 2 **Vol. 8** 500, 762
GOMORRON BILL ! **Vol. 5** 871, 923, 1043, 1080, 1164
GONDOLA DEL DIAVOLO, LA **Vol. 7** 1035, 1054
GONDOLA DELLE CHIMERE, LA **Vol. 7** 196, 1029
GONDOLE AUX CHIMERES, LA **Vol. 2** 153 **Vol. 7** 196, 453, 1029
GONDVISELES **Vol. 9** 103, 143, 238, 264
GONGEHOVDINGEN **Vol. 5** 2, 128, 179, 275, 304
GONZAGUE **Vol. 2** 388
GOOD-BYE AGAIN **Vol. 2** 824
GOOD-BYE EMMANUELLE **Vol. 2** 49, 457
GOOD BYE FIRENZE ! **Vol. 7** 1063
GOODBYE & AMEN **Vol. 7** 599
GOODBYE SEVILLA **Vol. 8** 437, 768
GORA DEL MULINO, LA **Vol. 7** 1189
GORACA LINIA **Vol. 1** 198, 237, 249, 381 **Vol. 6** *page 222, page 242*
GORACZKA **Vol. 6** *page 224, page 254*
GÖRANSSONS POJKE **Vol. 5** 840, 997
GORDIAN, DER TYRANN **Vol. 4** 79, 115, 119
GORDO DE NAVIDAD, EL **Vol. 8** 281, 563
GÖRE GABOR BIRO UR KALANDOZASAI BUDAPESTEN **Vol. 9** 248
GÖRE GABOR VISSZATER **Vol. 9** 13, 315
GORE NA CERESATA **Vol. 6** *page 40, page 64*
GORESTO PLADNE **Vol. 3** 83, 112 **Vol. 6** *page 41*
GORGO FASCINATORE, LA **Vol. 7** 266

GORGONA, LA Vol. 7 491, 552, 725, 777
GORGOPOTAMOS Vol. 3 192
GORILAS A TODO RITMO Vol. 8 297
GORILLA, LA Vol. 7 540
GORILLE VOUS SALUE BIEN, LE Vol. 2 185, 259, 644
GORILLES, LES Vol. 2 97, 309
GORKE TRAVE Vol. 3 349, 359, 391
GORKI DEO REKE Vol. 3 404, 405
GORODI FOGOLY, A Vol. 9 26, 117
GORRIONES DE PATIO, LOS Vol. 8 723
GORY O ZMIERZCHU Vol. 1 148
GOSCINNY WYSTEP Vol. 1 144
GOSPODIN NIKOJ Vol. 3 69, 75
GOSPODJA MINISTARKA Vol. 3 337, 349
GOSPODJICA Vol. 3 437, 454
GOSSE DE LA BUTTE, UN Vol. 2 303
GOSSE DE RICHE Vol. 2 595
GOSSE EN OR, UN Vol. 2 404, 514, 722
GOSSE SENSAS, UNE Vol. 2 243, 354, 664, 727
GOSSELINE, LA Vol. 2 169, 330, 613
GOSSES MENENT L'ENQUETE, LES Vol. 2 661, 671
GOSSETTE Vol. 2 192, 805
GÖSTA BERLINGS SAGA Vol. 5 889, 891, 964, 997, 999
GÖTA KANAL ELLER Vol. 5 915, 917, 1048, 1098, 1168
GOTO, L'ILE D'AMOUR Vol. 2 275
GOTT ALLEIN DIE EHRE Vol. 4 256, 605
GOTT DER RACHE, DER Vol. 4 300
GOTT MIT UNS Vol. 3 437 Vol. 7 585
GÖTTER, MENSCHEN UND TIERE Vol. 4 212, 571, 643, 726, 753
GOTTES MÜHLEN, DIE Vol. 10 256, 412
GOTTES WEGE SIND WUNDERBAR Vol. 4 514
GÖTTIN, DIRNE UND WEIB Vol. 4 366
GÖTTLICHE JETTE, DIE Vol. 4 49, 124, 339
GÖTZ VON BERLICHINGEN ZUBENANNT MIT DER EISERNEN HAND Vol. 4 125, 291, 340
GÖTZENDÄMMERUNG Vol. 4 5, 424
GOUALEUSE, LA Vol. 2 48, 722, 726
GOUBBIAH, MON AMOUR Vol. 2 402
GOUBIJAH Vol. 3 378
GOULUES, LES Vol. 2 565
GOULVE, LA Vol. 2 795
GOUPI MAINS ROUGES Vol. 2 124, 558, 601
GOURMANDES DE SEXE, LES Vol. 2 291
GOURMANDINES, LES Vol. 2 120
GOUT DE LA VIOLENCE, LE Vol. 2 248, 548, 738
GOUTS ET COULEURS Vol. 2 871
GOUVERNEUR, DER Vol. 4 445, 713
GOUVERNEUR DES TODES, DER Vol. 4 720, 799
GOYA Vol. 1 9, 17, 48 Vol. 3 59 Vol. 8 299, 348
GOYA QUE VUELVE Vol. 8 423
GOYESCAS Vol. 8 298, 484, 611
G.P.U. Vol. 4 435, 491, 740
GRA Vol. 1 233, 321 Vol. 6 *page 212*
GRAA DAME, DEN Vol. 5 97
GRABBARNA I 57 : AN Vol. 5 1205
GRABESTE – TE INCET Vol. 6 *page 278, page 282*
GRABMAL EINER GROSSEN LIEBE, DAS Vol. 4 94, 474, 634
GRABUGE, LE Vol. 2 49
GRACKA Vol. 6 *page 14, page 16, page 17*
GRADIVA Vol. 7 743
GRADUADA, LA Vol. 8 331, 553
GRAENSEFOLKET Vol. 5 265
GRAF CHARGON Vol. 4 256, 599
GRAF COHN Vol. 4 219, 560
GRAF GALLAS Vol. 4 634
GRAF MICHAEL Vol. 4 565
GRAF SYLVAINS RACHE Vol. 4 423, 780
GRAF VON CAGLIOSTRO, DER Vol. 4 288, 389
GRAF VON CHAROLAIS, DER Vol. 4 211, 247, 479
GRAF VON ESSEX, DER Vol. 4 240, 247, 266, 574, 673, 729
GRÄFIN ANKARSTRÖM Vol. 4 646
GRÄFIN DONELLI Vol. 4 646, 734
GRÄFIN HEYERS, DIE Vol. 6 *page 194*
GRÄFIN MARIZA Vol. 4 85, 158, 192, 626
GRÄFIN MARUSCHKA Vol. 4 159
GRÄFIN PLÄTTMAMSELL Vol. 4 113, 162
GRÄFIN VON MONTE CHRISTO, DIE Vol. 4 266, 527, 574
GRÄFIN WALEWSKA Vol. 4 423, 681
GRAIN DE SABLE, LE Vol. 2 368 Vol. 6 *page 80, page 121* Vol. 8 83
GRAINE AU VENT Vol. 2 25, 104, 330, 440
GRAINS DE BEAUTE Vol. 2 122, 197
GRAJSKI BIKI Vol. 3 375, 387
GRAMADA Vol. 3 80
GRAMMOS Vol. 3 156
GRAN AMOR DEL CONDE DRACULA, EL Vol. 8 626, 627
GRAN AVENTURA DE LOS PARCHIS, LA Vol. 8 680, 770
GRAN BARRERA, LA Vol. 8 614
GRAN BOLITTO Vol. 7 817
GRAN COARTADA, LA Vol. 8 470, 492
GRAN COMEDIA, LA Vol. 8 575
GRAN CRUCERO, EL Vol. 8 311, 465
GRAN FAMILIA, LA Vol. 8 524, 725
GRAN GALEOTO, EL Vol. 8 214, 449, 484

GRAN GOLPE AL SERVICIO DE SU MAJESTAD BRITANNICA **Vol. 8** 306, 708
GRAN GOLPE DE LOS SIETE HOMBRES DE ORO, EL **Vol. 8** 749
GRAN GOLPE DE NIZZA, EL **Vol. 8** 598
GRAN MENTIRA, LA **Vol. 8** 214
GRAN MOGOLLON, EL **Vol. 8** 749, 762
GRAN PREMIO **Vol. 7** 309, 639, 938
GRAN SENORA, UNA **Vol. 8** 207, 214
GRAN VARIETA **Vol. 7** 257
GRANA, LA **Vol. 7** 798
GRANATIERE DI POMERANIA, IL **Vol. 7** 559
GRANATIERE ROLAND, IL **Vol. 7** 1189
GRAND AMOR, LE **Vol. 2** 110, 525
GRAND AMOUR DE BEETHOVEN, UN **Vol. 2** 197, 309, 488
GRAND BAZAR, LE **Vol. 2** 114, 285
GRAND BIDULE, LE **Vol. 2** 598, 835
GRAND BLOND AVEC UNE CHAUSSURE NOIR, LE **Vol. 2** 564, 588
GRAND BLUFF, LE **Vol. 2** 16, 824
GRAND CARNAVAL, LE **Vol. 6** *page 75, page 86*
GRAND CEREMONIAL, LE **Vol. 2** 215, 570
GRAND CHEF, LE **Vol. 2** 402
GRAND CIRQUE, LE **Vol. 2** 205, 562, 698
GRAND COMBAT, LE **Vol. 2** 48, 560
GRAND DADAIS, LE **Vol. 2** 377, 873
GRAND DELIRE, LE **Vol. 2** 517
GRAND DEPART, LE **Vol. 2** 301
GRAND ELAN, LE **Vol. 2** 16, 344, 397, 687, 718
GRAND EMBOUTEILLAGE, LE **Vol. 6** *page 76*
GRAND ESCOGRIFFE, LE **Vol. 2** 196
GRAND ESCROC, LE **Vol. 2** 208
GRAND FANFARON, LE **Vol. 2** 513
GRAND GALA **Vol. 2** 322, 501
GRAND GUIGNOL **Vol. 7** 1113
GRAND HOTEL... **Vol. 4** 115, 359
GRAND HOTEL EXCELSIOR **Vol. 7** 30, 383
GRAND HOTEL PARADIS **Vol. 7** 108
GRAND JEU, LE **Vol. 2** 59, 308, 427, 573, 803
GRAND MEAULNES, LE **Vol. 2** 525
GRAND-MERE **Vol. 2** 688
GRAND PARDON, LE **Vol. 6** *page 75, page 100, page 121*
GRAND PATRON, UN **Vol. 2** 368, 615
GRAND PAVOIS, LE **Vol. 2** 36, 615
GRAND-PERE **Vol. 2** 858, 871
GRAND-PERE, LE **Vol. 6** *page 75, page 100*
GRAND PRIX **Vol. 7** 525
GRAND REFRAIN, LE **Vol. 2** 205, 479, 526, 644, 803
GRAND RENDEZ-VOUS, LE **Vol. 2** 96, 826
GRAND RESTAURANT, LE **Vol. 2** 128
GRAND RISQUE, LE **Vol. 2** 279, 680
GRAND SABORDAGE, LE **Vol. 2** 721
GRAND SECRET, LE **Vol. 2** 266, 372, 668
GRANDE AMIE, LA **Vol. 2** 381, 384, 413
GRANDE APPELLO, IL **Vol. 7** 472, 730, 1036, 1116
GRANDE ATTACCO, IL **Vol. 7** 1202
GRANDE AUDACIA, LA **Vol. 7** 825
GRANDE AVVENTURA, LA **Vol. 7** 958
GRANDE BAISE, LA **Vol. 2** 294
GRANDE BOUFFE, LA **Vol. 2** 142
GRANDE COLPO DI SURCOUF **Vol. 8** 426, 732
GRANDE DUELLO, IL **Vol. 7** 1194
GRANDE ELIAS, O **Vol. 8** 10, 77, 122, 131
GRANDE EPREUVE, LA **Vol. 2** 339
GRANDE FILLE TOUTE SIMPLE, UNE **Vol. 2** 388, 440
GRANDE FRIME, LA **Vol. 2** 349
GRANDE FROUSSE, LA **Vol. 2** 598, 642, 778
GRANDE GIORNO, IL **Vol. 7** 858
GRANDE GOLPO DEI SETTE UOMINI D'ORO, IL **Vol. 8** 749
GRANDE, GRANDE ERA A CIDADE **Vol. 8** 36
GRANDE GUERRA, LA **Vol. 7** 498, 514, 988
GRANDE ILLUSION, LA **Vol. 2** 124, 125, 526, 566, 719, 863
GRANDE JAVA, LA **Vol. 2** 878
GRANDE LECHE, LA **Vol. 2** 291
GRANDE LESSIVE, LA **Vol. 2** 840, 866
GRANDE LEVETTRE, LA **Vol. 2** 291
GRANDE LUCE, LA **Vol. 7** 101, 493, 902
GRANDE MAFFIA, LA **Vol. 2** 106, 781
GRANDE MAGUET, LA **Vol. 2** 32, 322
GRANDE MARNIERA, LA **Vol. 7** 53
GRANDE MARNIERE, LA **Vol. 2** 122, 722
GRANDE MEUTE, LA **Vol. 2** 330, 807
GRANDE MOMENTO **Vol. 7** 728
GRANDE MOUILLE, LA **Vol. 2** 291
GRANDE NOUBA, LA **Vol. 2** 740
GRANDE OLIMPIADE, LA **Vol. 7** 1044
GRANDE OMBRA, LA **Vol. 7** 1152
GRANDE PASSION, LA **Vol. 2** 180 **Vol. 6** *page 185*
GRANDE PASSIONE, LA **Vol. 7** 48, 520
GRANDE PAULETTE, LA **Vol. 2** 372
GRANDE RACKET, IL **Vol. 7** 728
GRANDE REVOLTE, LA **Vol. 2** 81
GRANDE RINUNCIA, LA **Vol. 7** 891
GRANDE SAUTERIE, LA **Vol. 2** 349
GRANDE SAVANA, LA **Vol. 7** 249
GRANDE SILENZIO, IL **Vol. 7** 998
GRANDE SPERANZA, LA **Vol. 7** 86

GRANDE STRADA AZZURRA, LA Vol. 3 387 Vol. 7 522, 797, 798
GRANDE TROUILLE, LA Vol. 2 814
GRANDE VADROUILLE, LA Vol. 2 21, 719
GRANDE VERGOGNA, LA Vol. 7 277
GRANDE VIE, LA Vol. 2 398, 496
GRANDE VOLIERE, LA Vol. 2 88, 719
GRANDES AMIGOS Vol. 8 402, 438
GRANDES FAMILLES, LES Vol. 2 644, 722
GRANDES GUEULES, LES Vol. 2 110, 772
GRANDES JOUISSEUSES, LES Vol. 2 291
GRANDES MANOEUVRES, LES Vol. 2 59, 232, 423, 488, 738 Vol. 6 *page 115*
GRANDES PERSONNES, LES Vol. 2 208, 285
GRANDES VACANCES, LES Vol. 2 97
GRANDEUR ET DECADENCE DE TRISTAN BERNARD Vol. 2 42
GRANDEUR NATURE Vol. 2 234, 837
GRANDHOTEL NEVADA Vol. 10 256, 586
GRANDI MAGAZZINI Vol. 7 196, 453
GRANDS CHEMINS, LES Vol. 2 21, 873
GRANDS, LES Vol. 2 402, 584, 697, 871
GRANDS MOYENS, LES Vol. 2 293, 746
GRANDS SENTIMENTS FONT LES BONS GUEULETONS, LES Vol. 2 7, 8
GRANDUCHESA SI DIVERTE, LA Vol. 7 16, 119, 236, 480, 512, 603, 788, 1060
GRANGES BRULEES, LES Vol. 2 377, 854
GRANICA Vol. 1 206, 282, 346
GRANICATA Vol. 3 31, 65 Vol. 6 *page 41*
GRANO DE MOSTAZA, EL Vol. 8 207, 214
GRANUJAS DECENTES, UNOS Vol. 8 762
GRAPHIQUE DE BOSCOP, LE Vol. 2 124, 177
GRAPPOLO D'ORO Vol. 7 1124
GRÄSANKLINGAR Vol. 5 917, 947, 1168
GRÄSET SJUNGER Vol. 5 851, 1198
GRASHAUS ODER DIE AUFTEILUNG VON 350000 FRAUEN AUF ZWEI MANN, DAS Vol. 1 57, 95, 104
GRAT ELSKEDE MANN Vol. 5 682, 787, 814
GRATIS MIDDAG Vol. 5 97
GRATITUD Vol. 8 238
GRATTACIELI Vol. 7 792, 1063
GRAUE NACHT, DIE Vol. 4 212, 571, 652, 726
GRAUEN, DAS Vol. 4 208, 232
GRAUSAME FREUNDIN, DIE Vol. 2 397
GRAUSIGE NÄCHTE Vol. 4 124, 663
GRAV ZARKAS BANDE Vol. 5 81
GRAVITACIJA ILI FANTASTICNA MLADOST CINOVNIKA BORISA HORVATA Vol. 3 437
GRAZIA, LA Vol. 7 730, 792, 1140
GRAZIE TANTE E ARRIVEDERCI Vol. 7 1026
GRAZIELLA Vol. 2 608 Vol. 7 106, 332, 621
GRAZIELLA LA GITANE Vol. 2 798
GREIFER, DER Vol. 4 192, 462
GRELUCHON DELICAT, LE Vol. 2 676, 824
GRENZFEUER Vol. 4 85, 286, 344, 541, 703
GRENZGÄNGER Vol. 4 25
GRENZJÄGER, DER Vol. 6 *page 184*
GRETA, MONIKA ET SUZELLE Vol. 2 721
GRETCHEN SCHUBERT Vol. 4 223, 479, 644
GRETEL ZIEHT DAS GROSSE LOS Vol. 4 239, 270
GREVARNA PA SVANSTA Vol. 5 996
GREVE DES DOMESTIQUES, LA Vol. 2 798
GREVE SVENSSON Vol. 5 840, 1091
GREVEN FRAN GRÄNDEN Vol. 5 878, 1173
GREVINDE CLARA Vol. 5 183
GREVINDE HJERTELOS Vol. 5 58
GREVINDE X Vol. 5 97
GREVINDENS AERE Vol. 5 183
GREY CONTRE X Vol. 2 152, 744
GRIBICHE Vol. 2 237, 308, 573
GRIBOUILLE Vol. 2 84, 427, 824, 837
GRIDO DELL AQUILA, IL Vol. 7 493
GRIDO DELLA TERRA, IL Vol. 7 1036, 1054
GRIDO D'OLTRE TOMBA, IL Vol. 7 837
GRIDO, IL Vol. 7 458, 705, 763, 854, 1167
GRIDO NELLA FORESTA, IL Vol. 7 843
GRIDO NELLA NOTTE, UN Vol. 7 1189
GRIFFARD Vol. 7 1189
GRIFFARD II Vol. 7 1189
GRIFFE ET LA DENT, LA Vol. 2 77, 853
GRILLE, DIE Vol. 4 603
GRILLON DU FOYER, LE Vol. 2 402
GRINGALET Vol. 2 128
GRINGO Vol. 8 311, 350
GRISJAKTEN Vol. 5 1093, 1170
GRISOU Vol. 2 191, 413, 595
GRITO NA NOITE, UM Vol. 8 44, 106
GRITOS ... A RITMO FUERTE Vol. 8 733
GRITOS DE ANSIEDAD Vol. 8 536
GRITOS EN LA NOCHE Vol. 8 600, 725
GROCK Vol. 4 186, 212, 479
GROF MEFISTO Vol. 9 73
GROFNÖ BETÖRÖI, A Vol. 9 136
GROG Vol. 7 405, 664
GROMADA Vol. 1 133, 227
GRÖNA HISSEN Vol. 5 863, 1032
GRONKOBINGS GLADE GAVTYVE Vol. 5 24, 83
GRONNE BILLE, DEN Vol. 5 265

GRONNE ELEVATOR, DEN **Vol. 5** 14, 281
GRONNE HEISEN, DEN **Vol. 5** 710, 778, 824
GROS BRAS, LES **Vol. 2** 348, 815
GROS CALIN **Vol. 2** 76, 130
GROS COUP, LE **Vol. 2** 348, 395
GROS MALINS, LES **Vol. 2** 728
GROSS PARIS **Vol. 2** 702
GROSSALARM **Vol. 4** 184, 325, 713
GROSSE ABENTEUER, DAS **Vol. 4** 89, 149, 660
GROSSE ABENTEUER DES KASPAR SCHMECK, DAS **Vol. 6** *page 159, page 174, page 176*
GROSSE ABENTEUERIN, DIE **Vol. 4** 246, 321, 752
GROSSE ATTRAKTION, DIE **Vol. 4** 306
GROSSE BLUFF, DER **Vol. 4** 79, 115, 132
GROSSE CAISSE, LA **Vol. 2** 644, 650
GROSSE CHANCE, DIE **Vol. 4** 124, 422, 703
GROSSE CHEF, DER **Vol. 4** 159, 759
GROSSE FALL, DER **Vol. 4** 292
GROSSE FREIHEIT NR. 7 **Vol. 4** 365, 381, 445
GROSSE GEHEIMNIS, DAS **Vol. 4** 98, 112
GROSSE GELEGENHEIT, DIE **Vol. 4** 339, 367, 613 **Vol. 6** *page 181*
GROSSE KÖNIG, DER **Vol. 4** 424, 462, 788
GROSSE LEIDENSCHAFT, DIE **Vol. 6** *page 185*
GROSSE LICHT, DAS **Vol. 4** 660, 704
GROSSE LIEBE, DIE **Vol. 4** 235, 742
GROSSE LOS, DAS **Vol. 4** 85, 410
GROSSE NUMMER, DIE **Vol. 4** 97, 553, 580, 608
GROSSE OPFER, DAS **Vol. 6** *page 194*
GROSSE PAUSE, DIE **Vol. 4** 213, 626
GROSSE PREIS, DER **Vol. 4** 124, 292, 640
GROSSE PREIS VON EUROPA, DER **Vol. 4** 329, 774
GROSSE REISE DER AGATHE SCHWEIGERT, DIE **Vol. 1** 34, 89, 124
GROSSE SCHATTEN, DER **Vol. 4** 15, 149, 188
GROSSE SEHNSUCHT, DIE **Vol. 4** 149, 219, 660
GROSSE SPIEL, DAS **Vol. 4** 285, 615, 678
GROSSE SPRUNG, DER **Vol. 4** 10, 15, 54, 115, 450, 481, 620
GROSSE TETE, UNE **Vol. 2** 185, 427
GROSSE UNBEKANNTE, DER **Vol. 4** 424, 527, 779
GROSSE UNBEKANNTE, DIE **Vol. 4** 133, 622
GROSSE UND DIE KLEINE PEST, DIE **Vol. 4** 522
GROSSE UND DIE KLEINE WELT, DIE **Vol. 4** 133, 302, 344, 647, 689
GROSSE WURF, DER **Vol. 4** 291
GROSSERENS OVERORDNEDE **Vol. 5** 81
GROSSINDUSTRIELLE, DER **Vol. 4** 356, 729
GROSSREINEMACHEN **Vol. 4** 257, 432, 647
GROSSTADT BEI NACHT, WIE SIE LACHT UND WEINT, DIE **Vol. 4** 183
GROSSTADTJUGEND **Vol. 4** 287, 384, 458
GROSSTADTKINDER **Vol. 4** 124, 125, 291
GROSSTADTKINDER ZWISCHEN SPREE UND PANKE **Vol. 6** *page 193*
GROSSTADTMÄDELS **Vol. 4** 593
GROSSTADTMELODIE **Vol. 4** 24, 525, 738
GROSSTADTNACHT **Vol. 4** 29, 40 **Vol. 6** *page 72*
GROSSTADTPIRATEN **Vol. 4** 495, 646, 733
GROSSTADTSCHMETTERLINGE **Vol. 4** 30, 192, 270, 614
GROSSTADTZAUBER **Vol. 4** 18, 558
GRÖSSTE GAUNER DES JAHRHUNDERTS, DER **Vol. 4** 384, 584, 700
GRÖSSTE LIEBE, DIE **Vol. 4** 340, 747
GROV SPOG, EN **Vol. 5** 97
GROZNICA LJUBAVI **Vol. 6** *page 310, page 317, page 319*
GRUBE MORGENROT **Vol. 1** 25, 28, 123 **Vol. 4** 188
GRUBEEJERENS DOD **Vol. 5** 228
GRÜN IST DIE HEIDE **Vol. 4** 119, 600
GRÜNE DOMINO, DER **Vol. 4** 30, 298, 567, 608
GRÜNE KAISER, DER **Vol. 4** 325, 774
GRÜNE LATERNE, DIE **Vol. 4** 270, 634
GRÜNE MANUELA, DIE **Vol. 4** 89, 316, 536, 546
GRÜNE MONOKEL, DAS **Vol. 4** 208, 479, 492
GRÜNE SALON, DER **Vol. 4** 361, 445
GRÜNE UNGEHEUER, DAS **Vol. 1** 72, 86, 112
GRUNT ! **Vol. 7** 599
GRUPNI PORTRET SA DAMON **Vol. 3** 356
GRUPPO DI FAMIGLIA IN UN INTERNO **Vol. 7** 498, 516, 736, 1025, 1081, 1142
GRÜSS MIR DAS BLONDE KIND AM RHEIN **Vol. 4** 242
GRÜSS' MIR DIE LORE NOCH EINMAL **Vol. 4** 124, 422, 703
GRUSS UND KUSS VERONIKA **Vol. 4** 135, 462
GRYR I NORDEN **Vol. 5** 670, 760
GRZECH ANTONIEGO GRUDY **Vol. 1** 131, 216 **Vol. 6** *page 257*
GRZECHY DZIECINSTWA **Vol. 6** *page 206, page 213, page 225*
GRZESZNA MILOSZ **Vol. 1** 176, 384

GRZESZNY ZYWOT FRANCISZKA BULY Vol. 6 *page 217, page 231, page 239, page 246*
G'SCHICHTEN AUS DEM WIENERWALD Vol. 4 239, 270, 769
GUAGLIONE Vol. 7 693
GUAI AI VINTI ! Vol. 7 161, 924, 1027
GUNAT MBI TELA Vol. 6 *page 15, page 17*
GUANTO, IL Vol. 7 1189
GUAPA, RICA Y ... ESPECIAL Vol. 8 286, 688
GUAPO HEREDERO BUSCA ESPOSA Vol. 8 297, 331
GUAPOS, LOS Vol. 8 281, 657
GUAPPARIA Vol. 7 275, 1170
GUARANY Vol. 7 231, 638, 639, 796
GUARDAMI NUDA Vol. 7 1213
GUARDATELE MA NON TOCCATELE Vol. 7 852
GUARDIA DEL CORPO, LA Vol. 7 639, 756
GUARDIA, GUARDIA SCELTA BRIGADIERE E MARESCIALLO Vol. 7 528
GUARDIA, LADRO E CAMERIERA Vol. 7 852
GUARDIAMARINAS, LOS Vol. 8 524, 725
GUARDIAN DEL PARAISO, EL Vol. 8 616, 630, 753
GUARDIE E LADRI Vol. 7 106, 781
GUAZZABUGLIO, IL Vol. 7 968
GUBBEN KOMMER Vol. 5 840, 909
GUD FADER OCH TATTAREN Vol. 5 1003, 1073
GUDERNES YNDLING Vol. 5 58
GUDRUN Vol. 5 23, 248
GUENDALINA Vol. 7 725, 1065
GUEPIER, LE Vol. 2 860
GUEPIOT, LE Vol. 2 49 Vol. 6 *page 115, page 130*
GUERILLERA, LA Vol. 6 *page 122*
GUERILLERO ET CELUI N'Y CROYAIT PAS, LE Vol. 2 172
GUERISSEUR, LE Vol. 2 368, 615, 857
GUERNICA Vol. 7 834, 1025 Vol. 9 167, 185, 259
GUERRA DE DIOS, LA Vol. 8 214, 400
GUERRA DE LOS LOCOS, LA Vol. 8 672
GUERRA DE LOS NINOS, LA Vol. 8 373, 466
GUERRA DEL FERRO, LA Vol. 7 426, 504
GUERRA DO MIRANDUM Vol. 8 10, 34, 60, 152, 153
GUERRA E IL SOGNO DI MOMI, LA Vol. 7 305
GUERRA E LA MODA, LA Vol. 7 943
GUERRA E PACE Vol. 7 299, 750, 1136
GUERRA EMPIEZA EN CUBA, LA Vol. 8 276, 298, 603
GUERRA IN TEMPO DI PACE Vol. 7 131
GUERRA, LA Vol. 7 943
GUERRA O PACE Vol. 7 796
GUERRA REDENTRICE Vol. 7 621
GUERRA SUCIA Vol. 8 321, 524
GUERRE DES BOUTONS, LA Vol. 2 47, 819
GUERRE DES GOSSES, LA Vol. 2 67, 360
GUERRE DES POLICES, LA Vol. 2 806 Vol. 6 *page 78*
GUERRE DES VALSES, LA Vol. 2 399
GUERRE DU FEU, LA Vol. 6 *page 69*
GUERRE EST FINIE, LA Vol. 2 770, 854
GUERRE SECRETE Vol. 2 112, 322, 661
GUERRES CIVILES EN FRANCE Vol. 2 405
GUERRIERI DELL'ANNO 2072, I Vol. 7 1165
GUERRILLA, LA Vol. 8 207, 214
GUERRILLERA DE VILLA, LA Vol. 8 395
GUERRILLEROS, LOS Vol. 8 311, 400
GUEULE COMME LA MIENNE, UNE Vol. 2 118, 875
GUEULE D'AMOUR Vol. 2 733
GUEULE D'ANGE Vol. 2 407, 433, 557
GUEULE DE L'AUTRE, LA Vol. 2 400, 564 Vol. 6 *page 125*
GUEULE DE L'EMPLOI, LA Vol. 2 74, 843
GUEULE DE LOUP, LA Vol. 6 *page 87, page 95*
GUEULE EN OR, UNE Vol. 2 80, 197, 488, 726
GUEULE OUVERTE, LA Vol. 2 5, 17
GUEUX AU PARADIS, LES Vol. 2 309
GUICHETS DU LOUVRE, LES Vol. 2 783, 835
GUIGNOLO, LE Vol. 2 22, 224
GUIGNON GUERISSEUR, LE Vol. 2 249
GUILLAUME LE CONQUERANT Vol. 6 *page 281*
GUILLOTINE Vol. 4 472, 565
GUINGUETTE Vol. 2 722
GUITARE AU POING Vol. 2 351
GUITARE ET LE JAZZ-BAND, LA Vol. 2 136
GUITARRA DE CARLOS GARDEL, LA Vol. 8 359, 396, 436, 704
GÜL BABA Vol. 9 164, 196
GULA BILEN, DEN Vol. 5 954, 1216
GULA DIVISIONEN Vol. 5 1073, 1164, 1207
GULA KLINIKEN Vol. 5 923, 982, 1205
GÜLBABA Vol. 9 234
GULD OG GRONNE SKOVE Vol. 5 31, 248, 272
GULD TIL PRAERIENS SKRAPPE DRENGE Vol. 5 35, 195, 215

GULDGRÄVARNA Vol. 5 1041, 1142
GULDKALVEN Vol. 5 183
GULDSPINDELN Vol. 5 925, 937
GULE DJAEVEL, DEN Vol. 5 97
GULLASCH GROSSERENS VOGTER Vol. 5 83
GULLFJELLET Vol. 5 676, 683, 714, 726, 738
GULNARA O STORIA DELL'INDIPENDENZA GRECA 1820 – 1830 Vol. 7 1189
GULOCKY Vol. 10 37, 542, 544
GUMMI TARZAN Vol. 5 185, 270, 283
GUNAN IL GUERRIERO Vol. 7 371, 414
GUNNAR HEDES SAGA Vol. 5 937, 997
GUNS Vol. 6 *page 105, page 111, page 133, page 138*
GÜNSTLING VON SCHÖNBRUNN, DER Vol. 4 49, 316
GUSAR Vol. 3 444
GUSTAF WASA Vol. 5 846, 925, 927
GUSTAV MOND... DU GEHST SO STILLE Vol. 4 115, 407
GUSTUL SI CULOAREA FERICIRII Vol. 3 284, 302, 308
GUTE ALTE ZEIT, DIE Vol. 4 85, 410
GUTE PARTIE, DIE Vol. 6 *page 189*
GUTE REISE, HERR MEYER Vol. 4 18
GUTE RUF, DER Vol. 4 208, 270
GUTE SIEBEN, DIE Vol. 4 378, 445
GUTEN ABEND, GUTE NACHT Vol. 4 115, 354
GUTEN TAG, LIEBER TAG Vol. 1 13, 47, 120
GUTEN TAG, SCHWIEGERMAMA Vol. 4 116, 291
GUTGEHENDES GESCHÄFT ZU VERKAUFEN Vol. 4 703
GUVERNORENS DATTER Vol. 5 9
GUXIMTARET Vol. 3 11, 29 Vol. 6 *page 18*
GUY DE MAUPASSANT Vol. 6 *page 84, page 95, page 141*
GVOZDENI KRST Vol. 6 *page 312*
GVOZDENI ORAO Vol. 3 426, 428
GWENDOLINE Vol. 6 *page 77, page 83*
GWIAZDY PORANNE Vol. 6 *page 225, page 241, page 252*
GWIAZDZISTA ESKADRA Vol. 1 278, 380, 404,
GYALOG A MENNYORSZAGBA Vol. 9 67, 121, 165
GYANU Vol. 9 117
GYARMAT A FÖLD ALATT Vol. 9 83, 91, 274
GYAVASAG Vol. 9 26, 117, 133
GYCKLARNAS AFTON Vol. 5 871, 1042, 1053, 1080
GYEREKBETEGSEGEK Vol. 9 215, 263, 287
GYERMEKRABLAS A PALANK UTCABAN Vol. 9 22, 50, 185
GYERMEKSZIV Vol. 9 220
GYERTEK EL A NEVNAPOMRA Vol. 9 134, 152, 215, 316
GYILKOS A HAZBAN VAN, A Vol. 9 152, 259, 339
GYIMESI VADVIRAG Vol. 9 2, 274, 315
GYLDNE DAGE, DE Vol. 5 303
GYLDNE HORN, DET Vol. 5 103
GYLDNE SMIL, DET Vol. 5 79, 183
GYLDNE UNGDOM Vol. 5 800, 831
GYMKATA Vol. 6 *page 312*
GYMNASIEPIGEN Vol. 5 2, 14
GYMNE EMERA Vol. 3 170
GYMNE STO DROMO Vol. 3 144, 205
GYMNI STO HIONI Vol. 3 196
GYONAS SZENTSEGE, A Vol. 9 323
GYP Vol. 7 1190
GYRISMOS TES MEDEIAS Vol. 3 149
GYURKOVICS LANYOK Vol. 9 77
GYURKOVICSARNA Vol. 5 891, 925
GYS OG GAEVE TANTER Vol. 5 199, 225, 304
GYULA VITEZ TELEN – NYARON Vol. 9 67, 84, 121
GYURKOVICS FIUK Vol. 9 166, 236

H

H – 8... Vol. 3 358, 453
HA EGYSZER HUSZ EV MULVA Vol. 9 218, 241, 332
HA ENTRADO UN LADRON Vol. 8 493, 515, 721
HA FATTO 13 Vol. 7 359, 712
HA FATTO UNA SIGNORA, L' Vol. 7 792
HA LLEGADO UN ANGELO Vol. 8 214, 261
HA MEGNJÖN JOZSEF Vol. 9 20, 154, 318
HA PASADO UN HOMBRE Vol. 8 614, 725
HAANDEN Vol. 5 97
HAAVISTON LEENI Vol. 5 549, 592
HÄÄYÖ Vol. 5 604, 657
HAB' MICH LIEB Vol. 4 8, 373
HABANERA Vol. 8 583, 772
HABANERA, LA Vol. 4 8, 736, 742
HABEN SIE FRITZCHEN NICHT VERGESSEN ? Vol. 4 251
HABIBI Vol. 8 237, 738
HABIT VERT, L' Vol. 2 279, 407
HABITACION PARA TRES Vol. 8 628, 630
HABITANTES DE LA CASA DESHABITADA, LOS Vol. 8 394, 396
HABLAMOS ESTA NOCHE Vol. 8 232, 770
HACE CIEN ANOS Vol. 8 298, 611

HACHA PARA LA LUNA DE MIEL, UN Vol. 8 464
HADDA PADDA Vol. 5 9
HADI JED Vol. 10 139, 425, 567
HAEGT MIG I RYGGEN Vol. 5 183
HAENDELIGT UHELD Vol. 5 16, 61, 272
HAENDTE I KOBENHAVN, DET Vol. 5 180, 199, 304
HAERVAERK Vol. 5 164, 259, 273
HAEVEN Vol. 5 97
HAEVENS NAT Vol. 5 9, 54, 167
HAEVNEREN Vol. 5 9, 197
HAEVNET Vol. 5 97
HAFENBARON, DER Vol. 4 329, 774
HAFENBRAUT, DIE Vol. 4 232, 774
HÄGRINGEN Vol. 5 1054
HAHN IM KORB, DER Vol. 4 321, 323, 430, 461, 546
HAHO, A TENGER Vol. 9 66, 83, 329
HAHO, ÖCSI ! Vol. 9 83, 259, 329
HAI-TANG Vol. 4 192, 462
HAIDARI: 3.30 Vol. 3 155
HÄIDEN VIETTO KARJALAN RUNOMAILLA Vol. 5 501, 603
HAIDUCII Vol. 3 216, 244, 254, 329, 335
HAIDUCII LUI SAPTECAI Vol. 3 254, 285, 335
HAINA DE PIELE Vol. 3 311
HAINE Vol. 2 219
HAJDUCKA VREMENA Vol. 3 421
HAJDUK Vol. 3 367, 421 Vol. 9 154, 318
HAJDUSKA KLETVA Vol. 3 44, 63, 69, 75
HAJKA Vol. 3 368, 404, 415
HAJNALI HAZTETÖK Vol. 9 141, 167, 340
HALADEK Vol. 9 20, 167, 255
HALAL UTAN, A Vol. 9 7
HALALCSENGÖ Vol. 9 89
HALALOS CSOK Vol. 9 58, 71
HALALOS TAVASZ Vol. 9 79, 133, 196, 270, 271
HALALTANC Vol. 9 196, 274
HALB UND HALB Vol. 4 361, 558
HALBBLUT Vol. 3 444 Vol. 6 *page 189*
HALBE UNSCHULD Vol. 4 159, 669
HALBE WEG, DER Vol. 4 517, 558, 796
HALBSEIDE Vol. 4 558
HALBWÜCHSIGEN, DIE Vol. 4 221, 340, 608
HALCON DE CASTILLA, EL Vol. 8 583, 772
HALCON DEL DESIERTO, EL Vol. 8 327, 772
HALHATATLAN LEGIOS, A Vol. 9 19, 152, 259
HALKA Vol. 1 134, 174, 194, 278, 317, 404
HALL ALLA DÖRRAR ÖPNA Vol. 5 361
HÄLLEBÄCKS GARD Vol. 5 871
HALLIN JANNE Vol. 5 468, 634
HALLO ! AFRIKA FORUDE ! Vol. 5 24, 92
HALLO, AMERIKA Vol. 9 77
HALLO BABY Vol. 5 955, 1020, 1027, 1160
HALLO, BUDAPEST ! Vol. 9 237, 319
HALLO CAESAR ! Vol. 4 149, 407, 660
HALLO, DU ALTES SPREEATHEN Vol. 1 60
HALLO HALLO ! HIER SPRICHT BERLIN ! Vol. 4 32, 95, 115, 373
HALLO JANINE Vol. 4 8, 184, 713
HALLO SZPICBRODKA CZYLI OSTATNI WYSTEP KROLA KASIARZY Vol. 1 227, 233
HALLO, TAXI ! Vol. 1 76
HALLOJ I HIMMELSENGEN Vol. 5 31, 39, 100, 195, 272
HALLO SZPICBRODKA Vol. 6 *page 207, page 231*
HALLUCINATIONS SADIQUES Vol. 2 721
HALO TAKSI Vol. 6 *page 310, page 134, page 318*
HÄLSINGAR Vol. 5 840, 841, 870, 878
HALT, MEINE UHR ! Vol. 4 25, 608
HALTA LENA OCH VINDÖGDA PER Vol. 5 996
HALTA LENA OCH VINDÖGDE PER Vol. 5 930
HALTA LOTTAS KROG Vol. 5 841, 923
HALTE...POLICE ! Vol. 2 88, 514
HALTEROFLIC Vol. 6 *page 69*
HALUAN RAKASTAA, PETER Vol. 5 494, 516
HALVBLOD Vol. 5 996
HALVEKSITTU Vol. 5 356, 428
HAMBURGER KRANKHEIT, DIE Vol. 2 618 Vol. 6 *page 90*
HAMELIN Vol. 8 600, 616
HAMENOI AGGELOI Vol. 3 157
HAMIDA Vol. 1 37
HAMIS IZABELLA, A Vol. 9 67, 84, 121
HAMLET Vol. 2 206 Vol. 4 42, 113, 189, 213, 722 Vol. 5 97
HAMMARFORSENS BRUS Vol. 5 928, 1052
HAMMERSLAGET Vol. 5 295
HAMNSTAD Vol. 5 940, 1173
HAMPELMANN, DER Vol. 4 49, 113, 391, 623
HAN ASPARUH Vol. 6 *page 46, page 61, page 63*
HAN ER MORMON Vol. 5 83
HAN GLÖMDE ALDRIG Vol. 5 923, 1205
HAN – HON Vol. 5 372
HAN, HON OCH PENGARNA Vol. 5 840, 841
HAN, HUN, DIRCH OG DARIO Vol. 5 104, 199, 201
HAN, HUN HOG HAMLET Vol. 5 24, 83

HAN MATADO UN CADAVER **Vol. 8** 217, 470
HAN ROBADO UNA ESTRELLA **Vol. 8** 495, 602
HAN SKAL DUELLERE **Vol. 5** 295
HAN SPILLER FODBOLD **Vol. 5** 83
HAN TRODDE DET VAR HAN **Vol. 5** 923
HÄN TULI IKKUNASTA **Vol. 5** 471, 566
HÄN VARASTI ELÄMÄN **Vol. 5** 516, 627
HAN VIL TIL FILMEN **Vol. 5** 83
HANBA **Vol. 6** *page 192* **Vol. 10** 275, 345
HAND DES WÜRGERS, DIE **Vol. 4** 584, 593
HÄNDE AUS DEM DUNKEL **Vol. 4** 49, 306
HÄNDE HOCH **Vol. 4** 634
HÄNDE HOCH, HIER EDDY POLO **Vol. 4** 211, 407
HANDELEN MED MENNESKELIV **Vol. 5** 183
HANDFULL KÄRLEK, EN **Vol. 5** 1052, 1067, 1068, 1096
HANDICAP DER LIEBE, DAS **Vol. 4** 186, 479
HANDLIRI **Vol. 10** 503, 532
HANDVOLL HOFFNUNG, EINE **Vol. 1** 18, 37, 92
HANDVOLL NOTEN, EINE **Vol. 1** 34, 37, 88
HÄNGEZÖPFCHEN **Vol. 4** 366
HANGYABOLY **Vol. 9** 134, 259, 265
HANIA **Vol. 1** 278, 346
HANICKO, CO S TEBOU BUDE ? **Vol. 10** 453, 481
HANKA **Vol. 1** 372, 407 **Vol. 3** 383, 440
HANKA A JINDRA **Vol. 10** 45, 345
HANNA D. - LA RAGAZZA DEL VONDELPARK **Vol. 7** 335
HANNA I SOCIETEN **Vol. 5** 1032, 1181
HANNA K. **Vol. 6** *page 71, page 83*
HANNE UND ELISABETH **Vol. 4** 51
HANNELES HIMMELFAHRT **Vol. 4** 28, 194, 247, 340, 544, 546
HANNEMANN, ACH HANNEMANN **Vol. 4** 216, 558
HANNERL UND IHRE LIEBHABER **Vol. 4** 190
HANNERL VOM ROLANDSBOGEN, DAS **Vol. 4** 221
HANNES SCHARF **Vol. 1** 74, 83
HANNIBAL TANAR UR **Vol. 9** 4, 222, 285
HANNO RAPITO UN UOMO **Vol. 7** 792, 796
HANNO RUBATO UN TRAM **Vol. 7** 106, 1035
HANS BEDRE HALVDEL **Vol. 5** 83
HANS BEIMLER, KAMERAD **Vol. 1** 27, 38, 59
HANS ENGELSKA FRU **Vol. 5** 891, 909, 997
HANS GODE GENIUS **Vol. 5** 183
HANS HUSTRUS FÖRFLUTNA **Vol. 5** 925
HANS IM GLÜCK **Vol. 4** 82, 266, 481, 574
HANS IN ALLEN GASSEN **Vol. 4** 527
HANS KAERESTE **Vol. 5** 183
HANS KONES MAND **Vol. 5** 83
HANS KONES VENINDE **Vol. 5** 83
HANS KUNGLIG HÖGHET SHINGLAR **Vol. 5** 891, 966, 1031, 1191
HANS LE MARIN **Vol. 2** 124, 279
HANS LIVS MATCH **Vol. 5** 332, 840, 878, 909
HANS MAJESTÄT FAR VÄNTA **Vol. 5** 329, 332, 362, 837, 840, 870, 878, 1173
HANS MAJESTÄTS RIVAL **Vol. 5** 332, 840, 997
HANS NADS TESTAMENTE **Vol. 5** 840, 909, 937, 996
HANS NIELSEN HAUGE **Vol. 5** 674, 724, 797
HANS OFFICIELLA FÄSTMÖ **Vol. 5** 332, 840, 997
HANS ONSDAGS - VENINDE **Vol. 5** 89, 136
HANS RIGTIGE KONE **Vol. 5** 58
HANS RÖCKLE UND DER TEUFEL **Vol. 1** 15, 79, 127
HANS STORE AFTEN **Vol. 5** 89, 161, 224
HANS STORE CHANCE **Vol. 5** 183
HANS TRUTZ IM SCHLARAFFENLAND **Vol. 4** 186, 202
HANS UND LEA GRUNDIG **Vol. 1** 7
HANS UNGDOMSBRUD **Vol. 5** 83
HANS VANSKELIGSTE ROLLE **Vol. 5** 9
HANS WESTMAR **Vol. 4** 149, 660, 742
HANSEATEN **Vol. 4** 461, 720
HÄNSEL UND GRETEL **Vol. 4** 401
HANSEN OG CO. **Vol. 5** 97
HANSEN OG HANSEN **Vol. 5** 676, 683, 714, 723
HANUSSEN **Vol. 9** 167, 259, 303
HANY AZ ORA VEKKER UR ? **Vol. 9** 6, 215, 316
HANYATT - HOMLOK **Vol. 9** 191, 265, 285
HAPPENING **Vol. 2** 249
HAPPY BOB PAA KEGLEBANEN **Vol. 5** 97
HAPPY BOB PAA ROTTEJAGT **Vol. 5** 97
HAPPY BOB SOM BOKSER **Vol. 5** 97
HAPPY BOB SOM FRIER **Vol. 5** 97
HAPPY BOB SOM TJENER **Vol. 5** 97
HAPPY DAY **Vol. 3** 160, 183
HAPPY-END **Vol. 1** 19, 68
HAPPY END **Vol. 10** 388, 425, 503
HAPPY ROAD, THE **Vol. 2** 423

HAPURA, DIE TOTE STADT **Vol. 4** 558, 704
HÄPY ENDÖ ? ELI KUINKA UUNO TURHAPURO SAI NIIN KAUNIIN JA RIKKAAN VAIMON **Vol. 5** 319, 345
HÄR BÖRJAR ÄVENTYRET **Vol. 5** 329, 345, 853
HAR DE IKKE SET CECILIE ? **Vol. 5** 83
HAR DU SET ALICE ? **Vol. 5** 66, 125, 185
HÄR HAR DU DITT LIV **Vol. 5** 881, 1081, 1192
HÄR KOMMER BÄRSÄRKARNA **Vol. 5** 862, 1058
HÄR KOMMER PIPPI LANGSTRUMP **Vol. 5** 862, 1149
HÄR KOMMER VI ! **Vol. 5** 928
HARA-KIRI **Vol. 2** 37, 308, 330
HARACZ SZAREGO DNIA **Vol. 6** *page 212, page 225, page 244*
HARAG NAPJA, A **Vol. 9** 117, 215, 312
HARAKIRI **Vol. 4** 159
HARALD HANDFASTE **Vol. 5** 357, 923, 939, 948, 1015
HARANGOK ROMABA MENTEK, A **Vol. 9** 120, 261, 273
HARAPOS FERJ, A **Vol. 9** 178, 237
HARCMODOR **Vol. 9** 167, 238
HARD KLANG **Vol. 5** 1042, 1181
HARD LOVE **Vol. 2** 76
HARD SENSATIONS **Vol. 7** 742
HARD SEXE ET PORNO PARTY **Vol. 2** 766
HARDA LEKEN, DEN **Vol. 5** 432, 1052
HARDA VILJOR **Vol. 5** 925
HARDI LES GARS ! **Vol. 2** 277, 648
HARDI PARDAILLAN **Vol. 2** 615, 659
HAREMSEVENTYR, EN **Vol. 5** 92
HAREMSMYSTIK **Vol. 5** 83
HAREMY KOUZLA ZBAVENE **Vol. 10** 126, 167
HARKA ! **Vol. 8** 400
HARLEKIN ES SZERELMESE **Vol. 9** 67, 152, 180
HARLEM **Vol. 7** 196, 453
HARMADIK NEKEFUTAS **Vol. 9** 316, 339
HÄRMÄSTÄ POIKIA KYMMENEN **Vol. 5** 372, 379
HARMINCADIK, A **Vol. 9** 26, 117, 133
HARMINCKET NEVEM VOLT **Vol. 9** 67, 84, 134
HARMLOSER FALL, EIN **Vol. 4** 76
HARMONIKA **Vol. 10** 345, 412
HARNA – ASKEL **Vol. 5** 369, 543, 544
HARO **Vol. 2** 543, 628
HAROM ARVA, A **Vol. 9** 7
HAROM CSENGÖ, A **Vol. 9** 26, 117
HAROM CSILLAG **Vol. 9** 36, 119, 122
HAROM GALAMB **Vol. 9** 132
HAROM HET **Vol. 9** 245
HAROM SARKANY **Vol. 9** 178, 237
HAROMSZAZEZER PENGÖ AZ UTCAN **Vol. 9** 94, 164, 208
HARRISON ES BARRISON **Vol. 9** 200
HARRY HILL AUF WELLE 1000 **Vol. 4** 223, 329
HARRY HILL IM BANNE DER TODESSTRAHLEN **Vol. 4** 223, 232
HARRY HILLS JAGD AUF DEN TOD **Vol. 4** 125, 321
HARRY MUNTER **Vol. 5** 869, 1068
HARRY OG KAMMERTJENEREN **Vol. 5** 2, 14
HART AM WIND **Vol. 1** 2, 34, 45, 92
HART INS LEBEN **Vol. 4** 110
HARY JANOS **Vol. 9** 14, 24, 87, 196, 199, 297
HASANAGINICA **Vol. 3** 356, 391
HASARD ET LA VIOLENCE, LE **Vol. 2** 244, 285
HASCHISCH, DAS PARADIES DER HÖLLE **Vol. 4** 219, 479
HASEMANNS TÖCHTER **Vol. 4** 603
HASENHÜTER, DER **Vol. 1** 34, 48, 49
HASENKLEIN KANN NICHTS DAFÜR **Vol. 4** 292, 558
HASENPFOTE, DIE **Vol. 4** 115, 354
HASKOVY POVIDKY ZE STAREHO MOCNARSTVI **Vol. 10** 70, 142, 532, 621
HASLO „KORN" **Vol. 1** 156, 282 **Vol. 6** *page 246*
HASS **Vol. 4** 424
HÄSSLICHE ENTLEIN, DAS **Vol. 4** 185
HÄSSLICHE MÄDCHEN, DAS **Vol. 4** 28, 558
HAST DU GELIEBT AM SCHÖNEN RHEIN **Vol. 4** 258, 544
HASTA LA ULTIMA GOTA DE SANGRE **Vol. 8** 306, 602
HÄSTHANDLARENS FLICKOR **Vol. 5** 840, 884
HAT HET BOLDOGSAG **Vol. 9** 196, 322
HATARI **Vol. 2** 604
HATASVADASZOK **Vol. 9** 281, 282, 316
HÄTAVARA **Vol. 5** 552, 592
HATHOLDAS ROZSAKERT **Vol. 9** 134, 215, 259
HATIFA **Vol. 1** 25, 92, 103
HATTEN ER SAT **Vol. 5** 2, 229
HATTEN MED SKATTEN **Vol. 5** 83
HATTESTPARANCSNOK, A **Vol. 9** 323
HATTYUDAL **Vol. 9** 241, 332
HAUPT DES JUAREZ, DAS **Vol. 4** 734, 747
HAUPTMANN FLORIAN VON DER MÜHLE **Vol. 1** 11, 63, 79, 85
HAUPTMANN VON KÖLN, DER **Vol. 1** 9, 54, 88

HAUPTMANN VON KÖPENICK, DER **Vol. 4** 102, 119, 135, 247, 308, 356, 373, 514, 516, 626
HAUPTSACHE GLÜCKLICH **Vol. 4** 524, 544, 623
HAURI **Vol. 10** 233, 313, 327
HAUS AM KRÖGEL, DAS **Vol. 4** 495, 704
HAUS AM MEER, DAS **Vol. 4** 395, 560, 735
HAUS DER LEIDENSCHAFTEN, DAS **Vol. 4** 124
HAUS DER LÜGE, DAS **Vol. 4** 132, 214
HAUS DES VERBRECHENS, DAS **Vol. 4** 217, 570, 780
HAUS GEGENÜBER, DAS **Vol. 4** 113
HAUS IN DER DRAGONERGASSE, DAS **Vol. 4** 208, 400
HAUS NUMMER 17 **Vol. 4** 292, 752
HAUS OHNE LACHEN, DAS **Vol. 4** 124, 779
HAUS OHNE MÄNNER, DAS **Vol. 4** 242, 558
HAUS OHNE TÜR, DAS **Vol. 4** 646
HAUS ZUM MOND, DAS **Vol. 4** 288, 479, 779
HÄUSCHEN IN GRINZING, DAS **Vol. 10** 168, 586
HAUSTYRANN, DER **Vol. 4** 323, 461
HAUT COMME TROIS POMMES **Vol. 2** 37
HAUT LE VENT **Vol. 2** 558, 595
HAVARIE **Vol. 10** 70, 283, 408
HAVASI MAGDOLNA **Vol. 9** 28
HAVASI NAPSÜTES **Vol. 9** 196, 312
HAVASI SZERELEM **Vol. 9** 315
HAVET OG MENNESKENE **Vol. 5** 4
HAVETS MELODI **Vol. 5** 870, 943
HAVETS SON **Vol. 5** 871, 1003, 1052
HAVI 200 FIX **Vol. 9** 288, 319, 334
HAVIS AMANDA - HELSINGIN KAUNOTAR **Vol. 5** 400, 404, 511, 542, 570
HAVKONGENS DATTER **Vol. 5** 97
HAVRANIA CESTA **Vol. 10** 273, 459
HÄXAN **Vol. 5** 197, 838, 848, 961, 1047
HÄXNATTEN **Vol. 5** 958, 1132
HAY ALGUIEN DETRAS DE LA PUERTA **Vol. 8** 214, 400
HAY QUE DESHACER LA CASA **Vol. 8** 219, 771
HAY QUE EDUCAR A PAPA **Vol. 8** 681, 749
HAY UN CAMINO A LA DERECHA **Vol. 8** 515, 753
HAY UN FANTASMA EN MI CAMA **Vol. 8** 413, 627
HAYDEE **Vol. 7** 1005
HAYDNS LETZTER BESUCHER **Vol. 4** 517, 558, 796
HAZ A SZIKLAK ALATT **Vol. 9** 67, 134, 326
HAZA OLTARA, A **Vol. 9** 172
HAZAFELE **Vol. 9** 274, 315
HAZAI PALYA **Vol. 9** 83, 183, 259
HAZAJARO LELEK **Vol. 9** 71, 274
HAZARDZISCI **Vol. 1** 216, 262 **Vol. 6** *page 222*
HAZASSAG **Vol. 9** 14, 68, 71
HAZASSAG A LIPOTVAROSBAN **Vol. 9** 136
HAZASSAG SZABADNAPPAL **Vol. 9** 152, 227, 259
HAZASSAGBOL ELEGSEGES **Vol. 9** 195, 305, 326
HAZASSAGGAL KEZDÖDIK **Vol. 9** 26, 117
HAZUGSAG NELKÜL **Vol. 9** 71, 97, 114, 115, 237
H.C.ANDERSENS SAGOR **Vol. 5** 1219
HE, DU ! **Vol. 1** 5, 60, 102
HE ! TU M'ENTENDS **Vol. 6** *page 133, page 135*
HECATE **Vol. 6** *page 104*
HECHO VIOLENTO, UN **Vol. 8** 261, 749
HECTOR **Vol. 8** 573, 672, 730
HEDDA GABLER **Vol. 4** 81, 179, 673 **Vol. 7** 1132
HEDELMÄTÖN PUU **Vol. 5** 356, 549
HEGYEK ALJAN **Vol. 9** 220
HEGYEK LEANYA, A **Vol. 9** 35, 196
HEI, RILLUMAREI ! **Vol. 5** 426
HEIDENNOVELLE **Vol. 4** 280
HEIDERÖSLEIN **Vol. 4** 201
HEIDESCHULMEISTER UWE KARSTEN **Vol. 4** 89, 298, 608
HEILIGE BERG, DER **Vol. 4** 10, 73, 397, 620
HEILIGE FLAMME, DIE **Vol. 4** 324
HEILIGE HASS, DER **Vol. 4** 212, 571, 726
HEILIGE LÜGE, DIE **Vol. 4** 247, 339
HEILIGE ODER DIRNE **Vol. 4** 132, 230 **Vol. 6** *page 192*
HEILIGE SCHWEIGEN, DAS **Vol. 4** 25
HEILIGE UND IHR NARR, DER **Vol. 4** 14, 186, 232, 323, 370
HEILIGE ZIEL, DAS **Vol. 4** 15
HEILIGEN DREI BRUNNEN, DIE **Vol. 4** 15, 204, 558, 774
HEIMAT **Vol. 4** 626, 742
HEIMAT AM RHEIN **Vol. 4** 232, 239, 270
HEIMAT UND FREMDE **Vol. 4** 634
HEIMATERDE **Vol. 4** 323, 329
HEIMATLAND **Vol. 4** 525, 736
HEIMKEHR **Vol. 4** 13, 85, 567, 574, 638
HEIMKEHR DES JOACHIM OTT, DIE **Vol. 6** *page 168, page 171, page 174*
HEIMKEHR DES ODYSSEUS, DIE **Vol. 4** 773

HEIMKEHR IN EIN FREMDES LAND Vol. 1 22, 27, 127
HEIMKEHR INS GLÜCK Vol. 4 462, 626
HEIMLICHE EHEN Vol. 1 34, 35, 58
HEIMLICHE GRÄFIN, DIE Vol. 4 614, 774
HEIMLICHE SÜNDER Vol. 4 25, 545
HEIMWEH Vol. 4 82, 306, 340, 466, 724
HEINITZER, DIE Vol. 1 1, 63, 100
HEINRICH AMÜSIERT SICH Vol. 4 97, 230, 239
HEINRICH, WO IST DIE HOSE ? Vol. 6 *page 186*
HEINZ HUSTET Vol. 4 339, 525
HEINZ IM MOND Vol. 4 132, 626
HEIRATEN - ABER WEN ? Vol. 4 515
HEIRATSANNONCEN Vol. 4 367, 373
HEIRATSBÜRO FORTUNA Vol. 4 13, 546
HEIRATSFALLE, DIE Vol. 4 71, 644, 717
HEIRATSFIEBER Vol. 4 204, 384, 391, 449
HEIRATSINSTITUT IDA & CO. Vol. 4 79, 115, 119
HEIRATSNEST, DAS Vol. 4 287, 565, 646
HEIRATSSCHWINDLER Vol. 4 167, 219, 339, 560, 727
HEISSE SPUREN Vol. 1 44, 45, 114
HEISSER SOMMER Vol. 1 19, 34, 40, 115
HEISSES BLUT Vol. 4 82, 124, 329, 370, 646
HEITERES UND ERNSTES UM DEN GROSSEN KÖNIG Vol. 4 361, 558
HEJ RUP ! Vol. 10 167, 300, 611
HEJ STINE ! Vol. 5 113, 284
HEJA ROLAND Vol. 5 1096
HEJRENAES Vol. 5 2, 23, 228
HEKSEN OG CYKLISTEN Vol. 5 97
HEKSENE FRA DEN FORSTENEDE SKOG Vol. 5 749
HEKSENETTER Vol. 5 738, 831
HEKUS LETTEM Vol. 9 214, 316
HELD ALLER MÄDCHENTRÄUME, DER Vol. 4 132, 479, 779
HELD DES TAGES, DER Vol. 4 389, 780
HELD EINER NACHT Vol. 4 666 Vol. 10 256, 524
HELD I UHELD Vol. 5 81
HELD WIDER WILLEN Vol. 4 536
HELDENTUM UND TODESKAMPF UNSERER EMDEN Vol. 4 344
HELDIN VON ST.HONOREE, DIE Vol. 4 124, 247
HELE, ON LETI ! Vol. 10 116, 183, 570
HELENA Vol. 4 118, 509, 535, 571, 726
HELENA, DAS LEBENDE TORPEDO Vol. 4 628
HELENA Y FERNANDA Vol. 8 426, 732
HELENE Vol. 2 8, 153, 158
HELGA FILLE ESCLAVE Vol. 2 393
HELIGA LÖGNEN, DEN Vol. 5 907, 923
HELLE Vol. 2 21, 719
HELLE FOR HELENE Vol. 5 31, 39, 248, 272 Vol. 6 *page 31, page 39, page 248, page 272*
HELLE FOR LYKKE Vol. 5 199, 201, 275
HELLES MÄDCHEN, EIN Vol. 4 342
HELLO AND GOOD-BYE Vol. 2 224
HELLYYS Vol. 5 606
HELMIKUUN MANIFEST Vol. 5 349, 481, 485
HELSINGIN KUULUISIN LIIKEMIES Vol. 5 576, 645
HELT FRA 64, EN Vol. 5 121
HELTEMOD FORSONER Vol. 5 97
HELTEN FRA OSTAFRIKA Vol. 5 81
HELUNAN HÄÄMATKA Vol. 5 549, 579
HELVEDES DATTER Vol. 5 97
HELVEDESMASKINEN Vol. 5 295
HELYET AZ ÖREGEKNEK Vol. 9 71, 322
HEM FRAN BABYLON Vol. 5 843, 923
HEMAT I NATTEN Vol. 5 327, 390, 851
HEMBRA Vol. 8 627, 749
HEMD DES GLÜCKS - EIN KURZWEILIG SPIEL VON EINEM, DER AUSZOG, DEN GLÜCKLICHEN MENSCHEN ZU SUCHEN, DAS Vol. 4 110, 517
HEMLIGA SVENSSON Vol. 5 1205
HEMLIGHETEN Vol. 5 1039
HEMMELIGHEDSFULDE X, DET Vol. 5 54, 65
HEMMELIGHETSFULLE LEILIGHETEN, DEN Vol. 5 676, 729
HEMPAS BAR Vol. 5 955, 1002, 1096, 1190
HEMSLAVINNOR Vol. 5 71, 928, 1210, 1215
HEMSÖBORNA Vol. 5 863, 925, 976, 1032, 1042, 1047, 1087, 1100, 1217
HEMTREVNAD I KASERN Vol. 5 864, 923
HENDES FORTID Vol. 5 9, 197, 228
HENDES HELT Vol. 5 183
HENDES MANDS FORLOVEDE Vol. 5 83
HENDES MODERS LOFTE Vol. 5 58
HENDES STILLE SVAERMERI Vol. 5 295
HENDES STORE AFTEN Vol. 5 14, 146, 180, 199
HENDES UNGDOMSFORELSKELSE Vol. 5 295
HENKER, FRAUEN UND SOLDATEN Vol. 4 344, 647, 689
HENKER VON SANKT MARIEN, DER Vol. 4 307, 325, 700, 720
HENNES LILLA MAJESTÄT Vol. 5 891, 928, 1031, 1156
HENNES MEGET KONGELIGE HOYHET Vol. 5 712, 718, 754, 799
HENNES MELODI Vol. 5 863, 1032, 1042
HENRETTELSEN Vol. 5 72
HENRIETTA Vol. 5 858, 1051, 1163

HENRIETTE JACOBY **Vol. 4** 159
HENRYS BAKVAERELSE **Vol. 5** 728, 762, 780, 799
HER STORMS FORSTE MONOKEL **Vol. 5** 97
HERBERG FOR HJEMLOSE **Vol. 5** 81
HERBSTMANÖVER **Vol. 4** 79, 115, 239, 270, 323, 644, 743
HERBSTZEIT AM RHEIN **Vol. 4** 287, 704
HERCULE **Vol. 2** 8, 427, 775
HERCULES **Vol. 2** 678 **Vol. 7** 1086
HERCULES CONTRA LOS HIJOS DEL SOL **Vol. 8** 598, 772
HEREDERO EN APUROS **Vol. 8** 390
HEREJE, EL **Vol. 8** 311, 611
HERENCIA EN PARIS, UNA **Vol. 8** 611
HERIA **Vol. 3** 186
HERIDA LUMINOSA, LA **Vol. 2** 775 **Vol. 8** 634, 717
HERITIER DU BAL TABARIN, L' **Vol. 2** 197, 381, 387
HERITIER, L' **Vol. 2** 588, 655
HERITIERS, LES **Vol. 2** 309
HERKULES **Vol. 4** 110
HERKULES MAIER **Vol. 4** 85, 208, 638
HERKULESFÜRDÖI EMLEK **Vol. 9** 67, 152, 253
HERMANA ALEGRIA, LA **Vol. 8** 261, 616, 630
HERMANA SAN SULPICIO, LA **Vol. 8** 261, 270, 298, 449, 606
HERMANN **Vol. 7** 295, 899
HERMÄNNCHEN, DAS **Vol. 4** 132
HERMANNS ERZÄHLUNGEN **Vol. 4** 565, 667
HERMANO BASTARDO DE DIOS, EL **Vol. 8** 372, 765
HERMANO MENOR **Vol. 8** 214, 630, 719
HERMELINMANTEL, DER **Vol. 4** 124
HERMINE UND DIE SIEBEN AUFRECHTEN **Vol. 4** 124, 422, 742
HERMIONE **Vol. 7** 655, 1113
HERODIADE **Vol. 2** 6
HEROE A LA FUERZA **Vol. 2** 646 **Vol. 8** 607
HEROE DE CASCORRO, EL **Vol. 8** 516, 542
HEROES A LA FUERZA **Vol. 8** 583
HEROES DE BLANCO **Vol. 8** 583
HEROES DE LA LEGION, LOS **Vol. 8** 606
HEROES DEL AIRE **Vol. 8** 495, 603, 752
HEROES DEL BARRIO, LOS **Vol. 8** 263
HEROES DEL PATIBULO, LOS **Vol. 8** 552
HEROES MILLONARIOS, LOS **Vol. 8** 306
HEROIN **Vol. 1** 13, 26, 68
HEROINE DU TRIANGLE D'OR, L' **Vol. 2** 301
HEROINES DU MAL, LES **Vol. 2** 215, 642
HEROIQUE MONSIEUR BONIFACE, L' **Vol. 2** 197, 309
HEROIS DO MAR **Vol. 8** 96, 106, 131, 167
HEROISME DE PADDY, L' **Vol. 2** 153
HEROS DE LA MARNE, LES **Vol. 2** 152, 273
HEROS DEL 95 **Vol. 8** 359, 611
HEROS N'ONT PAS FROID AUX OREILLES, LES **Vol. 2** 292, 628
HEROS SONT FATIGUES, LES **Vol. 2** 13, 615, 709
HERR ARNES PENGAR **Vol. 5** 912, 997
HERR ARNES PENNINGAR **Vol. 5** 909, 954, 1052
HERR AUF BESTELLUNG, DER **Vol. 4** 208, 479
HERR AUS DEM ZUCHTHAUS, DER **Vol. 4** 124, 192, 213
HERR BÜROVORSTEHER, DER **Vol. 4** 527, 584
HERR COLLINS ÄVENTYR **Vol. 5** 841, 923
HERR DER BESTIEN, DER **Vol. 4** 288, 626, 779
HERR DER LIEBE, DER **Vol. 4** 329
HERR DER NACHT, DER **Vol. 4** 287, 291, 565
HERR DER UNTERWELT, DER **Vol. 4** 747
HERR DER WELT, DER **Vol. 4** 119, 124, 135, 232, 270, 503 **Vol. 6** *page 194*
HERR DES TODES **Vol. 4** 190, 479, 700
HERR DES TODES, DER **Vol. 4** 124, 247
HERR FINANZDIREKTOR, DER **Vol. 4** 379, 734
HERR GENERALDIREKTOR, DER **Vol. 4** 190, 479, 733
HERR GRAF, DER **Vol. 4** 514
HERR HUSASSISTENTEN **Vol. 5** 840, 841
HERR IM HAUS, DER **Vol. 4** 18, 135, 517, 536, 558, 615, 660
HERR IWAN **Vol. 1** 49, 114
HERR KOBIN GEHT AUF ABENTEUER **Vol. 4** 82, 298
HERR MAHLER IN 1000 NÖTEN **Vol. 4** 361, 558
HERR MEISTER UND FRAU MEISTERIN **Vol. 4** 223, 373, 558
HERR ODER DIENER **Vol. 4** 361, 558
HERR OHNE WOHNUNG, DER **Vol. 4** 87, 115, 603
HERR PAPA, DER **Vol. 4** 18, 367
HERR PETERSENS DEBUT **Vol. 5** 256
HERR PUNTILA OCH HANS DRÄNG MATTI **Vol. 5** 434, 499, 613, 622, 1011, 1057, 1169, 1179
HERR SANDERS LEBT GEFÄHRLICH **Vol. 4** 298, 331, 727, 729
HERR SENATOR, DER **Vol. 4** 135, 461, 536

HERR ÜBER LEBEN UND TOD Vol. 4 476, 522
HERR VOM FINANZAMT, DER Vol. 4 221, 340
HERRA HUU - JESTAPA JEPULIS, PENIKAT SIPULIKS Vol. 5 377
HERRA JA YLHÄISYYS Vol. 5 552, 579
HERRA JOHTAJAN „HARHA - ASKEL" Vol. 5 346, 363
HERRA LAHTINEN LÄHTEE LIPETTIIN Vol. 5 363, 625
HERRA PUNTILA JA HÄNEN RENKINSÄ MATTI Vol. 5 434, 499, 613, 622
HERRA SOTAMINISTERI Vol. 5 443, 643
HERRAT OVAT HERKÄUSKOISIA Vol. 5 334, 417
HERRAT TÄYSIHOIDOSSA Vol. 5 359, 485
HERRE I KJOLE OG HVIDT, EN Vol. 5 55, 196
HERRE MED BART, EN Vol. 5 676, 683, 697, 723, 800
HERRE MED PORTFÖLJ Vol. 5 841, 923
HERREGAARDSMYSTERIET Vol. 5 228
HERREN OG HANS TJENERE Vol. 5 672, 724
HERREN VOM MAXIM, DIE Vol. 4 132, 338, 340, 546
HERRENPARTIE Vol. 3 347, 371, 374
HERRGOTT AM WEGE, DER Vol. 4 652
HERRGOTTSCHNITZER VON OBERAMMERGAU, DER Vol. 4 25, 571
HERRIN DER WELT, DIE Vol. 4 89, 298, 307, 325, 727
HERRIN UND IHR KNECHT, DIE Vol. 4 49, 626
HERRIN VON ATLANTIS, DIE Vol. 4 349, 452, 629
HERRIN VON MONBIJOU, DIE Vol. 4 219
HERRN FILIP COLLINS ABENTEUER Vol. 4 162, 359
HERROJEN EEVA Vol. 5 392
HERRSCHER, DER Vol. 4 13, 89, 266
HERZ DER KÖNIGIN, DAS Vol. 4 235, 531, 742
HERZ DER LILIAN THORLAND, DAS Vol. 4 93, 125, 291, 495, 575, 626, 674
HERZ DES CASANOVA, DAS Vol. 4 113, 780
HERZ EINER GATTIN, DAS Vol. 4 247
HERZ GEHT VOR ANKER Vol. 4 30, 135
HERZ IST TRUMPF Vol. 4 108, 115, 774
HERZ MODERN MÖBLIERT Vol. 4 373, 424, 788
HERZ MUSS SCHWEIGEN, DAS Vol. 4 13, 483
HERZ OHNE HEIMAT Vol. 4 79, 115, 525
HERZ SCHLÄGT FÜR DICH, EIN Vol. 4 135, 619, 647
HERZ VOM HOCHLAND, DAS Vol. 4 288
HERZEN OHNE ZIEL Vol. 4 344, 544
HERZENKNICKER Vol. 4 135
HERZENSDIEB Vol. 4 634
HERZENSFREUD - HERZENSLEID Vol. 4 247, 517, 558
HERZENSPHOTOGRAPH, DER Vol. 4 14, 186, 605
HERZOG FERRANTES ENDE Vol. 4 178, 546
HERZOG VON ALERIA, DER Vol. 4 71, 424
HERZOG VON REICHSTADT, DER Vol. 4 527
HERZTRUMPF Vol. 4 124, 247
HESTETYVEN Vol. 5 97
HET ÄR MIN LÄNGTAN Vol. 5 840, 1033
HET SNÖ Vol. 5 835, 919
HET TONNA DOLLAR Vol. 9 134, 138, 312
HETASTE LIGGEN Vol. 5 866
HETE SEPTEMBERDAGER Vol. 5 738, 826
HETEN MINT A GONOSZOK Vol. 9 274, 315
HETENKENT EGYSZER LATHATOM Vol. 9 319, 322
HETKIA YÖSSÄ Vol. 5 474, 584
HETS Vol. 5 840, 878
HETSZILVAFA Vol. 9 26, 79, 132
HEURE EXQUISE, L' Vol. 6 *page 122*
HEURES CHAUDES Vol. 2 703
HEUREUSE AVENTURE, L' Vol. 2 395
HEUREUX QUI, COMME ULYSSE... Vol. 2 294, 532
HEUT' KOMMT'S DRAUF AN Vol. 4 339, 462
HEUT' SPIELT DER STRAUSS Vol. 4 124, 527
HEUT' TANZT MARIETT Vol. 4 14, 354, 431, 788
HEUT WAR ICH BEI DER FRIEDA Vol. 4 124, 221
... HEUTE ABEND BEI MIR Vol. 4 115, 774
HEUTE ABEND IM ASTORIA Vol. 4 18
HEUTE NACHT - EVENTUELL Vol. 4 246, 527
HEVOSHUIJARI Vol. 5 349, 481, 493
HEXE, DIE Vol. 4 232, 673
HEXE VON LOLARUH, DIE Vol. 4 767
HEXEN Vol. 1 22, 34, 105
HEXEN VON SALEM, DIE Vol. 1 26, 35 Vol. 2 719, 801
HEXENGOLD Vol. 4 232, 673
HEXENLIED, DAS Vol. 4 565
HEXENZAUBER Vol. 4 187
HEXER, DER Vol. 2 397

HIBERNATUS **Vol. 2** 368, 457
HIC SUNT LEONES **Vol. 7** 1201
HIDEG NAPOK **Vol. 9** 183, 285, 332
HIER BIN ICH, MEIN VATER **Vol. 3** 395
HIER IRRT SCHILLER **Vol. 4** 115, 354
HIERRO DULCE **Vol. 8** 376, 672
HIEV UP **Vol. 1** 17, 60, 128
HIGGYETEK NEKEM ! **Vol. 9** 18, 50, 51
HIHAS DEL CID, LAS **Vol. 8** 523
HIIPIVÄ VAARA **Vol. 5** 472, 485, 510
HIJA DE JUAN SIMON, LA **Vol. 8** 239, 270, 517, 749
HIJA DEL CORREGIDOR, LA **Vol. 8** 572, 657
HIJA DEL MAR, LA **Vol. 8** 396, 732
HIJA DEL PENAL, LA **Vol. 8** 522, 751
HIJAS DE ELENA, LAS **Vol. 8** 450, 602
HIJAS DEL CID, LAS **Vol. 8** 732
HIJO DE JESSE JAMES, EL **Vol. 8** 311
HIJO DE LA NOCHE, EL **Vol. 8** 515, 721
HIJO DEL CAPITAN BLOOD, EL **Vol. 8** 214, 762
HIJO DEL CURA, EL **Vol. 8** 266, 762
HIJO DEL PISTOLERO, EL **Vol. 8** 276, 558
HIJO ES MIO, EL **Vol. 8** 681
HIJOS DE LA NOCHE, LOS **Vol. 8** 716
HIJOS DE PAPA **Vol. 8** 207, 343
HIJOS DE POBRES PERO DESHONESTOS PADRES...LES LLAMABAM CALAMIDAD **Vol. 8** 317, 505
HIJOS DE SCARAMOUCHE, LOS **Vol. 8** 641
HIJOS DEL TRABAJO, LOS **Vol. 8** 281
HILDE PETERSEN POSTLAGERND **Vol. 4** 97, 132, 230
HILDE UND DIE 4 PS **Vol. 4** 110, 135, 232, 323
HILDE WARREN UND DER TOD **Vol. 4** 113
HILFE, ERPRESSER ! **Vol. 4** 352, 700
HILFE, ICH ERBE **Vol. 4** 25
HILFE ! ÜBERFALL ! **Vol. 4** 132
HILJA, MAITOTYTTÖ **Vol. 5** 443, 549
HILMANPÄIVÄT **Vol. 5** 443, 549
HILMAR THATE SINGT DESSAU, EISLER, HOSALLA **Vol. 1** 84
HIMLASPELET **Vol. 5** 840, 956, 1114
HIMLEN ER BLA **Vol. 5** 23
HIMLENS STRAF **Vol. 5** 203
HIMMEL AUF ERDEN, DER **Vol. 4** 752, 796
HIMMEL OCH PANNKAKA **Vol. 5** 878, 1052, 1177
HIMMEL OG HAV **Vol. 5** 674, 799
HIMMEL OG HELVETE **Vol. 5** 682, 800, 814
HIMMEL, WIR ERBEN EIN SCHLOSS **Vol. 4** 2, 243
HIMMELBLAUE ABENDKLEID, DAS **Vol. 4** 65, 340, 536
HIMMELHUNDE **Vol. 4** 24, 354
HIMMELSKIBET **Vol. 5** 136, 183
HIMMELURET **Vol. 5** 760, 788
HIN OCH SMALÄNNINGEN **Vol. 5** 924, 966, 1052, 1080
HINCHA, EL **Vol. 8** 583, 772
HINGABE **Vol. 6** *page 196, page 198, page 199*
HINTER KLOSTERMAUERN **Vol. 4** 344, 544
HINTERTREPPE **Vol. 4** 211, 247, 396
HINTONJARO SZERELEM **Vol. 9** 4, 11, 199
HIPNOSIS **Vol. 8** 438, 719
HIRE, VALET DE COEUR, LA **Vol. 2** 169, 330, 613
HIRONDELLE ET LA MESANGE, L' **Vol. 2** 153
HIROSHIMA MON AMOUR **Vol. 2** 664, 854
HIRT VON MARIA SCHNEE **Vol. 4** 211, 247, 396
HIRTTÄMÄTTÖMÄT **Vol. 5** 563, 606
HIRURZI **Vol. 3** 101, 116 **Vol. 6** *page 53*
HISTOIRE D'ADELE H., L' **Vol. 2** 17, 434
HISTOIRE D'ADRIEN **Vol. 6** *page 116, page 117, page 122*
HISTOIRE D'AMOUR, UNE **Vol. 2** 185, 644, 646
HISTOIRE DE CHANTER **Vol. 2** 615
HISTOIRE DE PAUL **Vol. 2** 45
HISTOIRE DE RIRE **Vol. 2** 388, 402
HISTOIRE DE WAHARI **Vol. 2** 105
HISTOIRE D'EAU, UNE **Vol. 2** 465
HISTOIRE D'O **Vol. 2** 315, 682
HISTOIRE D'O, CHAPITRE II, L' **Vol. 8** 274
HISTOIRE D'UN PIERROT **Vol. 7** 965
HISTOIRE D'UN POISSON ROUGE, L' **Vol. 2** 781
HISTOIRE ENTRE MILLE, UNE **Vol. 2** 708
HISTOIRE IMMORTELLE, UNE **Vol. 2** 441, 675
HISTOIRE SIMPLE, UNE **Vol. 2** 515
HISTOIRE TRES BONNE ET TRES JOYEUSE DE COLINOT TROUSSE-CHEMISE, L' **Vol. 2** 190, 674
HISTOIRES D'A **Vol. 2** 754
HISTOIRES D'UN PEUPLE **Vol. 2** 405
HISTOIRES EXTRAORDINAIRES **Vol. 2** 21, 615, 644, 719, 776
HISTORIA DE AMOR, UNA **Vol. 8** 492, 772
HISTORIA DE BIENVENIDO, LA **Vol. 8** 214, 586
HISTORIA DE O, II PARTE **Vol. 8** 274
HISTORIA DE UN DURO, LA **Vol. 8** 604
HISTORIA DE UN TORERO, LA **Vol. 8** 281

HISTORIA DE UNA CHICA SOLA **Vol. 8** 232
HISTORIA DE UNA ESCALERA **Vol. 8** 294, 677
HISTORIA DE UNA NOCHE **Vol. 8** 298, 546
HISTORIA DE UNA PASION **Vol. 8** 396
HISTORIA DE UNA TRAICION **Vol. 8** 261, 282
HISTORIA EN DOS ALDEAS **Vol. 8** 359, 611
HISTORIA, JAKICH WIELE **Vol. 1** 174, 319
HISTORIA JEDNEGO MYSLIWCA **Vol. 1** 149, 321 **Vol. 6** *page 218*
HISTORIA PERVERSA, UNA **Vol. 8** 306, 762
HISTORIA WSPOLCZESNA **Vol. 1** 198, 303, 381
HISTORIA ZOLTEJ CIZEMKI **Vol. 1** 180, 211, 267, 320, 321, 377
HISTORIAS DE LA FERIA **Vol. 8** 207, 435
HISTORIAS DE LA RADIO **Vol. 8** 261, 438
HISTORIAS DE LA TELEVISION **Vol. 8** 524, 725
HISTORIAS DE MADRID **Vol. 8** 600
HISTORIAS DE MUJERES **Vol. 8** 300, 738
HISTORIAS SELVAGENS **Vol. 8** 7
HISTORIEN OM BARBARA **Vol. 5** 16, 37, 61, 195, 232
HISTORIEN OM EN GUT **Vol. 5** 798
HISTORIEN OM EN MODER **Vol. 5** 9
HISTORIEN OM HJORTHOLM **Vol. 5** 180, 199, 252
HITLER, CONNAIS PAS ! **Vol. 2** 668
HITLER IZ NASEG SOKAKA **Vol. 3** 349, 374, 441
HITLERJUNGE QUEX **Vol. 4** 17, 232, 713
HIT"R PET"R **Vol. 3** 66, 87 **Vol. 6** *page 46*
HITTEBARNET **Vol. 5** 58
HJAELPEN **Vol. 5** 295
HJÄLPSAMA HERRN **Vol. 5** 840, 1158
HJÄLTAR MOT SIN VILJA **Vol. 5** 2, 229, 1082
HJÄRTATS TRIUMF **Vol. 5** 891, 997, 1031
HJÄRTER KNEKT **Vol. 5** 871, 1042, 1052
HJÄTAR I GULT OCH BLATT **Vol. 5** 928, 1156
HJELP, VI FAR LEILIGHET **Vol. 5** 674, 698
HJEM GAR VI IKKE **Vol. 5** 738
HJERTE AF GULD, ET **Vol. 5** 9
HJERTEBETVINGEREN **Vol. 5** 83
HJERTEFEJLEN **Vol. 5** 81 **Vol. 6** *page 81*
HJERTEKNUSEREN **Vol. 5** 295
HJERTENES KAMP **Vol. 5** 9
HJERTENKRIGEN PAA RAVNSHOLT **Vol. 5** 295
HJERTER ER TRUMF **Vol. 5** 125, 261, 283
HJERTERKONGE **Vol. 5** 83
HJERTESTORME **Vol. 5** 256
HJERTETYVEN **Vol. 5** 83
HJORTENS FLUGT **Vol. 5** 97
HLADACI SVETLA **Vol. 10** 210, 273
HLAVNI VYHRA **Vol. 10** 359, 380, 433
HLEDA SE PAN TAU **Vol. 10** 173, 184, 564
HLEDA SE TATA **Vol. 10** 211, 235, 385
HLEDAM DUM HOLUBI **Vol. 10** 180, 299, 515
HLIDAC **Vol. 10** 369, 505
HLIDAC C. 47 **Vol. 10** 92, 122
HLJAB **Vol. 3** 75, 119
HNEV **Vol. 10** 369, 515, 550
HO ! **Vol. 2** 110, 770
HO AMATO UNA DIVA **Vol. 7** 587
HO FATTO SPLASH **Vol. 7** 38, 103
HO PERDUTO LA CHIAVE **Vol. 7** 1189
HO PERDUTO MIO MARITO ! **Vol. 7** 47, 902, 1116
HO SCELT L'AMORE **Vol. 7** 16, 792
HO SOGNATO IL PARADISO **Vol. 7** 1054
HO TANTA VOGLIA DI CANTARE **Vol. 2** 807
HO VISTO BRILLARE LE STELLE **Vol. 7** 687, 962, 978
HOA BINH **Vol. 2** 519
HOCH KLINGT DAS LIED VOM U-BOOT-MANN **Vol. 4** 192
HOCHBAHNKATASTROPHE, DIE **Vol. 4** 598
HOCHMUT KOMMT VOR DEM KNALL **Vol. 1** 34, 35, 43
HOCHSTAPLER **Vol. 4** 548, 646
HOCHSTAPLER-TRIO, DAS **Vol. 4** 135
HOCHSTAPLER WIDER WILLEN **Vol. 4** 119, 571, 652, 726
HOCHSTAPLERIN, DIE **Vol. 4** 124, 149, 292, 436, 640, 660
HOCHTOURIST, DER **Vol. 4** 135, 270, 344, 409, 615, 713, 751
HOCHVERRAT **Vol. 4** 270, 743 **Vol. 6** *page 190*
HOCHZEIT AM 13. **Vol. 4** 115, 774
HOCHZEIT AM WOLFGANGSEE **Vol. 4** 373, 734
HOCHZEIT AUF BÄRENHOF **Vol. 4** 13, 235
HOCHZEIT IM EXCENTRICCLUB, DIE **Vol. 6** *page 189*
HOCHZEIT IN WELTZOW **Vol. 1** 15, 79, 85
HOCHZEIT MIT HINDERNISSEN **Vol. 4** 79, 115, 564
HOCHZEIT VON LÄNNEKEN, DIE **Vol. 1** 44, 60, 101
HOCHZEITSFACKEL, DIE **Vol. 4** 87
HOCHZEITSHOTEL, DAS **Vol. 4** 184, 713

HOCHZEITSMORGEN, EIN **Vol. 4** 604, 652
HOCHZEITSNACHT **Vol. 4** 18, 188, 354, 558, 608
HOCHZEITSNACHT IM REGEN **Vol. 1** 33, 34, 48
HOCHZEITSREISE, DIE **Vol. 4** 13, 24, 574
HOCHZEITSREISE ZU DRITT **Vol. 4** 292, 517, 558
HOCHZEITSTRAUM, EIN **Vol. 4** 49, 234, 734
HOCKEY FEBER **Vol. 5** 684, 746, 767
HOCU ZIVIJETI **Vol. 6** *page 321, page 322*
HODIME SE K SOBE, MILACKU ...? **Vol. 10** 139, 593
HODINA MODRYCH SLONU **Vol. 10** 250, 263
HODINA PRAVDY **Vol. 10** 144, 516, 595
HODINA ZIVOTA **Vol. 10** 254, 313, 327
HODINAROVA SVATEBNI CESTA KORALOVYM MOREM **Vol. 10** 87, 144, 428
HODINY **Vol. 10** 82, 228, 459
HODY **Vol. 10** 73, 83, 459
HOFFMANNS ERZÄHLUNGEN **Vol. 1** 68, 74, 100 **Vol. 4** 366, 488
HOFFNUNGSLOSER FALL, EIN **Vol. 4** 729, 738
HOFINTRIGE, EN **Vol. 5** 9
HOFKONZERT, DAS **Vol. 4** 437, 738, 742
HOG OVER HOG **Vol. 5** 92
HOGO FOGO HOMOLKA **Vol. 10** 70, 402
HÖGRE ÄNDEMAL **Vol. 5** 891, 976, 1110
HÖGSTA VINSTEN **Vol. 5** 1112
HOGUERAS EN LA NOCHE **Vol. 8** 652
HOGY ALLUNK FIATALEMBER ? **Vol. 9** 183, 259, 285
HOGY SZALADNAK A FAK ... **Vol. 9** 183, 285, 316
HOGYAN FELEJTSÜK EL ELETÜNK LEGNAGYOBB SZERELMET...? **Vol. 9** 6, 152, 229
HOHE LIED DER LIEBE, DAS **Vol. 4** 729
HOHEIT TANZT WALZER **Vol. 4** 292
HOHELIED DER KRAFT, DAS **Vol. 4** 224, 329, 346, 508, 682
HÖHENFIEBER **Vol. 4** 71, 329, 654
HÖHERE BEFEHL, DER **Vol. 4** 28, 149, 660
HÖHERE TÖCHTER **Vol. 4** 84, 329
HÖJDHOPPAR'N **Vol. 5** 915, 1093, 1160, 1201
HOJT PAA EN KVIST **Vol. 5** 24, 92
HOJT SPIL **Vol. 5** 9
HOKUSPOKUS **Vol. 4** 266, 288, 574
HOL ALSZUNK VASARNAP ? **Vol. 9** 71
HOLA MUCHACHO **Vol. 8** 470, 477, 495
HOLA ... SENOR DIOS **Vol. 8** 405, 629
HOLD DA HELT FERIE **Vol. 5** 179, 281
HOLD FINGRENE FRA MOR **Vol. 5** 2, 228
HOLD-UP EN MUSIQUE **Vol. 2** 294, 401, 514
HOLDUDVAR **Vol. 9** 20, 154, 302
HÖLGY EGY KISSE BOGARAS, A **Vol. 9** 28, 34, 315, 322
HÖLGYEK ELÖNYBEN **Vol. 9** 132, 164
HOLKA NA ZABITI **Vol. 10** 18, 144, 359
HOLKA NEBO KLUK **Vol. 10** 256, 460
HOLKY Z PORCELANU **Vol. 10** 18, 144, 359
HÖLLE DER JUNGFRAUEN, DIE **Vol. 4** 270, 644, 725
HÖLLE DER LIEBE **Vol. 4** 230, 366
HÖLLE VON MANITOBA, DIE **Vol. 8** 493
HÖLLE VON MONTMARTRE, DIE **Vol. 4** 344, 544
HÖLLENMASCHINE, DIE **Vol. 4** 503
HÖLLENREITER, DER **Vol. 4** 277, 598
HÖLLENTEMPO **Vol. 4** 82, 239, 323, 558
HÖLLISCHE MACHT, DIE **Vol. 4** 36, 772
HOLNAP KEZDÖDIK AZ ELET **Vol. 9** 172
HOLNAP LESC FACAN **Vol. 9** 140, 263, 318
HOLOCAUST 2000 **Vol. 7** 763
HOLT VIDEK **Vol. 9** 152, 259, 339
HOLT VOLT, HOL NEM VOLT **Vol. 9** 82, 253, 259
HOLTAK VISSZAJARNAK, A **Vol. 9** 16, 231, 305
HOLUBICE **Vol. 10** 41, 87, 425
HOLZAPFEL WEISS ALLES **Vol. 4** 192, 462, 584
HOLZAUKTION, DIE **Vol. 4** 329, 365
HÖLZERNE KÄLBCHEN, DAS **Vol. 1** 22, 26, 82
HOMBRE ACOSADO **Vol. 8** 311, 583
HOMBRE COMO LOS DEMAS, UN **Vol. 8** 524, 616
HOMBRE DE CARACAS, EL **Vol. 8** 280, 450
HOMBRE DE LA DILIGENCIA, EL **Vol. 8** 583, 772
HOMBRE DE LA ISLA, EL **Vol. 8** 214, 611
HOMBRE DE LA LEGION, EL **Vol. 7** 359
HOMBRE DE LOS MUNECOS, EL **Vol. 8** 396
HOMBRE DE MARRAKECH, EL **Vol. 8** 749
HOMBRE DE MODA, EL **Vol. 8** 459
HOMBRE DE MUNDO, EL **Vol. 8** 294, 677
HOMBRE DE NEGOCIOS, UN **Vol. 8** 214, 357, 436
HOMBRE DEL GOLPE PERFECTO, EL **Vol. 8** 493
HOMBRE DEL PUNO DE ORO, EL **Vol. 8** 732
HOMBRE DEL RIO MALO, EL **Vol. 8** 558, 762
HOMBRE DEL SUR, EL **Vol. 8** 603

HOMBRE DEL VALLE MALIDITO, EL **Vol. 8** 400
HOMBRE LLAMADO NOON, UN **Vol. 8** 303
HOMBRE OCULTO, EL **Vol. 8** 739
HOMBRE QUE LAS ENAMORARA, EL **Vol. 8** 396
HOMBRE QUE MATO A BILLY EL NINO, EL **Vol. 8** 407, 551
HOMBRE QUE PERDIO EL TREN, EL **Vol. 8** 495, 509, 600
HOMBRE QUE SE QUISO MATAR, EL **Vol. 8** 207, 437
HOMBRE QUE SE REIA DEL AMOR, EL **Vol. 8** 285, 516, 648
HOMBRE QUE VEIA LA MUERTE, EL **Vol. 8** 390, 436
HOMBRE QUE VIAJABA DESPACITO, EL **Vol. 8** 600
HOMBRE QUE VINO DE UMMO, EL **Vol. 8** 331, 600
HOMBRE QUE VINO DEL ODIO, EL **Vol. 8** 396
HOMBRE SOLO, UN **Vol. 8** 298, 524
HOMBRE ... UNA CIUDAD, UN **Vol. 8** 539, 690
HOMBRE VA POR EL CAMINO, UN **Vol. 8** 276, 298
HOMBRE VINO A MATAR, UN **Vol. 8** 598
HOMBRE Y UN COLT, UN **Vol. 8** 396, 775
HOMBRES CONTRA HOMBRES **Vol. 8** 648
HOMBRES LAS PREFIEREN VIUDAS, LOS **Vol. 8** 306, 461
HOMBRES QUE RUGEN **Vol. 8** 471
HOMBRES SIN HONOR **Vol. 8** 640
HOMEN AS DIREITAS, UM **Vol. 8** 37, 44, 172
HOMEN DO DIA, O **Vol. 8** 130, 171
HOMEN DO RIBATEJO, UM **Vol. 8** 44, 67, 122
HOMICIDIO AL LIMITE DE LA LEY **Vol. 8** 611, 770
HOMICIDIOS EN CHICAGO **Vol. 8** 396
HOMME A ABATTRE, L' **Vol. 2** 197, 339, 655
HOMME A ABATTRE, UN **Vol. 8** 732
HOMME A FEMMES, L' **Vol. 2** 137, 552, 835
HOMME A LA BUICK, L' **Vol. 2** 347, 835
HOMME A L'HISPANO **Vol. 2** 62, 466, 824, 863
HOMME A L'IMPERMEABLE, L' **Vol. 2** 185, 401, 402
HOMME A L'OREILLE CASSEE, L' **Vol. 2** 99, 153
HOMME A MA TAILLE, UN **Vol. 6** *page 80, page 88, page 139*
HOMME ATLANTIQUE, L' **Vol. 6** *page 89, page 127*
HOMME AU CERVEAU GREFFE, L' **Vol. 2** 74, 674
HOMME AU CHAPEAU ROND, L' **Vol. 2** 833 ,863
HOMME AUX CLEFS D'OR, L' **Vol. 2** 47, 279, 386
HOMME AUX MAINS D'ARGILE, L' **Vol. 2** 254, 309
HOMME AUX TROIS MASQUES, L' **Vol. 2** 286
HOMME BLESSE, L' **Vol. 6** *page 91, page 104*
HOMME DE BALEARES, L' **Vol. 2** 48, 346
HOMME DE DESIR, L' **Vol. 2** 124
HOMME DE FEU, L' **Vol. 2** 114, 259
HOMME DE JOIE, L' **Vol. 2** 332, 427
HOMME DE LA JAMAIQUE, L' **Vol. 2** 270, 419
HOMME DE LA NUIT, L' **Vol. 2** 475, 529
HOMME DE L'INTERPOL, L' **Vol. 2** 800
HOMME DE LONDRES, L' **Vol. 2** 676
HOMME DE MA VIE, L' **Vol. 2** 629
HOMME DE MARRAKECH, L' **Vol. 8** 749
HOMME DE MYKONOS, L' **Vol. 2** 139, 570
HOMME DE NULLE PART, L' **Vol. 2** 47, 623
HOMME DE RIO, L' **Vol. 2** 602, 781
HOMME DE TROP, UN **Vol. 2** 835
HOMME DU BLED, L' **Vol. 2** 322, 501
HOMME DU FLEUVE, L' **Vol. 2** 600
HOMME DU JOUR, L' **Vol. 2** 402, 439
HOMME DU LARGE, L' **Vol. 2** 44, 330, 528
HOMME DU NIGER, L' **Vol. 2** 16, 153, 332, 832
HOMME DU TRAIN 117, L' **Vol. 2** 838
HOMME EN COLERE, L' **Vol. 2** 110
HOMME EN OR, UN **Vol. 2** 153
HOMME EST MORT, UN **Vol. 2** 457
HOMME ET L'ENFANT, L' **Vol. 2** 16, 395
HOMME ET UNE FEMME, UN **Vol. 2** 196, 532, 689
HOMME FRAGILE, L' **Vol. 2** 600 **Vol. 6** *page 89, page 137*
HOMME HEUREUX, UN **Vol. 2** 451, 688
HOMME INUSABLE, L' **Vol. 2** 42, 387, 825
HOMME LIBRE, UN **Vol. 2** 315
HOMME MARCHE DANS LA VILLE, UN **Vol. 2** 194, 395 **Vol. 7** 580
HOMME MASQUE, L' **Vol. 2** 167
HOMME NOIR, L' **Vol. 2** 50
HOMME ORCHESTRE, L' **Vol. 2** 598, 700
HOMME PRESSE, L' **Vol. 2** 175, 400
HOMME QUI AIMAIT LES FEMMES, L' **Vol. 2** 17, 434

HOMME QUI ASSASSINA, L' **Vol. 2** 206, 237
HOMME QUI CHERCHE LA VERITE, L' **Vol. 2** 32, 412, 688
HOMME QUI DORT, UN **Vol. 2** 879
HOMME QUI JOUE AVEC LE FEU, L' **Vol. 2** 330, 407
HOMME QUI ME PLAIT, UN **Vol. 2** 196
HOMME QUI NE SAIT PAS DIRE NON, L' **Vol. 2** 206
HOMME QUI PLEURE, L' **Vol. 2** 792, 864
HOMME QUI REVIENT DE LOIN, L' **Vol. 2** 595
HOMME QUI TRAHIT LA MAFFIA, L' **Vol. 2** 570, 689
HOMME QUI VALAIT DES MILLIARDS, L' **Vol. 2** 503, 791
HOMME QUI VENDIT SON AME, L' **Vol. 2** 124, 558
HOMME QUI VIENT DE LA NUIT, L' **Vol. 2** 738
HOMME SANS COEUR, L' **Vol. 2** 425, 698
HOMME SANS NOM, UN **Vol. 2** 399, 595, 698
HOMME SANS VISAGE, L' **Vol. 2** 169, 330, 613
HOMME SE PENCHE SUR SON PASSE, UN **Vol. 2** 354, 740 **Vol. 7** 982
HOMME TRAQUE, L' **Vol. 2** 475
HOMMES DE JOIE POUR FEMMES VICIEUSES **Vol. 2** 139
HOMMES DE PROIE, LES **Vol. 2** 152, 249, 601, 812
HOMMES EN BLANC, LES **Vol. 2** 185, 559, 661
HOMMES, LES **Vol. 2** 175, 567
HOMMES NE PENSENT QU'A CA, LES **Vol. 2** 515, 795, 856
HOMMES NOUVEAUX, LES **Vol. 2** 134, 217, 257, 488
HOMMES PREFERENT LES GROSSES, LES **Vol. 2** 534 **Vol. 6** *page 93*
HOMMES-RECLAME **Vol. 2** 678
HOMMES SANS NOM, LES **Vol. 2** 191, 279, 425, 595
HOMMES SANS PEUR, LES **Vol. 2** 249
HOMMES VEULENT VIVRE, LES **Vol. 2** 598, 856
HOMO SUM **Vol. 4** 212
HOMOLKA A TOBOLKA **Vol. 10** 70, 398
HOMUNCULUS **Vol. 4** 124, 288
HON DANSADE EN SOMMAR **Vol. 5** 1042, 1164
HON, DEN ENDA **Vol. 5** 891, 997
HON ELLER INGEN **Vol. 5** 863, 1032
HON GLÖMDE HENNE ALDRIG **Vol. 5** 843
HON, HAN OCH ANDERSSON **Vol. 5** 870, 891
HON KOM SOM EN VIND **Vol. 5** 1003, 1073
HON NA KOCKU **Vol. 10** 359, 430, 571
HON TRODDE DET VAR HAN **Vol. 5** 843
HONFOGLALAS **Vol. 9** 122, 261, 318
HONG-KONG UN ADDIO **Vol. 7** 300, 1136
HONNEUR D'UN CAPITAINE, L' **Vol. 6** *page 87, page 126, page 136*
HONNEURS DE LA GUERRE, LES **Vol. 2** 190, 515, 517
HONNING MANE **Vol. 5** 46, 66
HONORABLE CATHERINE, L' **Vol. 2** 58, 59, 205, 254, 601
HONORABLE SOCIETE, L' **Vol. 2** 806
HONORABLE STANISLAS AGENT SECRET, L' **Vol. 2** 348, 372
HONORABLES SINVERGUENZAS **Vol. 8** 233, 630
HONORE DE MARSEILLE **Vol. 2** 548, 875
HONRADEZ DE LA CERRADURA, EL **Vol. 8** 298, 396, 515
HONSEMINISTERENS BESOG **Vol. 5** 92
HONTE DE LA FAMILLE, LA **Vol. 2** 808
HONZA MALEM KRALEM **Vol. 10** 41, 283, 505
HONZIKOVA CESTA **Vol. 10** 299, 532
HOOK, LINE AND SINKER **Vol. 8** 196
HOP – A JE TU LIDOOP **Vol. 10** 172, 496, 529
HOPEAA RAJAN TAKAA **Vol. 5** 337, 634
HOPEAKIHLAJAISET **Vol. 5** 446, 552
HOPLA PA SENGEKANTEN **Vol. 5** 156, 218, 231, 296
HOPPSAN ! **Vol. 5** 871, 953, 1073
HOR, VAR DER IKKE EN, SOM LO ? **Vol. 5** 125, 179, 247
HORA BRUJA, LA **Vol. 8** 338, 358
HORA CERO: OPERACION ROMMEL **Vol. 8** 553
HORA DE AMOR, UMA **Vol. 8** 54, 160, 176
HORA DEL CORAJE, LA **Vol. 8** 762
HORA OT DALECE **Vol. 3** 51 **Vol. 6** *page 37*
HORACE 62 **Vol. 2** 368, 376
HORAS DE MARIA, AS **Vol. 8** 169
HORAS INCIERTAS **Vol. 8** 215, 551
HORAS PROHIBIDAS **Vol. 8** 397
HORCH, HORCH, DIE LERCH IM ÄTHERBLAU **Vol. 4** 79, 115, 354
HORDUBAL **Vol. 10** 82, 173, 544
HORDUBALOVE **Vol. 10** 34, 256
HOREA **Vol. 6** *page 268, page 274, page 275*
HORI, MA PANENKO **Vol. 10** 70, 398
HORIS METERA **Vol. 3** 164

HORIS SYNEIDESE **Vol. 3** 142
HORIZON, L' **Vol. 2** 208
HORIZONS SANS FIN **Vol. 2** 47, 322
HORIZONT **Vol. 9** 67, 152, 339
HORIZONTE TE HAPURA **Vol. 3** 6, 22
HORIZONTES DE LUZ **Vol. 8** 221, 461
HORKA KASE **Vol. 10** 28, 208, 344
HORKA LASKA **Vol. 10** 345, 580
HORKA ZIMA **Vol. 10** 325, 366, 394
HORKY PODZIM S VUNI MANGA **Vol. 10** 394, 496, 517
HORKY VZDUCH **Vol. 10** 299
HORLOGE, L' **Vol. 2** 357
HORLOGER DE SAINT-PAUL, L' **Vol. 2** 351, 548
HORMOONIT VALLOILLAAN **Vol. 5** 372, 379
HORNANKOSKI **Vol. 5** 310, 634
HOROSCOPE **Vol. 2** 97, 815
HOROSCOPO, EL **Vol. 8** 681, 749
HOROSKOP **Vol. 3** 353, 401
HOROUCI SRDCE **Vol. 10** 362, 564, 621
HORRIDO **Vol. 4** 149, 660, 704
HORROR **Vol. 8** 725, 762
HORROR STORY **Vol. 8** 552, 732
HORSE, LA **Vol. 2** 770, 875
HORSKE VOLANI SOS **Vol. 10** 169, 257, 524, 533
HORTOBAGY **Vol. 9** 124, 266
HORUCKA **Vol. 10** 82, 273, 548
HORVATOV IZBOR **Vol. 6** *page 312*
HOSE, DIE **Vol. 4** 132, 558, 626
HOSEKRAEMMEREN **Vol. 5** 179
HOSEN DES RITTERS VON BREDOW, DIE **Vol. 1** 18, 46, 128
HOSENKNÖPF, DIE **Vol. 4** 18, 230
HOSPEDE DO QUARTO 13, O **Vol. 8** 26, 55, 76
HOSPITAL DE URGENCIA **Vol. 8** 217, 470
HOSPITAL GENERAL **Vol. 8** 602, 772
HOSSZU AZ UT HAZAIG **Vol. 9** 134, 183, 314
HOSSZU VAGTA **Vol. 9** 215, 253, 259
HOST DO DOMU **Vol. 10** 109, 154
HOSTESS **Vol. 1** 76, 100, 104
HOSTINEC U KAMENNEHO STOLU **Vol. 10** 503, 580, 612
HOSTSONATEN **Vol. 5** 850
HOSZAKADAS **Vol. 9** 259, 263, 318
HOTEL „CENTRAL" **Vol. 6** *page 33, page 37, page 64*
HOTEL DE LA PLAGE, L' **Vol. 2** 335
HOTEL DER TOTEN GÄSTE **Vol. 8** 461
HOTEL DES AMERIQUE **Vol. 6** *page 84, page 90*
HOTEL DES ETUDIANTS **Vol. 2** 174, 626, 656
HOTEL DU LIBRE-ECHANGE, L' **Vol. 2** 388, 402, 573, 837
HOTEL DU NORD **Vol. 2** 626, 824, 837
HOTEL IMPERIAL **Vol. 9** 77
HOTEL KIKELET **Vol. 9** 218, 319, 322
HOTEL KLASKY LUX **Vol. 6** *page 214, page 251*
HOTEL MODRA HVEZDA **Vol. 10** 345, 412
HOTEL PARADIS **Vol. 5** 55, 101, 208, 295
HOTEL PARADISO **Vol. 2** 457
HOTEL POLAN UND SEINE GÄSTE **Vol. 6** *page 160, page 166, page 176*
HOTEL POUR JEUNES FILLES **Vol. 6** *page 129*
HOTEL PRO CIZINCE **Vol. 10** 289, 395, 505
HOTEL SACHER **Vol. 4** 82, 391, 631
HOTELBOY ED MARTIN **Vol. 1** 22, 82
HOTELGEHEIMNISSE **Vol. 4** 377, 444, 452
HOTELL KAKBRINKEN **Vol. 5** 928, 1042
HOTELL PARADISETS HEMLIGHET **Vol. 5** 899, 979, 1010
HOTELRATTEN **Vol. 4** 212, 571, 726, 753
HOTESSE TRES SPECIALE, UNE **Vol. 6** *page 93*
HOTESSES DU SEXE, LES **Vol. 2** 294
HOTESSES EN CHALEUR **Vol. 2** 291
HOTREAL **Vol. 9** 51, 61, 191
HOUKUTUSLINTU **Vol. 5** 472, 634
HOULA-HOULA **Vol. 2** 250, 659
HOUSATA **Vol. 10** 149, 250, 383
HOUSE OF FEAR **Vol. 2** 526
HOUSE OF MIRTH **Vol. 2** 579
HOUSLE A SEN **Vol. 10** 256, 412, 612
HOVDINGEN **Vol. 5** 750, 780, 784
HOVEDJAEGERNE **Vol. 5** 130, 212, 299
HOVMOD STAAR FOR FALD **Vol. 5** 203
HOY COMO AYER **Vol. 8** 450, 553
HOY NO PASAMOS LISTA **Vol. 8** 611, 704
HOYSOMMER **Vol. 5** 717, 804
HOZ Y EL MARTINEZ, LA **Vol. 8** 407, 414
H.P.HAENGER FAA'EN **Vol. 5** 81
HR. PETIT **Vol. 5** 89, 248
HR. TELL OG SON **Vol. 5** 24
HRA BEZ PRAVIDEL **Vol. 10** 70, 352, 433
HRA NA TELO **Vol. 10** 108, 263, 269
HRA O JABLKO **Vol. 10** 144, 299, 588
HRA O KRALOVNU **Vol. 10** 226, 317, 522
HRA O ZIVOT **Vol. 10** 296, 388, 433
HRABENKA Z PODSKALI **Vol. 10** 167, 453
HRABINA COSEL **Vol. 1** 202, 206, 227, 232, 302
HRAC **Vol. 10** 232, 296, 347
HRANJENIK **Vol. 3** 437, 451
HRATKY S CERTEM **Vol. 10** 369, 433, 503, 524
HRDINA JEDNE NOCI **Vol. 10** 256, 524

HRDINA MA STRACH **Vol. 10** 359, 385, 460
HRDINNY KAPITAN KORKORAN **Vol. 10** 256, 586
HRDINOVE MLCI **Vol. 10** 154, 256, 514
HRICH **Vol. 10** 167, 453
HRICH LASKY **Vol. 10** 138, 167
HRICH MLADI **Vol. 10** 61, 354
HRICH V MANZELSTVI **Vol. 10** 243, 345
HRICHY LASKY **Vol. 6** *page 188, page 197*
HRIECH KATARINY PADYCHOVEJ **Vol. 10** 72, 82, 287
HRISNA **Vol. 6** *page 199*
HRISNA KREV **Vol. 10** 586
HRISOMALLOUSA **Vol. 3** 206
HRISTE **Vol. 10** 233, 313, 398
HRISTINA **Vol. 3** 204
HROCH **Vol. 10** 227, 317, 567
HROMATA TES IRIDOS **Vol. 3** 159
HRONIKA LUDOSTI **Vol. 6** *page 304*
HRONIKA NA CUVSTVATA **Vol. 3** 47, 105, 121
HRONIKA OTPORA **Vol. 6** *page 303*
HRONIKA ZLOCINA **Vol. 6** *page 316*
HROZBA **Vol. 10** 139, 227, 325
HRY A SNY **Vol. 10** 433, 532, 621
HRY LASKY SALIVE **Vol. 10** 425, 503, 516
HRY PRO MIRNE POKROCILE **Vol. 10** 139, 233, 511
HRYSOMALOUSA **Vol. 3** 181
HU DAGMAR **Vol. 5** 692, 760
HUBAL **Vol. 1** 321, 389 **Vol. 6** *page 210*
HUDBA KOLONAD **Vol. 10** 41, 47, 327
HUDBA SRDCI **Vol. 10** 256, 586
HUDBA Z MARSU **Vol. 10** 313, 352, 569
HUELLA DE LUZ **Vol. 8** 214, 400
HUESO, EL **Vol. 8** 219, 299
HUESPED DE LAS TINIEBLAS, EL **Vol. 8** 276, 298
HUESPED DEL CUARTO NUMERO 13, EL **Vol. 8** 26, 55, 76, 359, 422, 725
HUESPED DEL SEVILLANO, EL **Vol. 8** 752
HUFEISEN FINDEN BRINGT GLÜCK **Vol. 4** 514
HUGO OCH JOSEFIN **Vol. 5** 869, 1068
HUGOS NACHTARBEIT **Vol. 4** 361, 409, 692
HÜHNERAUGENOPERATEUR, DER **Vol. 4** 634
HUHTIKUU ON KUUKAUSISTA JULMIN **Vol. 5** 590
HUHTIKUU TULEE **Vol. 5** 347, 379
HUIDA AL SUR **Vol. 8** 684
HUIDA, LA **Vol. 8** 405, 493
HUIJARIEN HUVITTAVAT HUIPUTTAJAT **Vol. 5** 440, 441, 562
HUIS CLOS **Vol. 2** 194, 423, 738
HUIT HOMMES DANS UN CHATEAU **Vol. 2** 560, 595
HUITIEME JOUR, LE **Vol. 2** 314, 386
HULLU KESÄ **Vol. 5** 507, 518, 521, 537, 568
HÜLYESEG NEM AKADALY **Vol. 9** 10, 160, 205
HUMAIN TROP HUMAIN **Vol. 2** 74
HUMAN **Vol. 2** 350, 590
HUMEUR VAGABONDE, L' **Vol. 2** 35, 49, 875
HUMITIDES SECRETES POUR MOUILLEUSES PRECOCES **Vol. 6** *page 88*
HUMORESKA **Vol. 10** 2, 460
HUN FIK HAM IKKE **Vol. 5** 83
HUN SKRIVER PAA MASKINE **Vol. 5** 295
HUND VON BASKERVILLE, DER **Vol. 4** 89, 123, 178, 186, 340, 608, 727, 774
HUNDERT MARK SIND WEG, DIE **Vol. 4** 311, 461
HUNDERT TAGE **Vol. 4** 382 **Vol. 7** 919, 1159
HUNDRA **Vol. 8** 303, 742
HUNDRA DRAGSPEL OCH EN FLICKA **Vol. 5** 841, 993
HUNDSTAGE **Vol. 4** 84, 666
HUNGER IN WALDENBURG **Vol. 6** *page 181*
HUNGERNDE MILLIONÄRE **Vol. 4** 277, 747
HUNGRY HEART, THE **Vol. 2** 24
HUR MARIE TRÄFFADE FREDERIK, ASNAN REBUS, KÄNGURUN PLOJ OCH... **Vol. 5** 874
HURA ZA NIM ! **Vol. 10** 161, 255, 508
HURAGAN **Vol. 1** 252, 278, 309, 364
HURRA FOR ANDERSENS **Vol. 5** 718, 762, 799
HURRA FOR DE BLA HUSARER **Vol. 5** 199, 201, 232, 236, 248
HURRA ! ICH BIN PAPA **Vol. 4** 51, 108, 623
HURRAH – EIN JUNGE ! **Vol. 4** 208, 246, 298
HURRAH ! ENDLICH ALARM **Vol. 4** 8, 742
HURRAH ! ICH LEBE ! **Vol. 4** 105, 113, 339, 340, 442
HUSAREN HERAUS **Vol. 4** 132, 340, 466
HUSAREN IN BERLIN **Vol. 1** 37, 55, 102
HUSARENFIEBER **Vol. 4** 124, 622
HUSARENLIEBE **Vol. 4** 232, 270, 471
HUSASSISTENTEN **Vol. 5** 58, 262
HUSET NR. 17 **Vol. 5** 840, 843
HUSMANDSTOSEN **Vol. 5** 44, 89
HUSSARDS, LES **Vol. 2** 124, 185, 226
HUSTLER, THE **Vol. 2** 778

HUSTRUER Vol. 5 767, 783, 797
HUSZ EVRE EGYMASTOL Vol. 9 36, 39, 318
HUSZ ORA Vol. 9 134, 259, 265
HUT AB, WENN DU KÜSST! Vol. 1 15, 75, 79
HÜTET EUCH VOR LEICHTEN FRAUEN Vol. 4 221, 340, 608
HUYENDO DE SI MISMO Vol. 8 524, 537
HUYENDO DEL HALCON Vol. 8 276, 558
HVAD MED OS ? Vol. 5 179, 281
HVAD VIL DE HA ? Vol. 5 23
HVEM EIER TYSSEDAL ? Vol. 5 757
HVEM ER GENTLEMANTYVEN ? Vol. 5 58
HVEM ER HUN ? Vol. 5 154
HVEM HAR BESTEMT...? Vol. 5 765, 784
HVEM MYRDER HVEM ? Vol. 5 159, 251
HVEM VAR FORBRYDEREN ? Vol. 5 9
HVEM VAR HUN ? Vol. 5 81
HVEZDA Vol. 10 173, 283, 577
HVEZDA JEDE NA JIH Vol. 10 299, 385, 412
HVEZDA PADA VZHURU Vol. 10 41, 144, 229
HVEZDA Z POSLEDNI STACE Vol. 10 532, 621
HVEZDA ZVANA PELYNEK Vol. 10 111, 564, 621
HVIDE DAME, DEN Vol. 5 58
HVIDE DJAEVEL, DEN Vol. 5 58
HVIDE HINGST, DEN Vol. 5 76, 257, 280
HVIDE RYTTERSKE, DEN Vol. 5 256
HVIDE SLAVEHANDEL I, DEN Vol. 5 97, 188
HVIDE SLAVEHANDEL II, DEN Vol. 5 97
HVIDE SLAVINDE, DEN Vol. 5 97, 178
HVIS ER BARNET ? Vol. 5 83
HVIS LILLE PIGE ER DU ? Vol. 5 31, 39, 248, 272
HVO SOM ELSKER SIN Vol. 5 58
HVOR ER FAR ? Vol. 5 2, 228
HVOR ER PELLE ? Vol. 5 97
HVOR ERE LIGET, MOLLER ? Vol. 5 35, 195
HVOR SORGERHE GLEMMES Vol. 5 58
HVORDAN JEG KOM TIL FILMEN Vol. 5 295
HVORLEDES LILLE PETER TAENKER SIG COOKS NORDPOLSFAERD Vol. 5 97
HYMENEE Vol. 2 404, 717, 722
HYPERION Vol. 3 239, 255
HYPOTHESE DU TABLEAU VOLE, L' Vol. 2 854
HYPPOLIT A LAKAY Vol. 9 71, 118, 128
HYRDEPIGENS KAERLIGHED Vol. 5 83
HYÜKYAALTOJA Vol. 5 351, 356

I

I BOG STVORI KAFANSKU PEVACICU Vol. 3 354
I BRIST PA BEVIS Vol. 5 923, 1022, 1078, 1205
I ... COMME ICARE Vol. 6 *page 147*
I DAG BEGYNDER LIVET Vol. 5 7
I DAG BÖRJAR LIVET Vol. 5 928, 1156
I DAG GIFTER SIG MIN MAN Vol. 5 878, 929, 1173
I DE GODE GAMLE DAGE Vol. 5 24, 136, 228
I DE LYSE NAETTER Vol. 5 199, 304
I DE UNGE AAR Vol. 5 256
I DEN GRONNE SKOV Vol. 5 16, 37, 195, 232
I DIMMA DOLD Vol. 5 940, 1052
I DIN FARS LOMME Vol. 5 113, 247, 259
I DOD MANS SPAR Vol. 5 919
I DODENS BRUDESLOR Vol. 5 121
I DÖDENS VÄNTRUM Vol. 5 871, 1042
I DOJDE DENJAT Vol. 3 110 Vol. 6 *page 34, page 40*
I DUR OCH SKUR Vol. 5 871, 1073
I EKDROME Vol. 3 142
I FANTASTICI TRE SUPERMEN Vol. 3 395
I FUEGO ! Vol. 8 53, 75
I GAAR OG I MORGEN Vol. 5 2, 228
I GABESTOKKEN Vol. 5 2, 179, 228
I GULT OCH BLATT Vol. 5 843, 923
I HUVET PA EN GAMMAL GUBBE Vol. 5 1200
I JOMFRUENS TEGN Vol. 5 127, 179
I KALPIKE LIRA Vol. 3 182, 198, 203
I KAMMERHERRENS KLAEDER Vol. 5 58
I KANTONNEMENT Vol. 5 184
I KONGENS KLAER Vol. 5 678
I KONGENS KLAE'R Vol. 5 14, 146, 180, 199
I KRAUGE Vol. 3 180
I LE LLAMABAN EL HALCON Vol. 8 626
I LIDELSENS FAMN Vol. 5 310
I LIVETS BRAENDING Vol. 5 58
I LIVETS VAR Vol. 5 1070
I LOVE YOU, JE T'AIME Vol. 2 351 Vol. 6 *page 82, page 146*
I LOVENS TEGN Vol. 5 11, 127, 257
I LUST OCH NÖD Vol. 5 569, 900
I MERCENARI MUOIONO ALL'ALBA Vol. 8 177
I MORALENS NAVN Vol. 5 760
I MORGEN - MIN ELSKEDE Vol. 5 10, 22, 162
I MÖRKASTE SMALAND Vol. 5 927, 1078
I NATT ELLER ALDRIG Vol. 5 840, 909
I NÖD OCH LUST Vol. 5 863, 930, 980, 1032

I OPIUMMETS MAGT **Vol. 5** 295
I PARADIS... **Vol. 5** 907, 923, 982
I PROSTATES **Vol. 3** 206
I RÖK OCH DANS **Vol. 5** 878, 953, 1173
I ROSLAGENS FAMN **Vol. 5** 928, 1042
I SEDSTE SEKUND **Vol. 5** 106
I SKYTTENS TEGN **Vol. 5** 127, 257
I SLIK EN NATT **Vol. 5** 724, 738
I SOM HÄR INTRÄDEN... **Vol. 5** 907, 923, 1044
I STAEVNEMODETS TIME **Vol. 5** 81
I STJERNERNE STAAR DET SKREVET **Vol. 5** 183
I TETI NE BRONX **Vol. 3** 20, 24 **Vol. 6** *page 17*
I TJENSTLIGT OJEMED **Vol. 5** 92
I TO CE PROCI **Vol. 6** *page 307, page 314*
I TRE SENTIMENTALI **Vol. 2** 32
I TVILLINGERNES TEGN **Vol. 5** 11, 127, 257
I TY ZOSTANIESZ INDIANEM **Vol. 1** 223, 321, 323
I TYRENS TEGN **Vol. 5** 11, 127, 257
I UNGDOMMENS MAKT **Vol. 5** 749, 765
I ZELIARA **Vol. 3** 161
IA IN OBERBAYERN **Vol. 4** 344, 647, 689
IABRA VERDE DE ACASA **Vol. 3** 242, 279, 311
IADES **Vol. 3** 212, 304
IANCU JIANU **Vol. 3** 315
IANCU JIANU, ZAPCIUL **Vol. 3** 225, 276, 293, 309
IANU JIANU, HAIDUCUL **Vol. 3** 225, 276, 293, 309
IARNA BOBOCILOR **Vol. 3** 284, 289, 301
IBA DEN **Vol. 10** 27, 315, 419, 420, 544
IBERICAS F.C., LAS **Vol. 8** 524
IBIS ROUGE, L' **Vol. 2** 866
IBRIS **Vol. 7** 802
ICH – AXEL CAESAR SPRINGER **Vol. 1** 22, 26, 46, 59, 70
ICH BEI TAG UND DU BEI NACHT **Vol. 4** 49, 298
ICH BIN 18 **Vol. 1** 87
ICH BIN DU **Vol. 4** 13, 574
ICH – BIN – DU ... **Vol. 4** 124, 219, 339
ICH BIN GLEICH WIEDER DA **Vol. 4** 28, 329
ICH BIN SEBASTIAN OTT **Vol. 4** 265, 288, 410, 614
ICH BITTE UM VOLLMACHT **Vol. 4** 135, 619, 647
ICH BLEIB' BEI DIR **Vol. 4** 536, 788
ICH BRAUCHE DICH **Vol. 4** 344, 445, 660
ICH – DANN EINE WEILE NICHTS **Vol. 1** 8, 53, 104
ICH DZIEN POWSZEDNI **Vol. 1** 189, 377, 403
ICH FÜR DICH – DU FÜR MICH **Vol. 4** 626, 634
ICH GEH' AUS UND DU BLEIBST DA **Vol. 4** 208, 270, 409
ICH GLAUB AN DICH **Vol. 4** 623
ICH GLAUB' NIE MEHR AN EINE FRAU **Vol. 4** 115, 373, 450
ICH HAB' IM MAI VON DER LIEBE GETRÄUMT **Vol. 4** 384, 584, 725
ICH HAB' MEIN HERZ IM AUTOBUS VERLOREN **Vol. 4** 43
ICH HAB MEIN HERZ IN HEIDELBERG VERLOREN **Vol. 4** 344, 545
ICH HAB' VON DIR GETRÄUMT **Vol. 4** 49, 149
ICH HATT' EINEN KAMERADEN **Vol. 4** 36, 124, 514
ICH HATTE EINST EIN SCHÖNES VATERLAND **Vol. 4** 565, 663
ICH HEIRATE MEINE FRAU **Vol. 4** 28, 424, 544
ICH HEIRATE MEINEN MANN **Vol. 4** 504
ICH KENN' DICH NICHT UND LIEBE DICH **Vol. 4** 246, 462, 638
ICH KENNE KEINE PARTEIEN MEHR **Vol. 4** 5
ICH KLAGE AN **Vol. 4** 49, 232, 300, 335, 416, 437
ICH KÜSSE IHRE HAND, MADAME **Vol. 4** 132, 479, 779
ICH LASSE DICH NICHT **Vol. 4** 232, 673
ICH LEBE FÜR DICH **Vol. 4** 316, 339, 691 **Vol. 6** *page 183*
ICH LIEBE ALLE FRAUEN **Vol. 4** 49, 234, 734
ICH LIEBE DICH **Vol. 4** 49, 113, 558, 738, 788
ICH MÖCHTE KEIN MANN SEIN **Vol. 4** 560, 663
ICH SEHE HELL, ICH SEHE DUNKEL **Vol. 4** 370
ICH SEHNE MICH NACH DIR **Vol. 4** 115, 774
ICH SING' MICH IN DEIN HERZ HINEIN **Vol. 4** 298, 323, 608
ICH TANKE, HERR FRANKE **Vol. 4** 361, 558
ICH UND DIE KAISERIN **Vol. 4** 49, 266, 574
ICH VERTRAUE DIR MEINE FRAU AN **Vol. 4** 270, 774
ICH VERWEIGERE DIE AUSSAGE **Vol. 4** 97, 430, 437
ICH WAR 19 **Vol. 1** 7, 9, 48
ICH WAR IN HONOLULU – WETTEN ? **Vol. 1** 52
ICH WAR JACK MORTIMER **Vol. 4** 373, 626

ICH WAR ZU HEIDELBERG STUDENT Vol. 4 340, 634
ICH WARTE AUF DICH Vol. 6 *page 199*
ICH WERDE DICH AUF HÄNDEN TRAGEN Vol. 4 331, 608, 729
ICH WILL DICH LIEBE LEHREN Vol. 4 113, 373, 490
ICH WILL EUCH SEHEN Vol. 1 8, 85, 91
ICH WILL NICHT WISSEN, WER DU BIST Vol. 4 49, 208, 255, 626
ICH ZWING DICH ZU LEBEN Vol. 1 1, 7, 14
ICI CEMO U GRAD Vol. 3 367
ICI ET AILLEURS Vol. 2 527
ICI L'ON PECHE Vol. 2 413, 529
I...COMME ICARE Vol. 2 668, 770
IDA Vol. 4 119, 517, 558
IDA, DIE PERLE Vol. 4 354, 367
IDA REGENYE Vol. 9 71, 322
IDEA FISSA, L' Vol. 7 1147
IDEAL HUSBAND, AN Vol. 2 656
IDEAL SEPTIMY Vol. 10 524, 621
IDEALE WOHNUNG, EINE Vol. 4 82, 115
IDEALER GATTE, EIN Vol. 4 17, 232, 634
IDEALIST Vol. 3 348, 387
IDEE FOLLE, UNE Vol. 2 37
IDEGEN ARCOK Vol. 9 140, 259
IDEGEN UTAKON Vol. 9 26, 117
IDEIGLENES PARADICSOM Vol. 9 134, 215, 316
IDEL ÄDEL ADEL Vol. 5 840, 1173, 1181
IDEMO DALJE Vol. 6 *page 318*
IDENTIFICAZIONE DI UNA DONNA Vol. 7 340, 363, 853, 895, 1168
IDENTITE JUDICIAIRE Vol. 2 231, 581
IDI MI, DODJI MI Vol. 6 *page 317, page 318*
IDILIO EN MALLORCA Vol. 8 417, 475
IDILLIO A BUDAPEST Vol. 7 16, 662
IDILLIO INTERROTTO Vol. 7 755
IDILLIO TRAGICO Vol. 7 490
IDIOT A PARIS, UN Vol. 2 532, 700
IDIOT, L' Vol. 2 59, 282, 566
IDIOTA, L' Vol. 7 1103
IDL MITN FIDL Vol. 1 194
IDÖ ABLAKAI, AZ Vol. 9 121, 149, 259
IDÖ VAN Vol. 9 61, 240, 243
IDÖK KEZDETEN, AZ Vol. 9 152, 316, 339
IDOL Vol. 6 *page 205, page 226, page 260*
IDOLE DES JEUNES, L' Vol. 2 685
IDOLE, L' Vol. 2 205, 330
IDOLES, LES Vol. 2 768
IDOLO DEL DOTTORE, L' Vol. 7 984
IDOLO INFRANTO Vol. 7 965
IDOLOS Vol. 2 646 Vol. 8 607
IDRAETSFILMENS INTERNATIONALE Vol. 5 242
IDU DANI Vol. 3 451
IDYLLE AU CAIRE Vol. 2 399
IDZIEM DO CIEBIE POLSKO, MATKO NASZA Vol. 1 174, 215
IERI, OGGI, DOMANI Vol. 7 483
IFIGENEIA Vol. 3 151
IFJU SZIVVEL Vol. 9 134, 168, 183
IGELFREUNDSCHAFT Vol. 1 68
IGEN Vol. 9 183, 259, 285
IGEN VAGY NEM ? Vol. 9 71, 322
IGERET FÖLDJE, AZ Vol. 9 121, 149, 332
IGLOI DIAKOK Vol. 9 71, 265, 288, 337
IGMANSKI MARS Vol. 6 *page 308, page 314*
IGNACE Vol. 2 67, 197, 488
IGNOTA, L' Vol. 7 288
IGNOTO, L' Vol. 7 168
IGRA NA LJUBOV Vol. 3 42, 109 Vol. 6 *page 37*
IGRASZKI PIENIADZA Vol. 1 372
IGRE NA SKELAMA Vol. 3 420, 435
IGREK 17 Vol. 3 46 Vol. 6 *page 42*
IGUANA DALLA LINGUA DI FUOCO, L' Vol. 7 585
IGY JÖTTEM Vol. 9 36, 273
IH, DU FORBARMENDE Vol. 5 179, 281
IHANA SEIKKAILU Vol. 5 443, 639
IHMEMIES Vol. 5 319, 320, 500, 585, 589
IHMESET SUVIYÖSSSÄ Vol. 5 372, 379
IHR DUNKLER PUNKT Vol. 4 132, 486, 584
IHR ERSTES ERLEBNIS Vol. 4 28, 82, 608
IHR FEHLTRITT Vol. 4 135
IHR GRÖSSTER ERFOLG Vol. 4 149, 373, 660
IHR GUTER RUF Vol. 4 183
IHR JUNGE Vol. 4 247 Vol. 10 168
IHR LEIBHUSAR Vol. 4 139, 723
IHR LETZTES LIEBESABENTEUER Vol. 4 43, 796
IHR PRIVATSEKRETÄR Vol. 4 30, 215, 416
IHR SCHLECHTER RUF Vol. 4 673
IHR SPORT Vol. 4 190, 319
IHR STRANDGUT Vol. 4 514
IHR TRICK Vol. 4 239
IHRE DURCHLAUCHT, DIE VERKÄUFERIN Vol. 4 527, 614
IHRE FREUNDIN Vol. 4 514
IHRE HOHEIT Vol. 4 183
IHRE HOHEIT BEFIEHLT Vol. 4 325, 567, 713
IHRE HOHEIT DIE TÄNZERIN Vol. 4 5, 231, 584
IHRE LETZTE DUMMHEIT Vol. 4 208, 424
IHRE MAJESTÄT DIE LIEBE Vol. 4 14, 321, 325
IHRE TOLLSTE NACHT Vol. 10 150, 164, 168
I+I Vol. 5 427, 439, 447

IKARIE XB 1 **Vol. 10** 229, 289, 498, 547, 621
IKARUS **Vol. 1** 1, 14, 27
IKARUS. IM HÖHENFLUG DER LEIDENSCHAFTEN **Vol. 4** 87, 232
IKONOSTAS"T **Vol. 3** 109, 114 **Vol. 6** *page 53*
IL EST MINUIT, DOCTEUR SCHWEITZER **Vol. 2** 226, 419, 698
IL ETAIT UNE FOIS **Vol. 2** 21, 55, 332
IL ETAIT UNE FOIS UN FLIC **Vol. 2** 293
IL ETAIT UNE FOIS...LA CHATTE MOUILLEE **Vol. 2** 291
IL FAUT TUER BIRGITT HAAS **Vol. 6** *page 82, page 92*
IL FAUT VIVRE DANGEREUSEMENT **Vol. 2** 49
IL NE FAUT JAMAIS S'ATTACHER AUX CHAUSSURES **Vol. 2** 272
IL NE FAUT PAS JOUER AVEC LE FEU **Vol. 2** 257
IL N'Y A PAS DE FUMEE SANS FEU **Vol. 2** 185, 293
IL PLEUT DANS MON VILLAGE **Vol. 3** 350, 408
IL PLEUT SUR SANTIAGO **Vol. 2** 60
IL PLEUT TOUJOURS OU C'EST MOUILLE **Vol. 2** 754
IL SUFFIT D'AIMER **Vol. 2** 270, 866
IL SUFFIT D'UNE FOIS **Vol. 2** 330, 566
IL VECCHIO TESTAMENTO **Vol. 3** 395
IL Y A LONGTEMPS QUE JE T'AIME **Vol. 2** 163 **Vol. 6** *page 86*
ILD OG JORD **Vol. 5** 2, 228
ILE AUX FEMMES NUES, L' **Vol. 2** 727 **Vol. 7** 968
ILE D'AMOUR, L' **Vol. 2** 254, 603, 826, 845
ILE DE LA MORT, L' **Vol. 2** 257
ILE DE LA SOLITUDE, L' **Vol. 2** 332
ILE DES VEUVES, L' **Vol. 2** 360
ILE DU BOUT DU MONDE, L' **Vol. 2** 249, 501, 709
ILE ENCHANTEE, L' **Vol. 2** 410, 688, 845
ILE SANS NOM, L' **Vol. 2** 474, 822
ILEGALET **Vol. 3** 10, 15
ILES ENCHANTEES, LES **Vol. 2** 700
ILES, LES **Vol. 6** *page 84, page 102, page 126, page 135, page 136*
ILHA DOS AMORES **Vol. 8** 7, 139, 163, 169, 198
ILHAS ENCANTADAS, AS **Vol. 8** 22, 39, 161, 189
ILIE IN LUNA DE MIERE **Vol. 3** 219, 224, 272
ILJUZIJA **Vol. 3** 60 **Vol. 6** *page 51, page 63*
I'LL NEVER FORGET YOU **Vol. 2** 656
ILLATOS UT A SEMMIBE **Vol. 9** 19, 67
ILLUSION **Vol. 4** 365, 381, 445
ILLUSION ZU DRITT **Vol. 4** 267
ILLUSIONE, L' **Vol. 7** 731
ILLUSIONS **Vol. 2** 346
ILLUSTRE ATTRICE CICALA FORMICA, L' **Vol. 7** 560
ILLUSTRE MAURIN, L' **Vol. 2** 152, 330, 437
ILOINEN LINNANMÄKI **Vol. 5** 374, 403, 579
ILONA **Vol. 4** 400, 733
ILS **Vol. 2** 326, 689
ILS APPELLENT CA UN ACCIDENT **Vol. 6** *page 109, page 118, page 123, page 132*
ILS ETAIENT CINQ **Vol. 2** 197, 501
ILS ETAIENT CINQ PERMISSIONNAIRES **Vol. 2** 6, 249
ILS ETAIENT NEUF CELIBATAIRES **Vol. 2** 32, 48, 197
ILS ONT TUE JAURES **Vol. 2** 546
ILS ONT VINGT ANS **Vol. 2** 48, 137, 475
ILS SONT FOUS CES SORCIERS **Vol. 2** 12, 224, 334
ILS SONT GRANDS CES PETITS **Vol. 2** 51, 746
ILUMINACJA **Vol. 1** 204, 245, 249
ILUSTRATE CU FLORI DE CIMP **Vol. 3** 297, 311, 326
ILUSTRE FREGONA, LA **Vol. 8** 744, 784
ILUZIJA **Vol. 3** 361, 415, 437
IM 100-KILOMETER-TEMPO **Vol. 4** 165
IM ABGRUND DES HASSES **Vol. 4** 634, 780
IM BAHNWÄRTERHÄUSL **Vol. 4** 603
IM BANN DES EULENSPIEGELS **Vol. 4** 354, 424, 437
IM BANNE DER BERGE **Vol. 4** 25
IM BANNE DER SUGGESTION **Vol. 6** *page 186*
IM BANNE DES ANDERN **Vol. 4** 98, 112
IM BANNE DES MONTE MIRACOLO **Vol. 4** 54, 145, 719
IM BEWUSSTSEIN DER SCHULD **Vol. 6** *page 194*
IM DIENSTE DER LIEBE **Vol. 4** 11, 192
IM FASCHING DER SINNE **Vol. 4** 385
IM GEHEIMDIENST **Vol. 4** 266, 288, 574
IM GLÜCK VERGESSEN **Vol. 4** 190
IM GLUTRAUSCH DER SINNE **Vol. 4** 93, 626
IM HAUSE DES KOMMERZIENRATS **Vol. 4** 146
IM HIMMEL IST DOCH JAHRMARKT **Vol. 1** 33, 75, 79
IM IRRWEGE DER LIEBE **Vol. 4** 308
IM KAMPF MIT DEM BERGE **Vol. 4** 10, 156

IM KAMPF MIT DER UNTERWELT Vol. 4 255, 361, 692
IM KAMPF UM DIAMANTENFELDER Vol. 4 98, 340
IM KAMPF UM DIE HEIMAT Vol. 4 340
IM LANDHAUS VON CHICAGO Vol. 4 167, 339
IM LETZTEN AUGENBLICK Vol. 4 212
IM LUXUSZUG Vol. 4 10, 14, 431, 788
IM NAMEN DES KAISERS Vol. 4 36, 270
IM NAMEN DES KÖNIGS Vol. 4 258, 390, 536
IM NAMEN DES VOLKES Vol. 4 564, 660, 728
IM PHOTOATELIER Vol. 4 754
IM PRATER BLÜHN WIEDER DIE BÄUME Vol. 4 85, 747
IM RAUSCHE DER LEIDENSCHAFT Vol. 4 384, 472, 565
IM RÜCKEN DES FEINDES Vol. 6 *page 172*
IM SCHATTEN DER MOSCHEE Vol. 4 223, 239, 610, 626, 763
IM SCHATTEN DER SCHULD Vol. 4 135, 503
IM SCHATTEN DES BERGES Vol. 4 85, 285, 541, 615, 623
IM SCHATTEN DES GLÜCKS Vol. 4 385
IM SCHLARAFFENLAND Vol. 1 8, 85
IM SONDERAUFTRAG Vol. 1 13, 17, 48
IM SPANNUNGSFELD Vol. 1 42, 53, 128
IM SPIEGEL Vol. 6 *page 166, page 168, page 174*
IM STAUB DER STERNE Vol. 1 47, 53, 113
IM TAUMEL EINER GROSSEN LIEBE Vol. 6 *page 188, page 191*
IM TOTEN WINKEL Vol. 1 21, 24, 120
IM TROMMELFEUER DER WESTFRONT Vol. 4 13
IM WEISSEN RÖSSL Vol. 4 558, 724
IM WIRBEL DES LEBENS Vol. 4 113, 780
IM ZEICHEN DER SCHLANGE Vol. 4 747
IM ZEICHEN DER SCHULD Vol. 4 160, 695
IMA LJUBAVI, NEMA LJUBAVI Vol. 3 368
IMAGE, L' Vol. 2 87, 153, 295, 648
IMAGES A PROPOS DE „ENLUMINURES AUTOUR DES MINUTES DU PROCES DE GILLES DE RAIS" Vol. 2 349
IMAM DVIJE MAME I DVA TATE Vol. 3 430, 437
IMBARCO A MEZZANOTTE Vol. 2 13 Vol. 7 447
IMBIANCHINO, L' Vol. 7 1189
IMBOSCATA, L' Vol. 7 369
IMBRANATO, L' Vol. 7 257
IMBROGLIONI, GLI Vol. 7 1027 Vol. 8 311
IMMACOLATA E CONCETTA, L'ALTRA GELOSIA Vol. 7 139, 933
IMMAGINE DELL'ALTRA, L' Vol. 7 843
IMMAGINE FATALE, L' Vol. 7 511
IMMAGINI DI UN CONVENTO Vol. 7 742
IMMENSEE Vol. 4 424, 462, 788
IMMER AM WEG DEIN GESICHT Vol. 1 7, 74, 92
IMMER NUR DU Vol. 4 123, 149, 354, 541
IMMOLAZIONE Vol. 7 369
IMMORALE, L' Vol. 2 294
IMMORALITA Vol. 7 852
IMMORTALITA Vol. 7 1173
IMMORTELLE, L' Vol. 2 57
IMPARIAMO AD AMARCI – GUIDA ALL' EDUCAZIONE SESSUALE Vol. 7 289
IMPASSE Vol. 2 70, 205
IMPASSE DES DEUX ANGES Vol. 2 279, 719
IMPATTO MORTALE Vol. 7 461
IMPECCABLE HENRI, L' Vol. 2 330, 595
IMPERATIV, DER Vol. 6 *page 239*
IMPERIAL REGIO CAPESTRO Vol. 7 64
IMPERO DEL SOLE, L' Vol. 7 129, 359, 712, 949
IMPIEGATA DI PAPA, L' Vol. 7 756, 796
IMPIEGATE STRADALI, LE Vol. 7 1179
IMPIEGATI Vol. 7 96, 947, 1043, 1134
IMPIEGATO, L' Vol. 7 401, 853
IMPORTANT C'EST D'AIMER, L' Vol. 2 35, 434
IMPORTANTE E NON FARSI NOTARE, L' Vol. 7 631, 763
IMPOSIBILA IUBIRE Vol. 6 *page 266, page 267*
IMPOSIBLE PARA UNA SOLTERONA Vol. 8 237, 603
IMPOSSIBLE MONSIEUR PIPELET, L' Vol. 2 158, 205, 516, 857
IMPOSSIBLE OBJET, L' Vol. 2 719, 837
IMPOSSIBLE...PAS FRANCAIS Vol. 2 368
IMPOSTOR, THE Vol. 2 526
IMPOSZTOROK Vol. 9 134, 265, 316
IMPRECATEUR, L' Vol. 2 588, 873
IMPRESSIONS ROMAINES Vol. 2 114
IMPRESSZARIO, AZ Vol. 9 337
IMPREVISTO, L' Vol. 7 266, 476, 514, 792
IMPREVU, L' Vol. 2 343, 598, 798 Vol. 7 514
IMPRONTO DELLA PICCOLA MANO, L' Vol. 7 490
IMPROVVISO Vol. 7 98, 383, 710
IMPRUDENZA FATALE Vol. 7 1189
IMPURES, LES Vol. 2 13, 185, 611, 709
IMPUSCATURI PE PORTATIV Vol. 3 268, 269, 323

IMPUSCATURI SUB CALR DE LUNA Vol. 3 330
IMPUTATO ALZATEVI ! Vol. 7 443, 493
IN 40 MINUTEN Vol. 4 323, 546
IN AMORE SI PECCA IN DUE Vol. 7 792, 1121
IN CAMPAGNA E CADUTA UNA STELLA Vol. 7 16, 193, 480
IN CERCA DI FELICITA Vol. 7 476, 756
IN DEM GROSSEN AUGENBLICK Vol. 4 124, 646
IN DEN STERNEN STEHT ES GESCHRIEBEN Vol. 4 255, 344
IN DER HEIMAT, DA GIBT'S EIN WIEDERSEHEN ! Vol. 4 356, 408, 424
IN DER SOMMERFRISCH'N Vol. 4 764, 779
IN DER TIEFE DES SCHACHTS Vol. 4 239
IN DUE SI SOFFRE MEGLIO Vol. 7 443
IN EINEM KÜHLEN GRUNDE Vol. 4 514
IN EINER KLEINEN KONDITOREI Vol. 4 344, 544
IN FLAGRANTI Vol. 4 344, 441
IN GEHEIMER MISSION Vol. 4 51, 108, 623
IN GINOCCHIO DA TE Vol. 7 743
IN HOC SIGNO VINCES Vol. 7 23, 825
IN ITALIA SI CHIAMA AMORE Vol. 7 537, 1152
IN JENA SIND ALLE MÄDELS SO BLOND Vol. 4 82
IN LETZTER MINUTE Vol. 4 344, 422
IN MANO AL DESTINO Vol. 7 1038, 1116
IN MAREMMA Vol. 7 796
IN NACHT UND EIS Vol. 4 239, 634, 780, 798
IN NOME DEL PAPA RE Vol. 7 383, 778
IN NOMINE DELLA LEGGE Vol. 7 86, 808
IN PUNTA DI PIEDI Vol. 7 610, 821
IN SACHEN HERDER CONTRA BRANDT Vol. 4 232, 378
IN SACHEN TIMPE Vol. 4 230, 232, 239
IN SAT LA NOI Vol. 1 278 Vol. 3 294
IN TERRA SARDA Vol. 7 878
IN TREUE STARK Vol. 4 339, 377
IN VIAGGIO CON PAPA Vol. 7 74, 834, 868
IN WERDER BLÜHEN DIE BÄUME Vol. 4 208, 560
IN WIEN HAB' ICH EINMAL EIN MÄDEL GELIEBT Vol. 4 270, 409, 796
INA LASKA Vol. 10 83, 114, 155
INACCESIBLE, LA Vol. 8 246, 657
INAFERRABILE DODICI, L' Vol. 7 16, 1136
INAFERRABILE, L' Vol. 7 1116
INAINTE DE TACERE Vol. 3 264, 319
INCANTESIMO Vol. 7 796, 1113
INCANTESIMO TRAGICO Vol. 7 923
INCANTEVOLE NEMICA Vol. 7 86
INCANTO DI MEZZANOTTE Vol. 7 830, 923, 951, 1111, 1177
INCATENATA Vol. 7 212
INCATENATA DAL DESTINO Vol. 7 17
INCENDIE, L' Vol. 2 678
INCENDIO DI ROMA, L' Vol. 7 33
INCENSURATIGLI Vol. 7 796
INCERTITUMBRE Vol. 8 285, 649, 651
INCLINACION SEXUAL AL DESNUDO Vol. 8 471
INCOGNITA, L' Vol. 7 1169
INCOGNITO Vol. 2 427, 615, 637
INCOMPRENSIBILE, L' Vol. 7 968
INCOMPRESSA, L' Vol. 7 1189
INCONNU DE HONG-KONG, L' Vol. 2 368
INCONNU DE MONACO, L' Vol. 2 173
INCONNU DE MONTE CARLO, L' Vol. 2 318, 727
INCONNU DE MONTREAL, L' Vol. 2 8, 158
INCONNU DES SIX JOURS, L' Vol. 2 712
INCONNU D'UN SOIR, L' Vol. 2 290
INCONNU N.13, L' Vol. 2 368
INCONNUE DE MONTE-CARLO, L' Vol. 7 822, 968
INCONNUE, L' Vol. 6 *page 79*
INCONNUS DANS LA MAISON, LES Vol. 2 332, 440
INCONSTANTE, L' Vol. 2 355
INCONTRI DI NOTTE Vol. 7 487, 1054
INCONTRI MOLTI RAVVICINATI DEL QUARTO TOPI Vol. 7 556
INCONTRO NELL'ULTIMO PARADISO Vol. 7 124
INCORRIGIBLE, L' Vol. 2 655
INCREDEREA Vol. 6 *page 268*
INCREDIBILE ATTESTA, L' Vol. 7 506
INCREVABLE, L' Vol. 2 318, 347, 807
INCUBO, L' Vol. 7 971
INCUBO SULLA CITTA CONTAMINATA Vol. 7 504
INDAGINE SU UN CITTADINO AL DI SOPRA DI OGNI SOSPETTO Vol. 7 599
INDAME ACCUSA Vol. 7 201
INDE AU FEMININ, L' Vol. 2 177
INDE FANTOME, L' Vol. 2 74, 546
INDECENCE Vol. 2 442
INDEKS Vol. 6 *page 232, page 233, page 246*
INDEPENDENTA ROMANIEI Vol. 3 234
INDERPIGEN Vol. 5 295
INDESIRABLE, L' Vol. 2 603
INDIA Vol. 7 1136 Vol. 8 7
INDIA SONG Vol. 2 633
INDIANA Vol. 7 409, 764
INDIANER MINE Vol. 3 218, 323

INDIANI Vol. 7 1189
INDIANI Z VETROVA Vol. 10 90, 134, 570
INDIANO, EL Vol. 8 614, 616, 630
INDIANO EN MORATILLA, UN Vol. 8 495, 600
INDIC, L' Vol. 6 *page 77, page 121, page 134*
INDIFFERENTI, GLI Vol. 7 1035, 1167
INDISCHE DIAMANT, DER Vol. 4 747
INDISCHE GRABMAL - DER TIGER VON ESCHNAPUR, DAS Vol. 4 108, 119, 270, 298, 307, 325, 536, 620, 633, 727, 733
INDISCHE GRABMAL - DIE SENDUNG DES JOGHI, DAS Vol. 4 89, 298, 307, 325, 536, 727, 733
INDISCHE RACHE, DIE Vol. 4 186, 560
INDISCHE TOD, DER Vol. 4 11, 208
INDISCRET, L' Vol. 2 635, 715, 814
INDISCRETES, LES Vol. 2 344, 475
INDISCRETION, L' Vol. 6 *page 78, page 87*
INDISKRETE FRAU, DIE Vol. 4 132, 626
INDISPENSABILE CAMILLO, L' Vol. 7 26
INDIZIENBEWEIS Vol. 4 212, 255, 359
INDOMPTABLE ANGELIQUE Vol. 2 347, 659
INDOVINA CHI VIENE A MERENDA ? Vol. 7 1027
INDULTO, EL Vol. 8 298, 611
INES DE CASTRO Vol. 8 55, 75, 178, 449
INEVITABLE MONSIEUR DUBOIS, L' Vol. 2 205, 698
INFAMIE Vol. 4 494, 612
INFANTICIDA, L' Vol. 7 1189
INFANZIA, VOCAZIONE E PRIME ESPERIENZE DI GIACOMO CASANOVA, VENEZIANO Vol. 7 869
INFEDELI, LE Vol. 7 1136
INFERMIERA DELLA MUTUA, GLI Vol. 7 333
INFERMIERA DI MIO PADRE, L' Vol. 7 311
INFERMIERA DI NOTTE, L' Vol. 7 1194
INFERMIERA, L' Vol. 7 567
INFERMIERA NELLA CORSIA DEI MILITARI, L' Vol. 7 1202
INFERMIERE DI TATA, L' Vol. 7 420
INFERNALES PENETRATIONS DANS LES 69 POSITIONS Vol. 2 692
INFERNO Vol. 7 13, 98, 623
INFERNO ADDOSSO, L' Vol. 7 501
INFERNO GIALLO Vol. 7 487, 1107, 1172
INFERNO IN DIRETTA Vol. 7 1086
INFIDELITES Vol. 2 294
INFIERNO DE LAS MUJERES, EL Vol. 8 237, 536
INFINIE TENDRESSE, UNE Vol. 2 60
INFIRMIERS PERVERSES Vol. 2 291
INGANNI Vol. 7 502, 653
INGANNO Vol. 7 16
INGE LARSEN Vol. 4 316, 319, 397, 415
INGE UND DIE MILLIONEN Vol. 4 149, 288, 660
INGEBORG HOLM Vol. 5 996
INGEN MANS KVINNA Vol. 5 909, 1052
INGEN MORGONDAG Vol. 5 356, 374, 1042
INGEN ROSER...TAKK ! Vol. 5 701, 712, 750
INGEN TID TIL KAERTEGN Vol. 5 14
INGEN VÄG TILLBAKA Vol. 5 871, 1156
INGENJÖR ANDREES LUFTFÄRD Vol. 5 852, 1192
INGENU, L' Vol. 2 300, 815
INGENUA, L' Vol. 7 1047
INGENUE LIBERTINE, L' Vol. 2 516
INGENUO, L' Vol. 7 965, 1032, 1071
INGHITITORUL DE SABII Vol. 6 *page 265, page 266, page 275*
INGIUSTA CONDANNA, L' Vol. 7 1121
INGMARSARVET Vol. 5 891, 997
INGMARSSÖNERNA Vol. 5 937, 997
INGORGO, L' Vol. 7 299, 340, 567
INGOVANY, AZ Vol. 9 237
INGRATI, GLI Vol. 7 1189
INHUMAINE, L' Vol. 2 44, 162, 492, 547, 569, 741, 798
INITIATION PERVERSE Vol. 2 291
INKOGNITO Vol. 4 28, 108 Vol. 5 24
INLED MIG I FRESTELSE Vol. 5 857
INMORALES, LOS Vol. 8 317
INNA Vol. 1 198, 213
INNAMORATA, L' Vol. 7 48, 640
INNAMORATI, GLI Vol. 7 1012
INNAMORATO PAZZO Vol. 7 30, 383
INNO AL SOLE, L' Vol. 7 304
INNOCENT, L' Vol. 2 360, 413, 538, 871
INNOCENTE, L' Vol. 7 498, 1025, 1142, 1189
INNOCENTI PAGANO, GLI Vol. 7 1121
INNOCENTS AUX MAINS SALES, LES Vol. 2 700
INOCENTE, EL Vol. 8 517, 614
INOCENTES, LOS Vol. 8 371, 578
INONDATION, L' Vol. 2 346
INQUIETUDES DE SHANTI ANDIA, LAS Vol. 8 276
INQUIETUDINE Vol. 7 372
INQUILINA DEL PIANO DI SOPRA, L' Vol. 7 869
I.N.R.I. Vol. 4 213, 373, 407, 452
INS BLAUE LEBEN Vol. 4 13, 713 Vol. 7 35
INSAISISSABLE FREDERIC, L' Vol. 2 67, 128
INSEGNANTE AL MARE CON TUTTA LA CLASSE Vol. 7 226, 426

INSEGNANTE BALLA ... CON TUTTA LA CLASSE, L' **Vol. 7** 383
INSEGNANTE, L' **Vol. 7** 426
INSEGNANTE VIENE A CASA, L' **Vol. 7** 426
INSEL DER DÄMONEN **Vol. 4** 605
INSEL DER GEZEICHNETEN, DIE **Vol. 4** 270, 634
INSEL DER GLÜCKLICHEN, DIE **Vol. 4** 291, 329, 634, 734
INSEL DER SCHWÄNE **Vol. 6** *page 161, page 166, page 169*
INSEL DER SELIGEN, DIE **Vol. 4** 743
INSEL DER SILBERREIHER, DIE **Vol. 1** 126
INSEL DER TRÄUME, DIE **Vol. 4** 113, 546
INSEL DER VERBOTENEN KÜSSE, DIE **Vol. 4** 626, 634, 720, 734
INSEL DER VERSCHOLLENEN, DIE **Vol. 4** 124, 239
INSEL, DIE **Vol. 4** 17, 232, 713
INSEMINACION ARTIFICIAL **Vol. 8** 491, 543
INSIRA - TE MARGARITE **Vol. 3** 236
INSOLENT, L' **Vol. 2** 251, 769
INSOLITA Y GLORIOSA HAZANA DEL CIPOTE DE ARCHIDONA, LA **Vol. 8** 245, 438
INSOLITO EMBARAZO DE LOS MARTINEZ, EL **Vol. 8** 249, 749
INSOUMIS, L' **Vol. 2** 285, 719
INSPECTEUR AIME LA BAGARRE, L' **Vol. 2** 128
INSPECTEUR CONNAIT LA MUSIQUE, L' **Vol. 2** 243, 717
INSPECTEUR GREY **Vol. 2** 217, 259, 273
INSPECTEUR LA BAVURE **Vol. 2** 224 **Vol. 6** *page 92*
INSPECTEUR SERGIL **Vol. 2** 131, 533
INSPEKCJA PANA ANATOLA **Vol. 1** 179, 232 **Vol. 6** *page 218*
INSPEKTOR **Vol. 3** 356, 436
INSPEKTOR WARREN WIRD BEMÜHT **Vol. 4** 18, 558
INSPEKTOR"T I NOSTTA **Vol. 3** 38, 68, 69
INSTANTA AMINA PRONUNTAREA **Vol. 3** 213, 225
INSTINCT, L' **Vol. 2** 138, 339, 413
INSTRUMENT LI E GAJDATA ? **Vol. 3** 30, 119 **Vol. 6** *page 43*
INSULA COMORILOR **Vol. 3** 235, 244, 329
INSULA SERPILOR **Vol. 3** 216
INTER RAIL **Vol. 5** 1050, 1051
INTERDIT AU PUBLIC **Vol. 2** 368, 857
INTERDIT AUX MOINS DE 13 ANS **Vol. 6** *page 95, page 122*
INTERDIT DE SEJOUR **Vol. 2** 194, 344
INTERESSANTE FALL, DER **Vol. 4** 329, 774
INTERMEZZO **Vol. 4** 119, 323, 385, 738, 788 **Vol. 5** 840, 909
INTERNATIONAL PROSTITUTION **Vol. 2** 49, 486
INTERNO BERLINESE **Vol. 7** 964, 1093, 1176
INTERNO DI UN CONVENTO **Vol. 7** 1145
INTERPOL CONTRE X **Vol. 2** 469, 727
INTERRABANG **Vol. 7** 184
INTILL HELVETETS PORTAR **Vol. 5** 878, 1115
INTILNIREA **Vol. 6** *page 265, page 270, page 277*
INTIMITÄTEN **Vol. 4** 147, 270, 278
INTIMNI OSVETLENI **Vol. 10** 70, 227, 398, 532
INTOARCE - TE SI MAI PRIVESTE O DATA **Vol. 6** *page 269, page 278*
INTOARCERE LA DRAGOSTEA DINTII **Vol. 6** *page 268, page 274, page 279*
INTOARCEREA **Vol. 3** 254, 260, 313
INTOARCEREA DIN IAD **Vol. 6** *page 265, page 268, page 271*
INTOARCEREA LUI MAGELLAN **Vol. 3** 218, 286, 326
INTOARCEREA LUI VODA LAPUSNEANU **Vol. 3** 250, 268, 301
INTOARCEREA VLASINILOR **Vol. 6** *page 274, page 280*
INTOHIMON VALASSA **Vol. 5** 310, 374
INTOLERANS 1920 **Vol. 3** 437
INTR-O DIMINEATA **Vol. 3** 319
INTRE OGLINZI PARALELE **Vol. 3** 209, 254, 255
INTREPIDE, L' **Vol. 2** 97
INTRIGA **Vol. 8** 484, 611
INTRIGA EN EL ESCENARIA **Vol. 8** 524, 772
INTRIGANTE, L' **Vol. 2** 55, 87
INTRIGANTES, LES **Vol. 2** 427, 722, 832
INTRIGO A LOS ANGELES **Vol. 7** 17
INTRIGUEN DER MADAME DE LA POMMERAYE, DIE **Vol. 4** 266, 288, 574
INTRODUCTIONS PERVERSES **Vol. 2** 766 **Vol. 6** *page 79*
INTRUSA, L' **Vol. 7** 404, 442, 726
INTUNERICUL ALB **Vol. 6** *page 276, page 280, page 282*
INUTILE ATTESA **Vol. 7** 1067
INUTILI PRECAUZIONI **Vol. 7** 1189
INVASION **Vol. 8** 606
INVASION DE LOS ZOMBIES ATOMICOS, LA **Vol. 8** 297, 299
INVASORE, L' **Vol. 7** 481
INVASORES DEL ESPACIO, LOS **Vol. 8** 219, 406

INVASORI, GLI Vol. 7 537
INVENCIBILI SETTE, GLI Vol. 8 450, 541
INVENCIBLE HOMBRE INVISIBLE, EL Vol. 8 331, 762
INVENCIBLES, LOS Vol. 8 450, 541
INVENTIAMO L'AMORE Vol. 7 955, 1116
INVERNADERO, EL Vol. 8 670
INVERNO DI MALATO Vol. 7 95, 910
INVEROSIMILE, L' Vol. 7 1092
INVESTIGACION CRIMINAL Vol. 8 261, 768
INVIATI SPECIALI Vol. 7 161, 827
INVICIBILE CAVALIERE MASCHERATO, L' Vol. 7 1121
INVICIBILI FRATELLI MACISTE, GLI Vol. 7 501
INVICIBILI TRE, GLI Vol. 7 587
INVIDIA, L' Vol. 7 441, 701
INVIDIA, L' – I SETTE PECCATI CAPITALI Vol. 7 1062
INVIERNO EN MARBELLA Vol. 8 700
INVINGATORUL Vol. 6 *page 266, page 275, page 282*
INVIOABILE, L' Vol. 7 561
INVITATION AU VOYAGE, L' Vol. 2 79, 380, 785 Vol. 6 *page 90, page 99*
INVITE DE LA ONZIEME HEURE, L' Vol. 2 368, 863
INVITE DU MARDI, L' Vol. 2 185, 644
INVITEE, L' Vol. 2 377
INZYNIER SZERUDA Vol. 1 225, 241, 290, 308
IO AMLETO Vol. 7 1036
IO AMO,... TU AMI Vol. 7 283
IO, CATERINA Vol. 7 587
IO, CHIARA E LO SCURO Vol. 7 290, 593
IO CON TE NON CI STO PIU Vol. 7 596, 819, 925
IO E CATERINA Vol. 7 80, 834
IO E LUI Vol. 7 817
IO HO PAURA Vol. 7 599
IO LA CONOSCEVO BENE Vol. 7 817
IO, MAMMETA E TU Vol. 7 739
IO, MONACA, TRE BASTARDI E SETTE PECCATRICI Vol. 7 492
IO NON SCAPPO ... FUGGO Vol. 7 817
IO PIACCIO Vol. 7 196
IO SO CHE TU SAI CHE IO SO Vol. 7 80, 834, 868
IO SONO CON GLI IPPOPOTAMI Vol. 7 869
IO SONO MIA Vol. 7 698, 908
IO, SUO PADRE Vol. 7 725, 1054
IO T'HO INCONTRATA A NAPOLI Vol. 7 1121
IO TI UCCIDO ! Vol. 7 107
IO TIGRO, TU TIGRI, EGLI TIGRA Vol. 7 221, 514
IO ZOMBO, TU ZOMBI, LEI ZOMBA Vol. 7 692
IOANNES O BIAIOS Vol. 3 134, 135, 183
ION – BLESTEMUL PAMINTULUI, BLESTUMUL IUBIRII Vol. 3 284, 286
IONEL SI MARIOARA Vol. 3 216
IPHIGENEIA Vol. 3 135
IPOLOHAGOS NATASSA Vol. 3 205
IPPOCAMPO, L' Vol. 7 86, 443
IRA, L' Vol. 7 442, 701
IRATAN E IRACEMA Vol. 8 4, 32, 47
IRCA V HNIZDECKU Vol. 10 58, 453
IRCIN ROMANEK Vol. 10 53, 243, 256, 310, 460
IRENA DO DOMU Vol. 1 281, 334, 370 Vol. 6 *page 247*
IRENE, IRENE Vol. 7 819
IRGENDWO IN BERLIN Vol. 1 23, 61
IRIGAWAY Vol. 2 445
IRIS Vol. 7 843
IRIS OCH LÖJTNANTSHJÄRTA Vol. 5 332, 606, 840
IRMELI, SEITSENTOISTAVUOTIAS Vol. 5 374, 592
IRON KISS, THE Vol. 2 526
IRONIA DEL DINERO, LA Vol. 8 298
IRONIE DELLA VITA Vol. 7 131
IRONIE DU SPORT, L' Vol. 2 513, 642, 688
IRREDENTI, GLI Vol. 7 314
IRRENDE SEELEN Vol. 4 213, 266, 574
IRREPARABILE, L' Vol. 7 69
IRRER DUFT VON FRISCHEM HEU, EIN Vol. 1 1, 69
IRRESISTIBLE CATHERINE, L' Vol. 2 92, 740 Vol. 7 982
IRRESISTIBLE, L' Vol. 2 522
IRRESISTIBLE REBELLE, L' Vol. 2 37
IRRFAHRT INS GLÜCK Vol. 4 288
IRRGARTEN DER LEIDENSCHAFT Vol. 4 544, 721
IRRLICHT Vol. 4 113, 780
IRRLICHT UND FEUER Vol. 1 13, 68, 100
IRRTUM DES HERZENS Vol. 4 97, 124, 232
IRRUNGEN Vol. 4 190
IRRWEGE DER LIEBE Vol. 4 340, 779 Vol. 6 *page 194*
ISABEL AUF DER TREPPE Vol. 6 *page 153, page 170, page 175*
ISABEL DE SOLIS Vol. 8 516
ISABELLA, DUCHESSA DEI DIAVOLO Vol. 7 1213
ISABELLE A PEUR DES HOMMES Vol. 2 404, 479
ISADORA Vol. 3 437
ISÄN VANHA JA UUSI Vol. 5 374, 443
ISÄNTÄ SOITTAA HANURIA Vol. 5 374, 443

ISÄPAPPA JA KELTANOKKA Vol. 5 443, 549
ISASKAR KETURIN IHMEELLISET SEIKKAILUT Vol. 5 374, 443
ISBJORNEJAGT Vol. 5 97
ISFUGLE Vol. 5 11, 185, 283
ISIDRO EL LABRADOR Vol. 8 298, 546
ISKAM DA ZIVEJA Vol. 3 76
ISKANJA Vol. 3 395, 421
ISKELMÄKARUSELLI PYÖRII Vol. 5 374, 469
ISKELMÄKETJU Vol. 5 613, 634
ISLA DE LA MUERTE, LA Vol. 8 311, 611
ISLA DEL TESORO, LA Vol. 8 215
ISMERETLEN ELLENFEL Vol. 9 196, 317
ISMERI A SZANDI MANDIT ? Vol. 9 67, 273
ISO VAALEE Vol. 5 319, 324, 448
ISOLA DEGLI UOMINI PESCE, L' Vol. 7 426
ISOLA DEL GABBIANO, L' Vol. 7 697, 817, 818
ISOLA DEL TESORO, L' Vol. 7 858
ISOLA DELLA FELICITA, L' Vol. 7 1169
ISOLA DELLE SVEDESI, L' Vol. 7 742
ISOLA DI ARTURO, L' Vol. 7 688
ISOLA DI PROGNE, L' Vol. 7 529
ISOLA, L' Vol. 7 819, 908
ISOLA MISTERIOSA E IL CAPITANO NEMO, L' Vol. 7 1062
ISOLA SCOMPARSA, L' Vol. 7 654
ISOLE INSANGUINATE, LE Vol. 7 39
ISOVIHA Vol. 5 420, 497, 609
ISPETTORE VARGAS, L' Vol. 7 40, 476
ISPRAVI SE, DELFINA Vol. 3 339, 384
ISRAEL Vol. 7 37
ISSIDORA Vol. 3 192
IST EDDY POLO SCHULDIG ? Vol. 4 211, 407
IST MEIN MANN NICHT FABELHAFT ? Vol. 4 323, 788
ISTANBUL-MASCHE Vol. 1 33, 57, 70
ISTEN ES EMBER ELÖTT Vol. 9 67, 183, 297
ISTEN FIA, ÖRDÖG FIA Vol. 9 234, 337
ISTEN HOZTA, ÖRNAGY UR Vol. 9 134, 259, 265
ISTEN ÖSZI CSILLAGA Vol. 9 121, 265, 314
ISTEN RABJAI Vol. 9 28, 76, 236
ISTEN VELETEK, BARATAIM ! Vol. 9 6, 53, 152
ISTID Vol. 5 812, 824
ISTINTO, L' Vol. 7 730
ISTITUTRICE DI SEI BAMBINE, L' Vol. 7 168
ISTORIA MIAS ZOES Vol. 3 149
ISTREL Vol. 3 434
ISTRUTTORIA, L' Vol. 7 369
ISTVAN, AKIRALY Vol. 9 60, 51, 61, 99, 111, 154, 210, 232, 279, 295
ISZONY Vol. 9 117, 149, 332
ITALIA A MANO ARMATA Vol. 7 1213
ITALIA IN PIGIAMA, L' Vol. 7 1137
ITALIA KZ Vol. 7 373, 416
ITALIA, PAESE DI GRIGANTI ? Vol. 7 369
ITALIA PICCOLA Vol. 7 897, 1027
ITALIA S'E DESTA, L' Vol. 7 829
ITALIA VIVA, L' Vol. 7 99, 610
ITALIAN BOYS Vol. 7 318
ITALIANE E L'AMORE - IL MATRIMONIO ASSURDO, LE Vol. 7 86, 827
ITALIANE E L'AMORE - LA FRENESIA DEL SUCCESSO, LE Vol. 7 86, 502
ITALIANE E L'AMORE - LA VEDOVA BIANCA, LE Vol. 7 86, 898
ITALIANE E L'AMORE, LE - IL PREZZO DELL'AMORE, LE Vol. 7 86, 1203
ITALIANE E L'AMORE - LE RAGAZZE MADRI, LE Vol. 7 86, 408
ITALIANE E L'AMORE - L'EDUCAZIONE SESSUALE DEI FIGLI, LE Vol. 7 86, 348
ITALIANE E L'AMORE - L'INFEDELTA CONIUGALE, LE Vol. 7 86, 502
ITALIANE E L'AMORE - VIAGGIO DI NOZE, LE Vol. 7 86, 502
ITALIANE E L'AMORE - LA SEPARAZIONE LEGALE, LE Vol. 7 86, 408
ITALIANE E L'AMORE - LE ADOLESCENTI E L'AMORE, LE Vol. 7 86, 408
ITALIANI BRAVA GENTE Vol. 7 690, 1056
ITALIEN DES ROSES, L' Vol. 2 301
ITALIENISCHES CAPRICCIO Vol. 1 8, 35, 125
ITALIENOS ESTAN LOCOS, LOS Vol. 8 517
ITEL A BALATON Vol. 9 71, 201
ITELEM Vol. 3 307, 329
ITELET Vol. 9 263, 318
ITENDENTI Vol. 3 10, 27 Vol. 6 *page 12*
ITENERAIRE BIS Vol. 6 *page 111, page 143*
IT'S HAPPENED TOMORROW Vol. 2 778
ITTO Vol. 2 8, 37, 514, 648, 714
IVALJO Vol. 3 36, 61, 99, 101 Vol. 6 *page 46*
IVAN Vol. 10 32, 72 Vol. 10 32, 72
IVAN IL FIGLIO DEL DIAVOLO BIANCO Vol. 7 798, 1054
IVAN KONDAREV Vol. 3 59, 121 Vol. 6 *page 37, page 46, page 56*
IVAN SISMAN Vol. 3 82
IVAN SUSANIN Vol. 3 66
IVANA V UTOKU Vol. 10 234, 250

IVANNA Vol. 8 332, 770
IVANOV Vol. 10 245, 273, 459
IVAR HVIDTFELDT Vol. 5 97
IVONNE LA BELLA DANZATRICE Vol. 7 264, 701
IVRIGE KORTSPILLERE Vol. 5 97
IWNA, LA PERLA DEI GANGE Vol. 7 1038
IWONKA Vol. 1 174, 176
IZA NENI Vol. 9 71, 322
IZBAVITELJ Vol. 3 430, 446
IZBIRACICA Vol. 3 359, 371, 413
IZDAJNIK Vol. 3 362, 418
IZDIRVA SE ... Vol. 6 *page 62*
IZGORI, ZA DA SVETIS Vol. 6 *page 62, page 63*
IZGUBLJENI ZAVICAJ Vol. 3 352
IZKUPLENIE Vol. 3 56
IZKUSTVENATA Vol. 3 72
IZKUSTVENATA PATICA Vol. 6 *page 29, page 54*
IZPIT Vol. 3 63, 110
IZPITI PO NIKOE VREME Vol. 3 101, 116 Vol. 6 *page 53*

J

J 3, LES Vol. 2 32, 439
... JA ALLA OLI TULINEN JÄRVI Vol. 5 354, 580
... JA HELENA SOITTAA Vol. 5 478, 559, 634
JA, JA, DIE FRAUN SIND MEINE SCHWACHE SEITE Vol. 4 221, 340 Vol. 6 *page 198*
JA JSEM STENA SMRTI Vol. 10 82, 173, 325
JA MILUJEM, TY MILUJES Vol. 10 82, 155, 228, 402
JA NEJSEM JA Vol. 10 23, 41, 208
JA, SPRAVEDLNOST Vol. 10 283, 369, 577
"JA TO TEDY BERU, SEFE ...! Vol. 10 139, 144, 593
JA, TREU IST DIE SOLDATENLIEBE Vol. 4 255, 292
JA TRUCHLIVY BUH Vol. 10 283, 369, 371
JA TU RZADZE Vol. 1 176, 278, 308
JA UZ BUDU HODNY, DEDECKU ! Vol. 10 139, 430, 593
JÄÄKÄRIN MORSIAN Vol. 5 315, 420, 471, 566, 608
J'ACCUSE Vol. 2 152, 153, 308, 402, 539
JACHT DER SIEBEN SÜNDEN, DIE Vol. 4 132, 584, 796
JACHTELE GALBENE Vol. 3 301, 320 Vol. 6 *page 272*
JACHYME, HOD DO STROJE ! Vol. 10 139, 292, 425
JACK Vol. 2 37 Vol. 5 858, 975, 1160, 1201
JACK CUOR DI LEONE Vol. 7 168
JACK, EL DESTRIPADOR DE LONDRES Vol. 8 732, 760
JACK EL NEGRO Vol. 8 298, 745
JACK FORBIS CONTRO ROBINET Vol. 7 668
JACK PERRYS EHEGLÜCK ODER DAS GESTÄNDNIS AM HOCHZEITSABEND Vol. 4 217, 356
JACKE WIE HOSE Vol. 1 56, 67, 88
JACOB SMITAREN Vol. 5 922, 1006, 1108, 1145
JACOBS STEGE Vol. 5 840, 909
JACOPO HORTIS Vol. 7 731
JACQUES ET JACOTTE Vol. 2 395, 858
JAD Vol. 3 384, 419
JADA, GOSCIE, JADA Vol. 1 143, 220, 243, 313, 401
JADZIA Vol. 1 176, 205
JAFET, DER SOGER SIG EN FADER Vol. 5 181
JAG Vol. 5 696
JAG ÄLSKAR DIG, ARBIGGA Vol. 5 1042, 1181
JAG ÄLSKAR DIG, KARLSSON Vol. 5 865, 949
JAG ÄLSKAR, DU ÄLSKAR Vol. 5 660, 1051
JAG ÄR ELD OCH LUFT Vol. 5 1042, 1125
JAG AR MARIA Vol. 5 858, 931, 933
JAG ÄR MARKIS - MED UPPDRAG ATT ÄLSKA Vol. 5 1106
JAG ÄR MED BARN Vol. 5 1051, 1071, 1093, 1097
JAG ÄR MED EDER... Vol. 5 1173
JAG ÄR NYFICKEN - BLA Vol. 5 1204
JAG ÄR NYFICKEN - GUL Vol. 5 1204
JAG DRÄPTE Vol. 5 1042, 1164
JAG - EN ÄLSKARE Vol. 5 1033
JAG - EN KVINNA Vol. 5 833, 839
JAG - EN KVINNA 2 Vol. 5 832, 860, 936
JAG EN MARKIS - MED UPPDRAG ATT ÄLSKA Vol. 5 839
JAG, EN OSKYLD Vol. 5 1024
JAG GIFTA MIG - ALDRIG Vol. 5 928
JAG HETER STELIOS Vol. 5 917
JAG RÄDNAR Vol. 5 946
JAG VILL LIGA MED DIN ÄLSKARE MAMMA... Vol. 5 1040
JAGAD Vol. 5 876, 923
JAGD AUF DICH Vol. 6 *page 181, page 195, page 198*
JAGD AUF MENSCHEN Vol. 4 255, 384, 725

JAGD AUF UNBEKANNT Vol. 3 395
JAGD NACH DEM GLÜCK, DIE Vol. 4 202, 556, 729
JAGD NACH DEM STIEFEL, DIE Vol. 1 7, 49, 92
JAGD NACH DEM TODE, DIE Vol. 4 115, 291, 734
JAGD NACH DER BRAUT, DIE Vol. 4 626, 634, 734
JAGD NACH DER FRAU, DIE Vol. 4 535, 584
JAGD NACH DER HUNDERTPFUNDNOTE ODER DIE REISE UM DIE WELT Vol. 4 217, 356
JAGD NACH DER MILLION, DIE Vol. 4 287, 384, 644, 646
JAGD NACH DER WAHRHEIT, DIE Vol. 4 211, 247
JÄGER AUS KURPFALZ, DER Vol. 4 354, 644, 738
JAGER SPAGETTI Vol. 6 *page 239*
JÄGER VON DER RISS, DER Vol. 4 571, 742
JÄGER VON FALL, DER Vol. 4 25, 344, 370, 495, 544
JAGODE U GRLU Vol. 6 *page 305, page 323*
JAGUAR Vol. 2 752 Vol. 9 37
JAGUAROV SKOK Vol. 6 *page 305, page 318*
JAHR VOLL MUSIK, EIN Vol. 1 52
JAHRE VERGEHEN, DIE Vol. 4 49
JAHRESZEITEN DES LEBENS Vol. 4 251
JAHRGANG 21 Vol. 1 82, 125
JAHRMARKT DES LEBENS Vol. 4 213, 424
JAI - ALAI Vol. 8 357, 752
J'AI ARROSE MES GALONS Vol. 2 273, 404
J'AI DIX-SEPT ANS Vol. 2 48, 627
J'AI EPOUSE UNE OMBRE Vol. 6 *page 100, page 132*
J'AI LE FEU AUX FESSES Vol. 6 *page 79*
J'AI TOUT DONNE Vol. 2 635
J'AI TUE Vol. 2 159, 237
J'AI TUE RASPOUTINE Vol. 2 59
J'AI VOULU RIRE COMME LES AUTRES Vol. 2 769, 817
JÄINEN SAARI Vol. 5 524
JAK BASNICI PRICHAZEJI O ILUZE Vol. 10 389, 577, 622
JAK BASNIKUM CHUTNA ZIVOT Vol. 10 226, 389, 577
JAK BYC KOCHANA Vol. 1 254, 305 Vol. 6 *page 216*
JAK DALEKO STAD, JAK BLISKO Vol. 1 187, 298 Vol. 6 *page 253*
JAK NAPALIT ADVOKATA Vol. 10 208, 286, 522
JAK RODI CHLAP Vol. 10 56, 139, 180, 181, 341, 344, 415
JAK ROZPETALEM II WOJNE SWIATOWA Vol. 1 198, 206, 313, 343
JAK SE BUDI PRINCEZNY Vol. 10 41, 425, 567
JAK SE DELA SMICH Vol. 10 389, 550
JAK SE KRADE MILION Vol. 10 153, 227, 296
JAK SE TOCI ROZMARYNY Vol. 10 144, 344, 496
JAK SE ZBAVIT HELENKY Vol. 10 299, 498, 564
JAK SVET PRICHAZI O BASNIKY Vol. 10 299, 496, 622
JAK TO SIE ROBI Vol. 1 311, 355
JAK UTOPIT DR. MRACKA ANEB KONEC VODNIKU V CECHACH Vol. 10 41, 388, 425
JAK VASICEK K NOHAM PRISEL Vol. 10 198
JAK ZYC Vol. 6 *page 221*
JAKKO Vol. 4 79, 124, 566
JAKO JED Vol. 10 149, 258, 497
JAKO ZAJICI Vol. 10 173, 383, 466
JAKOB DER LÜGNER Vol. 1 17, 48, 73
JAKOU BARVU MA LASKA Vol. 10 229, 313, 425
JAKTEN Vol. 5 766, 807, 826, 953, 1033
JAKUB Vol. 10 233, 398, 428
JALEA REAL Vol. 8 245, 368
JALISCO CANTA EN SEVILLA Vol. 8 298, 463
JÄLKEEN SYNTIINLANKEEMUKSEN Vol. 5 443, 643
JALL ALLA DÖRRAR ÖPNA Vol. 5 869
JALOUSIE DE BARBOUILLE, LA Vol. 2 9, 688
JALOUSIENS MAGT Vol. 5 83
JALOUX COMME UN TIGRE Vol. 2 156, 303
JALTA Vol. 6 *page 315*
JAMAIS AVANT LE MARIAGE Vol. 6 *page 83, page 96*
JAMAIS BATTU Vol. 6 *page 123*
JAMAIS DEUX SANS TROIS Vol. 2 460, 627
JAMAIS PLUS TOUJOURS Vol. 2 60
JAMBES EN L'AIR A BANGKOK Vol. 2 484
JAMBES EN L'AIR, LES Vol. 2 258
JAMBON D'ARDENNE Vol. 2 66, 251
JAN BIBIJAN Vol. 6 *page 47, page 62*
JAN CIMBURA Vol. 10 92, 569, 621
JAN GAR TIL FILMEN Vol. 5 150, 248, 272
JAN HUS Vol. 10 154, 503, 547, 561
JAN ROHAC Z DUBE Vol. 10 142, 524, 621

JAN UND DIE SCHWINDLERIN **Vol. 4** 737, 742
JAN VYRAVA **Vol. 10** 256, 564, 612
JAN ZIZKA **Vol. 10** 154, 503, 547, 561
JANA **Vol. 3** 437 **Vol. 10** 34, 453
JANA, DAS MÄDCHEN AUS DEM BÖHMERWALD **Vol. 10** 34, 453
JANDARMII **Vol. 3** 304
JANDRO **Vol. 8** 298, 426
JANE, MI PEQUENA SALVAJE **Vol. 8** 269, 681
JANIKA **Vol. 9** 71, 312
JÄNIKSEN VUOSI **Vol. 5** 500, 512, 541
JANKO MUZYKANT **Vol. 1** 176, 384
JANKO ZWYCIEZCA **Vol. 1** 404
JANNE MODIG **Vol. 5** 1183
JANNE VÄNGMAN I FARTEN **Vol. 5** 840, 928
JANNE VÄNGMAN OCH DEN STORA KOMETEN **Vol. 5** 840, 884
JANNE VÄNGMAN PA NYA ÄVENTYR **Vol. 5** 840, 1091
JANNE VÄNGMANS BRAVADER **Vol. 5** 840, 843
JANOS VITEZ **Vol. 9** 136, 178, 237
JANOSIK **Vol. 1** 179, 293 **Vol. 6** *page 215, page 245* **Vol. 10** 256, 412, 490, 536, 537
JANOSIK I.,II. **Vol. 10** 210, 217, 252, 576
JANSSONS FRESTELSE **Vol. 5** 891, 923, 1031, 1181
JANUSKOPF **Vol. 1** 5, 9, 14, 18
JANUSKOPF, DER **Vol. 4** 178, 288, 558
JAQUE MATE **Vol. 8** 426, 732
JAR ÄR ELD OCH LUFT **Vol. 5** 929, 1078
JAR AR MED EDER... **Vol. 5** 909
JARA CIMRAN, LEZICI, SPICI **Vol. 10** 144, 344, 571
JARA GOSPODA **Vol. 3** 390, 439
JARCIN PROFESOR **Vol. 10** 92, 580, 621
JARDIN DE LAS DELICIAS, EL **Vol. 8** 348
JARDIN DES SUPPLICES, LE **Vol. 2** 495, 599
JARDIN QUI BASCULE, LE **Vol. 2** 739
JARDIN SECRETO, EL **Vol. 8** 210, 570
JARDINIER D'ARGENTEUIL, LE **Vol. 2** 128, 875
JARDINIER, LE **Vol. 6** *page 80, page 116, page 125, page 126, page 142*
JARKA A VERA **Vol. 10** 460, 621
JARNI PISEN **Vol. 10** 92, 503, 569
JARNI POVETRI **Vol. 10** 313, 387
JARNI VODY **Vol. 10** 197, 569, 621
JAROSLAW DABROWSKI **Vol. 1** 321, 369 **Vol. 6** *page 259*
JARRAPELLEJOS **Vol. 8** 219, 610
JARRETELLES ROSES SUR BAS NOIRS **Vol. 6** *page 93, page 129*
JARVANY, A **Vol. 9** 107, 150, 167, 229
JARZEBINA CZERWONA **Vol. 1** 355, 391 **Vol. 6** *page 258*
JASNE LANY **Vol. 1** 225, 305, 397
JASNIE PAN SZOFER **Vol. 1** 278, 308, 404
JASTRZAB **Vol. 1** 326
JATEK A MUZEUMBAN **Vol. 9** 83, 183, 332
JATEK A SORSSAL **Vol. 9** 175, 220
JATEK A SZERELEMMEL **Vol. 9** 39, 261
JATSZANI KELL **Vol. 9** 6, 46, 51, 188, 316
JAULA, LA **Vol. 8** 232, 297, 770
JAULA NO TIENE SECRETOS, UNA **Vol. 8** 578, 635
JAVA DES OMBRES, LA **Vol. 6** *page 92*
JAVA, UNE **Vol. 2** 180, 193, 413
J'AVAIS SEPT FILLES **Vol. 2** 173, 318, 347, 807
JAVOR A JULIANA **Vol. 10** 82, 273, 548
JAZOL **Vol. 6** *page 316*
JAZZ GOSSEN **Vol. 5** 878, 1052, 1166
JE CHANTE **Vol. 2** 566
JE CRIE, JE JOUIS **Vol. 2** 291
JE FAUS ... OU ON ME DIT **Vol. 6** *page 93*
JE N'AI QUE TOI AU MONDE **Vol. 2** 217
JE N'AIME QUE TOI **Vol. 2** 247
JE PARLE D'AMOUR **Vol. 2** 278 **Vol. 6** *page 36*
JE PLAIDE NON COUPABLE **Vol. 2** 501, 671
JE PRENDS LA CHOSE DU BON COTE **Vol. 2** 600
JE REVIENDRAI A KANDARA **Vol. 2** 322, 601, 743
JE SAIS RIEN MAIS JE DIRAI TOUT **Vol. 2** 517, 783
JE SERAI SEULE APRES MINUIT **Vol. 2** 167, 412
JE SUIS A PRENDRE **Vol. 2** 2
JE SUIS AVEC TOI **Vol. 2** 9, 395
JE SUIS NEE POUR BAISER **Vol. 6** *page 120*
JE SUIS PIERRE RIVIERE **Vol. 2** 600, 854
JE SUIS TIMIDE...MAIS JE ME SOIGNE **Vol. 2** 7, 787
JE SUIS UN MOUCHARD **Vol. 2** 9, 833
JE SUIS UN SENTIMENTAL **Vol. 2** 194, 501
JE SUIS UNE BELLE SALOPE **Vol. 2** 294
JE T'AI ECRIT UNE LETTRE D'AMOUR **Vol. 2** 190
JE T'AIME, JE T'AIME **Vol. 2** 110, 263
JE T'AIME, MOI NON PLUS **Vol. 2** 441, 502, 588
JE TE CONFIE MA FEMME **Vol. 2** 646
JE TE TIENS, TU ME TIENS PAR LA BARBICHETTE **Vol. 2** 534

JE, TU, ELLES **Vol. 2** 344
JE VAIS CRAQUER **Vol. 2** 252, 739
JE VOUS AIME **Vol. 2** 74 **Vol. 6** *page 83*
JE VOUS AIMERAI TOUJOURS **Vol. 2** 29 **Vol. 7** 48, 466, 756
JE VOUS FERAI AIMER LA VIE **Vol. 2** 814
JE VOUS SALUE MAFIA **Vol. 2** 21, 208
JEAN CHOUAN **Vol. 2** 216
JEAN D'AGREVE **Vol. 2** 731, 805
JEAN DE LA CIRQUE **Vol. 2** 322
JEAN DE LA LUNE **Vol. 2** 407, 427, 573, 656
JEANNE **Vol. 2** 42, 526, 803
JEANNE DORE **Vol. 7** 48, 725, 955
JEANNETTE BOURGOGNE **Vol. 2** 404
JEANNOU **Vol. 2** 254, 595
JEANS E UNA MAGLIETTA, UN **Vol. 7** 123, 222
JEANS TONIC **Vol. 6** *page 121*
JEDAN DRUGI PUT **Vol. 6** *page 313*
JEDANAESTA ZAPOVIJED **Vol. 3** 421, 437
JEDE FRAU HAT EIN GEHEIMNIS **Vol. 4** 28, 424, 544
JEDE FRAU HAT ETWAS **Vol. 0** *page 83* **Vol. 4** 229
JEDE WOCHE HOCHZEITSTAG **Vol. 1** 19, 114, 122
JEDEN DEN PRE STARU PANIU **Vol. 10** 72, 459, 575
JEDEN STRIBRNY **Vol. 10** 82, 172, 577
JEDEN Z 36 – CIU **Vol. 1** 346, 371
JEDEN Z NICH JE VRAH **Vol. 10** 299, 531
JEDENACTE PRIKAZANI **Vol. 10** 125, 167, 256, 345, 612
JEDER FRAGT NACH ERIKA **Vol. 4** 49
JEDER STIRBT FÜR SICH ALLEIN **Vol. 1** 29, 47, 72
JEDINI IZLAZ **Vol. 3** 373, 425
JEDNA Z MILIONU **Vol. 10** 256, 524
JEDNOOKI VOJNICI **Vol. 3** 350
JEES JA JUST **Vol. 5** 488, 592
„... JEES OLYMPIALAISET", SANOI RYHMY **Vol. 5** 379, 579
JEFF **Vol. 2** 814
JEFTHAS DOTTER **Vol. 5** 1012
JEG DREPTE **Vol. 5** 726, 760
JEG ELSKER BLAT **Vol. 5** 128, 199, 201, 275
JEG ELSKER DIG **Vol. 5** 241, 248
JEG ELSKER EN ANDEN **Vol. 5** 89, 224
JEG – EN ELSKER **Vol. 5** 209, 286
JEG – EN KVINDE **Vol. 5** 2, 6
JEG – EN KVINDE II **Vol. 5** 1, 6, 76
JEG – EN MARKI **Vol. 5** 6, 248, 307
JEG ER SGU MIN EGEN **Vol. 5** 37, 272, 281, 292
JEG HAR ELSKET OG LEVET **Vol. 5** 31, 32, 55, 70, 101, 208
JEG MAA TILGIVE **Vol. 5** 97
JEG MODTE EN MORDER **Vol. 5** 89, 136
JEGO EKSELENCJA SUBIEKT **Vol. 1** 278, 346
JEGO OSTATNI CZYN **Vol. 1** 215, 326
JEGO WIELKA MILOSC **Vol. 1** 176, 278, 308
JEHOVA **Vol. 9** 2
JEJ PORTRET **Vol. 6** *page 217*
JEJI HRICH **Vol. 10** 472, 564
JEJI LEKAR **Vol. 10** 256, 524
JEJI PASTORKYNE **Vol. 10** 34, 256, 453, 509
JEJI PRINC **Vol. 10** 45
JEJICH SVATEBNI NOC **Vol. 10** 167, 310
JELBESZED **Vol. 9** 154, 259, 327
JELENIDÖ **Vol. 9** 339
JELEZBAL **Vol. 9** 236, 315
JEMNE UMENI OBRANY **Vol. 10** 250, 342, 508
JEN HO NECHTE, AT SE BOJI **Vol. 10** 41, 144, 153
JEN SI TAK TROCHU PISKNOUT **Vol. 10** 172, 383, 466
JENARO, EL DE LOS CATORCE **Vol. 8** 681, 749
JENGI **Vol. 5** 383, 639
JENNY **Vol. 2** 279, 402
JENNY UND DER HERR IM FRACK **Vol. 4** 285, 615, 688
JENNYS BUMMEL DURCH DIE MÄNNER **Vol. 4** 332, 422, 462
JENS DAGLYKKE **Vol. 5** 81
JENS LANGKNIV **Vol. 5** 142
JENS MANSSON I AMERIKA **Vol. 5** 1091, 1159
JENSEITS DER STRASSE **Vol. 4** 49, 234, 602
JENSEITS DES STROMES **Vol. 4** 634
JENSEITS VON GUT UND BÖSE **Vol. 4** 208, 390
JENSEN LAENGE LEVE **Vol. 5** 2, 179
JENTESPRANGET **Vol. 5** 762
JEPPE PA BJERGET **Vol. 5** 45, 195, 273
JEPPE PAA BJERGET **Vol. 5** 760, 788
JERGUS LAPIN **Vol. 10** 273, 320, 377
JERICHO **Vol. 2** 96, 156, 719
JERK A ISTANBUL **Vol. 2** 293, 475
JERNBANENS DATTER **Vol. 5** 97
JEROME PERREAU **Vol. 2** 113, 174, 330, 402
JEROMIN **Vol. 8** 611, 616, 630
JERRY, PESCADOR DE AGUAS TURVAS **Vol. 8** 196
JERUZSALEM **Vol. 9** 234
JE'SKU' TALE MED JOR'NSEN **Vol. 5** 83

JESLI SERCE MASZ BIJACE **Vol. 6** *page 241, page 247, page 248*
JESLI SIE ODNAJDZIEMY **Vol. 6** *page 213, page 245, page 252*
JESSE E LESTER DUE FRATELLI IN UN POSTO CHIAMATO TRINITA **Vol. 7** 780
JESSICA **Vol. 7** 923
JESTE SVATBA NEBYLA **Vol. 10** 407, 460, 612
JESTRAB KONTRA HRDLICKA **Vol. 10** 142, 154, 296
JESTRABI MOUDROST **Vol. 10** 74, 209, 389, 496
JESZCZE SLYCHAC SPIEW I RZENIE KONI **Vol. 1** 183, 360
J'ETAIS UNE AVENTURIERE **Vol. 2** 8, 59, 427, 657
JETPILOTER **Vol. 5** 235, 248
JETTATORE **Vol. 4** 270, 570
JETTATURA **Vol. 2** 42, 547
JETTCHEN GEBERT **Vol. 4** 159
JETZT UND IN DER STUNDE MEINES TODES **Vol. 1** 17, 49, 92
JEU **Vol. 2** 691
JEU AVEC LE FEU, LE **Vol. 2** 450, 536
JEU BRUTAL, UN **Vol. 6** *page 87*
JEU DE LA VERITE, LE **Vol. 2** 21, 248, 566
JEU DE MASSACRE **Vol. 2** 306, 738
JEU DU SOLITAIRE, LE **Vol. 2** 240
JEUNE COUPLE, UN **Vol. 2** 294
JEUNE FILLE ASSASSINEE, LA **Vol. 2** 351
JEUNE FILLE ET L'ENFER, LA **Vol. 8** 296, 302
JEUNE FILLE ET UN MILLION, UNE **Vol. 2** 424
JEUNE FILLE POUR PARTOUZES **Vol. 2** 52
JEUNE FILLE SAVAIT, UNE **Vol. 2** 344, 722
JEUNE FOLLE, LA **Vol. 2** 402, 837
JEUNE HOMME HONORABLE, UN **Vol. 2** 224
JEUNE MARIE, LE **Vol. 6** *page 71, page 78*
JEUNES COUPLES DECHAINES **Vol. 2** 814
JEUNES DANDISES AU PAIR **Vol. 6** *page 94*
JEUNES FILLES A MARIER **Vol. 2** 388, 833
JEUNES FILLES DE PARIS **Vol. 2** 249, 581, 644
JEUNES FILLES EN DETRESSE **Vol. 2** 23, 427
JEUNES FILLES IMPUDIQUES **Vol. 2** 721
JEUNES LOUPS, LES **Vol. 2** 738
JEUNES MARIES **Vol. 2** 232, 279, 488
JEUNESSE **Vol. 2** 308, 775, 803
JEUNESSE D'ABORD **Vol. 2** 48, 388
JEUNESSE, UNE **Vol. 6** *page 85, page 103*
JEUX D'ADULTES POUR GAMINES EXPERTES **Vol. 2** 291
JEUX DANGEREUX, LES **Vol. 2** 368, 503, 662
JEUX DE FEMMES **Vol. 2** 368, 627
JEUX DE LA COMTESSE D'OLINGEN DE GRATZ, LES **Vol. 2** 527 **Vol. 6** *page 138, page 142*
JEUX DE L'AMOUR, LES **Vol. 2** 285, 517, 655, 770
JEUX DE LANGUE **Vol. 2** 294
JEUX DE MAINS **Vol. 6** *page 79, page 106*
JEUX DE MINETTE **Vol. 2** 872
JEUX INTERDITS **Vol. 2** 96, 423
JEUX POUR COUPLES INFIDELES **Vol. 2** 649
JEUX SONT FAITS, LES **Vol. 2** 248, 566, 676
JEZDEC FORMULE RISK **Vol. 10** 98, 369, 505
JEZERO **Vol. 3** 374, 432
JEZIORO OSOBLIWOSCI **Vol. 1** 187, 202, 227
JIM BOUGNE BOXEUR **Vol. 2** 388
JIM CORWEY IST TOT **Vol. 4** 213
JIM LA HOULETTE **Vol. 2** 193, 460, 807
JIMMY, EIN SCHICKSAL VON MENSCH UND TIER **Vol. 4** 320, 423, 454
JINDRA **Vol. 10** 34
JINDRA, HRABENKA OSTROVINOVA **Vol. 10** 167, 243, 345, 453
JINETES DEL TERROR, LOS **Vol. 8** 598, 772
JINY VZDUCH **Vol. 10** 2, 524
J'IRAI COMME UN CHEVAL FOU **Vol. 2** 60, 823, 843
J'IRAI CRACHER SUR VOS TOMBES **Vol. 2** 118, 309
JIZDNI HLIDKA **Vol. 10** 167, 168, 256, 412
JMENEM JEHO VELICENSTVA **Vol. 10** 564
JO **Vol. 2** 97, 224
JO EJT, MUKI **Vol. 9** 28
JO ESTET NYAR, JO ESTET SZERELEM **Vol. 9** 41, 137, 305
JO-HO-HO **Vol. 6** *page 56, page 57, page 58*
JO LA ROMANCE **Vol. 2** 193, 197
JO, PAPA ! **Vol. 8** 276
JO UTAT AUTOBUSZ **Vol. 9** 241, 314
JOAN LUI **Vol. 7** 80, 347, 698
JOAO RATAO **Vol. 8** 122
JOAQUIN MURIETA **Vol. 8** 214, 551
JOB LAZADASA **Vol. 9** 55, 88, 137, 221, 279
JOBB ERKÖLCSÖKET **Vol. 9** 136
JOBBRA EN, BALRA TE **Vol. 9** 337
JOCASTE **Vol. 2** 452, 498, 805
JOCELYN **Vol. 2** 247, 402, 511, 524, 627

JOCKEY EST EN RETARD, LE **Vol. 2** 159, 286, 330
JOCKS **Vol. 7** 414
JOCONDES, LES **Vol. 6** *page 71*
JOE CALIGULA **Vol. 2** 74
JOE HILL **Vol. 5** 852, 917, 1096
JOE IL ROSSO **Vol. 7** 639, 756, 1116
JOE NAVIDAD **Vol. 8** 461, 495
JOE O TROMEROS **Vol. 3** 198, 203, 205
JOFFRE LA MIZIL **Vol. 3** 212
JOFROI **Vol. 2** 99, 871, 872
JOGHI, DER **Vol. 4** 202
JOGO DE MAO **Vol. 8** 15, 66
JOHAN **Vol. 5** 937, 996
JOHAN, CARNET INTIME HOMOSEXUEL **Vol. 2** 2, 30
JOHAN NYT ON MARKKINAT **Vol. 5** 392, 639
JOHAN PA SNIPPEN **Vol. 5** 933
JOHAN PA SNIPPEN TAR HEM SPELET **Vol. 5** 1003, 1073
JOHAN ULFSTJERNA **Vol. 5** 840, 891, 925, 997
JOHANN **Vol. 4** 110, 485, 619, 647
JOHANN BAPTISTE LINGG **Vol. 4** 35, 208, 223, 695
JOHANN STRAUSS, K. UND K. HOFBALLMUSIKDIREKTOR **Vol. 4** 132, 391, 424
JOHANNES GOTH **Vol. 4** 648, 747
JOHANNES KEPLER **Vol. 1** 18, 37, 55, 92
JOHANNISFEUER **Vol. 4** 79, 115, 774
JOHANNISNACHT **Vol. 4** 28, 340, 544
JOHANNISTRAUM **Vol. 4** 288
JOHANSSON OCH VESTMAN **Vol. 5** 840, 909
JOHN ERICSSON – SEGRARE VID HAMPTON ROADS **Vol. 5** 840, 909, 927, 1189
JOHN OG IRENE **Vol. 5** 180, 199, 304
JOHN TRAVOLTA ... DA UN INSOLITO DESTINO **Vol. 7** 1086
JOHNNY BANCO **Vol. 2** 279, 427
JOHNNY LARSEN **Vol. 5** 11, 46, 175
JOHNNY RATON **Vol. 8** 438, 627
JOHNNY WEST **Vol. 8** 306, 475
JÖJJÖN ELSEJEN **Vol. 9** 71, 133, 270
JOKER, LE **Vol. 2** 355
JOKEREN **Vol. 5** 70, 183
JOKIN IHMISESSÄ **Vol. 5** 462, 516
JOLANDA, LA FIGLIA DEL CORSARO NERO **Vol. 7** 91, 283, 333, 522, 781, 1195
JOLANTA – DEN GÄCKANDE SUGGAN **Vol. 5** 997, 1165
JOLI COEUR, LE **Vol. 6** *page 81, page 98*
JOLI MONDE **Vol. 2** 362, 488
JOLIE BRETONNE, LA **Vol. 2** 731
JOLLY **Vol. 7** 390
JOLLY, DER TEUFELSKERL **Vol. 4** 135, 334
JOMADAR, A **Vol. 9** 196, 270
JOMFRU TROFAST **Vol. 5** 770, 788
JON **Vol. 5** 404, 410, 511, 522
JÖN A ROZSON AT **Vol. 9** 220, 323
JONAS A MELICHAROVA **Vol. 10** 258, 394, 535
JONAS II ANEB JAK JE DULEZITE MITI MELICHAROVOU **Vol. 10** 139, 278, 330
JONE OR GLI ULTIMI GIORNI DI POMPEI **Vol. 7** 1055
JONNY ROOVA **Vol. 5** 990, 1163
JONNY STIEHLT EUROPA **Vol. 4** 119, 340
JÖNSSONLIGAN & DYNAMITHARRY **Vol. 5** 869, 947, 1212
JÖNSSONLIGAN FAR GULDFEBER **Vol. 5** 869, 882, 1038
JORAS PROHIBIDAS **Vol. 8** 537
JORDEN ER FLAD **Vol. 5** 13, 42, 273
JORDEN RUNT MED FANNY HILL **Vol. 5** 1093, 1127
JÖRG RATGEB, MALER **Vol. 1** 34, 37, 124
JOROBADO DE LA MORGUE, EL **Vol. 8** 626, 627
JORUND SMED **Vol. 5** 722, 723
JOS OISI VALTAA... **Vol. 5** 349, 485
JOS OVAJ PUT **Vol. 6** *page 308, page 315, page 321*
JOSE **Vol. 8** 270
JOSE DO TELHADO **Vol. 8** 18, 20
JOSE MARIA **Vol. 8** 753, 768
JOSEF KAJETAN TYL **Vol. 10** 167, 345
JOSEF UND SEINE BRÜDER **Vol. 1** 21, 32, 120
JOSEPH **Vol. 7** 980, 1002
JOSEPHA **Vol. 6** *page 70, page 96*
JOSETTE **Vol. 2** 533, 775
JOSSAIN ON RAILO **Vol. 5** 363, 379
JOU-JOU **Vol. 7** 26, 369, 701
JOUET, LE **Vol. 2** 74, 285
JOUEUR D'ECHECS, LE **Vol. 2** 9, 81, 158, 339, 538, 623, 657, 826, 871
JOUEUR, LE **Vol. 2** 226, 251, 625
JOUEUSE D'ORGUE, LA **Vol. 2** 192, 363, 412, 603
JOUIR ! **Vol. 2** 721
JOUIR JUSQU'AU DELIRE **Vol. 6** *page 79*
JOUISSEUR, LE **Vol. 2** 750
JOUISSEUSES, LES **Vol. 2** 291
JOULUKSI KOTIIN **Vol. 5** 651
JOUR AVEC VOUS, UN **Vol. 2** 514
JOUR DE FETE **Vol. 2** 318, 581, 615
JOUR DE GLOIRE, LE **Vol. 2** 154, 368
JOUR ET L'HEURE, LE **Vol. 2** 224, 285
JOUR, LA FETE, UN **Vol. 2** 196, 570
JOUR SE LEVE ET LES CONNERIES COMMENCEMENTS, LE **Vol. 6** *page 71*

JOUR SE LEVE, LE **Vol. 2** 8, 47, 100, 206, 837
JOURNAL D'UN CURE DE CAMPAGNE, LE **Vol. 2** 153, 170, 423
JOURNAL D'UN FOU, LE **Vol. 2** 570
JOURNAL D'UN SUICIDE, LE **Vol. 2** 301
JOURNAL D'UNE FEMME DE CHAMBRE, LE **Vol. 2** 294, 863
JOURNAL D'UNE FEMME EN BLANC, LE **Vol. 2** 251, 427
JOURNAL D'UNE MAISON DE CORRECTION **Vol. 2** 692
JOURNAL EROTIQUE D'UNE THAILANDAISE **Vol. 6** *page 134*
JOURNAL INTIME D'UNE JEUNE FILLE EN CHALEUR **Vol. 6** *page 129*
JOURNAL INTIME D'UNE NYMPHOMANE **Vol. 2** 139
JOURNAL TOMBE A CINQ HEURES, LE **Vol. 2** 122, 657
JOURNALISTEN SOM SJOVER **Vol. 5** 97
JOURNEE AU CAMP AVEC S.D.F., UNE **Vol. 2** 703
JOURNEE BIEN REMPLIE **Vol. 2** 527
JOURS GRIS, LES **Vol. 2** 359, 441
JOURS HEUREUX, LES **Vol. 2** 122,698
JOUSIAMPUJA **Vol. 5** 492, 511
JOVANA LUKINA **Vol. 3** 404
JOVANKA E LE ALTRE **Vol. 7** 988
JOVEN CASADA, LA **Vol. 8** 297, 299
JOVEN Y LA TENTACION, LA **Vol. 8** 296, 302
JOVENES AMANTES, LOS **Vol. 8** 348, 383
JOVENES AMIGUITAS BUSCAN PLACER **Vol. 8** 471
JOWITA **Vol. 1** 202, 227, 233
JOY **Vol. 6** *page 90, page 147*
JOYAS DEL DIABLO, LOS **Vol. 8** 583, 772
JOYEUSE PRISON, LA **Vol. 2** 595, 627, 726
JOYEUSES COLONIES DE VACANCES, LES **Vol. 2** 600
JOYEUX LURONS, LES **Vol. 2** 600
JOYEUX PELERINS, LES **Vol. 2** 15, 173, 778
J.R. CONTRAATACA **Vol. 8** 299, 756
JSEM DEVCE S CERTEM V TELE **Vol. 10** 167, 256, 524
JSI FALESNY HRAC **Vol. 10** 233, 250
JUAN Y JUNIOR EN UN MUNDO DIFERENTE **Vol. 8** 299, 681
JUANA LA LOCA ... DE VEZ EN CUANDO **Vol. 8** 250, 407
JUANILLO, PAPA Y MAMA **Vol. 8** 517, 732
JUANITA **Vol. 2** 122, 249, 644
JUANITO **Vol. 8** 306, 752
JUBILÄUM **Vol. 4** 123
JUBILÄUMSPREIS, DER **Vol. 6** *page 189*
JUBILEJ GOSPODINA IKLA **Vol. 3** 399, 444
JUD SÜSS **Vol. 4** 298, 462, 727
JUDAS **Vol. 9** 28
JUDAS FIAI **Vol. 9** 3
JUDAS ... TOMA TUS MONEDAS **Vol. 8** 317, 732
JUDAS VON TIROL, DER **Vol. 4** 306, 410, 774
JUDEL GRA NA SKRZYPCACH **Vol. 1** 194, 384
JUDEX **Vol. 2** 314, 330, 347, 375, 432, 613, 708
JUDEX 34 **Vol. 2** 807
JUDITH **Vol. 2** 183, 222
JUDITH THERPAUVE **Vol. 2** 517, 653
JUDITH TRACHTENBERG **Vol. 4** 284, 779
JUEGO DE AMOR PROHIBIDO **Vol. 8** 349
JUEGO DE HOMBRES **Vol. 8** 551, 630
JUEGO DE LA OCA, EL **Vol. 8** 298, 299, 402
JUEGO DE LA VERDAD, EL **Vol. 8** 524, 725
JUEGO DE NINOS **Vol. 8** 273, 477, 772
JUEGO DE PODER **Vol. 8** 466, 739
JUEGO DEL ADULTERIO, EL **Vol. 8** 626, 627
JUEGO SUCIO EN CASABLANCA **Vol. 8** 731, 734
JUEGOS DE SOCIEDAD **Vol. 8** 546, 549
JUERGAS DE EL SENORITO, LAS **Vol. 8** 317, 732
JUEVAS MILAGRO, LOS **Vol. 8** 214, 307, 378, 719
JUGANDO A MORIR **Vol. 8** 677, 772
JUGANDO A PAPAS **Vol. 8** 317, 552
JUGANDO CON LA MUERTE **Vol. 8** 297
JUGE ET L'ASSASSIN, LE **Vol. 2** 351, 746
JUGE FAYARD DIT „LE SHERIF", LE **Vol. 2** 522
JUGE, LE **Vol. 2** 97 **Vol. 6** *page 81, page 96*
JUGEMENT DE DIEU, LE **Vol. 2** 9, 226, 388, 402
JUGEMENT DE MINUIT, LE **Vol. 2** 318
JUGEMENT DERNIER, LE **Vol. 2** 558, 833
JUGEND **Vol. 4** 192, 234, 462, 734
JUGEND UND TOLLHEIT **Vol. 4** 646
JUGENDGELIEBTE, DIE **Vol. 4** 84, 85, 186
JUGENDLIEBE **Vol. 4** 49
JUGENDRAUSCH **Vol. 4** 115, 288, 663, 668
JUGENDSTÜRME **Vol. 4** 300
JUGENDSÜNDE, DIE **Vol. 4** 344, 647, 689
JUGENDSÜNDEN **Vol. 4** 221, 340, 608
JUGENDTRAGÖDIE **Vol. 4** 255, 379
JUHA **Vol. 5** 356, 363, 374, 453, 461, 514, 608

JUHLA MEREN RANNALLA **Vol. 5** 335, 420
JUICIO CONTRA UN ANGEL **Vol. 8** 517
JUICIO DE FALDAS **Vol. 8** 396, 749
JUICIO FINAL **Vol. 8** 614, 725
JUIF ERRANT, LE **Vol. 2** 42
JUIF POLONAIS, LE **Vol. 2** 205, 279, 488
JUKE-BOX - ATTRAVERSO LA LUCE **Vol. 7** 385, 859, 892
JUKE-BOX - IL VOLO **Vol. 7** 385, 562, 859
JUKE-BOX - LA CIFRA **Vol. 7** 385, 562, 859
JUKE-BOX - LA RICERCA **Vol. 7** 385, 562, 892
JUKE-BOX - NEI DINTORNI DI MEZZANOTTE **Vol. 7** 385, 859, 892
JUKE-BOX - URLI D'AMORE **Vol. 7** 1147
JULEFROKOSTEN **Vol. 5** 68, 195, 306
JULEK **Vol. 10** 18, 283, 299, 570
JULES ET JIM **Vol. 2** 208
JULIA **Vol. 9** 288
JULIA, ANNA, GENOWEFA **Vol. 1** 198, 213, 303
JULIA JUBILERAR **Vol. 5** 846
JULIA, JULIA **Vol. 5** 667, 749, 765
JULIA KISASSZONY **Vol. 9** 93
JULIA LEBT **Vol. 1** 9, 12, 34, 50
JULIA Y EL CELACANTO **Vol. 8** 492
JULIE DE CARNEILHAN **Vol. 2** 8, 615
JULIE ETAIT BELLE **Vol. 2** 133
JULIE LA ROUSSE **Vol. 2** 293, 294, 383
JULIE PAR DEVANT PAR DERRIERE **Vol. 2** 291
JULIE POT DE COLLE **Vol. 2** 457, 564, 565
JULIENNE ET ISAAC **Vol. 2** 74
JULIETA ENGANA A ROMEO **Vol. 8** 396
JULIETA Y ROMEO **Vol. 8** 475
JULIETTA **Vol. 2** 13, 279, 709
JULIETTE DU COTE DES HOMMES **Vol. 6** *page 89*
JULIETTE ET JULIETTE **Vol. 2** 196, 326
JULIETTE ET L'AIR DU TEMPS **Vol. 2** 51
JULIETTE OU LA CLE DES SONGES **Vol. 2** 13, 156, 402, 571, 837
JULIJA VREVSKAJA **Vol. 3** 59
JULISTEIDEN LIIMAAJAT **Vol. 5** 422
JULOT, DER APACHE **Vol. 4** 270, 384, 725
JUMALAN MYRSKY **Vol. 5** 430, 608
JUMALAN TUOMIO **Vol. 5** 349, 485
JUMEAUX DE BRIGHTON, LES **Vol. 2** 9, 205, 824
JUMENT VERTE, LA **Vol. 2** 226, 251, 625
JUMP FOR GLORY **Vol. 2** 32
JUMRUCI V PR"STTA **Vol. 6** *page 40, page 53, page 64*
JUNG UND ALT **Vol. 4** 183
JUNGE ADLER **Vol. 4** 541
JUNGE BARON NEUHAUS, DER **Vol. 4** 49, 266, 574
JUNGE DAME AUS GUTER FAMILIE **Vol. 4** 217
JUNGE ENGLÄNDER, DER **Vol. 1** 22, 56, 82
JUNGE FRAU VON 1914 **Vol. 1** 7, 36, 50
JUNGE GRAF, DER **Vol. 4** 123, 257, 432, 788
JUNGE HERZEN **Vol. 4** 631, 736
JUNGE, HEUT' IST FREITAG **Vol. 1** 19, 34, 114
JUNGE MAMA **Vol. 4** 89, 298
JUNGEN VOM KRANICHSEE, DIE **Vol. 1** 67, 129
JUNGENS **Vol. 4** 28, 246
JUNGER MANN **Vol. 1** 19, 128
JUNGES BLUT **Vol. 4** 535, 658, 752
JUNGES GEMÜSE **Vol. 1** 13, 48
JUNGFER, SIE GEFÄLLT MIR **Vol. 1** 37, 85
JUNGFERN VOM BISCHOFSBERG, DIE **Vol. 4** 2, 243
JUNGFRAU GEGEN MÖNCH **Vol. 4** 270, 462
JUNGFRAU VOM KYNAST, DIE **Vol. 4** 467
JUNGFRUKÄLLAN **Vol. 5** 1052, 1080, 1196
JUNGFRUN PA JUNGFRUSUND **Vol. 5** 1003, 1042
JUNGLA NERA **Vol. 7** 234, 359
JÜNGLING AUS DER KONFEKTION, DER **Vol. 4** 292, 391
JÜNGSTE GERICHT, DAS **Vol. 4** 379, 391
JUNINATT **Vol. 5** 940, 1052, 1053
JUNINATTEN **Vol. 5** 840, 909
JUNIO 44: DESEMBARCAREMOS EN NORMANDIA **Vol. 8** 600
JUNOST KOMANDIROW **Vol. 1** 298
JUOKSE KUIN VARAS **Vol. 5** 376, 383, 549
JUPITER **Vol. 2** 185, 600
JURAME **Vol. 8** 551, 749
JURASEK **Vol. 10** 460, 621
JÜRGENS RIECHT LUNTE **Vol. 4** 339, 525
JURRA FOR ANDERSENS **Vol. 5** 682
JURTZENKA : UN INVIERNO EN MALLORCA **Vol. 8** 348
JUS PRIMAE NOCTIS **Vol. 7** 585
JUSQU'A PLUS SOIF **Vol. 2** 686
JUSQU'AU BOUT DU MONDE **Vol. 2** 457, 795
JUSQU'AU DERNIER **Vol. 2** 279, 559, 661
JUSSI PUSSI **Vol. 5** 606
JUST LIKE AMERICA **Vol. 9** 61, 82, 98, 254, 276
JUSTE AVANT LA NUIT **Vol. 2** 520, 700
JUSTE DROIT, LE **Vol. 2** 684
JUSTICE D'ABORD **Vol. 2** 122
JUSTICE EST FAITE **Vol. 2** 124, 197

JUSTICES Vol. 2 49
JUSTICIA DE DIOS Vol. 8 422
JUSTICIA DEL COYOTE, LA Vol. 8 749, 752
JUSTICIERE, LA Vol. 2 136, 388, 646, 871
JUSTIN DE MARSEILLE Vol. 2 84, 193, 573
JUSTINE Vol. 2 814
JUSTINE & JULIETTE Vol. 5 946, 989, 1133
JUSTITIA Vol. 7 1064
JUSTOCOEUR Vol. 6 *page 113*
JUSTUS JÄRJESTÄÄ KAIKEN Vol. 5 516, 613
JUTALOMUTAZAS Vol. 9 138, 167, 305
JUTRO Vol. 3 423, 433
JUTRO MEKSYK Vol. 1 187, 196, 216 Vol. 6 *page 234*
JUTRO PREMIERA Vol. 1 249, 352, 373, 382
JUURAKON HULDA Vol. 5 349, 386
JUWELENMARDER, DER Vol. 4 116, 242
JUX WILL ER SICH MACHEN, EINEN Vol. 4 116, 291
JUXBARON, DER Vol. 4 213, 681
JUZGADO PERMANENTE Vol. 8 493, 732
JUZNA PATEKA Vol. 6 *page 316, page 320*
JUZNA POSTA Vol. 10 82, 84, 274
J'Y SUIS...J'Y RESTE Vol. 2 128, 595

K

K 1 GREIFT EIN ! Vol. 4 232
K. O. Vol. 3 213, 235, 286, 321
K. UND K. FELDMARSCHALL Vol. 10 167, 540, 586
KAASUA, KOMISARIO PALMU ! Vol. 5 516, 613
KABALA Vol. 9 221, 253
KABALE UND LIEBE Vol. 1 50, 90, 107
KABARETTPROGRAMM NR. 1 – 6 Vol. 4 89, 162, 536
KABINETT DES DR. LARIFARI, DAS Vol. 4 257, 292, 558
KACZMAREK Vol. 4 124, 221
KAD BUDEM MRTAV I BEO Vol. 3 367, 368
KAD CUJES ZVONA Vol. 3 437, 451
KAD GOLUBOVI POLETE Vol. 3 379, 436
KADAR KONTRA KEREKES Vol. 9 196, 270
KADDISCH Vol. 4 190
KADE PO DOZDOT Vol. 3 339, 384
KADETSZERELEM Vol. 9 196, 270
KADETTEN Vol. 4 13, 255, 292, 574 Vol. 6 *page 183*
KADETTKAMRATER Vol. 5 840, 878
KADRA SAFA Vol. 4 124, 646, 702
KAELEDAEGGEN Vol. 5 83
KAEMPEDAMENS BORTFORELSE Vol. 5 81
KAERE FAMILIE, DEN Vol. 5 31, 248, 272
KAERE HUSFRED, DEN Vol. 5 83
KAERE IRENE Vol. 5 46
KAERE KOBENHAVN, DET Vol. 5 2, 7
KAERE LEGETOJ, DET Vol. 5 74, 125, 257
KAERESTE FOR MEGET, EN Vol. 5 83
KAERLIGHED I ORIENTEN Vol. 5 97
KAERLIGHED OG FUGLEKVIDDER Vol. 5 83
KAERLIGHED OG KONTANTER Vol. 5 92
KAERLIGHED OG LOTTERI Vol. 5 83
KAERLIGHED OG MOBILISERING Vol. 5 81
KAERLIGHED OG OVERTRO Vol. 5 83
KAERLIGHED OG PAEDAGOGIK Vol. 5 83
KAERLIGHED OG SKRAEDDERGAELD Vol. 5 83
KAERLIGHED OVERFINDER ALT Vol. 5 242
KAERLIGHED PR. FLYTTEOMNIBUS Vol. 5 92
KAERLIGHED VED HOFFET Vol. 5 203
KAERLIGHEDENS ALMAGT Vol. 5 183
KAERLIGHEDENS FIRKLOVER Vol. 5 256
KAERLIGHEDENS MELODI Vol. 5 89, 281
KAERLIGHEDSDIGTEREN Vol. 5 83 Vol. 55 83
KAERLIGHEDSDOKTOREN Vol. 5 2, 23, 228
KAERLIGHEDSLAENGSEL Vol. 5 265
KAERLIGHEDSLEG Vol. 5 242
KAERLIGHEDSOEN Vol. 5 228
KAERLIGHEDSOFFER, ET Vol. 5 295
KAERLIGHEDSPEKULANTEN Vol. 5 83
KAERLIGHEDSPROVE, EN Vol. 5 183
KAERLIGHEDSVALSEN Vol. 5 291
KAHDEKSAN SURMANLUOTIA Vol. 5 318, 477, 596, 601, 650
KAHDEKSAS VELJES Vol. 5 606
KAHDEN LADUN POIKKI Vol. 5 443, 643
KAHDEN TANSSIN VÄLILLÄ Vol. 5 344, 356
KAHN DER FRÖHLICHEN LEUTE, DER Vol. 1 35, 67, 107
KAI E GYNE NA FOBEITAI TON ANDRA Vol. 3 163
KAI OI PENTE ESSAN KOLASMENES Vol. 3 131
KAI XANA PROS TE DOXA TRABA Vol. 3 176
KAIKENLAISIA KARKULAISIA Vol. 5 314, 560
KAIKENLAISIA VIERAITA Vol. 5 356, 428
KAIKKI RAKASTAVAT Vol. 5 349, 485

KAIKKIEN NAISTEN MONNI **Vol. 5** 451, 478
KAIN **Vol. 4** 529
KAINOVO ZNAMENI **Vol. 10** 45, 74, 261, 283, 345, 571
KAINUU 39 **Vol. 5** 391, 511, 606, 618
KAISER VON KALIFORNIEN, DER **Vol. 4** 54, 215, 311
KAISERIN ELISABETH VON ÖSTERREICH **Vol. 4** 282, 628
KAISERLIEBCHEN **Vol. 4** 774
KAISERWALZER **Vol. 4** 373
KAIVOPUISTON KAUNIS REGINA **Vol. 5** 471, 481, 549
KAJA GAR TILL SJÖSS **Vol. 5** 1042, 1164
KAJA, UBIT CU TE ! **Vol. 3** 444, 451
KAJASTUS **Vol. 5** 344, 356, 428
KAKAV DEDA TAKAV UNUK **Vol. 6** *page 318*
KAKKIENTRUPPEN **Vol. 7** 1213
KAKO SAM SISTEMATSKI UNISTEN OD IDIOTA **Vol. 6** *page 304, page 305*
KAKO SU SE VOLELI ROMEO I JULIJA **Vol. 3** 366
KAKS' TAVALLISTA LAHTISTA **Vol. 5** 549, 592
KAKSEN PA OVERLAND **Vol. 5** 668, 671, 700
KAKSI HAUSKAA VEKKULIA **Vol. 5** 462, 634
KAKSI KIVAA KAVERIA **Vol. 5** 441, 562
KAKSI VANHAA TUKKIJÄTKÄÄ **Vol. 5** 443, 646
KAKSI VIHTORIA **Vol. 5** 334, 636
KAKUK MARCI **Vol. 9** 253, 259, 265
KALA **Vol. 3** 375, 439
KALABALIKEN I BENDER **Vol. 5** 1048, 1093, 1096
KALABRYTA **Vol. 3** 149
KALAMITA **Vol. 10** 9, 428, 505
KALBSRAGOUT MIT CHAMPIGNON **Vol. 4** 28
KALDE SPOR **Vol. 5** 717, 800
KALEMERA ATHENA **Vol. 3** 168
KALI-YUG, LA DEA DELLA VENDETTA **Vol. 7** 1136
KALIBER FÜNF KOMMA ZWEI **Vol. 4** 560, 743
KALIFENS EVENTYR **Vol. 5** 97
KALIN OREL"T **Vol. 3** 56, 92 **Vol. 6** *page 40, page 44, page 57*
KALLE AALTOSEN MORSIAN **Vol. 5** 443, 549
KALLE KARLSSON FRAN JULARBO **Vol. 5** 924, 1052
KALLE – KUSTAA KORKIN SEIKKAILUT **Vol. 5** 367, 552
KALLE PA SPANGEN **Vol. 5** 930, 1032
KALLE STROPP, GRODAN BOLL OCH DERAS VÄNNER **Vol. 5** 840, 923, 1158, 1161
KALLE UTTER **Vol. 5** 1175, 1210
KALOJAN **Vol. 3** 31, 89, 111 **Vol. 6** *page 46*
KALOS ERTHE TO DOLLARIO **Vol. 3** 133
KALOSZE SZCZESCIA **Vol. 1** 148, 247, 286, 323, 397
KALOTASZEGI MADONNA **Vol. 9** 35, 71, 164
KALTE FÜSSE **Vol. 4** 167
KALTE HERZ, DAS **Vol. 1** 39, 77, 107 **Vol. 4** 192, 780
KALTE MAMSELL, DIE **Vol. 4** 232, 239, 270
KALVARIA **Vol. 9** 134, 149, 259
KAM CERT NEMUZE **Vol. 10** 359, 469
KAM DOSKACE RANNI PTACE **Vol. 10** 233, 371, 389
KAM NIKDO NESMI **Vol. 10** 106, 299, 327
KAM, PANOVE, KAM JDETE ? **Vol. 10** 87, 208, 313
KAM S NIM ? **Vol. 10** 45, 345, 532, 550
KAM ZMIZEL KURYR **Vol. 10** 74, 181, 226
KAMARAD DO DESTE **Vol. 10** 159, 495, 511
KAMARADSKE MANZELSTVI **Vol. 10** 275, 345
KAMARATKA SUSKA **Vol. 10** 228, 399, 592
KAMARATKY **Vol. 10** 269, 273, 548
KAMELDAMEN **Vol. 5** 238
KAMELEONTERNA **Vol. 5** 835, 919
KAMELIADAMEN **Vol. 5** 97
KAMENI HORIZONTI **Vol. 3** 341
KAMERAD HEDWIG **Vol. 4** 378
KAMERADEN **Vol. 4** 71, 302, 344, 479, 524, 544
KAMERADEN AUF SEE **Vol. 4** 517, 558, 620
KAMERADSCHAFT **Vol. 4** 28, 452, 727, 729
KAMERADSCHAFTSEHE **Vol. 6** *page 192*
KAMERADSKE MANZELSZVI **Vol. 6** *page 192*
KAMIENNE NIEBO **Vol. 1** 171, 281, 313, 382 **Vol. 6** *page 247*
KAMIENNE TABLICE **Vol. 6** *page 239, page 241, page 252*
KAMIONDZIJE OPET VOZE **Vol. 6** *page 313, page 321*
KAMION"T **Vol. 3** 114 **Vol. 6** *page 27, page 36*
KAMMERMUSIK **Vol. 4** 213, 626
KAMMERPIGEN **Vol. 5** 295
KAMMERSÄNGER, DER **Vol. 4** 660, 704
KAMOURASKA **Vol. 2** 135

KAMPEN MOD URETTEN Vol. 5 31, 150, 248
KAMPEN OM KVINDEN Vol. 5 183
KAMPEN OM NAESBYGAARD Vol. 5 23, 89, 281
KAMPEN OM TUNGTVANNET Vol. 5 679, 716, 722, 723, 820
KAMPF Vol. 4 232, 270, 774
KAMPF DER GESCHLECHTER, DER Vol. 4 339, 424, 377
KAMPF DER TERTIA, DER Vol. 4 306
KAMPF DES DONALD WESTHOF, DER Vol. 4 113, 479, 567
KAMPF GEGEN BERLIN, DER Vol. 4 316, 437, 725
KAMPF MIT DEM DRACHEN, DER Vol. 4 344, 647
KAMPF MIT DEM DRACHEN ODER DIE TRAGÖDIE DES UNTERMIETERS, DER Vol. 4 270, 567, 713
KAMPF UM ANASTASIA Vol. 4 8, 774
KAMPF UM BLOND Vol. 4 558, 634
KAMPF UM DAS ERBE, DER Vol. 4 247
KAMPF UM DEN BÄR, DER Vol. 4 232, 239, 270
KAMPF UM DEN MANN, DER Vol. 4 373, 558
KAMPF UM DIAMANTENFELDER Vol. 4 98
KAMPF UM DIE EHE, DER Vol. 4 270, 290
KAMPF UM DIE FRAU Vol. 4 125
KAMPF UM DIE HEIMAT, DER Vol. 4 290
KAMPF UM DIE SCHOLLE Vol. 4 49, 287, 458, 644
KAMPF UM ROM Vol. 3 307
KAMPF UMS ICH, DER Vol. 4 412 Vol. 6 *page 187*
KAMPF UMS MATTERHORN, DER Vol. 4 10, 558, 774
KÄMPFENDE GEWALTEN ODER WELT OHNE KRIEG Vol. 4 113, 780
KÄMPFENDE HERZEN Vol. 4 183, 306, 321, 444
KAMPFGESCHWADER LÜTZOW Vol. 4 65, 311, 361, 461, 580
KAMRATER I VAPENROCKEN Vol. 5 903, 958, 1156
KAN DISSE OJNE LYVE ? Vol. 5 228
KAN DOKTORN KOMMA ? Vol. 5 1042, 1078
KAN KAERLIGHED KURERES ? Vol. 5 83
KAN KVINDER FEJLE ? Vol. 5 157, 183
KANAL Vol. 1 242, 250 Vol. 6 *page 256*
KANARIFUGLEN Vol. 5 718, 767
KANAVAN LAIDALLA Vol. 5 372, 552
KANCSUKA HAZAJABAN, A Vol. 9 77
KANKA DO POHADKY Vol. 10 233, 395, 505
KANKKULAN KAIVOLLA Vol. 5 443, 639
KANNST DU PFEIFEN, JOHANNA ? Vol. 4 329, 751
KANON SAME ZLATO Vol. 10 250
KANSKE EN DIKTARE Vol. 5 1210
KANSKE EN GENTLEMAN Vol. 5 913, 923, 1205
KANTOR IDEAL Vol. 10 167, 453
KAOS Vol. 7 199, 607, 1112
KAPASZKODJ A FELLEGEKBE ! Vol. 9 12, 23, 36, 285
KAPEDANI Vol. 3 22, 23, 24
KAPELUSZ PANA ANATOLA Vol. 1 179, 198, 232
KAPERGASTEN Vol. 5 121
KAPETAN LESI Vol. 3 356, 366
KAPETAN MIKULA MALI Vol. 3 355, 374
KAPETAN SOROKADAS Vol. 3 203
KAPI, VODE, RATNICI Vol. 3 418
KAPITAN DABAC Vol. 10 287, 501, 576
KAPITÄN HANSENS ABENTEUER Vol. 4 240
KAPITAN KORDA Vol. 10 255, 286
KAPITÄNE BLEIBEN AN BORD Vol. 1 24, 26, 50, 73
KAPITAN"T Vol. 3 78, 126 Vol. 6 *page 62*
KAPLANDDIAMANT, DER Vol. 4 110, 246
KAPO Vol. 3 403, 436 Vol. 7 115, 522
KAPOTE KLAINE KAI OI DYNATOI Vol. 3 164
KAPRYSY LAZARZA Vol. 1 148
KAPTAJN KLYDE OG HANS VENNER VENDER TILBAGE Vol. 5 249, 270, 296
KAPUT LAGER GLI ULTIMI GIORNI DELLE SS Vol. 7 201
KÄPY SELÄN ALLA Vol. 5 516
KAR A BENZINERT Vol. 9 67, 117, 218
KAR – LEK SO GÖR VI. BREV TILL INGE OCH STEN Vol. 5 946, 1213, 1217
KARA BOZA Vol. 1 317
KÄRA FAMILJEN, DEN Vol. 5 867, 873, 1106, 1150
KÄRA LEKEN, DEN Vol. 5 534, 545, 709
KARA PLNA BOLESTI Vol. 10 108, 145, 474
KÄRA SLÄKTEN Vol. 5 840, 909
KARABINY Vol. 6 *page 251, page 253, page 258*
KARAMBOL Vol. 3 121 Vol. 3 38 Vol. 9 134, 215, 259
KARATE, EL COLT Y EL IMPOSTOR, EL Vol. 8 243, 762
KARATE PO POLSKU Vol. 6 *page 232, page 236, page 238*
KARBID UND SAUERAMPFER Vol. 1 48, 73, 98, 100

KARCZMA NA ROZDROZU Vol. 1 317, 371
KARD, A Vol. 9 152, 203, 339
KARD ES KOCKA Vol. 9 83, 326, 332
KARDIA ENOS ALETE Vol. 3 161
KARDIOGRAM Vol. 1 148, 202, 227, 289
KÅRE JOHN Vol. 5 880, 933, 1104
KAREL A JA Vol. 10 122, 580
KAREL HAVLICEK BOROVSKY Vol. 10 138, 167, 453, 586
KAREL HYNEK MACHA Vol. 10 154, 167, 256, 412, 532
KAREN, MAREN OG METE Vol. 5 31, 150, 248
KARGUS Vol. 8 229, 237, 529, 759
KARHANOVA PARTA Vol. 10 313, 532
KARHOZAT Vol. 9 209, 243
KARIERA Vol. 1 250, 381 Vol. 6 *page 255* Vol. 10 132, 165, 296, 428, 552, 564, 567
KARIERA MATKY LIZALKY Vol. 10 460, 621
KARIERA PAVLA CAMRDY Vol. 10 453, 524
KARIERTE WESTE, DIE Vol. 4 167
KARIN INGMARSDOTTER Vol. 5 879, 937, 996, 1085
KARIN MANSDOTTER Vol. 5 1042, 1080
KARINO Vol. 1 202, 233, 341
KARJOLSTEINEN Vol. 5 762, 766, 777
KARL ALS RADIOFACHMANN Vol. 4 97, 230, 239
KARL FÖR SIN HATT Vol. 5 871, 928, 1156
KARL FREDERIK REGERAR Vol. 5 840, 878, 909
KARL I KÖKET, EN Vol. 5 871, 1073
KARL MARX Vol. 6 *page 165, page 173*
KARL RENOVIERT SEINE WOHNUNG Vol. 4 97, 230, 239
KARL VALENTIN, DER SONDERLING Vol. 4 212, 571
KARL XII Vol. 5 891, 925, 927, 1105
KARLCHEN AUF DER BRAUTSCHAU Vol. 4 247, 340
KARLCHEN, DURCHHALTEN ! Vol. 6 *page 161, page 170, page 175*
KÄRLEK Vol. 5 909, 1173
KÄRLEK 1 - 1000 Vol. 5 1054
KÄRLEK 65 Vol. 5 1033
KÄRLEK EFTER NOTER Vol. 5 840, 878
KÄRLEK MASTE VI HA Vol. 5 1129
KÄRLEK OCH ALLSANG Vol. 5 864
KÄRLEK OCH DYNAMIT Vol. 5 893, 925
KÄRLEK OCH KASABRIST Vol. 5 840, 997
KÄRLEK OCH LANDSTORM Vol. 5 923, 925
KÄRLEK OCH STÖRTLOPP Vol. 5 871, 1042
KÄRLEK PA TURNE Vol. 5 1140, 1207
KÄRLEK, SOLSKEN OCH SANG Vol. 5 1003, 1156
KÄRLEKEN Vol. 5 851, 895
KÄRLEKEN SEGRAR Vol. 5 909, 944, 1173
KÄRLEKENS BRÖD Vol. 5 880, 1130, 1181
KÄRLEKENS DECIMALER Vol. 5 878, 1052, 1177
KÄRLEKENS KORS Vol. 5 310
KÄRLEKENS ÖGON Vol. 5 925
KÄRLEKENS XYZ Vol. 5 869
KÄRLEKS SOMMAR Vol. 5 869, 1109
KARLEKSEXPRESSEN Vol. 5 870, 923
KÄRLEKSHISTORIA Vol. 5 1096
KÄRLEKSLIVETS OFFER Vol. 5 928, 1032
KÄRLEKSNATT VID ÖRESUND, EN Vol. 5 928
KARLINE MANZELSTVA Vol. 10 178, 263, 592
KARLOVARSTI PONICI Vol. 10 283, 296, 577
KARMANKOLON KUNIGAS Vol. 5 398, 421
KARNAVALET Vol. 3 14, 22 Vol. 6 *page 18*
KARNEVAL Vol. 4 779 Vol. 5 97, 869, 1052, 1053, 1130
KARNEVAL DER LIEBE Vol. 4 51, 108, 139, 633
KARNEVALOVY PASTEL Vol. 10 335
KARNEVALSJAEVELEN Vol. 5 9
KARO-KÖNIG Vol. 4 525, 546
KAROLINA RIJECKA Vol. 3 391, 425
KAROSSZEK Vol. 9 164, 196
KARPACCY GORALE Vol. 1 188
KARPATHY ZOLTAN Vol. 9 122, 181, 332
KARRIÄR Vol. 5 928, 1132
KARRIERE Vol. 1 7, 14, 114 Vol. 4 255, 379, 409
KARRIERE IN PARIS Vol. 1 67, 101, 107
KARTANON NAISET Vol. 5 480, 579
KARTHAUSI Vol. 9 337
KARTKA Z PODROZY Vol. 6 *page 208, page 224, page 230*
KARTONAZENE Vol. 9 122, 199, 312
KARTYAVAR Vol. 9 122, 265, 332
KARUSELLEN Vol. 5 921, 997
KARUSELLEN GAR Vol. 5 843, 923
KARUSELLEN I FJÄLLEN Vol. 5 840, 1033
KARUSSELL Vol. 4 28, 232, 727
KARUSSELL DES LEBENS Vol. 4 133
KARUSSELL DES TODES, DAS Vol. 4 149, 239, 660
KARUZELA ZYCIA Vol. 1 151, 299, 394
KARVAT Vol. 5 606
KASARI Vol. 10 299, 433, 469, 532

KASARMIN TYTÄR Vol. 5 643
KASARMIN TYTTÖ Vol. 5 443
KASERNENZAUBER Vol. 4 270, 646
KASHIMA PARADISE Vol. 2 502
KASKADE RÜCKWÄRTS Vol. 6 *page 154, page 161, page 176*
KASPAREK KOUZELNIKEM Vol. 10 345, 524
KASSEN STEMMER Vol. 5 61, 125, 195
KASSENREVISION, DIE Vol. 6 *page 194*
KASSERER JENSEN Vol. 5 674
KASSIANE Vol. 3 165
KASTRULLRESAN Vol. 5 880, 1181
KASVOT KUVASTIMESSA Vol. 5 605
KASZEBE Vol. 1 136, 283
KATA A KROKODYL Vol. 10 235, 572
KATAKOMBY Vol. 10 154, 621
KATAPULT Vol. 10 139, 144, 516
KATARA TES MANAS Vol. 3 153
KATARAMENE AGAPE Vol. 3 130
KATARIINA JA MUNKKINIEMEN KREIVI Vol. 5 349, 481, 493
KATARINA KAUNIS LESKI Vol. 5 328, 356, 481, 549
KATASTROFA Vol. 1 179, 211 Vol. 6 *page 255*
KATASTROFA W GIBRALTARZE Vol. 6 *page 237, page 258*
KATEGORO TE ZOE Vol. 3 142
KATER LAMPE Vol. 1 21, 111 Vol. 4 132, 230
KATERINA A JEJI DETI Vol. 10 87, 296, 385
KATHARINA DIE GROSSE Vol. 4 178, 202, 651
KATHARINA KARASCHKIN Vol. 4 695
KATHARINA KNIE Vol. 4 247, 479, 600
KATHE KATERGARES STON PAGO TOU Vol. 3 147, 193
KATI NA KAIEI Vol. 3 144, 205
KATIA Vol. 2 34, 100, 226, 279, 332, 427, 488
KATINA Vol. 3 42 Vol. 6 *page 27, page 28*
KATINKA Vol. 4 98
KATIUSCIA Vol. 7 616, 1004
KATJUSCHA MASLOWA Vol. 4 704
KATKA Vol. 10 62, 349, 407
KATO BELJAZANI ATOMI Vol. 3 60, 109, 120
KATO PESEN Vol. 3 30, 116 Vol. 6 *page 47*
KATOAVIA TIMANTTEJA ELI HERRASMIESVARAS MOREL VASTUSTAJANAAN ETSIVÄ FRANCK Vol. 5 414
KATRINA Vol. 5 840, 997
KATTORNA Vol. 5 839
KATUPEILIN TAKANA Vol. 5 356, 634
KATYI Vol. 9 71, 270
KATZ' IM SACK, DIE Vol. 4 247
KATZ UND MAUS Vol. 1 360
KATZENPRINZ, DER Vol. 6 *page 173*
KATZENSPRUNG, EIN Vol. 1 46, 87
KATZENSTEG, DER Vol. 4 247, 343, 396, 423, 424, 771
KAUFHAUS HERZ Vol. 1 119
KAUFMANN VON VENEDIG, DER Vol. 4 213, 436, 734
KAULE Vol. 1 26, 98, 130
KAUNIS KAARINA Vol. 5 356, 646
KAUNIS MUISTO Vol. 5 611
KAUNIS VEERA ELI BALLAADI SAIMAALTA Vol. 5 374, 592
KAUTSCHUK Vol. 4 141, 184, 736, 774
KÄUTZCHENKUHLE Vol. 1 29, 57, 62, 87
KAVALIER VOM WEDDING, DER Vol. 4 339, 377, 437
KAVALIERE Vol. 4 35, 212
KAVALIERE VOM KURFÜRSTENDAMM Vol. 4 361, 558, 796
KAVARNA NA HLAVNI TRIDE Vol. 10 352, 569, 621
KAVEZ Vol. 3 436
KAVIARPRINZESSIN, DIE Vol. 4 257, 558
KÄYNTIKORTTINI... Vol. 5 467, 536
KAZDA KORUNA DOBRA Vol. 10 41, 87, 433
KAZDEMU JEHO NEBE Vol. 10 226, 317, 522
KAZDEMU WOLNO KOCHAC Vol. 1 176, 278
KAZDY DEN ODVAHU Vol. 10 22, 87, 395
KAZDY MLADY MUZ Vol. 10 184, 505
KAZDY TYZDEN SEDEM DNI Vol. 10 273, 459, 464
KAZIMIERZ WIELKI Vol. 1 227, 311 Vol. 6 *page 205*
KDE ALIBI NESTACI Vol. 10 532, 621
KDE REKY MAJI SLUNCE Vol. 10 197, 296, 569
K"DE ZIVEES ? Vol. 6 *page 26, page 49, page 61*
KDO CHCE ZABIT JESSII ? Vol. 10 371, 428, 433
KDO HLEDA ZLATE DNO Vol. 10 428, 516
KDO PRICHAZI PRED PULNOCI Vol. 10 371, 394, 517
KDO SE BOJI, UTIKA Vol. 10 44, 129, 149
KDY BRECI MUZI Vol. 10 299, 577
KDYBY TISIC KLARINETU Vol. 10 289, 359, 522
KDYBYCH BYL TATOU Vol. 10 524, 621
KDYZ MA SVATEK DOMINIKA Vol. 10 358, 359, 524

KDYZ ROZVOD, TAK ROZVOD Vol. 10 18, 129, 313
KDYZ SRDCE PROMLUVI Vol. 10 628
KDYZ STRUNY LKAJI Vol. 10 168, 586
KDYZ V RAJI PRSELO Vol. 10 344, 496, 497
KDYZ VALCIK ZNI Vol. 10 45, 345
KEAN Vol. 2 122, 361, 521, 623, 833 Vol. 4 319 Vol. 5 97 Vol. 7 915, 1167
KEAN, GENIO E SREGOLATEZZA Vol. 7 706, 1113
KEAN, GLI AMORI DI UN ARTISTA Vol. 7 472, 725, 1116, 1158
KEBY SOM MAL DIEVCA Vol. 10 273, 483, 548
KEBY SOM MAL PUSKU Vol. 10 15, 273, 548
KEDER SIG PAA LANDET, DE Vol. 5 83
KEDLUBNOVY KAVALIR V RAJI Vol. 10 45, 167, 345
KEDVES SZOMSZED, A Vol. 9 20, 311, 339
KEGYELMES UR ROKONA, A Vol. 9 26, 132
KEHRE WIEDER, AFRIKA Vol. 6 *page 183*
KEHRE ZURÜCK! ALLES VERGEBEN! Vol. 4 213, 558
KEIMENDES LEBEN Vol. 4 130, 560, 663
KEIN ÄRGER MIT CLEOPATRA Vol. 1 22, 35, 125
KEIN HÜSUNG Vol. 1 40, 56, 88
KEIN MANN FÜR CAMP DETRICK Vol. 1 2, 127
KEIN MANN FÜR ZWEI Vol. 6 *page 161, page 170, page 171*
KEIN PLATZ FÜR LIEBE Vol. 1 23, 108 Vol. 4 631
KEIN WORT VON LIEBE Vol. 4 2, 515 Vol. 10 2, 412
KEINE ANGST VOR LIEBE Vol. 4 28, 558
KEINE FEIER OHNE MEYER Vol. 4 212, 239
KEINEN TAG OHNE DICH Vol. 4 79, 97, 115, 132
KEINUMORSIAN Vol. 5 379, 443
KEITTIOKAVAL JEERIT Vol. 5 523, 634
KEJSAREN Vol. 5 882, 990, 1002, 1190
KEJSARN AV PORTUGALLIEN Vol. 5 840, 1114, 1125, 1173
KEJSER NERO PAA KRIGSSTIEN Vol. 5 97
KEK BALVANY, A Vol. 9 71, 322
KEKCEVE UKANE Vol. 3 395, 449
KEKEC Vol. 3 390, 406
KELLNERIN ANNA, DIE Vol. 4 340, 582
KELLO Vol. 5 434
KELLY'S HEROES Vol. 6 *page 312*
KEM, A Vol. 9 7
KEMENYKALAP ES KRUMPLIORR Vol. 9 78, 252, 332
KENGURU, A Vol. 9 316, 339
KENNEN SIE URBAN ? Vol. 1 5, 53, 80
KENNST DU DAS KLEINE HAUS AM MICHIGAN-SEE ? Vol. 6 *page 200*
KENNWORT MACHIN Vol. 4 79, 115, 365
KENRAALIN MORSIAN Vol. 5 374, 461, 592
KENTAURI Vol. 10 307, 375, 394, 488
KENTAUROK Vol. 9 229
KENYER ES CIGARETTA Vol. 9 107, 140
KEOMA Vol. 7 869
KEPVADASZOK Vol. 9 282, 286, 316
KERCENIMI Vol. 6 *page 14, page 18*
KEREK FERKO Vol. 9 330, 196
KERESZTELÖ Vol. 9 36, 121, 318
KERESZTUTON Vol. 9 58, 132, 270
KERKERSZENE AUS „FAUST" Vol. 4 634
KERMESSE DU SEXE, LA Vol. 2 721
KERMESSE HEROIQUE, LA Vol. 2 81, 573, 644, 803, 826, 837, 863
KERMESSE ROUGE Vol. 2 538, 595
KERSANTILLEKO EMMA NAUROI ? Vol. 5 592, 654
KERTES HAZAK UTCAJA Vol. 9 122, 207, 305
KERTOKAA SE HÄNELLE Vol. 5 383, 639
KESÄILLAN VALSSI Vol. 5 372, 566
KESÄKAPINA Vol. 5 515
KESÄLLÄ KELLO 5 Vol. 5 467, 536
KESÄN MAKU Vol. 5 465
KESÄYÖN LAPSET Vol. 5 535, 605
KESELYÜERCEG, A Vol. 9 7, 175
KESERÜ IGAZSAG Vol. 9 117, 314, 326
KESHILLTARET Vol. 3 10, 15
KESÖ Vol. 9 14, 68, 71
KESYTTÄMÄT VELJEKSET Vol. 5 516
KET ARVA, A Vol. 9 77
KET BAJTHAY, A Vol. 9 35, 196, 274, 336
KET ELHATAROZAS Vol. 9 279
KET EMELET BOLDOGSAG Vol. 9 83, 149, 314
KET FELIDÖ A POKOLBAN Vol. 9 75, 285
KET FOGOLY Vol. 9 178, 237
KET LANY AZ UTCAN Vol. 9 315, 322
KET PONT KÖZÖTT A LEGRÖVIDEBB GÖRBE Vol. 9 225, 265, 285
KET TÖRTENET A FELMULTBOL - TEGLAFAL MÖGÖTT; PHILEMON ES BAUCIS Vol. 9 51, 660, 67, 193
KET VALLOMAS Vol. 9 67, 165, 241
KETARCU ASSZONY Vol. 9 175
KETEZERPENGÖS FERFI, A Vol. 9 219
KETLELKÜ ASSZONY Vol. 9 200
KETSOS, MINI KAI SOUBLAKIA Vol. 3 174

KETSZER KETTÖ Vol. 9 196, 315
KETSZIVÜ FERFI, A Vol. 9 323
KETTE KLIRRT, DIE Vol. 4 390, 735
KETTEN HALTAK MEG Vol. 9 67, 83, 183
KETTENEKÜ DOB, A Vol. 9 253, 259, 318
KETTESBEN Vol. 9 117
KETTEVALT MENNYEZET Vol. 9 139, 311, 316
KETUNHÄNTÄ KAINALOSSA Vol. 5 334, 417
KEUSCHE GELIEBTE, DIE Vol. 4 232, 370, 678
KEUSCHE JOSEPH, DER Vol. 4 255, 774
KEUSCHE KOKOTTE, DIE Vol. 4 344, 544
KEUSCHE SÜNDERIN, DIE Vol. 4 135, 619, 647
KEUSCHE SUSANNE, DIE Vol. 4 192, 584
KI BESZEL ITT SZERELEMRÖL ? Vol. 9 6, 202
KI LATOTT ENGEM ? Vol. 9 259, 265, 285
KI VAN A TOJASBAN ? Vol. 9 121, 215, 335
KIALTAS ES KIALTAS Vol. 9 20, 152, 154
KICMA Vol. 3 350, 366, 400
KID RODELO Vol. 8 546, 626
KIDNAPNING Vol. 5 16, 111, 195 Vol. 7 845
KIDNAPPED Vol. 5 116, 142
KIEDY MILOSC BYLA ZBRODNIA – RASSENSCHANDE Vol. 1 220, 282, 403
KIELLETTY KIRJA Vol. 5 452
KIENTOPP IM KINO Vol. 4 291, 644
KIERION Vol. 3 142, 183
KIERUNEK BERLIN Vol. 1 211, 281 Vol. 6 *page 248*
KIFF TEBBY Vol. 7 92, 730
KIHAJOLNI VESZELYES Vol. 9 152, 253, 259
KIHLAUS Vol. 5 334, 406, 547, 625, 640
KIHLAUSKYLPYLÄ ELI DRITTELIN HERRASVÄKI KESÄÄ VIETTÄMÄSSÄ Vol. 5 368, 523
KIINNI ON JA PYSYY Vol. 5 443, 646
KIKI Vol. 2 397 Vol. 4 164, 257 Vol. 7 196, 989
KIKLOP Vol. 6 *page 320, page 321*
KILAS, O MAU DA FITA Vol. 8 15, 56, 118
KILENC HONAP Vol. 9 20, 154, 265
KILJUSEN HERRASVÄKI Vol. 5 320, 410, 546, 585, 647
KILL Vol. 2 728
KILLE OCH EN TJEJ, EN Vol. 5 1009, 1097
KILLEMANN HAT'N KLAPS Vol. 4 603
KILMA, REINA DE LAS AMAZONAS Vol. 8 515, 708
KILOMETRO 12 Vol. 8 233, 615
KILROY SEN TEKI Vol. 5 372, 552, 600
KIMEN Vol. 5 772
KINCSKERESÖ KISKÖDMÖN Vol. 9 67, 285, 318
KIND DER STRASSE, DAS Vol. 4 293
KIND – EIN HUND, EIN Vol. 4 550, 780
KIND, EIN HUND, EIN VAGABUND, EIN Vol. 4 354, 437, 738
KIND, ICH FREU MICH AUF DEIN KOMMEN Vol. 4 30, 325, 445, 567
KIND IST VOM HIMMEL GEFALLEN, EIN Vol. 4 558, 747
KIND MUSS DOCH 'NEN VATER HABEN, DAS Vol. 6 *page 186*
KIND RUFT, DAS Vol. 4 178, 213, 651
KINDER DER FINSTERNIS Vol. 4 178, 396, 397
KINDER DER LANDSTRASSE Vol. 4 270, 695
KINDER DER LIEBE Vol. 4 583
KINDER DER STRASSE Vol. 4 247, 424
KINDER DES GENERALS, DIE Vol. 4 646
KINDER DES GLÜCKS Vol. 4 505, 572
KINDER VON HEUTE Vol. 4 287, 293
KINDER VOR GERICHT Vol. 4 495, 747
KINDER, WIE DIE ZEIT VERGEHT Vol. 4 171
KINDERARZT DR. ENGEL Vol. 4 132, 738, 788
KINDERSEELEN KLAGEN EUCH AN Vol. 4 208, 558
KINDERTRAGÖDIE Vol. 4 25, 344
KINEK A TÖRVENYE ? Vol. 9 215, 221, 252
KING IN NEW YORK, A Vol. 2 656
KING-KONG-GRIPPE Vol. 1 68, 98
KING LEAR Vol. 5 8, 179, 293
KINNHAKEN, DER Vol. 1 54, 104
KINOAUTOMAT Vol. 10 359, 516, 543
KIPPARIKVARTETTI Vol. 5 592, 646
KIRA – KIRALINA Vol. 6 *page 197*
KIRALNYE HUSZARJA, A Vol. 9 28, 34, 237
KIRALYGYILKOSSAG Vol. 9 50, 51, 177, 263
KIRALYLANY ZSAMOLYA, A Vol. 9 8, 122, 259
KIRJE Vol. 5 387, 427
KIRKASTETTU SYDÄN Vol. 5 446, 549
KIRKASTUVA SÄVEL Vol. 5 356, 357, 395, 549
KIRKE OG ORGEL Vol. 5 31, 55, 101, 208
KIRSCHEN IN NACHBARS GARTEN Vol. 4 167, 430, 558
KIS HÖS, A Vol. 9 28
KIS KATALIN HAZASSAGA Vol. 9 71, 237
KIS LORD, A Vol. 9 200
KIS RONGYOS, A Vol. 9 71
KIS VALENTINO, A Vol. 9 107, 147

KISA **Vol. 3** 389
KISERTES **Vol. 9** 37, 68, 71, 215, 316
KISERTET LUBLON **Vol. 9** 152, 259, 273
KISERTETEK VONATA **Vol. 9** 71, 288
KISKRAJCAR **Vol. 9** 134, 183, 312
KISMET **Vol. 4** 281 **Vol. 6** *page 182*
KISPUS **Vol. 5** 29, 31, 241, 248
KISS KISS BANG BANG **Vol. 8** 523, 732
KIT **Vol. 3** 75, 105, 121 **Vol. 6** *page 52*
KIT & CO **Vol. 1** 46, 96, 102
KITÖRES **Vol. 9** 339
KITRA FIORE DELLA NOTTE **Vol. 7** 529
KITSCH **Vol. 4** 467, 522
KITTY SCHWINDELT SICH INS GLÜCK **Vol. 4** 132, 270, 409
KITTY UND DIE WELTKONFERENZ **Vol. 4** 445, 774
KIUSAUS **Vol. 5** 334, 528
KIVANDORLO, A **Vol. 9** 323
KIZIA-MIZIA **Vol. 1** 176
KJAERE LILLE NORGE **Vol. 5** 762
KJAERE MAREN **Vol. 5** 755, 780, 787, 824
KJAERLEIKENS FERGEREISER **Vol. 5** 728, 767, 801
KJAERLIGHET OG VENNSKAP **Vol. 5** 726, 738
KJAERLIGHET PA PINNE **Vol. 5** 741
KLAANI **Vol. 5** 404, 591
KLABAUTERMANDEN **Vol. 5** 16, 271, 272
KLABAUTERMANN, DER **Vol. 4** 287, 373, 471, 646
KLABAUTERMANNEN **Vol. 5** 797, 1148, 1150
KLADIVO NA CARODEJNICE **Vol. 10** 197, 254, 289, 503
KLAINE KI OI DYNATOI **Vol. 3** 152
KLAKIER **Vol. 6** *page 224, page 228, page 251*
KLAKSON **Vol. 3** 362, 416
KLAMCZUCHA **Vol. 6** *page 228, page 243, page 247*
KLAMSTWO KRYSTYNY **Vol. 1** 278, 308
KLÄNNINGEN **Vol. 5** 1042, 1080, 1130
KLAPKA LEGIO **Vol. 9** 101, 215
KLAPPERSTORCHVERBAND, DER **Vol. 4** 660
KLAPZUBOVA JEDENACTKA **Vol. 10** 524, 621
KLARA LUST **Vol. 5** 869
KLARAR BANANEN BIFFEN **Vol. 5** 883
KLART TILL DRABBING **Vol. 5** 840, 878, 997
KLASSKAMRATER **Vol. 5** 933, 1156
KLASYKA EUROPEJSKA **Vol. 1** 242, 254, 403
KLATOVSTI DRAGOUNI **Vol. 10** 256, 532
KLATSCH **Vol. 4** 340, 674
KLAUN FERDINAND A RAKETA **Vol. 10** 30, 229, 621
KLAVERVIRTOUSEN **Vol. 5** 83
KLEBOLIN KLEBT ALLES **Vol. 4** 514
KLEC PRO DVA **Vol. 10** 70, 153, 283
KLEIDER MACHEN LEUTE **Vol. 4** 119, 266, 740
KLEIN ABER MEIN **Vol. 4** 18
KLEIN DORRIT **Vol. 4** 123, 257, 787, 788
KLEINE AUS DER KONFEKTION, DIE **Vol. 4** 316, 385
KLEINE DAGMAR, DIE **Vol. 4** 213, 390
KLEINE DETEKTIV, DER **Vol. 4** 87
KLEINE FRÄULEIN TRÄUMT, DAS **Vol. 4** 24, 525
KLEINE FREUNDIN BRAUCHT EIN JEDER MANN, EINE **Vol. 4** 270, 774
KLEINE GRENZVERKEHR, DER **Vol. 4** 574, 631
KLEINE HERZOG, DER **Vol. 4** 291
KLEINE HOFKONZERT, DAS **Vol. 4** 149, 188, 729
KLEINE INGE UND IHRE DREI VÄTER, DIE **Vol. 4** 344, 544, 545, 775
KLEINE KOMMANDEUR, DER **Vol. 1** 34, 76, 85, 126
KLEINE KUNO, DER **Vol. 1** 34, 74, 101
KLEINE MÄDCHEN - GROSSE SORGEN **Vol. 4** 65, 121, 340
KLEINE MUCK, DER **Vol. 4** 735
KLEINE NACHTKOMÖDIE **Vol. 4** 339, 525
KLEINE NACHTMUSIK, EINE **Vol. 4** 15, 339
KLEINE PIT, DER **Vol. 4** 123, 247
KLEINE PRINZ, DER **Vol. 1** 34, 48, 73
KLEINE RESIDENZ **Vol. 4** 579, 647, 660, 678
KLEINE SCHWINDLERIN, DIE **Vol. 4** 132, 149, 660
KLEINE SEITENSPRUNG, DER **Vol. 4** 89, 266, 574, 614
KLEINE SKLAVIN, DIE **Vol. 4** 292, 788
KLEINE SOMMERMELODIE, EINE **Vol. 4** 453, 461
KLEINE UND DAS GROSSE GLÜCK, DAS **Vol. 1** 9, 35, 100
KLEINE UND DER GROSSE KLAUS, DER **Vol. 1** 26, 38, 48
KLEINE UND DIE GROSSE LIEBE, DIE **Vol. 4** 49, 738, 788
KLEINE UND IHR KAVALIER, DIE **Vol. 4** 208, 270
KLEINE VOM BUMMEL, DIE **Vol. 4** 239, 560
KLEINE VOM VARIETE, DIE **Vol. 4** 113, 306
KLEINE ZAUBERER UND DIE GROSSE FÜNF, DER **Vol. 1** 12, 55, 79

KLEINEN VERWANDTEN, DIE **Vol. 4** 329, 373
KLEINER GOLDENER RING, EIN **Vol. 4** 110, 230, 458, 796
KLEINER IRRTUM, EIN **Vol. 4** 558, 747
KLEINER MANN, GANZ GROSS **Vol. 4** 28, 124, 232, 475
KLEINER MANN – WAS NUN ? **Vol. 1** 29, 30, 47 **Vol. 4** 119
KLEINER REINFALL, EIN **Vol. 4** 261, 323
KLEINER VORSCHUSS AUF DIE SELIGKEIT, EIN **Vol. 4** 270, 323, 409, 462
KLEINES BEZIRKSGERICHT **Vol. 4** 323, 517, 558
KLEINES INTERMEZZO **Vol. 4** 517, 558, 796
KLEINES MÄDEL – GROSSES GLÜCK **Vol. 4** 28, 97, 437, 458, 525
KLEINHOFF HOTEL **Vol. 7** 910
KLEINSTADTPOET, DER **Vol. 4** 188, 608, 742
KLEINSTADTSÜNDER **Vol. 4** 329, 608, 742
KLEISTO PARATHYRO **Vol. 3** 166
KLEJNOT WOLNEGO SUMIENIA **Vol. 6** *page 224, page 246, page 256*
KLETT UND DATSCH, DIE PECHVÖGEL **Vol. 6** *page 181*
KLETTERMAXE **Vol. 4** 25, 255, 643
KLIC **Vol. 10** 193, 296, 569
KLIMAKS **Vol. 5** 712
KLIMBUSCH MACHT WOCHENENDE **Vol. 4** 18, 558
KLINCZ **Vol. 6** *page 231, page 239, page 252*
KLIPPEN DER EHE **Vol. 4** 340, 440, 608 **Vol. 6** *page 181*
KLIPPET **Vol. 5** 975
KLISURA **Vol. 3** 383, 436
KLJUC **Vol. 3** 421, 437, 445
KLK AN PTX – DIE ROTE KAPELLE **Vol. 1** 1, 26, 42
KLOCKAN PA RÖNNEBERGA **Vol. 5** 878, 1173
KLOCKORNA I GAMLA STAN **Vol. 5** 1032, 1139
KLOGE MAND, DEN **Vol. 5** 2, 7, 63
KLOGE OG VI GALE **Vol. 5** 89, 224
KLOKA GUBBEN **Vol. 5** 846
KLOKKER I MANESKINN **Vol. 5** 724, 772
KLOPKA ZA GENERALA **Vol. 3** 401
KLOPOTLIWY GOSC **Vol. 1** 144, 283, 323
KLOSTER VON DENSOMIR, DAS **Vol. 4** 747
KLOSTERFRIEDE **Vol. 4** 309
KLOSTERJÄGER, DER **Vol. 4** 25, 329, 370, 527
KLOSTERT I SENDOMIR **Vol. 5** 854, 937, 996
KLOTZ AM BEIN **Vol. 1** 35, 36, 100
KLOTZIGER JUNGE, EIN **Vol. 4** 8, 774
KLOVNEN **Vol. 5** 47, 136, 157, 228
KLUB KAWALEROW **Vol. 1** 211, 307
KLUB SZACHISTOW **Vol. 1** 313
KLUB TRI **Vol. 10** 460, 472
KLUCI NA RECE **Vol. 10** 122, 580
KLUCI Z BRONZU **Vol. 10** 70, 246, 408
KLUCZNIK **Vol. 6** *page 243, page 258*
KLUGE MARIANNE, DIE **Vol. 4** 614, 620, 685
KLUGE SCHWIEGERMUTTER, DIE **Vol. 4** 347, 354, 370
KLUGEN DINGE, DIE **Vol. 1** 29, 34, 92
KLUGEN FRAUEN, DIE **Vol. 4** 687
KLUK ZA DVE PETKY **Vol. 10** 139, 473, 570
KNABE EROS, DER **Vol. 4** 240, 494
KNABE IN BLAU, DER **Vol. 4** 270, 288
KNALLEFFEKT, DER **Vol. 4** 73, 371, 684
KNIGGE UND WIR **Vol. 4** 18, 558
KNIGHT WITHOUT ARMOUR **Vol. 2** 573, 803
KNIVEN I HJERTET **Vol. 5** 46, 249
KNJAZ"T **Vol. 3** 40, 93, 105 **Vol. 6** *page 46*
KNOCK **Vol. 2** 185, 387, 719
KNOCK OU LE TRIOMPHE DE LA MEDECINE **Vol. 2** 9, 122, 193
KNOCK OUT **Vol. 4** 123, 257, 745, 787, 788
KNUT FORMOS SISTE JAKT **Vol. 5** 712, 713, 799
KNUTSEN OG LUDVIGSEN **Vol. 5** 712, 777, 824
K.O. **Vol. 9** 152, 259, 339
KO PRIDE LEV... **Vol. 3** 395, 417
KO PRIDE LJUBEZEN **Vol. 3** 390
KO PUCA OTVORICE MU SE **Vol. 3** 367, 418
KO TO TAMO PEVA **Vol. 3** 347, 350, 407
KO ZORIJU JAGODE **Vol. 3** 421
KOBBERBRYLLUP **Vol. 5** 101, 183, 208
KOBENHAVN – KALUNDBORG OG – ? **Vol. 5** 24, 116
KOBENHAVN VED NAT **Vol. 5** 203
KOBENHAVNERE **Vol. 5** 24
KOBENHAVNERLIV **Vol. 5** 203
KOBIETA **Vol. 1** 174, 215, 326
KOBIETA I KOBIETA **Vol. 6** *page 219, page 235*
KOBIETA, KTORA GRZECHU PRAGNIE **Vol. 1** 174, 346
KOBIETA, KTORA WIDZIALA SMIERC **Vol. 1** 268
KOBIETA W KAPELUSZU **Vol. 6** *page 237, page 258*

KOBIETY NAD PRZEPASCIA **Vol. 1** 278, 308, 404
KOCAR DO VIDNE **Vol. 10** 197, 235, 289
KOCHA, LUBI, SZANUJE... **Vol. 1** 278, 308, 346
KOCHAJ ALBO RZUC **Vol. 6** *page 237, page 238, page 251*
KOCHAJ TYLKO MNIE **Vol. 1** 308, 372
KOCHAJMY SIE **Vol. 1** 262, 300
KOCHAJMY SYRENKI **Vol. 1** 163, 323 **Vol. 6** *page 240*
KOCHANKA SZAMOTY **Vol. 1** 193, 278, 346
KOCHANKOWIE Z MARONY **Vol. 1** 180, 311 **Vol. 6** *page 222*
KOCICI PRINC **Vol. 10** 18, 299, 425, 485
KOCKA **Vol. 10** 108, 148, 273
KOCKARI **Vol. 3** 350
KOCKY NEBEREM **Vol. 10** 250, 462
KOENIGSMARK **Vol. 2** 9, 32, 75, 102, 270, 410, 579, 661
KOEPUTKIAKUINEN JA SIMON ENKELIT **Vol. 5** 345, 424, 532, 652
KOFFER DES HERRN O.F., DIE **Vol. 4** 32, 115, 373
KOFFER MIT DYNAMIT **Vol. 1** 7
KOHE E LARGET **Vol. 6** *page 11, page 16, page 17*
KOHLHIESELS TÖCHTER **Vol. 4** 28, 321, 347, 608, 626, 663, 773
KOHO JSEM VCERA LIBAL **Vol. 10** 379, 453, 524, 586
KOHOUT NEZASPIEVA **Vol. 10** 73, 82, 399
KOHOUT PLASI SMRT **Vol. 10** 433, 564, 621
KOHTALO JOHTAA MEITÄ **Vol. 5** 363, 425, 481, 549
KOHTALO TEKEE SIIRRON **Vol. 5** 374, 443
KOJAK BUDAPESTEN **Vol. 9** 180
KOKKINA FANARIA **Vol. 3** 153, 163
KOKO, LE GORILLE QUI PARLE **Vol. 2** 17
KOKS I KULISSEN **Vol. 5** 26, 46
KOLBERG **Vol. 4** 424, 462, 531, 788
KÖLCSÖNADOTT ELET **Vol. 9** 117
KÖLCSÖNKERT CSECSEMÖK, A **Vol. 9** 28
KÖLCSÖNKERT FERJEK **Vol. 9** 196, 274
KÖLCSÖNKERT KASTELY **Vol. 9** 319, 322
KOLDUSGROF, A **Vol. 9** 337
KOLERAEN **Vol. 5** 295
KOLIESKO **Vol. 10** 72, 107, 459
KOLIK SLOV STACI LASCE **Vol. 10** 229, 313, 519
KOLKOTO SINAPENO Z"RNO **Vol. 6** *page 53, page 55*
KOLLEGE KOMMT GLEICH **Vol. 4** 124, 292, 640
KOLMAS LAUKAUS **Vol. 5** 523
KOLMASTOISTA KOPUTUS **Vol. 5** 549, 579, 600
KOLME MIESTÄ **Vol. 5** 467, 545, 648
KOLMIAPILA **Vol. 5** 451, 462, 634
KOLNATI SME, IRINA **Vol. 3** 434
KOLONIE, DIE **Vol. 6** *page 161, page 167*
KOLONIE LANFIERI **Vol. 10** 249, 286, 327, 425
KOLONNE X **Vol. 4** 208, 457 **Vol. 6** *page 195*
KOLOROWE PONCZOCHY **Vol. 1** 232, 281, 323 **Vol. 6** *page 209, page 247*
KÖLYÖK **Vol. 9** 67, 183, 241
KOM TVA MÄN, DET **Vol. 5** 842
KOMADIR"T NA OTRJADA **Vol. 3** 69, 101, 109
KOMBINA **Vol. 6** *page 28, page 36, page 50, page 64*
KOMEDI I HÄGERSKOG **Vol. 5** 909
KOMEDIA Z POMYLEK **Vol. 1** 305, 311, 388
KOMEDIANT **Vol. 10** 139, 283, 325
KOMEDIANTI **Vol. 10** 184, 522, 556
KOMEDIANTKA **Vol. 10** 122, 167
KOMEDIANTSKA PRINCEZNA **Vol. 10** 256, 412
KOMEDIANTY **Vol. 1** 169, 249, 305, 359
KOMEDIE POMYLEK **Vol. 6** *page 249, page 257*
KOMEDIE S KLIKOU **Vol. 10** 170, 286, 564
KOMENDANT **Vol. 1** 248, 357, 375
KOMISARI I DRITES **Vol. 3** 1, 4, 6 **Vol. 6** *page 17*
KOMISARIO PALMUN EREHDYS **Vol. 5** 443, 639
KOMM EN GÄST, DET **Vol. 5** 878, 1173
KOMM' ZU MIR ZUM RENDEZVOUS **Vol. 4** 292, 321
KOMM ZU MIR ZURÜCK **Vol. 4** 2, 228, 582
KOMMANDANT, DER **Vol. 3** 437
KOMMER DU, ELSA ? **Vol. 5** 696, 726, 760, 800
KOMMISSAR X - DIE DREI GELBEN KATZEN **Vol. 3** 395
KOMMISSAR X - IN DEN KLAUEN DES GOLDENEN DRACHENS **Vol. 3** 359, 395
KOMÖDIANTEN **Vol. 4** 85, 124, 211, 247, 251, 294, 678, 788
KOMÖDIANTEN DES LEBENS **Vol. 4** 113, 319
KOMÖDIANTEN - EMIL **Vol. 6** *page 160, page 167*
KOMÖDIANTENKIND, DAS **Vol. 4** 192, 247, 367
KOMÖDIE DES HERZENS **Vol. 4** 266, 574, 663
KOMÖDIE DES LEBENS **Vol. 4** 300

KOMPARSITA Vol. 3 30, 69 Vol. 6 *page 38*
KOMPLOTT IM BANKVIERTEL, DAS Vol. 4 428, 626
KOMPPANIAN NEROPATIT Vol. 5 443, 549, 620
KOMPTOIRISTKA Vol. 10 26, 115, 126, 198
KOMTESS URSEL Vol. 4 183
KOMTESSEN Vol. 5 2, 23
KOMTESSEN PAA STEENHOLT Vol. 5 101, 181, 208
KOMU TANCI HAVANA Vol. 10 154, 235
KOMUNISTE Vol. 10 228, 356, 504
KONCERT Vol. 3 399, 452 Vol. 6 *page 229* Vol. 10 139, 325, 425
KONCERT NA KONCI LETA Vol. 10 139, 254, 325
KONCERT NE VITIN Vol. 3 11, 26
KONCERT PRE POZOSTALYCH Vol. 10 82, 273, 459
KONCERT ZA FLEJTA I MOMICE Vol. 3 31 Vol. 6 *page 47, page 56*
KONE SOGES, EN Vol. 5 92
KONEC AGENTA W4C PROSTREDNICT-VIM PSA PANA FOUSTKY Vol. 10 567, 597
KONEC CESTY Vol. 10 193, 299, 385
KONEC JASNOVIDCE Vol. 10 564, 621
KONEC SRPNA V HOTELU OZON Vol. 10 41, 325
KONEC STARYCH CASU Vol. 10 129, 172, 516
KONEC STRASIDEL Vol. 10 122, 386, 503
KONECNA STANICE Vol. 10 173, 284, 466
KONECNE SAMI Vol. 10 532, 621
KONECNE SI ROZUMIME Vol. 10 308, 398, 428
KONFETTI Vol. 5 83
KONFIRMANDEN PAA FREDERIKSBERG Vol. 5 34
KONFLIKT Vol. 5 840, 878
KONFLIKTER Vol. 5 97
KONG LEAR Vol. 5 8, 179, 293
KONGELIGHT BESOG Vol. 5 31, 241, 248, 272
KONGEN AF PELIKANIEN Vol. 5 24, 55
KONGEN BOD Vol. 5 55, 208, 290
KONGEN BOD A.K.A. STAVNSBAANDS-FILMEN Vol. 5 101
KONGENS FOGED Vol. 5 81
KONGO-EXPRESS Vol. 4 347, 491, 736
KONGRESS TANZT, DER Vol. 2 399 Vol. 4 266, 574
KONIEC NASZEGO SWIATA Vol. 1 198, 237, 377, 381
KONIEC NOCY Vol. 1 155, 321, 391, 398
KONIEC WAKACJI Vol. 1 136, 213 Vol. 6 *page 232*
KÖNIG, DER Vol. 4 87, 734
KÖNIG DER MANEGE, DER Vol. 4 270, 384, 725
KÖNIG DER MITTELSTÜRMER, DER Vol. 4 359, 422, 643, 648
KÖNIG DROSSELBART Vol. 1 29, 34, 62, 87 Vol. 6 *page 181, page 190*
KÖNIG KRAUSE Vol. 4 87
KÖNIG NICOLO Vol. 4 496, 529
KÖNIG UND DIE KLEINEN MÄDCHEN, DER Vol. 4 223, 270
KÖNIG VON PARIS, DER Vol. 4 95, 113, 479, 600, 780
KÖNIGIN DER NACHT Vol. 4 183
KÖNIGIN DES VARIETES, DIE Vol. 4 212, 571, 726
KÖNIGIN DES WELTBADES, DIE Vol. 4 321, 584
KÖNIGIN EINER NACHT, DIE Vol. 4 359
KÖNIGIN KAROLINE VON ENGLAND Vol. 4 291
KÖNIGIN LUISE Vol. 4 306, 724
KÖNIGIN SEINES HERZENS, DIE Vol. 4 219, 479, 725
KÖNIGLICHE BETTLER Vol. 4 159
KÖNIGSGRENADIERE Vol. 4 571, 726
KÖNIGSKINDER Vol. 1 17, 48, 73
KÖNIGSLIEBCHEN Vol. 4 321, 340, 734
KÖNIGSTIGER Vol. 4 372, 558, 703
KÖNIGSTOCHTER VON TRAVANKORE, DIE Vol. 4 288
KÖNIGSWALZER Vol. 4 266, 531, 574, 713
KONJUH PLANINOM Vol. 3 346, 372
KONJUKTURRITTER Vol. 4 340, 354, 466
KONKURRENCEN Vol. 5 295
KONKURRENZ PLATZT, DIE Vol. 4 287, 458, 646
KONKURS Vol. 10 398
KÖNNEN GEDANKEN TÖTEN ? Vol. 4 423, 479
KONNIK"T Vol. 3 31, 70 Vol. 6 *page 46*
KÖNNYÜ MUZSA Vol. 9 71, 312
KÖNNYÜ TESTI SERTES Vol. 9 100, 291
KONOPIELKA Vol. 6 *page 220, page 250*
KONSTEN ATT ÄLSKA Vol. 5 909, 1173
KONSTGJORDA SVENSSON Vol. 5 891, 925
KONSTRUKCJA W PROCESIE Vol. 6 *page 233*
KONTAKT ! Vol. 5 738
KONTRAKT Vol. 6 *page 236, page 238, page 259*
KONTRYBUCJA Vol. 1 148, 232 Vol. 6 *page 246*
KONTSERTO ME POLYBOLO Vol. 3 159
KONZERT, DAS Vol. 4 331, 555
KONZERT FÜR BRATPFANNE UND ORCHESTER Vol. 1 15, 34, 92

KONZERT IN TIROL Vol. 4 97
KOOMIKKO Vol. 5 324, 525, 532
KOPASZKUTYA Vol. 9 109, 171, 240
KOPERNIK Vol. 1 5, 208, 237, 254, 321, 406 Vol. 6 *page 242*
KOPF DER PHARAONENTOCHTER, DER Vol. 4 356
KOPF DES GONZALES, DER Vol. 6 *page 192, page 197*
KOPF HOCH, CHARLY ! Vol. 4 213, 361, 681
KOPF HOCH, JOHANNES ! Vol. 4 49, 246
KÖPFCHEN KAMERAD Vol. 1 2, 7, 65
KOPFJÄGER VON BORNEO, DIE Vol. 4 15, 671, 708
KOPFÜBER INS GLÜCK Vol. 4 19, 536
KOPJASOK Vol. 9 83, 183, 259
KOPLJI POD BREZO Vol. 3 345, 406
KÖPMANSHUS I SKÄRGARDEN, ETT Vol. 5 1112
KOPORTOS Vol. 9 20, 238, 318
KOPPANYI AGA TESTAMENTUMA, A Vol. 9 57, 203, 265
KOPRETINY PRO ZAMECKOU PANI Vol. 10 250, 383, 462
KOPYTEM SEM, KOPYTEM TAM Vol. 10 44, 127, 172, 251
KORA TERRY Vol. 4 325, 531, 713
KORALLENPRINZESSIN, DIE Vol. 4 124, 232, 354
KORD KAMPHUES, DER RICHTER VON COESFELD Vol. 4 118, 367
KORENITE NA IZGRJAVASTOTO SL"NCE Vol. 3 69
KORITSIA STON ELIO Vol. 3 153
KÖRKARLEN Vol. 5 854, 937, 997, 1042, 1053, 1217
KORKEIN VOITTO Vol. 5 356, 428
KORMANYZO, A Vol. 9 323
KORNBLUMENBLAU Vol. 4 132, 270
KORNSPEKULANTENS FORBRYDELSE Vol. 5 295
KORPEN FLYGER Vol. 5 856, 946
KORPINPOLSKA Vol. 5 465
KORT ÄR SOMMAREN Vol. 5 880, 940
KORT ÄR SOMMARNATTEN Vol. 5 959
KORTE SOMMER, DEN Vol. 5 35, 179, 273
KORVETTENKAPITÄN, DER Vol. 4 186, 287, 384, 458, 644
KORYATOVIC Vol. 10 167, 453
KOSAVA Vol. 3 367, 368
KOSCIUSZKO POD RACLAWICAMI Vol. 1 174, 278, 308, 372, 404
KOSENIE JASTRABEJ LUKY Vol. 10 273, 548, 592
KOSILE STASTNEHO CLOVEKA Vol. 10 45
KOSKENKYLÄN LAULU Vol. 5 372, 552
KOSKENLASKIJAN MORSIAN Vol. 5 344, 349, 356, 406, 428, 476
KOSMETIKK – REVOLUSJONEN Vol. 5 694, 780
KOSTERVALSEN Vol. 5 1164
KÖSZIVÜ EMBER FIAI, A Vol. 0
KOTA 905 Vol. 3 421, 444
KÖTHINTA Vol. 9 75, 117, 165
KOTRMELEC Vol. 10 179, 226, 566
KOUZELNE DOBRODRUZSTVI Vol. 10 144, 292, 359
KOUZELNIKUV NAVRAT Vol. 10 227, 359, 535, 567
KOUZELNY DEN Vol. 10 211, 359, 433
KOUZELNY DUM Vol. 10 460, 621
KOVAA PELIÄ POHJOLASSA Vol. 5 443, 639
KOVANAAMA Vol. 5 347, 634
KOVARNATA PRINCESA TURANDOT Vol. 3 71
KÖYHÄ LAULAJA Vol. 5 549, 592
KOZACI Vol. 3 409
KOZARA Vol. 3 371, 436
KOZIE MLIEKO Vol. 10 349, 407
KOZIJAT ROG Vol. 3 48, 69 Vol. 6 *page 52*
KOZLONOH Vol. 10 92
KÖZÖS BÜN, A Vol. 9 67, 285, 318
KRAB I JOANNA Vol. 6 *page 206, page 248, page 255*
KRACH IM FORSTHAUS Vol. 4 230, 232, 239
KRACH IM HINTERHAUS Vol. 4 462, 734
KRACH IM VORDERHAUS Vol. 4 123, 135, 361
KRACH OPERACJI I TERROR Vol. 6 *page 241, page 251, page 253*
KRACH UM JOLANTHE Vol. 4 373, 626
KRACH UND GLÜCK UM KÜNNEMANN Vol. 4 135, 232, 354
KRADEC"T NA PRASKOVI Vol. 3 63, 82, 112 Vol. 6 *page 41*
KRAFT DES MICHAEL ARBOGAST, DIE Vol. 6 *page 189*
KRAFT OG SKONHED Vol. 5 55, 83
KRAFT UND STOFF Vol. 6 *page 187, page 188*
KRAFTMEIER, DER Vol. 4 560
KRAJ RATA Vol. 6 *page 310, page 315, page 321*
KRAJAT NA EDNA VAKANCIJA Vol. 3 63, 75, 89
KRAJAT NA PESENTA Vol. 3 30, 110 Vol. 6 *page 38*
KRAJAT NA P"TJA Vol. 3 37, 51 Vol. 6 *page 32*
KRAJINA S NABYTKEM Vol. 10 325, 344, 383

KRAJOBRAZ PO BITWIE Vol. 1 314, 360
KRAKATIT Vol. 10 21, 154, 621
KRAKONOS A LYZNICI Vol. 10 165, 299, 496
KRAL DROZDIA BRADA Vol. 10 145, 178, 283
KRAL KOMIKU Vol. 10 166, 428
KRAL KRALU Vol. 10 41, 193, 289
KRAL SUMAVY Vol. 10 70, 197, 433
KRAL ULICE Vol. 10 256, 586
KRALICI VE VYSOKE TRAVE Vol. 10 182, 184, 593
KRALJ PETROLEJA Vol. 3 371
KRALJEVSKI VOZ Vol. 6 *page 315, page 318*
KRALLE, DIE Vol. 4 385
KRALOVNA STRIBRNYCH HOR Vol. 10 99, 585
KRALOVSKA POLOVACKA Vol. 10 14, 273
KRALOVSKY OMYL Vol. 10 139, 154, 254
KRALOVSTVI ZA KYTARU Vol. 10 183, 289, 466
KRAMBAMBULI Vol. 4 85, 326, 410
KRANES KONDITORI Vol. 5 722, 723, 738, 776
KRANIOU TOPOS Vol. 3 134, 135
KRANKE SEELEN Vol. 4 87
KRASAVICE KATA Vol. 10 167, 584
KRASNA VYZVEDACKA Vol. 10 58, 345
KRASSE FUCHS, DER Vol. 4 208, 339
KRATKO SL"NCE Vol. 3 102 Vol. 6 *page 50, page 58*
KRATKY ZIVOT Vol. 10 72, 155, 390
KRAUSE UND KRUPP Vol. 1 113
KRAWATTENMACHER Vol. 4 720, 799
KRAZBATA V EKSPRESA Vol. 3 106
KRB BEZ OHNE Vol. 10 256, 460
KRC Vol. 3 386, 395
KREBSZ, AZ ISTEN Vol. 9 121, 183, 259
KREHKE VZTAHY Vol. 10 144, 299, 327
KREIVI Vol. 5 515, 546
KREJCOVSKA POVIDKA Vol. 10 524, 566
KREUTZEROVA SONATA Vol. 10 167, 453
KREUTZERSONATE, DIE Vol. 4 30, 291, 298, 608
KREUZ AM TEUFELSFELSEN, DAS Vol. 6 *page 186*
KREUZER EMDEN Vol. 4 344, 545, 775
KREUZIGET SIE ! Vol. 4 560, 663
KREUZLSCHREIBER, DIE Vol. 4 49, 638
KREUZZUG DES WEIBES Vol. 4 124, 733, 747
KREVATI I PERANDORIT Vol. 3 14, 26
KR"GOVE NA OBICTA Vol. 3 48 Vol. 6 *page 42, page 59*
KRIEGSMANS ERINAN Vol. 5 840, 940
KRIG OG KARELIGHED Vol. 5 58
KRIGENS FJENDE Vol. 5 58
KRIGERNES BORN Vol. 5 11, 165, 259
KRIGET ÄR SLUT Vol. 5 1131, 1194
KRIGSKORRESPONDENTER Vol. 5 154
KRIGSMILLIONAEREN Vol. 5 83
KRIK Vol. 10 41, 292, 385
KRIMINALASSISTENT BLOCH Vol. 5 304
KRIMINALFALL ERICH LEMKE Vol. 4 71, 370
KRIMINALGAADEN I KINGOSGADE Vol. 5 183
KRIMINALKOMMISSAR EYCK Vol. 4 184, 678
KRIMINALREPORTER HOLM Vol. 4 164, 462
KRIMINALSAGEN TOVE ANDERSEN Vol. 5 14, 180, 199
KRIMINEL HISTORIE, EN Vol. 5 83
KRIS Vol. 5 840, 1114
KRISCHNA Vol. 4 746
KRISS ROMANI Vol. 2 49, 137
KRISTALI Vol. 6 *page 35, page 54, page 57*
KRISTALL ODER PORZELLAN Vol. 4 18, 230
KRISTIAN Vol. 10 412, 621
KRISTIANE AF MARSTAL Vol. 5 180, 199, 201
KRISTIN KOMMENDERAR Vol. 5 909, 1173
KRISTINE VALDRESSDATTER Vol. 5 770
KRISTINUS BERGMAN Vol. 5 2, 31, 252
KRISTOFFERUS HUS Vol. 5 852, 896, 935
KRISTOVE ROKY Vol. 10 9, 323, 576
KRIZ U POTOKA Vol. 10 167, 412, 536, 621
KRIZNO OBDOBJE Vol. 6 *page 316*
KRIZOVA TROJKA Vol. 10 321, 580, 612
KRIZOVATKY Vol. 10 122, 193, 299, 335
KRKOLOMNA JIZDA Vol. 10 502
KROBLINGEN Vol. 5 97
KROK DO TMY Vol. 10 167, 256
KROL MACIUS PIERWSZY Vol. 1 305, 323, 355, 397 Vol. 6 *page 239*
KROLOWA PRZEDMIESCIA Vol. 1 278, 308, 347
KROLOWA PSZCZOL Vol. 6 *page 251*
KRONANS KÄCKA GOSSAR Vol. 5 843, 923
KRONANS KAVALJERER Vol. 5 891, 925, 997
KRONANS RALLARE Vol. 5 879
KRONBLOM Vol. 5 928
KRONBLOM KOMMER TILL STAN Vol. 5 923, 928, 1181
KRONDIAMANTEN Vol. 5 242
KRONE DER KAISERIN VON INDIEN, DIE Vol. 4 135
KRONE DES LEBENS, DIE Vol. 4 288

KRONE UND FESSEL **Vol. 4** 514
KRONIKA JEDNOG ZLOCINA **Vol. 3** 371, 424
KRONIKA ZHAVEHO LETA **Vol. 10** 70, 154, 385
KRONPRISEN **Vol. 5** 664, 732
KRONZEUGIN, DIE **Vol. 4** 28, 298
KROPKA NAD I **Vol. 1** 170, 346
KROTKIE ZYCIE **Vol. 1** 197 **Vol. 6** *page 248*
KROZ GRANJE NEBO **Vol. 3** 423, 455
KRST RAKOC **Vol. 3** 346, 351
KRTINY **Vol. 10** 144, 292, 359
KRUDT MED KNALD **Vol. 5** 24, 184
KRUDT OG KLUNKER **Vol. 5** 89, 150, 281
KRUH **Vol. 10** 70, 211
KRÜMELCHEN GEHT JAGEN **Vol. 6** *page 186*
KRÜMELCHEN IN DER SOMMERFRISCHE **Vol. 6** *page 186*
KRÜMELCHENS REISEABENTEUER **Vol. 6** *page 186*
KRUPP UND KRAUSE **Vol. 1** 13, 17, 68, 113
KRUT OCH KÄRLEK **Vol. 5** 546
KRUTA LUBOST **Vol. 10** 82, 273, 459
KRVAVA BAJKA **Vol. 1** 398 **Vol. 3** 356, 415
KRVAVA KOSULJA **Vol. 3** 337, 349, 403
KRVAVI PUT **Vol. 3** 356, 374
KR"VTA OSTAVA **Vol. 3** 77 **Vol. 6** *page 28, page 29*
KRWAWA DOLA **Vol. 1** 174, 317
KRWAWY WSCHOD **Vol. 1** 278, 372
KRYBSKYTTEN **Vol. 5** 97
KRYBSKYTTERNE PA NAESBYGAARD **Vol. 5** 195, 281
KRYENGRITJE NE PALLAT **Vol. 3** 5, 23, 24
KRYPSKYTTERE **Vol. 5** 701, 764, 767
KRYPTONIM „NEKTAR" **Vol. 6** *page 215, page 247*
KRYPTONYM „NEKTAR" **Vol. 1** 206, 281, 389
KRYSTA **Vol. 1** 174, 176, 326
KRYSTAL Z ISTANBULU **Vol. 10** 210, 576
KRYSTALLO **Vol. 3** 130
KRYSZTAL **Vol. 1** 220
KRZYK **Vol. 1** 255 **Vol. 6** *page 226, page 239, page 260*
KRZYK W NOCY **Vol. 1** 176, 235
KRZYZ WALECZNYCH **Vol. 1** 250, 398 **Vol. 6** *page 242*
KRZYZACY **Vol. 1** 187, 250 **Vol. 6** *page 233, page 242*
K"SCE NEBE ZA TRIMA **Vol. 3** 38, 126
KSIAZATKO **Vol. 1** 176, 308
KSIAZE JOZEF PONIATOWSKI **Vol. 1** 174, 215, 326
KSIEZNA LOWICKA **Vol. 1** 176, 278
K"STATA **Vol. 3** 72 **Vol. 6** *page 53, page 55*
K"STI BEZ OGRADI **Vol. 3** 30, 119 **Vol. 6** *page 43*
KTO ODCHADZA V DAZDI ... **Vol. 10** 263, 287, 425
KTO SI BEZ VINY **Vol. 10** 72, 287
KTO WIERZY W BOCIANY **Vol. 1** 216, 283 **Vol. 6** *page 235*
KTOKOLWIEK WIE ... **Vol. 1** 237, 249, 283, 321 **Vol. 6** *page 242*
KU WYZYNOM **Vol. 1** 290, 380
KUBINKE, DER BARBIER UND DIE DREI DIENSTMÄDCHEN **Vol. 4** 247, 461
KUCA **Vol. 3** 421, 437
KUCA NA OBALI **Vol. 3** 338
KUCA NA PJESKU **Vol. 6** *page 307, page 308*
KUCE V CEKMEDZE **Vol. 6** *page 31, page 49, page 64*
KUCKUCKS, DIE **Vol. 1** 3, 97, 117
KUD PUKLO DA PUKLO **Vol. 3** 454
KUDY KAM ? **Vol. 10** 41, 182, 524, 621
KUHHAVY DABEL **Vol. 10** 96, 172, 380
KUHLE WAMPE **Vol. 4** 234, 359
KÜHNE SCHWIMMER, DER **Vol. 4** 135, 270, 354
KUIN UNI JA VARJO **Vol. 5** 356, 485
KUISMA JA HELINÄ **Vol. 5** 340, 372, 379, 420, 481
KUKACKA V TEMNEM LESE **Vol. 10** 428, 434, 473, 513
KUKONLAULUSTA KUKONLAULUUN **Vol. 5** 443, 643
KÜLDETES EVIANBA **Vol. 9** 65, 215, 279
KULE, KTORE NIE TRAFIAJA **Vol. 1** 226
KULKURIN MASURKKA **Vol. 5** 462, 634
KULKURIN TYTTÖ **Vol. 5** 481, 579
KULKURIN VALSSI **Vol. 5** 357, 471, 481
KULLA - GULLA **Vol. 5** 871, 1073
KÜLÖNÖS HAZASSAG **Vol. 9** 117, 222, 237
KÜLÖNÖS ISMERTETÖJEL **Vol. 9** 165, 168, 241
KULORTE SLAVEHANDLER, DEN **Vol. 5** 81
KULOVY BLESK **Vol. 10** 359, 430, 505
KULT CIALA **Vol. 1** 139, 165, 266, 364 **Vol. 2** 100 **Vol. 4** 157
KULTAINEN KYNTTILÄNJALKA **Vol. 5** 356, 634
KULTAINEN VASIKKA **Vol. 5** 613, 634
KULTAMITALIVAIMO **Vol. 5** 356, 357, 620
KÜLVAROSI LEGENDA **Vol. 9** 71, 265, 314

KÜLVAROSI ÖRSZOBA Vol. 9 164, 196, 274
KUM SANDHEDEN Vol. 5 261
KUMMITUSKIEVARI Vol. 5 347, 634
KUN ENTIGGER Vol. 5 203
KUN HUNTTALAN MATTI SUOMEN OSTI Vol. 5 442, 516, 587
KUN ISÄ TAHTOO... Vol. 5 435, 608
KUN ISÄLLÄ ON HAMMASSÄRKY Vol. 5 356, 406
KUN ON TUNTEET Vol. 5 334
KUN ONNI PETTÄÄ Vol. 5 351
KUN SANDHEDEN Vol. 5 125, 283
KUN TAIVAS PUTOAA... Vol. 5 541
KUN TUOMI KUKKII Vol. 5 443, 549
KUNDSKABENS TRAE Vol. 5 219, 296
KUNG-FU Vol. 6 *page 233, page 243, page 257*
KUNG FU WU SU Vol. 2 272, 296
KUNGAJAKT Vol. 5 840, 909
KUNGEN AV DALARNA Vol. 5 840, 994, 1066
KUNGEN KOMMER Vol. 5 840, 841, 927
KUNGLIGA JOHANSSON Vol. 5 840, 997
KUNGLIGA PATRASKET Vol. 5 929, 954, 1032, 1142
KUNGLIGT ÄVENTYR, ETT Vol. 5 1073, 1080
KUNGSGATAN Vol. 5 993, 1119
KUNGSLEDEN Vol. 5 1216
KUNIGAS, JOLLA EI OLLUT SYDÄNTÄ Vol. 5 516, 644, 647
KUNIGAS KULKUREITTEN Vol. 5 347, 513, 516, 634
KUNNE VAERT DEG, DET Vol. 5 800, 831
KUNNIOITTAEN Vol. 5 462, 643
KÜNSTLERLIEBE Vol. 4 614, 634
KUNSTNERKONFLIKTER Vol. 5 97
KUNSTNERS KAERLIGHED, EN Vol. 5 265
KUOLLUT MIES KUMMITTELEE Vol. 5 356, 643
KUOLLUT MIES RAKASTUU Vol. 5 446, 488
KUOLLUT MIES VIHASTUU Vol. 5 549, 592
KUPELNOVY HRAC Vol. 10 205, 281, 288
KUR HIQEN MASKAT Vol. 3 12, 20
KUR PO XHIROHEJ NJE FILM Vol. 6 *page 11, page 16*
KUR PRANVERA VONOHET Vol. 3 368 Vol. 6 *page 305*
KUR ZBARDHI NJE DITE Vol. 3 10, 14, 23
KURATA NA CESTACH Vol. 10 193, 428, 597
KURFÜRSTENDAMM Vol. 4 131, 213, 288
KURIER DES ZAREN, DER Vol. 2 833 Vol. 4 119, 747
KURIER VON LISSABON, DER Vol. 4 192
KURITON SUKUPOLVI Vol. 5 347, 356, 428, 516
KURRAGÖMMA Vol. 5 880, 933, 1104
KURS AUF DIE EHE Vol. 4 234
KURT OG VALDE Vol. 5 16, 111, 195
KUSH VDES NE KEMBE Vol. 6 *page 17, page 18*
KUSS IM SCHNEE Vol. 10 154, 345, 460
KUSS IN DER SOMMERNACHT, EIN Vol. 4 344, 647
KÜSSCHEN UND DER GENERAL Vol. 1 32, 102
KÜSSE, DIE MAN NICHT VERGISST Vol. 4 158, 626
KÜSSE, DIE MAN STIEHLT IM DUNKELN Vol. 4 309
KÜSSE, DIE TÖTEN Vol. 4 99
KÜSSEN IST KEINE SÜND' Vol. 4 239, 585, 754
KUSTENS GLADA KAVALJERER Vol. 5 1181
KUTTEL Vol. 1 26, 43, 65
KUTYA EJI DALA Vol. 9 10, 116
KUUDES KÄSKY Vol. 5 356, 386
KUUMA KISSA ? Vol. 5 516
KUUMAT KUNDIT Vol. 5 336, 342
KUUN ON VAARALLINEN Vol. 5 443, 639
KUZIS STARI MOJ Vol. 3 341 Vol. 6 *page 304*
KVAR Vol. 3 367, 418
KVARNEN Vol. 5 925, 976
KVARTERET KORPEN Vol. 5 1033
KVARTERETS OLYCKSFAGEL Vol. 5 1181
KVARTETTEN SOM SPRÄNGDES Vol. 5 863, 909, 1032, 1042, 1125, 1173
KVET ZE SUMAVY Vol. 10 167, 453
KVICK SOM BLIXTEN Vol. 5 1034
KVINDE AF FOLKET, EN Vol. 5 97 Vol. 53 97
KVINDE ER OVERFLODIG, EN Vol. 5 31, 248, 272
KVINDEHADERENS FALD Vol. 5 81
KVINDELIGE DAEMON, DEN Vol. 5 9
KVINDELIGE SPION FRA BALKAN, DEN Vol. 5 203
KVINDELIST OG KAERLIGHED Vol. 5 104, 199, 304
KVINDEN HAN MODTE Vol. 5 183
KVINDEN MED DE SMUKKE OJNE Vol. 5 256
KVINDEREGIMENTET Vol. 5 183
KVINDESIND Vol. 5 46, 177, 218
KVINDESKAEBNE, EN Vol. 5 97, 203
KVINNA I LEOPARD Vol. 5 840, 923
KVINNA I VITT Vol, 5 909, 1173
KVINNA OM BORD Vol. 5 871, 1042, 1156
KVINNA UTAN ANSIKTE Vol. 5 840, 909, 1173

KVINNAN BAKOM ALLT Vol. 5 378, 395, 616, 928, 1042, 1171
KVINNAN GÖR MED GALEN Vol. 5 923, 928
KVINNAN SOM FÖRSVANN Vol. 5 840
KVINNAN TAR BEFALLET Vol. 5 863, 1032, 1181
KVINNAS ANSIKTE, EN Vol. 5 840, 909
KVINNAS MORGONDAG, EN Vol. 5 1021
KVINNENE Vol. 5 684, 787
KVINNENS PLASS Vol. 5 738
KVINNODRÖM Vol. 5 871, 965
KVINNOHUSET Vol. 5 1003, 1073
KVINNOLEK Vol. 5 909, 1142
KVINNOR I FANGENSKAP Vol. 5 1042, 1078
KVINNOR I VÄNTRUM Vol. 5 334, 875, 1042
KVINNORNA KRING LARSSON Vol. 5 863, 1042
KVINNORS VÄNTAN Vol. 5 940, 1173
KVOCNA Vol. 10 256, 564
KVOCNY A KRAL Vol. 10 144, 369, 496
KWANNON VON OKADERA, DIE Vol. 4 444, 637
KWESTIA SUMIENIA Vol. 1 302
KWIAT PAPROCI Vol. 1 220
KWIECIEN Vol. 1 227, 354 Vol. 6 *page 205*
KYLÄ Vol. 5 415
KYLÄRAITTIEN KUNINGAS Vol. 5 472, 634
KYLLÄ KAIKKI SELVIÄÄ Vol. 5 476
KYM SA SKONCI TATO NOC Vol. 10 459, 575, 576
KYNEGEMNE PROSFYGOPOULA Vol. 3 167
KYÖKIN PUOLELLA Vol. 5 330, 592
KYPROS Vol. 3 156
KYRIAKATIKO XYPNEMA Vol. 3 202
KYRITZ – PYRITZ Vol. 4 242, 270, 471
KYRKOHERDEN Vol. 5 869
KYS, KLAP OG KOMMERS Vol. 5 24, 55
KYS TIL HOJRE OG VENSTRE Vol. 5 232, 259
KYSKE LEVEMAND, DEN Vol. 5 179, 195, 210, 257
KYSS HENNE ! Vol. 5 840, 878
KYSSEN PA KRYSSEN Vol. 5 909, 1173
K.Z. 9 LAGER DI STERMINIO Vol. 7 311

L

LA CAPATUL LINIEI Vol. 6 *page 269, page 277*
LA ELVA LEVE ! Vol. 5 666
LA' OS VAERE Vol. 5 84
LA PORTILE PAMINTULUI Vol. 3 230, 291
LA RASCRUCEA MARILOR FURTUNI Vol. 3 242, 297, 302 Vol. 6 *page 273*
LA ROSA ROSSA Vol. 3 395
LA UN PUNCT DE AGITATIE Vol. 3 330
LA VIRSTA DRAGOSTEI Vol. 3 270, 303, 330
LABAKAN Vol. 10 296, 412, 569
LABBRA BAGNATE Vol. 7 742
LABBRA DI LURIDO BLU Vol. 7 910
LABBRA E IL CUORE, LE Vol. 7 825
LABBRA ROSSE Vol. 7 1027
LABBRA SERRATE Vol. 7 196, 443
LABBRA VOGLIOSE Vol. 7 742
LABERINTO DE PASIONES Vol. 8 223, 373
LABETE THESEIS Vol. 3 190
LABIRINTUL Vol. 3 246, 300, 309
LABIRINTUS Vol. 9 154, 316
LABYRINT SRDCE Vol. 10 503, 522, 582
LABYRINTH DER LIEBE, DAS Vol. 4 270, 309
LAC AUX DAMES Vol. 2 440, 573, 726
LAC D'ARGENT, LE Vol. 2 136
LAC DES MORTS VIVANTS, LE Vol. 6 *page 134*
LACENITE OBUVKI NA NEZNAJNIJAT VOJN Vol. 3 110 Vol. 6 *page 30, page 55*
LACHDOKTOR, DER Vol. 4 115, 617
LACHE, BAJAZZO Vol. 4 87, 347, 370, 486, 729
LACHE IN HAREM Vol. 3 315
LACHE-MOI LES VALSEUSES Vol. 2 721
LÄCHELN DER KLEINEN BEATE, DAS Vol. 4 385, 495
LACHENDE DRITTE, DER Vol. 4 167, 647
LACHENDE EHEMANN, DER Vol. 4 239, 458, 584
LACHENDE ERBEN Vol. 4 17, 292
LACHENDE GRAUEN, DAS Vol. 6 *page 188*
LACHENDE GRILLE, DIE Vol. 4 14, 166, 186, 743
LACHENDE HERZEN Vol. 4 216
LACHENDE TOD, DER Vol. 4 270, 309
LACHES VIVENT D'ESPOIR, LES Vol. 2 196
LACHTAUBEN WEINEN NICHT Vol. 6 *page 160, page 161, page 169*
LACOMBE LUCIEN Vol. 2 847
LACRIMA SUL VISO, UNA Vol. 7 743
LACRIME D'AMORE Vol. 7 1121
LACRIME DI SANGUE Vol. 7 359, 709
LACRIME DI SPOSA Vol. 7 1063
LACRIME NAPULITANE Vol. 7 222, 623, 1194
LADA CU ZESTRE Vol. 3 285, 295
LADAN Vol. 5 834, 1083, 1200

LADENPRINZ, DER **Vol. 4** 14, 186
LADIES & GENTLEMEN **Vol. 7** 544, 719, 1184
LADRA DI BRILLANTI **Vol. 7** 1189
LADRA DI FANCIULLI, LA **Vol. 7** 510
LADRA, LA **Vol. 7** 1027, 1189
LADRAO PRECISA - SE ! ... **Vol. 8** 44, 93, 168, 184
LADRI DI BICICLETTE **Vol. 7** 796, 1148
LADRI, I **Vol. 7** 448
LADRO DE QUEM SE FALA, O **Vol. 8** 130, 131
LADRO DI BAGDAD, IL **Vol. 7** 333
LADRO DI DAMASCO, IL **Vol. 3** 395
LADRO DI DONNE **Vol. 7** 583, 822
LADRO DI VENEZIA, IL **Vol. 7** 1035, 1054
LADRO, IL **Vol. 7** 941, 1030, 1169
LADRO IN PARADISO, UN **Vol. 7** 1044
LADRO SONO IO !, IL **Vol. 7** 792, 1036
LADRON DE CHATARRA **Vol. 8** 773
LADRON DE GUANTE BLANCO, UN **Vol. 8** 515, 721
LADRONA PARA UNA ESPIA, UNO **Vol. 8** 426, 732
LADRONE, IL **Vol. 7** 426
LADRONES SOMOS GENTE HONRADA, LOS **Vol. 8** 214, 276, 396, 493
LADY FOOTBALL **Vol. 7** 1133
LADY GODIVA **Vol. 4** 131
LADY HAMILTON **Vol. 4** 131, 288, 396, 720
LADY HARRINGTON **Vol. 2** 727, 745
LADY L. **Vol. 2** 282
LADY LIBERTINE **Vol. 6** *page 115, page 129*
LADY OSCAR **Vol. 2** 655
LADY PANAME **Vol. 2** 27, 279, 488
LADY VIOLETTA **Vol. 9** 71
LADY WINDERMERES FÄCHER **Vol. 4** 531, 544, 623
LAEGEN **Vol. 5** 9
LAEGENS BRYLLUPSAFTEN **Vol. 5** 295
LAEGENS OFFER **Vol. 5** 97
LAEREAAR, ET **Vol. 5** 9
LAF U SRCU **Vol. 6** *page 306, page 314, page 320*
LAGARTERANOS, LOS **Vol. 8** 655
LAGARTIJA CON PIEL DE MUJER, UNA **Vol. 8** 306
LAGER SSADIS KASTRAT KOMMANDATUR **Vol. 7** 289
LAGOR I DUNKLET **Vol. 5** 871, 1156
LAGRIMAE RERUM O NEL GOGRO DELLA VITA **Vol. 7** 264, 701
LAGRIME **Vol. 7** 264
LAGUNA NEGRA, LA **Vol. 8** 207, 214
LÄHELLÄ SYNTIÄ **Vol. 5** 397, 471
LAIA **Vol. 8** 232, 405
LAILA **Vol. 5** 689, 761, 899, 954, 1080
LAIN MUKAAN **Vol. 5** 347, 634
LAISSE ALLER...C'EST UNE VALSE **Vol. 2** 279, 293
LAISSE BETON **Vol. 6** *page 81, page 122*
LAISSE TOMBER TA CULOTTE **Vol. 6** *page 120*
LAISSEZ-LES VIVRE **Vol. 2** 881
LAISSEZ TIRER LES TIREURS **Vol. 2** 475, 659
LAITAKAUPUNGIN LAULU **Vol. 5** 356, 357, 395
LAITURI **Vol. 5** 432, 584
LAIVAN KANNELLA **Vol. 5** 322, 490, 516, 592, 640
LAIVASTON MONNIT MAISSA **Vol. 5** 516, 592
LAJLA **Vol. 5** 55
LAKEUKSIEN LUKKO **Vol. 5** 443, 643
LALIE POLNE **Vol. 10** 273, 401, 493
LALKA **Vol. 1** 216, 254, 279, 281, 294, 320 **Vol. 6** *page 248*
LAMIEL **Vol. 2** 513
LAMJATA **Vol. 3** 31, 60, 109
LÄMNA MIG INTE ENSAM **Vol. 5** 947, 975, 990, 1024
LAMPAANSYÖJÄT **Vol. 5** 495, 606
LAMPADA ALLA FINESTRA, LA **Vol. 7** 115, 574, 727, 792
LAMPADA DELLA NONNA, LA **Vol. 7** 1189
LAMPAS MALEHO PLAVCIKA **Vol. 10** 108, 188, 228
LAN MEG DIN KONE **Vol. 5** 7, 674, 797, 831
LANCELOT DU LAC **Vol. 2** 170
LAND DER LIEBE **Vol. 4** 82, 544
LAND DES LÄCHELNS, DAS **Vol. 4** 306, 373
LAND OHNE FRAUEN, DAS **Vol. 4** 96, 149, 321, 660, 703
LANDLIG USKYLDSIGHED, EN **Vol. 5** 83
LANDLIGGER - IDYL OG VANDGANG **Vol. 5** 83
LANDMANDSLIV **Vol. 5** 16, 31, 272
LANDPOMERANZE, DIE **Vol. 4** 711
LANDRU **Vol. 2** 700, 770
LANDSBYLAEGEN **Vol. 5** 23
LANDSFLYGTIGE, DE **Vol. 5** 937, 996
LANDSHÖVDINGENS DÖTTRAT **Vol. 5** 997
LANDSKAMP **Vol. 5** 840, 841
LANDSTORMENS LILLA ARBIGGA **Vol. 5** 840, 997
LANDSTORMENS LILLA LOTTA **Vol. 5** 840, 878
LANDSTRASSE UND GROSSTADT **Vol. 4** 288, 329

LANDSTREICHER, DIE **Vol. 4** 79, 123, 796
LANDSTRYGERLIV **Vol. 5** 97
LANDSVÄGSRIDDARE **Vol. 5** 941
LANG - LASSE I DELSBO **Vol. 5** 1052, 1080
LANGA VÄGEN, DEN **Vol. 5** 841, 993
LANGE GRETE, DIE **Vol. 4** 185, 558
LANGE RITT ZUR SCHULE, DER **Vol. 6** *page 154, page 161, page 162, page 166*
LANGE STRASSE, DIE **Vol. 6** *page 162, page 163, page 165*
LANGOK **Vol. 9** 71, 322
LANGSAME TOD, DER **Vol. 4** 329, 724
LANGT BORTA OCH NÄRA **Vol. 5** 423, 895, 1007, 1201
LANGTA EFTER KARLEK **Vol. 5** 430
LÄNGTAN TILL HAVET **Vol. 5** 1013, 1111
LANGTURSCHAUFFOR **Vol. 5** 26, 259
LANGUE DE VELOURS **Vol. 2** 735 **Vol. 6** *page 94*
LANGUES S... **Vol. 2** 294
LANGUIDI BACI ... PERFIDE CAREZZE **Vol. 7** 546
LÄNNEN LOKARIN VELI **Vol. 5** 443, 549
LANTERNA CIECA, LA **Vol. 7** 470, 790
LANTERNA CU AMINTIRI **Vol. 3** 258
LANTERNA DEL DIAVOLO, LA **Vol. 7** 196, 508, 1029
LANTERNA DI DIOGENE, LA **Vol. 7** 267, 701
LANYARCOK TÜKÖRBEN **Vol. 9** 152, 316, 339
LAOS KAI KOLONAKI **Vol. 3** 204
LAP VIRAGA, A **Vol. 9** 26, 117, 133
LAPATOSSU **Vol. 5** 356, 485
LAPATOSSU JA VINSKI OLYMPIAKUU-MEESSA **Vol. 5** 485, 615
LAPBLOB **Vol. 5** 904, 945, 1174
LÄPI USVAN **Vol. 5** 310, 634
LÄPIMURTO **Vol. 5** 387, 404
LAPSENI ON MINUN **Vol. 5** 349, 428
LAPUALAISBALLADI **Vol. 5** 313, 338
LAPUALAISMORSIAN **Vol. 5** 374, 563
LAQUELLE ? **Vol. 2** 798
LARGA AGONIA DE LOS PECES, LA **Vol. 8** 523
LARGA NOCHE DE LOS BASTONES BLANCOS, LA **Vol. 8** 274, 626
LARGO ALLE DONNE ! **Vol. 7** 454
LARGO DIA AGUILA, EL **Vol. 8** 331, 762
LARGOS DIAS DE LA VENGANZA, LOS **Vol. 8** 298, 523
LÄRINNA PA VIFT **Vol. 5** 840, 878
LÄRM UM WEIDEMANN **Vol. 4** 115, 471, 566
LARME DANS L'OCEAN, UNE **Vol. 2** 481
LARS HARD **Vol. 5** 924, 1052
LARS' HOVEDSTADSREJSE **Vol. 5** 230
LARS I PORTEN **Vol. 5** 699, 701, 749, 765
LARSSON I ANDRA GIFTET **Vol. 5** 928
LASCIA CANTARE IL CUORE **Vol. 7** 101
LASCIATE FARE A NINICHE **Vol. 7** 569
LASCIATE OGNI SPERANZA **Vol. 7** 792, 796
LASISYDÄN **Vol. 5** 374, 477, 516
LASKA **Vol. 10** 394, 408, 425
LASKA A DREVAKY **Vol. 10** 45
LASKA A LIDE **Vol. 10** 122, 403, 628
LASKA JE UTRPENIM **Vol. 10** 34, 198
LASKA NA DRUHY POHLED **Vol. 10** 41, 129, 153
LASKA NEBESKA **Vol. 10** 24, 286
LASKA NELASKAVA **Vol. 10** 72, 216, 459
LASKA S VUNI PRYSKYRICE **Vol. 10** 226, 286, 577
LASKA SI NEDA POROUCET **Vol. 10** 310, 393
LASKA SLECNY VERY **Vol. 10** 243, 536
LASKA Z PASAZE **Vol. 10** 113, 495, 571
LASKY JEDNE PLAVOVLASKY **Vol. 10** 70, 398
LASKY KACENKY STRNADOVE **Vol. 10** 167, 345
LASKY MEZI KAPKAMI DESTE **Vol. 10** 87, 172, 517
LASSATOK FELEIM ! **Vol. 9** 149, 259, 285
LASSE - MAJA **Vol. 5** 863, 1032, 1056
LASSE MANSSON FRA SKAANE **Vol. 5** 136, 157, 183
LASSE OG GEIR **Vol. 5** 765, 784
LASST BLUMEN SPRECHEN **Vol. 4** 361, 367
LAST, DIE **Vol. 4** 560
LASTER DER MENSCHHEIT **Vol. 4** 124, 407
„LATA DWUDZIESTE ..., LATA TRZYDZIESTE ..." **Vol. 6** *page 207, page 214, page 228, page 235, page 245*
LATA LENA OCH BLAÖGDA PER **Vol. 5** 1080, 1156
LATERNA FTOHEIA KAI FILOTIMO **Vol. 3** 168, 169, 205
LATIDOS DE PANICO **Vol. 8** 296, 770
LATITANTE, IL **Vol. 7** 943, 1169
LATSZAT CSAL, A **Vol. 9** 117
LATTERMASKEN **Vol. 5** 97
LATTJO MED BOCCACCIO **Vol. 5** 871, 1042, 1140
LAUANTAILEIKIT **Vol. 5** 452
LAUDE DELLA VITA E LA LAUDE DELLA MORTE, LA **Vol. 7** 399
LAUFENDE BERG, DER **Vol. 4** 329, 347, 378
LÄUFER VON MARATHON, DER **Vol. 4** 452, 629, 788
LAUFFEUER, DAS **Vol. 4** 171, 230

LAUKAS TEHTAALLA Vol. 5 516
LAUKAUS KYPROKSESSA Vol. 5 549
LAULAVA SYDÄN Vol. 5 356, 549
LAULU TULIPUNAISESTA KUKASTA Vol. 5 310, 374, 582
LAUNEN EINER WELTDAME, DIE Vol. 4 99
LAURA Vol. 3 170 Vol. 9 31, 238, 259
LAURA ... A 16 ANNI MI DICESTI SI Vol. 7 222, 474
LAURA NUDA Vol. 7 246, 1203
LAURA OU LES OMBRES D'ETE Vol. 2 215, 787
LAURE Vol. 7 406
LAURETTE OU LE CACHET ROUGE Vol. 2 864
LAUT UND LEISE IST DIE LIEBE Vol. 1 50, 53, 98
LAUTA MANCIA Vol. 7 853
LAUTER LIEBE Vol. 4 51, 108, 119
LAUTER LÜGEN Vol. 4 132, 270
LAVADEIRAS DE PORTUGAL Vol. 8 62, 111
LAVANDIERAS DE PORTUGAL, LAS Vol. 8 298
LAVANDIERES DU PORTUGAL, LES Vol. 2 137, 293, 294, 383 Vol. 8 62, 111, 298
LAVEATA TIETÄ Vol. 5 476, 631
LAVINA Vol. 6 *page 37, page 47, page 60* Vol. 9 288 Vol. 10 142, 154, 621
LAVINEN Vol. 5 183
LAVORO, IL Vol. 7 498
LAZ Vol. 9 87, 117, 199
LAZAR Vol. 6 *page 309, page 310, page 313*
LAZARILLO DE TORMES, EL Vol. 8 246, 276, 749
LAZNI CAR Vol. 3 451
LAZNI PRINC Vol. 6 *page 309*
LAZNIVKA Vol. 3 348, 387
LAZY SA POHLI Vol. 10 349, 351, 520
LEA Vol. 7 796
LEAC PENTRU SOACRE, UN Vol. 3 234
LEANDRAS, LAS Vol. 8 299, 538
LEANYBECSÜLET Vol. 9 93
LEANYFURFANG Vol. 9 77
LEANYVARI BOSZORKANY, A Vol. 9 132
LEANYVASA Vol. 9 26, 117, 337
LEAO DA ESTRELA, O Vol. 8 122, 140, 178, 184
LEBADA IARNA, O Vol. 6 *page 274, page 279*
LEBBROSO, IL Vol. 7 153, 987
LEBEN BEGINNT, DAS Vol. 1 81, 101, 120
LEBEN KANN SO SCHÖN SEIN, DAS Vol. 4 373
LEBEN LANG, EIN Vol. 4 265, 614, 620
LEBEN MIT UWE Vol. 1 34, 80, 126
LEBEN RUFT, DAS Vol. 4 8, 491
LEBEN UM LEBEN Vol. 4 112, 192
LEBEN UND LÜGE Vol. 4 300
LEBEN WIE IM PARADIES, EIN Vol. 3 395
LEBEN ZU ZWEIT Vol. 1 32, 44, 104
LEBENDE BRÜCKE, DIE Vol. 4 514
LEBENDE BUDDHAS Vol. 4 287, 373, 533, 571, 575, 577, 646, 652
LEBENDE FACKEL, DIE Vol. 4 634
LEBENDE LEICHNAM, DER Vol. 4 159, 565, 657 Vol. 6 *page 181*
LEBENDE MASKE – HEINRICH DER VIERTE, DIE Vol. 2 32
LEBENDE PROPELLER, DER Vol. 4 5, 584
LEBENDE TOTE, DIE Vol. 4 190, 773
LEBENDE WARE Vol. 1 17, 21, 46
LEBENSHUNGER Vol. 4 729
LEBENSKÜNSTLER, EIN Vol. 4 316, 437
LEBENSLIED, DAS Vol. 4 25, 544
LECCIONES DE BUEN AMOR Vol. 8 214, 400
LECHE-MOI PARTOUT Vol. 2 2
LECON D'AMOUR , UNE Vol. 2 798
LECON DE CONDUITE Vol. 2 8, 128
LECON PARTICULIERE, LA Vol. 2 732
LEDERSTRUMPF Vol. 4 90
LEDIGE MÜTTER Vol. 4 270, 724
LEDIGE TÖCHTER Vol. 4 379, 752
LEDOVE MORE VOLA Vol. 10 131, 621
LEDOVI MUZI Vol. 10 299, 577
LEENA Vol. 5 549, 579
LEGATI DA TENERA AMICIZIA Vol. 7 239, 241, 498, 738
LEGATO Vol. 9 134, 316, 318
LEGENDA Vol. 1 179, 325, 362 Vol. 6 *page 245*
LEGENDA A NYULPAPRIKASROL Vol. 9 134, 152, 259
LEGENDA A VONATON Vol. 9 83, 332
LEGENDA CELOR DOUA CRUCI Vol. 3 333
LEGENDA O ENEJI Vol. 3 403
LEGENDA O LASCE Vol. 10 296, 412, 569
LEGENDA ZA LJUBOVTA Vol. 3 95 Vol. 6 *page 26*
LEGENDA ZA PAISIJ Vol. 3 39, 56 Vol. 6 *page 29*
LEGENDE DE SOEUR BEATRIX, LA Vol. 2 167
LEGENDE DER HEILIGEN SIMPLICIA, DIE Vol. 4 89, 307, 325
LEGENDE VON PAUL UND PAULA, DIE Vol. 1 14, 71, 79
LEGENY A GATON Vol. 9 117
LEGERE ET COURT VETUE Vol. 2 344, 615
LEGGE DEI GANGSTERS, LA Vol. 7 474

LEGGE DELLA MONTAGNA, LA Vol. 7 298
LEGGE DI GUERRA Vol. 7 246, 1045
LEGGE E LA LEGGE, LA Vol. 7 915
LEGGE, LA Vol. 7 725
LEGGE VIOLENTA DELLA SQUADRA ANTICRIMINE Vol. 7 1194
LEGGENDA AZZURRA Vol. 7 902, 923, 1063
LEGGENDA DEI COSTAMALA, LA Vol. 7 131
LEGGENDA DEI TRE FIORI, LA Vol. 7 327
LEGGENDA DEL CASTELLO, LA Vol. 7 81, 755, 837
LEGGENDA DEL PIAVE, LA Vol. 7 424, 792, 891
LEGGENDA DEL RUNINO MALESE, LA Vol. 7 694, 875
LEGGENDA DELLA CROCE, LA Vol. 7 1189
LEGGENDA DELLA PRIMAVERA, LA Vol. 7 670, 673, 1192, 1193
LEGGENDA DELLE DOLOMITI, LA Vol. 7 993
LEGGENDA DI GENOVEFFA, LA Vol. 7 372
LEGGENDA DI PIERRETTE, LA Vol. 7 454
LEGGENDA DI SIMBARAB, LA Vol. 7 727
LEGGENDA DI WALLY, LA Vol. 7 493, 963
LEGGENDA MEDIOEVALE Vol. 7 1189
LEGION CONDOR Vol. 4 13
LEGION DE HEROES Vol. 8 524
LEGION DEL SILENCIO, LA Vol. 8 207, 298, 719
LEGION SAUTE SUR KOLWEZI, LA Vol. 2 208, 327, 519
LEGION ULICY Vol. 1 346
LEGIONAR Vol. 10 45, 536
LEGIONE DEI DANNATI, LA Vol. 8 762
LEGIONE DELLA MORTE, LA Vol. 7 420
LEGIONE STRANIERA Vol. 7 359, 1054
LEGIONES DE CLEOPATRA, LAS Vol. 8 602, 725
LEGIONI DI CLEOPATRA, LE Vol. 7 845 Vol. 8 602, 725
LEGIONS D'HONNEUR Vol. 2 193, 279, 386, 566, 719
LEGITIME VIOLENCE Vol. 6 *page 98, page 111*
LEGSZEBB FERFIKOR, A Vol. 9 138, 259, 339
LEGSZEBB KALAND, A Vol. 9 77
LEGUIGNON GUERISSEUR Vol. 2 832, 833
LEGY JO MINDHALALIG Vol. 9 71, 164, 165, 241, 259, 322
LEHMANN LERNT ZAUBERN Vol. 4 514
LEI NON SA CHI SONO IO Vol. 7 414
LEI O NESSUNA ! Vol. 7 965
LEIBA ZIBAL Vol. 3 315
LEIBEIGENEN, DIE Vol. 4 192, 462, 584
LEICHENFUND IM JAGEN 14 Vol. 1 111
LEICHENSACHE ZERNIK Vol. 1 59, 80, 110
LEICHTE ISABELL, DIE Vol. 4 97, 149, 208, 660
LEICHTE KAVALLERIE Vol. 4 79, 82, 115, 239, 340, 486, 703
LEICHTE MÄDCHEN, DAS Vol. 4 347, 370, 491
LEICHTE MUSE Vol. 4 270, 740, 774
LEICHTES BLUT Vol. 4 347, 370, 580
LEICHTSINN Vol. 4 167, 339
LEICHTSINN UND GENIE Vol. 4 760
LEICHTSINNIGE JUGEND Vol. 4 386
LEIDEN DES JUNGEN WERTHERS, DIE Vol. 1 18, 36, 50
LEIDENDES LAND Vol. 4 128, 144
LEIDENSCHAFT Vol. 4 517, 536, 558
LEIDENSCHAFT. DIE LIEBSCHAFTEN DER HELLA VON GILSA Vol. 4 220, 239, 584, 780
LEIDENSWEG DER EVA GRUNWALD, DER Vol. 4 5, 231, 584
LEIDENSWEG DER INGE KRAFFT, DER Vol. 4 298, 733
LEIDENSWEG EINER ACHTZEHNJÄHRIGEN, DER Vol. 4 162, 291
LEIKKIKALUGANGSTERI Vol. 5 606
LEINEN AUS IRLAND Vol. 4 124, 232, 620
LEJLIGHED TIL LEJE Vol. 5 2, 228
LEJON PA STAN Vol. 5 878, 881
LEJONET OCH JUNGFRUN Vol. 5 933, 1052
LEJONSOMMAR Vol. 5 919
LEJTÖN Vol. 9 157
LEK Vol. 5 792
LEK PA REGNBAGEN Vol. 5 940, 965, 1052
LEKARSTWO NA MILOSC Vol. 1 202, 400 Vol. 6 *page 240*
LEKCE Vol. 10 299, 385, 531
LEKCJA MARTWEGO JEZYKA Vol. 6 *page 251, page 253, page 258*
LEKCJA MILOSCI Vol. 1 195, 298
LEKTION, EN Vol. 5 97
LEKTION I KÄRLEK, EN Vol. 5 878, 1052
LELEK ORASA, A Vol. 9 220
LELEKIDOMAR, A Vol. 4 166 Vol. 9 71, 80
LELEKLATO SUGAR, A Vol. 9 288
LELICEK VE SLUZBACH SHERLOCKA HOLMESE Vol. 10 117, 138, 167, 524
LELKI KLINIKA Vol. 9 26, 133, 157, 290
LEMKES SEL. WITWE Vol. 4 247, 424

LEMMY POUR LES DAMES **Vol. 2** 598, 824
L'ENCLOS **Vol. 3** 387
LENE SPIELT LOTTERIE **Vol. 4** 323, 367
LENGYELVER **Vol. 9** 175, 220
LENIN, DIN GAVTYV **Vol. 5** 22, 278
LENIN W POLSCE **Vol. 1** 179, 233, 406
LENKENE BRYTES **Vol. 5** 670, 760
LENOR **Vol. 8** 214
LENTÄVA KALAKUKKO **Vol. 5** 461, 592
LENTÄVÄT LUUPÄÄT **Vol. 5** 345, 408, 424
LENZ **Vol. 9** 60, 216, 289
LEO ALS BAZILLENTRÄGER **Vol. 4** 598
LEO, DER AUSHILFSKELLNER **Vol. 4** 598
LEO SAPPERLOTER **Vol. 4** 598
LEO SKAL I POSTKASSEN **Vol. 5** 83
LEO UND SEINE DREI BRÄUTE **Vol. 4** 598
LEOFOROS TES PRODOSIAS **Vol. 3** 149
LEON DE DAMASCO, EL **Vol. 7** 2, 1116
LEON MORIN PRETRE **Vol. 2** 224, 376
LEONA DE CASTILLA, LA **Vol. 8** 298, 400
LEONCE **Vol. 2** 798
LEONE DI OMAR, IL **Vol. 7** 296,1005
LEONE DI TEBE, IL **Vol. 7** 439
LEONE MANSUETO, IL **Vol. 7** 490
LEONI AL SOLE **Vol. 7** 368, 853
LEONI LEO **Vol. 9** 288
LEONOR **Vol. 2** 11 **Vol. 7** 1145
LEONTINES EHEMÄNNER **Vol. 4** 166
LEON...TOUT COURT **Vol. 2** 65, 855
LEOPARD, A **Vol. 9** 288
LEOPARD, LE **Vol. 6** *page 75, page 97*
LEOPARDA FERITA, LA **Vol. 7** 108
LEOPARDI DI CHURCHILL, I **Vol. 8** 551
LEOPARDOS DE CHURCHILL, LOS **Vol. 8** 551
LEOPOLD LE BIEN-AIME **Vol. 2** 6, 99, 726
LEPA PARADA **Vol. 3** 366
„LEPAIN", DER KÖNIG DER UNSCHULDIGEN **Vol. 4** 135, 503
LEPAIN, DER KÖNIG DER VERBRECHER **Vol. 4** 628
LEPOTA POROKA **Vol. 6** *page 311, page 317*
LEPTIROV OBLAK **Vol. 3** 367, 368
LEPURI ME PESE KEMBE **Vol. 6** *page 309*
LEQUEL ? **Vol. 2** 159, 286, 330
LERPAR **Vol. 8** 154
LERSOENS KONGE **Vol. 5** 203
LES FETES GALANTES **Vol. 3** 218, 286
LESBO **Vol. 7** 780
LESETINSKY KOVAR **Vol. 10** 115, 198
LESMONEMENA PROSOPA **Vol. 3** 179
LEST WE FORGET **Vol. 2** 384, 579
LESTIA STEN ATHENA **Vol. 3** 159
LESZ, AMI LESZ **Vol. 9** 14, 71, 126
LET MRTVE PTICE **Vol. 3** 368
LET, PTAKU, LET ! **Vol. 10** 173, 222, 577
LETA ODLOCITVE **Vol. 6** *page 305, page 316*
LETACI VELIKOG NEBA **Vol. 3** 350, 451
LETALOTO **Vol. 6** *page 27, page 30, page 55*
LETISTE NEPRIJIMA **Vol. 10** 352, 359
LETNY STROM RADOSTI **Vol. 10** 269, 324, 399
LETO **Vol. 10** 407, 524
LETO JE KRIVO ZA SVE **Vol. 3** 366, 455
LETO S KOVBOJEM **Vol. 10** 286, 352, 425
LETOKRUHY **Vol. 10** 32, 82, 263
LETOS V ZARI **Vol. 10** 313, 572
LETSINDIGHEDENS LON **Vol. 5** 295
LETSINDING KONGE, EN **Vol. 5** 97
LETTERA ALL'ALBA, UNA **Vol. 7** 1177
LETTERA CHIUSA, LA **Vol. 7** 278, 971
LETTERA NAPOLETANA **Vol. 7** 603
LETTERE AL SOTTOTENENTE **Vol. 7** 250, 1054
LETTERE DAL CAMPO **Vol. 7** 1189
LETTERE DI UNA NOVIZIA **Vol. 7** 300, 514
LETTI SELVAGGI **Vol. 7** 817, 1007
LETTINO VUOTO, IL **Vol. 7** 168
LETTO A TRE PIAZZE **Vol. 7** 693, 1054
LETTOMANIA **Vol. 7** 1024
LETTRE CHARGE, LA **Vol. 2** 159, 286, 330
LETTRE, LA **Vol. 2** 384
LETTRE OUVERTE **Vol. 2** 124, 611
LETTRES D'AMOUR **Vol. 2** 8, 270, 439, 786
LETTRES D'AMOUR PERDUES **Vol. 6** *page 128*
LETTRES DE MON MOULIN, LES **Vol. 2** 347, 548, 871
LETTRES DE STALINGRAD, LES **Vol. 2** 139
LETTRICE DELLA DUCHESSA, LA **Vol. 7** 1189
LETTURA POCO INTERESSANTE **Vol. 7** 1189
LETZTE BRÜCKE, DIE **Vol. 3** 378
LETZTE CHANCE, DIE **Vol. 1** 7, 37, 85
LETZTE DROSCHKE VON BERLIN, DIE **Vol. 4** 176, 626
LETZTE EINES ALTEN GESCHLECHTES, DER **Vol. 4** 124, 514
LETZTE FAHRT DER SANTA MARGARETA **Vol. 4** 247, 340, 466
LETZTE FORT, DAS **Vol. 4** 85, 643, 729
LETZTE GALAVORSTELLUNG DES ZIRKUS WOLFSOHN, DIE **Vol. 4** 270, 291, 367, 486, 599, 725
LETZTE HEUER, DIE **Vol. 1** 35, 56, 90
LETZTE HOHENHAUS, DER **Vol. 4** 87
LETZTE KOMPAGNIE, DIE **Vol. 4** 14, 359

LETZTE LIEBESNACHT DER INGE TOLMEN, DIE **Vol. 4** 270, 695
LETZTE MANN, DER **Vol. 4** 178, 266, 574
LETZTE MASKE, DIE **Vol. 4** 320
LETZTE MOHIKANER, DER **Vol. 8** 523, 732
LETZTE NACHT, DIE **Vol. 1** 58, 120 **Vol. 4** 321, 752
LETZTE RUNDE, DIE **Vol. 4** 124, 232, 361
LETZTE SCHUSS, DER **Vol. 4** 135
LETZTE SONNENSOHN, DER **Vol. 4** 113
LETZTE SOUPER, DAS **Vol. 4** 85, 359
LETZTE STUNDE, DIE **Vol. 4** 500, 724
LETZTE TAG, DER **Vol. 4** 87, 734
LETZTE WALZER, DER **Vol. 4** 132, 246, 458, 546, 570, 663
LETZTE WORT, DAS **Vol. 1** 18, 47, 60
LETZTE ZEUGE, DER **Vol. 4** 131
LETZTEN GRÜSSE VON MARIE, DIE **Vol. 4** 13, 24
LETZTEN KOLCZAKS, DIE **Vol. 4** 132, 773
LETZTEN MENSCHEN, DIE **Vol. 4** 178, 237
LETZTEN TAGE VOR DEM WELTBRAND, DIE **Vol. 4** 219
LETZTEN VIER VON SANTA CRUZ, DIE **Vol. 4** 460, 713
LETZTES FACH UNTEN RECHTS **Vol. 1** 34, 81
LEUCHTE ASIENS, DIE **Vol. 4** 327, 587, 775
LEUCHTER DES KAISERS, DIE **Vol. 4** 614
LEUCHTFEUER **Vol. 1** 3, 107, 129
LEUCHTFEUER VON LUBACZOW, DAS **Vol. 4** 99
LEUR DERNIERE NUIT **Vol. 2** 8, 59, 540
LEUTE MIT DEM SONNENSTICH, DIE **Vol. 4** 339, 410, 422
LEUTE MIT FLÜGELN **Vol. 1** 9, 47, 54
LEUTNANT AUF BEFEHL **Vol. 4** 560, 663
LEUTNANT IHRER MAJESTÄT, DER **Vol. 4** 270, 292, 769
LEUTNANT MUCKI **Vol. 4** 366
LEUTNANT VOM SCHWANENKIETZ, DER **Vol. 1** 38, 59
LEUTNANT WARST DU EINST BEI DEN HUSAREN **Vol. 4** 186, 287, 644
LEUTNANT YORCK VON WARTENBURG, DER **Vol. 6** *page 164, page 165, page 171*
LEV FARLIGT **Vol. 5** 907, 1073
LEV LIVET LET **Vol. 5** 151, 304
LEV S BILOU HRIVOU **Vol. 10** 74, 87, 139, 144
LEV TOLSTOJ **Vol. 10** 82, 123, 228
LEVA PA „HOPPET“ **Vol. 5** 840, 843
LEVAGOTT KEZ **Vol. 9** 93
LEVE KRIDLO **Vol. 10** 233, 286, 577
LEVE SITT LIV **Vol. 5** 667, 737, 749
LEVEMANDEN PAA VARIETE **Vol. 5** 97
LEVENDE HITTEGODS, DET **Vol. 5** 83
LEVENTOPAIDO **Vol. 3** 135
LEVER DE MLLE LOUISE WILLY DE L'OLYMPIA A PARIS **Vol. 2** 678
LEVIATAN **Vol. 8** 237, 414
LEVIATHAN **Vol. 2** 395, 571
LEVINS MÜHLE **Vol. 6** *page 166, page 167, page 176*
LEVOTON VERI **Vol. 5** 310, 480
LEVRE DE SANG **Vol. 2** 739
LEVRES BRULANTES **Vol. 2** 349
LEVRES GLOUTONNES **Vol. 6** *page 106*
LEVRETTE, LA **Vol. 2** 721
LEVY ET COMPAGNIE **Vol. 2** 6, 48, 55, 193, 381
LEVY S'ENVOLE **Vol. 2** 728
LEY DE UNA RAZA, LA **Vol. 8** 402
LEY DEL COLT, LA **Vol. 8** 495, 541
LEY DEL FORASTERO, LA **Vol. 8** 438, 546
LEY DEL MAR **Vol. 8** 640
LEYENDA DE FERIA **Vol. 8** 437
LEYENDA DE NAVIDAD **Vol. 8** 502, 721
LEYENDA DEL TAMBOR, LA **Vol. 8** 245, 749
LEYENDA ROTA, LA **Vol. 8** 268
LEZIONE DI VIOLONCELLO CON TOCCATA E FUGA **Vol. 7** 910
LEZIONI PRIVATE **Vol. 7** 834
LI-HANG LE CRUEL **Vol. 2** 257
LI-PAO, MANDARINO **Vol. 7** 940
LIA **Vol. 3** 216
LIAISONS DANGEREUSES, LES **Vol. 2** 368, 383, 503, 581
LIANA, LA SCHIAVA BIANCA **Vol. 7** 796
LIANBRON **Vol. 5** 877, 1080
LIANE, DIE WEISSE SKLAVIN **Vol. 7** 796
LIANTES, LOS **Vol. 8** 237, 773
LIBERA, AMORE MIO **Vol. 7** 524
LIBERA USCITA **Vol. 7** 891
LIBERAZIONE **Vol. 7** 701, 725
LIBERI, ARMATI, PERICOLOSI **Vol. 7** 763
LIBERTE **Vol. 2** 131, 688
LIBERTE 1 **Vol. 2** 615, 857
LIBERTE EN CROUPE, LA **Vol. 2** 208
LIBERTE SURVEILLEE **Vol. 2** 680, 709, 770
LIBERTY BELLE **Vol. 6** *page 101, page 136*
LIBERXINA 90 **Vol. 8** 232, 732
LIBRO DELLA VITA, IL **Vol. 7** 1067
LICEALE AL MARE CON L'AMICA DI PAPA **Vol. 7** 811, 1202
LICEALE IL DIAVOLO E L'ACQUASANTA, LA **Vol. 7** 1202
LICEALE, LA **Vol. 7** 426
LICEALE NELLA CLASSE DEI RIPETENTI, LA **Vol. 7** 1202
LICEALE SEDUCE I PROFESSORI, LA **Vol. 7** 1202

LICEM U LICE Vol. 3 341, 444
LICENZA PREMIO Vol. 7 587
LICHT AUF DEM GALGEN, DAS Vol. 1 80, 96, 102, 103
LICHT DER SCHWARZEN KERZE, DAS Vol. 1 62, 91, 111
LICHT DES LEBENS, DAS Vol. 6 *page 194*
LICHT UM MITTERNACHT, DAS Vol. 4 356, 612
LICNE STVARI Vol. 6 *page 310*
LICZE NA WASZE GRZECHY Vol. 1 180, 211, 294
LIDE JAKO TY Vol. 10 193, 566
LIDE NA KRE Vol. 10 256, 412
LIDE POD HORAMI Vol. 10 256, 412, 532
LIDE V BOURI Vol. 10 40, 431
LIDE Z MARINGOTEK Vol. 10 524, 612, 621
LIDE Z METRA Vol. 10 87, 233, 395
LIDELSE Vol. 5 907, 1073
LIDERCNYOMAS Vol. 9 146, 148
LIEB, EIN DIEB, EIN WARENHAUS, EIN Vol. 4 82
LIEB HEIMATLAND Vol. 4 109, 125
LIEBE Vol. 4 52, 161, 424, 609, 724, 734
LIEBE AUF BEFEHL Vol. 4 691
LIEBE AUF DEM LETZTEN BLICK Vol. 1 13, 50
LIEBE AUF DEN ERSTEN BLICK Vol. 4 228, 471, 747
LIEBE AUF DEN ERSTEN TON Vol. 4 626
LIEBE AUGUSTIN, DER Vol. 4 234, 373, 738
LIEBE BEI HOF Vol. 4 527
LIEBE BESUCH, DER Vol. 4 119, 517, 558
LIEBE DER BRÜDER ROTT, DIE Vol. 4 14, 527
LIEBE DER JEANNE NEY, DIE Vol. 4 298, 379, 729
LIEBE DER MARION BACH, DIE Vol. 4 603
LIEBE DER SKLAVIN, DIE Vol. 4 300
LIEBE DES MAHARADSCHA, DIE Vol. 4 16 Vol. 6 *page 182* Vol. 7 48, 620
LIEBE DES VAN ROYK, DIE Vol. 4 476, 522
LIEBE DUMME MAMA Vol. 4 344, 647
LIEBE EINER KÖNIGIN, DIE Vol. 4 51, 213, 651
LIEBE GEHT SELTSAME WEGE Vol. 4 361, 373, 424, 560
LIEBE GEHT-WOHIN SIE WILL Vol. 4 658
LIEBE, HASS UND GELD Vol. 4 232, 673
LIEBE IM DREIVIERTELTAKT Vol. 4 582
LIEBE IM KUHSTALL Vol. 4 340, 535, 608
LIEBE IM RAUSCH Vol. 4 626, 634, 720, 734
LIEBE IM RING Vol. 4 149, 158, 660
LIEBE IM SCHNEE Vol. 4 287, 384, 458, 646
LIEBE IN UNIFORM Vol. 4 239, 558
LIEBE IST DER FRAUEN MACHT, DIE Vol. 4 467, 522, 644, 752
LIEBE IST ZOLLFREI Vol. 4 97, 391, 557
LIEBE KANN LÜGEN Vol. 4 197, 354, 738, 788
LIEBE, LEIDENSCHAFT UND LEID Vol. 4 121, 273, 791
LIEBE MACHT BLIND Vol. 4 89, 306
LIEBE MIT SECHZEHN Vol. 1 1, 52, 121
LIEBE MUSS VERSTANDEN SEIN Vol. 4 17, 30, 713
LIEBE NACH NOTEN Vol. 4 620
LIEBE PILGERFAHRT, DIE Vol. 4 773
LIEBE, SCHERZ UND ERNST Vol. 4 255, 292, 703
LIEBE SIEGT, DIE Vol. 4 119
LIEBE STRENG VERBOTEN Vol. 4 292, 329
LIEBE, TOD UND TEUFEL Vol. 4 298, 531, 608, 729
LIEBE UND ALLTAG Vol. 4 318
LIEBE UND CHAMPAGNER Vol. 4 219, 544
LIEBE UND DER CO-PILOT, DIE Vol. 1 58, 72, 129
LIEBE UND DIE ERSTE EISENBAHN, DIE Vol. 4 132, 230
LIEBE UND DIEBE Vol. 4 535, 626
LIEBE UND LEBEN Vol. 4 212
LIEBE UND TROMPETENBLASEN Vol. 4 220, 239, 560
LIEBE UND ZAHNWEH Vol. 4 329, 774
LIEBE VOM ZIGEUNER STAMMT..., DIE Vol. 4 71, 172
LIEBELEI Vol. 4 292, 517, 584
LIEBELEI UND LIEBE Vol. 4 108, 270, 796
LIEBEN MITMENSCHEN, DIE Vol. 1 44, 46, 47
LIEBES-KORRIDOR, DER Vol. 4 240, 329
„LIEBES MÄNNCHEN, FOLGE MIR..." Vol. 4 514
LIEBESARZT, DER Vol. 4 49, 86, 270, 409, 774
LIEBESBAROMETER, DAS Vol. 4 356, 779
LIEBESBRIEF DER KÖNIGIN, DER Vol. 4 319
LIEBESBRIEFE Vol. 4 688, 728
LIEBESBRIEFE AUS DEM ENGADIN Vol. 4 154, 437, 531, 536, 541, 564, 638
LIEBESBRIEFE DER BARONIN S..., DIE Vol. 4 186, 407, 725
LIEBESBRIEFE EINER VERLASSENEN, DIE Vol. 4 329, 385
LIEBESERKLÄRUNG AN G. T. Vol. 1 8, 79, 92

LIEBESERWACHEN Vol. 4 79, 115, 634
LIEBESEXPRESS, DER Vol. 4 132, 544
LIEBESFALLEN Vol. 1 9, 62, 79
LIEBESFEUER Vol. 4 113, 546
LIEBESFILIALE, DIE Vol. 4 191, 242, 270, 367, 471
LIEBESFRÜHLING Vol. 4 323, 634
LIEBESGESCHICHTEN Vol. 4 84, 119, 264, 491, 574, 734
LIEBESHANDEL Vol. 4 316, 729
LIEBESHEIRAT Vol. 4 110, 740
LIEBESKÄFIG, DER Vol. 4 192, 560
LIEBESKARNEVAL Vol. 4 479, 663, 743
LIEBESKLEEBLATT Vol. 4 84, 270, 409
LIEBESKOMMANDO Vol. 4 14, 124, 208
LIEBESKOMÖDIE Vol. 4 110, 340, 638
LIEBESLAUBE, DIE Vol. 4 473, 575, 626, 795
LIEBESLEUTE Vol. 4 49, 340, 466
LIEBESLIED Vol. 4 30, 347, 445, 567
LIEBESMARKT, DER Vol. 4 340, 535, 608
LIEBESNÄCHTE Vol. 4 255, 599
LIEBESNEST Vol. 4 202
LIEBESOPFER Vol. 4 212
LIEBESPREMIERE Vol. 4 266, 582
LIEBESPROBE, DIE Vol. 4 342
LIEBESREIGEN Vol. 4 584, 646
LIEBESSCHULE Vol. 4 28, 188, 608
LIEBESTAUMEL Vol. 4 320, 454
LIEBESTRÄUME Vol. 4 139
LIEBESWAAGE, DIE Vol. 4 529
LIEBESWALZER Vol. 4 89, 325, 713
LIEBET DAS LEBEN Vol. 4 327, 344, 509, 545
LIEBFRAUMILCH Vol. 4 340, 535, 608
LIEBLING DER GÖTTER Vol. 4 325, 567, 713
LIEBLINGSFRAU DES MAHARADSCHA, DIE Vol. 4 359, 400
LIEBSCHAFTEN DER KÄTE KELLER, DIE Vol. 4 660
LIEBSCHAFTEN DES HEKTOR DALMORE, DIE Vol. 4 131, 321
LIED, DAS MEINE MUTTER SANG, DAS Vol. 4 565, 704
LIED DER MATROSEN, DAS Vol. 1 40, 47, 72, 74
LIED DER MUTTER, DAS Vol. 4 288
LIED DER NACHTIGALL, DAS Vol. 4 110, 660
LIED DER NATIONEN, DAS Vol. 4 359
LIED DER NORNEN, DAS Vol. 4 165, 270, 584
LIED DER SCHWARZEN BERGE, DAS Vol. 4 123, 354, 481
LIED DER SONNE, DAS Vol. 4 292, 340
LIED DER WÜSTE, DAS Vol. 4 298, 727, 742
LIED, EIN KUSS, EIN MÄDEL, EIN Vol. 4 124, 208, 422
LIED EINER NACHT, DAS Vol. 4 28, 614, 729
LIED FÜR DICH, EIN Vol. 4 321, 614, 703
LIED GEHT UM DIE WELT, EIN Vol. 4 373, 626
LIED IST AUS, DAS Vol. 4 208, 479
LIED KLAGT AN, EIN Vol. 4 247, 339, 461
LIED VOM GLÜCK, DAS Vol. 4 246, 361
LIED VOM LEBEN, DAS Vol. 4 32, 710
LIED VOM TROMPETER, DAS Vol. 1 17, 42, 85
LIED VON LIEBE, EIN Vol. 4 525
LIENS DE SANG, LES Vol. 2 700
LIEUX SAINTS Vol. 2 740
LIEVRE ET LE RENARD, LE Vol. 2 581
LIFE AND DEATH OF COLONEL BLIMP, THE Vol. 2 656
LIFTING SHADOWS Vol. 2 384
LIGA NO ES COSA DE HOMBRES, LA Vol. 8 261, 687
LIGABUE Vol. 7 262, 514, 1165
LIGE FOR LIGE Vol. 5 92
LIGERAMENTE VIUDAS Vol. 8 249, 626
LIGHT AT THE EDGE OF THE WORLD, THE Vol. 8 214, 352, 611
LIGNE DE DEMARCATION, LA Vol. 2 520, 700
LIGNE DROITE, LA Vol. 2 288, 669
LIGUE STORY Vol. 8 749
LIGUERO MAGICO, EL Vol. 8 249, 368
LIJO A SU ALCANCE, UN Vol. 8 297, 577, 749
LIKA INFÖR LAGEN Vol. 5 1062
LIKAINEN PUOLITUSINA Vol. 5 314
LIKEROVA PRINCEZNICKA Vol. 10 34, 68
LILA AKAC Vol. 9 71, 90, 122, 152, 221, 237
LILA TEST, SARGA SAPKA Vol. 9 77
LILI Vol. 9 288
LILIKA Vol. 3 362, 418
LILIOM Vol. 2 195, 626, 722 Vol. 9 288
LILIOMFI Vol. 9 77, 241, 312, 326
LILITH, DAS MÄDCHEN VOM SEE Vol. 4 514
LILJANA Vol. 3 67
LILLA HELGONET Vol. 5 1042, 1164
LILLA MÄRTA KOMMER TILLBACKA Vol. 5 1042, 1181
LILLE CHAUFFOR, DEN Vol. 5 92
LILLE CLAUS OG STORE CLAUS Vol. 5 203
LILLE DANSERINDE, DEN Vol. 5 183
LILLE DON JUAN, DEN Vol. 5 83
LILLE DORRIT Vol. 5 136, 183, 228
LILLE FRIDOLF BLIR MORFAR Vol. 5 1042, 1217

LILLE FRIDOLF OCH JAG **Vol. 5** 1042, 1217
LILLE HANNE **Vol. 5** 97
LILLE HELT, EN **Vol. 5** 97
LILLE HORNBLAESER, DEN **Vol. 5** 277
LILLE HOTEL, DET **Vol. 5** 2, 14
LILLE LISE LET-PAA-TAA **Vol. 5** 24, 83
LILLE MAND, PAS PA **Vol. 5** 274, 276
LILLE NAPOLEON **Vol. 5** 878
LILLE PUS, DET **Vol. 5** 295
LILLE SPEJL **Vol. 5** 125, 195, 273
LILLE TEDDY **Vol. 5** 256
LILLE TILFAELDIGHED, EN **Vol. 5** 2, 24
LILLE VIRGIL OG ORLA FROSNAPPER **Vol. 5** 125, 152, 283
LILLE VIRTUOS, DEN **Vol. 5** 300
LILLEBROR OCH JAG **Vol. 5** 840, 909
„LILLI" UND „LILLIS EHE" **Vol. 4** 131, 423
LILLI'S ERSTE LIEBE **Vol. 6** *page 186*
LILLY A LILLETTE O L'ARTE DI FARSI AMARE **Vol. 7** 53
LILLY PUSSY **Vol. 7** 1057
LILY, AIME-MOI **Vol. 2** 12, 240
LILY VERTU **Vol. 2** 37
L'IMAGE **Vol. 4** 166
LIMELIGHT **Vol. 2** 526
LIMEUSES, LES **Vol. 2** 349
LIMONADOVY JOE **Vol. 10** 388, 503, 569
LIMOSNA DE AMORES **Vol. 8** 463, 722
LIMPAN **Vol. 5** 917, 1093, 1209
LIMPOPO **Vol. 8** 130
LIMUZYNA DAIMLER-BENZ **Vol. 6** *page 216, page 219, page 234, page 244*
LINA DE AGOSTO **Vol. 8** 759
LINCEUL N'A PAS DE POCHES, UN **Vol. 2** 866
LINCOLN ALS DECKENLÄUFER **Vol. 4** 135, 503
LINDA BEATRIZ, LA **Vol. 8** 516
LINDA DI CHAMOUNIX **Vol. 7** 865
LINDENWIRTIN AM RHEIN, DIE **Vol. 4** 36, 340
LINDENWIRTIN, DIE **Vol. 4** 32, 255, 321
LINE **Vol. 5** 717, 742, 800, 826
LINEA DEL CIELO, LA **Vol. 8** 373
LINEA DEL FIUME, LA **Vol. 7** 1205
LINGERIES INTIMES **Vol. 6** *page 94*
LINGUA DI ERIKA, LA **Vol. 7** 153
LINIA **Vol. 1** 198, 391 **Vol. 6** *page 239*
LINISTEA DIN ADINCURI **Vol. 6** *page 273, page 277*
LINJE SEX **Vol. 5** 843
LINKS DER ISAR – RECHTS DER SPREE **Vol. 4** 344, 347, 370, 544, 678
LINNAISTEN VIHREÄ KAMARI **Vol. 5** 379, 579
LINTUMIES **Vol. 5** 550, 551
LINUS ELLER TEGELHUSETS HEMLIGHET **Vol. 5** 858, 946, 1053
LIO DE PAPA, EL **Vol. 8** 639
LION DES MOGOLS, LE **Vol. 2** 100, 122, 521, 623
LION EN EL LABORATORIO **Vol. 8** 719
LIONCEAUX, LES **Vol. 2** 128, 232, 488
LIONS SONT LACHES, LES **Vol. 2** 185, 566
LIOS DE ESTEFANIA, LOS **Vol. 8** 546
LIQUIRIZIA **Vol. 7** 605, 939, 1025
LIRI A VDEKJE **Vol. 3** 2, 14
LIS DE MER, LE **Vol. 2** 781
LISA FLEURON **Vol. 7** 264, 701
LISACI-MYSACI A SIBENICAK **Vol. 10** 139, 496
LISBOA **Vol. 8** 10, 114
LISBOA, CRONICA ANEDOTICA **Vol. 8** 105
LISBON **Vol. 8** 10, 114
LISCA **Vol. 6** *page 266, page 268, page 277*
LISE KOMMER TIL BYEN **Vol. 5** 89, 224
LISELOTTE VON DER PFALZ **Vol. 4** 373, 626
LISICE **Vol. 3** 414
LISSISTRATE **Vol. 3** 180
LISSY **Vol. 1** 9, 46, 47, 72
LIST DO MATKI **Vol. 1** 278, 308, 346
LISTY CZYTELNIKOW **Vol. 1** 144
LIT A COLONNES, LE **Vol. 2** 601, 676, 786
LIT A DEUX PLACES, LE **Vol. 2** 722
LIT DE LA VIERGE, LE **Vol. 2** 312
LITA PA MEJ, ÄLSKLING ! **Vol. 5** 880, 1028, 1080, 1088
LITAN **Vol. 6** *page 74, page 94*
LITEN IDA **Vol. 5** 701, 858, 1201
LITET BO **Vol. 5** 1042, 1217
LITO VILOVITO **Vol. 3** 367, 418
LITRI Y SU SOMBRA, EL **Vol. 8** 214, 336, 484, 748
LITTLE FOX **Vol. 9** 71, 80
LITTLE FRAID LADY, THE **Vol. 2** 84
LIV **Vol. 5** 692, 767, 770, 787, 826
LIV OG DOD **Vol. 5** 765, 784
LIVE AND LET LIVE **Vol. 2** 84
LIVET ÄR STENKUL **Vol. 5** 1094
LIVET GAR VIDARE **Vol. 5** 871, 1156
LIVET I FINNSKOGARNA **Vol. 5** 334, 875, 1052
LIVET MASTE LEVAS **Vol. 5** 907, 923
LIVET PA EN PINNE **Vol. 5** 840, 997
LIVET PA FORSBYHOLM **Vol. 5** 997, 1173
LIVET PA LANDET **Vol. 5** 888, 891, 997, 1032, 1181
LIVET PAA HEGNGAARD **Vol. 5** 7, 290
LIVETS BAAL **Vol. 5** 97
LIVETS GOGLESPIL **Vol. 5** 58
LIVETS KARNEVAL **Vol. 5** 265

LIVETS KONFLIKTER Vol. 5 997
LIVETS VAR Vol. 5 992, 1064
LIVLIG DROM, EN Vol. 5 83
LIVLIGE STATUER, DE Vol. 5 83
LIVLIGT PENSIONAT, ET Vol. 5 83
LIVREURS, LES Vol. 2 97, 512
LIVSFORSIKRINGSAGENTEN Vol. 5 83
LIVSSLAVEN Vol. 5 97
LIVSTRAETTE TEODOR, DEN Vol. 5 81
LIZIN LET DO NEBE Vol. 10 168, 585
LIZINO STESTI Vol. 10 256, 564
LJUBAV I MODA Vol. 3 371, 374
LJUBAV I PONEKA PSOVKA Vol. 3 452
LJUBAVI BIJES Vol. 3 363, 442
LJUBAVNA PISMA S PREDUMISLJAJEM Vol. 6 *page 310, page 321*
LJUBAVNI SLUCAJ ILI TRAGEDIJA SLUBENICE PTT Vol. 3 359, 381, 418
LJUBAVNI ZIVOT BUDIMIRA TRAJKOVICA Vol. 3 336, 454 Vol. 6 *page 307*
LJUBEZEN Vol. 6 *page 305, page 319*
LJUBEZEN NA ODORU Vol. 3 348
LJUBI LJUBI, AL' GLAVU NE GUBI Vol. 6 *page 318, page 321*
LJUBICA Vol. 3 437, 454
LJUBIMEC 13 Vol. 3 38, 51, 82 Vol. 6 *page 26*
LJUBIOVTA NA MIRON Vol. 6 *page 50, page 52*
LJUBOV I PREST"PLENIE Vol. 3 33
LJUBOVNICATA NA GRAMIGNA Vol. 3 31, 40, 109
LJUBOVTA E LUDOST Vol. 3 98
LJUBOVTA NA SEMKARJA Vol. 3 118
LJUSET FRAN LUND Vol. 5 909, 1052
LJUSNANDE FRAMTID, DEN Vol. 5 840, 997
LJUTKO Vol. 3 356, 388
LJUVLIG ÄR SOMMARNATTEN Vol. 5 840, 923, 946, 959
LLAMA UN TAL ESTEBAN Vol. 8 444, 503
LLAMABAN J.R., LE Vol. 8 299
LLAMADA DEL LOBO, LA Vol. 8 208, 749
LLAMADA DEL MAR, LA Vol. 8 422
LLAMADA, LA Vol. 8 450, 708
LLAMAN DE JAMAICA, MISTER WARD Vol. 8 331
LLANERO, EL Vol. 8 396
LLANTO POR UN BANDIDO Vol. 8 214, 258
LLEGA SARTANA Vol. 8 598
LLEGADA DE NOCHE Vol. 8 207, 488
LLEGAR A MAS Vol. 8 298, 747
LLEGARON DOS HOMBRES Vol. 8 400
LLEGARON LOS FRANCESES Vol. 8 221, 541
LLEGARON LOS MARCIANOS Vol. 8 438
LLEGARON SIETE MUCHACHAS Vol. 8 630, 719
LLEGO, VEO DISPARO Vol. 8 306, 762
LLOVIDOS DEL CIELO Vol. 8 233
LLUVIA DE HIJOS Vol. 8 417, 606
LO, DIE KOKETTE Vol. 6 *page 186*
LO PAIS Vol. 2 161, 342, 600
LO QUE NUNCA MUERE Vol. 8 493, 732
LOA VUDU Vol. 8 749
LOBO, EL Vol. 8 604
LOBO NEGRO, EL Vol. 8 465, 549
LOBOS DA SERRA Vol. 8 67, 172
LOBOS, OS Vol. 8 105
LOCA DE LA CASA, LA Vol. 8 228
LOCA HISTORIA DE LOS TRES MOSQUETEROS, LA Vol. 8 368, 728
LOCA JUVENTUD Vol. 8 298, 569
LOCA POR EL CIRCO Vol. 8 237, 627
LOCANDA DELLA MALADOLESCENZA, LA Vol. 7 713, 780
LOCANDA DELLE OMBRE, LA Vol. 7 48
LOCANDIERA, LA Vol. 7 95, 453, 493, 827, 1060, 1088, 1194
LOCAS VACACIONES Vol. 8 302
LOCATAIRE, LE Vol. 2 377, 597, 634, 704, 787
LOCKENDE GEFAHR, DIE Vol. 4 535, 644
LOCKENDE ZIEL, DAS Vol. 4 306, 373
LOCKENDES GIFT Vol. 4 626, 720
LOCKFAGELN Vol. 5 1080
LOCKVOGEL Vol. 4 232, 437, 618, 713
LOCSOLOKOCSI, A Vol. 9 20, 154, 318
LOCURA DE AMOR Vol. 8 207, 298
LOCURAS DE BARBARA, LAS Vol. 8 493, 732
LOCURAS DE PARCHIS, LAS Vol. 8 368, 773
LODER, DER Vol. 4 213
LODERNDE LEIDENSCHAFTEN Vol. 4 514
LODSENS DATTER Vol. 5 97
LOFFE BLIR POLIS Vol. 5 1052, 1080, 1164
LOFFE PA LUFFEN Vol. 5 843, 923
LOFFE SOM MILJONÄR Vol. 5 1073, 1181
LOGN OG LOVEBROL Vol. 5 7, 20, 294
LOGNEREN Vol. 5 16, 61, 195
LOHENGRIN Vol. 7 460, 493, 792, 1035
LOHNBUCHHALTER KREMKE Vol. 4 28, 329, 344
LOI C'EST LA LOI, LA Vol. 2 683 Vol. 7 915
LOI DES HOMMES, LA Vol. 2 738, 836
LOI DES RUES, LA Vol. 2 185, 559, 661
LOI DU NORD, LA Vol. 2 279, 402
LOI DU PRINTEMPS, LA Vol. 2 566, 698
LOI DU SURVIVANT, LA Vol. 2 60
LOI, LA Vol. 2 347, 748 Vol. 7 725, 915
LÖJEN OCH TARAR Vol. 5 857, 1157

LOJTNANTENS VANDGANG **Vol. 5** 92
LÖJTNANTSHJÄRTAN **Vol. 5** 840, 997
LOJZICKA **Vol. 10** 345, 460
LOKAY **Vol. 1** 174, 404
LOKIS **Vol. 1** 254, 337, 403
LOKOMOTIVENBRAUT, DIE **Vol. 4** 13, 24
LOLA **Vol. 2** 72, 208, 285 **Vol. 8** 328, 605
LOLA DE TRIANA **Vol. 8** 606
LOLA, DICEN QUE NO VIVE SOLA, LA **Vol. 8** 221, 402
LOLA, DIE APACHENBRAUT **Vol. 4** 495, 575, 593
LOLA, ESPEJO OSCURO **Vol. 8** 331, 546
LOLA LA PICONERA **Vol. 8** 298, 607, 748
LOLA MONTES **Vol. 2** 27, 248, 279, 282, 386, 566 **Vol. 8** 359, 484, 725
LOLA MONTEZ, DIE TÄNZERIN DES KÖNIGS **Vol. 4** 400, 724
LOLA NOS LLEVA AL HUERTO, LA **Vol. 8** 728, 770
LOLA SE VA A LOS PUERTOS, LA **Vol. 8** 207, 214
LOLA TORBELLINO **Vol. 8** 326, 463
LOLETTE **Vol. 7** 807, 1156
LOLITA **Vol. 7** 965
LOLOS DE LOLA, LES **Vol. 2** 455, 817
LOMA **Vol. 5** 500, 512, 541
LONDRA CHIAMA POLO NORD **Vol. 7** 634, 910
LONG DES TROTTOIRS, LE **Vol. 2** 47, 279
LONG NIGHT, THE **Vol. 2** 526
LONG PLAY **Vol. 8** 402, 749
LONG SHIPS, THE **Vol. 6** *page 312, page 323*
LONGUE MARCHE, LA **Vol. 2** 154, 742
LONTANA DA DOVE **Vol. 7** 13, 904
LONTANO, LONTANO, LONTANO **Vol. 7** 1162
LOOPING **Vol. 1** 75, 116
LOOPING THE LOOP **Vol. 4** 266, 288, 574
LOPERJENTEN **Vol. 5** 728, 746, 784
LOPOTT BOLDOGSAG **Vol. 9** 297, 332
LOPPEN **Vol. 5** 295
LORD AM ALEXANDERPLATZ, EIN **Vol. 1** 14, 34, 114
LORD BLUFF **Vol. 7** 776
LORD REGINALDS DERBYRITT **Vol. 4** 35, 192
LORD ROBINET **Vol. 7** 1189
LÖRDAGSKVÄLLAR **Vol. 5** 1042, 1205
LORE **Vol. 4** 24, 365
LORELEI, DIE **Vol. 4** 223, 270
LORENZACCIO **Vol. 7** 476, 843
LORENZINO DE MEDICI **Vol. 7** 48, 1060
LORETANSKE ZVONKY **Vol. 10** 58, 345
LORRY **Vol. 3** 437
LORSQUE L'ENFANT PARAIT **Vol. 2** 368, 615, 627
LOS AMIGOS **Vol. 7** 222, 333
LOS NR. 13013 **Vol. 4** 361, 558
LOSE FALTER, DER **Vol. 4** 8, 774
LOSER TAKES ALL **Vol. 2** 656
LOTACAO ESGOTADA **Vol. 8** 54
LOTERIE DU BONHEUR, LA **Vol. 2** 118, 826
LOTNA **Vol. 1** 242, 313, 401
LOTS WEIB **Vol. 1** 26, 74, 130
LOTTA PER LA VITA, LA **Vol. 7** 196
LOTTCHENS GEBURTSTAG **Vol. 4** 115, 774
LOTTE **Vol. 4** 535, 626
LOTTE D'ANIMA **Vol. 7** 1189
LOTTE D'ELEMENTI – RAFFICHE D'ANIME **Vol. 7** 531
LOTTE DI GIGANTI **Vol. 7** 774, 1116
LOTTE IN WEIMAR **Vol. 1** 18, 36, 50
LOTTE LORE **Vol. 4** 232, 673
LOTTE NELL'ARIA **Vol. 7** 525
LOTTE NELL'OMBRA **Vol. 7** 595, 902
LOTTERIESCHWEDE, DER **Vol. 1** 47, 74
LOTTERISEDDEL NO. 22 152 **Vol. 5** 9
LOTTERISEDLEN **Vol. 5** 83
LOTTES ERSTE LIEBE **Vol. 6** *page 186*
LOTTI EZREDESEI **Vol. 9** 301
LOTTOVOITTAJA UKK TURHAPURO **Vol. 5** 606
LOUCO, O **Vol. 8** 109
LOUISE **Vol. 2** 47, 206, 539, 863
LOUISE DE LAVALLIERE **Vol. 4** 178, 195, 287, 294, 342
LOULOU **Vol. 2** 351 **Vol. 6** *page 73, page 86*
LOUP DES MALVENEUR, LE **Vol. 2** 538, 601
LOUP GAROU, LE **Vol. 2** 357
LOUPEZNICI NA CHLUMU **Vol. 10** 58, 167, 345
LOUPEZNIK **Vol. 10** 34, 138, 168, 586
LOUPIOTE, LA **Vol. 2** 37
LOUPS CHASSENT LA NUIT, LES **Vol. 2** 501, 615
LOUPS DANS LA BERGERIE, LES **Vol. 2** 581
LOUPS ENTRE EUX, LES **Vol. 2** 197, 339, 688
LOUSIANE **Vol. 6** *page 74*
LOUSTRAKOS **Vol. 3** 141
LOUTRE, LA **Vol. 2** 875
LOUVE, LA **Vol. 2** 205, 538
LOUVE SOLITAIRE, LA **Vol. 2** 686, 722
LOUVES, LES **Vol. 2** 118, 423, 738
LOV NA JELENE **Vol. 3** 355, 368
LOV NA MAMUTA **Vol. 10** 197, 254, 296

LOV U MUTNOM **Vol. 6** *page 313, page 317, page 320*
LOVAGIAS ÜGY **Vol. 9** 178, 237
LOVE FEROZ, EL **Vol. 8** 245
LOVE IS A BALL **Vol. 2** 781
LOVE IS WAR **Vol. 5** 718, 725
LOVEC SENZACI **Vol. 10** 254, 327, 568
LOVEJAGTEN **Vol. 5** 97
LOVEMAKER **Vol. 7** 854
LOVER'S OATH, A **Vol. 2** 84
LOVES BRULANTES **Vol. 2** 202
LOVIISA - NISKAVUOREN NUORI EMÄNTÄ **Vol. 5** 372, 379
LOVIND O PASARE DE PRADA **Vol. 6** *page 275*
LÖWE VON VENEDIG, DER **Vol. 4** 278, 327, 344, 545
LÖWENBRAUT, DIE **Vol. 4** 124, 247
LUCA IL CONTRABBANDIERI **Vol. 7** 222, 1015
LUCCIOLA **Vol. 7** 456, 668, 1190
LUCE CHE TORNA, LA **Vol. 7** 1104
LUCE DEL MONDO, LA **Vol. 7** 587, 606
LUCE DI DIO **Vol. 7** 1189
LUCE NELLE TENEBRE **Vol. 7** 118, 253, 493, 1054
LUCE NELL'OMBRA **Vol. 7** 1032
LUCERNA **Vol. 10** 167, 198, 453, 460, 621
LUCERTOLA CON LA PELLE DI DONNA, UNA **Vol. 8** 306
LUCES DE BOHEMIA **Vol. 8** 580, 759
LUCETTE **Vol. 2** 169, 330, 613
LUCI DEL VARIETA **Vol. 7** 214, 725
LUCI SOMMERSE **Vol. 7** 453, 796
LUCIA DI LAMMERMOOR **Vol. 7** 16
LUCIA, LUCI ... **Vol. 7** 39
LUCIANA, LA **Vol. 7** 603
LUCIANO SERRA PILOTA **Vol. 7** 48, 359, 476, 756, 941, 1121, 1136
LUCIE **Vol. 5** 772, 780, 824 **Vol. 10** 70, 182, 387
LUCIE A ZAZRAKY **Vol. 10** 370, 380, 395
LUCIE POSTRACH ULICE **Vol. 10** 70, 577
LUCIELLA **Vol. 7** 829
LUCIJA **Vol. 3** 345, 385
LUCKY, EL INTREPIO **Vol. 8** 775
LUCKY JO **Vol. 2** 481, 552
LUCKY LUCIANO **Vol. 7** 363, 1025
LUCRECE **Vol. 2** 566, 698
LUCRECE ADOLESCENTE, CURIEUSE **Vol. 6** *page 79*
LUCRECE BORGIA **Vol. 2** 248, 388, 402, 425, 566, 579, 722
LUCREZIA BORGIA **Vol. 4** 132, 178, 186, 287, 479, 720 **Vol. 7** 2, 552, 725, 938
LUCY UND DER ANGLER VON PARIS **Vol. 1** 8, 34, 47
LUDA KUCA **Vol. 6** *page 306*
LUDAS MATYI **Vol. 9** 87, 117, 199
LUDE, DER **Vol. 6** *page 167, page 168, page 173*
LUDE GODINE **Vol. 3** 356, 397, 402, 415
LUDI DANI **Vol. 6** *page 308*
LUDWIG **Vol. 7** 299, 817, 1051, 1142
LUDWIG DER ZWEITE, KÖNIG VON BAYERN **Vol. 4** 215, 681, 691
LUDZIE BEZ JUTRA **Vol. 1** 174, 176, 326
LUDZIE DZIESIEJSI **Vol. 1** 278, 407
LUDZIE MROKU **Vol. 1** 346
LUDZIE WISLY **Vol. 1** 241
LUDZIE Z POCIAGU **Vol. 1** 179, 382 **Vol. 6** *page 242*
LUEUR DANS LES TENEBRES, LA **Vol. 2** 871
LUFFAR - PETTER **Vol. 5** 1076
LUFFAREN OCH RASMUS **Vol. 5** 977
LUFTBUREN **Vol. 5** 869
LUFTENS VAGABOND **Vol. 5** 1205
LUFTPIRATEN, DIE **Vol. 4** 535
LUFTSCHIFF, DAS **Vol. 6** *page 159, page 161, page 171*
LUGAR DO MORTO, O **Vol. 8** 166
LUGAR LLAMADO GLORY, UN **Vol. 8** 214, 493
LÜGEN AUF RÜGEN **Vol. 4** 584, 646
LUGINA E PUSHKATAREVE **Vol. 3** 3, 5
LUI E PEGGIO DI ME **Vol. 7** 408, 771, 1127
LUIGI XI. RE DI FRANCIA **Vol. 7** 1189
LUIS CANDELAS **Vol. 8** 287, 357, 665
LUIS CANDELAS, EL LADRON DE MADRID **Vol. 8** 611
LUISA SANFELICE **Vol. 7** 50, 709, 923, 1177
LUISE, KÖNIGIN VON PREUSSEN **Vol. 4** 49
LUISE MILLERIN **Vol. 4** 165, 179, 266, 385, 400, 574, 660
LUK KRALOVNY DOROTKY **Vol. 10** 139, 325, 425
LUKAS **Vol. 10** 13, 250, 508
LUKKET AVDELING **Vol. 5** 712, 766, 817
LUKSUSCHAUFFOREN **Vol. 5** 295
LULEKUQET MBI MURE **Vol. 3** 9, 20
LULU **Vol. 2** 23 **Vol. 4** 14, 186, 560 **Vol. 6** *page 106, page 130* **Vol. 7** 1063 **Vol. 8** 602, 630 **Vol. 9** 93
LULU DE NOCHE **Vol. 8** 232, 694
LUME DELL'ALTRA CASA, IL **Vol. 7** 296
LUME DRITE **Vol. 3** 10, 14 **Vol. 6** *page 13*
LUME FARA CER, O **Vol. 6** *page 280*
LUMIERE **Vol. 2** 35
LUMIERE D'EN FACE, LA **Vol. 2** 194, 644
LUMIERE D'ETE **Vol. 2** 59, 251, 644, 837
LUMIERE ET L'AMOUR, LA **Vol. 2** 798

LUMIERES DE PARIS Vol. 2 23, 206, 719
LUMIERES DU SOIR Vol. 2 131, 259, 344
LUMIKKI JA SEITSEMÄN JÄTKÄÄ Vol. 5 322, 451, 592
LUMINA DE IULIE Vol. 3 260, 272, 285
LUMINA LA ETAJUL ZECE, O Vol. 6 *page 278*
LUMINA PALIDA A DURERII Vol. 3 209, 309, 328
LUMINI SI UMBRE Vol. 6 *page 268, page 270*
LUMISTEN METSIEN MIES Vol. 5 355, 621
LUMISTEN METSIEN TYTTÖ Vol. 5 443, 643
LUMPACIVAGABUNDUS Vol. 5 203
LUMPEN UND SEIDE Vol. 4 219, 560, 634
LUMPENBALL Vol. 4 212, 340, 471, 608
LUMPENBARON, DER Vol. 4 135
LUMPENKAVALIERE Vol. 4 247
LUNA Vol. 7 1152
LUNA DE AGOSTO Vol. 8 593
LUNA DE MIEL Vol. 8 634, 749
LUNA DE SANGRE Vol. 8 396, 515
LUNA DE VERANO Vol. 8 546
LUNA DI MIELE Vol. 7 639, 1054
LUNA DI MIELE IN TRE Vol. 7 1086
LUNA, LA Vol. 7 677, 1075, 1101, 1112
LUNA NOVA Vol. 7 16
LUNA NUOVA Vol. 7 424
LUNA ROSSA Vol. 7 250
LUNA VALE UN MILLON, LA Vol. 8 207
LUNATYCY Vol. 1 227, 389 Vol. 6 *page 215*
LUNDA-INDIANER Vol. 5 1112
LUNDA - TERRA DE DIAMANTES Vol. 8 122
LUNDRIMI I PARE Vol. 6 *page 14, page 19*
LUNE DANS LE CANIVEAU, LA Vol. 6 *page 87, page 95*
LUNE DE MIEL Vol. 2 47, 404
LUNEGARDE Vol. 2 13, 96, 158, 402, 440
LUNGA NOTTE DEL' 43, LA Vol. 7 401, 853
LUNGA STRADA SENZA POLVERE, LA Vol. 7 177
LUNGHI GIORNI DELLA VENDETTA, I Vol. 8 299, 523
LUNGI DAL NIDO Vol. 7 1038
LUOTTAMUS Vol. 5 339, 477, 504, 506
LUPA, LA Vol. 7 283, 401, 1136 Vol. 8 214, 276
LUPA MANNARA, LA Vol. 7 247
LUPENI'29 Vol. 3 220, 298, 306, 312
LUPI NELL'ABISSO Vol. 7 1147
LUPIC LEGENDA Vol. 10 317, 425, 567
LUPO DELLA FRONTIERA, IL Vol. 7 257
LUPO DELLA SILA, IL Vol. 7 1136
LUPO E L'ANGELO, IL Vol. 7 32, 817
LUPO, IL Vol. 7 107
LUPUL MARILOR Vol. 3 269, 281
LUSSURIA Vol. 7 742
LUSSURIA, LA Vol. 7 441, 701
LUST FOR LIFE Vol. 2 669
LUSTGARDEN Vol. 5 940, 1052, 1053, 1130
LUSTIGE EHEMANN, DER Vol. 4 560, 663
LUSTIGE KLEEBLATT, DAS Vol. 4 239, 329, 340
LUSTIGE WITWENBALL, DER Vol. 4 230, 458, 796
LUSTIGE WITWER, DER Vol. 4 132, 479, 779
LUSTIGEN MUSIKANTEN, DIE Vol. 4 287, 644, 646
LUSTIGEN VAGABUNDEN, DIE Vol. 4 51, 108, 292
LUSTIGEN WEIBER, DIE Vol. 4 13, 123, 788
LUSTIGEN WEIBER VON WIEN, DIE Vol. 4 14, 208, 517
LUSTIGEN WEIBER VON WINDSOR, DIE Vol. 1 23, 58, 78, 90
LUSTRASCARPE DEL RETTIFILO, IL Vol. 7 314
LUTE 2, EL Vol. 8 219
LUTE 2, EL - MANANA SERRE LIBRE Vol. 8 684
LUTE, EL - CAMINA O REVIENTA Vol. 8 219
LUTHER Vol. 4 155, 266, 548, 574, 614, 733
LÜTT MATTEN UND DIE WEISSE MUSCHEL Vol. 1 38
LUTTE POUR LA VIE, LA Vol. 2 731
LÜTZOWER Vol. 1 7, 63, 69, 114
LÜTZOWS WILDE VERWEGENE JAGD Vol. 4 119, 681
LUX, DER KÖNIG DER VERBRECHER Vol. 4 124, 221
LUXEMBURG GROFJA Vol. 9 7
LUXURE Vol. 2 294
LUXUSBAD, DAS Vol. 4 366
LUXUSWEIBCHEN Vol. 4 192, 584, 780
LUZ DEL FIN DEL MUNDO, LA Vol. 8 214, 352, 611
LVI SALONU Vol. 10 190, 318, 531
LYCEENNES PERVERSES Vol. 2 294
LYCKAN KOMMER Vol. 5 907, 929, 1184
LYCKANS GULLGOSSAR Vol. 5 840, 909
LYCKANS OST Vol. 5 1024, 1168
LYCKLIGA SKITAR Vol. 5 918, 933, 1068
LYCKLIGA VESTKÖPING Vol. 5 849, 958, 1121
LYCKOBARNEN Vol. 5 996
LYCKODRÖMMEN Vol. 5 869, 1052, 1130
LYCKORIDDARE, EN Vol. 5 891, 925, 976
LYDA SSANIN Vol. 4 390, 704

LYDIA Vol. 5 58
LYFTET Vol. 5 858, 869, 947
LYKKE PAA REJSEN Vol. 5 2, 228
LYKKEHJULET Vol. 5 24
LYKKELANDET Vol. 5 58
LYKKELIG SKILSMISSE, EN Vol. 5 93, 179, 244
LYKKEN Vol. 5 58
LYKKEN KOMMER Vol. 5 89, 136
LYKKEN SVUNDEN OG GENVUNDEN Vol. 5 183
LYKKENS BLAENDVAERK Vol. 5 58
LYKKENS BUDBRINGER Vol. 5 256
LYKKENS GALOSCHER Vol. 5 83, 97, 183
LYKKENS LUNEFULDE SPIL Vol. 5 81
LYKKENS MUSIKANTER Vol. 5 2, 7
LYKKEPER, EN Vol. 5 9, 197, 228
LYNET Vol. 5 101, 183, 208
LYNFOTOGRAFEN Vol. 5 228
LYNTOGET Vol. 5 14, 180, 199
LYON LEA Vol. 9 337
LYS, LES Vol. 2 798
LYSET OG LIVET Vol. 5 295
LYSSKY TRANSPORT GENNEM DANMARK Vol. 5 180, 199, 304
LYSTEM STYRET Vol. 5 58
L"ZOVNI ISTORII Vol. 3 36, 75

M

M Vol. 4 246, 727, 729
„M“ COMME MATHIEU Vol. 2 517
M. SCRUPULE GANGSTER Vol. 2 496
MA BLONDE ENTENDS-TU DANS LA VILLE Vol. 2 35
MA CHERIE Vol. 2 265 Vol. 6 *page 118*
MA COUSINE DE VARSOVIE Vol. 2 206
MA FEMME EST FORMIDABLE Vol. 2 158, 205, 516
MA FEMME EST UNE PANTHERE Vol. 2 566, 607, 875
MA FEMME, MA VACHE ET MOI Vol. 2 131, 419
MA FEMME, MON GOSSE ET MOI Vol. 2 875
MA FEMME S'APPELLE REVIENS Vol. 6 *page 80, page 132*
MA FEMME VOUS PLAIT...J'ADORE LA VOTRE Vol. 2 309, 588, 721
MA FEMME...HOMME D'AFFAIRES Vol. 2 152, 688, 775
MA ITALIDA STEN ELLADA Vol. 3 204
MA L'AMOR MIO NON MUORE Vol. 7 1038
MA LASKA S JAKUBEM Vol. 10 254, 428, 466
MA MEMOIRE COURTE Vol. 6 *page 87*
MA MERE ME PROSTITUE Vol. 6 *page 69*
MA MUJER DEL JUEZ Vol. 8 770
MA NON E UNA COSA SERIA Vol. 7 277, 756, 1116
MA NUIT CHEZ MAUD Vol. 2 17, 537, 701
MÄ OKSALLA YLIMÄLLÄ Vol. 5 328, 443, 481, 566
MA PETITE FOLIE Vol. 2 128, 595, 726
MA PETITE MARQUISE Vol. 2 395
MA POMME Vol. 2 13, 279
MA SOEUR DE LAIT Vol. 2 775
MA TANTE D'HONFLEUR Vol. 2 193, 237, 387, 388, 475
MA TANTE DICTATEUR Vol. 2 601, 698, 833
MA, TEGNAP, HOLNAP Vol. 9 126, 196
MAA ON SYNTINEN LAULU Vol. 5 381, 545, 606, 613
MAAILMAN KAUNEIN TYTTÖ Vol. 5 579
MAANEPRINSESSEN Vol. 5 58
MAARET – TUNTURIEN TYTTÖ Vol. 5 372, 379
MACABRO Vol. 7 333, 335, 394
MACADAM Vol. 2 36, 279, 644
MACAK POD SLJEMOM Vol. 3 343, 346, 401
MACAO, L'ENFER DU JEU Vol. 2 395, 460, 583, 676, 709, 858
MACARENA Vol. 8 276
MACARIO CONTRO ZAGOMAR Vol. 7 891
MACAU Vol. 8 159
MACBETH Vol. 6 *page 189*
MACCHERONI Vol. 7 285, 950, 964
MACCHIA ROSA, UNA Vol. 7 1145
MACCHIA ROSSA, LA Vol. 7 493
MACCHINA AMMAZZACATTIVI, LA Vol. 7 140, 1027
MACH' MICH GLÜCKLICH Vol. 4 28, 298
MACH MIR DIE WELT ZUM PARADIES Vol. 4 291
MACHINE, LA Vol. 2 720, 804
MACHINES A SOUS, LES Vol. 2 109
MACHITA Vol. 9 58, 132, 270
MACHO DES HALLES, LE Vol. 6 *page 69*
MACHT DER FINSTERNIS, DIE Vol. 4 14, 208, 420, 558
MACHT DER VERSUCHUNG Vol. 4 186, 773
MACHT DES GOLDES, DIE Vol. 4 124, 646
MACISTE Vol. 7 100
MACISTE ALLA CORTE DEL GRAN KHAN Vol. 7 165, 443, 852
MACISTE ALL'INFERNO Vol. 7 48, 363, 852, 1084, 1116
MACISTE ALPINO Vol. 7 100, 305, 469, 473, 1132
MACISTE ATLETA Vol. 7 1132

MACISTE CONTRO GLI UOMINI Vol. 7 1152
MACISTE CONTRO I CACCIATORI DI TESTE Vol. 7 1036
MACISTE CONTRO IL VAMPIRO Vol. 7 915
MACISTE CONTRO LO SCEICCO Vol. 7 196, 729, 850
MACISTE E IL NIPOTE D'AMERICA Vol. 7 196, 548
MACISTE E LA REGINA DI SAMAR Vol. 7 1152
MACISTE, GLADIATORE SI SPARTA Vol. 7 877
MACISTE IMPERATORE Vol. 7 1116
MACISTE IN VACANZA Vol. 7 878
MACISTE INNAMORATO Vol. 7 1078
MACISTE MEDIUM Vol. 7 1132
MACISTE NELLA GABBIA DEI LEONI Vol. 7 196, 640, 1116
MACISTE NELLA VALLE DEI RE Vol. 7 852
MACISTE NELL'INFERNO DI GENGIS KHAN Vol. 7 739
MACISTE POLIZIOTTO Vol. 7 1132
MACISTE SALVATO DALLE ACQUE Vol. 7 878
MACISTE UND DIE CHINESISCHE TRUHE Vol. 4 221, 356, 407, 545
MACKAN Vol. 5 917, 1027, 1195
MACKO UR KALANDJAI Vol. 9 73, 239
MACOCHA Vol. 1 317 Vol. 10 45
MACSKAJATEK Vol. 9 259, 297, 303
MAD LOVER Vol. 2 84, 579
MADALENA Vol. 3 172
MADAM SOUSSOU Vol. 3 146
MADAMA, LA Vol. 7 19
MADAME BESOKER OSLO Vol. 5 770
MADAME BOVARY Vol. 2 48, 346, 388, 526, 863 Vol. 4 247, 461
MADAME BUTTERFLY Vol. 7 988
MADAME CLAUDE Vol. 2 315, 783
MADAME CLAUDE 2 Vol. 6 *page 80, page 115*
MADAME CORALIE & C. Vol. 7 168
MADAME DE... Vol. 2 27, 248, 279, 566
MADAME DE THEBES Vol. 5 997
MADAME D'ORA Vol. 4 681
MADAME DU BARRY Vol. 2 248, 388, 566
MADAME DUBARRY Vol. 4 294, 424, 560, 663
MADAME ET LE MORT Vol. 2 407, 615
MADAME ET SON AUTO Vol. 2 131, 309
MADAME ET SON FLIRT Vol. 2 194
MADAME FLIRT Vol. 7 150
MADAME GOLVERY Vol. 10 34, 243, 272
MADAME HAT AUSGANG Vol. 4 157, 158
MADAME INCOGNITO Vol. 6 *page 186*
MADAME L'AMBASSADRICE Vol. 7 110
MADAME NE VEUT PAS D'ENFANTS Vol. 2 355
MADAME POMPADOUR Vol. 6 *page 194*
MADAME POTIPHOR Vol. 4 247
MADAME PUTIFARRE Vol. 7 1189
MADAME RECAMIER Vol. 4 270, 634
MADAME SANS GENE Vol. 2 6, 102, 193, 279, 343, 407, 439, 579, 707 Vol. 5 97
MADAME SANS-GENE Vol. 8 595
MADAME TALLIEN Vol. 7 490
MADAME WÜNSCHT KEINE KINDER Vol. 4 6, 28, 208, 663, 752
MADAME X, DIE FRAU FÜR DISKRETE BERATUNG Vol. 4 73, 747
MADAME X UND DIE „SCHWARZE HAND" Vol. 6 *page 186*
MADAMIGELLA DE MAUPIN Vol. 8 311
MADARKAK Vol. 9 67, 253, 318
MÄDCHEN AM KREUZ Vol. 4 158, 232, 565
MÄDCHEN AUF DEM BRETT, DAS Vol. 1 1, 36, 53
MÄDCHEN AUS DEM DSCHUNGEL, DAS Vol. 1 34, 74, 104
MÄDCHEN AUS DER ACKERSTRASSE, DAS Vol. 4 113, 223, 428, 495, 626, 651
MÄDCHEN AUS DER FREMDE, DAS Vol. 4 132, 513, 584, 597
MÄDCHEN AUS FRISCO, DAS Vol. 4 232, 747
MÄDCHEN AUS SCHNEE, EIN Vol. 1 5, 34
MÄDCHEN CHRISTINE, DAS Vol. 1 39, 58
MÄDCHEN, DAS WARTETE, DAS Vol. 4 208, 390
MÄDCHEN DER STRASSE, DAS Vol. 4 19, 149, 213, 660
MÄDCHEN, DIE MAN NICHT HEIRATET Vol. 4 119, 726
MÄDCHEN FÜR ALLES Vol. 4 79, 115, 132
MÄDCHEN GEHT AN LAND, EIN Vol. 4 234, 365, 608
MÄDCHEN, HÜTET EUCH ! Vol. 4 240, 565
MÄDCHEN IM VORZIMMER Vol. 4 24, 247, 736
MÄDCHEN IN UNIFORM Vol. 4 373, 437, 742, 772
MÄDCHEN IN WEISS Vol. 4 115, 239
MÄDCHEN IRENE, DAS Vol. 4 28, 544, 546
MÄDCHEN JOHANNA, DAS Vol. 4 266, 359, 574 Vol. 6 *page 183, page 192*
MÄDCHEN JUANITA, DAS Vol. 4 437
MÄDCHEN KRÜMEL, DAS Vol. 1 5, 27, 29
MÄDCHEN MALLE, DAS Vol. 1 22
MÄDCHEN MIT DEM GUTEN RUF, DAS Vol. 4 265, 614, 729

MÄDCHEN MIT DEM SCHWEFELHÖLZCHEN, DAS **Vol. 4** 646
MÄDCHEN MIT DEN FÜNF NULLEN, DAS **Vol. 4** 359, 558
MÄDCHEN MIT DER PROTEKTION, DAS **Vol. 4** 318, 359, 537
MÄDCHEN MIT PROKURA, EIN **Vol. 4** 246, 638, 646
MÄDCHEN OHNE GEWISSEN, DAS **Vol. 4** 309, 565
MÄDCHEN OHNE HEIMAT, DAS **Vol. 4** 211, 471
MÄDCHEN OHNE VATERLAND, DAS **Vol. 4** 646
MÄDCHEN STÖRTEBECKER, DAS **Vol. 6** *page 164, page 167, page 172*
MÄDCHEN UND DER JUNGE, DAS **Vol. 6** *page 159, page 166, page 169*
MÄDCHEN VOM MOORHOF, DAS **Vol. 4** 329, 774
MÄDCHEN VON DER HEILSARMEE, DAS **Vol. 4** 125, 565
MÄDCHEN VON FANÖ, DAS **Vol. 4** 288, 524, 544, 619
MÄDCHEN VON GESTERN NACHT, DAS **Vol. 4** 28, 329
MÄDCHEN VON SECHZEHNEINHALB, EIN **Vol. 1** 56, 81, 115
MÄDCHEN VON ST. COEUR, DAS **Vol. 4** 13, 24 **Vol. 6** *page 183*
MÄDCHEN ZU VERSCHENKEN, EIN **Vol. 4** 5
MÄDCHEN ZUM HEIRATEN **Vol. 4** 306, 321, 703
MÄDCHENHANDEL **Vol. 4** 255, 291
MÄDCHENHIRT, DER **Vol. 4** 225, 565, 783
MÄDCHENJAHRE EINER KÖNIGIN **Vol. 4** 30, 234, 462, 734
MÄDCHENPENSIONAT **Vol. 4** 723
MÄDCHENRÄUBER **Vol. 4** 361, 558
MÄDCHENSCHICKSALE **Vol. 4** 213, 422, 643, 788
MÄDCHENSCHIFF, DAS **Vol. 6** *page 193, page 199*
MADDALENA **Vol. 2** 719 **Vol. 7** 827, 1054
MADDALENA AL DESERTO **Vol. 7** 182, 1156
MADDALENA FERAT **Vol. 7** 264, 701
MADDALENA ZERO IN CONDOTTA **Vol. 7** 16, 756
MADE IN SWEDEN **Vol. 5** 940
MADE IN USA **Vol. 2** 208
MÄDEL AUF DER SCHAUKEL, DAS **Vol. 4** 85, 219
MÄDEL AUS DEM VOLKE, EIN **Vol. 4** 287, 565, 646
MÄDEL AUS DER HÖLLE, DAS **Vol. 4** 291, 390
MÄDEL AUS DER PROVINZ, DAS **Vol. 4** 124, 204
MÄDEL AUS GUTER FAMILIE, EIN **Vol. 4** 558, 644, 796
MÄDEL AUS U.S.A., DAS **Vol. 4** 257, 558
MÄDEL MIT DER MASKE, DAS **Vol. 4** 735
MÄDEL MIT DER PEITSCHE, DAS **Vol. 4** 257, 558
MÄDEL MIT TEMPERAMENT, EIN **Vol. 4** 149, 660, 725
MÄDEL UND DREI CLOWNS, EIN **Vol. 4** 158, 626
MÄDEL VOM BALLETT, DAS **Vol. 4** 560, 663
MÄDEL VOM BALLETT, EIN **Vol. 4** 123, 727, 774
MÄDEL VOM MONTPARNASSE, DAS **Vol. 4** 89
MÄDEL VON DER REEPERBAHN, EIN **Vol. 4** 292, 370, 489
MÄDEL VON NEBENAN, DAS **Vol. 4** 288
MÄDEL VON PICCADILLY, DAS **Vol. 4** 113
MÄDEL VON PONTECUCULI, DAS **Vol. 4** 367, 634
MÄDEL WIRBELT DURCH DIE WELT, EIN **Vol. 4** 132, 437, 458
MADELEINE **Vol. 4** 247, 385, 780
MADELON, LA **Vol. 2** 318, 347, 516, 807
MÄDELS RAN AN DIE FRONT ! **Vol. 4** 366
MÄDELS VON HEUTE **Vol. 4** 132, 270, 336
MADEMOISELLE **Vol. 2** 770
MADEMOISELLE ANGE **Vol. 2** 173, 279
MADEMOISELLE BEATRICE **Vol. 2** 254, 339
MADEMOISELLE DE LA FERTE **Vol. 2** 36, 231
MADEMOISELLE DE LA SEIGLIERE **Vol. 2** 153, 339
MADEMOISELLE DE MAUPIN **Vol. 8** 311
MADEMOISELLE DE PARIS **Vol. 2** 598, 626, 824
MADEMOISELLE DOCTEUR **Vol. 2** 13, 27, 688, 778
MADEMOISELLE DON QUICHOTTE **Vol. 7** 531
MADEMOISELLE ET SON GANG **Vol. 2** 347, 777, 807
MADEMOISELLE JOSETTE, MA FEMME **Vol. 2** 48, 247, 279, 407, 627
MADEMOISELLE MA MERE **Vol. 2** 153, 439
MADEMOISELLE MOZART **Vol. 2** 67, 249, 488

MADEMOISELLE PAS-CHIC Vol. 7 37
MADEMOISELLE S'AMUSE Vol. 2 807
MADEMOISELLE SE MAQUILLE Vol. 7 1113
MADEMOISELLE SOREL DANS LA LOGE Vol. 2 678
MADEMOISELLE STRIP-TEASE Vol. 2 97, 205, 516, 808
MADEMOISELLE SWING Vol. 2 270, 395
MADEMOISELLE X Vol. 2 156, 566, 863
MADICKEN PA JUNIBACKEN Vol. 5 1001, 1052, 1093, 1096
MADKAERESTEN Vol. 5 97
MADLA Z CIHELNY Vol. 10 256, 524
MADLA ZPIVA EVROPE Vol. 10 168, 564
MADLY Vol. 2 60, 457
MADO Vol. 2 110, 377
MADO MAVROGENOUS Vol. 3 163
MADONE DE L'ATLANTIQUE, LA Vol. 2 528
MADONE DES SLEEPINGS, LA Vol. 2 6, 153, 173, 671
MADONNA AM PORTAL, DIE Vol. 4 321, 558
MADONNA, CHE SILENZIO C'E STASERA Vol. 7 290, 303
MADONNA DELLA ROBBIA, LA Vol. 7 606
MADONNA DELLE ROSE Vol. 7 603
MADONNA DI NEVE Vol. 7 569, 574
MADONNA ERRANTE, LA Vol. 7 796
MADONNA IM FEGEFEUER Vol. 4 186, 306
MADONNINA Vol. 7 560
MADONNINA DEI MARINARI, LA Vol. 7 569
MADRAGOA Vol. 8 44, 130, 131
MADRE ALEGRIA Vol. 8 648
MADRE FOLLE, LA Vol. 7 569
MADRE GUAPA, LA Vol. 8 437
MADRE IGNORATA Vol. 7 1189
MADRE IN JAPAN Vol. 8 314, 627
MADRE, LA Vol. 7 428, 1103, 1162
MADRE MARTIRE Vol. 7 548
MADRE RITORNA, UNA Vol. 7 938
MADRES SOLTERAS Vol. 8 299, 738
MADRI PERICOLOSE Vol. 7 739
MADRID, COSTA FLEMING Vol. 8 762, 770
MADRID DE MIS SUENOS Vol. 7 332, 587, 968 Vol. 8 674
MADRID SE DIVORCIA Vol. 8 357
MADRIGUERA, LA Vol. 8 348, 772
MADRUGADA Vol. 8 78, 214
MADSALUNDE Vol. 5 181
MAELKEMANDENS HEST Vol. 5 83
MAERKELIG KAERLIGHED, EN Vol. 5 206
MAES Vol. 3 156
MAESTRA D'AMORE Vol. 7 1211
MAESTRA DI SCI, LA Vol. 7 536
MAESTRINA, LA Vol. 7 48, 196, 307, 359, 454, 756, 1054, 1060
MAESTRO DI DON GIOVANNI, IL Vol. 7 587
MAESTRO DI VIGEVANO, IL Vol. 7 263, 725
MAESTRO DI VIOLINO, IL Vol. 7 1015
MAESTRO E MARGHERITA, IL Vol. 7 514
MAESTRO, EL Vol. 8 233, 515, 546
MAESTRO, IL Vol. 8 233, 515, 546
MAESTRO LANDI Vol. 7 1159
MAESTRO, LE Vol. 2 409, 642
MAFIA Vol. 5 97
MAFIA DU PLAISIR, LA Vol. 2 769
MAFIA, UNA LEGGE CHE NON PERDONA Vol. 7 540
MAFIAEN – DET ER OSSE MIG Vol. 5 125, 195, 247
MAFIOSI, I Vol. 7 71
MAFIOSO Vol. 7 401, 817, 1017
MAG Vol. 10 139, 325, 434
MAGA E IL GRIFO, LA Vol. 7 824
MAGA LESZ A FERJEM Vol. 9 178, 237
MAGAS DIPLOMACIA Vol. 9 220
MAGASISKOLA Vol. 9 152, 253, 259
MAGDA Vol. 2 24
MAGDALENA Vol. 1 134, 257, 287 Vol. 10 125, 198, 536
MAGDALENE Vol. 5 97
MAGDAT KICSAPJAK Vol. 9 71, 322
MAGDOLNA Vol. 9 164, 196, 337
MAGIA Vol. 9 172
MAGIC BLACK Vol. 7 1054
MAGIC MOMENTS Vol. 7 290, 291, 405, 908
MAGICIENNES, LES Vol. 2 279, 566
MAGICIENS, LES Vol. 2 443, 700
MAGICZNE OGNIE Vol. 6 *page 224, page 251, page 253*
MAGIKE POLE Vol. 3 182, 198, 203
MAGISKA CIRKELN, DEN Vol. 5 973, 1042, 1096
MAGISTARNA PA SOMMARLOV Vol. 5 928, 1156
MAGISTRADO, EL Vol. 8 298, 546
MAGISTRALA Vol. 3 40 Vol. 6 *page 29, page 40*
MAGISTRATO, IL Vol. 7 781, 910 Vol. 8 298, 546
MAGLIARI, I Vol. 7 1167
MAGNAS MISKA Vol. 9 71, 237, 277, 323
MAGNATE, IL Vol. 7 649
MAGNIFICO AVENTURERO, EL Vol. 8 598, 772
MAGNIFICO AVVENTURIERO, IL Vol. 7 368, 739, 839 Vol. 8 598, 772

MAGNIFICO CORNUTO, IL **Vol. 7** 817
MAGNIFICO ROBIN HOOD, EL **Vol. 8** 317, 732
MAGNIFICO TEXANO, IL **Vol. 8** 677
MAGNIFICO TONY CARRERA, EL **Vol. 8** 562, 732
MAGNIFIQUE, LE **Vol. 2** 457, 564
MAGO PER FORZA, IL **Vol. 7** 16, 1170
MAGOT DE JOSEFA, LE **Vol. 2** 251, 625
MAGYAR FELTAMADAS **Vol. 9** 132, 237, 270
MAGYAR KIVANSAGHANGVERSENY **Vol. 9** 35, 164
MAGYAR RAPSZODIA **Vol. 9** 20, 154, 318
MAGYAR SASOK **Vol. 9** 132, 146, 164, 244, 247
MAGYAR UGARON, A **Vol. 9** 134, 215, 316
MAGYARENFÜRSTIN, DIE **Vol. 4** 329, 729
MAGYAROK **Vol. 9** 134, 152, 316
MAHARADJAHENS YNDLINGSHUSTRU **Vol. 5** 295
MAHARADJAHENS YNDLINGSHUSTRU II **Vol. 5** 295
MAHARADJAHENS YNDLINGSHUSTRU III **Vol. 5** 136, 157, 183
MAHARAJAENS YNDLINGSFLAMME **Vol. 5** 83
MAHLIA LA METISSE **Vol. 2** 538, 566
MAHNUNG, DIE **Vol. 6** *page 155, page 159*
MAHOVINA NA ASFALTU **Vol. 6** *page 317, page 319*
MAHULIENA, ZLATA PANNA **Vol. 10** 145, 178, 209
MAI 68 **Vol. 2** 640, 713
MAI CON LE DONNE **Vol. 7** 137, 362, 524
MAI LANYOK **Vol. 9** 319, 322
MAI PRESUS DE ORICE **Vol. 3** 279, 291
MAIBOWLE **Vol. 1** 68, 74
MAIGRET DIRIGE L'ENQUETE **Vol. 2** 270
MAIGRET ET L'AFFAIRE SAINT-FIACRE **Vol. 2** 644, 722
MAIGRET TEND UN PIEGE **Vol. 2** 644, 722, 832
MAIGRET VOIT ROUGE **Vol. 2** 197, 644
MAIJA LÖYTÄÄ SÄVELEN **Vol. 5** 374, 443
MAILLOT A POINTES D'ACIER, LE **Vol. 2** 159, 286, 330
MAIN A COUPER, LA **Vol. 2** 368
MAIN A FRAPPE, UNE **Vol. 2** 412, 709
MAIN CHAUDE, LA **Vol. 2** 185, 856
MAIN DE FER **Vol. 2** 798
MAIN DE FER CONTRE GANTS BLANCS **Vol. 2** 798
MAIN DU DIABLE, LA **Vol. 2** 23, 824
MAIN, LA **Vol. 2** 92, 854
MAIN MYSTERIEUSE OU LES DOIGTS QUI ETRANGLENT, LA **Vol. 2** 159, 286, 330
MAIN NOIRE, LA **Vol. 2** 503
MAINS D'ORLAC, LES **Vol. 2** 173, 233, 501, 671
MAINS FLETRIES, LES **Vol. 2** 6
MAINS LIEES, LES **Vol. 2** 419, 848
MAINS SALES, LES **Vol. 2** 48, 270, 722
MAIORUL MURA **Vol. 3** 315
MAIORUL SI MOARTEA **Vol. 3** 261, 291, 319
MAIS NE NOUS DELIVREZ PAS DU MAL **Vol. 2** 4
MAIS OU EST DONC ORNICAR **Vol. 2** 45, 94
MAIS OU EST DONC PASSEE LA SEPTIEME COMPAGNIE ? **Vol. 2** 368, 376
MAIS QU'EST-CE QUE J'AI FAIT AU BON DIEU POUR AVOIR UNE FEMME... ? **Vol. 2** 293
MAIS QU'EST-CE QUE'ELLES VEULENT ? **Vol. 2** 739
MAIS TOI TU ES PIERRE **Vol. 2** 347
MAISON AUX IMAGES, LA **Vol. 2** 644
MAISON BONNADIEU, LA **Vol. 2** 194, 395
MAISON DANS LA DUNE **Vol. 2** 124, 194, 566, 626, 824
MAISON DE CAMPAGNE, LA **Vol. 2** 97, 503
MAISON DE DANSES **Vol. 2** 32, 152, 388
MAISON D'EN FACE, LA **Vol. 2** 533, 698
MAISON DES BORIES, LA **Vol. 2** 190, 674
MAISON DES CONFIDENCES, LA **Vol. 2** 818
MAISON DES FILLES PERDUES, LA **Vol. 2** 139
MAISON DES HOMMES VIVANTS **Vol. 2** 136, 316, 381
MAISON DES SEPT JEUNES FILLES, LA **Vol. 2** 48, 676
MAISON DU BONHEUR, LA **Vol. 2** 13
MAISON DU MALTAIS, LA **Vol. 2** 33, 47, 194, 206, 652, 697, 805, 863
MAISON DU MYSTERE, LA **Vol. 2** 47, 122, 388, 404, 521, 623, 831, 833
MAISON DU PRINTEMPS, LA **Vol. 2** 118, 193
MAISON, LA **Vol. 2** 74
MAISON SOUS LA MER, LA **Vol. 2** 401, 719
MAISON SOUS LES ARBRES, LA **Vol. 2** 21
MAITRE APRES DIEU **Vol. 2** 185, 644
MAITRE BOLBEC ET SON MARI **Vol. 2** 279, 708, 807

MAITRE DE FORGES, LE **Vol. 2** 48, 347, 530, 722, 803
MAITRE D'ECOLE **Vol. 6** *page 90, page 146*
MAITRE NAGEUR, LE **Vol. 2** 150, 301
MAITRE-NAGEURS, LES **Vol. 2** 67, 131, 856
MAITRESSE **Vol. 2** 17
MAITRESSE DE VACANCES, LES **Vol. 2** 721
MAITRESSES DU VICE **Vol. 6** *page 81*
MAITRESSES POR COUPLE **Vol. 6** *page 94*
MAJ PA MALÖ **Vol. 5** 1042, 1080
MAJA DEL CAPOTE, LA **Vol. 8** 606, 674, 725
MAJA DESNUDA, LA **Vol. 7** 988
MAJA ZWISCHEN ZWEI EHEN **Vol. 4** 379, 391, 530
MAJALIS **Vol. 9** 117
MAJD HOLNAP ... **Vol. 9** 20, 253, 318
MAJESTÄT SCHNEIDET BUBIKÖPFE **Vol. 4** 712
MAJKA KATINA **Vol. 3** 378
MAJOR BAUK **Vol. 3** 365, 378
MAJORATSHERR, DER **Vol. 4** 230, 329, 373
MAJORDOME, LE **Vol. 2** 128
MAJORENS OP-PASSER **Vol. 5** 207, 294
MAJOVE HVEZDY **Vol. 10** 52, 193, 538, 566
MAJSTER KAT **Vol. 10** 210, 484, 576
MAJSTOR I MARGARITA **Vol. 3** 347, 356
MAJSTORI, MAJSTORI **Vol. 3** 438
MAJURI MAANTIELTÄ **Vol. 5** 443, 646
MAJUSI FAGY **Vol. 9** 39, 216, 261
MAKACS KATA **Vol. 9** 157
MAKE LIKE A THIEF **Vol. 5** 376, 383
MAKEDONSKA KRVAVA SVADBA **Vol. 3** 339, 384
MAKEDONSKIOT DEL OD PEKOLOT **Vol. 3** 339, 384
MAKKHETES **Vol. 9** 26, 28, 157
MAKRA **Vol. 9** 152, 259, 339
MAKRANCOS HÖLGY **Vol. 9** 196, 274
MAL AMADO, O **Vol. 8** 4, 185
MAL DE ESPANHA **Vol. 8** 105
MAL ESTUDIANTE **Vol. 8** 655
MAL PARTIS, LES **Vol. 2** 377, 728
MALA MANEKYNKA **Vol. 10** 287, 576
MALA MORSKA VILA **Vol. 10** 110, 165, 292, 394
„MALA“ ORDINA, LA **Vol. 7** 1179
MALA PASQUA **Vol. 7** 108
MALA PIANTA, LA **Vol. 7** 1189
MALA PLJACKA VLAKA **Vol. 6** *page 313, page 315*
MALABARS SONT AU PARFUM, LES **Vol. 2** 570, 815
MALABESTIA **Vol. 7** 828
MALADONNA **Vol. 7** 414, 631
MALAFEMMINA **Vol. 7** 852
MALAMORE **Vol. 7** 425, 549, 599
MALAREN **Vol. 5** 922, 1006, 1108, 1145
MALARIA **Vol. 2** 270, 595 **Vol. 4** 202, 743
MÄLARPIRATER **Vol. 5** 881, 889, 909, 1031
MALATO IMMAGINARIO, IL **Vol. 7** 817, 1035, 1142, 1175
MALAVITA ATTACCA... LA POLIZIA RISPONDE, LA **Vol. 7** 1022
MALCASADA, LA **Vol. 8** 422
MALDICION DE LA BESTIA **Vol. 8** 515, 641
MALDICION DE LOS KARNSTEIN, LA **Vol. 8** 598, 772
MALDICION GITANA **Vol. 8** 261, 298
MALDITAS PISTOLAS DE DALLAS, LA **Vol. 8** 206, 231
MALDONE **Vol. 2** 58, 475, 566, 656
MALDONNE **Vol. 2** 241
MALE DRAMATY **Vol. 6** *page 247, page 256*
MALE DRAMATY-KARUZELA **Vol. 1** 149, 280
MALE DRAMATY-UPADEK MILIONERA **Vol. 1** 211, 280
MALE DU SIECLE, LE **Vol. 2** 68, 142
MALE GRETE **Vol. 6** *page 315*
MALE LETNI BLUES **Vol. 10** 87, 173, 283
MALE STVARI **Vol. 3** 383, 401
MALEDETTA EURIDICE ! **Vol. 7** 643
MALEDETTI, VI AMERO **Vol. 7** 901
MALEDETTO IMBROGLIO, UN **Vol. 7** 86, 401
MALEFICES **Vol. 2** 128, 368
MALENCONTRE **Vol. 2** 37
MALENKA **Vol. 8** 495
MALER UND SEIN MODELL, DER **Vol. 4** 190, 479
MALET ME BLERIM MBULUAR **Vol. 3** 3, 20 **Vol. 6** *page 18*
MALETA PARA UN CADAVER, UNA **Vol. 8** 396
MALEVIL **Vol. 2** 655 **Vol. 6** *page 77*
MALHEURCHEN NO. 8 **Vol. 4** 356, 779
MALHEURS D'ALFRED, LES **Vol. 2** 110, 588
MALHEURS DE SOPHIE, LES **Vol. 2** 407, 538, 837 **Vol. 6** *page 92, page 95, page 119, page 148*
MALHEURS D'OCTAVIE, LES **Vol. 6** *page 78, page 93*
MALI COVEK **Vol. 3** 349, 413, 448
MALI MEDVEDARI **Vol. 10** 70, 154, 179
MALI VOJNICI **Vol. 3** 346, 450
MALIA **Vol. 7** 1136

MALIA, VERGINE E DI NOME MARIA Vol. 7 46
MALIBRAN, LA Vol. 2 48, 122, 579
MALIN PLAISIR, LE Vol. 2 175
MALINCONICO AUTUNNO Vol. 8 762
MALINOVY KOKTEJL Vol. 10 116, 337, 508
MALIZIA Vol. 7 1101
MALIZIA EROTICA Vol. 7 540
MALKATA Vol. 3 61, 75 Vol. 6 *page 41*
MALKATA RUSALKA Vol. 3 105, 121 Vol. 6 *page 56*
MALKEPIGEKOMTESSEN Vol. 5 295
MALKI TAJNI Vol. 3 75
MALLIA Vol. 7 264, 701
MALLORCAS SODE LIV Vol. 5 135
MALMEQUER Vol. 8 105
MALOM A POKOLBAN Vol. 9 20, 198, 265
MALOMBRA Vol. 7 414, 560, 750, 756, 1116
MALOSTRANSTI MUSKETYRI Vol. 10 122, 453, 586
MALPERTUIS: HISTOIRE D'UNE MAISON MAUDITE Vol. 2 154, 299
MALQUERIDA, LA Vol. 2 646 Vol. 8 607, 692
MALTESES, BURGUESES E AS VEZES ... Vol. 8 8, 130, 184
MALUQUINHA DE ARROIOS, A Vol. 8 33, 54, 145, 151
MALVA Vol. 4 36, 81, 270
MALVADO CARABEL, EL Vol. 8 323, 437, 749, 752
MALVALOCA Vol. 8 306, 357, 400, 437, 616, 630
MALVILLE: ETAT DE SIEGE Vol. 2 684
MALY Vol. 1 220, 310, 355
MALY BOBES Vol. 10 131, 359, 569
MALY BOBES VE MESTE Vol. 10 131, 359, 569
MALY MARYNARZ Vol. 1 241, 384
MALY PARTYZAN Vol. 10 184, 388
MALY VELKY HOKEJISTA Vol. 10 160, 161, 255
MALYISCHE DSCHONKE, DIE Vol. 4 99, 255
MALZENSTWO Z ROZSADKU Vol. 1 148, 196, 305
MAM TU SWOJ DOM Vol. 1 220, 281, 389 Vol. 6 *page 247*
MAMA Vol. 3 226, 284, 325, 332
MAMA CUMPLE CIEN ANOS Vol. 8 269, 358
MAMA DRACULA Vol. 2 441
MAMA, ICH LEBE Vol. 1 9, 48
MAMA, LEVANTATE Y ANDA Vol. 8 414
MAMAIA Vol. 2 689
MAMAN COLIBRI Vol. 2 180, 339, 387, 392, 698, 824, 826 Vol. 7 264, 701
MAMAN ET LA PUTAIN, LA Vol. 2 517
MAMAN NE CONNAIT RIEN A L'AMOUR Vol. 2 604
MAMAN POUPEE Vol. 7 569, 574, 1004
MAMBO Vol. 7 338
MAMBRU SE A LA GUERRA Vol. 8 219, 368
MAMI Vol. 9 71, 322
MAMINO SRDCE Vol. 10 58, 423
MAMITO Vol. 2 210
MAMMA Vol. 5 858, 1163, 1201 Vol. 7 493, 1054
MAMMA DORME, LA Vol. 7 1189
MAMMA EBE Vol. 7 13, 351, 957
MAMMA GIFTER SIG Vol. 5 923, 1205
MAMMA LONTANA Vol. 7 10
MAMMA MIA, CHE IMPRESSIONE ! Vol. 7 701, 796
MAMMA PERDONAMI Vol. 7 201
MAMMA ROMA Vol. 7 333, 781
MAMMA SCONOSCIOTA Vol. 7 17
MAMMA TAR SEMESTER Vol. 5 1073, 1164
MAMMASANTISSIMA, IL Vol. 7 474
MAMMY Vol. 2 254, 309
MAMORIAS DE LETICIA VALLE Vol. 8 316
MAMSELL NITOUCHE Vol. 2 397 Vol. 4 257
MAM'ZELLE BONAPARTE Vol. 2 332, 440
MAM'ZELLE EXTRA Vol. 7 168, 436
MAM'ZELLE NITOUCHE Vol. 2 279, 402, 626, 780, 797, 824
MAN ABOUT THE HOUSE, A Vol. 2 656
MAN BRAUCHT KEIN GELD Vol. 4 85, 208
MAN CALLED NOON, THE Vol. 8 303
MAN COULD GET KILLED, A Vol. 8 156
MAN ELSKER KUN EEN GANG Vol. 5 29, 30, 31, 55, 161
MAN FÖR MYCKET, EN Vol. 5 841, 923
MAN FROM RED GULCH, THE Vol. 2 84
MAN FROM THE OTHER SIDE, THE Vol. 5 909
MAN GLÖMMER INGENTING Vol. 5 923, 997
MÄN KAN INTE VALDTAS Vol. 5 327, 358, 522, 851, 1053, 1074, 1178
MÄN MASTE JU LEVA... Vol. 5 1126
MAN NEHME Vol. 4 124, 634
MAN OCH KVINNA Vol. 5 879
MAN QUE MAN Vol. 6 *page 134, page 135*
MAN REDE MIR NICHT VON LIEBE Vol. 4 491, 660

MAN SCHENKT SICH ROSEN, WENN MAN VERLIEBT IST **Vol. 4** 221, 340, 608
MAN SKU' VAERE NOGET VED MUSIKKEN **Vol. 5** 113, 179, 247
MAN SOLL ES NICHT FÜR MÖGLICH HALTEN ODER MACISTE UND DIE JAVANERIN **Vol. 4** 242, 560, 725
MAN SPIELT NICHT MIT DER LIEBE ! **Vol. 4** 379, 492, 646, 752
MAN SPRICHT ÜBER JAQUELINE **Vol. 4** 97, 108, 270
MAN STEIGT NACH **Vol. 4** 25, 452
MAN WHO LOVED REDHEADS, THE **Vol. 2** 656
MAN WHO WOULD BE KING, THE **Vol. 2** 837
MANADA DE LANA AZUL CELESTE, UNA **Vol. 8** 470, 492
MANANA **Vol. 8** 217, 470, 492
MANANA COMO HOY **Vol. 8** 21, 79, 153, 298, 311, 436
MANANA CUANDO AMANEZCA **Vol. 8** 613
MANANA DE DOMINGO **Vol. 8** 631
MANANA SERA OTRO DIA **Vol. 8** 348, 732
MANASSE **Vol. 3** 257
MANCHA QUE LIMPIA **Vol. 8** 572, 657
MANCHAS DE SANGRE EN UN COCHE NUEVO **Vol. 8** 276, 445
MANCHE ET LA BELLE, UNE **Vol. 2** 279, 566
MAND AF BETYDNING, EN **Vol. 5** 55, 101, 208
MAND MOD MAND **Vol. 5** 256
MANDAGARNA MED FANNY **Vol. 5** 869, 1109
MANDARINE, LA **Vol. 2** 457, 481
MANDARINO PER TEO, UN **Vol. 7** 1167
MANDAT D'AMENER **Vol. 2** 344, 627
MANDAT, LE **Vol. 2** 795
MANDEN, DER GOER **Vol. 5** 83
MANDEN, DER SEJREDE **Vol. 5** 228
MANDEN DER TAENKTE TING **Vol. 1** 238 **Vol. 5** 125, 187, 253
MANDEN MED ARRET **Vol. 5** 242
MANDEN MED DE NI FINGRE **Vol. 5** 242
MANDEN MED KAPPEN **Vol. 5** 295
MANDEN PA SVANEGARDEN **Vol. 5** 16, 61, 272
MANDEN UDEN FREMTID **Vol. 5** 58
MANDEN UDEN SMIL **Vol. 5** 58
MANDENS OVERMAND **Vol. 5** 83
MANDERLEY **Vol. 8** 455, 605
MANDINGA **Vol. 7** 289
MANDLIGE HUSSASSISTENT, DEN **Vol. 5** 7, 174
MANDOLINATA **Vol. 3** 48 **Vol. 6** *page 59*
MANDRAGOLA, LA **Vol. 7** 33, 388, 401
MANDRIN **Vol. 1** 305 **Vol. 2** 48, 193, 339, 475, 584, 697, 871
MANDRIN BANDIT GENTILHOMME **Vol. 2** 309, 377
MANDS VILJE **Vol. 5** 295
MANEGE **Vol. 4** 49, 73, 246, 354, 743
MANEGENRAUSCH **Vol. 4** 232, 300
MANEGENS BORN **Vol. 5** 183
MANEGES **Vol. 2** 124, 156
MANEVRI NA PETIJA ETAZ **Vol. 6** *page 44, page 62*
MANEWRY MILOSNE **Vol. 1** 278, 308, 404
MANGA SÄNGARNA, DE **Vol. 5** 887
MANGANTES, LOS **Vol. 8** 311
MANGIATI VIVI ! **Vol. 7** 504, 1202
MANHA SUBMERSA **Vol. 8** 169
MANHATTAN BABY **Vol. 7** 623, 694
MANI **Vol. 5** 39, 89, 153
MANI DI FATA **Vol. 7** 221, 771
MANI DI UNA DONNA SOLA **Vol. 7** 692
MANI DI VELLUTO **Vol. 7** 30, 347, 1008
MANI RAPACI, LE **Vol. 7** 932
MANI SUGLI OCCHI, LE **Vol. 7** 963
MANI SULLA CITTA, LE **Vol. 7** 238, 1167
MANIA **Vol. 4** 300, 396
MANIACI, I **Vol. 7** 852
MANICOMIO **Vol. 8** 611, 614, 749
MANICUREDAMEN MED DET STORE HJERTE **Vol. 5** 203
MANIGUA SIN DIOS, LA **Vol. 8** 276
MANINA, LA FILLE SANS VOILE **Vol. 2** 740 **Vol. 7** 982
MANINA, RAGAZZA SENZA VELI **Vol. 7** 982
MANJA VALEWSKA **Vol. 4** 391
MANN AN DER KETTE, DER **Vol. 4** 25
MANN AN DER WAND, DER **Vol. 4** 89
MANN AUF ABWEGEN, EIN **Vol. 4** 344, 430, 437
MANN AUF DEM KOMETEN, DER **Vol. 4** 223, 270, 725
MANN AUF DER FLASCHE, DER **Vol. 4** 424
MANN AUS DEM JENSEITS, DER **Vol. 4** 232, 535
MANN AUS KANADA, DER **Vol. 1** 38, 59
MANN AUS ZELLE 19, DER **Vol. 4** 135
MANN, DER DEN MORD BEGING, DER **Vol. 4** 113, 556, 734
MANN, DER NACH DER OMA KAM, DER **Vol. 1** 15, 92, 122
MANN, DER NICHT LIEBT, DER **Vol. 4** 558, 774
MANN, DER NICHT NEIN SAGEN KANN, DER **Vol. 4** 82, 517, 558

MANN, DER SEINEN MÖRDER SUCHT, DER Vol. 4 266, 574, 713
MANN, DER SHERLOCK HOLMES WAR, DER Vol. 4 298, 608, 729
MANN, DER SICH VERKAUFT, DER Vol. 4 242, 479
MANN, DER STERBEN MUSS, EIN Vol. 1 111, 121, 127
MANN DER TAT, DER Vol. 4 186, 560
MANN FÜR MANN Vol. 4 28, 298, 727
MANN FÜR MEINE FRAU, EIN Vol. 4 18, 391, 557
MANN GEGEN MANN Vol. 1 36, 55 Vol. 4 119, 270, 779
MANN GEHÖRT INS HAUS, EIN Vol. 4 620
MANN IM DUNKEL, DER Vol. 4 340, 608, 634
MANN IM EIS, DER Vol. 4 5
MANN IM FEUER, DER Vol. 4 59, 89, 287
MANN IM FRACK, DER Vol. 4 18, 339
MANN IM HINTERGRUND, DER Vol. 4 93, 570, 772
MANN IM KELLER, DER Vol. 4 159
MANN IM MONDE, DER Vol. 6 *page 194*
MANN IM SATTEL, DER Vol. 4 119, 232, 704
MANN IM SCHRANK Vol. 4 18, 558
MANN IM SCHRANK, DER Vol. 4 366, 418
MANN IM SPIEGEL, DER Vol. 4 178, 319
MANN IM STEINBRUCH, DER Vol. 4 790
MANN IN DER FALLE, DER Vol. 4 125, 626
MANN MIT DEM AFFEN, DER Vol. 4 115, 774
MANN MIT DEM LAUBFROSCH, DER Vol. 4 247, 461
MANN MIT DEM OBJEKTIV, DER Vol. 1 38, 92, 103
MANN MIT DEM PLAN, DER Vol. 4 339, 354
MANN MIT DEM PSST, DER Vol. 4 171, 230
MANN MIT DEM RING IM OHR, DER Vol. 6 *page 157, page 164, page 172*
MANN MIT DEN DREI FRAUEN, DER Vol. 4 390
MANN MIT DER EISERNEN MASKE, DER Vol. 4 124, 512, 747
MANN MIT DER FALSCHEN BANKNOTE, DER Vol. 4 124, 566
MANN MIT DER PRANKE, DER Vol. 4 232, 292, 558
MANN MIT GRUNDSÄTZEN ?, EIN Vol. 4 265, 266, 373
MANN MIT HERZ, EIN Vol. 4 208
MANN OHNE KOPF, DER Vol. 4 212, 255, 373, 384
MANN OHNE NAMEN, DER Vol. 4 186, 560
MANN OHNE NERVEN, DER Vol. 4 356, 471, 779
MANN OHNE SCHLAF, DER Vol. 4 85, 242
MANN SEINER FRAU, DER Vol. 4 329, 527
MANN ÜBER BORD Vol. 4 211, 247
MANN UM MITTERNACHT, DER Vol. 4 208, 316
MANN UND DAS MÄDCHEN, DER Vol. 1 37, 92, 121
MANN VON DER CAP ARCONA, DER Vol. 6 *page 159, page 161, page 168*
MANN WIE MAXIMILIAN, EIN Vol. 4 65, 110, 422
MANN WILL NACH DEUTSCHLAND, EIN Vol. 4 614, 729
MANNAJA Vol. 7 1202
MANNEKÄNG I RÖTT Vol. 5 871, 1042, 1053
MANNEN FRAN ANDRA SIDAN Vol. 5 909
MANNEN FRAN MALLORCA Vol. 5 1089, 1168
MÄNNEN I MÖRKER Vol. 5 1042, 1217
MANNEN I SKUGGAN Vol. 5 946, 1053, 1138
MANNEN PA TAKET Vol. 5 787, 852, 952, 1008, 1065, 1086, 1126, 1160, 1172
MANNEN SOM ALLA VILLE MÖRDA Vol. 5 960, 1005, 1042, 1117
MANNEN SOM BLEV MILJONÄR Vol. 5 933, 985, 1093, 1209
MANNEN SOM GICK UPP I RÖK Vol. 5 846 Vol. 9 6
MANNEN SOM IKKE KUNNE LE Vol. 5 664, 787
MANNEN SOM SLUTADE RÖKA Vol. 5 862, 1027, 1093
MANNEQUIN ASSASSINE, LE Vol. 2 9, 368, 786
MANNEQUINS Vol. 2 205, 279, 488
MANNEQUINS DE PARIS Vol. 2 158, 205, 516, 709
MÄNNER DER SYBILL, DIE Vol. 4 213, 390
MÄNNER MÜSSEN SO SEIN Vol. 4 119, 270
MÄNNER OHNE BART Vol. 1 7, 80, 128
MÄNNER OHNE BERUF Vol. 4 119, 479, 779
MÄNNER SOLL MAN NICHT ALLEINE LASSEN Vol. 4 25, 367
MÄNNER UM LUCIE, DIE Vol. 4 687
MÄNNER VOR DER EHE Vol. 4 108, 132, 270

MÄNNERWIRTSCHAFT Vol. 4 8, 713
MÄNNES WERDEGANG Vol. 4 520, 667
MÄNNISKOR MÖTS OCH LJUV MUSIK UPPSTAR HJÄRTAT Vol. 5 1016, 1052, 1153
MÄNNISKORS RIKE Vol. 5 1042, 1164
MANNSBILD MUSS HER, EIN Vol. 4 18, 558
MANO DE UN HOMBRE MUERTO, LA Vol. 8 600, 768
MANO DELLA MORTA, LA Vol. 7 196
MANO DELLO STRANIERO, LA Vol. 7 1035, 1062
MANO GUANTATA DI BIANCO, LA Vol. 7 81
MANO NEGRA, LA Vol. 8 373
MANO NERA, LA Vol. 7 131
„MANO NERA", PRIMA DELLA MAFIA, PIU DELLA MAFIA, LA Vol. 7 852
MANO TAGLIATA, LA Vol. 7 52, 715
MANOIR DE LA PEUR, LE Vol. 2 50
MANOLENKO Vol. 8 449
MANOLESCU Vol. 4 266, 295, 574 Vol. 6 *page 190*
MANOLESCU, DER FÜRST DER DIEBE Vol. 4 306, 634
MANOLESTA Vol. 7 208, 426
MANOLO, GUARDIA URBANO Vol. 8 214, 607
MANOLO LA NUIT Vol. 8 681, 749
MANON Vol. 2 251, 824
MANON 70 Vol. 2 674, 728
MANON DES SOURCES Vol. 2 231, 871
MANON LESCAUT Vol. 4 232, 396, 663 Vol. 7 196, 453, 621, 968, 1000
MANOS TORPES Vol. 8 551, 626
MANOVRE D'AMORE Vol. 7 1060
MANOVRE DEGLI ALPINI AL COLLE DELLA RANZOLA, LE Vol. 7 837
MANQUE DE MEMOIRE, LE Vol. 2 563
MANS KVINNA Vol. 5 1114, 1173
MANSARDA Vol. 1 198, 223, 249, 254, 321, 368
MANSION DE LA NIEBLA, LA Vol. 8 626
MANTELLI E SPADE INSANGUINATE Vol. 7 877
MANTELLO ROSSO, IL Vol. 7 17
MANTENUTO, IL Vol. 7 1044
MANTILLA DE BEATRIZ, LA Vol. 8 44, 88, 106, 178, 207, 475
MANUEL Y CLEMENTE Vol. 8 610, 756
MANULESCUS MEMOIREN Vol. 4 131, 423
MANYA, DIE TÜRKIN Vol. 4 779
MANZELE PANI MILENY Vol. 10 167, 536
MANZELKA NECO TUSI Vol. 10 345, 585
MANZELSTVI NA UVER Vol. 10 345, 580
MAR ABIERTO Vol. 8 207, 625
MAR BRAVA Vol. 8 465, 549
MAR NEM OLYAN IDÖKET ELÜNK Vol. 9 67, 84, 241
MARA Vol. 8 276
MARA WEST Vol. 7 1189
MARAKATUMBA ... MA NON E UNA RUMBA Vol. 7 978
MARATON Vol. 10 154, 283, 394, 503
MARATON POLSKI Vol. 1 380
MARATONA Vol. 7 796, 970
MARATONCI TRCE POCASNI KRUG Vol. 6 *page 311, page 314, page 318*
MARAVILLA Vol. 8 253, 298, 396, 458
MARAVILLAS Vol. 8 358, 580
MARBELLA, UN GOLPE DE CINCO ESTRELLAS Vol. 8 245, 338
MARC MATO, AGENTE S 077 Vol. 8 306, 603
MARCA DEL HOMBRE LOBO, LA Vol. 8 383, 396
MARCCO, DER BEZWINGER DES TODES Vol. 4 25, 255
MARCCO, DER RINGER DES MIKADO Vol. 4 25
MARCCO UNTER GAUKLERN UND BESTIEN Vol. 4 135
MARCCOS TOLLSTE WETTE Vol. 4 344, 544
MARCELINO PAN Y VINO Vol. 8 449, 725
MARCELLA Vol. 7 150, 453, 569, 701, 725
MARCELO, LA CHAVALA Y EL TROMBONGL Vol. 8 276
MARCHA O MUERE Vol. 8 214, 611
MARCHAND D'AMOUR Vol. 2 438, 775
MARCHAND DE PLAISIRS, LE Vol. 2 44, 367
MARCHAND DE SABLE, LE Vol. 2 6
MARCHANDES D'ILLUSIONS Vol. 2 294, 475, 559
MARCHANDS DE FILLES Vol. 2 581, 627
MARCHE NUPTIALE, LA Vol. 2 6, 29, 180
MARCHE OU CREVE Vol. 2 294, 475
MARCHE PAS SUR MES LACETS Vol. 2 294
MÄRCHEN VOM GLÜCK Vol. 4 187
MÄRCHENBRUNNEN Vol. 1 52
MÄRCHENSCHLOSS, DAS Vol. 1 90, 99
MARCHES DI LAUTEN, IL Vol. 7 1189
MARCHESA D'ARMINIANI, DIE Vol. 4 560, 663
MARCHESE DEL GRILLO, IL Vol. 7 80, 542, 834
MARCHESE DI RUVOLITO, IL Vol. 7 476, 750, 1060
MARCHIO DI KRIMINAL, IL Vol. 8 396
MARCIA NUZIALE, LA Vol. 7 48, 756
MARCIA O CREPA Vol. 8 214, 611
MARCIA TRIONFALE Vol. 7 411, 524
MARCIUSI MESE Vol. 9 71, 322

MARCO **Vol. 7** 1003
MARCO ANTONIO Y CLEOPATRA **Vol. 8** 215, 221, 603
MARCO POLO **Vol. 7** 877, 964, 1009, 1025
MARCO VISCONTI **Vol. 7** 16, 443, 501, 794, 796, 830, 888
MARE **Vol. 7** 601, 923, 941, 1111
MARE DI GUAI, UN **Vol. 7** 16, 756, 1027
MARE DI NAPOLI, IL **Vol. 7** 1057
MARE DU DIABLE, LA **Vol. 2** 387
MARE, IL **Vol. 7** 567, 979
MARE NOSTRUM **Vol. 7** 978 **Vol. 8** 400
MARECHIARO **Vol. 7** 476
„MARECKU, PODEJTE MI PERO !“ **Vol. 10** 173, 325, 425
MARELE SINGURATIC **Vol. 3** 250, 268, 330
MARENCO **Vol. 5** 738
MARGARITA SE LLAMA MI AMOR **Vol. 8** 207, 214
MARGARITKA, MAMA I AZ **Vol. 3** 48, 112 **Vol. 6** *page 33, page 41*
MARGE, LA **Vol. 2** 215, 642, 861
MARGET ET BENEDICT **Vol. 2** 798
MARGHERITA DA CORTONA **Vol. 7** 86, 709
MARGHERITA FRA I TRE **Vol. 7** 639, 1097
MARGHERITELLA **Vol. 7** 316
MARGINADAS, LAS **Vol. 8** 261
MARGINAL, LE **Vol. 6** *page 90, page 142*
MARGOT **Vol. 2** 308
MARGOT, LA PUPA DELLA VILLA ACCANTO **Vol. 7** 492
MARGOTON DU BATAILLON, LA **Vol. 2** 70, 193
MARGUERITE: 3 **Vol. 4** 24, 85, 620
MARGUERITE DE LA NUIT **Vol. 2** 251, 540, 625
MARI A PRIX FIXE, UN **Vol. 2** 503
MARI C'EST UN MARI, UN **Vol. 2** 214
MARI-JUANA **Vol. 8** 411, 437
MARI REVE, LE **Vol. 2** 397
MARIA **Vol. 4** 558 **Vol. 5** 869, 882, 1042, 1164 **Vol. 8** 123
MARIA ANTONIA LA CARAMBA **Vol. 8** 272, 359, 752
MARIA CHAPDELAINE **Vol. 2** 439, 440, 656, 824
MARIA DE LA NUIT **Vol. 2** 330, 528
MARIA DE LA O **Vol. 8** 298, 524
MARIA DE LOS REYES **Vol. 8** 357
MARIA DI MAGDALA **Vol. 7** 782, 796, 1113
MARIA, DIE GESCHICHTE EINES HERZENS **Vol. 4** 159, 374, 535
MARIA, DIE MAGD **Vol. 4** 82
MARIA DOLORES **Vol. 8** 583
MARIA DU BOUT DU MONDE **Vol. 2** 254, 309
MARIA FERNANDA LA JEREZANA **Vol. 8** 359, 579
MARIA ILONA **Vol. 4** 119, 266
MARIA KET EJSZAKAJA **Vol. 9** 196, 219, 236
MARIA MAGDALENA **Vol. 4** 131, 423
MARIA MALIBRAN **Vol. 7** 196, 709
MARIA MARUSJKA **Vol. 5** 767, 799
MARIA MATRICULA DE BILBAO **Vol. 8** 449, 558, 725
MARIA MIRABELA **Vol. 6** *page 272*
MARIA MORENA **Vol. 8** 276, 405
MARIA – NAP **Vol. 9** 20, 232, 281
MARIA NÖVER **Vol. 9** 71, 236, 322
MARIA PA KVARNGARDEN **Vol. 5** 843, 923, 1044
MARIA PAPOILA **Vol. 8** 18, 79, 199
MARIA PENTAGIOTISSA **Vol. 3** 157
MARIA PILAR **Vol. 2** 224
MARIA ROSA **Vol. 8** 214, 611
MARIA SONETTA, DAS FINDELKIND **Vol. 4** 187
MARIA STUART **Vol. 4** 124, 377
MARIA TEREZIA **Vol. 9** 288
MARIA TUDOR **Vol. 4** 131
MARIA ... VIENE A MARCELLO **Vol. 7** 732
MARIA Y LA OTRA **Vol. 8** 371, 769
MARIA ZEF **Vol. 7** 87
MARIA ZEFF **Vol. 7** 627
MARIAGE **Vol. 2** 196, 491
MARIAGE A LA MODE, LE **Vol. 2** 234
MARIAGE A RESPONSABILITE LIMITEE **Vol. 2** 153, 388, 466
MARIAGE D'AMOUR **Vol. 2** 332, 440
MARIAGE DE CHIFFON, LE **Vol. 2** 8, 407, 439, 758
MARIAGE DE FIGARO, LE **Vol. 2** 13, 453
MARIAGE DE KETTY, LE **Vol. 2** 798
MARIAGE DE L'APOTHICAIRE, LE **Vol. 2** 159, 286, 330
MARIAGE DE MADEMOISELLE BEULEMANS, LE **Vol. 2** 309
MARIAGE DE MINUIT **Vol. 2** 205, 723, 798
MARIAGE DE M.LLE BEULEMANS, LE **Vol. 2** 65, 387, 615, 824
MARIAGE DE RAISON **Vol. 2** 798
MARIAGE DE RAMUNTCHO, LE **Vol. 2** 92, 191
MARIAGE DE ROSINE, LE **Vol. 2** 197, 584
MARIAGE DE SUZIE, LE **Vol. 2** 798
MARIAGE DE VERENA, LE **Vol. 2** 42
MARIAGE DU FROTTEUR, LE **Vol. 2** 159, 286, 330
MARIAGES DE MLLE LEVY, LES **Vol. 2** 152, 330, 812

MARIAN Vol. 8 219, 450
MARIANELA Vol. 2 646 Vol. 8 607, 692, 717, 743
MARIANNE Vol. 5 840, 884
MARIANNE DE MA JEUNESSE Vol. 2 153, 279, 778
MARIANNE PA SYKEHUS Vol. 5 722, 738
MARIANNES BRYLLUP Vol. 5 180, 304
MARIBEL Y LA EXTRANA FAMILIA Vol. 8 207, 214
MARICRUZ Vol. 8 387
MARIDO A PRECIO FIJO, UN Vol. 8 436
MARIDO BARATO, UN Vol. 8 437
MARIDO DE IDA Y VUELTA, UN Vol. 8 214, 400
MARIDO SOLTEIRO, UM Vol. 8 54, 171, 172, 195
MARIDOS NO CENAN EN CASA, LOS Vol. 8 214, 396
MARIE Vol. 10 428, 433
MARIE-ANTOINETTE Vol. 2 81, 601, 722
MARIE ANTOINETTE Vol. 4 138, 159, 496, 529, 558
MARIE-CHANTAL CONTRE LE DOCTEUR KAH Vol. 2 700
MARIE DES ANGOISSES Vol. 2 84, 279, 407
MARIE DES ISLES Vol. 2 207, 661
MARIE DU PORT, LA Vol. 2 13, 156, 837
MARIE DU REGIMENT, LE Vol. 2 771
MARIE LA GAIETE Vol. 2 127
MARIE LA MISERE Vol. 2 558, 601
MARIE LEGENDE HONGROISE Vol. 2 676
MARIE LUISE Vol. 4 339
MARIE-MARTINE Vol. 2 407, 657
MARIE-OCTOBRE Vol. 2 232, 488, 863
MARIE-POUPEE Vol. 2 198
MARIE POUR MEMOIRE Vol. 2 312
MARIE SALOPE Vol. 2 291
MARIE-SOLEIL Vol. 2 72, 223
MARIEE DU REGIMENT, LA Vol. 2 187
MARIEE EST TROP BELLE, LA Vol. 2 273, 279, 465, 644
MARIEE ETAIT EN NOIR, LA Vol. 2 208, 377
MARIENWUNDER, DAS Vol. 4 634
MARIES DE L'AN II, LES Vol. 2 282, 400, 719, 837 Vol. 3 218, 286, 313
MARIETA DE L'ULL VIV Vol. 8 228
MARIJKA NEVERNICE Vol. 10 34, 120, 154
MARIKA Vol. 9 71, 208, 322
MARIKENS BRYLLUP Vol. 5 661, 762, 816
MARIN DE GIBRALTAR, LE Vol. 2 208
MARINA PIN PON Vol. 7 99
MARINAI, DONNE E GUAI Vol. 7 333 Vol. 8 749, 762
MARINAI SENZA STELLE Vol. 7 501
MARINARE LA SCUOLA Vol. 7 1189
MARINELLA Vol. 2 249, 644, 871 Vol. 7 1055
MARINEROS, NO MIREIS A LAS MUCHACHAS Vol. 8 749, 762
MARINO DE LOS PUNOS DE ORO, EL Vol. 8 207, 214
MARION Vol. 7 701
MARION, ARTISTA DI CAFFE – CONCERTO Vol. 7 264, 701
MARION, DAS GEHÖRT SICH NICHT Vol. 4 255, 323
MARIONA REBULL Vol. 8 433, 711
MARIONETTE Vol. 7 453, 493
MARIONETTEN Vol. 4 366
MARIONETTEN DER LEIDENSCHAFT Vol. 4 476, 522
MARIONETTEN DES TEUFELS Vol. 4 471, 686
MARIONETTES, THE Vol. 2 24
MARIONS-NOUS Vol. 2 384
MARIPOSA QUE VOLO SOBRE EL MAR, LA Vol. 8 583
MARIS DE LEONTINE, LES Vol. 2 6, 627
MARIS DE MA FEMME, LES Vol. 2 270
MARISOL RUMBA A RIO Vol. 8 214, 261
MARITI A CONGRESSO Vol. 7 915
MARITI ALLEGRI, I Vol. 7 131
MARITI, I Vol. 7 1060
MARITI IN CITTA Vol. 7 791, 817
MARITI IN PERICOLO Vol. 7 1179
MARITI – TEMPESTA D'AMORE, I Vol. 7 639, 750, 1054
MARITO DELL'AMICO, IL Vol. 7 886, 1079
MARITO DI ELENA, IL Vol. 7 1208
MARITO E MOGLIE Vol. 7 877, 1062
MARITO FELICE Vol. 7 1189
MARITO, IL Vol. 7 514
MARITO IN CAMPAGNA, IL Vol. 7 454
MARITO IN COLLEGIO, IL Vol. 7 869
MARITO IN VACANZA, IL Vol. 7 384, 909
MARITO, MOGLIE E... Vol. 7 554
MARITO PACIFICO, IL Vol. 7 1189
MARITO PER ANNA ZACCHEO, UN Vol. 7 725
MARITO PER IL MESE DI APRILE, UN Vol. 7 792, 948, 1036
MARITO PERDUTO, IL Vol. 7 250, 655
MARITO POVERO, IL Vol. 7 16
MARITZA Vol. 7 1189
MARIUS Vol. 2 646
MARIUTE Vol. 7 264, 701
MARIZZA, GENANNT DIE SCHMUGGLERMADONNA Vol. 4 178, 558
MARJA PIENI ! Vol. 5 391, 606
MARK COLPISCE ANCORA Vol. 7 1194

MARK IN POLIZIOTTO SPARA PER PRIMO **Vol. 7** 502, 1202
MARKAN TÄHDEN **Vol. 5** 330, 471
MARKENS GRODE **Vol. 5** 794
MARKETA LAZAROVA **Vol. 10** 24, 394, 425
MÄRKISCHE CHRONIK **Vol. 6** *page 158, page 162, page 164*
MÄRKISCHE FORSCHUNGEN **Vol. 6** *page 158, page 160, page 170*
MARKNADSAFTON **Vol. 5** 871, 1042, 1052
MARKO POLO **Vol. 3** 356, 380 **Vol. 6** *page 312*
MARKURELLS I WADKÖPING **Vol. 5** 891, 997
MARKUS – SETÄ KERTOO **Vol. 5** 401
MARLENE **Vol. 2** 128, 344
MARMAILLE, LA **Vol. 2** 131
MARMELADUPPRÖRET **Vol. 5** 1080, 1093
MARODÖRER **Vol. 5** 982
MARQUES DE SALAMANCA, EL **Vol. 8** 276, 298
MARQUES, EL **Vol. 8** 298, 600
MARQUES, O **Vol. 8** 143
MARQUESONA, LA **Vol. 8** 298, 716
MARQUIS D'EON, DER SPION DER POMPADOUR **Vol. 4** 81, 544, 545, 729
MARQUIS, DER **Vol. 8** 143
MARQUIS D'OR, DER **Vol. 4** 131, 385
MARQUISE D'O, LA **Vol. 2** 17, 595
MARQUISE VON O..., DIE **Vol. 4** 17, 496, 529
MARQUISE VON POMPADOUR, DIE **Vol. 4** 239, 479
MARQUISE VON PORNO **Vol. 2** 52
MARQUITTA **Vol. 2** 6, 48, 330
MARRAINE DE CHARLEY, LA **Vol. 2** 67, 118, 488, 875
MARRAINE DU REGIMENT, LA **Vol. 2** 366, 412
MARRIAGE DE CHIFFON, LE **Vol. 7** 390
MARS NA DRINU **Vol. 3** 349, 391
MARSCHALL VORWÄRTS **Vol. 4** 124, 204, 289, 422
MARSEILLAISE, LA **Vol. 2** 18, 59, 124, 248, 524, 540, 657, 863
MARSEILLE MES AMOURS **Vol. 2** 47, 514, 722
MARTA **Vol. 8** 390, 448, 692 **Vol. 9** 301
MARTER DER LIEBE **Vol. 4** 19, 149, 219, 669
MARTES Y TRECE **Vol. 8** 258, 311
MARTES Y TRECE NI TE CASES NI TE EMBARQUES **Vol. 8** 681
MARTHA **Vol. 4** 354, 437, 738 **Vol. 5** 16, 61, 272
MARTHE RICHARD AU SERVICE DE LA FRANCE **Vol. 2** 67, 488, 657
MARTHU CHE HA VISTO IL DIAVOLO **Vol. 7** 48
MARTIEN A PARIS, UN **Vol. 2** 131, 199
MARTIN **Vol. 5** 719, 778, 780
MARTIN A CERVENE SKLICKO **Vol. 10** 299, 380
MARTIN A DEVET BLAZNU **Vol. 10** 299, 380
MARTIN ET LEA **Vol. 2** 739
MARTIN LUTHER **Vol. 4** 119, 126, 127, 768
MARTIN ROUMAGNAC **Vol. 2** 402, 863
MARTIN SOLDAT **Vol. 2** 263, 481
MARTIN U OBLACIMA **Vol. 3** 341, 444
MARTIN XIII **Vol. 6** *page 165, page 173, page 175*
MARTINGALA **Vol. 8** 752
MARTINO IL TROVATELLO **Vol. 7** 469, 493
MARTIRE, LA **Vol. 7** 264
MARTIRI DI BELFIORE, I **Vol. 7** 64
MARTIRI, D'ITALIA, I **Vol. 7** 48, 198, 491, 1004, 1116
MARTWA FALA **Vol. 1** 179, 237, 344 **Vol. 6** *page 242*
MARTWY WEZEL **Vol. 1** 174, 380
MARTYR DE BOUGIVAL, LE **Vol. 2** 193, 254
MARTYRE **Vol. 2** 311
MARTYRE DE L'OBESE, LE **Vol. 2** 58, 388, 402, 528
MARTYRE DE SAINTE MAXENCE, LE **Vol. 2** 311
MARTYRIES **Vol. 3** 159
MARTYRIUM, DAS **Vol. 4** 729, 773
MARUXA **Vol. 8** 228
MARUZZELLA **Vol. 7** 792, 1121
MARWILLE DETEKTIVEM **Vol. 10** 243, 536
MARY ANN **Vol. 9** 200
MARY, LA STELLA DEL VARIETE **Vol. 7** 442
MARY LOU **Vol. 4** 14, 431
MARY WOOD, DIE TOCHTER DES STRÄFLINGS **Vol. 4** 447
MARYNIA **Vol. 6** *page 211, page 230, page 282*
MARYS GROSSES GEHEIMNIS **Vol. 4** 124, 157, 535
MARYSA **Vol. 10** 92, 122
MARYSIA I KRASNOLUDKI **Vol. 1** 243, 360
MARYSIA I NAPOLEON **Vol. 1** 305, 388, 408
MARZIANI HANNO DODICI MANI, I **Vol. 7** 347 **Vol. 8** 438
MARZY DEL VASTO MONDO **Vol. 7** 1073
M.A.S. **Vol. 7** 1027

MAS ALLA DE LA MUERTE Vol. 8 351, 426
MAS ALLA DE LAS MONTANAS Vol. 8 338, 538
MAS ALLA DE RIO MINO Vol. 8 306, 677
MAS ALLA DEL TERROR Vol. 8 237, 286
MAS BONITA QUE NINGUNA Vol. 8 749, 762
MAS FABULOSO GOLPE DEL FAR WEST, EL Vol. 8 297, 501, 552
MASAJISTA VOACIONAL, LA Vol. 8 249, 346
MASAMOD Vol. 9 73, 200
MASCA DE ARGINT Vol. 6 *page 270, page 282*
MASCARA DE SCARAMOUCHE, LA Vol. 8 214, 762
MASCHERA CHE RIDE, LA Vol. 7 1004
MASCHERA DEL DEMONIO, LA Vol. 7 106, 537
MASCHERA DEL MALE Vol. 7 48
MASCHERA DEL VIZIO, LA Vol. 7 829
MASCHERA DELLA COLPA Vol. 7 266, 701
MASCHERA DELLA FEMMINA, LA Vol. 7 940
MASCHERA DELL'AMORE Vol. 7 399
MASCHERA DELL'EROE, LA Vol. 7 277
MASCHERA DI CESARE BORGIA, LA Vol. 7 67, 910, 1063, 1096, 1107, 1172
MASCHERA DI VENERE, LA Vol. 7 503
MASCHERA E IL DESTINO, LA Vol. 7 548, 774
MASCHERA E IL VOLTO, LA Vol. 7 603, 994, 1054, 1060
MASCHERA FOLLE, LA Vol. 7 1038
MASCHERA, LA Vol. 7 654, 918
MASCHERA SUL CUORE Vol. 7 580, 676
MASCHIACCIO Vol. 7 456
MASCHIO FEMMINA FIORE FRUTTO Vol. 7 315, 834
MASCHIO RUSPANTE, IL Vol. 7 1086
MASCOTTCHEN Vol. 4 149, 219, 660
MASCOTTE Vol. 1 174, 278, 372 Vol. 4 186, 560
MASCOTTE, LA Vol. 2 339, 688, 775
MASCULINUM - FEMININUM Vol. 5 510
MASFEL MILLIO Vol. 9 67, 117, 327
MASIK EMBER, A Vol. 9 22, 141, 159
MASKARADA Vol. 3 417
MASKE DES TODES, DIE Vol. 4 477
MASKE FÄLLT, DIE Vol. 4 281
MASKE I MORGEN Vol. 5 257
MASKE IN BLAU Vol. 4 51, 108, 139
MASKE KU' VI Vol. 5 12, 48
MASKEN Vol. 4 219, 266, 359, 574, 734
MASKENFEST DES LEBENS, DAS Vol. 4 773
MASKENSCHERZ, EIN Vol. 4 183
MASKIERTE LIEBE Vol. 4 183
MASKIERTE SCHRECKEN, DER Vol. 4 718
MASKOTTI Vol. 5 334, 472, 529
MASKOVANA MILENKA Vol. 10 412, 621
MASNADIERA, LA Vol. 7 554
MASOCH Vol. 7 1034
MASOVO CUDO Vol. 6 *page 36, page 47*
MASQUE DE FER, LA Vol. 2 128, 661
MASQUE QUI TOMBE, LE Vol. 7 756, 796
MASQUERADER, THE Vol. 2 84
MASSACRE EN DENTELLES Vol. 2 158, 368
MASSEBA Vol. 10 172, 241, 409
MASSIMAMENTE FOLLE Vol. 7 153
MASSINELLI IN VACANZA Vol. 7 341
MASTER LOVE Vol. 2 187, 254
MÄSTERDETEKTIVEN BLOMKVIST Vol. 5 933, 1042
MÄSTERDETEKTIVEN LEVER FARLIGT Vol. 5 469
MÄSTERDETEKTIVEN OCH RASMUS Vol. 5 1207
MÄSTERKATTEN I STÖVLAR Vol. 5 944, 966, 976
MÄSTERMAN Vol. 5 937, 997
MASTODONTUL Vol. 3 294, 300, 317, 320
MASTRO LANDI Vol. 7 16
MATA-HARI Vol. 4 130, 316, 377
MATA-HARI AGENT H. 21 Vol. 2 427, 674
MATADOR Vol. 8 240, 373, 570, 684
MATAF, LE Vol. 2 20, 781
MATALOS Y VUELVE Vol. 8 626, 762
MATAR AL NANI Vol. 8 245, 765
MATAR PARA VIVIR Vol. 8 673
MATAR SAUDADES Vol. 8 56
MATASSA DI SETA, LA Vol. 7 133
MATCH CONTRE LA MORT Vol. 2 347, 407
MATCH DEI 100.000 DOLLARI, IL Vol. 7 1195
MATCH NULLO Vol. 7 316
MATCHLESS Vol. 7 408, 931, 1001, 1142
MATE DOMA LVA ? Vol. 10 55, 428, 593
MATECZKA Vol. 1 278, 308, 346
MATEJI, PROC TE HOLKY NECHTEJI ? Vol. 10 172, 570, 571
MATER AMATISSIMA Vol. 8 621, 669
MATER DEI Vol. 7 606
MATER DOLOROSA Vol. 2 153, 388, 402, 528 Vol. 4 423 Vol. 7 443, 639, 830, 1030, 1189
MATERIA - DER CLUB DER TOTEN Vol. 4 366
MATERNELLE, LA Vol. 2 37, 63, 67, 137, 195, 217, 758
MATERNITA Vol. 7 131
MATERNITE Vol. 2 187, 566

MATERNITE CLANDESTINE **Vol. 2** 404
MATHIAS SANDORF **Vol. 2** 194, 222, 648, 867 **Vol. 8** 611
MATIN ROUGE, UN **Vol. 6** *page 80, page 117*
MATIOUETTE OU L'ARRIERE-PAYS, LA **Vol. 6** *page 118, page 123, page 131*
MATKA JOANNA OD ANIOLOW **Vol. 1** 398, 403
MATKA KRACMERKA **Vol. 10** 256, 586
MATKALLA SEIKKAILUUN **Vol. 5** 356, 472, 634
MATKINA SPOVED **Vol. 10** 256, 532
MATOMENA HRISTOUGENNA **Vol. 3** 154
MATOMENE GE **Vol. 3** 130, 155
MATOMENO ELIOBASILEMA **Vol. 3** 136
MATOUS SONT ROMANTIQUES, LES **Vol. 6** *page 76, page 112*
MATRIARHAT **Vol. 3** 102, 105 **Vol. 6** *page 50*
MATRICULE 33 **Vol. 2** 579, 646
MATRIMONIO AL DESNUDO **Vol. 8** 245, 749
MATRIMONIO ALL'ITALIANA **Vol. 7** 514, 769
MATRIMONIO DI OLIMPIA, IL **Vol. 7** 48
MATRIMONIO D'INTERESSE **Vol. 7** 26
MATRIMONIO, IL **Vol. 7** 915, 1177
MATRIMONIO IN 27 MINUTI **Vol. 7** 26
MATRIMONIO MULIN, IL **Vol. 7** 1189
MATRIMONIO SEGRETO, IL **Vol. 7** 587
MATRIMONIOS SEPARADOS **Vol. 8** 553, 749
MATROSE PERUGINO, DER **Vol. 4** 439, 704
MATTATORE, IL **Vol. 7** 372
MATTEO AL MERCATO **Vol. 7** 1189
MATTINO DI PRIMAVERA **Vol. 7** 210
MATTO, IL **Vol. 7** 123, 242
MATURANTI **Vol. 6** *page 318*
MATURITA ZA SKOLOU **Vol. 10** 87, 139, 415, 568
MATYAS RENDET CSINAL **Vol. 9** 132, 164
MATYOLAKODALOM **Vol. 9** 73
MAUD ROCKEFELLERS WETTE **Vol. 4** 179, 673
MAUDITS, LES **Vol. 2** 13, 96
MAULKORB, DER **Vol. 4** 266, 373
MAULWÜRFE **Vol. 4** 291, 648
MAUNU KURKVAARAN KUJANJUOKSU **Vol. 5** 452
MAUPRAT **Vol. 2** 277, 426
MÄUSCHEN **Vol. 6** *page 186*
MAUSI **Vol. 4** 361, 558
MAUVAIS COUPS, LES **Vol. 2** 49, 170
MAUVAIS FILS, UN **Vol. 2** 110 **Vol. 6** *page 70*
MAUVAIS GRAINE **Vol. 2** 205, 228
MAUVAISES RENCONTRES, LES **Vol. 2** 232, 251
MAX, DER VIELGEPRÜFTE **Vol. 4** 131, 535
MAX ET LES FERRAILLEURS **Vol. 2** 377, 564
MAX UND SIEBENEINHALB JUNGEN **Vol. 6** *page 160, page 167, page 173*
MAXIME **Vol. 2** 185, 226, 566
MAY MORNING **Vol. 7** 649
MAYA **Vol. 2** 59, 826
MAYERLING **Vol. 2** 13, 23, 27, 194, 407, 676, 824, 863
MAYORES CON REPAROS **Vol. 8** 207, 595, 602
MAZ POD LOZKIEM **Vol. 1** 196, 305, 323
MAZ SWOJEJ ZONY **Vol. 1** 148, 189, 227
MAZEL TOV OU LE MARIAGE **Vol. 2** 190, 515
MAZEPA **Vol. 1** 179, 187 **Vol. 6** *page 250*
MAZLICEK **Vol. 10** 256, 586
MAZURKA **Vol. 4** 234, 713, 734
MAZURKA DER LIEBE **Vol. 1** 90, 107, 129
MAZURKA DI PAPA, LA **Vol. 7** 200, 460, 493, 1060
MAZURKA PA SENGEKANTEN **Vol. 5** 21, 223, 306
MAZZETTA, LA **Vol. 7** 599
MBROSTA STO THEO **Vol. 3** 203
ME **Vol. 5** 443, 639
ME APUESTO UN MILLON **Vol. 8** 357
ME DEBES UN MUERTO **Vol. 8** 331, 396
ME ENVENENO DE AZULES **Vol. 8** 219
ME FAIRE CA A MOI **Vol. 2** 168
ME FOBO KAI PATHOS **Vol. 3** 144
ME HACE FALTA UN BIGOTE **Vol. 8** 327, 756
ME HAS HECHO PERDER EL JUICIO **Vol. 8** 331, 396
ME OLVIDE DE VIVIR **Vol. 8** 214
ME QUIERO CASAR CONTIGO **Vol. 8** 382, 394, 493, 732
ME TE LAMPSE STA MATIA **Vol. 3** 142
ME TULEMME TAAS **Vol. 5** 374, 462
MEANDRE **Vol. 3** 274, 317
MECANISMO INTERIOR **Vol. 8** 688, 697
MECAVA **Vol. 3** 371, 421
MECTATEL **Vol. 6** *page 43, page 61, page 65*
MED DIG I MINA ARMAR **Vol. 5** 871, 1156
MED FARA FÖR LIVET **Vol. 5** 1041, 1142
MED FOLKET FÖR FOSTERLANDET **Vol. 5** 840, 841, 927
MED FULD MUSIK **Vol. 5** 24
MED GLORIAN PA SNED **Vol. 5** 878, 953, 1052

MED KAERLIG HILSEN Vol. 5 35, 257, 273
MED KRUT I NÄVARNA Vol. 5 914
MED LILL-KLAS I KAPPSÄCKEN Vol. 5 40, 217, 297, 874, 1212
MED LILLE KLAS I KUFFERTEN Vol. 5 40, 217, 297
MED LIVET SOM INSATS Vol. 5 863, 1032
MED OG UDEN KONE Vol. 5 83
MED STRAHOM IN DOLZNOSTJO Vol. 3 395, 417
MEDAGLIA E IL ROVESCIO, LA Vol. 7 133
MEDAGLIONE, IL Vol. 7 442
MEDAGLIONE INSANGUINATO, IL Vol. 7 335
MEDAILLE, DIE Vol. 4 82, 329
MEDAILLON DER LADY SINGTON, DAS Vol. 4 162, 291
MEDALJON SA TRI SRCA Vol. 3 337, 356, 413
MEDALLA DEL TORERO, LA Vol. 8 572, 657
MEDAN PORTEN VAR STÄNGD Vol. 5 863, 1032
MEDAN STADER SOVER Vol. 5 878, 1173
MEDEA Vol. 7 567, 1109, 1122, 1142
MEDEA DI PORTAMEDINA Vol. 7 829, 1082
MEDECIN DU BLED, LE Vol. 2 811
MEDENA VEZA Vol. 10 72, 287
MEDENI MJESEC Vol. 6 *page 308, page 321*
MEDIAS Y CALCETINES Vol. 8 280, 768
MEDICO A PALOS, EL Vol. 8 604
MEDICO DEI BAMBINI, IL Vol. 7 1078
MEDICO DEI PAZZI Vol. 7 852
MEDICO DELLE PAZZE, IL Vol. 7 668, 1151
MEDICO E LO STREGONE, IL Vol. 7 1147
MEDICO ... LA STUDENTESSA, IL Vol. 7 666
MEDICO PER FORZA, IL Vol. 7 48, 361, 796, 1116
MEDIKUS, A Vol. 9 77
MEDIUM, DAS Vol. 4 159, 232, 627
MEDIUM, IL Vol. 7 1016
MEDLODRAMMORE Vol. 7 901
MEDUSA VELATA Vol. 7 277
MEDVED A STRASIDLA Vol. 10 193, 569, 621
MEER, DAS Vol. 4 377, 479
MEER RUFT, DAS Vol. 4 452, 481, 774, 788
MEER SCHWEIGT, DAS Vol. 4 514
MEERE RUFEN, DIE Vol. 1 35, 54, 109
MEFIEZ-VOUS DES BLONDES Vol. 2 57, 158
MEFIEZ-VOUS FILLETTES Vol. 2 185, 423, 738
MEFIEZ-VOUS MESDAMES ! Vol. 2 615
MEFTURI 1900 Vol. 3 244
MEG KER A NEP Vol. 9 20, 154, 318
MEGALE APOFASE Vol. 3 184
MEGALEXANDROS Vol. 3 135, 166, 206
MEGALL AZ IDÖ Vol. 9 167, 169, 313
MEGALMODTALAK Vol. 9 196, 274
MEGALOS EROTIKOS Vol. 3 135
MEGATON YE-YE Vol. 8 406, 679
MEGBELYEGZETT, A Vol. 9 77
MEGBÜVOLTEK, A Vol. 9 172, 175
MEGEN KAERLIGHED, DEN Vol. 5 83
MEGFAGYOTT GYERMEK, A Vol. 9 2
MEGFELELÖ EMBER, A Vol. 9 117, 314, 326
MEGFELELÖ EMBER KENYES FELADATRA Vol. 9 10, 139, 286
MEGLIO SOLI CHE MALE ACCOMPAGNATI Vol. 7 1189
MEGÖLTEK EGY LEANYT Vol. 9 67, 134, 165
MEGSZALLOTTAK Vol. 9 134, 215, 332
MEGVEDTEM EGY ASSZONYT ! Vol. 9 94, 274, 338
MEHRI TO PLOIO Vol. 3 175, 183
MEIDÄN POIKAMME Vol. 5 356, 428
MEIDAN POIKAMME ILMASSA – ME MAASSA Vol. 5 356, 428, 538, 608
MEIDÄN POIKAMME MERELLÄ Vol. 5 359, 428
MEILLEURE FACON DE MARCHER, LA Vol. 2 536, 633
MEILLEURE PART, LA Vol. 2 13, 156
MEILLURE FACON DE S'ENVOYER EN L'AIR, LA Vol. 6 *page 108*
MEILTÄHÄN TÄMÄ KÄY Vol. 5 324, 540, 639
MEIN BLAUER VOGEL FLIEGT Vol. 1 52, 116, 124
MEIN FREUND, DER CHAUFFEUR Vol. 4 49, 287, 329, 644
MEIN FREUND, DER MILLIONÄR Vol. 4 292, 544
MEIN FREUND HARRY Vol. 4 287, 384, 458, 646
MEIN HEIDELBERG, ICH KANN DICH NICHT VERGESSEN Vol. 4 344, 544, 775
MEIN HERZ IST EINE JAZZBAND Vol. 4 14, 186, 709
MEIN HERZ RUFT NACH DIR Vol. 4 49, 614
MEIN HERZ SEHNT SICH NACH LIEBE Vol. 4 287, 644, 646
MEIN LEBEN Vol. 4 267
MEIN LEBEN FÜR DAS DEINE Vol. 4 246, 321, 752

MEIN LEBEN FÜR IRLAND **Vol. 4** 15, 123, 149
MEIN LEBEN FÜR MARIA ISABELL **Vol. 4** 149, 354, 660
MEIN LEOPOLD **Vol. 4** 87, 115, 208, 598, 603, 626, 771
MEIN LETZTER HAUCH **Vol. 4** 135, 514
MEIN LIEBER MANN UND ICH **Vol. 1** 11, 114
MEIN LIEBER ROBINSON **Vol. 1** 32, 53, 124
MEIN MANN DARF ES NICHT WISSEN **Vol. 4** 340, 566
MEIN SOHN, DER HERR MINISTER **Vol. 4** 13, 574
MEIN VATER ALFONS **Vol. 6** *page 162, page 164, page 176*
MEIN VATER IST EIN DIEB **Vol. 6** *page 158, page 163, page 168*
MEIN WILLE IST GESETZ **Vol. 4** 476, 522
MEINE COUSINE AUS WARSCHAU **Vol. 4** 113
MEINE FRAU, DIE FILMSCHAUSPIELERIN **Vol. 4** 560, 663
MEINE FRAU, DIE HOCHSTAPLERIN **Vol. 4** 149, 536, 629, 660
MEINE FRAU, DIE PERLE **Vol. 4** 558, 796
MEINE FRAU, DIE SCHÜTZENKÖNIGIN **Vol. 4** 97, 230
MEINE FRAU MACHT MUSIK **Vol. 1** 20, 54, 58, 88 **Vol. 4** 134
MEINE FRAU TERESA **Vol. 4** 215, 292, 640
MEINE FREUNDIN BARBARA **Vol. 4** 97, 340, 466
MEINE FREUNDIN JOSEFINE **Vol. 4** 123, 486, 489, 678
MEINE FREUNDIN SYBILLE **Vol. 1** 21, 79, 111
MEINE HERREN SÖHNE **Vol. 4** 347, 370, 438, 729
MEINE SCHWESTER UND ICH **Vol. 4** 52, 166, 186
MEINE STUNDE NULL **Vol. 1** 40, 44, 68, 106
MEINE TANTE – DEINE TANTE **Vol. 4** 8, 535, 626, 681, 742
MEINE TOCHTER LEBT IN WIEN **Vol. 4** 97, 391, 557
MEINE TOCHTER TUT DAS NICHT **Vol. 4** 97, 115
MEINE VIER JUNGENS **Vol. 4** 369, 539, 557
MEINEID **Vol. 4** 14, 208
MEINEIDBAUER, DER **Vol. 4** 97, 391, 491, 557, 585, 734
MEINES VATERS STRASSENBAHN **Vol. 6** *page 158, page 167, page 173*
MEIR EZOFOWICZ **Vol. 1** 174, 188
MEISEKEN **Vol. 4** 124, 232, 471, 566
MEISTER DER WELT, DER **Vol. 4** 85, 219
MEISTER PETZ IM WINTERSPORT **Vol. 4** 367
MEISTER VON NÜRNBERG, DER **Vol. 4** 38, 213, 536
MEISTERBOXER, DER **Vol. 4** 247, 461
MEISTERDETEKTIV, DER **Vol. 4** 340, 344, 525, 638, 647
MEISTERDIEB, DER **Vol. 1** 75
MEISTERSCHUSS, DER **Vol. 4** 98, 112
MEISTERSPRINGER VON KÜRNBERG, DIE **Vol. 4** 258
MEJ OCH DEJ **Vol. 5** 946
MEJOR DEL MUNDO, EL **Vol. 8** 602, 749
MEKANIKPIGEN **Vol. 5** 83
MEKO KAZANO **Vol. 6** *page 28, page 62*
MEKTOUB ? **Vol. 2** 139, 796
MELANCOLICAS, LAS **Vol. 8** 251, 346
MELCHIAD KOLOMAN **Vol. 10** 115
MELIMONTANT **Vol. 2** 21, 431, 722
MELISSA **Vol. 2** 721
MELLEM DE YDERSTE SKAER **Vol. 5** 242
MELLEM MUNTRE MUSIKANTER **Vol. 5** 83
MELLEM VENNER **Vol. 5** 135
MELO **Vol. 2** 440, 726
MELOCOTON EN ALMIBAR **Vol. 8** 524
MELODI OM VAREN, EN **Vol. 5** 840, 997, 1042, 1164
MELODIAS DE AMOR **Vol. 8** 429, 478
MELODIAS DE HOY **Vol. 8** 461, 762, 772
MELODIAS PROHIBIDAS **Vol. 8** 422
MELODIE DER LIEBE **Vol. 4** 49, 97, 255
MELODIE DER WELT **Vol. 4** 291, 373, 393, 538
MELODIE DES HERZENS **Vol. 4** 325, 567, 620
MELODIE DUSZY **Vol. 1** 174, 176, 326
MELODIE EN SOUS-SOL **Vol. 2** 185, 644
MELODIE ETERNE **Vol. 7** 196, 453, 1000
MELODIE IMMORTALI **Vol. 7** 528
MELODIE POUR MANUELA **Vol. 6** *page 81*
MELODIE POUR TOI **Vol. 2** 156, 173, 318, 863
MELODIEN FRAN GAMLA STAN **Vol. 5** 843, 923
MELODII LA COSTINESTI **Vol. 6** *page 271, page 273, page 280*
MELODII, MELODII **Vol. 3** 239, 245, 262
MELODRAMA **Vol. 3** 151, 156
MELODRAME **Vol. 2** 806
MELODRAMMA **Vol. 7** 48, 756
MELTOSAGOS KISASSZONY **Vol. 9** 288, 319
MELTOSAGOS RABASSZONY **Vol. 9** 323

MELUKA, DIE ROSE VON MARAKESCH Vol. 4 304
MELUSINE Vol. 4 15, 266
MEME LES MOMES ONT DU VAGUE A L'AME Vol. 2 352, 765
MEMED Vol. 6 *page 312*
MEMENTO Vol. 3 384, 396
MEMOIRE COMMUNE Vol. 2 26
MEMOIRE COURTE, LA Vol. 6 *page 86, page 97*
MEMOIREN DER TRAGÖDIN THAMAR, DIE Vol. 4 124, 232 Vol. 6 *page 194*
MEMOIREN EINES KAMMERDIENERS Vol. 4 223, 270
MEMOIRES DE LA VACHE YOLANDE, LES Vol. 2 191, 475
MEMOIRES D'UN FLIC Vol. 2 158, 205, 777
MEMOIRES D'UNE PETITE CULOTTE, LES Vol. 6 *page 120*
MEMORIA DEL DIAVOLO, LA Vol. 7 1038
MEMORIAS DE LETICIA VALLE Vol. 8 738
MEMORIAS DEL GENERAL ESCOBAR Vol. 8 511, 700
MEMORIE DEL DIAVOLO, LE Vol. 7 454
MEMORIE DELL'ALTRO, LE Vol. 7 1038
MEMORIE DI UN PAZZO, LE Vol. 7 843
MEMORIE DI UNA ISTITUTRICE, LE Vol. 7 1073
MEN ERÖTEUESE TO SABBATO Vol. 3 149
MEN, ETT LEJON, EN Vol. 5 840, 997
MEN OLSEN - BANDEN VAR IKKE DOD ! Vol. 5 762, 766, 815
MENACE DE MORT Vol. 2 595, 698
MENACE, LA Vol. 2 167, 248, 351, 388, 410, 434, 613, 856
MENACES Vol. 2 248, 395, 397, 413, 826
MENDIANTE DE ST.SULPICE, LA Vol. 2 440, 708
MENDIANTS ET ORGUEILLEUX Vol. 2 732
MENDICANTE DI SASSONIA, LA Vol. 7 398
MENE ME, MAMO, ZMEJ LJUBI Vol. 3 34
MENEKÜLÖ EMBER, A Vol. 9 71, 76
MENESGAZDA, A Vol. 9 67, 167, 332
MENESTYKSEN MAKU Vol. 5 541
MENGJEZE LUFTE Vol. 3 3, 5, 18
MENGJEZE TE REJA Vol. 3 7, 19, 22
MENNEISYYDEN VARJO Vol. 5 549, 592
MENNESKEDYRET Vol. 5 70
MENNESKER MODES OG SOD MUSIK OPSTAR I HJERTET Vol. 5 179, 200, 273
MENNESKESKAEBNER Vol. 5 121
MENNYEI SEREGEK Vol. 9 105, 167, 171
MENS PESTEN RASER Vol. 5 58
MENS PORTEN VAR LUKKET Vol. 5 115, 180, 199, 304
MENS SAGFOREREN SOVER Vol. 5 2, 228
MENSAJEROS DE PAZ Vol. 8 583, 772
MENSCH AM WEGE, DER Vol. 4 239, 561
MENSCH, BEZAHLE DEINE SCHULDEN Vol. 4 646
MENSCH GEGEN MENSCH Vol. 4 89, 316, 752
MENSCH OHNE NAMEN Vol. 2 399 Vol. 4 266, 288, 574
MENSCH UND DIE BESTIE, DER Vol. 3 371, 374
MENSCHEN Vol. 4 277, 425, 456
MENSCHEN AM MEER Vol. 4 151, 290
MENSCHEN AM SONNTAG Vol. 4 629
MENSCHEN HINTER GITTERN Vol. 4 750
MENSCHEN IM FEUER Vol. 4 119, 479
MENSCHEN IM KÄFIG Vol. 4 68, 177, 785
MENSCHEN IM RAUSCH Vol. 4 552, 718
MENSCHEN IM STURM Vol. 4 292
MENSCHEN IN DEN BERGEN Vol. 10 2, 412, 532
MENSCHEN OHNE VATERLAND Vol. 4 325, 713
MENSCHEN, TIERE, SENSATIONEN Vol. 4 181, 247, 339, 703
MENSCHEN UND MASKEN Vol. 4 471, 560, 734, 779
MENSCHEN UND TIERE Vol. 1 17, 68
MENSCHEN UNTEREINANDER Vol. 4 247, 461
MENSCHEN VOM VARIETE Vol. 4 124, 139, 232
MENSCHEN VON HEUTE Vol. 4 660, 704
MENSCHEN ZWEITER GÜTE Vol. 4 742
MENSCHENFEIND, DER Vol. 4 223
MENSCHENLEBEN IN GEFAHR Vol. 4 124, 725
MENSONGE DE NINA PETROVNA, LE Vol. 2 27, 67, 206, 332, 676
MENSONGES Vol. 2 197, 339
MENTEURES, LES Vol. 2 598, 824
MENTIRA DE LA GLORIA, LA Vol. 8 692
MENTIRA TIENE CABELLOS ROJOS, LA Vol. 8 492, 732
MENTIROSA Vol. 8 311, 546
MENTRE IL PUBLICO RIDE Vol. 7 168
MENUETT, DAS Vol. 4 118, 367
MENUETT DES BOCCHERINI, DAS Vol. 4 119, 558
MENYASSZONY GYÖNYÖRÜ VOLT, A Vol. 9 74, 154, 235
MENZOGNA Vol. 7 638, 1054
MENZOGNA FATALE Vol. 7 1189
MENZOGNA, LA Vol. 7 266, 456, 668, 1038
MEO PATACCA Vol. 7 997
MEPHISTO Vol. 9 105, 167, 259

MEPRIS, LE Vol. 2 208
MER COUPLEUR DE LARMES, LA Vol. 2 368
MERA SPORED MERA Vol. 6 *page 55*
MERA UR KÄRLEKENS SPRAK Vol. 5 361, 869
MERAVIGLIOSA Vol. 7 1045
MERAVIGLIOSE AVVENTURE DI GUERRIN MESCHINO Vol. 7 174, 1170
MERCADO PROHIBIDO Vol. 8 396, 503
MERCANTE DI EMOZIONI, IL Vol. 7 390
MERCANTE DI SCHIAVE, IL Vol. 7 67, 1107, 1136, 1172
MERCANTE DI VENEZIA, IL Vol. 7 493
MERCEDES Vol. 8 422
MERCENARI, I Vol. 7 364
MERCENARIO, EL Vol. 8 424
MERCENARIO, IL Vol. 8 770
MERCENARIOS DEL AIRE Vol. 8 551, 732
MERCI NATERCIA Vol. 2 160
MERE ROSII Vol. 3 264, 279, 288
MEREN JA LEMMEN AALLOT Vol. 5 315, 608
MEREN JUHLAT Vol. 5 452
MEREN KASVOJEN EDESSÄ Vol. 5 356, 406
MERENYLET Vol. 9 26, 122, 179
MERES TOU'36 Vol. 3 135, 166
MERII SALBATICI Vol. 3 213, 251, 287
MERIKOI TO PROTIMOUN KRYO Vol. 3 144
MERKŐZES, A Vol. 9 17, 263, 321
MERLE BLANC, LE Vol. 2 217, 558
MERLUSSE Vol. 2 38, 485
MERMOZ Vol. 2 100, 322, 496
MERRY GO ROUND Vol. 2 527
MERSEKELT EGÖV Vol. 9 20, 154, 318
MERVEILLEUSE ANGELIQUE Vol. 2 615, 659
MERVEILLEUSE JOURNEE, LA Vol. 2 6, 9
MERVEILLEUSE JOURNEE, UNE Vol. 2 409 Vol. 6 *page 78*
MERVEILLEUSE TRAGEDIE DE LOURDES, LA Vol. 2 84
MERVEILLEUSE VISITE, LA Vol. 2 475, 728
MERVEILLEUX PARFUM D'OSEILLE Vol. 2 808
MES ENNLA DE LA PASSIO Vol. 8 455
MES NUITS AVEC...ALICE, PENELOPE, ARNOLD, MAUD ET RICHARD Vol. 2 294
MES PETITES AMOUREUSES Vol. 2 17
MES TANTES ET MOI Vol. 2 395, 858
MESALLIANCE Vol. 5 97
MESE A 12 TALALATROL Vol. 9 71, 312
MESE DI ONESTA, UN Vol. 7 606
MESE HABBAL Vol. 9 166, 215, 339
MESEAUTO Vol. 9 15
MESEK AZ IROGEPRÖL Vol. 9 323
MESEORSZAG Vol. 9 175
MESIC NAD REKOU Vol. 10 412, 503, 569
MESIM PER LINDEN Vol. 3 3, 5
MESON DE GITANO, EL Vol. 8 595
MESONJETORJA Vol. 3 22 Vol. 6 *page 13*
MESSA E FINITA, LA Vol. 7 137, 411, 524, 801
MESSAGER, LE Vol. 2 309, 440, 451, 526
MESSAGGIO TRA I FLUTTI, UN Vol. 7 1189
MESSALINA Vol. 2 32 Vol. 7 53, 196, 621, 756, 785, 1044
MESSALINA, MESSALINA Vol. 7 738
MESSE DOREE, LA Vol. 2 600
MESSIA, IL Vol. 7 137, 798
MESSIEURS LES RONDS-DE-CUIR Vol. 2 67, 137, 344, 722
MESSIEURS LUDOVIC Vol. 2 96, 156, 501
MESTECKO NA DLANI Vol. 10 109, 122, 580
MESTER – HYPNOTISOREN Vol. 5 83
MESTERDETEKTIVERNE Vol. 5 83
MESTIZA, LA Vol. 8 495, 503, 600
MESTIZO Vol. 8 306, 708
MESTO ME NADEJE Vol. 10 255, 337, 462
MESTO NIC NEVI Vol. 10 41, 517, 531
META 40 MERES Vol. 3 194
METAMORFÖSIS Vol. 3 132
METAMORFOSIS Vol. 8 317
METAMORPHOSE DES CLOPORTES, LA Vol. 2 395, 770
METELLO Vol. 7 567
METELY, A Vol. 9 77
METEORA Vol. 7 442
METIER DE FOUS Vol. 2 153, 158
METODE I GALSKABEN Vol. 5 83
METRACEK Vol. 10 47, 255, 286
METROPOLI Vol. 7 1105
METROPOLIS Vol. 4 178, 298, 325, 567, 727, 771
METROPOLITAIN Vol. 2 270, 395, 583, 698, 858
METTETE L'AVVOCATO Vol. 7 829
METTI LO DIAVOLO TUO NE MIO INFERNO Vol. 7 1022
METY WARSZAWY Vol. 1 317
MEU CASO, O Vol. 8 15, 19, 118
MEU NOME E ..., O Vol. 8 60
MEURTRE A ETE COMMIS, UN Vol. 2 152, 812
MEURTRE EN 45 TOURS Vol. 2 548, 866
MEURTRE EST UN MEURTRE, UN Vol. 2 368
MEURTRES Vol. 2 57, 128, 344
MEURTRIER, LE Vol. 2 251, 625

MEUS AMIGOS Vol. 8 7, 103
MEXIKO Vol. 5 97
MEZCZYZNA Vol. 1 174, 215
MEZCZYZNA NIEPOTRZEBNY ! Vol. 6 *page 217, page 225, page 228*
MEZCZYZNI NA WYSPIE Vol. 1 206, 227, 381
MEZDU DVAMATA Vol. 3 76, 126
MEZDU RELSITE Vol. 3 38, 81, 100 Vol. 6 *page 32*
MEZEI PROFETA Vol. 9 196, 250
MEZEK Vol. 10 180, 246, 389
MEZI NAMI KLUKY Vol. 10 161, 255, 462
MEZI NAMI ZLODEJI Vol. 10 296, 433, 522
MEZI ZEMI A NEBEM Vol. 10 359, 380, 385
MEZTELEN DIPLOMATA Vol. 9 67, 121, 326
MEZTELEN VAGY Vol. 9 67, 154, 318
MEZZANOTTE Vol. 7 796
MEZZO DESTRO MEZZO SINISTRO - 2 CALCIATORI SENZA PALLONE Vol. 7 1202
MEZZO MILIONE ED UN MARITO Vol. 7 941
MGLA Vol. 1 131, 142, 318 Vol. 6 *page 211, page 244*
MI ADORABLE ESCLAVA Vol. 8 495, 762
MI ADORABLE SECRETARIA Vol. 8 628
MI ADORADO JUAN Vol. 8 486, 732
MI AMIGO EL VAGABUNDO Vol. 8 214, 286
MI BUDAPESTÜNK, A Vol. 9 35
MI CALLE Vol. 8 207, 214
MI CANCION ES PARA TI Vol. 8 311, 396
MI CHIAMANO MIMI Vol. 7 569
MI ENEMIGO EL DOCTOR Vol. 8 572
MI ENEMIGO Y YO Vol. 8 396
MI FACCIA CAUSA Vol. 7 257
MI FACCIO LA BARCA Vol. 7 381, 599
MI FANTASTICA ESPOSA Vol. 8 475, 625
MI FÖLDÜNK, A Vol. 9 120, 195, 261
MI GENERAL Vol. 8 358
MI HANNO RUBATO LA MOGLIE Vol. 7 1189
MI HIJO NO ES LO QUE PARECE Vol. 8 667
MI LESZ VELED ESZTERKE ? Vol. 9 121, 149, 332
MI MANDA PICONE Vol. 7 318, 908
MI MARIDO Y SUS COMPLEJOS Vol. 8 553, 749
MI MUJER ES DOCTOR Vol. 8 276
MI MUJER ES MUY DECENTE DENTRO DE LO QUE CABE Vol. 8 438, 681
MI MUJER ME GUSTA MAS Vol. 8 221
MI NOCHE DE BODAS Vol. 8 207, 363, 630
MI PERMETTE, BABBO ! Vol. 7 1027
MI PROFESORA PARTICULAR Vol. 8 348
MI QUERIDA SENORITA Vol. 8 348, 534
MI TIO JACINTO Vol. 8 449, 725
MI ULTIMO TANGO Vol. 8 261
MI VIDA EN TUS MANOS Vol. 8 417, 674
MIA CANZONE AL VENTO, LA Vol. 7 493, 1054
MIA ELLENIDA STO PARISI Vol. 3 204
MIA FIA Vol. 7 493
MIA FIGLIA Vol. 7 173, 967
MIA GYNAIKA FEUGEI Vol. 3 132
MIA GYNAIKA, MIA AGAPE, MIA ZOE Vol. 3 165
MIA GYNAIKA STEN ANTISTASE Vol. 3 158, 163
MIA KYRIA STA MPOUZOUKIA Vol. 3 144, 205
MIA MOGLIE E UNA STREGA Vol. 7 30, 347, 1008
MIA MOGLIE SI DIVERTE Vol. 7 639, 756, 796
MIA MOGLIE SI E FIDANZATA Vol. 7 53
MIA MOGLIE TORNA A SCUOLA Vol. 7 383
MIA NIPOTE ... CLEMENTINA Vol. 7 131
MIA NONNA POLIZIOTTO Vol. 7 891
MIA PARTHENA GIA DOLOMA Vol. 3 132
MIA SIGNORA, LA Vol. 7 725
MIA VITA E'TUA, LA Vol. 7 448
MIA VITA SEI TU, LA Vol. 7 506, 725
MIA ZOE TEN EHOUME Vol. 3 169
MIAMI SUPERCOPS - I POLIZIOTTI DELL'8 A STRADA Vol. 7 585
MIARKA, LA FILLE A L'OURSE Vol. 2 395
MIASTECZKO Vol. 1 155, 281, 389 Vol. 6 *page 218, page 247*
MIASTO NIEUJARZMIONE Vol. 1 184, 250 Vol. 2 407
MICH DÜRSTET Vol. 1 17, 47, 56, 74
MICHAEL Vol. 4 178, 244
MICHE Vol. 2 41, 579
MICHEL STROGOFF Vol. 2 59, 100, 153, 488, 521, 574, 593, 833
MICHELE STROGOFF Vol. 7 619
MICHELINE Vol. 2 584
MICI NENI KET ELETE Vol. 9 117, 149
MICROFONO DEL SELENIO, IL Vol. 7 91
MICROFONO E VOSTRO, IL Vol. 7 16, 86
MICSODA EJSZAKA Vol. 9 83, 165, 314
MIDAS KIRALY Vol. 9 288, 315
MIDI-MINUIT Vol. 2 870
MIDNATSSOLEN Vol. 5 295
MIDNATSSOLENS SON Vol. 5 845, 1156
MIDNIGHT PARTY Vol. 2 139
MIDSOMMARDANSEN Vol. 5 1033
MIDT I EN JAZZTID Vol. 5 179, 199, 273

MIDT OM NATTEN **Vol. 5** 16, 111, 165, 195, 217
MIEDO **Vol. 8** 600, 749
MIEDO A SALIR DE NOCHE **Vol. 8** 331, 414
MIEDZY BRZEGAMI **Vol. 1** 179, 354
MIEHEKE **Vol. 5** 349, 485
MIEHEN KUNNIA **Vol. 5** 549, 592
MIEHEN KYLKILUU **Vol. 5** 349, 476
MIEHEN TIE **Vol. 5** 334, 625
MIEHEN VANKINA **Vol. 5** 476
MIEJSCE DLA JEDNEGO **Vol. 1** 135, 202, 227, 354
MIEJSCE NA ZIEMI **Vol. 1** 171, 179, 294 **Vol. 6** *page 256*
MIEKAN TERÄLLÄ **Vol. 5** 420, 476
MIEL DEL DIABLO, LA **Vol. 8** 302, 762
MIEL, LA **Vol. 8** 297, 438
MIELE DEL DIAVOLO, IL **Vol. 8** 302, 762
MIELÖTT BEFEJEZI RÖPTET A DENEVER **Vol. 9** 92, 147, 240
MIENTRAS LA ALDEA DUERME **Vol. 8** 604
MIERT ? **Vol. 9** 68, 71
MIERT ROSSZAK A MAGYAR FILMEK ? **Vol. 9** 67, 117, 149
MIES ES MUCHA, LA **Vol. 8** 752
MIES JA HÄNEN OMATUNTONSA **Vol. 5** 328
MIES, JOKA EI OSANNUT SANOA EI **Vol. 5** 500, 541
MIES MARSELLESTA **Vol. 5** 344, 359
MIES TÄLTÄ TÄHDELTA **Vol. 5** 549, 579
MIESMALLI **Vol. 5** 349, 566, 578
MIESTÄ EI VOI RAISKATA **Vol. 5** 327, 358, 522, 619
MIETER SCHULZE GEGEN ALLE **Vol. 4** 626, 747
MIEUX VAUT ETRE RICHE ET BIEN PORTANT QUE FAUCHE ET MAL FOUTU **Vol. 2** 294
MIEZE VON BOLLE, DIE **Vol. 4** 342
MIEZUL FIERBINTE AL PIINII **Vol. 6** *page 277, page 280, page 283*
MIG OG CHARLY **Vol. 5** 12, 125, 179, 231
MIG OG DIG **Vol. 5** 88
MIG OG MAFIAEN **Vol. 5** 35, 94, 195
MIG OG MIN LILLEBROR **Vol. 5** 1, 76, 261
MIG OG MIN LILLEBROR OG BOLLE **Vol. 5** 236, 306
MIG OG MIN LILLEBROR – OG STORSMUGLERNE **Vol. 5** 16, 61, 272
MIGLIARI, I **Vol. 7** 94
MIGNON **Vol. 4** 274, 377 **Vol. 7** 622
MIGNONNE **Vol. 2** 159, 286, 330
MIGOVE V KIBRITENA KUTIJKA **Vol. 3** 121 **Vol. 6** *page 38*
MIGUELIN **Vol. 8** 311, 402
MIGUELON **Vol. 8** 357, 427, 542, 604
MIHAI VITEAZUL **Vol. 3** 230, 254, 325, 329
MIHAIL, CITNE DE CIRC **Vol. 3** 256, 292, 317
MIHAIL STROGOV **Vol. 3** 31, 109
MIJLOACAS LA DESCHIDERE **Vol. 3** 240, 269
MIKÄ YÖ ! **Vol. 5** 493, 566, 592
MIKOLA A MIKOLKO **Vol. 10** 72, 73, 155, 188
MIKOLAS ALES **Vol. 10** 412, 503, 569
MIKOR A SZÖLÖ ERIK **Vol. 9** 28
MIKOSCH RÜCKT EIN **Vol. 4** 239, 444, 452
MIKRES AFRODITES **Vol. 3** 155, 204
MIL GRITOS TIENE LA NOCHE **Vol. 8** 237, 524
MIL MILLONES PARA UNA RUBIA **Vol. 8** 450, 719
MILACEK PLUKU **Vol. 10** 138, 586
MILAGRO DEL CANTE, EL **Vol. 8** 249, 626
MILAGRO DEL CRISTO DE LA VEGA, EL **Vol. 8** 628
MILAGRO DEL SACRISTAN, EL **Vol. 8** 551, 749
MILAGRO EN LA CIUDAD **Vol. 8** 537, 753
MILAK, DER GRÖNLANDJÄGER **Vol. 4** 10, 15, 55
MILAN RASTISLAV STEFANIK **Vol. 10** 58, 296, 453, 580
MILANO '83 **Vol. 7** 835
MILANO MILLARDARIA **Vol. 7** 333
MILANO ROVENTE **Vol. 7** 221
MILANO VIOLENTA **Vol. 7** 1022
MILCZACE SLADY **Vol. 1** 149, 303, 355 **Vol. 6** *page 205*
MILCZENIE **Vol. 1** 237, 298, 408 **Vol. 6** *page 242*
MILENCI V ROCE JEDNA **Vol. 10** 87, 172, 233
MILENKY STAREHO KRIMINALNIKA **Vol. 10** 167, 345
MILIEU DU MONDE **Vol. 2** 284
MILIJUNI NA OTOKU **Vol. 3** 341, 452
MILION DE MADIGAN, EL **Vol. 8** 681
MILION ZA LAURE **Vol. 1** 142, 401 **Vol. 6** *page 253*
MILIONAR PENTRU O ZI **Vol. 3** 212
MILIONARIO, O **Vol. 8** 131, 185
MILIONE DI KADWING, IL **Vol. 7** 1118
MILIONE, IL **Vol. 7** 168
MILIONER **Vol. 1** 203, 206, 318
MILIONI DELLA GITANA, I **Vol. 7** 1090
MILIONI DELLA ZINGARA, I **Vol. 7** 91, 980

MILIONI DI SAETTA, I Vol. 7 965
MILITANTI Vol. 6 *page 12, page 17*
MILITARE E MEZZO, UN Vol. 7 1027
MILITÄRISCHE DISZIPLIN Vol. 4 646
MILIZIA TERRITORIALE Vol. 7 709, 730
MILJONÄÄRIMONNI Vol. 5 478, 613, 634
MILJONÄR FÖR EN DAG Vol. 5 1034
MILJOONALIIGA Vol. 5 452
MILJOONAVAILLINKI Vol. 5 443, 549
MILKA Vol. 5 381, 465, 651
MILLARD DANS UN BILLARD, UN Vol. 2 128, 481
MILLE CHILOMETRI AL MINUTO Vol. 7 1036
MILLE E UNA NOTTE ALL'ITALIANA, LE Vol. 7 852
MILLE ET DEUXIEME NUIT, LA Vol. 2 100, 339
MILLE ET UNE NUITS, LES Vol. 2 623
MILLE ET UNE PERVERSIONS DE FELICIA, LES Vol. 2 294
MILLE, I Vol. 7 1189
MILLE LIRE AL MESE Vol. 7 812, 1054
MILLE, MARIE OG MIG Vol. 5 101, 208, 228
MILLE MILLIARDS DE DOLLARS Vol. 6 *page 91, page 96*
MILLE PATTES FAIT DES CLAQUETTES, LE Vol. 2 97, 815
MILLEPIEDI, I Vol. 7 654
MILLERS DOKUMENT Vol. 5 925, 937
MILLIARDENSOUPER, DAS Vol. 4 511, 584
MILLIARDI, CHE FOLLIA ! Vol. 7 756, 1027, 1177
MILLIARDI, I Vol. 7 833
MILLIEME FENETRE, LA Vol. 2 515, 655
MILLION DOLLAR DOLLIES Vol. 2 384
MILLION, LE Vol. 2 81, 344, 573, 656, 708
MILLIONAER FOR EN AFEN Vol. 5 791, 800, 826
MILLIONAERDRENGEN Vol. 5 24, 58, 136
MILLIONEN DER YVETTE, DIE Vol. 1 72, 81, 129
MILLIONEN-MIENE, DIE Vol. 4 734
MILLIONENMÄDEL, DAS Vol. 4 222, 560
MILLIONENRAUB IM RIVIERAEXPRESS, DER Vol. 4 119, 573, 734
MILLIONENTESTAMENT, DAS Vol. 4 246, 462
MILLIONNAIERS D'UN JOUR Vol. 2 158, 368
MILLIONS DE L'ONCLE JAMES, LES Vol. 2 50
MILLIPILLERI Vol. 5 479, 606
MILLON EN LA BASURA, UN Vol. 8 524
MILLON POR TU HISTORIA, UN Vol. 8 471
MILLONA, LA Vol. 8 726
MILLONARIO POR UN DIA Vol. 8 214, 762
MILLONES DE POLICHINELA, LOS Vol. 8 436
MILLY, MARIA OCH JAG Vol. 5 899
MILORD L'ARSOUILLE Vol. 2 9, 226
MILOSC CI WSZYSTKO WYBACZY Vol. 6 *page 230, page 231, page 239, page 245*
MILOSC DWUDZIESTOLATKOW Vol. 1 148
MILOSC I SMIERC Vol. 1 317
MILOSC PRZEZ OGIEN I KREW Vol. 1 174, 404
MILOSRDNY CAS Vol. 10 228, 470, 483
MILOST Z ZIVITE Vol. 6 *page 37, page 42, page 64*
MILOVANI ZAKAZANO Vol. 10 460, 621
MILUJEME Vol. 10 179, 345
MIMI BLUETTE ... FIORE DEI MIO GIARDINO Vol. 7 347
MIMI E GLI STRACCIONI Vol. 7 731
MIMI PINSON Vol. 2 145, 385, 475, 743, 866
MIMOSA-SAN, BAMBOLA GIAPPONESE Vol. 7 441
MIMOZA LLASTICA Vol. 3 5, 23, 24
MIN ÄLSKADE Vol. 5 850, 946, 1093
MIN DATTER NELLY Vol. 5 89, 281
MIN FARMORS HUS Vol. 5 175, 177, 185
MIN FRU HAR EN FÄSTMAN Vol. 5 868, 870
MIN KÄRA ÄR EN ROS Vol. 5 833, 940
MIN KONE ENA PARIS Vol. 5 39
MIN KONE ER HUSAR Vol. 5 55, 101, 208
MIN KONE ER USKYLDIG Vol. 5 7, 89, 153
MIN KONE FRA PARIS Vol. 5 89, 281
MIN KONES FERIE Vol. 5 96, 108, 275, 304
MIN MARION Vol. 5 797, 800, 824
MIN OG MIN FAMILIE Vol. 5 180, 199, 201
MIN ONKEL GENERALKONSULEN Vol. 5 83
MIN SON PETER Vol. 5 2, 23, 228
MIN SOSTERS BORN Vol. 5 199, 201
MIN SOSTERS BORN NAR DE ER VAERST Vol. 5 199, 236, 261
MIN SOSTERS BORN PA BRYLLUP SREJSE Vol. 5 96, 199, 201, 275
MIN SOSTERS BORN VAELTER BYEN Vol. 5 128, 199, 261, 275
MIN SVÄRMOR – DANSÖSEN Vol. 5 928, 1032
MIN SVIGERINDE FRA AMERIKA Vol. 5 83
MIN SYSTER OCH JAG Vol. 5 871, 1156

MIN VÄN BALTHAZAR Vol. 5 898, 901
MIN VÄN OSCAR Vol. 5 1063, 1188
MIN VEN, LEVY Vol. 5 58
MIN VEN PRIVATDETEKTIVEN Vol. 5 157, 228
MINA DRÖMMARS STAD Vol. 5 1052, 1086, 1093
MINÄ ELÄN Vol. 5 552, 600
MINÄ JA MIEHENI MORSIAN Vol. 5 379, 600
MINÄ JA MINISTERI Vol. 5 359, 485
MINÄ JÄTÄN SINUT Vol. 5 370, 634
MINA, LA Vol. 7 1044
MINÄ SOITAN SINULLE ILLALLA Vol. 5 374, 443
MINÄKÖ ISÄ ! Vol. 5 379, 392
MINDEN KEZDET NEHEZ Vol. 9 134, 141, 259
MINDEN SZERDAN Vol. 9 20, 84, 238
MINDENKI ARTATLAN ? Vol. 9 121, 149, 314
MINDENKI MAST SZERET Vol. 9 163, 164, 196
MINDENNAP ELÜNK Vol. 9 83, 149, 332
MINDENT A NÖERT Vol. 9 71
MINDERJÄHRIGE, DIE Vol. 4 219, 672
MINDRIE Vol. 3 218, 281, 285, 310
MINE KAERE KONER Vol. 5 2, 228
MINE TOSSEDE DRENGE Vol. 5 241, 248
MINESTRONE, IL Vol. 7 434, 1093
MINETTES A TOUT FAIRE Vol. 2 814
MINETTES BRULANTES, LES Vol. 2 14
MING, RAGAZZI ! Vol. 7 1147
MINGEA Vol. 3 261, 263, 277
MINI TIA, LA Vol. 8 317, 768
MINIA Vol. 3 239, 254, 255
MINISTERN Vol. 5 918, 933, 1093
MINISZTER BARATJA, A Vol. 9 71, 274
MINKITURKKI Vol. 5 443, 639
MINNA VON BARNHELM Vol. 1 26, 90, 104, 115
MINNA VON BARNHELM ODER DAS SOLDATENGLÜCK Vol. 1 2
MINNESOTA CLAY Vol. 8 207, 311
MINOUCHE L'INSATIABLE Vol. 6 *page 101, page 118*
MINUIT CHAMPS-ELYSEES Vol. 2 243, 376
MINUIT, PLACE PIGALLE Vol. 2 217, 413, 528, 603, 727
MINUIT QUAI DE BERCY Vol. 2 401, 402, 480
MINULOST JANY KOSINOVE Vol. 10 154, 345
MINUTA ZA UMOR Vol. 3 345, 387
MINUTE DE VERITE, LA Vol. 2 488, 676
MINUTE PAPILLON Vol. 2 344, 376, 743
MINUTE VOR ZWÖLF, EINE Vol. 4 535
MIO AMANTE, IL Vol. 7 196
MIO AMICO JEKILL, IL Vol. 7 1147
MIO AMORE, IL Vol. 7 1189
MIO ANTROPOFAGO, IL Vol. 7 34, 437
MIO CADAVERE, IL Vol. 7 399, 929
MIO CARCERE, IL Vol. 7 968
MIO CARO ASSASSINO Vol. 8 311, 681
MIO DIARIO DI GUERRA, IL Vol. 7 301, 304
MIO FIGLIO NERONE Vol. 7 106
MIO FIGLIO PROFESSORE Vol. 7 756, 796
MIO GREGGE, IL Vol. 7 782, 783
MIO ZIO BRABASSOUS Vol. 7 1071
MIOCHE, LE Vol. 2 8, 427, 644, 775
MIODOWE MIESIACE Z PRZESZKODAMI Vol. 1 263
MIQUETTE Vol. 2 41, 152
MIQUETTE ET SA MERE Vol. 2 237, 332, 357, 412, 658, 824, 863
MIR NACH, CANAILLEN Vol. 1 46, 92
MIRABILE VISIONE, LA Vol. 7 254
MIRACLE DE LOURDES, LE Vol. 2 111
MIRACLE DES LOUPS, LE Vol. 2 65, 339, 368, 515, 547, 657
MIRACLES N'ONT LIEU QU'UNE FOIS, LES Vol. 2 156, 407, 837
MIRACOLI A MILANO Vol. 7 21, 85, 299, 453, 1177
MIRACOLINI Vol. 7 102, 263
MIRACOLO A VIGGIU Vol. 7 1036
MIRACOLO D'AMORE Vol. 7 1005
MIRACOLO DELLA MADONNA DI POMPEI, IL Vol. 7 829
MIRACOLO DELL'AMORE, IL Vol. 7 984
MIRACOLO, IL Vol. 7 107, 1136
MIRAGE DE PARIS, LE Vol. 2 167
MIRAGES Vol. 2 153
MIRAGES DE PARIS Vol. 2 23, 48, 55, 64
MIRAGGIO, IL Vol. 7 559
MIRAKEL DER LIEBE Vol. 4 77
MIRAKELTJERNEN Vol. 5 83
MIRAKLET Vol. 5 996
MIRANDA Vol. 7 31, 142, 585, 590
MIRCHA Vol. 9 71, 322
MIRE MEGVENÜLÜNK Vol. 9 172
MIREASA DIN TREN Vol. 3 218, 275, 319
MIREASMA PLOILOR TIRZII Vol. 6 *page 271, page 283*
MIRELE EFROS Vol. 1 240, 317
MIRIAM Vol. 5 443, 643
MIRIS DUNJA Vol. 6 *page 311, page 322*
MIRIS POLJSKOG CVECA Vol. 3 349, 454
MIRIS ZEMLJE Vol. 6 *page 222, page 238*
MIRISI, ZLATO I TAMJAN Vol. 3 375, 446
MIRKO I SLAVKO Vol. 3 370, 424
MIRNO LETO Vol. 3 419, 443
MIROIR Vol. 2 92, 402, 863
MIROIR A DEUX FACES, LE Vol. 2 197

MIRTIL Vol. 7 493
MIRYAM Vol. 7 326, 796
MIS Vol. 6 *page 208, page 237, page 239*
MIS RELACIONES CON ANA Vol. 8 414
MIS STON Vol. 3 440, 443
MISE A SAC Vol. 2 163, 517
MISERABILI „CACCIA ALL'UOMO", I Vol. 7 283, 453, 638
MISERABILI „TEMPESTA SU PARIGI", I Vol. 7 283, 453, 638
MISERABLES, LES Vol. 2 42, 158, 195, 248, 282, 307, 440, 452, 584, 625, 657, 676, 697, 832 Vol. 6 *page 85, page 88, page 94, page 139, page 145*
MISERIA E NOBILTA Vol. 7 665, 725, 993, 1102, 1147
MISERIE DEL SIGNOR TRAVET, LE Vol. 7 443, 756, 830, 1116
MISION ARENAS ARDIENZES Vol. 8 708
MISION BLANCA Vol. 2 427 Vol. 8 298, 311, 484, 628
MISION EN EL ESTRECHO Vol. 8 311, 611
MISION EN GINEBRA Vol. 8 562, 732
MISION EN MARRUECOS Vol. 8 611, 772
MISION ESPECIAL EN CARACAS Vol. 8 298
MISION EXTRAVAGANTE Vol. 8 515, 753
MISION LISBOA Vol. 8 725
MISION SECRETA EN EL CARIBE Vol. 8 515
MISS AND MRS. SWEDEN Vol. 5 918, 940, 1042, 1053
MISS ARIZONA Vol. 9 10, 72, 82, 251, 253, 254, 279
MISS BERYLL... DIE LAUNE EINES MILLIONÄRS Vol. 4 208, 390
MISS CARIBE Vol. 8 703, 736
MISS CATASTROPHE Vol. 2 293, 294, 401
MISS CUPLE Vol. 8 290, 493
MISS DEMONIO Vol. 7 577, 806
MISS DOLLAR Vol. 7 250, 655
MISS DOROTHY Vol. 7 277
MISS EDITH, DUCHESSE Vol. 2 392
MISS ELYETT Vol. 2 727
MISS EUROOPPAA METSÄSTÄMÄSSÄ Vol. 5 334, 438, 535, 572, 605
MISS EVELYNE, DIE BADEFEE Vol. 4 106 Vol. 6 *page 184*
MISS FLUFFY RUFFLES Vol. 7 196
MISS HELYETT Vol. 2 533
MISS LILLY ... PARDON Vol. 7 1206
MISS MUERTE Vol. 8 338, 762
MISS PIGALLE Vol. 2 703
MISS ROBINSON Vol. 7 454
MISS ROVEL Vol. 2 584
MISS SARA SAMPSON Vol. 4 429, 598
MISS TROUDUC Vol. 6 *page 120*
MISSION A TANGER Vol. 2 158, 368
MISSION SPECIALE Vol. 2 61, 131, 595, 758
MISSION SPECIALE A CARACAS Vol. 2 475, 661 Vol. 8 298
MISSIONE SPECIALE LADY CHAPLIN Vol. 8 438, 762
MISSIONNAIRE, LE Vol. 2 173
MISSIONNAIRE, UN Vol. 2 559, 671, 719
MISTENKA BEZ NAVRATU Vol. 10 154, 433, 503
MISTER DINAMITA, MANANA OS BESARA LA MUERTE Vol. 8 426, 768
MISTER DYNAMIT – MORGEN KÜSST EUCH DER TOD Vol. 8 426, 768
MISTER FLOW Vol. 2 339, 824
MISTER FREEDOM Vol. 2 263, 517
MISTER RADIO Vol. 4 223, 270, 384
MISTERELE BUCURESTILOR Vol. 6 *page 265, page 275*
MISTERI DELLA GIUNGLA NERA, I Vol. 7 694
MISTERI DELLA JUNGLA NERA, I Vol. 7 372
MISTERI DI MEZZANOTTE, I Vol. 7 295, 899
MISTERI DI PARIGI, I Vol. 7 493, 1037
MISTERI DI TERRA E DI MARE Vol. 7 1174
MISTERI DI VENEZIA, I Vol. 7 1035
MISTERIO DE LA PUERTA DEL SOL, EL Vol. 8 357
MISTERIO DE LA VIDA, EL Vol. 8 732
MISTERIO DELLE 12 E 35 Vol. 7 37
MISTERIO EN LA ISLA DE LOS MONSTRUOS Vol. 8 274
MISTERIO EN LA MARISMA Vol. 2 646 Vol. 8 607, 628, 725
MISTERIOS DE LA IMPERIAL TOLEDO, LOS Vol. 8 516
MISTERIOS DE TANGER, LOS Vol. 8 246
MISTERIOSA DAMA DALLE BIANCHE ROSE, LA Vol. 7 993
MISTERIOSO SENOR VAN EYCK, EL Vol. 8 450
MISTERO DEI BAULI NERI, IL Vol. 7 980
MISTERO DEL CASTELLO DI NANCY, IL Vol. 7 198
MISTERO DEL GIRL'S BAR Vol. 7 81
MISTERO DEL GRANDE ESPERO, IL Vol. 7 91
MISTERO DEL MORCA, IL Vol. 7 8, 178, 819
MISTERO DEL TEMPIO INDIANO, IL Vol. 7 1136
MISTERO DELL AMERICANO, IL Vol. 7 980
MISTERO DELLA CASA DIRIMPETTO, IL Vol. 7 196

MISTERO DELLA CASA N. 30, IL **Vol. 7** 796
MISTERO DELLA CORONA, IL **Vol. 7** 704
MISTERO DELLA DAMA VELATA, IL **Vol. 7** 81
MISTERO DELLA MANO, IL **Vol. 7** 1186
MISTERO DELLA SINDONE, IL **Vol. 7** 422
MISTERO DELL'ARTIDE, IL **Vol. 7** 1162
MISTERO DELL'ASSO DI PICCHE, IL **Vol. 7** 1104
MISTERO DELL'AUTO IN FIAMME, IL **Vol. 7** 1186
MISTERO DELL'UOMO GRIGIO, IL **Vol. 7** 91, 980
MISTERO DI BELLAVISTA, IL **Vol. 7** 383, 627
MISTERO DI BERNARDO BROWN, IL **Vol. 7** 751
MISTERO DI JACK HILTON, IL **Vol. 7** 1118
MISTERO DI OBERWALD, IL **Vol. 7** 112, 346, 564, 1039, 1145
MISTERO DI OSIRIS, IL **Vol. 7** 531
MISTERO DI QUELLA NOTTE, IL **Vol. 7** 264
MISTERO DI ROCCABRUNA **Vol. 7** 26
MISTERO DI UNA NOTTE DI PRIMAVERA, IL **Vol. 7** 22
MISTERO DI VILLA NIRVANA, IL **Vol. 7** 932
MISTERO IN CASA DEL DOTTORE **Vol. 7** 159
MISTERO O FOLLIA ? **Vol. 7** 327
MISTERUL LUI HERODOT **Vol. 3** 276, 289, 321
MISTICO, EL **Vol. 8** 572
MISTIGRI **Vol. 2** 803
MISTO V HOUFU **Vol. 10** 87, 131, 197, 299, 394, 517
MISTRAL, LE **Vol. 2** 205, 698
MIT CSINALT FELSEGED 3 – TOL 5 – IG? **Vol. 9** 134, 215, 312
MIT DEN AUGEN EINER FRAU **Vol. 4** 553, 566, 608
MIT DIR DURCH DICK UND DÜNN **Vol. 4** 344, 647, 792
MIT FAEDRELAND, MIN KAERLIGHED **Vol. 5** 295
MIT LIV ER MUSIK **Vol. 5** 89
MIT MEINEN AUGEN **Vol. 4** 65, 422, 619
MIT MIR NICHT, MADAM ! **Vol. 1** 48, 57, 60
MIT UNS DAS VOLK **Vol. 4** 761
MIT VERSIEGELTER ORDER **Vol. 4** 124, 232, 354
MIT VOLLGAS IN DIE EHE **Vol. 4** 410, 558
MITAD DEL CIELO, LA **Vol. 8** 219, 771
MITÄS ME SANKARIT **Vol. 5** 314, 560
MITÄS ME TAITEILIJAT **Vol. 5** 461, 592
MITGIFTJÄGER **Vol. 4** 436
MITICA POPESCU **Vol. 6** *page 277, page 281, page 282*
MITREA COCOR **Vol. 3** 217, 221, 228, 287, 299
MITSOU **Vol. 2** 131, 368, 516
MITT FOLK ÄR DITT **Vol. 5** 1052, 1164
MITT HEM ÄR COPACABANA **Vol. 5** 1167
MITTERNACHT **Vol. 4** 159, 686
MITTERNACHTS-TAXE, DIE **Vol. 4** 119, 339, 354, 437
MITTERNACHTSZUG, DER **Vol. 4** 273, 340, 420
MIUDO DA BICA, O **Vol. 8** 54
MIZAR **Vol. 7** 113, 915
MIZZICA ... MA CHE E PROIBITISSIMO **Vol. 7** 713
MJARKA ZA NEOTKLOENIE **Vol. 6** *page 39, page 40, page 57*
MLAD I ZDRAV KAO RUZA **Vol. 3** 380
MLADA LETA **Vol. 10** 412, 503, 569
MLADE SRDCIA **Vol. 10** 179, 273, 287, 345
MLADE VINO **Vol. 10** 23, 41, 577
MLADI GODINI **Vol. 6** *page 23, page 62*
MLADY MUZ A BILA VELRYBA **Vol. 10** 41, 496, 517
M"LCALIVITE P"TEKI **Vol. 3** 38, 73
MLCENI MUZU **Vol. 10** 250, 369
MLHY NA BLATECH **Vol. 10** 412, 530, 569, 621
MLODOSC CHOPINA **Vol. 1** 250, 307, 366
MLODOSC ZWYCIEZA **Vol. 1** 174, 404
MLODY LAS **Vol. 1** 278, 308, 404
MLYNAR A JEHO DITE **Vol. 10** 215, 452
MNICHOVO SRDCE **Vol. 10** 122, 257, 605
MNIEJSZE NIEBO **Vol. 6** *page 217, page 241, page 244*
MNIEJSZY SZUKA DUZEGO **Vol. 1** 254, 355 **Vol. 6** *page 242*
MOANA ISOLA DEL SOGNO **Vol. 7** 113
MOARA CU NOROC **Vol. 3** 244, 260, 329, 331
MOARA LUI CALIFAR **Vol. 6** *page 275*
MOARTEA LUI JOE-INDIANUL **Vol. 3** 260, 317, 323
MOBILA STIL **Vol. 6** *page 271, page 280*
MÖBLIERTE ZIMMER **Vol. 4** 255, 599
MOCNE UDERZENIE **Vol. 1** 211, 305, 388
MOCNY CZLOWIEK **Vol. 1** 175, 309
MOD LYSET **Vol. 5** 295
MOD MIG PA CASSIOPEIA **Vol. 5** 31, 39, 241, 248, 272
MOD OG MANDSHJERTE **Vol. 5** 2, 23
MODÄRNA FRUAR **Vol. 5** 899, 979, 1010
MODCHE A REZI **Vol. 10** 257, 453
MODE, ET **Vol. 5** 97

MODELL BIANCA Vol. 1 54, 97
MODELL LOU, DER LEBENSWEG EINES HUTES Vol. 4 517, 525, 558
MODELLA DI TIZIANO, LA Vol. 7 306
MODELLA, LA Vol. 7 266
MODELLE DI VIA MARGUTTA, LE Vol. 7 1027
MODELO Vol. 3 170
MODERATO CANTABILE Vol. 2 21, 824
MODERCA ZOSTAWIA SLAD Vol. 6 *page 216*
MODEREN Vol. 5 295
MODERN SALOME, A Vol. 2 384 Vol. 2 579
MODERNE AEGTESKAB Vol. 5 81
MODERNE CASANOVA, DER Vol. 4 287, 384, 458, 646
MODERNE DON JUAN, EN Vol. 5 81
MODERNE EHEN Vol. 4 31, 581
MODERNE LANDEVEJSRIDDER, EN Vol. 5 83
MODERNE MITGIFT Vol. 4 246, 298, 321, 703
MODERNE PIRATEN Vol. 4 166, 208
MODERNE SKLAVEN Vol. 4 385
MODERNE SOHELT, EN Vol. 5 97
MODERNE TÖCHTER Vol. 4 424
MODERNI MAGDALENA Vol. 10 167, 536
MODERNI MOSCHETTIERI, I Vol. 7 428
MODERS KAMP MED DODEN, EN Vol. 5 97
MODERSKAPETS KVAL OCH LYCKA Vol. 5 876
MODIFICATION, LA Vol. 2 860
MODIGA MINDRE MAN Vol. 5 862
MODO DI ESSERE DONNA, UN Vol. 7 514
MODRA PLANETA Vol. 10 18, 369, 517
MODRE Z NEBE Vol. 10 395, 458, 577
MODRY DEMANT Vol. 10 58
MODRY ZAVOJ Vol. 10 168, 564, 621
MOFTURI 1900 Vol. 3 231, 329
MOGILA NIEZNANEGO ZOLNIERZA Vol. 1 174, 273, 380
MOGLI E LE ARANCE, LE Vol. 7 254, 1004
MOGLI PERICOLOSE Vol. 7 817
MOGLIAMANTE Vol. 7 567
MOGLIE AMERICANA, UNA Vol. 7 478, 502
MOGLIE BELLE, LA Vol. 7 390
MOGLIE CHE SI GETTO DALLA FINESTRA, LA Vol. 7 436, 569
MOGLIE DEL MIO CLIENTE, LA Vol. 7 1189
MOGLIE DELL'AMICO... E SEMPRE PIU BUONA, LA Vol. 7 141
MOGLIE DI CLADIO, LA Vol. 7 369
MOGLIE DI MIO MARITO, LA Vol. 7 877
MOGLIE DI MIO PADRE, LA Vol. 7 1179
MOGLIE DI SUA ECCELENZA, LA Vol. 7 288
MOGLIE, DUE AMICI, QUATTRO AMANTI, UNA Vol. 7 245 Vol. 8 624, 770
MOGLIE E BUOI ... Vol. 7 634, 910
MOGLIE E UGUALE PER TUTTI, LA Vol. 7 798
MOGLIE IN BIANCO ... L'AMANTE, AL PEPE, LA Vol. 7 770
MOGLIE IN CASTIGO, LA Vol. 7 890
MOGLIE IN PERICOLO, UNA Vol. 7 196, 431, 902
MOGLIE IN VACANZA ... L'AMANTE IN CITTA, LA Vol. 7 116, 426
MOGLIE POSTICCIA, LA Vol. 7 1189
MOGLIE SCACCIATA, LA Vol. 7 984, 1079
MOGLIE VERGINE, LA Vol. 7 1213
MOHICANI DI PARIGI, I Vol. 7 1005
MOHR UND DIE RABEN VON LONDON Vol. 1 8, 53, 71
MOI ET LE COLONEL Vol. 2 190
MOI ET LES HOMMES DE QUARANTE ANS Vol. 2 322, 503
MOI, FLEUR BLEUE Vol. 2 198, 746
MOI, PIERRE RIVIERE, AYANT EGORGE MA MERE, MA SOEUR ET MON FRERE... Vol. 2 220, 321, 842
MOI RODZICE ROZWODZA SIE Vol. 1 194, 278
MOI, Y EN A VOULOIR DES SOUS Vol. 2 110, 263
MOINE, LE Vol. 2 251, 854
MOINEAUX DE PARIS Vol. 2 395
MOIS LE PLUS BEAU, LE Vol. 2 288, 504
MOISE ET SALOMON PARFUMEURS Vol. 2 113, 152, 330, 437
MOISSON SERA BELLE, LA Vol. 2 811
MOJ Vol. 4 660, 704
MOJ DOM Vol. 6 *page 253*
MOJ DRUGI OZENEK Vol. 1 149, 232, 303, 355
MOJ STARY Vol. 1 321, 398 Vol. 6 *page 246*
MOJ TATA NA ODREDJENO VREME Vol. 6 *page 317, page 318*
MOJA LUDA GLAVA Vol. 3 370, 408
MOJA STRANA SVIJETA Vol. 3 351, 371, 450
MOJA WOJNA, MOJA MILOSC Vol. 1 156, 325 Vol. 6 *page 250*
MOLIERE Vol. 2 789, 879
MOLINOS DE VIENTO Vol. 8 516
MOLJAC Vol. 6 *page 306, page 309, page 321*
MOLLEN Vol. 5 2, 97, 228, 290
MOLLENARD Vol. 2 13, 778, 837
MOLLERENS DATTER Vol. 5 203

MOLLY-FAMILJE FLICKMAN **Vol. 5** 989
MOLN ÖVER HELLESTA **Vol. 5** 933, 1073
MOLO **Vol. 1** 179, 220, 408
MOLOKAI **Vol. 8** 214, 276
MOLSKIS, SANOI EEMELI, MOLSKIS **Vol. 5** 639
MOLTI SOGNI PER LE STRADE **Vol. 7** 190, 1136
MOMCETO SI OTIVA **Vol. 3** 102 **Vol. 6** *page 26*
MOME AUX BOUTONS, LA **Vol. 2** 354
MOME PIGALLE, LA **Vol. 2** 118, 309
MOME VERT-DE-GRIS, LA **Vol. 2** 501, 615, 709
MOMENTO DE LA VERDAS, EL **Vol. 8** 450
MOMENTO DELLA VERITA, IL **Vol. 7** 869, 1025, 1167 **Vol. 8** 450
MOMENTO DELL'AVVENTURA, IL **Vol. 7** 909, 1084
MOMENTO PIU BELLO, IL **Vol. 7** 1147
MOMENTS **Vol. 2** 450
MOMIA NACIONAL, LA **Vol. 8** 346, 627
MOMICETO I ZMEJAT **Vol. 6** *page 61*
MOMINA SKALA **Vol. 3** 67
MOMMILAN VERITEOT 1917 **Vol. 5** 416, 434, 649
MON AMANT L'ASSASSIN **Vol. 2** 81, 308, 687
MON AMI LE CAMBRIOLEUR **Vol. 2** 67, 131, 856
MON AMI SAINFOIN **Vol. 2** 226, 231, 247, 615
MON AMI TIM **Vol. 2** 217, 727
MON AMI VICTOR ! **Vol. 2** 330, 407, 656
MON AMOUR EST PRES DE TOI **Vol. 2** 23, 67
MON AMOUR, MON AMOUR **Vol. 2** 441
MON BEGUIN **Vol. 2** 731, 775
MON COEUR AU RALENTI **Vol. 2** 538
MON COEUR BALANCE **Vol. 2** 84
MON COEUR EST ROUGE **Vol. 2** 324, 633
MON COQUIN DE PERE **Vol. 2** 247, 401, 517, 857
MON COUSIN ALBERT **Vol. 2** 871
MON CURE, CHAMPION DU REGIMENT **Vol. 2** 717
MON CURE CHEZ LE THAILANDAISES **Vol. 6** *page 73, page 79*
MON CURE CHEZ LES NUDISTES **Vol. 6** *page 73*
MON CURE CHEZ LES PAUVRES **Vol. 2** 137, 153, 245, 702
MON CURE CHEZ LES RICHES **Vol. 2** 32, 137, 197, 245, 528
MON DEPUTE ET SA FEMME **Vol. 2** 308, 366
MON FRANGIN DU SENEGAL **Vol. 2** 128, 247
MON GOSSE DE PERE **Vol. 2** 37, 153, 193, 197, 254, 424
MON MARI EST MERVEILLEUX **Vol. 2** 158, 368
MON ONCLE **Vol. 2** 124, 743, 776, 805
MON ONCLE BENJAMIN **Vol. 2** 339, 457, 513, 731
MON ONCLE D'AMERIQUE **Vol. 2** 770, 854 **Vol. 6** *page 74*
MON ONCLE DU TEXAS **Vol. 2** 294, 344
MON ONCLE ET MON CURE **Vol. 2** 249, 318, 726, 871
MON PERE AVAIT RAISON **Vol. 2** 84
MON PHOQUE ET ELLES **Vol. 2** 615, 833
MON POTE LE GITAN **Vol. 2** 475, 740
MON PREMIER AMOUR **Vol. 2** 879
MONA JA PALAVAN RAKKAUDEN AIKA **Vol. 5** 492, 511, 519
MONACA DI MONZA, LA **Vol. 7** 360
MONACA, LA **Vol. 7** 910
MONACHE DI S. ARCANGELO, LE **Vol. 7** 1007
MONACO DI MONZA, IL **Vol. 7** 1054
MONASTERO DI SANTA CHIARA **Vol. 7** 923
MÖNCH VON SANTAREM, DER **Vol. 4** 377, 681
MÖNCH VON ST.BARTHOLOMÄ, DER **Vol. 4** 25, 571
MONDE DU SILENCE, LE **Vol. 2** 781
MONDE EN ARMES, LE **Vol. 2** 122, 585
MONDE ETAIT PLEIN DE COULEURS, LE **Vol. 2** 742
MONDE NOUVEAU, UN **Vol. 2** 110, 251
MONDE OU L'ENNUIE, LE **Vol. 2** 122, 626
MONDE TREMBLERA, LE **Vol. 2** 59, 488, 657
MONDO BALDORIA **Vol. 7** 783
MONDO CANE **Vol. 7** 325, 478
MONDO CANE OGGI - L'ORRORE CONTINUA **Vol. 7** 743
MONDO DEI MIRACOLI, IL **Vol. 7** 1121
MONDO DEI SENSI DI EMY WONG, IL **Vol. 7** 694
MONDO DI NOTTE, IL **Vol. 7** 333
MONDO DI NOTTE OGGI **Vol. 7** 478
MONDO DI YOR, IL **Vol. 7** 738
MONDO LE CONDANNA, IL **Vol. 7** 196, 299
MONDO NUOVO - LA NUIT DE VARENNES, IL **Vol. 7** 434, 817, 873
MONDO SARA NOSTRO, IL **Vol. 7** 1027
MONDO SENZA VELI **Vol. 7** 18
MONDOLINATA A MARE **Vol. 7** 829
MONELLA DI STRADA **Vol. 7** 277
MONEMBASIA **Vol. 3** 193

MONEMVASIA **Vol. 3** 169
MONEY MANIAC, THE **Vol. 2** 384
MONEY, MONEY **Vol. 2** 732
MONGOLI, I **Vol. 7** 207, 476, 528, 1054
MONICA STOP **Vol. 8** 214, 602
MONICA VOGELSANG **Vol. 4** 190, 773
MONIKA **Vol. 4** 340, 466, 796
MONIQUE **Vol. 7** 654, 679, 918
MONIQUE ET JULIE, DEUX COLLEGIENNES EN PARTOUZE **Vol. 2** 293
MONISMANIE 1995 – EN FILM OM FRIHET **Vol. 5** 1001, 1080
MONJA ALFEREZ, LA **Vol. 8** 728, 749
MONNA VANNA **Vol. 4** 80, 300, 545, 584 **Vol. 9** 136
MONNAIE DE SINGE **Vol. 2** 781, 863
MONOCLE NOIR, LE **Vol. 2** 118
MONOCLE RIT JAUNE, LE **Vol. 2** 118, 294
MONOCOLI DELLA VERITA, IL **Vol. 7** 1189
MONOSABIO, EL **Vol. 8** 245, 331
MONOVRE D'AMORE **Vol. 7** 476
MONSEIGNEUR **Vol. 2** 8, 439
MONSIEUR **Vol. 2** 548, 644
MONSIEUR A LA TETE LOURDE, UN **Vol. 2** 159, 286, 330
MONSIEUR ALBERT **Vol. 2** 41, 190, 326, 563, 579
MONSIEUR BEGONIA **Vol. 2** 152, 273, 812
MONSIEUR BRELOQUE A DISPARU **Vol. 2** 48, 273
MONSIEUR BROTONNEAU **Vol. 2** 84, 143
MONSIEUR CHASSE **Vol. 2** 6, 475
MONSIEUR COCCINELLE **Vol. 2** 48, 100, 454
MONSIEUR DE CINQ HEURES, LE **Vol. 2** 249, 871
MONSIEUR DE COMPAGNIE, UN **Vol. 2** 208, 274
MONSIEUR DE FALINDOR **Vol. 2** 6, 627
MONSIEUR DE MINUIT, LE **Vol. 2** 330, 563, 573, 603, 658
MONSIEUR DE POURCEAUGNAC **Vol. 2** 330, 623
MONSIEUR DES LOURDINES **Vol. 2** 8, 9
MONSIEUR FABRE **Vol. 2** 226, 347, 719
MONSIEUR GREGOIRE S'EVADE **Vol. 2** 401, 833
MONSIEUR HECTOR **Vol. 2** 187, 538
MONSIEUR KLEIN **Vol. 2** 274, 299, 351, 837
MONSIEUR LA SOURIS **Vol. 2** 32, 439
MONSIEUR LE DIRECTEUR **Vol. 2** 614
MONSIEUR LE DUC **Vol. 2** 37, 332
MONSIEUR LE MARECHAL **Vol. 2** 397
MONSIEUR LE PRESIDENT DIRECTEUR GENERAL **Vol. 2** 97, 781
MONSIEUR LEQUIGNON, LAMPISTE **Vol. 2** 128, 496
MONSIEUR, MADAME ET BIBI **Vol. 2** 355
MONSIEUR PAPA **Vol. 2** 486, 781
MONSIEUR PELCHE FAIT L'OUVERTURE **Vol. 2** 159, 286, 330
MONSIEUR PERSONNE **Vol. 2** 533, 775
MONSIEUR SADE **Vol. 2** 738
MONSIEUR SANS GENE **Vol. 2** 439, 646
MONSIEUR SUZUKI **Vol. 2** 131, 581, 808
MONSIEUR TAXI **Vol. 2** 158, 205
MONSIEUR VERDOUX **Vol. 2** 206 **Vol. 2** 226, 719, 722
MONSTRUM Z GALAXIE ARKANA **Vol. 10** 173, 325, 381
MONTA IN SELLA, FIGLIO DI ... **Vol. 8** 243, 249
MONTAGNA DEL DIO CANNIBALE, LA **Vol. 7** 426
MONTAGNA DI LUCE, LA **Vol. 7** 649
MONTAGNE EST VERTE, LA **Vol. 2** 661
MONTAGNE INFIDELE, LA **Vol. 2** 378
MONTANA SIN LEY, LA **Vol. 8** 405, 677
MONTANAN REBELDE, LA **Vol. 8** 461, 495
MONTATORJA **Vol. 3** 22, 24
MONTDRAGON **Vol. 2** 513
MONTE-CHARGE, LE **Vol. 2** 47, 548
MONTE CRISTO **Vol. 2** 55, 100, 438, 521, 731
MONTE MIRACOLI **Vol. 7** 792, 1162, 1163
MONTECARLO **Vol. 7** 988
MONTECASSINO O MONTECASSINO NEL CERCHIO DI FUOCO **Vol. 7** 923
MONTEE VERS L'ACROPOLE, LA **Vol. 2** 614, 725
MONTENEGRO ELLER PÄRLOR OCH SVIN **Vol. 5** 886, 1093, 1102
MONTEVERGINE **Vol. 7** 639 **Vol. 73** 639
MONTIHO CARDAS **Vol. 10** 78, 399, 436
MONTMARTRE **Vol. 2** 731
MONTMARTRE-SUR-SEINE **Vol. 2** 270, 395
MONTPARNASSE 19 **Vol. 2** 27, 279, 549, 566
MONUMENTI **Vol. 3** 14, 20
MONUMENTO, EL **Vol. 8** 243, 426
MOONRAKER **Vol. 2** 835
MOORHUND, DER **Vol. 1** 74, 92, 103
MOR BAG RATTET **Vol. 5** 16, 195, 272
MOR-JAG HAR PATIENTER **Vol. 5** 305
MOR SKAL GIFTES **Vol. 5** 2, 14
MOR-VRAN **Vol. 2** 136, 378
MORA **Vol. 6** *page 129, page 131*
MORAENEN **Vol. 5** 136, 157, 183
MORAL **Vol. 4** 132, 232, 300, 361, 424, 461, 681

MORAL DER BANDITEN, DIE **Vol. 1** 12, 91, 126
MORAL DER GASSE, DIE **Vol. 4** 158, 626
MORAL UM MITTERNACHT **Vol. 4** 186, 558
MORAL UND LIEBE **Vol. 4** 373
MORAL UND SINNLICHKEIT **Vol. 4** 560
MORALISTE, IL **Vol. 7** 693
MORALKA PANI DULSKE **Vol. 10** 311, 388, 503
MORALNOSC PANI DULSKIEJ **Vol. 1** 175, 348
MORBO **Vol. 8** 214, 738
MORD AM MONTAG **Vol. 1** 48, 60, 102
MORD AN RATHENAU **Vol. 1** 86, 100, 111
MORD DER NIE VERJÄHRT, DER **Vol. 1** 21, 37
MORD FOR ABENT TAEPPE **Vol. 5** 257
MORD IM MÄRKISCHEN VIERTEL **Vol. 1** 70, 113
MORD IN DER GREENSTREET, DER **Vol. 4** 486, 648
MORD IN GATAWAY **Vol. 1** 9, 85
MORD IN RIVERPORT **Vol. 1** 2, 75
MORD LILLA VÄN **Vol. 5** 871, 1073
MORD OHNE SÜHNE **Vol. 1** 60, 104
MORD OHNE TÄTER, DER **Vol. 4** 124, 247
MÖRDAREN - EN HELT VANLIG PERSON **Vol. 5** 869, 881, 1017
MORDBRÄNNERSKAN **Vol. 5** 966
MORD...DIE TRAGÖDIE DES HAUSES GARRICK **Vol. 4** 192, 480
MÖRDER AUF URLAUB **Vol. 1** 9, 50 **Vol. 3** 346
MÖRDER DIMITRI KARAMASOFF, DER **Vol. 4** 49, 558, 712
MÖRDER SIND UNTER UNS, DIE **Vol. 1** 6, 51, 58
MÖRDER SITZT IM WEMBLEY-STADION, DER **Vol. 1** 10, 49, 96
MÖRDER UNTER UNS **Vol. 4** 246, 727, 729
MORDERCA ZOSTAWIA SLAD **Vol. 1** 179, 382
MORDEREN **Vol. 5** 81
MORDEREN UTEN ANSIKT **Vol. 5** 726, 822
MORDET PAA FYN **Vol. 5** 97
MORDETS MELODI **Vol. 5** 31, 55, 196
MORDI I FUGGI **Vol. 7** 1145
MORDOVA ROKLE **Vol. 10** 92, 345
MORDPROZESS MARY DUGAN **Vol. 4** 653
MORDS PAS, ON T'AIME **Vol. 2** 110
MORDSACHE HOLM **Vol. 4** 430, 461, 796
MORDSKAB **Vol. 5** 23, 25
MORDSMÄDEL, EIN **Vol. 4** 340, 359
MORDUS, LES **Vol. 2** 475
MORDVAPEN TILL SALU **Vol. 5** 914, 1142
MORE ZACINA ZA VSI **Vol. 10** 180, 495, 570
MORENA CLARA **Vol. 8** 295, 449, 496, 616, 630, 751
MORENA Y UNA RUBIA, UNA **Vol. 8** 516
MORETO **Vol. 3** 60, 78
MORFALOUS, LES **Vol. 6** *page 96*
MORGAN IL PIRATA **Vol. 7** 333, 915
MORGANE ET SES NYMPHES **Vol. 2** 600
MORGANE LA SIRENE **Vol. 2** 41, 153, 528
MORGEN BEGINNT DAS LEBEN **Vol. 4** 340, 354, 466
MORGEN WERDE ICH VERHAFTET **Vol. 4** 247, 266
MORGENROT **Vol. 4** 266, 288, 574
MORGENRÖTE **Vol. 4** 97, 367, 373
MORGIANA **Vol. 10** 144, 172, 292
MORGONDAGENS MELODI **Vol. 5** 923
MORIANERNA **Vol. 5** 1142, 1217
MORIR DE MIEDO **Vol. 8** 551
MORIR POR AMAR **Vol. 8** 346, 677
MORIRAI A MEZZANOTTE **Vol. 7** 99
MORIRE A ROMA **Vol. 7** 1145
MORITAT VOM BIEDERMANN **Vol. 4** 110, 246
MORITURUS **Vol. 4** 385, 495
MORITZ MACHT SEIN GLÜCK **Vol. 4** 643, 663
MORKE PUNKT, DET **Vol. 5** 9, 58
MORMONENONKEL, DER **Vol. 6** *page 186*
MORMONENS OFFER **Vol. 5** 97
MORMOR OG DE ATTE UNGENE I BYEN **Vol. 5** 701, 767, 824
MORMOR OG DE ATTE UNGENE I SKOGEN **Vol. 5** 701, 767, 824
MOROS Y CRISTIANOS **Vol. 8** 422, 728
MOROSO DE LA NONA, EL **Vol. 7** 493
MORPHIUM **Vol. 4** 213, 558
MORROR SAFARI **Vol. 7** 461
MORS AUX DENTS, LE **Vol. 2** 513, 682 **Vol. 6** *page 119*
MORS HUS **Vol. 5** 684, 718, 808, 824
MORSA DELLA MORTE, LA **Vol. 7** 442
MORSA, LA **Vol. 7** 369
MORSIAN YLLÄTTÄ **Vol. 5** 379, 600
MORSIUSSEPPELE **Vol. 5** 471
MORSKA PANNA **Vol. 10** 168, 259, 345, 412
MORT DE BELLE, LA **Vol. 2** 185, 668
MORT DE L'UTOPIE, LA **Vol. 2** 157
MORT DU CYGNE, LA **Vol. 2** 153, 158, 627, 832
MORT DU MILLIARDAIRE, LA **Vol. 2** 159, 286, 330
MORT DU SOLEIL, LA **Vol. 2** 648
MORT D'UN GUIDE **Vol. 2** 546, 849
MORT D'UN POURRI **Vol. 2** 224
MORT D'UN TUEUR **Vol. 2** 110, 457

MORT EN DIRECT, LA **Vol. 2** 351 **Vol. 6** *page 139*
MORT EN FRAUDE **Vol. 2** 128, 540, 781
MORT EN FUITE, LE **Vol. 2** 407, 657
MORT NE RECOIT PLUS, LE **Vol. 2** 460, 581, 611, 863
MORT N'EST PAS A VENDRE, LA **Vol. 2** 469
MORT OU EST TA VICTOIRE ? **Vol. 2** 661
MORT OU VIF **Vol. 2** 57, 627
MORT QUI FROLE, LA **Vol. 2** 159, 286, 330
MORT QUI RAMPE OU LE COLLIER VIVANT, LA **Vol. 2** 159, 286, 330
MORT SAISON DES AMOURS, LA **Vol. 2** 770, 854
MORT SANS IMPORTANCE, UNE **Vol. 2** 676, 727 **Vol. 7** 968
MORT TROUBLE, LA **Vol. 2** 721
MORTE A VENEZIA **Vol. 7** 1025, 1042, 1142
MORTE BIANCA, LA **Vol. 7** 148
MORTE BUSSA DUE VOLTE, LA **Vol. 7** 946
MORTE CAMMINA CON I TACCHI ALTI **Vol. 8** 245, 732
MORTE CHE ASSOLVE **Vol. 7** 69
MORTE CHE NON UCCIDE **Vol. 7** 1189
MORTE CIVILE, LA **Vol. 7** 250, 701, 796, 1054, 1060
MORTE DEL DUCA D'OFENIA, LA **Vol. 7** 254, 796, 1002, 1162
MORTE DI UN AMICO **Vol. 7** 1056, 1171
MORTE DI UN BANDITO **Vol. 7** 1044
MORTE IN AGGUATO, LA **Vol. 7** 1189
MORTE IN VATICANO **Vol. 7** 770 **Vol. 8** 549, 762
MORTE NON CONTA I DOLLARI, LA **Vol. 7** 443, 910
MORTE PIANGE, RIDE E POI ... S'ANNOIA, LA **Vol. 7** 369, 573
MORTE ROSSA, LA **Vol. 7** 843
MORTE SOSPETTA DI UNA MINORENNE **Vol. 7** 426
MORTE SULL'ALTA COLLINA, LA **Vol. 7** 839
MORTE VIENE DALLO SPAZIO, LA **Vol. 7** 106, 793
MORTELLE RANDONNEE **Vol. 6** *page 84, page 86*
MORTI RITORNANO, I **Vol. 7** 454
MOSCA D'ORO, LA **Vol. 7** 1071
MOSCHETTIERE FANTASMA, IL **Vol. 7** 665, 978
MOSCHETTIERI DEL MARE, I **Vol. 7** 915
MOSE **Vol. 7** 502
MOSEKONGEN **Vol. 5** 2, 89
MOSELFAHRT MIT MONIKA **Vol. 4** 369
MOSKAU-SHANGHAI **Vol. 4** 108, 270, 742
MOSQUITA EN PALACIO **Vol. 8** 276
MOST **Vol. 3** 444
MOST NA TU STRANU **Vol. 10** 320, 576
MOSTER FRA MOLS **Vol. 5** 89, 246
MOSTRI, I **Vol. 7** 347, 884
MOSTRO DELL'ISOLA, IL **Vol. 7** 792, 1121
MOSTRO DI FIRENZE, IL **Vol. 7** 32, 318
MOSTRO DI FRANKENSTEIN, IL **Vol. 7** 1078
MOSTRO ... E MEZZO, UN **Vol. 7** 1027
MOSTRO, IL **Vol. 7** 1067, 1194
MOSTRO INDIANO **Vol. 7** 1040, 1041
MOT DE CAMBRONNE, LE **Vol. 2** 84, 388
MOT FORESTILLING **Vol. 5** 664, 712, 718
MOT HÄRLIGA TIDER **Vol. 5** 869, 947
MOT NYA TIDER **Vol. 5** 843, 923, 928
MOTARDS, LES **Vol. 2** 224, 562
MÖTE I NATTEN **Vol. 5** 1032, 1091
MÖTE MED DJÄVULEN **Vol. 5** 1124
MÖTE MED LIVET **Vol. 5** 878, 1173
MÖTEN I SKYMNINGEN **Vol. 5** 940, 965
MOTIV PRO VRAZDU **Vol. 10** 165, 395, 415
MOTIVO IN MASCHERA **Vol. 7** 642
MOTIVOS DE BERTA, LOS **Vol. 8** 330, 454, 710
MOTODRAMA **Vol. 1** 216, 400 **Vol. 6** *page 235*
MOTORBRAUT, DIE **Vol. 4** 192, 220, 584
MOTORCYKLISTEN **Vol. 5** 97
MOTORINO, IL **Vol. 7** 311
MOTORIZADOS, LOS **Vol. 8** 311, 517
MOTORIZZATI, I **Vol. 8** 311, 517
MOTORKAVALJERER **Vol. 5** 871, 1052
MOTORVEJ PA SENGEKANTEN **Vol. 5** 21, 175, 223, 261, 304
MOTS POUR LE DIRE, LES **Vol. 6** *page 80*
MOTTE FLOG ZUM LICHT, EINE **Vol. 4** 467
MOTYLE **Vol. 1** 323, 391
MOTYLEM JESTEM CZYLI ROMANS CZTERDZIESTOLATKA **Vol. 1** 156, 187 **Vol. 6** *page 212*
MOUCHARDE, LA **Vol. 2** 57, 131
MOUCHETTE **Vol. 2** 190, 377
MOULIN DANS LE SOLEIL, LE **Vol. 2** 249, 708
MOULIN ROUGE **Vol. 2** 6, 413, 812
MOUMOU **Vol. 2** 67, 70
MOUNE ET SON NOTAIRE **Vol. 2** 533
MOUREZ, NOUS FERONS LE RESTE **Vol. 2** 401, 402, 419, 777
MOURIR A MADRID **Vol. 2** 60
MOURIR A TUE-TETE **Vol. 2** 135
MOURIR D'AIMER **Vol. 2** 185, 293
MOURIR D'AMOUR **Vol. 2** 503

MOUSQUETAIRES DU ROI, LES **Vol. 2** 581
MOUSSAILLON, LE **Vol. 2** 270, 395
MOUTARDE ME MONTE AU NEZ, LA **Vol. 2** 142, 151, 224
MOUTON A CINQ PATTES, LE **Vol. 2** 185, 626, 824
MOUTON ENRAGE, LE **Vol. 2** 481
MOUTON, LE **Vol. 2** 16, 875
MOUTON NOIR, LE **Vol. 2** 272 **Vol. 6** *page 130*
MOUTONNET **Vol. 2** 48, 217, 676
MOUTONS DE PANURGE, LES **Vol. 2** 512
MOUTONS DE PRAXOS, LES **Vol. 2** 196
MOVIE RUSH – LA FEBBRE DEL CINEMA **Vol. 7** 383
MOVIMENTO DAS COISAS, O **Vol. 8** 40
MOW DO MNIE JESZCZE **Vol. 1** 404
MOZA DEL CANTARO, LA **Vol. 2** 646 **Vol. 8** 298, 306, 607
MOZE BI FREGATA **Vol. 3** 47 **Vol. 6** *page 33, page 56*
MOZIKLIP **Vol. 9** 92, 275, 294
MOZITÜNDER, A **Vol. 9** 77
MP – MINÄ PELKÄÄN **Vol. 5** 439, 447
MPLOKO **Vol. 3** 183
MPOUMPOULINA **Vol. 3** 180, 204
MPROSTA STIN ANHONE **Vol. 3** 130
MR. UNIVERSE **Vol. 9** 100
MR. WU **Vol. 4** 213
MRAVENCI NESOU SMRT **Vol. 10** 385, 394, 408
MRAVNOST NADE VSE **Vol. 10** 256, 412
MRKACEK CIKO **Vol. 10** 180, 283, 299
MRTVA LADJA **Vol. 3** 417
MRTVI UCIA ZIVYCH **Vol. 10** 72, 148, 548
MRTVI ZIJI **Vol. 10** 126, 167
MRTVIM VSTOP PREPOVEDAN **Vol. 3** 383
MRTVY MEZI ZIVYMI **Vol. 10** 319, 461, 503
M'SIEUR LA CAILLE **Vol. 2** 131, 740
MSTITEL **Vol. 10** 226, 524, 621
MUCEDNICI LASKY **Vol. 10** 41, 289, 398
MUCHACHA DE LA PLAZA DE SAN PEDRO, LA **Vol. 8** 276
MUCHACHA DE LAS BRAGAS DE ORO, LA **Vol. 8** 219, 684
MUCHACHA DEL NILO, LA **Vol. 8** 583, 772
MUCHACHAS DE AZUL, LAS **Vol. 8** 440, 546
MUCHACHAS EN VACACIONES **Vol. 8** 583, 772
MUCHACHITA DE VALLADOLID, UNA **Vol. 8** 207, 214
MUCKMEIERS SOMMERFRISCHE **Vol. 4** 675
MUDAR DE VIDA **Vol. 8** 142, 169
MÜDE THEODOR, DER **Vol. 4** 292, 558
MÜDE TOD, DER **Vol. 4** 266, 462, 486, 574, 593, 729, 734
MUDLARK, THE **Vol. 2** 656
MUDUNU **Vol. 7** 730
MUERE UNA MUJER **Vol. 8** 243, 562
MUERTE BUSCA UN HOMBRE, LA **Vol. 8** 332, 737, 770
MUERTE CAMINA CON TACON ALTO, LA **Vol. 8** 245, 732
MUERTE CUMPLE CONDENA, LA **Vol. 8** 603, 626
MUERTE DE IMIKEL, LA **Vol. 8** 210, 764
MUERTE DE UN CICLISTA **Vol. 8** 214, 400
MUERTE DE UN PRESIDENTE, LA **Vol. 8** 252
MUERTE DE UN QUINQUI **Vol. 8** 346, 551
MUERTE DEL ESCORPION, LA **Vol. 8** 348
MUERTE EN EL VATICANO **Vol. 8** 549, 762
MUERTE EN PRIMAVERA **Vol. 8** 317
MUERTE ESPERA EN ATENAS, LA **Vol. 8** 258, 438
MUERTE INCIERTA **Vol. 8** 552, 732
MUERTE LLAMA OTRA VEZ, LA **Vol. 8** 492, 732
MUERTE LLEGA ARRASTRANDOSE, LA **Vol. 8** 407, 603
MUERTE SILBA UN BLUE, LA **Vol. 8** 524, 768
MUERTE VIAJA DEMASIADO, LA **Vol. 8** 438
MUERTOS NON PERDONAN, LOS **Vol. 8** 221, 681
MUFLES, LES **Vol. 2** 116, 145, 805
MÜHLE IM SCHWARZWALD, DIE **Vol. 4** 25
MÜHLE VON SANSSOUCI **Vol. 4** 14, 84, 186, 340, 715
MÜHLE VON WERBELLIN, DIE **Vol. 4** 186
MUHOKSEN MIMMI **Vol. 5** 443, 461
MUJ BRACHA MA PRIMA BRACHU **Vol. 10** 165, 380, 394
MUJ HRISNY MUZ **Vol. 10** 139, 233, 325
MUJ PRITEL FABIAN **Vol. 10** 296, 412, 569
MUJER CELOSA, LA **Vol. 8** 218, 687
MUJER CUALQUIERA, UNA **Vol. 2** 646 **Vol. 8** 214, 607
MUJER DE CABARET, UNA **Vol. 8** 667
MUJER DE NADIE, LA **Vol. 8** 295, 614
MUJER DE OTRO, LA **Vol. 8** 207, 214
MUJER DE TU PROXIMO, LA **Vol. 8** 356, 467
MUJER DEL DESIERTO, LA **Vol. 8** 317, 470
MUJER DEL JUEZ, LA **Vol. 8** 762
MUJER EN PELIGRO, UNA **Vol. 8** 268

MUJER EN UN TAXI, UNA **Vol. 8** 396
MUJER ES COSA DE HOMBRES, LA **Vol. 8** 438, 681
MUJER ESPANOLA, UN **Vol. 8** 357
MUJER PERDIDA, LA **Vol. 8** 214, 538
MUJERES ACORRALADAS **Vol. 8** 222
MUJERES AL BORDE DE UN ATAQUE DE NERVIOS **Vol. 8** 219, 340
MUJERES DE JEREMIAS, LAS **Vol. 8** 245
MUJERES LOS PREFIEREN TONTOS, LAS **Vol. 8** 371, 578
MUJERIEGO, EL **Vol. 8** 271, 753
MUKA OBRAZNOSTI **Vol. 10** 204, 389, 496
MUKE PO MATI **Vol. 3** 360, 437
MUKI **Vol. 9** 132
MULATTO, IL **Vol. 7** 113
MULHERES PA BEIRA **Vol. 8** 105
MULINO DEGLI SPETTRI, IL **Vol. 7** 53
MULINO DEL PO, IL **Vol. 7** 214, 517, 750, 1136
MULINO DELLE DONNE DI PIETRA, IL **Vol. 7** 877
MULINO, IL **Vol. 7** 26
MÜLLER CONTRA MÜLLER **Vol. 4** 18, 430
MÜLLER UND SEIN KIND, DER **Vol. 4** 135
MULLIN MALLIN **Vol. 5** 403, 579
MULT MAI DE PRET E IUBIREA **Vol. 6** *page 269, page 271, page 280*
MUMIA KOZBESZOL, A **Vol. 9** 145, 180, 316
MUMIENS HALSBAAND **Vol. 5** 136
MÜNCHHAUSEN **Vol. 4** 230, 246, 365, 713
MÜNCHHAUSENS NEUESTES ABENTEUER **Vol. 4** 361, 558
MÜNCHNERINNEN **Vol. 4** 688, 740
MUNDO PARA MI, UN **Vol. 8** 400, 515
MUNDO SEXUAL DE LA PAREJA, EL **Vol. 8** 456, 660
MUNDO SIGUE, EL **Vol. 8** 311, 396
MUNECA ROTA, LA **Vol. 8** 516, 724
MUNECAS **Vol. 8** 723
MUNECAS DE TRAPO **Vol. 8** 265, 465
MUNEQUITA **Vol. 8** 422
MUNKENS FRISTELSER **Vol. 5** 9, 197
MUNKSBROGREVEN **Vol. 5** 840, 909
MUNTELE ASCUNS **Vol. 3** 246, 283, 301
MUNTER KLINIK, EN **Vol. 5** 83
MUNTII IN FLACARI **Vol. 3** 242, 297, 302 **Vol. 6** *page 273*
MUNTRA MUSIKANTER **Vol. 5** 1042, 1205
MUR DE L'ATLANTIQUE, LE **Vol. 2** 377, 513, 514
MUR, LE **Vol. 2** 188
MURA DI MALAPAGA, LE **Vol. 7** 443, 517
MURAGLIA CINESE, LA **Vol. 7** 877
MURAILLES DU SILENCE, LES **Vol. 2** 286, 379
MURALLA FELIZ, LA **Vol. 8** 753
MURALLA, LA **Vol. 8** 611, 732
MURDER OBSESSION **Vol. 7** 384, 909
MURDEROCK UCCIDE A PASSO DI DANZA **Vol. 7** 901
MURHEENKRYYNIN POIKA **Vol. 5** 443, 639
MURIAS **Vol. 2** 183
MURIEL OU LE TEMPS D'UN RETOUR **Vol. 2** 770, 854
MURIO HACE QUINCE ANOS **Vol. 8** 214, 400
MURS MONT DES OREILLES, LES **Vol. 2** 97, 808
MURTOVARKAUS **Vol. 5** 344, 640
MUSA DEL PIANTO E QUELLA DEL SORISSO, LA **Vol. 7** 428, 1085
MUSA TRAGICA **Vol. 7** 279
MUSA Y EL FENIX, LA **Vol. 8** 498
MUSCHETARII IN VA CANTA **Vol. 6** *page 267, page 270, page 278*
MUSCHETARUL ROMAN **Vol. 3** 254, 256, 306
MUSEN **Vol. 5** 97
MUSEUMSMYSTERIET **Vol. 5** 97
MUSICA DE AYER **Vol. 8** 298, 611
MUSICA IN PIAZZA **Vol. 7** 493
MUSICA, LA **Vol. 2** 194, 854
MUSICA PARA SI **Vol. 8** 169
MUSICA PROFANA **Vol. 7** 266
MUSICA PROIBITA **Vol. 7** 518, 1031, 1063
MUSICA VIVA **Vol. 1** 37, 72
MUSICIENS DU CIEL, LES **Vol. 2** 13, 23, 688, 778
MUSICISTA, IL **Vol. 7** 1189
MUSIK AUS DEM FF **Vol. 1** 44, 74, 92
MUSIK DRINGEND GESUCHT **Vol. 1** 17, 65, 73
MUSIK I MÖRKER **Vol. 5** 1052, 1164
MUSIK IM BLUT **Vol. 4** 49, 734, 738
MUSIK IN SALZBURG **Vol. 4** 361
MUSIK OMBORD **Vol. 5** 871
MUSIK UND LIEBE IN TIROL **Vol. 3** 395
MUSIK ZU ZWEIN **Vol. 4** 571, 796
MUSIKANT VON DORNBUSCH, DER **Vol. 4** 110, 517
MUSIKANTER **Vol. 5** 724, 787
MUSIKFEIND, DER **Vol. 4** 167, 339
MUSIQUE EN TETE **Vol. 2** 270, 661
MUSJÖH LEBT GEFÄHRLICH **Vol. 1** 19, 122, 126
MUSKARCI **Vol. 3** 356, 359, 428
MUSKETIER MEIER III **Vol. 4** 323, 422, 435
MUSKI IZLET **Vol. 3** 347, 371, 374
MUSODURO **Vol. 7** 391

MUSOTTE **Vol. 7** 529
MUSS MAN SICH GLEICH SCHEIDEN LASSEN ? **Vol. 4** 321, 339, 703
MUSTA LUMIKKI **Vol. 5** 607
MUSTA RAKKAUS **Vol. 5** 443, 566
MUSTAA VALKOISELLA **Vol. 5** 516
MUSTALAISHURMAAJA **Vol. 5** 315, 608, 645
MUSTASUKKAISUUS **Vol. 5** 316
MUSTAT SILMÄT **Vol. 5** 476, 645
MUSTERGATTE, DER **Vol. 4** 82, 230, 458
MUSTERKNABEN **Vol. 1** 25, 107, 115
MUTA DI PORTICI, LA **Vol. 7** 493, 693, 1189
MUTIGE SEEFAHRER, DER **Vol. 4** 13, 230, 458
MUTINES DE L'ELSENEUR, LES **Vol. 2** 47, 70, 566, 623
MUTTER **Vol. 7** 493
MUTTER, DEIN KIND RUFT ! **Vol. 4** 202, 258, 318
MUTTER DER KOMPAGNIE, DIE **Vol. 4** 344, 545
MUTTER ERDE **Vol. 4** 217
MUTTER KRAUSENS FAHRT INS GLÜCK **Vol. 4** 234, 602 **Vol. 6** *page 181*
MUTTER UND DAS SCHWEIGEN, DIE **Vol. 1** 46, 85
MUTTER UND KIND **Vol. 4** 149, 190, 535, 536, 626, 660
MUTTER UND SOHN **Vol. 4** 327
MÜTTER, VERZAGET NICHT ! **Vol. 4** 190
MUTTERFREUDEN DES LANDWEHR-MANNES, DIE **Vol. 4** 514
MUTTERLIEBE **Vol. 4** 265, 340, 514, 536, 608, 614, 620
MUTTERLIED **Vol. 4** 97, 517, 558 **Vol. 7** 902, 1116
MUUAN SULHASMIES **Vol. 5** 372, 397
MUURAHAISPOLKU **Vol. 5** 67, 545
MUURMANIN PAKOLAISET **Vol. 5** 344, 640
MUZ BEZ SRDCE **Vol. 10** 11, 167, 453
MUZ, KTERY ROZDAVAL SMICH **Vol. 10** 327, 621
MUZ, KTERY STOUPL V CENE **Vol. 10** 289, 313, 522
MUZ, KTORY LUZE **Vol. 10** 273, 323
MUZ, KTORY SA NEVRATIL **Vol. 10** 210, 273
MUZ NA DRATE **Vol. 10** 165, 495, 505
MUZ NA UTEKU **Vol. 10** 283, 286, 447
MUZ NIE JE ZIADUCI **Vol. 10** 228, 491, 544
MUZ PRES PALUBU **Vol. 10** 139, 339, 473
MUZ S ORLEM A SLEPICI **Vol. 10** 286, 408, 517
MUZ V POVETRI **Vol. 10** 547, 564, 621
MUZ Z LONDYNA **Vol. 10** 41, 385, 473
MUZ Z NEZNAMA **Vol. 10** 345, 460
MUZ Z PRVNIHO STOLETI **Vol. 10** 289, 388, 621
MUZEUM ZAZRAKU **Vol. 10** 166, 289, 428
MUZI BEZ KRIDEL **Vol. 10** 503, 524
MUZI NESTARNOU **Vol. 10** 412, 569, 621
MUZI V OFFSIDU **Vol. 10** 586
MUZIKA PRO DVA **Vol. 10** 41, 385, 531
MUZIKANSTKA LIDUSKA **Vol. 10** 345, 460
MUZIKANT **Vol. 10** 193, 503, 569
MY DAUGHTER JOY **Vol. 2** 656
MY LITTLE BABY **Vol. 7** 264, 701
MY LITTLE DOG **Vol. 7** 53
MY NAME IS PETERSEN **Vol. 5** 2, 228
MY TRI A PES Z PETIPES **Vol. 10** 173, 283, 380
MY Z 9. A **Vol. 10** 268, 451
MY ZTRACENY HOLKY **Vol. 10** 369, 425, 505
MYÄHÄSTYNYT HÄÄYÖ **Vol. 5** 443, 566
MYGLAREN **Vol. 5** 981
MYKOIN PH 510 **Vol. 10** 172, 179, 433
MYKONOS, OU LE PAYS GAY **Vol. 6** *page 97*
MYLORD L'ARSOUILLE **Vol. 2** 339, 419, 731
MYRSKLYLUODON KALASTAJA **Vol. 5** 344, 356, 406
MYS DOBRE NADEJE **Vol. 10** 153, 173, 308
MYSAFIRI **Vol. 3** 10, 15
MYSTERE **Vol. 7** 340, 664
MYSTERE A SHANGHAI **Vol. 2** 70, 727, 968
MYSTERE BARTON, LE **Vol. 2** 96, 153
MYSTERE DE LA CHAMBRE JAUNE, LE **Vol. 2** 47, 58, 153, 251, 413, 573, 833
MYSTERE DE LA TOUR EIFFEL, LE **Vol. 2** 387, 824
MYSTERE DES ROCHES DE KADOR, LE **Vol. 2** 798
MYSTERE IMBERGER, LE **Vol. 2** 407, 540
MYSTERE SAINT VAL, LE **Vol. 2** 32, 197
MYSTERES DE L'OMBRE, LES **Vol. 2** 798
MYSTERES DE PARIS, LES **Vol. 2** 58, 59, 124, 153, 173, 192, 273, 368, 407, 515, 540, 708
MYSTERIET I LOUVRE **Vol. 5** 203
MYSTERIET I SOVEVOGNEN **Vol. 5** 228
MYSTERIEUSE LADY, LA **Vol. 2** 395
MYSTERIEUX MONSIEUR SYLVAIN, LE **Vol. 2** 15, 339
MYSTERIUM **Vol. 4** 680
MYSTISKE FODSPOR, DE **Vol. 5** 242
MYSTISKE FREMMEDE, DEN **Vol. 5** 58
MYSTISKE SELSKABSDAME, DEN **Vol. 5** 242

MYSTISKE TJENER, DEN **Vol. 5** 242
MYSTISKE Z - STRAALER, DE **Vol. 5** 265
MYTEN ELLER HAN SNODDE EN BLOMMA OCH FICK SPRINGA FÖR LIVET **Vol. 5** 1094
M"ZE **Vol. 3** 36, 78, 112 **Vol. 6** *page 41*
M"ZE BEZ RABOTA **Vol. 3** 52 **Vol. 6** *page 46*
M"ZE V KOMANDIROVKA **Vol. 3** 112 **Vol. 6** *page 41, page 51*
M"ZKI VREMENA **Vol. 3** 110 **Vol. 6** *page 23, page 34*

N

N. N. A HALAL ANGYALA **Vol. 9** 149, 316, 339
NA...! **Vol. 2** 663, 769
NA BIALYM SZLAKU **Vol. 1** 179, 206, 232 **Vol. 6** *page 215*
NA CISTO **Vol. 3** 119 **Vol. 6** *page 23, page 51*
NA DOBRE STOPE **Vol. 10** 122, 179
NA DRUHOM BREHU SLOBODA **Vol. 10** 155, 230, 324, 474
NA ISTARSKI NACIN **Vol. 6** *page 304, page 312, page 318*
NA JASNYM BRZEGU **Vol. 1** 404
NA KLANCU **Vol. 3** 348
NA KOHO TO SLOVO PADNE **Vol. 10** 18, 283, 428
NA KOLEJICH CEKA VRAH **Vol. 10** 296, 532
NA KOMETE **Vol. 10** 173, 283, 404, 422, 522
NA KONCI DIALNICE **Vol. 10** 108, 148, 228
NA KONCI MESTA **Vol. 10** 70, 566, 580
NA KONCI SVETA **Vol. 10** 286, 352, 425
NA KRAWEDZI **Vol. 1** 330, 401 **Vol. 6** *page 226*
NA LANE **Vol. 10** 353, 394, 517
NA LETNIM BYTE **Vol. 10** 213, 453
NA MALKIJA OSTROV **Vol. 3** 69, 84
NA MESTO GRADJANINE POKORNI **Vol. 3** 359, 383, 391
NA NIEBIE I NA ZIEMI **Vol. 1** 136, 325 **Vol. 6** *page 258*
NA NOCHE SECRETA DE LUCRECIA BORGIA **Vol. 8** 299
NA PAPIRNATIH AVIONIH **Vol. 3** 387, 449
NA POCHODE SA VZDY NESPIEVA **Vol. 10** 273, 287
NA POMOC DOHODE **Vol. 10** 213, 393
NA PYTLACKE STEZCE **Vol. 10** 184, 352, 517
NA RUZICH USTLANO **Vol. 10** 34, 256
NA SAMOTE U LESA **Vol. 10** 172, 516, 529
NA SLUNECNI STRANE **Vol. 10** 34, 120, 524
NA SMYCZY **Vol. 1** 298, 349
NA STARTU JE DELFIN **Vol. 10** 139, 516, 581
NA STRAZY SWEJ STAC BEDE **Vol. 6** *page 224, page 240, page 250, page 260*
NA STRIBRNEM ZRCADLE **Vol. 10** 186
NA SVATEM KOPECKU **Vol. 10** 34, 256
NA SVOJI ZEMLJI **Vol. 3** 377, 390, 406
NA SYBIR **Vol. 1** 166, 174, 176, 308 **Vol. 2** 290 **Vol. 4** 158
NA TIHIJA BRJAG **Vol. 3** 54, 109 **Vol. 6** *page 59*
NA T"MEN KR"STOP"T **Vol. 3** 49
NA TROPACH BARTKA **Vol. 6** *page 206, page 221, page 225*
NA TY LOUCE ZELENY **Vol. 10** 167, 256
NA VALOVIH MURE **Vol. 3** 406, 449
NA VELIKE RECE **Vol. 10** 139, 325, 425
NA VSEKI KILOMET"R **Vol. 6** *page 30, page 46, page 60, page 63, page 65*
NA VYSOKE STRANI **Vol. 10** 257, 613
NA WLASNA PROSBE **Vol. 6** *page 222, page 236*
NA WUNDERBAR **Vol. 4** 409, 536
NA WYLOT **Vol. 1** 164, 355 **Vol. 6** *page 246*
NA ZAZORJAVANE **Vol. 3** 75, 76, 103, 109, 114
NA ZIVOT I SM"RT **Vol. 6** *page 53, page 62, page 63*
NA ZIZKOVE VALECNEM VOZE **Vol. 10** 299, 380, 385
NAAR BONDER ELSKER **Vol. 5** 2, 7, 290
NAAR DJAEVLE ER PAA SPIL **Vol. 5** 97
NAAR FRUEN GAAR PAA EVENTYR **Vol. 5** 9
NAAR HAEVNGLODEN SLUKKES **Vol. 5** 295
NAAR HJERTERNE KALDER **Vol. 5** 183
NAAR HJERTET SAELGES **Vol. 5** 228
NAAR JULEKLOKKERNE RINGER **Vol. 5** 9
NAAR KATTEN ER UDE **Vol. 5** 89, 224
NAAR MAN KUN ER UNG **Vol. 5** 98, 119, 168
NAAR MOBLER FLYTTES **Vol. 5** 97
NAAR UNGDOMMEN RASER **Vol. 5** 83
NABADAILE CLEOPATREI **Vol. 3** 212, 305
NABOERNE **Vol. 5** 23, 100, 267
NACALOTO NA DENJA **Vol. 3** 60, 101 **Vol. 6** *page 53*
NACALOTO NA EDNA VAKANCIJA **Vol. 3** 83, 112 **Vol. 6** *page 41*
NACERADEC, KRAL KIBICU **Vol. 10** 138, 586

NACH DEM GESETZ **Vol. 4** 423, 780
NACHFOLGER, DER **Vol. 1** 1, 26, 90
NÄCHSTE BITTE, DER **Vol. 4** 340, 608, 691
NACHT AN DER AUTOBAHN, DIE **Vol. 1** 74, 120, 130
NACHT AN DER DONAU, EINE **Vol. 4** 246, 774
NACHT ASIENS, DIE **Vol. 4** 545
NACHT AUF DEM GOLDENHALL, DIE **Vol. 4** 178, 479
NACHT DER 12, DIE **Vol. 4** 285, 344
NACHT DER EINBRECHER, DIE **Vol. 4** 330, 527, 545, 745
NACHT DER ENTSCHEIDUNG, DIE **Vol. 4** 124, 232, 424, 536, 694
NACHT DER GROSSEN LIEBE, DIE **Vol. 4** 246, 729
NACHT DER KÖNIGIN ISABEAU, DIE **Vol. 4** 239, 772
NACHT DER TAUSEND SEELEN, DIE **Vol. 4** 212, 330, 509, 571
NACHT DER TOTEN, DIE **Vol. 4** 159, 217, 512
NACHT DER VERSUCHUNG, DIE **Vol. 4** 247, 544, 545
NACHT DER VERWANDLUNG, DIE **Vol. 4** 437, 462, 738
NACHT DES SCHRECKENS, DIE **Vol. 4** 219, 340, 608
NACHT GEHÖRT UNS, DIE **Vol. 4** 373, 450, 626
NACHT IM FORSTHAUS, DIE **Vol. 4** 247, 409
NACHT IM GRANDHOTEL, EINE **Vol. 4** 321, 452
NACHT IM GRENZWALD, DIE **Vol. 1** 15, 44, 75
NACHT IM MAI, EINE **Vol. 4** 28, 215, 445
NACHT IM PARADIES, EINE **Vol. 2** 397 **Vol. 4** 149, 257, 660 **Vol. 6** *page 83*
NACHT IN BERLIN ODER DIE LÖWEN SIND LOS, EINE **Vol. 4** 135
NACHT IN LONDON, EINE **Vol. 4** 28, 178, 734
NACHT IN VENEDIG, DIE **Vol. 4** 49, 65
NACHT IN VENEDIG, EINE **Vol. 4** 82
NACHT IN YOSHIWARA, EINE **Vol. 4** 43, 240
NACHT MIT DEM KAISER, DIE **Vol. 4** 234, 462, 734
NACHT MIT HINDERNISSEN, EINE **Vol. 4** 340, 466, 796
NACHT OHNE ABSCHIED **Vol. 4** 8, 373
NACHT OHNE MORGEN, DIE **Vol. 4** 211, 247
NACHT OHNE PAUSE, DIE **Vol. 4** 437, 517, 720
NACHT VOR DEM TODE, DIE **Vol. 6** *page 190, page 196*
NACHTASYL **Vol. 4** 747
NACHTBESUCH **Vol. 4** 167, 339
NACHTBESUCH IN DER NORTHERNBANK **Vol. 4** 132, 211, 729
NÄCHTE DES CORNELIS BROUWER, DIE **Vol. 4** 124, 219
NACHTFALTER **Vol. 4** 124, 646
NACHTGESTALTEN **Vol. 4** 131, 158, 288, 626
NACHTKOLONNE **Vol. 4** 186, 270, 409
NACHTSPIELE **Vol. 1** 122, 126
NACHTSTÜRME **Vol. 4** 159, 529, 558
NACHTWACHE, DIE **Vol. 4** 329, 774
NACHTWACHE IM PARADIES **Vol. 4** 799, 517, 703
NACIDO PARA LA MUSICA **Vol. 8** 616, 630, 762
NACIMENTO DE SALOME, IL **Vol. 7** 750, 796, 902, 1060
NACIMIENTO DE SALOME, EL **Vol. 8** 568
NACIONAL III **Vol. 8** 240, 738
NACIONALNA KLASA DO 785 SQUARE METERS **Vol. 6** *page 307*
NACIONALNA KLASA DO 785 X 3 **Vol. 3** 404, 454
NACISTE NELLE MINIERE DI RE SALOMONE **Vol. 7** 1147
NACKT UNTER WÖLFEN **Vol. 1** 48, 73, 102
NACKTE MANN AUF DEM SPORTPLATZ, DER **Vol. 1** 9, 10, 48
NACKTE SPATZ, DER **Vol. 4** 230, 458, 796
NACKTEN, DIE **Vol. 4** 270, 328
NAD NAMI SVITA **Vol. 10** 388, 503
NAD NIEMNEN **Vol. 1** 169, 308, 397
NAD PROPASTI **Vol. 10** 167, 536
NAD SANTJAGO VALI **Vol. 3** 48, 76
NADA **Vol. 2** 700 **Vol. 8** 276, 298, 311
NADA MENOS QUE TODO UN HOMBRE **Vol. 8** 207, 214
NADA MENOS QUE UN ARCANGEL **Vol. 8** 524, 595
NADEJE **Vol. 10** 235, 385, 577
NADIA, LA FEMME TRAQUEE **Vol. 2** 152, 248
NADIA LA JOUISSEUSE **Vol. 2** 291
NADIE LO SABRA **Vol. 8** 298, 484
NADIE OYO GRITAR **Vol. 8** 402, 749
NADLIDE **Vol. 10** 503, 564
NADVEREN **Vol. 5** 105
NADZOR **Vol. 6** *page 216, page 235, page 244*
NAESBYGAARDS ARVING **Vol. 5** 195, 281
NAESTLE STOP PARADIS **Vol. 5** 99, 249, 270
NAFTA **Vol. 1** 179, 354 **Vol. 6** *page 216*

NAFTALIN Vol. 9 71, 213
NAFTALINA Vol. 8 211, 341
NAGANA Vol. 2 96, 781, 856
NAGANIACZ Vol. 1 227, 254 Vol. 6 *page 224*
NAGON ATT ÄLSKA Vol. 5 1024, 1217
NAGRODY I ODZNACZENIA Vol. 1 156, 311 Vol. 6 *page 225*
NAGY GENERACIO, A Vol. 9 9, 31, 171, 253
NAGY KEK JELZES, A Vol. 9 253, 265, 316
NAGYMAMA, A Vol. 9 28, 71, 237
NAGYROZSDASI ESET, A Vol. 9 4, 285, 326
NAHA PASTYRKA Vol. 10 296, 433, 550
NAHIRA Vol. 4 319
NAHODOU JE PRIMA ! Vol. 10 28, 70, 208
NAHOTA Vol. 10 139, 227, 567
NAI MEN, ALLA Vol. 3 190
NAICA Vol. 3 313
NAIF AUX QUARANTE ENFANTS, LE Vol. 2 47, 471
N'AIMER QUE TOI Vol. 2 58, 407, 726
NAIMISIIN PÄIVÄKSI Vol. 5 415, 600
NAINEN ON VALTTIA Vol. 5 349, 386
NAIS Vol. 2 347, 807, 875
NAISENKUVIA Vol. 5 434
NAISET Vol. 5 452
NAISET, JOTKA MINULLE ANNOIT Vol. 5 443, 516
NAISKOHTALOITA Vol. 5 356, 549
NAIVKO Vol. 3 354
NAJ-D"LGATA NOST Vol. 3 96, 101 Vol. 6 *page 41*
NAJ-DOBRIJAT COVEK, KOGOTO POZNAVAM Vol. 3 96, 105 Vol. 6 *page 30*
NAJ-TEZKIJAT GRJAH Vol. 6 *page 45, page 56, page 65*
NAJ-VRARNATA STRAZA Vol. 3 49, 115
NAJDOLGIOT PAT Vol. 3 339, 384
NAJLEPSZE W SWIECIE Vol. 1 179, 408 Vol. 6 *page 210*
NÄKEMIIN, HELENA Vol. 5 478, 480, 634, 655
NÄKÖRADIOMIEHEN IHMEELLISET SIEKAILUT Vol. 5 318, 454, 477, 606, 614
NAKOVALNJA ILI CUK Vol. 3 109, 114 Vol. 6 *page 53*
NALEN Vol. 5 115, 281
NALLES BORNEHAVE Vol. 5 83
NAMENSHEIRAT Vol. 4 124, 204
NAMESTIE SVATEJ ALZBETY Vol. 10 36, 273, 320
NANA Vol. 2 44, 48, 203, 248, 282, 330, 388, 566 Vol. 5 203, 936 Vol. 6 *page 71*
NANAS DE ESPINAS Vol. 8 370
NANETTE Vol. 4 234, 373, 738 Vol. 6 *page 187*
NANETTE MACHT ALLES Vol. 4 242, 752
NANINE, FEMME D'ARTISTE Vol. 2 798
NANO ROSSO, IL Vol. 7 376, 523
NANON Vol. 4 325, 400, 605, 713, 724
NANTAS Vol. 2 217, 245, 723 Vol. 7 1132 Vol. 9 220
NANU, LA CUGINA D'ALBANIA Vol. 7 978
NANU, SIE KENNEN KORFF NOCH NICHT Vol. 4 97, 108, 270
NAO CAPITANA, LA Vol. 8 276, 298
NAO HA RAPAZES MAUS ! Vol. 8 21, 72, 106, 131
NAP LELKE, A Vol. 9 5, 237
NAPASTA Vol. 3 305, 333 Vol. 6 *page 274, page 275, page 278*
NAPFENY A JEGEN Vol. 9 83, 199, 259, 324
NAPLEMENTE DELBEN Vol. 9 20, 84, 273
NAPLES AU BAISER DE FEU Vol. 2 308, 332, 488
NAPLO GYERMEKEIMNEK Vol. 9 139, 152, 202
NAPLO SZERELMEIMNEK Vol. 9 139, 152, 202
NAPOLEON Vol. 2 83, 153, 247, 402, 440, 526, 528, 626, 672, 722, 775
NAPOLEON AUF ST. HELENA Vol. 4 28, 407, 526, 548, 729, 738, 743, 788
NAPOLEON BONAPARTE Vol. 2 138, 402, 579, 623, 722
NAPOLEON BONAPARTE, EMPEREUR DES FRANCAIS Vol. 2 706
NAPOLEON II – L'AIGLON Vol. 2 294, 383
NAPOLEON IST AN ALLEM SCHULD Vol. 4 49, 246
NAPOLEON OG HANS LILLE TROMPETIST Vol. 5 97
NAPOLEON UND DIE KLEINE WÄSCHERIN Vol. 4 131
NAPOLEONCINA Vol. 7 390
NAPOLEONE Vol. 7 476
NAPOLETANI A MILANO Vol. 7 86
NAPOLI CHE CANTA Vol. 7 48, 162, 796, 1116
NAPOLI CHE NON MUORE Vol. 7 196, 472
NAPOLI D'ALTRI TEMPI Vol. 7 196, 756, 1060
NAPOLI E SEMPRE NAPOLI Vol. 7 424, 500
NAPOLI E UNA CANZONE Vol. 7 569

NAPOLI ... LA CAMORRA SFIDA LA CITTA RISPONDE Vol. 7 1147
NAPOLI MILIONARIA Vol. 7 522, 1136
NAPOLI, PALERMO, NEW YORK, IL TRIANGOLO DELLA CAMORRA Vol. 7 222, 474
NAPOLI PIANGE E RIDE Vol. 7 16, 171
NAPOLI SERENATA CALIBRO NOVE Vol. 7 474
NAPOLI SI RIBELLA Vol. 7 1003
NAPOLI SIRENA DELLA CANZONE Vol. 7 829
NAPOLI SOLE MIO ! Vol. 7 798
NAPOLI TERRA D'AMORE Vol. 7 693, 829, 1054
NAPOLI VERDE - BLU Vol. 7 369
NAPOLI VIOLENTA Vol. 7 287, 1213
NAPOLYT LATNI ES ... Vol. 9 67, 190, 215
NAPPALI SÖTETSEG Vol. 9 75, 265, 297
NAPRAWDE WCZORAJ Vol. 1 206, 305
NAPULE ... E NIENTE CCHIU Vol. 7 569
NAPULE E SURRIENTO Vol. 7 424
NÄR ÄNGARNA BLOMMAR Vol. 5 841, 1052
NÄR BENG OCH ANDERS BYTTE HUSTRUR Vol. 5 1183 Vol. 5 871, 928, 1042
NAR ENDEN ER GO' Vol. 5 257, 281
NÄR KÄRLEKEN KOM TILLE BYN Vol. 5 907, 1052, 1181
NÄR KONSTNÄRER ÄLSKA Vol. 5 997
NÄR MILLIONERNA RULLA... Vol. 5 861, 986
NÄR MÖRKRET FALLER Vol. 5 871, 959, 1042
NÄR SEKLET VAR UNGT Vol. 5 1032, 1181
NÄR SYRENERNA BLOMMA Vol. 5 1052, 1080
NÄR UNGDOMEN VAKNAR Vol. 5 930, 1142
NAR VILLDYRET VAKNER Vol. 5 686, 799, 817
NÄRA LIVET Vol. 5 1042, 1217
NARAYANA Vol. 2 483
NARCISSE Vol. 2 6, 9, 32, 205
NARKO... EN FILM OM KAERLIGHED Vol. 5 195
NARKOS Vol. 5 871, 993, 1173
NARKOSE Vol. 4 85, 100, 359
NÄRKRINGARNA Vol. 5 870, 889
NARODNI POSLANIK Vol. 3 342, 343, 346
NAROZENINY REZISERA Z.K. Vol. 10 70, 299, 474, 544
NARR SEINER LIEBE, DER Vol. 4 14, 527
NARR UND DIE ANDEREN, DER Vol. 4 125, 329
NARREN IM SCHNEE Vol. 4 124, 232, 620
NARRENTANZ DER LIEBE Vol. 4 681
NARRIEN ILLAT Vol. 5 353, 437
NÄRRISCHE GLÜCK, DAS Vol. 4 287, 384, 458, 646
NÄRRISCHE WETTE DES LORD ALDINI, DIE Vol. 4 35, 385, 558
NAS AVTO Vol. 3 348, 387
NAS CLOVEK Vol. 6 *page 316*
NAS DEDEK JOSEF Vol. 10 197, 369, 385
NAS SKOLNIK Vol. 10 268, 459
NASA ZEMJA Vol. 3 53, 100, 109
NASCITA DI SOLOME Vol. 7 750, 796, 902, 1060
NASE BLAZNIVA RODINA Vol. 10 299, 524
NASE JEDENACTKA Vol. 10 256, 412
NASI FURIANTI Vol. 10 92, 345
NASI PRED BRANAMI Vol. 10 32, 263, 483
NASI SE PUTOVI RAZILAZE Vol. 3 358, 451
NASIEL SOM SI KAMARATA Vol. 10 548
NASIJAT SOSKANINI Vol. 6 *page 42, page 49, page 56*
NASILJE NA TRGU Vol. 3 436, 455
NASRESTNO DVIZENIE Vol. 6 *page 36, page 42, page 63*
NASTOJANJE Vol. 6 *page 312*
NASTRADIN HODZA I HIT'R PET'R Vol. 3 113 Vol. 6 *page 55*
NASTUP Vol. 10 154, 558, 566, 627
NASVIDENJE V NASLEDNJI VOJNI Vol. 3 387, 421
NASZINDULO Vol. 9 236
NASZINDULO FELARON Vol. 9 14, 132
NASZUT FELARON Vol. 9 71, 288
NAT - EKSPRESSEN P. 903 Vol. 5 2, 7
NAT PINKERTON Vol. 5 97
NATA DI MARZO Vol. 7 257, 913
NATA E PARE E LIRISE Vol. 6 *page 14, page 16*
NATACHA Vol. 7 148
NATALE Vol. 7 1189
NATALE AL CAMPO Vol. 7 756
NATALE DEL PERDONO, IL Vol. 7 1189
NATALE DI PIERINO, IL Vol. 7 1189
NATALE IN CASA D'APPUNTAMENTO Vol. 7 366
NATALE TRAGICO Vol. 7 1189
NATALIA Vol. 5 324, 516
NATALIZIO DELLA NONNA, IL Vol. 7 965
NATASSA Vol. 3 170
NATCAFEENS DRONNING Vol. 5 203
NATERCIA Vol. 2 854 Vol. 8 201
NATHALIE Vol. 2 232, 388, 557 Vol. 7 619
NATHALIE AGENT SECRET Vol. 2 185, 232, 488
NATHALIE GRANGER Vol. 2 190

NATHALIE RESCAPEE DE L'ENFER **Vol. 2** 393
NATHAN DER WEISE **Vol. 4** 212, 424, 448, 535, 628, 726, 753
NATLOGI BETALT **Vol. 5** 2, 23
NATT, EN **Vol. 5** 840, 879, 909, 997
NATT I HAMN **Vol. 5** 840, 940
NATT PA GLIMMINGEHUS, EN **Vol. 5** 933
NATT PA SMYGEHOLM, EN **Vol. 5** 840, 997
NATTBARN **Vol. 5** 1073, 1080
NATTEN FOR CHRISTIANS FODSELSDAG **Vol. 5** 97
NATTENS BARN **Vol. 5** 944
NATTENS GAADE **Vol. 5** 183
NATTENS LJUS **Vol. 5** 909, 1052
NATTENS MYSTERIUM **Vol. 5** 58
NATTEVANDREREN **Vol. 5** 58
NATTLEK **Vol. 5** 880, 933, 973
NATTLIGA TORNER **Vol. 5** 966
NATTMARA **Vol. 5** 1052, 1217
NATTVAKTENS HUSTRU **Vol. 5** 876, 923
NATTVARDSGÄSTERNA **Vol. 5** 1052, 1053, 1080
NATURBARNET **Vol. 5** 97
NATÜRLICH DIE NELLY **Vol. 1** 37, 92
NAUFRAGA DELLA VITA, LA **Vol. 7** 369
NAUFRAGES DES L'ILE DE LA TORTUE, LES **Vol. 2** 618
NAUFRAGEURS, LES **Vol. 2** 190, 517, 578, 743
NAUFRAGHI DELL'ONORE, I **Vol. 7** 454
NAUFRAGHI, I **Vol. 7** 783, 971, 1004, 1189
NAUFRAGIO **Vol. 7** 420, 525
NAUFRAGO, IL **Vol. 7** 1189
NAUKA LATANIA **Vol. 1** 183, 328 **Vol. 6** *page 246*
NAVAJEROS **Vol. 8** 349, 580
NAVE BIANCA, LA **Vol. 7** 170, 250
NAVE DEI LEONI, LA **Vol. 7** 1189
NAVE DEI MORTI, LA **Vol. 7** 519, 1092, 1118
NAVE DELLE DONNE MALEDETTE, LA **Vol. 7** 283, 443, 1136
NAVE FANTASMA, LA **Vol. 7** 131
NAVE, LA **Vol. 7** 668, 1038
NAVIDADES EN JUNIO **Vol. 8** 207, 214
NAVIRE NIGHT, LE **Vol. 2** 517
NAVRAT DOMU **Vol. 10** 92, 503
NAVRAT JANA PETRU **Vol. 10** 73, 269, 399
NAVRAT ZTRACENEHO SYNA **Vol. 10** 428, 517, 567
NAVRATY **Vol. 10** 18, 139, 227
NAVSTEVA Z OBLAK **Vol. 10** 184, 522, 582
NAZARE **Vol. 8** 77, 100, 130
NAZARE, PRAIA DE PESCADORES **Vol. 8** 105
NDERGJEGJA **Vol. 3** 2, 11, 29
NE 45.000 **Vol. 5** 356, 359, 428, 485
NE BOUGEZ PLUS **Vol. 2** 332, 824
NE CAKAJ NA MAJ **Vol. 3** 375, 395
NE CDO STINE **Vol. 3** 9, 23 **Vol. 6** *page 12*
NE DE PERE INCONNU **Vol. 2** 330, 719
NE DIRAJ U SRECU **Vol. 3** 356, 436
NE DITET TE PUSHIMIT **Vol. 6** *page 14, page 17, page 18*
NE FILLIM TE VERES **Vol. 3** 3, 26
NE JOCI, PETER **Vol. 3** 387, 390
NE JOUEZ PAS AVEC LES MARTIENS **Vol. 2** 642
NE KERDEZD, KI VOLTAM **Vol. 9** 196, 274
NE KUFI TE DY LEGJENDAVE **Vol. 6** *page 12, page 13, page 15*
NE LE CRIEZ PAS SUR LES TOITS **Vol. 2** 153, 347
NE ME PARLEZ PLUS JAMAIS D'AMOUR **Vol. 6** *page 148*
NE NAGINJI SE VAN **Vol. 3** 341, 452
NE NOUS FACHONS PAS **Vol. 2** 548
NE OKRECI SE, SINE **Vol. 3** 341, 452
NE PLEURE PAS **Vol. 2** 480
NE PRAG TE LIRISE **Vol. 6** *page 14, page 15, page 18*
NE PYJET ME BORE KA JETE **Vol. 3** 20, 24
NE SE OBR"STAJ NAZAD **Vol. 3** 63, 114 **Vol. 6** *page 31*
NE SHTEPINE TONE **Vol. 3** 12, 23 **Vol. 6** *page 12*
NE SI OTIVAJ **Vol. 3** 102, 105 **Vol. 6** *page 26*
NE SOIS PAS **Vol. 7** 619
NE SOIS PAS JALOUSE **Vol. 2** 488
NE UBIJ ! **Vol. 3** 447
NE VINIM NGA LUFTA **Vol. 3** 12, 23
NE ZNAM, NE CUH, NE VIDJAH **Vol. 6** *page 30, page 45*
NEA **Vol. 2** 285, 353, 873
NEA MARIN MILIARDA **Vol. 3** 233, 256, 271, 280
NEAMUL SOIMARESTILOR **Vol. 3** 230, 307, 312, 317, 332
NEBANTSVIRAG **Vol. 9** 315
NEBBIA AZZURRA **Vol. 7** 1104
NEBBIE SUL MARE **Vol. 7** 1177
NEBE A DUDY **Vol. 10** 412, 621
NEBEL **Vol. 1** 33, 40, 101
NEBELNACHT **Vol. 1** 71, 89, 102
NEBENBUHLER, DIE **Vol. 4** 223
NEBESKI ODRED **Vol. 3** 403, 428
NEBESTI JEZDCI **Vol. 10** 70, 283, 371

... NEBO BYT ZABIT Vol. 10 70, 283, 567
NEBOJSA Vol. 10 72, 108, 419
NECHCI NIC SLYSET Vol. 10 395, 398, 529
NECHTE TO NA MNE Vol. 10 184, 522
NECO JE VE VZDUCHU Vol. 10 144, 197, 369
NECROPHAGUS Vol. 8 583
NED MED KAERLIGHEDEN Vol. 5 83
NED MED VAABNENE Vol. 5 58
NEDALEKO DO NEBA Vol. 10 82, 145, 399, 544
NEDBRUDTE NERVER Vol. 5 183, 228
NEDELE VE VSEDNI DEN Vol. 9 134, 187 Vol. 10 313, 517, 582
NEDELJNI RUCAK Vol. 6 *page 311, page 314, page 320*
NEDELNITE MACOVE Vol. 3 43 Vol. 6 *page 48*
NEDJELJA Vol. 3 424
NEDTUR Vol. 5 701, 750, 767
NEFFE AUS AMERIKA, DER Vol. 4 424, 704
NEFNUKEJ, VEVERKO ! Vol. 10 74, 180, 233
NEGRE BLANC, LE Vol. 2 122, 380
NEGRO QUE TENIA ALMA BLANCA, EL Vol. 8 214, 401, 522, 548
NEGY LANY EGY UDVARBAN Vol. 9 117, 261, 265
NEGYEDIZIGLEN Vol. 9 196
NEGYEN AZ ARBAN Vol. 9 83, 265, 314
NEGYLOVAS HINTO Vol. 9 35, 58, 71
NEHEZ APANAK LENNI Vol. 9 178, 237
NEHEZ EMBEREK Vol. 9 304
NEHEZ KESZTYÜK Vol. 9 36, 120, 307
NEHEZ SZEREP Vol. 9 77
NEI LABIRINTI DELL'ANIMA Vol. 7 148
NEI LABIRINTI DI UN'ANIMA Vol. 7 1204
NEI LACCI DEL DESTINO Vol. 7 1189
NEIGE Vol. 6 *page 73, page 87*
NEIGE ETAIT SALE, LA Vol. 2 47, 615, 625
NEIGE SUR LES PAS, LA Vol. 2 84, 347
NEIGES BRULANTES Vol. 6 *page 88, page 98*
NEIGUNGSEHE Vol. 4 28, 235
NEINFRICATII Vol. 3 223
NEITI TALONMIES Vol. 5 443, 461
NEITI TUITTUPÄÄ Vol. 5 379, 446
NEJISTA SEZONA Vol. 10 344, 571, 606
NEJKRASNEJSI VEK Vol. 10 70, 402
NEJLEPSI CLOVEK Vol. 10 345, 503, 532, 569
NEJLEPSI TIP Vol. 10 247
NEJLEPSI ZENSKA MEHO ZIVOTA Vol. 10 460, 621
NEKA DALEKA SVJETLOST Vol. 3 351
NEKA DRUGA ZENA Vol. 6 *page 308, page 314, page 317*
NEKLIDNOU HLADINOU Vol. 10 182, 387, 428, 593
NEKONECNA - NEVYSTUPOVAT Vol. 10 255, 269, 462
NEKRE POLITEIA Vol. 3 168, 205
NEL BLU DIPINTO DI BLU Vol. 7 1167
NEL GORGO DEL PECCATO Vol. 7 1054, 1121
NEL GORGO DELLA SVENTURA Vol. 7 454
NEL MARE DELLA VITA - O - PEPPENIELLO - O - LA BELLA MORTE Vol. 7 420
NEL MONDO DEGLI AGGUATI Vol. 7 938
NEL PIU ALTO DEI CIELI Vol. 7 1133
NEL SEGNO DEL LEONE Vol. 7 863
NEL SEGNO DI ROMA Vol. 7 830, 1054, 1147
NEL SILENZIO DELL'ANIMA Vol. 7 1078
NELJÄ NAISTA Vol. 5 372, 379, 600
NELJÄ RAKKAUTTA Vol. 5 378, 395, 549, 616
NELJÄSTOISTA VIERAS Vol. 5 443, 620
NELKEN IN ASPIK Vol. 1 7, 42, 71
NELLA CITTA L'INFERNO Vol. 7 86, 1054
NELLA MISURA IN CUI ... Vol. 7 406
NELLA MORSA DELLA COLPA Vol. 7 574, 727
NELLA MORSA DI UN SOGNO Vol. 7 1156
NELLA RAGNA Vol. 7 1064
NELLA STRETTA DEL MISTERO Vol. 7 91
NELL'ANNO DEL SIGNORE Vol. 7 585
NELLE PIEGHE DELLA CARNE Vol. 8 602, 749
NELLINA Vol. 7 264, 701
NELLY Vol. 6 *page 120*
NELLY, DIE BRAUT OHNE MANN Vol. 4 437, 704
NELLY LA GIGOLETTE Vol. 7 264, 701
NELLYS FORLOVELSE Vol. 5 81
NELLYS RIDDERE Vol. 5 83
NEM Vol. 9 183, 259, 285
NEM AMANTES, NEM AMIGOS Vol. 8 124
NEM ELHETEK MUZSIKASZO NELKÜL Vol. 9 6, 15, 28, 259, 265
NEM ER A NEVEM Vol. 9 134, 183, 332
NEM LOPTAM EN ELETEMBEN Vol. 9 132, 164
NEM SZOKTAM HAZUDNI Vol. 9 36, 84, 304
NEM VAROK HOLNAPIG Vol. 9 199, 241, 332
NEMA BARIKADA Vol. 10 193, 621
NEMA DOSSZIE, A Vol. 9 67, 84, 122
NEMA KOLOSTOR Vol. 9 71, 236

NEMA MALIH BOGOVA Vol. 3 391
NEMA PROBLEMA Vol. 6 *page 305, page 306, page 318*
NEMES ROZSA Vol. 9 26, 133, 157
NEMESI DANZANTE, LA Vol. 7 91, 1005
NEMESIS Vol. 7 436, 569
NEMICA, LA Vol. 7 316, 796
NEMICO DI MIA MOGLIE, IL Vol. 7 1167
NEMICO, IL Vol. 7 792, 1063
NEMIR Vol. 6 *page 304*
NEMIRNI Vol. 3 362, 416
NEMOCNY BILY SLON Vol. 10 149, 251, 505
NEMODLENEC Vol. 10 125, 345
NEMOZNA Vol. 10 128, 328
NEMURITORII Vol. 3 226, 235, 244, 254, 334
NENAS DEL MINI-MINI, LAS Vol. 8 249
NENE Vol. 7 1025
NENI SIROTEK JAKO SIROTEK Vol. 10 70, 283, 550
NENNELLA Vol. 7 39
NENTORI I DYTE Vol. 6 *page 14, page 16*
NEOBHODIMIJAT GRESNIK Vol. 3 59, 102 Vol. 6 *page 31*
NEOBYCEJNA TRIDA Vol. 10 250, 286, 433
NEOHLIZEJ SE, JDE ZA NAMI KUN ! Vol. 10 286, 385, 408
NEP"LNOLETIE Vol. 6 *page 31, page 48*
NEPOCESTNA ZENA Vol. 10 34, 168
NEPORAZENA ARMADA Vol. 10 256, 460
NEPORAZENI Vol. 10 313, 352
NEPOTII GORNISTULUI Vol. 3 217, 221, 228, 260, 299
NEPRIMIRIMITE Vol. 3 93, 109, 126, 159
NERIKEJ HOCHU, ZE MAS MNE RAD Vol. 10 243
NERIKEJ MI MAJORE ! Vol. 10 180, 233, 286
NERINA Vol. 7 69
NERO Vol. 7 978
NERO CROCE, I Vol. 7 91
NERO VENEZIANO Vol. 7 347
NERONE Vol. 7 728, 796, 920, 978, 1189
NERONE E AGRIPPINA Vol. 7 1038
NERONE E POPPEA Vol. 7 311, 761
NERVEN Vol. 4 397
NESCHOVAVEJTE SE, KDYZ PRSI Vol. 10 41, 87, 433
NESI TES AFRODITES Vol. 3 159
NESPOKOEN DOM Vol. 3 36, 99, 121 Vol. 6 *page 46*
NESPOKOEN P"T Vol. 3 93, 100, 109
NESSUNO E PERFETTO Vol. 7 208, 347
NESSUNO HA TRADITO Vol. 7 938
NESSUNO TORNA INDIETRO Vol. 7 453, 1111, 1177
NESTO IZMEDJU Vol. 6 *page 305, page 322*
NESTOR BURMA DETECTIVE DE CHOC Vol. 6 *page 94, page 146*
NETTER BESUCH, EIN Vol. 4 361, 558
NETTES PFLÄNZCHEN, EIN Vol. 4 342
NETZWERK Vol. 1 7, 80, 92
NEU NOME E ..., O Vol. 8 34
NEUE, DER Vol. 1 7, 89, 104
NEUE FIMMEL, DER Vol. 1 42, 100, 101
NEUE GENERALDIREKTOR, DER Vol. 4 762
NEUE GOUVERNANTE, DIE Vol. 4 247
NEUE SCHIFFSJUNGE, DER Vol. 4 18, 558
NEUES AUS DER FLORENTINER 73 Vol. 1 11
NEUF DE TREFLE Vol. 2 270, 387
NEUF GARCONS, UN COEUR Vol. 2 9, 67
NEULAND Vol. 4 258, 397, 663, 735
NEUPLNE ZATMENI Vol. 10 395, 397, 496
NEUTRALIDAD Vol. 8 276
NEUVAINE DE COLETTE, LA Vol. 2 64
NEVE NEL BICCHIERE, LA Vol. 7 75, 227, 703
NEVERA Vol. 10 41, 580
NEVERA PO SLOVENSKY Vol. 10 51, 228, 548
NEVEROJATNA ISTORIJA Vol. 3 36, 82, 107 Vol. 6 *page 26*
NEVESINJSKA PUSKA Vol. 3 349, 419
NEVESTA Vol. 10 172, 327
NEVESTA HOL Vol. 10 82, 273, 544
NEVESTA K ZULIBANI Vol. 10 327, 344, 570, 577
NEVESTA S NEJKRASNEJSIMA OCIMA Vol. 10 139, 325, 425
NEVETÖ SZASZKIA, A Vol. 9 292, 337
NEVEU SILENCIEUX, UN Vol. 2 157, 645 Vol. 6 *page 147*
NEVIDELI JSTE BOBIKA ? Vol. 10 92, 503
NEVIDINI BATALJON Vol. 3 345, 385
NEVINATKA Vol. 10 122, 524
NEVINNA Vol. 10 168, 564
NEVINOST BEZ ZASTITE Vol. 3 405, 416
NEVITE O BYTE ? Vol. 10 413, 461
NEVJERA Vol. 3 373, 389, 425
NEVTELEN ASSZONY, A Vol. 9 77
NEVTELEN VAR Vol. 4 166 Vol. 9 35, 71, 80, 193, 312, 327
NEW YORK CITY INFERNO Vol. 2 2
NEW YORK EXPRESSKABEL Vol. 9 71
NEW YORK NE REPOND PLUS Vol. 2 814
NEZ DE CUIR Vol. 2 282, 402, 863
NEZABRAVIMIJAT DEN Vol. 3 38, 72 Vol. 6 *page 26*
NEZAV"RSENI IGRI Vol. 3 75, 124 Vol. 6 *page 23*

NEZBEDNY BAKALAR **Vol. 10** 142, 460, 569, 621
NEZLOB, KRISTINO ! **Vol. 10** 385, 522, 621
NEZLOBTE DEDECKA **Vol. 10** 167, 256
NEZNAMA KRASKA **Vol. 10** 322
NEZNAME MATKY **Vol. 10** 121, 122
NEZNY BARBAR **Vol. 10** 129, 172, 511
'NFAMA ! **Vol. 7** 829
NGA MEZI I ERRESIRES **Vol. 3** 14, 22 **Vol. 6** *page 12*
NGADHNJIM MBI VDEKJEN **Vol. 3** 11, 26 **Vol. 6** *page 15*
NI LIV **Vol. 5** 717, 800
NI LJUGER **Vol. 5** 918
NI POBRE NI RICO, SINO TODO LO CONTRARIO **Vol. 8** 614
NI SE LO LLEVO EL VIENTO NI PUNETERA FALTA QUE HACIA **Vol. 8** 592
NI VU, NI CONNU **Vol. 2** 510, 515, 549, 557
NIBELUNGEN, DIE **Vol. 4** 227, 288, 298, 325, 567, 591, 727, 771
NIC NIE STOI NA PRZESZKODZIE **Vol. 6** *page 242, page 243*
NICHOLAS NICKLEBY **Vol. 2** 656
NICHT LANGE TÄUSCHTE MICH DAS GLÜCK **Vol. 4** 309
NICHT SCHUMMELN, LIEBLING **Vol. 1** 40, 100, 113, 114
NICHT VERZAGEN, TRUDCHEN FRAGEN **Vol. 6** *page 160, page 170, page 173*
NICHTE AUS USA, DIE **Vol. 4** 18, 261
NICHTS ALS SÜNDE **Vol. 1** 33, 48, 53
NICK CARTER **Vol. 2** 24, 415
NICK CARTER CONTRE PAULIN BROQUET **Vol. 2** 24
NICK CARTER ET LE TREFLE ROUGE **Vol. 2** 72, 674
NICK CARTER ROI DES DETECTIVES **Vol. 2** 711
NICK CARTER VA TOUT CASSER **Vol. 2** 9, 129
NICK, DER KÖNIG DER CHAUFFEURE **Vol. 4** 158, 724
NICK WINTER NEGY UJ KALANDJA **Vol. 9** 308
NICKI **Vol. 6** *page 156, page 169*
NICOLE ET SA VERTU **Vol. 2** 488, 731
NIDO DE ESPIAS **Vol. 8** 243, 461
NIDO DE ZINGARA **Vol. 7** 825
NIDO DI FALASCO, IL **Vol. 7** 709, 1121
NIDO, EL **Vol. 8** 358
NIE **Vol. 10** 263, 483, 493
NIE BEDE CIE KOCHAC **Vol. 1** 144, 276, 391
NIE BYLO SLONCA TEJ WIOSNY **Vol. 6** *page 224, page 238, page 254*
NIE DAMY ZIEMI, SKAD NASZ ROD **Vol. 1** 317, 404
NIE JE ADAM AKO ADAM **Vol. 10** 268, 320
NIE LUBIE PONIEDZIALKU **Vol. 1** 156, 187, 303
NIE MA MOCNYCH **Vol. 1** 216, 306, 335 **Vol. 6** *page 226, page 246*
NIE MA POWROTU JOHNNY **Vol. 1** 180, 381 **Vol. 6** *page 215*
NIE MA ROZY BEZ OGNIA **Vol. 1** 276, 306 **Vol. 6** *page 209*
NIE MIALA BABA KLOPOTU **Vol. 1** 278, 372
NIE WIEDER LIEBE **Vol. 4** 28, 527, 574, 614 **Vol. 6** *page 189*
NIE ZAZNASZ SPOKOJU **Vol. 1** 212, 262, 341
NIEBEZPIECZNY KOCHANEK **Vol. 1** 174, 188
NIEBEZPIECZNY ROMANS **Vol. 1** 278, 364
NIEBIESKIE JAK MORZE CZARNE **Vol. 1** 212, 216, 283 **Vol. 6** *page 223*
NIEBLA Y SOL **Vol. 8** 415, 524, 755
NIECH CIE ODLECI MARA **Vol. 6** *page 210, page 252*
NIECIEKAWA HISTORIA **Vol. 6** *page 211, page 214, page 241*
NIEDALEKO WARSZAWY **Vol. 1** 223, 280, 377, 381 **Vol. 6** *page 247*
NIEDIELA SPRAWIEDLIWOSCI **Vol. 6** *page 237*
NIEDORAJDA **Vol. 1** 241, 278, 308
NIEDZIELA SPRAWIEDLIWOSCI **Vol. 1** 180, 232
NIEKOCHANA **Vol. 1** 202, 227, 283, 406
NIELS PIND OG HANS DRENG **Vol. 5** 7
NIEMAND WEISS ES **Vol. 4** 522
NIEMANDSLAND **Vol. 1** 127 **Vol. 4** 382, 639, 684
NIENASYCONE ZADZE **Vol. 1** 404
NIET INEJ CESTY **Vol. 10** 210, 312, 441, 576
NIEWIARYGODNE PRZYGODY MARKA PIEGUSA **Vol. 1** 313
NIEWINNI CZARODZIEJE **Vol. 1** 373, 391 **Vol. 6** *page 226*
NIEWOLNICA MILOSCI **Vol. 1** 174, 176
NIEWOLNICA ZMYSLOW **Vol. 1** 174, 215
NIEWOZNICY ZYCIA **Vol. 1** 226
NIEZNANY **Vol. 1** 354, 401 **Vol. 6** *page 234*
NIIN SE ON, POIJAAT **Vol. 5** 493
NIJE BILO UZALUD **Vol. 3** 358, 453
NIJE LAKO S MUSKARCIMA **Vol. 6** *page 304, page 314, page 323*

NIJE NEGO **Vol. 3** 409, 424 **Vol. 6** *page 306*
NIKDO NIC NEVI **Vol. 10** 532, 621
NIKDO SE NEBUDE SMAT **Vol. 10** 41, 289, 371
NIKLAS OCH FIGUREN **Vol. 5** 900, 1093, 1096
NIKODEM DYZMA **Vol. 1** 171, 305, 323 **Vol. 6** *page 220*
NIKOLA SUHAJ **Vol. 10** 92, 122, 350
NIKOLETINA BURSAC **Vol. 3** 341, 364
NIKOS MERTES **Vol. 3** 194
NIKT NIE WOLA **Vol. 1** 355, 398 **Vol. 6** *page 242*
NILS HOLGERSSINS UNDERBARA RESA **Vol. 5** 1042, 1130, 1196, 1217
NINA **Vol. 2** 279, 318, 807
NINA DE LA VENTA, LA **Vol. 8** 214, 276
NINA DE LUTO, LA **Vol. 8** 298, 402
NINA DE LUZMELA, LA **Vol. 8** 515, 721
NINA DEL PATIO **Vol. 8** 331, 396
NINA ESTA LOCA, LA **Vol. 8** 422
NINA JA ERIK **Vol. 5** 443, 549
NINA, NON FAR LA STUPIDA **Vol. 7** 650, 756
NINCS IDÖ **Vol. 9** 259, 263, 318
NINCSENEK VELETLENEK **Vol. 9** 315, 322
9-ES KORTEREM, A **Vol. 9** 134, 312
9 RAPAZES E 1 CAO **Vol. 8** 54
999. NACHT, DIE **Vol. 4** 681
19 DJEVOJAKA I MORNAR **Vol. 3** 351
1918 – MIES JA HÄNEN OMATUNTONSA **Vol. 5** 356, 566
1958 **Vol. 5** 701, 716, 746
1914, DIE LETZTEN TAGE VOR DEM WELTBRAND **Vol. 4** 219, 626
1919 – CRONICA DEL ALBA **Vol. 8** 580, 691
1990, I GUERRIERI DEL BRONX **Vol. 7** 623, 1015
1922 **Vol. 3** 159, 189
NINETTE Y UN SENOR DE MURCIA **Vol. 8** 438, 631
99 **Vol. 9** 71
99 DISGRAZIE DI DON CAMILLO, LE **Vol. 7** 701
99 MUJERES **Vol. 8** 595, 623
91 : A KARLSSON SLÄR KNOCK OUT **Vol. 5** 1041
91 : AN KARLSSON **Vol. 5** 997
91 : AN KARLSSON MUCKAR **Vol. 5** 1041
91 : AN KARLSSON RYCKER IN **Vol. 5** 840, 1041
91 : AN KARLSSONS BRAVADER **Vol. 5** 862, 1073
91 : AN KARLSSONS PERMIS **Vol. 5** 841, 1181
91 : AN OCH GENERALERNAS FNATT **Vol. 5** 858, 862
92 MINTTER AF I GAR **Vol. 5** 46, 134, 145
NINGUEM DUAS VEZES **Vol. 8** 7
NINGUNO DE LOS TRES SE LLAMABA TRINIDAD **Vol. 8** 261, 687
NINI FALPALA **Vol. 7** 493, 665
NINICHE **Vol. 4** 190, 584 **Vol. 7** 442
NINNOLA **Vol. 7** 886
NINO DE LAS COLES, EL **Vol. 8** 606
NINO DE LAS MONJAS, EL **Vol. 8** 217, 648, 655
NINO DE ORO, EL **Vol. 8** 422
NINO DE SU MAMA, EL **Vol. 8** 299, 728
NINO ES NUESTRO, EL **Vol. 8** 219, 221
NINO, NON FAR LA STUPIDA **Vol. 7** 493, 1060
NINO Y EL MURO, EL **Vol. 8** 298, 400
NINON DE LENCLOS **Vol. 4** 125
NINOS DEL HOSPICIO, LOS **Vol. 8** 650
NINOTSCHKA SUCHT DEN FRÜHLING **Vol. 1** 63, 74
90 MINUTEN AUFENTHALT **Vol. 4** 123, 182, 617, 727
NIPOTE D'AMERICA, LE **Vol. 7** 26, 701
NIPOTE SABELLA, LA **Vol. 7** 1177
NIPOTI DI ZORRO, I **Vol. 7** 1027
NIRVANA **Vol. 4** 565, 570
NISKAVUOREN AARNE **Vol. 5** 356, 643
NISKAVUOREN HETA **Vol. 5** 356, 643
NISKAVUOREN NAISET **Vol. 5** 349, 379, 386,392
NISKAVUORI **Vol. 5** 410, 511
NISKAVUORI TAISTELEE **Vol. 5** 374, 443
NITCHEVO **Vol. 2** 48, 167, 388, 676, 863
NITTEN RODE ROSER **Vol. 5** 52, 156, 218
NIXCHEN **Vol. 4** 223, 287, 565
NIZVODNO OD SUNCA **Vol. 3** 404, 426, 441
NJAMA NISTO PO-HUBAVO OT LOSOTO VREME **Vol. 3** 48, 69 **Vol. 6** *page 25*
NJE „GENERAL" ZIHET ROB **Vol. 3** 12
NJE GJENERAL KAPET ROB **Vol. 6** *page 11*
NJE NATE PA DRITE **Vol. 6** *page 15, page 17*
NJE NDODHI NE PORT **Vol. 3** 24 **Vol. 6** *page 16*
NJE SHOQE NGA FSHATI **Vol. 3** 2, 21
NJE UDHETIM I VESHTIRE **Vol. 3** 12 **Vol. 6** *page 14*
NJE VONESE E VOGEL **Vol. 6** *page 12, page 15*
NJERIU I MIRE **Vol. 6** *page 13, page 16, page 17*
NJERIU ME TOP **Vol. 3** 12, 18
NJERIU PREJ DHEU **Vol. 6** *page 306*

NJESITI GUERRIL **Vol. 3** 11, 22 **Vol. 6** *page 19*
NJIH DVOJICA **Vol. 3** 337, 349
NJR EMER MIDIS NJERZVE **Vol. 6** *page 11, page 13, page 14*
NJU **Vol. 4** 213, 273, 373, 563
NO ALLA VIOLENZA **Vol. 7** 949
NO CONTROL **Vol. 2** 84
NO DESEARA AL VECINO DEL QUINTO **Vol. 8** 297, 299
NO DESEARAS LA MUJER DEL PROJIMO **Vol. 8** 524, 725
NO DESEARAS LA MUJER DEL VECINO **Vol. 8** 291, 297
NO DISPARES CONTRA MI **Vol. 8** 470, 492
NO DISPONIBLE **Vol. 8** 214, 232
NO ENCONTRE ROSAS PARA MI MADRE **Vol. 8** 214
NO ES BUENO QUE EL HOMBRE ESTE SOLO **Vol. 8** 261, 306, 577
NO ES NADA MAMA, SOLO UN JUEGO **Vol. 8** 235, 762
NO FIRMES MAS LETRAS, CIELO **Vol. 8** 215, 267, 627
NO GRAZIE, IL CAFFE MI RENDE NERVOSO **Vol. 7** 351, 363, 947
NO HAGO LA GUERRA ... PREFIERO EL AMOR **Vol. 8** 306
NO HIGHWAY IN THE SKY **Vol. 2** 656
NO IMPORTA MORIR **Vol. 8** 602
NO LE BUSQUES TRES PIES **Vol. 8** 524
NO MATARAS **Vol. 8** 577, 627, 749
NO MR. JOHNSON **Vol. 3** 168
NO QUIERO, NO QUIERO ! **Vol. 8** 422
NO QUIERO PERDER LA HONRA **Vol. 8** 249, 566, 595
NO SOMOS DE PIEDRA **Vol. 8** 402, 406
NO SOMOS NI ROMEO NI JULIETA **Vol. 8** 517
NO TE NIEGUES A VIVIR **Vol. 8** 661
NO TEMAS A LA LEY **Vol. 8** 493
NOBEL FODSELSDAGSGAVE, EN **Vol. 5** 83
NOBELEZA DE CORAZONES **Vol. 8** 238
NOBELPRISTAGAREN **Vol. 5** 592 **Vol. 6** *page 293*
NOBENO SONCE **Vol. 6** *page 305, page 307*
NOBLEZA BATURRA **Vol. 8** 298, 422, 449, 493
NOBODY MACHT ALLES **Vol. 4** 747
NOC KLAVIRISTY **Vol. 10** 369, 577
NOC NA KARLSTEJNE **Vol. 10** 144, 292, 310, 359, 393
NOC NEVESTY **Vol. 10** 197, 235, 289
NOC ORANZOVYCH OHNU **Vol. 10** 139, 233, 496
NOC POSLIJE SMRTI **Vol. 6** *page 319, page 321*
NOC POSLUBNA **Vol. 1** 321, 400 **Vol. 5** 604, 657
NOC S MACKOU **Vol. 10** 210, 236
NOC SMARAGDOVEHO MESICE **Vol. 10** 254, 299, 327
NOC TRIKRALOVA **Vol. 10** 243, 536
NOCCIOLINE A COLAZIONE **Vol. 7** 1145
NOCE I DNIE **Vol. 1** 200, 243, 253 **Vol. 6** *page 249, page 258*
NOCES DE PORCELAINE, LES **Vol. 2** 110, 142
NOCES DE SABLE **Vol. 2** 16, 47
NOCES ROUGES, LES **Vol. 2** 700
NOCES VENITIENNES, LES **Vol. 2** 615
NOCHE DE ALBORADAS **Vol. 8** 246
NOCHE DE CELOS **Vol. 8** 548, 752
NOCHE DE LA FURIA, LA **Vol. 8** 346, 708
NOCHE DE LA IRA, LA **Vol. 8** 414, 698
NOCHE DE LAS GAVIOTAS, LA **Vol. 8** 237, 708
NOCHE DE LOS ASESINOS, LA **Vol. 8** 633
NOCHE DE LOS BRUJOS, LA **Vol. 8** 708
NOCHE DE REYES **Vol. 8** 396
NOCHE DE VERANO **Vol. 8** 492, 581, 686
NOCHE DE VINO TINTO **Vol. 8** 317, 470
NOCHE DE WALPURGIS, LA **Vol. 8** 596, 773
NOCHE DECISIVA **Vol. 8** 692
NOCHE DEL MARTES, LA **Vol. 8** 515, 640
NOCHE DEL SABADO, LA **Vol. 8** 214, 484, 638
NOCHE DEL TERROR CIEGO, LA **Vol. 8** 288, 677
NOCHE EN BLANCO, UNA **Vol. 8** 449
NOCHE FANTASTICA **Vol. 8** 436
NOCHE MAS HERMOSA, LA **Vol. 8** 738, 771
NOCHE SIN CIELO **Vol. 8** 677, 732
NOCHE Y EL ALBA, LA **Vol. 8** 221, 611, 616
NOCHES DE CASABLANCA **Vol. 8** 214
NOCHES DEL ESCORPION, LAS **Vol. 8** 317
NOCI I JUTRA **Vol. 3** 444
NOCNE UDERZENIE **Vol. 1** 305, 388
NOCNI HOST **Vol. 10** 503, 564
NOCNI IZLET **Vol. 3** 348, 387
NOCNI JEZDCI **Vol. 10** 72, 82, 567
NOCNI MOTYL **Vol. 10** 412, 569, 621
NOCTURNE, LE **Vol. 2** 169, 330, 613
NOCTURNO 29 **Vol. 8** 348
NOCTURNO DE CHOPIN, EL **Vol. 8** 604
NODDEBO PRAESTEGAARD **Vol. 5** 35, 55, 101, 203, 208, 247, 306
NODLANDING **Vol. 5** 738, 748
NODO, IL **Vol. 7** 701, 725

NOEL DU POILU, LE **Vol. 2** 798
NOEL EN ALSACE **Vol. 2** 678
NOEMI **Vol. 7** 1169
NOGLEHULLET **Vol. 5** 46
NOGLEN TIL PARADIS **Vol. 5** 195, 205
NOI CANNIBALI **Vol. 7** 528, 1035
NOI DUE SOLI **Vol. 7** 798
NOI DURI **Vol. 7** 1147
NOI E L'AMORE ... COMPORTAMENTO SESSUALE VARIANTE **Vol. 7** 289
NOI GANGSTERS **Vol. 7** 583
NOI NON SIAMO ANGELI **Vol. 7** 694
NOI PECCATORI **Vol. 7** 798, 1054
NOI SEITSEMÄN VELJESTÄ **Vol. 5** 400, 558, 606
NOI SIAM COME LE LUCCIOLE **Vol. 7** 666
NOI SIAMO DUE EVASI **Vol. 7** 798
NOI SIAMO LE COLONNE **Vol. 7** 877
NOI TRE **Vol. 7** 96, 947, 1043, 1091
NOI VIVI **Vol. 7** 3, 117, 151, 170, 250, 552
NOIDAN KIROT **Vol. 5** 344, 356
NOIR PRINTEMPS DES JOURS, LE **Vol. 2** 684
NOIRE ET CALINE **Vol. 2** 13
NOITA PALAA ALÄMÄÄN **Vol. 5** 347, 634
NOIVO DAS CALDAS, O **Vol. 8** 25, 130
NOIVO PARA DOS HERMANAS, UN **Vol. 8** 395
NOJO AOS CAES **Vol. 8** 169
NOKAUT **Vol. 3** 421 **Vol. 6** *page 303*
NOKEA JA KULTAA **Vol. 5** 357, 395
NOLLA FÖR MYCKET, EN **Vol. 5** 871, 880, 1104
NOMME LA ROCCA, UN **Vol. 2** 190, 517, 664
NOMOS 4000 **Vol. 3** 144
NON CANTARE BACIAMI **Vol. 7** 603
NON CANTO PIU **Vol. 7** 481, 1172
NON C'E AMORE PIU GRANDE **Vol. 7** 106
NON C'E BISOGNO DI DENARO **Vol. 7** 16, 621, 665
NON C'E DUE SENZA QUATTRO **Vol. 7** 585
NON C'E PACE TRA GLI ULIVI **Vol. 7** 401, 923
NON CI RESTA CHE PIANGERE **Vol. 7** 29, 482, 988
NON CONTATE SU DI NOI **Vol. 7** 1105
NON COUPABLE **Vol. 2** 15, 501
NON E' MAI TROPPO TARDI **Vol. 7** 257
NON E'VERO, MA CI CREDO **Vol. 7** 1063
NON FACCIO LA GUERRA, FACCIO L'AMORE **Vol. 8** 306
NON FUMO E ... SENTO ODOR DI FUMO **Vol. 7** 131
NON HO PAURA DI VIVERE **Vol. 7** 113
NON HO TEMPO **Vol. 7** 1173
NON ME LO DIRE ! **Vol. 7** 443, 1136
NON MI MUOVO ! **Vol. 7** 792, 1063
NON MI SPOSO PIU **Vol. 7** 862, 1061
NON PERDIAMO LA TESTA **Vol. 7** 910
NON SCHERZARE CON LE DONNE **Vol. 7** 910
NON SI DEVE PROFANARE IL SONNO DEI MORTI **Vol. 7** 627
NON SI SCRIVE SUI MURI A MILANO **Vol. 7** 741
NON SI SEVIZIA UN PAPERINO **Vol. 7** 834
NON SIAMO SPOSATI **Vol. 7** 1177
NON SIN GELOSA **Vol. 7** 508, 756, 796
NON, SON PIU GUAGLIONE **Vol. 7** 739
NON SONO SUPER-STIZIOSO...MA ! **Vol. 7** 443, 796
NON SPARATE SUI BAMBINI **Vol. 7** 289
NON STOP **Vol. 6** *page 275, page 282*
NON TI CONOSCO PIU **Vol. 7** 493, 709
NON TI CONOSCO PIU AMORE **Vol. 7** 381, 599
NON TI PAGO ! **Vol. 7** 638, 792
NON TUTTA IO MORRO ! **Vol. 7** 428
NON UCCIDERE ! **Vol. 7** 428
NON VOGLIAMO MORIRE **Vol. 7** 657
NONA **Vol. 3** 31, 72 **Vol. 6** *page 46*
NŐNEK MINDIG SIKERÜL, A **Vol. 9** 315, 322
NONNA FELICITA **Vol. 7** 443, 902
NONNA SABELLA, LA **Vol. 7** 333
NONNE UND TÄNZERIN **Vol. 4** 270, 570
NORA **Vol. 4** 246, 742
NORD-ATLANTIQUE **Vol. 2** 99, 174, 402
NORDHAVETS MAEND **Vol. 5** 7
NORDISK KVADRILLE **Vol. 5** 257
NORDLICHT **Vol. 4** 329, 378, 567
NORDPOL-AHOI ! **Vol. 4** 15, 51, 437
NORGE FOR FOLKET **Vol. 5** 670, 760
NORIS **Vol. 7** 369
NORMANDIE-NIEMEN **Vol. 2** 398, 625
NORMANNERNE **Vol. 5** 125, 159, 283
NORMANNI, I **Vol. 7** 1044
NORRLÄNNINGAR **Vol. 5** 870, 891
NORRTULLSIGAN **Vol. 5** 1210
NORSKE BYGGERLOSSER **Vol. 5** 718, 740
NOS INTENTIONS SONT PACIFIQUES **Vol. 2** 351
NOS MONDAINES **Vol. 2** 678
NOS POR CA TODOS BEM **Vol. 8** 118, 185
NOSFERATU **Vol. 4** 214, 729
NOSOTROS DOS **Vol. 8** 637, 749
NOSOTROS EN PARTICULAR **Vol. 8** 245, 324
NOSOTROS LOS DETECTIVES **Vol. 8** 438, 546, 577
NOSTALGHIA **Vol. 7** 363, 607, 1112

NOSTALGIA DEL CARCERE **Vol. 7** 1189
NOSTALGIE **Vol. 2** 27, 67, 488, 676, 863
NOSTALGIE DE LA PUREE, LA **Vol. 2** 159, 330
NÖSTENYFARKAS **Vol. 9** 245
NOSTNITE BDENIJA NA POP VECERKO **Vol. 3** 42, 109 **Vol. 6** *page 40*
NOSTRI ANNI PIU BELLI **Vol. 7** 798
NOSTRI BUONI VILLICI, I **Vol. 7** 26, 701
NOSTRI SOGNI, I **Vol. 7** 827, 1054, 1157
NOSTRO CAMPIONE, IL **Vol. 7** 1152
NOSTRO PROSSIMO, IL **Vol. 7** 1027, 1054
NOSTTA SRESTU TRINADESETI **Vol. 3** 93, 126 **Vol. 6** *page 27*
NÖSÜLNI TILOS **Vol. 9** 3
NOSZTY FIU ESETE TOTH MARIVAL, A **Vol. 9** 71, 149, 164, 241, 259, 322
NOTE MIT DEM SILBERKREUZ, DIE **Vol. 4** 604, 652
NOTGEMEINSCHAFT HINTERHAUS **Vol. 4** 325, 742
NOTRE-DAME AU PARADIS **Vol. 2** 475
NOTRE-DAME D'AMOUR **Vol. 2** 84, 584
NOTRE-DAME DE LA MOUISE **Vol. 2** 122, 698
NOTRE-DAME DE PARIS **Vol. 2** 81, 409, 427, 722
NOTRE FILLE **Vol. 2** 214
NOTRE HISTOIRE **Vol. 6** *page 78, page 91*
NOTRE PAUVRE COEUR **Vol. 2** 798
NOTSCHREI HINTER GITTERN **Vol. 4** 366, 747
NOTTA PORTA CONSIGLIU, LA **Vol. 7** 1136
NOTTE AL CONFINE, UN **Vol. 7** 160
NOTTE BRAVA, LA **Vol. 7** 401, 817
NOTTE CHE DORMI I SOTTO LE STELLE, LA **Vol. 7** 704
NOTTE CON TE, UNA **Vol. 7** 848
NOTTE DEI SERPENTI, LA **Vol. 7** 1194
NOTTE DEL GRANDE ASSALTO, LA **Vol. 7** 877
NOTTE DELL'ALTA MAREA, LA **Vol. 7** 478
NOTTE DELL'ANIMA, LA **Vol. 7** 196
NOTTE DELLE BEFFE, LA **Vol. 7** 665, 941
NOTTE DI FIAMMA **Vol. 7** 40, 734
NOTTE DI FORTUNA **Vol. 7** 639, 994, 1177
NOTTE DI NEVE **Vol. 7** 621
NOTTE DI SAN LORENZO, LA **Vol. 7** 524, 1034, 1112
NOTTE DI TEMPESTA **Vol. 7** 493, 731, 1116
NOTTE DI TENTAZIONE **Vol. 7** 390, 569, 1004
NOTTE DOPO L'OPERA, UNA **Vol. 7** 460, 734
NOTTE E FATTA PER RUBARE, LA **Vol. 8** 306
NOTTE INFERNALE, UNA **Vol. 7** 525
NOTTE, LA **Vol. 7** 1167, 1214
NOTTE ROMANTICA DI DOLLY, LA **Vol. 7** 508
NOTTE ROSSA DEL FALCO, LA **Vol. 7** 1024
NOTTE SENZA DOMANI, UNA **Vol. 7** 598
NOTTE TRAGICA **Vol. 7** 1187
NOTTI BIANCHE, LE **Vol. 7** 299, 498, 988, 1142
NOTTI DEI TEDDY-BOYS, LE **Vol. 7** 1054, 1062
NOTTI DEL TERRORE, LE **Vol. 7** 678
NOTTI DI CABIRIA, LE **Vol. 7** 522, 725, 1136
NOTTI DI LUCREZIA, LE **Vol. 7** 372
NOTTI DI RASPUTIN, LE **Vol. 7** 17
NOTTI EROTICHE DEI MORTI VIVENTI **Vol. 7** 742
NOTTI PECCAMINOSE DI PIETRO L'ARETINO, LE **Vol. 7** 742
NOTTI PORNO NEL MONDO, LE **Vol. 7** 152
NOTTI ROSSE **Vol. 7** 390, 1206
NOTTI SEGRETE DI LUCREZIA BORGIA, LE **Vol. 7** 609, 761 **Vol. 8** 299
NOTTURNI **Vol. 7** 1156
NOTTURNO **Vol. 7** 1213
NOTTURNO CON GRIDA **Vol. 7** 478
NOTTURNO DI CHOPIN, UN **Vol. 7** 731
NOTTURNO TRAGICO **Vol. 7** 266, 702
NOUS AURONS TOUTE LA MORT POUR DORMIR **Vol. 2** 600
NOUS AUTRES A CHAMPIGNOL **Vol. 2** 309, 318, 548
NOUS AVONS TOUS FAIT LA MEME CHOSE **Vol. 2** 131, 514
NOUS DEUX ? ON NE PENSE Q'A'CA **Vol. 6** *page 134*
NOUS ETIONS TOUS DES NOMS D'ARBRES **Vol. 6** *page 88, page 130*
NOUS ETIONS UN SEUL HOMME **Vol. 2** 2 **Vol. 6** *page 71*
NOUS IRONS A MONTE-CARLO **Vol. 2** 347, 807
NOUS IRONS A PARIS **Vol. 2** 627, 807
NOUS IRONS ADEAUVILLE **Vol. 2** 738
NOUS IRONS TOUS AU PARADIS **Vol. 2** 434, 564
NOUS LES GOSSES **Vol. 2** 9, 48
NOUS MAIGRIRONS ENSEMBLE **Vol. 2** 60
NOUS NE SOMMES PAS MARIES **Vol. 2** 194 **Vol. 7** 1177
NOUS NE SOMMES PLUS DES ENFANTS **Vol. 2** 488, 775, 803

NOUS N'IRONS PLUS AU BOIS **Vol. 2** 234, 570
NOUS NOUS SOMMES RENCONTRES DANS UN REVE **Vol. 2** 450
NOUS SOMMES TOUS DES ASSASSINS **Vol. 2** 124, 197
NOUVEAU JOURNAL D'UNE FEMME EN BLANC, LE **Vol. 2** 251, 427
NOUVELLE MISSION DE JUDEX, LA **Vol. 2** 330, 432
NOUVELLES RENCONTRES **Vol. 6** *page 93*
NOUVELLES ROUTES DE L'OPIUM, LES **Vol. 2** 301
NOVANTA GIORNI, I **Vol. 7** 108
NOVANTANOVE DISGRAZIE DEL SIG. CAMILLO **Vol. 7** 26
NOVE OSPITI PER UN DELITTO **Vol. 7** 1003
NOVE STELLE, LE **Vol. 7** 108
NOVEAU TESTAMENT, LE **Vol. 2** 48, 262, 454, 726
NOVEAUX ARISTOCRATES, LES **Vol. 2** 322, 738
NOVEAUX CONTES IMMORAUX, LES **Vol. 2** 215
NOVEAUX EXPLOITS DE NICK CARTER, LES **Vol. 2** 711
NOVEAUX MAITRES, LES **Vol. 2** 469, 858
NOVEAUX MESSIEURS, LES **Vol. 2** 237, 573, 656
NOVEAUX RICHES, LES **Vol. 2** 84, 197, 526, 726
NOVEAUX ROMTANTIQUES, LES **Vol. 2** 294, 806
NOVECENTO **Vol. 7** 484, 1101
NOVELLE GALEOTTE D'AMORE DAL „DECAMERON" **Vol. 7** 694
NOVELLE LICENZIOSE DI VERGINI VOGLIOSE **Vol. 7** 742
NOVENTA MINUTOS **Vol. 8** 298, 524
NÖVEREK **Vol. 9** 301
NOVIA DE JUAN LUCERO, LA **Vol. 8** 306, 600
NOVICES, LES **Vol. 2** 481
NOVICIA REBELDE, LA **Vol. 8** 261, 299
NOVINAR **Vol. 3** 371, 421
NOVIO A LA VISTA **Vol. 8** 233, 261, 298, 551, 611, 614
NOVIO DE MAMA, EL **Vol. 8** 606
NOVIO PARA DOS HERMANAS, UN **Vol. 8** 750
NOVIOS 68 **Vol. 8** 524, 725
NOVIOS, LOS **Vol. 8** 598, 626
NOVIZIA, LA **Vol. 7** 1179
NOWY **Vol. 1** 146, 147 **Vol. 6** *page 246*
NOZ **Vol. 3** 366, 426
NOZ W WODZIE **Vol. 1** 198, 242, 303
NOZZE DI SANGUE **Vol. 7** 16, 40, 1060, 1136
NOZZE VAGABONDE **Vol. 7** 4, 196, 640, 963
NTANTADES **Vol. 3** 189
NU BÖRJAR LIVET **Vol. 5** 909, 1173
NU FILMAM SA NE AMUZAM **Vol. 3** 250, 268, 269, 280
NU GAR DEN PA DAGMAR **Vol. 5** 125, 169, 283
NU STIGER DEN **Vol. 5** 23, 273, 281
NU VOTO A MAMME SCHIAVONA **Vol. 7** 314
NU VREAU SA MA INSOR **Vol. 3** 230, 278, 286, 317
NUAGE ENTRES LES DENTS, UN **Vol. 2** 142, 779
NUAGE, UN **Vol. 2** 798
NUBES DE VERANO **Vol. 8** 143, 692
NUCCIA LA PECORAIA **Vol. 7** 1189
NUCLEAIRE DANGER IMMEDIAT **Vol. 2** 502, 628
NUCLEO ZERO **Vol. 7** 233, 515, 610
NUDE PER L'ASSASSINO **Vol. 7** 335
NUDO DI DONNA **Vol. 7** 80, 383, 1008
NUDO DI ZINGARA **Vol. 7** 824
NUDO E CRUDELE **Vol. 7** 18, 694
NUESTRA NATACHA **Vol. 8** 522, 548
NUESTRO AGENTE EN CASABLANCA **Vol. 8** 214
NUEVA CENICIENTA, LA **Vol. 8** 214, 261
NUEVAS AMISTADES **Vol. 8** 221, 258
NUEVE CARTAS A BERTA **Vol. 8** 406, 747
NUEVO EN ESTA PLAZA **Vol. 8** 524, 725
NUEVOS CURANDEROS, LOS **Vol. 8** 213, 740
NUEVOS ESPANOLES, LOS **Vol. 8** 438, 681
NUEVOS EXTRATERRESTRES, LOS **Vol. 8** 524, 689
NUFARUL ROSU **Vol. 3** 224, 244, 322, 329
NUIT A LAS VEGAS, UNE **Vol. 2** 224
NUIT A MEGEVE, UNE **Vol. 2** 475, 706, 795
NUIT A TABARIN, UNE **Vol. 2** 67, 92
NUIT AMERICAINE, LA **Vol. 2** 351, 459
NUIT AU MOULIN ROUGE, UNE **Vol. 2** 243
NUIT AU PARADIS, UNE **Vol. 2** 397 **Vol. 6** *page 83*
NUIT AUX BALEARES, UNE **Vol. 2** 671, 848
NUIT BLANCHE **Vol. 2** 128, 488
NUIT BLANCHE ET ROUGE A LEVRES **Vol. 2** 293, 294
NUIT CLAIRE, LA **Vol. 2** 654
NUIT DE DECEMBRE **Vol. 2** 248, 279, 407, 579

NUIT DE FOLLIES, UNE Vol. 2 308, 419, 533
NUIT DE LA MORT, LA Vol. 2 198 Vol. 6 *page 113*
NUIT DE LA REVANCHE, LA Vol. 2 145
NUIT DE NOCES Vol. 2 67
NUIT DE NOCES, UNE Vol. 2 55, 475, 688
NUIT DE SAINT-GERMAIN-DES-PRES, LA Vol. 2 450
NUIT DE SYBILLE, LA Vol. 2 124, 615
NUIT DES ADIEUX, LA Vol. 2 398, 427
NUIT DES BULGARES, LA Vol. 2 675, 854
NUIT DES ESPIONS, LA Vol. 2 248, 598, 738
NUIT DES GENERAUX, LA Vol. 2 224
NUIT DES SUSPECTS, LA Vol. 2 131
NUIT DES TRAQUES, LA Vol. 2 270, 559, 661
NUIT D'OR Vol. 2 515
NUIT DU CARREFOUR, LA Vol. 2 37, 160, 533
NUIT D'UN VENDREDI 13, LA Vol. 2 277
NUIT EST A NOUS, LA Vol. 2 197, 395, 559, 585
NUIT EST MON ROYAUME, LA Vol. 2 8, 540, 598, 615
NUIT FANTASTIQUE, LA Vol. 2 538, 601, 615
NUIT INFIDELE, LA Vol. 2 172
NUIT MERVEILLEUSE, LA Vol. 2 566
NUIT S'ACHEVE, LA Vol. 2 727 Vol. 7 968
NUIT TOUS LES CHATS SONT GRIS, LA Vol. 2 434, 633
NUITS Vol. 2 27
NUITS ANDALOUSES Vol. 2 395, 627
NUITS BLANCHES DE SAINT-PETERSBOURG, LES Vol. 2 427, 574
NUITS BLANCHES SUR FOND D'OR NOIR Vol. 2 553
NUITS BRULANTES DE LINDA, LES Vol. 2 139
NUITS CHAUDES DE JUSTINE, LES Vol. 2 594
NUITS D'ALERTE Vol. 2 67, 698
NUITS DE FEU Vol. 2 332, 526, 626, 824
NUITS DE MONTMARTRE, LES Vol. 2 131, 777
NUITS DE PARIS Vol. 2 279, 488
NUITS DE PIGALLE Vol. 2 475, 658
NUITS DE PRINCES Vol. 2 122, 153, 775
NUITS MOSCOVITES, LES Vol. 2 23, 626, 679
NUITS PORNOGRAPHIQUES Vol. 2 349
NUITS ROUGES Vol. 2 532
NUITS SANS FIN Vol. 2 88, 514
NUKKEKAUPPIAS JA KANIS LILITH Vol. 5 443, 566
NUMERO 121, IL Vol. 7 277
NUMERO DEUX Vol. 2 527
NUMMISUUTARIT Vol. 5 356, 379, 392, 406, 428
... NUN E CARMELA MIA ! Vol. 7 424
NUN HATS GESCHNAPPT Vol. 6 *page 186*
NUNCA ES DEMASIADO TARDE Vol. 8 537, 753
NUNCA PASA NADA Vol. 8 258, 311
NUNTA DE PIATRA Vol. 3 223, 237, 316, 324
NUORA GIOVANE, LA Vol. 7 1022
NUORENA NUKKUNUT Vol. 5 334, 528
NUORI IHMISIÄ Vol. 5 349, 493, 566
NUORI LUOTSI Vol. 5 428, 640
NUORI MYLLÄRI Vol. 5 379, 392, 564
NUORUUS SUMUSSA Vol. 5 386, 395
NUORUUS VAUHDISSA Vol. 5 312, 463, 579
NUOVA CAMERIERA Vol. 7 1189
NUOVE BARBARI, I Vol. 7 1184, 1213
NUOVI ANGELI, I Vol. 7 128, 333, 781
NUOVI FURTO DI ROBINET, UN Vol. 7 1189
NUOVI MOSTRI, I Vol. 7 333
NUR AM RHEIN Vol. 4 149, 157, 219, 660
NUR DU Vol. 4 239, 479
NUR EIN DIENER Vol. 4 113, 780
NUR EIN VIERTELSTÜNDCHEN Vol. 4 379
NUR EINE FRAU Vol. 1 56, 81, 129
NUR EINE TÄNZERIN Vol. 4 527, 626
NUR NICHT WEICH WERDEN, SUSANNE ! Vol. 4 215, 646, 727
NUSJA Vol. 3 15, 26 Vol. 6 *page 14*
NUSJA DHE SHETRRETHIMI Vol. 3 10, 15
NUTRICE, LA Vol. 7 993
NUVOLA DI POLVERE ... UN GRIDO DI MORTE ... ARRIVA SARTANA Vol. 8 598
NXENESIT E KLASES S'IME Vol. 6 *page 17*
NY DAG GRYER, EN Vol. 5 151, 252
NY HAT TIL MADAMEN, EN Vol. 5 97
NY HUSASSISTENT, DEN Vol. 5 71
NY ROCAMBOLE, DEN Vol. 5 295
NYA HYSS AV EMIL LÖNNEBERGA Vol. 5 862, 1027, 1093
NYAR A HEGYEN Vol. 9 145, 316, 339
NYARON EGYSZERÜ Vol. 9 306, 332
NYBYGGARNA Vol. 5 1052, 1153, 1192
NYCKELN OCH RINGEN Vol. 5 863, 1032
NYDELIG ONKEL, EN Vol. 5 92
NYDELIGE NELLIKER Vol. 5 682, 718, 762, 799
NYDELIGT TREKLOVER, ET Vol. 5 9
NYE BUD, DET Vol. 5 97
NYE KOKKEPIGE, DEN Vol. 5 92

NYE LEGEN, DEN **Vol. 5** 683, 723, 760, 800
NYE LENSMANNEN, DEN **Vol. 5** 669, 806
NYE LYSEGRAA, DE **Vol. 5** 81
NYGIFTA **Vol. 5** 843, 923
NYGIFTE, DE **Vol. 5** 9, 81, 83
NYHAVN 17 **Vol. 5** 55, 101, 208
NYHTA HORIS XEMEROMA **Vol. 3** 179
NYHTERINE PERIPETEIA **Vol. 3** 179
NYHTES TES ATHENAS **Vol. 3** 204
NYHTOPERPATEMATA **Vol. 3** 171
NYITOTT ABLAK **Vol. 9** 134
NYLONOVY MESIAC **Vol. 10** 32, 273, 464
NYMPHETTES, LES **Vol. 2** 259
NYMPHOMANE LUBRIQUE, LA **Vol. 2** 291
NYMPHOMANE PERVERSE, LA **Vol. 2** 52
NYMPHOMANES ET SEXES GLOUTONS **Vol. 6** *page 112*
NYOM NELKÜL **Vol. 9** 107, 246, 280
NYORDNING PA SJÖGARDA **Vol. 5** 864, 1032
NYT LEGETOJ **Vol. 5** 11, 64, 247
NYT TYVEALERMERINGSAPPARAT, ET **Vol. 5** 97
NYUGATI ÖVEZET **Vol. 9** 71, 312
NYUZAK A RUHATARBAN **Vol. 9** 67, 84, 121

O

O AGAPETIKOS TES BOSKOPOULAS **Vol. 3** 169, 203
O AGNOSTOS EKENIS TES NYHTAS **Vol. 3** 165
O AGNOSTOS TES NYHTAS **Vol. 3** 204
O AHTYPETOS HTYPETHEKE **Vol. 3** 158
O AISIODOXOS **Vol. 3** 163
O ALTE BURSCHENHERRLICHKEIT **Vol. 4** 132, 321, 424, 788
O ANEFOROS TOU GOLGOTHA **Vol. 3** 157
O ANTHROPOS TOU TRENDU **Vol. 3** 168
O ANTROPOS ME TO GARYFALO **Vol. 3** 159, 207
O ANTROPOS ROLOI **Vol. 3** 201
O APITHONAS **Vol. 3** 149
O APOHAIRETISMOS **Vol. 3** 180
O-BI, O-BA **Vol. 6** *page 216, page 244*
O CACOULHA DO BARULHO **Vol. 7** 639
O CUPPE D'A MORTE **Vol. 7** 829
O CZYM MARZA KOBIETY **Vol. 1** 384, 404
O CZYM SIE NIE MOWI **Vol. 1** 174, 176, 278
O CZYM SIE NIE MYSLI **Vol. 1** 174, 176
O DEVCICU **Vol. 10** 53
O DIESE TANTE **Vol. 1** 76, 103, 127
O DIRLADAS **Vol. 3** 141
O DRAKOS **Vol. 3** 198, 207
O DROMOS TOU LAMORE **Vol. 3** 194, 206
O DWOCH TAKICH, CO UKRADLI KSIEZYC **Vol. 1** 232, 305, 377
O ELIOS TOU THANATOU **Vol. 3** 151, 196
O EMPORAKOS **Vol. 3** 165
O, EN SA'N NATT **Vol. 5** 840, 841
O FOBOS **Vol. 3** 153, 163
O GENITAROS **Vol. 3** 203
O HORA LA TARA **Vol. 3** 236
O IPPOKRATES KAI E DEMOKRATIA **Vol. 3** 166
O IPPOLYTOS KAI TO BIOLI TOU **Vol. 3** 192
O KOSMOS TRELATHEKE **Vol. 3** 182
O KYRIOS STATHMARHES **Vol. 3** 167
O LA BORSA O LA VITA **Vol. 7** 652, 756, 796
O LACRIMA DE FATA **Vol. 3** 209, 237, 281
O LAMBIRES ENANTIA STOUS PARANOMOUS **Vol. 3** 128
O MÄDCHEN, MEIN MÄDCHEN, WIE LIEB' ICH DICH ! **Vol. 4** 49, 287, 644
O MADIANA **Vol. 2** 46
O MANOLIOS STEN EUROPE **Vol. 3** 163
'O MARENARIELLO **Vol. 7** 424
O MEGALOS ENOHOS **Vol. 3** 149, 207
O MEGALOS EROTIKOS **Vol. 3** 160, 183
O MENOUSSES **Vol. 3** 130, 132
O METHYSTAKAS **Vol. 3** 157
O MICA INTIMPLARE **Vol. 3** 278, 285, 332
O MIHAEL DEN EHI PSILA **Vol. 3** 157
O MODISTROS **Vol. 3** 142
O MORAVSKE ZEMI **Vol. 10** 197, 369, 517
O MORFES MERES **Vol. 3** 141
'O MUNACIELLO **Vol. 7** 829
'O MUNASTERIO **Vol. 7** 314
O NANOS KAI OI EFTA HIONATES **Vol. 3** 163
O NECEM JINEM **Vol. 10** 67, 87, 299
O NIKITES **Vol. 3** 142
O NOAPTE **Vol. 3** 269
O NOAPTE DE POMINA **Vol. 3** 215
O NOAPTE FURTUNOASA **Vol. 1** 278 **Vol. 3** 294
O PALATOS TES ZOES **Vol. 3** 178
O POKOJNIKU SVE NAJLEPSE **Vol. 6** *page 308, page 314, page 320*
O POVESTE CA-N BASME **Vol. 3** 259, 286, 312
O POVESTE OBISNUITA **Vol. 3** 159, 286, 312
O PRINCEZNE JASNENCE A LETAJICIM SEVCI **Vol. 10** 44, 129, 395
O PSEUTES **Vol. 3** 144, 205
O PYRGOS TON IPPOTON **Vol. 3** 198

O SALTO **Vol. 2** 234
'O SCHIAFFO **Vol. 7** 39
O SCRISOARE PIERDUTA **Vol. 3** 248, 330
O SEVCI MATOUSOVI **Vol. 10** 154, 524, 612, 621
O SLAVE A TRAVE **Vol. 10** 82, 228, 548
O SLAVNOSTI A HOSTECH **Vol. 10** 41, 289, 516
O SNEHURCE **Vol. 10** 299, 425, 496
O SOLE MIO ! **Vol. 7** 52
O STATECNEM KOVARI **Vol. 10** 165, 462
O' SURDATO 'NNAMURATO **Vol. 7** 311, 648
O THANATOS TOU ALEXANDROU **Vol. 3** 168
O THIASOS **Vol. 3** 135, 166, 188
O TOIHOS **Vol. 3** 196
O VECECH NADPRIROZENYCH **Vol. 10** 311, 388, 469, 503, 532, 621
O VELKOU CENU **Vol. 10** 26, 257
O ZATOULANE PRINCEZNE **Vol. 10** 144, 325, 395
O ZAZRACENEM KVETU **Vol. 10** 480
O ZESTOS MENAS AUGOUSTOUS **Vol. 3** 185
O ZIVEJ VODE **Vol. 10** 72, 108, 129
OASIS **Vol. 2** 171, 280, 402, 576
OASIS DU DJERID **Vol. 2** 811
OASPETI DE SEARA **Vol. 3** 208, 225, 286
OAZA **Vol. 10** 283, 428, 577
OBCAN BRYCH **Vol. 10** 229, 503
OBCHOD NA KORZE **Vol. 10** 388, 503
OBECANA ZEMIJA **Vol. 6** *page 310, page 314*
OBEN FÄHRT DER GROSSE WAGEN **Vol. 1** 7, 15, 85
OBEN GEBLIEBEN IST NOCH KEINER **Vol. 6** *page 162, page 164, page 170*
OBERDAN **Vol. 7** 369
OBERSTEN OG SKLIDVAGTEN **Vol. 5** 97
OBERWACHTMEISTER SCHWENKE **Vol. 4** 626, 634
OBIC **Vol. 3** 59, 60 **Vol. 6** *page 27*
OBJECTIF 500 MILLIONS **Vol. 2** 513
OBJECTIVO BI - KI - NI ! **Vol. 8** 331, 553
OBJEV NA STRAPATE HURCE **Vol. 10** 369, 524, 569
OBJIZDKA **Vol. 10** 193, 296, 433
OBOK PRAWDY **Vol. 1** 305, 330 **Vol. 6** *page 240*
OBRACENI FERDYSE PISTORY **Vol. 10** 34, 138, 524, 599
OBRACUN **Vol. 3** 419, 443
OBRAZKI Z ZYCIA **Vol. 6** *page 215, page 218*
OBRONCY LWOWA **Vol. 1** 404
OBSESION **Vol. 8** 219, 276
OBSESIONES DE ARMANDO, LAS **Vol. 8** 667
OBSESSION **Vol. 2** 601, 722, 832
OBSESSION CHARNELLES **Vol. 6** *page 93*
OBSESSIONS PORNOS **Vol. 2** 294
OBSITOS, AZ **Vol. 9** 337
OBSTACULO, EL **Vol. 8** 614
OBUSKU, Z PYTELE VEN ! **Vol. 10** 345, 532, 550, 562
OBYCAJNY SPAS **Vol. 10** 93, 155, 188
OBZALOVANY **Vol. 10** 172, 352, 359, 512
OCA ALLA COLBERT, L' **Vol. 7** 1189
OCAKVANE **Vol. 3** 46 **Vol. 6** *page 23*
OCALENIE **Vol. 1** 144, 182, 281 **Vol. 6** *page 248*
OCALIC MIASTO **Vol. 1** 156, 178, 351, 355, 393, 396
OCASO DE UN PISTOLERO **Vol. 8** 551, 626
OCCASIONI DI ROSA, LE **Vol. 7** 933, 1105
OCCHEI, OCCHEI **Vol. 7** 817
OCCHI CHE ACCUSANO, GLI **Vol. 7** 1073
OCCHI CONSACRATI **Vol. 7** 806
OCCHI DALLE STELLE **Vol. 7** 763
OCCHI DELLA MORTA, GLI **Vol. 7** 1116
OCCHI DI VETRO **Vol. 7** 81
OCCHI DIPINTI **Vol. 7** 1225
OCCHI FREDDI DELLA PAURA, GLI **Vol. 8** 261, 306
OCCHI, LA BOCCA, GLI **Vol. 7** 96, 607, 801, 1043
OCCHI LUCENTI **Vol. 7** 320, 429
OCCHI SENZA LUCE, GLI **Vol. 7** 891
OCCHIO ALLA PENNA **Vol. 7** 524, 875
OCCHIO ALLA VEDOVA ! **Vol. 7** 1147
OCCHIO DIETRO LA PARETE, L' **Vol. 7** 909
OCCHIO MALOCCHIO PREZZEMOLO E FINOCCHIO **Vol. 7** 426, 504
OCCHIO NERO, OCCHIO BIONDO, OCCHIO FELINO ... **Vol. 7** 875, 1105
OCCHIO RIVELATORE O L'OCCHIO DELLA MORTE, L' **Vol. 7** 1116
OCCIDENT, L' **Vol. 2** 6, 248, 249, 388, 724
OCCIDENTE Y SABOTAJE **Vol. 8** 477, 495
OCCULTISMO **Vol. 7** 1211
OCCUPE-TOI D'AMELIE **Vol. 2** 47, 251, 833
... OCH ALLA DESSA KVINNOR **Vol. 5** 843, 923, 1044
... OCH EFTER SKYMNING KOMMER MÖRKER **Vol. 5** 970
OCHI DE URS **Vol. 6** *page 265, page 269, page 279*
„OCHRANA" WARZAWSKA I JEJ TAJEMNICE **Vol. 1** 326
OCHSE VON KULM, DER **Vol. 1** 58, 88
OCHSENKRIEG, DER **Vol. 4** 329, 378, 527

OCHSENMENUETT, DAS Vol. 4 239, 340, 466
OCI PRO PLAC Vol. 10 41, 254, 531
OCOVSKE PASTORALE Vol. 10 82, 228, 399
OCSKAY BRIGADEROS Vol. 9 200
OCTAVO MANDAMIENTO, EL Vol. 8 652
OCTOBRE A MADRID Vol. 2 391
OD KOBIETY DO KOBIETY Vol. 1 371, 407
OD PETKA DO PETKA Vol. 6 *page 320, page 321*
OD VRAZDY JENOM KROK KE LZI Vol. 10 41, 157, 531
OD ZITRKA NECARUJI Vol. 10 173, 425, 577
OD ZLATA JABUKA Vol. 6 *page 312, page 319, page 322*
ODDECHOVY CAS Vol. 10 222, 415, 428
ODDS 777 Vol. 5 101, 208
ÖDERMARKSPRÄSTEN Vol. 5 1052, 1164
ÖDESDIGERT INKOGNITO, ETT Vol. 5 998
ÖDESDIGRA KLOCKAN, DEN Vol. 5 1137
ODESSA IN FIAMME Vol. 7 196, 453
ÖDETS MAN Vol. 5 966
ÖDETS REDSKAP Vol. 5 1199
ODETTE Vol. 2 623 Vol. 7 264, 701, 1136
ODEUR DES FAUVES, L' Vol. 2 570, 808
ODHALENIE ALZBETY BATHORYCKY Vol. 10 268, 442, 451
ODIO Vol. 8 575, 640
ODIO LE BIONDE Vol. 7 263, 514, 585
ODIO MI CUERPO Vol. 8 221, 708
ÖDIPUS Vol. 1 34, 73, 104
ODISEA E TIFOZAVE Vol. 3 12, 22
ODISSEA DI SAN GIOVANNI, L' Vol. 7 316
ODISSEA NUDA Vol. 7 408
ODPLATA Vol. 10 167
ODVAZNA SLECNA Vol. 10 522, 621
ODVETA Vol. 10 263, 269, 399
ODWET Vol. 6 *page 258, page 260*
ODWIEDZINY PREZYDENTA Vol. 1 321, 377, 400
ODYGENS BELÖNING Vol. 5 863, 1032
ODYSSEUS A HVEZDY Vol. 10 110, 139, 144, 408
OEDIPUS ORCA Vol. 7 544
OEIL DE LYNX, DETECTIVE Vol. 2 41, 425
OEIL DU MAITRE, L' Vol. 2 71 Vol. 6 *page 108*
OEIL DU MALIN, L' Vol. 2 700
OEIL DU MONOCLE, L' Vol. 2 118, 293
OEIL EN COULISSES, L' Vol. 2 595, 791
OEIL, L' Vol. 6 *page 93*
OEIL POUR OEIL Vol. 2 197, 248,566
OESTE NEVADA JOE Vol. 8 768
OEUF, L' Vol. 2 263
OEUFS BROUILLES, LES Vol. 2 51
OEUFS DE L'AUTRUCHE, LES Vol. 2 128, 559, 661
OFFICIER DE POLICE SANS IMPORTANCE, UN Vol. 2 219, 787
OFFICINA DEL GRIGIONE, L' Vol. 7 968
OFIARY FANATYZMU Vol. 1 174, 188
OFICIOS DE CANDIDO, LOS Vol. 8 298, 318
OG SA ER DER BAL BAGEFTER Vol. 5 52, 243, 294
OGGETTI SMARRITI Vol. 7 142, 1105
OGGI, DOMANI, DOPODOMANI - L'UOMO DEI CINQUE PALLONI Vol. 7 1136
OGGI SPOSI Vol. 7 48, 852
OGIFT FADER SÖKES Vol. 5 933, 1042
OGNENA DIRJA Vol. 3 56, 90
OGNI GIORNI E DOMENICA Vol. 7 827
OGNIE SA JESZCZE ZYWE Vol. 1 156, 233 Vol. 6 *page 258*
OGNIOMISTRZ KALEN Vol. 1 187, 206, 281 Vol. 6 *page 247*
OGRADA Vol. 3 387
OGRO Vol. 7 54, 502, 855 Vol. 8 610
OH ANGELINA Vol. 7 600
OH DIESE MÄNNER Vol. 4 292, 391, 557
OH DOLCI BACI E LANGUIDE CAREZZE Vol. 7 257
OH ! MIA BELLA MATRIGNA Vol. 7 501
OH ! QUE MAMBO Vol. 2 131, 205
OH ! QUEL BOTTONE ! Vol. 7 1189
OH, SERAFINA ! Vol. 7 221, 345, 931
OHA, DIESE EHEMÄNNER Vol. 4 365, 608
OHI Vol. 3 149, 151, 185
OHLEDNUTI Vol. 10 395, 505, 517
OHM KRÜGER Vol. 4 49, 531, 536, 626, 729
OHNE FLEISS KEIN PREIS Vol. 4 367, 744
OHNE GELD DURCH DIE WELT Vol. 4 796
OHNE KAMPF KEIN SIEG Vol. 1 38, 53, 59
OHNE PASS IN FREMDEN BETTEN Vol. 1 38, 86
OHNEM A MECEM Vol. 10 115
OHNIVE KRIZOVATKY Vol. 10 155, 270, 273
OHNIVE LETO Vol. 10 92, 621
OHO, SANOI EEMELI Vol. 5 443, 549
OHRFEIGE, DIE Vol. 1 49, 100, 104
OI, AIKA VANHA KULTAINEN... Vol. 5 356, 386
OI ANDRES XEROUN NA AGAPOUN Vol. 3 161
OI APAHEDES TON ATHENON Vol. 3 154
OI ASSI TOU GEPEDOU Vol. 3 198

OI DYO TYHEROI **Vol. 3** 140
OI EGOISTES **Vol. 3** 163
OI EROES **Vol. 3** 200
OI GENAIOI PENTHAINOUN DYO FORES **Vol. 3** 153, 205
OI GERMANOI XANARHONTAI **Vol. 3** 157
OI GYNAIKES SEMERA **Vol. 3** 189
OI, KALLIS SUOMENMAA **Vol. 5** 356, 386
OI KLERONOMOI TOU KARAMPOUMPOUNA **Vol. 3** 165
OI KYINIGOI **Vol. 3** 135
OI MNESTERES TES PENELOPES **Vol. 3** 149
OI, MUISTATKOS... **Vol. 5** 392, 397
OI NEOI THELOUN NA ZESOUN **Vol. 3** 170
OI PEREFANOI **Vol. 3** 170
OI PROSTATES **Vol. 3** 191
OI SFAIRES DEN GYRIZOUN PISO **Vol. 3** 144
OI TELEUTAIOI TOU ROUPEL **Vol. 3** 151, 185
OI TEMPELIDES TES EUFORES KOILADAS **Vol. 3** 151, 181
OI THALLASIES OI HADRES **Vol. 3** 135, 205
OI TREIS PSEUTES **Vol. 3** 165
OIKOGENEIA HORAFA **Vol. 3** 141
OISEAU DE PARADIS, L' **Vol. 2** 503
OISEAU RARE, L' **Vol. 2** 17, 163
OISEAU RARE, UN **Vol. 2** 48, 67, 439
OISEAUX DE NUIT, LES **Vol. 2** 464
OISEAUX VONT MOURIR AU PEROU, LES **Vol. 2** 141
OJ OJ OJ ELLER SANGEN OM DEN ELDRÖDA HUMMERN **Vol. 5** 835, 940, 1081
OJCIEC KROLOWEJ **Vol. 6** *page 240, page 245, page 249*
OJEBLIKKET **Vol. 5** 11, 38, 165
OJO DE CRISTAL, EL **Vol. 8** 217, 501
OJO DEL HURACAN, EL **Vol. 8** 533, 762
OJOS AZUES DE MUNECA ROTA, LOS **Vol. 8** 506, 608
OJOS FRIOS DEL MIEDO, LOS **Vol. 8** 306
OJOS PERDIDOS, LOS **Vol. 8** 541, 598
OJOS VERDES **Vol. 8** 477, 495
ÖK KETTEN **Vol. 9** 20, 154, 280
O.K. NERONE **Vol. 7** 283, 453, 798
O.K. PATRON **Vol. 2** 293
O.K.: YEVTUSHENKO **Vol. 8** 249
OKENKO **Vol. 10** 256, 524
OKENKO DO NEBE **Vol. 10** 168, 532
O.K.FILE **Vol. 3** 142
OKLAHOMA JOHN **Vol. 8** 732
OKLAMANY HYPNOTISER SWENGALI **Vol. 10** 115, 260
OKNO **Vol. 6** *page 228, page 238, page 252*
OKO PROROKA **Vol. 6** *page 255, page 258*
OKOLICE SPOKOJNEGO MORZA **Vol. 6** *page 221, page 249, page 252*
OKOS MAMA, AZ **Vol. 9** 94, 126, 164, 288
OKOUZLENA **Vol. 10** 412, 569, 621
OKOVANI SOFERI **Vol. 6** *page 309*
OKOVY **Vol. 10** 198, 345
OKRAGLY TYDZIEN **Vol. 1** 391
OKRUTNY OJCIEC **Vol. 1** 317
OKSAT POIS **Vol. 5** 374, 443, 549
OKTOBER – DAGE **Vol. 5** 112, 179, 237
OKTOBER – ROSER **Vol. 5** 161, 304
OKTOBERI VASARNAP **Vol. 9** 192, 259, 311
OKUPACIJA U 26 SLIKA **Vol. 3** 360, 370, 415
OKURKOVY HRDINA **Vol. 10** 550, 626
OLA & JULIA **Vol. 5** 940
OLANDA PITTORESCA, L' **Vol. 7** 711
OLAVI **Vol. 9** 27, 93
OLD FIREHAND **Vol. 3** 444
OLD SETERHEND **Vol. 6** *page 312*
OLD SHATTERHAND **Vol. 6** *page 312*
OLD SUREHAND **Vol. 3** 444
OLDAS ES KÖTES **Vol. 9** 261, 273, 318
OLDRICH A BOZENA **Vol. 10** 139, 283, 325
OLE DOLE DOFF **Vol. 5** 1192
OLE KREIVITÄR, EN **Vol. 5** 328, 471, 481, 566
OLE OPFINDERS OFFER **Vol. 5** 24, 83
OLE, TORERO ! **Vol. 2** 440 **Vol. 8** 484, 625
OLEMME KAIKKI SYYLLISIÄ **Vol. 5** 516, 592, 634
OLENKA MINÄ TULLUT HAAREMIIN ? **Vol. 5** 356, 359, 428, 485
OLET MENNYT MINUN VEREENI **Vol. 5** 316, 458
OLGA AGAPE MOU **Vol. 3** 144, 205
OLI **Vol. 7** 454
OLIMPIADA 40 **Vol. 6** *page 221*
OLIMPIADI DEI MARITI, LE **Vol. 7** 693
OLIN NAHJUKSEN VAIMO **Vol. 5** 443, 639
OLIVIA **Vol. 2** 279, 516, 566
OLIVIER MESSIAEN ET LES OISEAUX **Vol. 2** 818
OLKUSKENS DROM **Vol. 5** 97
OLLE HENRY **Vol. 6** *page 161, page 171*
OLLIN OPPIVUODET **Vol. 5** 344, 356
OLOCAUSTO, L' **Vol. 7** 837
OLOVENY CHLEB **Vol. 10** 154, 313, 388, 514
OLOVNA BRIGADA **Vol. 3** 384, 434
OLSEN BANDEN **Vol. 5** 16, 61, 152, 195
OLSEN – BANDEN **Vol. 5** 718, 762
OLSEN – BANDEN DERNDA' **Vol. 5** 16, 61, 195

OLSEN - BANDEN FLUGT OVER PLANHKEVAERKET **Vol. 5** 11, 61, 111, 169
OLSEN - BANDEN FOR FULL MUSIKK **Vol. 5** 727, 762, 777
OLSEN - BANDEN GAR AMOK **Vol. 5** 16, 61, 272
OLSEN - BANDEN GAR I KRIG **Vol. 5** 16, 61, 131, 152
OLSEN - BANDEN GIR SEG ALDRI ! **Vol. 5** 703, 728, 762
OLSEN - BANDEN MOTER KONGEN OG KNEKTEN **Vol. 5** 707, 762, 824
OLSEN - BANDEN OG DATA - HARRY SPRENGER VERDENSBANKEN **Vol. 5** 703, 755, 762
OLSEN - BANDEN OG DYNAMITT - HARRY **Vol. 5** 718, 762, 819
OLSEN - BANDEN OG DYNAMITT - HARRY GAR AMOK **Vol. 5** 762, 819
OLSEN - BANDEN OG DYNAMITT - HARRY MOT NYE HOYDER **Vol. 5** 703, 728, 762
OLSEN - BANDEN OG DYNAMITT - HARRY PA SPORET **Vol. 5** 703, 712, 777
OLSEN - BANDEN OVER ALLE BJERGE **Vol. 5** 16, 61, 111, 169
OLSEN - BANDEN OVERGIVER SIG ALDREG **Vol. 5** 16, 61, 131, 195
OLSEN - BANDEN PA SPANDEN **Vol. 5** 16, 61, 272
OLSEN - BANDEN PA SPORET **Vol. 5** 16, 61, 195
OLSEN - BANDEN SER RODT **Vol. 5** 11, 16, 61, 125, 179, 195
OLSEN - BANDEN TAR GULL **Vol. 5** 762
OLSEN - BANDEN Y JYLLAND **Vol. 5** 16, 61, 272
OLSEN - BANDENS ALLER SISTE KUPP **Vol. 5** 762, 766, 815
OLSEN - BANDENS SIDSTE BEDRIFTER **Vol. 5** 16, 61, 179
OLSEN - BANDENS SISTE BEDRIFTER **Vol. 5** 712, 824
OLSEN - BANDENS STORE KUP **Vol. 5** 16, 61, 272
OLSNIENIE **Vol. 1** 159, 163, 275, 276
OLTRAGGIO AL PUDORE **Vol. 7** 1179
OLTRE I CONFINI DELL'ANIMA **Vol. 7** 108
OLTRE LA PORTA **Vol. 7** 434, 1142, 1145
OLTRE L'AMORE **Vol. 7** 196, 453, 671, 1000
OLVIDA LOS TAMBORES **Vol. 8** 207, 346
OLYAN, MINT OTTHON **Vol. 9** 20, 167, 280
OLYCKSFAGELN **Vol. 5** 864
OLYCKSFAGELN NR. 13 **Vol. 5** 923
OLYMPIA **Vol. 4** 117 **Vol. 7** 798
OM IN LODEN, UN **Vol. 3** 269, 328
OM KAP MED DODEN **Vol. 5** 295
OM KJAERLIGHETEN SYNGER DE **Vol. 5** 670, 760
OM SJU FLICKOR **Vol. 5** 917, 1027, 1093
OM TILLA **Vol. 5** 724, 800
OMBRA DEL BUON FORZATO, L' **Vol. 7** 454
OMBRA DEL SOGNO, L' **Vol. 7** 230
OMBRA DELLA COLPA, I **Vol. 7** 441 **Vol. 73** 441
OMBRA DI UN TRONO, L' **Vol. 7** 569
OMBRA FATALE, L' **Vol. 7** 1078
OMBRA IMPLACABILE, L' **Vol. 7** 545, 598
OMBRA, L' **Vol. 7** 48, 264, 333, 399, 701
OMBRA MISTERIOSA, L' **Vol. 7** 934
OMBRE **Vol. 7** 941
OMBRE BIANCHE **Vol. 7** 1136
OMBRE DE LA TERRE, L' **Vol. 6** *page 98*
OMBRE DECHIREE, L' **Vol. 2** 511
OMBRE DES CHATEAUX, L' **Vol. 2** 12, 517
OMBRE DU LOUP, L' **Vol. 2** 234
OMBRE DU PECHE, L' **Vol. 2** 871
OMBRE D'UN JEU, L' **Vol. 6** *page 140*
OMBRE D'UNE CHANCE, L' **Vol. 2** 730, 866
OMBRE ET LA NUIT, L' **Vol. 2** 13, 643 **Vol. 6** *page 103*
OMBRE ET LUMIERE **Vol. 2** 124, 598
OMBRE, L' **Vol. 2** 57
OMBRE ROUGE, L' **Vol. 6** *page 87, page 103, page 131, page 147*
OMBRE SU TRIESTE **Vol. 7** 558
OMBRE UMANE **Vol. 7** 1116
OMBYTE AV TAG **Vol. 5** 954, 1164
OMBYTE FÖRNÖJER **Vol. 5** 840, 841
OMEGA, OMEGA ... **Vol. 9** 20, 198
OMENA PUTOAA **Vol. 5** 379, 579
OMICIDIO PERFETTO A TERMINE DI LEGGE, UN **Vol. 8** 611
OMORFES MERES **Vol. 3** 165
OMRINGADE HUSET, DET **Vol. 5** 937, 1031
OMRINGET **Vol. 5** 672, 724
OMSTRIDTE JORD, DEN **Vol. 5** 58
OMUL CARE NE TREBUIE **Vol. 3** 235, 271, 289
OMUL DE LINGA TINE **Vol. 3** 263, 285, 332
OMUL SI UMBRA **Vol. 6** *page 266, page 269, page 281*
ÖN **Vol. 5** 869, 959, 1052
ON A JEHO SESTRA **Vol. 10** 117, 167
ON A RETROUVE LA SEPTIEME COMPAGNIE **Vol. 2** 154, 368

ON A TROUVE UNE FEMME NUE **Vol. 2** 538, 623
ON A VOLE LA CUISSE DE JUPITER **Vol. 2** 616, 779 **Vol. 6** *page 125, page 128*
ON A VOLE UN HOMME **Vol. 2** 193, 384
ON AURA TOUT VU **Vol. 2** 21, 293
ON DEMANDE LE COLONEL **Vol. 2** 131, 604, 872
ON DEMANDE UN ASSASSIN **Vol. 2** 173, 191, 475
ON DEMANDE UN MENAGE **Vol. 2** 131, 533
ON DEMANDE UN MODELE **Vol. 2** 678
ON EFFACE TOUT **Vol. 2** 119, 534
ON EST TOUJOURS TROP BON AVEC LES FEMMES **Vol. 2** 457
ON EST VENU LA POUR S'ECLATER **Vol. 2** 294
ON LAVTALLA PIENOINEN KAHVILA **Vol. 5** 443, 620
ON N'AIME QU'UNE FOIS **Vol. 2** 205
ON N'ARRETE PAS LE PRINTEMPS **Vol. 2** 351
ON NE BADINE PAS AVEC L'AMOUR **Vol. 2** 42, 246, 452
ON NE MEURT PAS COMME CA **Vol. 2** 70, 875
ON NE ROULE PAS ANTOINETTE **Vol. 2** 425, 579, 722
ON NE TRICHE PAS AVEC LA VIE **Vol. 2** 270, 309
ON N'ENTERRE PAS LE DIMANCHE **Vol. 2** 835
ON N'EST PAS DES ANGES ... ELLES NON PLUS **Vol. 2** 335 **Vol. 6** *page 81*
ON N'EST PAS SERIEUX QUAND ON A 17 ANS **Vol. 2** 720
ON, ONA, ONI **Vol. 6** *page 221, page 243*
ON PEUT LE DIRE SANS SE FACHER **Vol. 2** 151
ON PREND LA PILULE ET ON S'ECLATE **Vol. 6** *page 143*
ON PURGE BEBE **Vol. 2** 402, 780, 797
ON S'EN FOUT ... NOUS ON S'AIME **Vol. 6** *page 89*
ON S'EST TROMPE D'HISTOIRE D'AMOUR **Vol. 2** 7, 8
ON THE BEACH **Vol. 7** 988
ON VOIT BIEN QUE N'EST PAS TOI **Vol. 2** 108
ONARATA FAMIGLIA, L' **Vol. 7** 124
ONCE MORE, WITH FEELING ! **Vol. 2** 656
ONCE PARES DE BOTAS **Vol. 8** 515, 753
ONCE UPON A TIME IN AMERICA **Vol. 7** 203, 333, 893, 1077
ONCLE DE PEKIN, L' **Vol. 2** 70, 84, 193
ONDA CIRKELN, DEN **Vol. 5** 869, 1052
ONDA ÖGON **Vol. 5** 928, 1073
ONDATA DI CALORE **Vol. 7** 19
ONDATA DI PIACERE, UNA **Vol. 7** 247
ONDE VEJE, DE **Vol. 5** 203
ONDEES BRULANTES **Vol. 2** 294
ONDINO, L' **Vol. 7** 341, 361
ONDSKANS VÄRDSHUS **Vol. 6** *page 294*
ONE NIGHT WITH YOU **Vol. 2** 826
1-2-3 CORONA **Vol. 1** 3, 117
ONE, TWO, TWO: 122, RUE DE PROVENCE **Vol. 2** 315, 457
ONE WAY **Vol. 7** 763
100 000 **Vol. 3** 179
100 DE LEI **Vol. 3** 272, 285, 331
111 – ES, A **Vol. 9** 71, 80, 218, 322
141 PERC A BEFEJEZETLEN MONDABOL **Vol. 9** 134, 303, 316, 327
113, EL **Vol. 8** 329
120 – AS TEMPO **Vol. 9** 319
120, RUE DE LA GARE **Vol. 2** 698, 832
120, RUE MONTMARTRE **Vol. 2** 16, 118, 501
127 MILLONES LIBRES DE IMPUESTOS **Vol. 8** 438, 762
ONEIRA KORITSION **Vol. 3** 157
ONEIRO TES KYRIAKES **Vol. 3** 141
ONESIME **Vol. 2** 159, 286, 330
ONESTA CHE UCCIDE **Vol. 7** 965
ONESTA DEL PECCATO, L' **Vol. 7** 1132
ONESTO MONDO, L' **Vol. 7** 83
ONKEL BILL FRA NEW YORK **Vol. 5** 180, 199, 304
ONKEL BRÄSIG **Vol. 4** 97, 124
ONKEL FRIDOLIN **Vol. 4** 228, 471
ONKEL JOAKIMS HEMMELIGHED **Vol. 5** 23, 76, 307
ONKEL OG NEVO **Vol. 5** 97
ONKRAJ **Vol. 3** 356, 418
ONNELLINEN MINISTERI **Vol. 5** 386, 417, 471, 549
ONNELLISET **Vol. 5** 397, 471
ONNELLISET LEIKIT – JUULIA **Vol. 5** 434
ONNELLISET LEIKIT – TIKKU **Vol. 5** 434, 623
ONNEN – PEKKA **Vol. 5** 356, 357
ONNEN SAARI **Vol. 5** 363
ONNENPELI **Vol. 5** 541
ONNENPOTKU **Vol. 5** 356, 428
ONNI ETSII ASUNTOA **Vol. 5** 443, 599
ONNI PYÖRII **Vol. 5** 386, 471
ONOMASTICO DI ROBINET, L' **Vol. 7** 1189
ONORATA SOCIETA **Vol. 7** 1177
ONORE CASTIGLIANO **Vol. 7** 1189
ONORE DEL CASATO **Vol. 7** 1189
ONORE DELLA FAMIGLIA, L' **Vol. 7** 250
ONORE E GUAPPARIA **Vol. 7** 666
ONORE E SANGUE **Vol. 7** 1121

ONORE VOLE ANGELINA, L' **Vol. 7** 359
ONOREVOLE CON L'AMANTE SOTTO IL LETTO, L' **Vol. 7** 811, 953, 1202
ONORI DELLA GUERRA, GLI **Vol. 7** 277
ONORIFICENZA **Vol. 7** 1189
ONSDAGSVÄNINNAN **Vol. 5** 843, 1044, 1073
ONZE HEURES SONNAIENT **Vol. 2** 59
ONZE MILLE VERGES, LES **Vol. 2** 674
OP MED HODET **Vol. 5** 669, 675, 788, 805, 831
OP MED HUMORET **Vol. 5** 89
OP MED LILLE MARTHA **Vol. 5** 151, 304
OP OG NED LANGS KYSTEN **Vol. 5** 29, 180, 199, 304
OPADLY LISCIE Z DRZEW **Vol. 1** 298, 403 **Vol. 6** *page 225*
OPAK COVEK **Vol. 3** 116 **Vol. 6** *page 40, page 46*
OPALE DE FEU, L' **Vol. 2** 396
OPASEN CAR **Vol. 6** *page 50, page 65*
OPASEN POLET **Vol. 3** 101 **Vol. 6** *page 41, page 59*
OPASNI PUT **Vol. 3** 341, 452
OPASNI TRAG **Vol. 6** *page 308, page 314, page 317*
OPATICA I KOMESAR **Vol. 3** 346, 428
OPEN SEASON **Vol. 8** 245, 616
OPERA DE QUAT'SOUS, L' **Vol. 2** 23
OPERA-MUSETTE **Vol. 2** 9, 623
OPERA PRIMA **Vol. 8** 335, 373
OPERA VE VINICI **Vol. 10** 139, 144, 496
OPERABRANDEN **Vol. 5** 9
OPERACAO DINAMITE **Vol. 8** 54
OPERACAO ESTUPEFACIENTES **Vol. 8** 54
OPERACE ME DCERY **Vol. 10** 208, 325, 541
OPERACIJA BEOGRAD **Vol. 3** 356, 366, 441
OPERACIJA TICIJAN **Vol. 3** 374
OPERACION CABARETERA **Vol. 8** 553, 626
OPERACION CONTRAESPIONAJE **Vol. 8** 732
OPERACION DALILA **Vol. 8** 602, 725
OPERACION EMBAJADA **Vol. 8** 609, 725
OPERACION GOLDMAN **Vol. 8** 732
OPERACION LADY CHAPLIN **Vol. 8** 438, 762
OPERACION MANTIS **Vol. 8** 214, 296
OPERACION MATA HARI **Vol. 8** 553, 749
OPERACION MOGADOR **Vol. 8** 258, 495
OPERACION OGRO **Vol. 8** 610
OPERACION PLUS ULTRA **Vol. 8** 524, 725
OPERACION POKER **Vol. 8** 583, 772
OPERACION SECRETARIA **Vol. 8** 600, 749
OPERACION SILENCIO **Vol. 8** 732
OPERACIONI „ZJARRI" **Vol. 3** 13, 14, 22 **Vol. 6** *page 16*
OPERASJON COBRA **Vol. 5** 772, 780, 824
OPERASJON KANO **Vol. 5** 811
OPERASJON LOVSPRETT **Vol. 5** 718, 754, 762, 799
OPERASJON SJOSPROYT **Vol. 5** 718, 754, 762, 799
OPERASJON V FOR VANVIDD **Vol. 5** 718, 762, 799, 817
OPERATERI **Vol. 10** 628
OPERATEUR TENACE, L' **Vol. 2** 159, 286, 330
OPERATION BETON **Vol. 2** 687
OPERATION KIRSEBAERSTEN **Vol. 5** 16, 120
OPERATION LADY MARLENE **Vol. 2** 154, 368
OPERATION LEO **Vol. 5** 882, 946, 1093
OPERATION MACEDOINE **Vol. 2** 659
OPERATION MAGALI **Vol. 2** 70, 191
OPERATION TONNERRE **Vol. 2** 131, 243, 832
OPERATIUNEA „MONSTRUL" **Vol. 3** 250, 268, 291
OPERAZIONE GOLDMAN **Vol. 8** 732
OPERAZIONE KAPPA: SPTARATE A VISTA **Vol. 7** 311
OPERAZIONE NOTTE **Vol. 7** 373
OPERAZIONE POKER **Vol. 8** 583, 772
OPERENE STINY **Vol. 10** 168, 257, 524
OPERETTE **Vol. 4** 265, 614, 620
OPERNBALL **Vol. 4** 266, 774
OPERNREDOUTE **Vol. 4** 321, 738, 788
OPETANIE **Vol. 1** 216, 343 **Vol. 6** *page 257*
OPETTAJATAR SEIKKAILEE **Vol. 5** 443, 639
OPFER, DAS **Vol. 4** 183
OPFER DER ELLEN LARSEN, DAS **Vol. 4** 122, 560, 735
OPFER DER ISIS, DIE **Vol. 4** 498
OPFER DER LIEBE **Vol. 4** 9, 232, 247, 344, 545, 628, 673
OPFER UM OPFER **Vol. 4** 98
OPFERGANG **Vol. 4** 462
OPFINDERS SKAEBNE, EN **Vol. 5** 97
OPHELIA **Vol. 2** 700
OPINIONE PUBBLICA **Vol. 7** 171, 1170
OPIS OBYCZAJOW **Vol. 1** 198, 254
OPIUM **Vol. 4** 397
OPIUMKERINGÖ **Vol. 9** 26, 117
OPIUMSDROMMEN **Vol. 5** 58
OPKLADA **Vol. 3** 367, 449
OPLOB PAA FREDERKSBERG **Vol. 5** 34
OPOWIESC ATLANTYCKA **Vol. 1** 298, 305, 397

OPOWIESC W CZERWIENI Vol. 1 201, 230, 324, 408 Vol. 6 *page 239*
OPPAT MED GRÖNA HISSEN Vol. 5 930, 1042
OPPRIMES, LES Vol. 2 37
OPRI Vol. 5 356, 643
OPSADA Vol. 3 358, 445
OPSTANDELSE Vol. 5 97
OPSTANDELSE, EN Vol. 5 58
OPSTINSKO DETE Vol. 3 353, 436
OPTIMIST, DER Vol. 4 292
OPTIMISTE AU XXE SIECLE, L' Vol. 2 249
OPTIMISTISCHE TRAGÖDIE Vol. 1 98
OPVAKTE JONFRU, DEN Vol. 5 89, 118
OR DANS LA MONTAGNE, L' Vol. 2 57, 363, 595
OR DANS LA RUE, L' Vol. 2 388, 427
OR DE SAMORY, L' Vol. 2 158
OR DES MERS, L' Vol. 2 566
OR DES PHARAONS, L' Vol. 2 224
OR DU CRISTOBAL, L' Vol. 2 395, 526, 871
OR DU DUC, L' Vol. 2 386, 702
OR ET LE PLOMB, L' Vol. 2 502
OR, L' Vol. 2 733
ORA DELLA MORTE, L' Vol. 7 971
ORA „H" Vol. 3 267, 277
ORA TERRIBILE, L' Vol. 7 730
ORA ZERO Vol. 3 240, 272
ORAGE Vol. 2 626, 676, 824
ORAGE D'ETE Vol. 2 8, 15
ORAIA AIGIOTISSA Vol. 3 195
ORANZOVY MESIC Vol. 10 41, 388
ORAO U KAVEZU Vol. 3 444
ORASUL VAZUT DE SUS Vol. 3 281, 288, 291
ORATOTES MEDEN Vol. 3 144
ORBITA MORTAL Vol. 8 546, 626
ORCA, LA Vol. 7 544
ORCHESTRION, DAS Vol. 4 619, 647
ORCHIDEA FATALE, L' Vol. 7 1190
ORCHIDEA FATATA Vol. 7 1189
ORCHIDEEN Vol. 4 21, 607
ORDEN COMICO, EL Vol. 8 324, 728
ORDEN: FX 18 DEBE MORIR Vol. 8 258
ORDET Vol. 5 2, 23, 148, 840, 1114
ORDINATEUR DES POMPES FUNEBRES, L' Vol. 2 782
ORDINI SONO ORDINI, GLI Vol. 7 853
ÖRDÖG HEGEDÜSE, AZ Vol. 9 337
ÖRDÖG MATKAJA, AZ Vol. 9 3, 71
ÖRDÖG NEM ALSZIK, AZ Vol. 9 26, 95, 117, 133
ÖRDÖGLOVAS Vol. 9 26, 117, 133, 274
ORDONNANCE, L' Vol. 2 81, 122, 626, 676
ORDYNAT MICHOROWSKI Vol. 1 278, 308, 346
ORE 9 LEZIONE DI CHIMICA Vol. 7 86, 443, 1096
ORE DELL'AMORE, LE Vol. 7 62, 763
ORE NUDE, LE Vol. 7 853
ORECCHINI DELLA NONNA, GLI Vol. 7 26
ÖREG, AZ Vol. 9 259, 265, 285
ÖREGRUND – ÖSTHAMMAR Vol. 5 857, 1157
ORES AGAPES, ORES POLEMOU Vol. 3 161, 207
ORFANA DEL GHETTO, L' Vol. 7 209, 235
ORFANELLA DI MESSINA, L' Vol. 7 1189
ORFANI DEL PONTE DI NOSTRA SIGNORA, GLI Vol. 7 493
ORFEU NEGRO Vol. 2 124
ORFEUSZ ES EURYDIKE Vol. 9 96, 177, 263, 331
ORGAN Vol. 10 273, 396, 548
ORGASMO ESOTICO Vol. 7 311
ORGASMO INFERNALE Vol. 7 742
ORGASMO NERO Vol. 7 1086
ORGASMO NON STOP Vol. 7 311
ORGE Vol. 3 148, 153, 182
ORGIA, LA Vol. 8 641
ORGIA NOCTURNA DE LOS VAMPIROS, LA Vol. 8 237, 261
ORGIES ADOLESCENTES Vol. 2 735
ORGOGLIO Vol. 7 796, 902
ORGOGLIO O LA SUPERBIA, L' Vol. 7 442, 701
ORGOLII Vol. 6 *page 275, page 277, page 284*
ORGUEILLEUX, LES Vol. 2 156, 345
ORGULLA Vol. 8 214, 524
ORGULLA DE ALBACETE, EL Vol. 8 228
ORIENT Vol. 4 207, 400, 550, 724
ORIENT EKSPRES Vol. 6 *page 57, page 58*
ORIENT EXPRESS Vol. 7 1136, 1138 Vol. 9 76, 132
ORIENTALI, LE Vol. 7 528
ORIENTEXPRESS Vol. 4 37, 344, 424, 536, 544
ORILLA, LA Vol. 8 261, 749
ORINOCO, PARAISO DEL SEXO Vol. 8 237, 536
ORIONS BELTE Vol. 5 699, 778
ORIZZONTE DI SANGUE Vol. 7 16, 149
ORIZZONTI DEL SOLE Vol. 7 1048
ORKESTAR JEDNE MLADOSTI Vol. 6 *page 313, page 318*
ORKLONENIE Vol. 3 105
ORLANDO E I PALADINI DE FRANCIA Vol. 7 106
ORLANDO FURIOSO Vol. 7 1101, 1205
ORLE Vol. 1 160, 380, 404
ORLIE PIERKO Vol. 10 72, 287, 483

ORLOFF Y EL HOMBRE INVISIBLE **Vol. 8** 399, 537
ÖRLOGSMÄN **Vol. 5** 1042, 1164
ORLOVI RANO LETE **Vol. 3** 349, 374
ORLOW, DER **Vol. 4** 292, 584
ORMA, L' **Vol. 7** 209
ORME SULLA NEVE, LE **Vol. 7** 1189
ORMEN **Vol. 5** 839, 1052
ORNIERE, L' **Vol. 2** 614
ÖRNUNGAR **Vol. 5** 923, 1205
ORO **Vol. 7** 807
ORO DEGLI AZTEKI, L' **Vol. 7** 344
ORO DI NAPOLI – I GIOCATORI, L' **Vol. 7** 709
ORO DI NAPOLI – IL FUNERALINO, L' **Vol. 7** 709
ORO DI NAPOLI – IL GUAPPO, L' **Vol. 7** 709
ORO DI NAPOLI – IL PROFESSORE, L' **Vol. 7** 709
ORO DI NAPOLI, L' **Vol. 7** 725, 796
ORO DI NAPOLI – PIZZE A CREDITO, L' **Vol. 7** 709
ORO DI NAPOLI – TERESA, L' **Vol. 7** 709, 756
ORO DI ROMA, L' **Vol. 7** 763
ORO, DONNE E MARACAS **Vol. 7** 451
ORO MALDITO **Vol. 8** 626
ORO NERO **Vol. 7** 101, 603
ORO PER I CESARI **Vol. 7** 739, 1054
ORO ROJO **Vol. 8** 214
ORO VIL **Vol. 8** 298, 625
ORO Y MARFIL **Vol. 8** 394, 422
OROGLIO A CUCU, L' **Vol. 7** 472, 941, 1060
ÖRÖK CSEND TORNYA, A **Vol. 9** 28
ÖRÖK TITOK, AZ **Vol. 9** 94, 270, 271, 334
ÖRÖKBEFOGADAS **Vol. 9** 20, 167, 318
ÖRÖKÖS, AZ **Vol. 9** 83, 183, 259
ÖRÖKSEG **Vol. 9** 152, 253, 259
OROLIGT BLOD **Vol. 5** 310
OROSIA **Vol. 8** 449
OROSZLAN UGRANI KESZÜL, AZ **Vol. 9** 122, 134, 183
ORPHEE **Vol. 2** 279, 282, 395
ORPHELIN DE PARIS, L' **Vol. 2** 169, 330, 613
ORPHELINE, L' **Vol. 2** 169, 330, 613
ORPHELINS DE SAINT-VAAST, LES **Vol. 2** 404
ORPHEUS IN DER UNTERWELT **Vol. 1** 18, 37, 48
ORPHEUS UND EURYDIKE **Vol. 1** 1, 73, 79
ORPOPPJAN VALSSI **Vol. 5** 549, 592
ORRIBILE REALTA, L' **Vol. 7** 454
ORRIBILE SEGRETO DEL DR. HITCHOCK, L' **Vol. 7** 486, 739
ORRIZONTE DIPINTO, L' **Vol. 7** 196, 228
ORRIZONTE INFUOCATO **Vol. 7** 201
ÖRSEGVALTAS **Vol. 9** 26, 117, 133
ÖRSZEM, AZ **Vol. 9** 220
ORTALIONOWY DZIADEK **Vol. 1** 249
ORVOS HALALA, AZ **Vol. 9** 67, 183, 297
ORVOS VAGYOK **Vol. 9** 279
ORVOSOK **Vol. 9** 77
ORZEL **Vol. 1** 225, 305, 323 **Vol. 6** *page 315*
ORZEL I RESZKA **Vol. 1** 142, 203 **Vol. 6** *page 210*
ORZOWEI, IL FIGLIO DELLA SAVANA **Vol. 7** 528
OS BANDEIRANTES **Vol. 2** 368, 503
OSADA HAVRANU **Vol. 10** 139, 325, 425
OSADA MLADYCH SNU **Vol. 10** 34
OSALENA KOMTESA ZUZANA **Vol. 10** 201, 212
OSAM KILA SRECE **Vol. 3** 393, 400
ÖSASSZONY **Vol. 9** 28
OSCAR **Vol. 2** 503, 863
OSCAR STRIBOLT SOM JAEGER **Vol. 5** 97
OSCAR WILDE **Vol. 2** 656
OSCEOLA **Vol. 1** 46, 96, 102, 103
OSCURE VICENDE **Vol. 7** 843
OSCUROS SUENOS DE AGOSTO **Vol. 8** 214, 232
OS"DENI DUSI **Vol. 3** 117 **Vol. 6** *page 33*
OSEKA **Vol. 3** 391 **Vol. 6** *page 309*
ÖSEMBER, AZ **Vol. 9** 288
OSENI **Vol. 10** 197, 296, 569
OSHETIME NE BREGDET **Vol. 3** 10, 16 **Vol. 6** *page 15, page 17*
OSINDA **Vol. 3** 235, 272, 317
OSKAR **Vol. 5** 23, 248
OSKLIVA SLECNA **Vol. 10** 296, 522
OSM SRDCI V PLAMENECH **Vol. 10** 58, 125
OSMA VRATA **Vol. 3** 356, 440
OSMIJAT **Vol. 3** 51, 76 **Vol. 6** *page 49*
OSMNACTILETA **Vol. 10** 122, 524
OSMY DZIEN TYGODNIA **Vol. 6** *page 250, page 256, page 259*
OSO YPARHEI EROTAS **Vol. 3** 167
OSPITE DI UNA NOTTE, L' **Vol. 7** 574, 1004
OSPITE PERICOLOSO, UN **Vol. 7** 938
OSPITE SCONOSCIUTA, L' **Vol. 7** 441
OSS **Vol. 5** 701, 808, 824
OSS 117 N'EST PAS MORT **Vol. 2** 557, 866, 873
OSS 117 SE DECHAINE **Vol. 2** 503, 515
OSS ATOMFORSKERE IMELLON **Vol. 5** 800
OSS BARONER EMELLAN **Vol. 5** 1156, 1205

OSS EMELLAN **Vol. 5** 910
OSS TJUVAR EMELLAN ELLER EN BURG ANANAS **Vol. 5** 840, 878
OSSERVAZIONI DI ROBINET SUL BALLO **Vol. 7** 1189
OSSESSIONE **Vol. 7** 472, 750, 1036, 1136
OSSI HAT DIE HOSEN AN **Vol. 4** 103, 212, 247, 424, 739
OSSIS TAGEBUCH **Vol. 4** 560
ÖSSZEESKÜVÖK, AZ **Vol. 9** 288
OSTACOLO, L' **Vol. 7** 730
OSTAGGIO, L' **Vol. 7** 1189
OSTATNI DNI **Vol. 1** 211, 281
OSTATNI DZIEN LATA **Vol. 1** 233
OSTATNI ETAP **Vol. 1** 250, 270
OSTATNI KURS **Vol. 1** 400, 401 **Vol. 6** *page 218*
OSTATNI PO BOGU **Vol. 1** 179, 343 **Vol. 6** *page 257*
OSTATNI STRZAL **Vol. 1** 232, 305
OSTATNI SWIADEK **Vol. 1** 202, 227, 400
OSTATNIA BRYGADA **Vol. 1** 278, 308, 404
OSTATNIA ESKAPADA **Vol. 1** 290, 336
OSTATNIE DNI **Vol. 6** *page 215, page 225, page 248*
OSTERIA DI MOZZADITA, L' **Vol. 7** 730
ÖSTERMANS TESTAMENTE **Vol. 5** 1033
OSTIA **Vol. 7** 689
OSTPREUSSEN UND SEIN HINDENBURG **Vol. 4** 5
OSTRE SLEDOVANE VLAKY **Vol. 10** 41, 96, 516
OSTROV STRIBRNYCH VOLAVEK **Vol. 10** 87, 428, 517, 615
OSTROZNIE, YETI ! **Vol. 1** 155, 189, 321 **Vol. 6** *page 218*
OSTRVA **Vol. 3** 405
OSTRZE NA OSTRZE **Vol. 6** *page 214, page 229, page 240, page 247*
OSUD JMENEM KAMILA **Vol. 10** 172, 197, 385
OSUD TRI SRDCI **Vol. 10** 242
OSUDNA CHVILE **Vol. 10** 58, 345
OSUDNE NOCI **Vol. 10** 275, 345
OSVOBOZDENIE **Vol. 1** 401
OSVOBOZENI PRAHY **Vol. 10** 18, 184, 283, 313
OSYNLIGA MUREN, DEN **Vol. 5** 840, 909
ÖSZI ALMANACH **Vol. 9** 101, 147, 238, 243
ÖSZI VIHAR **Vol. 9** 234
OT DRUGATA STRANA NA OGLEDALOTO **Vol. 3** 48 **Vol. 6** *page 51*
OT NISTO NESTO **Vol. 3** 51 **Vol. 6** *page 40, page 54*
ÖT ORA 40 **Vol. 9** 71, 322
OTAC I VRTLOG **Vol. 3** 372
OTAC NA SLUZBENOM PUTU **Vol. 6** *page 303, page 306*
OTAGES **Vol. 6** *page 120*
OTAGES, LES **Vol. 2** 79, 488, 657
OTAGES SEQUESTRES ET VIOLETS **Vol. 6** *page 120*
OTALIA DE BAHIA **Vol. 2** 244, 667
OTAN E POLE PETHENEI **Vol. 3** 135
OTAN EMOUN 16 HRONON **Vol. 3** 131
OTAN OI GYNAIKES AGAPOUN **Vol. 3** 195
OTCHLAN POKUTY **Vol. 1** 176, 177, 317, 350
OTE – TOI DE MON SOLEIL **Vol. 6** *page 73, page 129*
OTEC **Vol. 10** 32, 72, 592
OTEC KONDELIK A ZENICH VEJVARA **Vol. 10** 92, 122, 453, 586
OTEC MA ZDERIE TAK CI TAK **Vol. 10** 72, 82, 210
OTELLO **Vol. 7** 21, 110, 196, 264, 1035
OTELLO Z M-2 **Vol. 1** 249
OTELUL RAZBUNA **Vol. 3** 234
OTHELLO **Vol. 1** 74, 95, 115 **Vol. 2** 16 **Vol. 4** 247, 356, 424, 511 **Vol. 5** 97, 228 **Vol. 7** 21, 196, 1035
OTHELLO GYULAHAZAN **Vol. 9** 57, 152, 203
OTK'DE SE ZNAEM ? **Vol. 6** *page 23, page 46*
OTK'DE ZNAEM **Vol. 3** 59
OTKLONENIE **Vol. 3** 112 **Vol. 6** *page 38, page 41*
OTKOGA TE CAKAM **Vol. 6** *page 31, page 37, page 57*
OTKRADNATIJAT VLAK **Vol. 3** 43, 75
OTNOVO V ZIVOTA **Vol. 3** 57, 86, 100
ÖTÖDIK OSZZALY, AZ **Vol. 9** 175, 220
ÖTÖDIK PECSET, AZ **Vol. 9** 134, 152, 316
OTONALES, LAS **Vol. 8** 692
OTPISANI **Vol. 3** 350, 391
OTRA ALCOBA, LA **Vol. 8** 577
OTRA IMAGEN, LA **Vol. 8** 492, 768
OTRA MUJER, LA **Vol. 8** 276, 611
OTRA ORILLA, LA **Vol. 8** 470, 492
OTRA RESIDENCIA, LA **Vol. 8** 681
OTRA SOMBRA, LA **Vol. 8** 298, 524
OTRA VEZ ADIOS **Vol. 8** 610, 738
OTRA VIDA DEL CAPITAN CONTRERAS, LA **Vol. 8** 214, 400
OTRA VUELTA DE TUERCA **Vol. 8** 274, 426
OTRAVENE SVETLO **Vol. 10** 167, 536
OTRAZENIJA **Vol. 6** *page 28, page 33, page 59*
OTRO ARBOL DE GUERNICA, EL **Vol. 8** 524
OTRO FU-MANCHU, EL **Vol. 8** 272, 744

OTTE AKKORDER **Vol. 5** 2, 228
OTTO ALS DIENSTMANN **Vol. 4** 603
OTTO E MEZZO **Vol. 7** 522, 1025, 1167
OTTO ER ET NAESEHORN **Vol. 5** 16, 61, 185, 299
OTTO HEIRATET **Vol. 4** 603
OTTO MILIONI DI DOLLARI **Vol. 7** 264, 701
OTTOKAR DER WELTVERBESSERER **Vol. 1** 15, 79, 85
OTV"D HORIZONTA **Vol. 3** 56, 88
OU EST PASSE TOM ? **Vol. 2** 163, 351
OU KLEPSES **Vol. 3** 192
OUBLIE, L' **Vol. 2** 79, 380
OUBLIE-MOI MANDOLINE **Vol. 2** 151, 815
OURANOS **Vol. 3** 141, 204
OURS ET LA POUPEE, L' **Vol. 2** 481, 674
OURS, L' **Vol. 2** 722, 781, 856
OURS, UN **Vol. 2** 277, 708
OUT OF THE FOG **Vol. 2** 579
OUT: SPELTRE **Vol. 2** 351
OUTONO **Vol. 8** 19, 34
OUTSIDER **Vol. 10** 206, 250, 508
OUVERT CONTRE X **Vol. 2** 57, 158
OVCAR **Vol. 3** 450
OVER ALLE GRAENSER **Vol. 5** 23, 29
OVER ALLE HAV **Vol. 5** 829
OVER GADEN - NEDEN VANDET **Vol. 5** 77, 282
ÖVERGREPPET **Vol. 5** 880, 933, 1068
OVERKLASSENS HEMMELIGE SEXGLAEDER **Vol. 5** 27, 124, 186
OVNI IN MAMUTI **Vol. 6** *page 313*
OVOCE STROMU RAJSKYCH JIME **Vol. 10** 289, 292, 299
OVRAOI EINAI DIKOI MAS **Vol. 3** 198
OXALA **Vol. 8** 166
OXFORD CONTRE MARTIGUES **Vol. 2** 159, 286, 330
OXYGEN **Vol. 3** 356, 449
OYE FOR OYE **Vol. 5** 699, 716, 728
OYE PA HVER FINGER, ET **Vol. 5** 674
OYEBLIKKET **Vol. 5** 748, 787
OZNAMUJE SE LASKAM VASIM **Vol. 10** 144, 313, 571
ÖZVEGY ES A SZAZADOS, AZ **Vol. 9** 67, 83, 183
ÖZVEGY MENYASSZONYOK **Vol. 9** 67, 83, 149

P

P & B **Vol. 5** 882, 1093, 1096
P. COMME PENETRATION **Vol. 2** 365
P... RESPECTUEUSE, LA **Vol. 2** 194, 778
P. S. **Vol. 1** 79, 80, 124
P... SENTIMENTALE, LA **Vol. 2** 404
PA D'ANGEL **Vol. 8** 297, 684
PA DESSA SKULDROR **Vol. 5** 1042, 1164
PA EN BÄNK I EN PARK **Vol. 5** 878, 1052, 1104
PA' EN IGEN AMALIE **Vol. 5** 35, 94, 195
PA FARLIGA VÄGAR **Vol. 5** 840, 878, 993
PA HEDER OCH SKOJ **Vol. 5** 871, 1052
PA HITTIGA JOHANSSON **Vol. 5** 1181
PA KRYSS MED ALBERTINA **Vol. 5** 923, 1205
PA KRYSS MED BLIXTEN **Vol. 5** 1037
PA LIV OCH DÖD **Vol. 5** 1042, 1116
PA LIVETS ÖDEVÄGAR **Vol. 5** 997
PA RYMNEN MED PIPPI LANGSTRUMP **Vol. 5** 862, 1027
PA SLAGET ATTE **Vol. 5** 703, 738
PA SOLSIDAN **Vol. 5** 840, 909
PA SOLSIDEN **Vol. 5** 674, 831
PA TOKT MED „TERNA“ **Vol. 5** 659, 756
PA TRO OG LOVE **Vol. 5** 31, 150, 248, 272
PAA DE VILDE VOVER **Vol. 5** 262
PAA LIVETS SKYGGESIDE **Vol. 5** 203
PÄÄ PYSTYN, HELENA **Vol. 5** 347, 451
PAA SLAGET 12 **Vol. 5** 183, 228
PAA SYNDENS TAERSKEL **Vol. 5** 81
PÄÄMAJA **Vol. 5** 418, 516, 635
PACALA **Vol. 3** 230, 291, 297
PACALA SI TINDALA LA BUCURESTI **Vol. 3** 304
PACAT **Vol. 3** 257
PACHA, LE **Vol. 2** 279, 293
PACHIN **Vol. 8** 306, 583
PACHIN ALMIRANTE **Vol. 8** 306, 396
PACHO, HYBSKY ZBOJNIK **Vol. 10** 82, 273, 280
PACIENTKA DOKTORA HEGLA **Vol. 10** 2, 412
PACIORKI JEDNEGO ROZANCA **Vol. 6** *page 249, page 250, page 260*
PACO L'INFAILLIBLE **Vol. 6** *page 72, page 107*
PACSIRTA **Vol. 9** 134, 199, 312
PACTO DE SILENCIO **Vol. 2** 440 **Vol. 8** 221, 486, 681
PAD ITALIJE **Vol. 6** *page 318, page 322*
PADELEK **Vol. 10** 503, 580
PADELLA CALIBRO 38 **Vol. 7** 959
PADRE **Vol. 7** 304, 305
PADRE COPILLAS, EL **Vol. 8** 499, 590
PADRE DE LA CRIATURA, EL **Vol. 8** 681, 749
PADRE MANOLO, EL **Vol. 8** 311, 402
PADRE NO HAY MAS QUE DOS **Vol. 8** 770, 773
PADRE NUESTRO **Vol. 8** 214, 232
PADRE PADRONE **Vol. 7** 741, 1034, 1112
PADRE PITILLO, EL **Vol. 8** 207, 298
PADRI E FIGLI **Vol. 7** 86

PADRINA, LA Vol. 7 290
PADRINO Y SUS AHIJADAS, EL Vol. 8 261, 450
PADRONA E SERVITA, LA Vol. 7 546
PADRONE DEL MONDO, IL Vol. 7 678, 692
PADRONE DEL VAPORE, IL Vol. 7 161, 333
PADRONE DELLE FERRIERE, IL Vol. 7 369, 798
PADRONE E L'OPERAIO, IL Vol. 7 599
PADRONE SONO ME Vol. 7 1054, 1147
PADRONI DELLA CITTA, I Vol. 7 763
PADUREA INDRAGESTITILOR Vol. 3 243
PADUREA NEBUNA Vol. 6 *page 275, page 280, page 281*
PADUREA PIERDUTA Vol. 3 269, 281, 319
PADUREA SPINZURATILOR Vol. 3 260, 297, 330
PAESE DEI CAMPANELLI, IL Vol. 7 1147
PAESE DEL SESSO SELVAGGIO, IL Vol. 7 852
PAESE DELLA PAURA, IL Vol. 7 277
PAESE SENZAPACE, IL Vol. 7 333
PAGANINI Vol. 4 412, 479
PAGANINI IN VENEDIG Vol. 4 95
PAGE VOM DALMASSE-HOTEL Vol. 4 323, 437, 458
PAGELLA, LA Vol. 7 311
PAGINA D'AMORE, UNA Vol. 7 441
PAGINA IGNOTA, LA Vol. 7 277
PAGINAS EN NEGRO, UNAS Vol. 8 399
PAGINE CHIUSE Vol. 7 1095
PAGINE SPARSE Vol. 7 131
PAGLIACCI Vol. 7 1054, 1189
PAGO CARA SU MUERTE Vol. 8 396
PAGODE, DIE Vol. 4 258, 412
PÄHKÄHULLU SUOMI Vol. 5 374, 558, 606
PAI TIRANO, O Vol. 8 11, 172
PAIDI RES MAMAS Vol. 3 161, 207
PAIDI TES ATHENAS Vol. 3 146
PAIDI TOU DROMOU Vol. 3 165
PAIDIA TES MADALENAS Vol. 3 186
PAIDIA TOU LIMANIOU Vol. 3 195
PAIMEN, PIIKA JA EMÄNTÄ Vol. 5 346, 363
PAIN DE JULES, LE Vol. 2 131, 514
PAIN VIVANT, LE Vol. 2 559, 776
PAIS, S.A. Vol. 8 219, 237
PAISA Vol. 7 725
PAISIBLE RETRAITE Vol. 2 703
PÄIVÄÄ, HERRA KIVI Vol. 5 387, 585, 589
PAIX SUR LE RHIN Vol. 2 273, 623, 858
PAJA I JARE - KAMIONDZIJE Vol. 3 359, 282, 391
PAJARITOS, LOS Vol. 8 368, 773
PAJAROS DE BADEN-BADEN, LOS Vol. 8 297, 385
PAJAROS DE CIUDAD Vol. 8 549
PAJASTA PALATSIIN Vol. 5 308, 370
PAKKETUR TIL PARADIS Vol. 5 732
PAKLENI OTOK Vol. 3 407, 420
PAKLIC Vol. 10 122, 154, 256
PAKSUNAHKA Vol. 5 462, 634
PAL UTCAI FIUK, A Vol. 9 134, 259, 265, 337
PAL-UTCAI FIUK, A Vol. 9 71, 220
PALABRAS DE AMOR Vol. 8 232, 732
PALAC Vol. 6 *page 217, page 219, page 240*
PALAC NA KOLKACH Vol. 1 285, 290, 372
PALACE HOTEL Vol. 6 *page 206, page 227, page 238*
PALACES Vol. 2 6, 381, 413
PALADINI, STORIA D'ARMI E D'AMORI, I Vol. 7 282, 283, 964, 1093
PALAVA ENKELI Vol. 5 487, 573
PALAZZO DEI SOGNI, IL Vol. 7 1189
PALCO ALL'OPERA, UN Vol. 7 481
PALEC BOZY Vol. 1 141, 180, 182 Vol. 6 *page 208*
PALETA LASKY Vol. 10 505, 569, 621
PALICOVA DCERA Vol. 10 45, 46, 412, 621
PALIKA Vol. 9 77
PALIO Vol. 7 196, 211, 508, 997, 1006
PALIOPAIDA Vol. 3 137
PALMA MEDJU PALMANA Vol. 3 357, 423
PALMER HA MUERTO Vol. 8 523, 537
PALMIRA Vol. 8 489, 560
PALOMA AZUL, LA Vol. 8 520, 560
PALOMA DE MIS AMORES Vol. 8 357, 522, 542
PALOMA, LA Vol. 4 132, 230
PALOMBIERE, LA Vol. 6 *page 116, page 122*
PALOMOS, LOS Vol. 8 396, 495
PALO'S BRUDEFAERD Vol. 5 263, 288
PALUDISME Vol. 2 811
PAMELA Vol. 2 339, 698, 786 Vol. 6 *page 139*
PAMIETNIK PANI HANKI Vol. 1 179, 237, 283 Vol. 6 *page 242*
PAMPA SALVAJE Vol. 8 276, 616
PAN Vol. 4 121, 124, 422, 703 Vol. 5 665, 795 Vol. 9 146, 148
PAN A HVEZDAR Vol. 10 287, 576
PAN A SLUHA Vol. 10 256, 412
PAN AMOR Y ANDALUCIA Vol. 8 261
PAN ANATOL SZUKA MILIONA Vol. 1 179, 232 Vol. 6 *page 236*
PAN DEBAJO DEL BRAZO, EL Vol. 8 728, 770
PAN DODEK Vol. 1 144, 232, 321
PAN HABETIN ODCHAZI Vol. 10 193, 296
PAN KONSUL POMERANC Vol. 1 174, 274

PAN MACHULA VO VESMIRE Vol. 10 136, 343, 521
PAN NA ROZTRHANI Vol. 10 300, 524, 611
PAN NOVAK Vol. 10 122, 580
PAN OTEC KARAFIAT Vol. 10 256, 586
PAN PROFESOR, NEPRITEL ZEN Vol. 10 201
PAN REDAKTOR SZALEJE Vol. 1 205, 372
PAN SI NEZELAL NIC Vol. 10 228, 459, 575
PAN TADEUSZ Vol. 1 174, 255, 380
PAN TAU Vol. 10 165, 172, 496
PAN TAU A CESTA KOLEM SVETA Vol. 10 173, 184, 283, 564
PAN TAU A CLAUDIE Vol. 10 173, 184, 283, 564
PAN TAU A NEDELE Vol. 10 173, 184, 283, 564
PAN TAU A SAMA VODA Vol. 10 184, 283, 564
PAN TAU JDE DO SKOLY Vol. 10 184, 283, 564
PAN TAU NA HORACH Vol. 10 173, 184, 283, 564
PAN TAU NADELUJE Vol. 10 173, 283, 564
PAN TAU PRICHAZI Vol. 10 184, 283, 564
PAN TWARDOWSKI Vol. 1 278, 308, 317, 346
PAN VOK ODCHAZI Vol. 10 226, 317, 522
PAN WOLODYJOWSKI Vol. 1 180, 202, 227, 242, 406
PANAGULIS VIVE Vol. 7 474, 952, 1046
PANAME Vol. 2 402, 440, 833
PANCEROVE AUTO Vol. 6 *page 196* Vol. 10 147, 364
PANCHO SE ZENI Vol. 10 256, 412, 566, 612
PANDILLA DE LOS ONCE, LA Vol. 8 311, 493
PANE ALTRUI, IL Vol. 7 559
PANE, AMORE E ... Vol. 7 756, 988
PANE, AMORE E FANTASIA Vol. 7 493, 756
PANE, AMORE E GELOSIA Vol. 7 756, 796
PANE, BURRO E MARMELLATA Vol. 7 514
PANE E CIOCCOLATA Vol. 7 1145
PANE, VY JSTE VDOVA ! Vol. 10 41, 154, 425
PANELKAPCSOLAT Vol. 9 216, 238
PANELSTORY ANEB JAK SE RODI SIDLISTE Vol. 10 51, 516, 595
PANENKA Vol. 10 256, 412
PANENSTVI Vol. 10 256, 460
PANENSTVI A KRIMINAL Vol. 10 154, 296, 391
PANG I BYGGET Vol. 5 909, 910, 1142
PANI BOVARY TO JA Vol. 1 138, 195, 201
PANI EDISONI Vol. 10 258, 497, 594
PANI KACKA ZASAHUJE Vol. 10 154, 621
PANI KATYNKA Z VAJECNEHO TRHU Vol. 10 257, 453
PANI KLUCI Vol. 10 359, 425, 496
PANI MINISTER TANCZY Vol. 1 174, 404
PANI MORALKA KRACI MESTEM Vol. 10 168, 412
PANI SA ZABAVAJU Vol. 10 36, 72, 210
PANI WALEWSKA Vol. 1 174, 188
PANIC BUTTON Vol. 7 1062
PANICO Vol. 8 347
PANICO EN EL TRANSIBERIANO Vol. 8 438, 762
PANIENKA Z OKIENKA Vol. 1 169, 303, 321, 359
PANIENKA Z POSTE-RESTANTE Vol. 1 308, 404
PANIER A CRABES, LE Vol. 2 250, 251, 374, 559, 661
PANIK Vol. 4 115, 119, 167, 263, 536, 779 Vol. 5 841, 1156
PANIK I FAMILIEN Vol. 5 89, 224
PANIK I PARADIS Vol. 5 110, 211, 258
PANIK IN CHIKAGO Vol. 4 208, 479
PANIKOS Vol. 3 135
PANIQUE Vol. 2 676
PANNA Vol. 10 92, 345
PANNA A NEVTOR Vol. 10 144, 299, 327
PANNA PO WOJNIE Vol. 1 268
PANNA ZAZRACNICA Vol. 10 273, 357, 548
PANNY Z WILKA Vol. 6 *page 241, page 254*
PANSER – BASSE Vol. 5 24
PANTALASKAS Vol. 2 97, 309, 808
PANTALLA DIABOLICA, LA Vol. 8 283, 296
PANTATA BEZOUSEK Vol. 10 34, 122, 167, 453
PANTELEJ Vol. 3 110 Vol. 6 *page 53, page 55*
PANTER CEKA V 17.30 Vol. 10 144, 286, 408
PANTERA NERA, LA Vol. 7 149, 606, 861
PANTERAS SE COMEN A LOS RICOS, LAS Vol. 8 207, 331
PANTHER Vol. 7 442
PANTHERBRAUT, DIE Vol. 4 560, 743
PANTINS D'AMOUR Vol. 2 191, 273, 595, 698
PANTOFFELHELD, DER Vol. 4 113, 752
PANTOFFELHELDEN Vol. 4 514
PANTOFUL CENUSARESEI Vol. 3 227, 244, 329, 332

PANTOMIMA DELLA MORTE, LA **Vol. 7** 1038
PANZERAUTO, DAS **Vol. 6** *page 196*
PANZERGEWÖLBE, DAS **Vol. 4** 159, 162, 396, 535
PAO, AMOR E ... TOTOBOLA ! **Vol. 8** 23, 129, 150, 159
PAO NOSSO ... **Vol. 8** 50, 172, 202
PAOLO E FRANCESCA **Vol. 7** 332, 796, 798, 1054
PAOLO IL CALDO **Vol. 7** 333
PAP' OCCHIO, IL **Vol. 7** 801, 952, 1145
PAPA **Vol. 7** 965 **Vol. 10** 243, 526
PAPA DIVENTA MAMMA **Vol. 7** 106
PAPA ECCELLENZA **Vol. 7** 844
PAPA KANN'S NICHT LASSEN **Vol. 4** 132, 479
PAPA LA STRADA **Vol. 7** 279
PAPA LEBONNARD **Vol. 7** 48, 168, 1159
PAPA, MAMAN, LA BONNE ET MOI **Vol. 2** 185, 309
PAPA, MAMAN, MA FEMME ET MOI **Vol. 2** 185, 309, 743
PAPA MIO, MI PIACCION TUTTI ! **Vol. 7** 569
PAPA PACIFICO **Vol. 7** 16
PAPA PER UNA NOTTE **Vol. 7** 701, 725
PAPA SANS LE SAVOIR **Vol. 2** 603, 672
PAPA SIE ZENI **Vol. 1** 278, 308, 346, 347
PAPA TI RICORDO **Vol. 7** 935
PAPAFLESSAS **Vol. 3** 151, 185
PAPAS FEHLTRITT **Vol. 4** 24, 525
PAPAS NEUE FREUNDIN **Vol. 1** 24, 85
PAPESSE, LA **Vol. 2** 777
PAPIERS DU MORT, LES **Vol. 2** 159, 286, 330
PAPIERSPUR, DIE **Vol. 4** 247
PAPILIO **Vol. 10** 226, 389, 577
PAPIRFUGLEN **Vol. 5** 684, 808
PAPPA BOM **Vol. 5** 878, 1173
PAPPA E CICCIA **Vol. 7** 882, 1086, 1153
PAPPA SÖKES **Vol. 5** 909, 1173
PAPPA TAR GULL **Vol. 5** 724, 800
PAPPA VARFÖR ÄR DU ARG – DU GJORDE LIKADANT SJÄLV NÄR DU VAR UNG **Vol. 5** 914, 1142
PAPPAS POJKE **Vol. 5** 840, 997
PAPPI **Vol. 4** 354, 734
PAPRIKA **Vol. 2** 563, 626 **Vol. 4** 340, 373
PAPUCSHÖS, A **Vol. 9** 132, 237
PAPY FAIT DE LA RESISTANCE **Vol. 6** *page 83, page 101*
PAQUEBOT TENACITY, LE **Vol. 2** 395, 439, 566, 824, 871
PAR AMOUR **Vol. 2** 798
PAR DE ASESINOS, UN **Vol. 8** 626
PAR DE HUEVOS, UN **Vol. 8** 297, 656
PAR DE ZAPATOS DEL 32, UN **Vol. 8** 306, 577, 600
PAR-DESSUS LE MUR **Vol. 2** 309, 376
PAR-DESSUS LES MURS **Vol. 2** 298
PAR HABITUDE **Vol. 2** 103, 699
PAR ICI LA MONNAIE **Vol. 2** 808
PAR ICI LA SORTIE **Vol. 2** 354
PAR LA FENETRE **Vol. 2** 57, 627
PAR LEPES A HATAR **Vol. 9** 117, 183, 332
PAR ORDRE DU TZAR **Vol. 2** 204, 395, 559, 698
PAR UN BEAU MATIN D'ETE **Vol. 2** 175, 863 **Vol. 8** 258
PARA NAD HRNCEM **Vol. 10** 193, 621
PARA SIEMPRE **Vol. 8** 396, 725
PARA TI ES EL MUNDO **Vol. 8** 516
PARABENS, SENHOR VICENTE ! **Vol. 8** 143
PARABOLA DI UNA VITA, LA **Vol. 7** 52
PARACELSUS **Vol. 4** 285, 531, 615, 678
PARADA REZERWISTOW **Vol. 1** 278, 380
PARADA WARSZAWY **Vol. 1** 174, 225, 241, 346
PARADE **Vol. 5** 853, 940, 947, 1093, 1128
PARADE DU RIRE **Vol. 2** 194, 496
PARADE EN SEPT NUITS **Vol. 2** 96, 566, 676
PARADIES DER DIRNEN, DAS **Vol. 4** 232
PARADIES DER JUNGGESELLEN **Vol. 4** 132, 270
PARADIES IM SCHNEE, DAS **Vol. 4** 113, 175, 272, 319, 622
PARADIS DE SATAN, LE **Vol. 2** 395, 526, 583, 807
PARADIS DES PILOTES PERDUS, LE **Vol. 2** 279, 824
PARADIS DES RICHES, LES **Vol. 2** 296, 543
PARADIS DES VOLEURS, LE **Vol. 2** 413, 533
PARADIS FAGELN **Vol. 5** 937
PARADIS PERDU **Vol. 2** 125, 423, 539, 566, 652
PARADIS POUR TOUS **Vol. 6** *page 95, page 132*
PARADIS RETUR **Vol. 5** 31, 272, 275
PARADISET **Vol. 5** 967, 1033
PARADISO **Vol. 2** 754 **Vol. 7** 48, 219, 238, 466
PARADISO BIANCO **Vol. 7** 1113, 1116
PARADISO, IL **Vol. 7** 53
PARADISTORG **Vol. 5** 850, 946, 1093
PARADO EN MOVIMIENTO, UN **Vol. 8** 474
PARAGGELIA **Vol. 3** 176, 207
PARAGON, GOLA **Vol. 1** 243, 276 **Vol. 6** *page 246*
PARAGRAF 224 **Vol. 10** 139, 227, 325

PARAGRAPH 182, MINDERJÄHRIG Vol. 4 495, 704
PARAISO PERDIDO, O Vol. 8 59, 166
PARAISO RECOBRADO, EL Vol. 8 268, 394
PARAISO SIN EVA Vol. 8 692
PARAISOS PERDIDOS, LOS Vol. 8 219, 749
PARALLELLE LIG, DET Vol. 5 179, 249, 299
PARAMEDICO, IL Vol. 7 46, 263
PARANINFO Vol. 7 493, 665
PARANOICO, EL Vol. 8 286, 407
PARAPIOGGIA DI FRICOT, IL Vol. 7 1189
PARAPLUIES DE CHERBOURG, LES Vol. 2 285, 700
PARAPLYEN Vol. 5 81
PARASUTISTII Vol. 3 218, 254, 335
PARATASE GIA ENAN ROLLO Vol. 3 166
PARAVENTO Vol. 7 1189
PARBAJ SEMMIERT Vol. 9 26, 132
PARBESZED Vol. 9 134, 149, 259
PARCE PLAVOG NEBA Vol. 3 342
PARCE QUE JE T'AIME Vol. 2 9
PARCE QUE JE VOUS AIME Vol. 2 719
PARCHIS ENTRA EN ACCION Vol. 8 250, 368
PARDON, TEVEDTEM ! Vol. 9 71
PARDONNEE Vol. 2 69
PARDONNEZ NOS OFFENSES Vol. 2 423, 548, 738
PAREJA DISTINTA, UNA Vol. 8 553, 770
PARENTHESE Vol. 3 142, 199
PARENTI DI BOUTALIN, I Vol. 7 1189
PARENTS NE SONT PAS SIMPLES CETTE ANNEE, LES Vol. 6 *page 99*
PARENTS TERRIBLES, LES Vol. 2 85, 332, 427
PARES Y NONES Vol. 8 691, 758
PARFUM DE LA DAME EN NOIR, LE Vol. 2 47, 251, 573, 644, 656, 775
PARFÜM DER MRS. WORRINGTON, DAS Vol. 4 176, 544, 726
PARI E DISPARI Vol. 7 599
PARIA, LE Vol. 2 340 Vol. 8 426, 732
PARIAS DE LA GLOIRE, LES Vol. 2 463, 475, 790 Vol. 8 493, 732
PARIAS DE LA GLORIA, LOS Vol. 8 493, 732
PARIGI MISTERIOSA Vol. 7 264, 701
PARIGI, O CARA Vol. 7 338, 853
PARIGINA A ROMA, UNA Vol. 7 587, 1054
PARIS Vol. 2 131, 257, 367, 623, 826
PARIS AU MOIS D'AOUT Vol. 2 285, 719
PARIS-BEGUIN Vol. 2 626, 676
PARIS BIEN VALE UNA MOZA Vol. 8 681, 749
PARIS BLUES Vol. 2 566
PARIS BRULE-T-IL ? Vol. 2 368, 400
PARIS CHANTE TOUJOURS Vol. 2 601, 824
PARIS CLANDESTIN Vol. 2 131, 777, 872
PARIS-COQUIN Vol. 2 401, 501, 801
PARIS-DEAUVILLE Vol. 2 488, 623
PARIS EN CINQ JOURS Vol. 2 197
PARIS EST TOUJOURS PARIS Vol. 2 13, 469
PARIS ESTAMBUL SIN REGRESO Vol. 8 258, 438
PARIS-GIRLS Vol. 2 42, 122, 197
PARIS, JE T'AIME Vol. 2 834
PARIS LA NUIT Vol. 2 237, 792
PARIS-MEDITERRANEE Vol. 2 48, 193, 197, 424, 830
PARIS, MES AMOURS Vol. 2 249, 395
PARIS MUSIC-HALL Vol. 2 376, 709
PARIS-NEW YORK Vol. 2 23, 402, 871
PARIS-NEW YORK-PARIS Vol. 2 402
PARIS N'EXISTE PAS Vol. 2 263, 364
PARIS NOUS APPARTIENT Vol. 2 101
PARIS-PALACE-HOTEL Vol. 2 8, 279, 540
PARIS QUI DORT Vol. 2 237, 313, 379
PARIS QUI S'AMUSE Vol. 2 381, 731
PARIS-SECRET Vol. 2 436, 686
PARIS-SOLEIL Vol. 2 339
PARIS VU PAR... Vol. 2 17, 74, 115, 513, 572, 700
PARISER UNTERWELT Vol. 4 340, 608, 634
PARISERINNEN Vol. 4 266, 574, 729
PARISETTE Vol. 2 169, 330, 613
PARISIENNE, UNE Vol. 2 21, 368, 503
PARISIENNES – ANTONIA, LES Vol. 2 21
PARISIENNES – ELLA, LES Vol. 2 21
PARISIENNES – FRANCOISE, LES Vol. 2 21
PARISIENNES, LES Vol. 2 13, 824
PARISIENNES – SOPHIE, LES Vol. 2 21
PARISISKOR Vol. 5 997, 1069
PARIZIANA Vol. 3 153
PARIZS KIRALYA Vol. 9 234
PARKETTSESSEL 47 Vol. 4 407, 734
PARKSTRASSE 13 Vol. 4 292, 517, 558
PÄRLEMOR Vol. 5 1042, 1130, 1216
PARLEZ-MOI D'AMOUR Vol. 2 67, 527, 579, 722, 727, 787
PARLOG Vol. 3 411
PARMIGIANA, LA Vol. 7 817, 1035
PARNASIE Vol. 10 243, 338, 345
PAROHY Vol. 10 49, 92, 566
PAROLA DI LADRO Vol. 7 401, 514, 522
PAROLA DI UN FUORILEGGE E LEGGE, LA Vol. 7 852
PAROLATA Vol. 3 51, 58 Vol. 6 *page 53*

PAROLE E SANGUE **Vol. 7** 383, 533
PAROLI DI LADO **Vol. 7** 514
PAROXISMUS **Vol. 7** 649
PARQUE DAS ILUSOES, O **Vol. 8** 23, 96, 150, 159
PARQUE DE MADRID **Vol. 8** 517
PARSIFAL **Vol. 7** 1038 **Vol. 8** 301, 611
PART DE L'OMBRE, LA **Vol. 2** 402, 676
PART DES LIONS, LA **Vol. 2** 219
PART DU FEU, LA **Vol. 2** 163, 175
PARTA HIC **Vol. 10** 41, 385, 473
PARTEA TA DE VINA **Vol. 3** 251, 307, 319
PARTENZA IMPOSSIBILE **Vol. 7** 1189
PARTENZA OR SETTE **Vol. 7** 796
PARTI SANS LAISSER D'ADRESSE **Vol. 6** *page 136, page 1474*
PARTIE DE CAMPAGNE, UNE **Vol. 2** 124, 523, 719
PARTIE DE PLAISIR, UNE **Vol. 2** 520, 700
PARTIE KRASNEHO DRAGOUNA **Vol. 10** 70, 158, 547
PARTIES CARREES CHAMPAGNARDES **Vol. 6** *page 79*
PARTIES CARREES TRES SPECIALES **Vol. 6** *page 93*
PARTIES CHAUDES **Vol. 2** 291
PARTIES FINES **Vol. 2** 721
PARTIES RAIDES **Vol. 2** 349
PARTIR **Vol. 2** 55, 84, 197
PARTIRE **Vol. 7** 196, 756
PARTITA A SCACCHI, UNA **Vol. 7** 1189
PARTITA DI BOXE, UNA **Vol. 7** 1189
PARTITA DOPPIA **Vol. 7** 1189
PARTITA NA INSTRUMENT DREWNIANY **Vol. 1** 179, 183 **Vol. 6** *page 246*
PARTIZANI **Vol. 3** 381, 398
PARTIZANI I VOGEL VELO **Vol. 3** 3, 26
PARTIZANSKA ESKADRILA **Vol. 3** 431 **Vol. 6** *page 305*
PARTIZANSKE PRICE **Vol. 3** 347, 373, 376, 423, 455
PARTOUZES DE MADAME PAULE, LES **Vol. 2** 291
PARTOUZES SUEDOISES **Vol. 2** 52
PARTYZANSKA STEZKA **Vol. 10** 560
PARYZ-WARSZAWA BEZ WIZY **Vol. 1** 202, 227, 243
PAS BESOIN D'ARGENT **Vol. 2** 153
PAS DE CAVIAR POUR TANTE OLGA **Vol. 2** 185
PAS DE COUP DUR POUR JOHNNY **Vol. 2** 243, 475, 559
PAS DE GRISBI POUR RICARDO **Vol. 2** 872
PAS DE LA MULE, LE **Vol. 2** 136
PAS DE PANIQUE **Vol. 2** 259, 586
PAS DE PITIE POUR LES CAVES **Vol. 2** 627, 872
PAS DE PITIE POUR LES FEMMES ! **Vol. 2** 401
PAS DE PROBLEME **Vol. 2** 293, 334
PAS DE SOURIS DANS LE BIZNESS **Vol. 2** 131, 871, 872
PAS DE VACANCES POUR M. LE MAIRE **Vol. 2** 496, 615
PAS DE WEEKEND POUR NOTRE AMOUR **Vol. 2** 247, 279
PAS GJURMEVE **Vol. 3** 14, 22 **Vol. 6** *page 12*
PAS KOJI JE VOLEO VOZOVE **Vol. 3** 367, 418
PAS PA RYGGEN, PROFESSOR **Vol. 5** 165, 261
PAS PAA PIGERNE **Vol. 5** 24, 228
PAS PAA SVINGET I SOLBY **Vol. 5** 7
PAS PARYGGEN, PROFESSOR **Vol. 5** 11
PAS PERDUS, LES **Vol. 2** 348, 481
PAS QUESTION LE SAMEDI **Vol. 2** 124
PAS SI BETE **Vol. 2** 319, 460, 467, 627
PAS SI MECHANT QUE CA **Vol. 2** 284, 665
PAS SUR LA BOUCHE **Vol. 2** 708, 753, 775 **Vol. 7** 796
PAS UN MOT A LA REINE-MERE **Vol. 2** 93, 871
PAS VDEKJES **Vol. 3** 28 **Vol. 6** *page 11, page 13*
PASA LA LUNA **Vol. 8** 583
PASACEK Z DOLINY **Vol. 10** 458, 541, 567
PASADO AMENAZA, EL **Vol. 8** 493, 732
PASAJE A VENEZUELA **Vol. 8** 217, 470
PASAK HOLEK **Vol. 10** 40, 168, 586
PASAPORTE PARA UN ANGEL **Vol. 8** 749, 753
PASAREA FURTUNII **Vol. 3** 248, 261, 286, 306
PASAZERKA **Vol. 1** 391, 403 **Vol. 6** *page 237*
PASCUALIN **Vol. 8** 495, 696
PASEILLO, EL **Vol. 8** 477, 495
PASI SPRE LUNA **Vol. 3** 218, 265, 286
PASIANS **Vol. 10** 286, 408, 614
PASION BAJO EL SOL **Vol. 8** 405, 492
PASION PROHIBIDA **Vol. 8** 368, 627
PASIONES DESENFRENADAS **Vol. 8** 249, 368
PASJA **Vol. 1** 219, 360, 398
PASKARZE **Vol. 1** 174, 176
PASKVALINO **Vol. 6** *page 312*
PASLA KONE NA BETONE **Vol. 10** 273, 491, 548
PASO AL FRENTE, UN **Vol. 8** 510, 719
PASODOBLE **Vol. 8** 245
PASODOBLE PRE TROCH **Vol. 10** 83, 459, 542
PASOS **Vol. 8** 401, 772

PASOS DE MUJER, UNOS **Vol. 8** 268, 449
PASOS LARGOS **Vol. 8** 466
PASOTA CON CORBATA, UN **Vol. 8** 527
PASQUALINO SETTEBELLEZZE **Vol. 6** *page 312* **Vol. 7** 333, 592
PASSA GIOVENTU **Vol. 7** 454
PASSA IL DRAMMA A LILIPUT **Vol. 7** 390
PASSA LA GUERRA **Vol. 7** 969
PASSA LA RUINA **Vol. 7** 824
PASSA SARTANA ... E L'OMBRA DELLA TUA MORTE **Vol. 7** 1179
PASSADO E O PRESENTE, O **Vol. 8** 7, 142
PASSAGE DE DEFILE **Vol. 2** 678
PASSAGE DE VENUS, LE **Vol. 2** 158, 205
PASSAGE DU RHIN, LE **Vol. 2** 185, 293, 294
PASSAGEM DE NIVEL **Vol. 8** 13, 150
PASSAGEM OU A MEIO CAMINHO **Vol. 8** 7
PASSAGER CLANDESTIN, LE **Vol. 2** 615
PASSAGER DE LA PLUIE, LE **Vol. 2** 377, 873
PASSAGER, LE **Vol. 2** 388, 663
PASSAGERE, LA **Vol. 2** 328, 496, 661
PASSAGERS, LES **Vol. 2** 39
PASSAGGERA, LA **Vol. 7** 369
PASSANO GLI UNNI **Vol. 7** 1038
PASSANTE DU SANS-SOUCI, LA **Vol. 6** *page 75, page 91, page 128*
PASSANTE, LA **Vol. 2** 501
PASSAPORTO PER LA PACE **Vol. 3** 380
PASSAPORTO ROSSO **Vol. 7** 48, 453, 1000
PASSARINHO DA RIBEIRA, O **Vol. 8** 69, 130, 155
PASSAROS DE ASAS CORTADAS **Vol. 8** 130, 175
PASSATO DI KASEIRA, IL **Vol. 7** 1189
PASSATORE, IL **Vol. 7** 796, 1054
PASSE DU DIABLE, LA **Vol. 2** 208
PASSE-MONTAGNE **Vol. 2** 281, 495
PASSE-MURAILLE, LA **Vol. 2** 347, 807
PASSE SIMPLE, LE **Vol. 2** 810
PASSE TON BAC D'ABORD **Vol. 2** 351, 411, 450
PASSEGGIATA, LA **Vol. 7** 85, 915, 1177, 1178
PASSEPORT 13.444 **Vol. 2** 339
PASSEPORT POUR LE MONDE **Vol. 2** 47
PASSER PASSER PIGER **Vol. 5** 199, 304
PASSEURS D'HOMMES **Vol. 2** 37, 131, 744
PASSEURS D'OR **Vol. 2** 228
PASSI FURTIVI IN UNA NOTTE BOIA **Vol. 7** 1024
PASSION **Vol. 2** 581, 611
PASSION DE FEMME **Vol. 2** 259
PASSION DE JEANNE D'ARC, LA **Vol. 2** 403, 438, 563, 852, 865
PASSION, EN **Vol. 5** 1052, 1053, 1080, 1144
PASSIONE CHE TRAVOLGE **Vol. 7** 1089
PASSIONE D'AMORE **Vol. 7** 893, 950, 1059
PASSIONE DEL GIOCO, LA **Vol. 7** 1189
PASSIONE DI POPOLO **Vol. 7** 316, 1013
PASSIONE DI ROBINET PER IL DIRIGIBILE **Vol. 7** 1189
PASSIONE FATALE **Vol. 7** 606, 1189
PASSIONE SALVA **Vol. 7** 1189
PASSIONI **Vol. 7** 250
PASSIONS **Vol. 6** *page 120*
PASSO FALSO, IL **Vol. 7** 605, 686
PAST **Vol. 10** 154, 184
PAST NA KACHNU **Vol. 10** 113, 172, 567
PASTEL DE SANGRE **Vol. 8** 348, 768
PASTEUR **Vol. 2** 48, 302, 388, 726
PASTO DE FIERAS **Vol. 8** 331, 681
PASTOR ANGELICUS **Vol. 7** 1162, 1163
PASTOR FIDO, IL **Vol. 7** 69, 1052
PASTOR JARMANN VENDER HJEM **Vol. 5** 672
PASTORA QUE SUPO AMAR, EL **Vol. 8** 227, 364, 516
PASTORALE HEROICA **Vol. 6** *page 229, page 241*
PASTORI JUSSILAINEN **Vol. 5** 374, 443
PASTRICI **Vol. 3** 395, 449
PAT CORNER **Vol. 5** 97
PAT, UNA DONNA PARTICOLARE **Vol. 7** 289
PATARAC **Vol. 7** 652, 796, 876
PATAROUVA **Vol. 5** 371, 452
PATATA BOLLENTE, LA **Vol. 7** 633
PATATE **Vol. 2** 251
PATATES, LES **Vol. 2** 251, 427
PATE KOLO U VOZU **Vol. 10** 41, 182, 524
PATE ODDELENI **Vol. 10** 87, 226, 369
PATEK NENI SVATEK **Vol. 10** 41, 165, 197
PATENTKUNSTSCHLOSS **Vol. 4** 525, 546
PATER SEBASTIAN **Vol. 9** 93
PATER VOJTECH **Vol. 10** 167, 412, 453
PATERA KATSE FRONEMA **Vol. 3** 137
PATERNITA **Vol. 7** 1211
PATIENCE **Vol. 4** 288, 396
PATIMA **Vol. 3** 230, 290, 297, 321
PATIO DAS CANTIGAS, O **Vol. 8** 11, 172
PATIO DE LOS NARANJOS, EL **Vol. 8** 516
PATISSIER, PETIT BLEU ET RAMONEUR **Vol. 2** 678
PATRES DU DESORDRE, LES **Vol. 2** 110, 382
PATRIA CHICA, LA **Vol. 8** 716, 752, 772
PATRIA DEL RATA **Vol. 8** 237, 373
PATRICIA **Vol. 2** 187, 698
PATRICIA MIA **Vol. 8** 276, 494
PATRICIA, VALERIA, ANNA ET LES AUTRES **Vol. 6** *page 98*

PATRICIO MIRO A UNA ESTRELLA **Vol. 8** 262, 515
PATRICK VIVE ANCORA **Vol. 7** 1179
PATRIE **Vol. 2** 27, 395, 615
PATRIMONIO NACIONAL **Vol. 8** 240, 570, 738
PATRIOT, DER **Vol. 6** *page 190*
PATRIOTE, LE **Vol. 2** 100, 290, 574, 626, 824
PATRIOTEN **Vol. 4** 13, 574
PATRONNE, LA **Vol. 2** 137, 344, 488
PATRONS-TELEVISION **Vol. 2** 161, 265, 600, 779
PATROUILLE BLANCHE **Vol. 2** 413, 529
PATROUILLE DE CHOC **Vol. 2** 875
PATROUILLE DES SABLES, LA **Vol. 2** 9, 496
PATRU PASI DE INFINIT, LA **Vol. 3** 281, 295, 330
PATRULLA, LA **Vol. 8** 276, 704
PATSCHULI-MÄUSCHEN, DAS **Vol. 4** 366
PATTERNS **Vol. 2** 425
PATTES BLANCHES **Vol. 2** 8, 59
PATTES DE VELOURS **Vol. 2** 53
PATTO COL DIAVOLO **Vol. 7** 453, 796
PATTO GIURATO **Vol. 7** 1162
PATTUGLIA DELL'AMBA ALAGI, LA **Vol. 7** 1115
PATTUGLIA SPERDUTA, LA **Vol. 7** 234, 352, 773
PATYOLAT AKCIO **Vol. 9** 122, 149, 332
PAUL **Vol. 2** 713
PAUL ET MICHELLE **Vol. 2** 377, 719
PAUL ET VIRGINIE **Vol. 2** 286, 371
PAUL FLIEGT **Vol. 4** 192 **Vol. 6** *page 186*
PAUL UND PAULINCHEN **Vol. 3** 395
PAUL UND PAULINE **Vol. 4** 124, 422, 634
PAUL WANGS SKAEBNE **Vol. 5** 97
PAULI **Vol. 7** 1189
PAULINA S'EN VA **Vol. 2** 359
PAULINE **Vol. 4** 734
PAULINE A LA PLAGE **Vol. 6** *page 70*
PAULINE ET L'ORDINATEUR **Vol. 2** 754
PAULO ROBERTO COTECHINO, CENTRAVANTI DI SFONDAMENTO **Vol. 7** 761, 1202
PAULOS MELAS **Vol. 3** 199
PAURA D'AMARE **Vol. 2** 100 **Vol. 7** 149, 910, 1054
PAURA FA 90, LA **Vol. 7** 1027
PAURA IN CITTA **Vol. 7** 123
PAURA, LA **Vol. 7** 257
PAURA NELLA CITTA DEI MORTI VIVENTI **Vol. 7** 504, 1015
PAUSTIANS UHR **Vol. 5** 7
PAUVRE VILLAGE, LE **Vol. 2** 216
PAVE DE PARIS, LE **Vol. 2** 9, 559, 661
PAVILJON 6 **Vol. 3** 347, 350, 368
PAVILLON BRULE, LE **Vol. 2** 601, 676
PAVILLONES HEMMELIGHED **Vol. 5** 183
PAVILON SELIM **Vol. 10** 178, 210, 454, 578
PAVLE PAVLOVIC **Vol. 3** 421, 426
PAVLINKA **Vol. 10** 394, 408, 425
PAVUCINA **Vol. 10** 42, 297
PAW **Vol. 5** 23, 53
PAWEL I GAWEL **Vol. 1** 241, 278
PAX **Vol. 8** 422
PAX AETERNA **Vol. 5** 58
PAX DOMINE **Vol. 2** 731
PAX VOBISCUM **Vol. 9** 27, 144
PAYS BLEU, LE **Vol. 2** 781, 791
PAYS DE COCAGNE, LE **Vol. 2** 504
PAYS D'OU JE VIENS, LE **Vol. 2** 8, 249, 540, 544
PAYS SANS ETOILES, LE **Vol. 2** 273, 388, 401, 402, 644
PAYSAGE AVEC UN PRINCE CHARMANT **Vol. 6** *page 85*
PAYSANNES **Vol. 2** 161
PAYSANS NOIRS **Vol. 2** 36
PAZ **Vol. 8** 359, 400, 704
PAZ EMPIEZA NUNCA, LA **Vol. 8** 221, 752
PAZAC"T NA KREPOSTTA **Vol. 3** 42, 63 **Vol. 6** *page 33*
PAZI STA RADIS **Vol. 6** *page 318*
PAZZA DI GIOIA **Vol. 7** 196, 756
PAZZARELLA DELLA PIANA MORTA **Vol. 7** 1078
PAZZO D'AMORE **Vol. 7** 332, 838
PAZZO, IL **Vol. 7** 268, 847
PE AICI NU SE TRECE **Vol. 3** 307, 321
PE FIRUL APEI **Vol. 3** 239, 242
PE MALUL STING AL DUNARII ALBASTRE **Vol. 6** *page 274, page 275, page 278, page 283*
PE RASPUNDEREA MEA **Vol. 3** 287, 299, 330
PEAU D'ANE **Vol. 2** 190, 263
PEAU DE BANANE **Vol. 2** 700, 863
PEAU DE L'OURS, LA **Vol. 2** 137, 496
PEAU DE TORPEDO, LA **Vol. 2** 722, 781
PEAU D'ESPION **Vol. 2** 348, 503
PEAU DOUCE, LA **Vol. 2** 208
PEAU D'UN HOMME, LA **Vol. 2** 8, 698
PEAU ET LES OS, LA **Vol. 2** 97, 480
PECADO DE AMOR **Vol. 8** 214, 261
PECADOR IMPECABLE, EL **Vol. 8** 232
PECADOR Y LA BRUJA, EL **Vol. 8** 298, 493
PECADOS CONYUGALES **Vol. 8** 299, 719
PECADOS DE UNA CHICA CASI DECENTE, LOS **Vol. 8** 261, 438
PECCATI A VENEZIA **Vol. 7** 1179
PECCATI DI GIOVANI MOGLI **Vol. 7** 713

PECCATI DI GIOVENTU Vol. 7 666
PECCATI DI UNA GIOVANE MOGLIE Vol. 7 713
PECCATO CHE SIA UNA CANAGLIA Vol. 7 528
PECCATO DI ANNA, IL Vol. 7 17
PECCATO DI CASTITA Vol. 7 196
PECCATO DI LOLA, IL Vol. 7 414, 631, 1165
PECCATO DI ROGLIA SANCHEZ, IL Vol. 7 453, 587, 1096
PECCATO DI UNA NOTTE, IL Vol. 7 932
PECCATO, IL Vol. 7 1121
PECCATO ORIGINALE Vol. 7 780
PECCATO SENZA MALIZIA Vol. 7 139
PECCATORI DI PROVINCI Vol. 7 347
PECCATORI, I Vol. 7 827
PECCATRICE DELL'ISOLA, LA Vol. 7 87
PECCATRICE, LA Vol. 7 1159, 1177
PECCATRICE MODERNA, LA Vol. 7 1067
PECCATRICE SENZA PECCATO Vol. 7 751
PECES ROJOS, LOS Vol. 8 616, 630, 719
PECHE DE JEUNESSE Vol. 2 97, 205
PECHES DE JEUNESSE Vol. 2 332, 824
PECHEUR D'ISLANDE Vol. 2 167, 208, 336, 402, 469, 503
PECHMARIE Vol. 4 246, 638, 774
PECORELLA, LA Vol. 7 277
PECORELLA SMARRITA Vol. 7 131
PEDER TORDENSKJOLD Vol. 5 34, 102
PEDIGUENOS, LOS Vol. 8 719, 749
PEDON MERKKI Vol. 5 400, 410, 500, 511, 651
PEDRO EL CRUEL Vol. 8 523, 749
PEDRO GOLFO Vol. 8 630
PEDRO SO Vol. 8 169
PEDRO SOLL HÄNGEN Vol. 4 424, 462, 788
PEER GYNT Vol. 4 159, 288, 727, 734
PEERLESS WORLD Vol. 2 24
PEGGY PA VIFT Vol. 5 907, 923, 1044
PEHAVY MAX A STRASIDLA Vol. 10 108, 228, 336
PEJZAZ HORYZONTALNY Vol. 1 277 Vol. 6 *page 249*
PEJZAZ Z BOHATEREM Vol. 1 2761 277, 303
PEJZAZI U MAGLI Vol. 6 *page 310*
PEKKA JA PÄTKA KETJUKOLARISSA Vol. 5 443, 599, 646
PEKKA JA PÄTKA LUMIMIEHEN JÄLJILLÄ Vol. 5 374, 443, 643
PEKKA JA PÄTKÄ MESTARIMAALAREINA Vol. 5 383, 443, 639
PEKKA JA PÄTKÄ MILJONÄÄREINA Vol. 5 443, 639
PEKKA JA PÄTKÄ NEEKEREINA Vol. 5 374, 443, 639
PEKKA JA PÄTKÄ PAHASSA PULASSA Vol. 5 443, 646
PEKKA JA PÄTKÄ PUISTOTÄTEINA Vol. 5 443, 646
PEKKA JA PÄTKÄ SALAPOLIISEINA Vol. 5 443, 461, 599
PEKKA JA PÄTKÄ SAMMAKKOMIEHINÄ Vol. 5 443, 599, 639
PEKKA JA PÄTKÄ SUEZ'ILLA Vol. 5 443, 599, 639
PEKKA PUUPÄÄ Vol. 5 461, 592
PEKKA PUUPÄÄ KESÄLAITUMILLA Vol. 5 443, 461
PELESKEI NOTARIUS, A Vol. 9 323
PELICAN, LE Vol. 2 335, 701
PELIKAN MA ALIBI Vol. 10 345, 532
PELLE DI BANDITO Vol. 7 742
PELLE, LA Vol. 7 434, 817, 884, 1142
PELLE SOTTO GLI ARTIGLI, LA Vol. 7 714
PELLE VIVA Vol. 7 496, 663, 1144
PELLEGRINI D'AMORE Vol. 7 171
PELLEGRINI, IL Vol. 7 1189
PELLICCIA DI VISONE, UNA Vol. 7 257
PELNIA Vol. 6 *page 205, page 216*
PELNIA NAD GLOWAMI Vol. 6 *page 208, page 240, page 253*
PELOTON D'EXECUTION Vol. 2 627, 833
PELUSA Vol. 8 602, 725
PENA DE MUERTE ? Vol. 8 217, 470
PENA DEL TAGLIONE, LA Vol. 7 1189
PENDRAGON LEGENDA, A Vol. 9 134, 259, 265
PENDULE A SALOMON, LA Vol. 2 344
PENELOPA Vol. 10 82, 273, 548
PENETRATIONS PORNOGRAPHIQUE Vol. 2 464
PENETRATIONS VICIEUSES Vol. 2 291
PENETREZ-MOI PAR LE PETITE TROU Vol. 2 294
PENGAR – EN TRAGIKOMISK SAGA Vol. 5 878, 1173
PENGAR FRAN SKYN Vol. 5 982
PENGE Vol. 5 228
PENGE SOM GRAES Vol. 5 17, 101, 150, 208
PENGENE ELLER LIVET Vol. 5 50, 113, 195
PENGENES MAGT Vol. 5 183
PENICKA A PARAPLICKO Vol. 10 70, 158, 547
PENIZE NEBO ZIVOT Vol. 10 453, 586
PENNE NERE Vol. 7 978
PENSACI GIACOMINO Vol. 7 792, 796, 954
PENSEUR, LE Vol. 2 798
PENSIERO D'AMORE Vol. 7 996
PENSION BOULANKA Vol. 1 46, 92, 100
PENSION DU LIBRE AMOUR Vol. 2 508

PENSION ELISE NOTTEBOHM Vol. 4 13, 558
PENSION JONAS Vol. 2 193, 676
PENSION LAUTENSCHLAG Vol. 4 385
PENSION MIMOSA Vol. 2 402, 573
PENSION PRO SVOBODNE PANY Vol. 10 286, 289, 352
PENSION SCHÖLLER Vol. 4 255, 471, 774
PENSION TRUOCHEN Vol. 6 *page 186*
PENSIONAT PARADISET Vol. 5 840, 864
PENSIONE PAURA Vol. 7 699
PENSIONNAT DES PETITES SELOPES, LE Vol. 6 *page 111*
PENSIONNAT ET SES INTIMITES, LE Vol. 2 810
PENSJA PANI LATTER Vol. 6 *page 258, page 259*
PENTE GYNAIKES GIA EINAN ANDRA Vol. 3 141
PENTEK ESTE Vol. 9 35, 239
PENTEK REZI Vol. 9 71, 322
PENTIMENTO Vol. 7 827
PENTITENCIER DE FEMMES Vol. 7 311
PENTITO, IL Vol. 7 585, 1153
PENTRU CA SE IUBESC Vol. 3 281, 321, 331
PENTRU PATRIE Vol. 3 211, 233, 235, 254, 256, 290
PENZ ALL A HAZHOZ Vol. 9 132, 270
PENZ BESZEL Vol. 9 26, 132
PENZCSINALO, A Vol. 9 83, 158, 265
PEPA DONCEL Vol. 8 261, 299
PEPANEK NEZDARA Vol. 10 115, 536
PEPE CONDE Vol. 2 646 Vol. 8 607, 692, 717
PEPE-HILLO Vol. 8 516
PEPE LE MOKO Vol. 2 309, 439, 440
PEPE, NO ME DES TORMENTO Vol. 8 368, 754
PEPEES AU SERVICE SECRET, LES Vol. 2 344, 475
PEPEES FONT LA LOI, LES Vol. 2 229, 479, 813
PEPEES LA BAISE Vol. 6 *page 114*
PEPI, LUCY, BOM Y OTRAS CHICAS DEL MONTON Vol. 8 372
PEPINA REJHOLCOVA Vol. 10 248, 453, 524
PEPITA JIMENEZ Vol. 8 228, 339
PEPITA KABAT Vol. 9 26, 117
PEPOTE Vol. 7 571
PEPPERMINT FRAPPE Vol. 8 348, 714
PEPPINO E LA NOBILE DAMA Vol. 7 16
PEPPINO LE MODELLE E...CHELLA LA Vol. 7 196
PEQUENA MOVIDA, UNA Vol. 8 297, 580
PEQUENECES Vol. 2 646 Vol. 8 298, 607
PEQUENO CORONEL, EL Vol. 8 298, 400
PER Vol. 5 46, 125, 283
PER AMORE Vol. 7 524
PER AMORE DI CESARINA Vol. 7 1114
PER AMORE DI POPPEA Vol. 7 426
PER AVER VISTO Vol. 7 399
PER DOLLARO DI GLORIA Vol. 8 396
PER ESSERE PIU LIBERO Vol. 7 131
PER FARE SUA CONOSCENZA Vol. 7 1189
PER FAVORE, OCCUPATI DI AMELIA Vol. 7 19, 20, 32 Vol. 8 219, 299
PER GUADAGNARE CENTO MILIONI Vol. 7 196
PER IL GUSTO DE UCCIDERE Vol. 8 406
PER IL PASSATO Vol. 7 654
PER LA PATRIA ! Vol. 7 965
PER LA SUA BOCCA Vol. 7 493
PER LA SUA FELICITA Vol. 7 196
PER LE ANTICHE SCALE Vol. 7 576
PER LE VIE DEL SOGNO Vol. 7 932
PER LE VIE DELLA CITTA Vol. 7 905
PER L'ONORE Vol. 7 369, 1189
PER MILLE DOLLARI AL GIORNO Vol. 8 602, 749
PER PIACERTI DI PIU Vol. 7 1211
PER POCCHI DOLLARI ANCORA Vol. 8 603
PER QUESTA NOTTE Vol. 7 199, 564, 1145
PER SALVARE IL PORCELLINO Vol. 7 369
PER SALVARTI HO PECCATO Vol. 7 196
PER TE, AMORE ! Vol. 7 1169
PER TUTTA LA VITA Vol. 7 369
PER UN BACIO Vol. 7 1125
PER UN FIGLIO Vol. 7 168
PER UN FIORE Vol. 7 493
PER UN PO DI GLORIA Vol. 7 866, 940
PER UN PUGNO DI DOLLARI Vol. 7 372, 612, 1077 Vol. 8 298, 493
PER UNA LAGRIMA Vol. 7 729
PER UN'ORA D'AMORE Vol. 7 1189
PER UOMINI SOLI Vol. 7 796
PER VIVERE MEGLIO DIVERTITEVI CON NOI Vol. 7 524, 599
PERBALLIMI Vol. 3 12 Vol. 6 *page 13*
PERCEVAL LE GALLOIS Vol. 2 17, 434
PERCHE FRICOT FU MESSO IN COLLEGIO Vol. 7 1189
PERCHE MI SONO FATTO GUARDINGO Vol. 7 1189
PERCHE NO ? Vol. 7 978
PERCHE NON FACCIAMO L'AMORE ? Vol. 8 373
PERCUSION Vol. 8 373, 580
PERDEU-SE UM MARIDO Vol. 8 85, 130
PERDIDO POR CEM ... Vol. 8 166
PERDIDOS EN EL TIEMPO Vol. 8 590

PERDIZIONE **Vol. 7** 2, 48, 170, 552
PERDONAME AMOR **Vol. 8** 738
PERDONAMI **Vol. 7** 106
PERDUTAMENTE TUO ... MI FIRMO MACALUSO CARMELO FU GIUSEPPE **Vol. 7** 1114
PERE DE MADEMOISELLE, LE **Vol. 2** 194, 423
PERE GORIOT, LE **Vol. 2** 32, 346, 722
PERE LAMPION, LE **Vol. 2** 425
PERE LEBONNARD, LE **Vol. 7** 48, 1159
PERE NOEL A LES YEUX BLEUS, LE **Vol. 2** 818
PERE NOEL EST UNE ORDURE, LE **Vol. 6** *page 101*
PERE PREMATURE, LE **Vol. 2** 648
PERE SERGE, LE **Vol. 2** 6, 330, 574, 869
PERE TRANQUILLE, LE **Vol. 2** 158, 719
PEREJE **Vol. 10** 168, 532
PERFECT STRANGERS **Vol. 2** 656
PERFEKT GENTLEMAN, EN **Vol. 5** 891, 1031
PERFEKTE SEKRETÄRIN, DIE **Vol. 4** 329, 365
PERFETTA EBBREZZA, LA **Vol. 7** 390
PERFETTA, LA **Vol. 7** 45
PERFIDO INCANTO **Vol. 7** 399, 929
PERFIL DE SATANAS, EL **Vol. 8** 207, 497
PERGOLA DE LAS FLORES, LA **Vol. 8** 467, 578
PERGOLESI **Vol. 7** 48, 196, 508, 1036, 1060, 1088
PERGÖTÜZBEN **Vol. 9** 34, 237
PERHEEN MUSTA LAMMAS **Vol. 5** 356, 386
PERINBABA **Vol. 10** 145, 165, 493
PERIPEZIE DELL'EMULO DI FORTUNELLO & C., LE **Vol. 7** 940
PERIPLANISE **Vol. 3** 181
PERISCOPE, LE **Vol. 2** 153
PERLA DEL CINEMA, LA **Vol. 7** 264, 701
PERLA DI CLEOPATRA, LA **Vol. 7** 196
PERLA DI POSILLIPO, LA **Vol. 7** 1174
PERLA INSANGUINATA, LA **Vol. 7** 1124
PERLA W KORONIE **Vol. 1** 198, 243, 303
PERLE DES ORIENTS, DIE **Vol. 4** 190, 560
PERLE, LA **Vol. 2** 646
PERLE, UNE **Vol. 2** 798
PERLEHALSBANDET **Vol. 5** 81
PERLEN BEDEUTEN TRÄNEN **Vol. 4** 329, 535
PERLEN DER LADY HARRISON, DIE **Vol. 4** 124, 240
PERLEN DES DR. TALMADGE, DIE **Vol. 4** 99, 255
PERLES DE LA COURONNE, LES **Vol. 2** 81, 309, 440, 657
PERLICKY NA DNE **Vol. 10** 22, 41, 292
PERLOVY NAHRDELNIK **Vol. 10** 72, 250
PERMIS DE CONDUIRE, LE **Vol. 2** 97, 810
PERMISSION, LA **Vol. 2** 427, 664
PERNIKOVA CHALOUPKA **Vol. 10** 58, 256, 345, 524
PERO ... EN QUE PAI VIVIMOS ? **Vol. 8** 400, 438
PERRAS CALLEJERAS **Vol. 8** 550, 762
PERRO DE ALAMBRE **Vol. 8** 297, 316
PERRO EN ORBITA, UN **Vol. 8** 342, 772
PERRO GOLFO **Vol. 8** 396
PERROI VERSHUES **Vol. 6** *page 305*, *page 309*
PERROQUET VERT, LE **Vol. 2** 158, 279
PERROS CALLEJEROS II **Vol. 8** 426, 473
PERSE BIE KJO DAULLE **Vol. 3** 4, 14 **Vol. 6** *page 19*
PERSECUCION A UN ESPIA **Vol. 8** 214
PERSECUCION EN MADRID **Vol. 8** 396, 503, 677
PERSECUCION HASTA VALENCIA **Vol. 8** 749
PERSEVERANCE MENE A BIEN **Vol. 2** 724
PERSONA **Vol. 5** 1042, 1053, 1080
PERTEJ MUREVE TE GURTA **Vol. 3** 3, 26 **Vol. 6** *page 12*
PERTILE DIMINETII **Vol. 3** 213
PERTUS **Vol. 2** 251, 427
PERÜCKE, DIE **Vol. 4** 397, 546
PERVENCHE **Vol. 2** 876
PERVERSA CARICIA DE SATAN, LA **Vol. 8** 629
PERVERSA SENORA, LA **Vol. 8** 396
PERVERSION **Vol. 8** 249, 626
PERVERSION PORNO **Vol. 2** 294
PERVERSIONS **Vol. 2** 442
PERVERSIONS SEXUELLES **Vol. 2** 396
PERVERSITES SUEDOISES **Vol. 2** 735
PESCADOR DE COPLAS, EL **Vol. 8** 524, 749
PESCATORE **Vol. 7** 1189
PESCATORE DEL RHONE, IL **Vol. 7** 488
PESCATORE 'E POSILLECO **Vol. 7** 603
PESCATORI **Vol. 7** 756
PESCE D'APRILE **Vol. 7** 1189
PESCENI GRAD **Vol. 3** 375
PE'SEMPRE **Vol. 7** 713
PESEN ZA COVEKA **Vol. 3** 62, 85, 91, 100, 109
PESENTA NA BALKANA **Vol. 3** 86
PESMA **Vol. 3** 374, 403
PESMA SA KUMBARE **Vol. 3** 356, 383, 388
PESO DEL DISONORE, IL **Vol. 7** 1189
PESSI JA ILLUSIA **Vol. 5** 505, 526, 532, 539, 563, 566, 569, 579, 589, 610
PEST IN FLORENZ, DIE **Vol. 4** 239, 574, 634, 734

PESTE, LA **Vol. 7** 1189
PESTI HAZTETÖK **Vol. 9** 67, 117
PESTI MESI **Vol. 9** 71, 322
PESTI VE TME **Vol. 10** 283, 344, 571
PET HOLEK NA KRKU **Vol. 10** 87, 235, 254
PET HRISNIKU **Vol. 10** 153, 296
PET MILIONU SVEDKU **Vol. 10** 154, 313, 569
PET MINUTA RAJA **Vol. 3** 338, 342, 387
PET MUZU A JEDNO SRDCE **Vol. 10** 41, 254, 505
PET PLJUS EDNO **Vol. 6** *page 30, page 45*
PET Z MILIONU **Vol. 10** 87, 182, 296
PETA ZASEDA **Vol. 3** 345, 385
PETER **Vol. 9** 71, 322
PETER ANDERSEN **Vol. 5** 2, 7
PETER DER GROSSE **Vol. 4** 4, 113, 131, 239
PETER, DER MATROSE **Vol. 4** 73, 186
PETER, FRANCK ET LES AUTRES **Vol. 2** 795
PETER LUMP **Vol. 4** 397
PETER, PAUL UND NANETTE **Vol. 4** 323, 734, 738
PETER SPIELT MIT DEM FEUER **Vol. 4** 20, 323, 424
PETER UND DAS EINMALEINS MIT DER SIEBEN **Vol. 1** 35, 74, 125
PETER VAN HEEREN **Vol. 5** 738, 831
PETER VOSS, DER MILLIONENDIEB **Vol. 4** 49, 124, 292, 544, 545, 640
PETERLE **Vol. 4** 135, 544, 619
PETERMANN IST DAGEGEN **Vol. 4** 79, 110, 115
PETERS BABY **Vol. 5** 104, 199, 201
PETERS ERBSCHAFT **Vol. 4** 423
PETERS HELD **Vol. 5** 97
PETERS JUGEND **Vol. 6** *page 163, page 166, page 173*
PETERS LANDLOV **Vol. 5** 104, 199, 304
PETERSBURGER NÄCHTE **Vol. 4** 49, 115
PETERSSON - SVERIGE **Vol. 5** 840, 909
PETHAINO KATHE XEMEROMA **Vol. 3** 153, 205
PETIC DE CER, UN **Vol. 6** *page 273, page 275, page 278*
PETIMATA OT MOBI DIK **Vol. 6** *page 41*
PETIMATA OT MOBY DICK **Vol. 3** 41, 45
PETISTOVKA **Vol. 10** 460, 503
PETIT A PETIT **Vol. 2** 752
PETIT ANGE ET SON PANTIN **Vol. 2** 216, 388
PETIT BAIGNEUR, LE **Vol. 2** 21, 835
PETIT BARBOUIN, LE **Vol. 2** 626
PETIT BOUGNAT, LE **Vol. 2** 514
PETIT CAFE, LE **Vol. 2** 264
PETIT CHAPERON ROUGE, LE **Vol. 2** 533
PETIT CHOSE, LE **Vol. 2** 99, 402, 584
PETIT GARCON DE L'ASCENSEUR, LE **Vol. 2** 395, 770
PETIT JASQUES, LE **Vol. 2** 64, 153, 243, 279, 282, 412, 475
PETIT JOSEPH **Vol. 6** *page 72, page 105*
PETIT LOUIS D'OR **Vol. 7** 1189
PETIT MARCEL, LE **Vol. 2** 359
PETIT MATIN, LE **Vol. 2** 220, 263, 549
PETIT MOINEAU DE PARIS, LE **Vol. 2** 136
PETIT MONDE DE DON CAMILLO, LE **Vol. 2** 395, 550
PETIT MONSTRE, LE **Vol. 2** 570
PETIT POUCET, LE **Vol. 2** 335, 347
PETIT PROF', LE **Vol. 2** 388, 395
PETIT RADJAH, LE **Vol. 2** 54
PETIT ROI, LE **Vol. 2** 9, 62, 824
PETIT SOLDAT, LE **Vol. 2** 208, 465
PETITE BANDE, LA **Vol. 6** *page 85, page 123, page 136*
PETITE CHOCOLADIERE, LA **Vol. 2** 371, 402, 452, 627, 656, 780,
PETITE DAME DU WAGON-LIT, LA **Vol. 2** 308
PETITE DUCHESSE, LA **Vol. 2** 798
PETITE FEMME DU FLORIDA, LA **Vol. 2** 37
PETITE FILLE EN VELOURS BLEU, LA **Vol. 2** 457
PETITE FILLE QUI CHERCHE LE PRINTEMPS, LA **Vol. 2** 7
PETITE FONCTIONNAIRE, LA **Vol. 2** 805
PETITE LISE, LA **Vol. 2** 48, 193, 332
PETITE MARCHANDE D'ALLUMETTES, LA **Vol. 2** 48
PETITE PENSIONNAIRE, LA **Vol. 2** 872
PETITE PESTE **Vol. 2** 131, 514, 595
PETITE SAUVAGE, LA **Vol. 2** 526, 581, 646
PETITE SIRENE, LA **Vol. 6** *page 92, page 101, page 124*
PETITE SOEUR DES PAUVRES **Vol. 2** 418
PETITE VERTU, LA **Vol. 2** 532, 700
PETITES ALLIEES, LES **Vol. 2** 153
PETITES ANNONCES TRES SPECIALES **Vol. 6** *page 94*
PETITES CARDINAL, LES **Vol. 2** 254, 368, 516
PETITES CURIEUSES, LES **Vol. 2** 214
PETITES DU QUAI AUX FLEURS, LES **Vol. 2** 13, 96, 156
PETITES FILLES DOCILES **Vol. 6** *page 120*
PETITES FILLES IMPUDIQUES **Vol. 2** 294
PETITES FILLES, LES **Vol. 2** 2
PETITES FILLES MODELES, LES **Vol. 2** 769
PETITES FILLES POUR GRANDS VICIEUX **Vol. 6** *page 79*

PETITES MARIONETTES, LES Vol. 2 432
PETITES PARISIENNES, LES Vol. 6 *page 79*
PETITES SAINTES Y TOUCHENT, LES Vol. 2 818
PETITES VICIEUSES FONT LES GRANDES EMMERDEUSES, LES Vol. 2 750
PETITES VICIEUSES PRENNENT LEUR PIED, LES Vol. 6 *page 79*
PETITS CALINS, LES Vol. 2 323, 781
PETITS CHATS, LES Vol. 2 824
PETITS DESSOUS DES GRANDS ENSEMBLES, LES Vol. 2 575, 592, 721
PETITS MATIN, LES Vol. 2 651
PETITS RIENS, LE Vol. 2 347, 871
PETKA S HVEZDICKOU Vol. 10 163, 250, 508
PETLA Vol. 1 148, 187, 401 Vol. 6 *page 208*
PETÖFI Vol. 9 239
PETÖFI '73 Vol. 9 154, 316
PETÖFI - DALCIKLUS Vol. 9 323
PETOMANE, IL Vol. 7 258, 283, 347
PETRIJIN VENAC Vol. 3 404, 421
PETROLE ! PETROLE ! Vol. 2 495 Vol. 6 *page 78*
PETROLEJOVE LAMPY Vol. 10 144, 172, 493
PETROLEUSES, LES Vol. 2 659
PETRONELLA Vol. 4 242, 357
PETRUSKA Vol. 7 621
PETTER FRA RUSKOY Vol. 5 672, 674
PETTERSSON & BENDEL Vol. 5 840, 909
PETTERSSON & BENDELS NYA AFFÄRER Vol. 5 841, 993
PETTINATRICE DI MARECHIARO, LA Vol. 7 314
PEU, BEAUCOUP, PASSIONNEMENT..., UN Vol. 2 110
PEU D'AMOUR, UN Vol. 2 206
PEU DE SOLEIL DANS L'EAU FROIDE, UN Vol. 2 49
PEUR, LA Vol. 2 626, 824
PEUR SUR LA VILLE Vol. 2 21, 655
PEVNOST NA RYNE Vol. 10 197, 296, 433
PEZZENTE GENTILUOMO, IL Vol. 7 133, 963
PEZZO, CAPOPEZZO E CAPITANO Vol. 7 910
PFAD DER SÜNDE, DER Vol. 4 124
PFARRER VON KIRCHFELD, DER Vol. 4 585, 725, 754
PFENNIGSCHLACHT, DIE Vol. 4 186, 367
PFERDEKUR, DIE Vol. 1 34
PFERDEMÄDCHEN, DAS Vol. 1 15 Vol. 6 *page 160, page 167, page 170*
PFINGSTORGEL, DIE Vol. 4 634, 647
PFLICHT, DIE Vol. 4 514
PFLICHT ZU SCHWEIGEN, DIE Vol. 4 255, 527
PHAEDRA Vol. 2 625
PHALENES, LES Vol. 2 472
PHALLOCRATES, LES Vol. 6 *page 149*
PHANTOM Vol. 4 115, 213, 501, 547, 734
PHANTOME DES GLÜCKS Vol. 4 158, 164, 584
PHANTOME DES LEBENS Vol. 4 329
PHARE DU BOUT DU MONDE, LE Vol. 2 224
PHEDRE Vol. 2 59, 282, 427
PHENOMENA Vol. 7 13, 51, 1084, 1153
PHILEMON ES BAUCIS Vol. 9 51, 66, 67, 193
PHILHARMONIKER Vol. 4 49
PHILIBUS Vol. 7 744
PHILINE Vol. 4 110
PHILIPP, DER KLEINE Vol. 1 52, 85, 121
PHOTO SCANDALE Vol. 2 735
PHOTOGRAPHE TRES SPECIALE, UNE Vol. 6 *page 112, page 129*
PHROSO Vol. 2 874
PI PI PIL... Vol. 5 314
PIA DE'TOLOMEI Vol. 7 50, 192, 493, 704, 709, 1027
PIACERE, IL Vol. 7 399, 1161
PIACERI DEL SABATO NOTTE, I Vol. 7 1044
PIACERI DELLO SCAPOLO Vol. 7 693
PIACEVOLE CONFRONTO Vol. 7 1023
PIAF Vol. 2 475, 781
PIAN DELLE STELLE Vol. 7 1054, 1136
PIANETA EROS Vol. 7 728
PIANGE ... IL TELEFONO Vol. 7 1015
PIANGE PIERROT Vol. 7 829
PIANISTA DI HAYNES Vol. 7 53
PIANKE Vol. 6 *page 166, page 168, page 173, page 176*
PIANOFORTE Vol. 7 405, 434, 817
PIANOFORTE SILENZIOSO Vol. 7 1189
PIANOS MECANICOS, LOS Vol. 8 214, 644
PIANOS MECANIQUES, LES Vol. 2 11
PIATKA Z ULICY BARSKIEJ Vol. 1 149, 305, 307, 366, 377
PIAZZA S. SEPOLCRO Vol. 7 1159
PIC ET PIC ET COLEGRAM Vol. 2 721
PICA SUL PACIFICO, LA Vol. 7 891
PICADOR, LE Vol. 2 308, 656, 687
PICARA MOLINERA, LA Vol. 8 214, 261, 295, 298
PICASSOS ÄVENTYR Vol. 5 882, 946, 947, 1160
PICCOLA AMICA, LA Vol. 7 574, 1004
PICCOLA FIORAIA, LA Vol. 7 760
PICCOLA FONTE, LA Vol. 7 264, 701
PICCOLA MANON Vol. 7 508, 574
PICCOLA MIA Vol. 7 730, 792

PICCOLA MODISTA, LA **Vol. 7** 1189
PICCOLA MOGLIE, UNA **Vol. 7** 1054, 1136
PICCOLA MONELLA **Vol. 7** 232, 849
PICCOLA PARROCHIA, LA **Vol. 7** 48
PICCOLA POSTA **Vol. 7** 333, 634
PICCOLA RIBELLE, LA **Vol. 7** 1189
PICCOLA SANTA **Vol. 7** 500
PICCOLE LABBRA **Vol. 7** 692
PICCOLI FUOCHI **Vol. 7** 542, 819
PICCOLI MARTIRI **Vol. 7** 209
PICCOLI NAUFRAGHI **Vol. 7** 360, 493, 1136
PICCOLO ALPINO **Vol. 7** 493, 1054, 1121
PICCOLO BURATTINAIO, IL **Vol. 7** 1189
PICCOLO FIGURINAIO **Vol. 7** 1189
PICCOLO HOTEL **Vol. 7** 639, 966
PICCOLO LUSTRASCARPE, IL **Vol. 7** 1189
PICCOLO MONDO ANTICO **Vol. 7** 261, 328, 493, 750, 756, 796,
PICCOLO RE, IL **Vol. 7** 985, 1063
PICCOLO RIBELLE **Vol. 7** 372
PICCOLO SALTATORE **Vol. 7** 1189
PICCOLO SANTO, IL **Vol. 7** 1032
PICCOLO SPAZZACAMINO, IL **Vol. 7** 1189
PICCOLO VANDEANO, IL **Vol. 7** 1189
PICCOLO VETRAIO, IL **Vol. 7** 234
PICCOLO VOM GOLDENEN LÖWEN, DER **Vol. 4** 212, 424
PICIU **Vol. 6** *page 265, page 276, page 280*
PICKPOCKET **Vol. 2** 153, 170
PICO 2, EL **Vol. 8** 210, 684
PICO, EL **Vol. 8** 297, 684
PICPUS **Vol. 2** 23, 67
PIEC I POL BLADEGO JOZKA **Vol. 1** 249 **Vol. 6** *page 240, page 249*
PIECIU **Vol. 1** 179, 303, 391
PIECZONE GOLABKI **Vol. 1** 198, 220, 343 **Vol. 6** *page 208*
PIED !, LE **Vol. 2** 76, 106
PIEDALU A PARIS **Vol. 2** 193, 475
PIEDALU DEPUTE **Vol. 2** 70, 131, 193
PIEDALU FAIT DES MIRACLES **Vol. 2** 193, 475
PIEDONE D'EGITTO **Vol. 7** 207, 599, 1010
PIEDONE L'AFRICANO **Vol. 7** 1086
PIEDRA DE TOQUE **Vol. 8** 298, 461
PIEDRAS VIVAS **Vol. 8** 396, 772
PIEDS DANS LE PLATRE, LES **Vol. 2** 60, 137, 664
PIEDS NICKELES, LES **Vol. 2** 514
PIEGE A CONS, LE **Vol. 2** 866 **Vol. 6** *page 93*
PIEGE A FILLES **Vol. 2** 57
PIEGE A HOMMES **Vol. 2** 254, 529
PIEGE A PUCELLES **Vol. 2** 514
PIEGE, LE **Vol. 2** 190, 650, 781
PIEGE POUR CENDRILLON **Vol. 2** 185, 824
PIEGES **Vol. 2** 194, 427, 581, 863
PIEKLO I NIEBO **Vol. 1** 265, 382, 403
PIEL QUEMADA, LE **Vol. 8** 217
PIENI LUUTATYTTÖ **Vol. 5** 443, 634
PIERINO AIUTANTE MESSO COMMUNALE ... PRATICAMENTE SPIONE **Vol. 7** 311
PIERINO COLPISCE ANCORA **Vol. 7** 811, 1202
PIERINO CONTRO TUTTI **Vol. 7** 811, 1202
PIERINO IL FICHISSIMO **Vol. 7** 672
PIERINO LA PESTE ALLA RISCOSSA **Vol. 7** 694
PIERINO MEDICO DELLA SAUB **Vol. 7** 226, 1202
PIERNA CRECIENTE, FALDA MENGUANTE **Vol. 8** 318, 331
PIERNAS CRUZADAS **Vol. 8** 344, 543
PIERNAS DE LA SERPIENTE, LAS **Vol. 8** 317, 768
PIERO E TERESA **Vol. 7** 266, 545
PIERPIN **Vol. 7** 587
PIERRE DANS LA BOUCHE, UNE **Vol. 6** *page 70, page 102, page 104*
PIERRE ET JEAN **Vol. 2** 23, 67, 226
PIERRE ET PAUL **Vol. 2** 220, 480
PIERROT **Vol. 5** 97 **Vol. 7** 965
PIERROT LA TENDRESSE **Vol. 2** 559, 795
PIERROT LE FOU **Vol. 2** 208, 377
PIERROT-PIERRETTE **Vol. 2** 169, 330, 613
PIERROTS LETZTES ABENTEUER **Vol. 4** 124, 681
PIERSCIEN KSIEZNEJ ANNY **Vol. 1** 202, 227, 400 **Vol. 6** *page 230*
PIERWSZA MILOSC KOSCIUSZKI **Vol. 1** 174, 292
PIERWSZE DNI **Vol. 1** 171, 227
PIERWSZY DZIEN WOLNOSCI **Vol. 1** 136, 389, 403, 406
PIERWSZY START **Vol. 1** 250
PIESEN O SIVOM HOLUBOVI **Vol. 10** 210, 576
PIESN O WIELKIM RZEZBIARZU **Vol. 1** 372, 384
PIESNIARZ WARSZAWY **Vol. 1** 278, 346
PIETA PER CHI CADE **Vol. 7** 487
PIETNO **Vol. 6** *page 208, page 234, page 256*
PIETRO, DER KORSAR **Vol. 4** 214, 436, 621, 729
PIETRO MICCA **Vol. 7** 360, 638, 639
PIETRO WYZEJ **Vol. 1** 278, 308, 346
PIGA BLAND PIGOR, EN **Vol. 5** 889, 923
PIGALLE-SAINT-GERMAIN-DES-PRES **Vol. 2** 627, 807
PIGE MED PEP, EN **Vol. 5** 32, 55, 101, 181, 208
PIGE UDEN LIGE, EN **Vol. 5** 89, 136
PIGEN FRA EGBORG **Vol. 5** 192

PIGEN FRA HAVET Vol. 5 46, 95
PIGEN FRA PALLS Vol. 5 9
PIGEN FRA SYDHAVSOEN Vol. 5 228
PIGEN I SOGELYSET Vol. 5 14, 281
PIGEN OG DROMMESLOTTET Vol. 5 298, 306
PIGEN OG GREVEN Vol. 5 179, 240
PIGEN OG MILLIONAEREN Vol. 5 23, 248
PIGEN OG PRESSEFOTOGRAFEN Vol. 5 20, 179
PIGEN OG VANDPYTTEN Vol. 5 248, 272
PIGER I TROJEN Vol. 5 175, 247, 306
PIGER I TROJEN 2 Vol. 5 298, 306
PIGER TIL SOS Vol. 5 158, 231, 306
PIGERNES JENSER Vol. 5 203
PIGEROVOREN Vol. 5 203
PIGESKOLEN PAA UDFLUGT Vol. 5 216
PIGESPEJDEREN Vol. 5 83
PIGGEN FRA KLUBBEN Vol. 5 58
PIGULKI DLA AURELII Vol. 1 227, 354
PIILOPIRTTI Vol. 5 362, 466, 602
PIKAJUNA POHJOISEEN Vol. 5 310, 472, 634
PIKKU ILONA JA HÄNEN KARITSANA Vol. 5 443, 646
PIKKU MATTI MAAILMALLA Vol. 5 357, 395, 620
PIKKU MYYJÄTÄR Vol. 5 356, 359, 485
PIKKU PELIMANNAISTA VIULUN KUNINKAAKSI Vol. 5 334, 349, 356, 566
PIKKU PELIMANNI Vol. 5 349, 566
PIKKU PIETARIN PIHA Vol. 5 312, 392
PIKKU SUORASUU Vol. 5 356, 481, 549, 639
PIKNIK Vol. 10 271, 313, 353
PIKNIK U TOPOLI Vol. 6 *page 314*
PIKRAGAPEMENE Vol. 3 168
PILAR GUERRA Vol. 8 246, 422
PILE OU FACE Vol. 2 815 Vol. 6 *page 81*
PILLANATNYI PENZZAVAR Vol. 9 94, 164
PILLOLE D'ERCOLE, LE Vol. 7 763, 915
PILLUELO DE MADRID, EL Vol. 8 606
PILOTA RITORNA, UN Vol. 7 709, 1063
PILOTI Vol. 10 74, 149, 283, 426, 550
PILSKA JULIA PA BRÖLLOPSRESA Vol. 5 866
PILVILINNA Vol. 5 391
PIM, PAM, PUM, ... FUEGO ! Vol. 8 245, 338
PIMEÄN PIRTIN HÄVITYS Vol. 5 372, 552, 579
PIMENTILLA Vol. 8 276
PIMERA ENTREGA, LA Vol. 8 338, 521
PIMPERNEL SVENSSON Vol. 5 840, 843
PINGWIN Vol. 1 220, 254, 401
PINK FLOYD Vol. 2 441
PINOT SIMPLE FLIC Vol. 6 *page 135, page 143*
PINSIÖN PAROONI Vol. 5 443, 639
PINTEA Vol. 3 284
PIOGGIA DEI DIAMANTI, LA Vol. 7 1132
PIOGGIA D'ORO Vol. 7 1189
PION, LE Vol. 2 495
PIOSENKA O MILOSCI Vol. 1 194, 205
PIOUPIOU ET NOUNOU Vol. 2 678
PIOVRA, LA Vol. 7 441, 701
PIOVUTO DAL CIELO Vol. 7 792, 1177
PIPA DEL NONNO, LA Vol. 7 1189
PIPE DE MASTER POUITTE, LA Vol. 2 159, 286, 330
PIPIN DER KURZE Vol. 4 239, 340, 466
PIPPI LANGSTRUMP Vol. 5 1003, 1052
PIPPI LANGSTRUMP PA DE SJU HAVEN Vol. 5 862
PIQUE-ASSIETTE, LES Vol. 2 671, 686
PIQUE DAME Vol. 4 132, 486, 626
PIRAT MIT HINDERNISSEN Vol. 1 49, 87, 121
PIRATA BOCANEGRA, EL Vol. 8 207
PIRATA DELLA SPARVIERO NERO, IL Vol. 7 1063
PIRATA SONO IO, IL Vol. 7 443, 1136
PIRATAS DE MALESIA, LOS Vol. 8 493, 732
PIRATE, LA Vol. 6 *page 90*
PIRATEN AUF DER PFERDEINSEL Vol. 1 12, 65, 103
PIRATEN DER OSTSEEBÄDER, DIE Vol. 4 270, 599
PIRATENE Vol. 5 701, 750, 767
PIRATES DU RAIL, LES Vol. 2 533, 775
PIRATI DEL GOLFO, IL Vol. 7 359
PIRATI DEL MARE, I Vol. 7 1189
PIRATI DELLA COSTA, I Vol. 7 1121
PIRATI DELLA MALESIA, I Vol. 7 649, 750, 792, 796 Vol. 8 493, 732
PIRATI DELL'ISOLA VERDE Vol. 7 846
PIRATII DIN PACIFIC Vol. 3 235, 244, 329
PIRI MINDENT TUD Vol. 9 71, 322
PIROS BUGYELLARIS Vol. 9 94, 274
PIROSBETÜS HETKÖZNAPOK Vol. 9 134, 187, 326
PIRUETAS JUVENILES Vol. 7 945, 1014, 1062 Vol. 8 720
PISCANALISTA PER SIGNORA Vol. 2 807
PISCINE, LA Vol. 2 814
PISEN DROZKARE Vol. 10 345, 460
PISEN LASKY Vol. 10 310, 393, 564, 621
PISEN NEMILOVANEHO Vol. 10 283, 408, 428
PISEN O STROMU A RUZI Vol. 10 41, 144, 153
PISEN O VELKE LASCE Vol. 10 34, 453
PISEN ZIVOTA Vol. 10 167, 453

PISICA DE MARE Vol. 3 272, 278, 285
PISITO, EL Vol. 8 719
PISMAK Vol. 6 *page 214*
PISMO – GLAVA Vol. 6 *page 313*
PISNE BY NEMELY UMIRAT Vol. 10 139, 165, 181, 340, 355
PISNE VEZNENEHO Vol. 10 45, 453
PISNICKA ZA GROS Vol. 10 184, 532
PISNICKAR Vol. 10 92, 453
PISO DE SOLTEROS Vol. 8 523, 732
PISO PISELLO Vol. 7 215, 425, 607
PISTA TEKINTETES UR Vol. 9 68, 196
PISTE DES GEANTS, LA Vol. 2 24
PISTE DU BONHEUR, LA Vol. 2 703
PISTE DU NORD, LA Vol. 2 174, 402
PISTE DU SUD Vol. 2 566, 719
PISTOLA PARA RINGO, UNA Vol. 8 523
PISTOLA PER RINGO, UNA Vol. 8 523
PISTOLAS NO DISCUTEN, LAS Vol. 8 598, 626
PISTOLE NON DISCUTONO, LA Vol. 8 598, 626
PISTOLEN Vol. 5 1014, 1187
PISTOLEROS DE ARIZONA Vol. 8 280
PISTOLEROS DE CASA GRANDE, LOS Vol. 8 207, 311, 546
PISTOLEROS DE PASO BRAVO, LOS Vol. 8 583, 626
PISTONNE, LE Vol. 2 234, 642
PISTRUIATUL Vol. 3 239, 295, 332
PITIE POUR LES VAMPS Vol. 2 118, 309, 743
PITKÄJÄRVELÄISET Vol. 5 356, 620
PITTORE MALEDETTO, IL Vol. 7 1132
PITTY Vol. 4 25, 367
PIU BELLA DONNA DEL MONDO, LA Vol. 7 525
PIU BELLA SERATA DELLA MIA VITA, LA Vol. 7 318
PIU BELLO DI COSI SI MUORE Vol. 7 347, 363
PIU CHE IL SOLE ! Vol. 7 751
PIU COMICO SPETTACOLO DEL MONDO, IL Vol. 7 978
PIU DOLCE CORONA, LA Vol. 7 454
PIU FORTE AMORE, IL Vol. 7 168
PIU FORTE DEL DESTINO Vol. 7 730
PIU FORTE DELLA VERITA Vol. 7 1116
PIU FORTE, LA Vol. 7 1189
... PIU FORTE, RAGAZZI ! Vol. 7 738
PIYAYO, EL Vol. 8 611, 616, 630
PIZZA CONNECTION Vol. 7 358, 383, 1153
PJEGAVA DJEVOJKA Vol. 3 343, 351
PJEVAM DANJU PJEVAM NOCU Vol. 3 438
PLACE DE LA CONCORDE Vol. 2 48, 248
PLACE DE LA REPUBLIQUE Vol. 2 74
PLACHE PRIBEHY Vol. 10 299, 505, 570
PLACIDO Vol. 8 719, 768
PLACIER DE DEMAIN, LE Vol. 2 159, 286, 330
PLACOWKA Vol. 6 *page 210, page 245, page 252*
PLAGE DES ENCULEES, LA Vol. 6 *page 112*
PLAGE TE VJETRA Vol. 3 3, 20 Vol. 6 *page 17*
PLAGEAANDER Vol. 5 97
PLAISIR A TROIS Vol. 2 139
PLAISIR, LE Vol. 2 8, 27, 279, 386
PLAISIRS DE PARIS Vol. 2 226, 279, 438, 488, 863
PLAISIRS FOUS, LES Vol. 2 349
PLAISIRS SECRETS Vol. 2 814
PLAISIRS SOLITAIRES, LES Vol. 2 294
PLAKU DHE HASMI Vol. 6 *page 11, page 12, page 16*
PLAMENY ZIVOTA Vol. 10 53, 167, 310
PLAN DER DREI, DER Vol. 4 125, 626
PLANETA KRAWIEC Vol. 6 *page 220, page 236*
PLANICIE HEROICA Vol. 8 141, 191
PLANINA NA GNEVOT Vol. 3 384, 396
PLANQUE, LA Vol. 2 207, 866
PLANQUE N.1-SEX PORNO, LA Vol. 2 259
PLANQUES DU REGIMENT, LES Vol. 6 *page 85, page 101, page 124*
PLANTAGENSTRASSE 19 Vol. 6 *page 168, page 174, page 175*
PLASMO ... DISTRUSSE ... Vol. 7 531, 1079
PLAT ELLER KRONE Vol. 5 24
PLATERO Y YO Vol. 8 311, 719
PLATZ ODER SIEG ? Vol. 6 *page 155, page 171*
PLAVCIK A VRATKO Vol. 10 82, 228, 459
PLAVECKY MARIAS Vol. 10 122, 412
PLAVENI HRIBAT Vol. 10 385, 395, 473
PLAVI 9 Vol. 3 358, 445, 453
PLAYA DE FORMENTOR Vol. 8 396, 741
PLAYA DE LOS PERROS, LA Vol. 8 612
PLAYA PROHIBIDA Vol. 8 637, 749
PLAYBOY EN PARO Vol. 8 407, 762
PLAYTIME Vol. 2 49, 315, 747, 873
PLAZA DE ORIENTE Vol. 8 603, 749
PLAZA DEL DIAMANTE, LA Vol. 8 250, 684
PLAZO PARA MORIR Vol. 8 598, 772
PLECAREA VLASINILOR Vol. 6 *page 274, page 279, page 280*
PLEIN AUX AS Vol. 2 388, 407, 603
PLEIN DU SUPER, LE Vol. 2 739
PLEIN LES POCHES POUR PAS UN ROND Vol. 2 210
PLEIN SOLEIL Vol. 2 96, 224
PLEIN SUD Vol. 6 *page 87* Vol. 8 684

PLEINS FEUX SUR L'ASSASSIN **Vol. 2** 137, 314, 540
PLEINS FEUX SUR STANISLAS **Vol. 2** 348, 372
PLEITO DE SANGRE **Vol. 8** 493, 515
PLENENO JATO **Vol. 3** 32, 109 **Vol. 6** *page 29*
PLENNIK"T **Vol. 3** 71
PLES V DEZJU **Vol. 3** 375, 395
P.L.M. O L'ASSASSINIO DELLA PARIS – LYON – MEDITERRANEE **Vol. 7** 503
PLOMIENIE **Vol. 1** 354, 355 **Vol. 6** *page 231*
PLOMIENNE SERCA **Vol. 1** 194, 205
PLOMO SOBRE DALLAS **Vol. 8** 495, 773
PLONGEURS SOUDANAIS **Vol. 2** 678
PLOUM-PLOUM, TRA-LA-LA **Vol. 2** 70, 871
PLUIE D'OR **Vol. 2** 708, 855
PLUKOVNIK SVEC **Vol. 10** 247, 345
PLUMARD EN FOLIE, LE **Vol. 2** 154, 503
PLUME AU VENT **Vol. 2** 427
PLUME LA POULE **Vol. 2** 193
PLURIELLES **Vol. 2** 600
PLUS BEAU QUE MOI TU MEURS **Vol. 6** *page 98, page 148*
PLUS BELLE DES VIES, LA **Vol. 2** 249, 875
PLUS BELLE FILLE DU MONDE, LA **Vol. 2** 401, 413
PLUS BELLES ESCROQUERIES DU MONDE-PARIS: „L'HOMME QUI VENDIT LA TOUR EIFFEL" **Vol. 2** 700
PLUS CA VA, MOINS CA VA **Vol. 2** 60
PLUS DE VACANCES POUR LE BON DIEU **Vol. 2** 32, 615
PLUS DE WHISKY POUR CALLAGHAN **Vol. 2** 354, 740 **Vol. 7** 982
PLUS HEUREUX DES HOMMES, LE **Vol. 2** 368, 615
PLUS JOLI PECHE DU MONDE, LE **Vol. 2** 15, 368
PLUSZ-MINUSZ EGY NAP **Vol. 9** 134, 215, 316
PO' DI CIELO, UN **Vol. 7** 634, 877
PO DIRJATA NA BEZSLEDNO IZCEZNALITE **Vol. 3** 103 **Vol. 6** *page 49, page 51*
PO DRODZE **Vol. 6** *page 230, page 238*
PO ISTI POTI SE NE VRACAJ **Vol. 3** 395, 447
PO NOCI DEN **Vol. 10** 70, 503, 532, 557
PO STOPACH KRVE **Vol. 10** 227, 313, 496
PO TROTOARA **Vol. 3** 75, 99, 124
POBEL **Vol. 5** 716, 824
POBOCNIK JEHO VYSOSTI **Vol. 10** 118, 167, 169
POBOJOWISKO **Vol. 6** *page 237*
POBRE GARCIA, EL **Vol. 8** 719, 749
POBRE RICO, EL **Vol. 8** 396
POBRE VALBUENA, EL **Vol. 8** 572, 657
POCESTNE PANI PARDUBICKE **Vol. 10** 122, 142, 154
POCHARDE, LA **Vol. 2** 37, 131, 538, 661
POCHISCENIE „SAVOJI" **Vol. 6** *page 240*
POCHODNE **Vol. 10** 296, 311, 522
POCIAG **Vol. 1** 233, 298 **Vol. 6** *page 233*
POCKAM, AZ ZABIJES **Vol. 10** 385, 428, 550
POCTI LJUBOVNA ISTORIJA **Vol. 6** *page 34, page 48*
POCTI REVIZIJA **Vol. 6** *page 60, page 62*
POCZMISTRZ **Vol. 1** 237
POD BANDERA MILOSCI **Vol. 1** 278, 371
POD EDNO NEBE **Vol. 6** *page 26, page 38, page 61*
POD GWIAZDA FRYGIJSKA **Vol. 1** 227
POD IGOTO **Vol. 3** 66, 95, 108
POD ISTO NEBO **Vol. 3** 396, 443
POD JEDNOU STRECHOU **Vol. 10** 345, 524
POD JEZEVCI SKALOU **Vol. 10** 184, 352, 517
POD KAMENNYM NEBOM **Vol. 5** 813
POD NOHAMA NEBE **Vol. 10** 18, 286, 570
POD ORLOVOTO GNEZDO **Vol. 3** 35
POD STAROTO NEBE **Vol. 3** 67
POD SUMNJOM **Vol. 3** 346, 435
POD TWOJA OBRONE **Vol. 1** 174, 215, 404
PODEJ STESTI PRST, ONO TI DA RUKU **Vol. 10** 243
PODER DE BARBA AZUL, EN **Vol. 8** 262
PODER DEL DESEO, EL **Vol. 8** 214, 426
PODEROSA INFLUJO DE LA LUNA, EL **Vol. 8** 297, 299
PODEROSO CABALLERO **Vol. 8** 718
PODEZRENI **Vol. 10** 25, 172, 229
PODFUK **Vol. 10** 233, 344, 571
PODHALE W OGNIU **Vol. 1** 227, 305, 323 **Vol. 6** *page 213*
PODIVNA PRATELSTVI HERCE JESENIA **Vol. 10** 226, 317, 522
PODIVNY HOST **Vol. 10** 4, 100, 171
PODIVNY VYLET **Vol. 10** 144, 173, 380
PODNE **Vol. 3** 373, 423, 433
PODOBIZNA **Vol. 10** 314, 460, 621
PODRIAS CON 5 CHICAS A LA VEZ ? **Vol. 8** 471
PODROZ DO ARABII **Vol. 6** *page 206, page 229, page 235, page 256*
PODROZ ZA JEDEN USMIECH **Vol. 1** 198, 243 **Vol. 6** *page 240*
PODROZE PANA KLEKSA **Vol. 6** *page 210*
PODSKALAK **Vol. 10** 58, 586
POEMA A SVEDOMI **Vol. 10** 82, 228, 320
POETA E LA PRINCIPESSA, IL **Vol. 7** 390

POETEN OG LILLEMOR Vol. 5 31, 39, 241, 248, 272
POETEN OG LILLEMOR I FORARSHUMOR Vol. 5 31, 248, 272
POETEN OG LILLEMOR OG LOTTE Vol. 5 31, 248, 272
POET"T I DJAVOL"T Vol. 6 *page 33, page 37, page 53*
POFONOK VÖLGYE Vol. 9 102
POGANY MADONNA, A Vol. 9 67, 84, 122
POGANYOK Vol. 9 94, 288
POGLED U NOC Vol. 3 442
POGLED U ZJENICU SUNCA Vol. 3 361, 371
POGON B Vol. 3 391, 392, 448
POGON ZA ADAMEM Vol. 1 231, 323
POGRZEB SWIERSZCZA Vol. 1 163, 355 Vol. 6 *page 227*
POHADKA MAJE Vol. 10 412, 453, 586, 621
POHADKA O MALICKOVI Vol. 10 74, 239, 333
POHADKA O PUTOVANI Vol. 10 124, 539, 554
POHADKA O STARE TRAMVAJI Vol. 10 299, 380, 385
POHADKA SVATOJANSKE NOCI Vol. 10 187, 250, 462
POHISTENIE V Z"LTO Vol. 6 *page 37*
POHJALAISIA Vol. 5 344, 356,428, 640
POHJAN TÄHTEET Vol. 5 606
POHLAD KOCCE USI Vol. 10 103, 250, 299, 508, 522
POHLED DO OCI Vol. 10 385, 388, 503
POHOD Vol. 3 418
POHORSKA VESNICE Vol. 10 58, 453
POIA EINAI E MARGARITA Vol. 3 192
POIGNARD MALAIS, LE Vol. 2 122, 193, 197
POIKA ELI KESÄÄNSÄ Vol. 5 347, 634
POIKAMIESPAPPA Vol. 5 349, 386
POIKAMIESTEN HOLHOKKI Vol. 5 552, 592
POIL DE CAROTTE Vol. 2 9, 17, 67, 142, 158, 231, 824, 864
POINT DE CHUTE Vol. 2 241
POINT DE MIRE, LE Vol. 2 142, 224
POINT DOULOUREUX, LE Vol. 2 450
POINT DU JOUR, LE Vol. 2 47, 96
POINT FINAL A LA LIGNE Vol. 6 *page 108, page 133*
POINTE COURTE, LA Vol. 2 801
POIOS THANASES Vol. 3 185
POIR UN SOU D'AMOUR Vol. 2 58, 205, 279
POISON, LA Vol. 2 48, 270
POISSON D'AVRIL Vol. 2 197, 259, 309
POJAT Vol. 5 383, 481, 639
POJDZIESZ PONAD SADEM Vol. 1 141, 163, 233, 401 Vol. 6 *page 244*
POJKARNA PA STORHOLMEN Vol. 5 840, 841
POJKEN I TRADET Vol. 5 880, 940, 1167
POKANI, PÄÄKONSULI Vol. 5 552, 592
POKER Vol. 5 1003, 1052
POKER DE ASES Vol. 8 311, 716, 752
POKFOCI Vol. 9 221, 253, 303
POKHALO Vol. 9 132, 288, 318, 334, 339
POKKERS PIGEBARN, ET Vol. 5 203
POKKERS UNGER, DE Vol. 5 44, 51, 98, 252
POKLAD BYZANTSKEHO KUPCE Vol. 10 425, 503, 522
POKLAD HRABETE CHAMARE Vol. 10 41, 44, 129
POKLAD RYTIRE MILOTY Vol. 10 116, 265, 337
POKOJ NR. 13 Vol. 1 174, 215, 326
POKOJ Z WIDOKIEM NA MORZE Vol. 1 136, 204 Vol. 6 *page 230*
POKOLENIE Vol. 1 242, 250, 321 Vol. 6 *page 255*
POKOLREV Vol. 9 29, 67, 180
POKORENE RIEKY Vol. 10 268, 459
POKRIV Vol. 3 41 Vol. 6 *page 27, page 29*
POKUS O VRAZDU Vol. 10 70, 154, 385
POKUSENI PANI ANTONIE Vol. 10 256, 586
POLAR Vol. 6 *page 93, page 112, page 145*
POLARE Vol. 5 975, 1024, 1201
POLARSTÜRME Vol. 4 774
POLE NEORANE Vol. 10 320, 345, 520
POLE POPPENSPÄLER Vol. 1 40, 56, 129 Vol. 4 110, 135
POLEMONTA Vol. 3 189
POLENBLUT Vol. 4 257, 432 Vol. 10 167, 168, 256
POLENOV PRAH Vol. 3 343, 450
PÖLHÖLÄ Vol. 5 409, 424
POLIBEK VE SNEHU Vol. 10 154, 345, 460
POLIBEK ZE STADIONU Vol. 10 256, 460
POLICARPO, UFFICIALE DI SCRITTURA Vol. 7 333, 781, 988, 1142
POLICE JUDICIAIRE Vol. 2 131, 595
POLICE MONDAINE Vol. 2 279, 407
POLICEJNI HODINA Vol. 10 311, 564, 621
POLICHE Vol. 2 402, 573, 803
POLICIA DETIENE, LA LEY JUZGA, LA Vol. 8 695, 762
POLICIER PAS COMME LES AUTRES, UN Vol. 2 872
POLICMAJSTER TAGIEJEW Vol. 1 174, 346
POLIKUSKA Vol. 7 1049

POLIS PAULIS' PASKASMÄLL **Vol. 5** 891, 1031
POLITICA CU DELICATESE **Vol. 3** 218
POLITISCHE TEPPICH, DER **Vol. 4** 612
POLIZEIAKTE 909 **Vol. 4** 192, 398, 638
POLIZEIBERICHT MELDET, DER **Vol. 4** 49, 270
POLIZEIBERICHT ÜBERFALL **Vol. 4** 84
POLIZEIFUNK MELDET, DER **Vol. 4** 51, 108, 132
POLIZEIRUF 110 – AUSKÜNFTE IN BLINDENSCHRIFT **Vol. 6** *page 163, page 166, page 170*
POLIZEISPION 77 **Vol. 4** 239, 486, 546
POLIZIA E SCONFITTA, LA **Vol. 7** 738
POLIZIA INCRIMINA, LA LEGGE ASSOLVE, LA **Vol. 8** 695, 762
POLIZIA INTERVIENE: ORDINE DI UCCIDERE ! **Vol. 7** 852
POLIZIOTTA A NEW YORK, LA **Vol. 7** 426, 504
POLIZIOTTA DELLA SQUADRA DEL BUONCOSTUME, LA **Vol. 7** 426
POLIZIOTTA DILETTANTE **Vol. 7** 1189
POLIZIOTTA FA CARRIERA, LA **Vol. 7** 426
POLIZIOTTA, LA **Vol. 7** 1189
POLIZIOTTO SCOMODO, UN **Vol. 7** 1003
POLIZIOTTO SENZA PAURA **Vol. 7** 852
POLIZIOTTO, SOLITUDINE E RABBIA **Vol. 7** 315, 1022
POLIZIOTTO SPRINT **Vol. 7** 335, 852
POLIZIOTTO SUPERPIU **Vol. 7** 381, 585
POLIZON A BORDO **Vol. 8** 449, 548
POLKA DES MENOTTES, LA **Vol. 2** 273, 423, 738
POLLO PERA, EL **Vol. 8** 604
POLNISCHE WIRTSCHAFT **Vol. 4** 208, 212, 560, 599
POLNOCNA OMSA **Vol. 10** 72, 503, 522, 582
POLOCAS STESTI **Vol. 10** 344, 496, 517
POLONIA-EXPRESS **Vol. 1** 24, 101, 120
POLONIA RESTITUTA **Vol. 6** *page 237, page 249, page 258*
POLOWANIE NA MUCHY **Vol. 1** 136, 314 **Vol. 6** *page 258*
POLSKA APROVISUJE **Vol. 10** 92, 363
POLSKA KREV **Vol. 10** 167, 168, 256
POLTERABEND **Vol. 4** 292, 517, 558
POLTERABEND, EIN **Vol. 1** 3, 107 **Vol. 4** 202
POLUDNIK ZERO **Vol. 1** 198, 282 **Vol. 6** *page 222*
POLVERE ALLEGRA **Vol. 7** 1189
POLVERE BIRBONA **Vol. 7** 1189
POLVO ERES **Vol. 8** 331, 627
POLVORILLA **Vol. 8** 551, 749
POLY ARGA GIA DAKRYA **Vol. 3** 205
POLYCE PYTHON 357 **Vol. 2** 74, 434
POLYGAME OU LE SEXE A L'ORIENTALE, LE **Vol. 2** 215
POLYKARPOVO ZIMNI DOBRODRUZSTVI **Vol. 10** 92
POLYTEEKKARIFILMI **Vol. 5** 406, 576
POMATA MERAVIGLIOSA, LA **Vol. 7** 1189
POMERANCOVY KLUK **Vol. 10** 233, 369, 505
POMICIONE, IL **Vol. 7** 714
POMLADNI VETER **Vol. 3** 417
POMME D'AMOUR **Vol. 2** 113, 626, 644, 656
POMOCNIK **Vol. 10** 82, 263, 493
POMPIERI **Vol. 7** 32, 633, 882
POMPIERI DI VIGGIU, I **Vol. 7** 1136
POMSTA MORE **Vol. 10** 92, 104, 392, 601
POMSTA MRTVYCH RYB **Vol. 10** 16, 93, 320
PONAD SNIEG **Vol. 1** 278, 404
PONEDELJAK ILI UTORAK **Vol. 3** 421
PONT DE SINGE, LE **Vol. 2** 350
PONT DU NORD, LE **Vol. 6** *page 87*
PONT VERD LE SOLEIL **Vol. 2** 669, 866
PONTCARRAL, COLONEL D'EMPIRE **Vol. 2** 27, 566, 676
PONTE DE LOS SUSPIROS, EL **Vol. 8** 603
PONTE DEI SOSPIRI, IL **Vol. 7** 2, 491, 825, 846, 879, 1116, 1147 **Vol. 8** 603
PONTE DELLE STREGHE, IL **Vol. 7** 1189
PONTE DI VETRO, IL **Vol. 7** 2, 48, 701
PONTE SULL'INFINITO, IL **Vol. 7** 487
PONTEVAT POMMARIPOJAT **Vol. 5** 374, 592
PONTON, EMBARCADERE **Vol. 2** 678
PONZIO PILATO **Vol. 7** 1054
POP CIRA I POP SPIRA **Vol. 3** 349, 374
POPCORN E PATATINE **Vol. 7** 474
POPE JOAN **Vol. 3** 254, 269
POPELKA **Vol. 10** 338, 345
POPESCU LO IN CONTROL **Vol. 3** 261, 263
POPIOL I DIAMENT **Vol. 1** 148, 250, 398
POPIOLY **Vol. 1** 206, 242, 305, 337 **Vol. 6** *page 256*
POPLACH V OBLACICH **Vol. 10** 173, 425, 577
POPOLVAR NAJVACSI NA SVETE **Vol. 10** 228, 459, 548, 592
POPPEA UNA PROSTITUTA AL SERVIZIO DELL'IMPERO **Vol. 7** 1179
POPPERS **Vol. 8** 376, 672
POPRASK NA SILNICI E 4 **Vol. 10** 70, 246, 380
POPSY POP **Vol. 2** 570, 814
POR EL GRAN PREMIO **Vol. 8** 298, 449

POR FAVOR OCUPATE DE AMELIA Vol. 8 219
POR FIN SE CASA ZAMORA Vol. 8 357
POR MIL DOLARES AL DIA Vol. 8 602, 749
POR QUE NO HACEMOS EL AMOR ? Vol. 8 373
POR QUE PECAMOS A LOS 40 Vol. 8 524
POR QUE SEGUIR MATANDO Vol. 8 732
POR QUE TE ENGANA TU MARIDO ? Vol. 8 299, 426
POR QUE VIVIR TRISTES ? Vol. 8 716
POR UN AMOR Vol. 8 246
POR UN MILAGRO DE AMOR Vol. 8 228
POR UN PUNADO DE DOLARES Vol. 8 298, 598
PORADNIK MATRYMONIALNY Vol. 1 180, 389 Vol. 6 *page 220*
PORCA MISERIA Vol. 7 1170
PORCA SOCIETA Vol. 7 1213
PORCA VACCA ! Vol. 7 593, 599
PORCI CON LA P 38 Vol. 7 949
PORCI CON LE ALI Vol. 7 854
PORCILE Vol. 7 333, 388, 817, 1007
PORCO MONDO Vol. 7 191
PORETTA ELI KEISARIN UUDET PISTEET Vol. 5 379, 552, 555, 561, 600
PORI Vol. 6 *page 184*
PORNO EROTIC LOVE Vol. 7 742
PORNO HOLOCAUST Vol. 7 742
PORNO INVESTIGATRICI, LE Vol. 7 742
PORNO KILLERS, LE Vol. 7 152
PORNO LUI EROTICA LEI Vol. 7 311
PORNO PARTY Vol. 2 291
PORNO STORY Vol. 2 291
PORNO VIDEO Vol. 7 742
PORNOCHATTES, LES Vol. 2 120
PORNOCRATES, LES Vol. 2 294
PORNOGRAPHIE CHEZ MADAME SAINT-CLAUDE Vol. 2 692
PORNOGRAPHIE SPECIALE Vol. 2 766
PORNOGRAPHIE SUEDOISE Vol. 2 82
PORNOSHOP DELLA SETTIMA STRADA, IL Vol. 7 742
PORNOTISSIMO Vol. 2 52
PORN'S GIRL Vol. 2 484
PORQUE TE VI LLORAR Vol. 8 357, 400
PORT-ARTHUR Vol. 2 397
PORT ARTHUR Vol. 4 361, 413, 442 Vol. 10 167, 168, 256, 275, 564
PORT D'ATTACHE Vol. 2 9, 339
PORT DU DESIR, LE Vol. 2 9,13
PORTA CHIUSA, LA Vol. 7 837
PORTA DEI SOGNI, LA Vol. 7 171, 978
PORTA DEL CIELO, LA Vol. 7 40, 1136
PORTA DEL MONDO, LA Vol. 7 968
PORTA FORTUNA Vol. 7 1189
PORTA UN BACIONE A FIRENZE Vol. 7 16
PORTALETTERE, IL Vol. 7 1189
PORTASI Vol. 10 142, 345, 460
PORTATRICE DI PANE, LA Vol. 7 196, 796
PORTE DES LILAS Vol. 2 59, 226, 232, 488
PORTE D'ORIENT Vol. 2 496, 836
PORTE DU LARGE, LA Vol. 2 332, 626, 824
PORTE-VEINE, LE Vol. 2 273, 460, 807
PORTES CLAQUENT, LES Vol. 2 383, 540, 686
PORTES DE FEU, LES Vol. 2 503
PORTES DE LA NUIT, LES Vol. 2 8, 47, 571, 837
PORTEUSE DE PAIN, LA Vol. 2 224, 227, 347, 516, 526, 581, 606, 614, 803
PORTFEL Vol. 1 220
PORTICO DE LA GLORIA, EL Vol. 8 298, 400
PORTIERE DI NOTTE, IL Vol. 7 347
PORTILE ALBASTRE ALE ORASULUI Vol. 3 281, 306, 331
PORTILE DIMINETII Vol. 3 316, 327 Vol. 6 *page 273*
PORTINAIA BURLATA Vol. 7 1189
PORTO Vol. 7 196, 640, 796
PORTO DE ABRIGO Vol. 8 44, 50
PORTO FRANCO Vol. 3 231, 287, 330
PORTRAIT DE DORIAN GRAY, LE Vol. 2 763
PORTRAIT DE GROUPE AVEC DAME Vol. 2 351
PORTRAIT DE MARIANNE, LE Vol. 2 60, 142
PORTRAIT DE SON PERE, LE Vol. 2 595, 627
PORTRAIT D'UN ASSASSIN Vol. 2 402, 698
PORTRAIT ROBOT Vol. 2 97, 856
PORTRETTET Vol. 5 690, 762
PORTRETUL UNEI GENERATII Vol. 3 211, 229, 245
PORTRETUL UNUI NECUNOSCUT Vol. 3 263
PORTUGUESA EM PARIS, UMA Vol. 8 201
PORUCIK ALEXANDER RJEPKIN Vol. 10 256, 564
POSALJI COVEK U POLA DVA Vol. 3 418
POSAUNE, DIE Vol. 4 367, 744
POSAUNIST, DER Vol. 4 184, 713
POSCIG Vol. 1 169, 179 Vol. 6 *page 240*
POSEBAN TRETMAN Vol. 3 367, 418
POSEL USVITU Vol. 10 412, 503, 569
POSELEDNATA DUMA Vol. 3 60, 75
POSELEDNIJAT ERGEN Vol. 3 38, 73

POSELEDNIJAT VOJVODA Vol. 3 36, 101, 126
POSELEDNJI MOST Vol. 3 378
POSICION AVANZADA Vol. 8 611, 749
POSLEDNA BOSORKA Vol. 10 146, 193, 273
POSLEDNATA DUMA Vol. 6 *page 51*
POSLEDNI BOHEM Vol. 10 138, 524
POSLEDNI ETAPA Vol. 10 299, 353, 380
POSLEDNI LEC Vol. 10 389, 466, 534
POSLEDNI MEJDAN Vol. 10 35
POSLEDNI MOHYKAN Vol. 10 413, 503, 532
POSLEDNI MUZ Vol. 10 34, 256
POSLEDNI PLES NA ROZNOVSKE PLOVARNE Vol. 10 87, 425, 568
POSLEDNI PODSKALAK Vol. 10 345, 564
POSLEDNI POLIBEK Vol. 10 126, 167
POSLEDNI PROPADNE PEKLU Vol. 10 18, 70, 254
POSLEDNI RADOST Vol. 10 243, 310
POSLEDNI RUZE OD CASANOVY Vol. 10 154, 359, 498
POSLEDNI VLAK Vol. 10 106, 139, 511
POSLEDNI VYSTREL Vol. 10 296, 412
POSLEDNI VYSTREL DAVIDA SANDELA Vol. 10 327, 425, 590
POSLEDNI ZELANIJA Vol. 6 *page 47, page 64*
POSLEDNIJAT ERGEN Vol. 6 *page 59*
POSLEDNIJAT RUND Vol. 3 69, 125 Vol. 6 *page 44*
POSLEDNIJAT VOJVODA Vol. 6 *page 41*
POSLEDNJA POSTAJA Vol. 3 449
POSLEDNJA TRKA Vol. 3 367, 402, 415
POSLEDNJI DAN Vol. 3 388, 425
POSLEDNJI KOLOSEK Vol. 3 349, 376
POSLEDNJI OBRACUN Vol. 6 *page 314*
POSLEDNJI PODVIG DIVERZANTA OBLAKA Vol. 3 371, 407
POSLEDNO LJATO Vol. 3 42 Vol. 6 *page 34, page 36*
POSLEDNOTO PRIKLJUCENIE Vol. 6 *page 38, page 60*
POSLEDNY NAVRAT Vol. 10 320, 576
POSLICEK LASKY Vol. 10 256, 412
POSLIJEPODNE JEDNOG FAZANA Vol. 3 451
POSLIZG Vol. 1 144, 245, 314
POSLUSNE HLASIM Vol. 10 132, 460, 503
POSSEDEES, LES Vol. 2 190, 204, 650, 781
POSSESSION, LA Vol. 2 413, 603, 633
POST RESTANT Vol. 3 218, 256, 276
POSTA IN GUERRA, LA Vol. 7 341
POSTAV DOM, ZASAD STROM Vol. 10 101, 129, 178, 228
POSTAVA K PODPIRANI Vol. 10 41, 87
POSTAVENI MIMO HRU Vol. 10 344, 408, 425
POSTILLON VON LONJUMEAU, DER Vol. 4 391
POSTLAGERND XYZ Vol. 4 329, 774
POSTMEISTER, DER Vol. 4 265, 614, 620
POSTO ALL'INFERNO, UN Vol. 7 743
POSTO, IL Vol. 7 221, 635
POSTOJ Vol. 10 542, 607
POSTOVSKY PANACEK Vol. 10 250
POSTRIZINY Vol. 10 172, 425, 516
POSZUKIWANY - POSZUKIWANA Vol. 1 218, 233 Vol. 6 *page 209*
POT-BOUILLE Vol. 2 59, 282, 427
POTE DE SE XEHASA Vol. 3 180
POTE DEN EINAI ARGA Vol. 3 149
POTEM NASTAPI CISZA Vol. 1 237, 249, 355, 398 Vol. 6 *page 242*
POTERE SOVRANO, IL Vol. 7 369, 640, 965
POTERE TEMPORALE Vol. 7 965
POTMULI ODJECI Vol. 6 *page 315*
POTOP Vol. 1 202, 227, 260, 398 Vol. 6 *page 230, page 256*
POTOTO, BOLICHE Y COMPANIA Vol. 8 272
POTPOURRI Vol. 4 79, 115, 239
POTRAGA Vol. 3 337, 373, 448
POTRAGA ZA ZMAJEM Vol. 3 345, 387
POTRAZI VANDU KOS Vol. 3 376, 448
POTSDAM, DAS SCHICKSAL EINER RESIDENZ Vol. 4 85, 219
POTTEPLANTEN Vol. 5 265
POUCE Vol. 2 190, 674
POUCETTE Vol. 2 237
POUDRE D'ESCAMPETTE Vol. 2 564
POUIC-POUIC Vol. 2 97, 309
POULE, LA Vol. 2 41, 460, 469, 579
POULE SUR UN MUR, UNE Vol. 2 871, 872
POUPEE D'AMOUR Vol. 2 873
POUPEE, LA Vol. 2 208, 436
POUPEE ROUGE, LA Vol. 2 214, 513
POUR CENT BRIQUES, T'AS PLUS RIEN Vol. 6 *page 75, page 117*
POUR CLEMENCE Vol. 2 754
POUR DES FUSILS PERDUS Vol. 2 736
POUR DON CARLOS Vol. 2 212, 216
POUR ETRE AIME Vol. 2 708
POUR LA PEAU D'UN FLIC Vol. 6 *page 89, page 99*
POUR LE CORPS D'UNE FEMME Vol. 2 720
POUR LE MAILLOT JAUNE Vol. 2 153, 247, 273, 412, 601
POUR LE MERITE Vol. 4 13, 311, 574
POUR QUELQUE CHOSE DE PLUS Vol. 2 76

POUR REGNER **Vol. 2** 37
POUR UN AMOUR LOINTAIN **Vol. 2** 363, 779
POUR UN BARIL DE PETROLE **Vol. 2** 396
POUR UN SOURIRE **Vol. 2** 860
POUR UNE NUIT D'AMOUR **Vol. 2** 173, 501, 671
POUR UNE POIGNEE DE CACAHUETES OU LE RETOUR DE SCRATCH DANS LE PLUS EGALE MOINS **Vol. 2** 806
POUR VIVRE HEUREUX **Vol. 2** 646
POURQOUI **Vol. 2** 2, 30
POURQOUI J'AI TUE MON ENFANT **Vol. 2** 864
POURQUOI ISRAEL **Vol. 2** 527, 618
POURQUOI PARIS ? **Vol. 2** 194
POURQUOI PAS ! **Vol. 2** 739
POURQUOI PAS NOUS ? **Vol. 6** *page 82, page 91*
POURQUOI VIENS-TU SI TARD ? **Vol. 2** 9, 566
POURVU QU'ON AIT L'IVRESSE **Vol. 2** 810
POUSSEZ PAS GRAND-PERE DANS LES CACTUS **Vol. 2** 878
POUSSIERE D'EMPIRE **Vol. 6** *page 91, page 103, page 140*
POUTA **Vol. 10** 70, 193, 433
POUTNICI **Vol. 10** 63, 297, 384
POVERA PICCOLA **Vol. 7** 968
POVERE AMORE **Vol. 7** 371, 609
POVERI FIGLIUOLI **Vol. 7** 1211
POVERI, MA BELLI **Vol. 7** 333, 443
POVERI MILLIONARI **Vol. 7** 333
POVERO MARITO **Vol. 7** 1189
POVERO PIERO, IL **Vol. 7** 306
POVERO RICCO, UN **Vol. 7** 29, 524
POVEST O DOBRIH LJUDEH **Vol. 3** 449
POVEST O STRIBRNE JEDLI **Vol. 10** 286, 567
POVESTA CALATORIILOR **Vol. 6** *page 280, page 282*
POVESTE SENTIMENTALA **Vol. 3** 221, 268, 286
POVESTE TRISTA **Vol. 3** 243
POVESTEA DRAGOSTEI **Vol. 3** 265, 270, 301
POVIDKY O DETECH **Vol. 10** 182, 299, 369, 577
POVODEN **Vol. 10** 154, 566
POVRATAK **Vol. 3** 367, 371, 418, 421
POVRATAK OTPISANIH **Vol. 3** 381, 408
POWRANIE SAVOI **Vol. 6** *page 240*
POWROT **Vol. 1** 174, 176, 211, 250, 326, 355
POWROT NA ZIEMIE **Vol. 1** 198, 243 **Vol. 6** *page 222*
POZARY A SPALENISTE **Vol. 10** 233, 408, 428
POZDNI LASKA **Vol. 10** 154, 453, 460
POZDNI MAJ **Vol. 10** 256, 460
POZEGNANIA **Vol. 1** 187, 401 **Vol. 6** *page 216*
POZEGNANIE Z DIABLEM **Vol. 1** 179, 381 **Vol. 6** *page 239*
POZNAJETE LI PALVA PLESA **Vol. 3** 367, 415, 438
POZNANSKIE SLOWIKI **Vol. 1** 143, 243 **Vol. 6** *page 222*
POZNE POPOLUDNIE **Vol. 1** 249, 382, 403
POZNEJ SVEHO MUZE **Vol. 10** 345, 460
POZO DE LOS ENAMORADOS, EL **Vol. 8** 628
POZOR, IDE JOZEFINA ... **Vol. 10** 82, 263, 548
POZOR, STRASI ... ! **Vol. 10** 256, 412
POZOR, VIZITA ! **Vol. 10** 87, 106, 395
POZORISNA VEZA **Vol. 3** 336, 398, 426
POZZO CHE PARLA, IL **Vol. 7** 1189
POZZO DEI MIRACOLI, IL **Vol. 7** 1189
PRACE **Vol. 10** 197, 433, 621
PRACH A BROKY **Vol. 10** 257, 453
PRADLENKA JEHO JASNOSTI **Vol. 6** *page 200* **Vol. 10** 167
PRAERIENS SKRAPPE DRENGE **Vol. 5** 10, 35, 195
PRAESIDENTEN **Vol. 5** 291
PRAESTEN FRA HAVET **Vol. 5** 9
PRAESTEN I VEJLBY **Vol. 5** 101, 125, 183, 195, 208, 273
PRAESTENS DATTER **Vol. 5** 58
PRAESTENS SONNER **Vol. 5** 97
PRAHA NULTA HODINA **Vol. 10** 299, 382, 522, 616 **Vol. 10** 299, 382, 522, 616
PRAIRIE PIRATE, THE **Vol. 2** 84
PRAMEN LASKY **Vol. 10** 243, 345, 524 **Vol. 10** 243, 345, 524
PRANA **Vol. 8** 438
PRANKE, DIE **Vol. 7** 361, 796
PRANVERE NE GJIROKASTER **Vol. 3** 2, 14
PRÄSIDENT BARRADA **Vol. 4** 113, 780
PRÄSIDENT, DER **Vol. 4** 157, 219, 479
PRÄSTÄNKÄN **Vol. 5** 1135
PRÄSTEN **Vol. 5** 925
PRÄSTEN I UDDARBO **Vol. 5** 1042, 1217
PRÄSTEN, SOM SLOG OUT **Vol. 5** 840, 878
PRATELE BERMUDSKEHO TROJUHELNIKU **Vol. 10** 149, 180, 383
PRATELE NA MORI **Vol. 10** 95, 179, 407, 433
PRATER **Vol. 4** 149, 219, 535, 660
PRATO, IL **Vol. 7** 524, 1034, 1112
PRAVE ZACINAME **Vol. 10** 142, 256, 532

PRAVIDLA KRUHU Vol. 10 149, 163, 511
PRAVO NA HRICH Vol. 10 34, 256, 453, 524
PRAVO NA MINULOST Vol. 10 16, 72, 548, 565
PRAVO STANJE STVARI Vol. 3 421, 437
PRAWDZIE W OCZY Vol. 1 156, 303, 354
PRAWDZIWY KONIEC WIELKIEJ WOJNY Vol. 1 242, 250, 313
PRAWO I PIESC Vol. 1 242, 294
PRAWO JEST PRAWEM Vol. 6 *page 247, page 249*
PRAZDNINY PRO PSA Vol. 10 18, 385, 595
PRAZDNINY S MINKOU Vol. 10 250, 286
PRAZDNINY V OBLACICH Vol. 10 502, 532
PRAZNIK Vol. 3 104, 418 Vol. 6 *page 51*
PRAZNOVANJE POMLADI Vol. 3 395, 449
PRAZSKA PETKA Vol. 10 409, 467, 495, 574, 610
PRAZSKA SVADLENKA Vol. 10 45
PRAZSKE BLUES Vol. 10 296, 371, 519
PRAZSKE DETI Vol. 10 53, 453
PRAZSKE NOCI Vol. 10 96, 229, 352, 425, 503, 547, 567
PRAZSKE SVADLENKY Vol. 10 58, 167, 345
PRAZSKE TAJEMSTVI Vol. 10 76, 139, 304, 551
PRAZSKY FLAMENDR Vol. 10 2, 257, 453, 585, 612
PRAZSKY KAT Vol. 10 453, 509
PRAZSKY MONTE CHRISTO Vol. 10 137, 532
PRAZSTI ADAMITE Vol. 10 92, 363
PRE ISTINE Vol. 3 367, 418
PRE MNA NEHRA BLUES Vol. 10 478, 548
PRE RATA Vol. 3 388, 409
PREA CALD PENTRU LUNA MAI Vol. 6 *page 273, page 275, page 278*
PREA MIC PENTRU UN RAZBOI ATIT DE MARE Vol. 3 287, 326, 330
PREA TINERI PENTRU RIDURI Vol. 6 *page 273, page 279, page 280*
PREBROJAVANE NA DIVITE ZAJCI Vol. 3 46 Vol. 6 *page 34*
PRECIEUSES RIDICULES, LES Vol. 2 488
PRECIO DE UN ASESINO Vol. 8 562, 768
PRECIO DE UN HOMBRE, EL Vol. 8 311
PRECIO DEL ABORTO, EL Vol. 8 523
PRECLIK Vol. 10 70, 550
PRED MATURITOU Vol. 10 120, 167, 453, 586
PRED OTECESTOVO DA ZABRAVIM OMRAZATA SI Vol. 3 106
PREDA, LA Vol. 7 442, 490, 1169
PREDATORI DI ATLANTIDE, I Vol. 7 406
PREDCASNE LETO Vol. 10 16, 72, 470
PREDICITIONS POUR 1914 Vol. 2 159, 286, 330
PREDJARIE Vol. 10 273, 320
PREDNOSTA STANICE Vol. 10 168, 564
PREDONE DELL'ARIA Vol. 7 1073
PREDONI DEL SAHARA, I Vol. 7 591
PREDSTAVA HAMLETA U SELU MRDUSA DONJA Vol. 3 414, 437
PREDTUCHA Vol. 10 524, 621
PREDUPREZDENIETO Vol. 6 *page 63, page 65*
PREERIA Vol. 3 218
PREFERISCO L'INFERNO ! Vol. 7 454
PREFETTO DI FERRO, IL Vol. 7 585
PREHLIDCE VELIM JA Vol. 10 41, 153, 569
PREISLIED, DAS Vol. 4 652
PREKOBROJNA Vol. 3 341, 365
PRELUDES A LA GLOIRE Vol. 2 611, 719
PRELUDIO 11 Vol. 1 42, 48, 102
PRELUDIO D'AMORE Vol. 7 410, 923
PRELUDIUM Vol. 10 92, 621
PRELUDY POUSTE Vol. 10 325, 425, 518
PREMEDITATION Vol. 2 559, 627, 661
PREMIER BAL Vol. 2 402, 722
PREMIER DE CORDEE Vol. 2 8, 9
PREMIER MAI Vol. 2 16, 368, 503, 861
PREMIER RENDEZ-VOUS Vol. 2 488, 657
PREMIER VOYAGE Vol. 2 527
PREMIERA Vol. 3 245, 306, 307
PREMIERE BRIGADE CRIMINELLE Vol. 2 287
PREMIERE DER BUTTERFLY Vol. 6 *page 74*
PREMIERE FÄLLT AUS, DIE Vol. 1 24, 34, 35
PREMIERE FOIS, LA Vol. 2 177
PREMIERE GIFLE, LA Vol. 2 724
PREMIERE I HELVEDE Vol. 5 2, 201, 279
PREMIERE IM ADMIRALSPALAST Vol. 1 29, 47, 103
PREMIERES ARMES Vol. 2 318, 776
PREMIERS DESIRS Vol. 6 *page 77, page 97, page 110, page 140*
PREMIERS OUTRAGES, LES Vol. 2 259, 404
PREMIERS PAS Vol. 2 244
PRENDS DIX MILLE BALLES ET CASSE -TOI Vol. 6 *page 148*
PRENDS TON PASSE-MONTAGNE, ON VA A LA PLAGE Vol. 6 *page 76*
PRENEZ GARDE A LA PEINTURE Vol. 2 167, 388, 744
PRENEZ LA QUEUE COMME TOUT LE MONDE Vol. 2 294, 804
PREPAREZ VOS MOUCHOIRS Vol. 2 616, 655

PREPOTENTI, I **Vol. 7** 883
PREPOTENTI PIU DI PRIMA **Vol. 7** 910
PRERUSENA HRA **Vol. 11** 72, 280, 592
PRERUSENA PIESEN **Vol. 10** 268, 326, 340
PRESA DELLA BASTIGLIA, LA **Vol. 7** 454
PRESA DI ROMA **Vol. 7** 14
PRESA DI TROIA, LA **Vol. 7** 186
PRESAGIO **Vol. 8** 732
PRESAGIO, IL **Vol. 7** 456, 668, 1038
PRESANTE QUINCE DIAS **Vol. 8** 249
PRESENTAT'ARMI ! **Vol. 7** 390
PRESIDENT HAUDECOEUR, LE **Vol. 2** 339, 698
PRESIDENT, LE **Vol. 2** 644
PRESIDENTE DELLA BA. CE. CRE. MI., IL **Vol. 7** 48, 466
PRESIDENTE, LA **Vol. 2** 48, 197, 722
PRESIDENTESSA, LA **Vol. 7** 86, 709
PRESIDIO, EL **Vol. 8** 396
PRESSENS MAGT **Vol. 5** 9
PRESTAME QUINCE DIAS **Vol. 8** 306
PRESTAME TU MUJER **Vol. 8** 249, 409
PRESTAMI TUA MOGLIE **Vol. 7** 383, 811
PRESTIDIGITATEUR **Vol. 2** 678
PRESTOP **Vol. 6** *page 305, page 307*
PRESUDA **Vol. 3** 384, 434 **Vol. 6** *page 308*
PRESUNTOS, LOS **Vol. 8** 536
PRETE SCOMODO, UN **Vol. 7** 46
PRETORA, LA **Vol. 7** 1147
PRETRES INTERDITS **Vol. 2** 304, 702
PREUSSISCHE LIEBESGESCHICHTE **Vol. 4** 28 **Vol. 6** *page 187*
PREVIERKA LASKY **Vol. 10** 349, 576
PREZIL JSEM SVOU SMRT **Vol. 10** 292, 313
PREZZO DEL POTERE, IL **Vol. 7** 743 **Vol. 8** 252
PREZZO DELLA FELICITA, IL **Vol. 7** 399
PREZZO DELLA GLORIA, IL **Vol. 7** 476, 1035
PREZZO DELL'ONORE, IL **Vol. 7** 201
PRI NIKOGO **Vol. 3** 76, 116 **Vol. 6** *page 59*
PRIBEH '88 **Vol. 10** 176, 448, 495
PRIBEH JEDNOHO DNE **Vol. 10** 167, 345
PRIBEH LASKY A CTI **Vol. 10** 283, 313, 398
PRIBEH ONDREJA GREGORA **Vol. 10** 548
PRIBLIZAVANJE **Vol. 6** *page 303*
PRICA O FABRICI **Vol. 3** 388
PRICHAZEJI Z TMY **Vol. 10** 193, 566
PRICHOZI Z TEMNOT **Vol. 10** 167, 272
PRIEHRADA **Vol. 10** 287, 351
PRIETENI FARA GRAI **Vol. 3** 265, 321
PRIGIONE BIANCA **Vol. 7** 443, 941
PRIGIONE, LA **Vol. 7** 40, 639
PRIGIONE SOTTO LA NEVE, LA **Vol. 7** 1125
PRIGIONIERA DELLA TORRE DEL FUOCO, LA **Vol. 7** 1152
PRIGIONIERA DI AMALFI, LA **Vol. 7** 693
PRIGIONIERE DELL'ISOLA DEL DIAVOLO, LE **Vol. 7** 1131
PRIGIONIERI DEL MALE **Vol. 7** 257
PRIGIONIERI DELLE TENEBRE **Vol. 7** 1027
PRIGIONIERO DI SANTA CRUZ, IL **Vol. 7** 101, 638, 1060
PRIHODY PANA PRIHODY **Vol. 10** 226, 317, 522
PRIJDU HNED **Vol. 10** 412, 621
PRIJEKI SUD **Vol. 3** 430, 437
PRIJELA K NAM POUT **Vol. 10** 18, 359, 425
PRIKLADY TAHNOU **Vol. 10** 564, 621
PRIKLJUCENIE V POLUNOST **Vol. 3** 75, 99, 124
PRIKO SINJEG MORA **Vol. 3** 380, 389, 414
PRILIS VELKA SANCE **Vol. 10** 18, 172, 233
PRILIV NA NEZNOST **Vol. 6** *page 24, page 41*
PRIM **Vol. 8** 281
PRIMA ANGELICA, LA **Vol. 8** 584
PRIMA BICICLETTA DI ROBINET, LA **Vol. 7** 1189
PRIMA CHE SIA TROPPO PRESTO **Vol. 7** 363, 1202
PRIMA COMUNIONE **Vol. 7** 73, 332, 359
PRIMA CORSA AUTOMOBILISTICA SUSA-MONCENISIO, LA **Vol. 7** 837
PRIMA DELLA RIVOLUZIONE **Vol. 7** 1045
PRIMA DI SERA **Vol. 7** 407, 796
PRIMA DONNA CHE PASSA, LA **Vol. 7** 253, 1027, 1054
PRIMA DONNA, LA **Vol. 7** 827
PRIMA MELODIE **Vol. 3** 272
PRIMA NOTTE DI NOZZE **Vol. 7** 649
PRIMA NOTTE DI QUIETE, LA **Vol. 7** 854
PRIMA NOTTE, LA **Vol. 7** 1167
PRIMA VERAS SAGA OLAV DEN HELLIGE **Vol. 5** 695, 746, 778
PRIMADONNA, LA **Vol. 7** 481, 862, 1000
PRIMANERLIEBE **Vol. 4** 208, 788
PRIMAVERA **Vol. 7** 731
PRIMAVERA MORTAL **Vol. 3** 382, 427
PRIMER AMOR **Vol. 8** 628
PRIMER CUARTEL, EL **Vol. 8** 317, 768
PRIMER DIVORCIO, EL **Vol. 8** 346, 407, 627
PRIMER TORERO PORNO, EL **Vol. 8** 242, 420
PRIMERA AVENTURA, LA **Vol. 8** 207, 214

PRIMEROSE Vol. 2 193, 573, 826 Vol. 7 168
PRIMI CALORI Vol. 7 1189
PRIMO AMORE Vol. 7 333, 453, 1131, 1177
PRIMO APPLAUSO Vol. 7 792, 1121
PRIMO BASILIO, O Vol. 8 10, 84, 129, 131
PRIMO ED ULTIMO BACIO Vol. 7 1004
PRIMO PREMIO MARIAROSA Vol. 7 476
PRIMPRINETTE Vol. 7 148
PRIMULA BIANCA, LA Vol. 7 796
PRIMULE INSANGUINATE, LE Vol. 7 72
PRIMULE, LA Vol. 7 72
PRIN CENUSA IMPERIULUI Vol. 3 246, 307, 326
PRINC A VECERNICE Vol. 10 197, 313, 425
PRINC BAJAJA Vol. 10 50, 325, 369
PRINCE BOUBOULE Vol. 2 131, 604, 871
PRINCE CHARMANT, LE Vol. 2 31, 197, 440, 446, 521, 833
PRINCE DE MINUIT Vol. 2 193, 197
PRINCE DE MON COEUR Vol. 2 131, 152, 812
PRINCE JEAN, LE Vol. 2 318, 722, 727
PRINCE OF FOXES, THE Vol. 7 978
PRINCES DE LA CRAVACHE Vol. 2 62
PRINCES, LES Vol. 6 *page 86, page 109*
PRINCESA DE LOS URSINOS, LA Vol. 8 207, 214, 400
PRINCESA Y VAGABUNDA Vol. 8 443, 532
PRINCESSE AUX CLOWNS, LA Vol. 2 41, 614, 845
PRINCESSE CZARDAS Vol. 2 399
PRINCESSE DE CLEVES, LA Vol. 2 13, 282, 669, 722
PRINCESSE MANDANE Vol. 2 79, 380, 762, 785
PRINCESSE MASHA, LA Vol. 2 547
PRINCESSE TAM-TAM Vol. 2 84, 573, 775
PRINCEZNA SE ZLATOU HVEZDOU Vol. 10 296, 460, 547
PRINCEZNA Z CHALUPY Vol. 10 363
PRINCIPE APACHE, IL Vol. 7 967
PRINCIPE COM ORELHAS DE BURRO, O Vol. 8 169
PRINCIPE DALLA MASCHERA ROSSA Vol. 7 1054, 1063
PRINCIPE DEI DOLLARI, IL Vol. 7 454
PRINCIPE DELLE VOLPI, IL Vol. 7 978
PRINCIPE DELL'IMPOSSIBILE, IL Vol. 7 48, 464
PRINCIPE DI HOMBURG Vol. 7 11, 819
PRINCIPE DI KAYTAN, IL Vol. 7 531
PRINCIPE ENCADENADO, EL Vol. 8 298, 762
PRINCIPE FOLLE, IL Vol. 7 1116
PRINCIPE FUSTO, IL Vol. 7 1012
PRINCIPE IDIOTA, IL Vol. 7 418, 1103
PRINCIPE MASCHERATO, IL Vol. 7 994
PRINCIPE ZILAH, LA Vol. 7 886
PRINCIPESSA BARBARA, LA Vol. 7 456, 1132
PRINCIPESSA BEBE, LA Vol. 7 559, 679
PRINCIPESSA D'AZZURRO, LA Vol. 7 493
PRINCIPESSA DEL SOGNO, LA Vol. 7 639, 966, 1074
PRINCIPESSA DELLE CANARIE, LA Vol. 7 1062
PRINCIPESSA DI BAGDAD Vol. 7 150
PRINCIPESSA GIORGIO, LA Vol. 7 264, 701
PRINCIPESSA MARIA, LA Vol. 7 250, 929
PRINCIPESSA MISTERIOSA, LA Vol. 7 121, 168, 441
PRINCIPESSA NADINE O IPNOTISMO TRAGICO, LA Vol. 7 160
PRINCIPESSA NERA, LA Vol. 7 548, 774
PRINCIPESSA STRANIERA, LA Vol. 7 965
PRINCIPESSA TARAKANOVA, LA Vol. 7 453, 476, 487, 1116
PRINCIPESSA ZOE, LA Vol. 7 168
PRINCIPESSINA Vol. 7 86
PRINCIPIO DA SABEDORIA, O Vol. 8 6, 104, 169
PRINGKEPAS TON ALETON Vol. 3 178
PRINS FOR EN DAG Vol. 5 81
PRINS GUSTAF Vol. 5 1042, 1164
PRINS HATT UNDER JORDEN Vol. 5 880, 1027, 1080, 1096
PRINS PIWI Vol. 5 226
PRINSESSA RUUSUNEN Vol. 5 328, 356, 468, 549
PRINSESSAN Vol. 5 839, 938
PRINSESSE ELENA Vol. 5 58
PRINSESSE FOR EN DAG Vol. 5 14, 20, 179
PRINSESSEN SOM INGEN KUNNE MALBINDE Vol. 5 760
PRINSESSENS HJERTE Vol. 5 183
PRINSESSENS TILBEDER Vol. 5 256
PRINTEMPS Vol. 2 390
PRINTEMPS A PARIS Vol. 2 243, 354, 776
PRINTEMPS D'AMOUR Vol. 2 193
PRINTEMPS DU COEUR, LE Vol. 2 798
PRINTEMPS, L'AUTOMNE ET L'AMOUR, LE Vol. 2 197, 824
PRINTRE COLINELE VERZI Vol. 3 272, 276, 291
PRINZ HINTER DEN SIEBEN MEEREN, DER Vol. 6 *page 160, page 163, page 167*
PRINZ KUCKUCK Vol. 4 288, 396
PRINZ LOUIS FERDINAND Vol. 4 132, 636, 788
PRINZ SAMI Vol. 4 242, 560

PRINZ UND DIE TÄNZERIN, DER Vol. 4 192, 560
PRINZENRAUB, DER Vol. 4 124, 247
PRINZESSCHEN Vol. 4 660, 704
PRINZESSIN OLALA Vol. 4 208, 479
PRINZESSIN SISSY Vol. 4 97
PRINZESSIN SUWARIN, DIE Vol. 4 30, 115, 359
PRINZESSIN TATJANA Vol. 4 213, 414
PRINZESSIN TRULALA Vol. 4 239, 560
PRINZESSIN TURANDOT Vol. 4 266, 574, 729
PRINZESSIN UND DER GEIGER, DIE Vol. 4 663
PRINZESSIN VON NEUTRALIEN, DIE Vol. 4 178, 319
PRIPAD BARNABAS KOS Vol. 10 32, 253, 576
PRIPAD DANIELA Vol. 10 225, 369
PRIPAD DR. KOVARE Vol. 10 352, 559
PRIPAD JESTE NEKONCI Vol. 10 211, 296
PRIPAD KRASNEJ NERESTNICE Vol. 10 32, 263, 483
PRIPAD LUPINEK Vol. 10 502, 524
PRIPAD MRTVEHO MUZE Vol. 10 517, 531, 568
PRIPAD MRTVYCH SPOLUZAKU Vol. 10 144, 531, 568
PRIPAD PRE OBHAJCU Vol. 10 287, 483, 576
PRIPAD PRO ZACINAJICIHO KATA Vol. 10 229, 369, 425
PRIPAD Z – 8 Vol. 10 296, 552, 564
PRIRODNA GRANICA Vol. 3 362, 418
PRISCILLAS FAHRT INS GLÜCK Vol. 4 724, 734
PRISE DU POUVOIR PAR LOUIS XIV Vol. 2 480
PRISIONERO EN LA CIUDAD Vol. 8 331, 719
PRISNE TAJNE PREMIERY Vol. 10 498, 550, 621
PRISON SANS BARREAUX Vol. 2 566, 719, 863
PRISON WITHOUT BARS Vol. 2 656
PRISONNIER DE MON COEUR Vol. 2 37, 387, 573
PRISONNIERE, LA Vol. 2 770, 873
PRISONNIERS DE LA BROUSSE Vol. 2 740 Vol. 7 982
PRISONNIERS DE MAO Vol. 2 283
PRISONS AUSSI..., LES Vol. 2 502, 691
PRISONS DE FEMMES Vol. 2 347, 407, 581, 698
PRIS"STVIE Vol. 3 46, 101 Vol. 6 *page 27, page 50*
PRISTE BUDEME CHYTREJSI, STAROUSKU ! Vol. 10 18, 233, 428
PRITA Vol. 3 11, 22 Vol. 6 *page 18*
PRITEL LHAR Vol. 10 380, 407
PRITELKYNE PANA MINISTRA Vol. 10 345, 412
PRIVATDETEKTIVENS OFFER Vol. 5 97
PRIVATE LIFE OF DON JUAN, THE Vol. 2 656
PRIVATE LIFE OF HENRY VIII, THE Vol. 2 656
PRIVATSEKRETÄRIN, DIE Vol. 4 257, 298, 373
PRIVILEGIO DELL'AMORE Vol. 7 209
PRIV"RZANIJAT BALON Vol. 3 121 Vol. 6 *page 35*
PRIX DE BEAUTE Vol. 2 388, 563, 626
PRIX DU DANGER, LE Vol. 6 *page 78, page 82*
PRIZNANI Vol. 10 122, 532
PRIZNANIE Vol. 3 63, 126
PRIZORI Vol. 3 75, 78 Vol. 6 *page 59*
PRIZOVANIJAT NE SE JAVI Vol. 3 42, 121 Vol. 6 *page 26*
PRKOSNA DELTA Vol. 3 431
PRO HUBICKU DO AFRIKY Vol. 10 201, 584
PRO KAMARADA Vol. 10 168, 580, 621
PRO PATRIA Vol. 4 194, 247 Vol. 5 9
PROBA DE MICROFON Vol. 3 284 Vol. 6 *page 265*
PROBA OGNIA I WODY Vol. 1 186, 216, 284 Vol. 6 *page 233*
PROBABILITA ZERO Vol. 7 1136
PROBAUT Vol. 9 134, 215, 314
PROBIO, EL Vol. 8 647
PROBLEMA RESUELTO Vol. 8 526
PROBLEMATICKY GENTLEMAN Vol. 10 243, 335
PROBLEMATISCHE NATUREN Vol. 4 183
PROBLEME PERSONALE Vol. 6 *page 266, page 272, page 273*
PROBUZENE SVEDOMI Vol. 10 167, 536
PROBUZENI Vol. 10 202, 388, 503
PROC ? Vol. 10 44, 149, 251
PROC NEVERIT NA ZAZRAKY Vol. 10 395, 505
PROC SE NESMEJES ? Vol. 10 122, 167
PROCES AU VATICAN Vol. 2 419, 698
PROCES DE JEANNE D'ARC, LE Vol. 2 153, 170
PROCES, LE Vol. 2 548, 728
PROCESI K PANENCE Vol. 10 41, 292
PROCESO A JESUS Vol. 8 339, 348
PROCESO A UNA ESTRELLA Vol. 8 450, 482
PROCESO DE CONCIENCIA Vol. 8 425, 467
PROCESO DE GIBRALTAR Vol. 8 517

PROCESO DE LAS BRUJAS, EL **Vol. 8** 546, 706
PROCESSO ALLA CITTA **Vol. 7** 750, 1062, 1131
PROCESSO ALL'AMORE **Vol. 7** 1121
PROCESSO CLEMENCEAU, IL **Vol. 7** 264, 701
PROCESSO CONTRO IGNOTI **Vol. 7** 603, 1054
PROCESSO DELLE ZITELLE, IL **Vol. 7** 1036
PROCESSO D'ESPARBES, IL **Vol. 7** 1162
PROCESSO DI VERONA, IL **Vol. 7** 86, 353
PROCESSO E MORTE DI SOCRATE **Vol. 7** 48, 701
PROCESSO MONTEGU, IL **Vol. 7** 621
PROCES'T **Vol. 3** 51, 75, 105
PROCESUL ALB **Vol. 3** 254, 306, 312
PROCITNUTI ZENY **Vol. 10** 243, 335, 345
PROCUREUR HALLERS, LE **Vol. 2** 290
PRODANA NEVESTA **Vol. 10** 92, 115, 167, 256, 283, 363, 536, 568, 586
PRODAVAC HUMORU **Vol. 10** 318, 466, 535
PRODE ANSELMO E IL SUO SCUDIERO, IL **Vol. 7** 1136
PRODIGA, LA **Vol. 8** 214, 400
PRODLOUZENY CAS **Vol. 10** 139, 144, 292
PRODOSIA **Vol. 3** 153
PROF ENSEIGNE SANS PRESERVATIFS, LA **Vol. 6** *page 97, page 146*
PROF. GUIDO TERSILLI, PRIMARIO DELLA CLINICA VILLA CELESTE..., IL **Vol. 7** 6
PROFANAZIONE **Vol. 7** 53, 107
PROFESIONALES DE LA MUERTE, LOS **Vol. 8** 523, 732
PROFESOR DE MI MUJER, EL **Vol. 8** 483
PROFESOR EROTICUS **Vol. 8** 266, 681
PROFESOR WILCZUR **Vol. 1** 176, 308, 404
PROFESORI ZA SKOLOU **Vol. 10** 180, 184, 369, 550
PROFESSEUR CUPIDON, LE **Vol. 10** 167, 453, 586
PROFESSEUR RASPOUTINE, LE **Vol. 6** *page 95*
PROFESSION: AVENTURIERS **Vol. 2** 294
PROFESSIONE FIGLIO **Vol. 7** 1194
PROFESSIONE P... ATTRICE **Vol. 7** 540
PROFESSIONE REPORTER **Vol. 7** 27, 913, 1100, 1145
PROFESSIONNEL, LE **Vol. 6** *page 77*
PROFESSOR BUCHS REJSEEVENTYR **Vol. 5** 97
PROFESSOR GATTI E I SUOI GATTINI, IL **Vol. 7** 432
PROFESSOR KRANZ TEDESCO DI GERMANIA **Vol. 7** 383
PROFESSOR LAROUSSE **Vol. 4** 219
PROFESSOR MAMLOCK **Vol. 1** 7, 9, 50
PROFESSOR PETERSENS PLEJEBORN **Vol. 5** 24, 83
PROFESSOR POPPES PRILLIGA PRILLERIER **Vol. 5** 843, 923
PROFESSORE MIOPE, IL **Vol. 7** 1189
PROFESSORESSA DI LINGUE **Vol. 7** 666
PROFESSORI MASA **Vol. 5** 374, 443
PROFESSORI UUNO D. G. TURHAPURO **Vol. 5** 345, 606
PROFETA VELATO, IL **Vol. 7** 1189
PROFETA VOLTAL SZIVEM **Vol. 9** 145, 223, 285, 316
PROFETUL, AURUL SI ARDELENII **Vol. 3** 269, 280, 283
PROFUMO E BALOCCHI **Vol. 7** 649
PROGNOZA POGODY **Vol. 6** *page 214, page 247*
PROGON **Vol. 6** *page 310, page 314, page 317*
PROHIBIDO ENAMORARSE **Vol. 8** 214, 400
PROHIBIDO SONAR **Vol. 8** 342, 749
PROIBITO **Vol. 7** 522, 1136
PROIBITO RUBARE **Vol. 7** 522, 1136
PROIE DU VENT, LA **Vol. 2** 65, 148, 357, 573, 753
PROIE POUR L'OMBRE, LA **Vol. 2** 368, 770
PROJECTION PRIVEE **Vol. 2** 49, 851
PROJEKT AQUA **Vol. 1** 38, 59
PROKA **Vol. 6** *page 309*
PROKLETI DOMU HAJNU **Vol. 10** 183, 204, 283, 511
PROKLJATIETO **Vol. 6** *page 43, page 63*
PROKURATOR ALICJA HORN **Vol. 1** 308, 404
PROLETARDRENGEN **Vol. 5** 228
PROM **Vol. 1** 283, 294 **Vol. 6** *page 257*
PROM DO SZWECJI **Vol. 6** *page 215, page 223, page 237*
PROMENONS – NOUS DANS LE BOIS **Vol. 6** *page 69*
PROMESSE A L'INCONNUE **Vol. 2** 84, 347
PROMESSE DE L'AUBE, LA **Vol. 2** 837
PROMESSE DI MARINAIO **Vol. 7** 373
PROMESSE, LA **Vol. 2** 263, 593
PROMESSES **Vol. 2** 761
PROMESSES DANGEREUSES, LES **Vol. 2** 404
PROMESSI SPOSI, I **Vol. 7** 196, 389, 756, 1055, 1060, 1189 **Vol. 8** 598 626
PROMETEJ S OTOKA VISEVICE **Vol. 3** 421, 437
PROMETHEAS SE DEUTERO PROSOPO **Vol. 3** 156, 187
PROMETHEUS I LAENKER I-II **Vol. 5** 193

PROMETHEUS IN MÄRKISCH-BUCHHOLZ Vol. 1 49
PROMOCE Vol. 10 335
PRONTO AD UCCIDERE Vol. 7 406
PRONTO, CHI PARLA ? Vol. 7 796
PRONTO ... LUCIA Vol. 7 817
PROPANTOS PSYHRAIMIA Vol. 3 179
PROPRE DE L'HOMME, LE Vol. 2 110
PROPRIETA NON E PIU UN FURTO, LA Vol. 7 599
PROPRIETARII Vol. 3 246, 284, 301
PROSCRITO DEL RIO COLORADO, EL Vol. 8 626
PROSIM NEBUDIT Vol. 10 182, 286, 460
PROSIM, PANE PROFESORE Vol. 10 168, 564
PROSIT NEUJAHR 1910 ! Vol. 4 646
PROSOPO ME PROSOPO Vol. 3 199
PROSTITUTION Vol. 2 294
PROSTITUTION CLANDESTINE Vol. 2 139
PROSTITUTION, DIE Vol. 4 159, 404
PROSTITUTION, LA Vol. 2 287, 581, 727
PROTEA Vol. 2 24
PROTECTEUR, LE Vol. 2 808
PROTEGIDAS, LAS Vol. 8 407, 738
PROTEKTIONSKIND, DAS Vol. 4 329, 355
PROTEST Vol. 3 430
PROTI VSEM Vol. 10 154, 503, 547, 561
PROTIV KINGA Vol. 3 397
PROTUVA Vol. 3 437
PROUDY Vol. 10 126, 167
PROVA D'ORCHESTRA Vol. 7 434, 893, 988
PROVA TRIGICA Vol. 7 1189
PROVAZ Z OBESENCE Vol. 10 524
PROVDAM SVOU ZENU Vol. 10 532, 621
PROVERENO NEMA MINA Vol. 3 357
PROVIDENCE Vol. 2 35, 770
PROVINCIA VIOLENTA Vol. 7 492
PROVINCIALE, LA Vol. 7 21, 332, 459, 522, 781, 1036
PROVINCIALE, LE Vol. 6 *page 74, page 95*
PROVINCIALES EN CHALEUR Vol. 6 *page 94*
PROVINSEN KALDER Vol. 5 24
PROVINZONKEL, DER Vol. 4 85, 321
PROVOCATION, LA Vol. 2 60
PROXENIO TIS ANNAS Vol. 3 159
PROXIMA ESTACION, LA Vol. 8 219, 368
PROXIMA OTONO, EL Vol. 8 747
PROXIMA VEZ QUE VIVAMOS, LA Vol. 8 276, 298
PROZESS WIRD VERTAGT, DER Vol. 1 25, 88, 107
PROZESS WORTH Vol. 4 165
PROZOREC"T Vol. 3 117 Vol. 6 *page 34, page 43*
PROZVAN JE I V3 Vol. 3 356
PRSELO JIM STESTI Vol. 10 229, 369
PRSTYNEK Vol. 10 122, 154, 256, 580
PRÜFSTEIN, DER Vol. 4 339, 525
PRULOM Vol. 10 122, 142, 256, 412
PRUNCUL, PETROLUL SI ARDELENII Vol. 3 275, 320 Vol. 6 *page 270*
PRVA LJUBAV Vol. 3 415, 416 Vol. 6 *page 305, page 314*
PRVAK Vol. 10 451, 576
PRVI GRADJANIN MALE VAROSI Vol. 3 347, 356, 389
PRVI SPLITSKI ODRED Vol. 3 366, 371
P"RVI UROK Vol. 3 44, 69, 84
P"RVIJAT KURIER Vol. 3 82 Vol. 6 *page 46*
PRVNI A POSLEDNI Vol. 10 70, 179, 433
PRVNI DEN MEHO SYNA Vol. 10 184, 550
PRVNI PARTA Vol. 10 229, 503
PRVNI POLIBENI Vol. 10 256, 586
PRZEBUDZENIE Vol. 1 384, 397
PRZECIWKO BOGOM Vol. 1 168, 321, 330 Vol. 6 *page 237*
PRZED ODLOTEM Vol. 6 *page 229, page 243, page 249*
PRZEDSWIATECZNY WIECZOR Vol. 1 202, 227, 254 Vol. 6 *page 246*
PRZEDWIOSNIE Vol. 1 252, 309
PRZEJSCIE PODZIEMNE Vol. 1 144
PRZEKLADANIE Vol. 1 136
PRZEKLETA ZIEMA Vol. 6 *page 213, page 249, page 252*
PRZEKLETE OKO PROROKA Vol. 6 *page 246, page 258*
PRZEMYTNICY Vol. 6 *page 249*
PRZEOR KORDECKI Vol. 1 290, 372
PRZEPRASZAM, CZY TU BIJA ? Vol. 1 147, 264, 327, 349
PRZEPROWADZKA Vol. 6 *page 248, page 257*
PRZERAZLIWE LOZE Vol. 1 198, 313, 354
PRZERWANY LOT Vol. 1 171, 227, 237
PRZESADY Vol. 1 174, 319
PRZESLUCHANIE Vol. 6 *page 222, page 239, page 253*
PRZESTEPCY Vol. 1 174, 215, 326
PRZEWODNIK Vol. 6 *page 229*
PRZEZ LZY DO SZCZESCIA Vol. 1 278, 347
PRZEZ PIEKLO Vol. 1 404
PRZEZNACZENIE Vol. 1 290, 371 Vol. 6 *page 234, page 244, page 254*
PRZYBLEDA Vol. 1 278, 404
PRZYBYLI DO WSI ZOLNIERZE Vol. 1 191, 290
PRZYGODA JEDNEJ NOCY - 9,25 Vol. 1 278, 371
PRZYGODA NA MARIENSZTACIE Vol. 1 173, 225, 250, 298, 377

PRZYGODA NOWOROCZNA Vol. 1 237, 305, 399
PRZYGODA Z PIOSENKA Vol. 1 196, 253 Vol. 6 *page 232*
PRZYGODY PANA PIORUNKIEWICZA Vol. 1 241
PRZYGODY PSA CYWILA Vol. 1 144, 294
PRZYSPIESZENIE Vol. 6 *page 214, page 244, page 247*
PRZYSTAN Vol. 1 187, 206, 220, 401
PSELA TA HERIA HITLER ! Vol. 3 180
PSEVDONIM : LUKAC Vol. 9 154
PSI A LIDE Vol. 10 41, 289, 292
PSICHA, DIE TÄNZERIN KATHARINA DER GROSSEN Vol. 4 656
PSOHLAVCI Vol. 10 138, 142, 248, 296, 453, 524, 586
PSOMI GIA ENAN DRAPETE Vol. 3 130
PSST, ICH BIN TANTE EMMA Vol. 4 365, 608
PSY Vol. 2 779 Vol. 6 *page 89*
PSYCHE Vol. 9 10, 122, 169, 215
PSYCHEDELICA BLUES Vol. 5 715, 747, 802, 804
PTAKI PTAKOM Vol. 1 293, 327, 390
P"TESESTVIE Vol. 6 *page 47, page 48, page 59*
P"TESESTVIE MEZDU DVA BRJAGA Vol. 6 *page 62*
PTICI DOLITAT Vol. 6 *page 39, page 54*
PTICI I HR"TKI Vol. 3 41 Vol. 6 *page 23, page 34*
P"TISTA Vol. 3 31, 51, 55, 93, 100 Vol. 6 *page 41*
P'TIT CON Vol. 6 *page 70, page 96*
P'TIT PARIGOT, LE Vol. 2 55, 613
P'TITE, LA Vol. 2 745
P'TITE LILI, LA Vol. 2 1
P'TITES TETES, LES Vol. 6 *page 141*
P"TJAT K"M SOFIJA Vol. 6 *page 28, page 45, page 56, page 60, page 63*
P"TJAT MINAVA PREZ BELOVIR Vol. 3 31, 51, 55
P"TJAT NA BEZP"TNITE Vol. 3 71
P"TJAT NA MUZIKANITE Vol. 6 *page 55, page 60*
PUCANJ Vol. 3 352, 412, 437
PUCCINI Vol. 2 719 Vol. 7 756
PUCE ET LE PRIVE, LA Vol. 2 409 Vol. 6 *page 84*
PUCE, LA Vol. 2 678
PUCK Vol. 5 357, 471
PUCK HETER JAG Vol. 5 871, 1042
PUDR A BENZIN Vol. 10 453, 586, 624
PUEBLA DE LAS MUJERES Vol. 8 524, 749
PUENTAS DE INVIERNO Vol. 8 626, 728
PUENTE AEREO Vol. 8 297, 616
PUENTE DE COPLAS Vol. 8 495, 614
PUENTE DE LA PAZ, EL Vol. 8 214, 555, 607
PUENTE DEL DIABLO Vol. 8 614, 725
PUENTE SOBRE EL TIEMPO, UN Vol. 8 493, 749
PUERTA ABIERTA, LA Vol. 7 1062 Vol. 8 214, 378, 410, 721
PUGNI IN TASCA, I Vol. 7 646, 723
PUGNI, PUPE E MARINAI Vol. 7 1044
PUGNO DEL GIGANTE, IL Vol. 7 108
PUGOWITZA Vol. 6 *page 157, page 158, page 160*
PUITS AUX TROIS VERITES, LE Vol. 2 457, 540, 738
PUITS DE JACOB, LE Vol. 2 102, 579, 767
PUKOTINA RAJA Vol. 3 391, 392, 452
PUKOVNIKOVICA Vol. 3 418 Vol. 6 *page 309*
PUL DOMU BEZ ZENICHA Vol. 10 389, 517, 567
PULAKAPINA Vol. 5 465, 500
PULAPKA Vol. 1 163, 249, 298
PULCINELLA Vol. 2 136 Vol. 7 621
PULL-OVER ROUGE, LE Vol. 2 110
PULNOCNI KOLONA Vol. 10 41, 254, 505
PULSACIONES Vol. 8 265, 465
PULVERMÜHLE, DIE Vol. 4 247
PUMPARI OD ZLATE PODKOVY Vol. 10 41, 144, 153
PUNAHILKKA Vol. 5 594, 606
PUNAHOUSUT Vol. 5 330, 592
PUNAINEN VIIVA Vol. 5 483, 516, 613
PUNATUKKA Vol. 5 452
PUNGA CU LIBELULE Vol. 3 271, 272
PUNITION, LA Vol. 2 135, 260, 609, 787
PUNKS KOMMT AUS AMERIKA Vol. 4 132, 230
PUNKT WYJSCIA Vol. 1 249
PUNKTY ZA POCHODZENIE Vol. 6 *page 217, page 223, page 227*
PUNTA A CTYRLISTEK Vol. 10 296, 564
PUNTO NERO, UN Vol. 7 730
PUPATELLA Vol. 7 39
PUPILLA DEL MARINAIO, LA Vol. 7 1189
PUPILLA RIACCESA, LA Vol. 7 796
PUPILLE NELL'OMBRA Vol. 7 369
PUPILLE SPENTE Vol. 7 144
PUPPE, DIE Vol. 4 560, 663
PUPPE VOM LUNAPARK, DIE Vol. 4 291, 348, 626
PUPPEN FÜR DIE NACHT Vol. 6 *page 164, page 170, page 173*
PUPPENKÖNIGIN, DIE Vol. 4 207, 400, 550, 724
PUPPENMACHER VON KIANG-KING, DER Vol. 4 239, 333, 543
PURE CU'ME Vol. 7 314

PURGATORIO, IL **Vol. 7** 1189
PURITAIN, LE **Vol. 2** 67, 206, 579, 676
PURO, EL **Vol. 8** 261, 768
PURPUR UND WASCHBLAU **Vol. 4** 119
PURZEL-FILM-SERIE **Vol. 4** 514
PUSCIK JEDE DO PRAHY **Vol. 10** 60, 407, 579
PUSILLECO ADDIRUSO **Vol. 7** 829
PUSKAK ES GALAMBOK **Vol. 9** 183, 241, 259
PUSS & KRAM **Vol. 5** 1170
PUSTI SNOVI **Vol. 3** 349, 374
PUSTIUL **Vol. 3** 244, 278, 325, 329
PUSTOLOV PRED VRATIMA **Vol. 3** 341, 444
PUSTOTA **Vol. 6** *page 305, page 311, page 317*
PUSZCZA **Vol. 1** 278, 380
PUSZTAI KIRALYKISASSZONY, A **Vol. 9** 132, 237
PUSZTAI SZEL **Vol. 9** 178, 237
PUT **Vol. 3** 416
PUT OKO SVETA **Vol. 3** 349
PUT U RAJ **Vol. 3** 421, 437
PUTAINS D'HISTOIRE D'AMOUR **Vol. 6** *page 88*
PUTAPELA **Vol. 8** 455
PUTEREA SI ADEVARUL **Vol. 3** 239, 274, 319
PUTKINOTKO **Vol. 5** 347, 634
PUTNICI SA SPLENDIDA **Vol. 3** 376, 444
PUTOVANI JANA AMOSE **Vol. 10** 283, 313, 516
PUTOVANIE DO SAN JAGA **Vol. 10** 82, 273
PUTOVANJE NA MJESTO NESRECE **Vol. 3** 355, 418, 437
PUTSCHLIESEL **Vol. 4** 190, 560
PUTTANA GALERA ! **Vol. 7** 643
PUZZLE **Vol. 8** 297, 361
PYGMALION **Vol. 4** 246, 462
PYGMALION XII **Vol. 1** 2, 57, 127
PYHÄ PERHE **Vol. 5** 434
PYLLI I LIRISE **Vol. 3** 11, 18
PYRAMIDE FÜR MICH, EINE **Vol. 1** 1, 57, 63
PYRAMIDE HUMAINE, LA **Vol. 2** 546, 609
PYRETO STEN ASFALTO **Vol. 3** 135, 205
PYSNA PRINCEZNA **Vol. 10** 283, 407, 460
PYTLACI **Vol. 10** 389, 517, 567
PYTLAKOVA SCHOVANKA **Vol. 10** 92, 503, 582, 612

Q

Q **Vol. 2** 294, 309
QIVITOQ **Vol. 5** 31, 241, 248
QORTIMET E VJESHTES **Vol. 6** *page 11, page 16*
QUA LA MANO **Vol. 7** 426, 450, 1142
QUADRANTE D'ORO, IL **Vol. 7** 825
QUADRIFOGLIO D'ORO **Vol. 7** 899
QUADRILLE **Vol. 2** 488, 657
QUADRILLE D'AMOUR **Vol. 2** 402
QUADRILLE D'AUTREFOIS **Vol. 2** 678
QUADRILLE DE GOMMEUX **Vol. 2** 678
QUADRILLE DE TITIS **Vol. 2** 678
QUADRILLE DEL'AVENIR **Vol. 2** 678
QUADRO DI OSVALDO MARS, IL **Vol. 7** 196
QUAI DE GRENELLE **Vol. 2** 9, 368
QUAI DES BLONDES **Vol. 2** 158, 368, 857
QUAI DES BRUMES **Vol. 2** 309, 644, 778, 837
QUAI DES ILLUSIONS **Vol. 2** 717
QUAI DES ORFEVRES **Vol. 2** 251, 824
QUAI DU DESIR, LE **Vol. 2** 139
QUAI DU POINT DU JOUR **Vol. 2** 322, 835
QUAI NOTRE-DAME **Vol. 2** 686
QUALCOSA DI BIONDO **Vol. 7** 514, 893
QUALE DEI DUE ? **Vol. 7** 825
QUAND EL SOLEIL MONTERA **Vol. 2** 344
QUAND FEMMES ONT PRIS LA COLERE **Vol. 2** 186, 621, 736
QUAND LA FEMME S'EN MELE **Vol. 2** 279, 344
QUAND LA RAISON S'EN VA **Vol. 2** 757
QUAND LES FILLES SE DECHAINENT **Vol. 2** 484
QUAND L'OMBRE DESCEND **Vol. 2** 229, 648
QUAND MINUIT SONNERA **Vol. 2** 41, 47, 425
QUAND NOUS ETIONS DEUX **Vol. 2** 32, 138, 339
QUAND PASSENT LES FAISANS **Vol. 2** 503, 650
QUAND SONNERA MIDI **Vol. 2** 153, 249, 702
QUAND TE TUES-TU ? **Vol. 2** 43, 460, 717
QUAND TU LIRAS CETTE LETTRE **Vol. 2** 13, 322, 376, 832
QUAND TU SERAS DEBLOQUE FAIS-MOI SIGNE **Vol. 6** *page 95, page 134*
QUAND VIENT L'AMOUR **Vol. 2** 826 **Vol. 3** 390
QUANDO C'ERA LUI ... CARO LEI ! **Vol. 7** 1136
QUANDO GALLINA CANTA... GALLO TACE **Vol. 7** 26, 701
QUANDO GLI ANGELI PIANGONO **Vol. 7** 1147
QUANDO IL CUORE PARLA **Vol. 7** 1189
QUANDO LA COPPIA SCOPPIA **Vol. 7** 199, 599

QUANDO LA PRIMAVERA RITORNO Vol. 7 168
QUANDO L'AMORE E SENSUALITA Vol. 7 981
QUANDO LE DONNE SI CHIAMAVANO „MADONNE" Vol. 7 649
QUANDO TRAMONTA IL SOLE Vol. 7 1054, 1167
QUANTO E BELLO LU MURIRE ACCISO Vol. 7 381, 699
QUANTO VOLTE ... QUELLA NOTTE Vol. 7 975
QUARANTA MILIARDI E UNA CORONA Vol. 7 107
QUARANTÄNE Vol. 4 236, 321, 734
QUART D'HEURE AMERICAIN, LE Vol. 6 *page 77, page 80, page 141*
QUARTA PAGINA Vol. 7 838, 994
QUARTET Vol. 6 *page 75, page 86*
QUARTETT, DAS Vol. 4 329, 365
QUARTETT ZU FÜNFT Vol. 1 35, 41, 67, 129 Vol. 4 247
QUARTETTO BASILEUS Vol. 7 331, 1093, 1165
QUARTETTO PAZZO Vol. 7 48, 493, 1121
QUARTIER CHINOIS Vol. 2 193, 475
QUARTIER LATIN Vol. 2 32, 366, 368, 388, 727 Vol. 4 292, 626
QUARTIER SANS SOLEIL Vol. 2 229, 294, 802
QUARTIERE DEI LILLA Vol. 7 619
QUARTIERI ALTI Vol. 7 725, 756, 945, 1136
QUASI QUASI MI SPOSO Vol. 7 1114
QUATORZE JUILLET Vol. 2 573, 644, 656
QUATRE CAVALIERS DE L'APOCALYPSE, LES Vol. 2 385
QUATRE CENT COUPS, LES Vol. 2 224, 285
QUATRE CHARLOTS MOUSQUETAIRES, LES Vol. 2 251, 738
QUATRE DE LA LEGION Vol. 2 76
QUATRE HEURES DU MATIN Vol. 2 6, 722
QUATRE HOMMES AUX POINGS NUS Vol. 2 139
QUATRE JOURS A PARIS Vol. 2 368, 516, 627
QUATRE NUITS D'UN REVEUR Vol. 2 170, 517
QUATRE PAS DANS LES NUAGES Vol. 2 318, 395
QUATRE POUR CENT DE RISQUE Vol. 2 655
QUATRE SERGENTS DU FORT CARRE, LES Vol. 2 533
QUATRE VAGABONDS, LES Vol. 2 778
QUATRE VERITES, LES Vol. 2 59, 581, 824 Vol. 8 719, 766
QUATRE VEUVES Vol. 2 259
QUATRE-VINGT PREND DES VACANCES, LE Vol. 2 70, 309
QUATRE-VINGT-TREIZE Vol. 2 798
QUATRIEME SEXE, LE Vol. 2 198
QUATRO AGENTES SECRETOS, OS Vol. 8 62
QUATTRO DEL GETTO TONANTE, I Vol. 7 60, 1170
QUATTRO DELL' APOCALIPSE, I Vol. 7 1015
QUATTRO DONNE NELLA NOTTE Vol. 7 619
QUATTRO GIORNATE DI NAPOLI, LE Vol. 7 502, 915
QUATTRO INESORABILI, I Vol. 8 551, 626
QUATTRO MOSCHETTIERI, I Vol. 7 454, 937, 1027, 1189
QUATTRO PASSI FRA LE NUVOLE Vol. 7 709, 1177
QUATTRO RAGAZZE SOGNANO Vol. 7 792, 1063
QUATTRO ROSE ROSSE Vol. 7 359, 443
QUAX, DER BRUCHPILOT Vol. 4 311, 461
QUAX IN FAHRT Vol. 4 119, 270
QUE ARMAN LAS MUJERES, LO Vol. 8 249
QUE CONTENTA ESTOY ! Vol. 7 587 Vol. 8 475
QUE COSAS TIENE EL AMOR ! Vol. 8 249, 338
QUE EMPIEZAN A LOS 15 ANOS, LAS Vol. 8 471
QUE ESPEREN LOS CUERVOS Vol. 8 331
QUE FAMILIA Vol. 8 422
QUE GOZADA DE DIVORCIO ! Vol. 8 249, 368
QUE HACE UNA CHICA COMO TU EN UN SITIO COMO ESTE ? Vol. 8 210, 293
QUE HACEMOS CON LOS HIJOS ? Vol. 8 524, 725
QUE HE HECHO YO PARA MERECER ESTO ? Vol. 8 240, 373 ,570
QUE LA BETE MEURE Vol. 2 700
QUE LA FETE COMMENCE Vol. 2 351, 377
QUE LES GROS SALAIRES LEVENT LE DOIGT !!! Vol. 6 *page 72, page 92*
QUE LES HOMMES SONT BETES Vol. 2 137, 409,427
QUE NOCHE DE BODAS, CHICAS ! Vol. 8 378
QUE NOCHE, MUCHACHOS ! Vol. 8 306, 600
QUE NOS IMPORTA LA REVOLUCION ? Vol. 8 595, 762

QUE NOS QUITEN LO BAILAO **Vol. 8** 210
QUE PERSONNE NE SORTE **Vol. 2** 131
QUE TIA LA C.I.A. ! **Vol. 8** 728, 770
QUE TIENEN QUE SERVIR, LAS **Vol. 8** 331, 611
QUE TIO MAS GRANDE ! **Vol. 8** 606
QUE TOCAN EL PIANO, LOS **Vol. 8** 331, 681
QUE VERDE ERA MI DUQUE ! **Vol. 8** 299, 728
QUE VIENEN LOS SOCIALISTAS ! **Vol. 8** 368, 773
QUE VIVA CARRANCHO ! **Vol. 8** 732
QUEI DUE **Vol. 7** 453, 796, 1116
QUEIMADA **Vol. 7** 502, 1007
QUEL BANDITO SONO IO **Vol. 7** 106, 522
QUEL DISCOLO DI PAPA **Vol. 7** 1189
QUEL GALANTUOMO DI UN CAMERIERE **Vol. 7** 1189
QUEL GRAN PEZZO DELL'UBALDA, TUTTA NUDA E TUTTA CALDA **Vol. 7** 1027
QUEL MOVIMENTO CHE MI PIACE TANTO **Vol. 7** 514
QUEL PASTICCIACCIO BRUTTO DI VIA MERULANA **Vol. 7** 661, 1007, 1166
QUEL TESORO DI PAPA **Vol. 7** 603
QUELLA PESTE DI PIERINA **Vol. 7** 414
QUELLA PICCOLA DIFFERENZA **Vol. 7** 318
QUELLA PROVINCIA MALIZIOSA **Vol. 7** 1047
QUELLA STRANA VOGLIA D'AMARE **Vol. 7** 13
QUELLA VECCHIA CANAGLIA **Vol. 7** 196, 756
QUELLA VILLA ACCANTO AL CIMITERO **Vol. 7** 623, 1015
QUELLE DROLE DE GOSSE ! **Vol. 2** 488, 581, 803
QUELLE JOIE DE VIVERE ! **Vol. 7** 1214
QUELLE JOLIE DE VIVRE **Vol. 2** 224, 882
QUELLE SACREE SOIREE **Vol. 2** 294, 401
QUELLE STRANE OCCASIONI **Vol. 7** 1138
QUELLEN DER LIEBE **Vol. 4** 718
QUELLI DELLA CALIBRO 38 **Vol. 7** 910
QUELLI DELLA MONTAGNA **Vol. 7** 326, 359, 1160
QUELLO CHE NON MUORE **Vol. 7** 731
QUELLO STRANO DESIDERIO **Vol. 7** 141
QUELQUE PART, QUELQU'UN **Vol. 2** 60
QUELQUES MESSIEURS TROP TRANQUILLES **Vol. 2** 293
QUELQUES PAS DANS LA VIE **Vol. 2** 374
QUELQU'UN A TUE **Vol. 2** 42, 388, 402, 863
QUELQU'UN DANS L'OMBRE **Vol. 2** 64
QUELQU'UN DERRIERE LA PORTE **Vol. 2** 320, 517
QUEM ESPERA POR SAPATOS DE DEFUNTO MORRE DESALCO **Vol. 8** 7
QUEMA EL SUELO **Vol. 8** 207, 567
QUEREMOS UN HIJO TUYO **Vol. 8** 368, 773
QUERIDA, LA **Vol. 8** 214
QUERIDI PROFESOR **Vol. 8** 450, 708
QUERIDO LILAS, O **Vol. 8** 32
QU'EST-CE MAMAN COMPREND A L'AMOUR **Vol. 2** 604
QU'EST-CE QUE TU VEUX, JULIE ? **Vol. 2** 265
QU'EST-CE QUI FAIT CRAQUER LES FILLES ? **Vol. 6** *page 73*
QU'EST-CE QU'ON ATTEND POUR ETRE HEUREUX ! **Vol. 6** *page 80, page 81, page 143, page 147*
QU'EST QUI FAIT COURIR DAVID ? **Vol. 6** *page 97, page 101*
QUESTA E LA VITA **Vol. 7** 603, 897
QUESTA PAZZE, PAZZE, PAZZE DONNE **Vol. 7** 451
QUESTI FANTASMI **Vol. 7** 501
QUESTION, LA **Vol. 2** 513
QUESTO AMORE AI CONFINI DEL MONDO **Vol. 7** 889
QUESTO E QUELLO **Vol. 7** 381, 408
QUESTO SI CHE E AMORE **Vol. 7** 1194
QUI **Vol. 2** 124, 787
QUI ? **Vol. 2** 798
QUI COMINCIA L'AVVENTURA **Vol. 7** 854
QUI ETES-VOUS M. SORGE ? **Vol. 2** 857
QUI ETES-VOUS POLLY MAGOO ? **Vol. 2** 110, 285
QUICK **Vol. 4** 30, 325, 567
QUIEN ME COMPRA UN LIO ! **Vol. 8** 422
QUIEN ME QUEIRE AMI ? **Vol. 8** 270
QUIEN SOY YO ? **Vol. 8** 207, 732
QUIERE CASARSE CONMIGO ? **Vol. 8** 544, 578
QUIEREME CON MUSICA **Vol. 8** 400, 470
QUIERO MUERTO, LO **Vol. 8** 306
QUIERO SONAR **Vol. 8** 345
QUIESTI RAGAZZI **Vol. 7** 493, 756
QU'IL EST JOLI GARCON L'ASSASSIN DE PAPA ! **Vol. 2** 133 **Vol. 6** *page 110*
QUINCE HORCAS PARA UN ASESINO **Vol. 8** 299
QUINDICI FORCHE PER UN ASSASSINO **Vol. 8** 299
QUINIELA, LA **Vol. 8** 477, 717
QUINTA DEL PORRO, LA **Vol. 8** 554, 656
QUITTE OU DOUBLE **Vol. 2** 32, 131
QUO VADIS **Vol. 2** 206

QUO VADIS ? **Vol. 4** 113 **Vol. 7** 390, 436, 1088, 1189
QUO VADIS HOMO SAPIENS **Vol. 6** *page 272*
QUO VADIS SPIRINTION **Vol. 3** 140
QUO VADIS ZIVORADE **Vol. 3** 342, 357
QUOI DE NEUF, PUSSICAT ? **Vol. 2** 770
QYTETI ME I RI NE BOTE **Vol. 3** 2, 10, 14

R

R. A. S. **Vol. 2** 522, 642
RAB RABY **Vol. 9** 67, 199, 268
RABAGAS **Vol. 7** 490
RABAKA **Vol. 10** 128, 210, 578
RABANEK DJAVOLJI PIRAT **Vol. 3** 444
RABATTEUSE, LA **Vol. 2** 196
RABAUKEN-KABARETT, DAS **Vol. 1** 84, 85, 103
RABI LÖW **Vol. 10** 89
RABLELEK **Vol. 9** 301
RABLOLOVAG, A **Vol. 9** 293, 337
RABMADAR **Vol. 9** 28, 288, 325
RABNA VERDE, LA **Vol. 8** 493, 541
RABOLIOT **Vol. 2** 48, 57, 788
RABOUILLEUSE, LA **Vol. 2** 579
RACA **Vol. 8** 38, 54, 130
RACA, LASKA MOJA **Vol. 10** 32, 140, 228, 340, 373, 439
RACCOMANDATO DI FERRO, IL **Vol. 7** 257
RACCONTI D'ESTATE **Vol. 7** 1062
RACCONTI DI CANTERBURY, I **Vol. 7** 333, 388, 434
RACCONTI DI CARNEVALE, IL **Vol. 7** 490, 679
RACCONTI PROIBITI ... DI NIENTE VESTITI **Vol. 7** 1147
RACCONTI ROMANI **Vol. 7** 798, 1131
RACCONTI ROMANI DI PIETRO L'ARETINO, I **Vol. 7** 46
RACE D'EP, LA **Vol. 6** *page 134, page 143*
RACE DES SEIGNEURS, LA **Vol. 2** 377, 875
RACEK MA ZPOZDENI **Vol. 10** 345, 460
RACHE DER TOTEN, DIE **Vol. 4** 159
RACHE DES BANDITEN, DIE **Vol. 4** 718
RACHE DES BASTARDS, DIE **Vol. 4** 309
RACHE DES KAPITÄNS MITCHELL, DIE **Vol. 6** *page 160, page 166, page 176*
RACHE DES MARQUIS DOKAMA, DIE **Vol. 4** 135, 366
RACHE DES TITANEN, DIE **Vol. 4** 27
RACHE EINER FRAU, DIE **Vol. 4** 660, 735
RACHE FÜR EDDY **Vol. 4** 107, 221, 223, 367
RÄCHENDE LIEBE **Vol. 6** *page 194*
RÄCHER, DIE **Vol. 4** 232, 673
RACHUNEK SUMIENIA **Vol. 1** 220, 280, 281, 343 **Vol. 6** *page 248*
RACKARUNGE, EN **Vol. 5** 857
RAD NA ODREDJENO VREME **Vol. 3** 404, 408, 415
RADBIZOM A FELESEGEM **Vol. 9** 71, 322
RADIKALNI REZ **Vol. 10** 139, 226, 577
RADIO FOLLO, LA **Vol. 8** 248, 297
RADIO-HEIRAT, DIE **Vol. 4** 165, 563
RADIO TEKEE MURRON **Vol. 5** 374, 395, 646
RADIO TULEE HULLUKSI **Vol. 5** 374, 462
RADIO VIHOR ZOVE ANDJELIJU **Vol. 3** 354
RADIOAMATERI **Vol. 10** 363, 564
RADIOKILLER **Vol. 6** *page 157, page 159, page 169*
RADIOSTACIONI **Vol. 3** 11, 22 **Vol. 6** *page 18*
RADIUM **Vol. 4** 135, 503
RADIUM VENDICATORE, IL **Vol. 7** 292, 824
RADMIROV KATALIN **Vol. 9** 288
RADNI NASLOV **Vol. 6** *page 309, page 311, page 317*
RADOPOLJE **Vol. 3** 343, 436
RADOST AZ DO RANA **Vol. 10** 106, 197, 313, 583
RADUZ A MAHULENA **Vol. 10** 223, 299, 498, 569
RAFAGA DE PLOMO, UNA **Vol. 8** 732
RAFAL U NEBO **Vol. 3** 366, 378
RAFFICA DI PIOMBO, UNA **Vol. 8** 306
RAFFICA SULLA FELICITA **Vol. 7** 26, 701
RAFFICHE **Vol. 7** 1209
RAFFINIERTE FRAUEN **Vol. 4** 342
RAFFINIERTESTE FRAU BERLINS, DIE **Vol. 4** 25, 159, 255
RAFLE EST POUR CE SOIR, LA **Vol. 2** 47, 70, 318
RAFLES SUR LA VILLE **Vol. 2** 9, 368, 503
RAG. ARTURO DE FANTI, BANCARIO PRECARIO **Vol. 7** 770, 1003
RAGAN **Vol. 8** 299, 517
RAGANELLA **Vol. 7** 1004
RAGASZKODOM A SZERELEMHEZ **Vol. 9** 132, 164
RAGAZZA **Vol. 7** 756
RAGAZZA ALLA PARI **Vol. 7** 1022
RAGAZZA CHE DORME **Vol. 7** 136, 1198
RAGAZZA CHE SAPEVA TROPPO, LA **Vol. 7** 537
RAGAZZA CON LA VALIGIA, LA **Vol. 7** 781, 1027
RAGAZZA DALLA PELLE DI LUNA, LA **Vol. 7** 185

RAGAZZA DE PIAZZA SAN PIETRO, LA Vol. 8 276
RAGAZZA DEL PALIO, LA Vol. 7 988
RAGAZZA DEL VAGONE LETTO, LA Vol. 7 46
RAGAZZA DELLA SALINA, LA Vol. 7 1177
RAGAZZA DI BUBE, LA Vol. 7 522, 1167
RAGAZZA DI PASSAGGIO Vol. 7 1095
RAGAZZA DI PIAZZA S. PIETRO, LA Vol. 7 126, 792
RAGAZZA DI PRAGA, UNA Vol. 7 949
RAGAZZA DI TRIESTE, LA Vol. 7 29, 347
RAGAZZA DI VIA CONDOTTI, LA Vol. 7 247
RAGAZZA DI VIA MILLELIRE, LA Vol. 7 223, 854
RAGAZZA DI VIA VENETO, LA Vol. 7 796
RAGAZZA FUORISTRADA, LA Vol. 7 184
RAGAZZA IN PRESTITO, LA Vol. 7 1062
RAGAZZA IN VETRINA, LA Vol. 7 582, 725
RAGAZZA MERAVIGLIOSA, LA Vol. 8 438, 602
RAGAZZA SOTTO IL LENZUOLO, LA Vol. 7 1121
RAGAZZA VOGLIOSA Vol. 7 311
RAGAZZE DA MARITO Vol. 7 86, 118, 443
RAGAZZE DI PIAZZE DI SPAGNA, LE Vol. 7 498, 638, 1167
RAGAZZE DI SAN FREDIANO, LE Vol. 7 319, 1167
RAGAZZE D'OGGI Vol. 7 1062
RAGAZZE NON SCHERZATE Vol. 7 390
RAGAZZI CHE SI AMANO, I Vol. 7 255
RAGAZZI DEI PARIOLI, I Vol. 7 1044
RAGAZZI DEL JUKE BOX Vol. 7 796, 1054
RAGAZZI DEL MASSACRO, I Vol. 7 1179
RAGAZZI DELLA PERIFERIA SUD, I Vol. 7 474
RAGAZZI DI TORINO SOGNANO TOKYO E VANNO A BERLINO Vol. 7 356
RAGAZZO Vol. 7 1036, 1116
RAGAZZO CHE SAPEVA AMERE, IL Vol. 7 19
RAGAZZO CHE SORRIDE, IL Vol. 7 649
RAGAZZO DI BORGATA Vol. 7 546, 646, 1166
RAGAZZO DI CAMPAGNA Vol. 7 30, 358, 383
RAGAZZO DI EBAULUS Vol. 7 237
RAGAZZO E UNA RAGAZZA, UN Vol. 7 664, 957
RAGE AU CORPS, LA Vol. 2 9, 402
RAGE AU POING, LA Vol. 2 600
RAGE DU SEXE, LA Vol. 2 735
RAGENS RIKE Vol. 5 891, 978, 1052, 1080
RAGGARE ! Vol. 5 923
RAGGARGÄNGET Vol. 5 883, 1142
RAGGEN – DET AR JAG DET Vol. 5 928, 1042
RAGGIO DI SOLE Vol. 7 1189
RAGIONE DI STATO Vol. 7 817
RAGIONE PER VIVERE, UNA PER MORIRE, UNA Vol. 8 762
RAGISMENES KARDIES Vol. 3 203
RAGNO, IL Vol. 7 1189
RAINFOX Vol. 5 25, 111, 220
RAINHA SANTA Vol. 8 41, 72, 110
RAISINS DE LA MORT, LES Vol. 2 76
RAISON D'ETAT, LA Vol. 2 185
RAISON DU PLUS FOU, LA Vol. 2 635
RAISON DU PLUS PORT, LA Vol. 2 163, 175
RAJ NA ZIEMI Vol. 1 249, 280, 281, 298, 311 Vol. 6 *page 248*
RAKAS Vol. 5 452
RAKAS LURJUS Vol. 5 443, 643
RAKAS VARKAANI Vol. 5 443, 639, 643
RAKASTIN SINUA, HILDE Vol. 5 549, 579
RAKASTUNUT RAMPA Vol. 5 317, 385, 396, 499, 553, 606, 622, 647
RAKEV VE SNU VIDETI ... Vol. 10 70, 283, 550
RAKKAUDEN KAIKKIVALTA – AMOR OMNIA Vol. 5 356, 406
RAKKAUDEN RISTI Vol. 5 310, 552
RAKKAUS ALKAA AAMUYÄSTÄ Vol. 5 374, 383
RAKKAUS KAHLEISSA Vol. 5 443, 461
RAKKAUS ON NOPEAMPI PIIROISEN PÄSSIÄKIN Vol. 5 443, 620
RAKKAUSELOKUVA Vol. 5 434
RAKKAUSTENSA UHRI Vol. 5 334, 365, 640
RÄKNA DE LYCKELIGA STUNDERNA BLOTT Vol. 5 843, 923, 1022, 1044, 1146
RÄKNA MED BRAK Vol. 5 871, 1073
RAKOCZI HADNAGYA Vol. 9 11, 274, 326
RAKOCZI INDULO Vol. 9 71, 322
RAKOCZI NOTAJA Vol. 9 14, 68, 71, 199, 222
RAKOCZY-MARSCH Vol. 4 139, 208
RAKUUNA KALLE KOLLOLA Vol. 5 346, 348, 473
RALI Vol. 3 72 Vol. 6 *page 43*
RALIUL Vol. 6 *page 274, page 279, page 280*
RALLARBLOD Vol. 5 664, 732, 823
RALLARE Vol. 5 878, 1173
RALLYE DES JOYEUSES, LE Vol. 2 76
RAM ZA SLIKU MOJE DRAGE Vol. 3 340, 351, 415
RAMADAN Vol. 2 811

RAMASAGUL Vol. 6 *page 272, page 274, page 276, page 279*
RAMASJANG I KANTONNEMENTET Vol. 5 97
RAMIN CU TINE Vol. 6 *page 268, page 281, page 284*
RAMPER. DER TIERMENSCH Vol. 4 73, 354, 743
RAMUNTCHO Vol. 2 208, 526, 833
RANA JESEN Vol. 3 421, 444
RANCHO TEXAS Vol. 1 148, 179, 202, 223
RANCON DU BONHEUR, LA Vol. 2 798
RANCUNE, LA Vol. 2 59
RAND ROVER Vol. 7 713
RANGON ALUL Vol. 9 117, 165, 314
RANI RADOVI Vol. 3 360
RANI SNIJEG U MINHENU Vol. 6 *page 310, page 322*
RÄNNSTENSUNGAR Vol. 5 862, 923, 1045, 1142
RANNY W LESIE Vol. 1 233, 401
RANSON Vol. 5 701, 797
RANTASALMEN SULTTAANI Vol. 5 426, 443, 566
RANTASUON RAATAJAT Vol. 5 356, 386
RAPACE, LE Vol. 2 163
RAPACITE Vol. 2 407, 824
RAPARIGA NO VERAO, UMA Vol. 8 32, 47
RAPAX Vol. 2 236
RAPAZES DE TAXIS Vol. 8 54, 176
RAPHAEL-LE-TATOUE Vol. 2 6, 318, 775
RAPHAEL OU LE DEBAUCHE Vol. 2 481, 674
RAPIDE DE NUIT Vol. 2 67, 197
RAPIMENTI DI MISS ELLEN, IL Vol. 7 971
RAPINA AL QUARTIERE OVEST Vol. 7 1152
RAPIREA FECIOARELOR Vol. 3 248, 254, 335
RAPORTTI ELI BALLADI LAIVA-TYTÖISTÄ Vol. 5 452
RAPPEL IMMEDIAT Vol. 2 16, 338, 395, 583
RAPPESAGLIA Vol. 7 502
RAPPORTPIGEN Vol. 5 179
RAPSODIA BALTYKU Vol. 1 278, 308, 404
RAPSODIA DE SANGRE Vol. 8 468, 732
RAPSODIA SATANICA Vol. 7 965
RAPT Vol. 2 833
RAPT AU DEUXIEME BUREAU Vol. 2 309, 376
RAPTEME USTED Vol. 8 516, 716
RAPTO DE T.T., EL Vol. 8 747, 749
RAPTURE Vol. 2 368 Vol. 7 1021
RAPTUS Vol. 7 487
RAS DEL QUARTIERE, IL Vol. 7 340, 599
RAS LE COEUR Vol. 2 218, 784
RASARE SOARELE Vol. 3 217, 220, 221, 228, 299
RASCEL-FIFI Vol. 7 1167
RASCEL MARINE Vol. 7 1167
RASCOALA Vol. 3 218, 286, 319
RASKE RIVIERA REJSENDE Vol. 5 24, 83
RASKENSTAM Vol. 5 858, 869, 1053
RASKOLNIKOFF Vol. 2 23
RASKOLNIKOV Vol. 9 288, 337
RASKOLNIKOW Vol. 4 14, 208
RASMINES BRYLLUP Vol. 5 55, 79, 101, 208
RASMUS PA LUFFEN Vol. 5 933, 985, 1093, 1209
RASMUS PONTUS OCH TOKER Vol. 5 977
RASPOUTINE Vol. 2 27, 249, 479, 661
RASPUTIN Vol. 4 113, 340, 546
RASPUTINS LIEBESABENTEUER Vol. 4 230, 350, 599
RASUNA VALEA Vol. 1 278 Vol. 3 294, 299
RAT Vol. 3 361, 371
RAT DER GÖTTER, DER Vol. 1 6, 78, 101
RATACIRE Vol. 3 288, 324
RATAPLAN Vol. 8 422, 437
RATAS NO DUERMEN DE NOCHE, LA Vol. 8 732
RATATAA ELLER THE STAFFAN STOLLE STORY Vol. 5 909, 953, 1052
RATATAPLAN Vol. 7 38, 103
RATAVARTIJAN KAUNIS INKERI Vol. 5 471, 634
RATKAISUN PÄIVÄT Vol. 5 392, 397
RATNICI Vol. 6 *page 312*
RATS - NOTTE DI TERRORE Vol. 7 335
RÄTSEL DER NACHT Vol. 4 30
RÄTSEL DER SPHINX, DAS Vol. 4 131, 240
RÄTSEL EINER LIEBE Vol. 4 270
RÄTSEL EINER NACHT Vol. 4 471, 779
RÄTSEL IM MENSCHEN, DAS Vol. 4 385
RÄTSEL UM BEATE Vol. 4 340, 466, 623
RÄTSEL VON BANGALOR, DAS Vol. 4 396
RÄTSELHAFTE FRAU, DIE Vol. 4 87, 124
RÄTSELHAFTE TOD, DER Vol. 4 35, 240
RÄTSELHAFTER BLICK, EIN Vol. 4 71, 488
RATTE, DIE Vol. 4 423
RÄTTEN ATT ÄLSKA Vol. 5 840, 933
RATTEN DER GROSSTADT Vol. 4 495, 565, 593, 634
RATTEN, DIE Vol. 4 178, 479
RATTENFÄNGER VON HAMELN, DER Vol. 4 186, 202
RATTENLOCH, DAS Vol. 4 159
RATTENMÜHLE, DIE Vol. 4 135, 334

RATTENS MUSKETÖRER Vol. 5 864, 993
RATTO DELLE SABINE, IL Vol. 7 603
RAUB DER DOLLARPRINZESSIN, DER Vol. 4 135, 334
RAUB DER MONA LISA, DER Vol. 4 14, 124, 208
RAUB DER SABINERINNEN, DER Vol. 4 208, 373, 479, 626
RÄUBER, DIE Vol. 4 308
RÄUBERBANDE, DIE Vol. 4 321, 752
RÄUBERHAUPTMANN NULPE Vol. 4 646
RAUBTIER, DAS Vol. 1 45, 62, 100
RAUHREIF Vol. 1 37, 68
RAUM IST IN DER KLEINSTEN HÜTTE Vol. 4 135, 503
RAUSCH Vol. 4 178, 202
RAUSCHENDE MELODIEN Vol. 1 25, 35, 56
RAUTACIOSUL ADOLESCENT Vol. 3 218, 276, 313
RAUTAKAUPPIAS UUNO TURHAPURO Vol. 5 345, 424
RAUTAKYLÄN VANHA PAROONI Vol. 5 344, 356
RAVAGEUSES DE SEXE Vol. 2 808
RAVISSANTE Vol. 2 488, 598
RAVISSANTE IDIOTE, UNE Vol. 2 21, 873
RAVNOVESIE Vol. 6 *page 39, page 45, page 55*
RAYE DES VIVANTS Vol. 2 249,395
RAYO DE LUZ, UN Vol. 8 214, 276
RAYO DESINTEGRADOR, EL Vol. 8 298, 681
RAYO, EL Vol. 8 262
RAZA Vol. 8 298, 449, 611
RAZA DE HIDALGOS Vol. 8 246
RAZARAC ZAGREB Vol. 6 *page 319*
RAZBOIUL DOMNITELOR Vol. 3 273, 317
RAZBOIUL INDEPENDENTEI Vol. 3 230, 317
RAZBUNAREA Vol. 3 234, 269, 281
RAZBUNAREA HAIDUCILOR Vol. 3 218, 254, 335
RAZMEDJA Vol. 3 412, 437
RAZON PARA VIVIR Y UNA PARA MORIR, UNA Vol. 8 762
RAZSELJENA OSEBA Vol. 6 *page 305, page 316*
RAZZA SELVAGGIA Vol. 7 19, 240
RAZZA VIOLENTA Vol. 7 371, 514
RAZZIA Vol. 1 6, 51, 58 Vol. 2 407 Vol. 8 515, 552 Vol. 9 36, 119, 122, 312
RAZZIA IN ST. PAULI Vol. 4 608, 747
RAZZIA SUR LA CHNOUF Vol. 2 322, 601, 832
RDECE KLASJE Vol. 3 367, 368
RDECI BOOGIE, ALI KAJ TI JE DEKLICA Vol. 6 *page 303, page 313*
RDZA Vol. 6 *page 252, page 257*
RE BURLONE Vol. 7 196, 453, 570
RE DAVID Vol. 7 978
RE DEI CRIMINALI, IL Vol. 8 600, 626
RE DEL CIRCO, IL Vol. 7 725, 960
RE DELLA FORZA, IL Vol. 7 108
RE DELLA NOTTE, IL Vol. 7 390
RE DELL'ABISSO, IL Vol. 7 304
RE DELL'ATLANTICO, IL Vol. 7 731
RE DELLE BANANE Vol. 7 1064
RE DI DENARI Vol. 7 48, 640
RE D'INGHILTERRA NON PAGA, IL Vol. 7 163, 603, 826, 838, 1159
RE, LE TORRI, GLI ALFIERI, IL Vol. 7 84, 254, 796
RE SI DIVERTE, IL Vol. 7 48, 170, 421, 830
REALI DI FRANCIA, I Vol. 7 1121
REBELDES DE ARIZONA, LOS Vol. 8 495, 773
REBELDIA Vol. 8 214, 488, 719
REBELION DE LAS MUERTES, LA Vol. 8 237
REBELION DE LOS BUCANEROS, LA Vol. 8 332, 749
REBELION DE LOS ESCALVOS, LA Vol. 8 438, 611
REBELION DE LOS GLADIADORES, LA Vol. 8 602
REBELION DE LOS MUERTOS, LA Vol. 8 708
REBELION DE LOS PAJAROS, LA Vol. 8 362, 621
REBELL, DER Vol. 4 10, 54, 73, 208, 373, 437, 556
REBELLE, LE Vol. 2 537
REBELOTE Vol. 6 *page 106, page 123, page 133*
REBRO ADAMOVO Vol. 3 44, 100 Vol. 6 *page 45*
REBUS Vol. 1 198, 349 Vol. 6 *page 232* Vol. 7 927, 1055 Vol. 8 311, 611
RECADO, O Vol. 8 137
RECE DO GORY Vol. 6 *page 227, page 243, page 258*
RECHT AUF GLÜCK Vol. 4 558
RECHT AUF LIEBE, DAS Vol. 4 30, 124, 158, 232, 270
RECHT DER FREIEN LIEBE, DAS Vol. 4 423
RECHT DER UNGEBORENEN, DAS Vol. 4 306, 379
RECIF DE CORAIL, LE Vol. 2 440
RECLAME DI UN SARTO, LA Vol. 7 1189
RECLUSO, IL Vol. 7 1189
RECLUTA CON NINO Vol. 8 214, 546

RECOMENADO, EL Vol. 8 296, 770
RECOMINCIO DA ZERO Vol. 7 949
RECOMPENSA Vol. 8 108, 130
RECONRES AU MASCULIN Vol. 6 *page 97*
RECONSTITUIREA Vol. 3 267, 269, 332
RECOURS DE LA METHODE, LE Vol. 2 35
RECOURS EN GRACE Vol. 2 59, 204, 427
RECREATION, LA Vol. 2 655, 669
RED BULL, DER LETZTE APACHE Vol. 4 718
RED LANTERN, THE Vol. 2 579
REDENZIONE Vol. 7 402, 488, 733, 796
REDENZIONE D'ANIMA Vol. 7 56, 654
REDL EZREDES Vol. 9 167, 259
REDS Vol. 5 626
REFAIS – LE MOI ENCORE Vol. 6 *page 145*
REFIUTI DEL TEVERE, I Vol. 7 493
REFLUX, LE Vol. 2 97
REFORMES SE PORTENT BIEN, LES Vol. 2 76, 130
REFUGE, LE Vol. 2 339
REFUGIO ALL'ALBA, IL Vol. 7 824
REGAIN Vol. 2 143, 485, 583, 871
REGALO, IL Vol. 7 411
REGARD, LE Vol. 2 390, 654
REGASIRE Vol. 3 316, 320, 327
REGATES DE SAN FRANCISCO, LES Vol. 2 251, 824
REGEMENTETS ROS Vol. 5 840, 930
REGENTA, LA Vol. 8 581
REGENTRUDE, DIE Vol. 1 48, 49
REGGIA SUL FIUME, LA Vol. 7 101, 476
REGI IDÖK FOCIJA Vol. 9 152, 253, 259
REGI KERINGÖ Vol. 9 132, 274
REGI NYAR Vol. 9 26, 117, 133
REGIMENT DE LIGNE Vol. 2 678
REGIMENTSKOMMANDEUR, DER Vol. 1 19, 59, 121
REGIMENTSMUSIK Vol. 4 524, 544
REGIMENTSTOCHTER, DIE Vol. 4 229, 558
REGINA DEI CARAIBI, LA Vol. 7 91, 1195
REGINA DEI TARTARI, LA Vol. 7 347
REGINA DEL CARBONE, LA Vol. 7 37
REGINA DEL MERCATO, LA Vol. 7 49, 579, 831, 1119, 1120, 1183
REGINA DELLA NOTTE, LA Vol. 7 804
REGINA DELLA SCALA Vol. 7 822, 1177
REGINA DELLE AMAZZONI, LA Vol. 7 17, 1054
REGINA DELL'EMILIA: BOLOGNA, LA Vol. 7 1055
REGINA DI NAVARRA, LA Vol. 7 3, 50, 493, 792, 1000
REGINA DI NINIVE, LA Vol. 7 1189
REGINA DI SABA, LA Vol. 7 174, 798
REGINA JA MIEHET Vol. 5 434
REGINA MARGOT, LA Vol. 7 583
REGINE Vol. 4 89, 149, 660
REGINE, DIE TRAGÖDIE EINER FRAU Vol. 4 49, 316
REGINELLA Vol. 7 10
REGINETTA DELLE ROSE, LA Vol. 7 1002
REGINETTA DI GHIACCIO, LA Vol. 7 890
REGLE DU JEU, LA Vol. 2 18, 48, 251, 501, 526, 718, 719
REGLEMENTS DE COMPTE Vol. 2 259
REGNAR PA VAR KÄRLEK, DET Vol. 5 871, 1052, 1164
REGNEN HOLDT OP Vol. 5 2, 7
REGRESA UN DESCONOCIDO Vol. 8 470, 492
REGRESO DE CAPONE, EL Vol. 8 396
REGRESO DE LOS SIETE MAGNIFICOS Vol. 8 221
REI DAS BERLENGAS, O Vol. 8 35, 56
REIFE KIRSCHEN Vol. 1 8, 79, 92
REIFENDE JUGEND Vol. 4 373, 424, 461, 626
REIFENDER SOMMER Vol. 1 13, 47, 103
REIGEN, DER Vol. 4 131, 288
REINA DEL CHANTECLER, LA Vol. 8 214, 569
REINA DEL MATE, LA Vol. 8 219, 580
REINA DEL TABARIN Vol. 8 517, 772
REINA MORA, LA Vol. 8 522, 572, 595, 657
REINA SANTA Vol. 8 41, 72, 110, 214, 400
REINE DE BIARRITZ, LA Vol. 2 279, 533
REINE DES RESQUILLEUSES, LA Vol. 2 55, 332, 623
REINE MARGOT, LA Vol. 2 13, 194, 540, 709, 776
REINE SÜNDERIN, DIE Vol. 4 467
REISE IN DIE VERGANGENHEIT Vol. 4 435, 678, 740
REISE INS EHEBETT Vol. 1 34, 40, 95, 115
REISE INS GLÜCK, DIE Vol. 4 774
REISE NACH SUNDEVIT, DIE Vol. 1 14, 128
REISE NACH TILSIT, DIE Vol. 4 430, 437, 462
REISE UM DIE ERDE IN 80 TAGEN Vol. 4 159
REISEBEKANNTSCHAFT Vol. 4 85, 167, 367, 744
REISEN TIL HAVET Vol. 5 787
REISEN TIL JULESTJERNEN Vol. 5 694, 701, 772
REISEZIEL ERFURT Vol. 1 44, 86, 90, 112
REITER OHNE KOPF, DER Vol. 4 270, 471
REITER VON DEUTSCH-OSTAFRIKA, DIE Vol. 4 124, 422, 634
... REITET FÜR DEUTSCHLAND Vol. 4 298, 365, 485, 727

REIZENDE FAMILIE, EINE **Vol. 4** 51, 108, 631
REJS **Vol. 1** 282, 323 **Vol. 6** *page 225*
REJSEN TIL MAANEN **Vol. 5** 83
REJTÖZKÖDÖ, A **Vol. 9** 20, 152, 154
REKA **Vol. 10** 122, 524
REKA CARUJE **Vol. 10** 142, 154, 580, 612, 621
REKNEM SI TO PRISTI LETO **Vol. 10** 139, 516, 517
REKOLEKCJE **Vol. 1** 203, 207
REKOPIS ZNALEZIONY W SARAGOSSIE **Vol. 1** 187, 276, 320
REKRUT PETERSEN **Vol. 5** 180, 199, 201
REKTOR PA SENGEKANTEN **Vol. 5** 21, 175, 223, 261
REKVIEM ZA EDNA MR"SNICA **Vol. 6** *page 57, page 58*
REKVIEM ZA RYTIEROV **Vol. 10** 72, 79, 548
REKVIJEM **Vol. 3** 418, 427
RELACAO FIEL E VERDADEIRA **Vol. 8** 185
RELACION MATRIMONIAL Y OTRAS COSAS **Vol. 8** 492, 732
RELACIONES CASI PUBLICAS **Vol. 8** 400, 438
RELATO POLICIACO **Vol. 8** 405, 493
RELAXE-TOI CHERIE **Vol. 2** 347
RELEVO PARA UN PISTOLERO **Vol. 8** 495, 752
RELICARIO, EL **Vol. 8** 207, 214, 422
RELITTO, IL **Vol. 7** 923
REMBRANDT **Vol. 2** 656 **Vol. 4** 15, 574
REMENY JOGA, A **Vol. 9** 20, 339
REMENYKEDŐK **Vol. 9** 152, 259, 339
REMO E ROMOLO **Vol. 7** 728
REMONTONS LES CHAMPS-ELYSEES **Vol. 2** 48, 81, 158, 309, 530, 657, 722
REMORQUES **Vol. 2** 824, 837
REMOUS **Vol. 2** 402, 775
REMPART DES BEGUINES, LE **Vol. 2** 457, 873
REMPARTS D'ARGILE **Vol. 2** 873
REMPLISSEZ-MOI...LES TROIS TROUS **Vol. 2** 421
RENA **Vol. 1** 278, 404
RENA RAMAR SANNINGEN **Vol. 5** 840, 841
RENACER **Vol. 8** 691
RENATE IM QUARTETT **Vol. 4** 424, 774, 788
RENCONTRE A PARIS **Vol. 2** 197, 566
RENCONTRES **Vol. 2** 501, 663, 738
REND MIG I REVOLUTIONEN **Vol. 5** 16, 61, 272
REND MIG I TRADITIONERNE **Vol. 5** 16, 165, 294, 296
RENDEZ-MOI MA PEAU **Vol. 2** 39 **Vol. 6** *page 124*
RENDEZ-VOUS A BRAY **Vol. 2** 190, 674
RENDEZ-VOUS A GRENADE **Vol. 2** 128, 344
RENDEZ-VOUS A MELBOURNE **Vol. 2** 479, 514, 855
RENDEZ-VOUS A PARIS **Vol. 2** 698, 824
RENDEZ-VOUS AVEC LA CHANCE **Vol. 2** 247, 676
RENDEZ-VOUS CHAMPS-ELYSEES **Vol. 2** 273, 603, 871
RENDEZ-VOUS D'ANNA, LES **Vol. 2** 655
RENDEZ-VOUS DE JUILLET **Vol. 2** 330, 719
RENDEZ-VOUS DE MINUIT, LE **Vol. 2** 49, 285
RENDEZ-VOUS, LE **Vol. 2** 9
RENDEZ-VOUS ME MIAN AGNOSTE **Vol. 3** 151, 153
RENDEZVOUS MIT UNBEKANNT **Vol. 1** 76, 104
RENDITE DI ROBINET, LE **Vol. 7** 1189
RENDS-MOI LA CLE **Vol. 6** *page 92*
RENEGATE, LA **Vol. 2** 88, 514
RENGASMATKA **Vol. 5** 405, 579
RENNFIEBER **Vol. 4** 159
REPAS DE L'ELEPHANT **Vol. 2** 678
REPAS DES FAUVES, LE **Vol. 2** 279, 661
REPMANAD **Vol. 5** 952, 1024
REPOKER DE BRIBONES **Vol. 8** 243, 249
REPORTAGE 57 **Vol. 1** 43, 82, 125
REPORTAZ PSANA NA OPRATCE **Vol. 10** 153, 296, 385
REPORTER X **Vol. 8** 118, 185
REPORTEROS, LOS **Vol. 8** 212
REPOS DU GUERRIER, LE **Vol. 2** 21, 824
REPRIMIDO, EL **Vol. 8** 331
REPRODUCTION INTERDITE **Vol. 2** 9, 501
REPROUVES, LES **Vol. 2** 47, 125, 451, 566
REPTILICUS **Vol. 5** 104, 172, 199, 304
REPUBLICA DI MUSSOLINI – R.S.I., LA **Vol. 7** 242
REPUBLIK DER BACKFISCHE, DIE **Vol. 4** 149, 219, 339, 437, 660
REPUBLIKA VO PLAMEN **Vol. 3** 384, 396
REPUBLIKANEREN **Vol. 5** 9, 197
REPUBLIQUE EST MORTE A DIEN-BIEN-PHU, LA **Vol. 2** 640, 713
REQUIEM **Vol. 9** 134, 215, 316
REQUIEM FÜR HANS GRUNDIG **Vol. 1** 89
REQUIEM PARA EL GRINGO **Vol. 8** 602, 749
REQUIEM PER UN AGENTE SEGRETO **Vol. 8** 306, 772
REQUIEM POR UN CAMPESINO ESPANOL **Vol. 8** 250
REQUIEM POUR UN CAID **Vol. 2** 309, 347

REQUIEM POUR UN VAMPIRE **Vol. 2** 685
REQUIN, LE **Vol. 2** 153, 189, 573
REQUINS DE GIBRALTAR, LES **Vol. 2** 70, 173, 488
REQUINS DU PETROLE, LES **Vol. 2** 778
REQUISITOIRE, LE **Vol. 2** 803
RESA DI TITI, LA **Vol. 7** 1116
RESA I NATTEN **Vol. 5** 933, 1073
RESAN BORT **Vol. 5** 840, 878, 1173
RESAN TILL DEJ **Vol. 5** 954, 1073, 1164
RESEAU PARTICULIER **Vol. 6** *page 349*
RESEAU SECRET **Vol. 2** 814
RESERVEHELD, DER **Vol. 1** 26, 46, 85
RESERVESTATISTEN **Vol. 5** 83
RESERVIERT FÜR DEN TOD **Vol. 1** 7, 13, 115
RESIDENCIA, LA **Vol. 8** 276, 438
RESIDENCIA PARA ESPIAS **Vol. 8** 517, 725
RESISTANCE **Vol. 1** 104
RESPUESTA, LA **Vol. 8** 317, 361
RESTAURANT INTIM **Vol. 5** 840, 1091
RESTOS DEL NEUFRAGIO, LOS **Vol. 8** 611
RESURRECTIO **Vol. 7** 652, 756, 796
RESURRECTION **Vol. 2** 162, 798
RESURREZIONE **Vol. 7** 37, 360, 369, 910
RETALHOS DA VIDA DE UM MEDICO **Vol. 8** 131
RETEAUA S **Vol. 3** 245
RETENEZ-MOI ... OU JE FAIS UN MALHEUR **Vol. 6** *page 70, page 89, page 99*
RETEZ **Vol. 10** 51, 369, 516
RETFAERDIGES HUSTRU, DEN **Vol. 5** 242
RETORNO **Vol. 8** 721
RETORNO A LA VERDAD **Vol. 8** 401, 490, 772
RETORNO DE RINGO, EL **Vol. 8** 533, 732
RETORNO DE WALPURGIS, EL **Vol. 8** 237, 708
RETORNO DEL HOMBRE LOBO, EL **Vol. 8** 762, 770
RETORNO DI RINGO, IL **Vol. 8** 523
RETOUR A LA BIEN-AIMEE **Vol. 2** 517
RETOUR A LA VIE **Vol. 2** 15, 251, 402, 644, 866
RETOUR A L'AUBE **Vol. 2** 153, 676, 832
RETOUR AU BONHEUR **Vol. 2** 152, 538
RETOUR AU PARADIS **Vol. 2** 84, 775
RETOUR DE CHRISTOPHE COLON **Vol. 6** *page 114*
RETOUR DE DON CAMILLO, LE **Vol. 2** 140, 550
RETOUR DE FLAMME **Vol. 2** 48, 322
RETOUR DE MANIVELLE **Vol. 2** 128, 601, 832
RETOUR DE MARTIN GUERRE, LE **Vol. 6** *page 135*
RETOUR DES BIDASSES EN FOLIE, LE **Vol. 6** *page 73*
RETOUR DES VEUVES, LE **Vol. 2** 291
RETOUR DU GRAND BLOND, LE **Vol. 2** 564, 588
RETOUR DU PASSE, LE **Vol. 2** 798
RETOUR EN FORCE **Vol. 2** 450, 851
RETROSCENA **Vol. 7** 50, 756, 1177
RETTEN SEJRER **Vol. 5** 58
RETTENDE ENGEL, DER **Vol. 4** 25, 647, 787
RETTER DER MENSCHHEIT **Vol. 4** 495, 693
RETTILE DELLA METROPOLITAN BANK **Vol. 7** 732
RETTILI DELLA MINIERA, I **Vol. 7** 726
RETURN OF THE SEVEN **Vol. 8** 221
REVAILLE, DAS GROSSE WECKEN **Vol. 4** 377, 424
REVANCHE DE ROGER LA HONTE, LA **Vol. 2** 197, 226, 824
REVANCHE, LA **Vol. 6** *page 78, page 141*
REVANSA **Vol. 3** 211, 218, 256, 292
REVE, LE **Vol. 2** 167, 227, 346, 388
REVE PLUS LONG QUE LA NUIT, UN **Vol. 2** 879
REVEIL DE MADDALONE, LE **Vol. 2** 145
REVEILLE-TOI, CHERIE **Vol. 2** 348, 372
REVELACION **Vol. 8** 268, 298, 548
REVELATEUR, LE **Vol. 2** 312
REVELATION, LA **Vol. 2** 72
REVELY **Vol. 5** 944, 976
REVENANT, UN **Vol. 2** 558, 644 **Vol. 32** 558, 644
REVENANTE, LA **Vol. 2** 169, 330, 613
REVES D'AMOUR **Vol. 2** 388, 488, 571, 756, 837
REVES DE CLOWN **Vol. 2** 37, 712
REVISOR **Vol. 10** 256, 524
REVOLTE IM ERZIEHUNGSHAUS **Vol. 4** 14, 382, 492
REVOLTE, LE **Vol. 2** 57, 338, 339, 807
REVOLTEE, LE **Vol. 2** 388, 566
REVOLTES DE LOHMANACH, LES **Vol. 2** 248, 496, 676
REVOLTES DU DANAE, LES **Vol. 2** 562, 871
REVOLTOSA, LA **Vol. 8** 214, 228, 336, 396, 400, 523,
REVOLUCION MATRIMONIAL, LA **Vol. 8** 261, 338
REVOLUCNI ROK 1848 **Vol. 10** 142, 524, 569, 621
REVOLUTION DER JUGEND **Vol. 6** *page 198*
REVOLUTION D'OCTOBRE, LA **Vol. 2** 60

REVOLUTIONEN I VANDKANTEN **Vol. 5** 305
REVOLUTIONENS DATTER **Vol. 5** 711
REVOLUTIONSBRYLLUP, ET **Vol. 5** 9, 97
REVOLUTIONSHOCHZEIT **Vol. 4** 306, 313, 361, 424, 605
REVOLVER **Vol. 7** 1045
REVOLVER DES CORPORALS, DER **Vol. 1** 17, 22, 75
REVUE NA ZAKAZKU **Vol. 10** 144, 237, 325
REVUE UM MITTERNACHT **Vol. 1** 36, 56, 115
REWIZJA OSOBISTA **Vol. 1** 180, 217, 238 **Vol. 6** *page 236, page 257*
REY BALTASAR, EL **Vol. 8** 503, 611
REY DE AFRICA **Vol. 8** 598
REY DE LA CARRETERA, EL **Vol. 8** 280, 306
REY DE LAS FINANZAS, EL **Vol. 2** 775 **Vol. 8** 400
REY QUE RABIO, EL **Vol. 8** 516
REZA POR TU ALMA ... Y MUERRE **Vol. 8** 331
RHEINISCHE BRAUTFAHRT **Vol. 4** 340, 703
RHEINISCHES MÄDCHEN BEIM RHEINISCHEN WEIN, EIN **Vol. 4** 584, 646
RHEINLAND, DAS **Vol. 4** 270, 691
RIASZTOLÖVES **Vol. 9** 202, 255, 339
RIAUX, LES **Vol. 2** 46
RIAVANTI ... MARSCH ! **Vol. 7** 1003
RIBALTA, LA **Vol. 7** 1189
RIBATEJO **Vol. 8** 25, 146, 172, 184
RICAR BEZ BRONJA **Vol. 3** 59, 114 **Vol. 6** *page 31*
RICATTO DI UN PADRE, IL **Vol. 7** 587
RICCHEZZA SENZA DOMANI **Vol. 7** 493, 1030
RICCHI, RICCHISSIMI, PRATICAMENTE IN MUTANDE **Vol. 7** 116, 426, 1202
RICCHIAMO DEL CUORE, IL **Vol. 7** 978
RICCHIAMO DEL GHIACCIAIO, IL **Vol. 7** 475
RICCHIAMO, IL **Vol. 7** 301
RICHARD WAGNER **Vol. 4** 183
RICHEST GIRL, THE **Vol. 2** 579
RICHIAMO DEL LUPO, IL **Vol. 8** 208, 749
RICHTER VON ZALAMEA, DER **Vol. 1** 35, 72, 81 **Vol. 4** 444, 734, 747
RICHTHOFEN, DER ROTE RITTER DER LUFT **Vol. 4** 565, 603, 643
RICOMINCIO DA TRE **Vol. 7** 834, 882
RICORDAMI **Vol. 7** 1152
RICORDATI DI NAPOLI **Vol. 7** 1121
RICOTTA, LA **Vol. 7** 388
RID I NATT ! **Vol. 5** 840, 909, 1173
RIDDERSEN AF DEN BEDROVELIGE SKIKKELSE **Vol. 5** 83
RIDE BENE CHI RIDE ULTIMO **Vol. 7** 1114
RIDEAU ROUGE, LE **Vol. 2** 57, 544
RIDENDO E SCHERZANDO **Vol. 7** 347
RIDERE, RIDERE, RIDERE **Vol. 7** 373
RIDETI CA-N VIATA **Vol. 6** *page 268, page 276, page 277*
RIDI PAGLIACCIO ! **Vol. 7** 460, 910, 1096
RIDICOLO, IL **Vol. 7** 442
RIEN NE VA PLUS **Vol. 2** 151, 737, 879
RIEN QUE LA VERITE **Vol. 2** 579, 722
RIEN QUE LES HEURES **Vol. 2** 745
RIEN QUE PAR DERRIERE **Vol. 2** 10
RIFFLE BILL, LE ROI DE LA PRAIRIE **Vol. 2** 711
RIFFPIRATEN **Vol. 4** 309, 380
RIFIFI EN AMSTERDAM **Vol. 8** 495, 541
RIFIFI EN LA CIUDAD **Vol. 8** 600, 772
RIGE ENKE, DEN **Vol. 5** 90, 207
RIGENERATO **Vol. 7** 86
RIGOLBOCHE **Vol. 2** 344, 362, 533
RIGOLETTO **Vol. 7** 1189
RIGOLETTO E LA SUA TRAGEDIA **Vol. 7** 17
RIIHALAN VALTIAS **Vol. 5** 372, 397
RIKARD NORDRAAK **Vol. 5** 676, 696, 729, 781
RIKAS TYTTÖ **Vol. 5** 386, 430
RIKKI OG MAENDENE **Vol. 5** 89, 281
RIKOLLINEN NAINEN **Vol. 5** 374, 457
RIKOS JA RANGAISTUS **Vol. 5** 404, 591, 644
RIMORSO **Vol. 7** 603, 792
RIMSKE DEVICE **Vol. 3** 349
RINA **Vol. 10** 53, 453
RINA, L'ANGELO DELLE ALPI **Vol. 7** 209
RINALDO RINALDINI **Vol. 4** 239, 374, 384, 449 **Vol. 7** 606
RINCON PARA QUERERNOS, UN **Vol. 8** 629, 768
RINCONCITO MADRILENO **Vol. 8** 329
RING DER BAJADERE, DER **Vol. 4** 746
RING DES SCHICKSALS, DER **Vol. 4** 366
RINGARD, LE **Vol. 2** 721
RINGARDS, LES **Vol. 2** 275, 568
RINGENDE SEELEN **Vol. 4** 300
RINGO DE NEBRASKA **Vol. 8** 495
RINGUL **Vol. 6** *page 270, page 281, page 283*
RINTAMALOTTA **Vol. 5** 462, 634
RINUNZIA **Vol. 7** 168
RIO ABAJO **Vol. 8** 358, 746
RIO DE ORO, EL **Vol. 8** 434, 738
RIO GUADALQUIVIR **Vol. 8** 546, 749
RIO MALDITO **Vol. 8** 629, 768
RION YÖ **Vol. 5** 549, 592

RIPA DRACULUI Vol. 3 224, 231, 299, 330
RIPACSOK Vol. 9 10, 51, 253
RIPETENTE FA L'OCCHIETTO AL PRESIDE, LA Vol. 7 226, 1202
RIPOSO FESTIVO Vol. 7 1189
RIPUDIATA Vol. 7 71
RIRES DE PARIS Vol. 2 67, 131
RIRIKA Vol. 3 146
RISA E LACRIME NAPOLETANE Vol. 7 493
RISAIA, LA Vol. 7 294, 667, 1147
RISATE DI GIOIA Vol. 7 86
RISCATTO, IL Vol. 7 17, 277
RISCOSSA DELLE MASCHERE, LA Vol. 7 1116
RISIKO Vol. 1 2, 44, 59
RISO AMARO Vol. 7 401, 725
RISQUES DU METIER, LES Vol. 2 128
RISTI JA LIEKKI Vol. 5 328, 356, 481, 566
RISTIKON VARJOSSA Vol. 5 356, 566
RISTO VANARIN PIILOKAMERA Vol. 5 408, 424
RITA Vol. 8 117, 193
RITA DA CASCIA Vol. 7 443, 701, 938
RITAM ZLOCINA Vol. 6 *page 307, page 310*
RITORNANO QUELLI DELLI CALIBRO 38 Vol. 7 909
RITORNO Vol. 7 902
RITORNO AL NIDO Vol. 7 1054
RITORNO ALLA TERRA Vol. 7 369, 685
RITORNO ALL'ONDA Vol. 7 829
RITORNO DAI BAGNI, IL Vol. 7 1189
RITORNO DALL'OMBRA Vol. 7 53
RITORNO DI DON CAMILLO, IL Vol. 7 196, 709
RITORNO DI RINGO, IL Vol. 8 523, 732
RITRATTO DI BORGHESIA IN NERO Vol. 7 817
RITROVARSI ALL'ALBA Vol. 7 302
RITT IN DIE FREIHEIT Vol. 4 30, 265, 567, 614
RITT IN DIE SONNE Vol. 4 321, 546
RITTER BLAUBART Vol. 1 68, 74, 95, 100
RITTER DER NACHT Vol. 4 73, 186
RITTER WIDER WILLEN Vol. 4 329, 774
RIUL CARE URCA MUNTELE Vol. 3 218, 245, 281
RIVA DEI BRUTO, LA Vol. 7 978
RIVA DELLE TRE GIUNCHE, LA Vol. 7 983
RIVALE DI SHERLOCK HOLMES, UN Vol. 7 1189
RIVALE, IL Vol. 7 399, 914
RIVALE, LA Vol. 2 860 Vol. 7 257
RIVALEN Vol. 4 443, 471
RIVALEN AM STEUER Vol. 1 25, 73, 88
RIVALEN DER LUFT Vol. 4 115, 620
RIVALEN IM WELTREKORD Vol. 4 84, 452
RIVALER Vol. 5 97
RIVAUX D'ARNHEIM, LES Vol. 2 237
RIVE GAUCHE Vol. 2 579, 646, 803
RIVELAZIONI DI UN MANIACO SESSUALE AL CAPO DELLA SQUADRA MOBILE Vol. 7 996
RIVER, THE Vol. 2 526, 719
RIVIERE DES TROIS JONQUES, LA Vol. 2 124, 376, 740 Vol. 7 982
RIVINCITA DEL DOTTORE, LA Vol. 7 1189
RIVINCITA DEL PASSATO, LA Vol. 7 298
RIVINCITA DI IVANHOE, LA Vol. 7 501
RIVINCITA DI MACISTE, LA Vol. 7 878
RIVINCITA, LA Vol. 7 81, 304
RIVOLTA DEGLI SCHIAVI, LA Vol. 8 438, 611
RIVOLTA DEI GLADIATORI, LA Vol. 8 602
RIVOLTA DEI PRETORIANI, LA Vol. 7 877
RIVOLUZIONE DEI PESCIANI, LA Vol. 7 940
RIZES TOU TOPOU MAS Vol. 3 145
RO HESTE VINDER LOBET, DE Vol. 5 76
„ROAH-ROAH" DER SCHREI DER SEHNSUCHT Vol. 6 *page 184*
ROBE NOIRE POUR UN TUEUR, UNE Vol. 6 *page 83*
ROBE ROUGE, LA Vol. 2 727
ROBERT ET ROBERT Vol. 2 491, 616
ROBERT I BERTRAND Vol. 1 241, 278, 308
ROBERT KOCH, DER BEKÄMPFER DES TODES Vol. 4 416, 729, 740
ROBERT MAYER – DER ARZT AUS HEILBRONN Vol. 1 3, 56, 105
ROBERT UND BERTRAM Vol. 4 49, 204, 287, 423, 424, 458, 646, 788 Vol. 6 *page 184*
ROBERTE Vol. 2 94, 114
ROBERTO BURAT Vol. 7 454
ROBERTO EL DIABLO Vol. 8 546, 749
ROBIN HOOD E I PIRATI Vol. 7 739
ROBIN HOOD, EL ARQUERO INVENCIBLE Vol. 8 749
ROBIN HOOD NUNCA MUERE Vol. 8 534, 641
ROBINET AMA LA FIGLIA DEL GENERALE Vol. 7 1189
ROBINET AMA LA FIORAIA Vol. 7 1189
ROBINET ANARCHICO Vol. 7 1189
ROBINET ATTACCATO ALLA SELLA Vol. 7 1189
ROBINET AVIATORE Vol. 7 1189
ROBINET BOXEUR Vol. 7 1189

ROBINET CARICATURISTA Vol. 7 1189
ROBINET COCCHIERE Vol. 7 1189
ROBINET CONTRO ROBINET Vol. 7 1189
ROBINET CORTEGGIATORE Vol. 7 1189
ROBINET COSTRETTO A FARE IL LADRO Vol. 7 1189
ROBINET E BOUTALIN SI BATTONO Vol. 7 1189
ROBINET E IL BIGLIETTO DA CENTO LIRE Vol. 7 1189
ROBINET FINTO COW-BOY Vol. 7 1189
ROBINET GUARDIA CICLISTA Vol. 7 1189
ROBINET GUIDA PER AMORE Vol. 7 1189
ROBINET HA IL SONNO DURO Vol. 7 1189
ROBINET HA LA MANIA DEL SALVADANAIO Vol. 7 1189
ROBINET HA PAURA DEI MICROBI Vol. 7 1189
ROBINET HA PERSO IL TRENO Vol. 7 1189
ROBINET HA SONNO Vol. 7 1189
ROBINET HA UN ALLIEVO Vol. 7 1189
ROBINET IN BURLETTA Vol. 7 1189
ROBINET IN RITARDO Vol. 7 1189
ROBINET IN UN EDUCANDATO Vol. 7 1189
ROBINET IN VACANZA Vol. 7 1189
ROBINET INNAMORATO DI UNA CHANTEUSE Vol. 7 1189
ROBINET JOCKEY Vol. 7 1189
ROBINET OPERATORE Vol. 7 1189
ROBINET POLIZIOTTO Vol. 7 1189
ROBINET QUESTURINO Vol. 7 1189
ROBINET REPORTER Vol. 7 1189
ROBINET RICCO Vol. 7 1189
ROBINET SCIOPERANTE Vol. 7 1189
ROBINET SI ALLENA PER IL GIRO D'ITALIA Vol. 7 1189
ROBINET SI DA'ALLA MALA VITA Vol. 7 1189
ROBINET SI DEDICA AGLI SPORTS Vol. 7 1189
ROBINET SOGNO AL MARE Vol. 7 1189
ROBINET SOLDATO ALPINO Vol. 7 1189
ROBINET SPOSA A VAPORE Vol. 7 1189
ROBINET STUDENTE Vol. 7 1189
ROBINET, SUA MOGLIE ET IL CUGINO Vol. 7 1189
ROBINET TENORE Vol. 7 1189
ROBINET TIMIDO Vol. 7 1189
ROBINET TROPPO AMATO Vol. 7 1189
ROBINET VUOL LAVORARE Vol. 7 1189
ROBINET VUOL PIANTARE UN CHIODO Vol. 7 1189
ROBINETTE FEMINISTA Vol. 7 1189
ROBINSON CRUSOE Vol. 5 97 Vol. 7 838
ROBINSON, EIN Vol. 4 10, 54, 135, 154, 660
ROBINSON ET LE TRIPORTEUR Vol. 2 11, 554
ROBINSON I ROSLAGEN Vol. 5 933, 1156
ROBINSON I SKÄRGARDEN Vol. 5 976, 1110
ROBINSONKA Vol. 10 345, 394, 408, 425, 460, 562
ROBO DE DIAMANTES Vol. 8 493, 595
ROBOT GIRL NR. 1 Vol. 10 256, 412
ROCAMBOLE Vol. 2 153, 615, 708, 838 Vol. 7 980, 1054
ROCCIA DELLA MORTE, LA Vol. 7 1067
ROCCIA INCANTATA, LA Vol. 7 1063
ROCCO E I SUOI FRATELLI Vol. 7 498, 988, 1142
ROCHE AUX MOUETTES, LA Vol. 2 131, 404
ROCIO DE ALBAICIN Vol. 8 723
ROCIO DE LA MANCHA Vol. 8 214, 261
ROCKING SILVER Vol. 5 25, 48, 175
ROCK'N ROLL Vol. 7 310
ROCK'N TORAH Vol. 6 *page 71, page 140*
ROCKY CARAMBOLA Vol. 8 438, 681
ROCKY ROAD TO DULLIN Vol. 2 208
ROCNIK JEDENADVACET Vol. 10 229, 313, 374
RÖDA DAGEN Vol. 5 891, 909
RÖDA HÄSTARNA, DE Vol. 5 834
RÖDA KAPPAN, DEN Vol. 5 916, 1052, 1153
RODBLATT PARADIS Vol. 5 718, 808
RODE ALFABET, DET Vol. 5 67
RODE ENGE, DE Vol. 5 89, 224
RODE HESTE, DE Vol. 5 184, 248
RODE HESTE VINDER LOBET, DE Vol. 5 1, 234
RODE KAPPE, DEN Vol. 5 23, 200, 273
RODE RUBIN, DEN Vol. 5 199, 223, 306
RODE RUSLAND, DET Vol. 5 203
RODELKAVALIER, DER Vol. 4 560
RODEO Vol. 10 250, 395, 517
RODINNE TRAMPOTY OFICIALA TRISKY Vol. 10 460, 621
RODNA ZEM Vol. 10 273, 287, 501, 621
RODOLFI APACHE Vol. 7 1189
RODRIGUEZ AU PAYS DES MERGUEZ Vol. 2 76, 289
RODZINA Vol. 1 262
RODZINA LESNIEWSKICH Vol. 6 *page 206, page 237, page 247, page 254*
RODZINA MILCARKOW Vol. 1 198, 303, 381
RÖG, A Vol. 9 239
ROG BRZESKIEJ I CAPRI Vol. 6 *page 207, page 225*
ROGELIA Vol. 8 214, 484

ROGER-LA-HONTE **Vol. 2** 131, 346, 570, 603, 604, 722, 835
ROGOPAG **Vol. 2** 700
ROGOPAG – IL POLLO RUSPANTE **Vol. 7** 128, 781
ROGOPAG – ILLIBATEZZA **Vol. 7** 333, 781, 1147
ROGOPAG – LA RICOTTA **Vol. 7** 333, 388, 781
ROHANY VELEM ! **Vol. 9** 134, 215, 229
ROHY **Vol. 10** 210, 268, 618
ROI DE CAMARGUE **Vol. 2** 187, 833
ROI DE COEUR, LE **Vol. 2** 457, 517
ROI DE LA MONTAGNE, LE **Vol. 2** 798
ROI DE PARIS, LE **Vol. 2** 206, 217
ROI DES AULNES, LE **Vol. 2** 65, 672
ROI DES BRICOLEURS, LE **Vol. 2** 866
ROI DES CAMELOTS, LE **Vol. 2** 460, 627
ROI DES CHAMPS-ELYSEES, LE **Vol. 2** 41, 488
ROI DES CONS, LE **Vol. 6** *page 86, page 91*
ROI DES PALACES, LE **Vol. 2** 658, 676
ROI DES RESQUILLEURS, LE **Vol. 2** 55, 122, 193, 197, 330, 533
ROI DU BLA-BLA-BLA, LE **Vol. 2** 538, 661
ROI DU CAMEMBERT, LE **Vol. 2** 563
ROI DU CIRAGE, LE **Vol. 2** 48, 197
ROI DU VILLAGE, LE **Vol. 2** 259
ROI, LE **Vol. 2** 197, 226, 309, 332, 440, 488
ROI PANDORE, LE **Vol. 2** 627, 807
ROI SANS DIVERTISSEMENT, UN **Vol. 2** 19, 49
ROINILAN TALOSSA **Vol. 5** 356, 428
ROIS BIS, LE **Vol. 2** 397
ROIS DE LA FLOTTE, LES **Vol. 2** 6, 366, 719
ROIS DU SPORT, LES **Vol. 2** 32, 197, 423
ROJ **Vol. 3** 356, 391
ROJAL"T **Vol. 3** 77 **Vol. 6** *page 29, page 34*
ROJO Y NEGRO **Vol. 8** 400, 625, 674
ROK 1863 **Vol. 1** 404
ROK 1914 **Vol. 1** 234, 364, 372, 384
ROK NA DEDINE **Vol. 10** 32, 273
ROK PIERWSZY **Vol. 1** 227, 354 **Vol. 6** *page 205*
ROK SPOKOJNEGO SLONCA **Vol. 6** *page 206, page 239, page 253*
ROKKI DIGGARI **Vol. 5** 326, 350, 464
ROKONOK **Vol. 9** 71, 222, 312
ROLE EFFACE DE MARIE, LE **Vol. 2** 109
ROLF **Vol. 7** 311
ROLF HAT EIN GEHEIMNIS **Vol. 4** 410, 558
ROLLENDE HOTEL, DAS **Vol. 4** 423
ROLLENDE KORRIDOR, DER **Vol. 4** 583
ROLLENDE KUGEL, DIE **Vol. 4** 215, 219, 681, 749
ROLLENDE SCHICKSAL, DAS **Vol. 4** 255, 344
ROLLERBOY **Vol. 7** 406
ROLLO DE SEPTIEMBRE, EL **Vol. 8** 376, 728
ROLLS PARA HIPOLITO, UN **Vol. 8** 667, 728
ROMA **Vol. 7** 388, 664, 988
ROMA A MANO ARMATA **Vol. 7** 1202
ROMA CITTA APERTA **Vol. 7** 48, 757
ROMA, CITTA LIBERA **Vol. 7** 756, 1136
ROMA DROGATA: LA POLIZIA NON PUO INTERVENIRE **Vol. 7** 1024
ROMA, L'ALTRA FACCIA DELLA VIOLENZA **Vol. 7** 124
ROMA – L'ANTICA CHIAVE DEI SENSI **Vol. 7** 414, 691
ROMA O MORTE 1867 – 1870 **Vol. 7** 782
ROMA ORE 11 **Vol. 2** 59 **Vol. 7** 725
ROMA VIOLENTA **Vol. 7** 1213
ROMAN BOXERA **Vol. 10** 243, 310
ROMAN DE WERTHER, LE **Vol. 2** 122, 526, 688, 778, 802
ROMAN DER CHRISTINE VON HERRE **Vol. 4** 38, 178, 648
ROMAN DER LILIAN HAWLEY, DER **Vol. 4** 479, 634
ROMAN D'UN JEUNE HOMME PAUVRE, LE **Vol. 2** 388, 402
ROMAN D'UN MOUSSE, LE **Vol. 2** 798
ROMAN D'UN SPAHI, LE **Vol. 2** 279, 407, 824
ROMAN D'UN TRICHEUR, LE **Vol. 2** 386, 533, 579, 580
ROMAN EINER HALBWELTDAME, DER **Vol. 4** 93, 125, 367
ROMAN EINER JUNGEN EHE **Vol. 1** 23, 56, 90
ROMAN EINER NACHT **Vol. 4** 344, 647
ROMAN EINES ARZTES **Vol. 4** 292, 517, 558
ROMAN EINES DIENSTMÄDCHENS, DER **Vol. 4** 424, 663
ROMAN HLOUPEHO HONZY **Vol. 10** 53, 138, 510
ROMAN HOLIDAY **Vol. 2** 13
ROMAN I MAGDA **Vol. 1** 264, 306
ROMANA, LA **Vol. 7** 781, 1062
ROMANCE **Vol. 2** 739
ROMANCE A L'INCONNUE **Vol. 2** 113, 731
ROMANCE A TROIS **Vol. 2** 32, 439
ROMANCE DE PARIS **Vol. 2** 539, 566
ROMANCE DO LUACHIMO, O **Vol. 8** 122
ROMANCE PRO KRIDLOVKU **Vol. 10** 18, 289, 503

ROMANCE ZA KORUNU **Vol. 10** 139, 233, 496
ROMANETO **Vol. 10** 283, 344, 571
ROMANETTI **Vol. 2** 48, 367
ROMANIN **Vol. 7** 732
ROMANS **Vol. 5** 1032, 1181
ROMANS PANNY OPOLSKIEJ **Vol. 1** 174, 407
ROMANS TERESY HENNERT **Vol. 1** 207, 294, 333, 354
ROMANS Z INTRUZEM **Vol. 6** *page 225, page 242, page 250*
ROMANTICA AVVENTURA, UNA **Vol. 7** 493, 756, 1060
ROMANTICISMO **Vol. 7** 420
ROMANTICNA ISTORIJA **Vol. 6** *page 40, page 61*
ROMANTIK PA SENGEKANTEN **Vol. 5** 247, 304
ROMANTIKA **Vol. 9** 20, 154, 318
ROMANTISCHE BRAUTFAHRT **Vol. 4** 85, 714
ROMANTYCZNI **Vol. 1** 141, 237, 385, 391, 403 **Vol. 6** *page 215, page 242*
ROMANZA DEI RICORDI, LA **Vol. 7** 731
ROMANZA FINAL **Vol. 8** 299, 762
ROMANZE IN MOLL **Vol. 4** 97, 149, 188
ROMANZE MIT AMELIE **Vol. 6** *page 159, page 169, page 174*
ROMANZO A PASSO DI DANZA **Vol. 7** 945, 1014, 1062
ROMANZO D'AMORE **Vol. 7** 923
ROMANZO DEL DIAVOLI **Vol. 7** 668
ROMANZO DELLA MIA VITA, IL **Vol. 7** 501
ROMANZO DI FABIENNE, IL **Vol. 7** 131
ROMANZO DI UN ATLETA, IL **Vol. 7** 420
ROMANZO DI UN CHIRURGO, IL **Vol. 7** 968
ROMANZO DI UN CUORE, IL **Vol. 7** 1189
ROMANZO DI UN DERELITTO **Vol. 7** 837
ROMANZO DI UN GIOVANE POVERO, IL **Vol. 7** 369, 853, 1027, 1054, 1060, 1140
ROMANZO DI UNA GIOVANE POVERA, IL **Vol. 7** 730
ROMANZO DI UNA VESPA, IL **Vol. 7** 266
ROMANZO NERO E ROSA, IL **Vol. 7** 48, 520
ROME-EXPRESS **Vol. 2** 401
ROMEO I JULCIA **Vol. 1** 278, 380
ROMEO, JULIE A TMA **Vol. 10** 154, 202, 503
ROMEO UND JULIA **Vol. 1** 1, 73
ROMEO UND JULIA AUF DEM DORFE **Vol. 6** *page 157, page 163*
ROMEO UND JULIA IM SCHNEE **Vol. 4** 663
ROMERSKE MODEL, DEN **Vol. 5** 97
ROMLOTT EMBEREK KÖZÖTT **Vol. 9** 136
ROMOLA **Vol. 7** 978
RONDA ESPANOLA **Vol. 2** 646 **Vol. 8** 607, 725
RONDE DE NUIT **Vol. 2** 401 **Vol. 6** *page 82, page 92*
RONDE DE NUIT, LA **Vol. 2** 547
RONDE DES HEURES, LA **Vol. 2** 153, 279, 538, 871
RONDE, LA **Vol. 2** 27, 224, 279, 457, 557, 566
RONDINE **Vol. 7** 569
RONDINE, LA **Vol. 7** 606
RONDINI IN VOLO **Vol. 7** 792, 923
RONDINI NEL TURBINE **Vol. 7** 212
RONDO **Vol. 3** 421, 437
RONGYOSOK **Vol. 9** 26, 308
RONJA RÖVARDOTTER **Vol. 5** 827, 852, 933
RONNY **Vol. 4** 17, 28, 614, 729
ROQUEVILLARD, LES **Vol. 2** 488, 698, 786
ROSA AL VIENTO, UNA **Vol. 8** 639, 732
ROSA DE ALFAMA **Vol. 8** 44, 54, 130
ROSA DE LIMA **Vol. 8** 583, 772
ROSA DE LOS VIENTOS, LA **Vol. 8** 412, 530, 712
ROSA DE MADRID **Vol. 8** 676
ROSA DI FORTUNIO, LA **Vol. 7** 301
ROSA DI SANGUE **Vol. 2** 29 **Vol. 7** 48, 491, 701, 838
ROSA DIAMANT, DER **Vol. 4** 50, 142 **Vol. 6** *page 199*
ROSA, LA **Vol. 7** 529
ROSA MISTICA **Vol. 7** 731
ROSA PANTÖFFELCHEN, DAS **Vol. 4** 36, 366
ROSA ROJA, LA **Vol. 8** 551
ROSA ROSSA, LA **Vol. 7** 738
ROSA TRIKOT, DAS **Vol. 4** 223, 560
ROSAIRE, LE **Vol. 2** 48
ROSALBA **Vol. 7** 890, 1045
ROSALBA LA FANCIULLA DI POMPEI **Vol. 7** 501
ROSARIO DELLA COLPA, IL **Vol. 7** 196
ROSARIO LA CORTIJERA **Vol. 8** 572, 648, 657
ROSAROTE BRILLE, DIE **Vol. 4** 373
ROSAS DE OTONO **Vol. 8** 214, 357, 436
ROSAS Y ESPINAS **Vol. 8** 460
ROSCOVANUL **Vol. 3** 256, 281, 334
ROSE **Vol. 2** 833
ROSE BERND **Vol. 4** 37, 190, 773
ROSE DE FER, LA **Vol. 2** 721
ROSE DE LA MER, LA **Vol. 2** 128, 407
ROSE DEL MIRACOLO, LE **Vol. 7** 1004, 1103
ROSE DI DANZICA, LE **Vol. 7** 46, 132, 784
ROSE DI PASSIONE **Vol. 7** 250, 929

ROSE DI SANGUE **Vol. 7** 621
ROSE ECORCHEE, LA **Vol. 2** 294
ROSE EFFEUILLEE, LA **Vol. 2** 864
ROSE ET LE BLANC, LE **Vol. 6** *page 81, page 88, page 99, page 142*
ROSE ET LE RESEDA, LA **Vol. 2** 57
ROSE FATALI **Vol. 7** 569
ROSE NOIRE POUR UN TUEUR, UNE **Vol. 2** 779
ROSE ROUGE, LA **Vol. 2** 47, 194 **Vol. 7** 63
ROSE SCARLATTE **Vol. 7** 595, 756
ROSE VON DSCHIANDUR, DIE **Vol. 4** 565
ROSE VON STAMBUL, DIE **Vol. 4** 529, 681
ROSEANNA **Vol. 5** 881, 1017, 1080
ROSEMARIE WILL NICHT MEHR LÜGEN **Vol. 4** 517, 525, 558
ROSEN **Vol. 5** 97, 926, 932, 1051
ROSEN AUS DEM SÜDEN **Vol. 4** 123, 213, 626, 774, 788
ROSEN BLÜH'N AUF DEM HEIDEGRAB **Vol. 4** 535, 565 **Vol. 6** *page 185*
ROSEN, DIE DER STURM ENTBLÄTTERT **Vol. 4** 309
ROSEN IN TIROL **Vol. 4** 266
ROSEN PA TISTELÖN **Vol. 5** 1042, 1164
ROSEN UND LIEBE **Vol. 4** 323, 329
ROSENBORG HAVE **Vol. 5** 97
ROSENMONTAG **Vol. 4** 89, 266, 321, 574, 734
ROSER TIL MONICA **Vol. 5** 738
ROSES ROUGES ET PIMENTS VERTS **Vol. 2** 427
ROSICA **Vol. 3** 90
ROSIER DE MADAME HUSSON, LE **Vol. 2** 100, 347, 657, 807
ROSIERE DE PESSAC, LA **Vol. 2** 818
ROSIERE DES HALLES, LA **Vol. 2** 55, 332, 646
ROSINA FUMO VIENE IN CITTA PER FARSI IL CORREDO **Vol. 7** 1044
ROSITA **Vol. 4** 189
ROSNICKA **Vol. 10** 263, 269, 493
ROSOLINO PATERNO, SOLDATO **Vol. 7** 333
ROSOR VARJE KVÄLL **Vol. 5** 899, 979, 1010
ROSPIGGAR **Vol. 5** 928, 1042
ROSSA, LA **Vol. 7** 603
ROSSETTO, IL **Vol. 7** 73, 877
ROSSINI **Vol. 7** 16, 443, 830
ROSSO SEGNO DELLA FOLLIA, IL **Vol. 7** 975 **Vol. 8** 464
ROSSZEMBEREK **Vol. 9** 109, 152, 191
ROSTRO AL MAR **Vol. 8** 537, 614, 753
ROSTRO DEL ASESINO, EL **Vol. 8** 562, 772
ROSVO ROOPE **Vol. 5** 430, 481, 634
RÖTÄGG **Vol. 5** 907, 923
ROTAIE **Vol. 7** 48, 219, 238, 361, 1140
ROTATION **Vol. 1** 77, 101
ROTE FADEN, DER **Vol. 4** 247, 329
ROTE FALTER, DER **Vol. 6** *page 186*
ROTE GEIGE, DIE **Vol. 1** 79
ROTE HENKER, DER **Vol. 4** 223, 773
ROTE HEXE, DIE **Vol. 4** 240, 424
ROTE JULE, DIE **Vol. 4** 247
ROTE KAMILLE, DIE **Vol. 1** 7, 90
ROTE KREIS, DER **Vol. 4** 124, 186, 586
ROTE MARIANNE, DIE **Vol. 4** 98, 112
ROTE MAUS, DIE **Vol. 4** 407, 734
ROTE MÜHLE **Vol. 4** 422, 588, 774
ROTE MÜHLE, DIE **Vol. 4** 212
ROTE ORCHIDEEN **Vol. 4** 344, 738, 788
ROTE PLAKAT, DAS **Vol. 4** 232, 724
ROTE REDOUTE, DIE **Vol. 4** 71, 479
ROTE REITER, DER **Vol. 1** 45, 53 **Vol. 4** 27, 123, 323, 673, 674, 788
ROTE ROSEN – BLAUE ADRIA **Vol. 4** 582
ROTE SARAFAN, DER **Vol. 4** 159
ROTE SCHATTEN, DER **Vol. 4** 25, 645
ROTE STREIFEN, DER **Vol. 4** 186, 309
ROTFUCHS **Vol. 1** 22, 26, 71
ROTHAUSGASSE, DIE **Vol. 4** 340, 527, 608
ROTHSCHILD **Vol. 2** 332, 648, 744
ROTHSCHILDS, DIE **Vol. 4** 28, 123, 370
ROTKÄPPCHEN **Vol. 1** 8, 21, 107
RÖTMANAD **Vol. 5** 975, 1004, 1052
ROTSCHILD LEANY **Vol. 9** 28
ROTSCHLIPSE **Vol. 1** 29, 127
ROTTASOTA **Vol. 5** 452
ROTTENKNECHTE **Vol. 1** 7, 49, 50, 73
ROUE, LA **Vol. 2** 9, 152, 153, 277, 419, 559, 661
ROUES FOLLES **Vol. 2** 703
ROUGE EST MIS, LE **Vol. 2** 185, 644, 801
ROUGE ET LE NOIR, LE **Vol. 2** 251, 427, 625 **Vol. 7** 168
ROULE MOI UN PATIN, SALOPE **Vol. 6** *page 112*
ROULETABILLE AVIATEUR **Vol. 2** 122, 676
ROULETABILLE CHEZ LES BOHEMIENS **Vol. 2** 452, 764, 871
ROULETABILLE CONTRE LA DAME DE PIQUE **Vol. 2** 124
ROULETABILLE JOUE ET GAGNE **Vol. 2** 88, 124
ROUMANIE, TERRE D'AMOUR **Vol. 2** 308
ROUTE DE CORINTHE, LA **Vol. 2** 700
ROUTE DU BAGNE, LA **Vol. 2** 153, 698
ROUTE DU BONHEUR, LA **Vol. 2** 128, 247, 738
ROUTE ENCHANTEE, LA **Vol. 2** 84, 249
ROUTE EST BELLE, LA **Vol. 2** 751

ROUTE IMPERIALE, LA **Vol. 2** 388, 427, 626
ROUTE INCONNUE, LA **Vol. 2** 706
ROUTE, LA **Vol. 2** 283
ROUTE NAPOLEON, LA **Vol. 2** 153, 676
ROUTE SANS ISSUE **Vol. 2** 197
ROUTES DU SUD, LES **Vol. 2** 299, 351, 828
ROVANIEMEN MARKKINOILLA **Vol. 5** 443, 461, 643
ROVASTIN HÄÄMATKAT **Vol. 5** 356, 428
ROVERENS BRUD **Vol. 5** 97
ROVERHOVDINGENS FLUGT OG DOD **Vol. 5** 97
ROVERNE FRA ROLD **Vol. 5** 44, 51, 150
ROVERS DOD, DEN **Vol. 5** 97
ROVETO ARDENTE **Vol. 7** 968
ROYAL DERBY **Vol. 7** 268
ROZA **Vol. 1** 278, 308, 346
ROZANIEC Z GRANATOW **Vol. 1** 206
ROZDELENI **Vol. 10** 72, 210, 483
ROZINA SEBRANEC **Vol. 10** 142, 460, 569, 621
ROZKOSNY PRIBEH **Vol. 10** 256, 460
ROZMARING **Vol. 9** 315, 322
ROZMARNE LETO **Vol. 10** 41, 96, 516
ROZPOREK I SPOLKA **Vol. 1** 174, 176
ROZPUSTENY A VYPUSTENY **Vol. 10** 18, 172, 570
ROZPUSTILA NOC **Vol. 10** 61, 164, 168, 460, 532
ROZSAFABOT **Vol. 9** 196, 236
ROZSTANIE **Vol. 1** 254, 320
ROZSUDOK **Vol. 10** 273, 476
ROZTOMILY CLOVEK **Vol. 10** 154, 345
ROZTRHLA SE HUDACKOVI STRUNA **Vol. 10** 32, 273
ROZTRZENE FOTO **Vol. 10** 167, 536
ROZTRZKA **Vol. 10** 182, 193, 359
ROZVOD PANI EVY **Vol. 10** 345, 580
ROZWODOW NIE BEDZIE **Vol. 1** 156, 242 **Vol. 6** *page 237*
R.P.Z. APPELLE BERLIN **Vol. 2** 605 **Vol. 4** 148
RRUGA E LIRISE **Vol. 6** *page 12, page 18*
RRUGE TE BARDHA **Vol. 3** 12, 22, 23
RRUGICAT QE KERKONIN DIELL **Vol. 6** *page 17*
RRUGICAT QUE KERKONIN DIELL **Vol. 3** 22, 24
RUA SEM SOL **Vol. 8** 21, 30, 75, 147
RUBACUORI **Vol. 7** 756, 1116
RÜBEZAHLS HOCHZEIT **Vol. 4** 202, 425
RUBIA DEL BAR, LA **Vol. 8** 641
RUBLAK – DIE LEGENDE VOM VERMESSENEN LAND **Vol. 6** *page 153, page 155*
RUBLO DE DOS CARAS **Vol. 8** 276, 732
RUCHOME PIASKI **Vol. 1** 148, 156, 192, 311
RÜCKKEHR ALS TOTER **Vol. 1** 14, 106
RUDA ZARE NAD KLADNEM **Vol. 10** 142, 296, 433, 524
RUDE JOURNEE POUR LA REINE **Vol. 2** 188, 220, 468
RUE BARBARE **Vol. 6** *page 95, page 102*
RUE CASES NEGRES **Vol. 6** *page 109*
RUE DE LA PAIX **Vol. 2** 384
RUE DE L'ENFER, LA **Vol. 2** 472
RUE DE L'ESTRAPADE **Vol. 2** 279, 368
RUE DES BOUCHES PEINTES, LA **Vol. 2** 48, 131, 259
RUE DES PRAIRIES **Vol. 2** 644, 722
RUE DES SAUSSAIES **Vol. 2** 96, 124, 173, 743
RUE DU PAVE D'AMOUR, LA **Vol. 2** 584
RUE DU PIED-DE-GRUE **Vol. 2** 739
RUE SANS JOIE, LA **Vol. 2** 152, 273, 740, 812
RUE SANS LOI, LA **Vol. 2** 217, 254
RUE SANS NOM, LA **Vol. 2** 277, 623, 698
RUEDA DE LA VIDA, LA **Vol. 8** 268
RUEDA DE SOSPECHOSOS **Vol. 8** 298, 600
RUF AN DAS GEWISSEN **Vol. 4** 340, 354, 416
RUF AUS DEM JENSEITS, DER **Vol. 4** 118, 367
RUF DER LIEBE, DER **Vol. 4** 319
RUF DER WILDNIS **Vol. 8** 303, 442
RUF DES NORDENS, DER **Vol. 4** 84, 143, 558, 774
RUF DES SCHICKSALS, DER **Vol. 4** 112, 729
RUFFIAN, LE **Vol. 6** *page 83, page 93, page 96, page 99*
RUG SI FLACARA **Vol. 3** 214, 276, 293
RUHE IST DIE ERSTE BÜRGERPFLICHT **Vol. 4** 239, 517
RUHIGES HEIM MIT KÜCHENBENUTZUNG **Vol. 4** 255, 535, 788
RUIDO DEL SILENCIO, EL **Vol. 8** 396
RUINAS NO INTERIOR, AS **Vol. 8** 28, 169
RUISENOR DE LAS CUMBRES, EL **Vol. 8** 524
RUISSEAU, LE **Vol. 2** 8, 42, 427, 452, 722
RUKAVICKA **Vol. 10** 154, 168, 612, 621
RUKOJMI V BELLA VISTA **Vol. 10** 70, 153, 154, 385
RULLESKOJTERNE **Vol. 5** 97
RUMA ELSA **Vol. 5** 443, 549
RUMBO **Vol. 8** 214, 625
RUMBO A BELEN **Vol. 8** 681, 706
RUMBO AL CAIRO **Vol. 8** 522, 548, 663
RUMBURAK **Vol. 10** 369, 425, 496
RUMEURS **Vol. 2** 57, 131
RUN LIKE A THIEF **Vol. 8** 493, 595

RUND UM EINE MILLION Vol. 4 687
RUNDA 6-A Vol. 3 272, 290, 317
RUNOILIJA JA MUUSA Vol. 5 400, 500, 511, 522, 651
RUNOILOJA MUUTTAA Vol. 5 344
RUNON KUNINGAS JA MUUTTOLINTU Vol. 5 349, 481, 566
RUOTA DEL FALCO, LA Vol. 7 110, 815
RUOTA DEL VIZIO, LA Vol. 7 751
RUPE MALEDETTA, LA Vol. 7 1189
RUPE TARPEA, LA Vol. 7 796
RÜPEL Vol. 1 85, 103, 111
RUPTURE, LA Vol. 2 520, 700
RURALES DE TEXAS, LOS Vol. 8 311, 400
RUSALKA Vol. 10 74, 170, 223, 498, 524, 566
RUSE, LA Vol. 2 245, 257
RUSES DU DIABLE, LES Vol. 2 504
RUSH Vol. 7 124
RUSKAN JÄLKEEN Vol. 5 460, 477, 540, 639
RUSSIA Vol. 7 1189
RUSTLERS' RHAPSODY Vol. 8 219, 535
RUTA DE LOS NARCOTICOS, LA Vol. 8 217, 768
RUTA GLORIOSA Vol. 8 227
RUTH Vol. 5 97
RUTSCHBAHN Vol. 4 192, 266, 614
RUUSU JA KULKURI Vol. 5 552, 600
RUUSUJEN AIKA Vol. 5 323, 413, 449, 541, 595
RUY BLAS Vol. 2 282, 427, 863
RUZOVE KOMBINE Vol. 10 169, 524
RUZOVE SNY Vol. 10 82, 263, 493
RYBA NA SUCHU Vol. 10 412, 621
RYCERZ Vol. 6 *page 227, page 255, page 256*
RYCHLIK DO OSTRAVY Vol. 10 70, 154, 569
RYCK MEJ I SVANSEN, ÄLSKLING ! Vol. 5 869, 916, 1016
RYHMY JA ROMPPAINEN Vol. 5 334, 552, 579
RYKMENTIN MURHEENKRYYNI Vol. 5 356, 485
RYMD INVASION I LAPPLAND Vol. 5 871
RYNGRAF Vol. 1 380
RYS Vol. 6 *page 258, page 259*
RYSKA SNUVAN Vol. 5 840, 909, 969
RYSOPIS Vol. 1 261
RYTIR BLEDE RUZE Vol. 10 45, 526, 629
RYTMUS 1934 Vol. 10 172, 233, 466
RYTTARE I BLATT Vol. 5 871, 1042, 1053
RYWALE Vol. 1 346
RYYSYRANNAN JOOSEPPI Vol. 5 347, 634
RZECZPOSPOLITA BABSKA Vol. 1 303, 389
RZECZYWISTOSC Vol. 1 180, 211, 247, 389 Vol. 6 *page 237*

S

S-A FURAT O BOMBA Vol. 3 265, 297
S CERTY NEJSOU ZERTY Vol. 10 33, 172, 254, 516
S DECA NA MORE Vol. 3 73, 101
S D"H NA BADEMI Vol. 3 38, 114
S LJUBOV I NEZNOST Vol. 3 59
S MARGINAL, UM Vol. 8 32, 118, 124
S. O. S. NORONHA Vol. 2 224, 671, 856
S OSOBENO MNENIE Vol. 3 101, 109, 119
S PODELENA LJUBOV Vol. 6 *page 33*
S TEBOU ME BAVI SVET Vol. 10 344, 429, 434
S VYLOUCENIM VEREJNOSTI Vol. 10 92, 167, 168
S.A. L'AMORE Vol. 7 131
SA-MANN BRAND Vol. 4 344, 647
SA MEILLEURE CLIENTE Vol. 2 32, 197, 488
SA MORI RANIT DIN DRAGOSTE DE VIATA Vol. 6 *page 269, page 276, page 282*
SA MOTES VI I MORGEN Vol. 5 726, 800
SA TETE Vol. 2 136
SA TUKTAS EN ÄKTA MAN Vol. 5 863, 1032
SA TUKTAS KÄRLEKEN Vol. 5 1042, 1217
SAA MODES VI HOS TOVE ! Vol. 5 28, 89, 224
SAA TIL SOS Vol. 5 101, 183, 208
SAARISTON TYTTÖ Vol. 5 347, 634
SAAT GEHT AUF, DIE Vol. 4 323, 461, 646
SAATANAN RADIKAALIT Vol. 5 606
SABADO DOMINGO Y VIERNES Vol. 8 243, 762
SABADO EN LA PLAYA Vol. 8 348, 464
SABATO, DOMENICA E VENERDI Vol. 7 426, 1046 Vol. 8 243, 762
SABELA DE CAMBADOS Vol. 8 214, 484, 625
SABGERINDENS DIAMANTER Vol. 5 97
SABIAN DEMASIADO Vol. 8 630, 719
SABINA Vol. 4 324 Vol. 5 869
SABINA, LA Vol. 8 299
SABINE KLEIST, 7 JAHRE Vol. 6 *page 159, page 172, page 173*
SABINE WULFF Vol. 1 7, 9, 12, 126
SABINO ARANA Vol. 8 210, 654
SABLAZAN Vol. 6 *page 317, page 320*
SABLES Vol. 2 402, 440
SABLES MOUVANTS, LES Vol. 2 381

SABOR DE LA GLORIA, EL Vol. 8 516, 707
SABOR DE LA VENGANZA, EL Vol. 8 603, 626
SABOR DEL ODIO, EL Vol. 8 762
SABOTIER DU VAL DE LOIRE Vol. 2 504
SABRA Vol. 1 383, 384 Vol. 2 142, 513
SABU PRINCIPE LADRO Vol. 7 196
SAC DE BILLES, UN Vol. 2 450, 455
SACCO BELLO, UN Vol. 7 567, 1077
SACCO, DER HUNGERKÜNSTLER Vol. 4 135, 503
SACCO DI ROMA, IL Vol. 7 174, 333, 681
SACHE MIT DEM HERMELIN, DIE Vol. 4 8, 742
SACHE MIT DEM SCHORRSIEGEL, DIE Vol. 4 158, 306
SACHE MIT STYX, DIE Vol. 4 97, 347, 370
SACHTA POHRBENYCH IDEI Vol. 10 257, 536
SACRA BANDIERA Vol. 7 1189
SACRA BIBBIA, LA Vol. 7 316, 654, 937
SACRE LEONCE Vol. 2 579, 722, 872
SACREE JEUNESSE Vol. 2 318, 627, 875
SACRIFICIO SUBLIME Vol. 7 304
SACRIFIES, LES Vol. 6 *page 81, page 122*
SADAN DU VILLE HA MIG Vol. 5 310, 363
SADAN ER DE ALLE Vol. 5 16, 61, 272
SADAN ER JEG OSSE Vol. 5 162, 247, 296
SADAN MIEKAN MIES Vol. 5 379, 481, 579
SADEAIKA Vol. 5 464
SAETA DEL RUISENOR Vol. 8 524, 749
SAETA RUBIA Vol. 8 517, 749
SAETTA Vol. 7 304
SAETTA CONRO LA GHIGLIOTTINA Vol. 7 1067
SAETTA CONTRO L'ORCO DI MARCOUF Vol. 7 553
SAETTA E IL CLUB DEI CIUFFI Vol. 7 553
SAETTA E LE MOGLI DEL PASCIA Vol. 7 751
SAETTA IMPARA A VIVERE ! Vol. 7 1116
SAETTA PRINCIPE PER UN GIORNO Vol. 7 751
SAETTO SALVA LA REGINA Vol. 7 304
SAFARI DIAMANTS Vol. 2 873
SAFARI EXPRESS Vol. 7 318
SAFFO Vol. 7 531, 782
SAFFO VENERE DI LESBO Vol. 7 257
SÄG DET I TONER Vol. 5 891, 997
SÄG DET MED BLOMMOR Vol. 5 878, 1173
SAG' MIR, WER DU BIST Vol. 4 119, 149, 373, 660
SAGA DE LOS DRACULA, LA Vol. 8 237, 708
SAGAEN OM SELBUSTEINEN Vol. 5 673
S'AGAPO Vol. 3 193
SAGARMATHA Vol. 10 128, 269, 399
SAGEATA CAPITANULUI ION Vol. 3 265, 297
SÄGS PA STAN, DET Vol. 5 907, 923, 982
SAH LA REGE Vol. 3 250, 306
SAHARA Vol. 2 526 Vol. 8 703
SAHARA BRULE, LE Vol. 2 97, 309, 615
SAHARA CROSS Vol. 7 524
SAHARA D'AUJOURD'HUI Vol. 2 510
SAHIN Vol. 10 268, 544
SAI CHE I PAPAVERI..., LO Vol. 7 852
SAI COSA FACEVA STALIN ALLE DONNE ? Vol. 7 502
SAI L'ULTIMA SU ... I MATTI ?, LA Vol. 7 1022
„SAID". EIN VOLK IN KETTEN Vol. 4 297, 743
SAIGNEE, LA Vol. 2 185, 294
SAINT JOAN Vol. 2 656
SAINT MARTIN'S LANE Vol. 2 440
SAINT MENE LA DANSE, LE Vol. 2 402
SAINT PREND L'AFFUT, LE Vol. 2 548, 661
SAINT VOYOU, PRIEZ POUR NOUS Vol. 2 795
SAINTE FAMILLE, LA Vol. 2 854
SAINTES NITOUCHES, LES Vol. 2 671, 768
SAISON DANS LA VIE D'EMMANUEL, UNE Vol. 2 202
SAISON DE PAIX A PARIS, UNE Vol. 6 *page 314, page 317*
SAISON D'ETE Vol. 2 678
SAISON IN KAIRO Vol. 2 399 Vol. 4 288 Vol. 6 *page 189*
SAIT-ON JAMAIS Vol. 2 824
SAITRO LA RIBELLE Vol. 7 53
SAKAFLIAS Vol. 3 186
SÄKKIJÄRVEN POLKKA Vol. 5 480, 592
SAL GORDA Vol. 8 373, 580
SALAINEN ASE Vol. 5 428, 484, 529
SALAIRE DE LA PEUR, LE Vol. 2 423, 626, 722, 824
SALAIRE DU PECHE, LE Vol. 2 13, 128, 832
SALAKULJETTAJAN LAULU Vol. 5 451, 600
SALAMANDER Vol. 6 *page 181*
SALAMBO Vol. 7 923
SALAMDRE D'OR, LA Vol. 2 570, 795
SALAMMBO Vol. 2 153
SALARIO DEL CRIMEN, EL Vol. 8 298, 517
SALARIO PARA MATAR Vol. 8 762, 770

SALAS U MALOM RITU **Vol. 3** 341, 356, 441
SALAUDS VONT EN ENFER, LES **Vol. 2** 427, 676, 801
SALAVIINANPOLTTAJAT **Vol. 5** 351, 356, 476
SALE AFFAIRE, UNE **Vol. 2** 175 **Vol. 6** *page 129*
SALE HISTOIRE, UNE **Vol. 2** 164, 517, 716
SALE REVEUR **Vol. 2** 450, 588
SALICE PIANGENTE, IL **Vol. 7** 148
SALKA VALKA **Vol. 5** 1042, 1080, 1130
SALKATTEN **Vol. 5** 843, 942, 1073
SÄLLSKAPSLEK **Vol. 5** 954, 1217
SÄLLSKAPSRESAN **Vol. 5** 963, 1096
SALLY OCH FRIHETEN **Vol. 5** 946, 1093
SALO O LE 120 GIORNATE DI SODOMA **Vol. 7** 333, 388, 434
SALOME **Vol. 7** 741, 825
SALON DORA GREEN **Vol. 4** 340, 462
SALON KITTY **Vol. 7** 450, 585
SALONWAGEN E 417 **Vol. 4** 247, 424, 430, 437
SALSTÄNK OCH KRUTGUBBAR **Vol. 5** 1080
SALTA GUBBAR OCH SEXTANTER **Vol. 5** 852, 905
SALTIMBANC LA POLUL NORD, UN **Vol. 6** *page 267, page 276, page 279*
SALTIMBANCHI, I **Vol. 7** 321
SALTIMBANCII **Vol. 6** *page 276, page 279*
SALTIMBANCOS **Vol. 8** 10, 50, 77
SALTIMBANQUES, LES **Vol. 2** 355
SALTO **Vol. 1** 355, 382 **Vol. 6** *page 224*
SALTO A LA GLORIA **Vol. 8** 214, 600
SALTO DELLA MORTE, IL **Vol. 7** 454
SALTO MORTAL **Vol. 8** 495, 546
SALTO MORTALE **Vol. 4** 49, 157, 316, 437
SALTO NEL VUOTO **Vol. 7** 411, 607, 801
SALUT BERTHE **Vol. 2** 570, 815
SALUT GERMAIN **Vol. 1** 12, 91
SALUT I FORCA AL CANUT **Vol. 8** 310, 641
SALUT J'ARRIVE **Vol. 6** *page 82, page 105*
SALUT L'ARTISTE **Vol. 2** 588, 655
SALUT, LES FRANGINES **Vol. 2** 600
SALUT VOLEURS **Vol. 2** 72
SALUTARI DE LA AGIGEA **Vol. 6** *page 268, page 270, page 277*
SALUTI E BACI **Vol. 7** 796
SALUTO ITALICO **Vol. 7** 806
SALVAJE KURDISTAN, EL **Vol. 8** 523, 732
SALVAJES EN PUENTE SAN GIL, LOS **Vol. 8** 402
SALVATORE GIULIANO **Vol. 7** 241, 401, 1167
SALVATORI DI ROBINET, I **Vol. 7** 1189
SALVE SAUEGJETER **Vol. 5** 745
SAM **Vol. 3** 356, 436
SAM NA SAM **Vol. 1** 138, 325, 340
SAM POSROD MIASTA **Vol. 1** 171, 202, 211, 227 **Vol. 6** *page 207*
SAMAN **Vol. 9** 107, 253
SAMANKA **Vol. 6** *page 125*
SAMBA **Vol. 4** 98 **Vol. 8** 214, 538, 644
SAMBIZANGA **Vol. 2** 7 **Vol. 8** 3
SAMBO **Vol. 7** 1063
SAMEDI SOIR **Vol. 2** 198
SAMHÄLLETS DOM **Vol. 5** 997
SAMI SRED V"LCI **Vol. 6** *page 37, page 42, page 62*
SAMI SWOI **Vol. 1** 179, 254 **Vol. 6** *page 257*
SAMO JEDNOM SE LJUBI **Vol. 6** *page 312, page 320*
SAMO LJUDI **Vol. 3** 341, 444
SAMODIVSKO HORO **Vol. 3** 110 **Vol. 6** *page 24*
SAMORASTNIKI **Vol. 3** 348, 387
SAMOROST **Vol. 10** 389, 517, 550
SAMOTNOSC WE DWOJE **Vol. 1** 237, 243, 403 **Vol. 6** *page 242*
SAMOURAI, LE **Vol. 2** 224, 457
SAMPO **Vol. 5** 407, 431, 508, 637
SAMPO LAPPELILL **Vol. 5** 924, 1203
SAMRTNO PROLECE **Vol. 3** 382, 427, 441
SAMS **Vol. 5** 920, 946
SAMSON **Vol. 1** 156, 373, 398 **Vol. 2** 32, 193, 332 **Vol. 6** *page 205*
SAMSON, SEIN EIGENER MÖRDER **Vol. 4** 535
SAMVETSÖMMA ADOLF **Vol. 5** 840, 997
SAMVITTIGHEDSKVALER **Vol. 5** 9, 197
SAN **Vol. 3** 423
SAN ANTONIO NE PENSE QU'A CA **Vol. 6** *page 75, page 87*
SAN BABILA ORE 20: UN DELITTO INUTILE **Vol. 7** 928
SAN GIORGIO **Vol. 7** 574, 1004
SAN GIOVANNI DECCOLATO **Vol. 7** 26, 941, 978
SAN MICHELE AVEVA UN GALLO **Vol. 7** 741, 1034
SAN PASQUALE BAYLONNE, PROTETTORE DELLE DONNE **Vol. 7** 834
SAN REMO, LA GRANDE SFIDA **Vol. 7** 603, 1054
SAN STRALANDE DAG, EN **Vol. 5** 909, 1142
SAN – ZURKA – SAN **Vol. 7** 708
SANATORIUM POD KLEPSYDRA **Vol. 1** 294, 320, 325
SANCE **Vol. 10** 18, 359, 612
SANCTA MARIA **Vol. 7** 756, 796
SANDE ANSIGT, DET **Vol. 5** 89

SANDE KAERLIGHED, DEN **Vol. 5** 9
SANDENER KINDESMORDPROZESS, DER **Vol. 1** 49, 70
SANDERS OF THE RIVER **Vol. 2** 656
SANDGRÄFIN, DIE **Vol. 4** 149, 536, 660
SANDOKAN **Vol. 7** 738 **Vol. 8** 493, 732
SANDOKAN ALLA RISCOSSA **Vol. 7** 17
SANDOKAN CONTRO IL LEOPARDO DI SARAWAK **Vol. 7** 17
SANDOKAN IL MACISTE DELLA GIUNGIA **Vol. 7** 649
SANDOKAN, LA TIGRE DI MOMPRACEM **Vol. 8** 492, 732
SANDY **Vol. 6** *page 79, page 97*
SANFTE JACOB, DER **Vol. 4** 230, 232, 239
SANG A LA TETE, LE **Vol. 2** 118, 517, 826
SANG D'ALLAH, LE **Vol. 2** 440
SANG DES AUTRES, LE **Vol. 6** *page 107, page 110*
SANG DES FINOEL, LE **Vol. 2** 183, 222
SANG DES IMMORTELLES, LE **Vol. 2** 216
SANG DU FLAMBOYANTS, LE **Vol. 6** *page 84, page 142*
SANG D'UN POETE, LE **Vol. 2** 279, 656
SANG ET LUMIERES **Vol. 2** 57, 347, 454, 548
SANGEN OM DEN ELDRÖDA BLOMMAN **Vol. 5** 840, 909, 937, 996, 997, 1052
SANGEN OM RONDANE **Vol. 5** 760, 788
SANGEN OM STOCKHOLM **Vol. 5** 1156, 1164
SANGEN TIL LIVET **Vol. 5** 676, 696, 831
SANGEN TILL HENNE **Vol. 5** 840, 878
SÄNGERIN, DIE **Vol. 6** *page 187*
SÄNGKAMMARTJUVEN **Vol. 5** 840, 843, 923, 959, 1104
SANGRAAL LA SPADA DI FUOCO **Vol. 7** 371, 426
SANGRE EN CASTILLA **Vol. 8** 298
SANGRE EN EL RUEDO **Vol. 8** 214, 484
SANGRE EN LA CARIBE **Vol. 8** 344
SANGRE EN LA NIEVE **Vol. 8** 276
SANGRE EN MIS ZAPATOS **Vol. 8** 731
SANGRE FRIA, LA **Vol. 8** 732
SANGRE SOBRE TEXAS **Vol. 8** 457
SANGUE A CA'FOSCARI **Vol. 7** 1035
SANGUE BLEU **Vol. 7** 965
SANGUE DI SBIRRO **Vol. 7** 474
SANGUE DI ZINGARA **Vol. 7** 1036
SANGUE SUL SAGRATO **Vol. 7** 1036
SANGUE TOUREIRO **Vol. 8** 85, 88
SANKARI KUIN SANKARI **Vol. 5** 471, 634
SANKARIALOKAS **Vol. 5** 451, 462
SANKT ELISABETH IN UNSEREN TAGEN **Vol. 6** *page 184*
SANKT HANS FEST **Vol. 5** 726, 738
SANKT URBAN **Vol. 1** 63, 98
SANO SE SUOMEKSI **Vol. 5** 523, 588
SANS FAMILLE **Vol. 2** 48, 269, 322, 388, 423, 571, 688, 727, 738, 837
SANS LAISSER D'ADRESSE **Vol. 2** 309, 676
SANS LENDEMAIN **Vol. 2** 526, 778
SANS MOBILE APPARENT **Vol. 2** 475, 655
SANS SOMMATION **Vol. 2** 570, 600
SANS TAMBOUR, NI TROMPETTE **Vol. 2** 419, 510, 676
SANSONE **Vol. 7** 1057
SANSONE ACROBATA DEL KOLOSSAL **Vol. 7** 198
SANSONE BURLONE **Vol. 7** 1186
SANSONE CONTRO I PIRATI **Vol. 7** 1121
SANSONE E I RETTILI UMANI, IL **Vol. 7** 489, 525
SANSONE E LA LADRA DI ATLETI **Vol. 7** 525, 553
SANSONE MUTO **Vol. 7** 1186
SANSONETTE AMAZZONE DELL'ARIA **Vol. 7** 543, 684
SANSONETTE DANZATRICE DELLA PRATERIA **Vol. 7** 543, 684
SANSONETTE E I QUATTRO ARLECCHINI **Vol. 7** 543, 684
SANT HANDER INTE HÄR **Vol. 5** 940, 1173
SANTA CECILIA **Vol. 7** 429, 469, 1113
SANTA DERVIS, A **Vol. 9** 22, 152, 259, 262
SANTA ISABEL DE CERES **Vol. 8** 281
SANTA, LA **Vol. 7** 277
SANTA LUCIA LONTANA **Vol. 7** 336
SANTA LUCIA LUNTANA **Vol. 7** 1063
SANTA RELLINA **Vol. 7** 53, 108
SANTAJ **Vol. 3** 256, 281, 308
SANTANDER EN LLAMAS **Vol. 8** 268, 625
SANT'ELENA PICCOLA ISOLA **Vol. 7** 106, 117, 830, 1159
SANT'ILARIO **Vol. 7** 978, 1071
SANTO ANTERO **Vol. 8** 122
SANTO CONTRA EL DOCTOR MUERTE **Vol. 8** 600, 626
SANTO CONTRA LOS ASESINOS DE LA MAFIA **Vol. 8** 405, 492
SANTO DISONORE **Vol. 7** 1121
SANTO PATRONO, IL **Vol. 7** 1022
SANTOS INCOCENTES, LOS **Vol. 8** 297, 610
SANTUARIO NO SE RINDE, EL **Vol. 8** 225, 524, 693
SAPATO DE CETIM, O **Vol. 8** 35, 118, 169
SAPHO **Vol. 2** 32, 197, 279, 488, 738 **Vol. 4** 479, 724
SAPORE DI MARE **Vol. 7** 617, 664
SAPORE DI MARE 2 UN ANNO DOPO **Vol. 7** 544
SAPTAMINA NEBUNILOR **Vol. 3** 225, 226, 241, 254, 306, 335

SAPTE BAIETI SI O STRENGARITA **Vol. 3** 244, 286
SAPTE ZILE **Vol. 3** 255, 269
SARA FELTOM **Vol. 7** 454
SARA LÄR SIG FOLKVETT **Vol. 5** 840, 997
SARACINESCA **Vol. 7** 490, 796, 1052
SARACINESCA O DON GIOVANNI SARACINESCA **Vol. 7** 701
SARAH **Vol. 6** *page 95, page 111*
SARAH GROFNÖ **Vol. 9** 77
SARAJEVO **Vol. 9** 182, 315, 322
SARAJEVSKI ATENTAT **Vol. 3** 356, 359, 451
SARAJEVSKY ATENTAT **Vol. 10** 48, 87, 283, 428
SARANO **Vol. 7** 816
SARASATE **Vol. 8** 298, 400, 674
SARATI-LE-TERRIBLE **Vol. 2** 6, 273, 812, 874
SARATOGA-KOFFER, DER **Vol. 4** 747
SAREN SVJAT **Vol. 3** 73
SARGA ARNYEK **Vol. 4** 166 **Vol. 9** 71, 80
SARGA CSIKO **Vol. 9** 94, 217, 248, 322
SARGA KASZINO **Vol. 9** 132, 236
SARGA ROZSA **Vol. 9** 24
SARI BIRO **Vol. 9** 274, 315
SARIKA DRAGAM **Vol. 9** 20, 152, 253
SARILHO DE FRALDAS **Vol. 8** 24, 113, 176
SÄRKELÄ ITTE **Vol. 5** 357, 592, 620
SARRE **Vol. 2** 13
SARTINA E L'OPERAIO, LA **Vol. 7** 1189
SARUTUL **Vol. 3** 218, 227
SAS 181 ANTWORTET NICHT **Vol. 1** 13, 48, 120
SAS A SAN SALVADOR **Vol. 6** *page 86, page 98*
SASA **Vol. 3** 374
SASEK A KRALOVNA **Vol. 10** 293, 334, 467, 623
SASIEDZI **Vol. 1** 305, 323, 382 **Vol. 6** *page 257*
SASOM I EN SPEGEL **Vol. 5** 1052, 1053, 1080
SASTANAK SA NECASNIM **Vol. 3** 403
SATAN OPIUM **Vol. 4** 192
SATANA **Vol. 7** 1189
SATANAS **Vol. 4** 178, 681
SATANELLA **Vol. 7** 306
SATANICA **Vol. 7** 493
SATANSKETTEN **Vol. 4** 266, 385, 574
SATIRO, IL **Vol. 7** 1189
SATURNIN **Vol. 2** 366, 530, 531
SAUDADES PARA DONA GENCIANA **Vol. 8** 185
SAUL E DAVID **Vol. 8** 298, 697
SAUL E DAVIDE **Vol. 7** 738
SAUL Y DAVID **Vol. 8** 298, 697
SAURE WOCHEN, FROHE FESTE **Vol. 1** 25, 56, 117
SAUT DE L'ANGE, LE **Vol. 2** 110, 588
SAUT, LE **Vol. 2** 674
SAUTS D'OBSTACLES **Vol. 2** 678
SAUVAGE, LE **Vol. 2** 251, 517
SAUVE QUI PEUT LA VIE **Vol. 2** 527 **Vol. 6** *page 122, page 133*
SAUVEUR, LE **Vol. 2** 527
SAVAGE GUNS, THE **Vol. 8** 311, 400
SAVAMALA **Vol. 6** *page 313, page 322*
SAVANA VIOLENZA CARNALE **Vol. 7** 1024
SAVITRI **Vol. 7** 508
SAVITRY SATYVAN **Vol. 7** 630
SAVNET SIDEN MANDAG **Vol. 5** 676, 717, 729, 738, 831
SAVOIR-VIVRE DE CHEMIN DE FER **Vol. 2** 368
SAVOY-HOTEL 217 **Vol. 4** 266, 531, 574, 729
SAXOFONE **Vol. 7** 221
SAXOPHON-SUSI **Vol. 4** 257, 329
SAZKA NA TRINACTKU **Vol. 10** 18, 227, 313
SAZKA O HUBICKU **Vol. 10** 363
SB ZATVARA KRUG **Vol. 3** 367, 402
SBAGLIO DI ESSERE VIVO, LO **Vol. 7** 493
SBALLATO, GASATO, COMPLETAMENTE FUSO **Vol. 7** 599, 697
SBAMM ! **Vol. 7** 728, 894
SBANDATI, GLI **Vol. 7** 915, 1167
SBATTI IL MOSTRO IN PRIMA PAGINA **Vol. 7** 599, 763
SBERNE SUROVOSTI **Vol. 10** 41, 352
S"BESEDNIK PO ZELANIE **Vol. 6** *page 31, page 33, page 49, page 64*
SBIRULINO **Vol. 7** 692
SBOGOM, PRIJATELI **Vol. 3** 63, 114 **Vol. 6** *page 31*
SCACCO ALLA REGINA **Vol. 7** 514
SCACCO MATTO **Vol. 7** 293
SCACCO TUTTO MATTO, UNO **Vol. 8** 517
SCALA DI FRICOT, LA **Vol. 7** 1189
SCALA DI SETA, LA **Vol. 7** 508
SCALA, LA **Vol. 7** 219, 238, 361, 652, 796
SCALABRINO **Vol. 7** 292, 1118
SCALDALETTO, LO **Vol. 7** 993
SCALDINO, LO **Vol. 7** 48
SCAMPALO **Vol. 7** 954, 1027, 1054
SCAMPOLO **Vol. 2** 32 **Vol. 7** 1162
SCAMPOLO '53 **Vol. 7** 16
SCAMPOLO, EIN KIND DER STRASSE **Vol. 4** 113
SCANDALE **Vol. 2** 270, 309

SCANDALE AUX CHAMPS-ELYSEES **Vol. 2** 9, 866
SCANDALE AUX GALERIES, UN **Vol. 2** 273, 727
SCANDALE, LE **Vol. 2** 566, 598, 700, 775, 778
SCANDALO **Vol. 7** 1101
SCANDALO PER BENE **Vol. 7** 40, 493
SCANDALO PERBENE, UNO **Vol. 7** 258, 347, 450
SCANDALOSA GILDA **Vol. 7** 1194
SCANO BOA **Vol. 8** 517
SCAPA FLOW **Vol. 4** 558, 796
SCAPOLO, LO **Vol. 7** 1167
SCARABEI D'ORO, GLI **Vol. 7** 26
SCARLATINE, LA **Vol. 6** *page 81, page 101, page 144*
SCARLET WEST, THE **Vol. 2** 84
SCARPE AL SOLE, LE **Vol. 7** 487, 1116, 1172
SCARPE GROSSE **Vol. 7** 941, 1036
SCARPETTA, LA **Vol. 7** 1189
SCATOLA MACCHIATA DI SANGUE, LA **Vol. 7** 730
SCEICCO BIANCO, LO **Vol. 7** 118, 493, 983, 1128
SCELERATS, LES **Vol. 2** 248, 285, 738, 770
SCEMO DI GUERRA **Vol. 7** 102, 617, 697
SCENE GIAPPONESI **Vol. 7** 1189
SCENES BORN **Vol. 5** 203 **Vol. 5** 9, 197
SCENES DE LA VIE DROLE **Vol. 2** 375
SCENES DE MENAGE **Vol. 2** 279, 626, 824
SCENINGANG **Vol. 5** 1073, 1164
SCENY DZIECIECE Z ZYCIA PROWINCJI **Vol. 6** *page 239*
SCERIFFA, LA **Vol. 7** 891
SCERIFFO EXTRATERRESTRE ... POCO EXTRA E MOLTO TERRESTRE **Vol. 7** 524
SCHABERNACK **Vol. 4** 149, 462, 660
SCHACH DER EVA **Vol. 4** 340, 466, 787
SCHACHMATT **Vol. 4** 52, 115, 247, 774
SCHÄDEL DER PHARAONENTOCHTER, DER **Vol. 4** 240
SCHANDE **Vol. 6** *page 192*
SCHATTEN **Vol. 4** 214, 729
SCHATTEN AUS DEM TOTENREICH **Vol. 4** 159
SCHATTEN DER GABY LEED, DER **Vol. 4** 212, 545
SCHATTEN DER MANEGE **Vol. 4** 97, 124, 204, 422
SCHATTEN DER NACHT **Vol. 4** 239
SCHATTEN DER UNTERWELT **Vol. 4** 119, 124, 246
SCHATTEN DER VERGANGENHEIT **Vol. 4** 159
SCHATTEN DER WELTSTADT **Vol. 4** 149, 213, 660
SCHATTEN DES LEBENS **Vol. 4** 183
SCHATTEN DES MEERES **Vol. 4** 183
SCHATTEN DES TOTEN, DER **Vol. 4** 71, 434
SCHATTEN ÜBER DEN INSELN **Vol. 1** 56, 58, 129
SCHATTEN ÜBER ST.PAULI **Vol. 4** 79, 115, 729
SCHATZ, DER **Vol. 4** 266, 574, 704
SCHATZ DER AZTEKEN, DER **Vol. 6** *page 312*
SCHATZ DES ABDAR RAHMAN, DER **Vol. 4** 99, 124
SCHATZ IM BRAHMANENTEMPEL, DER **Vol. 4** 239
SCHATZ IM SILBERSEE, DER **Vol. 3** 366, 371
SCHATZ MACH' KASSE **Vol. 4** 479, 527
SCHATZKAMMER IM SEE, DIE **Vol. 4** 634
SCHATZSUCHER **Vol. 1** 63 **Vol. 6** *page 155, page 167*
SCHAUSPIELDIREKTOR, DER **Vol. 4** 13, 24
SCHEIDUNGSGRUND, DER **Vol. 4** 353, 515 **Vol. 10** 256, 412
SCHEIDUNGSREISE **Vol. 4** 110, 124, 232
SCHEIN TRÜGT, DER **Vol. 4** 323, 461
SCHEINE DES TODES **Vol. 4** 400, 536
SCHEINHEILIGE FLORIAN, DER **Vol. 4** 647, 660, 678, 728
SCHELETRO DI CASSIO, LO **Vol. 7** 843
SCHENK MIR DAS LEBEN **Vol. 4** 25, 270, 646
SCHERBEN **Vol. 4** 743
SCHERBEN BRINGEN GLÜCK **Vol. 4** 212
SCHERZANDO CON IL FUOCO **Vol. 7** 1189
SCHERZANDO CON LA MORTE **Vol. 7** 454
SCHERZI DA PRETE **Vol. 7** 869
SCHERZO DEL DESTINO IN AGGUATO DIETRO L'ANGOLO COME UN BRIGANTE DA STRADE **Vol. 7** 109, 592
SCHIAVA DEL PECCATO, LA **Vol. 7** 911, 1044, 1054
SCHIAVA DI ROMA, LA **Vol. 7** 1063
SCHIAVE BIANCHE - VIOLENZA IN AMMAZZONIA **Vol. 7** 585
SCHIAVE DELLA LEGGE **Vol. 7** 603
SCHIAVE DI CALIGOLA, LE **Vol. 7** 414, 691
SCHIAVE DI CARTAGINE, LE **Vol. 7** 17
SCHIAVI PIU FORTI DEL MONDO, GLI **Vol. 7** 694

SCHICKSAL Vol. 4 70, 265, 329, 527, 614, 620
SCHICKSAL AM STROM Vol. 4 243, 273, 791
SCHICKSAL DER CAROLA VAN GELDERN, DAS Vol. 4 660
SCHICKSAL DER RENATE LANGEN, DAS Vol. 4 186, 287, 644
SCHICKSAL EINER NACHT, DAS Vol. 4 219, 681, 749
SCHICKSAL EINER SCHÖNEN FRAU, DAS Vol. 4 208, 292
SCHICKSALS-DOLCH, DER Vol. 4 61, 126, 127, 271, 510
SCHICKSALSSTUNDE, DIE Vol. 4 276
SCHICKSALSTAG, DER Vol. 4 190, 479
SCHICKSALSWÜRFEL Vol. 4 474, 634 Vol. 6 *page 198, page 199*
SCHIEBER Vol. 4 424
SCHIEBERKÖNIG, DER Vol. 4 628
SCHIFF DER VERLORENEN MENSCHEN, DAS Vol. 2 290 Vol. 4 158, 626
SCHIFF IN NOT Vol. 4 82, 270, 725
SCHIFF IN NOT S.O.S. Vol. 4 19, 49, 164, 584, 697
SCHIFF OHNE HAFEN, DAS Vol. 4 119, 270
SCHIFFBRÜCHIGEN, DIE Vol. 4 135, 628
SCHIFFE UND MENSCHEN Vol. 4 35, 219
SCHIMMELKRIEG IN DER HOLLEDAU, DER Vol. 4 430, 461, 774
SCHIMMELREITER, DER Vol. 4 382, 517
SCHINDERHANNES Vol. 4 359, 558
SCHIRM MIT DEM SCHWAN Vol. 4 251
SCHLACHT VON BADEMÜNDE, DIE Vol. 4 232, 242, 471, 751
SCHLAFWAGENKONTROLLEUR, DER Vol. 4 247, 340, 466
SCHLAGENDE WETTER Vol. 4 211, 247
SCHLANGENTÄNZERIN Vol. 4 247
SCHLEICHENDE GIFT, DAS Vol. 4 192
SCHLEIERTÄNZERIN, DIE Vol. 4 97, 149, 359, 660
SCHLEMIHL, DER Vol. 4 558, 774
SCHLEPPZUG M 17 Vol. 4 234, 601, 747
SCHLOSS EINÖD Vol. 4 113
SCHLOSS HUBERTUS Vol. 4 25, 238, 329
SCHLOSS IM SÜDEN, DAS Vol. 4 246, 729
SCHLOSS IN FLANDERN, DAS Vol. 4 89, 246, 638
SCHLOSS VOGELÖD Vol. 4 24, 25, 370, 599, 729, 734 Vol. 6 *page 188*
SCHLÖSSER UND KATEN Vol. 1 48, 74, 103
SCHLOSSHERR VON HOHENSTEIN, DER Vol. 4 159
SCHLUCHT DES TODES, DIE Vol. 4 270, 400, 725
SCHLUSSAKKORD Vol. 4 28, 325
SCHLÜSSEL, DIE Vol. 1 18, 36, 50
SCHMETTERLINGSSCHLACHT, DIE Vol. 4 179, 673
SCHMIEDE DES GRAUENS, DIE Vol. 4 25
SCHMUCK DER HERZOGIN, DER Vol. 4 71
SCHMUCK DES RAJAH, DER Vol. 4 309
SCHMUGGLER VON RAJGROD, DIE Vol. 6 *page 163, page 165, page 176*
SCHMUGGLERBRAUT VON MALORCA, DIE Vol. 4 49, 270
SCHMUGGLERIN, DIE Vol. 4 125
SCHNEEMANN FÜR AFRIKA, EIN Vol. 1 33, 55
SCHNEESCHUHBANDITEN Vol. 4 56, 58, 359, 584
SCHNEEWEISSCHEN UND ROSENROT Vol. 1 34, 76, 114
SCHNEEWITTCHEN Vol. 4 190
SCHNEEWITTCHEN UND DIE SIEBEN ZWERGE Vol. 1 2, 72, 92
SCHNEIDER WIBBEL Vol. 4 49, 234, 239, 270, 424, 738
SCHNELLER ALS DER TOD Vol. 4 356, 471, 779
SCHNITZEL FLIEGT Vol. 4 370, 774
SCHÖN IST DIE MANÖVERZEIT Vol. 4 558, 774
SCHÖN IST ES, VERLIEBT ZU SEIN Vol. 4 361, 558
SCHÖN IST JEDER TAG, DEN DU MIR SCHENKST, MARIE LUISE Vol. 4 28, 339, 544
SCHÖN WARS DOCH Vol. 4 239, 424
SCHÖNE ABENTEUER, DAS Vol. 4 28, 614, 720, 729, 752
SCHÖNE FRÄULEIN SCHRAGG, DAS Vol. 4 24, 82, 370
SCHÖNE JOLAN, DIE Vol. 4 747
SCHÖNE LURETTE, DIE Vol. 1 36, 56, 115
SCHÖNE MÄDCHEN, DAS Vol. 4 329
SCHÖNE MISS LILIAN, DIE Vol. 4 232, 673
SCHÖNEN TAGE VON ARANJUEZ, DIE Vol. 4 49, 325, 445
SCHÖNER TAG, EIN Vol. 4 15, 426, 489
SCHÖNHEITSFLECKCHEN, DAS Vol. 4 373, 626
SCHÖNHEITSKONKURRENZ ODER : DAS URTEIL DES PARIS, DIE Vol. 6 *page 194*
SCHÖNSTE FRAU DER WELT, DIE Vol. 4 17, 123, 192, 220, 584
SCHÖNSTE FRAU VON PARIS, DIE Vol. 4 292, 626
SCHÖNSTE GESCHENK, DAS Vol. 4 560
SCHÖNSTE MANN IM STAATE, DER Vol. 4 212, 239, 424, 461

SCHÖNSTEN BEINE VON BERLIN, DIE **Vol. 4** 213, 681
SCHÖPFER, DER **Vol. 4** 250, 747
SCHOTTIS **Vol. 8** 716
SCHPOUNTZ, LE **Vol. 2** 485, 871
SCHRAMMELN **Vol. 4** 13, 614
SCHRECK IN DER ABENDSTUNDE **Vol. 4** 85, 410
SCHRECKEN DER GARNISON, DER **Vol. 4** 212, 232, 239, 270
SCHRECKEN DER WESTKÜSTE, DER **Vol. 4** 401, 674
SCHRECKEN DES MEERES, DER **Vol. 4** 255, 344
SCHRECKEN VOM HEIDEKRUG, DER **Vol. 4** 97, 230
SCHRECKENSNACHT AUF SCHLOSS DRAGENEDD, DIE **Vol. 4** 482, 598
SCHRECKENSNACHT IM HAUSE CLARQUE, DIE **Vol. 4** 67, 356
SCHRECKENSNACHT IM IRRENHAUS VON IVOY, DIE **Vol. 4** 598, 780
SCHRECKENSNACHT IN DER MENAGERIE, DIE **Vol. 4** 288, 779
SCHRECKLICHE FRAU, EINE **Vol. 1** 68, 125
SCHREI DES GEWISSENS, DER **Vol. 4** 232, 300
SCHREI IN DER WÜSTE, DER **Vol. 4** 25, 545
SCHREIN DER MEDICI, DER **Vol. 4** 583
SCHREITEN WIR DEN SONNENBOGEN AB **Vol. 1** 111
SCHRITT FÜR SCHRITT **Vol. 1** 13, 82, 125
SCHRITT VOM WEGE, DER **Vol. 4** 119, 469
SCHUBERT'S FRÜHLINGSTRAUM **Vol. 4** 208, 626
SCHÜCHTERNE CASANOVA, DER **Vol. 4** 123, 774, 788, 789
SCHUHE EINER SCHÖNEN FRAU, DIE **Vol. 4** 170, 320, 454
SCHUHPALAST PINKUS **Vol. 4** 560
SCHULD DER LAVINIA MORLAND, DIE **Vol. 4** 89, 307, 325, 396, 788
SCHULD, DIE **Vol. 4** 190, 742, 773
SCHULD UND SÜHNE **Vol. 4** 36, 646, 660
SCHULDIG **Vol. 4** 113, 183, 293, 546
SCHULDIGE, DIE **Vol. 4** 428, 626
SCHULEULE PAULE, DIE **Vol. 6** *page 161, page 165, page 170*
SCHÜRZEN-REGIMENT ODER DAS WEIBER-HEER IM GEFECHT !, DAS **Vol. 4** 308
SCHUSS AM NEBELHORN, DER **Vol. 4** 25, 664
SCHUSS, DER **Vol. 4** 228, 471
SCHUSS IM MORGENGRAUEN **Vol. 4** 82, 270, 409, 713
SCHUSS IM PAVILLON, DER **Vol. 4** 344, 544
SCHUSS IM TONFILMATELIER, DER **Vol. 4** 89, 270, 409
SCHUSS IM TRAUM **Vol. 4** 319
SCHUSS UM MITTERNACHT **Vol. 4** 416, 688, 740
SCHÜSSE AN DER GRENZE **Vol. 4** 340, 466, 741
SCHÜSSE DER ARCHE NOAH, DIE **Vol. 6** *page 160, page 167, page 175*
SCHÜSSE IN KABINE 7 **Vol. 4** 79, 115, 132
SCHÜSSE IN MARIENBAD **Vol. 1** 85
SCHÜSSE UNTERM GALGEN **Vol. 1** 14, 92
SCHUSTER LEHMANN IN DER REICHSTAGS-WÄHLERVERSAMMLUNG **Vol. 4** 514
SCHUT, DER **Vol. 3** 367
SCHÜTZENFEST IN SCHILDA **Vol. 4** 186, 340
SCHÜTZENKÖNIG, DER **Vol. 4** 344, 544
SCHÜTZENKÖNIG WIRD DER FELIX **Vol. 4** 422, 703
SCHÜTZENLIESEL **Vol. 4** 239, 374, 449
SCHUZKA SE STINY **Vol. 10** 165, 511, 541
SCHWÄCHE FÜR MUSIK, EINE **Vol. 1** 60
SCHWACHE STUNDE, DIE **Vol. 4** 451, 688
SCHWARZ AUF WEISS **Vol. 4** 136, 451, 714
SCHWARZ GEGEN BLOND **Vol. 4** 314, 375
SCHWARZE BOOT, DAS **Vol. 4** 383, 476
SCHWARZE DIAMANT, DER **Vol. 4** 135, 503
SCHWARZE DOMINO, DER **Vol. 4** 287, 384, 458, 646
SCHWARZE GALEERE, DIE **Vol. 1** 50, 90, 107
SCHWARZE GOTT, DER **Vol. 4** 132, 773
SCHWARZE GRAF, DER **Vol. 4** 240
SCHWARZE HANNE, DIE **Vol. 4** 366, 470
SCHWARZE HARLEKIN, DER **Vol. 4** 135, 509, 720
SCHWARZE HUSAR, DER **Vol. 4** 266, 527, 574, 628
SCHWARZE KATZE, DIE **Vol. 4** 423
SCHWARZE KUVERT, DAS **Vol. 4** 270, 443, 471
SCHWARZE LOCKE, DIE **Vol. 4** 251
SCHWARZE LOS, DAS **Vol. 4** 124, 247, 681
SCHWARZE MASKE, DIE **Vol. 4** 95
SCHWARZE MÜHLE, DIE **Vol. 1** 48, 73
SCHWARZE PANTHER **Vol. 1** 1, 102 **Vol. 6** *page 177*

SCHWARZE PANTHERIN, DIE Vol. 4 115, 306, 321
SCHWARZE PERLEN Vol. 4 113, 780
SCHWARZE PIERROT, DER Vol. 4 463, 471, 560, 634, 779
SCHWARZE ROBE, DIE Vol. 4 24, 485, 631
SCHWARZE ROSEN Vol. 4 325, 445, 729
SCHWARZE SCHACHDAME, DIE Vol. 4 356, 612
SCHWARZE SCHAF, DAS Vol. 4 243, 325
SCHWARZE SPINNE, DIE Vol. 4 192, 277
SCHWARZE WALFISCH, DER Vol. 4 149, 634, 660
SCHWARZER JÄGER JOHANNA Vol. 4 79, 382, 517
SCHWARZER SAMT Vol. 1 7, 13, 115
SCHWARZER ZWIEBACK Vol. 1 53, 106
SCHWARZES GOLD Vol. 6 *page 160, page 168, page 176*
SCHWARZFAHRT INS GLÜCK Vol. 4 132, 437
SCHWARZWALDMÄDEL Vol. 4 97, 119, 164, 270, 292, 584, 796
SCHWEBENDE JUNGFRAU, DIE Vol. 4 373, 626
SCHWEDISCHE NACHTIGALL, DIE Vol. 4 119, 266, 435, 740
SCHWEIGEN IM WALDE, DAS Vol. 4 25, 135, 230, 370, 517, 691
SCHWEIGENDE STERN, DER Vol. 1 40, 72, 298, 305
SCHWERE JUNGE, DER Vol. 4 424
SCHWERE JUNGENS - LEICHTE MÄDCHEN Vol. 4 212, 339
SCHWERER FALL, EIN Vol. 4 239, 681
SCHWERER JUNGE, EIN Vol. 4 437, 747
SCHWERT DES DAMOKLES, DAS Vol. 4 119, 517, 558
SCHWERT UND HERD Vol. 4 596
SCHWERVERBRECHER, DER Vol. 4 652
SCHWESTER AGNES Vol. 1 12, 26, 59
SCHWESTER VERONIKA Vol. 4 247, 461, 771
SCHWIERIG SICH ZU VERLOBEN Vol. 6 *page 164, page 166, page 171*
SCHWUR DER RENATE RABENAU, DER Vol. 6 *page 189*
SCHWUR DES PETER HERGATZ, DER Vol. 4 178
SCHWUR DES SOLDATEN POOLEY, DER Vol. 1 35, 111, 125
SCIANA CZAROWNIC Vol. 1 156, 220, 294, 391
SCIANCATO, LO Vol. 7 1189
SCIMITARRA DEL BARBAROSSA, LA Vol. 7 508, 529
SCIMITARRA DEL SARACENO, LA Vol. 7 1121
SCIMMIA DI ROBINET, LA Vol. 7 1189
SCIOPEN Vol. 7 387
SCIOPERO DEI MILIONI, LO Vol. 7 200, 487
SCIOPERO DELLA VIRTU, LO Vol. 7 569
SCIPIONE L'AFRICANO Vol. 7 55, 196, 1136
SCIUSCIA Vol. 7 101, 196, 641, 756
SCOGLIERA DEL PECCATO, LA Vol. 7 46, 792, 827
SCOGLIERE DELLA MORTE, LE Vol. 7 133, 963
SCOIATTOLO DEL MARE, LO Vol. 7 963
SCOIATTOLO, LO Vol. 7 780
SCOMMESSA ORIGINALE, UNA Vol. 7 1189
SCONCERTO ROCK Vol. 7 1146
SCONFITTA DELL'IDOLO, LA Vol. 7 180, 1067
SCONOSCIUTA, LA Vol. 7 493
SCOPONE SCIENTIFICO Vol. 7 1007
SCOPUL SI MIJLOACELE Vol. 6 *page 265, page 279, page 280*
SCORPION, LE Vol. 2 293, 475
SCORTICATELI VIVI Vol. 7 1024
SCOUMOUNE, LA Vol. 2 163, 873
SCOUT, DER Vol. 6 *page 163, page 166, page 172*
SCRATCH Vol. 6 *page 149*
SCRIGNO DEI MILIONI, LO Vol. 7 131
SCROLLINA Vol. 7 53
SCUGNI Vol. 7 320
SCUGNIZZO A NEW YORK, UNO Vol. 7 222, 426
SCUGNIZZO, LO Vol. 7 474
SCUOLA DEI TIMIDI, LA Vol. 7 792, 1063
SCUOLA D'EROI Vol. 7 168
SCUOLA ELEMENTARE Vol. 7 86, 283
SCUSA SE E POCO Vol. 7 208, 599
SCUSATE IL RITARDO Vol. 7 13
SCUSI, LEI E NORMALE ? Vol. 7 694
SE ACABO EL PETROLEO Vol. 8 474, 617
SE ALKOI OMENASTA Vol. 5 311, 613, 634
SE ALKOI SATEESSA Vol. 5 443, 643
SE ARMO EL BELEN ! Vol. 8 214, 484
S.E. EL MAYORDOMO Vol. 8 422
SE ESTAS MUERTO, POR QUE BAILAS ? Vol. 8 214, 681
SE. EXZELLENZ DER REVISOR Vol. 4 291
SE FOSSI RE ! Vol. 7 1189
SE HA FUGADO UN PRESO Vol. 8 648
SE HA PERDIDO UN CADAVER Vol. 8 430
SE INFIEL Y NO MIRES CON QUIEN Vol. 8 232, 684
SE IO FOSSI ONESTO Vol. 7 16, 196, 1054
S.E. L'AMBASCIATRICE Vol. 7 559

SE LE FUE EL NOVIO **Vol. 8** 396
SE NECESITA CHICO **Vol. 8** 686, 747
SE NON SON MATTI NON LI VOGLIAMO **Vol. 7** 50, 792, 1036
SE OPP FOR SPIONER ! **Vol. 5** 841, 843, 923
SE PARHAITEN NAURAA, JOKA VIIMEKSI NAURAA **Vol. 5** 344, 356
SE PERMETTETE PARLIAMO DI DONNE **Vol. 7** 408
SE QUELL'IDIOTA CI PENSASSE ... **Vol. 7** 595
SE QUIERES VIVIR ... DISPARA **Vol. 8** 396
SE SEI VIVO SPARA **Vol. 8** 626
SE TUTTO VA BENE SIAMO ROVINATI **Vol. 7** 241, 426
SE VENDE UN PALACIO **Vol. 8** 437
SE VINCESSI 100 MILIONI **Vol. 7** 978
SEBASTIAN **Vol. 7** 728
SEBASTIEN PARMI LES HOMMES **Vol. 2** 484
SEBECHLEBSKI HUDCI **Vol. 10** 209, 228, 399
SECCOTINE **Vol. 7** 1189
SECERNA VODICA **Vol. 6** *page 310, page 319*
SECHS MÄDCHEN SUCHEN NACHTQUARTIER **Vol. 4** 321, 752
SECHS TAGE HEIMATURLAUB **Vol. 4** 30, 79, 124
SECHSE KOMMEN DURCH DIE WELT **Vol. 1** 7, 32, 128
SECHZEHN TÖCHTER UND KEIN PAPA **Vol. 4** 290, 565
SECOND SOUFFLE, UN **Vol. 2** 537, 701
SECONDA B **Vol. 7** 756, 796, 1036, 1060
SECONDA MADRE, LA **Vol. 7** 1125
SECONDA MOGLIE, LA **Vol. 7** 560
SECONDE VERITE, LA **Vol. 2** 251, 661
SECONDO TRAGICO FANTOZZI, IL **Vol. 7** 763
SECRET D'ALTA ROCCA, LE **Vol. 2** 286
SECRET DE DELHIA, LE **Vol. 2** 41, 153, 579
SECRET DE L'EMERAUDE, LE **Vol. 2** 259, 273, 595
SECRET DE MADAME CLAPAIN, LE **Vol. 2** 48, 676
SECRET DE MAYERLING, LE **Vol. 2** 254, 282, 488
SECRET DE MONTE-CRISTO, LE **Vol. 2** 65
SECRET DE POLICHINELLE, LE **Vol. 2** 37, 84, 227, 243, 248
SECRET DE ROSETTE LAMBERT, LE **Vol. 2** 42, 547
SECRET DE SOEUR ANGELE, LE **Vol. 2** 279, 488
SECRET D'HELENE MARIMON, LE **Vol. 2** 9, 247, 709
SECRET DU CARGO, LE **Vol. 2** 792
SECRET DU CHEVALIER D'EON, LE **Vol. 2** 13, 194, 226, 368, 669, 837
SECRET DU FLORIDA, LE **Vol. 2** 217, 670, 671
SECRET D'UNE MERE, LE **Vol. 2** 222, 404, 864
SECRET D'UNE NUIT, LE **Vol. 2** 388, 407
SECRET, LE **Vol. 2** 74, 772
SECRET PROFESSIONNEL **Vol. 2** 318, 475, 875
SECRETAIRES, LES **Vol. 2** 484
SECRETAIRES TRES PARTICULIERES **Vol. 2** 721 **Vol. 6** *page 137*
SECRETARIA PARA TODO **Vol. 8** 400, 772
SECRETARIA PELIGROSA **Vol. 8** 692, 772
SECRETARIAS, LAS **Vol. 8** 524
SECRETAS INTENCIONES, LAS **Vol. 8** 348
SECRETISIMO **Vol. 8** 396, 775
SECRETO ... AMOR, EL **Vol. 8** 369, 767
SECRETO DE ANA MARIA, EL **Vol. 8** 437
SECRETO DE BILL NORTH, EL **Vol. 8** 214, 602
SECRETO DE LA MUJER MUERTA, EL **Vol. 8** 516
SECRETO DE LAS ESMERALDAS, EL **Vol. 8** 708
SECRETO DE LOS HOMBRES AZULES, EL **Vol. 8** 305
SECRETO DE MONICA, EL **Vol. 8** 578
SECRETO DE PAPA, EL **Vol. 8** 614, 772
SECRETO DE TOMY, EL **Vol. 8** 207, 298
SECRETO DEL CAPITAN O'HARA, EL **Vol. 8** 583, 772
SECRETO DEL DOCTOR ORLOFF, EL **Vol. 8** 450, 583
SECRETO INCONFESABLE DE UN CHICO BIEN, EL **Vol. 8** 261, 688
SECRETS **Vol. 2** 566, 657
SECRETS D'ALCOVE **Vol. 2** 153, 248, 566, 615, 722 **Vol. 7** 1062
SECRETS DE LA MER ROUGE, LES **Vol. 2** 9, 48
SECRETUL CIFRULUI **Vol. 3** 227, 261, 285
SECRETUL LUI BACHUS **Vol. 6** *page 278, page 279, page 282*
SECTION SPECIALE **Vol. 2** 251, 374, 503, 873
SECUESTRO A LA ESPANOLA **Vol. 8** 299, 546
SECUESTRO BAJO EL SOL **Vol. 8** 258
SECUESTRO EN LA CIUDAD **Vol. 8** 207, 558
SECVENTE **Vol. 6** *page 265, page 267, page 269, page 279, page 282*
SED, LA **Vol. 8** 546

SEDDOK Vol. 7 528
SEDEM JEDNOU RANOU Vol. 10 128, 188, 544
SEDIA ELETTRICA Vol. 7 834
SEDIM NA KONARI A JE MI DOBRE Vol. 10 129, 273, 274
SEDM HAVRANU Vol. 10 193, 286, 433
SEDM ZABITYCH Vol. 10 22, 197, 395
SEDMA VELMOC Vol. 10 34, 92, 256
SEDME NEBE Vol. 10 116, 462, 570
SEDMEHO DNE VECER Vol. 10 286, 352, 612
SEDMIKRASKY Vol. 10 289, 292, 313
SEDMINA Vol. 3 395, 449
SEDMY KONTINENT Vol. 10 182, 184, 572
SEDOTTA E ABBANDONATA Vol. 7 401, 868
SEDUCTEURS, LES Vol. 2 7
SEDUCTOR DE GRANADA, EL Vol. 8 261, 298
SEDUTTORE, IL Vol. 7 234, 915
SEDUTTORI DELLA DOMENICA, I Vol. 7 333
SEDZIOWIE Vol. 1 174, 317
SEE RUFT, DIE Vol. 4 72
SEEBADNIXE, DIE Vol. 6 *page 186*
SEEFAHRT, DIE IST LUSTIG, EINE Vol. 4 186, 340, 361, 466
SEEFAHRT IST NOT ! Vol. 4 36, 660
SEEKADETT, DER Vol. 4 211, 242
SEELEN IM STURM Vol. 4 300
SEELENVERKÄUFER Vol. 4 131, 476
SEELENVERKÄUFER, DER Vol. 4 522
SEFUL SECTORULUI SUFLETE Vol. 3 218, 256, 332
SEGER I MÖRKER Vol. 5 940, 1173
SEGITSEG ÖRÖKÖLTEM ! Vol. 9 319, 322
SEGNI PARTICOLARI: BELLISSIMO Vol. 7 30, 383
SEGNO DEI TEMPI, IL Vol. 7 209
SEGNO DI VENERE, IL Vol. 7 756, 796
SEGNO DI ZORRO, IL Vol. 8 517
SEGREDO DE LOS HOMBRES AZULES, EL Vol. 8 305, 450, 541
SEGRETARIA PELIGROSA Vol. 8 476
SEGRETARIA PER TUTTI, LA Vol. 7 477, 493
SEGRETARIA PRIVATA DI MIO PADRE Vol. 7 947
SEGRETARIA PRIVATA, LA Vol. 7 851, 1036, 1116
SEGRETE ESPERIENZE DI LUCA E FANNY, LE Vol. 7 540
SEGRETI SEGRETI Vol. 7 482, 868, 1105
SEGRETISSIMO Vol. 8 396, 775
SEGRETO DEI COBRA, IL Vol. 7 26
SEGRETO DEL GOBBO, IL Vol. 7 1189
SEGRETO DEL LAGO, IL Vol. 7 1189
SEGRETO DEL MORTO, IL Vol. 7 196
SEGRETO DELLA FIDANZATA, IL Vol. 7 1189
SEGRETO DELLA GROTTA AZZURRA, IL Vol. 7 390
SEGRETO DELLA MINIERA D'ORO, IL Vol. 7 1090
SEGRETO DELLE TRE PUNTE, IL Vol. 7 16
SEGRETO DI EMMA, IL Vol. 7 1189
SEGRETO DI JACK, IL Vol. 7 168
SEGRETO DI STATO Vol. 7 1189
SEGRETO DI SUOR ANGELA, IL Vol. 7 619
SEGRETO DI UN BANDITO, IL Vol. 7 81
SEGRETO DI VILLA PARADISO, IL Vol. 7 40, 595
SEGRETO, IL Vol. 7 260, 1209
SEGRETO INVIOLABILE, IL Vol. 7 360, 1177
SEGRETO NEL CHIOSTRO, UN Vol. 7 1005
SEGRETTO DEL DOTTORE, IL Vol. 7 978
SEGUNDA GUERRA DE LOS NINOS, LA Vol. 8 299, 373
SEGUNDO LOPEZ, AVENTURERO URBANO Vol. 8 477
SEHNSUCHT Vol. 4 288, 479
SEHNSUCHT NACH AFRIKA Vol. 4 18, 57
SEI BAMBINE E IL PERSEO Vol. 7 1136, 1159
SEI DONNE PER L'ASSASSINO Vol. 7 1177
SEI GEGRÜSST, DU MEIN SCHÖNES SORRENT Vol. 4 255, 548, 720
SEI GETREU BIS IN DEN TOD Vol. 6 *page 194*
SEI MIA ! Vol. 7 409
SEI MOGLI DI BARBABLU, LE Vol. 7 16
SEI TU FELICITA Vol. 7 1070
SEIKKAILU JALKAMATKALLA Vol. 5 344, 359
SEILERGASSE 8 Vol. 1 58, 129
SEIMILA CHILOMETRI DI PAURA Vol. 7 692
SEIN BESTER FREUND Vol. 4 13, 49, 119, 479, 738, 779, 788
SEIN GROSSER FALL Vol. 4 89, 306
SEIN GROSSER SIEG Vol. 1 56, 64, 115
SEIN GRÖSSTER BLUFF Vol. 4 270, 471, 779
SEIN IST DAS GERICHT Vol. 4 648, 663
SEIN LETZTER BERICHT: AUS DEM VATERHAUS VERSTOSSEN Vol. 4 300
SEIN LETZTER FALL Vol. 1 17, 98
SEIN LETZTER TRICK Vol. 4 98, 112
SEIN LETZTES EDELWEISS Vol. 4 25, 571
SEIN LETZTES MODELL Vol. 4 139, 723

SEIN SCHEIDUNGSGRUND Vol. 4 270, 409, 527, 751
SEIN SOHN Vol. 4 340, 582, 638
SEIN STÖRENFRIED Vol. 4 356, 779
SEIN WEIB Vol. 4 98, 686
SEINE BEICHTE Vol. 4 130, 131, 240
SEINE BESTE ERFINDUNG Vol. 4 329, 774
SEINE BESTE ROLLE Vol. 4 243, 273
SEINE ERSTE LIEBE Vol. 4 79, 115, 742
SEINE EXCELLENZ - DER REVISOR Vol. 6 *page 193*
SEINE EXZELLENZ VON MADAGASKAR Vol. 4 186, 479
SEINE FRAU, DIE UNBEKANNTE Vol. 4 186, 306
SEINE HOHEIT - GENOSSE PRINZ Vol. 1 63, 79, 115
SEINE KAMMERJUNGFER Vol. 4 603
SEINE KLEINE MADONNA Vol. 4 226
SEINE KOKETTE FRAU Vol. 4 5
SEINE LETZTE MASKE Vol. 4 159
SEINE SCHWERSTE STUNDE Vol. 4 18, 461
SEINE STÄRKSTE WAFFE Vol. 4 119, 270, 502, 779
SEINERZEIT ZU MEINER ZEIT Vol. 4 65, 287, 774
SEINES BRUDERS LEIBEIGENER Vol. 4 153, 209, 212, 705
SEINS DE GLACE, LES Vol. 2 21, 293
SEITENSPRUNG Vol. 6 *page 167, page 175, page 176*
SEITENSPRÜNGE Vol. 4 340, 379, 410, 485
SEITENSPRÜNGE DES HERRN BLOHM Vol. 4 247, 461
SEITSEMÄNVELJESTÄ Vol. 5 356, 428
SEKI SNIMA-PAZI SE ! Vol. 3 359, 383
SEKRET Vol. 1 202, 227, 289 Vol. 6 *page 225, page 257*
SEKRET ENIGMY Vol. 6 *page 240, page 241, page 245, page 260*
SEKRETÄRIN DES GESANDTEN, DIE Vol. 4 558
SEKSDAGESLOBET Vol. 5 31, 241, 248
SEKSMISJA Vol. 6 *page 219, page 245, page 253*
SEKSOLATKI Vol. 1 156, 239 Vol. 6 *page 240*
SEKSTET Vol. 5 2, 7
SELBST IST DER MANN Vol. 4 239, 437
SELBSTGERICHTET Vol. 4 583
SELCE E L'ACCIAIO, LA Vol. 7 971
SELIGE EXZELLENZ, DIE Vol. 4 89, 115, 135, 263, 361, 461
SELIKA Vol. 7 787, 899
SELIM NUNO A BÖRZECEZAR Vol. 9 234
SELJACKA BUNA 1573 Vol. 3 341, 350
SELJANIN"T S KOLELOTO Vol. 3 102, 105 Vol. 6 *page 26*
SELKOR Vol. 3 75, 124 Vol. 6 *page 39, page 56*
SELKSKABSDAMEN Vol. 5 58
SELKVINNEN Vol. 5 717, 738, 828
SELLA D'ARGENTO Vol. 7 834
SELLAISENA KUIN SINÄ MINUT HALUSIT Vol. 5 310, 363, 379, 419, 510, 634
SELLÖ A PECSETGYÜRÜN Vol. 9 152, 162, 203
SELO KRAJ ZAVOD Vol. 3 99, 124
SELSKOTO CUDOVISTE Vol. 3 34
SELTSAME GESCHICHTE DES BARONS TORELLI, DIE Vol. 4 159
SELTSAME KÖPFE Vol. 4 159
SELTSAME NACHT DER HELGA WANGEN, DIE Vol. 4 242, 461
SELTSAME VERGANGENHEIT DER THEA CARTER, DIE Vol. 4 626, 691 Vol. 6 *page 183*
SELTSAMEN ABENTEUER DES HERRN FRIDOLIN B., DIE Vol. 1 6, 23, 90
SELTSAMER GAST, EIN Vol. 4 135, 247, 461
SELV OM DE ER SMA Vol. 5 717, 738
SELVA BLANCA, LA Vol. 8 303, 442
SELVA ESTA LOCA, LOCA, LOCA, LA Vol. 8 455
SELVAGGIA, LA Vol. 7 19
SELVMORDERLUNDEN Vol. 5 97
SELVMORDSSKOLEN Vol. 5 179, 248
SEM HLADOVYCH Vol. 10 44, 149, 251
SEM SOMBRA DE PECADO Vol. 8 181
SEMAINE DE VACANCES, UNE Vol. 2 351, 682
SEMANA DE FELICIDAD, UNA Vol. 8 718
SEME DELL'UOMO, IL Vol. 7 1194
SEMINARISTE, LE Vol. 7 501, 1202
SEMMELWEIS Vol. 9 71, 237, 241, 299, 322, 326
SEMMELWEISS - RETTER DER MÜTTER Vol. 1 39, 58, 107
SEMNUL SARPELUI Vol. 6 *page 276*
SEMPRE AVANTI, SAVOIA ! Vol. 7 829
SEMPRE PIU DIFFICILE Vol. 7 603, 709
SEN FRATERA ONDREJE Vol. 10 331
SEN NOCI ... Vol. 10 174, 203, 229, 361, 568, 621
SENDA IGNORADA Vol. 8 207
SENDA TORCIDA Vol. 8 732, 753
SENDAS CRUZADAS Vol. 8 677, 732
SENDAS MARCADAS Vol. 8 217, 537
SENDERO DE ESPERANZA Vol. 8 590
SENDERO DEL ODIO, EL Vol. 8 276
SENECHAL LE MAGNIFIQUE Vol. 2 318, 347, 807
SENHORA DE FATIMA Vol. 8 94, 199

SENILITA Vol. 7 817, 1035
SENIORCHEF, DER Vol. 4 28, 381, 445
SENKYRKA U DIVOKE KRASY Vol. 10 122, 453, 586
SENNI JA SAVON SULTTAANI Vol. 5 443, 646
SENOR DE LA SALLE, EL Vol. 8 517, 749
SENOR DON JUAN TENORIO, EL Vol. 8 357
SENOR ESTA SERVIDO, EL Vol. 8 331, 627
SENOR ESTEVE, EL Vol. 8 276, 298
SENOR GALINDEZ, EL Vol. 8 373, 451
SENORA AMA Vol. 2 646 Vol. 8 607, 749
SENORA DE FATIMA, LA Vol. 8 214, 484
SENORA DOCTOR Vol. 8 627, 749
SENORA ESTUPENDA, UNA Vol. 8 406, 492
SENORA LLAMADA ANDERS, UNA Vol. 8 603, 626
SENORITA DE TREVELEZ, LA Vol. 8 244, 268
SENORITA INUTIL, LA Vol. 8 246, 657
SENORITAS DE MALA COMPANIA, LAS Vol. 8 261
SENORITO OCTAVIO, EL Vol. 8 634, 732
SENORITO Y LAS SEDUCTORAS, EL Vol. 8 207, 732
SENSATION IM WINTERGARTEN Vol. 4 149, 219, 660
SENSATIONS-PROZESS Vol. 4 247, 461
SENSATIONSPROZESS CASILLA Vol. 4 82, 329
SENSO Vol. 2 282 Vol. 7 21, 555, 597, 988, 1054, 1142
SENSUALIDAD Vol. 8 219, 338
SENSUALITA Vol. 7 1136
SENSUELA Vol. 5 638
SENTA AUF ABWEGEN Vol. 1 26, 50, 90
SENTENCE, LA Vol. 2 224, 285, 770
SENTENCIA CONTRA UNA MUJER Vol. 8 523, 732
SENTIERI DELLA VITA, I Vol. 7 279
SENTIERI DELL'ODIO, I Vol. 7 451 Vol. 8 276
SENTIERO DELLA GLORIA, IL Vol. 7 279
SENTINELLA DELLA PATRIA Vol. 7 621
SENTINELLA MORTA, LA Vol. 7 436, 559
SENTINELLA PIETRIFICATA Vol. 7 1189
SENTINELLE DI BRONZO Vol. 7 476, 906, 1116
SENTINELLE ENDORMIE, LA Vol. 2 270, 661
SENTINTA Vol. 3 307, 329
SENTIVANO ... IMO STRANO, ECCITANTE, PERICOLOSO PUZZO DI DOLLARI Vol. 7 692
SENZA AMORE Vol. 7 529
SENZA BANDIERA Vol. 7 792
SENZA BUCCIA Vol. 8 627
SENZA CIELO Vol. 2 100 Vol. 7 149, 174, 1177
SENZA COLPA Vol. 7 212, 560
SENZA FAMIGLIA Vol. 7 1054
SENZA FAMIGLIA NULLATENENTI, CERCANO AFFETTO ... Vol. 7 1194
SENZA MAMMA Vol. 7 39
SENZA NOME Vol. 7 390
SENZA PADRE Vol. 7 108
SENZA PECCATO Vol. 7 1162
SENZA PIETA Vol. 7 522, 1136
SENZA RAGIONE Vol. 7 1137
SENZA SAPERE NIENTE DI LEI Vol. 7 1025
SENZA SCRUPOLI Vol. 7 19, 770
SENZA SOLE Vol. 7 493, 508
SENZA UNA DONNA Vol. 7 443, 476, 1062
SENZA VELI Vol. 7 586, 938
SENZA VIA D'USCITA Vol. 8 611, 732
SENZI MAMA Vol. 10 268, 459, 464
SEPARACION MATRIMONIAL Vol. 8 284, 299, 681
SEPARATI IN CASA Vol. 7 11, 383, 667
SEPKAJUCI FANTOM Vol. 10 32, 263, 483
SEPOLCRO DEI RE, IL Vol. 7 196
SEPOLTA VIVA Vol. 7 1194
SEPPE MORIR E FU REDENTO Vol. 7 1162
SEPT FOIS FEMME Vol. 2 285
SEPT GARS ET UNE GARCE Vol. 2 347
SEPT GARS...UNE GARCE Vol. 3 244, 286
SEPT HOMMES...UNE FEMME Vol. 2 67, 722
SEPT MORTS SUR ORDONNANCES Vol. 2 163, 873
SEPT PARTOUZARDS DE L'OUEST, LES Vol. 2 692
SEPT PECHES CAPITAUX-LA COLERE, LES Vol. 2 655
SEPT PECHES CAPITAUX-LA GOURMANDISE, LES Vol. 2 655
SEPT PECHES CAPITAUX-LA LUXURE, LES Vol. 2 224
SEPT PECHES CAPITAUX-LA PARESSE, LES Vol. 2 224
SEPT PECHES CAPITAUX-L'AVARICE, LES Vol. 2 700
SEPT PECHES CAPITAUX, LES Vol. 2 47, 156, 251, 546, 826, 837, 863 Vol. 7 1062
SEPT PECHES CAPITAUX-L'ORGUEIL, LES Vol. 2 224
SEPTEMBERLIEBE Vol. 1 40, 48, 100
SEPTEMBRIE Vol. 3 292, 320

SEPTEMVRIJCI Vol. 3 56, 75, 104
Vol. 6 *page 58*
SEPTIEMBRE, EN Vol. 8 358, 580
SEPTIEME CIEL, LE Vol. 2 194, 232, 488
SEPTIEME COMMANDEMENT, LE Vol. 2 128, 344, 454
SEPTIEME COMPAGNIE AU CLAIR DE LUNE, LA Vol. 2 154, 368
SEPTIEME COMPAGNIE FAIT TOUT SAUTER, LA Vol. 2 368
SEPTIEME JOUR DE SAINT-MALO, LE Vol. 2 866
SEPTIEME JURE, LE Vol. 2 118
SEPTIEME PORTE, LA Vol. 2 322, 368
SEPTIMA PAGINA Vol. 8 311, 436
SEQUESTRATI DI ALTONA, I Vol. 7 483, 514
SER, EL Vol. 8 426
SERA C'INCONTRAMMO, UNA Vol. 7 221
SERAIL Vol. 2 35, 787
SERATA Vol. 3 239, 256, 274
SERBARILE GALANTE Vol. 3 218, 286
SERCE BATIARA Vol. 1 278, 404
SERCE MATKI Vol. 1 278, 308, 404
SERCE NA ULICE Vol. 1 346, 384
SERENAADI SOTATORVELLA Vol. 5 349, 566
SERENAADILUUTNANTTI Vol. 5 374, 443
SERENADA PENTRU ETAJUL XII Vol. 3 214, 331
SERENADE Vol. 2 194, 425, 652, 719, 863
Vol. 4 82, 265, 614
SERENADE AU BOURREAU Vol. 2 254, 309
SERENADE AU TEXAS Vol. 2 419, 598
SERENADE AUX NUAGES Vol. 2 67, 863
SERENATA A LA LUZ DE LA LUNA Vol. 8 256
SERENATA A MARIA Vol. 7 1121
SERENATA AL VENTO Vol. 7 587
SERENATA AMARA Vol. 7 476
SERENATA DELLA MORTE, LA Vol. 7 196, 563
SERENATA ESPANOLA Vol. 8 298, 311, 436
SERENATA ITALICA Vol. 7 493
SERENATA PER SEDICI BRONDE Vol. 7 603
SERENATA TRAGICA Vol. 7 1170
SERENATELLA SCIUE'SCIUE' Vol. 7 603
SERENIDADE Vol. 8 169
SERENISSIMUS UND DIE LETZTE JUNGFRAU Vol. 4 208, 560
SERGE PANINE Vol. 2 273, 338, 339, 820
SERGEANT BERRY Vol. 4 344, 430, 437, 531, 638
SERGENT X Vol. 2 615, 719, 873
SERGENT X, LE Vol. 2 187, 574, 833
SERGENTE D'ISPEZIONE Vol. 7 587
SERGIL CHEZ LES FILLES Vol. 2 471, 514
SERGIOS KAI ANNA Vol. 3 170
SERGIUS PANIN Vol. 9 77
SERGYL ET LE DICTATEUR Vol. 2 328, 496
SERIE COMIQUE Vol. 2 792
SERIE D'ART Vol. 2 792
SERIE D'ART ET D'HISTOIRE Vol. 2 792
SERIE DES BEBES Vol. 2 792
SERIE DRAMATIQUE Vol. 2 792
SERIE FILMS HISTORIQUES Vol. 2 792
SERIE NOIRE Vol. 2 158, 205, 351, 857
SERIEUX COMME LE PLAISIR Vol. 2 49
SERIF ZA MREZAMI Vol. 10 72, 544
SERIOUS CHARGE Vol. 2 656
SERPE CONTRO SERPE Vol. 7 442
SERPE, LA Vol. 7 264, 701
SERPENT, LE Vol. 2 719, 770
SERPENTE A SONAGLI, IL Vol. 7 460, 508, 1029
SERPIENTE DE MAR Vol. 8 407, 627
SERRA BRAVA Vol. 8 80, 96
SERTES Vol. 9 60, 10, 215
SERTORIO Vol. 8 174
SERVA PADRONA, LA Vol. 7 1189
SERVANTE ET MAITRESSE Vol. 2 810
SERVANTE, LA Vol. 2 808, 864
SERVANTES IZ MALOG MISTA Vol. 6 *page 311, page 315*
SERVICE DE NUIT Vol. 2 339, 615
SERVICIO EN LA MAR Vol. 8 311, 436
SERVIZIO POSTALE Vol. 7 1189
SESENTA HORAS EN EL CIELO Vol. 8 522, 640
SESION CONTINUA Vol. 8 368, 681
SESSO ALLEGRO Vol. 7 311
SESSO E VOLENTIERI Vol. 7 408, 697
SESSO NERO Vol. 7 152, 771
SESSO PROFONDO Vol. 7 1015, 1086
SESSOACERBO Vol. 7 742
SEST CERNYCH DIVEK ANEB PROC ZMIZEL ZAJIC ? Vol. 10 41, 227, 352
SEST DANA JUNA Vol. 6 *page 303, page 310*
SEST MEDCEDU S CIBULKOU Vol. 10 286, 388, 425
SEST MUSKETYRU Vol. 10 257, 453
SESTA BRZINA Vol. 6 *page 313, page 317, page 321*
SESTAPADESAT NEOMLUVENYCH HODIN Vol. 10 144, 173, 352
SESTRA ANGELIKA Vol. 10 92, 453
SESTRICKY Vol. 10 87, 139, 535
SET DEL PICERE, IL Vol. 7 414
SETEA Vol. 3 306, 310, 312
SETEA PAMINTULUI Vol. 3 269
SETIMO DE CABELLERIA, EL Vol. 8 541

SETKANI V CERVENCI **Vol. 10** 87, 222, 568
SETRELE PISMO **Vol. 10** 167
SETSNACTILETA **Vol. 10** 213, 331
SETTE ANNI DI FELICITA **Vol. 7** 443, 1117
SETTE CANZONE PER SETTE SORELLE **Vol. 7** 603
SETTE DELL'ORSA MAGGIORE, I **Vol. 7** 1136
SETTE FOLGORI DI ASSUR, LE **Vol. 7** 1027
SETTE GIORNI ALL'ALTRO MONDO **Vol. 7** 493, 756
SETTE GIORNI CENTO LIRE **Vol. 7** 196, 508, 756
SETTE NANI ALLA RISCOSSA **Vol. 7** 528
SETTE NOTE IN NERO **Vol. 7** 1015
SETTE ORE DI GUAI **Vol. 7** 1111
SETTE PECCATI CAPITALI **Vol. 7** 1062
SETTE PECCATI, I **Vol. 7** 532, 786, 938
SETTE PISTOLE PER I MACGREGOR **Vol. 8** 331, 626, 762
SETTE SCIALLI DI SETA GIALLA **Vol. 7** 694
SETTE SPADE DEL VENDICATORE, LE **Vol. 7** 739
SETTE UOMINI D'ORO **Vol. 7** 404 **Vol. 8** 626
SETTIMA DONNA, LA **Vol. 7** 909
SETTIMANA AL MARE, LA **Vol. 7** 315, 672
SETTIMANA BIANCA, LA **Vol. 7** 315, 672
SEUIL DU VIDE, LE **Vol. 2** 133, 325, 793
SEUL AMOUR, UN **Vol. 2** 158, 566, 676
SEUL DANS LA NUIT **Vol. 2** 388
SEUL DANS PARIS **Vol. 2** 231, 581
SEUL...A CORPS PERDU **Vol. 2** 3, 344
SEULS AU MONDE **Vol. 2** 615, 833
7 BALAS PARA SELMA **Vol. 8** 7
7 DI MARSA MATRUH, I **Vol. 7** 1024
7, HYDEN PARK – LA CASA MALEDETTA **Vol. 7** 99
SEVEN NIGHTS IN JAPAN **Vol. 2** 224
7 – 9 – 13 **Vol. 5** 24
7 ORE DI GUAI **Vol. 7** 638
7 UOMINI D'ORO NELLO SPAZIO **Vol. 7** 369, 474
1789 **Vol. 2** 478, 587, 879
70 VECES 7 **Vol. 8** 719
SEVERINO **Vol. 1** 46, 57, 96
SEVERNI PRISTAV **Vol. 10** 70, 352, 564, 582
SEVILLANE, LA **Vol. 2** 88, 871
SEX O-CLOCK U.S.A. **Vol. 2** 196
SEX-POWER **Vol. 2** 781
SEX SHOP **Vol. 2** 263, 517
SEXE A LA BOUCHE, LE **Vol. 2** 349
SEXE A TRAVERS LE MONDE, LE **Vol. 6** *page 118, page 119*
SEXE FAIBLE, LE **Vol. 2** 41, 469, 824
SEXE NU, LE **Vol. 2** 513
SEXLINGAR **Vol. 5** 840, 878
SEXO SANGRIENTO **Vol. 8** 767
SEXOMANIA **Vol. 3** 142
SEXORCISME **Vol. 2** 750
SEXTA-FEIRA, 13 **Vol. 8** 29
SEXTANERIN, DIE **Vol. 10** 256, 460
SEXTANKA **Vol. 10** 256, 348, 453, 460
SEXTETTEN KARLSSON **Vol. 5** 876, 923
SEXTO SENTIDO, EL **Vol. 8** 270, 600, 676
SEXTON BLAKE **Vol. 5** 121
SEXUAL ABERRATION – SESSO PERVERSO **Vol. 7** 311
SEXUELLA **Vol. 2** 795
SEXUELLA SEXY **Vol. 2** 795
SEXY CAT **Vol. 8** 583
SEXY GANG **Vol. 2** 126
SEXYRELLA **Vol. 2** 294
SEZONA MIRA U PARIZU **Vol. 6** *page 314, page 317*
SEZONOWA MILOSC **Vol. 1** 174, 215, 326, 345
SF-PARAATI **Vol. 5** 485, 615
SFIDA AL RE DI CASTIGLIA **Vol. 8** 523, 749
SFIDA ALLA MORTE **Vol. 7** 108
SFIDA, LA **Vol. 7** 688, 915, 1167
SFINGE, D'ACCIAIO, LA **Vol. 7** 1005
SFINGE D'ORO, LA **Vol. 8** 243, 555
SFINGE, LA **Vol. 7** 264, 701, 1113
SFINGE SORRIDE PRIMA DI MORIRE – STOP – LONDRA **Vol. 7** 1179
SFINTA TEREZA SI DIAVOLII **Vol. 3** 239, 256
SFINXENS HEMMELIGHED **Vol. 5** 136
SFIRSITUL NOPTII **Vol. 6** *page 265, page 269, page 276*
SFRATTATO CERCO CASA EQUO CANONE **Vol. 7** 882, 1202
SGARRO, LO **Vol. 7** 688
SHARK – ROSSO NELL'OCEANO **Vol. 7** 426
SHARON VESTIDA DE ROJO **Vol. 8** 249, 383
SHEHERAZADE **Vol. 2** 100, 521, 566, 833, 863 **Vol. 4** 66 **Vol. 8** 311
SHEMBJA E IDHUJVE **Vol. 3** 2, 11 **Vol. 6** *page 13*
SHEPHERD KING, THE **Vol. 7** 978
SHERIFF NO DISPARA, EL **Vol. 8** 772
SHERIFF TEDDY **Vol. 1** 81, 100, 115
SHERIFF TERRIBLE, EL **Vol. 8** 626, 752
SHERLOCK HOLMES **Vol. 4** 430, 558, 796 **Vol. 5** 97
SHETLANDSJENGEN **Vol. 5** 738, 762

SHINING SEX Vol. 2 139
SHIRAT E VJESHTES Vol. 6 *page 14, page 16*
SHOCK Vol. 7 1086
SHOCK CORRIDOR Vol. 2 526
SHOCKING Vol. 2 294
SHOCKING BOYS Vol. 6 *page 96*
SHOKET Vol. 6 *page 15, page 18, page 19*
SHOKU YNE TILI Vol. 6 *page 12, page 15, page 17*
SHPERTHIMI Vol. 3 11, 26
SHTEPIA JONE PERBASHKET Vol. 6 *page 11, page 13, page 17*
SHTIGJE TE LUFTES Vol. 3 3, 5, 23
SHYLOCK VON KRAKAU, DER Vol. 4 734
SI CA PEUT VOUS FAIRE PLAISIR Vol. 2 344, 401
SI CA VOUS CHANTE Vol. 2 224
SI C'ETAIT A REFAIRE Vol. 2 491, 616 Vol. 6 *page 85*
SI ELLE DIT QUI ... JE NE DIS PAS NON ! Vol. 6 *page 74, page 79*
SI FULANO FOSE MENGANO Vol. 8 331, 553
SI GENTILLE PETITE FILLE, UNE Vol. 2 814
SI GJITHE TE TJERET Vol. 6 *page 13, page 18*
SI ILIE FACE SPORT Vol. 3 310
SI J'ETAIS LE PATRON Vol. 2 48, 439
SI J'ETAIS UN ESPION Vol. 2 235, 666
SI JEUNESSE SAVAIT Vol. 2 137, 309
SI JOLI VILLAGE, UN Vol. 2 175
SI JOLIE PETITE PLAGE, UN Vol. 2 13, 194
SI J'TE CHERCHE J'ME TROUVE Vol. 6 *page 90, page 95*
SI LAS MUJERES MANDARAN Vol. 8 237, 297
SI LE ROI SAVAIT CA Vol. 2 863
SI ... LO VOGLIO ! Vol. 7 713
SI MA GUEULE VOUS PLAIT ... Vol. 6 *page 85, page 108, page 128, page 138*
SI PARIS M'ETAIT CONTE Vol. 2 540
SI PARIS NOUS ETAIT CONTE Vol. 2 8
SI PRES SI LOIN PLACE DE LA REUNION Vol. 2 647
SI PUO FARE ... AMIGO Vol. 7 1136 Vol. 8 749
SI QUIERES VIVIR ... DISPARA Vol. 8 685
SI RINGRAZIA LA REGIONE PUGLIA PER AVERCI FORNITO I MILANESI Vol. 7 1022
SI SALVE CHI VUOLE Vol. 7 623, 947
SI TE HUBIERAS CASADO CONMIGO Vol. 8 214, 400
SI TE VDISET Vol. 3 419 Vol. 6 *page 308*
SI TOUS LES GARS DU MONDE Vol. 2 388, 626, 824
SI TU REVIENS Vol. 2 419
SI TU VEUX Vol. 2 152, 330
SI VE VULESSE BENE Vol. 7 39
SI VERSAILLES M'ETAIT CONTE Vol. 2 601, 626, 722
SI VOLVEMOS A VERNOS Vol. 8 348, 772
SIAMO DONNE Vol. 7 157, 725, 910, 915, 1036, 1062
SIAMO FATTI COSI Vol. 7 474
SIAMO IN RITARDO Vol. 7 1189
SIAMO RICCHI E POVERI Vol. 7 528
SIAMO TUTTI IN LIBERTA PROVVISORIA Vol. 7 1044
SIAMO TUTTI INQUILINI Vol. 7 1044
SIAMO TUTTI MILANESI Vol. 7 891
SIAMO TUTTI NECESSARI Vol. 7 1058
SIAMO UOMINI O CAPORALI ! Vol. 7 1136
SIBIL Vol. 3 63, 119 Vol. 6 *page 28*
SIBIRSKA LEDI MAGBET Vol. 3 349, 436
SICARIO, IL Vol. 7 95, 877, 1184
SICILIANERBROVERENS BRYLLUP Vol. 5 300
SICILIEN, LE Vol. 2 118, 875
SIDI-BEL-ABBES Vol. 2 158, 407, 433
SIDI-BRAHIM Vol. 2 339, 368
SIDONIE PANACHE Vol. 2 84, 404
SIDSTE AF SLAEGTEN, DEN Vol. 5 157
SIDSTE DANS, DEN Vol. 5 157, 183
SIDSTE NAT, DEN Vol. 5 295
SIDSTE VINTER, DEN Vol. 5 2, 160
SIE KANNTEN SICH ALLE Vol. 1 35, 58, 125
SIE NANNTEN IHN AMIGO Vol. 1 9, 34, 101
SIE NANNTEN IHN GRINGO Vol. 8 438, 546
SIE ODER SIE Vol. 4 361, 558
SIE UND DIE DREI Vol. 4 123, 319, 397, 774, 788
SIEBEN AFFAIREN DER DONA JUANITA, DIE Vol. 1 7, 48, 73
SIEBEN BRIEFE Vol. 4 273, 688
SIEBEN JAHRE GLÜCK Vol. 4 758 Vol. 7 443, 1177
SIEBEN JAHRE PECH Vol. 4 375, 391, 558
SIEBEN OHRFEIGEN Vol. 4 325, 713
SIEBEN SOMMERSPROSSEN Vol. 1 52, 71, 121
SIEBEN TÖCHTER DER FRAU GYURKOVICS, DIE Vol. 4 101, 288 Vol. 5 988
SIEBENTE JAHR, DAS Vol. 1 1, 32, 100
SIEBENTE JUNGE, DER Vol. 4 255, 344, 524, 544, 623
SIEBENTE KUSS, DER Vol. 4 385

SIEBENTE NACHT, DIE **Vol. 4** 223, 799
SIEBZEHNJÄHRIGE, EINE **Vol. 4** 149, 354, 660
SIEBZEHNJÄHRIGEN, DIE **Vol. 4** 306, 743, 798
SIEDEM-SIEDEM-SIEDEM **Vol. 1** 206
SIEDMY KONTINENT **Vol. 10** 268, 287, 527
SIEG DER JUGEND, DER **Vol. 4** 648, 747, 762
SIEG DES HOSENROCKS, DER **Vol. 4** 646
SIEG DES MAHARADSCHA, DER **Vol. 4** 270, 725
SIEGA VERDE **Vol. 8** 214, 449, 611
SIEGER, DER **Vol. 4** 325, 567
SIEGERIN, DIE **Vol. 6** *page 184, page 198*
SIEGFRIED **Vol. 7** 1038
SIEGFRIEDS TOD **Vol. 4** 288, 567
SIEH DEN MENSCHEN **Vol. 1** 2, 65
SIEMPRE EN MI RECUERDO **Vol. 8** 497, 541
SIEMPRE ES DOMINGO **Vol. 8** 762
SIEMPRE MUJERES **Vol. 8** 400
SIEMPRE VUELVEN DE MADRUGADA **Vol. 8** 214, 486
SIERRA DE ARACENA, LA **Vol. 8** 357
SIERRA DE RONDA **Vol. 8** 648
SIERRA MAESTRA **Vol. 7** 502
SIERRA MALDITA **Vol. 8** 541, 614, 748, 749
SIESTA VETA **Vol. 10** 82, 454, 548
SIESTE, LA **Vol. 2** 678
SIETE BRAVISIMOS, LOS **Vol. 8** 396
SIETE CALLES **Vol. 8** 560, 782
SIETE CHICAS PELIGROSAS **Vol. 8** 643
SIETE CUCAS, LAS **Vol. 8** 507
SIETE DE PANCHO VILLA, LOS **Vol. 8** 583, 772
SIETE DIAS DE ENERO **Vol. 8** 549, 773
SIETE DOLARES AL ROJO **Vol. 8** 207
SIETE ESPARTANOS, LOS **Vol. 8** 541, 725
SIETE ESPIAS EN LA TRAMPA **Vol. 8** 207, 495
SIETE HOMBRES DE ORO **Vol. 8** 626
SIETE MAGNIFICAS, LAS **Vol. 8** 243
SIETE MINUTOS PARA MORIR **Vol. 8** 331
SIETE PISTOLAS PARA LOS MACGREGOR **Vol. 8** 331, 626, 762
SIETE PISTOLAS PARA TIMOTHY **Vol. 8** 562
SIETE VIDAS DE GATO, LAS **Vol. 8** 306, 681
SIFFLEURS, LES **Vol. 5** 584
SIGARO, IL **Vol. 7** 1189
SIGFRIDO **Vol. 7** 827, 1027
SIGGE NILSSON OCH JAG **Vol. 5** 893, 923, 1205
SIGNAL IN DER NACHT **Vol. 4** 410, 738, 788
SIGNAL ROUGE, LE **Vol. 2** 173, 191, 475
SIGNALE – EIN WELTRAUMABENTEUER **Vol. 1** 37, 62, 87, 102, 103, 401
SIGNALI NAD GRADOM **Vol. 3** 366, 444
SIGNE, ARSENE LUPIN **Vol. 2** 57, 185
SIGNE DU LION, LE **Vol. 2** 395, 517
SIGNE FURAX **Vol. 2** 244 **Vol. 6** *page 105*
SIGNE ILLISIBLE **Vol. 2** 254, 529
SIGNES EXTERIEURS DE RICHESSE **Vol. 6** *page 78, page 148*
SIGNOR CAMILLO, CACCIATORE D'ORSI, IL **Vol. 7** 131
SIGNOR CAMILLO IN FASCE, IL **Vol. 7** 131
SIGNOR DIOTISALVI, IL **Vol. 7** 943, 1169
SIGNOR MAX, IL **Vol. 7** 196, 756, 1060
SIGNOR MINISTRO LI PRESE TUTTI E SUBITO **Vol. 7** 130
SIGNOR ROBINSON – MOSTRUOSA STORIA D'AMORE E D'AVVENTURE, IL **Vol. 7** 502
SIGNORA ARLECCHINO, LA **Vol. 7** 369
SIGNORA DELLA CAMELIE, LA **Vol. 7** 264, 701, 807
SIGNORA DELLA NOTTE, LA **Vol. 7** 30, 564, 1007
SIGNORA DELLE MINIERE, LA **Vol. 7** 548
SIGNORA DELLE PERLE, LA **Vol. 7** 399
SIGNORA DELLE ROSE, LA **Vol. 7** 399
SIGNORA DELL'OVEST, UNA **Vol. 7** 2, 48, 170, 552
SIGNORA DI MONTECARLO, LA **Vol. 2** 727 **Vol. 7** 587, 707, 822, 968
SIGNORA DI TUTTI, LA **Vol. 7** 48, 244
SIGNORA HA FATTO IL PIENO, LA **Vol. 7** 1026
SIGNORA IN NERO, LA **Vol. 7** 487
SIGNORA INNAMORATA, LA **Vol. 7** 886, 1079
SIGNORA PARADISO, LA **Vol. 7** 48, 460, 792
SIGNORA ROBINET, LA **Vol. 7** 1189
SIGNORA SENZA CAMELIE, LA **Vol. 7** 915, 1045, 1062
SIGNORA SENZA PACE, LA **Vol. 7** 1103
SIGNORE A DOPPIO PETTO, IL **Vol. 7** 792, 1036
SIGNORE CHE PRANZA, UN **Vol. 7** 1189
SIGNORE CHE SOFFRE IL SOLLETICO **Vol. 7** 1189
SIGNORE DELLA TAVERNA, IL **Vol. 7** 2, 701, 725
SIGNORE DESIDERA ?, IL **Vol. 7** 792, 796
SIGNORE E SIGNORI, BUONANOTTE **Vol. 7** 80, 778, 950, 1084

SIGNORE, LE **Vol. 7** 113
SIGNORE METODICO, IL **Vol. 7** 1189
SIGNORE NON BEVE, IL **Vol. 7** 1189
SIGNORETTE **Vol. 7** 902, 1036
SIGNORI GIURATI ! ... **Vol. 7** 37
SIGNORI, IN CARROZZA ! **Vol. 7** 796
SIGNORI LADRI, I **Vol. 7** 1189
SIGNORI SI NASCE **Vol. 7** 693
SIGNORINA CHICCHIRICCHI **Vol. 7** 569
SIGNORINA CICLONE, LA **Vol. 7** 456, 792
SIGNORINA DELL'ALTRO MONDO, LA **Vol. 7** 825
SIGNORINA DELL'AUTOBUS, LA **Vol. 7** 665, 730
SIGNORINA IN LOTTERIA, UNA **Vol. 7** 1067
SIGNORINA, LA **Vol. 7** 332, 432, 436, 460, 598, 838
SIGNORINA ... MADRE DI FAMIGLIA, LA **Vol. 7** 569
SIGNORINE DELLA VILLA ACCANTO, LE **Vol. 7** 1027, 1140
SIGNORINE DELLO 04, LE **Vol. 7** 333
SIGNORINELLA **Vol. 7** 1136
SIGNORINO, IL **Vol. 7** 430
SIGNUM LAUDIS **Vol. 10** 172, 283, 567
SIKATOR **Vol. 9** 83, 183, 332
SIKKEN EN NAT **Vol. 5** 151, 304
SIKKE'N FAMILIE **Vol. 5** 89, 143, 281
S'IL VOUS PLAIT, LA MER **Vol. 2** 349
SILA DEL TSORO, LA **Vol. 8** 611
SILBERKÖNIG, DER **Vol. 4** 113
SILBERMÖWE, DIE **Vol. 4** 428, 626
SILBERNE FESSEL, DIE **Vol. 4** 232, 300
SILBERNE LÖFFEL, DER **Vol. 4** 79, 115, 239
SILENCE DE LA MER, LE **Vol. 2** 224
SILENCE DES ORGANES, LE **Vol. 2** 341
SILENCE EST D'OR, LE **Vol. 2** 59, 332, 824
SILENCE, LE **Vol. 2** 167
SILENCE...ON TOURNE **Vol. 2** 326, 810
SILENCIEUX, LE **Vol. 2** 196, 674
SILENE SMUTNA PRINCEZNA **Vol. 10** 359, 498, 572
SILENY KANKAN **Vol. 10** 82, 129, 299, 466
SILENZIO, IL **Vol. 7** 1169
SILENZIO ! SI GIRA ! **Vol. 7** 86
SILJA – NUORENA NUKKUNUT **Vol. 5** 549, 579
SILKESSTRUMPAN **Vol. 5** 941
SILLABE ARDENTI **Vol. 7** 144
SILLANKORVAN EMÄNTÄ **Vol. 5** 372
SILMÄT HÄMÄRÄSSÄ **Vol. 5** 403, 530
SILNA VODA **Vol. 3** 30, 122 **Vol. 6** *page 31*
SILNEJSI NEZ STRACH **Vol. 10** 286, 408
SILOM OTAC **Vol. 3** 349, 374
SILOUTES **Vol. 3** 199
SILTALAN PEHTOORI **Vol. 5** 359, 379, 392, 485
SILURAMENTO DELL'OCEANIA, IL **Vol. 7** 456, 668
SILURI UMANI **Vol. 7** 852
SILVESTERPUNSCH **Vol. 1** 54, 68, 90
SILVESTRE **Vol. 8** 7, 89
SILVIA AMA A RAQUEL **Vol. 8** 590
SILVIA ZULU **Vol. 7** 471, 1190
SILVIO E LO STRADIVARIUS **Vol. 7** 965
SILVIO PELLICO **Vol. 7** 843
SIMEN MUSTROENS BESYNDERLIGE OPLEVELSER **Vol. 5** 669, 671, 741
SI'MO DDICE 'ORE **Vol. 7** 39
SIMO HURTTA **Vol. 5** 359, 497
SIMON BOLIVAR **Vol. 8** 276, 749
SIMON, CONTAMOS CONTIGO **Vol. 8** 297, 331
SIMON I BACKABO **Vol. 5** 840, 878
SIMON JUDIT **Vol. 9** 337
SIMON MENYHERT SZÜLETESE **Vol. 9** 241, 274
SIMON SYNDAREN **Vol. 5** 1033
SIMONE **Vol. 2** 346
SIMONE BARBES OU LA VERTU **Vol. 2** 281 **Vol. 6** *page 104*
SIMOUN, LE **Vol. 2** 419, 775
SIMPATICO MASCALZONE, UN **Vol. 7** 292, 891, 1118
SIMPATICUL DOMN R **Vol. 3** 239, 272
SIMPLET **Vol. 2** 23, 824
SIMPLETTE **Vol. 2** 64
SIMPLON-TUNNEL **Vol. 1** 22, 56, 115
SIMUN **Vol. 7** 467, 574, 727
SIN ALIENTO **Vol. 8** 598
SIN LA SONRISA DE DIOS **Vol. 8** 493, 732
SIN NOVEDAD EN EL ALCAZAR **Vol. 8** 475, 540
SIN UN ADIOS **Vol. 8** 438, 627, 667
SIN UNIFORME **Vol. 8** 311, 436
SIN-VENTURA, LA **Vol. 2** 245
SINÄ OLET KOHTALONI **Vol. 5** 435, 634
SINAIS DE VIDA **Vol. 8** 2
SINAL VERMELHO **Vol. 8** 125, 149
SINDACALISTA, IL **Vol. 7** 763
SINDBAD CONTRO I SETTE SARACENI **Vol. 7** 867
SINFONIA D'AMORE – SCHUBERT **Vol. 7** 798, 1035
SINFONIA DEL HOGAR **Vol. 8** 677
SINFONIA DEL MARE **Vol. 7** 409, 1169
SINFONIA FATALE **Vol. 7** 1054
SINFONIA PASTORALE **Vol. 7** 971
SING, COWBOY, SING **Vol. 6** *page 164, page 172, page 173*
SING SING **Vol. 7** 381, 408

SINGAPUR DE CART Vol. 6 *page 267, page 275, page 278*
SINGAPUR, HORA CERO Vol. 8 396, 772
SINGE EN HIVER, UN Vol. 2 185, 644
SINGENDE, KLINGENDE BÄUMCHEN, DAS Vol. 1 56, 90, 129
SINGENDE STADT, DIE Vol. 4 113, 724
SINGENDE TOR, DER Vol. 4 89 Vol. 7 902
SINGENDEN BABIES, DIE Vol. 4 89
SINGOALLA Vol. 2 388, 566 Vol. 5 968, 1060
SINGUR PRINTRE PRIETENI Vol. 3 235, 252, 286
SINGURATATEA FLORILOR Vol. 3 245, 297, 311
SININEN VARJO Vol. 5 476
SININEN VARJO ELI KESKIYÖN MURHA Vol. 5 645
SININEN VIIKO Vol. 5 374, 443
SINJATA BEZPREDELNOST Vol. 6 *page 57, page 58*
SINJI GALEB Vol. 3 445, 452
SINNESRAUSCH Vol. 4 192
SIN"T NA MARIJA Vol. 6 *page 36, page 47*
SINTFLUT Vol. 4 99, 150, 173, 612
SINUT MINÄ TAHDON Vol. 5 312, 379, 579
SIPOLO MACSKAKÖ, A Vol. 9 6, 259
SIRENA Vol. 7 493, 730 Vol. 10 132, 256, 564
SIRENA DEL CANTABRICO, LA Vol. 8 516
SIRENA, LA Vol. 7 1189
SIRENA NEGRA, LA Vol. 8 753
SIRENE DES TROPIQUES, LA Vol. 2 205, 277, 547, 775
SIRENE DU MISSISSIPPI, LA Vol. 2 188, 674
SIRIUS Vol. 10 286, 567
SIRO LJATO Vol. 6 *page 37*
SIROCCO Vol. 2 87, 407
SIROKKO Vol. 9 20, 154, 318
SIROKO JE LISCE Vol. 6 *page 315, page 319*
SIROMA SAM AL'SAM BESAN Vol. 3 430
SIROMASKA RADOST Vol. 3 37, 93, 126 Vol. 6 *page 26*
SIROMASKO LJATO Vol. 3 38, 52
SIRON TUL Vol. 9 71
SIROTA MARIJA Vol. 3 368, 379
SISKA Vol. 5 869, 940, 1042, 1053
SISSIGNORA Vol. 7 750, 796, 862
SISSIT Vol. 5 516, 613
SISTA ÄVENTYRET, DET Vol. 5 975
SISTA BUDET Vol. 5 1019, 1071
SISTA LEKEN Vol. 5 321, 429, 475, 509
SISTA NATTEN Vol. 5 914
SISTA PARET UT Vol. 5 878, 954
SISTA RINGEN Vol. 5 1073, 1080
SISTA STEGEN, DE Vol. 5 1042, 1080
SISTE FLEKSNES, DEN Vol. 5 664, 709, 808
SISTEMA PELEGRIN, EL Vol. 8 503, 677
SISTO V Vol. 7 1189
SISULLA JA SYDÄMELLÄ Vol. 5 472, 634
SITIADOS EN LA CIUDAD Vol. 8 437, 768
SITTENRICHTER PARAGRAPH 218, DER Vol. 4 221, 340
SITUATION DELICATE D'UN CAMBRIOLEUR Vol. 2 159, 286, 330
SITUATION EST GRAVE...MAIS PAS DESESPEREE, LA Vol. 2 368
SIVOOKY DEMON Vol. 10 115, 260, 584
SIX CENT MILLE FRANCS PAS MOIS Vol. 2 833
SIX ET DEMI, ONZE Vol. 2 426, 656
SIX HEURES A PERDRE Vol. 2 332, 601
SIX PETITES FILLES EN BLANC Vol. 2 249, 460
6. TRAEKNING Vol. 5 55, 101, 208
SIXIEME ETAGE Vol. 2 99, 174, 402
65, 66 OCH JAG Vol. 5 840, 841
SIXTYNINE (69) Vol. 5 1155
69 : AN, SERGEANTEN OCH JAG Vol. 5 871, 1073
69 SIXTYNINE Vol. 5 606
63 7157, LA Vol. 7 1067
SIZILIANISCHE BLUTRACHE Vol. 4 131
SJAELEBYTNING Vol. 5 97
SJAELETYVEN Vol. 5 58
SJÄTTE BUDET Vol. 5 876, 1073
SJÄTTE SKOTTET Vol. 5 871, 954
SJECAS LI SE DOLLY BELL Vol. 6 *page 303, page 313*
SJENKA SLAVE Vol. 3 366, 371
SJÖCHARMÖRER Vol. 5 840, 1119
SJÖMAN GAR I LAND, EN Vol. 5 958, 1121
SJÖMAN I FRACK, EN Vol. 5 928, 1042, 1156
SJÖMAN TILL HÄST, EN Vol. 5 863, 1032
SJÖSLAVAR Vol. 5 1042, 1080
SJOV I GADEN Vol. 5 76, 195
SJOVE AR, DE Vol. 5 89, 281
SJU SVARTA BE-HA Vol. 5 1033
SJU VACKRA FLICKOR Vol. 5 933, 1073
SJUNDE INSEGLET, DET Vol. 5 940, 1036, 1052
SJUNDES HIMLEN Vol. 5 909, 1052
SJUTTON AR Vol. 5 933, 1073
SKA' RU' ME' PA FEST ? Vol. 5 914
SKA' VI LEGE SKJUL ? Vol. 5 16, 61, 195
SKADESKUTT Vol. 5 738, 831
SKAEBNE, EN Vol. 5 295
SKAEBNENS VEJE Vol. 5 58

SKAEBNESVANGRE OPFINDELSE, DEN Vol. 5 97
SKAEBNESVANGRE VILDFARELSER Vol. 5 9, 197
SKAF EN SENSATION ! Vol. 5 55, 101, 208
SKAF MIG EN KAERESTE Vol. 5 83
SKAL VI DANSE FORST Vol. 5 185, 270, 283
SKAL VI VAEDDE EN MILLION ? Vol. 5 55, 101, 208
SKALNI PLEMENO Vol. 10 122, 564
SKALNI SEVCI Vol. 10 138, 524
SKALNI V OFSAJDE Vol. 10 268, 368, 451
SKALPEL, PROSIM Vol. 10 465, 511, 541
SKALY A LUDIA Vol. 10 210, 268
SKAMMEN Vol. 5 1017, 1052, 1080
SKANDAALI TYTTÖKOULUSSA Vol. 5 443, 481, 549, 639
SKANDAL IM VIKTORIA-CLUB, DER Vol. 4 232, 673
SKANDAL IN BADEN-BADEN Vol. 4 49, 115
SKANDAL IN BUDAPEST Vol. 4 139
SKANDAL IN DER PARKSTRASSE Vol. 4 452, 471
SKANDAL OCH KÄRLEK I KRÄKLUNDA Vol. 5 1075
SKANDAL UM DEN HAHN Vol. 4 339, 422, 616
SKANDAL UM DIE FLEDERMAUS Vol. 4 115, 634
SKANDAL UM EVA Vol. 4 626, 729
SKANDAL V GRI-GRI BARU Vol. 10 226, 317, 567
SKANDALEMEGEREN Vol. 5 83
SKANDALEN PAA SORUPGAARD Vol. 5 183
SKANINGAR Vol. 5 1045
SKANÖR – FALSTERBO Vol. 5 933, 1032
SKAPANI W OGNIU Vol. 1 283, 298 Vol. 6 *page 246*
SKARABÄUS, DER Vol. 4 119, 517, 558
SKARB Vol. 1 225, 298, 305, 363
SKARB KAPITANA MARTENSA Vol. 1 232, 305 Vol. 6 *page 242*
SKARB TRZECH LOTROW Vol. 1 249
SKAREDA DEDINA Vol. 10 325, 394, 517
SKÄRGÄRDFLIRT Vol. 5 928
SKÄRGARDSKALJERER Vol. 5 870
SKÄRGARDSNATT, EN Vol. 5 1003
SKÄRSELD Vol. 5 890, 895
SKAZANY Vol. 1 314, 378 Vol. 6 *page 258*
SKELETHAANDEN Vol. 5 81
SKELETROS Vol. 7 961
SKELETT, DAS Vol. 4 46, 277, 695
SKELETT DES HERRN MARKUTIUS, DAS Vol. 4 560, 729
SKEPP KOMMER LASTAT, ETT Vol. 5 923, 1092
SKEPP OHOJ ! Vol. 5 840, 841, 878
SKEPP TILL INDIALAND Vol. 5 1052, 1164
SKEPPAR JANSSON Vol. 5 863, 1032
SKEPPARE I BALSVÄDER Vol. 5 878, 1173
SKEPPARGATAN 40 Vol. 5 891, 1154
SKEPPARKÄRLEK Vol. 5 840, 841, 878, 909
SKEPPSBRUTNE MAX Vol. 5 855, 923, 978, 1042, 1193
SKETE PA MOLLEGAARDEN, DET Vol. 5 89, 281
SKETERRE 43 Vol. 3 11, 20
SKIBET ER LADET MED Vol. 5 7, 76
SKICKA HEM N:R 7 Vol. 5 863, 928, 1032
SKIERRI – VAIVAISKOVUJEN MAA Vol. 5 465, 498
SKIES STEN AMMO Vol. 3 142
SKILSMISSE, EN Vol. 5 92, 228
SKILSMISSENS BORN Vol. 5 31, 32, 55, 101, 208
SKILT OG ATTER FORENET Vol. 5 97
SKIMEISTER VON MORGEN Vol. 1 13, 34, 101
SKINDOD AEGTEMAND, EN Vol. 5 92
SKINOUSSA, PAYSAGE AVEC LA CHUTE D'ICARE Vol. 2 272
SKINSYGEDIGTEREN Vol. 5 83
SKJAER I SJOEN Vol. 5 718, 762, 799
SKJAERGARDSFLIRT Vol. 5 692, 770
SKLAMALAM Vol. 1 176, 278
SKLAVA Vol. 3 165
SKLAVEN DER LIEBE Vol. 4 212, 509, 571, 652, 753
SKLAVEN DER SEELEN Vol. 4 584, 637
SKLAVEN FREMDEN WILLENS Vol. 4 270, 570
SKLENENY DUM Vol. 10 116, 595, 622
SKLENIKOVA VENUSA Vol. 10 73, 82, 399
SKOGEN ÄR VAR ARVEDEL Vol. 5 843, 923, 1022
SKOK Vol. 1 209, 237, 391 Vol. 6 *page 242*
SKOK DO TMY Vol. 10 229, 369, 569
SKÖKE TILL SALU Vol. 5 863, 1032
SKOLA HRISNIKU Vol. 10 130, 369, 577
SKOLA OTCU Vol. 10 286, 412, 433
SKOLA ZAKLAD ZIVOTA Vol. 10 256, 524
SKOLKA SKOLAN Vol. 5 871, 1156
S'KOMMT ANDERS Vol. 6 *page 186*
SKÖNA HELENA Vol. 5 871, 880, 954
SKÖNA JUVELER Vol. 5 917, 1048, 1065, 1093

SKÖNA SUSANNA OCH GUBBARNA Vol. 5 909, 1052
SKONHEDEN OG UDYRET Vol. 5 125, 249, 296
SKONNE EVELYN, DEN Vol. 5 83, 295
SKORPAN Vol. 5 878, 1052
SKORPIO, A Vol. 9 28
SKORPION, PANNA I LUCZNIK Vol. 1 187 Vol. 6 *page 220, page 231*
SKORPION SRESTU D"GA Vol. 3 63, 73 Vol. 6 *page 36*
SKORSTENSFEJEREN KOMMER I MORGEN Vol. 5 262
SKORTEJAEGEREN Vol. 5 83
SKORVRIDERGARDEN Vol. 5 7
SKOTTET Vol. 5 910
SKOVSOENS DATTER Vol. 5 70
SKOYTEKONGEN Vol. 5 674
SKRAA BRAEDDER, DE Vol. 5 295
SKRADZIONA KOLEKCJA Vol. 6 *page 225, page 240, page 244*
SKRAPHANDLERE Vol. 5 664, 808, 824
SKRATTBOMBEN Vol. 5 1003, 1073
SKREALLAN, RUSPRICK OCH KNORHANE Vol. 5 862
SKRIFT I SNE Vol. 5 712
SKRIGET FRA DJAEVLEKLOFTEN Vol. 5 256
SKRIVANCI NA NITI Vol. 10 41, 516
SKRIVANCI PISEN Vol. 10 256, 586
SKRIVANCI TICHO Vol. 10 283, 318, 571
SKRUEBRAEKKEREN Vol. 5 83
SKRYTY PRAMEN Vol. 10 82, 145, 273, 320
SKRZYDLATY ZWYCIEZCA Vol. 1 140, 193, 380
SKUD FOR MIDNAT, ET Vol. 5 2, 228
SKUD I MORKET, ET Vol. 5 97
SKUESPILLEREN Vol. 5 97
SKUGGAN Vol. 5 862, 1042
SKULLE DET DUKKE OPP FLERELIK, ER DET BARE A RINGE Vol. 5 703, 718, 762
SKUPLJACI PERJA Vol. 3 350, 421
SKYLDIG - IKKE SKYLDIG Vol. 5 9
SKYTTEN Vol. 5 11, 175, 261, 270
SLA FORST, FREDE ! Vol. 5 16, 61, 249, 272
SLADDER Vol. 5 9
SLADKA JOSEFINKA Vol. 10 167, 345
SLADKE HRY MINULEHO LETA Vol. 10 172, 493
SLADKE STAROSTI Vol. 10 269, 273, 493
SLADKO I GORCIVO Vol. 3 48, 72
SLADKY CAS KALIMAGDORY Vol. 10 36, 268, 544
SLADY WILCZYCH ZEBOW Vol. 6 *page 219*
SLAEGTEN Vol. 5 158, 175, 253, 261
SLAEGTERNES KAMP Vol. 5 67
SLAGSKÄMPEN Vol. 5 882, 1038, 1096
SLAGTERREGNINGEN Vol. 5 97
SLÄKTEN ÄR BÄST Vol. 5 928, 1042, 1164
SLÄKTEN ÄR VÄRST Vol. 5 840, 878
SLALAM UNDER HIMMELEN Vol. 5 674, 831
SLAMENY KLOBOUK Vol. 10 289, 292, 503
SLANA RUZE Vol. 10 74, 283, 408, 444
SLANE CUKRIKY Vol. 10 73, 148, 328
SLAP AF, FREDE ! Vol. 5 16, 61, 249, 272
SLÄPP FANGARNE LOSS - DET ÄR VAR ! Vol. 5 869, 953
SLASTI OTCE VLASTI Vol. 10 283, 550
SLAVIA L - BROX Vol. 10 600
SLAVICA Vol. 3 369 Vol. 6 *page 309*
SLAVKO, NEDEJ SE ! Vol. 10 167, 256, 412
SLAVNOST V BOTANICKEJ ZAHRADE Vol. 10 72, 79, 493
SLAVNOSTI SNEZENEK Vol. 10 172, 325, 535
SLECNA GOLEM Vol. 10 184, 197, 296, 391
SLECNA MATINKA Vol. 10 256, 460
SLECNA OD VODY Vol. 10 41, 433, 524
SLECNY PRIJDOU POZDEJI Vol. 10 70, 522
SLED POZARA NAD RUSIJA Vol. 3 115
SLEDITE OST RAVAT Vol. 3 66
SLEDITE OSTAVAT Vol. 3 37, 109
SLEDOVATELJAT I GORATA Vol. 3 30, 41 Vol. 6 *page 31*
SLEM DRENG, EN Vol. 5 92, 203
SLEMME DRENGE Vol. 5 97
SLEPICE A KOSTELNIK Vol. 10 460, 559
SLIKE IZ ZIVOTA UDARNIKA Vol. 3 338, 360
SLIM CALLAGHAN ... IL DURO Vol. 7 982
SLINGREVALSEN Vol. 5 46, 113, 175
SL"NCETO I SJANKATA Vol. 3 38, 47, 69
SL"NCEV UDAR Vol. 6 *page 27, page 45, page 55*
SLNKO V SIETI Vol. 10 72, 548
SLODYCZ GRZECHU Vol. 1 174, 188
SLOGAN Vol. 2 72
SLONA ROZA Vol. 6 *page 211, page 224*
SLONCE WSCHODZI RAZ NA DZIEN Vol. 1 355, 408 Vol. 6 *page 215*
SLOTTET Vol. 5 23, 248, 275
SLOVO ANDREJA VITUZNIKA Vol. 3 390
SLOVO DELA ZENU Vol. 10 70, 193, 582
SLUBOWANIE Vol. 1 241, 346, 384
SLUBY ULANSKIE Vol. 1 176, 308

SLUCAEN KONCERT Vol. 3 31, 61, 95 Vol. 6 *page 41*
SLUCAJAT PENLEVE Vol. 3 41, 63 Vol. 6 *page 41*
SLUCAJNI ZIVOT Vol. 3 430
SLUMRANDE TONDER FJÄRRAN UR TIDEN Vol. 5 1086
SLUMRANDE TONER Vol. 5 512, 629
SLUNCE, SENO A PAR FACEK Vol. 10 74, 129, 153
SLUNCE SENO JAHODY Vol. 10 41, 129, 153
SLUZBENI POLOZAJ Vol. 3 371, 421
SLUZEBNO POLOZENIE - ORDINAREC Vol. 3 110 Vol. 6 *page 25*
S.M. IL DANARO Vol. 7 131
SMAA LANDSTRYGERE, DE Vol. 5 97
SMAEKLAASEN Vol. 5 97
SMAGLIATURA, LA Vol. 7 1145
SMAK WODY Vol. 6 *page 221, page 224, page 244, page 247*
SMALÄNNINGAR Vol. 5 840, 841, 997
SMARAGD, DER Vol. 4 135
SMARKULA Vol. 1 220, 227, 408 Vol. 6 *page 242*
SMARRITA ! Vol. 7 277
SMÄRTGRÄNSEN Vol. 5 990
SMAZALNIA STORY Vol. 6 *page 229, page 241*
SMEDER PA LUFFEN Vol. 5 1052, 1164
SMEDESTRAEDE 4 Vol. 5 2, 228
SMEDJE OKO - ZLO OKO Vol. 3 343, 346, 374
SMEKER Vol. 6 *page 314, page 315*
SMEKMÄNAD Vol. 5 889, 1086
SMEMORATO, LO Vol. 7 48, 792
SMERTENS BORN Vol. 5 46, 66
SMERY ZIVOTA Vol. 10 345, 532, 580
SMESNY PAN Vol. 10 144, 408
SMIC, SMAC, SMOC Vol. 2 499
SMICH SE LEPI NA PATY Vol. 10 172, 254, 505
SMIERC PREZYDENTA Vol. 1 227, 244, 280, 325, 377, 393 Vol. 6 *page 248*
SMIERC W SRODKOWYM POKOJU Vol. 1 206
SMIERC ZA ZYCIE Vol. 1 278, 317
SMIL EMIL Vol. 5 130
SMIL MAND Vol. 5 120
SMIL OG TAARER Vol. 5 83
SMOG Vol. 7 248
SMOKE Vol. 5 919
SMOLIARI Vol. 10 210, 263, 494
SMORFIE DI PULCINELLA, LE Vol. 7 669
SMRT CERNEHO KRALE Vol. 10 70, 154, 547
SMRT GOLUZU Vol. 6 *page 309*
SMRT GOSPODINA GOLUZE Vol. 6 *page 309*
SMRT KRASNYCH SRNCU Vol. 10 144, 313, 511
SMRT MOUCHY Vol. 10 144, 292, 568
SMRT NA CERNO Vol. 10 59, 286, 408
SMRT NA CUKROVEM OSTROVE Vol. 10 229, 313, 489
SM'RT NJAMA Vol. 3 48, 63, 112 Vol. 6 *page 41*
SMRT PANA GOLUZU Vol. 10 214, 454, 548
SMRT PRICHADZA V DAZDI Vol. 10 36, 72, 451
SMRT SI RIKA ENGELCHEN Vol. 10 352, 359, 385
SMRT SI VYBIRA Vol. 10 41, 425, 522
SMRT SITA NA MIERU Vol. 10 72, 287, 474
SMRT STOPAREK Vol. 10 369, 425, 577
SMRT TALENTOVANEHO SEVCE Vol. 10 139, 165, 325
SMRT V SEDLE Vol. 10 41, 352, 385
SMRT ZA OPONOU Vol. 10 283, 369, 371
SM'RTTA NA ZAEKA Vol. 6 *page 28, page 33, page 48*
SMUGA CIENIA Vol. 1 136, 224, 325, 340, 405 Vol. 6 *page 254*
SMUGGLARKUNGEN Vol. 5 853, 1053
SMUGLERE Vol. 5 740, 799, 817
SMUGLERE I SMOKING Vol. 5 672
SMUGLEREN OG HANS BRUD Vol. 5 216
SMUKKE ARNE OG ROSA Vol. 5 199, 201, 275
SMULTRONSTÄLLET Vol. 5 940, 1166
SMUTECNI SLAVNOST Vol. 10 70, 227, 327
SMUTJES SCHWESTER Vol. 4 18, 558
SMUTNY LORD Vol. 10 332, 544
SMUTSIGA FINGRAR Vol. 5 919, 1052
SMYK Vol. 10 229, 503, 519
SNACKA GAR JU ... Vol. 5 917, 931, 932
SNADJI SE DRUZE Vol. 6 *page 315, page 319*
SNADNY ZIVOT Vol. 10 296, 382, 564
SNAHA Vol. 3 42, 53, 63, 66, 76 Vol. 6 *page 28*
SNAPPHANNAR Vol. 5 1032, 1181
SNART 17 Vol. 5 716, 728, 746
SNATKOVA KANCELAR Vol. 10 92, 138, 453
SNEDIGE SKRAEDDER, DEN Vol. 5 81
SNEH POD NOHAMI Vol. 10 82, 273, 399
SNEHOVA RUZENKA Vol. 10 480
SNEHULKA A SEDM TRPASLIKU Vol. 10 58, 256, 480
SNEZENKA Z TATER Vol. 10 119, 310, 393, 536

SNEZENKY Vol. 10 198, 536
SNEZENKY A MACHRI Vol. 10 149, 383, 571
SNIMKI ZA SPOMEN Vol. 6 *page 23, page 24, page 59*
SNOBS Vol. 2 866
SNÖSTORMEN Vol. 5 863, 1032
SNOVI, ZIVOT I SMRT FILIPA FILIPOVICA Vol. 3 347, 407
SNOW TREASSURE Vol. 5 674
SNURRIGA FAMILJEN Vol. 5 841
SNURRIGA FAMNEN Vol. 5 840
SNUSHANERNE Vol. 5 24
SNUSTOBAKSDAASEN Vol. 5 90
SNY I MARZENIA Vol. 6 *page 229*
SNY I MARZENIA. PART II : DO WIDZENIA, KOCHANI Vol. 6 *page 247*
SNY I MARZENIA - SEN O VIOLETCIE Vol. 6 *page 214, page 230*
SNY NA NEDELI Vol. 10 433, 460, 621
SNY O ZAMBEZI Vol. 10 70, 246, 380
SO Vol. 3 351, 370
SO EIN BIENCHEN Vol. 1 11, 122
SO EIN FLEGEL Vol. 4 115, 132
SO EIN FRÜCHTCHEN Vol. 4 337, 558 Vol. 6 *page 181*
SO EIN GLÜCK KANNST DU NUR HABEN Vol. 4 87
SO EIN HUNDELEBEN Vol. 10 256, 586
SO EIN MÄDEL VERGISST MAN NICHT Vol. 4 28, 600
SO EIN THEATER Vol. 4 786
SO ENDETE EINE LIEBE Vol. 4 527, 614
SO GEFÄLLST DU MIR Vol. 4 97, 391, 557
SO IST DAS LEBEN Vol. 6 *page 195, page 197*
SO KÜSST NUR EINE WIENERIN Vol. 4 340, 608, 720
SO LANG' NOCH EIN WALZER VON STRAUSS ERKLINGT Vol. 4 132, 608
SO LEBEN WIR ALLE TAGE Vol. 4 97, 230, 239
SO LIEBT UND KÜSST MAN IN TIROL Vol. 3 395
SO RÄCHT DIE SONNE Vol. 4 213
SO SIND DIE MÄNNER Vol. 4 307, 437, 622, 634
SO THIS IS PARIS Vol. 2 526
SO WEIT GEHT DIE LIEBE NICHT Vol. 4 344, 647, 689
SOBALVANY, A Vol. 9 67, 241, 326
SOBOL I PANNA Vol. 6 *page 225, page 259*
SOBOTA Vol. 10 412, 503
SOBOTKI Vol. 1 179 Vol. 6 *page 232*
SOBRE EL CIENO Vol. 8 357, 648
SOBRE LACRADO, EL Vol. 8 422
SOBRE VERDE, EL Vol. 8 207, 214
SOBRENATURAL Vol. 8 349, 749
SOBRESALIENTE Vol. 8 207, 772
SOBRINA DEL CURA, LA Vol. 8 604
SOBRINO DE BUFALO BILL, EL Vol. 8 475
SOCIAL HYPOCRITES Vol. 2 579
SOCIETY Vol. 2 526
SOCIO INVISIBLE, IL Vol. 7 2, 701, 1116
SOCRATE, LE Vol. 2 721
SOD LOGEREND, EN Vol. 5 83
SÖDER OM LANDSVÄGEN Vol. 5 863, 930, 1032
SÖDERKARAR Vol. 5 1205
SÖDERPOJKAR Vol. 5 870, 923
SODOMS ENDE Vol. 4 773
SÖDRANS REVY Vol. 5 836, 871, 924, 933, 1053, 1164, 1207
SODRASBAN Vol. 9 259, 263, 318
SOEURS BRONTE, LES Vol. 2 434, 633
SOEURS D'ARMES Vol. 2 254, 595
SOEURS HORTENSIAS, LES Vol. 2 193, 722, 727
SOFER DE MARE VITEZA Vol. 3 281
SOFIA E A EDUCACAO SEXUAL Vol. 8 35, 164, 185
SOFKA Vol. 3 349, 405
SOGNI D'ORO Vol. 7 524, 801, 1034
SOGNI MOSTRUOSAMENTE PROIBITI Vol. 7 882, 1086
SOGNI MUOIONO ALL'ALBA, I Vol. 7 1214
SOGNI NEL CASSETTO, I Vol. 7 86, 271
SOGNO D'AMORE Vol. 7 493
SOGNO DI BUTTERFLY, IL Vol. 7 196, 453, 487
SOGNO DI FIRULI, IL Vol. 7 1189
SOGNO DI MOMO, IL Vol. 7 305
SOGNO DI PRIMAVERA Vol. 7 1169
SOGNO DI RIRETTE, IL Vol. 7 934
SOGNO DI ROBINET, IL Vol. 7 1189
SOGNO DI TUTTI, IL Vol. 7 476, 792
SOGNO DI UN GIORNO, IL Vol. 7 730
SOGNO DI UN PADRE Vol. 7 1189
SOGNO DI UN TRAMONTO D'AUTUNNO Vol. 7 1189
SOGNO DI UNA NOTTE D'ESTATE Vol. 7 202, 425, 1093
SOGNO DI UNA NOTTE DI MEZZA SBORNIA, IL Vol. 7 693
SOGNO DI ZORRO, IL Vol. 7 453, 796, 798, 830, 1003
SOGNO D'UNA NOTTE D'ESTATE VENEZIA, IL Vol. 7 390
SOHA, SEHOL, SENKINEK Vol. 9 52, 185, 238
SOHA TÖBBE ... MINDÖRÖKKE Vol. 9 77
SOHN DER GÖTTER, DER Vol. 4 488, 681

SOHN DER HAGAR, DER **Vol. 4** 28, 325, 359, 663, 727
SOHN DER WEISSEN BERGE, DER **Vol. 4** 54, 479, 481, 527
SOHN DES HANNIBAL, DER **Vol. 4** 479, 527, 773
SÖHNE DER GROSSEN BÄRIN, DIE **Vol. 1** 68, 102
SÖHNE DER HÖLLE **Vol. 4** 651
SÖHNE DER NACHT **Vol. 4** 424
SOIF DES HOMMES, LA **Vol. 2** 322, 615, 866
SOIR A MARSEILLE, UN **Vol. 2** 413
SOIR A TIBERIADE, UN **Vol. 2** 661
SOIR AU FRONT, UN **Vol. 2** 413
SOIR DE BOMBE, UN **Vol. 2** 308, 698
SOIR DE RAFLE, UN **Vol. 2** 153, 833
SOIR DE REVEILLON, UN **Vol. 2** 833
SOIR SUR LA PLAGE, UN **Vol. 2** 153, 173
SOIR, UN TRAIN, UN **Vol. 2** 190, 674
SOIREE A LA COMEDIE FRANCAISE, UNE **Vol. 2** 32
SOIREE MONDAINE **Vol. 2** 528
SOIR...PAR HASARD, UN **Vol. 2** 514
SOIRS DE PARIS **Vol. 2** 344, 480
SOIS BELLE ET TAIS-TOI ! **Vol. 2** 156, 626, 824, 861
SOITA MINULLE, HELENA **Vol. 5** 471, 634
SOJKY V HLAVE **Vol. 10** 82, 454, 459
SOK HÜHO EMMIERT **Vol. 9** 26, 132
SOK HUSEG SEMMIERT **Vol. 9** 36, 83, 183
SOK OD SLJIVA **Vol. 6** *page 305, page 307, page 322*
SOKOLOVO **Vol. 10** 18, 283, 313, 318, 587
SOL DE VERANO **Vol. 8** 470, 551
SOL E TOIROS **Vol. 8** 18, 44, 88, 131
SOL EN EL ESPEJO **Vol. 8** 214, 276
SOL EN LA NIEVE **Vol. 8** 246
SOL NAD ZLATO **Vol. 10** 72, 425, 493
SOL OVER DANMARK **Vol. 5** 202, 256
SOL ÖVER KLARA **Vol. 5** 863, 888, 1032
SOL ÖVER SVERIGE **Vol. 5** 928, 1042, 1181
SOL SALE TODOS LOS DIAS, EL **Vol. 8** 207, 214
SOL ZIEMI CZARNEJ **Vol. 1** 198 **Vol. 6** *page 240, page 259*
SOLA ANTE EL TERROR **Vol. 8** 731
SOLAJA **Vol. 3** 342, 343
SOLAMENTE NERO **Vol. 7** 1194
SOLANGE LEBEN IN MIR IST **Vol. 1** 13, 17, 101
SOLCO DI PESCA, IL **Vol. 7** 141
SOLCO E LA SEMENTA, IL **Vol. 7** 529
SOLDADOS **Vol. 8** 219, 338
SOLDADOS DE PLOMO **Vol. 8** 580, 691
SOLDAT BOM **Vol. 5** 940, 1173
SOLDAT DER MARIE, DER **Vol. 4** 560, 663
SOLDAT DUROC, CA VA ETRE LA FETE **Vol. 2** 241
SOLDAT LAFORET, LE **Vol. 2** 804
SOLDAT TOUJOURS, TOUJOURS AUX COTES DU PEUPLE **Vol. 2** 95
SOLDATEN - KAMERADEN **Vol. 4** 340, 466, 703
SOLDATEN OG JENNY **Vol. 5** 304
SOLDATENLIEDER **Vol. 4** 329, 667
SOLDATENRAMMERATER PA VAGT **Vol. 5** 194, 201, 276, 304
SOLDATERKAMMERATER **Vol. 5** 14, 281
SOLDATERKAMMERATER PA BJORNETJENESTE **Vol. 5** 23, 76
SOLDATERKAMMERATER PA EFTERATSMANOVRE **Vol. 5** 207
SOLDATERKAMMERATER PA SJOV **Vol. 5** 207
SOLDATERKAMMERATER RYKKER UD **Vol. 5** 281, 304
SOLDATESSA ALLA VISITA MILITARE, LA **Vol. 7** 426
SOLDATESSA ALLE GRANDI MANOVRE, LA **Vol. 7** 426
SOLDATESSE, LE **Vol. 7** 333
SOLDATI FARA UNIFORMA **Vol. 3** 270, 297, 330
SOLDATINI DEL RE DI ROMA, I **Vol. 7** 1190
SOLDATO DI VENTURA, IL **Vol. 7** 738
SOLDATO D'ITALIA, IL **Vol. 7** 1073
SOLDIER, THE **Vol. 3** 391, 455
SOLE **Vol. 7** 250, 652, 756, 796, 838
SOLE DI MONTECASSINO, IL **Vol. 7** 796
SOLE E I PAZZI, IL **Vol. 7** 1070
SOLE NEGLI OCCHI, IL **Vol. 7** 915, 1036
SOLE NUDO **Vol. 7** 347
SOLE SORGE ANCORA, IL **Vol. 7** 494, 1136
SOLE TORNERA, IL **Vol. 7** 606
SOLEDAD **Vol. 3** 350, 359 **Vol. 7** 949
SOLEIL A TOUJOURS RAISON, LE **Vol. 2** 156, 644, 837, 863
SOLEIL DANS L'OEIL, LE **Vol. 2** 16, 419
SOLEIL DE MINUIT, LE **Vol. 2** 48, 203, 270
SOLEIL DES VOYOUS, LE **Vol. 2** 840, 875
SOLEIL EN FACE, LE **Vol. 6** *page 103*
SOLEIL ET OMBRE **Vol. 2** 216
SOLEIL NOIR, LE **Vol. 2** 59, 824
SOLEIL O **Vol. 2** 161
SOLEIL QUI RIT ROUGE, LE **Vol. 2** 692
SOLEILS DE L'ILE DE PAQUES, LES **Vol. 2** 196
SOLEILS ET L'OMBRE, LES **Vol. 2** 117
SOLEN, DER DRAEBTE **Vol. 5** 242

SOLI PER LE STRADE Vol. 7 501
SOLIGA SOLBERG Vol. 5 863, 1032
SOLISTIN ANNA ALT Vol. 4 97, 437, 701
SOLIST"T Vol. 6 *page 28, page 38, page 64*
SOLITA DE CORDOUE Vol. 2 6, 70
SOLITAIRE PASSE A L'ATTAQUE, LE Vol. 8 537, 551
SOLITAIRE, UN Vol. 2 219, 532
SOLITAIRES, LES Vol. 7 1189
SOLITARIO DELLA MONTAGNA, IL Vol. 7 48, 466, 756
SOLITARIO PASA AL ATAQUE, EL Vol. 8 537, 551
SOLITI IGNOTI, I Vol. 7 522
SOLITI IGNOTI ... VENT'ANNI DOPO, I Vol. 7 74, 1025
SOLITI RAPINATORI A MILANO Vol. 7 1044
SOLITUDE DU CHANTEUR DE FOND, LA Vol. 2 502, 517, 716
SOLITUDINE Vol. 7 1027, 1054
SOLL MAN HEIRATEN ? Vol. 4 720, 734
SOLL UND HABEN Vol. 4 124, 294, 733
SOLLAZZEVOLI STORIE DI MOGLI GAUDENTI E MARITI PENITENTI Vol. 7 742
SOLLERS ET GUEGAN ONT DEUX MOTS A SE DIRE Vol. 6 *page 117*
SOLO Vol. 2 866
SOLO AL MONDO Vol. 7 1189
SOLO ANTE EL STREAKING Vol. 8 407, 681
SOLO CONTRO ROMA Vol. 3 395 Vol. 7 585, 1147
SOLO DIO MI FERMERA Vol. 7 201
SOLO FÜR MARTINA Vol. 6 *page 161, page 162, page 172*
SOLO PARA HOMBRES Vol. 8 752
SOLO PER TE Vol. 7 902, 1116
SOLO PER TE, LUCIA Vol. 7 493
SOLO PRO STAROU DAMU Vol. 10 227, 428, 567
SOLO SUNNY Vol. 6 *page 153, page 159*
SOLO UN ATAUD Vol. 8 396, 772
SOLOS EN LA MADRUGADA Vol. 8 438, 681
SOLOS LOS DOS Vol. 8 214, 261
SOLSKINSBORNENE Vol. 5 295
SOLSKINSDALEN Vol. 5 136, 157
SOLSTIK Vol. 5 29, 89
SOLTANO UN BACIO Vol. 7 443, 638
SOLTERA Y MADRE EN LA VIDA Vol. 8 331, 681
SOLTERO DIFICIL, UN Vol. 8 486, 732
SOLTERO Y PADRE EN LA VIDA Vol. 8 249, 577, 749
SOLTEROS DE VERANO Vol. 8 276, 732
SOLTI IGNOTI, I Vol. 7 1167
SOLUNSKI ATENTATORI Vol. 3 419, 443
SOLVMUNN Vol. 5 684
SOLVYLLUPSDAG, EN Vol. 5 83
SOM DU VIL HA' MIG ! Vol. 5 2, 228
SOM DU VILLHA MEJ Vol. 5 843, 923
... SOM EN TJUV OM NATTEN Vol. 5 1042, 1181
SOM FOLK ÄR MEST Vol. 5 871, 1205
SOM HAN BÄDDER FAR HON LIGGA Vol. 5 843, 1142
... SOM HAVETS NAKNA VIND Vol. 5 869, 1017, 1142
SOM I DRÖMMAR Vol. 5 1033, 1206
SOM MAN BÄDDAR... Vol. 5 933, 1042
SOM MAN REDER ... Vol. 5 92
SOM NATT OCH DAG Vol. 5 1170
SOM SENDT FRA HIMLEN Vol. 5 115, 149, 281
SOMAEND OG SVIGERMODRE Vol. 5 89, 281
SOMAEND PA SENGEKANTEN Vol. 5 156, 296
SOMAND GAR I LAND, EN Vol. 5 89, 281
SOMAND HAN MAA DOJE, EN Vol. 5 83
SOMAND I KNIBE Vol. 5 89, 281
SOMANDSLIV Vol. 5 97
SOMBRA DEL JUDOKA CONTRA EL DOCTOR WONG, LA Vol. 8 222
SOMBRA EN EL COLEGIO, UNA Vol. 8 357
SOMBRA EN LA VENTANA, UNA Vol. 8 614
SOMBRE DIMANCHE Vol. 2 254, 658
SOMBRE ILUMINADA, LA Vol. 8 311, 524
SOMEND PA SENGEKANTEN Vol. 5 231
SOMMAR OCH SYNDARE Vol. 5 909, 959, 1042
SOMMARÄVENTYR, ETT Vol. 5 909
SOMMAREN MED MONIKA Vol. 5 940, 1052
SOMMARFLICKAN Vol. 5 1073, 1164
SOMMARLEK Vol. 5 940, 1173
SOMMARNATTENS LEENDE Vol. 5 940, 1052, 1053
SOMMARNÖJE SÖKES Vol. 5 878, 984, 1052
SOMME, DIE Vol. 4 77, 97, 204, 784
SOMMER I TYROL Vol. 5 31, 248, 272
SOMMER, SOL OG STUDINER Vol. 5 83
SOMMER, SONNE, ERIKA Vol. 4 373, 626
SOMMEREN JEG FYLTE 15, DEN Vol. 5 712, 755
SOMMERFUGLENE Vol. 5 739
SOMMERGIBILE PIU PAZZO DEL MONDO Vol. 7 335, 770, 1202
SOMMERGLAEDER Vol. 5 204, 228
SOMMERKRIG Vol. 5 257

SOMMERLIEBE **Vol. 1** 82, 94 **Vol. 4** 28, 483, 641
SOMMERNÄCHTE **Vol. 4** 378, 728
SOMMERNACHTSTRAUM, EIN **Vol. 4** 373, 452, 646
SOMMERNACHTSTRAUM IN UNSERER ZEIT, EIN **Vol. 4** 646
SOMMERTAG MACHT KEINE LIEBE, EIN **Vol. 1** 81, 88, 125
SOMMES-NOUS DEFENDUS ? **Vol. 2** 193
SOMNAMBUL **Vol. 4** 379, 565
SÖMNEN **Vol. 5** 858, 869, 1098
SON ALTESSE L'AMOUR **Vol. 2** 48, 206
SON AUTRE AMOUR **Vol. 2** 698
SON CONTENTO **Vol. 7** 239, 290, 593
SON CRIME **Vol. 2** 864
SON DERNIER NOEL **Vol. 2** 401, 402, 419
SON DERNIER ROLE **Vol. 2** 270, 595, 758
SON EXCELLENCE ANTONIN **Vol. 2** 388, 404
SON EXCELLENCE LE BOUIF **Vol. 2** 584
SON FATTA COSI ! ..., IO **Vol. 7** 268
SON NOM DE VENISE DANS CALCUTTA DESERT **Vol. 2** 633
SON OF A BITCH **Vol. 7** 818
SON ONCLE DE NORMANDIE **Vol. 2** 205, 698
SON TORNATE A FIORIRE LE ROSE **Vol. 7** 1114
SO'N WINDHUND ! **Vol. 4** 242, 270, 471
SONAMBULOS **Vol. 8** 581
SONATA A KREUTZER, LA **Vol. 7** 764
SONATA PRO ZRZKU **Vol. 10** 116, 295, 337
SONATAS **Vol. 8** 311, 387, 611
SONCNI KRIK **Vol. 3** 375, 387
SÖNDAG I SEPTEMBER, EN **Vol. 5** 833, 946, 1053
SONDAG PAA AMAGER, EN **Vol. 5** 55, 70, 101, 208
SONDAGSJAEGENS JAGTAEVENTYR **Vol. 5** 183
SONDERBARE LIEBE, EINE **Vol. 6** *page 156, page 161, page 176*
SONDIS KLEINE **Vol. 6** *page 186*
SONG **Vol. 4** 192, 270, 462
SONG OF SCHEHERAZADE, THE **Vol. 2** 526
SONGE DE BUTTTERFLY, LE **Vol. 6** *page 74*
SONHAR E FACIL **Vol. 8** 44, 131
SONHO DE AMOR **Vol. 8** 42, 50, 58
SONIDO PREHISTORICO, EL **Vol. 8** 276, 616
SONJA **Vol. 5** 871, 954
SONJA – 16 AR **Vol. 5** 76, 261
SONJAS RAPPORT **Vol. 6** *page 157, page 163, page 176*
SONNAMBULA, LA **Vol. 7** 187, 257, 734
SONNE ASIENS, DIE **Vol. 4** 252, 321
SONNE VON ST. MORITZ, DIE **Vol. 4** 390, 467
SONNEN **Vol. 5** 9, 31, 150, 248, 272
SONNEN AV NORGE **Vol. 5** 674, 717, 800
SONNEN FRA AMERIKA **Vol. 5** 2, 23
SONNEN FRA RULLEKAELDEREN **Vol. 5** 97
SONNEN FRA VINGARDEN **Vol. 5** 35, 247, 306
SONNENBRUCKS, DIE **Vol. 1** 67, 101
SONNENSUCHER **Vol. 1** 9, 72, 105
SONNER AV NORGE KJOPER BIL **Vol. 5** 682, 717, 800
SONNTAG DES LEBENS **Vol. 4** 555
SONNTAGSFAHRER **Vol. 1** 8, 68, 125
SONNTAGSKIND, DAS MANCHMAL SPINNT, EIN **Vol. 1** 15, 59, 79
SONNTAGSKINDER **Vol. 4** 18, 317, 558
SONO FOTOGENICO **Vol. 7** 29, 333, 841, 964
SONO IO L'ASSASSINO **Vol. 7** 1063
SONO SARTANA, IL VOSTRO BECCHINO **Vol. 7** 124
SONO STATO IO ! **Vol. 7** 347, 709, 931, 1177
SONO STATO UN AGENTE C.I.A. **Vol. 7** 763
SONO UN FENOMENO PARANORMALE **Vol. 7** 265, 381, 834
SONORA **Vol. 8** 732
SONRIA, POR FAVOR **Vol. 8** 603, 677, 725
SONS KAERLIGHED, EN **Vol. 5** 183
SOPHIE AIME LES SUCETTES **Vol. 2** 291
SOPHIE ET LE CRIME **Vol. 2** 16, 96, 427, 517, 801
SOPHIENLUND **Vol. 4** 120, 270, 774
SOPLA GAITAS, EL **Vol. 8** 368, 773
SOPOR **Vol. 5** 882, 947, 1160
SOPRAVVISSUTI DELLA CITTA MORTA, I **Vol. 7** 692
SOPRAVVISSUTO, IL **Vol. 7** 796
SOR ANGELICA **Vol. 8** 503, 653, 677
SOR CITROEN **Vol. 8** 524, 725
SOR INTERPIDA **Vol. 8** 214, 400
SOR YE – YE **Vol. 8** 395
SORAYA, REINA DEL DESIERTO **Vol. 8** 762
SORBOLE ... CHE ROMAGNOLA ! **Vol. 7** 556
SORCIER BLANC, LE **Vol. 2** 475, 479
SORCIER DU CIEL, LE **Vol. 2** 67, 131
SORCIERE, LA **Vol. 2** 9, 368
SORCIERES DE SALEM, LES **Vol. 2** 615, 719, 801
SORELLA DEL BANDITO, LA **Vol. 7** 1189
SORELLA DEL FORZATO, LA **Vol. 7** 37

SORELLA DEL MISSIONARIO, LA Vol. 7 1189
SORELLA DI URSULA, LA Vol. 7 130
SORELLA, LA Vol. 7 36
SORELLA MINORE, LA Vol. 7 497
SORELLE, LE Vol. 7 19
SORELLE MATERASSI Vol. 7 493, 1060, 1080
SOREN SONDERVOLD Vol. 5 196
SORENSEN OG RASMUSSEN Vol. 5 31, 32, 55, 70, 101, 208
SORGMUNTRE BARBER, DEN Vol. 5 70, 183
SORPASSO, IL Vol. 7 347, 884
SORPRESE DEL DIVORZIO, LE Vol. 7 2, 48, 1116, 1159
SORPRESE DEL VAGONE LETTO, LE Vol. 7 68, 476, 680
SORPRESE DELL'AMORE Vol. 7 257
SORPRESE DI DON CAMILLO, LE Vol. 7 26
SORRISI E CANZONI Vol. 7 322
SORRISI E SPASIMI DELLA MENZOGNA Vol. 7 1073
SORRISO DI SFINGE, IL Vol. 7 1174
SORRISSO DEL GRANDE TENTATORE, IL Vol. 7 1194
SORTE DOMINO, DEN Vol. 5 97
SORTE DROM, DEN Vol. 5 154, 245
SORTE FAMILIE, DEN Vol. 5 245
SORTE HAETTE, DEN Vol. 5 97
SORTE HERTUG, DEN Vol. 5 97
SORTE KUGLE, DEN Vol. 5 228
SORTE MASKE, DEN Vol. 5 97
SORTE SHARA Vol. 5 23
SORTE SKIB, DET Vol. 5 106
SORTIE DE BOITE Vol. 6 *page 69*
SORTIE DE SECOURS Vol. 2 475, 814
SORTILEGES Vol. 2 15, 388, 644
SORTILEGIO Vol. 8 270, 676
SORTILEGIO AFRICANO Vol. 8 80, 184
SORTÜZ EGY FEKETE BIVALYERT Vol. 9 20, 42, 154
S.O.S. ABUELITA Vol. 8 600, 749
S.O.S. DIE INSEL DER TRÄNEN Vol. 4 36, 89, 416, 546, 663, 720
S.O.S. EISBERG Vol. 4 15, 437, 521, 556, 620
S.O.S. INVASION Vol. 8 583
SOSESC PASARILE CALATOARE Vol. 6 *page 278, page 280, page 281*
SOSKEN PA GUDS JORD Vol. 5 716, 765
SOSKENDE Vol. 5 257, 261, 281
SOTAGULASHI KAIUN HÄIRITTY KESÄLOMA Vol. 5 344, 433, 649
SOTANA, EL Vol. 8 311, 752
SOTAPOJAN HEILAT Vol. 5 451, 613
SOTAPOLULLA Vol. 5 501, 533, 534
SOTLUGG OCH LINLUGG Vol. 5 923, 1181
SOTTO A CHI TOCCA Vol. 7 692
SOTTO DIECI BANDIERE Vol. 7 1136
SOTTO' E CANCELLE Vol. 7 424
SOTTO I PONTI DI PARIGI Vol. 7 774, 813, 972
SOTTO IL BACIO DEL FUOCO Vol. 7 993, 1129
SOTTO IL SOLE DI ROMA Vol. 7 1036
SOTTO IL VESTITO NIENTE Vol. 7 225, 258, 367, 664, 857
SOTTO LA CROCE DEL SUD Vol. 7 360, 493, 1136
SOTTO LA MASCHERA Vol. 7 1207
SOTTO L'ALA DELLA MORTE Vol. 7 399
SOTTO S. FRANCISCO Vol. 7 829
SOTTO ... SOTTO STRAPAZZATO DA ANOMALA PASSIONE Vol. 7 592, 1093
SOTTOMARINO N.27 Vol. 7 965
SOUBOJ LASKY Vol. 10 7, 34
SOUBOJ S BOHEM Vol. 10 257, 536, 605
SOUBRETTE PERVERSE, LA Vol. 2 218
SOUCOUPE DIRECTION VENUS Vol. 2 309
SOUD BOZI Vol. 10 532, 621
SOUDNY DEN Vol. 10 132, 133, 564
SOUFFLE AU COEUR, LE Vol. 2 35, 163
SOUHVEZDI PANNY Vol. 10 229, 369
SOUKROMA VICHRICE Vol. 10 41, 96, 371
SOULIER DE SATIN, LE Vol. 8 35, 118, 169
SOUMISSION Vol. 6 *page 79*
SOUMISSION PERVERSES Vol. 2 365
SOUND OF NÄVERLUR Vol. 5 1040
SOUPCONS Vol. 2 47, 70
SOUPE A LA GRIMACE, LA Vol. 2 475, 540, 866
SOUPE AU LAIT Vol. 2 401
SOUPE AUX CHOUX, LA Vol. 6 *page 73, page 94*
SOUPE AUX POULETS Vol. 2 475, 702
SOUPE FROIDE, LA Vol. 2 151, 275, 359
SOUPER A LA CASERNE Vol. 2 678
SOUPER UM MITTERNACHT, DAS Vol. 4 634, 780
SOUPIRANT, LE Vol. 2 514, 836
SOUPIRS PROFONDS Vol. 2 294
SOURCE DE BEAUTE, LA Vol. 2 153
SOURIANTE MADAME BEUDET, LA Vol. 2 308, 614, 648
SOURICIERE, LA Vol. 2 501, 535
SOURIRE DANS LA TEMPETE, UN Vol. 2 247, 557
SOURIRE VERTICAL, LE Vol. 2 685, 721
SOURIS CHEZ LES HOMMES, UNE Vol. 2 368, 457
SOURIS D'HOTEL Vol. 2 573
SOURIS, LA Vol. 2 678

SOUS-DOUES EN VACANCES, LES **Vol. 6** *page 73*
SOUS-DOUES, LES **Vol. 2** 114, 862 **Vol. 6** *page 73*
SOUS LA GRIFFE **Vol. 2** 159, 286, 330, 722, 871
SOUS LA TERRE **Vol. 2** 566
SOUS LE CASQUE DE CUIR **Vol. 2** 308, 775
SOUS LE CIEL DE PARIS **Vol. 2** 395, 562, 615
SOUS LE CIEL DE PROVENCE **Vol. 2** 347, 395, 548 **Vol. 7** 530, 580
SOUS LE SIGNE DE MONTE-CARLO **Vol. 2** 503, 515
SOUS LE SIGNE DU TAUREAU **Vol. 2** 185, 875
SOUS LES TOITS DE PARIS **Vol. 2** 573, 656, 708
SOUS LES YEUX D'OCCIDENT **Vol. 2** 8, 526, 626
SOUS-MARIN DE CRISTAL, LE **Vol. 2** 387, 824
SOUTHERNER, THE **Vol. 2** 24, 526
SOUVENIR D'ITALIE **Vol. 7** 1136
SOUVENIRS D'EN FRANCE **Vol. 2** 324, 633
SOUVENIRS NE SONT PAS A VENDRE, LES **Vol. 2** 70, 871,
SOUVENIRS PERDUS **Vol. 2** 388, 566
SOVENDE HUS, DET **Vol. 5** 183
SOVIEL LIEDER SOVIEL WORTE **Vol. 1** 47
SOVMUNN **Vol. 5** 808
SOVNIG BRUDGOM, EN **Vol. 5** 81
SOVRANETTA **Vol. 7** 968
SOVRAZNIK **Vol. 3** 367, 418
SOWIZDRZAL SWIETOKRZYSKI **Vol. 6** *page 240, page 244*
SOYAS TAGSTEN **Vol. 5** 29, 96, 199, 249, 275, 304
SOYEZ LES BIENVENUS **Vol. 2** 96, 193, 397, 460, 627
SPACE MEN **Vol. 7** 738
SPADA DI DAMOCLE, LA **Vol. 7** 886, 1079
SPADA E LA CROCE, LA **Vol. 7** 740
SPADA NORMANNA, LA **Vol. 8** 214
SPADLA S MESICE **Vol. 10** 235, 385, 522
SPAGHETTI A MEZZANOTTE **Vol. 7** 19, 116
SPAGHETTI HOUSE **Vol. 7** 546, 696, 1165
SPÄHTRUPP HALLGARTEN **Vol. 4** 51, 108, 292
SPALOVAC MRTVOL **Vol. 10** 96, 172, 353
SPANILA JIZDA **Vol. 10** 197, 296, 547
SPANISCHE FLIEGE, DIE **Vol. 4** 306, 410, 774
SPAREKASSEBOGEN **Vol. 5** 203
SPARKASSE MIT LIKÖR **Vol. 4** 247, 461
SPARTACO **Vol. 7** 634, 910
SPARTACUS E I DIECI GLADIATORI **Vol. 7** 1027
SPARVIERI DEL RE, GLI **Vol. 7** 877
SPARVIERO DEL NILO, LO **Vol. 7** 792, 1027
SPASENIETO **Vol. 6** *page 63*
SPASSVÖGEL **Vol. 4** 167, 461
SPÄTE LIEBE **Vol. 4** 614, 620
SPÄTSAISON **Vol. 1** 75, 89
SPAVALDI E INNAMORATI **Vol. 7** 588
SPECCHIO E LA MORTE, LO **Vol. 7** 192, 704
SPECIALIST PO VSICKO **Vol. 3** 51, 99 **Vol. 6** *page 41*
SPECIALISTI, GLI **Vol. 7** 854
SPECIJALNO VASPITANJE **Vol. 3** 454 **Vol. 6** *page 305, page 307*
SPEED CROSS **Vol. 7** 875, 1022
SPEED DRIVER **Vol. 8** 299, 728
SPEEDING VENUS, THE **Vol. 2** 84
SPEEDY GONZALES - NOIN 7 VELJEKSEN POIKA **Vol. 5** 606, 613, 632
SPEJLETS SPAADOM **Vol. 5** 183
SPELUNKE **Vol. 4** 560, 599
SPERANTA **Vol. 3** 246, 281
SPERDUTA DI ALLAH, LA **Vol. 7** 326, 796
SPERDUTI NEL BUIO **Vol. 7** 709, 984
SPERGIURA ! **Vol. 7** 1189
SPERIAMO CHE SIA FEMMINA **Vol. 7** 109, 450
SPETTRI, GLI **Vol. 7** 731
SPETTRO DI MEZZANOTTE, LO **Vol. 7** 279
SPETTRO, LO **Vol. 7** 299, 739
SPEZIALIST FÜR ALLES **Vol. 4** 239, 340, 466
SPHINX **Vol. 9** 220
SPIA PER VENDETTA **Vol. 7** 1189
SPIAGGIA, LA **Vol. 7** 283, 359, 1065
SPIE FRA LE ELICHE **Vol. 7** 753, 838
SPIEGEL DES GROSSEN MAGUS, DER **Vol. 6** *page 165, page 168, page 173*
SPIEGEL DES LEBENS **Vol. 4** 85, 89
SPIEL **Vol. 4** 85, 582
SPIEL AN BORD **Vol. 4** 115, 703
SPIEL AUF DER TENNE **Vol. 4** 424, 774
SPIEL DER LIEBE, DAS **Vol. 4** 384, 472, 565
SPIEL IM SOMMERWIND **Vol. 4** 367, 564
SPIEL IM SPIEL **Vol. 4** 33
SPIEL MIT DEM FEUER **Vol. 4** 325, 445, 729
SPIEL MIT DEM FEUER, DAS **Vol. 4** 266, 574, 729

SPIEL MIT DEM SCHICKSAL, DAS Vol. 4 124, 290
SPIEL MIT DEM WEIBE, DAS Vol. 4 132, 479
SPIEL MIT DER LIEBE, DAS Vol. 4 287, 379, 458, 646
SPIEL UM DEN MAMM Vol. 4 292, 716, 796
SPIEL ZU DRITT Vol. 1 54, 73
SPIELBANK-AFFÄRE Vol. 1 40, 47, 78
SPIELER, DER Vol. 4 30, 266, 740
SPIELER, DIE Vol. 4 290, 429
SPIELERBLUT ODER DAS ERBE DER VÄTER Vol. 4 514
SPIELEREIEN EINER KAISERIN Vol. 4 149, 219, 660
SPIELERIN, DIE Vol. 4 14, 166, 663
SPIELMANN, DER Vol. 4 535
SPIELZEUG DER ZARIN, DAS Vol. 4 747
SPIELZEUG SCHÖNER FRAUEN, DAS Vol. 4 85, 212, 652
SPIEWY PO ROSIE Vol. 6 *page 205, page 225, page 228, page 250*
SPIGA, LA NUBE E IL RE, LA Vol. 7 342
SPINA NEL CUORE, UNA Vol. 7 362, 599, 931
SPINNE, DIE Vol. 4 247
SPINNEN, DIE Vol. 4 634, 734
SPINOLAS LETZTES GESICHT Vol. 4 87, 124
SPION 503 Vol. 5 7
SPION, DER Vol. 4 734
SPIONE Vol. 4 298, 727, 729
SPIONE AM WERK Vol. 4 28, 729, 738, 788
SPIONE IM SAVOY-HOTEL Vol. 4 340, 373
SPIONEN Vol. 5 97
SPIONEN FRA TOKIO Vol. 5 97
SPIONUL Vol. 3 234
SPIRALA Vol. 1 138, 204, 403
SPIRALE DI NEBBIA, UNA Vol. 7 544
SPIRALE, LA Vol. 2 74, 161
SPIRINTION BEBES Vol. 3 140
SPIRINTION HAMAILEON Vol. 3 140
SPIRITI D'IMITAZIONE Vol. 7 1189
SPIRITISMO Vol. 7 441, 701
SPIRITISMUS Vol. 4 232
SPIRITISTEN Vol. 5 58
SPITZEN Vol. 4 316, 733
SPITZEN DER GESELLSCHAFT, DIE Vol. 4 342
SPITZENKLÖPPLERIN VON VALENCIENNES, DIE Vol. 4 297, 405
SPITZENTUCH DER FÜRSTIN WOLKOWSKA, DAS Vol. 4 124, 232 Vol. 6 *page 194*
SPLAV MEDUZE Vol. 3 360, 393 Vol. 6 *page 304*
SPOD TE UCKE GORI Vol. 3 366, 371
SPODNIE JASNIE PANA Vol. 1 188
SPOGELSESTOGET Vol. 5 179, 273, 299
SPOGELSESTYVEN Vol. 5 81
SPOGER I VILLAEN, DET Vol. 5 9
SPOGLIAMOCI COSI SENZA PUDOR... Vol. 7 426, 514
SPÖKAR, DET SPÖKAR, DET Vol. 5 930, 993
SPÖKBARONEN Vol. 5 870, 891
SPÖKE PA SEMESTER Vol. 5 1052, 1140, 1164
SPOKELSE FORELSKER SEG, ET Vol. 5 676
SPÖKET PA BRAGEHUS Vol. 5 903, 1121
SPÖKET PA JUNKERSHUS Vol. 5 944, 966, 976
SPOKOJNE LATA Vol. 6 *page 233, page 235, page 237, page 241*
SPÖKREPORTERN Vol. 5 928, 1156
SPOMEN Vol. 3 41 Vol. 6 *page 43*
SPOMEN ZA BLIZNACKATA Vol. 3 40, 47, 122 Vol. 6 *page 30*
SPORCKSCHEN JÄGER, DIE Vol. 4 339, 377, 724
SPORCK'SCHEN JÄGER, DIE Vol. 4 536, 558
SPORT SUPERSTAR Vol. 7 242, 1016
SPORTOWIEC MIMO WOLI Vol. 1 194, 228, 308, 397
SPORTSZERELEM Vol. 9 7, 153, 288, 334
SPOSA BELLA, LA Vol. 7 365, 988
SPOSA DEI RE, LA Vol. 7 485, 644, 1150
SPOSA, LA Vol. 7 501
SPOSA NON PUO ATTENDERE O ANSELMO HA FRETTA Vol. 7 796
SPOSA PERDUTA, LA Vol. 7 144, 815
SPOSATA IERI Vol. 7 619
SPOSERO SIMON LE BON – CONFESSIONE DI UNA SEDICENNE ... Vol. 7 664, 957
SPOSINA, LA Vol. 7 800
SPOSOB BYCIA Vol. 1 148, 202, 227, 242
SPOTKANIA Vol. 1 179, 302
SPOTKANIA NA ATLANTYKU Vol. 6 *page 212, page 244*
SPOTKANIA W MROKU Vol. 1 305, 382 Vol. 6 *page 218*
SPOTKANIE W „BAJCE" Vol. 1 187, 227, 377
SPOTKANIE ZE SZPIEGIEM Vol. 1 202, 377, 400
SPOZNIENIE PRZECHODNIE Vol. 1 187, 227, 237, 305
SPRAVCA SKANZENU Vol. 10 83, 454, 607
SPRAVNA TREFA Vol. 10 116, 465, 508
SPRAWA BARTIENIEWA Vol. 1 174, 215, 326

SPRAWA DO ZALATWIENIA Vol. 1 169, 232, 298, 305
SPRAWA GORGONOWEJ Vol. 1 314, 339, 340 Vol. 6 *page 244*
SPRAWA PILOTA MARESZA Vol. 1 225, 250, 313, 323, 334
SPRAWA SIE RYPLA Vol. 6 *page 213, page 218, page 229*
SPREEWALDMÄDCHEN, DAS Vol. 4 213, 242, 558
SPRENGBAGGER 1010 Vol. 4 14, 323, 354, 397, 643
SPRINGDYKKEREN Vol. 5 295
SPRINGER VON PONTRESINA, DER Vol. 4 10, 124, 422
SPRINGPOJKAR ÄR VI ALLIHOPA Vol. 5 840, 909
SPRUNG IN DEN ABGRUND Vol. 4 119, 232, 270
SPRUNG INS DUNKLE, DER Vol. 4 12, 159, 759
SPRUNG INS GLÜCK Vol. 2 32 Vol. 4 582
SPRUNG INS GLÜCK, DER Vol. 4 19, 49, 149, 660
SPRUNG INS LEBEN, DER Vol. 4 162, 729
SPRUNG INS NICHTS, DER Vol. 4 555
S.P.U.K. Vol. 6 *page 304, page 306*
SPUK IM HAUSE DES PROFESSORS, DER Vol. 4 159
SPUK IM HOCHHAUS Vol. 6 *page 164, page 170, page 174*
SPUK IM MUSEUM Vol. 4 517, 558, 796
SPUK IM SCHLOSS Vol. 4 416, 688, 740
SPUK IN VILLA SONNENSCHEIN Vol. 1 13, 26, 86
SPUK UNTERM RIESENRAD Vol. 1 26, 87, 113
SPUR DER STEINE Vol. 1 50, 72, 73
SPUR DES ERSTEN, DIE Vol. 4 251
SPUR DES FALKEN, DIE Vol. 1 37, 68, 102
SPUR DES VERMISSTEN, DIE Vol. 6 *page 162, page 163, page 165*
SPUR FÜHRT IN DEN SIEBENTEN HIMMEL, DIE Vol. 1 33, 34, 86
SPUR IN DIE NACHT Vol. 1 24, 92, 120
SPUREN IM SCHNEE Vol. 4 344, 544 Vol. 6 *page 200*
SPURENSUCHER Vol. 6 *page 162, page 164, page 169*
SPURVE UNDER TAGET Vol. 5 196, 304
SQUADRA ANTIFURTO Vol. 7 738
SQUADRA ANTIGANGSTERS Vol. 7 310, 738
SQUADRA ANTIMAFIA Vol. 7 738
SQUADRA ANTISCIPPO Vol. 7 287
SQUADRA ANTITRUFFA Vol. 7 738
SQUADRIGLIA BIANCA Vol. 3 210 Vol. 7 402, 756
SQUADRONE BIANCO Vol. 7 4, 196, 369, 453, 1116
SQUARTATORE DI NEW YORK, LO Vol. 7 599, 623
SQUAW TSCHAPAJEWS, DIE Vol. 1 15, 87
SQUEAKER, THE Vol. 2 656
SQUILLO Vol. 7 714
SRAMNO LETO Vol. 3 344
S"RCE COVESKO Vol. 3 75, 109, 114
SRDCE NA KOLEJICH Vol. 10 168, 580
SRDCE NA LANE Vol. 10 36, 72, 470
SRDCE V CELOFANU Vol. 10 168, 564
SRDCE V SOUMRAKU Vol. 10 168, 256, 460
SRDCE ZA PISNICKU Vol. 10 256, 412, 524
SRDECNY POZDRAV ZE ZEMEKOULE Vol. 10 144, 325, 369
SRECA DOLAZI U9 Vol. 3 366, 452
SRECA NA VRVICI Vol. 3 348, 395
SRECA U TORBI Vol. 3 383, 391
SRECNA NOVA '49 Vol. 6 *page 304, page 316, page 321*
SRECNA PORODICA Vol. 3 370
SRECNO, KEKEC ! Vol. 3 390
SREDNOSTNA SRESTA Vol. 3 48, 63, 100
SRESCEMO SE VECERAS Vol. 3 343, 375
SRESTA NA SILITE Vol. 6 *page 53*
SRESTU VJAT'RA Vol. 3 73 Vol. 6 *page 36*
SRETNI UMIRU DVAPUT Vol. 3 343, 346, 373
SRPNOVA NEDELE Vol. 10 170, 286, 543, 564
SS LAGER 5 - L'INFERNO DELLE DONNE Vol. 7 289
ST-TROPEZ BLUES Vol. 2 517
STA DIHTYA TES ARAHNES Vol. 3 142
STA SE ZGODI KAD SE LJUBAV RODI Vol. 6 *page 306, page 309, page 311*
STAATSANWALT BRIANDS ABENTEUER Vol. 4 270, 570
STAATSANWALT JORDAN Vol. 4 113, 636, 780, 788, 796
STAATSANWALT KLAGT AN, DER Vol. 4 428, 663, 712
STABSSERGENTENS SON Vol. 5 97
STABWECHSEL IN DER 7 C Vol. 6 *page 168, page 171, page 175*
STACKARS FERDINAND Vol. 5 840, 878
STACKARS LILLA SVEN Vol. 5 840, 1114
STACKARS MILJONÄRER Vol. 5 903, 1121
STADACCHINO MISTERIOSO, LO Vol. 7 476
STADIO Vol. 7 756, 796, 1036, 1116
STADT ANATOL Vol. 4 298, 536, 608

STADT DER MILLIONEN, DIE Vol. 4 232, 293
STADT DER TAUSEND FREUDEN, DIE Vol. 4 149, 660
STADT DER VERSUCHUNG, DIE Vol. 4 211, 294, 721
STADT IN SICHT Vol. 4 461, 779
STADT STEHT KOPF, EINE Vol. 4 202, 517, 527
STADT UND LAND Vol. 4 18, 367
STAERKERE MAGT, EN Vol. 5 183
STAFETA Vol. 9 78, 253
STAGIONE DEI SENSI, LA Vol. 7 408
STAGIONE DELLE PIOGGE, UNA Vol. 7 766
STÄHLERNDE STRAHL, DER Vol. 4 382, 517
STAHLTIER, DAS Vol. 4 794
STAJNIA NA SALWATORZE Vol. 1 305, 313, 408
STAKKELS KARIN Vol. 5 183
STAKKELS META Vol. 5 228
STAL Vol. 5 840, 909
STALLONE, LO Vol. 7 666
STALOWE SERCA Vol. 1 169, 250, 305
STAMMBAUM DES DR. PISTORIUS, DER Vol. 4 24, 28, 736
STAMPEN Vol. 5 940, 1052
STANCES A SOPHIE, LES Vol. 2 732
STANDESAMT 10, 15 UHR Vol. 4 79, 115, 354
STANDHAFTIG FRIER, EN Vol. 5 97
STANDHAFTIGE SPILLEMAND, DEN Vol. 5 83
STANDSCHÜTZE BRUGGLER Vol. 4 10, 24, 25, 370, 382
STANGATA NAPOLETANA – LA TRASTOLA Vol. 7 19
STANZA DEL VESCOVO, LA Vol. 7 524, 1035
STANZA SEGRETA, LA Vol. 7 1189
STAR MIT FREMDEN FEDERN Vol. 1 2, 94, 115, 120
STARCI NA CHMELU Vol. 10 41, 111, 524
STAREC"T Vol. 3 31, 69
STARINNATA MONETA Vol. 3 82 Vol. 6 *page 32*
STARK SYSTEM Vol. 7 13, 411
STÄRKARE ÄN LAGEN Vol. 5 933, 1052
STARKASTE, DEN Vol. 5 891, 1031
STARKE HERZEN Vol. 4 567
STÄRKER ALS DIE LIEBE Vol. 4 558, 703
STÄRKER ALS DIE NACHT Vol. 1 13, 54, 88, 90
STÄRKER ALS PARAGRAPHEN Vol. 4 97, 230, 458
STÄRKERE MACHT, DIE Vol. 4 494, 612
STÄRNSMÄLL I FRUKOSTKLUBBEN Vol. 5 1042, 1164
STARS OF GLORY Vol. 2 384
STARY HRICH Vol. 10 167, 345
STASERA ALLE UNDICI Vol. 7 453, 1177
STASERA NIENTE DI NUOVO Vol. 7 443, 1136
STASERA SCIOPERA Vol. 7 891
STASTIE PRIDE V NEDELU Vol. 10 210, 268
STASTNOU CESTE Vol. 10 412, 621
STATE BUONI SE POTETE Vol. 7 383, 778
STATIVERNE PAA MORGENKOMMERS Vol. 5 97
STATO INTERESSANTE Vol. 7 46
STATOCNY ZLODEJ Vol. 10 193, 273
STATUA DI CARNE, LA Vol. 7 48
STATUA VIVENTE, LA Vol. 7 994, 1136
STATUETTA DI NELLY Vol. 7 1189
STAV ZTROSKOTANI Vol. 10 56, 116, 508
STAVISKY Vol. 2 770, 854
STAVITEL CHRAMU Vol. 10 53, 613
STAVRAETOI STA METEORA Vol. 3 132
STAWKA WIEKSZA NIZ ZYCIE Vol. 1 249, 400
STAZIONE TERMINI Vol. 7 21, 43, 125, 555, 709
STE DOJDAT NOVI DNI Vol. 3 66
STEALERS, THE Vol. 2 84
STEAUA FARA NUME Vol. 3 298, 312
STECCHINI GIAPPONESI Vol. 7 209
STECKBRIEF 606 Vol. 4 365, 608
STECKBRIEF EINES UNERWÜNSCHTEN Vol. 1 38, 128
STEEPLE CHASE Vol. 10 392, 601
STEFAN LUCHIAN Vol. 6 *page 267, page 275, page 276*
STEFANIA Vol. 7 1113
STEH ICH IN FINSTRER MITTERNACHT Vol. 4 223, 560
STEINERNE REITER, DER Vol. 4 279, 288, 567, 727
STEINREICHER MANN, EIN Vol. 4 79, 115, 373
STEINZEITBALLADE Vol. 1 72, 101 Vol. 6 *page 164*
STEJAR EXTREMA URGENTA Vol. 3 225, 254, 284
STEK Vol. 10 64, 178, 188
STELE DE IARNA Vol. 3 289 Vol. 6 *page 269*
STELLA Vol. 2 721 Vol. 3 198, 202 Vol. 6 *page 82, page 92* Vol. 7 655, 1113
STELLA DEL CINEMA, LA Vol. 7 196, 508
STELLA DEL MARE Vol. 7 40, 419
STELLA DELL'INDIA, LA Vol. 7 379
STEMMERETSKVINDER Vol. 5 97

STEMTÖL SZEMBE **Vol. 9** 134
STEN STENSSON KOMMER TILL STAN **Vol. 5** 923, 1045
STEN STENSSON KOMMER TILLBAKA **Vol. 5** 905, 1142
STEN STENSSON STEEN FRAN ESLÖV **Vol. 5** 872
STEN STENSSON STEEN FRAN ESLÖV PA NYA ÄVENTYR **Vol. 5** 840, 997
STENANSIKTET **Vol. 5** 869, 1052
STENATA **Vol. 6** *page 41, page 47* **Vol. 10** 380, 385, 503
STENBROENS „HELTE" **Vol. 5** 15
STENHUGGEREN **Vol. 5** 97
STEPAN CEL MARE **Vol. 3** 269, 290, 307
STEPENICE HRABROSTI **Vol. 3** 374
STEPENICE ZA NEBO **Vol. 6** *page 310*
STEPPA, LA **Vol. 7** 388, 1035, 1062
STEPUTAT & CO. **Vol. 4** 215, 564
STERBEN IM WALDE, DAS **Vol. 4** 5
STERBENDE MODELL, DAS **Vol. 4** 309
STERBENDE VÖLKER **Vol. 4** 135, 247, 397
STERMINATOR VESEVO **Vol. 7** 34
STERN VON OLIVERA, DER **Vol. 4** 452
STERN VON RIO **Vol. 4** 424, 462, 788
STERN VON VALENCIA **Vol. 4** 89, 298, 536
STERNE **Vol. 1** 9, 21
STESSO MARE STESSA SPIAGGIA **Vol. 7** 610
STESTI PRO DVA **Vol. 10** 532, 621
STEUERLOS **Vol. 4** 356, 550, 724
STEUERMANN HOLK **Vol. 4** 36, 202, 660
STEVNEMOTE MED GLEMTE AR **Vol. 5** 804
STICENIK **Vol. 3** 408, 429
STICI PRE SVITANJA **Vol. 3** 381, 411
STICO **Vol. 8** 338, 358
STIER VON OLIVERA, DER **Vol. 4** 190, 452, 560
STIER VON SALANDA, DER **Vol. 4** 71, 488
STIGMATE, LE **Vol. 2** 169, 330, 538, 613
STIHAN A PODEZRELY **Vol. 10** 394, 517, 550
STIHOVE **Vol. 3** 76, 103 **Vol. 6** *page 45*
STIKA V RYBNICE **Vol. 10** 70, 247
STILETT, DAS **Vol. 4** 517, 525, 558
STILLA FLIRT, EN **Vol. 5** 840, 909
STILLE DAGE I CLICHY **Vol. 5** 130, 226, 289
STILLIGA AUGUSTA **Vol. 5** 841, 945
STIMME AUS DEM ÄTHER, DIE **Vol. 4** 132, 149, 660, 728
STIMME DER LIEBE, DIE **Vol. 4** 132, 437, 458
STIMME DES HERZENS **Vol. 4** 247, 391, 557
STIMME DES HERZENS, DIE **Vol. 4** 213, 344, 605, 626, 647, 689
STIMME DES TOTEN, DIE **Vol. 6** *page 189*
STIMME, DIE **Vol. 4** 131, 217
STIMULANTIA **Vol. 5** 880, 1052
STIMULANTIA – BIRGIT NILSSON **Vol. 5** 940
STIMULANTIA – DYGDENS BELÖNING **Vol. 5** 881, 933
STIMULANTIA – HAN-HON **Vol. 5** 940
STIMULANTIA – KONFRONTATIONER **Vol. 5** 1004
STIMULANTIA – NEGRESSEN I SKAPET **Vol. 5** 881, 933
STIMULANTIA – SMYCKET **Vol. 5** 940, 1130
STIN KAPRADINY **Vol. 10** 425, 541, 567
STIN LETAJICIHO PTACKA **Vol. 10** 172, 233, 327
STIN VE SVETLE **Vol. 10** 338, 345, 524
STINE OG DRENGENE **Vol. 5** 16, 61, 195
STINSEN PA LYCKAS **Vol. 5** 863, 1032
STINY **Vol. 10** 167, 536
STINY HORKEHO LETA **Vol. 10** 394, 425, 505
STIRPE, LA **Vol. 7** 1005
STJAALNE ANSIGT, DET **Vol. 5** 58
STJAALNE MINISTER, DEN **Vol. 5** 2, 228
STJAALNE NAVN, DET **Vol. 5** 81
STJENKA RASIN **Vol. 4** 340, 466, 742
STJERNESKUD **Vol. 5** 89
STO KONI DO STU BRZEGOW **Vol. 1** 280, 311 **Vol. 6** *page 217, page 248*
STO METROW MILOSCI **Vol. 1** 380
STO PET PROCENT ALIBI **Vol. 10** 503, 514, 532
STO PIECDZIESIAT NA GODZINE **Vol. 1** 180, 391
STO TONA STASTIE **Vol. 3** 116
STOCKHOLMSSOMMMAR **Vol. 5** 1160
STOD I AVISEN, DET **Vol. 5** 7, 20
STODDERPRINSESSEN **Vol. 5** 242
STOJAN MUTIKASA **Vol. 3** 440, 444
STOJNE U KOSTENURKA **Vol. 3** 90
STOLZ DER 3. KOMPAGNIE, DER **Vol. 4** 49, 479
STOLZ DER FIRMA, DER **Vol. 4** 743
STOLZ DER KOMPAGNIE, DER **Vol. 4** 149, 321, 660
STOLZE SCHWEIGEN, DAS **Vol. 4** 179, 673
STOLZENFELS AM RHEIN **Vol. 4** 584, 747
STOMPA & CO. **Vol. 5** 717, 754, 772
STOMPA FORELSKER SEG **Vol. 5** 674, 717
STOMPA, SELVFOLGELIG ! **Vol. 5** 682, 772
STON DASKALO MAS ME AGAPE **Vol. 3** 141

STOP – CADRU LA MASA **Vol. 6** *page 266*
STOP FLIRTING **Vol. 2** 84
STOP TYVEN ! **Vol. 5** 183
STOPAR **Vol. 10** 156, 192, 250
STOPP ! TÄNK PA NAGOT ANNAT **Vol. 5** 1032, 1181
STOPY NA SITNE **Vol. 10** 36, 72, 320
STORA ÄVENTYRET, DET **Vol. 5** 1167
STORA FAMNEN **Vol. 5** 840, 997
STORA HOPAREGRÄND OCH HIMMELRIKET **Vol. 5** 1042, 1164
STORA KÄRLEKEN, DEN **Vol. 5** 840, 878
STORA SKRÄLLEN **Vol. 5** 840, 878
STORCH STREIKT, DER **Vol. 4** 340, 527
STORE AMATÖREN, DET **Vol. 5** 878, 953, 1052
STORE ANSVAR, DET **Vol. 5** 2, 7
STORE BARNEDAPEN, DEN **Vol. 5** 669
STORE DAG, DEN **Vol. 5** 202
STORE FALD, DET **Vol. 5** 203
STORE FORVENTNINGER **Vol. 5** 136, 183, 228
STORE GAVTYV, DEN **Vol. 5** 23, 281
STORE GEVINST, DEN **Vol. 5** 83, 97, 203
STORE HJERTE, DET **Vol. 5** 228
STORE LOB, DET **Vol. 5** 44, 89
STORE MAGT, DEN **Vol. 5** 157, 228
STORE MIDDAG, DEN **Vol. 5** 9
STORE OPERATION, DEN **Vol. 5** 58, 295
STORE VARPET, DET **Vol. 5** 800
STÖRENFRIED, DER **Vol. 4** 115
STÖRENFRIEDE, DIE **Vol. 1** 23, 25
STORFOLK OG SMAFOLK **Vol. 5** 722, 734, 781, 800
STORGARDS – ANNAS FRIARE **Vol. 5** 986
STORIA AMBIGUA, UNA **Vol. 7** 414
STORIA D'AMORE E DI AMICIZIA **Vol. 7** 766, 1098, 1191
STORIA D'AMORE, UNA **Vol. 7** 493, 693, 756
STORIA DE FRATELLI E DE COLTELLI **Vol. 7** 1213
STORIA DEI TRE, LA **Vol. 7** 108
STORIA DEI TREDICI, LA **Vol. 7** 560
STORIA DEL PALETOT, LA **Vol. 7** 1189
STORIA DELLA DAMA DAL VENTAGLIO BIANCO **Vol. 7** 1162
STORIA DI CLO – CLO, LA **Vol. 7** 48
STORIA DI FIFA E DI COLTELLO **Vol. 7** 1213
STORIA DI LULU, LA **Vol. 7** 1189
STORIA DI PICCOLINO, LA **Vol. 7** 1189
STORIA DI PIETRA **Vol. 7** 405, 567, 628
STORIA DI UN DELITTO, LA **Vol. 7** 192, 704
STORIA DI UN PAIO DI STIVALI **Vol. 7** 1189
STORIA DI UN PECCATO, LA **Vol. 7** 1004
STORIA DI UNA CAPINERA, LA **Vol. 7** 16, 1162
STORIA DI UNA DONNA, LA **Vol. 7** 369
STORIA DI UNA MINORENNE **Vol. 7** 739
STORIA DI UNA SIGARETTA, LA **Vol. 7** 53, 903
STORIA MILANESE, UNA **Vol. 7** 221, 635
STORIA RUSSA **Vol. 7** 1189
STORIA VERA DELLA SIGNORA DELLE CAMELIE, LA **Vol. 7** 498, 567, 1142
STORIE DI ORDINARIA FOLLIA **Vol. 7** 333, 351, 405, 434
STORIE SCELLERATE **Vol. 7** 333
STORM ÖVER SKÄREN **Vol. 5** 930, 1032
STORM ÖVER TJURÖ **Vol. 5** 1073, 1080
STORMEN PAA KOBENHAVN DEN 11. FEBRUAR 1659 **Vol. 5** 198, 203
STORMENS BARN **Vol. 5** 870, 923
STORMFAGELN **Vol. 5** 997
STORMVARSEL **Vol. 5** 117, 199, 306
STORSTADSFAROR **Vol. 5** 966, 976
STORSTE I VERDEN, DET **Vol. 5** 58
STORSTE SPILLET, DET **Vol. 5** 718, 762, 769, 799
STÖRTEBEKER **Vol. 4** 288, 558
STORTJUVEN **Vol. 5** 844, 946, 1050
STOSSTRUPP 1917 **Vol. 4** 247, 652, 775, 787
STOT STAAR DEN DANSKE SOMAND **Vol. 5** 89, 248
STÖTEN **Vol. 5** 878, 1052
STOV FOR ALLE PENGENE **Vol. 5** 199, 246, 304
STOV PA HJERNEN **Vol. 5** 104, 199, 201, 304, 682, 717, 814
STOVER STADIG, DET **Vol. 5** 104, 199, 304
STOVSUGERBANDEN **Vol. 5** 85, 143, 281
STRAALEMESTEREN **Vol. 5** 83
STRACH **Vol. 1** 150, 204, 340 **Vol. 6** *page 236* **Vol. 10** 24, 182, 313
STRACHY **Vol. 1** 169, 342, 397
STRACONA NOC **Vol. 1** 144
STRADA BUIA, LA **Vol. 7** 1054
STRADA DEI GIGANTI, LA **Vol. 7** 1062
STRADA, LA **Vol. 7** 257, 294, 539, 718, 725, 956
STRADA LUNGA UN ANNO, LA **Vol. 7** 1044, 1200
STRADA SENZA USCITA **Vol. 7** 946
STRADIVARI **Vol. 4** 89, 246
STRADIVARIS SCHÜLERGEIGE **Vol. 4** 13, 24
STRÄFLING AUS STAMBUL, DER **Vol. 4** 164, 247, 584
STRÄFLING VON CAYENNE, DER **Vol. 4** 208 **Vol. 6** *page 193*
STRÄFLINGSKAVALIER, DER **Vol. 4** 223

STRAFSACHE VAN GELDERN **Vol. 4** 149, 634, 660
STRAGE DEGLI INNOCENTI, LA **Vol. 7** 806, 1189
STRAH **Vol. 3** 395, 421
STRAHIL VOJVODA **Vol. 3** 64, 86 **Vol. 6** *page 25*
STRAHLEN DES TODES, DIE **Vol. 4** 634
STRAINUL **Vol. 3** 213, 218, 281, 321, 331
STRAKA V HRSTI **Vol. 10** 74, 397, 466
STRAKATI ANDELE **Vol. 10** 111, 235, 427, 626
STRAKONICKY DUDAK **Vol. 10** 132, 193, 359
STRANA **Vol. 7** 806
STRANA PASSIONE, UNA **Vol. 7** 13
STRANDHUGG I SOMRAS **Vol. 5** 887, 1065
STRANDHUGG PA KAVRINGEN **Vol. 5** 770, 788
STRANEN DVUBOJ **Vol. 3** 41 **Vol. 6** *page 41, page 51*
STRANGE AFFAIR OF UNCLE HARRY, THE **Vol. 2** 526
STRANGER'S HAND, THE **Vol. 7** 1035, 1062
STRANGOLATORE MUTO, LO **Vol. 7** 91
STRANI CASI DI COLLERICCIO, GLI **Vol. 7** 212
STRANIERA, LA **Vol. 7** 978
STRANIERO, LO **Vol. 7** 498, 988, 1142
STRANO APPUNTAMENTO **Vol. 7** 1044
STRANO INVITO A PRANZO, UNO **Vol. 7** 1189
STRANO VIAGGIO DI PIN-POPO, LO **Vol. 7** 133, 963
STRANO VIZIO DELLA SIGNORA WARD, LO **Vol. 8** 396
STRASIDLA Z VIKYRE **Vol. 10** 255, 462, 570
STRASNA ZENA **Vol. 10** 309, 352, 621
STRASSE DER VERLORENEN SEELEN, DIE **Vol. 6** *page 198*
STRASSE DES VERGESSENS, DIE **Vol. 4** 49, 287, 644
STRASSE, DIE **Vol. 4** 211, 247, 441
STRASSENBEKANNTSCHAFT **Vol. 1** 16 **Vol. 4** 97
STRASSENMUSIK **Vol. 4** 344, 647
STRASZNY DWOR **Vol. 1** 278, 308, 404
STRATEGIA DEL RAGNO **Vol. 7** 524, 1101
STRATENA DOLINA **Vol. 10** 82, 273, 280
STRATIOTES HORIS STOLE **Vol. 3** 162
STRATTAGEMMA D'AMORE **Vol. 7** 1189
STRAZILE AU AMINTIRI **Vol. 3** 250, 317
STRAZNA NOC **Vol. 1** 278, 372
STREGA, LA **Vol. 7** 1189
STREGHE – LA STREGA BRUCIATA VIVA, LE **Vol. 7** 498, 913, 988
STREHA E RE **Vol. 3** 9 **Vol. 6** *page 18*
STREICHHÖLZER, KAUFT STREICHHÖLZER ! **Vol. 4** 743
STREIK ! **Vol. 5** 767
STREIK DER DIEBE, DER **Vol. 4** 423, 681
STREIT UM DEN KNABEN JO **Vol. 4** 28, 298, 608
STREIT UM DEN SERGEANTEN GRISCHA, DER **Vol. 1** 22
STREITBARE HERR KICKEL, DER **Vol. 4** 323
STREITHAMMEL, DER **Vol. 4** 97, 773
STREJKEN **Vol. 5** 997
STREJKEN PAA DEN GAMLE FABRIK **Vol. 5** 9
STRELLE NEL FOSSO, LE **Vol. 7** 333, 810
STREPY PRO EVU **Vol. 10** 106, 327, 344
STRESS ES TRES, TRES **Vol. 8** 348
STRETNUTIE **Vol. 10** 32, 263, 483
STRETTA E BAGNATA **Vol. 7** 742
STRETTA, LA **Vol. 7** 168
STRIBRNA OBLAKA **Vol. 10** 168, 524
STRIBRNY VITR **Vol. 10** 412, 503, 569
STRICH DURCH DIE RECHNUNG **Vol. 4** 82, 89, 270, 409
STRIDEN GAR VIDARE **Vol. 5** 840, 909
STRIDZA SPOD HAJA **Vol. 10** 20
STRIEBORNY FAVORIT **Vol. 10** 459, 576
STRINGIMI FORTE PAPA **Vol. 7** 426
STRIP-TEASE **Vol. 2** 503
STRIP-TEASEUSES, LES **Vol. 2** 814
STRIP-TIS A LA INGLESIA **Vol. 8** 702
STR"MNATA P"TEKA **Vol. 3** 31, 100, 101, 109
STROGIJAT OT KVARTAL „AKACIJA" **Vol. 6** *page 23, page 33, page 58*
STROHWITWER **Vol. 4** 186, 340, 644
STROM, DER **Vol. 4** 15, 147, 232
STROMBOLI, TERRA DI DIO **Vol. 7** 725
STROMER **Vol. 5** 158, 175, 261
STRUKTURA KRYSZTALU **Vol. 1** 254, 403 **Vol. 6** *page 222*
STRUPPS **Vol. 4** 124, 634
STRYCEK Z AMERIKY **Vol. 10** 365, 453, 524
STRZAL **Vol. 1** 176, 371
STUBLENSKITE LIPI **Vol. 3** 37, 89 **Vol. 6** *page 45*
STUD **Vol. 10** 96, 428, 473
STUD. CHEM. HELENE WILLFUER **Vol. 4** 255, 527
STUDENCI **Vol. 1** 174, 215
STUDENT SEIN, WENN DIE VEILCHEN BLÜHEN **Vol. 4** 124, 204
STUDENT VON PRAG, DER **Vol. 4** 124, 202, 234, 359, 462, 486, 559, 646, 734

STUDENTE, LO Vol. 7 311
STUDENTENS GLADE LIV Vol. 5 92
STUDENTER I PARIS Vol. 5 1162
STUDENTERKAMMERATERNE Vol. 5 83, 183
STUDENTERNA PA TRÖSTEHULT Vol. 5 1112
STUDENTESSA, LA Vol. 7 765
STUDENTMAGERSVENDEN Vol. 5 83
STUDENTSKA MAMA Vol. 10 256, 586
STUDIO LEGALE PER UNA RAPINA Vol. 7 364
STUDUJEME ZA SKOLOU Vol. 10 122, 532, 621
STUKAS Vol. 4 323, 455, 469, 580, 736
STULEN VALS, EN Vol. 5 870
STÜLKE UND LEHMANN Vol. 4 18, 558
STUMME GAST, DER Vol. 4 28, 246, 329, 376
STUMME VON PORTICI, DIE Vol. 4 89, 149, 660
STUNDE DER TÖCHTER Vol. 6 *page 160, page 170, page 173*
STUNDE DER VERSUCHUNG, DIE Vol. 4 247, 424, 461
STUNDE GLÜCK, EINE Vol. 4 34, 506, 691
STUNDE NACH MITTERNACHT, DIE Vol. 4 558
STUPNE PORAZENYCH Vol. 10 139, 304, 516
STUREC V UHOTO Vol. 3 119 Vol. 6 *page 30, page 51*
STURM Vol. 6 *page 189*
STURM AUF DREI HERZEN Vol. 4 97, 232
STURM AUF MARAKESCH Vol. 4 304
STÜRME Vol. 4 113
STÜRME DER LEIDENSCHAFT Vol. 4 325, 567
STÜRME DES LEBENS Vol. 4 366
STÜRME ÜBER DEM MONTBLANC Vol. 4 10, 15, 73, 620
STURMFLUT Vol. 4 344, 544
STURMFLUT DER LIEBE Vol. 4 339, 534, 590, 599
STURMFLUT DES LEBENS Vol. 4 735
STÜRMISCH DIE NACHT Vol. 4 478, 535
STURMTRUPPEN N. 2 Vol. 7 13, 134, 988
STURMVOGEL Vol. 4 99
STURZ, DER Vol. 6 *page 166, page 167, page 174*
STURZ IN DIE FLAMMEN, DER Vol. 4 192, 565
STURZ INS GLÜCK, DER Vol. 4 190, 287
STÜRZENDE GÖTTER Vol. 4 67, 179, 400, 522
STÜTZEN DER GESELLSCHAFT Vol. 4 132, 230
STVANI LIDE Vol. 10 91, 479, 524, 586
STVORYLKA Vol. 10 287, 306, 569
STVRTY ROZMER Vol. 10 66, 108, 454
STYRIDSATSTYRI Vol. 10 268, 320, 478, 576
STYRMAN KARLSSONS FLAMMOR Vol. 5 840, 864, 878, 891, 996
STYRMAND KARLSEN Vol. 5 180, 199, 201, 248
STYRMANDENS SIDSTE FART Vol. 5 97
SU 13, UNA Vol. 7 1007
SU DESCONSOLADO ESPOSA Vol. 8 515
SU E GIU - IL COLPO DEL LEONE Vol. 7 1136
SU E GIU - IL MARITO D'AGOSTO Vol. 7 1147
SU E GIU - QUESTIONE DI PRINCIPIO Vol. 7 1136
SU FA COSI Vol. 7 454
SU NOMBRE ES DAPHNE Vol. 8 249
SU PROPIO DESTINO Vol. 8 732
SU ULTIMA NOCHE Vol. 8 207, 400
SUA ALTEZZA HA DETTO : NO ! Vol. 7 16, 665
SUA FIGLIA Vol. 7 148
SUA STRADA, LA Vol. 7 636, 638, 1054
SUB CUPOLA ALBASTRA Vol. 3 276
SUB CUPOLA CIRCULUI Vol. 3 218, 256
SUBDESARROLLADOS, LOS Vol. 8 331, 681
SUBLIME RINUNCIA Vol. 7 890
SUBLOKATOR Vol. 1 303, 382, 403
SUBOTOM UVECE Vol. 3 356, 436
SUCCEDE A NAPOLI Vol. 7 863
SUCEDIO EN DAMASCO Vol. 2 646 Vol. 8 607, 717
SUCEDIO EN MI ALDEA Vol. 8 501, 753
SUCEDIO EN SEVILLA Vol. 8 214, 396
SUCESO DE ANOCHE, EL Vol. 8 423
SUCETTE MAGIQUE, LA Vol. 2 259
SUCEUSES, LES Vol. 2 201, 291
SUCHE NACH DEM WUNDERBUNTEN VÖGELCHEN, DIE Vol. 1 26, 29, 104
SUCHENDE SEELE, DIE Vol. 4 25, 70, 239
SUCRE, LE Vol. 2 505, 564
SUDAR NA PARALELAMA Vol. 3 389, 452
SUDARIO A LA MEDIA, UN Vol. 8 299
SUDBINE Vol. 3 362, 402
SUDIM TA LASKOU Vol. 10 108, 263, 269
SUENA EL CLARIN Vol. 8 233, 450
SUENO O REALIDAD ? Vol. 8 678
SUENOS DE MUJER Vol. 8 207, 732
SUENOS HUMEDOS DE PATRICIA, LOS Vol. 8 471
SUFFRAGETTE, DIE Vol. 4 634
SUGAR COLT Vol. 8 495, 762
SUGARKA Vol. 9 175, 220
SUGGESTIONATA, LA Vol. 7 949
SUICIDA ! Vol. 7 1189

SUICIDIO SUBLIME, IL **Vol. 7** 304
SUIJANJE **Vol. 6** *page 303*
SUIVEZ CET HOMME **Vol. 2** 124, 128
SUIVEZ-MOI JEUNE HOMME **Vol. 2** 57, 131
SUJET OU LE SECRETAIRE AUX MILLE ET UN TIROIRS, LE **Vol. 2** 806
SUKCESKOMPONISTEN **Vol. 5** 2, 23
SUKROMNA VOJNA **Vol. 10** 72, 82, 287
SUL LIMITE DEL NIRVANA **Vol. 7** 420
SUL LIMITE DELLA FOLLIA **Vol. 7** 1073
SUL PASSO ESTREMO **Vol. 7** 320
SUL PONTE DEI SOSPIRI **Vol. 7** 796
SUL TRANSATLANTICO **Vol. 7** 1189
SUL TRAPEZIO **Vol. 7** 965
SULEJMAN VELICANSTVENI **Vol. 3** 444
SULL' ALTRA, UNA **Vol. 8** 306, 762
SULLA STRADA MAESTRA **Vol. 7** 22
SULLE ALPI RICONSACRATA **Vol. 7** 341
SULLE ROVINE DELL' AMORE **Vol. 7** 69
SULLIVAN **Vol. 7** 701
SULT **Vol. 5** 2, 63, 179, 743, 797
SULTANA SAFIYE, LA **Vol. 7** 476
SULTANE DE L'AMOUR, LA **Vol. 2** 277, 333, 708
SULTANS, LES **Vol. 2** 722
SUMARIO SANGRIENTO DE LA PEQUENA STEFANIA **Vol. 8** 311, 681
SUMMER OF SILENCE **Vol. 5** 739
SUMMER STORM **Vol. 2** 778
SUMNJIVO LICE **Vol. 3** 349, 374
SUMURU **Vol. 8** 546
SUMURUN **Vol. 4** 294, 560, 663 **Vol. 6** *page 196, page 199*
SUNCE TUDJEG NEBA **Vol. 3** 351
SÜNDE AM WEIBE **Vol. 4** 230, 258, 523
SÜNDE DER LISSY KRAFFT, DIE **Vol. 4** 97, 495
SÜNDE EINER SCHÖNEN FRAU, DIE **Vol. 6** *page 188, page 197*
SÜNDE UND MORAL **Vol. 4** 565, 704
SÜNDEN DER ELTERN **Vol. 4** 270, 570
SÜNDEN DER VÄTER, DIE **Vol. 4** 646
SÜNDENBABEL **Vol. 4** 219, 316
SÜNDENBOCK, DER **Vol. 4** 340, 361
SÜNDENLUST **Vol. 4** 98
SÜNDERIN, DIE **Vol. 4** 339, 560, 725, 743
SÜNDIG UND SÜSS **Vol. 4** 257, 558
SÜNDIGE DORF, DAS **Vol. 4** 10, 524, 544
SÜNDIGE HOF, DER **Vol. 4** 344, 545, 787
SÜNDIGE LIEBE **Vol. 4** 646
SÜNDIGE VESTALIN, DIE **Vol. 4** 626
SUNNUNTAIMETSÄSTÄJÄT **Vol. 5** 556
SUNRISE **Vol. 4** 202, 716
SUO DESTINO, IL **Vol. 7** 639, 751, 902, 1211
SUO NOME E DONNA ROSA, IL **Vol. 7** 743
SUO PIU GRANDE AMORE, IL **Vol. 7** 17
SUOCERA DI ROBINET, LA **Vol. 7** 1189
SUOCERO DI SE STESSO **Vol. 7** 369
SUOMALAISTYTTÖJÄ TUKHOLMASSA **Vol. 5** 310, 634
SUOMISEN OLLI RAKASTUU **Vol. 5** 356, 386
SUOMISEN OLLI YLLÄTTÄÄ **Vol. 5** 356, 386
SUOMISEN OLLIN TEMPAUS **Vol. 5** 356, 386
SUOMISEN PERHE **Vol. 5** 471, 549
SUOMISEN TAITEILIJAT **Vol. 5** 356, 386
SUONATRICE D'ARPA, LA **Vol. 7** 493
SUONNO D'AMORE **Vol. 7** 603
SUOPURSU KUKKII **Vol. 5** 356, 549
SUOR BIANCA, LA **Vol. 7** 978
SUOR EMANUELLE **Vol. 7** 694
SUOR GIOVANE, LA **Vol. 7** 763
SUOR LETIZIA **Vol. 7** 634, 1167
SUOR MARIA **Vol. 7** 739
SUOR TERESA **Vol. 7** 164, 891, 965
SUOTORPAN TYTTÖ **Vol. 5** 349, 566
SUPE FÖR TVA **Vol. 5** 357, 871, 948, 1026
SUPER CLIMAX **Vol. 7** 742
SUPER HARD CORE **Vol. 7** 742
SUPER RAPINA A MILANO **Vol. 7** 820
SUPERANDY IL FRATELLO BRUTTO DI SUPERMAN **Vol. 7** 1015
SUPERARGO CONTRA DIABOLICUS **Vol. 8** 523, 732
SUPERARGO CONTRO DIABOLIKUS **Vol. 8** 523, 732
SUPERARGO EL GIGANTE **Vol. 8** 600, 626
SUPERCAMORRISTAS, LOS **Vol. 8** 550, 670
SUPERSEVEN CHIAMA CAIRO **Vol. 7** 1121
SUPPLENTE, LA **Vol. 7** 501
SUPPLENTE VA IN CITTA, LA **Vol. 8** 626, 627
SUPPLIZIO DEL SILENZIO, IL **Vol. 7** 1162
SUPREMA BELLEZZA **Vol. 7** 529
SUPREMO CONVEGNO, IL **Vol. 7** 175, 183, 1156, 1189
SUPREMO OLOCAUSTO **Vol. 7** 131
SUR, EL **Vol. 8** 219, 269
SUR LA ROUTE DE SALINA **Vol. 2** 279
SUR LE BANC **Vol. 2** 48, 231, 259
SUR LE PLANCHER DES VACHES **Vol. 2** 270
SUR LES RAILS **Vol. 2** 798
SUR UN ARBRE PERCHE **Vol. 2** 598, 781
SURCOS **Vol. 8** 488, 614
SURCOUF **Vol. 2** 216, 584
SURCOUF, LE TIGRE DES SEPT MERS **Vol. 6** *page 81*
SURCOUF, L'EROE DEI SETTE MARI **Vol. 8** 426, 732

SURDOUES DE LA 1RE COMPAGNIE, LES Vol. 6 *page 72*
SURIS IN PLINA VARA, UN Vol. 3 230, 287, 291
SURORILE Vol. 6 *page 281*
SURPRISE DU CHEF, LA Vol. 2 618
SURPRISE PARTY Vol. 6 *page 111*
SURPRISES DE LA RADIO, LES Vol. 2 193, 254, 339, 538
SURPRISES DU DIVORCE, LES Vol. 2 388
SURPRISES DU SLEEPING, LES Vol. 2 646, 722
SURPRISES D'UNE NUIT DE NOCES, LES Vol. 2 131, 205
SURRY, DER STEHER Vol. 4 239
SURSIS POR UN ESPION Vol. 2 259, 475, 738
SURVIVRE Vol. 2 65, 253, 308
SUS ANOS DORADAS Vol. 8 459
SUS OG DUS PA BY'N Vol. 5 718, 762, 772
SUSANA Vol. 8 553, 749
SUSANA TIENE UN SECRETO Vol. 8 285, 648
SUSANA Y YO Vol. 8 214, 762
SUSANNA TUTTA PANNA Vol. 7 333
SUSANNE Vol. 5 150, 208, 902
SUSANNE I BADET Vol. 5 81
SUSANNE IM BADE Vol. 4 13, 340, 466
SUSANNE MACHT ORDNUNG Vol. 4 82, 239, 270
SUSANNE PAA EVENTYR Vol. 5 81
SUSANNE UND DER ZAUBERRING Vol. 1 29, 34, 55
SUSE KERKSTRAATEN Vol. 4 62, 465, 690
SUSE, LIEBE SUSE Vol. 1 14, 92, 100
SUSI UND DER SCHWARZE MANN Vol. 4 329, 525
SUSPECTS, LES Vol. 2 131, 194, 344, 595, 726, 787, 815
SUSPENDIDO EN SINVERGUENZA Vol. 8 495, 546
SUSPENSE AU DEUXIEME BUREAU Vol. 2 279, 514
SUSPENSO EN COMUNISMO Vol. 8 546
SUSPIRIA Vol. 7 98, 1145
SUSPIROS DE TRIANA Vol. 8 261, 298, 496
SUSS GOTT Vol. 5 840, 1029
SÜSSE MÄDEL, DAS Vol. 4 321, 340, 734
SÜSSES GEHEIMNIS, EIN Vol. 4 132, 479
SUSSIE Vol. 5 907, 923
SUSSURRO NEL BUIO, UN Vol. 7 318
SUSU-MUSU Vol. 3 94
SUT Vol. 3 367
SÜT A NAP Vol. 9 94, 315, 322
SUTJESKA Vol. 3 420, 421, 444
SUTON Vol. 6 *page 317, page 320*
SUTRA CU BITI SLOBODNA Vol. 3 435
SUTURP - EINE LIEBESGESCHICHTE Vol. 6 *page 162, page 163, page 165*
SUURI SÄVELPARAATI Vol. 5 549, 656
SUURIN VOITTO Vol. 5 357, 471, 566
SUURSALON HÄÄT Vol. 5 344, 356, 571
SUVINEN SATU Vol. 5 406, 640
SUVIYÖN SALAISUUS Vol. 5 472, 510, 634
SUZANNE Vol. 2 833
SUZANNE ET SES BRIGANDS Vol. 2 158, 368
SUZANNE OG LEONARD Vol. 5 259, 270, 283
SUZANNE SIMONIN, LA RELIGIEUSE DE DIDEROT Vol. 2 288, 513
SVADLENKA Vol. 10 256, 412
SVÄLT Vol. 5 833, 916, 1016, 1148
SVANDA DUDAK Vol. 10 92, 154, 345
SVÄNGER PA SLOTTET, DET Vol. 5 940, 1052
SVANUCE Vol. 3 371
SVÄRMOR KOMMER Vol. 5 840, 997
SVARSKÖTT PASTORAT, ETT Vol. 5 840, 884, 923, 1033
SVART HAV Vol. 5 701, 824
SVARTA FAGLAR Vol. 5 892, 1186
SVARTA MASKERNA, DE Vol. 5 997
SVARTA PALMKRONOR Vol. 5 940, 974, 1035
SVARTA ROSOR Vol. 5 840, 843
SVARTE FUGLER Vol. 5 684, 746, 808
SVARTE RUDOLF Vol. 5 870, 891
SVARTERE ENN NATTEN Vol. 5 765, 784
SVATA HRISNICE Vol. 10 47, 59, 193, 286, 621
SVATBA Vol. 3 66
SVATBA BEZ PRSTYNKU Vol. 10 193, 296, 612
SVATBA JAKO REMEN Vol. 10 394, 532, 627
SVATBA S PODMINKOU Vol. 10 197, 395, 519
SVATBITE NA JOAN ASEN Vol. 3 46, 59 Vol. 6 *page 43, page 45, page 48*
SVATBY PANA VOKA Vol. 10 283, 313, 567
SVATEBNI CESTA Vol. 10 256, 460
SVATEBNI KOSILE Vol. 10 243
SVATEJ Z KREJCARKU Vol. 10 18, 70
SVATEK VERITELU Vol. 10 2, 532
SVATY VACLAV Vol. 10 143, 167, 168, 185, 257, 298, 524, 586
SVED, AKINEK NYOMA VESZETT, A Vol. 9 6, 202
SVEDECTVI MRTCYCH OCI Vol. 10 70, 153, 569
SVEDESI, LE Vol. 7 853
SVEDOMI Vol. 10 321, 522, 566

SVEDSKI KRALE Vol. 3 59, 114 Vol. 6 *page 31*
SVEHLAVICKA Vol. 10 345, 509
SVEJK NA FRONTE Vol. 10 167, 345
SVEJK V CIVILU Vol. 10 137, 453
SVEJK V RUSKEM ZAJETI Vol. 10 345, 586
SVEN TUSAN Vol. 5 840, 843
SVEN TUUVA Vol. 5 356, 374, 481, 639
SVEND DYRINGS HUS Vol. 5 97
SVENGALI Vol. 4 306, 724
SVENSK TIGER, EN Vol. 5 878, 1173, 1197
SVENSKA BILDER Vol. 5 878, 1052, 1083, 1130
SVENSKA FLICKOR I PARIS Vol. 5 1054
SVENSKA FLOYD Vol. 5 871, 880, 1028
SVENSKE RYTTAREN Vol. 5 909, 1173
SVENSSON ORDNAR ALLT Vol. 5 846
SVENTATELLA, UNA Vol. 7 369
SVERIGE AT SVENSKARNA Vol. 5 1014, 1068, 1187
SVET, KDE SE ZEBRA Vol. 10 256, 524
SVET NA KAJZARJU Vol. 3 390, 406
SVET NIC NEVI Vol. 10 183, 204, 283, 577
SVET OTEVRENY NAHODAM Vol. 10 283, 317, 567
SVET PATRI NAM Vol. 10 167, 256
SVETACI Vol. 10 144, 572
SVETI PESAK Vol. 3 382
SVETLO JEHO OCI Vol. 10 92, 345
SVETOZAR MARKOVIC Vol. 3 370, 418
SVI NA MORE Vol. 3 394, 455
SVIGERMODEREN Vol. 5 97
SVIRAC"T Vol. 3 69, 78, 82 Vol. 6 *page 46*
SVITALO CELOU NOC Vol. 10 139, 254, 292, 327
SVITANI Vol. 10 256, 524
SVITATA, LO Vol. 7 60, 817
SVIZDUK U OSAM Vol. 3 376
SVOBODA ILI SM"RT Vol. 3 48 Vol. 6 *page 42*
SVOGA TELA GOSPODAR Vol. 3 399, 452
SVUNDNE LYKKE, DEN Vol. 5 203
SWEDENHJELMS Vol. 5 840, 909
SWEET MOVIE Vol. 2 417, 455, 517
SWIADECTWO URODZENIA Vol. 1 243, 403 Vol. 6 *page 232*
SWIETA WOJNA Vol. 1 198, 330 Vol. 6 *page 222*
SWIETO KWIATKA W WARSZAWIE Vol. 1 317
SWING IT FRÖKEN ! Vol. 5 871, 1073
SWING IT, MAGISTERN ! Vol. 5 928, 1156
SWINGMEN IN EUROPE Vol. 2 700
SWITCH Vol. 7 1133
SWOOSIE Vol. 7 99
SYD FOR TANA RIVER Vol. 5 85, 141
SYGNALY Vol. 1 211, 241, 250, 308
SYKORKA Vol. 10 16, 27, 73
SYKSYLLÄ KAIKKI ON TOISIN Vol. 5 319, 399, 470, 648
SYLVESTER Vol. 4 124, 247, 559, 646
SYLVESTERNACHT AM ALEXANDER-PLATZ Vol. 4 49, 246
SYLVI Vol. 5 351, 356, 357, 481
SYLVIA, L'ENCHANTEE Vol. 2 648
SYLVIANE DE MES NUITS Vol. 2 407, 433, 475
SYLVIE ET LE FANTOME Vol. 2 8, 158, 439
SYMPHONIE DER LIEBE Vol. 4 666
SYMPHONIE DES BRIGANDS, LA Vol. 2 778
SYMPHONIE EINES LEBENS Vol. 4 149, 188, 288, 486
SYMPHONIE FANTASTIQUE, LA Vol. 2 23, 226, 824
SYMPHONIE PASTORALE, LA Vol. 2 27, 722, 824
SYMPHONIE POUR UN MASSACRE Vol. 2 59, 719
SYN HOR Vol. 10 167, 453
SYN SZATANA Vol. 1 346
SYND Vol. 5 891, 928, 997, 1077
SYND SKAL SONES Vol. 5 256
SYNDEN LOCKAR OCH STRAFFAR Vol. 5 357, 471
SYNDENS DATTER Vol. 5 90
SYNDENS SOLD Vol. 5 97
SYNDERE I SOMMERSOL Vol. 5 770, 803
SYNDERINDE, EN Vol. 5 97
SYNDIG KAERLIGHED Vol. 5 9
SYNNIN JÄLJET Vol. 5 471, 549
SYNNIN PUUMERKKI Vol. 5 552, 592
SYNNITÖN LANKEEMUS Vol. 5 357, 471, 493
SYNNÖVE SOLBAKKEN Vol. 5 928, 954, 1080
SYNOIKIA TO ONEIRO Vol. 3 193, 207
SYNOVE HOR Vol. 10 70, 305, 532
SYNTEZA Vol. 6 *page 210, page 211, page 212, page 252*
SYNTINYT TERVE TYTTÖ Vol. 5 446, 549
SYNTIPUKKI Vol. 5 347, 356, 428, 516
SYÖKSYKIERRE Vol. 5 319, 324, 606
SYSKONBÄDD 1782 Vol. 5 869, 1052, 1067
SYSMÄLÄINEN Vol. 5 310, 386, 592
SYSTEME DU DOCTEUR BITUME, LE Vol. 2 159, 286, 330
SYTTEN Vol. 5 96, 199, 201, 249, 275
SYV DAGER FOR ELISABETH Vol. 5 669, 771
SYYLLISIÄKÖ ? Vol. 5 356, 428
SZABAD LELEGZET Vol. 9 167
SZABADGYALOG Vol. 9 216, 238

SZABADITS MEG A GONOSZTOL Vol. 9 20, 152, 253
SZABONE Vol. 9 134, 312
SZABOTAZS Vol. 9 35, 132, 186, 284
SZAKADEK Vol. 9 26, 165, 241
SZAKALLAS EMBER, A Vol. 3 249 Vol. 9 77
SZAKITANI NEHEZ DOLOG Vol. 9 26, 117
SZALAS MILOSCI Vol. 1 226
SZALENCY Vol. 1 404
SZALENSTWA PANNY EWY Vol. 6 *page 226, page 228, page 239*
SZAMARKÖHÖGES Vol. 9 106, 204, 259
SZANSA Vol. 6 *page 219, page 235, page 241*
SZARADA Vol. 1 293, 390, 401
SZARNYAS DANDAR Vol. 9 28, 35, 127, 196
SZARNYAS ÜGYNÖK, A Vol. 9 56, 212, 240
SZARVASSA VALT FIUK Vol. 9 67, 123, 154
SZARZA CZYLI PRZYPOMNIENIE KANONU Vol. 6 *page 226, page 246*
SZATAN Z SIODMEJ KLASY Vol. 1 305, 308, 346, 400 Vol. 6 *page 242*
SZCZESCIARZ ANTONI Vol. 1 171, 305, 321 Vol. 6 *page 230*
SZCZESCIE PRZYCHODZKI KIEDY CHCE Vol. 1 194
SZCZESLIWA TRZYNASTIKA Vol. 1 372, 384
SZCZESLIWY WISIELEC Vol. 1 301
SZEGENY DZSONI ES ARNIKA Vol. 9 107, 210, 215
SZEGENY GAZDAGOK Vol. 9 71, 134, 199, 208, 332
SZEGENYLEGENYEK Vol. 9 20, 273, 318
SZELEBURDI CSALAD, A Vol. 9 78, 84, 147, 221
SZELEBURDI VAKACIO Vol. 9 51, 107, 134
SZELEHAMOSNÖ, A Vol. 9 83, 183, 259
SZEMÜNEGESEK Vol. 9 155
SZENT MIHALY Vol. 9 220
SZENT PETER ESERNYÖJE Vol. 9 34, 134, 172, 176, 183
SZENTJANOS FEJEVETELE Vol. 9 36, 265, 285
SZENTJOBI ERDÖ TITKA, A Vol. 9 28
SZENZACIO Vol. 9 132, 178, 237
SZEP CSILLAG Vol. 9 132, 274
SZEP ILONKA Vol. 9 7
SZEP LANYOK, NE SIRJATOK Vol. 9 20, 154
SZEP MAGYAR KOMEDIA Vol. 9 20, 154, 329
SZEP PONGRACZNE KRINOLINJA Vol. 9 71
SZEPEK ES BOLONDOK Vol. 9 152, 167, 229
SZEPTEMBER VEGEM Vol. 9 132, 236
SZERECSEM KIRALYFI, A Vol. 9 7
SZERELEM Vol. 9 149, 259, 297
SZERELEM CSÜTÖRTÖK Vol. 9 83, 149, 314
SZERELEM ELSÖ VERIG Vol. 9 6, 53
SZERELEM HALALTUSAJA, A Vol. 9 77
SZERELEM HATARAI, A Vol. 9 140, 152, 316
SZERELEM MASODIK VERIG Vol. 9 6, 82, 254
SZERELEM MINDENT LEGYÖZ, A Vol. 9 73
SZERELEM NEM SZEGYEN, A Vol. 9 196, 322
SZERELEMBÖL NÖSÜLTEM Vol. 9 178, 237
SZERELENCSES DANIEL Vol. 9 171, 253, 281
SZERELMEM, ELEKTRA Vol. 9 20, 154, 318
SZERELMES BICIKLISTAK Vol. 9 134, 145, 259
SZERELMESFILM Vol. 9 190, 215, 316
SZERELMI ALMOK Vol. 9 67, 71, 122, 199, 322
SZERELMI LAZ Vol. 9 186, 196
SZERENCSES FLOTAS Vol. 9 196, 270
SZERESSETEK ODOR EMILIAT Vol. 9 152, 316, 339
SZERESSÜK EGYMAST Vol. 9 26, 79
SZERESSÜK EJSZAKA Vol. 9 117
SZERETES, A Vol. 9 77
SZERETÖ FIA, PETER Vol. 9 26, 117, 133
SZERETÖK Vol. 9 20, 37, 206
SZEROKIEJ DROGI, KOCHANIE Vol. 1 179, 306 Vol. 6 *page 257*
SZERVUSZ, PETER Vol. 9 25, 132, 270
SZEVASZ, VERA Vol. 9 67, 149, 339
SZIAMI MACSKA Vol. 9 132, 164
SZIBERIA Vol. 9 323
SZIGET A SZARAZFÖLDÖN Vol. 9 20, 253, 318
SZIKRAZO LANYOK Vol. 9 316, 339
SZINDBAD Vol. 9 263, 303, 316
SZINES TINTAKROL AMODOM Vol. 9 167, 199, 312
SZINESZNÖ Vol. 9 7
SZINHAZ SZERELMESE, A Vol. 9 33, 76, 312
SZIRIUSZ Vol. 9 132, 164
SZIRMOK, VIRAGOK, KOSZORUK Vol. 9 169, 171, 253
SZIV SZAVA, A Vol. 9 153

SZIVZÜR Vol. 9 20, 140
SZKICE WARSZAWSKIE Vol. 1 355, 389 Vol. 6 *page 209*
SZKICE WEGLEM Vol. 1 148, 171, 202, 247, 302, 392
SZKLANA GORA Vol. 1 237, 305, 391
SZKLANA KULA Vol. 1 298, 391 Vol. 6 *page 256*
SZKODA TWOICH LEZ Vol. 6 *page 243, page 250, page 255*
SZLAKIEM HANBY Vol. 1 174
SZOBALANY, A Vol. 9 77
SZÖKEVENYK Vol. 9 1, 65, 110, 240
SZÖKÖTT KATONA, A Vol. 9 248
SZOMBATTOL HETFÖIG Vol. 9 259, 297, 326
SZOP CADRU LA MASA Vol. 3 314
SZÖRNYEK EVADJA Vol. 9 20, 154
SZOVATHY EVA Vol. 9 58, 132, 247, 270
SZPIEG Vol. 1 215
SZPIEG W MASCE Vol. 1 176, 308
SZPITAL PRZEMIENIA Vol. 1 144, 325, 403
SZTABS-KAPITAN GUBANIEW Vol. 1 278, 394
SZTANDAR WOLNOSCI Vol. 1 174, 380, 404
SZULAMIT Vol. 9 136
SZÜRKERUHAS HÖLGY, A Vol. 9 7
SZÜTS MARA HAZASSAGA Vol. 9 71, 270
SZÜZ ES A GÖDÖLYE, A Vol. 9 132, 274
SZYB L 23 Vol. 1 278, 380
SZYFRY Vol. 1 187, 216, 320

T

TA' BRILLER PAA ! Vol. 5 2, 7
TA CHVILE, TEN OKAMZIK Vol. 10 428, 473, 517
TA' DET SOM EN MAND, FRUE Vol. 5 87, 251
TA DOLLARIA TES ASPASIAS Vol. 3 142
TA HAND OM ULLA Vol. 5 923, 1022, 1079, 1205
TA HEIROCROTEMATA Vol. 3 157, 202
TA HROMATA TES IRIDOS Vol. 3 151
TA' HVAD DU VIL HA' Vol. 5 150, 248
TA KOURELIA TRAGOUDANE AKOMA Vol. 3 156
TA' LIDT SOLSKIN Vol. 5 199, 201, 275
TA MEJ I DALEN Vol. 5 887
TA NASE PISNICKA CESKA Vol. 10 289, 359, 572
TA PAHTUI KAUKANA Vol. 5 374, 443, 620
TA' PELLE MED Vol. 5 2, 23, 228
TA TESSERA SKALOPATIA Vol. 3 154
TA TRAGOUDIA TES FOTIAS Vol. 3 128
TA TRETI Vol. 10 41, 197, 283
TÄÄLÄ POHJANTÄHDEN ALLA Vol. 5 494, 594, 613, 635, 639
TÄÄLLÄ ALKAA SEIKKAILU Vol. 5 329
TÄÄLTÄ TULAAN ELÄMÄ ! Vol. 5 319, 320, 500
TAAPE TÄHTENÄ Vol. 5 613, 634
TAAS TAPAAMME SUOMISEN PERHEEN Vol. 5 443, 639
TAAS TYTTÖ KADOKSISSA Vol. 5 451, 592
TABARIN Vol. 2 419, 598
TABARNAC Vol. 2 551
TABEA, STEHE AUF ! Vol. 4 132, 773
TABLE AUX CREVES, LA Vol. 2 344, 615
TABLICZKA MARZENIA Vol. 1 276, 313, 381 Vol. 6 *page 227*
TABOR Vol. 2 871
TABU Vol. 5 844, 869, 1053 Vol. 8 406
TABUSSE Vol. 2 595
TAC TAC Vol. 8 407, 738
TADEL, DER Vol. 1 73
T'ADERFIA ORKISTEKAN EKDIKESE Vol. 3 141, 148
TAENK PA ET TAL Vol. 5 16, 61, 195
TAG 56 Vol. 5 841, 923
TAG AUF DEM MARS, EIN Vol. 4 113, 651
TAG DER ROSEN IM AUGUST ... DA HAT DIE GARDE FORTGEMUSST, EIN Vol. 4 208, 560
TAG DET SOM EN MAND ! Vol. 5 163, 228
TAG NACH DER SCHEIDUNG, DER Vol. 4 49, 738, 788
TAG TIL MARKED I FJORDBY Vol. 5 180, 199, 201, 304
TAG TIL RONNEBY KRO Vol. 5 89, 196, 264
TAGEBUCH COLLINS, DAS Vol. 4 192
TAGEBUCH DES DR. HART, DAS Vol. 4 288 Vol. 6 *page 189*
TAGEBUCH EINER KOKOTTE Vol. 4 212, 255
TAGEBUCH EINER VERLORENEN Vol. 4 10, 246, 452
TAGEBUCH EINER VERLORENEN, DAS Vol. 4 159, 565
TAGET GAR KLOCKAN 9 Vol. 5 840, 1119
TÄHDET KERTOVAT, KOMISARIO PALMU Vol. 5 613, 634
TÄHTIRAPRTTERIT TULEVAT Vol. 5 363, 481, 567
TÄHTISILMÄ Vol. 5 549, 579, 592, 639
TAHUR, EL Vol. 8 309
TAIKAPEILI Vol. 5 496, 546
TAIKAYÖ Vol. 5 461, 462

TAIS - TOI QUAND TU PARLES Vol. 7 450, 1194
TAISTELU HEIKKILÄN TALOSTA Vol. 5 334, 636
TAJEMNICA ALEI ROZ Vol. 1 174, 215, 326
TAJEMNICA DZIKIEGO SZYBU Vol. 1 179, 381
TAJEMNICA MEDALIONU Vol. 1 404
TAJEMNICA PANNY BRINX Vol. 1 278, 308, 346
TAJEMNICA PRZYSTANKU TRAMWAJOWEGO Vol. 1 174, 176, 371
TAJEMNICA SKRZYNKI POCZTOWEJ Vol. 1 134, 278, 309
TAJEMNICA STAREGO OGRODU Vol. 6 *page 219, page 243, page 245*
TAJEMNICA STAREGO RODU Vol. 1 174, 176
TAJEMNICE NALEWEK Vol. 1 154, 157, 161
TAJEMSTVI HRADOU V KARPATECH Vol. 10 144, 466, 523, 621
TAJEMSTVI KRVE Vol. 10 142, 296, 524
TAJEMSTVI OCELOVEHO MESTA Vol. 10 144, 369, 371
TAJEMSTVI STARE PUDY Vol. 10 189, 250, 462
TAJEMSTVI VELIKEHO VYPRAVECE Vol. 10 197, 289, 394
TAJEMSTVI ZLATEHO BUDDHY Vol. 10 369, 531, 569
TAJFUN Vol. 9 337
TAJFUNI S NEZNI IMENA Vol. 3 38 Vol. 6 *page 30, page 58*
TAJNA NIKOLE TESLE Vol. 3 350 Vol. 6 *page 321*
TAJNATA NA APOLONIJA Vol. 6 *page 59, page 63*
TAJNATA VECERJA NA SEDMACITE Vol. 3 37, 100 Vol. 6 *page 45*
TAJNE STAROG TAVANA Vol. 6 *page 322*
TAJNI DNEVNIK SIGMUNDA FROJDA Vol. 6 *page 315*
TAJVANSKA KANASTA Vol. 6 *page 305, page 308*
TAK BLIZKO U NEBE Vol. 10 286, 473
TAK FORDI DU KOM, NICK ! Vol. 5 2, 7
TAK LASKA ZACINA Vol. 10 425, 428, 516
TAKIAISPALLO Vol. 5 375
TAKKI JA LIIVIT POIS Vol. 5 356, 485
TAKOVA LASKA Vol. 10 154, 202, 503
TAKOVY JE ZIVOT Vol. 6 *page 195, page 197* Vol. 10 346, 477
TAKT OG TONE I HIMMELSENGEN Vol. 5 215, 257, 273
TAKT, TONE OG TOSSER Vol. 5 83
TAKTSTOCK RICHARD WAGNERS, DER Vol. 6 *page 194*
TAKZE AHOJ Vol. 10 70, 179, 279, 437
TAL DER SIEBEN MONDE, DAS Vol. 1 2, 62, 87
TAL DES TRAUMES, DAS Vol. 4 183
TAL DULCINEA, UNA Vol. 8 311, 546
TALENTPROBE Vol. 6 *page 186*
TALER DER TANTE SIDONIE, DER Vol. 4 329, 774
TALFAHRT DES SEVERIN HOYER, DIE Vol. 4 67, 152, 187, 698
TALIRE NAD VELKYM MALIKOVEM Vol. 10 87, 395, 517
TALISMAN Vol. 3 42, 76 Vol. 6 *page 49*
TALISMAN, DER Vol. 4 89
TALLONE D'ACHILLE, IL Vol. 7 1111
TALPALATNYI FÖLD Vol. 9 196, 237, 277
TALPRA, GYÖZÖ ! Vol. 9 51, 140, 259
TALPUK ALATT FÜTYÜL A SZEZ Vol. 9 253, 291, 318
TALTET - VEM TILHOR VARLDEN ? Vol. 5 1108
TAM KDE HNIZDI CAPI Vol. 10 226, 317, 567
TAM NA HORACH Vol. 10 257, 605, 613
TAM NA KONECNE Vol. 10 41, 433, 522
TAM-TAM Vol. 2 625
TAM TAM MAYUMBE NAPOLITANO Vol. 7 1027
TAM TAM NELL'OLTRE GIUBA Vol. 7 87
TAM ZA LESEM Vol. 10 184, 193, 227
TAMANGO Vol. 2 251, 781
TAMARA Vol. 8 246
TAMARA, COMMENT J'AI ENTERRE MA VIE DE JEUNE FILLE Vol. 2 198
TAMARA LA COMPLAISANTE Vol. 2 9, 395
TAMBARI Vol. 1 7, 37, 92
TAMBIEN HAY CIELO SOBRE EL MAR Vol. 8 677
TAMBIEN LOS ANGELES COMEN JUDIAS Vol. 8 523, 732
TAMBOR DEL BRUCH, EL Vol. 8 677
TAMBOR, EL Vol. 8 618
TAMBOR Y CASCAVEL Vol. 8 396
TAMBOUR BATTANT Vol. 2 538, 661
TAMBURIN UND KASTAGNETTEN Vol. 4 560, 743
TAMBURINO SARDO, IL Vol. 7 420
T'AMERO SEMPRE Vol. 7 48, 466, 493, 756
TAMPON DU CAPISTON, LE Vol. 2 48, 627
TAMTEN Vol. 1 317
TANA Vol. 3 17, 20
TANA A DVA PISTOLNICI Vol. 10 255, 462

TÄNÄÄN OLET TÄÄLÄ **Vol. 5** 452
TANAR UR KEREM ... **Vol. 9** 117, 312, 326
TANASE SCATIU **Vol. 3** 281, 283, 324
TANC, A **Vol. 9** 337
TANCZACY JASTRZAB **Vol. 1** 312, 378 **Vol. 6** *page 210*
TANDEM MIT KETTMANN **Vol. 1** 95
TANDLAEGE PA SENGEKANTEN **Vol. 5** 21, 257, 273
TANDPINENS KVALER **Vol. 5** 97
TANECEK PANNY MARINKY **Vol. 10** 29, 141, 256
TANECNIÇE **Vol. 10** 213, 331, 412, 569, 621
TANGO **Vol. 3** 63, 119 **Vol. 5** 55, 101, 208
TANGO DEI TRAPASSATI, IL **Vol. 7** 1116
TANGO DELLA GELOSIA, IL **Vol. 7** 102, 208
TANGO FÜR DICH, EIN **Vol. 4** 208, 479
TANGO NOTTURNO **Vol. 4** 79, 115, 729
TANGO PRE MEDVEDA **Vol. 10** 273, 478, 544
TANGO PTAKA **Vol. 6** *page 221, page 227, page 236*
TANGOFIEBER **Vol. 4** 743
TANITONÖ, A **Vol. 9** 71, 77, 112, 237
TANITVANYOK, A **Vol. 9** 147, 243
TANK DELLA MORTE, IL **Vol. 7** 64
TÄNK, OM JAG GIFTER MIG MED PRÄSTEN **Vol. 5** 915, 982
TANKMÄDEL, DAS **Vol. 4** 646
TANKOVA BRIGADA **Vol. 10** 87, 407
TANNENBERG **Vol. 4** 97, 124, 204
TANSSI YLI HAUTOJEN **Vol. 5** 356, 481, 549
TANT D'AMOUR PERDU **Vol. 2** 47, 270
TANT GRÖN, TANT BRUN OCH TANT GREDELIN **Vol. 5** 840, 940
TANT QUE JE VIVRAI **Vol. 2** 332, 566
TANT QU'IL Y AURA DES FEMMES **Vol. 2** 9, 153
TANT QU'ON A LA SANTE **Vol. 2** 110, 642
TANTE CLEMENTINE **Vol. 4** 18, 558
TANTE CRAMERS TESTAMENTE **Vol. 5** 2, 228
TANTE MARIECHEN **Vol. 4** 361, 558
TANTE POSE **Vol. 5** 714, 726, 738, 781
TANTE SIDONIES ERBE **Vol. 4** 329, 421, 649
TANTE TUT FRA PARIS **Vol. 5** 2, 23
TANTE ZITA **Vol. 2** 110, 770
TANTES FODSELSDAG **Vol. 5** 97
TANTO CALORE **Vol. 7** 414
TANTO VA LA GATTA AL LADRO ... **Vol. 7** 347
TANU, A **Vol. 9** 316, 339
TANULMANY A NÖKRÖL **Vol. 9** 67, 84, 122
TANZ AM SONNABEND – MORD ? **Vol. 1** 13, 82, 107
TANZ AUF DEM VULKAN **Vol. 4** 24, 119, 422 **Vol. 6** *page 187*
TANZ AUF DEM VULKAN, DER **Vol. 4** 270, 570
TANZ GEHT WEITER, DER **Vol. 4** 281
TANZ IN DEN ABGRUND, DER **Vol. 4** 270
TANZ INS GLÜCK, DER **Vol. 4** 85, 409
TANZ MIT DEM KAISER **Vol. 4** 325, 373
TANZ UM LIEBE UND GLÜCK **Vol. 4** 428, 626
TANZENDE HERZ, DAS **Vol. 4** 218, 497
TANZENDE TOD, DER **Vol. 4** 239, 287, 291, 560, 644
TANZENDE WIEN, DAS **Vol. 4** 14, 52, 186, 325
TÄNZER, DER **Vol. 4** 660, 704
TÄNZER MEINER FRAU, DER **Vol. 4** 158, 396
TÄNZERIN BARBARINA, DIE **Vol. 4** 219, 681
TÄNZERIN FÜR SÜD-AMERIKA GESUCHT **Vol. 4** 774
TÄNZERIN MARION, DIE **Vol. 4** 240, 644
TÄNZERIN TOD **Vol. 4** 192
TÄNZERIN VON SANSSOUCI, DIE **Vol. 4** 49, 73, 608
TANZHUSAR, DER **Vol. 4** 255, 292, 725
TANZMÄDEL, DAS **Vol. 4** 339, 544
TANZSTUDENT, DER **Vol. 4** 584, 796
TAO **Vol. 2** 192, 452, 871
TAPAGE NOCTURNE **Vol. 2** 28, 119, 231, 247 **Vol. 6** *page 102, page 121*
TAPEINOS KAI KATAFRONEMENOS **Vol. 3** 164
TAPFERE SCHNEIDERLEIN, DAS **Vol. 1** 3, 105, 107 **Vol. 4** 329
TAPFERE SCHULSCHWÄNZER, DER **Vol. 1** 65, 80
TAPINARII **Vol. 6** *page 282*
TAPPA INTE SUGEN **Vol. 5** 940, 1173
TAPPE DELLA GLORIA E DELL'ARDIRE ITALICI, LE **Vol. 7** 1004
TAPPRE JACOB, DEN **Vol. 5** 81
TAPPRE SOLDATEN JÖNSSON, DER **Vol. 5** 933, 1073, 1080
TAPRE SKRAEDDER, DEN **Vol. 5** 83
TAPRE SVIGERMOR, DEN **Vol. 5** 92
TARAKANOVA **Vol. 2** 23, 27, 41, 100, 206, 440, 657
TARALEZITE SE RAZDAT BEZ BODLI **Vol. 3** 73, 101
TARANTEL, DIE **Vol. 4** 190, 773
TARANTELLA NAPOLETANA **Vol. 7** 526
TARANTOS, LOS **Vol. 8** 350, 732

TARAS BULBA Vol. 4 67, 135, 545, 612
TARASS BOULBA Vol. 2 9, 23, 48, 626
TARDE DE TOROS Vol. 8 449, 725
TARJETA DE VISITA Vol. 8 752
TÄRNINGEN ÄR KASTAD Vol. 5 880, 933
TARNISHED REPUTATION Vol. 2 384
TAROTS Vol. 8 762, 770
TARPANY Vol. 1 179, 408 Vol. 6 *page 242*
TARPS ELIN Vol. 5 1042, 1217
TARTARI, I Vol. 7 506
TARTARIN DE TARASCON Vol. 2 158, 440, 475, 488, 657, 875
TARTARUGA DEL DIAVOLO, LA Vol. 7 277
TARTARUGA, LA Vol. 7 1113
TARTASSATI, I Vol. 7 1044
TARTÜFF Vol. 4 178, 266, 574
TARTUFFE Vol. 2 824
TARZAK CONTRO GLI UOMINI LEOPARDO Vol. 7 867
TARZAN DO 5. ESQUERDO, O Vol. 8 130, 155
TARZAN EN LA GRUTA DEL ORO Vol. 8 749
TARZAN Y EL MISTERIO DE LA SELVA Vol. 8 473
TARZANA, SESSO SELVAGGIO Vol. 7 1121
TARZANOVA SMRT Vol. 10 153, 296, 425
TASCHENDIEBE Vol. 4 414, 598
TASIO Vol. 8 219, 771
TASSE, LA Vol. 6 *page 69*
TASSINARO, IL Vol. 7 834, 868, 957
TAT MIT VORBEDACHT, EINE Vol. 4 18, 430
TATA DE DUMINICA Vol. 3 245, 311
TATA MIA Vol. 8 358, 675
TATAL RISIPITOR Vol. 3 272, 281, 321, 331
TATARJARAS Vol. 9 28
TÄTER GESUCHT Vol. 4 242, 270, 471
TÄTER IST UNTER UNS, DER Vol. 4 269, 678, 740
TATIANA, LA DANZATRICE POLACCA Vol. 7 1162
TATINEK NA NEDELI Vol. 10 153, 170, 503
TATJANA Vol. 4 132, 773
TATO, PRECTI TO ! Vol. 10 394, 572
TATO, SEZEN STENE Vol. 10 299, 380, 385
TATORT BERLIN Vol. 1 34, 74, 92
TATOUE, LE Vol. 2 185, 854
TATRANSKA ROMANCE Vol. 10 92, 453
TATUL Vol. 3 73 Vol. 6 *page 38, page 47*
TAUBE AUF DEM DACH, DIE Vol. 1 126
TAUBENJULE Vol. 6 *page 154, page 162, page 176*
TAUGENICHTS, DER Vol. 4 165
TAULANTI KERKON NJE MOTER Vol. 6 *page 12, page 16*
TAUSE FLERTALL, DER Vol. 5 765, 784
TAUSEND FÜR EINE NACHT Vol. 4 379 Vol. 10 58, 75
TAUSEND MARK BELOHNUNG Vol. 4 367
TAVARATALO LAPATOSSU JA VINSKI Vol. 5 471, 549
TAVASZ A TELBEN Vol. 9 200
TAVASZ A VIHARBAN Vol. 9 71
TAVASZI SZERELEM Vol. 9 175, 308
TAVASZI SZONATA Vol. 9 182, 270, 315
TAVASZI ZAPOR Vol. 9 71, 201
TAVERNA ROSSA Vol. 7 487, 1000, 1054
TAVERNE DU POISSON COURONNE, LA Vol. 2 558, 833
TAVERNE VERDE, LA Vol. 7 196
TAVOLA DEI POVERI, LA Vol. 7 652, 756, 796
TAXA K 1640 EFTERLYSES Vol. 5 89, 281
TAXI 13 Vol. 5 1003, 1073
TAXI 313 X 7 Vol. 2 298
TAXI DE LOS CONFLICTOS, EL Vol. 8 338, 396
TAXI GIRL Vol. 7 426
TAXI PARA TOBRUK, UN Vol. 8 753
TAXI POUR TOBROUK, UN Vol. 2 128, 368 Vol. 8 753
TAXI, ROULOTTE ET CORRIDA Vol. 2 97, 205
TAXIDI STON EROTA KAI STON THANATO Vol. 3 158
TÄYOSUMA Vol. 5 357, 471
TAZI HUBAVA ZRJALA V"ZRAST Vol. 6 *page 24, page 49, page 61*
TAZI KR"V TRJABVASE DA SE PROLEE Vol. 6 *page 41, page 62*
TAZNI PTACI Vol. 10 369, 524, 569
TAZZA DI THE, UNA Vol. 7 369
TCHAN Vol. 10 41, 385, 531
TCHAN KONDELIK A ZET VEJVARA Vol. 10 345, 586
TCHAO PANTIN Vol. 6 *page 90, page 99*
TCHYNE Vol. 10 153, 299
TE CSAK PIPALJ LADANYI ! Vol. 9 178, 237
TE DOY MI VIDA Vol. 8 399, 537
... TE LASSO ! Vol. 7 39
TE LO DIRO DOMANI Vol. 7 493
TE MARRE PAS ... C'EST POUR RIRE ! Vol. 6 *page 91*
TE NOCI Vol. 3 436, 455
TE POBEDIHA Vol. 3 71, 86
TE QUIERO PARA MI Vol. 8 449

TE QUIERO, TE QUIERO, TE QUIERO **Vol. 8** 592
TE, RONGYOS ELET ! **Vol. 9** 6, 215, 316
TE STO ASPETTANNO **Vol. 7** 1117
TE VAGY A DAL **Vol. 9** 14, 132, 133, 270
TEATERTOSSET **Vol. 5** 63, 89
TEATRO APOLO **Vol. 8** 214, 301, 401
TECH NEKOLIK DNU ... **Vol. 10** 96, 388, 621
TECHO DE CISTAL, EL **Vol. 8** 299, 402
TECNICA DE UNA ESPIA **Vol. 8** 583, 772
TECNICA E IL RITO, LA **Vol. 7** 93
TECNICA PARA UN SABOTAJE **Vol. 8** 438
TECNICA PER UN MASSACRO **Vol. 8** 438
TECUMSEH **Vol. 1** 15, 17, 96
TED L'INVISIBILE **Vol. 7** 548
TED ZAS MY **Vol. 10** 168, 412
TEDDY BOYS DELLA CANZONE, I **Vol. 7** 739
TEDDY BY KOURIL **Vol. 10** 449
TEE TYÖ OPI PELAAMAAN **Vol. 5** 366, 394
TEE ZU ZWEIEN **Vol. 4** 13, 24
TEERENPELIÄ **Vol. 5** 383, 598
TEGNAP **Vol. 9** 7, 117, 259
TEGNAPELÖTT **Vol. 9** 60, 20, 215
TEILNEHMER ANTWORTET NICHT **Vol. 4** 316, 527
TEKETORIA **Vol. 9** 167, 265
TEKOZLO FIU, A **Vol. 9** 200
TEL EST PRIS QUI CROYAIT PRENDRE **Vol. 6** *page 69*
TEL PERE ET FILS, UN **Vol. 2** 128, 226, 332, 440
TELA DE ARANA **Vol. 8** 484, 772
TELEFONATA DEL DIAVOLO, LA **Vol. 7** 968, 1079
TELEFONI BIANCHI **Vol. 7** 318
TELEFONISTA, LA **Vol. 7** 196, 466, 756
TELEFONO NEL MEDIO EVO **Vol. 7** 1189
TELEGRAME **Vol. 3** 231, 254, 306
TELEPHONE PUBLIC **Vol. 2** 495, 561
TELEPHONE ROSE, LE **Vol. 2** 389, 457
TELEVISIONE **Vol. 7** 978
TELEVIZE V BUBLICICH ANEB V BUBLICE TELEVIZI **Vol. 10** 233, 394, 398
TELJES GÖZZEL **Vol. 9** 117, 165, 274
TELO DIANY **Vol. 10** 398
TEMA 13 : BATRINETEA **Vol. 6** *page 273*
TEMA DI MARCO, IL **Vol. 7** 540
TEMI **Vol. 7** 796
TEMI IL LEONE **Vol. 7** 1174
TEMNE SLUNCE **Vol. 10** 184, 197, 283, 313, 398
TEMNO **Vol. 10** 132, 143, 296, 524
TEMOIN DANS LA VILLE, UN **Vol. 2** 224, 515
TEMOIN DE MINUIT, LE **Vol. 2** 249, 294
TEMOIN, LE **Vol. 2** 259
T'EMPECHES TOUT LE MONDE DE DORMIR ! **Vol. 6** *page 90, page 96*
TEMPEL DER LIEBE, DER **Vol. 4** 385
TEMPEL DES SATANS **Vol. 1** 2, 120
TEMPELDANSERINDENS ELSKOV **Vol. 5** 58
TEMPELRÄUBER, DER **Vol. 4** 558, 704
TEMPESTA IN UN CRANIO **Vol. 7** 1092, 1118
TEMPESTA, LA **Vol. 7** 299, 750, 1136
TEMPESTA NEL NIDO **Vol. 7** 993
TEMPESTA SUL GOLFO **Vol. 7** 16, 756, 830, 1140
TEMPESTAD EN EL ALMA **Vol. 8** 207, 704
TEMPESTAD EN LA LUNA **Vol. 8** 359
TEMPESTAD, LA **Vol. 8** 278
TEMPESTAIRE, LE **Vol. 2** 777
TEMPETE **Vol. 2** 8, 34
TEMPETE SUR L'ASIE **Vol. 2** 581, 646
TEMPETE SUR LES MAUVENTS **Vol. 2** 440
TEMPETE SUR PARIS **Vol. 2** 34
TEMPI DURI PER I VAMPIRI **Vol. 7** 1044
TEMPI NOSTRI **Vol. 7** 90, 283, 332, 453, 910
TEMPIO DEL SACRIFICIO, IL **Vol. 7** 525
TEMPO, CARLO, TEMPO **Vol. 4** 361
TEMPO DEGLI ASSASSINI, IL **Vol. 7** 1147
TEMPO DELLE BELVE, IL **Vol. 7** 383
TEMPO DI ROMA **Vol. 2** 368
TEMPO DI VILLEGGIATURA **Vol. 7** 60, 1012
TEMPO MASSIMO **Vol. 7** 756, 796
TEMPO PRVNI LASKY **Vol. 10** 229, 369
TEMPO SI E FERMATO, IL **Vol. 7** 113
TEMPO ! TEMPO ! **Vol. 4** 287, 384, 458, 646
TEMPORALE ROSY **Vol. 7** 80, 333, 542
TEMPOS DIFICEIS **Vol. 8** 169
TEMPS DE L'AMOUR, LE **Vol. 2** 247
TEMPS DE MOURIR, LE **Vol. 2** 441
TEMPS DE VIVRE, LE **Vol. 2** 527
TEMPS DES CERISES, LE **Vol. 2** 124, 863
TEMPS DES COPAINS, LE **Vol. 2** 19
TEMPS DES LOUPS, LE **Vol. 2** 241, 567
TEMPS DES OEUFS DURS, LE **Vol. 2** 247, 627, 832
TEMPS DES VACANCES, LE **Vol. 2** 154, 409
TEMPS DU GHETTO, LE **Vol. 2** 314
TEMPTATIONS OF SHOP GIRL **Vol. 2** 84
TEN OKRUTNY NIKCZEMNY CHLOPAK **Vol. 1** 202, 227, 408 **Vol. 6** *page 228*
TEN SVETR SI NESVLIKEJ ! **Vol. 10** 41, 254, 531

TENACIA ABBRUZZESE Vol. 7 34
TENACIDAD Vol. 8 647
TENDRE ENNEMIE, LA Vol. 2 193, 362, 778
TENDRE ET MALICIEUSE CHRISTINA Vol. 6 *page 97*
TENDRE ET VIOLENTE ELISABETH Vol. 2 185, 661
TENDRE ET VOYOU Vol. 2 759
TENDRE POULET Vol. 2 457, 779
TENDRE VOYOU Vol. 2 781, 863
TENDREMENT VACHE Vol. 2 734 Vol. 6 *page 137*
TENDRES COUSINES Vol. 2 215 Vol. 6 *page 97*
TENDRES SOUVENIRS Vol. 6 *page 113, page 137*
TENEBRE Vol. 7 98, 312, 730, 912, 1145
TENEBROSA MANO, LA Vol. 7 304
TENEMOS 18 ANOS Vol. 8 541, 749
TENENTE BERTH, IL Vol. 7 441, 442
TENENTE DEI CARABINIERI, IL Vol. 7 291, 482, 1127
TENENTE GIORGIO, IL Vol. 7 443, 796, 1147
TENERA FOLLIA, UNA Vol. 7 311
TENERO TRAMONTO, UN Vol. 7 318
TENGERPARTI RANDEVU Vol. 9 35, 164
TENGO 17 ANOS Vol. 8 438, 602
TENGO QUE ABANDONARTE Vol. 8 331, 493
TENKES KAPITANYA, A Vol. 9 36, 149, 214
TENKRAT O VANOCICH Vol. 10 70, 197
TENTACOLI Vol. 7 406
TENTATION DE BARBIZON, LA Vol. 2 15, 368, 388
TENTATION DE SAINT ANTOINE, LA Vol. 2 678
TENTATION D'UN HOMME VERTUEUX, LA Vol. 2 147
TENTATION, LA Vol. 2 84, 167, 249, 388, 407
TENTATIONS DE MARIANNE, LES Vol. 2 359
TENTAZIONE Vol. 7 476, 966
TEO EL PELIRROJO Vol. 8 672, 765
TEODORA Vol. 7 194, 1169, 1189
TEODORA, IMPERATRICE DI BISANZIO Vol. 7 332, 638, 799, 1033, 1159
TEODORO E SOCIO Vol. 7 965
TEOREMA Vol. 7 710, 939, 1007
TEPEPA Vol. 8 523
TEPPICH DES GRAUENS, DER Vol. 8 438, 600
TERAZ I W KAZDA GODZINE Vol. 1 237
TERCERA LUNA, LA Vol. 8 220, 536, 590
TERCIO DE QUITES Vol. 8 725, 735
TERESA Vol. 2 294, 532
TERESA CONFALONIERI Vol. 7 196, 453, 1060
TERESA DE JESUS Vol. 8 207, 749
TERESA ETIENNE Vol. 7 583
TERESA LA LADRA Vol. 7 854
TERESA RAQUIN Vol. 7 984, 1167
TERESA VENERDI Vol. 7 1054, 1063
TEREZA Vol. 10 184, 193, 226
TEREZU BYCH KVULI ZADNE HOLCE NEOPUSTIL Vol. 10 255, 462
TERIKE Vol. 9 172
TERJE VIGEN Vol. 5 937, 997
TERMETSZES Vol. 9 246
TERMINAL Vol. 7 177
TERMINSPEKULANTEN Vol. 5 203
TERORE NELLO SPAZZIO Vol. 7 537
TERPSICHORE Vol. 4 377
TERRA Vol. 7 1189
TERRA DE GIGANTES Vol. 8 749
TERRA DI FUOCO Vol. 7 453, 619
TERRA DI NESSUNO Vol. 7 639, 830, 978, 1111, 1121
TERRA D'INCANTI Vol. 7 984
TERRA INCOGNITA Vol. 1 49, 70, 100
TERRA MADRE Vol. 7 652, 796, 851, 1020, 1029
TERRA MELOPHON MAGAZIN NR.1 Vol. 4 115
TERRA NOVA MAR VELHO Vol. 8 60
TERRA SENZA TEMPO Vol. 7 46, 827
TERRA STRANIERA Vol. 7 1167
TERRA TREMA, LA Vol. 7 21, 1167
TERRA VISTA DALLA LUNA, LA Vol. 7 498, 913, 988, 1142, 1197
TERRAIN VAGUE Vol. 2 96, 719
TERRAZZA, LA Vol. 7 29, 964, 1025
TERRE AU VENTRE, LA Vol. 2 713
TERRE DE FEU Vol. 2 488 Vol. 7 453, 619
TERRE DU DIABLE, LA Vol. 2 440
TERRE, LA Vol. 2 153, 159, 339, 415
TERRE PROMISE, LA Vol. 2 440, 845
TERRE QUI MEURT, LA Vol. 2 37, 698
TERREUR Vol. 2 6, 102
TERREUR DE CES DAMES, LA Vol. 2 347
TERREUR DES DAMES, LA Vol. 2 318, 807
TERRIBILE TEODORO, IL Vol. 7 891
TERRIBILI SETTE, I Vol. 7 627, 738
TERRIBLE DE CHICAGO, EL Vol. 8 317, 768
TERRITOIRE DES AUTRES, LE Vol. 2 77, 853
TERRITOIRE, LE Vol. 8 5
TERRITORIO, O Vol. 8 5
TERROR Vol. 5 12, 156, 175
TERROR CANIBAL Vol. 8 396, 632

TERROR EN LA NOCHE Vol. 8 438, 600
TERRORE DEI BARBARI, IL Vol. 7 17
TERRORE DEI MANTELLI ROSSI, IL Vol. 8 598, 772
TERRORE DEL ANDALUSIA, IL Vol. 7 725
TERRORE DELLA MASCHERA ROSSA, IL Vol. 7 796
TERRORE SULLA CITTA Vol. 7 1167
TERRORISTA, IL Vol. 7 221, 347, 1039
TERRORISTAS Vol. 8 519
TERRORISTI A MADRID Vol. 7 476
TERVETULOA AAMUKAHVILLE Vol. 5 462
TERVETULOA AAMUKAHVILLE ELI TOTTAKO TOINEN PUOLI Vol. 5 481, 643
TERZA IPOTESI SU UN CASO DI PERFETTA STRATEGIA CRIMINALE Vol. 7 1179
TERZA LICEO Vol. 7 106
TERZO GODE, IL Vol. 7 1189
TERZO INCOMODO, IL Vol. 7 442
T'ES FOLLE OU QUOI ? Vol. 6 *page 89*
T'ES HEUREUSE ? MOI, TOUJOURS ... Vol. 6 72
TES MOIRAS TO APOPAIDI Vol. 3 157
TESCHIO D'ORO, IL Vol. 7 1092
TESEO CONTRO IL MINOTAURI Vol. 7 528
TESNA KOZA Vol. 6 *page 306, page 314, page 320*
TESORO DE LA DIOSA BLANCA, EL Vol. 8 222, 571
TESORO DE LAS CUATRO CORONAS, EL Vol. 8 549, 582
TESORO DE MAKUBA, EL Vol. 8 677
TESORO DEL BENGALA, IL Vol. 7 476
TESORO DEL CAPITAN TORNADO, EL Vol. 8 249
TESORO DEL CASTILLO, EL Vol. 8 214, 391
TESORO DEL SUD, IL Vol. 7 1139
TESORO DELLA CATTEDRALE, IL Vol. 7 1190
TESORO DELLE QUATTRO CORONE, IL Vol. 8 582
TESORO DI BELZEBU, IL Vol. 7 400
TESORO DI ROMMEL, IL Vol. 7 476
TESORO EN EL CIELO, UN Vol. 8 515, 753
TESOROMIO Vol. 7 406, 674, 841
TESS Vol. 2 190, 449, 710, 784 Vol. 6 *page 83*
TESSERIS ASSI Vol. 3 163
TEST PILOTA PIRXA Vol. 1 289, 324 Vol. 6 *page 257*
TEST PILOTA PIRXA - DOZNANIJE PILOTA PIRKSA Vol. 6 *page 248*
TESTA DELLA MEDUSA, LA Vol. 7 470, 790
TESTA GIU ... GAMBE IN ARIA Vol. 7 981
TESTA O CROCE Vol. 7 1213
TESTA O CROCE - IL FIGLIO DEL BEDUINO Vol. 7 841, 1153
TESTA O CROCE - LA PECORELLA SMARRITA Vol. 7 318, 841, 907
TESTAMENT Vol. 3 379, 418
TESTAMENT DES CORNELIUS GULDEN, DAS Vol. 4 208, 255
TESTAMENT DES DR. MABUSE, DAS Vol. 4 729
TESTAMENT DES IVE SIEVERS, DAS Vol. 4 424
TESTAMENT D'ORPHEE, LE Vol. 2 377, 686, 702
TESTAMENT DU DOCTEUR CORDELIER, LE Vol. 2 239, 480
TESTAMENT PROF. WILCZURA Vol. 1 278, 404
TESTAMENTET Vol. 5 97
TESTAMENTETS HEMMELIGHED Vol. 5 58
TESTAMENTO DEL VIRREY, EL Vol. 8 611
TESTAMENTO ORIGINALE Vol. 7 1189
TESTE ALATE Vol. 7 296
TESTE DI QUOIO Vol. 7 585
TESTIMONE, IL Vol. 7 40, 834, 1136
TESTÖR, A Vol. 9 200
TET D'UN HOMME, LA Vol. 2 824, 863
TETE A CA, LA Vol. 2 739
TETE A CLAQUE Vol. 6 *page 78, page 97*
TETE BLONDE Vol. 2 32, 332
TETE CONTRE LES MURS, LA Vol. 2 475, 778
TETE DU CLIENT, LA Vol. 2 322
TETE FOLLE Vol. 2 207, 407
TETE QUI TOURNE, LA Vol. 2 46
TETEMREHIVAS Vol. 9 28
TETES BRULEES, LES Vol. 2 740 Vol. 8 221
TETICKA Vol. 10 168, 412
TETOVANE CASOM Vol. 10 82, 263, 493
TETTES ISMERETLEN, A Vol. 9 26, 134, 165
TETTO, IL Vol. 7 756, 796
TETUAN, IL GALEOTTO DETECTIVE Vol. 7 209
TEUER GEKAUFTES GLÜCK Vol. 4 183
TEUFEL UND CIRCE Vol. 4 356, 575
TEUFEL UND DIE MADONNA, DER Vol. 4 131
TEUFEL VOM MÜHLENBERG, DER Vol. 1 81, 105, 107
TEUFELCHEN Vol. 4 385

TEUFELEI MIT CHIC **Vol.** 1 29
TEUFELSKREIS, DER **Vol.** 1 43, 82
TEUFELSLOCH, DAS **Vol.** 4 135, 503
TEUFELSREPORTER, DER **Vol.** 4 340, 608, 691
TEUFELSWALZER, DER **Vol.** 4 71, 488
TEURE HEIMAT **Vol.** 4 221, 255
TEX E IL SIGNORE DEGLI ABISSI **Vol.** 7 504, 802, 875
TEXAS 2000 **Vol.** 7 742
TEXAS KID **Vol.** 8 523, 732
TEXAS TEX **Vol.** 5 97
TEXAS TRAIL, THE **Vol.** 2 84
TEXICAN, THE **Vol.** 8 523, 732
TEZKY ZIVOT DOBRODRUHA **Vol.** 10 34, 345
THAIS **Vol.** 6 *page 207, page 211, page 251* **Vol.** 7 843, 929
THAMAR, DAS KIND DER BERGE **Vol.** 4 270, 733
THANASE PARE TO OPLO ! **Vol.** 3 207
THANASE SFIXE KI ALLO TO ZONARI **Vol.** 3 180
THANASES, GIULIETTA KAI TA LOUKANIKA **Vol.** 3 207
THAT DANGEROUS AGE **Vol.** 2 656
THEATERBESUCH, DER **Vol.** 4 344
THEATERPRINZ, DER **Vol.** 4 366
THEATRE DES MATIERES, LE **Vol.** 2 804
THEATRES EROTIQUES DE PARIS, LES **Vol.** 2 139
THEMA SYNEIDESEOS **Vol.** 3 205
THEMIS **Vol.** 4 673
THEMROC **Vol.** 2 455, 732
THEODOR KÖRNER **Vol.** 4 89, 192, 546
THEODORE ET COMPAGNIE **Vol.** 2 32, 197
THERESE DESQUEYROUX **Vol.** 2 168, 566
THERESE ETIENNE **Vol.** 2 128, 226, 402
THERESE MARTIN **Vol.** 2 131, 623
THERESE RAQUIN **Vol.** 2 23, 96, 402 **Vol.** 4 14, 186, 605, 788
THESARI **Vol.** 6 *page 14, page 15, page 18*
THIEF OF BAGDAD, THE **Vol.** 2 656
THIEF, THE **Vol.** 2 24
THIHIJAT BEGLEC **Vol.** 3 40
THINGS TO COME **Vol.** 2 656
THIRD WORLD – PRISONNIER DE LA RUE **Vol.** 2 292 **Vol.** 6 *page 84*
THIRRJA **Vol.** 3 2, 15
13 AAR **Vol.** 5 55, 101, 208
13 DNI **Vol.** 3 56, 75, 76, 105
13. JUL **Vol.** 6 *page 308, page 312, page 321*
13 STOLAR **Vol.** 5 572
13 – 13, EL **Vol.** 8 214, 357, 436
THIRTEENTH CHAIR, THE **Vol.** 2 384
13.000, EL **Vol.** 8 276
38 GRADI PARALLELO **Vol.** 7 891
39-ES DANDAR, A **Vol.** 9 241, 312, 326
36 ORE ALL'INFERNO **Vol.** 7 689
33.333 **Vol.** 5 840, 878, 891, 1031
THIS LAND IS MINE **Vol.** 2 526
THOMAS **Vol.** 2 527
THOMAS CHATTERTON **Vol.** 7 1189
THOMAS ER FREDLOS **Vol.** 5 130
THOMAS GRAALS BÄSTA BARN **Vol.** 5 937, 996
THOMAS GRAALS BÄSTA FILM **Vol.** 5 937, 996
THOMAS GRAALS MYNDLING **Vol.** 5 870
THOMAS L'IMPOSTEUR **Vol.** 2 155, 314, 674
THOMAS MÜNTZER **Vol.** 1 23, 54, 72, 81
THOMPSON 1880 **Vol.** 8 562, 732
THOR IL CONCUISTATORE **Vol.** 7 124
THORA VAN DEKEN **Vol.** 5 891, 925, 976
THORVALD OG LINDA **Vol.** 5 130, 131, 247
1000 CHILOMETRI AL MINUTO **Vol.** 7 443, 954
1000 TALAROW **Vol.** 6 *page 239*
1000 WORTE DEUTSCH **Vol.** 4 255, 379
3 **Vol.** 5 712
THREE CASES OF MURDER **Vol.** 2 656
3 CODONAS, DIE **Vol.** 4 49, 246
3 S, AGENTE ESPECIAL **Vol.** 8 438
3 SLAGS KAERLIGHED **Vol.** 5 6, 35
3 : 1 A SZERELEM JAVARA **Vol.** 9 71, 322
300 EVES EMBER, A **Vol.** 9 28
317. SECTION, LA **Vol.** 2 208
THRILLER **Vol.** 5 860
THRONFOLGER, DER **Vol.** 4 124, 247
THUMMELUMSEN **Vol.** 5 31, 32, 55, 101, 208
THUNDER **Vol.** 7 1015
THUNDERSTORM **Vol.** 8 276, 732
THYMESOU AGAPE MOU **Vol.** 3 153
TI ASPETTERO ALL'INFERNO **Vol.** 7 1147
TI CONOSCO, MASCHERINA ! **Vol.** 7 792, 1063
TI EKANES STON POLEMO THANASE **Vol.** 3 135, 207
TI GUTTER OG EN GJENTE **Vol.** 5 662, 711, 831
TI HO SEMPRE AMATO **Vol.** 7 333
TI – KOYO E IL SUO PESCECANE **Vol.** 7 877
TI LOVIS **Vol.** 3 345, 395
TI PROKLETI HEROJI **Vol.** 6 *page 315*
TI RITROVERO **Vol.** 7 1136
TI SPACCO IL MUSO, BIMBA **Vol.** 7 632
TI VOGLIO ... BENE ! **Vol.** 7 52
TIA DE CARLOS EN MINI FALDA, LA **Vol.** 8 492, 768

TIA DE CARLOS, LA **Vol. 8** 667, 681
TIA RAMONA, LA **Vol. 8** 640
TIA TULA, LA **Vol. 8** 243, 258
TIBETANA **Vol. 8** 546, 595
TICCHIO DELLA SIGNORA, IL **Vol. 7** 1189
TICHA RADOST **Vol. 10** 93, 148, 332, 544
TICHY AMERICAN V PRAZE **Vol. 10** 330, 369, 577
TICHY SPOLECNIK **Vol. 10** 223, 495, 570
TICKENDE BRATPFANNE, DIE **Vol. 1** 49
TIDENS BARN **Vol. 5** 228
TIE MALE VYLETY **Vol. 10** 82, 263, 493
TIE PIMEÄÄN **Vol. 5** 516
TIEFE FURCHEN **Vol. 1** 32, 100, 130
TIEFEN DER GROSSTADT **Vol. 4** 67, 135
TIEFLAND **Vol. 4** 54, 247, 363, 743
TIEMPO DE AMOR **Vol. 8** 258, 438
TIEMPO DE SILENCIO **Vol. 8** 232, 684
TIEMPO DE VALS **Vol. 8** 357, 504
TIEMPO DE VIOLENCIA **Vol. 8** 438, 631
TIEMPOS DE CHICAGO **Vol. 8** 396
TIEMPOS FELICES **Vol. 8** 401, 749
TIENDA DE ANTIGUEDADES, LA **Vol. 8** 357
TIENE USTED OJOS DE MUJER FATAL **Vol. 8** 649, 651
TIERCE A COEUR **Vol. 2** 558, 833
TIERRA BRUTAL **Vol. 8** 311, 400
TIERRA DE FUEGO **Vol. 8** 523, 732
TIERRA SEDIENTA **Vol. 8** 400
TIERRA Y CIELO **Vol. 8** 268
TIFOSO, L'ARBITRO E IL CALCIATORE, IL **Vol. 7** 1126, 1202
TIGAR **Vol. 3** 347, 362, 424
TIGER, DER **Vol. 4** 270, 288
TIGERIN, DIE **Vol. 4** 288, 626, 779
TIGGERSKEN **Vol. 5** 97
TIGRANA **Vol. 7** 621
TIG"RCETO **Vol. 3** 63 **Vol. 6** *page 50*
TIGRE AIME LA CHAIR FRAICHE, LE **Vol. 2** 700
TIGRE DE CHAMBERI, EL **Vol. 8** 214, 493
TIGRE DE LOS SIETE MARES, EL **Vol. 8** 426, 732
TIGRE DEL KYBER, EL **Vol. 8** 749
TIGRE DU CIEL, LE **Vol. 2** 299
TIGRE E ANCORA VIVA ... SANDOKAN ALLA RISCOSSA **Vol. 7** 738
TIGRE REALE **Vol. 7** 305, 1132
TIGRE SE PARFUME A LA DYNAMITE, LE **Vol. 2** 700 **Vol. 8** 280, 732
TIGRE SE PERFUMA CON DINAMITA, EL **Vol. 8** 280, 732
TIGRES DE MOMPRACEM, LOS **Vol. 8** 396
TIGRI DI MOMPRACEM, LE **Vol. 8** 396
TIGRIS **Vol. 7** 304
TIH-MINH **Vol. 2** 432
TIHIJAT BEGLEC **Vol. 6** *page 31, page 32*
TIL SAETERS **Vol. 5** 809
TILL EULENSPIEGEL **Vol. 1** 7, 47, 80
TILL EULENSPIEGEL: WIE EULENSPIEGEL EIN URTEIL SPRICHT **Vol. 4** 18
TILL EULENSPIEGEL: WIE EULENSPIEGEL SICH EINMAL ERBOT, ZU FLIEGEN **Vol. 4** 18
TILL EULENSPIEGELS LUSTIGE STREICHE **Vol. 1** 74, 92, 95, 100
TILL GLÄDJE **Vol. 5** 940, 1173
TILL ÖSTERLAND **Vol. 5** 891, 997
TILLSAMMANS MED GUNILLA MANDAG KVÄLL OCH TISDAG **Vol. 5** 869, 881, 1130
TILMAN RIEMENSCHNEIDER **Vol. 1** 58, 72, 129
TILOS A CSOK **Vol. 9** 220
TILOS A SZERELEM **Vol. 9** 68, 71, 83, 183, 332
TILTOTT TERÜLET **Vol. 9** 70, 149, 339
TIM PADEM **Vol. 10** 106, 173, 567
TIMANFAYA **Vol. 8** 501, 552, 577
TIME IS MONEY **Vol. 4** 192, 390
TIMEGLASSET **Vol. 5** 9, 197
TIMIDO, EL **Vol. 8** 551, 749
TIMOCKA BUNA **Vol. 6** *page 318, page 320, page 322*
TIMON **Vol. 3** 421, 446
TINE **Vol. 5** 2, 140, 179, 275
TINERETE FARA BATRINETE **Vol. 3** 226, 241, 251, 286, 306
TINGEL-TANGEL **Vol. 4** 270, 409, 462
TINGELTANGEL **Vol. 4** 239, 340
TINGUJT E LUFTES **Vol. 3** 14, 26
TINIEBLAS QUEDARON ATRAS, LAS **Vol. 8** 661
TINKO **Vol. 1** 2, 105, 120
T'INQUIETE PAS CE SE SOIGNE **Vol. 6** *page 58, page 106*
TINTENFISCHKLUB, DER **Vol. 4** 35, 212
TINTIN ET LE MYSTERE DE LA TOISON D'OR **Vol. 2** 19, 503, 819
TINTIN ET LES ORANGES BLEUES **Vol. 2** 49, 819
TINTO CON AMOR **Vol. 8** 219, 383
TINTOMARA **Vol. 5** 261
TIO, DE VERDAD VIENEN DE PARIS ? **Vol. 8** 237, 546, 667
TIP AUF AMALIA **Vol. 4** 340, 481
TIPI DA SPIAGGIA **Vol. 7** 852
TIR GROUPE **Vol. 6** *page 70, page 82*
TIRANA, LA **Vol. 8** 298, 611
TIRANNO DEL GARDA, IL **Vol. 7** 1063
TIRANNO DI PADOVA, IL **Vol. 7** 1035, 1054
TIRANNO DI SIRACUSA, IL **Vol. 7** 1136
TIRARSE AL MONTE **Vol. 8** 739

TIRE-AU-FLANC Vol. 2 1, 48, 381, 388, 722
TIRE-AU-FLANC 62 Vol. 2 208
TIRE PAS SUR MON COLLANT Vol. 2 818
TIRETTE DEVANT DERRIERE Vol. 2 721
TIREZ SUR LE PIANISTE Vol. 2 208, 578
TIRLITTAN Vol. 5 452
TIRO AL PICCIONE Vol. 7 401, 853
TIRO POR LA ESPALDA, UN Vol. 8 298, 762
TISCHGAST, DER Vol. 4 119
TISIC ZA JEDNU NOC Vol. 10 58, 75
TISICROCNA VCELA Vol. 10 82, 101, 228, 424
TISNOVE VOLANI Vol. 10 74, 153, 410, 428
TISTEGA LEPEGA DNE Vol. 3 385, 390
TISTTESSEG NEVEBEN Vol. 9 2
TISZAVIRAG Vol. 9 44, 274
TISZTA AMERIKA Vol. 9 61, 82, 98, 254, 276
TISZTELET A KIVETELNEK ! Vol. 9 153, 288, 334
TISZTITOTŰZ Vol. 9 77
TITANENKAMPF Vol. 4 159
TITANIA, TITANIA Vol. 9 6, 103, 316
TITANIC Vol. 4 49, 124, 269, 416, 437 Vol. 7 442
TITANIC VALS Vol. 1 278 Vol. 3 265, 286, 294
TITI I., ROI DES GOSSES Vol. 2 339, 731
TITIMEKUV NAHRDELNIK Vol. 10 115, 584
TITIN DES MARTIGUES Vol. 2 9, 122
TITOKZATOS IDEGEN Vol. 9 288, 319, 334
TIZ EV MULVA Vol. 9 107, 147, 256
TIZEDES MEG A TÖBBIEK, A Vol. 9 67, 149, 241
TIZENNEGYEDIK, A Vol. 9 175, 220
TIZEZER NAP Vol. 9 259, 263
TIZIO, CAIO, SEMPRONIO Vol. 7 792, 891
TJAEREHANDLEREN Vol. 5 130, 247
TJENESTEPIGE Vol. 5 97
TJOCKA SLÄKTEN Vol. 5 863
TJORVEN, BATSMAN OCH MOSES Vol. 5 862
TJORVEN OCH MYSAK Vol. 5 862
TJORVEN OCH SKRALLAN Vol. 5 862
TJUTJUN Vol. 3 48, 97
TKO PJEVA ZLO NE MISLI Vol. 3 430, 437
TO Vol. 5 227
TO ALLO GRAMMA Vol. 3 156
TO AMAXAKI Vol. 3 169, 205
TO BYL CESKY MUZIKANT Vol. 10 2, 412
TO BYLA NOC Vol. 10 154, 299
TO', E MORTA LA NONNA ! Vol. 7 599
TO ER MUNDO E DEMASIAO Vol. 8 349
TO FLUER I ETT SMEKK Vol. 5 736
TO FORAELDRELOSE Vol. 5 97
TO GENTLEMEN Vol. 5 97
TO GULDGRAVERE, DE Vol. 5 97
TO HOMA VAFTEKE KOKKINO Vol. 3 130, 144
TO JA ZABILEM Vol. 1 202, 343 Vol. 6 *page 257*
TO KANONI KAI TO AEDONI Vol. 3 142
TO KELLI MEDEN Vol. 3 194
TO KORITSI BOMBA Vol. 3 183
TO KORITSI KAI TO ALOGO Vol. 3 196
TO KORITSI ME TA MAVRA Vol. 3 172
TO KORITSI ME TA PARAMYTHIA Vol. 3 130
TO KORITSI TES GEITONIAS Vol. 3 157, 202
TO KORITSI TOU 17 Vol. 3 173, 175
TO KORITSI TOU LUNA PARK Vol. 3 137, 193
TO LEVENDE OG EN DOD Vol. 5 691
TO LIMANI TON DAKRYON Vol. 3 154
TO LIV Vol. 5 663, 676, 723
TO MINUTTER FOR SENT Vol. 5 31, 150, 248, 272
TO MISTENKELIGE PERSONER Vol. 5 676, 729
TO NESI TES AGAPES Vol. 3 180
TO NEZNATE HADIMRSKU Vol. 10 167, 453
TO PA TOPP Vol. 5 682, 717, 772
TO PIKRO PSOMI Vol. 3 157, 198, 203
TO PIO LAMPRO ASTERI Vol. 3 161
TO PLOIO TIS HARAS Vol. 3 141
TO POTAMI Vol. 3 204
TO PROPO KAI TA MPOUZOUKIA Vol. 3 171
TO RIVALER Vol. 5 295
TO SKORE HO' DER Vol. 5 239
TO SO GADI Vol. 3 390, 395
TO SOM ELSKER HINANDEN Vol. 5 39, 196, 304
TO SPITI STOUS BRAHOUS Vol. 3 129, 196
TO TAXIDI TOU MELITOS Vol. 3 181
TO TELEUTAIO PSEMA Vol. 3 133, 172, 202
TO TELEUTAOI FILI Vol. 3 131
TO THE VICTOR Vol. 2 368
TO THYMA Vol. 3 144
TO TRITO SPITI TES ANGELAS Vol. 3 132
TO TYLKO ROCK Vol. 6 *page 216, page 235, page 253*
TO VIOLINER, DET Vol. 5 97
TOA Vol. 2 61, 706
TOAMNA BOBOCILOR Vol. 3 284, 301
TOATE PINZELE SUS Vol. 3 218, 285, 286, 319

TOBE HRANA ZVONIT NEBUDE **Vol. 10** 105, 313, 550
TOBIAS BREMSER **Vol. 1** 116
TOBIAS BUNTSCHUH **Vol. 4** 400, 646
TOBIASNAETTER **Vol. 5** 2, 7, 39, 163
TOBOGAN **Vol. 10** 116, 440, 506
TOBOGGAN **Vol. 2** 451
TOCATA Y FUGA DE LOLITA **Vol. 8** 438, 681
TOCCA PRIMA A TERESA **Vol. 7** 369
TOCHTER DER GRÄFIN STACHOWSKA, DIE **Vol. 6** *page 189*
TOCHTER DES HENKERS, DIE **Vol. 4** 165
TOCHTER DES KUNSTREITERS, DIE **Vol. 4** 223, 287
TOCHTER DES MEHAMED, DIE **Vol. 4** 186, 560
TOCHTER DES REGIMENTS, DIE **Vol. 2** 397
TOCHTER DES SAMURAI, DIE **Vol. 4** 15, 564
TÖCHTER IHRER EXZELLENZ, DIE **Vol. 4** 89, 325, 445
TOCHTER NAPOLEONS, DIE **Vol. 4** 291 **Vol. 6** *page 193*
TOCKA P"RVA **Vol. 3** 56, 69, 104 **Vol. 6** *page 26*
TOD AM MISSISSIPPI **Vol. 1** 33
TOD AUS DEM OSTEN, DER **Vol. 4** 186
TOD DES ERASMUS, DER **Vol. 6** *page 189*
TOD EINES MILLIONÄRS **Vol. 1** 29, 55, 57
TOD HAT EIN GESICHT, DER **Vol. 1** 33, 40, 101
TOD IM PREIS INBEGRIFFEN **Vol. 1** 38, 53, 59
TOD IM SPIEGEL, DER **Vol. 4** 270, 360
TOD IN DEN BERGEN, DER **Vol. 4** 187
TOD IN DER KURVE **Vol. 1** 49, 85, 121
TOD IN SEVILLA, DER **Vol. 4** 646
TOD ÜBER SHANGHAI **Vol. 4** 239, 558
TOD UND DIE LIEBE, DER **Vol. 4** 320, 356
TOD UND DIE MUTTER, DER **Vol. 4** 124, 247
TODBRINGER, DER **Vol. 4** 36
TODESFAHRT IM WELTREKORD **Vol. 4** 73, 259, 535
TODESFLIEGER, DER **Vol. 4** 598, 644
TODESKARAWANE, DIE **Vol. 4** 340, 535
TODESREIGEN, DER **Vol. 4** 192, 743, 780
TODESRITT IM RIESENRAD, DER **Vol. 4** 527
TODESSPIELER, DER **Vol. 4** 718
TODESSTURZ, DER **Vol. 4** 247
TODESWEG AUF DIE BERNINA, DER **Vol. 4** 221, 495
TODFEIND, DER **Vol. 4** 35, 514
TODGEWEIHTEN, DIE **Vol. 4** 232, 300, 673
TÖDLICHER IRRTUM **Vol. 1** 11, 96, 102, 179
TODO ES POSIBLE EN GRANADA **Vol. 2** 646 **Vol. 8** 438, 607, 728, 770
TODO MODO **Vol. 7** 434, 599
TODO POR ELLAS **Vol. 8** 628
TODO UN HOMBRE **Vol. 8** 474, 577
TODOS AL SUELO **Vol. 8** 250
TODOS ERAN CULPABLES **Vol. 8** 306, 482
TODOS LOS COLORES DE LA OSCURIDAD **Vol. 8** 551
TODOS ME LLAMAN GATO **Vol. 8** 286
TODOS O NINGUNO **Vol. 8** 317, 732
TODOS PARA UNO, GOLPES PARA TODOS **Vol. 8** 603, 667, 695
TODOS SOMOS NECESARIOS **Vol. 8** 214, 719
TOFFELHELTEN **Vol. 5** 97, 295
TOFFELHELTENS FODSELSDAG **Vol. 5** 83
„TOFFLAND" – EN LYCKLIG KOMEDI **Vol. 5** 954, 1093, 1217
TOGGER **Vol. 4** 340, 373, 466
TOGLI LE GAMBE DAL PARABREZZA **Vol. 7** 779
TOI C'EST MOI **Vol. 2** 67, 722
TOI, LE VENIN **Vol. 2** 21, 423
TOILE D'ESCARGOT, LA **Vol. 2** 2
TOINE **Vol. 2** 338
TOIROS DE MARY FOSTER, OS **Vol. 8** 54, 96
TOIVELAULUJA **Vol. 5** 374, 579
TOKA E JONE **Vol. 6** *page 17*
TOKA JONE **Vol. 3** 1, 8
TOKAJI ASZU **Vol. 9** 312
TOKAJI RAPSZODIA **Vol. 9** 94, 184, 322
TOKE E PERGJAKUR **Vol. 3** 22, 24
TÖKELETES CSALAD **Vol. 9** 236, 315
TÖKELETES FERFI **Vol. 9** 132, 133, 219, 270
TOLLE BOMBERG, DER **Vol. 4** 132, 149, 660
TOLLE HEIRAT VON LALO, DIE **Vol. 4** 476, 522
TOLLE HERZOGIN, DIE **Vol. 4** 149, 660, 733
TOLLE KOMTESS, DIE **Vol. 4** 255, 323, 462
TOLLE LOLA, DIE **Vol. 4** 192, 462, 584
TOLLE NACHT **Vol. 4** 347, 370, 486, 774
TOLLE NACHT, EINE **Vol. 4** 321, 558, 796
TOLLER EINFALL, EIN **Vol. 4** 82, 85, 713
TOLLER SCHWIEGERSOHN, EIN **Vol. 6** *page 186*
TOLLER TAG, EIN **Vol. 6** *page 190*
TOLONC, A **Vol. 9** 77
TOM OG METTE PA SPORET **Vol. 5** 804, 807
TOMBE DU CIEL **Vol. 2** 67, 698

TOMBEUR, LE **Vol. 2** 475, 595, 726, 738
TOMBOLA **Vol. 7** 1005 **Vol. 8** 214, 261
TOMBOLA PARADISO NERO **Vol. 7** 923
TOMBOY - I MISTERI DEL SESSO **Vol. 7** 946
TOMI, A MEGFAGYOTT GYERMEK **Vol. 9** 21, 34, 237
TOMKA DHE SHOKET E TIJ **Vol. 3** 3, 22 **Vol. 6** *page 12*
TON SORT **Vol. 4** 93, 680
TONELLI **Vol. 4** 344, 524, 544
TONI **Vol. 2** 121, 143, 719
TONKA SIBENICE **Vol. 6** *page 187, page 190* **Vol. 10** 138, 431
TONNERRE DE DIEU, LE **Vol. 2** 185, 875
TONNERRE, LE **Vol. 2** 346
TONNY **Vol. 5** 772, 797, 830
TONTA DEL BOTE, LA **Vol. 8** 261, 331, 422
TONTO DE LAGARTERA, EL **Vol. 8** 606
TONTONS FLINGUEURS, LES **Vol. 2** 293, 548
TONY ARZENTA **Vol. 7** 585
TONY, L'ALTRA FACCIA DELLA TORINO VIOLENTA **Vol. 7** 59
TONY, TOBE PRESKOCILO **Vol. 10** 428, 499
TOPARTI LATOMAS **Vol. 9** 71, 278, 322
TOPAZE **Vol. 2** 8, 347, 469, 552, 722, 872
TÖPFCHEN, DAS **Vol. 4** 247
TOPHAR-MUMIE, DIE **Vol. 4** 637, 648
TOPICAL SPANISH **Vol. 8** 232, 768
TOPIEL **Vol. 1** 236
TOPLE GODINE **Vol. 3** 357, 368
TOPLO **Vol. 3** 40, 105 **Vol. 6** *page 33*
TOPOLINI RICONOSCENTI **Vol. 7** 1189
TOPPAR - BRITTA **Vol. 5** 950
TOR DER FREIHEIT, DAS **Vol. 4** 212
TORBELLINO **Vol. 8** 436
TORDENSKJOLD GAAR I LAND **Vol. 5** 31, 32, 55, 101, 208
TORDENSTENENE **Vol. 5** 55, 83
TÖREDEK AZ ELETRÖL **Vol. 9** 152, 221, 279
TORERO, EL **Vol. 8** 214
TORERO POR ALEGRIAS **Vol. 8** 677, 772
TÖRICHTE HERZ, DAS **Vol. 4** 113
TÖRICHTE JUNGFRAU, DIE **Vol. 4** 17, 28, 232
TORINO NERA **Vol. 7** 1025
TORMENTA **Vol. 8** 276, 732
TORMENTA DE ALMAS **Vol. 8** 396
TORMENTA, LA **Vol. 7** 569
TORMENTA SOBRE AL PACIFICO **Vol. 8** 426, 732
TORMENTO **Vol. 7** 417, 1027, 1054, 1072, 1190 **Vol. 8** 245
TORMENTO D'AMORE **Vol. 7** 468 **Vol. 8** 749
TORMENTO D'ANIME **Vol. 7** 201
TORMENTO DEL PASSATO **Vol. 7** 1027
TORMENTO GENTILE **Vol. 7** 965
TORNA ! **Vol. 7** 275, 443, 916, 1027
TORNA A SURRIENTO **Vol. 7** 39
TORNA, CARO IDEAL ! **Vol. 7** 750, 1027, 1054, 1060
TORNA, PICCINA, MIA ! **Vol. 7** 693
TORNADO **Vol. 7** 692
TORNAVARA **Vol. 2** 9, 826
TORNO DI FUOCO **Vol. 7** 694
TORO BRAVO **Vol. 8** 749
TORO DA MONTA, UN **Vol. 7** 1179
TOROCKOI MENYASSZONY **Vol. 9** 164, 178, 237
TÖRÖKFEJES KOPJA, A **Vol. 9** 57, 303, 332
TORPILLE AERIENNE, LA **Vol. 2** 159, 286, 330
TORPINI NASZ **Vol. 9** 71, 182, 322, 337
TORPPARALLI **Vol. 5** 469
TORQUATO TASSO **Vol. 7** 1189
TORRE DE LOS SIETE JOROBADOS, LA **Vol. 8** 268, 625
TORRE DELLE SENSAZIONI, LA **Vol. 7** 1090
TORREJON CITY **Vol. 8** 221, 461, 772
TORRENTE, IL **Vol. 7** 614, 822, 1121
TORRENTS **Vol. 2** 15, 339
TORREPARTIDA **Vol. 8** 298, 753
TORRES SNORTEVOLD **Vol. 5** 676, 738, 803
TÖRST **Vol. 5** 940, 1173
TÖRTENELMI MAGANÜGYEK **Vol. 9** 67, 84, 122
TORTILLARDS, LES **Vol. 2** 131, 433
TORTUE SUR LE DOS, LA **Vol. 2** 94, 633
TORTURA DEL SILENZIO, LA **Vol. 7** 144
TORTURA NADZIEI **Vol. 1** 302
TORTURADOS **Vol. 8** 207, 295
TOSCA **Vol. 2** 29 **Vol. 4** 16 **Vol. 7** 2, 48, 170, 476, 524, 552, 701, 988, 1060
TOSCA, LA **Vol. 5** 97
TÖSEN FRAN STRORMYRTORPET **Vol. 5** 878, 906, 937, 996, 1173
TOSITARKOITUKSELLA **Vol. 5** 379, 446
TOSSEDE KVINDFOLK, DE **Vol. 5** 83
TOSSEDE MANDFOLK, DE **Vol. 5** 83
TOSSEDE PARADIS, DET **Vol. 5** 23, 243, 248, 275
TOT ODER SCHEINTOT **Vol. 4** 232, 673
TOTAGAS **Vol. 9** 67, 83, 318
TOTE REDEN NICHT **Vol. 1** 29, 34
TOTEN AUGEN, DIE **Vol. 4** 309
TOTEN BLEIBEN JUNG, DIE **Vol. 1** 34, 42, 47

TOTEN ERWACHEN, DIE **Vol. 4** 159
TOTEN SCHWEIGEN, DIE **Vol. 4** 183
TÖTENDES SCHWEIGEN **Vol. 4** 747
TOTENINSEL **Vol. 4** 266, 574, 734, 747
TOTENKLAUS, DER **Vol. 4** 762, 780
TOTENKOPF, 50 000 MARK-PRÄMIENFILM, DER **Vol. 4** 756
TOTENKOPFHUSAR, DER **Vol. 4** 628
TOTENKOPFREITER, DIE **Vol. 4** 240
TOTENTANZ **Vol. 4** 239, 734
TOTENTANZ, DER **Vol. 4** 646
TÖTET NICHT MEHR ! **Vol. 4** 270, 522
TOTINEN TORVENSOITTAJA **Vol. 5** 349, 386
TOTO **Vol. 2** 193, 197 **Vol. 5** 434
TOTO A COLORI **Vol. 7** 333, 443
TOTO A PARIGI **Vol. 7** 693, 798
TOTO AL GIRO D'ITALIA **Vol. 7** 1027, 1028
TOTO ALL'INFERNO **Vol. 7** 1136
TOTO CERCA MOGLIE **Vol. 7** 16
TOTO CONTRO I QUATTRO **Vol. 7** 1027
TOTO CONTRO IL PIRATA NERO **Vol. 7** 693
TOTO D'ARABIA **Vol. 7** 814
TOTO DE ARABIA **Vol. 8** 732
TOTO E CAROLINE **Vol. 7** 1036
TOTO E CLEOPATRA **Vol. 7** 1054
TOTO E IL RE DI ROMA **Vol. 7** 603
TOTO E LE DONNE **Vol. 7** 333
TOTO E MARCELLINO **Vol. 7** 476, 1054
TOTO E PEPPINO DIVISI A BERLINO **Vol. 7** 1027
TOTO, EVA E IL PENNELLO PROIBITO **Vol. 7** 693 **Vol. 8** 276, 630
TOTO, FABRIZI E I GIOVANI D'OGGI **Vol. 7** 693
TOTO LASCIA O RADDOPPIA ? **Vol. 7** 451
TOTO LE MOKO **Vol. 7** 891
TOTO LETO DOMA **Vol. 10** 37, 72, 155
TOTO NELLA LUNA **Vol. 7** 1044
TOTO, PEPPINO E I ... FUORILEGGE **Vol. 7** 16
TOTO, PEPPINO E LA MALAFEMMINA **Vol. 7** 16
TOTO, PEPPINO E LE FANATICHE **Vol. 7** 196
TOTO SCEICCO **Vol. 7** 16
TOTO TARZAN **Vol. 7** 602
TOTO TERZO UOMO **Vol. 7** 333
TOTO, VITTORIO E LA DOTTORESSA **Vol. 7** 693 910 **Vol. 8** 276
TOTONNO SE NE VA **Vol. 7** 455
TOTOTE DI GYP **Vol. 7** 45, 390
TOTTE ET SA CHANCE **Vol. 2** 32 **Vol. 4** 19
TOTTISALMEN PERILLINEN **Vol. 5** 471, 552
TOTUL PENTRU FOTBAL **Vol. 3** 211, 245, 262, 272
TOTUUS ON ARMOTON **Vol. 5** 361, 463
TOUBIB, LE **Vol. 2** 719, 783
TOUCH PAS A MON BINIOU **Vol. 6** *page 88*
TOUCHE-A-TOUT **Vol. 2** 59
TOUCHE PAS A LA FEMME BLANCHE **Vol. 2** 74
TOUCHE PAS A MON COPAIN **Vol. 2** 742
TOUCHEZ PAS AU GRISBI **Vol. 2** 279, 601
TOUCHEZ PAS AUX BLONDES **Vol. 2** 318, 347, 581
TOUCHONS DU BOIS **Vol. 2** 833
TOUHA **Vol. 10** 182, 292, 313
TOUHA SHERLOCKA HOLMESE **Vol. 10** 369, 425, 473
TOUHA ZVANA ANADA **Vol. 10** 36, 388, 503
TOUKOKUUN TAIKA **Vol. 5** 549, 592
TOUR DE COCHON, UN **Vol. 2** 388, 863
TOUR DE FRANCE PAR DEUX ENFANTS, LE **Vol. 2** 387
TOUR DE NESLE, LA **Vol. 2** 118, 330, 538, 826
TOUR, PRENDS GARDE !, LA **Vol. 2** 197, 204, 282, 801
TOURBILLON **Vol. 2** 15, 226, 309
TOURBILLON DE PARIS, LE **Vol. 2** 217
TOURMENTE, LA **Vol. 2** 798
TOURMENTS **Vol. 2** 294, 475, 559
TOURNANT DANGEREUX, LE **Vol. 2** 243, 471
TOURNEE DES GRANDS DUCS, LA **Vol. 2** 661
TOURNOI DANS LA CITE, LE **Vol. 2** 547, 623
TOURNOI, LE **Vol. 2** 237, 533, 871
TOUS LE CHEMINS MENENT A ROME **Vol. 2** 59, 185, 566
TOUS LES DEUX **Vol. 2** 137, 153, 439
TOUS LES GARCONS S'APPELLENT PATRICK **Vol. 2** 465
TOUS PREUVENT ME TUER **Vol. 2** 322, 601
TOUS VEDETTES ! **Vol. 2** 285, 335
TOUT CA NE VAUT PAS L'AMOUR **Vol. 2** 9
TOUT CHANTE AUTOUR DE MOI **Vol. 2** 475, 777
TOUT DEPEND DES FILLES **Vol. 2** 561
TOUT EST A NOUS **Vol. 2** 352
TOUT EST PERMIS **Vol. 2** 291
TOUT FEU TOUT FLAMME **Vol. 6** *page 77, page 86, page 87*

TOUT LE MONDE, IL EST BEAU, TOUT LE MONDE, IL EST GENTIL Vol. 2 110, 263
TOUT LE MONDE PEUT SE TROMPER Vol. 6 *page 83, page 98*
TOUT L'OR DU MONDE Vol. 2 59, 661
TOUT PEUT ARRIVER Vol. 2 441
TOUT POUR JOUIR Vol. 2 721
TOUT POUR LE TOUT, LE Vol. 2 194, 873
TOUT POUR RIEN Vol. 2 55, 197
TOUT S'ARRANGE Vol. 2 469
TOUT VA BIEN Vol. 2 263, 551
TOUT VA TRES BIEN, MADAME LA MARQUISE Vol. 2 48, 193
TOUTE FOLLES DE LUI Vol. 2 781, 863
TOUTE LA FAMILLE ACCUSE Vol. 2 153, 270, 479
TOUTE LA FAMILLE ETAIT LA Vol. 2 6, 698
TOUTE UNE NUIT Vol. 6 *page 109*
TOUTE UNE VIE Vol. 2 196, 457
TOVA SE SLUCI NA ULICATA Vol. 3 53, 63, 65
TOVARITCH Vol. 2 9, 488
TOWN CALLES BASTARD, A Vol. 8 276, 558
TOYA Vol. 5 768, 800
TOYA OG HEIDI Vol. 5 756
TOZI HUBAV ZIVOT Vol. 3 30, 46 Vol. 6 *page 42*
TOZI INSTINSKI M"Z Vol. 3 31, 52, 109
TRA FUMI DI CHAMPAGNE Vol. 7 279
TRA I SORRISI DI NAPOLI Vol. 7 503
TRABAJO TRANQUILO, UN Vol. 8 426
TRACASSIN OU LES PLAISIRS DE LA VILLE, LE Vol. 2 309, 598
TRACCE MISTERIOSE Vol. 7 1189
TRACE, LA Vol. 6 *page 82, page 131, page 133*
TRAD Vol. 1 180, 283 Vol. 6 *page 246*
TRADATOR FARA VOIE Vol. 3 236
TRADIMENTO Vol. 7 222, 474
TRADIMENTO O A PASSATO CHE UCCIDE, IL Vol. 7 73, 1062
TRADITION DE MINUIT, LA Vol. 2 626, 698, 722, 824
TRADLÖST OCH KÄRLEKSFULLT Vol. 5 1090
TRAFIC Vol. 2 749, 866
TRAFIC SUR LES DUNES Vol. 2 404
TRAFIQUANTS DE LA MER, LES Vol. 2 460
TRÄFRACKEN Vol. 5 946, 1104, 1142
TRAGEDIA DI BAMBOLA Vol. 7 1004
TRAGEDIA DI UN UOMO RIDICOLO, LA Vol. 7 853, 1076, 1112
TRAGEDIA ROSJI I JEJ TRZY EPOKI Vol. 1 317
TRAGEDIA SU TRE CARTE Vol. 7 559
TRAGEDIA SULL'ISONZO Vol. 7 78
TRAGEDIA UBOGIEGO SKRZYPKA Vol. 1 268
TRAGEDIE DE CARMEN, LA Vol. 6 *page 90, page 100, page 118*
TRAGEDIE DE LOURDES, LA Vol. 2 827
TRAGEDIE IMPERIALE, LA Vol. 2 8, 81, 332, 427, 526
TRAGICA BEFFA Vol. 7 320
TRAGICA CONVEGNO Vol. 7 399
TRAGICA FIAMMA Vol. 7 1124
TRAGICA NOTTE Vol. 7 2, 170, 552, 725, 1060, 1116
TRAGICA NOTTE DI ASSISI, LA Vol. 7 938
TRAGICA VISIONE Vol. 7 488
TRAGÖDIE Vol. 4 213, 626
TRAGÖDIE DER ENTEHRTEN, DIE Vol. 4 424, 643
TRAGÖDIE DER LIEBE Vol. 4 294, 396, 528, 536, 733
TRAGÖDIE EINER EHE Vol. 4 219, 232, 681
TRAGÖDIE EINER LIEBE Vol. 4 18 Vol. 7 756
TRAGÖDIE EINER LIEBESNACHT Vol. 4 344, 545
TRAGÖDIE EINES GROSSEN, DIE Vol. 4 644
TRAGÖDIE EINES STAATSANWALTES, DIE Vol. 4 25, 424
TRAGÖDIE EINES STREIKS Vol. 4 183
TRAGÖDIE EINES VERLORENEN, DIE Vol. 4 242, 316
TRAGÖDIE IM HAUSE HABSBURG Vol. 4 157, 174, 558
TRAGÖDIE IM ZIRKUS ROYAL Vol. 4 270, 644, 796
TRAGOUDIA TES FOTIAS Vol. 3 142, 149, 153, 159, 168, 176
TRAGOVI CRNE DEVOJKE Vol. 3 367, 368
TRAICAO INVEROSIMIL Vol. 8 122
TRAIN D'AMOUR, LE Vol. 2 595
TRAIN DE 8 H 47, LE Vol. 2 332, 381, 864
TRAIN DE BERLIN S'EST ARRETE, LE Vol. 2 294
TRAIN DE PLAISIR Vol. 2 488
TRAIN D'ENFER Vol. 2 677 Vol. 8 517, 732
TRAIN, LE Vol. 2 400, 770, 875
TRAIN POUR VENISE, LE Vol. 2 279
TRAIN SANS YEUX, LE Vol. 2 745
TRAITE LES BLANCHES, LA Vol. 2 207, 479
TRAITEMENT DE CHOC Vol. 2 435, 738
TRAJA SVEDKOVIA Vol. 10 210, 484, 576
TRAJE BLANCO, UN Vol. 8 214, 611

TRAJE DE LUCES, EL Vol. 8 209, 268, 298, 311
TRAJE DE ORO, EL Vol. 8 732, 753
TRALLANDE JÄNTA, EN Vol. 5 840, 909, 1028
TRAMONTO Vol. 7 965
TRAMONTO DEI BLASONI Vol. 7 471, 1190
TRAMONTO DELL'UMANITA, IL Vol. 7 720
TRAMONTO DI FUOCO Vol. 7 796, 918
TRAMPA Vol. 3 109, 110 Vol. 6 *page 55*
TRAMPA BAJO EL SOL Vol. 8 517, 732
TRAMPA MORTAL Vol. 8 732, 753
TRAMPA PARA CATALINA Vol. 8 258, 311
TRAMPA PARA UN FORAJIODO Vol. 8 396
TRAMPA SEXUAL Vol. 8 367
TRAMPAS DEL MATRIMONIO, LAS Vol. 8 551
TRAMPEN NACH NORDEN Vol. 1 11, 62
TRAMPOSOS, LOS Vol. 8 546, 749
TRAMPOTY DIVADELNIHO REDITELE Vol. 10 57, 285
TRAMVIERA N. 47, LA Vol. 7 488
TRANDAFIRUL GALBEN Vol. 6 *page 271, page 273, page 281*
TRANQUILLE DONNE DI CAMPAGNE Vol. 7 633
TRANS EUROP EXPRESS Vol. 2 441
TRANSATLANTIQUES, LES Vol. 2 197
TRANSIT A SAIGON Vol. 2 856
TRANSIT CARLSBAD Vol. 10 96, 229, 369
TRANSPORT Z RAJE Vol. 10 87, 289, 503
TRANSZPORT, A Vol. 9 107, 227, 318
TRAPENI Vol. 10 70, 197, 433
TRAPERA, LA Vol. 8 238
TRAPIANTO, IL Vol. 7 257 Vol. 8 299, 517
TRAPPOLA D'AMORE Vol. 7 796, 830
TRAPPOLA DI FUOCO, LA Vol. 7 1036
TRAPPOLA, LA Vol. 7 53
TRAQUA, LA Vol. 2 719
TRAQUE, LE Vol. 2 96, 778
TRAQUENARDS Vol. 2 445
TRARA UM LIEBE Vol. 4 192, 270, 409, 462
TRASA Vol. 10 218, 296, 463
TRASPLANTE A LA ITALIANA Vol. 8 299, 517
TRASPLANTE DE UN CEREBRO Vol. 8 557, 706
TRATTA DELLE BIANCHE, LA Vol. 7 1147
TRATTATO SCOMPARSO, IL Vol. 7 756, 796
TRAU, SCHAU WEM Vol. 4 13, 24
TRAUM DER HERZOGIN, DER Vol. 4 267
TRAUM DES HAUPTMANN LOY, DER Vol. 1 68, 90, 107
TRAUM VOM GLÜCK, EIN Vol. 4 186, 400, 605, 662
TRAUM VOM RHEIN, DER Vol. 4 132, 230
TRAUM VON SCHÖNBRUNN Vol. 4 132, 270
TRÄUME SIND SCHÄUME Vol. 4 354, 367
TRÄUMENDE MUND, DER Vol. 4 368
TRÄUMEREI Vol. 4 28, 246
TRAUMMUSIK Vol. 4 49, 246 Vol. 7 902
TRÄUM'NICHT, ANNETTE ! Vol. 1 23, 56, 66
TRAUMSCHIFF, DAS Vol. 1 24, 100, 105
TRAUMULUS Vol. 4 373, 626
TRAV, HOPP OCH KÄRLEK Vol. 5 1045
TRAVAIL Vol. 2 167, 375
TRAVAIL C'EST LA LIBERTE, LE Vol. 2 347, 368, 503
TRAVAILLEURS DE LA MER Vol. 2 159
TRAVERSATA NERA Vol. 7 40, 595, 827
TRAVERSEE DE LA LOIRE, LA Vol. 2 479
TRAVERSEE DE PARIS, LA Vol. 2 251, 625
TRAVESIAS DE MORUCHA, LAS Vol. 8 768
TRAVESURAS DE MORUCHA, LAS Vol. 8 629
TRAVIATA '53 Vol. 7 493
TRAVIATA, LA Vol. 7 567, 756, 944, 1142
TRAVIESA MOLINERA, LA Vol. 2 440 Vol. 8 486, 516
TRAVOLTA DAGLI AFFETTI FAMILIARI Vol. 7 123
TRAVOLTI DA UN INSOLITO DESTINO NELL' AZZURRO MARE D'AGOSTO Vol. 7 592
TRE AMANTI, I Vol. 7 48
TRE AQUILOTTI, I Vol. 7 196, 443
TRE CORSARI, I Vol. 7 283, 333, 781
TRE DAR I BUREN Vol. 5 905
TRE DAR PA LUFFEN Vol. 5 843
TRE DESIDERI, I Vol. 7 453
TRE EGLE OG FEM LOVER Vol. 5 16, 195
TRE ENGLE OG FEM LOVER Vol. 5 111
TRE ESPERIMENTI DI ELIANA, I Vol. 7 392
TRE FINDER EN KRO Vol. 5 23, 281
TRE FRATELLI IN GAMBA Vol. 7 827, 1192
TRE GLADA TOKAR Vol. 5 841, 923
TRE ILLUSIONI, LE Vol. 7 560
TRE LADRI, I Vol. 7 501, 634
TRE LADRONI, I Vol. 7 1189
TRE MA MAN VAERE Vol. 5 14, 281

TRE MAND FREM FOR EN TROLD **Vol. 5** 38, 63, 179
TRE MENO DUE **Vol. 7** 899
TRE MILIONI DI DOTE **Vol. 7** 26, 701
TRE NOTTI D'AMORE - LA VEDOVA **Vol. 7** 483, 798
TRE NOTTI VIOLENTE **Vol. 8** 396, 775
TRE OMBRE, LE **Vol. 7** 510
TRE ÖNSKNINGAR **Vol. 5** 840, 843, 923, 959
TRE PASSI A NORD **Vol. 7** 756
TRE PASSI NEL DELIRIO - TOBY DAMMIT **Vol. 7** 664, 931, 988, 1142
TRE PECORE VIZIOSE **Vol. 7** 993
TRE PERSONE PER BENE **Vol. 7** 559
TRE PIGER FRA JYLLAND **Vol. 5** 180, 199, 201
TRE PIGER I PARIS **Vol. 5** 31, 43, 272
TRE PIGER OG EN GRIS **Vol. 5** 159
TRE RAGAZZE CERCANO MARITO **Vol. 7** 1036, 1172
TRE RAGAZZE VIENNES **Vol. 7** 756
TRE SENTIMENTALI, I **Vol. 7** 53
TRE SERGENTI DEL BENGALA, I **Vol. 7** 649 **Vol. 8** 493
TRE SIMPATICHE CAROGNE **Vol. 7** 1136
TRE SKOKIGA SKOJARE **Vol. 5** 841, 923
TRE SKOLEKAMMERATER, DE **Vol. 5** 2, 228
TRE SMA PIGER **Vol. 5** 16, 61, 195, 272
TRE SÖNER GICK TILL FLYGET **Vol. 5** 1042, 1078
TRE SORELLE, LE **Vol. 7** 731
TRE SORRISI DI UNA MONELLA, I **Vol. 7** 277
TRE SOTTO IL LENZUOLO **Vol. 7** 1003, 1022
TRE STORIE PROIBITE **Vol. 7** 21, 499
TRE STRANIERE A ROMA **Vol. 7** 817
TRE TIGRI CONTRO TRE TIGRI **Vol. 7** 502, 633
TRE UOMINI IN FRACK **Vol. 7** 730, 1189
TRE VAGABONDI, I **Vol. 7** 304
TRE VOLTI - AMANTI CELEBRI **Vol. 7** 1142
TRE VOLTI DELLA PAURA, I **Vol. 7** 537
TRE VOLTI, I **Vol. 7** 724, 725
TRE VOLTI - IL PROVINO, I **Vol. 7** 853, 1142
TRE VOLTI - LATIN LOVER, I **Vol. 7** 188, 1142
TREASURE ISLAND **Vol. 8** 611
TRECATOARELE IUBIRI **Vol. 3** 250, 268, 301
TRECE ONZAS DE ORO **Vol. 8** 437, 614
TRECENTO DELLA SETTIMA **Vol. 7** 615, 941
TRECI KLJUC **Vol. 6** *page 307, page 310*
TREDICESIMO COMMENSALE, IL **Vol. 7** 454
TREDICI UOMINI E UN CANNONE **Vol. 7** 16, 603, 1121
TREDIE MAGT, DEN **Vol. 5** 9
TREDOWATA **Vol. 1** 174, 176, 243, 278, 308, 333, 346, 360
TREFFPUNKT AIMEE **Vol. 1** 2, 35, 125
TREFFPUNKT GENF **Vol. 1** 36, 85
TREI SCRISORI SECRETE **Vol. 3** 269, 319
TREI ZILE SI TREI NOPTI **Vol. 3** 269, 272, 326 **Vol. 6** *page 269*
TREIBENDE KRAFT **Vol. 4** 306, 648, 747
TREINTA Y NUEVE CARTAS DE AMOR **Vol. 8** 752
TREIS DRAPETES TOU FREBOKOMIOU **Vol. 3** 203
TREIUS KOUKLES KAI EGO **Vol. 3** 165
TREIZE A TABLE **Vol. 2** 158, 205, 808
TREIZE JOURS EN FRANCE **Vol. 2** 196
TREIZIEME CAPRICE, LE **Vol. 2** 110, 377
TREIZIEME ENQUETE DE GREY, LA **Vol. 2** 42
TREKANTER **Vol. 5** 261
TRELLOS PALABOS KAI BEGGOS **Vol. 3** 197
TREMBLEMENTS DE CHAIR **Vol. 2** 484
TREN **Vol. 3** 381, 398
TREN DE LAS 8.45, EL **Vol. 8** 522, 640
TREN, EL **Vol. 8** 227, 364, 516
TREN EXPRESO, EL **Vol. 8** 752
TRENCK **Vol. 4** 97, 115, 132, 247
TRENCK, DER PANDUR **Vol. 4** 344, 415, 437
TRENER **Vol. 3** 336, 400
TRENO CROCIATO, IL **Vol. 7** 86, 170, 313
TRENO DELLE 21,15, IL **Vol. 7** 493, 730, 985
TRENO DI PIACERE **Vol. 7** 196
TRENO POPOLARE **Vol. 7** 196, 1036
TRENTA SECONDI D'AMORE **Vol. 7** 796
TRENTE ET QUARANTE **Vol. 2** 460, 475
TRENTESIMA PERLA, LA **Vol. 7** 344
TRENUTKI ODLOCITVE **Vol. 3** 387, 390
TREPTE PE CER **Vol. 3** 209, 307, 319
TRES ALCOBAS **Vol. 8** 780
TRES DA VIDA AIRADA, OS **Vol. 8** 44, 54, 131
TRES DE LA CRUZ ROJA **Vol. 8** 725, 762
TRES DIAS SEM DEUS **Vol. 8** 1, 42, 58, 194
TRES DOLARES DE PLOMO **Vol. 8** 461
TRES EN RAYA **Vol. 8** 551
TRES ESPEJOS **Vol. 8** 26, 75, 119, 311, 449
TRES ESPELHOS **Vol. 8** 26, 75, 119
TRES GORRIONES Y PICO **Vol. 8** 342, 450
TRES GRACIAS, LAS **Vol. 8** 16, 18, 43, 50, 82, 122, 162, 170, 180

TRES HERMANAS, LAS Vol. 8 640
TRES HOMBRES BUENOS Vol. 8 603
TRES HOMBRES VAN A MORIR Vol. 8 298
TRES INSUFFISANT Vol. 2 406 Vol. 6 *page 144*
TRES LADRONES IN CASA Vol. 8 214, 436
TRES MENOS EU Vol. 8 19, 34
TRES MUJERES DE HOY Vol. 8 414, 742
TRES NOCHES VIOLENTAS Vol. 8 396, 775
TRES PERFECTAS CASADAS, LAS Vol. 8 245, 406
TRES PERROS LOCOS, LOCOS, LOCOS Vol. 8 595, 773
TRES SARGENTOS BENGALES Vol. 8 493
TRES SUECAS PARA TRES RODRIGUEZ Vol. 8 450, 553
TRES SUPERMEN EN LA SELVA, LOS Vol. 8 342
TRESA Vol. 7 369
TRESOR DE CANTENAC, LE Vol. 2 593, 706, 722
TRESOR DE FEMME, UN Vol. 2 197, 407
TRESOR DES HOMMES BLEUS, LE Vol. 2 310, 383 Vol. 8 305, 450, 541
TRESOR DES PIEDS NICKELES, LE Vol. 2 217, 627
TRESORS DE LA MER ROUGE, LES Vol. 2 740
TRESTANCI NA SPILBERKU Vol. 10 267
TRETA SLED SL"NCETO Vol. 3 119 Vol. 6 *page 42, page 54*
TRETI PRINC Vol. 10 369, 434, 473
TRETI ROTA Vol. 10 138, 248, 453, 524
TRETI SARKAN Vol. 10 209, 273, 459
TRETI SKOBA PRO KOCOURA Vol. 10 255, 462
TRETI ZVONENI Vol. 10 167, 621
TRETOTO LICE Vol. 6 *page 37, page 38, page 60*
TRETTIO PINNAR MUCK Vol. 5 887, 1142
TREVE, LA Vol. 2 188
TREVO DE QUATRO FOLHAS, O Vol. 8 9, 16, 86, 96, 112, 122
TREVOGA Vol. 3 100, 127 Vol. 6 *page 52*
TR"GNI NA P"T Vol. 3 60, 105 Vol. 6 *page 29*
TRHAK Vol. 10 144, 359, 505
TRHANI Vol. 10 34, 122
TRI Vol. 3 381, 421, 429
TRI ANE Vol. 3 341, 444
TRI CETRTINE SONCA Vol. 3 395, 449
TRI CHLAPI NA CESTACH Vol. 10 139, 388, 425
TRI CHLAPI V CHALUPE Vol. 10 352, 621
TRI DCERY Vol. 10 107, 273, 548
TRI KAMARADI Vol. 10 142, 193, 256
TRI KORAKA U PRAZNO Vol. 3 383, 391, 392
TRI KROKY OD TELA Vol. 10 167, 300, 611
TRI MUZI VE SNEHU Vol. 10 256, 460
TRI NEVINNI Vol. 10 408, 425, 621
TRI OD MORE Vol. 10 432, 531, 619
TRI ORISKY PRO POPELKU Vol. 10 41, 197, 425, 555
TRI PRANI Vol. 10 41, 522, 549
TRI SATA ZA LJUBAV Vol. 3 451
TRI STO DVADCAT DVA Vol. 10 39, 273, 544
TRI TUNY PRACHU Vol. 10 296, 580
TRI VEJCE DO SKLA Vol. 10 256, 412
TRI VETERANI Vol. 10 144, 325, 369
TRI ZGODBE Vol. 3 345, 390, 406, 449
TRI ZLATE VLASKY DEDA VSEVEDA Vol. 10 131, 395, 569
TRIANGLE ECORCHE, LE Vol. 2 713
TRIANGOLO GIALLO, IL Vol. 7 277
TRIANGULITO Vol. 8 732
TRIANGULO Vol. 8 73, 179, 284, 406, 719
TRIANGULO CIRCULAR, O Vol. 8 83
TRIATRICET STRIBRNYCH KREPELEK Vol. 10 369, 371
TRIBLY Vol. 2 84
TRIBULATIONS D'UN CHINOIS EN CHINE, LES Vol. 2 475, 781
TRICET JEDNA VE STINU Vol. 10 22, 24, 296
TRICET PANEN A PYTHAGORAS Vol. 10 250, 303, 462, 500
TRICHEURS, LES Vol. 2 96, 598, 719, 873
TRICK 17B Vol. 1 30, 37, 50
TRICK-TRACK Vol. 4 646
TRICOCHE ET CACOLET Vol. 2 197, 626, 824
TRIDENTUL NU RASPUNDE Vol. 3 381, 320 Vol. 6 *page 266, page 270*
TRIESTE MIA Vol. 7 693
TRIESTE O I VENDICATORI DI OBERDAN Vol. 7 78
TRIESTE O L'IMPERO DELLA FORCA Vol. 7 114
TRIGO E O JOIO, O Vol. 8 122
TRIGO LIMPIO Vol. 8 217, 768
TRIJE PRISPEVKI K SLOVENSKI BLAZNOSTI Vol. 6 *page 303, page 304, page 316*
TRIKYMIA MIAS KARDIAS Vol. 3 168
TRILOGIA DI MACISTE, LA Vol. 7 1092
TRIMATA OT MORETO Vol. 6 *page 50, page 58*
TRIMATA OT ZAPASA Vol. 3 120, 124 Vol. 6 *page 45*

TRINACTA KOMNATA **Vol. 10** 151, 471, 503
TRINACTY REVIR **Vol. 10** 122, 142, 256, 524
TRINCA DEL AIRE, LA **Vol. 8** 214, 463
TRINE **Vol. 5** 723, 738
TRINGLEUSES, LES **Vol. 2** 795
TRINI **Vol. 1** 38, 62
TRINTA Y NUEVE CARTAS DE AMOR **Vol. 8** 438
TRIO ANGELOS **Vol. 10** 287, 439, 576
TRIO BIRBONE, UN **Vol. 7** 1189
TRIO DE DAMAS **Vol. 8** 546
TRIO INFERNAL, LE **Vol. 2** 163, 873
TRIOMPHE DE MICHEL STROGOFF, LE **Vol. 2** 722, 781
TRIOMPHE DU LUTTEUR, LE **Vol. 2** 159, 330
TRIONFO CRISTIANO **Vol. 7** 829
TRIONFO DEI DIECI GLADIATORI, IL **Vol. 7** 1027
TRIONFO DELLA VITA, IL **Vol. 7** 732
TRIONFO DI SANGUE **Vol. 7** 488
TRIPES AU SOLEIL, LES **Vol. 2** 207, 407, 777
TRIPLE CROSS **Vol. 2** 13, 722
TRIPLE ENQUETE **Vol. 2** 70, 152, 475
TRIPLE MORT DU TROISIEME PERSONNAGE, LA **Vol. 8** 219, 656
TRIPLE MUERTE DEL TERCER PERSONNAJE, LE **Vol. 8** 219, 656
TRIPLE P... **Vol. 2** 596
TRIPLEPATTE **Vol. 2** 42, 547
TRIPOLI, BEL SUOL D'AMORE **Vol. 7** 798
TRIPORTEUR, LE **Vol. 2** 250, 661
TRIPORTEUSES, LES **Vol. 2** 294
TRIQUE, GAMIN DE PARIS **Vol. 2** 266
TRISTAN ET ISEULT **Vol. 2** 633
TRISTAN ET YSEULT **Vol. 2** 708, 867
TRISTANA **Vol. 8** 207, 214
TRISTE AMORI **Vol. 7** 1162
TRISTE REALITA **Vol. 7** 314
TRISTE SKAEBNER **Vol. 5** 97
TRISTI AMORI **Vol. 7** 196, 682, 756, 1060
TRISTI PASSIONI, LE **Vol. 7** 1162
TRITE SM"RTNI GRJAHA **Vol. 3** 40 **Vol. 6** *page 29*
TRITTICO DELL'AMORE, IL **Vol. 7** 301, 1032, 1071
TRITTICO DI BONNARD, IL **Vol. 7** 369
TRIUMPH DES TODES, DER **Vol. 4** 239
TRIUMPHS OF A MAN CALLED HORSE **Vol. 8** 303, 374
TRIUNFO DE LOS DIEZ GLADIADORES, EL **Vol. 8** 523
TRIUNFO DE UN HOMBRE LLAMADO CAVALLO, EL **Vol. 8** 303, 374
TRIUNFO DEL AMOUR, EL **Vol. 2** 727
TRIX, DER ROMAN EINER MILLIONENERBIN **Vol. 4** 208, 390
TRNOVE POLE **Vol. 10** 13, 250, 462
T"RNOVSKATA CARICA **Vol. 6** *page 58, page 61*
TRNY A KVETY **Vol. 10** 167, 536
TRO, HAB OG KAERLIGHED **Vol. 5** 177, 219, 249, 296
TRO, HAB OG TROLDDOM **Vol. 5** 31, 241, 306
TRO OG VILLIG PIGE, EN **Vol. 5** 9
TRÖDLER VON AMSTERDAM, DER **Vol. 4** 14, 340, 471
TROEN, DER FRELSER **Vol. 5** 256
TROFAST KAERLIGHED **Vol. 5** 70
TROFEJ **Vol. 3** 411, 415
TROFEJ NEZNAMEHO STRELCA **Vol. 10** 36, 263, 459
TROIKA **Vol. 2** 833 **Vol. 4** 164, 584, 707
TROIKA SUR LA PISTE BLANCHE **Vol. 2** 9, 339
TROIS ARGENTINS A MONTMARTRE **Vol. 2** 6, 413
TROIS ARTILLEURS A L'OPERA **Vol. 2** 404, 722
TROIS ARTILLEURS AU PENSIONNAT **Vol. 2** 270, 676, 833
TROIS ARTILLEURS EN VADROUILLE **Vol. 2** 9, 833
TROIS BALLES DANS LA PEAU **Vol. 2** 254, 469
TROIS CENTS A L'HEURE **Vol. 2** 381
TROIS CHAMBRES A MANHATTAN **Vol. 2** 59, 778
TROIS COQUILLAGES DE TUNISIE **Vol. 2** 811
TROIS COURONNES DU MATELOT, LES **Vol. 6** *page 99*
TROIS COUSINES, LES **Vol. 2** 32, 347
TROIS DE LA CANEBIERE **Vol. 2** 131, 309, 743
TROIS DE LA MARINE **Vol. 2** 205, 309, 777
TROIS DE SAINT-CYR **Vol. 2** 322, 533, 538, 775
TROIS ENFANTS DANS LE DESORDRE **Vol. 2** 128, 659
TROIS ETC...DU COLONEL **Vol. 2** 124, 226
TROIS FEMMES, TROIS AMES **Vol. 2** 13, 47, 57, 398, 571
TROIS FILLES DANS LE VENT **Vol. 6** *page 134*
TROIS FILLES NUES DANS L'ILE AUX PECHES **Vol. 6** *page 120*
TROIS FILLES VERS LE SOLEIL **Vol. 2** 293
TROIS FONT LA PAIRE, LES **Vol. 2** 8, 322

TROIS GARCONS, UNE FILLE **Vol. 2** 158, 627, 727 **Vol. 7** 968
TROIS HOMMES A ABATTRE **Vol. 6** *page 74, page 99*
TROIS HOMMES EN HABIT **Vol. 7** 730, 1189
TROIS HOMMES SUR UN CHEVAL **Vol. 2** 570, 689
TROIS JEUNES FILLES NUES **Vol. 2** 9, 37
TROIS JOURS A VIVRE **Vol. 2** 137, 626, 824
TROIS JOURS DE BRINGUE A PARIS **Vol. 2** 404, 703, 717
TROIS JOURS DE PERM **Vol. 2** 249
TROIS LYCEENNES A PARIS **Vol. 6** *page 113*
TROIS LYCEENNES EN CHALEUR **Vol. 6** *page 113*
TROIS LYS, LES **Vol. 2** 595
TROIS MARINS DANS UN COUVENT **Vol. 2** 243, 717
TROIS MARINS EN BORDEE **Vol. 2** 717
TROIS MILLIARDS SANS ASCENSEUR **Vol. 2** 20, 835
TROIS MOUSQUETAIRES, LES **Vol. 2** 158, 232, 368, 402, 412, 466, 516, 547, 615, 824, 857
TROIS NOUVELLES DE TCHEKHOV **Vol. 2** 811
TROIS PASSIONS, LES **Vol. 2** 153
TROIS POUR CENT **Vol. 2** 332
TROIS TAMBOURS, LES **Vol. 2** 131
TROIS TELEGRAMMES **Vol. 2** 156, 395
TROIS VALSES **Vol. 2** 279, 322, 386, 778
TROIS VIEILLES FILLES EN FOLIE **Vol. 2** 243, 717
TROISIEME CHEMINEE GAUCHE **Vol. 2** 191
TROISIEME DALLE, LA **Vol. 2** 249, 533
TROISIEME WAGON, LE **Vol. 6** *page 133*
TROIS...SIX...NEUF **Vol. 2** 8, 59, 427
TROJE I LAS **Vol. 1** 202, 237, 400
TROJKA HULTAJSKA **Vol. 1** 241, 278, 308
TROLD KAN TAEMMES **Vol. 5** 58
TROLDEN I AESKEN **Vol. 5** 92
TROLL **Vol. 5** 933
TROLL-ELGEN **Vol. 5** 803, 823
TROLL I ORD **Vol. 5** 674
TROLLEBOKUNGEN **Vol. 5** 691, 1154
TROLLFLÖJTEN **Vol. 5** 935, 1080
TROLLFOSSEN **Vol. 5** 729, 738
TROLLSOMMAR **Vol. 5** 869, 1048
TROLOS **Vol. 5** 9
TROLOSE HUSTRU, DEN **Vol. 5** 183
TROMBITAS, A **Vol. 9** 253, 259, 303
TROMMELN ASIENS, DIE **Vol. 4** 330, 509, 527
TROMP-LA-MORT **Vol. 7** 108
TROMPETAS SEL APOCALIPSIS, LAS **Vol. 8** 681
TRONO E LA SEGGIOLA, IL **Vol. 7** 991
TRONO PARA CHRISTY, UN **Vol. 8** 261
TROP C'EST TROP **Vol. 2** 742
TROP JOLIES POUR ETRE HONNETES **Vol. 2** 808
TROP PETITE MON AMI **Vol. 2** 814
TROPENBLUT **Vol. 4** 681
TROPENNÄCHTE **Vol. 4** 229
TROPERNES DATTER **Vol. 5** 183
TROPICA DEL CANCRO **Vol. 7** 738
TROPPO BUON CUORE **Vol. 7** 1189
TROPPO FORTE **Vol. 7** 383, 625, 1166
TROPPO ONESTA **Vol. 7** 1189
TROPPO TARDI T'HO CONOSCIUTA **Vol. 7** 205, 402, 1148
TROPPO ZELO **Vol. 7** 1189
TROR DU JEG ER FODT I GAAR ? **Vol. 5** 89
TROR HAN **Vol. 5** 1041
TROSKABSVAESKEN **Vol. 5** 81
TROST I TAKLAMPA **Vol. 5** 698, 807, 810
TROTS **Vol. 5** 909, 1173
TRÖTTE TEODOR **Vol. 5** 840, 863, 909, 997, 1032
TROTZ ALLEDEM ! **Vol. 1** 1, 14, 17, 59
TROTZIGES BLUT **Vol. 4** 247
TROU DANS LE MUR, UN **Vol. 2** 294, 717
TROU, LE **Vol. 2** 190, 598
TROU NORMAND, LE **Vol. 2** 347, 807
TROUBLE FESSES, LE **Vol. 2** 219
TROUMPA'67 **Vol. 3** 185
TROVATA DELLO SPORTMANN, LA **Vol. 7** 163, 250
TROVATELLA DI MILANO, LA **Vol. 7** 493
TROVATELLA DI POMPEI, LA **Vol. 7** 827
T"RSI SE S"PRUG ZA MAMA **Vol. 6** *page 50, page 64*
TRST **Vol. 3** 377, 449
TRUANDS, LES **Vol. 2** 57, 454, 676
TRUANGOLITO, EL **Vol. 8** 611
TRÜBE WASSER **Vol. 1** 58
TRUC DU BRESILIEN, LE **Vol. 2** 84, 279
TRUCIDO E LO SBIRRO, IL **Vol. 7** 383, 599
TRUDE, DIE SECHZEHNJÄHRIGE **Vol. 4** 36, 124
TRUDNA LJUBOV **Vol. 3** 59, 121 **Vol. 6** *page 58*
TRUDNA MILOSC **Vol. 1** 171, 305 **Vol. 6** *page 255*
TRUET LYKKE **Vol. 5** 256
TRUHANES **Vol. 8** 245, 549
TRUHANES DE HONOR **Vol. 8** 524, 772
TRUITE, LA **Vol. 6** *page 70, page 99, page 135*
TRUST DER DIEBE **Vol. 4** 208, 211

TRUTZE VON TRUTZBERG, DIE Vol. 4 330, 527, 544
TRUXA Vol. 4 49, 626
TRYLLEKUNSTNERNEN Vol. 5 97
TRYLLESAEKKEN Vol. 5 97
TRYSIL - KNUT Vol. 5 676, 683, 711, 714, 723, 770
TRYTON, A Vol. 9 288
TRZEBA ZABIC TE CYWILA Vol. 1 294
TRZEBA ZABIC TE MILOSC Vol. 1 294, 314 Vol. 6 *page 236*
TRZECIA CZESC NOCY Vol. 1 136, 313, 324, 325
TRZECIA GRANICA Vol. 1 144
TRZY KOBIETY Vol. 1 133, 305 Vol. 6 *page 209*
TRZY KROKI PO ZIEMI Vol. 1 171, 401 Vol. 6 *page 246*
TRZY OPOWIESCI Vol. 1 133, 227
TRZY SERCA Vol. 1 278, 404
TRZY STARTY Vol. 1 179 Vol. 6 *page 209*
TRZY STARTY - NOWELA BOKSERSKA Vol. 1 354
TRZY STARTY - NOWELA KOLARSKA Vol. 1 400
TRZY STARTY - NOWELA PLYWACKA Vol. 1 152
TRZYDZIESCI KARATOW SZCZESCIA Vol. 1 278, 404
TSCHELJUSKIN Vol. 1 68, 79, 95
TSCHTSCHENSEN-RACHE Vol. 4 380, 385
TU CHE NE DICI ? Vol. 7 910
TU DIOS Y MI INFIERNO Vol. 8 237, 245, 577
TU ES PIERRE Vol. 2 540
TU ESTAS LOCO, BRIONES Vol. 8 373, 771
TU FOSA SERA LA EXACTA ... AMIGO Vol. 8 629
TU L'AS DANS LE CUL ! MARRAINE Vol. 6 *page 112*
TU M'APPARTIENS Vol. 2 722
TU MARIDO NOS ENGANA Vol. 8 396, 515
TU M'AS SAUVE LA VIE Vol. 2 61, 706
TU MI TURBI Vol. 7 405, 658, 1173
TU NE TUERAS POINT Vol. 2 250, 251, 625 Vol. 3 447
TU PERDONAS ... YO NO Vol. 8 770
TU SERAS DUCHESSE Vol. 2 41, 579
TU SERAS TERRIBLEMENT GENTILLE Vol. 2 259
TU SOLO Vol. 8 358
TU TEN KAMEN Vol. 10 167, 431, 536
TU Y YO SOMOS TRES Vol. 8 214, 449
TUA DONNA, LA Vol. 7 1177
TUA PER LA VITA ! Vol. 7 454, 1063
TUA, SOLTANTO TUA ! Vol. 7 806
TUA VITA E IN QUESTO SCUDO ! Vol. 7 1125
TUA VITA PER MIO FIGLIO, LA Vol. 7 474
TUAREG Vol. 8 237, 303
TUAREG - IL GUERRIERO DEL DESERTO Vol. 7 216
TUBE, LE Vol. 2 866
TÜCSÖK Vol. 9 67, 121, 149
TUDJA ZEMLJA Vol. 3 346, 390
TUDOR Vol. 3 227, 244, 254, 329
TUE-LA-MORT Vol. 2 205, 346
TUEUR, LE Vol. 2 142, 719
TUFA DE VENETIA Vol. 3 280, 284, 291
TUGTHUSFANGE NO. 97 Vol. 5 9
TUHLAAJAPOIKA Vol. 5 332, 414
TUHOOTU NUORUUS Vol. 5 386, 471
TUKKIJOELLA Vol. 5 310, 344, 356, 359, 478, 634
TUKKIPOJAN MORSIAN Vol. 5 356, 359, 428
TÜKÖRKEPEK Vol. 9 8, 140, 202
TUKUMA Vol. 5 46, 177, 231
TULAK Vol. 10 243
TULAK MACOUN Vol. 10 168, 564
TULAKOVO SRDCE Vol. 10 34, 45, 126
TULINEN JÄRVI Vol. 5 1119
TULIPÄÄ Vol. 5 445, 511, 531, 563, 606, 630, 641, 647
TULIPANI DI HAARLEM, I Vol. 7 1145
TULIPE NOIRE, LA Vol. 2 224, 598
TULIPUNAINEN KYYHKYNEN Vol. 5 443, 549
TULITIKKUJA LAINAAMASSA Vol. 5 341, 356, 412, 485, 540, 557, 579
TULL - BOM Vol. 5 878, 1173
TULLUVAPAA AVIOLIITTO Vol. 5 324, 450, 565, 575, 591
TUMBA DEL PISTOLERO, LA Vol. 8 551, 626
TUMBA PARA EL SHERIFF, UNA Vol. 8 598, 626
TUMBA PARA UN FORAJIDO Vol. 8 629, 768
TUNDRA Vol. 5 340
TUNELUL Vol. 3 218, 281
TUNISI TOP SECRET Vol. 7 817
TUNNEL Vol. 7 728
TUNNEL, DER Vol. 4 213, 288, 647, 727, 734
TUNTEITA - CONFITEOR Vol. 5 574, 618
TUNTEITA - KRISTINA Vol. 5 375
TUNTEITA - LIUKASTELUA Vol. 5 444
TUNTEITA - TOIVO, RAKKAUS... Vol. 5 444, 593
TUNTEMATON SOTILAS Vol. 5 374, 443, 583, 639, 643

TUNTEMATON YSTÄVÄ Vol. 5 324, 516
TUO VIZIO E UNA STANZA CHIUSA E SOLO IO NE HO LA CHIAVE, IL Vol. 7 426
TUOMARI MARTTA Vol. 5 357, 471
TUP AKKA LAKKO Vol. 5 345, 424, 652
TUPPEN Vol. 5 985, 1093, 1096, 1209
TÜR GEHT AUF, EINE Vol. 4 82, 89, 298
TUR SVENTON - PRIVATDETEKTIV Vol. 5 694, 821, 1160, 1201, 1042, 1148
TURBAMENTO Vol. 7 453, 493
TURBANTE BLANCO Vol. 8 640
TURBINA Vol. 10 2, 154, 569
TURBINE Vol. 7 792, 1063
TURBINE ROSSO Vol. 7 986
TURCO NAPOLETANO, UN Vol. 7 852, 1102
TURFPIRATEN Vol. 4 162, 550
TURI E I PALADINI Vol. 7 242
TURISMO ES UN GRAN INVENTO, EL Vol. 8 524, 725
TURISTAS Y BRIBONES Vol. 8 492, 732
TURKASEN TENAVA Vol. 5 443, 549
TURLIS ABENTEUER Vol. 1 15, 34, 42, 50
TURLUPINS, LES Vol. 2 522 Vol. 6 *page 103, page 124*
TURM DES SCHWEIGENS, DER Vol. 4 162, 567
TURNO DI NOTTE Vol. 7 856
TURNO, IL Vol. 7 95, 567, 778
TUSENIE Vol. 10 82, 228, 544
TUSET STREET Vol. 8 214, 762
TUSK Vol. 2 301
TÜSKE A KÖRÖM ALATT Vol. 9 52, 143, 177, 215
TUT OG KJOR Vol. 5 701, 762, 777, 797, 824
TUTTA COLPA DEL PARADISO Vol. 7 405, 482, 819, 1007
TUTTA DA SCOPRIRE OR L'AMANTE TUTTA DA SCOPRIRE Vol. 7 383
TUTTA LA CITTA CANTA Vol. 7 446, 479, 481
TUTTA LA VITA IN UNA NOTTE Vol. 7 40, 419
TUTTA LA VITA IN VENTIQUATTR'ORE Vol. 7 796
TUTTI A CASA Vol. 7 257, 401
TUTTI A SCUOLA Vol. 7 257
TUTTI DENTRO Vol. 7 834, 868, 957
TUTTI FIGLI DI „MAMMASANTISSIMA" Vol. 7 694
TUTTI FRATELLI NEL WEST ... PER PARTE DI PADRE Vol. 7 981
TUTTI GLI UOMINI DEL PARLAMENTO Vol. 7 946
TUTTI I COLORI DEL BUIO Vol. 7 426 Vol. 8 551
TUTTI INNAMORATI Vol. 7 634, 817
TUTTI PER UNO ... BOTTE PER TUTTI Vol. 7 846 Vol. 8 603, 667, 695
TUTTI POSSONO ARRICCHIRE TRANNE I POVERI Vol. 7 901
TUTTO ! Vol. 7 932
TUTTO A POSTO E NIENTE IN ORDINE Vol. 7 592
TUTTO NEL MONDO E BURLA Vol. 7 196
TUTTO PER LA DONNA Vol. 7 493, 756, 796, 1181
TUTTO PER TUTTO Vol. 8 762
TUTTO SUO PADRE Vol. 7 1194
TUTTOBENIGNI Vol. 7 1105
TUULINEN PÄIVÄ Vol. 5 584
TUVO LA CULPA ADAN Vol. 8 214, 357, 436
TÜZ Vol. 9 117, 274
TÜZ, A Vol. 9 288
TÜZ A HEGYEN Vol. 9 69
TÜZET KEREK Vol. 9 86
TÜZGÖMBÖK Vol. 9 134, 152, 335
TÜZKERESZTSEG Vol. 9 134, 274
TÜZOLTO UTCA 25. Vol. 9 215, 259, 263
TÜZPROBA Vol. 9 337
TVA AR I VARJE KLASS Vol. 5 840, 909
TVA HJÄRTAN OCH EN SKUTA Vol. 5 923, 978
TVA KILLAR OCH EN TJEJ Vol. 5 1051, 1093, 1209
TVA KONUNGAR Vol. 5 859, 879, 927, 997
TVA KVINNOR Vol. 5 843, 923
TVA LEVANDE OCH EN DÖD Vol. 5 940, 1042
TVA MAN OM EN ÄNKA Vol. 5 840, 997
TVA MÄNNISKOR Vol. 5 940, 1173
TVA SKÖNA JUVELER Vol. 5 871, 1053, 1073, 1164
TVA SOLKIGA BLONDINER Vol. 5 1050, 1108
TVA TRAPPOR ÖVER GARDEN Vol. 5 907, 1173
TVAR POD MASKOU Vol. 10 18, 283, 359
TVAR V OKNE Vol. 10 32, 253, 576
TVAR ZA SKLEM Vol. 10 395, 517, 567
TVÄRBALK Vol. 5 880, 933, 1093
TVILLINGEBRODRENE Vol. 5 9, 97
TVOJE SRDCE INKOGNITO Vol. 10 122, 412, 460
TWARZ ANIOLA Vol. 1 198, 243, 303
TWARZA W TWARZ Vol. 1 220
12 MINUTEN NACH 12 Vol. 4 184, 365
„28" - AS, A Vol. 9 26, 117
28 OKTOBRIOU ORA 5.30 Vol. 3 185
24 CASA D"ZD Vol. 6 *page 53, page 63*
24 IOULIOU 1974 Vol. 3 159
24 STUNDEN AUS DEM LEBEN EINER FRAU Vol. 4 49, 321, 626

24 TIMER Vol. 5 252
29 IRMAOS Vol. 8 54
2019 DOPO LA CADUTA DI NEW YORK Vol. 7 504
27 HORAS Vol. 8 210, 230
20.000 DOLARES POR UN CADAVER Vol. 8 495, 773
TWILIGHT TIME Vol. 6 *page 317, page 320*
TWIN PAWNS Vol. 2 384
2 AAR EFTER Vol. 5 39, 150, 153, 173
2 X 2 NEHA 5 Vol. 9 117, 274, 326
2 X 2 IM HIMMELBETT Vol. 5 31, 39, 100, 195, 272
250 GRAMMA Vol. 5 606, 647, 651
201 KANARINIA Vol. 3 141
2000 PENGÖS FERFI, A Vol. 9 26, 117, 133
TY, CO W OSTREJ SWIECISZ BRAMIE Vol. 1 278, 303, 404
TYDEN V TICHEM DOME Vol. 10 154, 612, 621
TYKSAKKERNES OLGILDE Vol. 5 97
TYLKO UMARLY ODPOWIE Vol. 1 179, 187, 216
TYÖMIEHEN PÄIVÄKIRJA Vol. 5 541
TYÖN SANKARILAULU Vol. 5 335, 420
TYPE COMME MOI NE DEVRAIT JAMAIS MOURIR, UN Vol. 2 60
TYPHON SUR NAGASAKI Vol. 2 13, 388
TYRANNEI DES TODES Vol. 4 240
TYRANNENS FALD Vol. 5 89
TYRMA Vol. 7 1062
TYSIAC TALAROW Vol. 1 330, 401
TYSTNADEN Vol. 5 1052, 1080, 1130, 1196
TYSTNADEN HUS Vol. 5 923, 982
TYTÖN HUIVI Vol. 5 549, 592
TYTTO ASTUU ELÄMÄÄN Vol. 5 356, 386
TYTTO JA HATTU Vol. 5 374, 443
TYTTÖ KUUNSILLALTA Vol. 5 443, 549
TYTTÖ LÄHTEE KASARMIIN Vol. 5 462, 516
TYTTÖ TULI TALLON... Vol. 5 549, 579
TYVEN Vol. 5 97
TZAVELLES Vol. 3 149
TZIGANA, PASSIONE Vol. 7 196

U

U 9 WEDDIGEN Vol. 4 208, 424, 626
U-BOOTE WESTWÄRTS Vol. 4 491, 660, 728
U GORI RASTE ZELEN BOR Vol. 3 444
U KRESU DROGI Vol. 1 278, 404
U MREZI Vol. 3 451
U NAS V KOCOURKOVE Vol. 10 34, 256
U NAS V MECHOVE Vol. 10 313, 569, 577
U NAS V VEVEREK Vol. 10 92, 621
U OLUJI Vol. 3 444, 451
U POKLADNY STAL Vol. 10 524, 621
U RALJAMA ZIVOTA Vol. 6 *page 312, page 320*
U RASKORAKU Vol. 3 356, 381, 402
U SNEDENEHO KRAMU Vol. 10 142, 256, 586
U SUKOBU Vol. 3 338, 342
U SVATEHO ANTONICKA Vol. 10 453, 586
U SVATEHO MATEJE Vol. 10 168, 412, 580
U SVATEHO MATEJE, KDYZ SE SLUNKO ZASMEJE Vol. 10 524
U TELEFONU MARTIN Vol. 10 299, 380, 433
UANSTAENDIGE, DE Vol. 5 66, 269, 299
ÜB' IMMER TREU UND REDLICHKEIT Vol. 4 246, 752, 796
UBAT 39 Vol. 5 1003
ÜBER ALLES – DAS RECHT Vol. 4 779
ÜBER ALLES IN DER WELT Vol. 4 365, 574
ÜBER GANZ SPANIEN WOLKENLOSER HIMMEL Vol. 1 44, 68, 76
ÜBERBLICKT MAN DIE JAHRE Vol. 6 *page 161, page 163, page 165*
ÜBERFALL Vol. 4 73, 428, 452
ÜBERFALL AUF DEN EUROPA-EXPRESS, DER Vol. 4 67, 356
ÜBERFALL IM KAUKASUS, EIN Vol. 4 254
ÜBERFLÜSSIGE MENSCHEN Vol. 4 14, 25, 321, 400
ÜBERGANG, DER Vol. 1 14
ÜBERLEBENDEN, DIE Vol. 1 29, 87
ÜBERMUT IM SALZKAMMERGUT Vol. 3 395
ÜBERRASCHENDE SÄUGLING, DER Vol. 4 517, 525, 558
UBETAENKSOMME ELSKER, DEN Vol. 5 26, 125, 185
UBICA NA ODSUSTVU Vol. 3 346
UBIJ ME NEZNO Vol. 3 348, 387
UBISTVO NA PODMUKAO I SVIREP NACIN I IZ NISKIH POBUDA Vol. 3 389, 441
UBOJ Vol. 1 317
UBRANIE PRAWIE NOWE Vol. 1 323, 389 Vol. 6 *page 220*
UCCELACCI E UCCELLINI Vol. 7 128, 333, 388, 1035
UCCELLI D'ITALIA Vol. 7 123, 350, 504, 1161
UCCELLO DALLE PIUME DI CRISTALLO, L' Vol. 7 1101
UCCELLO MIGRATORE, L' Vol. 7 567
UCCIDETE IL VITELLO GRASSO E ARROSTITELO Vol. 7 524

UCHO Vol. 10 197, 289, 394
UCIEC JAK NAJBLIZEJ Vol. 1 202, 204, 227 Vol. 6 *page 246*
UCITEL ORIENTALNICH JAZYKU Vol. 10 92, 363
UCNA LETA IZUMITELJA POLZA Vol. 6 *page 305, page 307*
UCZTA BALTAZARA Vol. 1 232, 250, 298, 331 Vol. 6 *page 207*
UD I DEN KOLDE SNE Vol. 5 24
UDAR"T Vol. 6 *page 24, page 61*
UDEN EN TRAEVIL Vol. 5 199, 225, 304
UDEN NATTEGN Vol. 5 97
UDHETIM NE PRANVERE Vol. 3 2, 9, 18
UDJI, AKO HOCES Vol. 3 351
UDOLI KRASNYCH ZAB Vol. 10 47, 286, 380
UDOLI VCEL Vol. 10 139, 425, 567
UDOLIE VECNYCH KARAVAN Vol. 10 273, 451, 603
UDSKEJELSE, EN Vol. 5 81
UDVARI LEVEGÖ Vol. 9 220
UF-ONI JSOU TADY Vol. 10 233, 325, 495
UFA-MÄRCHEN Vol. 4 265, 774
UFFICIALE NON SI ARRENDE MAI, NEMMENO DI FRONTE ALL'EVIDENZA... Vol. 7 1205
UGLER I MOSEN Vol. 5 672, 726
UHELDIG DEBUT, EN Vol. 5 83
UHELDIG SKYGGE, EN Vol. 5 92
UHELDIG TYVEKNAEGT, EN Vol. 5 83
UHOL POHLADU Vol. 10 145, 148, 459
UHRENLADEN, DER Vol. 4 329, 774
UJ FÖLDESUR, AZ Vol. 9 15, 71, 164, 322
UJ GILGAMES Vol. 9 26, 117, 183
UJ ROKON, AZ Vol. 9 71, 322
UJED ANDJELA Vol. 6 *page 313, page 314*
UJJASZÜLETES Vol. 9 7
UJRAELÖK Vol. 9 146, 148
UJSZÜLÖTT APA, AZ Vol. 9 136
UKA I BJESHKAVE TE NEMUNA Vol. 3 357, 366
UKJENT MANN Vol. 5 723, 776
UKJENTES MARKED, DE Vol. 5 664, 800
UKRADENA BITVA Vol. 10 152, 238, 286
UKRADENA VZDUCHOLOD Vol. 10 283, 286, 387, 404, 422
UKRIZOVANA Vol. 10 257, 536, 605
UKRYTY W SLONCU Vol. 6 *page 243, page 257, page 260*
ULAN KS. JOSEFA Vol. 1 176, 278, 308, 404
ULANI, ULANI CHLOPSCY MALOWANI Vol. 1 176, 278
ULICA GRANICZNA Vol. 1 366
ULICE BEZ MENA Vol. 10 93, 544, 598
ULICE ZPIVA Vol. 10 168, 564,
ULICKA HRICHU A LASKY Vol. 10 34, 272
ULICKA V RAJI Vol. 10 8, 167, 168
ULICNI BOZESTVA Vol. 3 71
ULICNICE Vol. 10 167, 256, 412
ULISSE Vol. 7 338, 781, 999, 1136, 1147
ULKOSAARELAISET Vol. 5 609
ULLA MIN ULLA Vol. 5 891, 997
ULLABELLA Vol. 5 2, 7
ULLI UND MAREI Vol. 4 13, 15, 167, 317
ULOGA MOJE PORODICE U SVJETSKOJ REVOLUCIJI Vol. 3 360, 409
ULOUPENA HRANICE Vol. 10 122, 460
ULTIMA AMICA, L' Vol. 7 1189
ULTIMA AVENTURA DEL ZORRO, LA Vol. 8 749
ULTIMA AVENTURA, L' Vol. 7 48, 466, 756, 1113
ULTIMA CANZONE, L' Vol. 8 537
ULTIMA CARROZZELLA, L' Vol. 7 443, 1027
ULTIMA CARTA, L' Vol. 7 1169
ULTIMA CITA, LA Vol. 8 640
ULTIMA DOGARESSA, L' Vol. 7 1154
ULTIMA DONNA, L' Vol. 2 142 Vol. 7 197, 1145
ULTIMA FALLA, LA Vol. 7 40, 1177 Vol. 8 778
ULTIMA FIABA, L' Vol. 7 1064
ULTIMA FIAMMA Vol. 7 40, 1177
ULTIMA FRONTIERA A MORTII Vol. 3 240, 272, 308
ULTIMA GIOVINEZZA Vol. 2 29 Vol. 7 48, 838
ULTIMA IMPRESA, L' Vol. 7 277
ULTIMA INVENZIONE, L' Vol. 7 81
ULTIMA JUGADA, LA Vol. 8 590
ULTIMA LIVREA, L' Vol. 7 1070
ULTIMA NEMICA, L' Vol. 7 16, 830, 851
ULTIMA NOAPTE A COPILARIEI Vol. 3 208, 261
ULTIMA NOAPTE A SINGURATATII Vol. 3 240, 269
ULTIMA NOAPTE DE DRAGOSTE Vol. 3 211, 256, 280
ULTIMA NOTTE D'AMORE, L' Vol. 7 1062
ULTIMA ORGIA DEL 3. REICH, L' Vol. 7 276
ULTIMA PRIMAVERA, L' Vol. 7 671
ULTIMA RECITA DI ANNA PARNELL, L' Vol. 7 843, 921
ULTIMA SENORA ANDERSON, LA Vol. 8 438, 521
ULTIMA SENTENZA, L' Vol. 7 603
ULTIMA SERA DI CARNEVALE, L' Vol. 7 1189
ULTIMA SERATA DEI DIAVOLI VOLANTI, L' Vol. 7 548

ULTIMA SPIAGGIA, L' Vol. 7 988
ULTIMA VIOLENZA, L' Vol. 7 443, 637, 638
ULTIMA VOLTA INSIEME, L' Vol. 7 311, 648
ULTIMA VOLTA, L' Vol. 7 909
ULTIMAS BANDERAS, LAS Vol. 8 449, 725
ULTIMAS DE FILIPINAS, LAS Vol. 8 403
ULTIMAS TARDES CON TERESA Vol. 8 245, 587, 594, 605
ULTIMATUM Vol. 2 273, 488, 581, 646 Vol. 6 *page 213, page 224, page 251*
ULTIME AVVENTURE DI GALAOR, LE Vol. 7 104
ULTIME LETTERE DI JACOPO ORTIS, LE Vol. 7 559
ULTIMELE ZILE ALE VERII Vol. 3 247, 300, 327
ULTIMI CINQUE MINUTI, GLI Vol. 7 796
ULTIMI DELLA STRADA, GLI Vol. 7 16, 809, 1155
ULTIMI FILIBUSTIERI, GLI Vol. 7 91, 809, 838, 897, 1195
ULTIMI GIORNI DI POMPEI, GLI Vol. 7 21, 53, 230, 219, 390, 837, 955, Vol. 8 261, 438
ULTIMI ZAR, GLI Vol. 7 198, 640, 1116
ULTIMO ADDIO Vol. 7 201
ULTIMO ADDIO, L' Vol. 7 1177
ULTIMO AMANTE, L' Vol. 7 1136
ULTIMO AMORE, L' Vol. 7 725
ULTIMO BALLO, L' Vol. 7 493, 792
ULTIMO CABALLO, EL Vol. 8 298, 401
ULTIMO CACCIATORE, L' Vol. 7 852, 1046
ULTIMO CANTO, L' Vol. 7 37
ULTIMO CAVALIERE, L' Vol. 7 304, 1073
ULTIMO COMBATTIMENTO, L' Vol. 7 941, 966
ULTIMO CONTRABANDISTA, EL Vol. 8 357, 427, 542, 604
ULTIMO CUPLE, EL Vol. 8 207, 298, 336
ULTIMO DEI BERGERAC, L' Vol. 7 217, 796
ULTIMO DEI BORGIA, L' Vol. 7 196
ULTIMO DEI COGNAC, L' Vol. 7 390, 1052
ULTIMO DEI FRONTIGNAC OVVERO IL ROMANZO DI UN GIOVANE POVERO, L' Vol. 7 1038
ULTIMO DEI VICHINGHI, L' Vol. 7 1162
ULTIMO DESEO Vol. 8 226, 551
ULTIMO DIA Vol. 2 646 Vol. 8 295, 311, 607
ULTIMO DIA DE LA GUERRA, EL Vol. 8 204
ULTIMO ENCUENTRO Vol. 8 348, 772
ULTIMO GUAPPO, L' Vol. 7 474
ULTIMO GUERRIERO, L' Vol. 7 694
ULTIMO INCONTRO, L' Vol. 7 196
ULTIMO KAMIKAZE, EL Vol. 8 214, 296
ULTIMO LORD, L' Vol. 7 53, 796
ULTIMO MERCENARIO, L' Vol. 8 426
ULTIMO MOHICANO, EL Vol. 8 523, 732
ULTIMO MONDO Vol. 7 738
ULTIMO PARADISO, L' Vol. 7 1044
ULTIMO REY DE LOS INCAS, EL Vol. 8 450, 524
ULTIMO RICORDO, L' Vol. 7 1189
ULTIMO ROMANZO DI GIORGIO BELFIORE, L' Vol. 7 940
ULTIMO SABADO, EL Vol. 8 470, 492
ULTIMO SAPORE DELL'ARIA, L' Vol. 7 318
ULTIMO SCUALO, L' Vol. 7 124, 1086, 1165
ULTIMO SCUGNIZZO, L' Vol. 7 792, 1036
ULTIMO SOGNO, L' Vol. 7 264, 701
ULTIMO SOLDADO, O Vol. 8 95, 117, 193
ULTIMO TANGO A PARIGI, L' Vol. 7 1042, 1101
ULTIMO TANGO A ZAGAROL Vol. 7 1147
ULTIMO TANGO EN MADRID Vol. 8 702
ULTIMO TRAVESTIMENTO, L' Vol. 7 796
ULTIMO UOMO DI SARA, L' Vol. 7 741
ULTIMO VERANO, EL Vol. 8 551
ULTIMO VIERJE, EL Vol. 8 552
ULTIMOS DE FILIPINAS, LOS Vol. 8 298, 311, 449
ULTIMOS DIAS DE POMPEYA, LOS Vol. 8 261, 438
ULTIMOS GOLPES DE EL TORETE, LOS Vol. 8 426
ULTIMUL CARTUS Vol. 3 218, 235, 272
ULTIMUL MOHICAN Vol. 3 269
ULVEJAEGERNE Vol. 5 24, 83
ULVETID Vol. 5 23, 125, 249
ULVOVA MYLLÄRI Vol. 5 410, 585, 651
ULYKKENS GALOSCHER Vol. 5 97
ULYSSE Vol. 2 778
ULZANA Vol. 1 8, 96
UM 500 000 MARK Vol. 4 192
UM 9 KOMMT HARALD Vol. 4 347, 370, 491
UM DAS GLÜCK BETROGEN Vol. 4 183
UM DAS MENSCHENRECHT Vol. 4 647, 652, 689, 787
UM DEN SOHN Vol. 4 208, 390
UM DES LEBENS GLÜCK BETROGEN Vol. 4 356, 779
UM EIN BISSCHEN GLÜCK Vol. 10 256, 524
UM EIN HAAR... Vol. 4 213, 626
UM EIN WEIB – IN EWIGER NACHT Vol. 4 514

UM EINE NASENLÄNGE **Vol. 4** 292, 298
UM EINES WEIBES EHRE **Vol. 4** 25, 329
UM HAARESBREITE **Vol. 4** 183
UM KOPF UND KRAGEN **Vol. 4** 25, 608
UM LIEBE UND EHRE **Vol. 4** 603
UM LIEBE UND THRON **Vol. 4** 344, 527, 756
UM MILLON EN LA BASURA **Vol. 8** 725
UM RECHT UND EHRE **Vol. 4** 558
UM RECHT UND LIEBE **Vol. 4** 67, 764, 787
UMANITA **Vol. 7** 796
UMANOIDE, L' **Vol. 7** 585
UMARLI RZUCAJA CIEN **Vol. 1** 159, 203, 221 **Vol. 6** *page 244*
UMBERTO D. **Vol. 7** 21, 709
UMILLE EROE **Vol. 7** 1189
UMIRAJ SAMO V KRAEN SLUCAJ **Vol. 3** 38 **Vol. 6** *page 26, page 58*
UMLCENE RTY **Vol. 10** 168, 585
UM'S TÄGLICHE BROT **Vol. 6** *page 181*
UMWEGE DES SCHÖNEN KARL, DIE **Vol. 4** 373, 626
UMWEGE ZU DIR **Vol. 4** 85, 714
UMWEGE ZUM GLÜCK **Vol. 4** 123, 365, 544
UN AUGUST IN FLACARI **Vol. 3** 222, 223, 225, 237, 265, 276, 288
UN, DEUX, TROIS **Vol. 2** 669, 863
UN, DEUX, TROIS, QUATRE **Vol. 2** 13, 184
UN, DOS, TRES AL ESCONDITE INGLES **Vol. 8** 348, 785
UN, DOS, TRES ... ENSAIMADES Y RES MES **Vol. 8** 232
UN-ERHÖRTE FRAU, DIE **Vol. 4** 246, 361
UNA **Vol. 6** *page 308, page 320*
UNAS PAGINAS EN NEGRO **Vol. 8** 537
UNBEDEUTENDE FRAU, EINE **Vol. 4** 232, 673
UNBEKANNTE, DIE **Vol. 4** 82, 382 **Vol. 6** *page 183*
UNBEKANNTE GAST, DER **Vol. 4** 452, 774
UNBEKANNTE MORGEN, DAS **Vol. 4** 733
UNBEKANNTE TÄTER, DER **Vol. 4** 115, 742
UNBERÜHRTE FRAU, DIE **Vol. 4** 211, 219
UNBESIEGBAREN, DIE **Vol. 1** 40, 56, 105
UNBEWOHNTE HAUS, DAS **Vol. 4** 217, 277, 503
UNCAS, EL FIN DE UNA RAZA **Vol. 8** 450
UND ALLES WEGEN MARIETTA **Vol. 6** *page 156, page 171, page 174*
...UND DAS IST DIE HAUPTSACHE **Vol. 4** 298, 321
... UND DEINE LIEBE AUCH **Vol. 1** 7, 84, 130
UND DENNOCH KAM DAS GLÜCK **Vol. 4** 461, 522
UND DENNOCH WARD ES MORGEN **Vol. 4** 91
UND DER MOND LACHT DAZU **Vol. 4** 135
...UND DIE MUSIK SPIELT DAZU **Vol. 4** 24, 30, 347
... UND DIE MUSIK SPIELT DAZU **Vol. 7** 101
UND DU, MEIN SCHATZ, FÄHRST MIT **Vol. 4** 354
...UND ES LEUCHTET DIE PUSZTA **Vol. 4** 536
...UND ES LOCKT EIN RUF AUS SÜNDIGER WELT **Vol. 4** 176, 211
UND JETZT WIRD GEARBEITET **Vol. 4** 382
UND NÄCHSTES JAHR AM BALATON **Vol. 6** *page 165, page 171, page 174*
UND SIE SINGT DOCH **Vol. 4** 361, 558
UND SO EIN GLÜCK KANNST DU NUR HABEN **Vol. 4** 87, 239
UND WENN DER GANZE SCHNEEE VERBRENNT **Vol. 3** 395
UND WENN'S NUR EINER WÄR **Vol. 1** 25, 56 **Vol. 6** *page 153, page 154*
UND WER KÜSST MICH ? **Vol. 4** 255, 323
UND WIEDER 48 ! **Vol. 1** 77, 101, 107
UNDENRIGSKORRESPONDENTEN **Vol. 5** 99
UNDER BLINKFYRETS **Vol. 5** 295
UNDER BYENS TAGE **Vol. 5** 2, 24
UNDER DEN GAMLE FANE **Vol. 5** 202
UNDER DITT PARASOLL **Vol. 5** 843, 1104, 1142
UNDER EN STEINHIMMEL **Vol. 5** 813
UNDER FALSK FLAG **Vol. 5** 81
UNDER FALSKA FLAGG **Vol. 5** 840, 841
UNDER KAERLIGHEDENS AAG **Vol. 5** 256
UNDER MINDERNES TRAE **Vol. 5** 58
UNDER SAVKLINGENS TAENER **Vol. 5** 58
UNDER SKAEBNEN HJUL **Vol. 5** 58
UNDER SÖDRA KORSET **Vol. 5** 1080
UNDER SVÄLLANDE SEGEL **Vol. 5** 1029
UNDER THE RED ROBE **Vol. 2** 656
UNDERBARA LÖGNEN, DEN **Vol. 5** 1080, 1156
UNDICESIMO COMANDAMENTO, L' **Vol. 7** 296
UNDSKYLD VI ER HER **Vol. 5** 11, 231, 259, 299
UNE CHANTE, L'AUTRE PAS, L' **Vol. 2** 45
UNE DE LA CAVALERIE **Vol. 2** 871
UNE ET L'AUTRE, L' **Vol. 2** 49, 288
UNEHELICHEN, DIE **Vol. 4** 247, 461

UNELMA KARJAMAJALLA Vol. 5 357, 379, 581
UNENDLICHE WEG, DER Vol. 4 344, 660
UNENTSCHULDIGTE STUNDE, DIE Vol. 4 437
UNERKANNTE, DER Vol. 4 593, 626
UNFUG DER LIEBE Vol. 4 166, 725
UNG DAM MED TUR Vol. 5 863, 1032
UNG FLUKT Vol. 5 674, 724
UNG FRUE FORSVUNNET Vol. 5 762, 831
UNG KAERLIGHED Vol. 5 18, 41
UNG LEG Vol. 5 7
UNG MAN SÖKER SÄLLSKAP Vol. 5 871, 1073
UNG OCH KÄR Vol. 5 840, 994
UNG OG FORELSKET Vol. 5 83
UNG SOMMAR Vol. 5 862, 1042
UNGA GREVEN TAR FLICKAN OCH PRISET Vol. 5 1210
UNGA HJÄRTAN Vol. 5 899, 979, 1010
UNGARISCHE RHAPSODIE Vol. 4 288, 325
UNGDOM Vol. 5 891, 1031
UNGDOM AV I DAG Vol. 5 863, 1032
UNGDOM I BOJOR Vol. 5 871, 923
UNGDOM I FARA Vol. 5 876, 1055, 1073
UNGE BLOD, DET Vol. 5 58
UNGE HJERTER Vol. 5 711, 825
UNGE PIGER FORSVINDER I KOBENHAVN Vol. 5 180, 199, 201
UNGE VILJER Vol. 5 662, 692
UNGEBETENE GAST, DER Vol. 6 *page 163, page 169, page 175*
UNGEKLÄRTER FALL, EIN Vol. 4 656
UNGEN Vol. 5 674, 692, 701, 770, 824
UNGETREUE ECKEHART, DER Vol. 4 149, 292, 517, 558, 609, 660
UNGEWÖHNLICHER TAG, EIN Vol. 1 2, 86
UNGKARL OG AEGTEMAND Vol. 5 262
UNGKARLELIV Vol. 5 83
UNGKARLSHOTELLET Vol. 5 1086
UNGKARLSPAPPAN Vol. 5 840, 841
UNGKARLSPARADISET Vol. 5 870, 983
UNGLÜCK IN DER KINDERSTUBE, EIN Vol. 4 514
UNGT BLOD Vol. 5 843, 923, 1044
UNHEILIGE SOPHIA, DIE Vol. 1 19, 85
UNHEIMLICHE CHINESE, DER Vol. 4 212, 545, 571
UNHEIMLICHE, DER Vol. 4 168, 321, 558
UNHEIMLICHE GESCHICHTEN Vol. 4 192, 288, 546, 626
UNHEIMLICHE HAUS, DAS Vol. 4 159, 488
UNHEIMLICHE HELENE, DIE Vol. 4 110, 517
UNHEIMLICHE LICHT, DAS Vol. 4 240
UNHEIMLICHE NACHT, DIE Vol. 4 608, 630
UNHEIMLICHE WANDLUNG DES ALEX ROSCHER, DIE Vol. 4 110, 269, 285
UNHEIMLICHEN WÜNSCHE, DIE Vol. 4 15, 339, 475
UNICA LEGGE IN CUI CREDO, L' Vol. 7 611
UNICO PECCATO, L' Vol. 7 133
UNICO TESTIGO, EL Vol. 8 723
UNIFORMES ET GRANDES MANOEUVRES Vol. 2 128, 514, 593
UNKNOWN LOVE Vol. 2 384
UNMÖGLICHE FRAU, DIE Vol. 4 135, 382, 517
UNMÖGLICHE HERR PITT, DER Vol. 4 149, 181, 536, 660
UNMÖGLICHE LIEBE Vol. 4 306, 462
UNMÖGLICHE WETTE, EINE Vol. 4 323
UNMORAL Vol. 4 119, 215, 681
ÜNNEPI VACSORA Vol. 9 26, 134, 183
ÜNNEPNAPOK Vol. 9 149, 259, 263
UNO A UNO SIN PIEDAD Vol. 8 396
UNO CONTRO L'ALTRO ... PRATICAMENTE AMICI Vol. 7 208, 256, 617
UNO, DOS, TRES ... DISPARA OTRA VEZ Vol. 8 407
UNO E DI TROPPO Vol. 7 1189
UNOS Vol. 10 313, 352
UNOS BANKERE FUXE Vol. 10 167, 453, 586
UNOS MORAVANKY Vol. 10 153, 344, 385
UNRUHE Vol. 6 *page 153*
UNS ET LES AUTRES, LES Vol. 2 110, 491 Vol. 6 *page 93*
UNSCHULD Vol. 4 85, 321
UNSCHULD OHNE KLEID, DIE Vol. 4 270, 599, 644
UNSCHULD VOM LANDE, DIE Vol. 4 232, 239
UNSER FRÄULEIN DOKTOR Vol. 4 234, 696, 738
UNSER KLEINER JUNGE Vol. 4 270, 582
UNSER KURZES LEBEN Vol. 6 *page 161, page 165, page 170*
UNSER STILLER MANN Vol. 1 18, 63, 124
UNSER TÄGLICH BROT Vol. 1 3, 117 Vol. 4 219, 424
UNSERE EMDEN Vol. 4 82, 119, 287, 544, 643, 775
UNSERE KLEINE FRAU Vol. 4 411, 464 Vol. 7 639, 756, 796
UNSICHTBARE FRONT, DIE Vol. 4 232, 270, 462
UNSICHTBARE GAST, DER Vol. 4 208
UNSICHTBARE MENSCH, DER Vol. 4 71, 780

UNSICHTBARE VISIER, DAS Vol. 1 12, 17, 62, 70, 98
UNSICHTBARE VISIER - INSEL DES TODES, DAS Vol. 6 *page 157, page 168, page 172*
UNSICHTBARER GEHT DURCH DIE STADT, EIN Vol. 4 119, 270
UNSTERBLICHE HERZ, DAS Vol. 4 435, 462, 734
UNSTERBLICHE LIED, DAS Vol. 4 620
UNSTERBLICHE LUMP, DER Vol. 4 266, 288, 574
UNSTERBLICHER WALZER Vol. 4 85, 620
UNTER AUSSCHLUSS DER ÖFFENTLICHKEIT Vol. 4 124, 247, 436, 461
UNTER BLUTSCHULD Vol. 4 21
UNTER DEM PANTOFFEL Vol. 4 186
UNTER DEN BRÜCKEN Vol. 4 491, 736
UNTER DER DORNENKRONE Vol. 4 366
UNTER DER LATERNE Vol. 4 247, 461
UNTER DER SCHWARZEN STURMFAHNE Vol. 4 132, 682
UNTER FALSCHER FLAGGE Vol. 4 298
UNTER FALSCHER MASKE Vol. 4 114
UNTER GESCHÄFTSAUFSICHT Vol. 10 167, 453
UNTER HEISSEM HIMMEL Vol. 4 266, 574, 729
UNTER RÄUBERN UND BESTIEN Vol. 4 288, 779
UNTER VIER AUGEN Vol. 4 329, 774
UNTERGANG DER EMMA, DER Vol. 1 45, 70
UNTERM BIRNBAUM Vol. 1 15, 55, 57, 59
UNTERNEHMEN MICHAEL Vol. 4 13, 574
UNTERSCHLAGUNG, DIE Vol. 4 247, 461
UNTERTAN, DER Vol. 1 3, 105, 107, 129
UNTERWEGS NACH ATLANTIS Vol. 1 18, 80, 128
UNTERWEGS ZU LENIN Vol. 1 14, 114
UNÜBERWINDLICHE, DER Vol. 4 287, 379, 384, 458
UNUS. DER FÜRST DER BERGE Vol. 4 270, 443, 471
UNUS. DER WEG IN DIE WELT Vol. 4 270, 443, 471
UNVERBESSERLICHE BARBARA, DIE Vol. 1 1, 69, 116
UNVOLLKOMMENE LIEBE, DIE Vol. 4 8, 491
UNWIDERSTEHLICHE, DER Vol. 4 89, 246
UNWILLKOMMENE ZAUNGÄSTE Vol. 4 635
UOMA DI PAGLIA, L' Vol. 7 86
UOME DEL SANGUE, L' Vol. 7 40
UOMINI, CHE MASCALZONI ..., GLI Vol. 7 756, 796, 1036, 1116
UOMINI E CIELI Vol. 7 106
UOMINI E LUPI Vol. 7 923, 1054
UOMINI E NOBILUOMINI Vol. 7 693
UOMINI GIALLI Vol. 7 144
UOMINI NERI, GLI Vol. 7 196
UOMINI NO Vol. 7 340, 524, 1034, 1112
UOMINI NON GUARDANO IL CIELO, GLI Vol. 7 501
UOMINI NON SONO INFRATI, GLI Vol. 7 40, 725
UOMINI OMBRA Vol. 7 113
UOMINI SENZA DOMANI Vol. 7 372
UOMINI SI NASCE POLIZIOTTI SI MUORE Vol. 7 694
UOMINI SONO NEMICI, GLI Vol. 7 891
UOMINI SUL FONDO Vol. 7 71, 106, 113, 154, 170, 250
UOMO CHE DORMI 130 ANNI Vol. 7 1190
UOMO CHE RIDEVA, L' Vol. 7 81
UOMO CHE SORRIDE, L' Vol. 7 453, 493
UOMO CHE VIDE LA MORTE, L' Vol. 7 878
UOMO DA BRUCIARE, UN Vol. 7 913, 1056
UOMO DA RISPETTARE, UN Vol. 7 333
UOMO DAI CALZONI CORTE, L' Vol. 7 796
UOMO DAL COLPO PERFETTO, L' Vol. 8 493
UOMO DAL GUANTO GRIGIO, L' Vol. 7 1054
UOMO DAL MANTELLO VERDE, L' Vol. 7 53
UOMO DALLA CROCE, L' Vol. 7 58, 636, 756, 1111
UOMO DALLA LINGUA MOZZA, L' Vol. 7 91
UOMO DALL'ARTIGLIO, L' Vol. 7 361, 796
UOMO DEL ROMANZO, L' Vol. 7 796
UOMO DELLA FORESTA, L' Vol. 7 108
UOMO DELLA LEGIONE, L' Vol. 7 359, 951, 1111
UOMO DELLA MANCHA, L' Vol. 7 988
UOMO DELLA MIA VITA, L' Vol. 7 257
UOMO DELLA ROSA Vol. 7 630
UOMO DELLA VALLE MALEDETTA, L' Vol. 7 468 Vol. 8 400
UOMO DI PAGLIA, L' Vol. 7 401
UOMO DOPO L'UOMO, L' Vol. 7 505
UOMO E IL DIAVOLO, L' Vol. 7 535
UOMO FACILE, UN Vol. 7 514
UOMO GIALLO, L' Vol. 7 1189
UOMO IN GINOCCHIO, UN Vol. 7 567
UOMO, LA BESTIA E LA VIRTU, L' Vol. 7 373
UOMO MECCANICO, L' Vol. 7 296
UOMO PAPPAGALLO, L' Vol. 7 131

UOMO PIU ALLEGRO DI VIENNA, L' **Vol. 7** 53, 219, 665
UOMO PUMA, L' **Vol. 7** 1194
UOMO, UN **Vol. 7** 347
UOMO VENUTO DAL MARE, L' **Vol. 7** 638, 756
UONI **Vol. 3** 73 **Vol. 6** *page 40, page 53*
UP FROM THE BEACH **Vol. 2** 875
UPAL **Vol. 1** 298, 337, 338, 408
UPIOR **Vol. 1** 237
UPIR Z FERATU **Vol. 10** 299, 425, 571
UPLNE VYRIZENY CHLAP **Vol. 10** 193, 296, 433
UPPAT IGEN **Vol. 5** 870, 923
UPPDRAG I KOREA **Vol. 5** 1208
UPPDRAGET **Vol. 5** 852, 896, 1153
UPPSAGD **Vol. 5** 840, 878
UPRCHLIK **Vol. 10** 3, 286, 309, 387
UR KÄRLEKENS SPRAK **Vol. 5** 1217
URA PRANE KESHTJELLES **Vol. 6** *page 11, page 15*
URAGANO **Vol. 7** 304, 1113
URAGANO AI TROPICI **Vol. 7** 830, 1054, 1136
URAMISTEN **Vol. 9** 191, 204, 242
URDEMA CANZONA MIA, L' **Vol. 7** 752
URGIA **Vol. 3** 209, 281, 335
URGIA CEREASCA **Vol. 3** 234
URI MURI **Vol. 9** 71, 119, 179, 237
URI VILAG **Vol. 9** 71, 208
URIEL ACOSTA **Vol. 4** 288, 558
URILANY SZOBAT KERES **Vol. 9** 178, 237
URLATORI ALLA SBARRA **Vol. 7** 1054
URLAUB AUF EHRENWORT **Vol. 4** 13, 574
URLAUB OHNE DICH **Vol. 1** 73, 86, 110
URLO DALLE TENEBRE, UN **Vol. 7** 289
URMARIREA **Vol. 3** 307, 311
URODA ZYCIA **Vol. 1** 278, 346
URODZINY MATYLDY **Vol. 1** 202, 210, 254
URODZINY MLODEGO WARSZAWIAKA **Vol. 6** *page 236, page 241, page 253*
UROK NA ISTORIJATA **Vol. 3** 53 **Vol. 6** *page 59*
URSULA **Vol. 1** 12, 18, 59
URSULA - FLICKAN I FINNSKOGARNA **Vol. 5** 871, 1003, 1042
URSULE ET GRELU **Vol. 2** 322, 814
URSUS **Vol. 7** 107, 759 **Vol. 8** 541, 725
URSUS NELLA TERRA DI FUOCO **Vol. 7** 1147
URSUS NELLA VALLE DEI LEONI **Vol. 7** 1027
URTAIN, EL REY DE LA SELVA ... O ASI **Vol. 8** 348
URTEIL DES ARZTES, DAS **Vol. 4** 423
URWALDSYMPHONIE **Vol. 4** 98, 141
U.S.A. VIOLACION Y VENGANZA **Vol. 8** 466, 536
USIJANJE **Vol. 3** 350, 424
USKYLDIG DOMT **Vol. 5** 97
USMEV DIABLA **Vol. 10** 32, 148, 542
USMEVAVA ZEM **Vol. 10** 54, 413
USMIECH LOSU **Vol. 1** 174, 176, 222
USMIECHY ZYCIA **Vol. 1** 322, 376
USPORENO KRETANJE **Vol. 3** 412, 437
USTED PUEDE SER UN ASESINO **Vol. 8** 524, 725
USTED TIENE OJOS DE MUJER FATAL **Vol. 8** 583, 630
USURAIO, L' **Vol. 7** 40, 476
USVOJENJE-FERIJE-PROBNI ROK **Vol. 3** 430
USYNLIGE HAER, DEN **Vol. 5** 2, 7, 89
UT AV MORKET **Vol. 5** 674, 831
UT MINE STROMTID **Vol. 4** 131, 628
UTAZAS A KOPONYAM KÖRÜL **Vol. 9** 134, 259, 265
UTAZAS JAKABBAL **Vol. 9** 152, 154
UTE BLASER SOMMARVIND **Vol. 5** 831, 843, 1042
UTEK **Vol. 10** 428, 473
UTEK DO VETRU **Vol. 10** 250, 462
UTEK S CEZAREM **Vol. 10** 233, 325, 541
UTEK ZE STINU **Vol. 10** 182, 211, 313
UTEKAJME, UZ IDE ! **Vol. 10** 16, 210, 454
UTEKY DOMU **Vol. 10** 395, 516, 517
UTIMAS HORAS, LAS **Vol. 8** 396, 772
UTKÖZBEN **Vol. 6** *page 230, page 238* **Vol. 9** 6, 202, 280
ÜTKOZET BEKEBEN **Vol. 9** 237, 241
UTOLSO DAL, AZ **Vol. 9** 236, 315
UTOLSO EJSZAKA, AZ **Vol. 9** 77
UTOLSO ELÖTTI EMBER, AZ **Vol. 9** 265, 285, 314
UTOLSO ELÖTTI ITELET **Vol. 9** 50, 169, 253
UTOLSO KEZIRAT, AZ **Vol. 9** 51, 297, 316
UTOLSO KÖR, AZ **Vol. 9** 36, 84, 285
UTOLSO VACSORA, AZ **Vol. 9** 67, 122
UTOLSO VERECZKEY, AZ **Vol. 9** 132, 133, 270
UTOPIA **Vol. 2** 74
UTOSZEZON **Vol. 9** 134, 259, 265
UTRO **Vol. 5** 29, 38
UTRO NAD RODINATA **Vol. 3** 31, 61, 66, 120, 123
UTROTO E NEPOVTORIMO **Vol. 3** 72, 120 **Vol. 6** *page 37*
UTRPENI MLADEHO BOHACKA **Vol. 10** 385, 522, 621
UTRPENI SEDE SESTRY **Vol. 10** 7
UTRPENIM KE SLAVE **Vol. 10** 198, 243, 526
UTSTÖTTA, DE **Vol. 5** 841

UTTÖRÖK **Vol. 9** 241
UTVANDRARNA **Vol. 5** 1052, 1153, 1192
UUNO TURHAPURO **Vol. 5** 570, 617
UUNO TURHAPURO ARMEJAN LEIVISSÄ **Vol. 5** 309, 345, 424
UUNO TURHAPURO MENETTÄÄ MUISTINSA **Vol. 5** 345, 424
UUNO TURHAPURON AVIOKRIISI **Vol. 5** 345, 409, 424
UUNO TURHAPURON MUISTI PALAILEE PÄTKITTÄIN **Vol. 5** 345, 424, 456, 546
UUTEEN ELÄMÄÄN **Vol. 5** 349, 493
UWAGA - SZPIEG **Vol. 1** 176, 308
UWIEDZIONA **Vol. 1** 364, 384, 407
UZ BENCE **Vol. 9** 71, 208
UZ SE NEBOJIM **Vol. 10** 13, 250, 508
UZ ZASE SKACU PRES KALUZE **Vol. 10** 197, 289
UZAVRELI GRAD **Vol. 3** 361, 371
UZAVRENY OKRUH **Vol. 10** 44, 226, 541
ÜZENET A VOLGARPARTROZ **Vol. 9** 76
UZICKA REPUBLIKA **Vol. 3** 389, 409, 424
UZROK SMRTI NE POMINJATI **Vol. 1** 398

V

V BLOUZNENI **Vol. 10** 345, 524
V BLUDISKU PAMATI **Vol. 10** 210, 419, 454
V CIZIM REVIRU **Vol. 10** 61, 150, 256, 460, 532
V GORI RASTE ZELEN BOR **Vol. 3** 451
V HLAVNI ROLI OLDRICH NOVY **Vol. 10** 129, 313, 327
V HODINE DVANASTEJ **Vol. 10** 287, 576
V HORACH DUNI **Vol. 10** 256, 345, 532
V KAZDEM POKOJI ZENA **Vol. 10** 144, 172, 327
V KAZDOM POCASI **Vol. 10** 32, 263, 282
V KRAJA LJATOTO **Vol. 6** *page 52*
V KRAJA NA LJATOTO **Vol. 3** 31, 70, 109, 124
V MESICI LASKY **Vol. 10** 45
V MESTE PLNOM DAZDNIKOV **Vol. 10** 19, 69, 542
V NOKTITE NA POROKA **Vol. 3** 33
V PANSKEM STAVU **Vol. 10** 53, 345
V PATEK RANO **Vol. 10** 359, 411
V PIATOK TRINASTEHO **Vol. 10** 268, 380
V PODSTATE JSME NORMALNI **Vol. 10** 41, 165, 550
V POKUSENI **Vol. 10** 256, 460
V PROUDECH **Vol. 10** 359, 388
V SEST RANO NA LETISTI **Vol. 10** 52, 179, 502
V TIHA VECER **Vol. 3** 38, 75, 82 **Vol. 6** *page 59*
V TOM DOMECKU POD EMAUZY **Vol. 10** 168, 586
V TRESTNEM UZEMI **Vol. 10** 313, 352
V ZACETKU JE BIL GREH **Vol. 3** 387
VA BANQUE **Vol. 4** 124, 208, 223, 479, 646
VA VOIR MAMAN, PAPA TRAVAILLE **Vol. 2** 655, 770
VAADESKUD **Vol. 5** 97
VAADESKUDDET **Vol. 5** 183
VAAKSA VAARAA **Vol. 5** 434, 489
VAÄPELI MYNKHAUSEN **Vol. 5** 443, 639
VÄÄPELIN KAUHU **Vol. 5** 462, 613, 634
VAARALLISTA VAPAUTTA **Vol. 5** 402, 524
VABANK **Vol. 6** *page 210, page 245, page 252*
VABANK II CZYLI RIPOSTA **Vol. 6** *page 224, page 245, page 254*
VACACIONES AL DESNUDO **Vol. 8** 627
VACACIONES PARA IVETE **Vol. 8** 524, 725
VACANCES **Vol. 2** 42
VACANCES DE MONSIEUR HULOT, LES **Vol. 2** 137, 581, 619, 776, 856
VACANCES DU DIABLE, LES **Vol. 2** 727
VACANCES EN ENFER **Vol. 2** 314, 318
VACANCES EXPLOSIVES **Vol. 2** 259, 309, 401
VACANCES FINISSENT DEMAIN, LES **Vol. 2** 460
VACANCES PAYEES **Vol. 2** 187
VACANCES PORTUGAISES **Vol. 2** 208
VACANCES ROMAINES **Vol. 2** 13
VACANCES ROYALES **Vol. 6** *page 101*
VACANCIERS, LES **Vol. 2** 600
VACANTA LA MARE **Vol. 3** 248, 277, 332
VACANTA TRAGICA **Vol. 3** 239, 289
VACANZA BESTIALE, UNA **Vol. 7** 408, 450
VACANZA DEL CACTUS, UNA **Vol. 7** 315, 672
VACANZA DEL DIAVOLO, LA **Vol. 7** 978
VACANZE AD ISCHIA **Vol. 7** 725
VACANZE COL GANGSTER **Vol. 7** 781, 923
VACANZE D'AMORE **Vol. 7** 739
VACANZE DEL SOR CLEMENTE, LE **Vol. 7** 638
VACANZE D'ESTATE **Vol. 7** 311
VACANZE DI NATALE **Vol. 7** 318, 841, 1059
VACANZE D'INVERNO **Vol. 7** 1136
VACANZE IN AMERICA **Vol. 7** 318
VACANZE IN ARGENTINA **Vol. 7** 556
VACANZE IN VAL TREBBIA **Vol. 7** 576, 836, 1173
VACANZE PER UN MASSACRO **Vol. 7** 514
VACCA E IL PRIGIONIERO, LA **Vol. 7** 583

VACHE ET LE PRISONNIER, LA Vol. 2 98, 168, 402 Vol. 4 65
VACKER DAG..., EN Vol. 5 880, 933
VACKERT MINNE, ETT Vol. 5 611
VACKRASTE PA JORDEN, DET Vol. 5 863, 1032
VAD KVINNAN VILL Vol. 5 1037
VAD VETA VÄL MÄNNEN ? Vol. 5 840, 997
VADDORZO, A Vol. 9 77
VADO A VIVERE DA SOLO Vol. 7 664, 697, 1010
VADO, VEDO, SPARO Vol. 8 306, 762
VADROZSA Vol. 9 94
VAEDDEKAMP, EN Vol. 5 97
VAEDDELOBEREN Vol. 5 83
VAERDIFULDE HUSASSISTENT, DEN Vol. 5 83
VAERELESE NO. 17 Vol. 5 83
VAGABOND BIEN-AIME, LE Vol. 2 23
VAGABOND, DELL'AMORE Vol. 7 807, 843
VAGABONDA, LA Vol. 7 569
VAGABONDE, LA Vol. 2 626
VAGABONDEN Vol. 5 2, 142
VAGABONDERNE PA BAKKEGAARDEN Vol. 5 89, 281
VAGABONDI DELLE STELLE, I Vol. 7 891
VAGABONDO, IL Vol. 7 838
VAGABONDS DU REVE, LES Vol. 2 131, 191
VAGABONDS MAGNIFIQUES, LES Vol. 2 277
VAGABONZII DE LA CARABUS Vol. 3 333
VAGABUND Vol. 4 437, 677 Vol. 6 *page 199*
VÄGEN GENOM SKA Vol. 5 907, 1073
VÄGEN TILL KLOCKRIKE Vol. 5 1042, 1080
VAGHE STELLE DELL'ORSA Vol. 7 195, 333, 498, 817
VAGON LI Vol. 3 356, 398
VAGY, A Vol. 9 77
VAHAVY STRELEC Vol. 10 70, 87
VAI ALLA GRANDE Vol. 7 102
VAI AVANTI TU, CHE MI VIEN DA RIDERE Vol. 7 102, 770
VAI COL LISCIO Vol. 7 544
VAI GORILLA Vol. 7 1194
VAIHDOKAS Vol. 5 344, 356
VAIMOKE Vol. 5 359, 485
VAIN KAKSI TUNTIA Vol. 5 310, 634
VAIN LAULAJAPOIKIA Vol. 5 374, 443, 549
VAIN NELJÄ KERTAA Vol. 5 382, 467, 545
VAIN SINULLE Vol. 5 357, 471
VAIVAISUKON MORSIAN Vol. 5 356, 566
VAJZAT ME KORDELE TE KUQE Vol. 3 2, 3
VAKTPOSTENE Vol. 5 724, 787
VAKVILAGBAN Vol. 9 17, 215, 238
VAL D'ENFER, LE Vol. 2 332, 824
VALAHOL EUROPABAN Vol. 9 30, 117, 237
VALAHOL MAGYARORSZAGON Vol. 9 32, 285, 296
VALAKI FIGYEL Vol. 9 107, 147, 174, 240
VALAMIT VISZ A VIZ Vol. 9 196
VALANGA, LA Vol. 7 993
VALBORGSMÄSSOAFTON Vol. 5 840, 878
VALCIK PRO MILION Vol. 10 154, 299, 385
VALD Vol. 5 909, 1052
VALDEMAR SEJR Vol. 5 121
VALDEZ, IL MEZZOSANGUE Vol. 7 817
VALECNE TAJNOSTI PRAZSKE Vol. 10 345, 586
VALEHTELIJA Vol. 5 612
VALENCIA Vol. 4 344, 544
VALENTIN DOBROTIVY Vol. 10 122, 154
VALENTINA Vol. 8 517, 580, 691, 749
VALENTINA RAGAZZA IN CALORE Vol. 7 657
VALERIA DENTRO E FUORI Vol. 7 946
VALERIA D'ISSOGNE Vol. 7 1189
VALERIE A TYDEN DIVU Vol. 10 87, 289, 395
VALET MAITRE, LE Vol. 2 187, 698
VALFANGARE Vol. 5 840, 878, 997
VALIDITA GIORNI DIECI Vol. 7 472, 639
VALIENTE Vol. 8 298, 546
VALIGIA DEI SOGNI, LA Vol. 7 1177
VALIGIA DELLE INDIE, LA Vol. 7 1189
VALISE, LA Vol. 2 293
VALKOINEN PEURA Vol. 5 334, 530
VALKOISEN NEILIKAN VELHO Vol. 5 552, 600
VALKOSIT RUUSUT Vol. 5 357, 471, 481
VALLALD ÖNMAGADAT Vol. 9 8, 190, 229
VALLE DE LAS ESPADAS, EL Vol. 8 450, 602, 603
VALLE DE LAS VIUDAS, EL Vol. 8 245, 331
VALLE DE LOS HOMBRES DE PIEDRA, EL Vol. 8 541, 626
VALLE DEL DIAVOLO, LA Vol. 2 807 Vol. 7 476, 941, 954
VALLE DELL'ECO TONANTE, LA Vol. 7 528
VALLE DI CAINO, LA Vol. 7 606
VALLE SANTA Vol. 7 493
VALLEE, LA Vol. 2 17
VALLFARTEN TILL KEVLAR Vol. 5 854, 1210

VALOR CIVILE Vol. 7 420
VALPARAISO, VALPARAISO Vol. 2 689
VALSE ARDENTE, LA Vol. 7 168
VALSE BLANCHE, LA Vol. 2 88, 339
VALSE BRILLANTE DE CHOPIN Vol. 2 679
VALSE BRILLANTE, LA Vol. 2 9, 153
VALSE DE L'ADIEU, LA Vol. 2 413, 614, 871
VALSE DE PARIS, LA Vol. 2 185, 566
VALSE DU GORILLE, LA Vol. 2 615, 719, 873
VALSE ETERNELLE Vol. 2 355
VALSE ROYALE Vol. 4 574
VALSE TRISTE Vol. 5 340
VALSEUSES, LES Vol. 2 163, 633
VALTER BRANI SARAJEVO Vol. 3 351, 370
VALTOZO FELHÖZET Vol. 9 67, 84, 122
VALURILE DUNARII Vol. 3 270, 330, 332
VAMMENTES HAZASSAG Vol. 5 324, 450, 575, 591 Vol. 9 253, 259, 311
VAMONOS BARBARA Vol. 8 219, 398, 684
VAMOS A CONTAR MENTIRAS Vol. 8 523, 725
VAMOS POR LA PAREJITA Vol. 8 517
VAMPIRE DE DÜSSELDORF, LE Vol. 2 457, 513 Vol. 8 214
VAMPIRE NUE, LA Vol. 2 721
VAMPIRES, LES Vol. 2 330
VAMPIRESAS 1930 Vol. 8 221, 541
VAMPIRI Vol. 3 435
VAMPIRI, I Vol. 7 106, 793
VAMPIRO DE LA AUTOPISTA, EL Vol. 8 518
VAMPIRO, IL Vol. 7 420
VAMPIRO PARA DOS, UN Vol. 8 541, 783
VAMPYR Vol. 2 563, 626, 865
VAMPYRDANSERINDEN Vol. 5 97
VANAERDE, DEN Vol. 5 183
VANDGANG, EN Vol. 5 92
VANDINY TRAMPOTY Vol. 10 412, 532, 621
VÄNDKORSET Vol. 5 907, 1146
VANDRIG I SOLEN, EN Vol. 5 439, 543
VANDRING MED MANEN Vol. 5 840, 1114
VANGELO SECONDO MATTEO, IL Vol. 7 333, 388, 1035
VANGELO SECONDO SIMONE E MATTEO, IL Vol. 7 287
VANGELO SECONDO, UN Vol. 7 1003
VANICE Vol. 10 70, 179, 182
VANINA Vol. 4 186, 546
VANINA VANINI Vol. 1 22, 72, 104 Vol. 7 1035, 1147
VANITA Vol. 7 603
VANOCE S ALZBETOU Vol. 10 235, 283, 408
VANSKELIGT VALG Vol. 5 58
VAQUILLA, LA Vol. 8 214
VAR EN GANG ..., DET Vol. 5 871, 993, 1182, 1217
VAR EN GANG EN SJÖMAN, DET Vol. 5 1003, 1042, 1053
VAR EN LORDAG AFTEN, DET Vol. 5 16, 61, 195, 272
VAR HERRE LUGGAR JOHANSSON Vol. 5 863, 1032
VAR HERRE TAR SEMESTER Vol. 5 1107
VAR I VARDEN, EN Vol. 5 840, 909
VAR PA RUNDETARN, DET Vol. 5 180, 199, 304
VAR SIN VÄG Vol. 5 863, 1142
VARA ROMANTICA Vol. 3 219, 263, 281, 317
VARAKOZOK Vol. 9 20, 154, 265
VARAN POJKE Vol. 5 863, 1042
VARASTETTU KUOLEMA Vol. 5 334, 363, 528, 624
VARAT GÄNG Vol. 5 1042, 1078
VARAVENTTIILI Vol. 5 372, 379
VARAZSKERINGÖ Vol. 9 128
VARAZSLO, A Vol. 9 83, 183, 259
VÄREITÄ Vol. 5 360, 586
VARELSERNA Vol. 5 1018, 1101
VARGTIMMEN Vol. 5 1017, 1053, 1080, 1196
VARHANIK U SVATEHO VITA Vol. 10 34, 138
VARIETE Vol. 4 19, 178, 316, 752 Vol. 6 *page 189*
VARIETE CSILLAGAI, A Vol. 9 64, 71, 104
VARIETES Vol. 2 32, 193, 676 Vol. 4 19 Vol. 8 538, 667
VARIOLA VERA Vol. 6 *page 308, page 310, page 314*
VARJOJA KANNAKSELLA Vol. 5 428, 472
VARJOSTETTUA VALOA Vol. 5 469
VARJU A TORONYORAN Vol. 9 178, 237
VÄRLDENS BÄSTA KARLSSON Vol. 5 869, 1027, 1093
VARLJIVO LETO '68 Vol. 6 *page 304, page 305, page 320*
VÄRMLÄNNINGARNA Vol. 5 840, 843, 870, 909, 911, 923, 1161, 1189
VARNATT Vol. 5 724, 772, 780
VARNING FÖR JÖNSSONLIGAN Vol. 5 947, 1160
VAROS ALATT, A Vol. 9 168, 241
VAROSBUJOCSKA Vol. 9 107, 174, 279
VARSOVAN LAULU Vol. 5 443, 549
VART HJÄRTA HAR SIN SAGA Vol. 5 840, 863
VARTIOITU KYLÄ 1944 Vol. 5 437, 541, 546, 589, 647
VARUJ ... ! Vol. 10 154, 177, 256, 287

VARUSKUUNNAN „PIKKU“ MORSIAN Vol. 5 472, 476
VAS-Y MAMAN Vol. 2 74
VASARNAP DELUTAN Vol. 9 258, 267
VASARNAPI SZÜLÖK Vol. 9 221, 253
VASENS HEMMELIGHED Vol. 5 9
VASER PERDE IL TRENO Vol. 7 1189
VASGYAROS, A Vol. 9 77
VASIKANA HÄNTÄ Vol. 5 389
VASKATA Vol. 6 *page 61*
VASO CERRADO Vol. 8 539
VASTAMYRKKY Vol. 5 415, 482
VASTAUS Vol. 5 373
VÄSTKUSTENS HJÄLTAR Vol. 5 846, 995
VASVIRAG Vol. 9 285, 312, 326
VATER GEHT AUF REISEN Vol. 4 212, 232, 239, 270
VATER UND SOHN Vol. 4 208, 479, 600
VÄTER UND SÖHNE Vol. 4 315
VATER VOSS Vol. 4 124, 321
VATER WERDEN IST NICHT SCHWER... Vol. 4 245, 584, 729
VATERLAND RUFT, DAS Vol. 4 5
VATI WILL NICHT HEIRATEN Vol. 1 75
VATSA SISÄÄN – RINTA UIOS Vol. 5 443, 639
VAUTRIN Vol. 2 205, 722 Vol. 7 108
VAXDOCKAN Vol. 5 909
VAZENI PRATELE, ANO Vol. 10 283, 577, 594
VCERA NEDELE BYLA Vol. 10 256, 412
VDAVALA SE JEDNA PANNA Vol. 10 243
VDAVKY NANYNKY KULICHOVY Vol. 10 167, 256, 412, 453
VDOVICKA SPADLA S NEBE Vol. 10 460, 621
VDOVSTVO KAROLINE ZASLER Vol. 3 395, 421
VE DVOU SE TO LEPE TAHNE Vol. 10 345, 586
VE SPARECH UPIRA Vol. 10 53, 453
VE ZNAMENI TYRKYSOVE HORY Vol. 10 59, 81, 197, 595
VEAU GRAS, LE Vol. 2 8, 197, 425
VECCHIA GIORNALAIO Vol. 7 1189
VECCHIA GUARDIA Vol. 7 166, 725
VECCHIA SIGNORA, LA Vol. 7 1189
VECCHIO NIDO, IL Vol. 7 1189
VECCHIO REGNO Vol. 7 234
VECEN KALENDAR Vol. 3 60, 105 Vol. 6 *page 52*
VECERNJA ZVONA Vol. 6 *page 309*
VECES ETAIENT FERMES DE L'INTERIEUR, LES Vol. 2 633, 642
VECINOS Vol. 8 373
VECNI VREMENA Vol. 6 *page 23, page 50, page 64*
VED FAENGSLETS PORT Vol. 5 97
VED KONGELUNDEN Vol. 5 180, 199, 201
VEDETTA LOMBARDA, LA Vol. 7 420
„VEDI COME SEI, LO ... LO VEDI COME SEI ?!“ Vol. 7 443, 639
VEDI NAPOLI E POI MUORI Vol. 7 443, 910
VEDI NAPULE E PO'MORI Vol. 7 53
VEDIAMOCI CHIARO Vol. 7 315, 383
VEDO NUDO Vol. 7 408, 763
VEDOVA ALLEGRO, IL Vol. 7 1136, 1189
VEDOVA DEL TRULLO, LA Vol. 7 1016
VEDOVA, LA Vol. 7 48, 1060, 1159
VEDOVO, IL Vol. 7 1147
VEGAS 500 MILLONES, LAS Vol. 8 338, 426, 732
VEGGENTE, LA Vol. 7 764, 767
VEGLIA D'ARMI DEL BOY – SCOUT, LA Vol. 7 442
VEGRE ... Vol. 9 71, 322
VEGRE, HETFÖ Vol. 9 6, 161
VEGSZO Vol. 9 71, 80
VEGÜL Vol. 9 167
VEIEN TILBAKE Vol. 5 745, 768
VEILCHEN VOM POTSDAMER PLATZ, DAS Vol. 4 340, 466, 471, 566
VEILCHENFRESSER, DER Vol. 4 14, 186, 363
VEILLE D'ARMES Vol. 2 167, 309, 388, 440
VEINARDS, LES Vol. 2 97, 268
VEINTE MIL DUROS Vol. 8 452
VEINTE PASOS PARA LA MUERTE Vol. 8 261, 768
VEJEN TIL BYEN Vol. 5 159, 287
VEJRHANEN Vol. 5 89
VELA INCANTATA, LA Vol. 7 593, 1173
VELBLOUD UCHEM JEHLY Vol. 10 29, 92, 256, 412
VELE AMMAINATE Vol. 7 756, 1036, 1116
VELENO DEL PIACERE, IL Vol. 7 277
VELHOS SAO OS TRAPOS Vol. 8 32, 66
VELIKA PRILEZITOST Vol. 10 503, 580
VELIKA TURNEJA Vol. 3 343, 346, 401
VELIKI DAN Vol. 3 356, 416
VELIKI I MALI Vol. 3 425, 436
VELIKI PLAVI PUT Vol. 3 387
VELIKI SKRETNICAR USKOTRACNOG KOLOSJEKA Vol. 6 *page 307, page 309*
VELIKI TRANSPORT Vol. 6 *page 312, page 319*
VELIKO SUDJENJE Vol. 3 338, 361
VELIKONOCNI DOVOLENA Vol. 10 173, 227, 577
VELKA FILMOVA LOUPEZ Vol. 10 139, 283, 496
VELKA NEZNAMA Vol. 10 283, 428, 516
VELKA NOC A VELKY DEN Vol. 10 82, 273, 548

VELKA PREHRADA **Vol. 10** 154, 168, 564
VELKA SAMOTA **Vol. 10** 313, 387, 412
VELKE DOBRODRUZSTVI **Vol. 10** 184, 311, 388, 582
VELKE PRANI **Vol. 10** 180, 286, 339
VELKE TRAPENI **Vol. 10** 173, 233, 380
VELKY PRIPAD **Vol. 10** 345, 380, 580
VELLEZER DHE SHOKE **Vol. 6** *page 11, page 12, page 15*
VELLUTO NERO **Vol. 7** 536
VELO DELLA COLPA, IL **Vol. 7** 730
VELO DELLA FELICITA, IL **Vol. 7** 1113
VEM DÖMER **Vol. 5** 854, 937, 997
VEN BAKANCSOS ES FIA, A HUSZAR, A **Vol. 9** 77
VEN GAZEMBER, A **Vol. 9** 71, 189, 272, 322
VEN I BOLIGNODEN, EN **Vol. 5** 128, 190, 199, 201
VENA D'ORO, LA **Vol. 7** 257, 730, 1140
VENCEDORES DE LA MUERTE, LOS **Vol. 8** 655
VENDANGES, LES **Vol. 2** 302
VENDAVAL **Vol. 8** 207, 298
VENDAVAL MARAVILHOSO **Vol. 8** 44, 88, 122, 184
VENDEDORA DE ROPA INTERIOR, LA **Vol. 8** 237, 414
VENDELINUV OCISTEC A RAJ **Vol. 10** 257
VENDEMIAIRE **Vol. 2** 330, 432
VENDETTA **Vol. 4** 239
VENDETTA ALSAZIANA **Vol. 7** 1189
VENDETTA DA FUTURO **Vol. 7** 426
VENDETTA DEI BARBARI, LA **Vol. 7** 891
VENDETTA DEI SERPI, LA **Vol. 7** 454
VENDETTA DEI TUGHS, LA **Vol. 7** 372
VENDETTA DEL CORSARO, LA **Vol. 7** 792, 910
VENDETTA DEL MARE, LA **Vol. 7** 1189
VENDETTA DEL MORTO, LA **Vol. 7** 1189
VENDETTA DEL PADRONE DELLE FERRIERE **Vol. 7** 965
VENDETTA DELLO SCEMO, LA **Vol. 7** 314
VENDETTA DI AQUILA NERA, LA **Vol. 7** 443, 481, 1053
VENDETTA DI CAMILLO, LA **Vol. 7** 26, 701
VENDETTA DI ERCOLE, LA **Vol. 7** 798
VENDETTA DI SPARTACUS **Vol. 7** 694
VENDETTA DI UNA PAZZA, LA **Vol. 7** 566, 603, 1054
VENDETTA DI URSUS, LA **Vol. 7** 1152
VENDETTA EN CAMARGUE **Vol. 2** 401, 419
VENDETTA FATALE **Vol. 7** 1189
VENDETTA, LA **Vol. 2** 347, 686 **Vol. 7** 845
VENDETTA NEL SOLE **Vol. 7** 1113
VENDETTA SARDA **Vol. 7** 16
VENDICAME ! **Vol. 7** 1103
VENDICATA **Vol. 7** 891
VENDICATORE, IL **Vol. 7** 1035
VENDIMI **Vol. 6** *page 15, page 17*
VENEDOR DE ILUSIONES, EL **Vol. 8** 249, 383
VENENOSA, LA **Vol. 2** 113, 413, 614, 871
VENERE **Vol. 7** 493, 531, 730, 781
VENERE DEI PIRATI, LA **Vol. 7** 739
VENERE DI CHERONEA, LA **Vol. 7** 638
VENERE NERA, LA **Vol. 7** 1057
VENERE PROPIZIA **Vol. 7** 1078
VENERI, NINFE E SIRENE **Vol. 7** 316, 806
VENEXIANA, LA **Vol. 7** 213, 263, 607
VENEZIA LA LUNA E TU **Vol. 7** 333
VENEZIANISCHE LIEBESABENTEUER **Vol. 4** 545
VENEZIANISCHE NACHT, EINE **Vol. 4** 743
VENGA A PRENDERE IL CAFFE ... DA NOI **Vol. 7** 221, 283, 931
VENGADOR DE CALIFORNIA, EL **Vol. 8** 626, 752
VENGANZA DE DON MENDO, LA **Vol. 8** 207, 595
VENGANZA DE LA MOMIA, LA **Vol. 8** 237, 708
VENGANZA DEL ZORRO, LA **Vol. 8** 221, 603
VENGANZA, LA **Vol. 8** 214, 602
VENGERRAK, A **Vol. 9** 337
VENNER **Vol. 5** 724, 800
VENOUSEK A STAZICKA **Vol. 10** 65, 126, 167, 168, 412
VENT DE L'ILLA, EL **Vol. 8** 432
VENT DEBOUT **Vol. 2** 731
VENT SE LEVE, LE **Vol. 2** 137, 626, 637, 824
VENTA DE VARGAS **Vol. 8** 311, 602
VENT'ANNI **Vol. 7** 891
VENTE A ALEMANIA, PEPE **Vol. 8** 331, 627
VENTE A LIGAR AL OESTE **Vol. 8** 267, 331, 627
VENTESIMO DUCA, IL **Vol. 7** 481, 1107, 1172
VENTO DEL SUD **Vol. 7** 1167
VENTO DELL'EST **Vol. 7** 1194
VENTO DI MILIONI **Vol. 7** 902, 1036
VENTO DI PRIMAVERA **Vol. 7** 1121
VENTOLERA **Vol. 8** 214, 261
VENTRE DI PARIGI, IL **Vol. 7** 52
VENTRILOQUO, IL **Vol. 7** 454
VENUS **Vol. 2** 41, 153
VENUS AVEUGLE **Vol. 2** 13, 96, 153, 539

VENUS DE L'OR, LA Vol. 2 270, 395
VENUS FRA VESTO Vol. 5 104, 199, 201, 246
VENUS IM FRACK Vol. 4 752, 788, 796
VENUS VOM MONTMARTRE, DIE Vol. 4 242, 340
VENUS VOR GERICHT Vol. 4 623, 647, 660, 728
VENUSTA ETSIMÄSSÄ ELI ERÄÄN NUOREN MIEHEN IHMEELLISET SEIKKAILUT Vol. 5 344
VERA LUKASOVA Vol. 10 266, 460, 621
VERA PANINA Vol. 4 309
VERA STORIA DELLA MONACA DI MONZA, LA Vol. 7 127
VERA, UN CUENTO CRUEL Vol. 8 219, 251
VERÄCHTER DES TODES, DER Vol. 4 270, 471
VERANEO EN ESPANA Vol. 8 217, 390
VERANO 70 Vol. 8 524
VERANO DE INFIERNO, UN Vol. 8 358, 549
VERANO PARA MATAR, UN Vol. 8 426, 732
VERBENA DE LA PALMOS, LA Vol. 8 762
VERBENA DE PALOMA, LA Vol. 8 214, 522, 548, 727
VERBOTENE FRUCHT, DIE Vol. 4 660, 704
VERBOTENE LIEBE Vol. 4 113, 340, 725, 780
VERBOTENE WEG, DER Vol. 4 743, 772
VERBRECHEN IN DER WALLSTREET Vol. 4 13
VERBRECHEN UND LIEBE Vol. 4 628
VERBRECHEN VON HOUNDSDITCH, DAS Vol. 4 356, 529
VERDACHT AUF EINEN TOTEN Vol. 1 33, 53, 75
VERDACHT AUF URSULA Vol. 4 85, 97
VERDAD, LA Vol. 8 319
VERDAD SOBRE EL CASO SAVOLTA, LA Vol. 8 255
... VERDAMMT, ICH BIN ERWACHSEN Vol. 1 113, 122, 124, 127
VERDE DONCELLA Vol. 8 207, 214
VERDE ETA, LA Vol. 7 1210
VERDE POR FORA, VERMELHO POR DENTRO Vol. 8 34, 60
VERDEN ER FULD AF BORN Vol. 5 11, 120, 175
VERDENS HERKULES Vol. 5 7
VERDENS RIGESTE PIGE Vol. 5 89, 281
VERDENS UNDERGANG Vol. 5 9, 183
VERDES ANOS, OS Vol. 8 127, 169
VERDES PRADERAS, LAS Vol. 8 245, 658
VERDES VACACIONES DE UNA FAMILIA BIEN, LAS Vol. 8 426
VERDI BANDIERE DI ALLAH, LE Vol. 7 1147
VERDICT Vol. 2 49, 185
VERDUGO, EL Vol. 8 203, 214, 334, 450, 614
VERDUN, SOUVENIRS D'HISTOIRE Vol. 2 595
VERDUN, VISION D'HISTOIRE Vol. 2 65, 595
VEREB IS MADAR, A Vol. 9 84, 122, 332
VEREDAS Vol. 8 7, 10
VERERBTE TRIEBE Vol. 4 186, 558
VERFEHLTE JUGEND Vol. 4 187
VERFLIXTE BANDE Vol. 1 89, 130
VERFLUCHTE, DER Vol. 4 344, 544
VERFÜHRBAREN, DIE Vol. 1 53, 114
VERFÜHRERIN, DIE Vol. 6 *page 184, page 191*
VERFÜHRTE, DER Vol. 4 124, 646
VERFÜHRTEN, DIE Vol. 4 660
VERFÜHRUNG AM MEER Vol. 3 405
VERGELOSE, DE Vol. 5 714, 726, 738
VERGESSEN SIE BITTE NICHT DIE ANTENNEN ZU ERDEN Vol. 4 76
VERGESST MIR MEINE TRAUDEL NICHT Vol. 1 2, 48, 54
VERGIFTETE STROM, DER Vol. 4 219
VERGIFTETES BLUT Vol. 4 11, 514
VERGINE DEI VELENI, LA Vol. 7 209
VERGINE DEL GIGLIO, LA Vol. 7 1189
VERGINE DEL MARE, LA Vol. 7 548
VERGINE DEL RONCADOR, LA Vol. 7 210
VERGINE DI BABILONIA, LA Vol. 7 1189
VERGINE DI BALI, LA Vol. 7 692
VERGINE FOLLE, LA Vol. 7 399
VERGINE, IL TORO E IL CAPRICORNO, LA Vol. 7 426
VERGINE IN FAMIGLIA, UNA Vol. 7 1024
VERGINE MODERNA, LA Vol. 7 798
VERGINELLA, LA Vol. 7 1152
VERGINITA Vol. 7 1054 Vol. 7 603
VERGISS MEIN NICHT ! Vol. 4 97, 354, 424, 437, 481, 703, 738
VERGNÜGTE HOCHZEITSREISE, EINE Vol. 6 *page 186*
VERGÖDÖ SZIVEK Vol. 9 77
VERHÄNGNISVOLLE ANDENKEN, DAS Vol. 4 309
VERHÄNGNISVOLLES GEIGENSOLO Vol. 4 302
VERHEXTE FISCHERDORF, DAS Vol. 1 46, 54, 92
VERHEXTE SCHEINWERFER, DER Vol. 4 786
VERHÖR, DAS Vol. 1 79, 124
VERI AZ ÖRDÖG A FELESEGET Vol. 9 84, 167, 259
VERIGATA Vol. 3 47, 109, 121

VERIRRTE JUGEND **Vol. 4** 213, 306 **Vol. 45** 213, 306
VERITA, LA **Vol. 7** 330, 1205
VERITA NUDA, LA **Vol. 7** 559
VERITAS VINCIT **Vol. 4** 126, 396, 423, 780
VERITE, LA **Vol. 2** 21, 824
VERITE SUR BEBE DONGE, LA **Vol. 2** 153, 249
VERITE SUR L'IMAGINAIRE PASSION D'UN INCONNU, LA **Vol. 2** 390
VERITES ET MENSONGES **Vol. 2** 635
VERJÜNGTE ADOLAR, DER **Vol. 4** 192, 270, 409
VERKANNTE EXISTENZEN **Vol. 4** 549
VERKANNTE LEBEMANN, DER **Vol. 4** 340, 466, 796
VERKAUFTE BRAUT, DIE **Vol. 4** 301, 344, 373, 600, 776
VERKAUFTE GROSSVATER, DER **Vol. 4** 135, 619, 647
VERKEHRTE BERLIN, DAS **Vol. 4** 514, 646
VERKLUNGENE MELODIE **Vol. 4** 24, 378, 445, 567
VERKLUNGENE TRÄUME **Vol. 4** 97, 599
VERKOMMEN **Vol. 4** 321, 644
VERKRACHTE EXISTENZEN **Vol. 4** 297, 549, 780
VERLEGENHEITSKIND, DAS **Vol. 4** 28, 329
VERLIEB' DICH NICHT AM BODENSEE **Vol. 4** 108, 270, 323
VERLIEBT UND VORBESTRAFT **Vol. 1** 36, 86, 120
VERLIEBTE BLASEKOPF, DER **Vol. 4** 239, 424
VERLIEBTE FIRMA, DIE **Vol. 4** 479, 536, 600
VERLIEBTE HERZEN **Vol. 10** 256, 412
VERLIEBTE HOTEL, DAS **Vol. 4** 123, 257, 775, 786, 788
VERLIEBTES ABENTEUER **Vol. 4** 361, 738, 788
VERLOBTE, DIE **Vol. 6** *page 158, page 161, page 162*
VERLOBUNG IN HOLLERBUSCH **Vol. 6** *page 169, page 174, page 175*
VERLOGENE MORAL **Vol. 4** 178, 479
VERLORENE BALL, DER **Vol. 1** 81, 92, 103
VERLORENE ENGEL, DER **Vol. 1** 72, 80, 92
VERLORENE LÄCHELN, DAS **Vol. 4** 110, 246
VERLORENE MELODIE, DIE **Vol. 4** 30, 567
VERLORENE NÄCHTE **Vol. 4** 256, 605
VERLORENE PARADIES, DAS **Vol. 4** 232, 536, 751
VERLORENE SCHATTEN, DER **Vol. 4** 178, 202, 560
VERLORENE SCHUH, DER **Vol. 4** 30, 38, 359, 770
VERLORENE SEELEN **Vol. 4** 159
VERLORENE SOHN, DER **Vol. 4** 54, 373, 437, 458, 531
VERLORENE TAL, DAS **Vol. 4** 306, 742
VERLORENE TÖCHTER **Vol. 4** 212, 270, 309
VERMÄCHTNIS DES INKA, DAS **Vol. 8** 450, 524
VERNI ZUSTANEME **Vol. 10** 94, 240, 283, 416, 418, 421, 475, 546, 595
VERNUNFT DES HERZENS, DIE **Vol. 4** 247
VERONICA **Vol. 3** 241, 286, 330
VERONICA SE INTOARCE **Vol. 3** 241, 286
VERONICAS SVEDEDUG **Vol. 5** 255
VERONIKA **Vol. 10** 139, 283, 325
VERONIQUE **Vol. 2** 615
VERONIQUE NIQUE NIQUE **Vol. 2** 294
VERONIQUE OU L'ETE DE MES TREIZE ANS **Vol. 2** 742
VERRAT AUF SCHLOSS TREUENFELS **Vol. 4** 232, 673
VERRÄTER **Vol. 4** 13, 311, 347, 445
VERRÄTERIN, DIE **Vol. 4** 124, 646
VERRUFENEN, DIE **Vol. 4** 247, 461
VERS L'ARGENT **Vol. 2** 871
VERS L'EXTASE **Vol. 2** 383, 566
VERS L'IMMORTALITE **Vol. 2** 159, 286, 330
VERSCHLEIERTE BILD VON GROSSKLEINDORF, DAS **Vol. 4** 159
VERSCHLEPPT **Vol. 4** 212, 495, 594
VERSCHWÖRUNG ZU GENUA, DIE **Vol. 4** 211, 247, 288, 452
VERSCHWUNDENE HAUS, DAS **Vol. 4** 270, 443, 471
VERSCHWUNDENE PRINZESSIN, DIE **Vol. 4** 113
VERSCHWUNDENE TESTAMENT, DAS **Vol. 4** 379, 558
VERSO LA GLORIA **Vol. 7** 965
VERSO L'ARCOBALENO **Vol. 7** 796
VERSPIELTE HEIMAT **Vol. 1** 44, 98, 128
VERSPRICH MIR NICHTS **Vol. 4** 49, 738, 788
VERSUNKENE FLOTTE, DIE **Vol. 4** 232, 321, 340, 703
VERSUNKENE WELTEN **Vol. 4** 192, 780
VERSZERZÖDES **Vol. 9** 242, 259, 279
VERT GALANT, LE **Vol. 2** 339, 731
VERTA KÄSISSÄMME **Vol. 5** 356, 461
VERTAGTE HOCHZEITSNACHT, DIE **Vol. 4** 742

VERTAUSCHTE BRAUT, DIE **Vol. 4** 17, 84, 123, 232, 257, 316, 432, 437
VERTAUSCHTE GESICHTER **Vol. 4** 132, 734
VERTAUSCHTE HUND, DER **Vol. 4** 367, 744
VERTAUSCHTE KÖNIGIN, DIE **Vol. 6** *page 168, page 172, page 173*
VERTAUSCHTE PALETOTS **Vol. 6** *page 186*
VERTE MOISSON, LA **Vol. 2** 795, 819
VERTEIDIGER HAT DAS WORT, DER **Vol. 4** 97, 416, 437
VERTIGE, LE **Vol. 2** 511, 547, 722, 727, 820
VERTIGE POUR UN TUEUR **Vol. 8** 331
VERTIGES **Vol. 2** 344, 401
VERTIGNE **Vol. 7** 150, 756, 1055
VERTIGNE BIANCA **Vol. 7** 1044
VERTIGNE D'AMORE **Vol. 7** 16
VERTIGO **Vol. 8** 214, 276, 336, 625
VERTIGO DEL CRIMEN, EL **Vol. 8** 708
VERTIGO EN LA PISTA **Vol. 8** 299, 728
VERTUDIEU **Vol. 2** 814
VERURTEILTE DORF, DAS **Vol. 1** 78, 90, 117
VERWANDLUNG, DIE **Vol. 4** 321
VERWANDTE SIND AUCH MENSCHEN **Vol. 4** 340, 361
VERWANDTE UND BEKANNTE **Vol. 1** 33, 70
VERWEHTE SPUREN **Vol. 4** 234, 462, 734
VERWIRRUNG DER LIEBE **Vol. 1** 8, 54, 88
VERWORRENE WEGE **Vol. 4** 300
VERWUNSCHENE PRINZESSIN, DIE **Vol. 4** 113
VERZAUBERTE TAG, DER **Vol. 4** 215, 361
VERZEIHUNG, SEHEN SIE FUSSBALL ? **Vol. 6** *page 154, page 164, page 169*
VES V POHRANICI **Vol. 10** 321, 580
VESELA BIDA **Vol. 10** 460, 621
VESELA B"LGARIJA **Vol. 3** 33
VESELE VANOCE PREJI CHOBOTNICE **Vol. 10** 139, 165, 496
VESELICA **Vol. 3** 390, 395
VESELO GOSTIVANJE **Vol. 6** *page 303, page 316*
VESELY SOUBOJ **Vol. 10** 70, 179
VESELYE MUZYKANTY **Vol. 1** 298
VESNA **Vol. 3** 387, 422, 447 **Vol. 3** *page 303*
VESNICKO MA STREDISKOVA **Vol. 10** 172, 516, 535
VESPERAS IMPERIALES **Vol. 8** 449, 484, 674
VESPRO SICILIANO **Vol. 7** 1036
VESSEL OF WRATH **Vol. 2** 440
VESTA DA SPOSA, LA **Vol. 7** 1189
VESTALE DU GANGE, LA **Vol. 2** 413, 688, 845
VESTEC **Vol. 10** 359, 580
VESTENS BORN **Vol. 5** 136, 242
VESTENS HYAENE **Vol. 5** 106
VESTER-VOV-VOV **Vol. 5** 55, 83
VESTERHAVSDRENGE **Vol. 5** 233, 241, 272
VESTIDA DE AZUL **Vol. 8** 358
VESTIDO COR DE FOGO, O **Vol. 8** 169
VESTIRE GLI IGNUDI **Vol. 7** 1062
VESTITO ALLA MODA **Vol. 7** 1189
VESZELYBEN A POKOL **Vol. 9** 220
VESZELYES JATEKOK **Vol. 9** 81, 140, 328
VETAR JE STAO PRED ZORU **Vol. 3** 349, 374
VETERAANIN VOITTO **Vol. 5** 443, 643
VETERAN VOTRUBA **Vol. 10** 364
VETRNA HORA **Vol. 10** 294, 313, 352
VETRNE MORE **Vol. 10** 197, 367, 397, 428
VETTER AUS DINGSDA, DER **Vol. 4** 123, 462, 788
VETTURALE DEL MONCENISIO, IL **Vol. 7** 48, 491, 640, 730
VETTURALE DEL SAN GOTTARDO, IL **Vol. 7** 476, 518, 966
VEUVE COUDERC, LA **Vol. 2** 770, 875
VEUVE EN OR, UNE **Vol. 2** 60, 279
VEUVE ET L'INNOCENT, LA **Vol. 2** 330, 826
VEUVE LUBRIQUE, LA **Vol. 2** 219
VEUVES EN CHALEUR **Vol. 2** 291
VEVERKA A KOUZELNA MUSLE **Vol. 10** 74, 180, 233
VEZ AL ANO SER HIPPY NO HACE DANO, UNA **Vol. 8** 331, 681
VEZEN NA BEZDEZI **Vol. 10** 244, 524
VI ARME SYNDERE **Vol. 5** 31, 150, 241, 248, 272
VI BEHÖVER VARAN **Vol. 5** 841, 871, 1052, 1164, 1181
VI BYGGER LANDET **Vol. 5** 670, 760
VI ER ALLESAMMEN TOSSEDE **Vol. 5** 248, 304
VI FIXAR ALLT **Vol. 5** 909
VI FLYGER PA RIO **Vol. 5** 907, 928, 1042
VI FLYR PA RIO **Vol. 5** 700
VI GAR LANDSVÄGEN **Vol. 5** 840, 997
VI GIFTER OSS **Vol. 5** 672, 729
VI HAR DET JO DEJLIGT **Vol. 5** 23, 248
VI HEMSLAVINNOR **Vol. 5** 928, 929, 1042
VI KUNDE NA' DET SAA RART ! **Vol. 5** 2, 228
VI MASTHUGGSPOJKAR **Vol. 5** 840, 878
VI PA SALTKRAKAN **Vol. 5** 862, 1042
VI PA SOLGLÄNTAN **Vol. 5** 863
VI PA VÄDDÖ **Vol. 5** 933, 1042

VI SOM GAR KEOKKENVEJEN Vol. 5 31, 150, 241, 248
VI SOM GAR KÖKSVÄGEN Vol. 5 840, 909
VI SOM GAR SCENVÄGEN Vol. 5 863, 1032
VI SOM GAR STJERNEVEJEN Vol. 5 23, 281, 301
VI SPILLOPPER Vol. 5 762
VI TRE Vol. 5 928, 1156
VI TRE DEBUTERA Vol. 5 940, 1173
VI TVA Vol. 5 928, 1042
VI VIL HA' ET BARN Vol. 5 89, 248
VI VIL LEVE Vol. 5 670, 760
VI VIL SKILLES Vol. 5 674, 722
VI VOKSNE Vol. 5 133
VIA COL ... PARAVENTO Vol. 7 603
VIA CRUCIS Vol. 5 183 Vol. 7 1078
VIA DEGLI SPECCHI Vol. 7 109, 199, 841
VIA DEL DOLORE, LA Vol. 7 621
VIA DEL PECCATO, LA Vol. 7 53, 560, 594, 665
VIA DEL SILENZIO, LA Vol. 7 1003
VIA DELLA DROGA, LA Vol. 7 124
VIA DELLA LUCE, LA Vol. 7 150
VIA DELLE CINQUE LUNE Vol. 7 453, 796, 1060
VIA DELLE LACRIME, LA Vol. 7 144
VIA DELLE STELLE, LA Vol. 7 198
VIA DOLOROSA, LA Vol. 7 792
VIA MACAO Vol. 8 130
VIA MACAU Vol. 8 130
VIA MALA Vol. 4 288, 574
VIA MARGUTTA Vol. 7 86
VIA PADOVA, 46 Vol. 7 21, 796
VIA PENOSA, LA Vol. 7 488
VIACCIA, LA Vol. 7 86, 781, 1142
VIADUKT, A Vol. 9 6, 191, 221, 265, 298
VIAGER, LE Vol. 2 400, 835
VIAGGIATORI DELLI SERA, I Vol. 7 134, 567
VIAGGIO CON ANITA Vol. 7 80, 333, 564
VIAGGIO DEI BERLUPON, IL Vol. 7 26
VIAGGIO DEL SIGNOR PERRICHON Vol. 7 196, 443
VIAGGIO DI NOZZE Vol. 7 1189
VIAGGIO DI PIACERE, UN Vol. 7 654, 701
VIAGGIO, IL Vol. 7 1116
VIAGGIO IN ITALIA Vol. 7 443, 1062
VIAGGIO NELLA LUNA, UN Vol. 7 172, 865, 872, 930
VIAGGIO NELL'IMPOSSIBILE, UN Vol. 7 369
VIAGGIO VERSO LA MORTE, UN Vol. 7 730, 986
VIAJE A BANGKOK, ATAUD INCLUIDO Vol. 8 731, 734
VIAJE A NINGUNA PARTE, EL Vol. 8 219, 368
VIAJE AL MAS ALLA Vol. 8 330, 351
VIAJE DE NOVIOS Vol. 8 311, 600
VIAJE DE NOVIOS A LA ITALIANA Vol. 8 311, 396
VIAJE SIN DESTINO Vol. 8 437
VIAJERO DEL CLIPPER, EL Vol. 8 396
VIALE DELLA SPERANZA Vol. 7 106, 781
VIATA INVINGE Vol. 3 231, 260
VIATA NU IARTA Vol. 3 224, 230, 244, 250, 268, 329
VIBRACII Vol. 6 *page 24, page 42, page 60*
VICA, A VADEVEZÖS Vol. 9 272, 322
VICARIO DE OLOT, EL Vol. 8 641, 684
VICE ET LA VERTU, LE Vol. 2 21, 368
VICENDE DELL'ILLUSIONE, LE Vol. 7 1067
VICENZELLA Vol. 7 39
VICIEUSES POUR HOMME SEUL Vol. 2 795
VICIO EN EL ALMA Vol. 8 773
VICIO Y LA VIRTUD, EL Vol. 8 407, 577, 738
VICIOSAS AL DESNUDOR Vol. 8 302, 639
VICOMTE DE BRAGELONNE, LE Vol. 2 204, 419, 559, 698
VICOMTE REGLE SES COMPTES, LE Vol. 2 249
VICTIMA DEL ODIO Vol. 8 246, 657
VICTIMAS, LAS Vol. 8 611, 732
VICTOIRE EN CHANTANT, LA Vol. 2 7, 251, 791
VICTOR Vol. 2 231, 419
VICTOR FRANKENSTEIN Vol. 5 946, 1024
VICTORIA Vol. 8 274, 275
VICTORIA 2 – LA DISBAUXA DEL 17 Vol. 8 274, 275
VICTORIA 3, LA RAZON Y EL ARREBATO Vol. 8 274, 275
VIDA ALFREDEDOR, LA Vol. 8 400, 749
VIDA DE FAMILIA Vol. 8 289, 479, 482, 608
VIDA DE UNA MODISTA, LA Vol. 8 655
VIDA, EL AMOR Y LA MUERTE, LA Vol. 8 728, 770
VIDA EMPIEZA A MEDIANOCHE Vol. 8 357, 436
VIDA EN SOMBRAS Vol. 8 380, 753
VIDA EN UN BLOC, LA Vol. 8 214, 611
VIDA EN UN HILO, LA Vol. 8 268
VIDA ENCADENADA, LA Vol. 2 646 Vol. 8 214, 336, 607
VIDA ES MAGNIFICA, LA Vol. 8 768
VIDA ES MARAVILLOSA, LA Vol. 8 276, 753
VIDA INTIMA DE UN SEDEUCTOR CINICO Vol. 8 249

VIDA NUEVA DE PEDRITO DE ANDIA, LA Vol. 8 214, 644
VIDA PARA DOIS, UMA Vol. 8 80, 96
VIDA PERRA Vol. 8 681
VIDA POR DELANTE, LA Vol. 8 752
VIDA PRIVADA DE FULANO DE TAL, LA Vol. 8 470, 493
VIDA SIGUE IGUAL, LA Vol. 8 438
VIDA SIN RISAS Vol. 8 616, 630, 762
VIDAS Vol. 8 7
VIDAS CONFUSAS Vol. 8 214, 752
VIDAS CRUZADAS Vol. 8 400
VIDAS ROTAS Vol. 8 436, 449, 751
VIDAS SEM RUMO Vol. 8 96, 122
VIDDENES FOLK Vol. 5 752, 794
VIDOCQ Vol. 2 833
VIDUNDERCIGAREN Vol. 5 97
VIDUNDERHUNDEN Vol. 5 92
VIDUNDERLIGE HAARELEXIR, DEN Vol. 5 97
VIE A DEUX, LA Vol. 2 173, 232, 322, 488
VIE A L'ENVERS, LA Vol. 2 348, 738
VIE AMOUREUSE DE L'HOMME INVISIBLE, LA Vol. 8 399, 537
VIE CHANTEE, LA Vol. 2 153, 270
VIE COMMENCE DEMAIN, LA Vol. 2 47, 460
VIE CONJUGALE, LA Vol. 2 185, 294
VIE CONTINUE, LA Vol. 6 *page 70, page 85*
VIE DE BOHEME, LA Vol. 2 601, 863
VIE DE CHATEAU, LA Vol. 2 517, 770
VIE DE CHIEN, UNE Vol. 2 347, 871
VIE DE GARCON, UNE Vol. 2 318, 627, 807
VIE DE PLAISIR, LA Vol. 2 67, 205, 332
VIE DEL CUORE, LE Vol. 7 603, 1054, 1060
VIE DEL PECCATO, LE Vol. 7 603
VIE DELL'IGNOTO, LE Vol. 7 1189
VIE DEVANT SOI, LA Vol. 2 17, 285
VIE D'UN HONNETE HOMME, LA Vol. 2 48, 70
VIE EN ROSE, LA Vol. 2 615, 644
VIE EST A NOUS, LA Vol. 2 124, 248, 407, 719
VIE EST BELLE, LA Vol. 2 185, 709, 857
VIE EST MAGNIFIQUE, LA Vol. 2 174, 402
VIE EST UN JEU, LA Vol. 2 191
VIE EST UN REVE, LA Vol. 2 137, 514
VIE EST UN ROMAN, LA Vol. 6 *page 90, page 96, page 105, page 128*
VIE FACILE, LA Vol. 2 804
VIE, L'AMOUR, LA MORT, LA Vol. 2 196, 704
VIE MIRACULEUSE DE THERESE MARTIN, LA Vol. 2 180, 387, 824
VIE NORMALE, LA Vol. 2 570
VIE PARISIENNE, LA Vol. 2 157, 197, 256, 407, 824
VIE PASSIONNEE DE CLEMENCEAU, LA Vol. 2 247
VIE PRIVEE Vol. 2 205, 224, 285, 322
VIE SANS JOIE, UNE Vol. 2 132
VIE TELLE QU'ELLE EST, LA Vol. 2 375
VIE T'EN AS QU'UNE, LA Vol. 2 342
VIE, UNE Vol. 2 96, 593, 719, 873
VIECHEREIEN Vol. 1 76, 127
VIEGE POUR ST. TROPEZ, UNE Vol. 2 396
VIEIL HOMME ET L'ENFANT, LE Vol. 2 515, 655, 664
VIEILLE DAME INDIGNE, LA Vol. 2 188, 599
VIEILLE FILLE, LA Vol. 2 517
VIEJA MUSICA, LA Vol. 8 297, 610
VIEL LÄRM UM NICHTS Vol. 1 2, 54, 75
VIEL LÄRM UM NIXI Vol. 4 373, 738 Vol. 7 862, 1061
VIELA Vol. 8 21, 30, 75, 147
VIELA RUA SEM SOL Vol. 8 79
VIELGELIEBTES STERNCHEN Vol. 1 9, 85
VIENI AVANTI CRETINO Vol. 7 763, 925
VIENS CHEZ MOI, J'HABITE CHEZ UNE COPINE Vol. 2 879 Vol. 6 *page 132*
VIENT DE PARAITRE Vol. 2 501, 698
VIENTO DE SIGLOS Vol. 8 276
VIENTO DEL NORTE Vol. 8 306, 396, 616, 630
VIENTO DEL SUR Vol. 8 551, 753
VIER GESELLEN, DIE Vol. 4 373, 626
VIER JAHRESZEITEN IM MENSCHLICHEN LEBEN, DIE Vol. 4 514
VIER MÄDEL UND EIN MANN Vol. 4 329, 774
VIER MUSKETIERE, DIE Vol. 4 124, 703
VIER TREPPEN RECHTS Vol. 4 246, 292
VIER VOM BOB 13, DIE Vol. 4 246, 298, 361, 692
VIERAS MIES TULI TALOON Vol. 5 349, 428
VIERAS NIES Vol. 5 392, 397
VIERGE AU PORTAIL, LA Vol. 2 424
VIERGE DU RHIN, LA Vol. 2 197, 259, 309
VIERGE ... FACON DE PARLER Vol. 6 *page 97*
VIERGE FOLLE, LA Vol. 2 55, 254, 446, 623, 775
VIERGES DE ROME, LES Vol. 2 309, 593
VIERGES, LES Vol. 2 778, 840
VIERNES Y TRECE Vol. 8 29
VIERTE GEBOT, DAS Vol. 4 247, 321
VIERTE KOMMT NICHT, DER Vol. 4 339, 422, 729
VIERTE VON RECHTS, DIE Vol. 4 213, 246

VIETATO AI MINORENNI Vol. 7 476
VIETNAM, SCENE DEL DOPOGUERRA Vol. 7 723
VIETTELYSTEN TIE Vol. 5 347, 379, 380
VIEUX DE LA VIEILLE, LES Vol. 2 118, 644, 832
VIEUX FUSIL, LE Vol. 2 74, 772
VIEUX LOUPS BENISSENT LA MORT, LES Vol. 2 640
VIEW FROM THE LEFT Vol. 3 437
VIFORNITA Vol. 3 246, 284, 285
VIGDIS Vol. 5 723, 760
VIGILE, IL Vol. 7 86, 781
VIGILI E VIGILESSE Vol. 7 909
VIGNES DU SEIGNEUR, LES Vol. 2 205, 279, 318, 347, 807
VIHAAN SINUA - RAKAS Vol. 5 374, 462
VIHAR Vol. 9 134, 168
VIHAR UTAN Vol. 9 68, 71
VIHARBRIGAD Vol. 9 76, 186, 230
VIHDOINKIN HÄÄYÖ Vol. 5 443, 643
VIHELTÄJÄT Vol. 5 584
VIHREÄ KULTA Vol. 5 386, 430, 592
VIHREÄ LESKI Vol. 5 434, 515
VIHTORI JA KLAARA Vol. 5 357, 581
VIIKON TYTTÖ Vol. 5 379, 600
VIIMEINEN KESÄ Vol. 5 321, 429, 475, 509
VIIMEINEN SAVOTTA Vol. 5 388, 477, 522, 540, 639
VIIMEINEN VIERAS Vol. 5 334, 592
VIISI VEKKULIA Vol. 5 374, 443
VIKEND BEZ RODICU Vol. 10 233, 286, 408
VIKEND ZA MILION Vol. 10 82, 128, 155
VIKI Vol. 9 178, 288, 334
VIKINGEBLOD Vol. 5 97
VIKINGO, EL Vol. 8 438, 667, 719
VIKTOR UND VIKTORIA Vol. 4 17, 232, 713
VIKTORIA Vol. 4 13, 614
VIKTORIA UND IHR HUSAR Vol. 4 373, 626
VIL DU SE MIN SMUKKE NALVE ? Vol. 5 259, 283
VIL SEDUCCION, LA Vol. 8 299, 719
VILAG CSAK HANGULAT, A Vol. 9 136
VILAGREM Vol. 9 77
VILD PA SEX Vol. 5 1124
VILDA JAKTEN PA LIKBILEN, DEN Vol. 5 1051
VILDANDEN Vol. 5 718, 754, 762, 799
VILDEMARKENS SANG Vol. 5 870, 949, 960, 987, 1005, 1049
VILDFAGEL, EN Vol. 5 891, 925, 937
VILDFAGLAR Vol. 5 878, 1052
VILDLEDT ELSKOV Vol. 5 97
VILDMANDEN Vol. 5 97
VILDMARKSSOMMAR Vol. 5 930, 1137
VILE, UN Vol. 7 1189
VILJELOS KAERLIGHED Vol. 5 9
VILL SA GÄRNA TRO Vol. 5 919
VILLA ALEGRE Vol. 8 551, 772
VILLA BORGHESE Vol. 7 106
VILLA DA VENDERE Vol. 7 476, 756
VILLA DELLE ANIME MALEDETTE, LA Vol. 7 59
VILLA DESTIN Vol. 2 44, 330, 506, 528
VILLA ELETTRICA, LA Vol. 7 1079
VILLA FALCONIERI Vol. 4 558, 725
VILLA IM TIERGARTEN, DIE Vol. 4 25, 255
VILLA „LES DUNES" Vol. 2 647
VILLA MEPHISTO Vol. 4 334
VILLA SANS-SOUCI, LA Vol. 2 496, 540, 722
VILLA VENNELY Vol. 5 135, 170
VILLAFRANCA Vol. 7 48, 1121, 1159
VILLAGE DE LA COLERE, LE Vol. 2 6, 475, 627
VILLAGE PERDU, LE Vol. 2 9, 368
VILLAGES QUI MEURENT Vol. 2 811
VILLAGREM Vol. 3 249
VILLAR STA GYNAIKEIA LOUTRA Vol. 3 157
VILLE ANDESONS ÄVENTYR Vol. 5 891, 925
VILLE BIDON, LA Vol. 2 190
VILLE DE SILENCES, LA Vol. 6 *page 128*
VILLE EST A NOUS, LA Vol. 2 157
VILLE MOYENNE, UNE Vol. 2 633
VILLEGGIATURA, LA Vol. 7 24
VILLERVALLE I SÖDERHAVET Vol. 5 885, 914
VILLI POHJOLA Vol. 5 462, 516
VILLIN POHJOLAN KULTA Vol. 5 383, 598
VILLIN POHJOLAN SALATTU LAAKSO Vol. 5 383, 598
VILLMARKENS LOV Vol. 5 662
VILNA ZONA Vol. 3 110 Vol. 6 *page 55, page 58*
VIM, ZE JSI VRAH Vol. 10 139, 425, 593
VIN CICLISTII Vol. 3 265
VINA VLADIMIRA OLMERA Vol. 10 229, 359, 433
VINATA Vol. 3 30, 122 Vol. 6 *page 31*
VINATOAREA DE VULPI Vol. 3 255, 272, 282
VINCENT, FRANCOIS, PAUL ET LES AUTRES Vol. 2 110, 588
VINCENT MIT L'ANE DANS UN PRE...ET S'EN VINT L'AUTRE Vol. 2 114
VINCITA DEL MILIONE Vol. 7 1189
VINCOLO SEGRETO Vol. 7 390
VINDICTA Vol. 2 169, 330, 613, 877

VINDINGEVALS **Vol. 5** 893, 1173
VINDUNDERDOKTOREN **Vol. 5** 97
VINELLA E DON PEZZOTTA **Vol. 7** 540
VINETA **Vol. 4** 71, 223, 291, 340
VINGAR KRING FYREN **Vol. 5** 928, 1156
VINGARNA **Vol. 5** 937, 997
VINGSLAG I NATTEN **Vol. 5** 862, 1042
VINGT ANS APRES **Vol. 2** 237
VINGT-CINQ ANS DE BONHEUR **Vol. 2** 67, 332
VINGT-CINQUIEME HEURE, LA **Vol. 2** 185
VINGT-HUIT JOURS DE CLAIRETTE, LES **Vol. 2** 152, 330
VINGT MILLE ANS A LA FRANCAISE **Vol. 2** 275
VINGT MILLE LIEUES SOUS LES MERS **Vol. 2** 679
VINGT-QUATRE HEURES DE LA VIE D'UN CLOWN **Vol. 2** 709
VINGT-QUATRE HEURES DE LA VIE D'UNE FEMME **Vol. 2** 173, 457, 875
VINGT-QUATRE HEURES DE PERM' **Vol. 2** 397, 863
VINGT-QUATRE HEURES D'UN AMERICAIN A PARIS **Vol. 2** 234
VINGT-SEPT RUE DE LA PAIX **Vol. 2** 9, 727
VINOBRANI **Vol. 10** 389, 517, 567
VINOVNA LI E ? **Vol. 3** 67
VINTERBORN **Vol. 5** 11, 38, 165
VINTERNAT, EN **Vol. 5** 705, 798
VINTI, I **Vol. 7** 135, 915, 1045, 1062
VIOLA AZ ALFÖLDI HARAMIA **Vol. 9** 27, 144
VIOLADOR VIOLADO, EL **Vol. 8** 551
VIOLANTA **Vol. 4** 8, 378
VIOLANTHA **Vol. 4** 213, 626
VIOLEE MAIS CONSENTANTE **Vol. 6** *page 69*
VIOLENCE ET PASSION **Vol. 7** 498, 736, 1025, 1142
VIOLENTAS, LAS **Vol. 8** 358
VIOLENTI DELLA ROMA BENE, I **Vol. 7** 728
VIOLENTS, LES **Vol. 2** 254, 407
VIOLENZA IN UN CARCERE FEMMINILE **Vol. 7** 311, 686
VIOLENZA SEGRETA **Vol. 7** 915, 1045
VIOLENZA SUL LAGO **Vol. 7** 1036
VIOLER ER BLA **Vol. 5** 125, 261
VIOLET **Vol. 4** 115, 291
VIOLETERA, LA **Vol. 8** 214, 261, 363
VIOLETTE ET FRANCOIS **Vol. 2** 21, 873
VIOLETTE NEI CAPELLI **Vol. 7** 443, 603
VIOLETTE NOZIERE **Vol. 2** 141, 700
VIOLETTES IMPERIALES **Vol. 2** 42, 59, 122, 388, 440, 566, 688
VIOLINES Y TROMPETAS **Vol. 8** 466, 728
VIOLINISTA NA KROVU **Vol. 6** *page 312*
VIOLINISTINDENS ROMAN **Vol. 5** 97
VIOLINO D'AUTORE **Vol. 7** 1189
VIOLINO DE JOAO, O **Vol. 8** 18, 24, 146
VIOLINO INCANTATO, IL **Vol. 7** 1189
VIOLONS DU BAL, LES **Vol. 2** 502, 527, 787
VIORICA **Vol. 3** 234
VIRAGES **Vol. 2** 131
VIRAGVASARNAP **Vol. 9** 130, 285, 316, 318
VIRAJ PERICULOS **Vol. 6** *page 270, page 273, page 277*
VIRE-VENT **Vol. 2** 581, 615
VIREE SUPERBE, LA **Vol. 2** 164, 220, 329, 640
VIRGEN DE CRISTAL, LA **Vol. 8** 228
VIRGEN DEL MAR, LA **Vol. 8** 723
VIRGILE **Vol. 2** 185, 248, 423, 738
VIRGINIE **Vol. 2** 347, 566
VIRGINITES A PRENDRE **Vol. 6** *page 144*
VIRILIDAD A LA ESPANOLA **Vol. 8** 677
VIRRAD **Vol. 9** 117, 332
VIRSTELE OMULUI **Vol. 3** 232, 272, 290
VIRTASET JA LAHTISET **Vol. 5** 371, 393
VIRTOUS MODEL **Vol. 2** 574
VIRUS, L'INFERNO DEI MORTI VIVENTI **Vol. 7** 216
VIS DE IANUARIE **Vol. 3** 281 **Vol. 6** *page 269, page 270*
VISA FÜR OCANTROS **Vol. 1** 8, 34, 104
VISA POUR L'ENFER **Vol. 2** 727
VISAGES DE FEMME **Vol. 2** 270, 722, 727
VISAGES D'ENFANTS **Vol. 2** 153, 297, 648
VISCONTE GIOVENTU E IL CONTE CENT'ANNI, IL **Vol. 7** 45, 1162
VISION, DIE **Vol. 4** 212
VISITA, LA **Vol. 7** 817, 1035
VISITA QUE NO TOCO EL TIMBRE, LA **Vol. 8** 562, 595
VISITEUR, LE **Vol. 2** 722, 826
VISITEURS DU SOIR, LES **Vol. 2** 402, 837, 863
VISKNINGAR OCH ROP **Vol. 5** 1080, 1196
VISNJA NA TASMAJDANU **Vol. 3** 381, 416
VISOKI NAPON **Vol. 6** *page 311, page 315*
VISPA TERESA, LA **Vol. 2** 807 **Vol. 7** 443
VISSZA AZ UTON **Vol. 9** 196, 322
VISSZAESÖK **Vol. 9** 20, 103, 154
VISSZASZAMLALAS **Vol. 9** 107, 238, 279, 280
VISUL LUI TANASE **Vol. 3** 257
VISUL UNEI NOPTI DE IARNA **Vol. 1** 278 **Vol. 3** 215, 294
VITA, A VOLTE, E MOLTO DURA, VERO „PROVVIDENZA" ? **Vol. 7** 408
VITA AGRA, LA **Vol. 7** 763

VITA DI SCORTA, LA **Vol. 7** 141
VITA DIFFICILE, UNA **Vol. 7** 86, 299, 1051
VITA E BELLA, LA **Vol. 7** 208, 599, 638, 756
VITA E FUMO, LA **Vol. 7** 1005
VITA E LA COMMEDIA, LA **Vol. 7** 45, 390
VITA E LA LEGGENDA, LA **Vol. 7** 843, 921
VITA FRUN **Vol. 5** 871, 880, 1028
VITA FUTURISTA **Vol. 7** 527
VITA IN GIOCO, LA **Vol. 7** 1145
VITA IN VENDITA, UNA **Vol. 7** 869
VITA KATTEN, DEN **Vol. 5** 1042, 1164
VITA NEJLIKAN **Vol. 5** 919, 933, 1054, 1213
VITA NUOVA **Vol. 9** 175, 220
VITA RICOMINCIA, LA **Vol. 7** 48, 756
VITA SELVAGGIA, SELVAGGIA TERRA **Vol. 7** 1137
VITA SEMPLICE, LA **Vol. 7** 250
VITA SENZA SCOPO, LA **Vol. 7** 608, 640
VITA TORNA, LA **Vol. 7** 187, 1062
VITA TRAVIATA **Vol. 7** 399
VITA VÄGGEN, DEN **Vol. 5** 858, 917
VITA VENDUTA, UNA **Vol. 7** 335, 852
VITA VIOLENTA, UNA **Vol. 7** 1035
VITAZ **Vol. 10** 82, 263, 544
VITE PERDUTE O LA LEGGE DEL MITRA **Vol. 7** 1136
VITEJI NEAMULUI **Vol. 3** 333
VITELLONI, I **Vol. 7** 257, 299, 539, 725, 1147, 1179
VITET E PARA **Vol. 3** 4, 16 **Vol. 6** *page 17*
VITEZNA KRIDLA **Vol. 10** 154, 184
VITEZNY LID **Vol. 10** 18, 184, 283, 313
VITR V KAPSE **Vol. 10** 389, 516, 545
VITRIOLDRAMA **Vol. 5** 97
VITTIMA DEL DOVERE **Vol. 7** 1189
VITTIMA DELLA PROPRIA ONESTA **Vol. 7** 1189
VITTIMA, LA **Vol. 7** 554
VITTIMA NASCOSTA, LA **Vol. 7** 69
VITTIME ! **Vol. 7** 973
VITUS THAVONS GENERALCOUP **Vol. 4** 49, 149, 660, 735
VIUDAS, LA **Vol. 8** 524, 602, 611, 725
VIUDITA NAVIERA, LA **Vol. 8** 214, 546
VIUDITA YE-YE, LA **Vol. 8** 261, 768
VIUUULENTEMENTE MIA **Vol. 7** 1059, 1086
VIV - HAH HAH - TAJA **Vol. 5** 324, 606
VIVA AMERICA ! **Vol. 8** 396
VIVA E ARENA **Vol. 7** 1063
VIVA EL PRESIDENTE **Vol. 2** 35, 787
VIVA IL CINEMA ! **Vol. 7** 257
VIVA LA AVENTURA ! **Vol. 8** 306, 426
VIVA LA CLASE MEDIA **Vol. 8** 214, 297
VIVA LA MUERTE **Vol. 2** 732
VIVA LA RIVISTA **Vol. 7** 257
VIVA LA VIDA ! **Vol. 8** 648
VIVA LA VIE **Vol. 6** *page 75, page 87*
VIVA L'ITALIA **Vol. 7** 710, 716, 1147
VIVA LO IMPOSIBLE **Vol. 8** 214, 338, 400
VIVA MADRID QUE ES MI PUEBLO **Vol. 8** 281, 364
VIVA MANOLIO ! **Vol. 3** 186
VIVA MARIA **Vol. 2** 224, 285
VIVA RENA **Vol. 3** 137
VIVAN LOS NOVIOS ! **Vol. 8** 338, 492
VIVANTE EPINGLE, LA **Vol. 2** 871
VIVE HENRI IV, VIVE L'AMOUR **Vol. 2** 226, 251, 625
VIVE LA COMPAGNIE **Vol. 2** 530, 531
VIVE LA FRANCE **Vol. 2** 112, 171
VIVE LA LIBERTE **Vol. 2** 579, 826
VIVE LA SOCIALE **Vol. 6** *page 75, page 89*
VIVE LA VIE **Vol. 2** 530, 531
VIVE LES FEMMES ! **Vol. 6** *page 99, page 104*
VIVE LES VACANCES **Vol. 2** 309, 611, 743
VIVEMENT DIMANCHE ! **Vol. 6** *page 70, page 87*
VIVENDO CANTANDO CHE MALE TI FO ? **Vol. 7** 1027
VIVERA A SBAFO **Vol. 7** 603
VIVERE ! **Vol. 7** 725, 1029
VIVERE ANCORA **Vol. 7** 1036
VIVERE IN PACE **Vol. 7** 101, 796, 798
VIVI O PREFERIBILMENTE MORTI **Vol. 8** 338, 681
VIVIENDO AL REVES **Vol. 8** 614, 640
VIVIR AL SOL **Vol. 8** 681, 741
VIVIR MANANA **Vol. 8** 601, 627
VIVIR UN LARGO INVIERNO **Vol. 8** 562, 768
VIVO RETRATO, EL **Vol. 8** 664, 703
VIVOS O PREFERIBILIMENTE MUERTOS **Vol. 8** 338, 681
VIVRE **Vol. 2** 388, 688
VIVRE ENSEMBLE **Vol. 2** 8
VIVRE LA NUIT **Vol. 2** 110, 597
VIVRE LIBRE **Vol. 2** 526
VIVRE LIBRE OU MOURIR **Vol. 2** 210
VIVRE SA VIE **Vol. 2** 208
VIZA ZA OKEANA **Vol. 3** 41 **Vol. 6** *page 27, page 43*
VIZA ZA ZLOTO **Vol. 3** 396, 443
VIZI MORBOSI DI UNA GOVERNANTE, I **Vol. 7** 1022
VIZI PRIVATI **Vol. 3** 437
VIZIETTO II, IL **Vol. 7** 375, 817, 1035
VIZIETTO, IL **Vol. 7** 817
VIZINHOS DO RES - DO - CHAO, OS **Vol. 8** 46, 88, 106

VIZIO ATAVICO **Vol. 7** 730
VIZIO DI FAMIGLIA, IL **Vol. 7** 1202
VIZIO E DELITTO **Vol. 7** 138
VIZIVAROSI NYAR **Vol. 9** 122, 259, 318
VJAT"RNATA MELNICA **Vol. 3** 32, 38, 120 **Vol. 6** *page 26, page 43*
VJAT"R"T NA P"TESESTVIJATA **Vol. 3** 48, 72, 78 **Vol. 6** *page 42*
VJERHANEN **Vol. 5** 118
VKUS NA BISER **Vol. 6** *page 54, page 55, page 62*
VLAD TEPES **Vol. 3** 254, 276, 297, 321
VLAK BEZ VOZNOG REDA **Vol. 3** 361, 371
VLAK DO STANICE NEBE **Vol. 10** 197, 394, 425
VLAK U SNIJEGU **Vol. 3** 430, 452
VLAKARI **Vol. 10** 128, 269, 493
VLAKOM PREMA JUGU **Vol. 6** *page 310*
VLASTNE SE NIC NESTALO **Vol. 10** 173, 325, 517
VLCI BOUDA **Vol. 10** 165, 495, 516
VLCI JAMA **Vol. 10** 154, 202, 503
V"LCICATA **Vol. 3** 47, 63, 69, 75
VLCIE DIERY **Vol. 10** 287, 503
VLEPE LOUKIANOS **Vol. 3** 183
VMV 6 **Vol. 5** 334, 349
VOCATION, LA **Vol. 2** 316, 608, 613
VOCE CHE UCCIDE, LA **Vol. 7** 905
VOCE DEL CUORE, LA **Vol. 7** 266
VOCE DEL SANGUE, LA **Vol. 7** 603
VOCE DEL SILENZIO, LA **Vol. 7** 910
VOCE D'ORO, LA **Vol. 7** 890, 1164
VOCE MONTANA, LA **Vol. 7** 244, 466, 796
VOCE SENZA VOLTO, LA **Vol. 7** 792, 1036
VOCE, UNA CHITARRA, UN PO'DI LUNA, UNA **Vol. 7** 333
VODKAA KOMISARIO PALMU **Vol. 5** 364, 443, 516
VOENNI DEJSTVIJA V MIRNO VREME **Vol. 3** 98
VOGEL IM KÄFIG, DER **Vol. 4** 603
VOGELHÄNDLER, DER **Vol. 4** 108, 119, 270
VOGLIA DI DONNA **Vol. 7** 509
VOGLIA DI GUARDARE **Vol. 7** 742
VOGLIA DI SESSO **Vol. 7** 742
VOGLIA, LA **Vol. 7** 742
VOGLIAMO I COLONNELLI **Vol. 7** 1086
VOGLIO BENE SOLTANTO A TE **Vol. 7** 938
VOGLIO MORTO, LO **Vol. 8** 306
VOGLIO TRADIRE MIO MARITO ! **Vol. 7** 196
VOGLIO VIVERE CON LETIZIA **Vol. 7** 453, 756, 1060
VOGLIO VIVERE COSI **Vol. 7** 443, 1136
VOGUE MON COEUR **Vol. 2** 42, 249, 360, 412
VOI ! **Vol. 7** 968, 1079
VOI JUKU MIKÄ LAUANTAI **Vol. 5** 560
VOI MEITÄ ! ANOPPI TULEE **Vol. 5** 356, 485
VOI VELJET, MIKÄ PÄIVÄ ! **Vol. 5** 403, 517
VOICI LE TEMPS DES ASSASSINS **Vol. 2** 388, 824
VOIE LACTEE, LA **Vol. 2** 377
VOIE SANS DISQUE, LA **Vol. 2** 451, 595
VOILE BLEU, LE **Vol. 2** 226, 339, 722
VOILE DU BONHEUR, LE **Vol. 2** 37, 134, 639
VOIR VENISE ET CREVER **Vol. 2** 344
VOISINE EST A DEPUCELER, LA **Vol. 6** *page 120*
VOIX DE SON MAITRE, LA **Vol. 2** 161, 265, 600
VOIX DU METAL, LA **Vol. 2** 648
VOIX DU REVE, LA **Vol. 2** 124
VOIX HUMAINE, LA **Vol. 2** 423
VOIX SANS VISAGE, LA **Vol. 2** 412, 778
VOJNARKA **Vol. 10** 61, 460, 621
VOJNATA NA TARALEZITE **Vol. 3** 59, 116 **Vol. 6** *page 53*
VOJNIK **Vol. 3** 391, 455
VOJNIKOVA LJUBAV **Vol. 3** 381, 398, 415
VOJNIK"T OT OBOZA **Vol. 3** 40, 120 **Vol. 6** *page 56*
VOJTECH, RECENY SIROTEK **Vol. 10** 41, 221
VOKSDAMEN **Vol. 5** 9
VOL, LE **Vol. 2** 145
VOLANI RODU **Vol. 10** 139, 325, 425
VOLANIE DEMONOV **Vol. 10** 72, 287, 478
VOLATA, LA **Vol. 7** 1057
VOLCA NOK **Vol. 3** 396, 419, 443
VOLDTEKT **Vol. 5** 664, 694, 767
VÖLEGENY **Vol. 9** 215, 259, 279
VŐLEGENY NYOLCKOR ERKEZIK, A **Vol. 9** 19, 138, 316
VOLEJTE MARTINA **Vol. 10** 299, 380
VOLETS CLOS, LES **Vol. 2** 234
VOLEUR DE BAGDAD, LE **Vol. 2** 81
VOLEUR DE CRIMES, LE **Vol. 2** 870
VOLEUR DE FEMMES, LE **Vol. 2** 174, 402 **Vol. 7** 583, 822
VOLEUR DE FEUILLES, LE **Vol. 6** *page 89, page 131, page 137*
VOLEUR DE TIBIDABO, LE **Vol. 8** 768
VOLEUR, LE **Vol. 2** 197, 206, 224, 770
VOLEUR SE PORTE BIEN, LE **Vol. 2** 92, 193
VOLEURS DE GLOIRE, LES **Vol. 2** 688
VOLEURS DE LA NUIT, LES **Vol. 6** *page 70, page 95*

VOLEUSE, LA **Vol. 2** 655, 773
VOLGA EN FLAMMES **Vol. 2** 23
VÖLGY, A **Vol. 9** 83, 183, 332
VOLHA V PLAMENECH **Vol. 10** 8, 256, 586, 602
VOLK IN NOT **Vol. 4** 208, 270
VOLKSFEIND, EIN **Vol. 4** 234, 536, 734
VOLKSGERICHT IM MITTELALTER ODER DIE ZEIT DES SCHRECKENS UND DES GRAUENS, EIN **Vol. 4** 308
VOLLDAMPF VORAUS **Vol. 4** 235, 247, 401, 626, 634, 720
VOLNA NOHA **Vol. 10** 43, 106, 541
VOLO DAL NIDO **Vol. 7** 626
VOLO DEGLI AIRONI, IL **Vol. 7** 301, 1032
VOLONTÄR, DER **Vol. 4** 734
VOLONTE **Vol. 2** 167
VOLOTARI PER DESTINAZIONE IGNOTA **Vol. 7** 426
VOLPONE **Vol. 2** 9, 58, 59, 386, 402, 657, 824
VOLT EGYSZER EGY CSALAD **Vol. 9** 152, 259, 273
VOLTA ALLA SETTIMANA, UNA **Vol. 7** 476, 756
VOLTATI EUGENIO **Vol. 7** 257, 340
VOLTI DELL'AMORE, I **Vol. 7** 569, 1140
VOLTO DEL PASSATO, IL **Vol. 7** 150
VOLTO DI MEDUSA, IL **Vol. 7** 45, 390
VOLUNTAD **Vol. 8** 723
VOLUTTA DI MORTE **Vol. 7** 131
VOLVER A EMPEZAR **Vol. 8** 582, 616
VOLVER A VIVIR **Vol. 8** 258
VOLVERE A NACER **Vol. 8** 245, 306
VOM KÖNIG MIDAS **Vol. 1** 29, 82, 125
VOM LEBEN GETÖTET **Vol. 4** 366, 367
VOM REGEN IN DIE TRAUFE **Vol. 4** 24, 525
VOM SCHICKSAL VERWEHT **Vol. 4** 361 **Vol. 7** 453
VOM TÄTER FEHLT JEDE SPUR **Vol. 4** 49, 584
VON BUTTIGLIONE STURMTRUP-PENFÜHRER **Vol. 7** 1015
VON MELODIE ZU MELODIE **Vol. 1** 73
VON MORGENS BIS MITTERNACHT **Vol. 4** 288, 479
VON STUFE ZU STUFE BIS IN DEN TOD **Vol. 4** 232
VONTADE MAIOR, UMA **Vol. 8** 122, 130
VOR FAELLES VEN **Vol. 5** 136, 228
VOR LIEBE WIRD GEWARNT **Vol. 4** 123, 727, 774
VOR TIDS DAME **Vol. 5** 97
VOR TIDS HELTER **Vol. 5** 798
VORACES, LES **Vol. 2** 241, 376
VORAGINE **Vol. 7** 442
VORBESTRAFTEN, DIE **Vol. 4** 124, 747
VORDERHAUS UND HINTERHAUS **Vol. 4** 558
VORDERTREPPE UND HINTERTREPPE **Vol. 4** 178, 213, 651
VORE VENNERS VINTER **Vol. 5** 24, 83
VORES FJERDE FAR **Vol. 5** 228
VORES LILLE BY **Vol. 5** 122
VORGLUTEN DES BALKANBRANDES **Vol. 4** 239
VORHANG FÄLLT, DER **Vol. 4** 325, 378, 567
VÖRÖS FÖLD **Vol. 9** 59, 238
VÖRÖS GROFNÖ **Vol. 9** 37, 238
VÖRÖS REKVIEM **Vol. 9** 20, 253, 318
VÖRÖS SAMSON, A **Vol. 9** 28
VÖRÖS TINTA **Vol. 9** 149, 259, 297
VORSICHT AM PLATZE **Vol. 4** 261, 323
VORTICE **Vol. 7** 638, 1054
VORUNTERSUCHUNG **Vol. 4** 30, 325, 713
VOS GUEULES LES MOUETTES **Vol. 2** 414, 835
VOTO, IL **Vol. 7** 333, 429, 968
VOTRE DEVOUE BLAKE **Vol. 2** 185, 501
VOTRE ENFANT M'INTERESSE **Vol. 2** 502
VOTRE SOURIRE **Vol. 2** 55, 84
VOULEZ-VOUS DANSER AVEC MOI ? **Vol. 2** 21, 232, 488
VOULEZ-VOUS FAIRE DU CINEMA ? **Vol. 2** 104
VOULEZ-VOUS UN BEBE NOBEL ? **Vol. 2** 275
VOULOIR **Vol. 2** 131
VOUS HABITEZ CHEZ VOS PARENTS ? **Vol. 6** *page 87, page 104*
VOUS INTERESSEZ-VOUS A LA CHOSE ? **Vol. 2** 335
VOUS N'AUREZ PAS L'ALSACE ET LA LORRAINE **Vol. 2** 7, 251
VOUS N'AVEZ RIEN A DECLARER ? **Vol. 2** 48, 249, 488, 523
VOUS NE L'EMPORTEREZ PAS AU PARADIS **Vol. 2** 779
VOUS PIGEZ ? **Vol. 2** 611, 702
VOUS SEULE QUE J'AIME **Vol. 2** 55, 70, 338, 339
VOYAGE A BIARRITZ **Vol. 2** 402
VOYAGE A TROIS **Vol. 2** 581
VOYAGE AU BOUT DU MONDE, LE **Vol. 2** 618
VOYAGE AU JARDIN DES MORTS **Vol. 2** 331
VOYAGE D'AMELIE, LE **Vol. 2** 164, 341
VOYAGE DANS LA LUNE, LE **Vol. 2** 182, 591
VOYAGE DE MONSIEUR PERRICHON, LE **Vol. 2** 362, 388, 863
VOYAGE DE NOCES **Vol. 2** 48, 351
VOYAGE DU PERE, LE **Vol. 2** 128, 835

VOYAGE EN AMERIQUE, LE **Vol. 2** 13, 776
VOYAGE EN BALLON, LE **Vol. 2** 294, 819
VOYAGE EN CAPITAL **Vol. 2** 406, 502
VOYAGE EN DOUCE, LE **Vol. 2** 31, 481
VOYAGE IMAGINAIRE, LE **Vol. 2** 87, 388, 610, 614
VOYAGE IMPREVU, LE **Vol. 2** 122, 674
VOYAGE SANS ESPOIR **Vol. 2** 388, 488
VOYAGE-SURPRISE **Vol. 2** 156, 786
VOYAGEUR DE LA TOUSSAINT, LE **Vol. 2** 615, 826
VOYAGEUR SANS BAGAGES, LE **Vol. 2** 439, 566
VOYAGEUSE INATTENDUE, LA **Vol. 2** 197, 488
VOYANTE, LA **Vol. 2** 6, 193, 346
VOYANTS, LES **Vol. 2** 872
VOYOU, LE **Vol. 2** 196
VRAGET **Vol. 5** 97
VRAGOVE **Vol. 3** 34
VRAH SKRYVA TVAR **Vol. 10** 22, 395, 593
VRAH ZO ZAHROBIA **Vol. 10** 72, 287, 478
VRAI COUPABLE, LE **Vol. 2** 9, 124
VRAIE HISTOIRE DE GERARD LECHOMEUR, LA **Vol. 6** *page 72*
VRAK **Vol. 10** 74, 129, 466, 625
VRANE **Vol. 3** 381
VRAT SE DO HROBU ! **Vol. 10** 88, 139, 293
VRATA OSTAJU OTVORENA **Vol. 3** 338, 375
VRATA RAJA **Vol. 1** 187, 337 **Vol. 3** 349
VRATICU SE **Vol. 3** 390, 444
VRAZDA ING. CERTA **Vol. 10** 289, 325, 359
VRAZDA PO NASEM **Vol. 10** 227, 313, 371
VRAZDA V HOTELU EXCELSIOR **Vol. 10** 70, 154, 547
VRAZDA V OSTROVNI ULICI **Vol. 10** 453, 586
VRAZEDNE POCHYBNOSTI **Vol. 10** 59, 286, 408
VRCHNI, PRCHNI **Vol. 10** 106, 395, 505
VREAU SA STIU DE CE AM ARIPI **Vol. 6** *page 269, page 271, page 280*
VRECKARI **Vol. 10** 32, 72, 107
VREDENS DAG **Vol. 5** 2, 7, 91, 285
VREME BEZ VOJNA **Vol. 3** 419
VREME LEOPARDS **Vol. 6** *page 319*
VREME LJUBAVI **Vol. 3** 416
VREME, VODI **Vol. 3** 384, 419 **Vol. 6** *page 308*
VREMEA ZAPEZILOR **Vol. 3** 297, 317, 319
VRHOVI ZELENGORE **Vol. 3** 374
VRIMEN TAMBIEN JUEGA, EL **Vol. 8** 311
VRTLOG **Vol. 3** 372
VSANOU NOVI BOJOVNICI **Vol. 10** 142, 296, 564
VSE PRO LASKU **Vol. 10** 365
VSECHNO NEBO NIC **Vol. 10** 106, 313, 325
VSEKI DEN, VSJAKA NOST **Vol. 3** 121 **Vol. 6** *page 50*
VSICHNI DOBRI RODACI **Vol. 10** 289, 292, 313
VSICHNI MAJI TALENT **Vol. 10** 149, 258, 570
VSICHNI MUSI BYT V PYZAMU **Vol. 10** 394, 505
VSICHNI PROTI VSEM **Vol. 10** 317, 517, 522
VSICKI I NIKOJ **Vol. 3** 122 **Vol. 6** *page 27, page 28*
VSICKO E LJUBOV **Vol. 3** 58 **Vol. 6** *page 38, page 56*
VSTUP ZAKAZAN **Vol. 10** 77, 87, 184, 286, 387
VSUDE ZIJI LIDE **Vol. 10** 30, 359, 532
VTACKOVIA, SIROTY A BLAZNI **Vol. 10** 9, 273, 323
VTAK NOCIAR **Vol. 10** 155, 228, 592
VU ARME SYNDERE **Vol. 5** 233
VU DU PONT **Vol. 2** 427, 507, 770
VUDU, LOS **Vol. 8** 590
VUELO 971 **Vol. 8** 541, 725
VUELTA, LA **Vol. 8** 492, 768
VUELVE SAN VALENTIN **Vol. 8** 214
VUES D'ICI **Vol. 2** 108
VUK SAMOTNJAK **Vol. 3** 374
VULA **Vol. 3** 48, 63
VULCANO **Vol. 2** 178
VULTURUL 101 **Vol. 3** 263, 266, 281
VUOKRASULHANEN **Vol. 5** 379, 600
VY NEZNATE ALBERTA ? **Vol. 10** 168, 412
VYBUCH BUDE V PET **Vol. 10** 255, 299, 337
VYDELECNE ZENY **Vol. 10** 256, 460
VYDERA **Vol. 10** 460, 621
VYHRAVAT POTICHU **Vol. 10** 233, 395, 511
VYHYBKA **Vol. 10** 210, 268, 451
VYJIMECNA SITUACE **Vol. 10** 28, 495, 570
VYKRIK DO SIBIRSKE NOCI **Vol. 10** 92, 256
VYLET DO MLADOSTI **Vol. 10** 31, 155, 454
VYLET PO DUNAJI **Vol. 10** 268, 276, 320
VYNALEZ ZKAZY **Vol. 10** 462, 550
VYSOKA MODRA ZED **Vol. 10** 193, 286, 296, 612, 614
VYSOKA ZED **Vol. 10** 235, 433, 577
VYSSI PRINCIP **Vol. 10** 503, 564, 582

VYSTRAHA Vol. 10 142, 460, 621
VYSTRELY V MARIANSKYCH LAZNICH Vol. 10 283, 318, 405, 550
VYZNAVACI SLUNCE Vol. 10 34, 453
VZBOURENI NA VSI Vol. 10 122, 552, 585
VZDUSNE TORPEDO 48 Vol. 10 34, 167, 256
V"ZDUSNIJAT COVEK Vol. 3 110 Vol. 6 *page 52, page 63*
VZDY MOZNO ZACAT Vol. 10 210, 253, 268
VZHURU NOHAMA Vol. 10 532, 621
VZORNY KINEMATOGRAF HASKA JAROSLAVA Vol. 10 311, 388, 503
VZTEKLY ZENICH Vol. 10 167, 198

W

W BIALY DZIEN Vol. 6 *page 210, page 253, page 255*
W KAZDA POGODE Vol. 1 220
W LA FOCA Vol. 7 315
W LASACH POLSKICH Vol. 1 170, 371
W OBRONIE WLASNEJ Vol. 6 *page 211, page 226*
W PUSTYNI I W PUSZCZY Vol. 1 210, 232, 251, 323, 355
W SRODKU LATA Vol. 1 190, 198, 295, 297 Vol. 6 *page 254*
W STARYM DWORKU, CZYLI NIEPODLEGLOSC TROJKATOW Vol. 6 *page 228, page 233, page 235*
W SZPONACH CARSKICH SIEPACZY Vol. 1 174, 176, 326
W TE DNI PRZEDWIOSENNE Vol. 1 198, 306 Vol. 6 *page 258*
WACHSFIGURENKABINETT, DAS Vol. 4 396, 397, 437, 681
WACUS Vol. 1 278, 308, 404
WAFFEN DER JUGEND, DIE Vol. 4 514
WAGON SHOW, THE Vol. 2 84
WAHLVERWANDTSCHAFTEN, DIE Vol. 1 80, 110 Vol. 6 *page 158*
WAHN DES PHILIPP MORRIS, DER Vol. 4 660, 704
WAHNSINN Vol. 4 270, 288
WAHRE JAKOB, DER Vol. 4 536
WAISE VOM WEDDING, DIE Vol. 4 232, 774
WAISE VON CAPRI, DIE Vol. 4 149, 535, 660
WAISE VON LOWOOD, DIE Vol. 4 329, 450, 733
WALDFRIEDEN Vol. 4 344, 647
WALDRAUSCH Vol. 4 323, 347, 370
WALDWINTER Vol. 4 30, 445, 567
WALET PIKOWY Vol. 1 198, 303, 323, 343
WALKOWER Vol. 1 202, 227, 283
WALLENSTEIN Vol. 4 124, 213
WALLY Vol. 7 48, 652, 756, 885
WALPURGISNACHT Vol. 4 89, 124, 258
WALZER FÜR DICH, EIN Vol. 4 115, 774
WALZER IM SCHLAFCOUPE, EIN Vol. 4 270, 409, 634
WALZER MIT DIR, EIN Vol. 4 340, 623, 638
WALZERKÖNIG, DER Vol. 4 208, 255
WALZERKRIEG, DER Vol. 2 399 Vol. 4 266, 288, 574
WALZERNACHT, EINE Vol. 4 560
WALZERPARADIES Vol. 4 97, 373, 479
WALZERTRAUM, EIN Vol. 4 38, 89 Vol. 6 *page 194*
WAMPIRY WARSZAWY Vol. 1 371, 380
WANDA, LA PECCATRICE Vol. 7 476, 796, 1054
WANDA SARACETO Vol. 7 1125
WANDA WARENINE Vol. 7 1073
WANDERING DAUGHTERS Vol. 2 84
WANDERNDE BILD, DAS Vol. 4 298, 325, 536, 646
WANDERNDE KOFFER, DER Vol. 4 428, 626
WANKENDE GLAUBE, DER Vol. 4 183
WAR AND PEACE Vol. 2 178 Vol. 7 299, 750, 1136
WAR ES DER IM 3. STOCK ? Vol. 4 340, 354, 466
WÄRE DIE ERDE NICHT RUND ... Vol. 6 *page 158, page 163, page 164*
WARE FÜR KATALONIEN Vol. 1 58, 100, 129
WARENHAUSPRINZESSIN, DIE Vol. 4 49, 424
WARSCHAUER ZITADELLE, DIE Vol. 4 270, 409, 462, 471, 734 Vol. 6 *page 187*
WARSZAWKA Vol. 1 174, 176, 326
WARSZAWSKA PREMIERA Vol. 1 133, 250
WARSZAWSKA SYRENA Vol. 1 187, 227, 232 Vol. 6 *page 239*
WARUM LÜGST DU, ELISABETH ? Vol. 4 72, 329
WARUM LÜGT FRÄULEIN KÄTHE ? Vol. 4 108, 115, 373
WARUM SICH SCHEIDEN LASSEN ? Vol. 4 85, 535
WARUM SO AUFGEREGT ? Vol. 4 361, 558
WAS BIN ICH OHNE DICH ? Vol. 4 340, 354, 546
WAS DER TOTENKOPF ERZÄHLT Vol. 6 *page 186*
WAS EINE FRAU IM FRÜHLING TRÄUMT Vol. 4 471, 560
WAS FRAUEN TRÄUMEN Vol. 4 208, 246

WAS FÜR EIN TAG Vol. 1 75, 100, 119
WAS GESCHAH IN DIESER NACHT ? Vol. 4 51, 108, 729
WAS HALTEN SIE VON MUSIK? Vol. 1 37, 85, 103
WAS IHR WOLLT Vol. 1 26, 90, 115
WAS IST LOS IM ZIRKUS BEELY ? Vol. 4 471, 560, 779
WAS IST LOS MIT NANETTE ? Vol. 4 212, 329, 567
WAS KINDER DEN ELTERN VERSCHWEIGEN Vol. 4 25, 329
WAS KOSTET DIE LIEBE ? Vol. 6 *page 197*
WAS SAGEN SIE DAZU ? Vol. 4 18, 558
WAS SAGT ONKEL EMIL DAZU ? Vol. 4 623
WAS STEINE ERZÄHLEN Vol. 4 124, 408
WAS TUN SYBILLE ? Vol. 4 24, 28, 445
WAS WÄRE, WENN...? Vol. 1 21, 36, 120
WAS WILL BRIGITTE ? Vol. 4 124, 344, 615
WAS WIRD HIER GESPIELT ? Vol. 4 49, 249, 340
WAS WISSEN DENN MÄNNER ? Vol. 4 247, 614
WASSER FÜR CANITOGA Vol. 4 123, 302, 344, 638
WASSERTEUFEL VON HIEFLAU, DIE Vol. 4 15, 354, 372, 424
WATERLOO Vol. 4 323, 544, 729, 775 Vol. 7 498, 817
WE WLASNE SIDLA Vol. 1 404
WEBER, DIE Vol. 4 14, 186, 743
WEDLE WYROKOW TWOICH ... Vol. 6 *page 238, page 241, page 250*
WEEK-END Vol. 2 208 Vol. 5 2, 24, 227
WEEK-END A ZUYDCOOTE Vol. 2 185, 224
WEEKEND IM PARADIES Vol. 4 208, 379, 479
WEEKEND Z DZIEWCZYNA Vol. 6 *page 246*
WEEKENDS MALEFIQUES DU COMTE ZAROFF, LES Vol. 2 818
WEEKENDY Vol. 1 313
WEEKENDY - AUTOBUSY JAK ZOLWIE Vol. 1 143, 243, 355
WEEKENDY - JULIA Vol. 1 143, 243, 355
WEEKENDY Z DZIEWCZYNA Vol. 1 283, 403
WEG DER ERÖSUNG, DER Vol. 6 *page 194*
WEG DER GRETE LESSEN, DER Vol. 4 660, 704
WEG DER TRÄNEN, DER Vol. 4 743
WEG, DER ZUR VERDAMMNIS FÜHRT, DER Vol. 4 239, 288
WEG DES LEBENS, DER Vol. 4 183
WEG DES TODES, DER Vol. 4 124
WEG DURCH DIE NACHT, DER Vol. 4 73, 97
WEG INS FREIE, DER Vol. 4 159, 235, 742
WEG NACH RIO, DER Vol. 4 149, 157, 208, 660
WEG ZU GOTT, DER Vol. 4 255, 743
WEG ZU ISABELL, DER Vol. 4 13, 119, 234, 738 Vol. 6 *page 187*
WEG ZUM LICHT, DER Vol. 4 344, 545
WEGE DER LIEBE Vol. 4 3, 634
WEGE, DIE INS DUNKLE FÜHREN Vol. 6 *page 186*
WEGE ZU KRAFT UND SCHÖNHEIT Vol. 4 95, 149, 293, 481, 511, 660, 743
WEGE ZUR GUTEN EHE Vol. 4 164, 287, 361
WEHE, WENN ER LOSGELASSEN Vol. 4 257
WEHE, WENN SIE LOSGELASSEN Vol. 4 113, 626
WEHRLOSE OPFER Vol. 4 270, 695
WEIB DES PHARAO, DAS Vol. 4 242, 294, 452, 560, 663, 681
WEIB, EIN TIER, EIN DIAMANT, EIN Vol. 4 89, 416, 546
WEIB IN FLAMMEN Vol. 4 166, 527
WEIBERKRIEG, DER Vol. 4 344, 565
WEIBERREGIMENT Vol. 4 445, 713
WEIBERWIRTSCHAFT Vol. 6 *page 155, page 163*
WEIHNACHTS-FREUD UND LEID Vol. 4 135, 503
WEIHNACHTSGANS AUGUSTE, DIE Vol. 1 99
WEIHNACHTSGLÜCK Vol. 4 646
WEIHNACHTSMANN HEISST WILLI, DER Vol. 1 14, 34, 128
WEIHNACHTSTRÄNEN Vol. 4 247
WEIL DU ES BIST Vol. 4 287, 373, 420, 644
WEIL ICH DICH LIEBE... Vol. 1 15, 71, 121
WEIN, WEIB, GESANG Vol. 4 95, 261, 277, 364
WEISSE ABENTEUER, DAS Vol. 2 871
WEISSE DÄMON, DER Vol. 4 85, 288
WEISSE HÖLLE VOM PIZ PALÜ, DIE Vol. 4 10, 15, 452, 620
WEISSE MAJESTÄT, DIE Vol. 4 432
WEISSE PFAU, DER Vol. 4 124, 247, 351, 396
WEISSE RAUSCH, DER Vol. 4 15, 73, 212, 481
WEISSE SKLAVEN Vol. 4 354, 422, 788
WEISSE SKLAVIN, DIE Vol. 4 149, 223, 247, 270, 535, 594, 660

WEISSE SPINNE, DIE Vol. 4 242, 357
WEISSE TEUFEL, DER Vol. 2 833 Vol. 4 66, 113, 413, 442, 707 Vol. 6 *page 192, page 195*
WEISSE TOD, DER Vol. 4 724
WEISSE TRAUM, DER Vol. 4 620, 685
WEISSE UNTER KANNIBALEN, EINE Vol. 4 98, 112
WEISSE WÄSCHE Vol. 4 215, 234, 292
WEISSE WÖLFE Vol. 1 11, 96, 102
WEISSEN ROSEN VON RAVENSBURG, DIE Vol. 4 158, 247, 340, 429, 734
WEISSER FLIEDER Vol. 4 79, 115, 564
WEISSER OTHELLO, EIN Vol. 6 *page 186*
WEISSES BLUT Vol. 1 36, 56, 115
WEITE STRASSEN - STILLE LIEBE Vol. 1 32, 114
WELCOME STRANGER Vol. 2 84
WELLE 4711 Vol. 4 329
WELLEN DER LEIDENSCHAFT Vol. 4 119, 234, 394
WELLINGTON CONTRO SE STESSO Vol. 7 433, 1067
WELT DES SCHEINS, DIE Vol. 4 471
WELT GEHT UNTER, DIE Vol. 4 514
WELT IN FLAMMEN, DIE Vol. 4 329
WELT OHNE HUNGER, DIE Vol. 4 321
WELT OHNE MASKE, DIE Vol. 4 119, 270
WELT OHNE WAFFEN, DIE Vol. 4 276, 634
WELT STEHT KOPF, DIE Vol. 4 747
WELT WILL BETROGEN SEIN, DIE Vol. 4 113, 297, 681, 733, 780
WELTBRAND Vol. 4 124, 423
WELTMEISTER, DER Vol. 4 113, 780
WELTREKORD IM SEITENSPRUNG Vol. 4 171, 517, 558, 647
WELTSPIEGEL, DER Vol. 4 476, 522
WEM DURCH DIE LIEBE LEID GESCHAH Vol. 6 *page 199*
WEN DIE GÖTTER LIEBEN Vol. 4 13, 85
WENDEE, LA CHIAVE DEL PIACERE Vol. 7 99
WENN AM SONNTAGABEND DIE DORFMUSIK SPIELT Vol. 4 247, 734
WENN COLOMBINE WINKT Vol. 4 300
WENN DAS HERZ DER JUGEND SPRICHT Vol. 4 149, 190, 208, 660
WENN DAS HERZ IN HASS ERGLÜHT Vol. 4 309
WENN DAS LEBEN NEIN SAGT Vol. 4 385
WENN DAS LEBEN RUFT Vol. 4 514
WENN DEM ESEL ZU WOHL IST Vol. 4 344, 544
WENN DER HAHN KRÄHT Vol. 4 373, 626
WENN DER JUNGE WEIN BLÜHT Vol. 4 65, 232, 354, 422, 535, 566
WENN DER WEISSE FLIEDER WIEDER BLÜHT Vol. 4 344, 544
WENN DIE ABENDGLOCKEN LÄUTEN Vol. 4 25
WENN DIE LAWINEN STÜRZEN Vol. 6 *page 189*
WENN DIE LIEBE MODE MACHT Vol. 4 85, 89
WENN DIE LIEBE NICHT WÄR' ! Vol. 4 36, 270
WENN DIE MASKE FÄLLT Vol. 4 646
WENN DIE MUSIK NICHT WÄR' ! Vol. 4 373, 437, 738
WENN DIE MUTTER UND DIE TOCHTER... Vol. 4 212, 424
WENN DIE SOLDATEN... Vol. 4 255, 292, 725
WENN DIE SONNE WIEDER SCHEINT Vol. 4 119, 265 Vol. 6 *page 189*
WENN DU EINE SCHWIEGERMUTTER HAST Vol. 4 132, 232, 339, 727
WENN DU EINE TANTE HAST Vol. 4 85, 242
WENN DU EINMAL DEIN HERZ VERSCHENKST Vol. 4 164, 576, 584, 729
WENN DU NOCH EINE HEIMAT HAST Vol. 4 124, 208
WENN DU ZU MIR HÄLTST Vol. 1 26, 42
WENN EIN KLEINES MÄDEL SPIELT Vol. 4 376, 633
WENN EIN MÄDCHEN HÜBSCH IST Vol. 4 232, 673
WENN EIN MÄDEL HOCHZEIT MACHT Vol. 4 97, 135
WENN EIN WEIB DEN WEG VERLIERT Vol. 4 213, 414
WENN EINER EINE REISE TUT Vol. 4 361
WENN FRAUEN LIEBEN UND HASSEN Vol. 4 366
WENN FRAUEN SCHWEIGEN Vol. 4 24, 30, 445
WENN ICH KÖNIG WÄR ! Vol. 4 234, 471, 566, 734
WENN IN WERDER BLÜHN DIE BÄUME Vol. 4 394
WENN MAN BUMMELN GEHT Vol. 4 514
WENN MÄNNER KOCHEN Vol. 4 97, 230, 239
WENN MÄNNER VERREISEN Vol. 4 340, 703
WENN MENSCHEN IRREN Vol. 4 36, 356
WENN MENSCHEN REIF ZUR LIEBE WERDEN Vol. 4 292, 340
WENN MUTTER NICHT ZUHAUSE IST Vol. 4 230, 232, 239
WENN TOTE SPRECHEN Vol. 4 124, 232 Vol. 6 *page 194*

WENN VIER DASSELBE TUN **Vol. 4** 560
WENN WIR ALLE ENGEL WÄREN **Vol. 4** 373, 626
WER BIST DU ? **Vol. 4** 119, 517, 558
WER DAS SCHEIDEN HAT ERFUNDEN **Vol. 4** 149, 219, 660
WER HAT ANGST VOR MARMADUKE ? **Vol. 4** 24, 365
WER HAT ROBBY GESEHEN ? **Vol. 4** 239, 340
WER IST DER TÄTER ? **Vol. 4** 779
WER KÜSST MADELEINE ? **Vol. 4** 564, 608
WER KÜSST MICH ? **Vol. 6** *page 189*
WER NIMMT DIE LIEBE ERNST ? **Vol. 4** 113, 558
WER REISST DENN GLEICH VORM TEUFEL AUS **Vol. 1** 15, 59, 79
WER SEINE FRAU LIEB HAT **Vol. 1** 22, 34
WER STIRBT SCHON GERN UNTER PALMEN **Vol. 3** 395
WER WAGT – GEWINNT **Vol. 4** 149, 634, 660
WER WAR ES ? **Vol. 4** 644, 724
WER WEISS ? **Vol. 4** 560
WER WIRD DENN WEINEN, WENN MAN AUSEINANDERGEHT **Vol. 4** 192, 270, 462, 614 **Vol. 6** *page 195*
WER WIRFT DEN ERSTEN STEIN ? **Vol. 4** 223, 270
WER ZAHLT HEUTE NOCH ? **Vol. 4** 270, 409, 536
WERFT ZUM GRAUEN HECHT, DIE **Vol. 4** 124, 742
WERTHER **Vol. 8** 297, 616, 699 **Vol. 10** 167, 453
WESELA NIE BEDZIE **Vol. 1** 189, 207, 212
WESELE **Vol. 1** 325, 403, 405
WESELI BIEDACY **Vol. 1** 169, 290
WEST-INDIES **Vol. 2** 161, 770
WEST TI VA STRETTO AMICO ... E ARRIVATO ALLELUJA !, IL **Vol. 7** 743
WESTERN DI COSE NOSTRE **Vol. 7** 178, 819, 908
WESTERPLATTE **Vol. 1** 227, 237, 398 **Vol. 6** *page 242*
WESTFRONT 1918 **Vol. 4** 450, 452, 542, 729
WETTE, DIE **Vol. 4** 354
WETTE UM DIE SEELE, DIE **Vol. 4** 594
WETTE UM EINEN KUSS **Vol. 4** 239, 340, 466
WETTERLEUCHTEN UM BARBARA **Vol. 4** 10, 517, 558
WETTERWART, DER **Vol. 4** 149, 165, 190, 589, 660, 704
WETTLAUF UMS GLÜCK **Vol. 4** 219, 558, 634, 763
WEZWANIE **Vol. 1** 311 **Vol. 6** *page 238, page 245*
WHAT A FLASH **Vol. 2** 685
WHAT'S A WIFE WORTH ? **Vol. 2** 84
WHEN THIEF MEETS THIEF **Vol. 2** 32
WHISKY Y VODKA **Vol. 8** 402, 749
WHISTLE, THE **Vol. 5** 76
WHITE „POP" JESUS **Vol. 7** 311
WHITE SISTER, THE **Vol. 7** 978
WHITECHAPEL **Vol. 4** 247, 558
WHY GET MARRIED ? **Vol. 2** 84
WIANO **Vol. 1** 179, 382 **Vol. 6** *page 228*
WIATR OD MORZA **Vol. 1** 278, 364
WIDE OPEN **Vol. 5** 1217
WIDZIADLO **Vol. 6** *page 214, page 226*
WIE BLEIBE ICH JUNG UND SCHÖN **Vol. 4** 239, 270, 644
WIE DER HASE LÄUFT **Vol. 4** 30, 424, 461
WIE DER VATER, SO DER SOHN **Vol. 4** 514
WIE DIE WILDEN **Vol. 1** 46, 86, 100
WIE DU MIR – SO ICH DIR **Vol. 4** 361, 558
WIE EIN DIEB IN DER NACHT **Vol. 4** 483, 714
WIE EIN EI DEM ANDERN **Vol. 4** 18, 558
WIE EIN WUNDER KAM DIE LIEBE ÜBER NACHT **Vol. 4** 323, 329
WIE EINST IM MAI **Vol. 4** 149, 213, 396, 660, 774
WIE EULENSPIEGEL ZU MARBURG DEN LANDGRAFEN MALTE **Vol. 4** 774
WIE FRITZCHEN SICH DIE REICHSTAGS-KÄMPFE UND NEU-WAHLEN DENKT **Vol. 4** 308
WIE FÜTTERT MAN EINEN ESEL **Vol. 1** 128
WIE HEIRATE ICH MEINEN CHEF **Vol. 4** 255, 527
WIE HEIRATET MAN EINEN KÖNIG **Vol. 1** 80, 92, 102
WIE IMPONIERE ICH MEINER FRAU ? **Vol. 4** 79, 115, 239
WIE KOMMEN DIE DIE LÖCHER IN DEN KÄSE ? **Vol. 4** 536
WIE KONNTEST DU, VERONIKA **Vol. 4** 329, 373
WIE MAN MILLIONÄR WIRD **Vol. 6** *page 186*
WIE SAG ICH'S MEINEM MANN **Vol. 4** 288, 614
WIE SAGEN WIR ES UNSEREN KINDERN **Vol. 4** 28, 728
WIE SICH DAS KIND RÄCHT **Vol. 4** 593

WIE SOLL SICH EINE FRAU ENTSCHEIDEN Vol. 1 73, 114, 122
WIE TAUSEND SONNEN Vol. 1 46
WIE WÄRS MIT UNS BEIDEN Vol. 6 *page 167, page 174, page 175*
WIE WERD ICH BLOSS DIE PERLE LOS Vol. 4 18, 461
WIE WERDE ICH ENERGISCH ? Vol. 4 30, 567
WIE WERDE ICH REICH UND GLÜCKLICH ? Vol. 4 306, 373
WIE WOHNE ICH GUT UND BILLIG Vol. 4 479, 536
WIECZNA MLODOSC Vol. 1 346, 357
WIECZNE PRETENSJE Vol. 1 164, 378 Vol. 6 *page 217*
WIEDERGEFUNDEN Vol. 4 514
WIEDERSEH'N MACHT FREUDE Vol. 4 247, 461
WIEGENLIED, DAS Vol. 4 270
WIELKA MAJOWKA Vol. 6 *page 218, page 235, page 249*
WIELKA MILOSC BALZAKA Vol. 1 303
WIELKA POLSKA Vol. 1 380
WIELKA, WIEKSZA I NAJWIEKSZA Vol. 1 198, 213, 323, 381 Vol. 6 *page 218*
WIELKI PODRYW Vol. 6 *page 236, page 260*
WIELKI PODRYW - OBECNOSC Vol. 6 *page 260*
WIELKI SZU Vol. 6 *page 243, page 255, page 257*
WIELKI UKLAD Vol. 1 182, 216, 264, 306
WIELKOMIEJSKI MROK Vol. 1 290, 329, 380
WIEN 1910 Vol. 4 620, 738
WIEN - BERLIN Vol. 4 242, 306
WIEN, DU STADT DER LIEDER Vol. 4 49, 291, 626
WIEN, WIE ES WEINT UND LACHT Vol. 4 132, 584
WIEN, WIEN, NUR DU ALLEIN Vol. 4 97, 232
WIENER BLUT Vol. 4 614, 666
WIENER G'SCHICHTEN Vol. 4 391, 557, 774
WIENER HERZEN Vol. 4 257 Vol. 6 *page 200*
WIENER MÄDELN Vol. 4 666
WIENER MODELL, DAS Vol. 4 329, 365
WIENERBARNET Vol. 5 2, 7, 228
WIERNA RZEKA Vol. 1 278, 308, 372, 404 Vol. 6 *page 255*
WIERNE BLIZNY Vol. 6 *page 252, page 255, page 258*
WIESENPIEPER Vol. 6 *page 161, page 174, page 175*
WIGILIA Vol. 6 *page 224*
WIKTORYNA Vol. 1 302
WILCZE ECHA Vol. 1 179, 243 Vol. 6 *page 245*
WILCZY BILET Vol. 1 143, 223 Vol. 6 *page 212*
WILCZYCA Vol. 6 *page 241, page 249*
WILD BEASTS - BELVE FEROCI Vol. 7 694
WILD WEST STORY Vol. 5 871, 877, 880
WILDDIEBE Vol. 4 24, 525
WILDE HUMMEL, EINE Vol. 6 *page 186*
WILDERER, DER Vol. 4 149, 364, 535, 660
WILDERER IM JAGEN 161 Vol. 4 186
WILDFANG, DER Vol. 10 256, 460
WILDKATZE, DIE Vol. 4 247
WILDNIS Vol. 4 219, 468, 749, 763
WILDSCHÜTZ DES ERZGEBIRGES, KARL STÜLPNER, DER Vol. 4 655
WILDSCHÜTZ JENNERWEIN Vol. 4 25, 571
WILDVOGEL Vol. 4 247, 340, 638
WILHELM MEISTERS THEATRALISCHE SENDUNG Vol. 6 *page 166, page 167, page 173*
WILHELM TELL Vol. 4 10, 124, 162, 422, 467, 646, 681, 742 Vol. 5 97
WILLY DRILL Vol. 9 71
WINCHESTER BILL Vol. 8 583, 598
WINDJAMMER Vol. 5 672
WINDSTÄRKE 9 Vol. 4 626, 676
WINDSTOSS, EIN Vol. 4 149, 167, 188
WINDSURF, IL VENTO NELLE MANI Vol. 7 544, 821
WINETOU I. Vol. 3 444
WINETOU I. LOVAC NA MEDVJEDE Vol. 3 444
WINETOU II. Vol. 3 444
WINETOU III. Vol. 3 444
WINETOU U DOLINI SMIRTI Vol. 3 444
WINTERFLIRT Vol. 4 187
WINTERNACHTSTRAUM Vol. 4 89, 246
WINTERSTÜRME Vol. 4 312, 321
WIOSNA, PANIE SIERZANCIE Vol. 1 179, 343 Vol. 6 *page 258*
WIR Vol. 6 *page 224*
WIR ARMEN, KLEINEN MÄDCHEN Vol. 4 743
WIR BITTEN ZUM TANZ Vol. 4 85, 234, 375
WIR FAHREN NACH AMERIKA Vol. 4 392
WIR GRATULIEREN Vol. 4 18
WIR HALTEN FEST UND TREU ZUSAMMEN Vol. 4 239, 255
WIR KAUFEN EINE FEUERWEHR Vol. 1 15, 34, 71
WIR LASSEN UNS SCHEIDEN Vol. 1 33, 71, 100
WIR MACHEN MUSIK Vol. 4 381, 445, 582

WIR PARKEN, WO ES UNS GEFÄLLT Vol. 4 643
WIR SCHALTEN UM AUF HOLLYWOOD Vol. 4 22, 562
WIR SEH'N UNS WIEDER Vol. 4 517
WIR SIND VOM K. UND K. INFANTERIEREGIMENT Vol. 4 558, 724
WIR TANZEN UM DIE WELT Vol. 4 354, 430, 437
WIR ZIEHEN UM Vol. 4 430, 558, 796
WIRD SCHULTE VERHAFTET ? Vol. 4 410, 558
WIRTIN ZUM WEISSEN RÖSS'L, DIE Vol. 4 124, 292, 580, 640
WIRTSHAUS IM SPESSART, DAS Vol. 4 119, 726
WISKOTTENS, DIE Vol. 4 339, 377, 437
WITWENBALL, DER Vol. 4 49, 626
WIZJA LOKALNA 1901 Vol. 6 *page 216, page 234, page 238, page 244*
WLASNIE O MILOSCI Vol. 6 *page 212, page 234*
WLOCZEGI Vol. 1 278, 404
WNIEBOWSTAPIENIE Vol. 1 187
WO ANDERE SCHWEIGEN Vol. 6 *page 159, page 164, page 171*
WO DER ZUG NICHT LANGE HÄLT Vol. 1 26, 33, 40, 86
WO DIE ALPENROSEN BLÜHN Vol. 4 25
WO DIE LERCHE SINGT Vol. 4 89, 303
WO DU HINGEHST... Vol. 1 22, 56, 101
WO IST COLETTI ? Vol. 4 87, 734
WOCHENENDBRAUT Vol. 4 221, 367
WOCHENENDBRAUT, DIE Vol. 4 292, 626
WOCHENENDFRIEDEN Vol. 4 517, 525, 558
WOCHENENDZAUBER Vol. 4 115, 239, 287, 384, 458, 646
WODZIREJ Vol. 1 136, 204, 264
WOJEWODA Vol. 1 317
WOJEWODZIANKU Vol. 1 188
WOJNA DOMOWA Vol. 1 237
WOJNA SWIATOW - NASTEPNE STULECIE Vol. 6 *page 211, page 216, page 251*
WOLF UNTER WÖLFEN Vol. 1 21, 37, 120
WÖLFIN, DIE Vol. 4 98, 112
WOLGA-WOLGA Vol. 4 14, 157, 255, 527, 788
WOLGABURLACKEN, DIE Vol. 6 *page 312, page 315*
WOLGAMÄDCHEN, DAS Vol. 4 700 Vol. 6 *page 193*
WOLNE CHWILE Vol. 6 *page 248, page 250, page 252*
WOLNE MIASTO Vol. 1 171, 227 Vol. 6 *page 242*
WOLZ Vol. 1 1, 14, 27
WOMAN AND WIFE Vol. 2 24
WOMAN HATER Vol. 2 826
WOMAN HE SCORNED, THE Vol. 6 *page 198*
WOMAN'S VENGEANCE, A Vol. 2 526
WOMIT SCHNURRT DIE KATZE ? Vol. 4 410, 558
WONDER MAN, THE Vol. 2 84
WONDERFUL COPENHAGEN Vol. 5 78
WOPYT Vol. 6 *page 153*
WOZZEK Vol. 1 77, 107 Vol. 4 734 Vol. 6 *page 157*
W.R. MISTERIJE ORGANIZMA Vol. 3 367, 415, 418, 424
WRAKI Vol. 1 149, 250, 323, 377, 401
WRZOS Vol. 1 194, 278, 308
WSCIEKLY Vol. 6 *page 248, page 252, page 257*
WSCIEKLY RYWAL Vol. 1 174, 215, 326
WSPOLNA POKOJ Vol. 1 227, 254, 320, 377
WSROD NOCNEJ CISZY Vol. 6 *page 235, page 249, page 255*
WSZYSCI I NIKT Vol. 1 311, 316, 318
WSZYSTKO SIE KRECI Vol. 1 176, 407
WUNDER DER MADONNA, DAS Vol. 4 124
WUNDER DER SCHÖPFUNG Vol. 4 95, 458, 625, 743
WUNDER DES FLIEGENS Vol. 4 124, 311, 620
WUNDER DES SCHNEESCHUHS, DAS Vol. 4 10, 156
WUNDERBARE LÜGE DER NINA PETROWNA, DIE Vol. 4 266, 288, 574
WUNDERKUR, DIE Vol. 4 110, 517
WUNDERLICHEN GESCHICHTEN DES THEODOR HUBER, DIE Vol. 4 52, 628
WUNDERSAME LÜGE DER NINA PETROWNA, DIE Vol. 6 *page 190*
WUNDERSCHIESSBUDE, DIE Vol. 4 774
WUNSCHKONZERT Vol. 4 13, 51, 108, 132, 742
WÜRGER, DER Vol. 4 321 Vol. 6 *page 200*
WÜRGER DER WELT, DER Vol. 4 124, 514
WÜSTENGRAB, DAS Vol. 4 558
WÜSTENKÖNIG VON BRANDENBURG, DER Vol. 1 15, 79, 127
WYCIECZKA W NIEZNANE Vol. 1 233, 237, 355
WYKLETA CORKA Vol. 1 317
WYKOLEJENI Vol. 1 174
WYROK Vol. 1 171, 321
WYROK SMIERCI Vol. 6 *page 241, page 243*
WYROK ZYCIA Vol. 1 176, 308

WYSCIE AWARYJNE Vol. 6 *page 252, page 257*
WYSOKIE LOTY Vol. 6 *page 241, page 251, page 253*
WYSPA ZLOCZYNCOW Vol. 1 148, 243, 280, 281 Vol. 6 *page 248*
WYSTRZAL Vol. 1 189, 253

X

X 25 JAVLJA Vol. 3 375, 387, 447
X - PARONI Vol. 5 411
XEHASMENOI EROES Vol. 3 153
XYPNA BASILE Vol. 3 153, 205
XYPNA KAEMENE PERIKLE Vol. 3 193

Y

Y A-T-IL UN FRANCAIS DANS LA SALLE Vol. 6 *page 84, page 133*
Y A-T-IL UN PIRATE SUR L'ANTENNE ? Vol. 6 *page 89*
Y AL TERCER ANO RESUCITO Vol. 8 207, 524
Y DEL SEGURO ... LIBRANOS, SENOR Vol. 8 232, 368
Y DESPUES DEL CUPLE Vol. 8 602, 772
Y EL CUERPO SIGUE AGUANTANDO Vol. 8 221, 258
Y EL PROJIMO ? Vol. 8 219, 616
Y ELIGIO EL INFIERNO Vol. 8 276, 603, 616, 630
Y EN A MARRE Vol. 2 118, 559, 661
Y LE LLAMABAN EL HALCON Vol. 8 551
Y SI NO, NOS ENFADAMOS Vol. 8 595, 659
YA SOY MUJER Vol. 8 219, 299
YA TENEMOS COCHE Vol. 8 515, 517
YACHT MISTERIOSO, IL Vol. 7 668
YAMATA Vol. 9 172, 200
YAMICE SOUS LES CEDRES Vol. 2 309
YANEKEE, EL Vol. 8 732
YANKEE Vol. 8 732
YAO Vol. 2 258
YASMINA Vol. 2 413, 845, 871
YBRIS Vol. 7 802
YEBALA Vol. 8 272
YEHUDI MENUHIN - CHEMIN DE LUMIERE Vol. 2 635
YELLOW HAIR Y PECOS KID Vol. 8 303, 742
„YELLOW: LE CUGINE" Vol. 7 1147
Y'EN A PAS DEUX COMME ANGELIQUE Vol. 2 113, 381
YENDO HACIA TI Vol. 8 245, 346
YERMA Vol. 9 141, 279, 316
YESTERDAY Vol. 6 *page 235*
YETI, IL GIGANZE DEL 20 SECOLO Vol. 7 692
YEUX CERNES, LES Vol. 2 110
YEUX DE L'AME, LES Vol. 2 103, 699
YEUX DE L'AMOUR Vol. 2 128, 661
YEUX DU COEUR, LES Vol. 2 731
YEUX FERMES, LES Vol. 2 351
YEUX NOIRS, LES Vol. 2 526, 626, 676, 824
YEUX SANS VISAGE, LES Vol. 2 156, 778
YHDEN MIEHEN SOTA Vol. 5 541
YHDEN YÖN HINTA Vol. 5 443, 643
YHTEINEN VAIMOMME Vol. 5 379, 392
YJE MBI DRIN Vol. 3 12, 18
YJET E NETEVE TE GJATA Vol. 3 10, 14 Vol. 6 *page 16*
YKS' TAVALLINEN VIRTANEN Vol. 5 443, 549
YKSITYISALUE Vol. 5 452
YLI RAJAN Vol. 5 488, 592
YLIJÄÄMAINEN Vol. 5 372, 634
YLIOPPILAS PÖLLÖVAARANKIHLAUS Vol. 5 344, 433, 640
YNGSSÖMORDET Vol. 5 369, 1052
YO CANTO PARA TI Vol. 8 357, 648
YO CREO QUE ... Vol. 8 439
YO, EL VAQUILLA Vol. 8 455, 550
YO FUI EL REY Vol. 8 534, 603
YO LA VI PRIMERO Vol. 8 221
YÖ MEREN RANNALLEN Vol. 5 511, 647
YO NO ME CASO Vol. 8 359, 475
YO NO SOY LA MATA-HARI Vol. 8 276, 298
YÖ ON PITKA Vol. 5 374, 530
YO SOY FULANA DE TAL Vol. 8 261, 385, 667
YO SOY MI RIVAL Vol. 7 796
YÖ VAI PÄIVÄ Vol. 5 463, 486, 503, 585, 595, 597
YOKMOK Vol. 1 180, 330, 332
YÖN SAALISTAJAT Vol. 5 491
YÖN SYLISSÄ Vol. 5 327, 390
YORICKOVA LEBKA Vol. 10 243, 526
YORK Vol. 4 266, 288, 574
YOSHIWARA Vol. 2 58, 59, 778
YOUNG SANCHEZ Vol. 8 562, 768
YÖVARTIJA VAIN ... Vol. 5 356, 485
YOYO Vol. 2 110, 322, 836
YPEROHE OPTASIA Vol. 3 132
YPOBRYHIO PAPANIKOLES Vol. 3 180
YPOLOHAGOS NATASSA Vol. 3 153, 193
YPOTRON Vol. 8 603
YRJÄNÄN EMÄNNAN SYNTI Vol. 5 356, 386
YUPPIES Vol. 7 70, 299, 599
YVETTE Vol. 2 1, 745 Vol. 4 738, 742, 788

YVETTE, DIE MODEPRINZESSIN **Vol. 4** 291, 390

Z

Z **Vol. 2** 208, 642
Z 7 OPERACION REMBRANDT **Vol. 8** 331
Z BLATA DO LOUZE **Vol. 10** 256, 524
Z CESKYCH MLYNU **Vol. 10** 125, 154, 198, 345, 621
Z CESKYCH MLYNU II **Vol. 10** 345, 524
Z DNIA NA DZIEN **Vol. 1** 278, 315
Z LASKY **Vol. 10** 453, 586
Z MEHO ZIVOTA **Vol. 10** 296, 412, 569
Z RAMION W RAMIONA **Vol. 1** 226
Z TAMTEJ STRONY TECZY **Vol. 1** 202, 227, 306 **Vol. 6** *page 253*
ZA CESKOSLOVENSKY STAT **Vol. 10** 600
ZA CEST VITEZU **Vol. 10** 198
ZA EDNA TROJKA ... **Vol. 6** *page 30, page 35, page 56, page 61*
ZA GLOSEM SERCA **Vol. 1** 145, 357
ZA GOSPOZICATA I NEJNATA M"ZKA KOMPANIJA **Vol. 6** *page 30, page 35, page 45*
ZA GRZECHY **Vol. 1** 132, 241, 290
ZA HUMNY JE DRAK **Vol. 10** 161, 255, 462
ZA-LA-MORT **Vol. 7** 369
ZA LJUBAV JE POTREBNO TROJE **Vol. 6** *page 304, page 323*
ZA OPONOU SMRTI **Vol. 10** 122, 167
ZA PET MINUT SEDM **Vol. 10** 55, 313, 593
ZA RADOVYMI DVERMI **Vol. 10** 92, 256
ZA RANNICH CERVANKU **Vol. 10** 34, 256
ZA RODINATA **Vol. 3** 94
ZA RODNOU HROUDU **Vol. 10** 524, 586
ZA ROK, ZA DZIEN, ZA CHWILE **Vol. 1** 201, 256, 262
ZA RZADOW CARSKICH **Vol. 1** 174, 176, 326
ZA SPITSKAMI **Vol. 5** 341, 412, 540, 557, 579
ZA SVOBODU NARODA **Vol. 10** 53, 167, 198, 201, 310, 363
ZA TICHYCH NOCI **Vol. 10** 122, 412
ZA TRNKOVYM KEREM **Vol. 10** 184, 352, 517
ZA TRZI SPOJRZENIA **Vol. 1** 235, 317
ZA VOLANTEM NEPRITEL **Vol. 10** 317, 517, 567
ZA WAMI POJDA INNI **Vol. 1** 305, 397
ZA WINY BRATA **Vol. 1** 255
ZA WINY NIEPOPELNIONE **Vol. 1** 176, 308
ZA ZASLONA **Vol. 1** 191, 290
ZABAWKA **Vol. 1** 308, 346
ZABEC **Vol. 10** 345, 412
ZABICIE CIOTKI **Vol. 6** *page 224*
ZABIJAKA **Vol. 1** 237
ZABIJCIE CZARNA OWCE **Vol. 1** 179, 293 **Vol. 6** *page 246*
ZABIL JSEM EINSTEINA, PANOVE **Vol. 10** 139, 425, 505, 565
ZABITA NEDELE **Vol. 10** 254, 299, 435, 596
ZABRAVETE TOZI SLUCAJ **Vol. 6** *page 55, page 64*
ZABUDNITE NA MOZARTA **Vol. 10** 454, 474, 493
ZABY A INE RYBY **Vol. 10** 269, 542, 544
ZACHARIUS **Vol. 6** *page 281*
ZACHEV STRACHU **Vol. 10** 283, 344, 571
ZACIATOK SEZONY **Vol. 10** 188, 454, 544
ZACIT ZNOVA **Vol. 10** 70, 211, 433
ZACNE GRZECHY **Vol. 1** 143, 249, 267, 320, 398
ZACZAROWANE KOLO **Vol. 1** 174, 215
ZACZAROWANE PODWORKO **Vol. 1** 198, 211 **Vol. 6** *page 210*
ZACZAROWANY ROWER **Vol. 1** 214, 250, 355, 377
ZADARSKI MEMENTO **Vol. 6** *page 315, page 319, page 320*
ZADEH TELA **Vol. 6** *page 317, page 319, page 320*
ZADUSZKI **Vol. 1** 355, 382 **Vol. 6** *page 242*
ZADZWONCIE DO MOJEJ ZONY **Vol. 1** 232, 302, 305
ZAGANELLA E IL CAVALIERE **Vol. 7** 264, 730, 1189
ZAGLIOSTRO **Vol. 7** 1054
ZAGROZENIE **Vol. 1** 186, 321, 335, 361
ZAGUBIONE UCZUCIA **Vol. 1** 179, 400 **Vol. 6** *page 239*
ZAHADA MODREHO POKOJE **Vol. 10** 256, 524
ZAHADNY PRIPAD **Vol. 10** 167, 310
ZAHADNY PRIPAD GALGINUV **Vol. 10** 243, 536
ZAHL BAR, WENN DU KANNST **Vol. 6** *page 164, page 175, page 176*
ZAHN UM ZAHN **Vol. 4** 247
ZAIDA, DIE TRAGÖDIE EINES MODELLS **Vol. 4** 733, 773
ZAIRA **Vol. 3** 157
ZAJEDNICKI STAN **Vol. 3** 349, 391
ZAJTRA BUDE NESKORO ... **Vol. 10** 82, 135, 273, 417, 544
ZAKAZANE PIOSENKI **Vol. 1** 149, 169, 174, 290, 305 **Vol. 6** *page 220*
ZAKAZANE UVOLNENIE **Vol. 10** 83, 459
ZAKAZANY VYLET **Vol. 10** 129, 197, 313
ZAKLETE REWIRY **Vol. 1** 144, 328, 403

ZAKOCHANI SA MIEDZY NAMI Vol. 1 149, 355 Vol. 6 *page 250*
ZAKON"T NA MORETO Vol. 3 45, 65 Vol. 6 *page 54, page 61*
ZAKRET Vol. 1 162, 163, 379
ZALACAIN EL AVENTURERO Vol. 8 207, 298, 336, 751, 763
ZALAMORT Vol. 4 240, 673
ZALAZAK SUNCA Vol. 6 *page 318*
ZALICZENIE Vol. 1 237
ZALOBNICI Vol. 10 380, 385, 503
ZALOGA Vol. 1 169, 229
ZAMACH Vol. 1 242, 305
ZAMACH STANU Vol. 6 *page 241, page 251, page 253, page 256*
ZAMARLE ECHO Vol. 1 169, 205, 228, 290, 397
ZAMBO, IL DOMINATORE DELLA FORESTA Vol. 7 1022
ZAMEK „NEKONECNO" Vol. 10 28, 180, 495, 570
ZAMEK PRO BARBORKU Vol. 10 227, 433, 503, 532
ZAMILOVANA TCHYNE Vol. 10 201
ZAMPA DI VELLUTO Vol. 7 940
ZAMPO Y YO Vol. 8 258, 438, 581
ZAMPOGNARO INNAMORATO Vol. 7 222, 817
ZAMPUGNARO INNAMORATO Vol. 7 298
ZAN, RE DELLA GIUNGLA Vol. 7 694
ZANCOS, LOS Vol. 8 338
ZANIK SAMOTY BERHOF Vol. 6 *page 219* Vol. 10 165, 359, 360, 511
ZANIM NADEJDZIE DZIEN Vol. 1 163, 323 Vol. 6 *page 257*
ZANNA BIANCA Vol. 7 763
ZAOSTRIT, PROSIM Vol. 10 411, 503
ZAPACH PSIEJ SIERSCI Vol. 6 *page 207, page 226, page 250*
ZAPACH ZIEMI Vol. 6 *page 222, page 238*
ZAPADLI VLASTENCI Vol. 10 122, 167, 256
ZAPALNICZKA Vol. 1 249
ZAPAMIETAJ IMIE SWOJE Vol. 1 210, 232
ZAPATAS BANDE Vol. 4 178, 213, 651
ZAPFENSTREICH Vol. 4 208, 339
ZAPFENSTREICH AM RHEIN Vol. 4 49, 527, 546
ZAPIS ZBRODNI Vol. 1 150, 314 Vol. 6 *page 236*
ZAPOMNIANA MELODIA Vol. 1 278, 308, 346
ZAPOR Vol. 9 117, 183, 332
ZAPOWIEDZ CISZY Vol. 6 *page 217, page 255*
ZAPOWIEDZ CISZY - DOM Vol. 6 *page 217*
ZAPOWIEDZ CISZY - ZWIASTOWANIE Vol. 6 *page 217*
ZAPPA Vol. 5 171, 175, 219, 296
ZAPPATORE Vol. 7 222, 474
ZAR UND ZIMMERMANN Vol. 1 40, 78, 129
ZARABANDA, BING, BING Vol. 8 311, 611
ZARAZA Vol. 1 148, 202, 227, 289, 296
ZAREVO NAD DRAVA Vol. 3 73, 101 Vol. 6 *page 48*
ZAREWITSCH, DER Vol. 4 270, 292, 536, 626, 703, 769
ZARIJOVE NOCI Vol. 10 292, 313
ZARKI Vol. 3 391
ZAROTA Vol. 3 390, 395
ZART TARGYALAS Vol. 9 26, 126, 132
ZÄRTLICHEN VERWANDTEN, DIE Vol. 4 49, 626
ZÄRTLICHKEIT Vol. 4 462
ZASEDA Vol. 3 367, 368
ZASIEKI Vol. 6 *page 241, page 252*
ZASTAVA Vol. 3 444, 445
ZASTIHLA ME NOC Vol. 10 129, 299, 466
ZATAH Vol. 10 70, 246, 408, 444
ZATOULANE DELO Vol. 10 369, 580
ZATYKAC NA KRALOVNU Vol. 10 70, 144, 531
ZAUBERGEIGE, DIE Vol. 4 391, 557, 623
ZAUBERHAFTE MELODIEN Vol. 1 2, 54, 92
ZAUBERMÄNNCHEN, DAS Vol. 1 2, 86, 125
ZAVAROS EJSZAKA Vol. 9 26
ZAVESCANE Vol. 1 298
ZAVET PODIVINOVA Vol. 10 198
ZAVET"T NA INKATA Vol. 3 38
ZAVORRA UMANA Vol. 7 1189, 1190
ZAVRAT Vol. 10 70, 385, 577
ZAV"RTETE VSICKI SFERI Vol. 6 *page 25, page 39, page 43*
ZAWILOSCI UCZUC Vol. 1 153, 254, 284, 298, 395
ZAWSZE W NIEDZIELE Vol. 1 213, 303, 403
ZAZA Vol. 2 70, 243 Vol. 7 5, 750, 756, 1116
ZAZDROSC Vol. 1 176, 177, 350
ZAZDROSC I MEDYCYNA Vol. 1 144, 403, 408 Vol. 6 *page 222*
ZAZIDANI Vol. 3 362, 418
ZAZIE DANS LE METRO Vol. 2 285, 702
ZAZRACNY HLAVOLAM Vol. 10 227, 235, 532
ZBABELEC Vol. 10 296, 532
ZBEHOVIA A PUTNICI Vol. 10 9, 200, 332, 576
ZBOGOM OSTAJ BUNKERU NA RECI Vol. 3 389

ZBOHOM, SLADKE DRIEMOTY **Vol. 10** 194, 224, 544
ZBOR PERICULOS **Vol. 6** *page 267, page 271, page 283*
ZBOR PLANAT **Vol. 3** 229 **Vol. 6** *page 272*
ZBOROV **Vol. 10** 248, 532, 621
ZBRANE PRO PRAHU **Vol. 10** 197, 359, 425
ZBRODNIA LORDA ARTURA SAVILLE **Vol. 1** 198, 313, 354
ZBRODNIARZ I PANNA **Vol. 1** 242, 321
ZBRODNIARZ, KTORY UKDRADL ZBRODNIE **Vol. 1** 283, 403
ZDE JSOU LVI **Vol. 10** 296, 564
ZDJECIA PROBNE **Vol. 1** 264, 288, 291, 411
ZE DO BURRO, O **Vol. 8** 64
ZE SOBOTY NA NEDELI **Vol. 10** 147, 586
ZE SVETA LESNICH SAMOT **Vol. 10** 122, 256, 586
ZE VSECH JEDINA **Vol. 10** 168, 564, 585
ZED **Vol. 3** 419, 443
ZEDER **Vol. 7** 96, 1043
ZEICHEN DER DREI, DAS **Vol. 4** 718
ZEIT DER EINSAMKEIT, DIE **Vol. 6** *page 280*
ZEIT DER STÖRCHE **Vol. 1** 10, 36, 59
ZEIT ZU LEBEN **Vol. 1** 1, 8
ZEJSCIE DO PIEKLA **Vol. 1** 232, 355 **Vol. 6** *page 224*
ZELENA KNIZKA **Vol. 10** 184, 524
ZELENA LETA **Vol. 10** 74, 153, 246
ZELENA VLNA **Vol. 10** 397, 425, 577
ZELENE OBZORY **Vol. 10** 154, 503, 517
ZELENITE POLJA **Vol. 6** *page 52, page 55*
ZELENY AUTOMOBIL **Vol. 10** 536
ZELENY DEDEK **Vol. 10** 132, 345, 532, 552
ZEMALJSKI DANI TEKU **Vol. 3** 367, 438
ZEMENT **Vol. 1** 19
ZEMIANSKA CEST **Vol. 10** 193, 273, 618
ZEMJA **Vol. 3** 66, 71, 104 **Vol. 6** *page 41*
ZEMJATA GORI **Vol. 3** 106
ZEMLJACI **Vol. 3** 388
ZEMRA QE NUK PLAKEN **Vol. 3** 2, 3
ZEMSTA **Vol. 1** 173, 250, 302, 321
ZENA, KTERA VI, CO CHCE **Vol. 10** 167, 300, 524, 611
ZENA NA ROZCESTI **Vol. 10** 168, 580
ZENA POD KRIZEM **Vol. 10** 256, 460
ZENA PRO TRI MUZE **Vol. 10** 473, 517, 595
ZENA Z VRCHOV **Vol. 10** 268, 273, 320, 501
ZENELÖ MALOM **Vol. 9** 71, 76, 186
ZENICA **Vol. 3** 383, 413
ZENTRALE RIO **Vol. 4** 270, 525
ZENU ANI KVETINOU NEUHODIS **Vol. 10** 359, 517, 572
ZENY U BENZINU **Vol. 10** 345, 412
ZENY V OFSAJDU **Vol. 10** 70, 387
ZEPPELIN **Vol. 5** 716, 780
ZERBAL **Vol. 7** 99
ZERBROCHENE KRUG, DER **Vol. 4** 266, 729
ZERBROCHENE PUPPE, DIE **Vol. 4** 251
ZERO **Vol. 2** 633 **Vol. 7** 196
ZERO DE CONDUITE **Vol. 2** 425 **Vol. 6** *page 143, page 147*
087 MISION APOCALIPSIS **Vol. 8** 598, 626
ZERO IN CONDOTTA **Vol. 7** 1202
091, POLICIA AL HABLA **Vol. 8** 524
07 CON EL 2 DELANTE **Vol. 8** 629, 768
07 ... TASSI **Vol. 7** 402, 665
001, OPERACION CARIBE **Vol. 8** 524, 749
00-2 AGENTI SEGRETESSIMI **Vol. 7** 17
ZERT **Vol. 10** 9, 87, 235
ZERTANZTEN SCHUHE, DIE **Vol. 1** 48, 49
ZERWANE CUMY **Vol. 6** *page 241, page 246*
ZERWANY MOST **Vol. 1** 237, 283, 298
ZESTOKE GODINE **Vol. 3** 343, 351
ZESTOS MENAS AUGOUSTOS **Vol. 3** 207
ZESTREA **Vol. 3** 319, 321
ZESTREA DOMNITEI RALU **Vol. 3** 254, 285, 306, 335
ZETEMA ZOES KAI THANATOU **Vol. 3** 193
ZEUGEN GESUCHT **Vol. 4** 18, 340, 558, 608, 691
ZEUGENDE TOD, DER **Vol. 4** 518, 779
ZEUS **Vol. 7** 531
ZEW MORZA **Vol. 1** 174, 346
ZEZOWATE SZCZESCIE **Vol. 1** 179, 198, 242, 313, 323, 391
ZGODBA KI JE NI **Vol. 3** 385, 449
ZIA BERTA, LA **Vol. 7** 1189
ZIA ... CAMILLO **Vol. 7** 264, 701
ZIA D'AMERICA VA A SCIARE, LA **Vol. 7** 891
ZIA DI CARLO, LA **Vol. 7** 460, 1062
ZIA SMEMORATA, LA **Vol. 7** 487, 756
ZIA SVEDESE, LA **Vol. 7** 311
ZIBALDONE N. 1 **Vol. 7** 283, 332, 634, 796, 910
ZIDUL **Vol. 3** 237, 264, 281
ZIEL IN DEN WOLKEN **Vol. 4** 620, 660, 728
ZIELONE LATA **Vol. 6** *page 240, page 243*
ZIEMIA **Vol. 1** 321, 400
ZIEMIA OBIECANA **Vol. 1** 163, 174, 176, 204, 216, 303
ZIG ZAG STORY **Vol. 6** *page 72, page 148*
ZIG-ZIG **Vol. 2** 68, 434

ZIGANO, DER BRIGANT VON MONTE DIAVOLO **Vol. 4** 340, 471, 779
ZIGEUNER DER NACHT **Vol. 4** 452, 629
ZIGEUNERBARON **Vol. 4** 30, 567, 614
ZIGEUNERBARON, DER **Vol. 2** 733 **Vol. 3** 356 **Vol. 4** 14, 166, 186, 743
ZIGEUNERBLOD **Vol. 5** 295
ZIGEUNERBLUT **Vol. 4** 124, 167, 234, 385, 462, 646, 734
ZIGEUNERENS HAEVN **Vol. 5** 97
ZIGEUNERLIEBCHEN **Vol. 4** 186
ZIGEUNERPRIMAS, DER **Vol. 4** 124, 221
ZIGEUNERPRINSESSEN **Vol. 5** 242
ZIGEUNERWEISEN **Vol. 4** 747
ZIGOMAR **Vol. 2** 24, 415
ZIGOMAR CONTRE NICK CARTER **Vol. 2** 24, 711
ZIGOMAR PEAU D'ANGUILLE **Vol. 2** 24, 711
ZIGOTO **Vol. 2** 159, 286, 330
ZIKINA DINASTIJA **Vol. 6** *page 306, page 314, page 318*
ZILE DE VARA **Vol. 3** 261, 285, 295
ZILE FIERBINTI **Vol. 3** 256, 272, 317
ZILLE TYPEN **Vol. 4** 82
ZILLE UND ICK **Vol. 6** *page 160, page 167, page 170*
ZIMBET PENTRU MAI TIRZIU, UN **Vol. 3** 245, 272, 317
ZIMMER 107 **Vol. 4** 208
ZIMMER ZU VERMIETEN **Vol. 4** 329, 774
ZIMMERMÄDCHEN..DREIMAL KLINGELN **Vol. 4** 239, 340
ZIMOVANJE U JAKOBSFELDU **Vol. 3** 341, 356, 441
ZIMOWY ZMIERZCH **Vol. 1** 179, 187 **Vol. 6** *page 240*
ZINARI, GLI **Vol. 7** 1189
ZINGARA **Vol. 7** 304, 1027
ZINGARA, LA **Vol. 7** 1189
ZINGARI **Vol. 7** 48, 254, 621
ZINKER, DER **Vol. 4** 164, 257
ZIO ADOLFO IN ARTE FÜHRER **Vol. 7** 426
ZIO SI SPOSA, LO **Vol. 7** 1090
ZIRAFA V OKNE **Vol. 10** 255, 462
ZIRKUS DES LEBENS **Vol. 4** 734
ZIRKUS RENZ **Vol. 4** 8, 36, 270, 774
ZIRKUSPRINZESSIN, DIE **Vol. 4** 71, 287, 329, 384, 458, 646
ZISKA **Vol. 2** 302, 311
ZITELLONE DELUSE **Vol. 7** 304
ZITELLONI, GLI **Vol. 7** 1027
ZITELLONI, I **Vol. 8** 276
ZITOUNTAI GAMPROI ME PROIKA **Vol. 3** 164
ZITRA SE BUDE TANCIT VSUDE **Vol. 10** 184, 193
ZITRA TO ROZTOCIME, DRAHOUSKU ... ! **Vol. 10** 41, 425, 593
ZITRA VSTANU A OPARIM SE CAJEM **Vol. 10** 229, 369, 425
ZITTO QUANDO PARLI **Vol. 7** 450, 1194
ZIUA Z **Vol. 6** *page 266, page 280*
ZIVA ISTINA **Vol. 3** 412
ZIVA VODA **Vol. 10** 72, 269, 399
ZIVATAR KEMENESPUSZTAN **Vol. 9** 266, 288
ZIVETI KAO SAV NORMALAN SVET **Vol. 6** *page 308, page 318*
ZIVI BILI PA VIDJELI **Vol. 3** 454
ZIVI HORA **Vol. 6** *page 60, page 65*
ZIVJECE OVAD NAROD **Vol. 3** 349, 399
ZIVJETI OD LJUBAVI **Vol. 3** 371, 430
ZIVOT BEZ KYTARY **Vol. 10** 289, 369, 371
ZIVOT JE KRASNY **Vol. 10** 345, 412
ZIVOT JE LEP **Vol. 6** *page 303, page 314, page 318*
ZIVOT JE MASOVNA POJAVA **Vol. 3** 351
ZIVOT JE NAS **Vol. 3** 369, 391, 432
ZIVOT JE PES **Vol. 10** 256, 586
ZIVOT NA UTEKU **Vol. 10** 82, 263, 459
ZIVOT PRO JANA KASPARA **Vol. 10** 313, 577
ZIVOT SEL KOLEM **Vol. 10** 45, 201
ZIVOT VOJENSKY, ZIVOT VESELY **Vol. 10** 453, 524
ZIVOTEM VEDLA JE LASKA **Vol. 10** 92, 453
ZIVY BIC **Vol. 10** 32, 273
ZIZANIE, LA **Vol. 2** 719
ZIZEN **Vol. 10** 133, 247
ZIZIS EN FOLIE, LES **Vol. 2** 735
ZIZKOVSKA ROMANCE **Vol. 10** 87, 296
ZIZNIVE MLADI **Vol. 10** 122, 580
ZKAZA JERUZALEMA **Vol. 10** 70, 580
ZKAZENA KREV **Vol. 10** 201
ZKOUSKA POKRACUJE **Vol. 10** 296, 385, 522
ZKROCENI ZLEHO MUZE **Vol. 10** 344, 429, 434
ZLA NOC **Vol. 10** 352, 395, 457
ZLALNA FRALKA **Vol. 3** 340
ZLATA KATERINA **Vol. 10** 256, 586
ZLATA RENETA **Vol. 10** 18, 503
ZLATA REVA **Vol. 10** 102, 263, 493
ZLATA SLEPICE **Vol. 10** 41, 144, 153
ZLATA SVATBA **Vol. 10** 41, 153, 254
ZLATA ZENA **Vol. 10** 198, 536
ZLATE CASY **Vol. 10** 82, 273, 548
ZLATE DNO **Vol. 10** 532, 621
ZLATE KAPRADI **Vol. 10** 24, 433, 503
ZLATE PTACE **Vol. 10** 256, 524
ZLATE RYBKY **Vol. 10** 153, 283, 395
ZLATE SRDECKO **Vol. 10** 363
ZLATNA PRACKA **Vol. 3** 371, 374

ZLATNATA REKA Vol. 6 *page 44, page 45, page 65*
ZLATNIJAT Z"B Vol. 3 100, 101 Vol. 6 *page 32*
ZLATO ZA ODVAZNE Vol. 6 *page 312*
ZLATY CLOVEK Vol. 10 345, 412
ZLATY KLICEK Vol. 10 257, 536, 605, 613
ZLATY KVET Vol. 10 17
ZLATY PAVOUK Vol. 10 433, 564, 566
ZLATY VLK Vol. 10 112
ZLE PARE Vol. 3 383, 435
ZLE PONDELI Vol. 10 299, 380, 385
ZLOCIN SLECNY BACILPYSKY Vol. 10 162, 263, 544
ZLOCIN U SKOLI Vol. 6 *page 321*
ZLOCIN V DIVCI SKOLE Vol. 10 41, 96, 153, 286, 352, 522
ZLOCIN V MODRE HVEZDE Vol. 10 227, 369, 505
ZLOCIN V SANTANU Vol. 10 41, 325, 517, 567
ZLODEJ Vol. 10 45
ZLODZIEJ I DZIEWCZYNKA Vol. 1 158, 371, 387
ZLOTA MASKA Vol. 1 278, 346
ZLOTE BAGNO Vol. 1 174, 215, 326
ZLOTO Vol. 1 254, 276, 320
ZLOTY KLUCZYK Vol. 1 298
ZLOZOR Vol. 10 72, 107, 320
ZMARTWYCHWSTANIE OFFLANDA Vol. 1 388
ZMLUVA S DIABLOM Vol. 10 210, 576
ZMORY Vol. 1 303, 378, 408 Vol. 6 *page 259*
ZNACHOR Vol. 1 278, 308, 404 Vol. 6 *page 236, page 238, page 241, page 256*
ZNAK Vol. 1 144
ZNAKI NA DRODZE Vol. 1 220, 276, 283
ZNAKI ZODIAKU Vol. 1 262, 409 Vol. 6 *page 210*
ZNAKOW SZCZEGOLNYCH BRAK Vol. 1 271, 321 Vol. 6 *page 244*
ZNAMENI KOTVY Vol. 10 256, 412, 569
ZNAMENI RAKA Vol. 10 24, 41, 96
ZNAMOST SESTRY ALENY Vol. 10 515, 568, 577
ZNAMOST Z INSERATU Vol. 10 335
ZNAMOST Z ULICE Vol. 10 167, 345
ZNAS ONEN MALY DOMEK U JEZERA ? Vol. 6 *page 200* Vol. 10 185, 608
ZNICZ OLIMPIJSKI Vol. 1 232, 298
ZNIKAD DONIKAD Vol. 1 198, 272, 327, 391 Vol. 6 *page 241*
ZNISZCZYC PIRATA Vol. 1 237
ZOBACZYMY SIE W NIEDZIELE Vol. 1 179, 354
ZOCELENI Vol. 10 503, 564
ZODIA PECIO AREI Vol. 3 268, 287, 298
ZOE Vol. 2 13, 248, 615
ZOE ENOS ENTHROPOU Vol. 3 185
ZOE HORIS HAMOGELO Vol. 3 131
ZOFIA Vol. 1 354, 379, 401
ZÖLDAR Vol. 9 121, 318, 332
ZOLNIERZ KROLOWEJ MADAGASKARU Vol. 1 169, 290, 307, 308, 377, 397
ZOLNIERZ ZWYCIESTWA Vol. 1 298, 305, 397 Vol. 6 *page 223*
ZOMBI 2 Vol. 7 875, 1015
ZOMBI 3 Vol. 7 1058
ZOMBI HOLOCAUST Vol. 7 1213
ZONA Vol. 1 215
ZONA DLA AUSTRALIJCZYKA Vol. 1 211, 220, 403
ZONA I NIE ZONA Vol. 1 176, 278
ZONE DE LA MORT, LA Vol. 2 153
ZONE FRONTIERE Vol. 2 404, 722
ZONE, LA Vol. 2 656
ZONGORA A LEVEGÖBEN Vol. 9 316, 339
ZONJA NGA QYTETI Vol. 3 2, 12
ZOO Vol. 2 633
ZOPF UND SCHWERT Vol. 4 132, 681
ZOPPO CHE FA STRADE Vol. 7 1189
ZORBA THE GREEK Vol. 3 150, 172
ZÖRGETNEK AZ ABLAKON Vol. 9 196
ZORICA Vol. 3 344, 383
ZORRA Y EL ESCORPION, LA Vol. 8 536
ZORRITA MARTINEZ Vol. 8 331, 360, 627
ZORRO ALLA CORTE DI SPAGNA Vol. 7 1152
ZORRO ALLA CORTE D'INGHILTERRA Vol. 7 1121
ZORRO CABALGA OTRA VEZ, EL Vol. 8 598, 772
ZORRO, CABALLERO DE LA JUSTICIA, EL Vol. 8 332, 770
ZORRO CONTRO MACISTE Vol. 7 1121
ZORRO DE MONTERREY, EL Vol. 8 557, 749
ZOSIA Vol. 1 242, 269, 325, 401
ZOUZOU Vol. 2 425, 427, 573, 581, 626
ZOYA Vol. 7 399
ZOZOS, LES Vol. 2 618
ZPEV ZLATA Vol. 10 167, 272
ZPIVAJICI FILM Vol. 10 173, 233, 577
ZPIVAJICI PUDRENKA Vol. 10 299, 380, 385
ZPIVALI JSME ARIZONU Vol. 10 70, 371, 433
ZPOVEDNICE Vol. 10 540
ZRALE VINO Vol. 10 41, 425, 496
ZRCADLENI Vol. 10 82, 172, 327
ZRCADLO PRO KRISTYNU Vol. 10 165, 313, 415
ZRELA MLADOST Vol. 10 16, 324, 399
ZSARNOK SZIVE, A Vol. 9 20, 154
ZSUZSANNA ES VENEK Vol. 9 71, 237

ZTRACENA STOPA **Vol. 10** 70, 292
ZTRACENA TVAR **Vol. 10** 55, 313, 593
ZTRACENI **Vol. 10** 41, 388, 457
ZTRATILA SE BILA PANI **Vol. 10** 61, 345, 585
ZU BEFEHL, HERR UNTEROFFIZIER **Vol. 4** 239
ZU JEDER STUNDE **Vol. 1** 2, 26, 86
ZU NEUEN UFERN **Vol. 4** 8, 437, 742 **Vol. 6** *page 197*
ZU STRASSBURG AUF DER SCHANZ' **Vol. 4** 135, 339, 774
ZU TODE GEHETZT **Vol. 4** 646
ZUCCHERO, MIELE E PEPERONCINO **Vol. 7** 116, 426
ZUCKERWASSER **Vol. 6** *page 310, page 319*
ZUFLUCHT **Vol. 4** 535, 626
ZUG FÄHRT AB, EIN **Vol. 4** 273, 285, 688
ZÜGELLOSES BLUT **Vol. 4** 309
ZUGVERKEHR UNREGELMÄSSIG **Vol. 1** 56, 64, 101
ZUGVOGEL AM SUND **Vol. 1** 57, 62, 111
ZUHANAS KÖZBEN **Vol. 9** 92, 232, 240
ZUM 5-UHR-TEE SPIELT BERNHARD ETTE BEI ADOLF GONDRELL **Vol. 4** 745
ZUM BEISPIEL JOSEF **Vol. 1** 12, 79, 128
ZUM PARADIES DER DAMEN **Vol. 4** 124, 190, 663, 779
ZUM ZUM ZUM N.2 **Vol. 7** 1213
ZÜND AN, ES KOMMT DIE FEUERWEHR **Vol. 1** 7, 19, 92
ZUR CHRONIK VON GRIESHUUS **Vol. 4** 132, 266, 486, 533, 574, 729
ZUR SEE **Vol. 1** 59, 60, 95
ZURIVY REPORTER **Vol. 10** 254, 327, 568
ZURÜCK AUS DEM WELTALL **Vol. 5** 333, 443, 455, 634
ZUTA **Vol. 3** 368, 444
ZUZANNA I CHLOPCY **Vol. 1** 198, 202, 400
ZVANI **Vol. 7** 1190
ZVEZDA PUTUJE NA JUG **Vol. 3** 444
ZVEZDE SU OCI RATNIKA **Vol. 3** 356, 415, 418
ZVEZDI **Vol. 6** *page 59*
ZVEZDI V KOSITE, SALZI V OCITE **Vol. 3** 42 **Vol. 6** *page 23, page 32, page 46, page 56*
ZVIRATA VE MESTE **Vol. 10** 149, 195, 497
ZVIZDUK U OSAM **Vol. 3** 376
ZVONY PRE BOSYCH **Vol. 10** 273, 459, 478
ZVONY Z RAKOSU **Vol. 10** 142, 411, 413
ZWANGSLIEBE IM FREISTAAT **Vol. 4** 226, 342, 446
ZWARIOWANA NOC **Vol. 1** 311, 321
ZWECK HEILIGT DIE MITTEL, DER **Vol. 4** 18, 558
ZWEI BLAUE JUNGS **Vol. 6** *page 189*
ZWEI BRIEFE AN POSPISCHIEL **Vol. 1** 7, 19, 92
ZWEI BRÜDER **Vol. 4** 157, 565
ZWEI FRAUEN **Vol. 4** 361, 738, 788
ZWEI FREUNDE **Vol. 4** 514
ZWEI GENIES **Vol. 4** 329, 713
ZWEI GLÜCKLICHE MENSCHEN **Vol. 4** 215, 536
ZWEI GLÜCKLICHE TAGE **Vol. 4** 246, 323, 584, 646
ZWEI GUTE KAMERADEN **Vol. 4** 124, 422, 646
ZWEI HERZEN IM 3/4 TAKT **Vol. 4** 95, 208, 479
ZWEI HERZEN UND EIN SCHLAG **Vol. 4** 17, 289, 614 **Vol. 6** *page 130*
ZWEI HIMMELBLAUE AUGEN **Vol. 4** 239
ZWEI IM SONNENSCHEIN **Vol. 4** 115, 132
ZWEI IN EINEM AUTO **Vol. 4** 29, 321, 703
ZWEI IN EINER GROSSEN STADT **Vol. 4** 79, 288, 486, 650
ZWEI IN EINER KLEINEN STADT **Vol. 1** 15
ZWEI KINDER **Vol. 4** 113, 424
ZWEI KRAWATTEN **Vol. 4** 158, 452, 608
ZWEI MENSCHEN **Vol. 4** 15, 73, 219, 400, 550, 724,
ZWEI MÜTTER **Vol. 1** 48, 74, 103
ZWEI PAAR FÜSSE **Vol. 4** 514
ZWEI REPARIEREN SICH DURCH **Vol. 4** 704
ZWEI ROTE ROSEN **Vol. 4** 14, 431
ZWEI SCHWARZE LATERNEN **Vol. 4** 205, 299, 467
ZWEI UND DIE DAME, DIE **Vol. 4** 461, 720
ZWEI UNTERM HIMMELSZELT **Vol. 4** 276, 363, 663
ZWEI VOM SÜDEXPRESS, DIE **Vol. 4** 301, 344, 544, 776
ZWEI WELTEN **Vol. 4** 219, 469, 525, 578
ZWEI ZEILEN KLEINGEDRUCKT **Vol. 6** *page 166, page 172, page 173*
ZWEIERLEI MORAL **Vol. 4** 247, 461, 603
ZWEIMAL HOCHZEIT **Vol. 4** 246, 774
ZWEIMAL LUX **Vol. 4** 270, 634
ZWEIMAL ZWEI IM HIMMELBETT **Vol. 4** 135, 370, 373
ZWEITE FRAU, DIE **Vol. 4** 159
ZWEITE GLEIS, DAS **Vol. 1** 26, 47, 111
ZWEITE LEBEN, DAS **Vol. 4** 213, 390
ZWEITE LEBEN DES FRIEDRICH WILHELM GEORG PLATOW, DAS **Vol. 1** 19, 110, 128

ZWEITE MUTTER, DIE Vol. 4 87, 163, 261, 603
ZWEITE SCHUSS, DER Vol. 4 228, 582, 740, 743
ZWEIUNDFÜNFZIG WOCHEN SIND EIN JAHR Vol. 1 36, 40, 47, 100
ZWIELICHT Vol. 4 24, 323, 329
ZWISCHEN 1/2 11 UND 11 Vol. 6 *page 186*
ZWISCHEN 12 UND 2 Vol. 4 519, 741
ZWISCHEN ABEND UND MORGEN Vol. 4 417, 729
ZWISCHEN DEN ELTERN Vol. 4 30, 123, 544
ZWISCHEN FLAMMEN UND FLUTEN Vol. 4 383, 424
ZWISCHEN FREITAG UND MORGEN Vol. 1 7, 9, 48
ZWISCHEN HAMBURG UND HAITI Vol. 4 8, 28
ZWISCHEN HIMMEL UND ERDE Vol. 4 28, 235, 344, 647
ZWISCHEN LIEBE UND PFLICHT Vol. 4 747
ZWISCHEN NACHT UND MORGEN Vol. 4 247, 373, 461, 661 Vol. 6 *page 186*
ZWISCHEN NACHT UND TAG Vol. 1 42, 68, 102
ZWISCHEN STROM UND STEPPE Vol. 4 89, 134
ZWISCHEN TOD UND LEBEN Vol. 4 681
ZWISCHEN VIERZEHN UND SIEBZEHN Vol. 4 208, 255
ZWISCHEN ZWEI HERZEN Vol. 4 124, 247, 703
ZWISCHEN ZWEI WELTEN Vol. 4 113
ZWISCHEN ZWÖLF UND VIERTEL EINS Vol. 4 95
ZWISCHENBILANZ Vol. 1 11, 65
ZWISCHENFALL IN BENDERATH Vol. 1 34, 43, 82
ZWÖLF UHR MITTAGS KOMMT DER BOSS Vol. 1 7, 49, 104
ZWYCIESTWO Vol. 1 211, 280 Vol. 6 *page 225, page 248*
ZYCIE RAZ JESZCZE Vol. 1 156, 180, 185, 237, 249, 398 Vol. 6 *page 242*
ZYCIE RODZINNE Vol. 1 325, 388, 403
ZYWOT MATEUSZA Vol. 1 217

FSC
www.fsc.org
MIX
Papier | Fördert
gute Waldnutzung
FSC® C083411

Kolibri 360 GmbH
Ferdinand-Jühlke-Straße 7
99095 Erfurt, Deutschland
produktsicherheit@kolibri360.de